APPLIED MATHEMATICS FOR BUSINESS,
ECONOMICS, LIFE AND SOCIAL SCIENCES

# APPLIED MATHEMATICS FOR BUSINESS, ECONOMICS, LIFE AND SOCIAL SCIENCES

## Edmond C. Tomastik

UNIVERSITY OF CONNECTICUT

*Saunders College Publishing*

**HARCOURT BRACE COLLEGE PUBLISHERS**

Fort Worth   Philadelphia   San Diego   New York   Orlando   Austin
San Antonio   Toronto   Montreal   London   Sydney   Tokyo

Text Typeface: 10/12 Times Roman
Compositor: CRWaldman Graphic Communications
Executive Editor: Jay Ricci
Developmental Editors: Anita M. Fallon and Anne Wightman
Managing Editor: Carol Field
Project Editor: Maureen Iannuzzi
Copy Editor: Merry Post
Manager of Art and Design: Carol Bleistine
Art Directors: Anne Muldrow and Susan Blaker
Art Assistant: Sue Kinney
Text Designer: Rebecca Lemna
Cover Designer: Jennifer Dunn
Text Artwork: Tech-Graphics
Director of EDP: Tim Frelick
Producer Manager: Carol Florence
Marketing Manager: Monica Wilson

**On the cover:** The detail on the cover is from a design adapted from a Turkish flatweave in the collection of the Textile Museum, Washington, D.C. (1978.19.1, gift of Mrs. Arthur S. Rudd). The original weave measures 160.5 cm x 229.5 cm and was purchased in 1958 in Crete. The weave is a supplementary-weft patterning.

Printed in the United States of America

Applied Mathematics for Business, Economics, Life and Social Sciences

ISBN: 0-03-094073-7

Library of Congress Catalog Card Number: 93-086849

3456 016 987654321

This book was printed on paper made from waste paper, containing 10% post-consumer waste and 40% pre-consumer waste, measured as a percentage of total fiber weight content.

*To Nancy*

# A Note About Custom Publishing

Courses in finite mathematics with calculus are structured in various ways, differing in length, content, and organization. To cater to these differences, Saunders College Publishing is offering *Applied Mathematics for Business, Economics, Life and Social Sciences* in a custom-published format. Instructors can rearrange, add, or cut chapters to produce a text that best meets their needs.

The diagram below shows the chapter dependencies in *Applied Mathematics for Business, Economics, Life and Social Sciences* which instructors should consider. Beyond these dependencies, instructors, with custom publishing, are free to choose the topics they want to cover in the order they want to cover them, thereby creating a text that follows their course syllabi.

Saunders College Publishing is working hard to provide the highest quality service and product for your courses. If you have any questions about custom publishing, please contact your local Saunders sales representative.

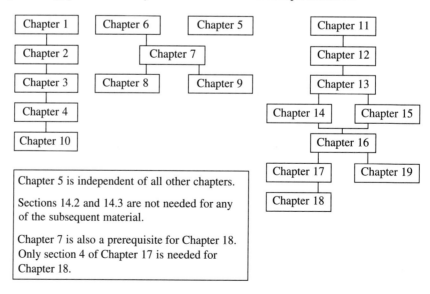

Chapter 5 is independent of all other chapters.

Sections 14.2 and 14.3 are not needed for any of the subsequent material.

Chapter 7 is also a prerequisite for Chapter 18. Only section 4 of Chapter 17 is needed for Chapter 18.

# Accuracy and "The Saunders Solution"

To ensure the accuracy of our math texts, Saunders is taking an additional step, "The Saunders Solution," in the proofreading process. At this time, we are the only publisher taking this extra step to ensure accuracy in our mathematics texts.

All Saunders mathematics textbooks, regardless of edition, are proofread in both the galley and pages stages of production. Instructor's Review Copies, identified as such on the book's cover, are then printed in the fall for adoption consideration only. The Instructor's Review Copy then undergoes "The Saunders Solution" accuracy check, completed independently by both the author and paid accuracy reviewers. Any inconsistencies or errors in the texts are corrected in the Student Edition, which is available in the spring.

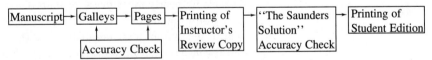

The implementation of "The Saunders Solution" ensures that all textbooks will be as clean as second printings and students will use textbooks that are as error-free and accurate as possible.

All publishers face the same concerns in dealing with the problem of errors in math texts. Only Saunders has taken an extra step to find a solution.

# Preface

*Applied Mathematics for Business, Economics, Life and Social Sciences* is designed to be used in a course that combines finite mathematics with calculus and is aimed for students majoring in business, management, economics, or the life or social sciences. The text can be understood by the average student with two years of high school algebra. A wide range of topics is included, giving the instructor considerable flexibility in designing a course. Optional graphing calculator material is included.

## Distinguishing Features

I have taught both the finite mathematics course and the introductory calculus course for nearly thirty years, examined countless texts for adoption, used a variety of different texts in my classes, and, for a number of reasons, have found all of the current texts unsatisfactory. Thus I decided to write this text, which distinguishes itself from the others in the following ways.

First, applications truly play a central and prominent role in the text. Second, and in keeping within emphasis on applications, I have stressed translating applied problems into mathematical equations or into an appropriate mathematical format for solution by calculators and computers. Third, and again in keeping with an emphasis on applications, I have presented probability, not as an abstract mathematical subject, but as a natural phenomenon that arises in business and science. By beginning with empirical probability I show how the abstract definition of probability arises, resulting in deep and practical insights into probability and statistics. Fourth, I have given an easily understood geometric presentation of why and how the simple method words, giving deep insights into this fundamentally important method. The calculus is presented by relying heavily on graphs and the geometry of curves. The idea is to have the student see geometrically all the concepts of the calculus. Furthermore, as in the finite mathematics part, a wide variety of timely applications are used to motivate the material.

Let me amplify on these points, one at a time. First, this text is written for *users* of mathematics. Thus, for example, a concrete applied problem is presented first as a motivation before developing a needed mathematical topic. After the mathematical topic has been developed, further applications are given so that the student understands the practical need for knowing the mathematics. This is done so consistently and thoroughly that after going through a number of chapters, the student should come to believe that mathematics is everywhere.

Second, I believe that no other skill is more important than the ability to translate a real-life problem into an appropriate mathematical format for finding the solution. Students often refer to this process as ''word problems.'' Whereas linear systems of equations, linear programming problems, and financial problems, for example, can easily be solved by computers and some calculators, no computer or calculator, now or in the foreseeable future, can translate these applied problems into the necessary mathematical language. Thus students, in their jobs, will most likely use their mathematical knowledge to translate applied problems into necessary mathematical formats for solution by computers.

To develop these needed skills many word problems, requiring the writing of one linear equation, are given in all of the sections in the first introductory chapter. This prepares the student for the many word problems that require creating systems of linear equations in the second chapter on linear systems and matrices. The word problems continue in the third chapter on linear programming, with the last section of this chapter devoted entirely to setting up linear programming problems for solution on computers. If desired, students can then use the software packages that are free to adopters of this text to solve the linear programming problems that have been written. Word problems then continue in the next chapter on the simplex method and beyond. In the calculus many timely applications are considered, such as the optimal management of renewable resources, the Laffer curve in federal tax policy, the logistic equation in population dynamics, to name just a few.

The third major area that distinguishes this text from others, is the discussion found on probability. Other finite mathematics texts present probability as a *mathematical* subject with an *abstract mathematical* definition of probability given to the reader to accept on faith. If, on occasion, a text attempts to explain why this definition is used, the explanation is always confined to the equally likely outcomes case.

Since this text is not meant for mathematics majors but for business, finance, and science majors, the approach is to first present probability as a *natural* phenomenon that arises in concrete problems in business, finance, and science. Thus empirical probability in a natural context is presented first. An experiment is performed $N$ times or a survey of $N$ people is taken. The frequency, $f(E)$, that some particular event occurs is noted. The significance of the fraction $\dfrac{f(E)}{N}$, which is called the empirical probability, is then easily explained. The student readily understands that changing $N$ will normally result in changing the empirical probability and that making $N$ large should result in this empirical probability being very close to some number, which is then defined as the actual probability. This, of course, is precisely how probability arises everyday in business and science. The basic properties of empirical probability are easily grasped, extended to probability, and then provide a basic motivation for the mathematical definition of probability. The mathematical definition of probability then arises in a natural context. Relating probability to relative frequencies is then used throughout the text to give deep insights into other subjects.

For example, in statistics, the text easily demonstrates that after $N$ trials of an experiment with outcomes $x_1, x_2, \ldots, x_n$, the *average* outcome is just

$$x_1 \frac{f(x_1)}{N} + x_2 \frac{f(x_2)}{N} + \cdots + x_n \frac{f(x_n)}{N},$$

where $x_1$ is the first outcome, $x_2$ the second, and so on, while $f(x_1)$ is the frequency of occurrence of the first outcome, $f(x_2)$ is the frequency of occurrence of the second, and so on. Now when $N$ is large, we expect the relative frequency $\frac{f(x_1)}{N}$ to be approximately the probability $p_1$ of the first outcome, $\frac{f(x_2)}{N}$ to be approximately the probability $p_2$ of the second outcome, and so on. Thus we have a natural definition of expected value as

$$x_1 p_1 + x_2 p_2 + \cdots + x_n p_n,$$

and the student sees the somewhat mysterious expected value arising from an everyday *average*. Additionally, these insights are valuable when studying continuous probability using the calculus.

Fourth, a problem with current texts in finite mathematics is their presentation of the simplex method as a mechanical process with only a hint as to why the method might work. In this text I give a geometrical presentation that my students find easy to follow and that shows precisely how the simplex method works. The students see precisely the roles played by the various steps, ratios, and pivots. With these insights, the students are able to learn the additional topics involved in the dual, the extended simplex method, and post-optimal analysis.

Finally, the calculus is presented in a very intuitive and geometric manner, with emphasis on a clear understanding of the fundamentals. Applications are woven into the development of the material and are also used to motivate and to maintain interest. For example, average and instantaneous velocity are introduced at the beginning. This is followed immediately with obtaining the slope of the tangent line as the limit of the slope of secant lines. These ideas are presented in expanded form over several sections. Thus the student is immediately presented with both a physical and geometric interpretation of the notion of the derivative before the derivative is even defined.

Great care is taken to present the reader with the basic assumptions, in order to avoid possible confusion. For example, we state clearly that we only deal with functions that have a finite number of critical points on any finite interval. Thus, a continuous function with a positive derivative at one point will be increasing on some interval that contains the point, a fact which need not be true for functions with an infinite number of critical points on a finite interval.

## Important Features

**Style.**  The text is interesting and can be understood by the average student with a minimum of outside assistance. Material on a variety of topics is presented in an interesting, informal, and student-friendly manner without compromising the mathematical content and accuracy. Concepts are developed gradually, always introduced intuitively, and culminate in a definition or result. Where possible, general concepts are presented only after particular cases have been presented. Scattered throughout the text, and set-off in boxes, are historical and anecdotal comments. The historical comments are not only interesting in themselves, but also indicate that mathematics is a continually developing subject. The anecdotal comments relate the material to contemporary problems.

**Applications.**  The text includes many meaningful applications drawn from a variety of fields. For example, nearly every section opens by posing an interesting and relevant applied problem using familiar vocabulary, which is then solved later in the section after the appropriate mathematics have been developed. Applications are given for all the mathematics that are presented and are used to motivate the student.

**Worked Examples.**  About 700 worked examples, including about 200 Self-Help Exercises mentioned below, have been carefully selected to take the reader progressively from the simplest idea to the most complex. All the steps needed for the complete solutions are included.

**Self-Help Exercises.**  Immediately preceding each exercise set is a set of Self-Help Exercises. These 200 exercises have been very carefully selected to bridge the gap between the exposition in the chapter and the regular exercise set. By doing these exercises and checking the complete solutions provided, students will be able to test or check their comprehension of the material. This, in turn, will better prepare them to do the exercises in the regular exercise set.

**Exercises.**  The book contains over 4600 exercises. Each set begins with drill-type problems to build skills, and then gradually increases in difficulty. The exercise sets also include an extensive array of realistic applications from diverse disciplines. Graphing calculator exercises are included. (See below.)

**Graphing Calculator Exercises.**  These optional exercises appear at the end of appropriate exercise sets under the heading of Graphing Calculator Exercises, marked with the icon ▦. They are provided for those instructors who wish to supplement and complement the material by using graphing calculators.

**Graphing Calculator Appendix.** Located at the end of the text, this appendix gives step-by-step explanations on how to use the TI-81 graphing calculator. Topics covered include: basic graphing, the Gauss–Jordan method, matrix inverses, and statistical calculations. Sample programs are also provided.

## Students Aids

- **Boldface** is used when defining new terms
- **Boxes** are used to highlight definitions, theorems, results, and procedures.
- **Remarks** are used to draw attention to important points that might otherwise be overlooked.
- **Warnings** alert students to common mistakes.
- **Titles** for worked examples help to identify the subject.
- **Chapter summary outlines**, at the end of each chapter, conveniently summarize all the definitions, theorems, and procedures in one place.
- **Review exercises** are found at the end of each chapter.
- **Answers** to odd-numbered exercises and to all the review exercises are provided in an appendix.
- A student's **solution manual** that contains completely worked solutions to all odd-numbered exercises and to all chapter review exercises is available.
- The software package **MathPath** by George Bergeman, Northern Virginia Community College, is available free to users of this text. It supplies graphical and computational support for many of the important topics in each chapter. Available for the IBM or IBM-compatible.
- **Graph 2D/3D**, a software package by George Bergeman, Northern Virginia Community College, is also available free to users. This software graphs functions in one variable and graphs surfaces of functions in two variables. It also provides computational support for solving calculus problems and investigating concepts. Available for the IBM or IBM-compatible.

## Instructor Aids

- An **instructor's manual** with completely worked solutions to the even-numbered exercises is available free to adopters. A **student's solution manual** is free to adopters and contains the completely worked solutions to all odd-numbered and Chapter Review exercises.
- The **printed test bank** contains short-answer and multiple-choice test questions for the instructor to use in their own format. Approximately 100 test questions per chapter are given. The printed test bank is available free to adopters.
- A **computerized test bank** allows instructors to quickly create, edit, and print tests or different versions of tests from the set of test questions accompanying the text. It is free to adopters and is available in IBM or Mac versions.
- The software package MathPath and Graph 2D/3D (see Student Aids) are free to adopters.

## Content Overview

**Chapter 1.** The first three sections contain review material on sets, Cartesian coordinates, and lines. An introduction to the theory of the firm with some necessary economics background is provided to take into account the students' diverse backgrounds. The fourth (optional) section on least squares provides an example of the use of linear equations.

**Chapter 2.** Linear systems and the Gauss–Jordan method are covered in the first two sections of this chapter. The next three sections cover the basic material on matrices. Although many applications are included in the first five sections, the sixth section of this chapter is entirely devoted to input-output analysis, which is an application of linear systems and matrices used in economics.

**Chapter 3.** The first section presents linear inequalities with an emphasis on translating applied problems into an appropriate mathematical format. Basic material on geometric linear programming is included in the second section. The third section stresses translating more complicated applied problems into the linear programming format. If computers are available, all of these problems can be solved on the computer using the softwave available free to adopters.

**Chapter 4.** The first two sections present the simplex method for standard maximization problems and give considerable geometric insight into why the method works. The third section shows how to solve minimization problems by solving the dual problem, while the fourth section considers more general linear programming problems. The last section presents some post-optimal analysis.

**Chapter 5.** This chapter on finance does not depend on any of the other material and can be covered at any point in the course.

**Chapter 6.** This chapter contains basic material on sets and counting, which is used extensively in the next chapter on probability.

**Chapter 7.** The basic material on probability is covered in this chapter. The second section introduces probability in a natural way by emphasizing empirical probability and the long-term behavior of the relative frequency of an event. This sets the background for the mathematical definition of probability given in the third section, motivates the definition of expected value and variance of a random variable given in the next chapter, and provides the background needed to understand the law of large numbers also given in the next chapter. Conditional probability, Bayes' theorem, and Bernoulli trials round out this chapter.

**Chapter 8.** This chapter contains the basic material on statistics. The expected value and variance are motivated using the notions of empirical probability and the long-term behavior of the relative frequency given in Section 7.2. Thus the discussion in Section 7.2 allows a deeper coverage of statistics than otherwise possible with other texts. The normal distribution and its approximation by the binomial distribution are covered in the fourth and fifth sections. The law of large numbers is found in Section 8.5.

**Chapter 9.**   The basic material on Markov processes, covering both regular and absorbing Markov processes, is presented in this chapter.

**Chapter 10.**   Game theory and its important connection to linear programming is presented in this chapter. This material gives the basics on the extensive interrelationship between linear programming and the celebrated theory of games developed by von Neumann and important in economic theory.

**Chapter 11.**   A review of some needed precalculus subjects makes up this chapter, with a primary emphasis on functions. The first section contains the definition of function, function notation, and the vertical line test. Quadratic and other elementary functions are considered in the second section. The third section considers combinations and composition of functions.

**Chapter 12.**   The study of calculus begins here. In the first section, limits are introduced by using both geometric and numerical examples. The two fundamental problems of determining the instantaneous velocity and slope of the tangent line are introduced also in order to motivate the need to study limits and to set the stage for the later material on derivatives. The rules of limits are also given. Continuity, rates of change, and the derivative are found in the next three sections. The last three sections contain the rules of differentiation, including the chain rule.

**Chapter 13.**   The first section of this chapter uses the sign of the first derivative to determine where functions are increasing and decreasing, while the second section considers relative extrema and the first derivative test. The third section discusses concavity and the second derivative test. These first sections also include a discussion of federal tax revenue and the Laffer curve and also the management of renewable natural resources. The fourth section contains the material on limits at infinity and some extended material on curve sketching. The fifth section considers absolute extrema on both infinite and finite intervals and the last section contains many applications of optimization problems.

**Chapter 14.**   Several additional topics on the derivative are included in this chapter. The first section contains the differential and the linear approximation using differentials, the second section covers implicit differentiation, while the third section includes both related rate problems and a discussion of elasticity of demand.

**Chapter 15.**   The first section of this chapter develops exponential functions and the second section develops logarithmic functions. The third section gives the derivative formulas for the exponential and logarithmic functions, while the last section gives extensive applications, including population growth, radioactive decay, and the logistic equation with derived estimates of the limiting human population of the earth.

**Chapter 16.** This chapter begins the theory of integration. The antiderivative and substitution are given in the first two sections. The definite integral as a limit of Riemann sums is the subject of the third section. The fundamental theorem of calculus and area between curves is covered in the last two sections. Extensive applications are given throughout this chapter, including applications to the theory of the firm and economics, average value, density, consumer's and producer's surplus, and Lorentz's curves.

**Chapter 17.** This chapter rounds out additional topics in integration. Integration by parts and by using tables is covered in the first two sections. An extensive section on numerical integration, including the trapezoid and Simpson's rule and errors, is given in the third section. Improper integrals, with applications, is contained in the fourth section and continuous money flow is the subject of the last section.

**Chapter 18.** Continuous probability distributions using calculus is the subject of the first section, with the mean and variance covered in the second. The normal distribution is covered in the last section. This last section overlaps the material covered in Section 8.4, giving two options of when to cover the normal distribution.

**Chapter 19.** The first section presents an introduction to functions of several variables, including cost and revenue curves, Cobb-Douglas production functions, and level curves. The second section then introduces partial derivatives with applications that include competitive and complementary demand relations. The third section gives the second derivative test for functions of several variables and applied applications on optimization. The fourth section is on Lagrange multipliers and carefully avoids algebraic complications. The differential and approximations is given in the fifth section. The last section on double integrals covers double integrals over general domains, Riemann sums, and applications to average value and density.

## Acknowledgments

Many people at Saunders have played important roles in the production of this text. I especially want to thank Elizabeth Widdicombe, Publisher, for her generous support from the beginning to the end of this project. I also especially thank Robert Stern, my Acquisitions Editor, who signed me on and patiently guided the manuscript through many revisions, and my Executive Editor, Jay Ricci, who capably guided the project to its completion. I greatly appreciate the very important start that my first Developmental Editor, Donald Gacewicz, provided, and also greatly appreciate the considerable help given by the Developmental Editors Anne Wightman and Anita Fallon and Marketing Manager, Monica Wilson. I also thank Maureen Iannuzzi, the Project Editor, for doing such a great job and Merry Post for her excellent copy editing.

I wish to express my sincere appreciation to each of the following reviewers for their many helpful suggestions: Paul Wayne Britt, *LSU–Baton Rouge*; Darrell Clevidence, *Carl Sandburg College*; Michael Friedberg, *University of Houston*; Carl T. Brezovec, *University of Kentucky*; Carol DeVille, *LA Tech University*; John Haverhals, *Bradley University*; Henry Mark Smith, *University of New Orleans*; Daniel Marks, *Auburn University*; Norman Martin, *Northern Arizona University*; Wayne Powell, *Oklahoma State University*; Debra Landre, *San Joaquin Delta College*; Caroline Woods, *Marquette University*; and Harvey Yarborough, *Brazosport College*.

I also wish to thank the following for checking the accuracy of the text. Eric Bibelnieks, *Normandale Community College*; Michael Friedberg, *University of Houston*; Sudhir Goel, *Valdosta State University*; Murli M. Gupta, *George Washington University*; Debra Landre, *San Joaquin Delta College*; John Haverhals, *Bradley University*; William R. Livingston, *Missouri Southern State College*; Frank Miles, *California State University, Dominguez Hills*; Dave Morestad, *University of North Dakota*; Diane Tischer, *Metropolitan Community College*; and Steve Winters, *University of Wisconsin, Oshkosh*.

At the University of Connecticut, there are many people that I would like to thank for playing an important role in this project. First I thank James Hurley for his continuous support, encouragement, and many suggestions. Also I thank Gerald Leibowitz and Joy Mark for using the manuscript in the classroom and providing many suggestions. I also thank Joseph McKenna for his interesting suggestions, Walter Lowrie for our discussions on linear programming, and Vince Giambalvo for his help in a variety of ways. I would like to also thank the teaching assistants Yong-Hong Chen, Yue Chen, Bill Gilmartin, Haiming Jen, Kelly Penfield, and Chunying Wang for their considerable help in accuracy reviewing. On a personal level, I am grateful to my wife, to whom I dedicate this book, for her love, patience, and support.

Edmond Tomastik

*November 1993*

# Contents

## 1
## CARTESIAN COORDINATES AND LINES   1

1.1   Sets and the Cartesian Coordinate System   1

1.2   Lines   15

1.3   Linear Models   28

1.4   Method of Least Squares   40

## 2
## SYSTEMS OF LINEAR EQUATIONS AND MATRICES   5

2.1   Introduction to Systems of Linear Equations   52

2.2   General Systems of Linear Equations   65

2.3   Introduction to Matrices   81

2.4   Matrix Multiplication   91

2.5   Inverse of a Square Matrix   108

2.6   Leontief Input-Output Models   122

## 3
## LINEAR PROGRAMMING: THE GRAPHICAL METHOD   121

3.1   Graphing Linear Inequalities   122

3.2   Graphical Solution of Linear Programming Problems   133

3.3   More Linear Programming   151

## 4
## LINEAR PROGRAMMING: THE SIMPLEX METHOD   185

4.1   Slack Variables and the Simplex Tableau   186

4.2   Simplex Method: Standard Maximization Problems   201

4.3   Minimization Problems and the Dual   219

4.4   Simplex Method: Nonstandard and Minimization Problems   239

4.5   Post-Optimal Analysis   253

## 5
## MATHEMATICS OF FINANCE   269

5.1   Simple Interest and Discount   269

5.2   Compound Interest   277

5.3   Annuities and Sinking Funds   288

5.4   Present Value of Annuities and Amortization   297

# 6
# SETS AND COUNTING   314

6.1   Sets   315

6.2   The Number of Elements In a Set   327

6.3   Multiplication Principle, Permutations   334

6.4   Combinations   343

6.5   Binomial Theorem, Partitions   351


# 7
# PROBABILITY   365

7.1   The Basis of Probability   366

7.2   Empirical and Intuitive Probability   373

7.3   Mathematical Probability   386

7.4   Spaces With Equally Likely Outcomes   398

7.5   Conditional Probability   406

7.6   Bayes' Theorem   421

7.7   Bernoulli Trials   430

Solution to Game Show Problem   439


# 8
# STATISTICS AND PROBABILITY
# DISTRIBUTIONS   443

8.1   Probability Distributions   444

8.2   Measures of Central Tendency   456

8.3   Measures of Dispersion   472

8.4   The Normal Distribution   489

8.5   Normal Approximation to the Binomial
Distribution   499


# 9
# MARKOV PROCESSES   515

9.1   Markov Processes   515

9.2   Regular Markov Processes   528

9.3   Absorbing Markov Processes   536


# 10
# GAME THEORY   551

10.1   Strictly Determined Games   552

10.2   Mixed Strategy Games   560

10.3   Linear Programming and Game Theory   572


# 11
# FUNCTIONS AND GRAPHS   589

11.1   Functions   589

11.2   Some Special Functions   605

11.3   Combinations of Functions   619


# 12
# THE DERIVATIVE   631

12.0   Introduction to Calculus   632

12.1   Limits   634

12.2   Continuity   656

12.3   Rates of Change   665

12.4   The Derivative   678

12.5   Derivatives of Constants, Powers,
and Sums   691

12.6   Derivatives of Products and Quotients   705

12.7   The Chain Rule   714


# 13
# APPLICATIONS OF THE
# DERIVATIVE   729

13.1   Increasing and Decreasing   729

13.2   Relative Extrema   744

13.3   The Second Derivative   759

13.4   Limits at Infinity; More Curve Sketching   775

13.5   Absolute Extrema   786

13.6   Optimization Problems   797

# 14
# MORE ON THE DERIVATIVE   815

14.1   The Linear Approximation and
       Differentials   815
14.2   Implicit Differentiation   824
14.3   Related Rates and Elasticity of Demand   830

# 15
# EXPONENTIAL AND LOGARITHMIC FUNCTIONS   841

15.1   Exponential Functions   841
15.2   Natural Logarithms   851
15.3   Derivatives of Exponential and Logarithmic
       Functions   860
15.4   Growth and Decay   872

# 16
# INTEGRATION   893

16.1   Antiderivatives   894
16.2   Substitution   904
16.3   The Definite Integral   913
16.4   The Fundamental Theorem of Calculus   931
16.5   Area Between Two Curves   948

# 17
# ADDITIONAL TOPICS IN INTEGRATION   971

17.1   Integration by Parts   971
17.2   Integration Using Tables   978
17.3   Numerical Integration   983
17.4   Improper Integrals   995
17.5   Continuous Money Flow   1003

# 18
# PROBABILITY AND CALCULUS   1017

18.1   Continuous Probability Density
       Functions   1017

18.2   Expected Value and Variance   1027
18.3   The Normal Distribution   1038

# 19
# FUNCTIONS OF SEVERAL VARIABLES   1053

19.1   Functions of Several Variables   1054
19.2   Partial Derivatives   1068
19.3   Extrema of Functions of Two Variables   1081
19.4   Lagrange Multipliers   1096
19.5   Total Differentials and Approximations   1105
19.6   Double Integrals   1113

# A
# REVIEW   A-1

A.1   Exponents and Roots   A-1
A.2   Polynomials and Rational Expressions   A-7
A.3   Equations   A-14
A.4   Inequalities   A-20

# B
# TABLES   B-1

B.1   Area Under a Normal Curve   B-1
B.2   Financial Tables   B-3
B.3   Basic Geometric Formulas   B-35
B.4   Table of Integrals   B-36

# C
# THE GRAPHING CALCULATOR   C-1

C.1   The Graphing Calculator   C-1
C.2   Programs   C-5

# ANSWERS   AN-1

# INDEX   I-1

# Applications Index

## Business

Adhesive tape manufacturing, 902
Advertising, 359, 396, 419, 508, 570, 582
  campaign, 967
  promotion, 465
Air pollutant concentration, 486
Airline scheduling, 106
Airline tours, 181–182
Alcohol, 485
Allocation of funds, 167
Allocation of machine time, 59
Amortization, 302
Amortization schedule, 303
Animal nutrition, 246
Animal ranching, 219
Annuity, 289
  amortization, 302
  amortization schedule, 303
  equity in sinking fund, 293
  future amount, 270
  future value, 276
  maturity value, 273
  ordinary, 289
  outstanding principal, 304–305
  payment toward principal, 305
  present value, 271, 284, 848
  sinking fund, 292, 312
  sinking fund payment, 292
  term of, 289
Apple orchard, 807
Appreciation, 741, 902
Auto-catalytic reaction, 757
Automobile, accidents, 487
  life, 1050
  selection, 534
Average amount of money, 946

Average cost, 653, 708, 812–813
  function, 783
  of long-distance call, 726–727
Average price, 946
Average profit, 947
Average rate of change, 666
Average revenue, 947
  function, 757
Awarding of contracts, 405, 420

Ball bearing quality control, 1046
Bath towel manufacturing, 977
Battery life, 1026
Battery reliability, 438
Bicycle sales, 943
Bidding on contracts, 393, 395
Blending, 170
Brick manufacturing, 982
Bus company revenue, 806, 813

Camera sales, 912
Capital expenditure, 308
  analysis, 301
Car loan, 302–303, 308
Car rental, 549
Car sales, 552
Cardboard box manufacturing, 911
Casino gambling, 381
Cereal packaging, 479
Cereal profits, 1055–1056
Chemical spill, 945–946
Chemical toxin, 654
  cost, 783
Chinese farming, 78
Chinese way of calculating with arrays, 78

Cigarettes, 485
Clock manufacturing, 911
Closed Leontief input-output model, 126, 129
Closet construction, 1093
Cobb-Douglass production function, 1057,
  1100, 1104, 1109, 1112, 1131,
  1138–1139
Coin appreciation, 287
Committee selection, 342, 344, 346–347,
  349, 359
Commuting time, 1025
Company contracts, 438
Comparing bank rates, 287–288
Comparing investments, 301, 468–469
Comparing present values, 307–308
Competitive demand relation, 1073–1075
Competitive pricing, 1094, 1139
Complementary demand relation,
  1073–1075
Component selection, 359
Compound interest, 278, 722
  effective annual yield, 283
  effective rate, 274
Compounding, continuous, 844, 859, 1079
Computer chips, 433
Computer failure, 415
Consumer advertising, 362
Consumer survey, 362, 382
Consumers' surplus, 955, 957, 961, 976,
  982
Container construction, 799–801, 806
Continuous compounding, 844, 859, 1079
Continuous money flow, 1003
  continuous reinvestment, 1007–1008
  on infinite intervals, 1008–1009

Continuous money flow (*continued*)
  present value, 1004–1005
  present value on infinite intervals,
    1008–1009
  rate, 1003
  total income, 1003
Contractor costs, 74
Copper mine, 912
Copper production, 945, 962
Corporate farming, 178
Corporate success determination, 372
Cost(s), 13, 29, 43–44, 1066, 1094, 1111
  airline, 175, 234–235, 251
  fencing, 615
  fixed, 29
  of long-distance call, 664
  oil refinery, 175, 234, 251
  postage, 628
  variable, 29
Cost equation, 676, 703, 822, 835–836
Credit card account, 270
Crude oil blending, 170
Customer selection, 350

Defective component, 438
Defective parts manufacturing, 419
Delivery fees, 603, 662
Demand, 36, 44, 48
  and supply, 79
  competitive, 1073–1075
  complementary, 1073–1075
  for cameras and film, 1066
  for electricity, 1037
Demand curve, 31
Demand equation, 28, 31, 703, 713, 721,
    756, 822, 828, 835, 863, 870, 960
Demand function, 741, 828
Density, 1128
  of pollutants, 947
Depreciation, 902–903, 912, 946, 967
Detergent box weight distribution, 1046
Diminishing returns, 772–773, 887
Discount, 272
Discounted note, 277
Disk manufacturing, 902
Disposable diaper market, 713
Distribution of production, 135
Doubling time, 276, 282, 312, 857, 859,
    884

Economics, 425, 428
Education fund, 296–297
Elastic demand, 833
Elasticity of demand, 832
Electric can opener manufacturing, 976
Electric utility average cost, 773
Electricity, current intensity, 727
  demand, 1026–1027
  tax, 663–664
Employee, absences, 484
  assignments, 350
  attendance, 468
  bonus, 496

productivity, 713, 912
  selection, 350
Entertainment, 333
Envelope manufacturing, 902
Equilibrium price, 32, 955
Equilibrium quantity, 32, 955
Equipment fund, 297
Equity in a sinking fund, 293
Excess resources, 166, 211
Executive option, 354
Executive scheduling, 341

Facility location, 49, 1094
Farming, 176–177, 181–182, 266–267
Federal employees, 469
Federal tax revenue, 736, 742
Fence construction, 780, 797, 805–806,
    808, 813, 1098, 1104
Fixed costs, 29
Flight control, 415
Flight safety, 415
Food survey, 332
Furniture manufacturing, 119, 147, 162,
    178, 182, 215–216, 260
Future amount, 270
Future value, 276
  of annuity, 289
  of ordinary annuity, 291

Gasoline blending, 170

Hair brush manufacturing, 912
Hand purse manufacturing, 911
Heating oil consumption, 1037
Homeowners survey, 438
Hotel profits, 807
Hotel revenue, 807
House down payment, 297
House selection, 341
Household survey, 328

Income maximization, corporate farming,
    178
Income survey, 372
Individual retirement account, 296
Industrial pollutants, 941, 943
Inflation, 962
Input-output analysis, closed Leontief
    model, 126, 129
  input-output matrix, 123
  open Leontief model, 123
Insurance, 467
Interest, 270, 278, 722
  compound, 278, 722
  doubling time, 276, 282, 312, 857, 859,
    884
  per period, 278
  present value, 271, 284
  simple, 270
  simple discount, 272, 277
  tripling time, 277, 312
Interest rate(s), 270

nominal, 283
  tables of, 1167–1198
Inventory, 36
Inventory control, 808, 813
Inventory cost, 802
Inventory cost model, 802
Investment(s), 25, 62–63
Investment advisor, 468
Investment mix, 115, 119, 173–174, 250,
    332
Investment return, 465, 468
Investment risk, 174–175, 235, 251, 477,
    483–484

Kennedy administration tax policy, 738
Keogh account, 312
Knife manufacturing revenue, 901

Labor allocation, 36
Labor relations, 363
Laboratory management, 363
Laffer, Arthur, 736
Laffer curve, 736
Leaf area, 742
License plate(s), number of, 339
  selection, 340
Life insurance, 511
Life insurance policy, 461
Life of a part, 1026
Light bulb life span, 1023–1024, 1034
Linear cost model, 29
Lipstick manufacturing, 912
Lloyd's of London, 374
London life table, 375, 382
Lot acceptance sampling, 504
Lottery, 381, 466, 511
Lottery prize, 312
Lotto game, 343, 349, 403
Louisiana purchase, 287

Machine allocation, 36
Make or buy decision, 37
Malfunctioning machine, 438
Management personnel selection, 363
Manufacturing, 147, 162–163, 254, 396,
    428–429, 438, 512
  adhesive tape, 902
  bath towel, 977
  brick, 982
  cardboard box, 911
  clock, 911
  defective parts, 419
  disk, 902
  electric can opener, 976
  envelope, 902
  furniture, 119, 147, 162, 178, 182,
    215–216, 260
  hair brush, 912
  hand purse, 911
  knife, 901
  lipstick, 912
  metal connectors, 145, 148, 157, 215,
    240

Manufacturing (*continued*)
  paper plate, 901
  park benches, 142, 182
  pen, 754
  piston rings, 498
  purses, 267, 911
  salt shaker, 901
  shoulder bag, 911
  spark plug, 902
  table cloth, 901
  tape, 1047
  toaster, 982
  wire, 1050
  wooden boat, 138
Marginal, analysis, 696
  average cost, 708
  cost, 698, 967
  productivity of capital, 1072
  productivity of labor, 1072
  profit, 698
  propensity to consume, 903
  propensity to save, 903, 994
  revenue, 698, 721, 967
Market competition, 570
Market share, 654, 783
Marketing, 396, 526–527, 535, 571, 582
Maturity value, 273
Meal selection, 340, 349
Measurement(s), 1139
Measurement conversions, 623, 625
Menu selection, 342
Metal connector manufacturing, 145, 148, 157, 215, 240
Mining, 182, 1014
  copper, 945
  platinum, 1001
Mixed strategy, 560
Mixture, 62, 262–263
  acid, 135
  alcohol, 116
  candy, 176
  fruit drinks, 180
  fruit juices, 262–263
  nuts, 147, 162, 178–179, 216
  tea, 119
Mortgage payments, 312
Mutual fund, 359, 405, 444, 452
  selection of, 347

Natural gas extraction, 937
Natural resource depletion, 945
Networks, 106
Newsletter copies, 1047
Newspaper readers, 331
Nominal rate, 283
Nuclear accident, 877
Nuclear reactors, 883

Octane rating of gasoline, 172
Office scheduling, 77
Oil drilling, 434
Oil pollution, 469–470
Oil production, 909, 944

Oil refinery, 170, 234, 251
  blending, 176
Oil slick, 12
Oil well income, 1004, 1010–1011
Oil well price, 1006
Open Leontief input-output model, 123
Ordinary annuity, 289
Outstanding principal, 304–305

Package construction, 799–801, 806
Packaging, 497, 603, 628, 1139
Paper plate manufacturing revenue, 901
Park bench manufacturing, 142, 182
Payment toward principal, 305
Pen manufacturer, 754
Pension fund investment selection, 77
Personnel, 435
Personnel retention, 549
Pet food manufacturer survey, 332
Petroleum consumption, 675
Pipe construction, 807
Piston ring manufacturing, 498
Plant fertilizer, 144, 156
Plant productivity, 783
Plant scheduling, 179
Portfolio management, 167
Postage cost, 628
Postal rates, 653
Present value, 271, 284, 848
  of ordinary annuity, 300
Price wars, 570, 581
Principal, 270
Prize selection, 350
Proceeds, 273
Process selection and capacity, 37
Producers' surplus, 957, 961, 982
Product, defects, 419
  life, 372, 498, 1037
  mix, 581
  reliability, 420
  selection, 340, 342, 349
  testing, 372, 383, 405
Production, 15, 23
  costs, 1139
  function, 1056
  line, 544–545
  matrix, 86, 89–90
  of perishable items, 77–78
  planning, 98, 101
  scheduling, 63, 77–78, 147, 163, 235, 237, 251–252
    mining, 175, 235–236, 251–252
Profit(s), 13, 29, 49, 237, 756, 1001, 1037, 1066, 1094–1095
  average, 947
  equation, 703, 822
  function, 690, 741, 870
  orchard, 615
  sleds, 216, 237
Promotion, 570, 581
Purchasing, flowers, 236
Purchasing decision, 73

Purchasing mix, 176
Purse manufacturing, 267, 911

Quality, 487
Quality control, 396, 405, 419, 428–429, 455, 468, 495, 498, 507, 512, 1046

Rate of sales, 974, 976
Reagan administration tax policy, 736
Real estate appreciation, 287
Refrigerator sales, 982
Retirement and taxes, 296
Retirement fund, 296
Retirement packages, comparing of, 308
Revenue, 13, 927, 962, 992, 1037, 1046, 1111
  average, 947
  bus company, 613, 618
  candy, 261–262
  equation, 676, 703, 822
  function, 625, 690, 756–757, 772, 870
  orchard, 615
  shoes, 603
Risk, in investments, 174–175, 235, 251, 477, 483–484

Safety, 507
Salary equation, 664
Salary incentives, 498
Sales, 43, 63, 396, 467, 484, 487, 497, 928, 945, 967, 992, 1001, 1046
  car dealership, 455
  commissions, 36, 603
  condominiums, 135
  equation, 676
  function, 690
  machine tools, 420
  of new product, 962
  records, 383
  restaurant, 135
  selection, 333
  sweaters, 77
  sweepers, 135
Salt shaker manufacturing, 901
Saving, for machinery, 288
Savings, 312
Savings certificate interest, 659–660
Scheduling, 26, 62–63
Scrap value, 33
Selections for promotion, 371
Setting prices, 570, 581
Shelf-life, 1037, 1050
Shipping, 147, 162
Shoulder bag manufacturing, 911
Simple discount, 272, 277
Simple interest, 270
Sinking fund, 292, 312
Sinking fund payment, 292
Site selection, 349
Soda production, 1047
Spare parts, planning, 105
  production, 999

Spark plug manufacturing, 902
Spread of technology, 764, 772, 784, 883, 886
Steel beam quality control, 1046
Steel production, 993
Stock(s), 483
Stock indices, 420
Stock selection, 349, 358, 405
Straight-line depreciation, 33, 36
Stumpage value, 881–882
Sulfur content of oil, 171
Suppliers, 419
Supply and demand, 891
  equilibrium, 955
Supply curve, 32
Supply equation, 32
Survey, 327

Table cloth manufacturing, 901
Tape manufacturing, 1047

Taxes, 605
Technological innovation, spread of, 764, 772, 784, 883, 886
Telephone company charges, 599
Term of an annuity, 289
Testing a product, 405
Theater revenue, 799
Threshold strength, 727, 784
Toaster manufacturing, 982
Toothpaste market, 783
Toxic emissions costs, 713
Toy sales, 980
Traffic flow, 78–79, 106
Transportation, 2, 11–12, 48, 174, 236, 244, 250
  air cargo, 147, 162
  charges, 12
  costs, 26, 168–169
  selection, 333
Tripling time, 277, 312, 859

Useful life, 487

Variable costs, 29
VCR life, 1026
VCR profits, 738

Waiting time, 1026, 1037
Warrantee period, 496
Water allocation, 178
Water consumption, 1026, 1037
Weight of top soil, 487
Wire manufacturing, 1050
Wooden boat manufacturing, 138
Work scheduling, 177–178
Work stoppages, 675

Yogurt selection, 334

# Life Sciences

Abortions, 484–485
AIDS, 511
Air pollutant concentration, 486
Alcohol, 485
Amount of pollutants, 1128
Animal nutrition, 246
Animal ranching, 219
Antarctic blue whale, 792
  carrying capacity, 792
  measuring, 37
  minimum viable population level, 792
Average air pollution, 947
Average amount of drug, 947
Average blood pressure, 947
Average concentration of pollutants, 1126
Average population, 1131

Bacteria, 703, 823, 838, 928, 993, 1112
Biology, 13, 527, 535, 615, 703–704, 714, 727, 742, 757, 773, 783–784, 983, 1001, 1066–1067, 1080
Biomass, 751
Birth rate, 487
Bladder cancer, 507
Blood flow, 742, 828
Blood pressure, average, 947
Blood type, 429
Bone measurement, 823
Bronchi, 757, 829

Cancer, 383, 419, 499, 625, 676, 816–817
  bladder, 507
  carcinogen, 383
  cervical, 499
  lung, 407
  thyroid, 507
Carbon-14, 878

Carbon dating, 878
Carbon monoxide, 486
Carcinogen, 383
Cardiac output, 1066
Carrying capacity, 751
  Antarctic blue whale, 792
  mice, 776
Cervical cancer, 499
Chemical spill, 945–946
Chemical toxin, 654
  cost, 783
Cigarettes, 485
Cocaine, 485, 507
Concentration of drug, 714, 727, 784
Concentration of pollutants, 1128, 1131

Death rate, 487
Deer, 714
  antlers, 26
  population, 714, 887
Density, 1128
  of pollutants, 947
Detection range of a school of fish, 722
Diet, 163, 173, 229
  health food mix, 183
  meal preparation, 177
  vitamin selection, 174
  vitamin supplements, 147, 162, 235
Disease(s), 1001
  cancer, 383, 419, 499, 625, 676, 816–817
    bladder, 507
    carcinogen, 383
    cervical, 499
    lung, 407
    thyroid, 507
  gonorrhea, 484

heart, 414, 420, 507
  meningitis, 486
  plague, 485
  polio, 485
  spread of, 886
  syphilis, 484
  tuberculosis, 424
  venereal, 484
  viral hepatitis, 486
Distance between two ships, 625
Distribution of population by height, 454–455
Douglas fir, 881–882
Drug(s), average amount, 947
  cocaine, 485, 507
  concentration, 714, 727, 784
  dosage, 654
  effectiveness, 508
  marijuana, 485, 507
  testing, 439
  use, 485

Eastern Pacific yellowfin tuna, 881

Facility location, 49, 1094
False negative, 424
False positive, 424
Fick method, 1066
Fishery, 163–164, 751, 791, 795, 886, 1079, 1139
  Antarctic blue whale, 792
  bass, 147, 163, 179
  biomass, 751
  carrying capacity, 752
  coastal shark, 754
  harvesting, 751
  maximum sustainable yield, 751

Fishery (*continued*)
  minimum viable biomass level, 791
  overfishing, 791
  Pacific halibut, 753
  Pacific salmon, 753
  renewable natural resources, 751
  shark, 754
  trout, 216
Flight control, 415
Forest fire, 623, 701
Forestry, 722, 881–882

Gompertz growth curve, 871
Gonorrhea, 484

Health, 485
Health food mix, 183
Heart disease, 414, 420, 507
Heart operations, 468
Highway speed measurement, 382

Insect temperature, 44
International Whaling Commission, 792
Isobar, 1061

Leaf area, 742
Limited growth, 878
Lung(s), 757, 829
Lung cancer, 407

Marijuana, 485, 507
Maximum sustainable yield, 751
Medical diagnosis, 429, 885
Medical operation time, 1047
Medicine, 37, 396, 419, 424–425, 428, 435,
      468, 625, 654, 676, 703–704, 714,
      727, 757, 784, 828–829, 836, 885,
      968, 1066–1067, 1112
Meningitis, 486
Mice, carrying capacity, 776
  population, 764, 1001
Minimum viable biomass level, 791
Monomolecular growth, 871

National Marine Fisheries Service, 754
Nitrogen dioxide, 486

Nuclear accident, 877
Nuclear reactors, 883
Nutrition, 13, 49, 62–63, 78
  meal selection, 234

Oil slick, 12, 831
Overfishing, 791
Ozone, 486

Photosynthesis, 742, 757, 783, 829
Pine disease, 628
Plagues, 485, 765
Plant fertilizer, 144, 156
Plant nutrition, 175, 235, 237, 251
Plant roots, 871, 1080
Plutonium-239, 883
Poiseuille's law, 704, 742, 828
Polio, 485
Pollutants, 982, 1001
Pollution, 512
Population theory, 902
  average, 968, 1131
  bacteria, 993
  biomass, 751
  birth rate, 487
  carrying capacity, 752
  death rate, 487
  deer, 887
  density, 942, 947
  Eastern Pacific yellowfin tuna, 881
  exploding, 764
  fishery, 751, 791
  growth, 874
  harvesting, 751
  lake trout, 891
  limited resources, 763
  logistics equation, 879
  maximum sustainable yield, 751
  mice, 764, 776, 1001
  minimum viable biomass level, 791
  predator, 615, 773
  prey, 615, 773
  renewable natural resources, 751, 791
  S-curve, 763
Predators, 615, 773
Prey, 615, 773
Public health, 885

Radioactive decay, 875, 885, 902
Radioactive tracers, 885
Radioactive waste, 1001
Recovery time, 1027, 1037
Renewable natural resources, 751, 791
Room temperature, 497

Schools of fish, 722, 1079
Sharks, 754
Site selection, 349
Smoking, 414, 420
Spherical tumor, 703
Spread of disease, 886
Spread of infestation, 628
Spread of technology, 764, 772, 784, 883,
      886
Squirrel weight distribution, 1047
Sulfur dioxide, 486
Survey, 327
Syphilis, 484

Threshold strength, 727, 784
Thyroid cancer, 507
Thyroid gland, 885
Tine test, 424
Tornadoes, 471
Trachea, 757, 829
Tree leaves, 1080
Tripling time, 277, 312, 859
Tuberculosis, 424
Tumor, 835
  cancerous, 625, 676
Twigs, 26

Venereal disease, 484
Viral hepatitis, 486
Vitamin supplements, 147, 162

Water consumption, 1026, 1037
Water vapor, 829
Weather records, 382
Weight of top soil, 487
Whales, 37
Wiess' law, 727, 742, 784
Windpipe, 757, 829

## Physical Sciences

Acceleration, 765, 773, 902
Asteroid, 40
Auto-catalytic reaction, 757
Average velocity, 640–641, 666

Ceres, 40
Chemistry, 714, 784

Earth satellite, 420
Einstein's relativity theory, 915

Free water vapor diffusion coefficient, 704

Light intensity, 757

Measurement(s), 1139
Measurement conversion, 623, 625
Median, 463, 1034

Navigation, 12, 603, 831–832, 836
Networks, 106
Newton's law of cooling, 878
Newton's law of heating, 886

Optical lens, 420

Physical chemistry, 871
Physics, 836, 838, 886

Salt water concentration, 714, 784
Space craft, 438
Sulfur content of oil, 171

Temperature, 37, 485

Velocity, 603, 615–616, 666, 703, 727,
      900, 902, 928, 993

# Social Sciences

Accidental deaths, 507
Anthropology, 885, 891
Arrests, 507
Automobile accidents, 487
Award selection, 341

Birthday problem, 402
Birthdays, 404–405

Carbon dating, 878
Coefficient of inequality, 958–959
College instruction, 439
Commuting time, 1025
Course registration, 331–333
Course selection, 349
Crime, 512

Death rate, 487
Divorce rate, 487–488, 493

Education, 333
Education fund, 296–297
Elections, 37
Entertainment, 333
Estes learning model, 544

Family organization, 350
Family planning, 403, 418
Flight safety, 415

Grade distribution, 383–384, 452
Grade point average, 465
Guessing on a test, 453

Health care unit location, 470
Highway speed measurement, 382
Hours studying, 452

Immigration, 507
Intelligence test (IQ) scores, 498, 1047
Itinerary selection, 342

Learning, 823, 1027, 1038
Learning time, 722, 1047
Lorentz curve, 958

Marriage rate, 487–488
Median age, 44, 730–731
Medical team rescue, 804
Memorization, 912
Menu selection, 342

Parental determination, 429
Politics, 526, 535
Population explosion, 883
Population of farmers, 730
Population of United States, 730
Population theory, 902
    average, 968, 1131
    birth rate, 487
    carrying capacity, 752
    death rate, 487
    deer, 887
    density, 942, 947
    exploding, 764
    growth, 874
    human, 874
    India, 890

Indonesia, 888
Japan, 884
limiting human population, 882
logistics equation, 879
Nigeria, 884, 887
of Northeast, 881
S-curve, 763
United States, 885–886
world, 884–885
Prize selection, 350
Psychology, 722
    and sales, 419
    of success, 415
Public expenditure on education, 471

Safety, 507
Scholarships, 359
Scholastic Aptitude Test scores, 486
School rezoning, 176, 236
Social interaction, 107
Social mobility, 549
Spread of news, 887
Spread of rumors, 615, 773, 886
Student ratings, 453
Successful marriage odds, 395
Supreme Court cases, 469
Supreme Court decisions, 350

Test grades, 498

Vacation selection, 527
Voter turnout, 469

Work stoppages, 675

The operation of a factory involves both fixed and variable costs.

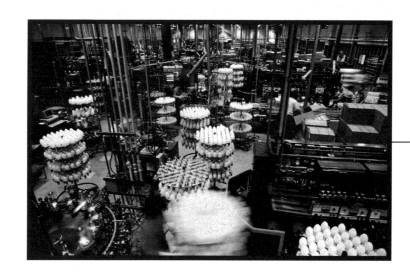

# Cartesian Coordinates and Lines

1.1  Sets and the Cartesian Coordinate System

1.2  Lines

1.3  Linear Models

1.4  Method of Least Squares

In this chapter we present some very basic material that will be used throughout this text. We first give a brief introduction to sets. Introduced second is the Cartesian coordinate system, which provides a geometric way of presenting data and relationships between variables. Lines are then introduced, followed by applications of linear models.

## 1.1  SETS AND THE CARTESIAN COORDINATE SYSTEM

▶ *Sets*

▶ *Cartesian Coordinate System*

▶ *Distance Between Two Points*

▶ *Creating Equations*

▶ *Graphs*

**René Descartes,**
**1596–1650**

It is sometimes said that Descartes's work *La Geometrie* marks the turning point between medieval and modern mathematics. He demonstrated the interplay between algebra and geometry, tying together these two branches of mathematics. Due to poor health as a child, he was always permitted to remain in bed as long as he wished. He maintained this habit throughout his life and did his most productive thinking while lying in bed in the morning. Descartes was a great philosopher as well as a mathematician and felt that mathematics should be a model for other branches of study. The following is a quote from his famous *Discours sur la Méthode*: "The long chain of simple and easy reasonings by means of which geometers are accustomed to reach conclusions of their most difficult demonstrations led me to imagine that all things, to the knowledge of which man is competent, are mutually connected in the same way, and that there is nothing so far removed from us as to be beyond our reach, or so hidden that we cannot discover it, provided only we abstain from accepting the false for the true, and always preserve in our thoughts the order necessary for the deduction of one truth from another."

**APPLICATION**

**Getting to the Church**
**on Time**

Figure 1.1

A prospective groom is located at point O in Figure 1.1 and must drive his car to his wedding taking place in 25 minutes at the church located at point C. The church at C is 13 miles east and 4 miles north of O. The only highway goes through a town located at B which is 10 miles due east of O. From B the highway goes straight to C. From O to B is a nice open stretch of highway, and he can average 50 miles per hour. The stretch from B to C has traffic lights and some congestion, and he can average only 25 miles per hour on this leg of the trip. Can he get to the church on time? See Example 5 for the answer.

## Sets

The notion of a *set* forms the foundation for a great deal of mathematics. We need to introduce a few basic terms about sets and to consider some notation that is used to represent sets.

---

**Sets**

A **set** is a collection of objects, specified in a manner that enables one to determine if a given object is or is not in the collection.

---

**EXAMPLE 1**    **Determining if a Collection is a Set**

Which of the following collections are sets?

- The numbers from 1 to 10

- The past presidents of the United States
- The solutions of the equation $x^2 = 4$
- Good players who have played for the New York Yankees
- Interesting books published in 1992

**Solution**  The first 3 collections are sets. The last two collections are not sets since we are not given the criteria to determine whether a player is good or a book is interesting.  ∎

Each object in a set is called a **member** or **element** of the set. We often describe a set by listing all the objects between braces. Thus the set consisting of the first three months of the year can be written in **roster notation** as {January, February, March}.

We also use the **set-builder notation**.

---

**Set-Builder Notation**

The "set of all elements $x$ such that $x$ has property $P$" is written in set-builder notation as

$$\{x \mid x \text{ has property } P\}.$$

---

In this notation the vertical bar is read "such that."

**E X A M P L E  2**  **Using Set-Builder Notation**

Write the set {1, 2, 3, 4} in set-builder notation.

**Solution**  One solution is

$$\{1, 2, 3, 4\} = \{x \mid x \text{ is an integer and } 0 < x < 5\}. \quad ∎$$

We also use the following notation.

---

| | | |
|---|---|---|
| $x \in A$ | means | "$x$ is an element in $A$." |
| $x \notin A$ | means | "$x$ is not an element in $A$." |

---

Thus, for example,

$$3 \in \{x \mid x \text{ is an integer and } 0 < x < 5\}, \text{ and}$$

$$6 \notin \{x \mid x \text{ is an integer and } 0 < x < 5\}.$$

---

**Equality of Sets**

Two sets $A$ and $B$ are **equal**, written $A = B$ if, and only if, they consist of precisely the same elements.

---

**E X A M P L E   3**     **Showing Two Sets Are Equal**

     a.    Determine if {1, 2, 3, 4} equals {4, 3, 2, 1}.
     b.    Determine if {1, 1, 1, 1, 2, 3, 4} equals {1, 2, 3, 4}.
     c.    Determine if {1, 2, 4} equals {1, 2, 3, 4}.

**Solutions**     a.    Since rearranging the order of the elements does not change the collection, {1, 2, 3, 4} = {4, 3, 2, 1}.
     b.    Since listing the same element several times does not add any elements to the set, {1, 1, 1, 1, 2, 3, 4} = {1, 2, 3, 4}.
     c.    Since $3 \notin \{1, 2, 4\}$, but $3 \in \{1, 2, 3, 4\}$, $\{1, 2, 4\} \neq \{1, 2, 3, 4\}$.   ∎

## Cartesian Coordinate System

It is very useful to have a geometric representation of the real numbers called the **number line**. The number line, shown in Figure 1.2, is a (straight) line with an arbitrary point selected to represent the number zero. This point is called the **origin**. A unit of length is selected, and, if the line is horizontal, the number one is placed an arbitrary distance to the right of the origin. The distance from the origin to the number one then represents the unit length. Each positive real number $x$ then lies $x$ units to the right of the origin, and each negative real number $x$ lies $-x$ units to the left of the origin. In this manner each real number corresponds to exactly one point on the number line, and each point on the number line corresponds to exactly one real number.

     We will use the following convenient notation.

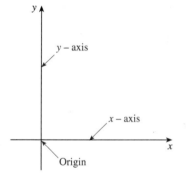

**Figure 1.2**

**Figure 1.3**

| Interval Notation | Definition | Geometric Representation |
|---|---|---|
| $(a, b)$ | $\{x \mid a < x < b\}$ | |
| $[a, b]$ | $\{x \mid a \leq x \leq b\}$ | |
| $(a, b]$ | $\{x \mid a < x \leq b\}$ | |
| $[a, b)$ | $\{x \mid a \leq x < b\}$ | |
| $[a, +\infty)$ | $\{x \mid a \leq x\}$ | |
| $(a, +\infty)$ | $\{x \mid a < x\}$ | |
| $(-\infty, b)$ | $\{x \mid x < b\}$ | |
| $(-\infty, b]$ | $\{x \mid x \leq b\}$ | |

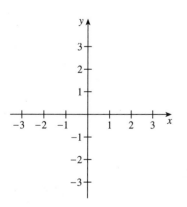

**Figure 1.4**

The Cartesian coordinate system (named after René Descartes) permits us to express data or relationships between two variables as geometric pictures. Representing numbers and equations as geometric pictures using the Cartesian co-

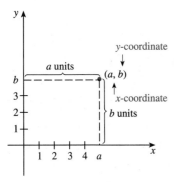

**Figure 1.5**

ordinate system yields deeper insight into the equations and data since we can see them graphically.

To describe a **Cartesian coordinate system** we begin with a horizontal line called the **x-axis** and a vertical line called the **y-axis**, both drawn in the plane. The point of intersection of these two lines is called the **origin**. The plane is then called the **xy-plane** or the **coordinate plane**. See Figure 1.3.

For each axis select a unit of length. The units of length for each axis need not be the same. Starting from the origin as zero, mark off the scales on each axis. For the x-axis, positive numbers are marked off to the right of the origin and negative numbers to the left. For the y-axis, positive numbers are marked off above the origin and negative numbers below the origin. See Figure 1.4.

Each point $P$ in the xy-plane is assigned a pair of numbers $(a, b)$ as shown in Figure 1.5. The first number $a$ is the horizontal distance from the point $P$ to the y-axis and is called the **x-coordinate**, and the second number $b$ is the vertical distance to the x-axis and is called the **y-coordinate**.

Conversely, every pair of numbers $(a, b)$ determines a point $P$ in the xy-plane with x-coordinate equal to $a$ and y-coordinate equal to $b$. Figure 1.6 indicates some examples.

We use the standard convention that $P(x, y)$ is the point $P$ in the plane with Cartesian coordinates $(x, y)$.

**W A R N I N G .** The symbol $(a, b)$ is used to denote both a point in the xy-plane and also an interval on the real line. The context will always indicate in which way this symbol is being used.

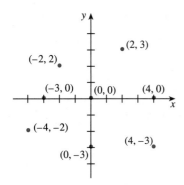

**Figure 1.6**

The x-axis and the y-axis divide the plane into four **quadrants**. The **first quadrant** is all points $(x, y)$ with both $x > 0$ and $y > 0$. The **second quadrant** is all points $(x, y)$ with both $x < 0$ and $y > 0$. The **third quadrant** is all points $(x, y)$ with both $x < 0$ and $y < 0$. The **fourth quadrant** is all points $(x, y)$ with both $x > 0$ and $y < 0$. See Figure 1.7.

## Distance Between Two Points

We wish now to find the distance $d$ between two given points $P_1(x_1, y_1)$ and $P_2(x_2, y_2)$ in the xy-plane shown in Figure 1.8.

Recall the Pythagorean theorem that says that $d^2 = a^2 + b^2$ in Figure 1.8. Then

$$d^2 = |x_2 - x_1|^2 + |y_2 - y_1|^2 = (x_2 - x_1)^2 + (y_2 - y_1)^2.$$

Taking square roots of each side gives the following.

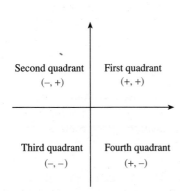

**Figure 1.7**

---

**Distance Formula**

The distance $d = d(P_1, P_2)$ between the two points $P_1(x_1, y_1)$ and $P_2(x_2, y_2)$ in the xy-plane is given by

$$d(P_1, P_2) = \sqrt{(x_2 - x_1)^2 + (y_2 - y_1)^2}.$$

Figure 1.8

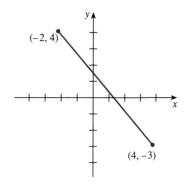

Figure 1.9

**E X A M P L E   4**     **Finding the Distance Between Two Points**

Find the distance between the two points $(-2, 4)$ and $(4, -3)$ shown in Figure 1.9.

**Solution**     We have $(x_1, y_1) = (-2, 4)$ and $(x_2, y_2) = (4, -3)$. Thus

$$d(P_1, P_2) = \sqrt{(x_2 - x_1)^2 + (y_2 - y_1)^2}$$
$$= \sqrt{([4] - [-2])^2 + ([-3] - [4])^2}$$
$$= \sqrt{(6)^2 + (-7)^2} = \sqrt{85}. \ \blacksquare$$

**E X A M P L E   5**     **Using the Distance Between Two Points**

Figure 1.10

A prospective groom is located at point O in Figure 1.10 and must get, by car, to his wedding located at the church at point C in 25 minutes. The church at C is located 13 miles east and 4 miles north of O. The only highway goes through a town located at B. From O to B is a nice open stretch of highway, and he can average 50 miles per hour. The stretch from B to C has traffic lights and some congestion, and he can average only 25 miles per hour on this leg of the trip. Can he get to the church on time?

**Solution**     The distance from O to B is 10 miles, while the distance from B to C is $\sqrt{(13 - 10)^2 + (4 - 0)^2} = 5$ miles. Distance is the product of velocity with time, $d = v \cdot t$. Thus $t = d/v$. Then the time to travel the given route is

$$\frac{10}{50} + \frac{5}{25} = \frac{2}{5}$$

of an hour, or 24 minutes. This is less than 25 minutes, so he can make his wedding on time. $\blacksquare$

### Creating Equations

In order to solve any applied problem, we must first take the problem and translate it into mathematics. Doing this requires creating equations.

E X A M P L E  6        **Creating an Equation**
_____

A small shop makes two styles of shirts. The first style costs $7 to make, and the second $9. The shop has $252 a day to produce these shirts. If $x$ is the number of the first style produced each day and $y$ is the number of the second style, find an equation that $x$ and $y$ must satisfy.

**Solution**        Create a table that includes the needed information. See Table 1.1.

Since $x$ are produced at a cost of $7 each, the cost in dollars of producing $x$ of these shirts is $7x$. Since $y$ are produced at a cost of $9 each, the cost in dollars of producing $y$ of these shirts is $9y$. This is indicated in Table 1.1. Since the cost of producing the first style plus the cost of producing the second style must total $252, the equation we are seeking is

$$7x + 9y = 252. \quad \blacksquare$$

**Table 1.1**

| *Style*           | *First* | *Second* | *Total* |
|-------------------|---------|----------|---------|
| Cost of each      | $7      | $9       |         |
| Number produced   | $x$     | $y$      |         |
| Cost              | $7x$    | $9y$     | 252     |

### Graphs

Given any equation in $x$ and $y$ (such as $y = 2x - 1$), the **graph of the equation** is the set of all points $(x, y)$ such that $x$ and $y$ satisfy the equation.

---
**Graph of an Equation**

The **graph of an equation** is the set

$$\{(x, y) | x \text{ and } y \text{ satsify the equation}\}.$$
---

For example the point $(0, -1)$ is on the graph of the equation $y = 2x - 1$, because when $x$ is replaced with 0 and $y$ replaced with $-1$ the equation is satisfied. That is, $(-1) = 2(0) - 1$.

E X A M P L E   7      **Determining a Graph of an Equation**

Sketch the graph of $y = 2x - 1$, that is, find the set $\{(x, y)\,|\,y = 2x - 1\}$.

**Solution**      We take a number of values for $x$, put them into the equation and solve for the corresponding values of $y$. For example if we take $x = -3$, then $y = 2(-3) - 1 = -7$. Table 1.2 summarizes this work

Table 1.2

| $x$ | $-3$ | $-2$ | $-1$ | $0$ | $1$ | $2$ | $3$ |
|---|---|---|---|---|---|---|---|
| $y$ | $-7$ | $-5$ | $-3$ | $-1$ | $1$ | $3$ | $5$ |

The points $(x, y)$ found in Table 1.2 are graphed in Figure 1.11. We see that the graph is a straight line.  ∎

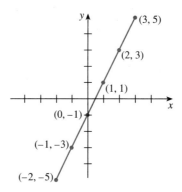

Figure 1.11

To find the graph of an equation, plot points until a pattern emerges. More complicated equations will require more points.

E X A M P L E   8      **Determining the Graph of an Equation**

Sketch the graph of $x = 2$, that is, find the set $\{(x, y)\,|\,x = 2\}$.

**Solution**      Points such as $(2, -1)$, $(2, 1)$, and $(2, 2)$ are all on the graph. The graph is a vertical line, as shown in Figure 1.12.  ∎

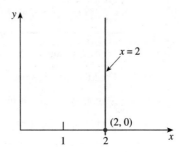

**Figure 1.12**

**E X A M P L E 9** **Determining a Graph of an Equation**

Sketch the graph of $y = x^2$, that is, find the set $\{(x, y) | y = x^2\}$.

**Solution** We take a number of values for $x$, put them into the equation and solve for the corresponding values of $y$. For example if we take $x = -2$, then $y = (-2)^2 = 4$. Table 1.3 summarizes this work.

The points $(x, y)$ found in Table 1.3 are graphed in Figure 1.13. ∎

**Table 1.3**

| $x$ | $-3$ | $-2$ | $-1$ | 0 | 1 | 2 | 3 |
|---|---|---|---|---|---|---|---|
| $y$ | 9 | 4 | 1 | 0 | 1 | 4 | 9 |

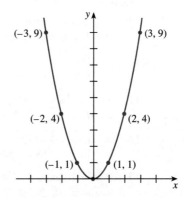

**Figure 1.13**

## SELF-HELP EXERCISE SET 1.1

1.  Plot the points $(-2, 5)$, $(-2, -4)$, and $(3, -2)$.

2.  Find the distance between the two points $(-4, -2)$ and $(3, -4)$.

3.  Sketch the graph of the equation $7x + 9y = 252$ found in Example 6.

## EXERCISE SET 1.1

In Exercises 1 through 4 which of the statements determine sets?

1.  **a.**  The past presidents of Harvard University
    **b.**  The past and future presidents of Harvard University

2.  **a.**  The good past presidents of Harvard University
    **b.**  The past presidents of Harvard University who were at least 6 feet tall

3.  **a.**  Individuals who played for the New York Yankees in 1991
    **b.**  All-around good New York Yankee players during 1991

4.  **a.**  Los Angeles Dodger players who had a batting average of at least .300 during 1991
    **b.**  The good-looking Los Angeles Dodger players of 1991

In Exercises 5 through 8 write the sets in set-builder notation.

5.  **a.**  $\{1, 3, 5, 7, 9, 11, 13, 15\}$
    **b.**  $\{3, 6, 9, 12, 15, 18, 21\}$

6.  **a.**  The positive integers less than 10
    **b.**  The solutions to $(x - 2)(x - 3) = 0$

7.  **a.**  $\{2, 4, 6, 8, 10, 12, 14, 16\}$
    **b.**  The odd positive integers

8.  **a.**  $\{5, 10, 15, 20, 25, 30\}$
    **b.**  $\{a, b, c, d, e\}$

In Exercises 9 through 12 write the sets in roster notation.

9.  **a.**  $\{x | x$ is an integer, $5 \leq x < 12\}$
    **b.**  The positive even integers less than or equal to 13

10. **a.**  $\{x | x$ is a letter in the word "mathematics"$\}$
    **b.**  $\{x | (x + 1)(x - 1) = 0\}$

11. **a.**  $\{x | x$ is an integer, $3 < x \leq 10\}$
    **b.**  The positive odd integers less than or equal to 13

12. **a.**  $\{x | x$ is a letter in the word "mathematics" that appears at least twice$\}$
    **b.**  $\{x | (x + 3)(x - 2) = 0\}$

13. If $A = \{u, v, y, z\}$, determine whether the following statements are true or false.
    **a.**  $w \in A$,  **b.**  $x \notin A$,  **c.**  $z \in A$,  **d.**  $v \notin A$

14. If $A = \{u, v, y, z\}$, determine whether the following statements are true or false.
    **a.**  $x \in A$,  **b.**  $y \notin A$,  **c.**  $y \in A$,  **d.**  $t \in A$

In Exercises 15 through 22 draw a number line and identify.

**15.** $(-1, 3)$        **16.** $[-2, 6]$

**17.** $(2, 5]$        **18.** $[-3, -1)$

**19.** $(-2, +\infty)$        **20.** $[2, +\infty)$

**21.** $(-\infty, 1.5]$        **22.** $(-\infty, -2)$

In Exercises 23 through 34 plot the given point in the $xy$-plane.

**23.** $(1, 2)$        **24.** $(-3, -4)$

**25.** $(-2, 1)$        **26.** $(2, -3)$

**27.** $(1, 0)$        **28.** $(-2, 0)$

**29.** $(0, -2)$        **30.** $(0, 4)$

**31.** $(-2, -1)$        **32.** $(2, 2)$

**33.** $(3, -4)$        **34.** $(-3, 0)$

In Exercises 35 through 46 find the set of points $(x, y)$ in the $xy$-plane that satisfies the given equation or inequality.

**35.** $x \geq 0$        **36.** $y \geq 0$

**37.** $x < 0$        **38.** $y < 0$

**39.** $x = 2$        **40.** $x = -1$

**41.** $y = 3$        **42.** $y = -2$

**43.** $xy > 0$        **44.** $xy < 0$

**45.** $xy = 0$        **46.** $x^2 + y^2 = 0$

In Exercises 47 through 54 find the distance between the given points.

**47.** $(2, 4), (3, 2)$        **48.** $(-2, 5), (2, -3)$

**49.** $(-2, 5), (-3, -1)$        **50.** $(1, 3), (1, 6)$

**51.** $(-3, -2), (4, -3)$        **52.** $(-5, -6), (-3, -2)$

**53.** $(a, b), (b, a)$        **54.** $(x, y), (0, 0)$

In Exercises 55 through 64 sketch the graphs.

**55.** $y = -2x + 1$        **56.** $y = -2x - 2$

**57.** $y = 3x - 2$        **58.** $y = 2x + 3$

**59.** $y = 3$        **60.** $x = 4$

**61.** $y = 3x^2$        **62.** $y = -x^2$

**63.** $y = x^3$        **64.** $y = \sqrt{x}$

## Applications

**65.** **Transportation.**   A truck travels a straight highway from O to A and then another straight highway from A to B. How far does the truck travel?

**66.** **Transportation.**   If the truck in the previous exercise travels 30 miles per hour from O to A and 60 miles per hour from A to B, how long does the trip take?

**67.** **Navigation.**   A ship leaves port and heads due east for 4 hours with a speed of 3 miles per hour, after which it turns due north and maintains this direction at a speed of 2.5 miles per hour. How far is the ship from port 6 hours after leaving?

**68.** **Navigation.**   A ship leaves port and heads due east for 1 hour at 5 miles per hour. It then turns north and heads in this direction at 4 miles per hour. How long will it take the ship to get 13 miles from port?

**69.** **Oil Slick.**   The shoreline stretches along the $x$-axis as shown with ocean consisting of the first and second quadrants. An oil spill occurs at the point (2, 4), and a coastal town is located at (5, 0) where the numbers represent miles. If the oil slick is approaching the town at a rate of one-half mile per day, how many days until the oil slick reaches the coastal town located at (5, 0)?

*(© Michel Follorov/Photo Researchers, Inc.)*

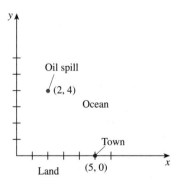

**70.** **Transportation Charges.**   A furniture store is located 5 miles east and 14 miles north of the center of town. You live 10 miles east and 2 miles north of the center of town. The store advertises free delivery within a 12-mile radius of the store. Do you qualify for free delivery?

**71. Profits.**   A furniture store sells chairs at a profit of $100 each and sofas at a profit of $150 each. Let $x$ be the number of chairs sold each week and $y$ the number of sofas sold each week. If the profit in one week is $3000, write an equation that $x$ and $y$ must satisfy. Graph this equation.

**72. Revenue.**   A restaurant has two specials, steak and chicken. The steak special is $13, and the chicken special is $10. Let $x$ be the number of steak specials served and $y$ the number of chicken specials. If the restaurant made $390 in sales on these specials, find the equation that $x$ and $y$ must satisfy.

**73. Nutrition.**   An individual needs 800 mg of calcium daily but is unable to eat any dairy products due to the large amounts of cholesterol in such products. This individual does enjoy eating canned sardines and steamed broccoli however. Let $x$ be the number of ounces of canned sardines consumed each day and let $y$ be the number of cups of steamed broccoli. If there are 125 mg of calcium in each ounce of canned sardines and 190 mg in each cup of steamed broccoli, what equation must $x$ and $y$ satisfy for this person to satisfy the daily calcium requirement from these two sources?

*(© Spencer Grant/Photo Researchers, Inc.)*

**74. Costs.**   A contractor builds ranch and split-level style homes. The ranch costs $130,000 to build and the split-level $150,000. Let $x$ be the number of ranch-style homes built and $y$ the number of split-level style homes. If the contractor has $1,360,000 to build these homes, find the equation that $x$ and $y$ must satisfy. Graph this equation.

**75. Biology.**   A fisheries biologist decides to set a gill net on his next trip to a remote mountain lake. The net must stretch between points A and B shown in the figure. Since he must backpack to the lake he wants to carry no more net than necessary. He walks off the distance $a$ from C to B and the distance $b$ from A to C, being careful that AC is perpendicular to AB. Find the distance $c$ from A to B in terms of $a$ and $b$.

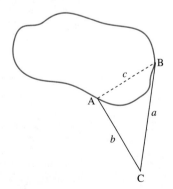

## Solutions to Self-Help Exercise Set 1.1

**1.** The points are plotted in the following figure.

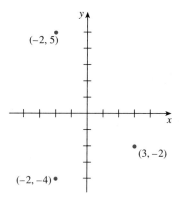

**2.** The distance between the two points $(-4, -2)$ and $(3, -4)$ is given by

$$\sqrt{[3 - (-4)]^2 + [-4 - (-2)]^2} = \sqrt{7^2 + (-2)^2} = \sqrt{53}$$

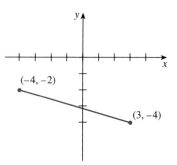

**3.** First solve for $y$ and obtain $y = -\frac{7}{9}x + 28$. Now take a number of values for $x$, put them into the equation, and solve for the corresponding values of $y$. For example if we take $x = 9$, then $y = -\frac{7}{9}(9) + 28 = 21$. The following table summarizes this work.

| $x$ | 0 | 9 | 18 | 27 | 36 |
|---|---|---|---|---|---|
| $y$ | 28 | 21 | 14 | 7 | 0 |

The points $(x, y)$ found in the table are graphed in the figure. We see that apparently the graph is a straight line.

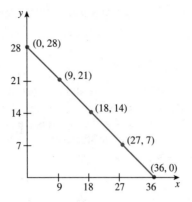

## 1.2 LINES

▶ *Slope*

▶ *Equations of Lines*

▶ *Applications*

**APPLICATION**

**Production of Vacuum Cleaners**

A firm produces two models of vacuum cleaners, regular and deluxe. The regular model requires 8 hours of labor to assemble while the deluxe requires 10 hours of labor. Let $x$ be the number of regular models produced and $y$ the number of deluxe models. If the firm wishes to use 500 hours of labor to assemble these cleaners, find an equation that the two quantities $x$ and $y$ must satisfy. If the number of regular models to be assembled is increased by 5, how many fewer deluxe models will be assembled? See Example 8 for the answer.

### Slope

In order to describe a straight line, one must first describe the "slant" or **slope** of the line. See Figure 1.14.

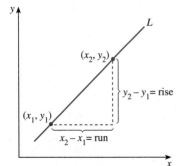

**Figure 1.14**

---

**Slope of a Line**

Let $(x_1, y_1)$ and $(x_2, y_2)$ be two points on a line $L$. If the line is nonvertical $(x_1 \neq x_2)$, the **slope** $m$ of the line $L$ is defined to be

$$m = \frac{y_2 - y_1}{x_2 - x_1}.$$

If $L$ is vertical $(x_2 = x_1)$, then the slope is said to be **undefined**.

---

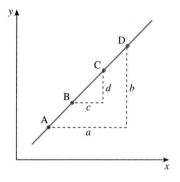

**Figure 1.15**

**REMARK.**     The term $x_2 - x_1$ is called the **run** and $y_2 - y_1$ is called the **rise**. See Figure 1.14. Thus the slope $m$ can be thought of as

$$m = \frac{y_2 - y_1}{x_2 - x_1} = \frac{\text{rise}}{\text{run}}.$$

The slope of the line does not depend on which two points on the line are chosen to use in the above formula. To see this, notice in Figure 1.15 that the two right triangles are similar since the corresponding angles are equal. Thus the ratio of the corresponding sides must be equal. That is,

$$\frac{b}{a} = \frac{d}{c},$$

where the ratio on the left is the slope using the points $A$ and $D$ and the ratio on the right is the slope using the points $B$ and $C$.

**E X A M P L E   1**     **Finding the Slope of Lines**

Find the slope (if it exists) of the line through each pair of points. Sketch the line.
(a) (1, 2), (3, 6),   (b) $(-3, 1)$, $(3, -2)$,   (c) $(-1, 3)$, $(2, 3)$,   (d) (3, 4), (3, 1)

**Solutions**     (a)   $(x_1, y_1) = (1, 2)$ and $(x_2, y_2) = (3, 6)$. So

$$m = \frac{y_2 - y_1}{x_2 - x_1} = \frac{(6) - (2)}{(3) - (1)} = \frac{4}{2} = 2.$$

See Figure 1.16. Notice that for each unit we move to the right, the line moves up $m = 2$ units.

(b)   $(x_1, y_1) = (-3, 1)$ and $(x_2, y_2) = (3, -2)$. So

$$m = \frac{y_2 - y_1}{x_2 - x_1} = \frac{(-2) - (1)}{(3) - (-3)} = \frac{-3}{6} = -\frac{1}{2}.$$

See Figure 1.17. Notice that for each unit we move to the right, the line moves down $\frac{1}{2}$ unit.

**Figure 1.16**

**Figure 1.17**

**Figure 1.18**

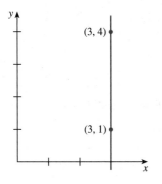

**Figure 1.19**

(c) $(x_1, y_1) = (-1, 3)$ and $(x_2, y_2) = (2, 3)$. So

$$m = \frac{y_2 - y_1}{x_2 - x_1} = \frac{(3) - (3)}{(2) - (-1)} = \frac{0}{3} = 0.$$

See Figure 1.18. Notice that this line is horizontal.

(d) $(x_1, y_1) = (3, 4)$ and $(x_2, y_2) = (3, 1)$. So

$$m = \frac{y_2 - y_1}{x_2 - x_1} = \frac{(1) - (4)}{(3) - (3)} = \frac{-3}{0} = \text{undefined}.$$

See Figure 1.19. Notice that this line is vertical. ∎

Consider any line $L$ that is not vertical and let $P_1(x_1, y_1)$ be one point on the line. We noted earlier that in determining the slope of a line, *any* two points on the line can be used. We will now take a second point $P_2(x_2, y_2)$ on the line where $P_2$ is chosen with $x_2 = x_1 + 1$. Thus we have run $= x_2 - x_1 = 1$, and, therefore,

$$m = \frac{y_2 - y_1}{x_2 - x_1} = \frac{y_2 - y_1}{1} = y_2 - y_1 = \text{rise}. \tag{1}$$

This says that if the run is taken as one, then the slope will equal the rise.

If now $m$ is positive, as in Figure 1.20a, then moving over one unit results in moving *up* $m$ units. Thus the line *rises* (moving from left to right) and the larger the value of $m$ the steeper the rise.

If $m$ is negative, as in Figure 1.20b, then moving over one unit results in moving *down* $|m| = -m$ units. Thus the line *falls* and the more negative the value of $m$, the steeper the fall.

If $m$ is zero, as in Figure 1.20c, then moving over one unit results in moving up *no* units. Thus the line is *horizontal*.

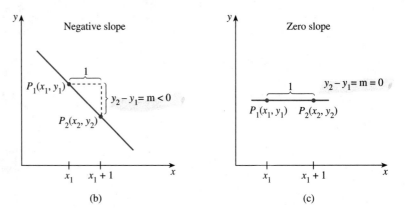

(a)  (b)  (c)

**Figure 1.20**

The first three parts of Example 1 give specific examples of these three general cases.

Figure 1.21 shows several lines through the same point with different slopes.

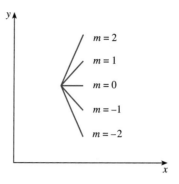

**Figure 1.21**

E X A M P L E  2      **Using the Slope of a Line**

Find the $y$-coordinate of point $P$ if $P$ has $x$-coordinate 5 and $P$ is on the line through (2, 7) with slope $-2$.

**Solution**     See Figure 1.22. Using the definition of slope, we have

$$-2 = \frac{y - 7}{5 - 2}$$

$$= \frac{y - 7}{3}$$

$$-6 = y - 7$$

$$y = 1 \quad \blacksquare$$

**Figure 1.22**

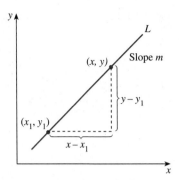

**Figure 1.23**

## Equations of Lines

Suppose we have a line through the point $(x_1, y_1)$ with slope $m$ as shown in Figure 1.23. If $(x, y)$ is any other point on the line, then we must have

$$m = \frac{y - y_1}{x - x_1},$$

or

$$y - y_1 = m(x - x_1).$$

This equation is the **point-slope equation**.

---

**The Point-Slope Equation of a Line**

The equation of the line through the point $(x_1, y_1)$ with slope $m$ is given by

$$y - y_1 = m(x - x_1).$$

---

**E X A M P L E   3**

### Finding the Point-Slope Equation of a Line

Find an equation of the line through $(-2, 3)$ with slope $m = -2$. Sketch the graph.

**Solution**    Since $(x_1, y_1) = (-2, 3)$, the point-slope equation is

$$y - y_1 = m(x - x_1)$$
$$y - 3 = -2[x - (-2)]$$
$$y = -2x - 1 \quad \blacksquare$$

**R E M A R K .**    One quick way of finding the graph is to notice that when $x = 0$, $y = -2(0) - 1 = -1$. Thus $(0, -1)$ is a second point on the graph. See Figure 1.24.

A point $(x, y)$ will be on the graph of a vertical line such as the one shown in Figure 1.25 if, and only if, $x = a$, where $(a, 0)$ is the point where the vertical line crosses the $x$-axis. Thus we have the following.

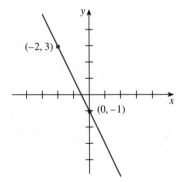

**Figure 1.24**

---

**Vertical Lines**

A vertical line has the equation

$$x = a$$

where $(a, 0)$ is the point at which the vertical line crosses the $x$-axis.

---

Given any line that is not vertical, then the line has a slope $m$ and by Figure 1.26 must cross the $y$-axis at some point $(0, b)$. The number $b$ is called the

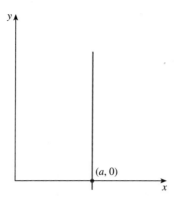

**Figure 1.25**                                    **Figure 1.26**

**y-intercept**. Thus we can use the point-slope equation where $(x_1, y_1) = (0, b)$ is a point on the line. This gives

$$y - b = m(x - 0)$$
$$y = mx + b$$

This is the **slope-intercept equation** of a nonvertical line.

---

**The Slope-Intercept Equation of a Line**

The equation of the line with slope $m$ and $y$-intercept $b$ is given by

$$y = mx + b.$$

---

**E X A M P L E   4**    **Finding the Slope-Intercept Equation of a Line**

Find the equation of a line with slope 3 and $y$-intercept $-2$. Draw a graph.

**Solution**    We have $m = 3$ and $b = -2$. Thus

$$y = mx + b = 3x - 2.$$

To draw a graph, we first notice that since the $y$-intercept is $-2$, the point $(0, -2)$ is on the graph. See Figure 1.27. Since the slope is 3, if we move over 1 unit to the right, the line will move up 3 units so that the point $(1, 1)$ is also on the line.  ■

We have already shown that every line has an equation of the general form

$$ax + by = c,$$

where $a$ and $b$ are not both zero. This is called a **linear equation**.

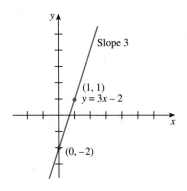

Figure 1.27

Suppose now we are given the equation $ax + by = c$, where $a$ and $b$ are not both zero. Then if $b \neq 0$,

$$y = -\frac{a}{b}x + \frac{c}{b},$$

which, according to the slope-intercept equation, is the equation of a line with slope $-a/b$ and $y$-intercept $c/b$. If $b = 0$, then $a \neq 0$ and $ax + by = c$ becomes $ax = c$ and thus $x = c/a$, which is the equation of a vertical line. Thus we have the following theorem.

---

**The General Form of the Equation of a Line**

The graph of a linear equation $ax + by = c$ is a straight line, and, conversely, every line is the graph of a linear equation.

---

We have already defined the $y$-intercept. We now need to define the $x$-intercept. Any line that is not horizontal must cross the $x$-axis at some point $(a, 0)$. The number $a$ is called the **$x$-intercept**. To find the $x$-intercept of a line, simply set $y = 0$ in the equation of the line and solve for $x$. Similarly, to find the $y$-intercept of a line, set $x = 0$ in the equation of the line and solve for $y$. The intercepts are very useful in graphing, as the next example illustrates.

E X A M P L E  5      **Graphing an Equation of a Line**

Sketch a graph of $2x + 3y = 6$.

**Solution**      From the preceding theorem we know that this is a line. Perhaps the easiest way to graph this line is to find the $x$- and $y$-intercepts. Setting $x = 0$ yields $3y = 6$ or $y = 2$ as the $y$-intercept. Setting $y = 0$ yields $2x = 6$ or $x = 3$ as the $x$-intercept. See Figure 1.28.  ∎

Since slope is a number that indicates the direction or slant of a line, the following theorem can be proven.

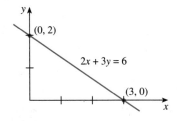

Figure 1.28

---

**Slope and Parallel Lines**

Two lines are parallel if and only if they have the same slope.

---

E X A M P L E  6      **Finding the Equation of a Line**

Let $L_1$ be the line that goes through the two points $(-3, 1)$ and $(-1, 7)$. Find the equation of the line $L$ through the point $(2, 1)$ and parallel to $L_1$. See Figure 1.29.

**Solution**    We already have the point $(2, 1)$ on the line $L$. In order to use the point-slope equation, we need to now find the slope $m$ of $L$. But since $L$ is parallel to $L_1$, $m$ will equal the slope of $L_1$. The slope of $L_1$ is

$$m = \frac{y_2 - y_1}{x_2 - x_1} = \frac{(7) - (1)}{(-1) - (-3)} = \frac{6}{2} = 3.$$

Now let us use the point-slope equation for $L$ where $(2, 1)$ is a point on the line $L$ with slope $m = 3$. Then we obtain

$$y - y_1 = m(x - x_1)$$
$$y - 1 = 3(x - 2)$$
$$y = 3x - 5. \quad \blacksquare$$

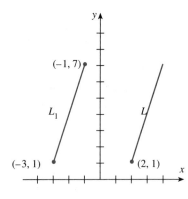

**Figure 1.29**

We also have the following theorem. (See Exercise 46.)

---

**Slope and Perpendicular Lines**

Two nonvertical lines with slope $m_1$ and $m_2$ are perpendicular if, and only if,

$$m_2 = -\frac{1}{m_1}.$$

---

**E X A M P L E  7**    **Finding the Equation of a Line**

Find the equation of the line $L$ through the point $(3, -1)$ and perpendicular to the line $L_1$ given by $2x + 3y = 6$.

**Solution**    See Figure 1.30. We rewrite $2x + 3y = 6$ as

$$y = -\tfrac{2}{3}x + 2.$$

This indicates that the slope of $L_1$ is $-\tfrac{2}{3}$. Thus the slope $m$ of $L$ will be the negative reciprocal or $m = \tfrac{3}{2}$. Now use the point-slope equation for $L$ to give

$$y - y_1 = m(x - x_1)$$
$$y - (-1) = \frac{3}{2}(x - 3)$$
$$y = \frac{3}{2}x - \frac{11}{2}. \quad \blacksquare$$

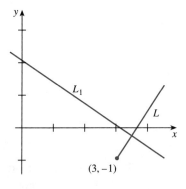

**Figure 1.30**

## Applications

**E X A M P L E   8**

### Production of Vacuum Cleaners

A firm produces two models of vacuum cleaners, regular and deluxe. The regular model requires 8 hours of labor to assemble while the deluxe requires 10 hours of labor. Let $x$ be the number of regular models produced and $y$ the number of deluxe models. If the firm wishes to use 500 hours of labor to assemble these cleaners, find an equation that the two quantities $x$ and $y$ must satisfy. If the number of the regular models that are assembled is increased by 5, how many fewer deluxe models will be assembled?

**Solution**

In order to find the required equation, construct a table with the given information. See Table 1.4.

**Table 1.4**

| Type | Regular | Deluxe | Total |
|------|---------|--------|-------|
| Work hours for each | 8 | 10 | |
| Number produced | $x$ | $y$ | |
| Work hours | $8x$ | $10y$ | 500 |

The number of work-hours needed to assemble all of the regular models is the number of work-hours required to produce one of these models times the number produced. This gives $8x$. In a similar fashion we see that the number of work-hours needed to produce all of the deluxe models is $10y$. The number of hours allocated to assemble the regular model plus the number allocated to assemble the deluxe must equal 500. That is,

$$8x + 10y = 500.$$

This is the required equation. Solving for $y$ yields

$$y = -\tfrac{4}{5}x + 50.$$

This is a line with slope $-\tfrac{4}{5}$. See Figure 1.31. Thus increasing $x$ by 5 will decrease $y$ by 4.  ■

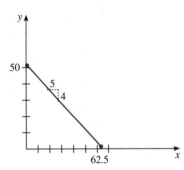

**Figure 1.31**

## SELF-HELP EXERCISE SET 1.2

1. Find the equation of the line through the two points $(3, 1)$ and $(-2, -3)$ and then find the $y$-intercept. Sketch.

2. Find the equation of the line through the point $(2, -1)$ and perpendicular to the line in the previous exercise. Sketch.

3. Sketch the graph of $2x + 3y - 12 = 0$. Find the slope of this line. Sketch.

## EXERCISE SET 1.2

In Exercises 1 through 8 find the slope (if it exists) of the line through the given pair of points.

1.  $(1, 3), (2, 7)$

2.  $(3, -2), (2, 5)$

3.  $(-2, -2), (-3, 1)$

4.  $(-3, 2), (5, 4)$

5.  $(-1, 3), (2, 3)$

6.  $(-1, 2), (-3, -4)$

7.  $(-3, -1), (-3, -7)$

8.  $(-5, -2), (-3, -4)$

In Exercises 9 through 20 find the slope (if it exists) of the line given by each of the equations. Also find the $x$- and $y$-intercepts if they exist.

9.  $y = 2x - 1$

10.  $y = -3x + 2$

11.  $x + y = 1$

12.  $x - 2y - 3 = 0$

13.  $-x + 2y - 4 = 0$

14.  $x + y = 0$

15.  $x = 3$

16.  $x = 0$

17.  $y = 4$

18.  $y = 0$

19.  $x = 2y + 1$

20.  $x = -3y + 2$

In Exercises 21 through 42 find the equation of the line given the indicated information.

21.  Through $(2, 1)$ with slope $-3$

22.  Through $(-3, 4)$ with slope 3

23.  Through $(-1, -3)$ with slope 0

24.  Through $(2, 3)$ with slope undefined

25.  Through $(-2, 4)$ and $(-4, 6)$

26.  Through $(-1, 3)$ and $(-1, 6)$

27.  Through $(-2, 4)$ and $(-2, 6)$

28.  Through $(-1, 4)$ and $(-3, 2)$

29.  Through $(a, 0)$ and $(0, b)$, $ab \neq 0$

30.  Through $(a, b)$ and $(b, a)$, $a \neq b$

31.  Through $(-2, 3)$ and parallel to the line $2x + 5y = 3$

32.  Through $(2, -3)$ and parallel to the line $2x - 3y = 4$

33.  Through $(0, 0)$ and parallel to the line $y = 4$

34.  Through $(0, 0)$ and parallel to the line $x = 4$

35.  Through $(1, 2)$ and perpendicular to the line $y = 4$

36. Through $(1, 2)$ and perpendicular to the $x$-axis

37. Through $(-1, 3)$ and perpendicular to the line $x + 2y - 1 = 0$

38. Through $(2, -4)$ and perpendicular to the line $x - 3y = 4$

39. Through $(3, -1)$ and with $y$-intercept 4

40. Through $(-3, -1)$ and with $y$-intercept 0

41. $y$-intercept 4 and parallel to $2x + 5y = 6$

42. $y$-intercept $-2$ and perpendicular to $x - 3y = 4$

43. Are the two lines $2x + 3y = 6$ and $3x + 2y = 6$ perpendicular?

44. Are the two lines $2x + 3y = 6$ and $3x - 2y = 6$ perpendicular?

45. A person has \$2.35 in change consisting entirely of dimes and quarters. If $x$ is the number of dimes and $y$ is the number of quarters, write an equation that $x$ and $y$ must satisfy.

46. Establish the following theorem: Two lines with slope $m_1$ and $m_2$ are perpendicular if, and only if,

$$m_2 = -\frac{1}{m_1}.$$

Give a proof in the special case shown in the figure in which the two lines intersect at the origin. (The general proof is very similar.) *Hint:* The two lines $L_1$ and $L_2$ are perpendicular if and only if the triangle $AOB$ is a right triangle. By the Pythagorean theorem this will be the case if, and only if,

$$[d(A, B)]^2 = [d(A, O)]^2 + [d(B, O)]^2$$

Now use the distance formula for each of the three terms in the previous equation and simplify.

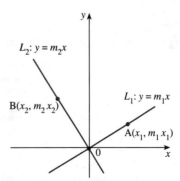

## Applications

47. **Investment.**  A person invests some money in a bond that yields 8% a year, and some money in a money market fund that yields 5% a year, and obtains \$576 of interest in the first year. If $x$ is the amount invested in the bond and $y$

the amount invested in the money market fund, find an equation that $x$ and $y$ must satisfy. If the amount of money invested in bonds is increased by $100, by how much must the money invested in the money market decrease if the total interest is to stay the same?

48. **Transportation Costs.**   New cars are transported from Detroit to Boston and also from Detroit to New York City. It costs $90 to transport each car to Boston and $80 to transport each car to New York City. Suppose $4000 is spent on transporting these cars. If $x$ is the number transported to Boston and $y$ is the number transported to New York City, find an equation that $x$ and $y$ must satisfy. If the number of cars transported to Boston is increased by 8, how many fewer cars must be transported to New York City if the total transportation costs remained the same?

49. **Scheduling.**   An attorney has two types of documents to process. The first document takes 3 hours and the second takes 4 hours to process. If $x$ is the number of the first type that she processes and $y$ is the number of the second type, find an equation that $x$ and $y$ must satisfy if she works 45 hours on these documents. If she processes 8 additional of the first type of document, how many fewer of the second type of document would she process?

50. **Size of Deer Antlers.**   The size of a deer's antlers depends primarily on the age of the deer. Mule deer in Northern Colorado have no antlers at age 10 months, but after that the weight of the antlers increases by .12 kilogram for every 10 months of additional age. Let $x$ be the age of these mule deer in months and $y$ the weight of their antlers in kilograms. Assuming a linear relationship, write an equation that $x$ and $y$ must satisfy. What will be the weight in kilograms of antlers carried by a 40-month-old mule deer?

*(© John O. Sumner /Photo
Researchers, Inc.)*

51. **Length of Twigs.**   If $d$ is the twig diameter in inches of bitter bush, then it is estimated that the length $l$ in inches of the twig is given by $l = 1.25 + 89.83d$. Find the slope of this line and interpret what the slope means in this instance.

## Graphing Calculator Exercises

52. On your graphing calculator graph the equations $y = 3 + 2x$, $y = 3 + x$, $y = 3 - x$, $y = 3 - 2x$ in the same viewing rectangle.

53. On your graphing calculator graph $y = 1.43x$, $y = 1.43x + 1$, and $y = 1.43x + 2$ in the same viewing rectangle. Are they parallel?

## Solutions to Self-Help Exercise Set 1.2

1. To find the equation of the line through the two points $(3, 1)$ and $(-2, -3)$, we first must find the slope of the line through the two points. This is

$$m = \frac{y_2 - y_1}{x_2 - x_1} = \frac{1 - (-3)}{3 - (-2)} = \frac{4}{5}.$$

We now have the slope of the line and *two* points on the line. Pick one of the points, say (3, 1), and use the point-slope form of the equation of the line and obtain

$$y - y_1 = m(x - x_1)$$

$$y - 1 = \frac{4}{5}(x - 3)$$

$$y = \frac{4}{5}x - \frac{7}{5}.$$

This last equation is in the form of $y = mx + b$ with $b = -\frac{7}{5}$. Thus the $y$-intercept is $-\frac{7}{5}$.

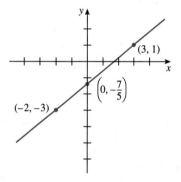

2.  The equation through the point (2, −1) and perpendicular to the line in the previous exercise must have a slope equal to the negative reciprocal of the slope of that line. Then the slope of the line whose equation we are seeking is $-\frac{5}{4}$. Using the point-slope form of the equation yields

$$y - y_1 = m(x - x_1)$$

$$y + 1 = -\frac{5}{4}(x - 2)$$

$$y = -\frac{5}{4}x + \frac{3}{2}.$$

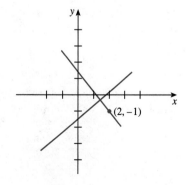

3.  Perhaps the easiest way to sketch the graph of $2x + 3y - 12 = 0$ is to find the $x$- and $y$-intercepts. To find the $y$-intercepts, set $x = 0$ and obtain $3y = 12$ or $y = 4$. To find the $x$-intercept, set $y = 0$ and obtain $2x = 12$ or $x = 6$. These intercepts are plotted in the figure, and the line is drawn through them. To find the slope, write the equation in the form $y = -\frac{2}{3}x + 4$. The slope is then seen to be $m = -\frac{2}{3}$.

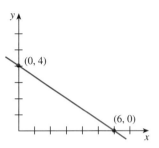

## 1.3    LINEAR MODELS

► *Cost, Revenue, and Profits*

► *Supply and Demand*

► *Straight-Line Depreciation*

**Augustin Cournot,**
**1801–1877**

*(Historical Pictures Collection/Stock Montage, Inc.)*

The first significant work dealing with the application of mathematics to economics was Cournot's "Researches into the Mathematical Principles of the Theory of Wealth," which was published in 1836. It was Cournot who originated the supply and demand curves that are discussed in this section. Irving Fisher, a prominent economics professor at Yale and one of the first exponents of mathematical economics in the United States, wrote that Cournot's book "seemed a failure when first published. It was far in advance of the times. Its methods were too strange, its reasoning too intricate for the crude and confident notions of political economy then current."

A P P L I C A T I O N

**Finding the Demand Equation**

Suppose that 1000 items of a certain product are sold per day by the entire industry at a price of $10 per item and that 2000 items can be sold per day by the same industry at a price of $8 per item. What is the demand equation, that is, the equation that gives the price $p$ of each item when given the number of items $x$ that are made and sold, if the graph of the demand equation is a straight line? See Example 3 for the answer.

### Cost, Revenue, and Profits

Any manufacturing firm has two types of costs—fixed costs and variable costs. **Fixed costs** are costs that do not depend on the amount of production. These costs include real estate taxes, interest on loans, some management salaries, certain minimal maintenance, and protection of plant and equipment. **Variable costs** depend on the amount of production. They include cost of material and labor. Total cost or simply **cost** is the sum of fixed and variable costs:

$$(\text{cost}) = (\text{variable cost}) + (\text{fixed cost}).$$

Let $x$ denote the number of units of a given product or commodity produced by a firm. In the *linear cost model* we assume that the cost $m$ of manufacturing one item is the same no matter how many items are produced. Thus the variable cost is the number of items produced times the cost of each item:

$$(\text{variable cost}) = (\text{cost per item}) \times (\text{number of items produced})$$

$$= mx.$$

If $b$ is the fixed cost and $C$ is the cost, we then have the following:

$$C = \text{cost}$$

$$= (\text{variable cost}) + (\text{fixed cost})$$

$$= mx + b.$$

A graph is shown in Figure 1.32. Note that the $y$-intercept is the fixed cost and the slope is the cost per item.

In the **linear revenue model** we assume that the price $p$ of an item sold by a firm is the same no matter how many items are sold. (This is a reasonable assumption if the number sold by the firm is small compared to the total number sold by the entire industry.) Thus the revenue is the price per item times the number of items sold. Let $x$ be the number of items sold. (We always assume that *the number of items sold equals the number of items produced.*) Then, if we denote the revenue by $R$,

$$R = \text{revenue}$$

$$= (\text{price per item}) \times (\text{number sold})$$

$$= px.$$

See Figure 1.33. Notice that this straight line goes through (0, 0), since nothing sold results in no revenue. The slope is the price per item.

No matter if our models of cost and revenue are linear or not, **profit** $P$ is always revenue less cost. Thus

$$P = \text{profit}$$

$$= (\text{revenue}) - (\text{cost})$$

$$= R - C.$$

**Figure 1.32**

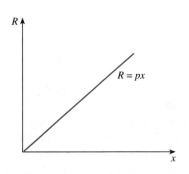

**Figure 1.33**

E X A M P L E   1     **Finding Cost, Revenue, and Profit Equations**

A firm has weekly fixed costs of $80,000 associated with the manufacture of dresses that cost $25 per dress to produce. The firm sells all the dresses it produces at $75 per dress. Find the cost, revenue, and profit equations if $x$ is the number of dresses produced per week.

**Solution**

a.   $C$ = (variable cost) + (fixed cost)

$$= mx + b$$

$$= 25x + 80,000.$$

b.   $R$ = (price per item) × (number sold)

$$= px$$

$$= 75x.$$

c.   $P$ = (revenue) − (cost)

$$= R - C$$

$$= (75x) - (25x + 80,000)$$

$$= 50x - 80,000.$$

These three equations are graphed in Figure 1.34.   ∎

**Figure 1.34**

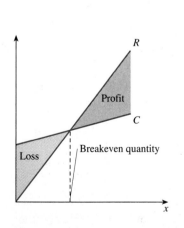

**Figure 1.35**

One can see in Figure 1.34c that, for smaller values of $x$, $P$ is *negative*, that is, the firm has losses. For larger values of $x$, $P$ turns positive and the firm has (positive) profits. The value of $x$ at which the profit is zero is called the **break-even quantity**. Figure 1.35 shows that the break-even quantity is the value of $x$ at which cost equals revenue. This is true since $P = R - C$, and $P = 0$ if, and only if, $R = C$.

E X A M P L E 2    **Finding the Break-Even Quantity**

Find the break-even quantity in the previous example.

Solution    Set $P = 0$ and solve for $x$.

$$0 = P$$

$$= 50x - 80{,}000$$

$$50x = 80{,}000$$

$$x = 1600$$

Thus the firm needs to manufacture and sell 1600 dresses for profits to be zero.  ■

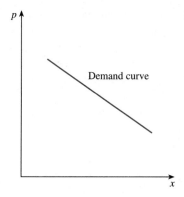

**Figure 1.36**

## Supply and Demand

In the previous discussion we assumed that the number of items produced and sold by the given firm was small compared to the number sold by the industry. Under this assumption it was then reasonable to conclude that the price was constant and did not vary with the number $x$ sold. But if the number of items sold by the firm represented a *large* percentage of the number sold by the entire industry, then trying to sell significantly more items could only be accomplished by *lowering* the price of each item. Thus, under these assumptions, the price $p$ of each item would depend on the number sold. If, in addition, the price is assumed to be linear, that is, $p = mx + b$, then the graph of this equation must slope down as shown in Figure 1.36.

We assume that $x$ is the number of a certain item produced and sold by the entire industry during a given time period and that $p = mx + b$ is the price of the item if $x$ are sold, that is, $p = mx + b$ is the price of the $x$th item sold. We call $p = mx + b$ the **demand equation** and the graph the **demand curve**.

E X A M P L E 3    **Finding the Demand Equation**

Suppose that 1000 units of a certain item are sold per day by the entire industry at a price of $10 per item and that 2000 units can be sold per day by the same industry at a price of $8 per item. Find the demand equation $p = mx + b$ assuming the demand curve is linear.

Solution    Figure 1.37 shows the two points (1000, 10) and (2000, 8) that lie on the demand curve. Since we are assuming that the demand curve is a straight line, we need only find the straight line through these two points. The slope $m$ is

$$m = \frac{8 - 10}{2000 - 1000} = -0.002.$$

Notice the slope is *negative*, indicating the line slants down moving from left to right. This is what we expect of a reasonable demand curve.

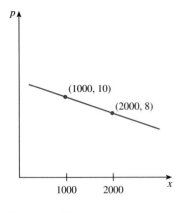

**Figure 1.37**

Now using the point-slope equation for a line with (1000, 10) as the point on the line, we have

$$p - 10 = m(x - 1000)$$
$$= -0.002(x - 1000)$$
$$p = -0.002x + 12. \quad \blacksquare$$

The **supply equation** $p = mx + b$ is the price $p$ that it will take for suppliers to make available $x$ items to the market. The graph of this equation is called the **supply curve**. Figure 1.38 shows a *linear* supply curve. A reasonable supply curve rises moving from left to right since the suppliers of an item will naturally want to sell more items if the price per item is higher.

The best known law of economics is the law of supply and demand. Figure 1.39 shows a demand equation and a supply equation that intersect. The point of intersection is the point at which supply equals demand, which is called the **equilibrium point**. The $x$-coordinate of the equilibrium point is called the **equilibrium quantity** and the $p$-coordinate is called the **equilibrium price**.

**Figure 1.38**

**Figure 1.39**

**E X A M P L E  4**      **Finding the Equilibrium Point**

Suppose the supply curve is given by $p = 0.008x + 5$ and the demand curve is the same as in the previous example. Find the equilibrium point.

**Solution**      The demand and supply equations are given by

$$\text{demand equation: } p = -0.002x + 12$$
$$\text{supply equation: } p = 0.008x + 5.$$

**Figure 1.40**

These are shown in Figure 1.40. The equilibrium point must be on *both* curves, and thus the coordinates of this point must satisfy both equations. Thus

$$-0.002x + 12 = p = 0.008x + 5$$

$$0.01x = 7$$

$$x = 700.$$

Then $x = 700$ is the equilibrium quantity, and when $x = 700$,

$$p = -0.002(700) + 12 = 10.6$$

is the equilibrium price. Thus the equilibrium point is (700, 10.6). ■

## Straight-Line Depreciation

Many assets such as machines or buildings have a *finite* useful life and further-more *depreciate* in value from year to year. For purposes of determining profits and taxes, various methods of depreciation can be used. In **straight-line depre-ciation** we assume that the value $V$ of the asset is given by a *linear* equation in time $t$, say, $V = mt + b$. The slope $m$ must be *negative* since the value of the asset *decreases* over time.

**E X A M P L E   5**     **Straight-Line Depreciation**

A company has purchased a new machine for $100,000 with a useful life of ten years, after which it is assumed that the scrap value of the machine is $5000. If we use straight-line depreciation, write an equation for the value $V$ of the machine where $t$ is measured in years. What will be the value of the machine after the first year? After the second year? After the ninth year?

**Solution**     We assume that $V = mt + b$, where $m$ is the slope and $b$ is the $V$-intercept. We then must find both $m$ and $b$. We are told that the machine is initially worth $100,000, that is, when $t = 0$, $V = 100,000$. Thus the point (0, 100,000) is on the line, and thus 100,000 is the $V$-intercept $b$. See Figure 1.41.

Since the value of the machine in 10 years will be $5000, this means that when $t = 10$, $V = 5000$. Thus (10, 5000) is also on the line. From Figure 1.41, the slope can then be calculated since we now know that the two points (0, 100,000) and (10, 5000) are on the line. Then

$$m = \frac{5000 - 100,000}{10 - 0} = -9500.$$

Thus

$$V = -9500t + 100,000.$$

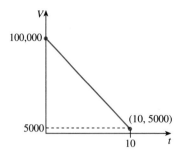

**Figure 1.41**                    **Figure 1.42**

Then

$$\text{When } t = 1, \ V = -9500(1) + 100{,}000 = 90{,}500.$$

$$\text{When } t = 2, \ V = -9500(2) + 100{,}000 = 81{,}000.$$

$$\text{When } t = 9, \ V = -9500(9) + 100{,}000 = 14{,}500. \quad \blacksquare$$

In general, suppose that the initial value of the machine is $I$, the scrap value is $S$, and the useful life is $n$ years. If we assume straight-line depreciation and $V = mt + b$ is the value after $t$ years, then, from Figure 1.42, $(0, I)$ and $(n, S)$ are on the graph. This is true since when $t = 0$, $V = I$, and when $t = n$, $V = S$. Thus the $y$-intercept is $I$. The slope $m$ is

$$m = \frac{S - I}{n - 0} = -\frac{I - S}{n}.$$

Thus

$$V = mt + b = -\frac{I - S}{n} t + I.$$

## SELF-HELP EXERCISE SET 1.3

1. A manufacturer has fixed costs of $10,000 per week and variable costs of $20 per item. The item is sold for $30. Let $x$ denote the number of items manufactured and sold.
   a.  Find the cost equation          b.  Find the revenue equation.
   c.  Find the profit equation.       d.  Find the break-even quantity.

2. A manufacturer notes that it can sell 10,000 items at $6 per item and 8,000 items at $8 per item. Find the demand curve, assuming it is a straight line.

3. Suppose the supply equation for the manufacturer in the previous exercise was $p = .002x + 4$. Find the equilibrium quantity.

# EXERCISE SET 1.3

In Exercises 1 through 4 you are given the cost per item and the fixed costs over a certain time period. Assuming a linear cost model, find the cost equation, where $C$ is cost and $x$ is the number produced.

1. Cost per item = $3, fixed cost = $10,000

2. Cost per item = $6, fixed cost = $14,000

3. Cost per item = $0.02, fixed cost = $1000

4. Cost per item = $0.1, fixed cost = $4000

In Exercises 5 through 8 you are given the price of each item assumed to be independent of the number of items sold. Find the revenue equation, where $R$ is revenue and $x$ is the number sold, assuming the revenue is linear in $x$.

5. Price per item = $5      $R = 5x$          6. Price per item = $10

7. Price per item = $0.1                      8. Price per item = $0.5

9. Using the cost equation found in Exercise 1 and the revenue equation found in Exercise 5, find the profit equation for $P$, assuming the number produced equals the number sold.

10. Using the cost equation found in Exercise 2 and the revenue equation found in Exercise 6, find the profit equation for $P$, assuming the number produced equals the number sold.

11. Using the cost equation found in Exercise 3 and the revenue equation found in Exercise 7, find the profit equation for $P$, assuming the number produced equals the number sold.

12. Using the cost equation found in Exercise 4 and the revenue equation found in Exercise 8, find the profit equation for $P$, assuming the number produced equals the number sold.

In Exercises 13 through 16 linear cost and revenue equations are given. Find the break-even quantity.

13. $C = 2x + 4, R = 4x$                 14. $C = 3x + 10, R = 6x$

15. $C = 0.1x + 2, R = 0.2x$             16. $C = 0.03x + 1, R = 0.04x$

In Exercises 17 through 20 you are given a demand equation and a supply equation. Find the equilibrium point.

17. Demand: $p = -x + 6$, supply: $p = x + 3$

18. Demand: $p = -3x + 12$, supply: $p = 2x + 5$

19. Demand: $p = -10x + 25$, supply: $p = 5x + 10$

20. Demand: $p = -0.1x + 2$, supply: $p = 0.2x + 1$

## Applications

21. **Straight-Line Depreciation.**   A new machine that costs $50,000 has a useful life of 9 years and a scrap value of $5000. Using straight-line depreciation, find the equation for the value $V$ in terms of $t$ where $t$ is in years. Find the value after one year. After 5 years.

22. **Straight-Line Depreciation.**   A new building that costs $1,100,000 has a useful life of 50 years and a scrap value of $100,000. Using straight-line depreciation, find the equation for the value $V$ in terms of $t$ where $t$ is in years. Find the value after one year. After two years. After 40 years.

23. **Demand.**   Suppose that 500 units of a certain item are sold per day by the entire industry at a price of $20 per item and that 1500 units can be sold per day by the same industry at a price of $15 per item. Find the demand equation for $p$, assuming the demand curve to be a straight line.

24. **Demand.**   Suppose that 10,000 units of a certain item are sold per day by the entire industry at a price of $150 per item and that 8000 units can be sold per day by the same industry at a price of $200 per item. Find the demand equation for $p$, assuming the demand curve to be a straight line.

25. **Machine Allocation.**   A furniture manufacturer makes bookcases and small desks, each requiring use of a cutting machine. Each bookcase requires 3 hours on the machine, and each desk requires 4 hours on the machine. The machine is available 120 hours each week.
    a.   If $x$ is the number of bookcases and $y$ the number of desks that are manufactured each week, find the equation that $x$ and $y$ must satisfy if all available time is used on the machine.
    b.   Find the slope of the equation in the previous part and interpret its meaning.

26. **Labor Allocation.**   A company manufacturers three-speed and ten-speed bikes. Each three-speed bike requires 6 hours of labor and each ten-speed bike requires 12 hours of labor. There are 240 hours of labor available each day.
    a.   If $x$ is the number of three-speed bikes and $y$ the number of ten-speed bikes that are manufactured each day, find the equation that $x$ and $y$ must satisfy if all available hours of labor are used.
    b.   Find the slope of the equation in the previous part and interpret its meaning.

27. **Sales Commissions.**   A salesperson is offered two salary plans. The first gives a base salary of $1000 a month and a 5% commission on all sales, while the second plan gives a base salary of $1500 a month and a 4% commission.
    a.   If $x$ is the amount of sales, and $y$ the monthly salary, find the relationship between $x$ and $y$ for each plan.
    b.   How much sales are required so that each plan yields the same total salary?

28. **Inventory.**   A store has 1000 units of a certain item in stock and sells 20 units a day.
    a.   If $y$ is the number of items in stock at the end of day $t$, find the relationship between $y$ and $t$.
    b.   How long until the inventory is exhausted?

29. **Medicine.** Two rules have been suggested by Cowling and by Friend to adjust adult drug dosage levels for young children. Let $a$ denote the adult dosage and let $t$ be the age (in years) of the child. The two rules are given respectively by

$$C = \frac{a}{24}(t + 1) \quad \text{and} \quad F = \frac{2}{25}at.$$

If, for a particular drug, $a = 100$ in the appropriate units, graph both functions. Find the age for which the two rules give the same dosage.

30. **Temperature.** Let $C$ denote the temperature measured on the Celsius scale and $F$ denote the temperature measured on the Fahrenheit scale. The freezing and boiling points of water are 0 degrees and 100 degrees, respectively, on the Celsius scale and 32 degrees and 212 degrees, respectively, on the Fahrenheit scale. If $F$ is a linear equation in $C$, find the equation that relates $F$ to $C$.

31. **Elections.** A candidate for the Senate of the United States obtained 41% of the vote in the previous election and spent $3 million in advertisements. The candidate's advisors estimate that each additional $1 million in advertisement expenditures will obtain an additional 6% of the vote for the next election. Write an equation that relates the advertisement expenditures in millions of dollars with the percent of the vote obtained, assuming a linear relationship. How much must be spent on advertising, according to this analysis, to obtain 50% of the vote in the next election?

32. **Weight of Whales.** It is difficult to weigh large whales, but easy to measure their length. The International Whaling Commission adopted the formula $w = 3.51l - 192$ as a reliable way of relating the length $l$ in feet of adult blue whales with their weight $w$ in British tons. Graph this equation. According to this formula, what must be the increase in weight of these whales for every 5-foot increase in their length?

33. **Make or Buy Decision.** A company includes a manual with each piece of software and is trying to decide whether to contract with an outside supplier or produce the manual in house. The lowest bid of any outside supplier is $.75 per manual. The company estimates that producing the manuals in house will require fixed costs of $10,000 and variable costs of $.50 per manual. Which alternative has the lower total cost if demand is 20,000 manuals?

34. **Make or Buy Decision.** Repeat the previous exercise for a demand of 50,000 manuals.

35. **Process Selection and Capacity.** A machine shop needs to drill holes in a certain plate. An inexpensive manual drill press could be purchased that will require large labor costs to operate, or an expensive automatic press can be purchased that will require small labor costs to operate. The following table summarizes the options.

| Machine | Annual Fixed Costs | Variable Labor Costs | Production Rate |
|---------|---------|---------|---------|
| Manual | $1000 | $16.00 per hour | 10 plates per hour |
| Automatic | $8000 | $2.00 per hour | 100 plates per hour |

If these are the only fixed and variable costs, determine which machine should be purchased by filling in the following table if the number produced is (a) 1000; (b) 10,000.

| Volume | Total Costs | |
|---|---|---|
| | *Manual* | *Automatic* |
| 1,000 | | |
| 10,000 | | |

36. **Costs Per Unit.**   If $C$ is the total cost to produce $x$ items, then the cost per unit is $\dfrac{C}{x}$. Fill in the table with the costs per unit for the indicated volumes for each of the machines in the previous exercise.

| Volume | Costs Per Unit | |
|---|---|---|
| | *Manual* | *Automatic* |
| 1,000 | | |
| 10,000 | | |
| 100,000 | | |

## Graphing Calculator Exercises

In Exercises 37 through 38 find the equilibrium point to two decimal places, using the zoom feature of your graphing calculator.

37.   Demand: $p = -0.07x + 5.15$, supply: $p = 0.03x + 0.65$.

38.   Demand: $p = -3.1x + 25.7$, supply: $p = 0.22x + 2.6$.

## Solutions to Self-Help Exercise Set 1.3

1. **a.**   A manufacturer has fixed costs of $10,000 per week and variable costs of $20 per item, and the number of items manufactured is $x$. The variable costs are the cost per item times the number of items, or $20x$. Thus the cost in dollars is

$$C = 20x + 10,000.$$

   **b.**   The item is sold for $30. Thus the revenue equation is

$$R = 30x.$$

   **c.**   Profit is always revenue less costs. Thus

$$P = R - C = 30x - (20x + 10,000) = 10x - 10,000.$$

   **d.**   The break-even quantity is the quantity for which the profits are zero. Setting $P = 0$ then gives

$$0 = 10x - 10,000$$

$$x = 1000$$

2. If a manufacturer sells 10,000 items at $6 per item and sells 8000 items at $8 per item and the demand curve is assumed to be a straight line, then the slope of this line is given by

$$m = \frac{6 - 8}{10{,}000 - 8000} = -.001$$

See the figure. Then the demand equation is given by the point-slope equation of the line as

$$p - 8 = -.001(x - 8000)$$

$$p = -.001x + 16.$$

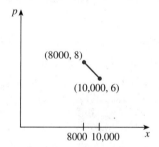

3. If the supply equation for the manufacturer in the previous exercise is $p = .002x + 4$, then to find the equilibrium quantity, we have

$$.002x + 4 = -.001x + 16$$

$$.003x = 12$$

$$x = 4000.$$

## 1.4   METHOD OF LEAST SQUARES*

*Carl Gauss, the Missing Planet, and the Method of Least Squares*

*(David Eugene Smith Collection/Rare Book and Manuscript Library/Columbia University.)*

Carl Gauss (1777–1855), one of the greatest mathematicians of all time, first formulated the method of least squares. He did this in 1794 at the age of 16! In 1781 Sir William Herschel discovered the planet Uranus, bringing the number of planets known at that time to seven. At that time astronomers were certain that an eighth planet existed between Mars and Jupiter, but the search for the missing planet had proved fruitless. On January 1, 1801, Giuseppe Piazzi, director of the Palermo observatory, announced the discovery of a new planet Ceres in this location. (It was subsequently realized that Ceres was actually a large asteroid.) Unfortunately, a few weeks later he lost sight of it. Gauss took up the challenge of determining the orbit from the few recorded observations. He had in his possession one remarkable mathematical tool—the method of least squares, which he had not bothered to publish. Using this method, he predicted the orbit. Astronomers around the world were astonished to find Ceres exactly where he said it would be. This incident catapulted him to fame.

We assumed in the last section that the demand equations for some products were linear. Sometimes we supposed that we were given two data points $(x_1, p_1)$ and $(x_2, p_2)$. That is, $x_1$ items were sold if the price was set at $p_1$, and $x_2$ items were sold if the price was set at $p_2$. If we then *assume* that the demand equation is linear, then of course there is only *one* straight line through these two points, and we can easily calculate the equation $y = ax + b$ of this line.

But suppose that we have more than two data points. Suppose, as in Table 1.5, we have five points available. The $p_i$ are the prices in dollars for a certain product, and the $x_i$ are the corresponding demands for the product in the number of thousands of items sold per day.

**Table 1.5**

| $x_i$ | 1 | 2 | 3 | 5 | 9 |
|-------|----|----|----|----|----|
| $p_i$ | 10 | 9 | 8 | 7 | 5 |

These are plotted in Figure 1.43 which is called a **scatter diagram**. If we eyeball the scatter diagram, we see clearly that the points do not lie on any straight line but seem to be scattered in a more or less linear fashion. Under such circumstances we might be justified in assuming that the demand equation was more or less a straight line. But what straight line? Any line that we draw will miss most of the points. We might then think to draw a line that is somehow closest to the

**Figure 1.43**

*Optional.

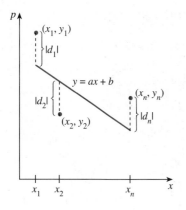

**Figure 1.44**

data points. To actually follow such a procedure, we need to state exactly how we are to measure this closeness.

In this section we will measure this closeness in a manner that will lead us to the **method of least squares**. To be given a nonvertical straight line is the same as to be given two numbers $a$ and $b$ with the equation of the straight line given as $y = ax + b$. Suppose we are given $n$ data points $(x_1, y_1), (x_2, y_2), \ldots, (x_n, y_n)$, and a line $y = ax + b$. We then define $d_1 = y_1 - (ax_1 + b)$ and note from Figure 1.44 that $|d_1|$ is just the vertical distance from the first data point $(x_1, y_1)$ to the line $y = ax + b$. Doing the same for all the data points, we then have

$$d_1 = y_1 - (ax_1 + b)$$
$$d_2 = y_2 - (ax_2 + b)$$
$$\vdots$$
$$d_n = y_n - (ax_n + b),$$

where $|d_2|$ is the vertical distance from the second data point $(x_2, y_2)$ to the line $y = ax + b$, and so on.

Now if all the data points were on the line $y = ax + b$, then all the distances $|d_1|, |d_2|, \ldots, |d_n|$, would be zero. Unfortunately, this will rarely be the case. We then use the sum of the squares of these distances

$$d = d_1^2 + d_2^2 + \cdots + d_n^2$$

as a measure of how close the set of data points is to the line $y = ax + b$. Notice that this number $d$ will be different for different straight lines—large if the straight line is far removed from the data points and small if the straight line passes close to all the data points. We then seek the line or, what is the same thing, the two numbers $a$ and $b$ that will make this sum of squares the least. Thus the name *least squares*.

It can be shown that the numbers $a$ and $b$ that determine the line $y = ax + b$ closest to the set of data points must satisfy the two linear equations

$$\left. \begin{array}{l} (x_1^2 + \cdots + x_n^2)a + (x_1 + \cdots + x_n)b = x_1y_1 + \cdots + x_ny_n \\ (x_1 + \cdots + x_n)a + \qquad\qquad\qquad nb = y_1 + \cdots + y_n \end{array} \right\} \quad (1)$$

We then have the following.

---

**Method of Least Squares**

The line $y = ax + b$ closest to the data points $(x_1, y_1), (x_2, y_2), \ldots, (x_n, y_n)$ can be found by solving the following two linear equations for $a$ and $b$:

$$(x_1^2 + \cdots + x_n^2)a + (x_1 + \cdots + x_n)b = x_1y_1 + \cdots + x_ny_n$$
$$(x_1 + \cdots + x_n)a + \qquad\qquad\qquad nb = y_1 + \cdots + y_n.$$

---

**E X A M P L E   1**      **Using the Method of Least Squares to Find a Demand Equation**

a. Find the best fitting line through the data points in Table 1.5 and thus find a linear demand equation.

b. Estimate the price if the demand is 6000.

**Solutions**      a. Create the following table.

| $x_i$ | $y_i$ | $x_i^2$ | $x_i y_i$ |
|---|---|---|---|
| 1 | 10 | 1 | 10 |
| 2 | 9 | 4 | 18 |
| 3 | 8 | 9 | 24 |
| 5 | 7 | 25 | 35 |
| 9 | 5 | 81 | 45 |
| Sum    20 | 39 | 120 | 132 |

Equation (1) then becomes

$$120a + 20b = 132$$
$$20a + 5b = 39.$$

One can solve by eliminating one of the variables. For example, to eliminate the variable $b$, multiply the second equation by $-4$ and add the result to the first, giving

first equation                  $120a + 20b = \quad 132$

$-4$ times second equation     $-80a - 20b = -156$

add                                   $40a = \quad -24$

This yields $a = -0.6$. Since $a = -0.6$ and $20a + 5b = 39$,

$$b = \frac{39 - 20a}{5} = 10.2.$$

Thus the equation of the best fitting straight line that we are seeking is

$$y = -0.6x + 10.2.$$

The graph is shown in Figure 1.45.

b. The answer to the second question is

$$p = y = (-0.6)(6) + 10.2 = 6.6,$$

that is, if 6000 items are to be sold, then the price should be $6.60.  ∎

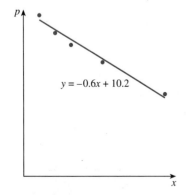

$y = -0.6x + 10.2$

**Figure 1.45**

## SELF-HELP EXERCISE SET 1.4

1. Using the method of least squares, find the best fitting line through the three points $(0, 0)$, $(2, 2)$, $(3, 2)$.

## EXERCISE SET 1.4

In Exercises 1 through 8 find the best fitting straight line to the given set of data, using the method of least squares.

1. $(0, 0)$, $(1, 2)$, $(2, 1)$

2. $(0, 1)$, $(1, 2)$, $(2, 2)$

3. $(0, 0)$, $(1, 1)$, $(2, 3)$, $(3, 3)$

4. $(0, 0)$, $(1, 2)$, $(2, 2)$, $(3, 0)$

5. $(1, 4)$, $(2, 2)$, $(3, 2)$, $(4, 1)$

6. $(0, 0)$, $(1, 1)$, $(2, 2)$, $(3, 4)$

7. $(0, 4)$, $(1, 2)$, $(2, 2)$, $(3, 1)$, $(4, 1)$

 8. $(0, 0)$, $(1, 2)$, $(2, 1)$, $(3, 2)$, $(4, 4)$

## Applications

9. **Sales.** A firm has just introduced a new product and the sales during the first five weeks are given by the following table.

| Week Number | $(x_i)$ | 1 | 2 | 3 | 4 | 5 |
|---|---|---|---|---|---|---|
| Units Sold | $(y_i)$ | 10 | 16 | 20 | 26 | 30 |

Find the best fitting straight line using the method of least squares. What do you think sales will be next week?

10. **Costs.** In the last four years a firm has produced varying amounts of an item at a certain plant and incurred varying total costs given by the following table where the number of items is in thousands per year and the total cost is in millions of dollars per year.

| Number Produced | $(x_i)$ | 2 | 3 | 1 | 3 |
|---|---|---|---|---|---|
| Total Cost | $(y_i)$ | 2 | 4 | 2 | 3 |

Find the best fitting straight line using the method of least squares. What do you think costs per year would be next year if production was set at 2500 units?

11. **Sales.**    A firm has undertaken four different radio advertising campaigns, and the sales in millions of dollars versus the amount spent on advertising in millions of dollars is given in the following table.

| Cost of Advertising | $(x_i)$ | 1 | 2.5 | 3.5 | 4 |
|---|---|---|---|---|---|
| Sales in Dollars | $(y_i)$ | 3.5 | 5 | 6 | 6.5 |

Find the best fitting straight line using the method of least squares. What do you think sales would be in the next campaign if $3 million dollars were spent on radio advertising?

12. **Insect Temperature.**    An insect cannot control its body temperature; the body temperature depends on the temperature of the surrounding air. The following data have been collected for a particular insect species.

| Air Temperature | $(x_i)$ | 18.2 | 23.0 | 25.6 | 30.4 |
|---|---|---|---|---|---|
| Insect Temperature | $(y_i)$ | 19.7 | 24.5 | 26.2 | 31.5 |

Find the best fitting straight line using the method of least squares.

13. **Median Age of U.S. Population.**    The following table gives the median age of the U.S. population.

| Year | $(x_i)$ | 1820 | 1880 | 1950 | 1990 |
|---|---|---|---|---|---|
| Median Age | $(y_i)$ | 17 | 20 | 30 | 33 |

Find the best fitting straight line using the method of least squares.

## Graphing Calculator Exercises

14. **Demand.**    A firm has set different prices for its product in different cities and has obtained the following data relating the price of the product to the number of thousands sold per week at that price.

| Price in Dollars | $(x_i)$ | 8 | 9 | 10 | 11 | 12 | 13 | 14 | 15 | 16 | 17 |
|---|---|---|---|---|---|---|---|---|---|---|---|
| Sales Volume in Thousands | $(y_i)$ | 10 | 8 | 7 | 6 | 5 | 4.5 | 4 | 3.6 | 3 | 2 |

Find the best fitting straight line using the method of least squares. Graph this function. What do you think sales per week would be if the price were set at $7?

## Solutions to Self-Help Exercise Set 1.4

**1.** Create the following table.

| | $x_i$ | $y_i$ | $x_i^2$ | $x_iy_i$ |
|---|---|---|---|---|
| | 0 | 0 | 0 | 0 |
| | 2 | 2 | 4 | 4 |
| | 3 | 2 | 9 | 6 |
| Sum | 5 | 4 | 13 | 10 |

This leads to the following two linear equations.

$$13a + 5b = 10$$
$$5a + 3b = 4$$

Multiplying the first equation by 3 and the second by $-5$ and adding them gives

| | | |
|---|---|---|
| Multiply first equation by 3 | $39a + 15b =$ | $30$ |
| Multiply second equation by $-5$ | $-25a - 15b =$ | $-20$ |
| Adding | $14a =$ | $10$ |

or $a = \dfrac{5}{7}$. Then since $5a + 3b = 4$,

$$b = \frac{4}{3} - \frac{5}{3}a = \frac{4}{3} - \frac{5}{3} \cdot \frac{5}{7} = \frac{1}{7}.$$

The best fitting line is then $y = \dfrac{5}{7}x + \dfrac{1}{7}$.

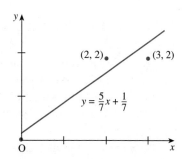

## SUMMARY OUTLINE OF CHAPTER 1

- A **set** is a collection of objects, specified in a manner that enables one to determine if a given object is or is not in the collection. Each object in a set is called a **member** or **element** of the set.

- The "set of all elements $x$ such that $x$ has property $P$" is written in **set-builder notation** as $\{x|x$ has property $P\}$.

- Two sets $A$ and $B$ are **equal**, written $A = B$, if, and only if, they consist of precisely the same elements.

- Construct a **Cartesian coordinate system** with a horizontal line called the **$x$-axis** and a vertical line called the **$y$-axis** both drawn in the plane. The point of intersection of these two lines is called the **origin**. The plane is then called the **$xy$-plane** or the **coordinate plane**. Each point $P$ in the $xy$-plane is assigned a pair of numbers $(a, b)$. The first number $a$ is the horizontal distance from the point $P$ to the $y$-axis and is called the **$x$-coordinate**. The second number $b$ is the vertical distance to the $x$-axis and is called the **$y$-coordinate**.

- The $x$-axis and the $y$-axis divide the plane into four **quadrants**. The **first quadrant** is all points $(x, y)$ with both $x > 0$ and $y > 0$. The **second quadrant** is all points $(x, y)$ with both $x < 0$ and $y > 0$. The **third quadrant** is all points $(x, y)$ with both $x < 0$ and $y < 0$. The **fourth quadrant** is all points $(x, y)$ with both $x > 0$ and $y < 0$.

- The distance $d = d(P_1, P_2)$ between the two points $P_1(x_1, y_1)$ and $P_2(x_2, y_2)$ in the $xy$-plane is given by

$$d(P_1, P_2) = \sqrt{(x_2 - x_1)^2 + (y_2 - y_1)^2}.$$

- The **graph of an equation** is the set $\{(x, y)|x$ and $y$ satisfy the equation$\}$.

- Let $(x_1, y_1)$ and $(x_2, y_2)$ be two points on a line $L$. If the line is nonvertical ($x_1 \neq x_2$), the **slope** $m$ of the line $L$ is defined to be

$$m = \frac{y_2 - y_1}{x_2 - x_1}.$$

If $L$ is vertical ($x_2 = x_1$), then the slope is said to be **undefined**.

- **The Point-Slope Equation of a Line.**   The equation of the line through the point $(x_1, y_1)$ with slope $m$ is given by

$$y - y_1 = m(x - x_1).$$

- **Equation of Vertical Line.**   A vertical line has the equation $x = a$ where $(a, 0)$ is the point at which the vertical line crosses the $x$-axis.

- **The Slope-Intercept Equation of a Line.**   The equation of the line with slope $m$ and $y$-intercept $b$ is given by $y = mx + b$.

- **The General Form of the Equation of a Line.**   The graph of a linear equation $ax + by = c$ is a straight line, and, conversely, every line is the graph of a linear equation.

- **Slope and Parallel Lines.**   Two lines are parallel if, and only if, they have the same slope.

- **Slope and Perpendicular Lines.**   Two nonvertical lines with slope $m_1$ and $m_2$ are perpendicular if, and only if,

$$m_2 = -\frac{1}{m_1}.$$

- **Linear Cost, Revenue, and Profit Equations.** Let $x$ be the number of items made and sold.

$$\text{variable cost} = (\text{cost per item}) \times (\text{number of items produced})$$

$$= mx$$

$$C = \text{cost} = (\text{variable cost}) + (\text{fixed cost})$$

$$= mx + b$$

$$R = \text{revenue} = (\text{price per item}) \times (\text{number sold})$$

$$= px.$$

$$P = \text{profit} = (\text{revenue}) - (\text{cost})$$

$$= R - C.$$

- The quantity at which the profit is zero is called the **break-even quantity**.

- Let $x$ be the number of items made and sold and $p$ the price of each item. A **linear demand equation**, which governs the behavior of the consumer, is of the form $p = mx + b$, where $m$ must be negative. A **linear supply equation**, which governs the behavior of the producer, is of the form $p = mx + b$, where $m$ must be positive.

- The point at which supply equals demand is called the **equilibrium point**. The $x$-coordinate of the equilibrium point is called the **equilibrium quantity** and the $p$-coordinate is called the **equilibrium price**.

- In **straight-line depreciation** we assume that the value $V$ of the asset is given by a *linear* equation in time $t$, say, $V = mt + b$, where the slope $m$ must be *negative*.

- The line $y = ax + b$ closest to the data points $(x_1, y_1), (x_2, y_2), \ldots, (x_n, y_n)$, can be found solving the following two linear equations for $a$ and $b$.

$$(x_1^2 + \cdots + x_n^2)a + (x_1 + \cdots + x_n)b = x_1 y_1 + \cdots + x_n y_n$$

$$(x_1 + \cdots + x_n)a + \qquad\qquad nb = y_1 + \cdots + y_n$$

## Chapter 1 Review Exercises

1. Write the set of odd positive integers in set-builder notation.

2. Write the following sets in roster notation.
   **a.** $\{x \mid x$ is a letter in the word Mississippi$\}$   **b.** $\{x \mid (x + 1)(x - 3) = 0\}$

In Exercises 3 through 6 plot the given point in the $xy$-plane.

3. $(2, 3)$  4. $(-2, -5)$  5. $(-3, 2)$  6. $(4, -5)$

7. Find the points $(x, y)$ in the $xy$-plane that satisfy $xy \leq 0$.

8. Find the distance between the two points $(2, -3)$ and $(-3, -6)$.

9. Sketch the graph of $y^2 = 9x$.

In Exercises 10 through 12 find the slope (if it exists) of the line through each pair of points.

**10.**  $(-2, 4), (1, -3)$   **11.**   $(-2, -3), (2, 3)$   **12.**   $(-2, 5), (-2, -3)$

In Exercises 13 through 15 find the slope (if defined) of the lines. Find the x- and y-intercepts if they exist.

**13.**   $y = 3x + 2$   **14.**   $y = -2$   **15.**   $x = 3$

**16.** Find the equation of the line through the point $(-2, 5)$ with slope $-2$.

**17.** Find the equation of the line through the two points $(-4, 2)$ and $(3, -5)$.

**18.** Find the equation of the line through the point $(-3, -5)$ and parallel to the line $y = 3x - 2$.

**19.** Find the equation of the line through the point $(-3, 7)$ and perpendicular to the line $y = -4x + 5$.

**20.** Assuming a linear cost model, find the equation for the cost $C$, where $x$ is the number produced, the cost per item is $6, and the fixed costs are $2000.

**21.** Assuming a linear revenue equation, find the revenue equation for $R$, where $x$ is the number sold and the price per item is $10.

**22.** Assuming the cost and revenue equations in the previous two problems, find the profit equation. Also find the break-even quantity.

**23.** Given that the cost equation is $C = 5x + 3000$ and the revenue equation is $R = 25x$, find the break-even quantity.

**24.** Given the demand equation $p = -2x + 4000$ and the supply equation $p = x + 1000$, find the equilibrium point.

**25. Transportation.**   A plane travels a straight path from O to A and then another straight path from A to B. How far does the plane travel? Refer to the figure.

**26. Transportation.**   If the plane in the previous exercise travels 200 miles per hour from O to A and 250 miles per hour from A to B, how long does the trip take?

**27. Demand.**   A company notes from experience that it can sell 100,000 pens at $1 each and 120,000 of the same pens at $0.90 each. Find the demand equation, assuming it is linear.

**28.** **Profit.**   It costs a publisher $2 to produce each copy of a weekly magazine. The magazine sells for $2.50 a copy, and the publisher obtains advertising revenue equal to 30% of the revenue from sales. How many copies must be sold to obtain a profit of $15,000?

**29.** **Nutrition.**   A certain woman needs 15 mg of iron in her diet each day but cannot eat fish or liver. She plans to obtain all of her iron from kidney beans (4.5 mg per cup) and soybeans (5.5 mg per cup). If $x$ is the number of cups of kidney beans and $y$ the number of cups of soybeans she consumes each day, what linear equation must $x$ and $y$ satisfy? Interpret the meaning of the slope of this equation.

**30.** **Facility Location.**   A company is trying to decide whether to locate a new plant in Houston or Boston. Information on the two possible locations is given below. The initial investment is in land, buildings, and equipment.

|  | *Houston* | *Boston* |
| --- | --- | --- |
| Variable Cost | $.25 per item | $.22 per item |
| Annual Fixed Costs | $4,000,000 | $4,200,000 |
| Initial Investment | $16,000,000 | $20,000,000 |

Suppose 10,000,000 items are produced each year.
  **a.**   Find which city has the lowest annual total costs, not counting the initial investment.
  **b.**   Which city has the lowest total cost over five years, counting the initial investment?

**31.** Using the method of least squares, find the best fitting line through the 4 points (0, 0), (2, 2), (3, 2), (4, 3).

Systems of equations
are used in the
management of
portfolios.
(© Jeffrey Sylvester/
FPG International
Corporation.)

# Systems of Linear Equations and Matrices

2.1   Introduction to Systems of Linear Equations

2.2   General Systems of Linear Equations

2.3   Introduction to Matrices

2.4   Matrix Multiplication

2.5   Inverse of a Square Matrix

2.6   Leontief Input–Output Analysis

In the last chapter on several occasions we encountered a system of two linear equations in two unknowns. In this chapter we wish to find an efficient method for solving a system with any number of linear equations and any number of unknowns. For example, the following is a system of three linear equations in the four unknowns $x$, $y$, $z$, and $u$:

$$2x + 5y - 7z + 3u = 7$$
$$3x - 4y + 8z + 9u = 4$$
$$6x + 3y - 2z - 4u = 3.$$

Such a system is referred to as a *system of linear equations*.

In the first two sections of this chapter we examine the Gauss–Jordan method, which is used extensively by computers to solve systems of linear equations with many equations and unknowns. Such systems arise in a wide variety of applications. We shall examine some of these applications.

**Then we shall study the basic ideas of matrix theory. We shall look at various applications of this theory, including how matrix theory is connected with solving systems of linear equations.**

## 2.1    INTRODUCTION TO SYSTEMS OF LINEAR EQUATIONS

▶ *Introduction*

▶ *Gauss–Jordan Method*

▶ *Applications*

**Carl Friedrich Gauss, 1777–1855 Wilhelm Jordan, 1842–1899**

Most historians place Gauss among the very top mathematicians of all time. Gauss was a child prodigy born of working-class parents. Fortunately, his mother, although uneducated herself, encouraged the young Gauss to pursue his education. Gauss made fundamental contributions to algebra and number theory and also to astronomy, geodesy, and electricity. He initiated a new branch of mathematics called differential geometry. To determine the orbits of asteroids, he developed a method of solving systems of linear equations by eliminating one variable at a time. This idea was later improved by Wilhelm Jordan, a prominent professor of geodesy, who needed to solve large systems of equations that were involved in triangulation techniques needed for his major geodetic surveys of Germany and the Libyan desert. His improved method is now called the Gauss–Jordan method. It is this method that will be considered in this chapter.

**A P P L I C A T I O N**
**Allocation of Machine Time**

A firm produces three products, a dress shirt, a casual shirt and a sport shirt, and uses a cutting machine, a sewing machine, and a packaging machine in the process. To produce each gross of dress shirts requires 3 hours on the cutting machine and 2 hours on the sewing machine. To produce each gross of casual shirts requires 5 hours on the cutting machine and 1 hour on the sewing machine. To produce each gross of sport shirts requires 7 hours on the cutting machine and 3 hours on the sewing machine. It takes 2 hours to package each gross of each type of shirt. If the cutting machine is available for 480 hours, the sewing machine 170 hours, and the packaging machine 200 hours, determine the number of gross of each shirt that should be produced to use all of the available time on the three machines. (A gross is a dozen dozen or 144.) See Example 3 for the answer.

## Introduction

We first write the equations that are needed to solve the above problem on the allocation of machine time. As usual we let the unknowns be denoted by letters. Let $x$ be the number of gross of dress shirts made, $y$ the number of gross of casual shirts, and $z$ the number of gross of sport shirts. As in the last chapter, create a table that summarizes all of the given information. See Table 2.1.

**Table 2.1**

|  | Dress Shirts | Casual Shirts | Sport Shirts | Total Available |
|---|---|---|---|---|
| Time on packaging machine | $2x$ | $2y$ | $2z$ | 200 |
| Time on cutting machine | $3x$ | $5y$ | $7z$ | 480 |
| Time on sewing machine | $2x$ | $y$ | $3z$ | 170 |

Since each gross of each style of shirt requires two hours on the packaging machine, and this machine is available for 200 hours, we must have

$$2x + 2y + 2z = 200.$$

Since the number of hours on the cutting machine is $3x + 5y + 7z$, while the total hours available is 480, we must have

$$3x + 5y + 7z = 480.$$

Looking at the time spent and available on the sewing machine gives

$$2x + y + 3z = 170.$$

The three equations together give a **system of linear equations**.

$$\left.\begin{array}{l} 2x + 2y + 2z = 200 \\ 3x + 5y + 7z = 480 \\ 2x + y + 3z = 170 \end{array}\right\} \tag{1}$$

In the previous problem there were three equations in three unknowns. In other applications one routinely encounters a system of dozens or even hundreds of linear equations with a similar number of unknowns. Such large systems cannot be solved by hand. They must be solved by a computer, and the computer needs a method to systematically solve such large systems efficiently. Such a method, called the Gauss–Jordan method, is introduced here. Our goal here is to see how this system works so that we can understand how the computer will solve such problems. This will enable us to see what types of problems can be solved by this method, to know what can be expected in the way of solutions (or if there even are solutions), and to be able to interpret our results. Another important goal will be to gain skill in converting certain applied problems into systems of linear equations.

We will solve by hand only small systems. Our goal is not so much to find the solution(s) to a system of linear equations as it is to learn how the Gauss–Jordan method works. We tried to limit the number of computations in the systems and, in particular, to avoid too many fractions.

In this section we will consider how to solve in a systematic manner a system of linear equations such as given in system (1). A **solution** of this system is a triplet of numbers $(a, b, c)$ such that replacing $x$ with $a$, replacing $y$ with $b$, and replacing $z$ with $c$ will yield equalities. Thus the ordered pair (30, 50, 20) is a solution of the above system since replacing $x$ with 30, replacing $y$ with 50, and replacing $z$ with 20 yields

$$2(30) + 2(50) + 2(20) = 200$$

$$3(30) + 5(50) + 7(20) = 480$$

$$2(30) + (50) + 3(20) = 170.$$

(We will see how to obtain this solution shortly.) The set of all solutions to a system is called the **solution set**. Also two systems of equations with the same solution set are said to be **equivalent**.

Let us now consider the following system of linear equations:

$$3x + 9y = 45$$

$$2x + y = 10.$$

The graphs of each of the equations in the above system are lines. For example, the graph of the first equation $3x + 9y = 45$ is a line, say $L_1$. This line is graphed in Figure 2.1. The graph of the second equation $2x + y = 10$ is also a line, say $L_2$. This line is also graphed in Figure 2.1. The reader can easily check that (3, 4) is a solution. The solution (3, 4) must be the point of intersection of the two lines since the ordered pair must satisfy both equations and thus lie on both lines.

Not all systems of linear equations have solutions. Consider the following system.

$$x + y = 2$$

$$x + y = 1$$

Since we cannot have two numbers $x$ and $y$ that add to 2 and also add to 1, there cannot be a solution to this system. The two lines are graphed in Figure 2.2, where we see that the two lines are parallel and not equal. Thus the lines do not intersect.

There is a third possibility. The system of linear equations can have an infinite number of solutions. Consider the following:

$$x - y = 2$$

$$2x - 2y = 4.$$

Since the second equation is simply twice the first, a pair of numbers $(x, y)$ will satisfy the second equation if, and only if, the pair satisfies the first. In other

**Figure 2.1**

**Figure 2.2**

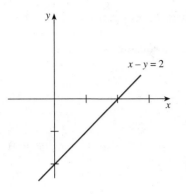

$x - y = 2$

**Figure 2.3**

words the second equation is superfluous. Thus any point on the line $x - y = 2$ will satisfy *both* equations, and there are an infinite number of such points. Geometrically both equations represent the *same* line, as indicated in Figure 2.3. In this case the solution set is the set of all points on the line $x - y = 2$.

In this section we will consider the case when a system has exactly one solution and the case when a system has no solutions. The case when a system has an infinite number of solutions will be left until the next section.

## Gauss–Jordan Method

We now will develop the **Gauss–Jordan method** to solve linear systems of equations. This method, which works for any number of equations, has the following considerable advantages.

---

**Advantages of the Gauss–Jordan Method**

1.  If a solution exists, this method will find it.
2.  If no solution exists, this method will so indicate.
3.  If an infinite number of solutions exist, this method will find all of them.
4.  The method is readily programmable for use on computers.

---

The method will be illustrated with the following example.

**E X A M P L E  1**  **Using the Gauss–Jordan Method**

Solve

$$2x - 2y - 2z = 2$$
$$2x - 3y + z = 10$$
$$x + y - 2z = 0$$

**Solution**  The first step is to place a 1 as the coefficient of $x$ in the first equation.

$$2x - 2y - 2z = 2$$
$$2x - 3y + z = 10$$
$$x + y - 2z = 0$$

STEP 1:
Want a 1 here.

To accomplish this multiply the first equation by $\frac{1}{2}$ to obtain

$$x - y - z = 1$$
$$2x - 3y + z = 10$$
$$x + y - 2z = 0.$$

STEP 2:
Want a 0 here.

Now, for the second step we wish to eliminate $x$ from the second and third equations. To eliminate $x$ from the second equation, add $-2$ times the first equation to the second equation and replace the second equation with the result.

$$\begin{array}{lrcr} -2 \text{ times first} & -2x + 2y + 2z & = & -2 \\ \text{second} & 2x - 3y + z & = & 10 \\ \hline \text{new second} & -y + 3z & = & 8 \end{array}$$

To eliminate $x$ from the third equation add $-1$ times the first equation to the third equation and replace the third equation with the result.

$$\begin{array}{lrcr} -1 \text{ times first} & -x + y + z & = & -1 \\ \text{third} & x + y - 2z & = & 0 \\ \hline \text{new third} & 2y - z & = & -1 \end{array}$$

This yields

$$\begin{array}{rcrcrcr} x & - & y & - & z & = & 1 \\ & & -y & + & 3z & = & 8 \\ & & 2y & - & z & = & -1. \end{array}$$

**STEP 3:** Want a 1 here.

Notice that we have now eliminated $x$ from the second and the third equations. For now we neglect the first equation and only consider the second and third equations, and repeat the application of the first two steps to these two equations. Our goal is to eliminate $y$ from the third equation. To prepare for this, we wish to have a one as the coefficient of $y$ in the second equation. Then for our third step we multiply the second equation by $-1$ and obtain

$$\begin{array}{rcrcrcr} x & - & y & - & z & = & 1 \\ & & y & - & 3z & = & -8 \\ & & 2y & - & z & = & -1. \end{array}$$

**STEP 4:** Want a 0 here.

Now to eliminate $y$ from the third equation, we add $-2$ times the second equation to the third equation and replace the third equation with the result.

$$\begin{array}{lrcr} -2 \text{ times second} & -2y + 6z & = & 16 \\ \text{third} & 2y - z & = & -1 \\ \hline \text{new third} & 5z & = & 15 \end{array}$$

This gives

$$\begin{array}{rcrcrcr} x & - & y & - & z & = & 1 \\ & & y & - & 3z & = & -8 \\ & & & & 5z & = & 15. \end{array}$$

**STEP 5:** Want a 1 here.

Now we can readily solve for $z$ by dividing the third equation by 5 and obtain

$$x - y - z = 1$$
$$y - 3z = -8$$
$$z = 3.$$

STEP 6:
Want a 0 here.

Now we wish to use the third equation to eliminate $z$ from the first and second equations. To do this we add the third equation to the first equation and replace the result with the first equation; we also add three times the third equation to the second equation and replace the result with the second equation. This then yields

$$x - y = 4$$
$$y = 1$$
$$z = 3.$$

STEP 7:
Want a 0 here.

Finally we wish to use the second equation to eliminate $y$ from the first equation. We accomplish this by adding the second equation to the first equation and replacing the result with the first equation giving

$$x = 5$$
$$y = 1$$
$$z = 3.$$

Thus the solution is $(5, 1, 3)$.  ∎

**REMARK.**    We should check to see if this is the correct answer by replacing $x$ with 5, $y$ with 1, and $z$ with 3 in the original system to obtain

$$2(5) - 2(1) - 2(3) = 2$$
$$2(5) - 3(1) + (3) = 10$$
$$(5) + (1) - 2(3) = 0.$$

This verifies the solution.

Sometimes it is necessary to interchange two equations. This is required, for example, if there is no $x$ in the first equation but there is an $x$ in the second equation. In such a case the two equations are interchanged. For example, in the following problem

$$3y = 6$$
$$2x + y = 1,$$

interchange the equations to obtain

$$2x + y = 1$$
$$3y = 6.$$

Then begin with Step 1 and follow the subsequent steps indicated above.

We have performed three basic operations on the system. These are called **elementary** operations; they are summarized as follows.

---

**Elementary Operations**

1. Interchange the $i$th equation with the $j$th equation ($E_i \leftrightarrow E_j$).
2. Replace the $i$th equation by a nonzero constant $k$ times the $i$th equation ($kE_i \rightarrow E_i$).
3. Replace the $i$th equation with the $i$th equation plus $k$ times the $j$th equation ($E_i + kE_j \rightarrow E_i$).

---

It can be proven that such elementary operations do not change the solution set to the system. Thus no solutions are introduced or lost when these operations are applied to the system.

---

**Elementary Operations Yield Equivalent Systems**

Elementary operations yield an equivalent system of equations; that is, elementary operations do not introduce or eliminate any solutions to the system.

---

Notice how in Example 1 we used elementary operations to go from the system

$$2x - 2y - 2z = 2$$
$$2x - 3y + z = 10 \qquad \text{to the system}$$
$$x + y - 2z = 0$$

$$x = 5$$
$$y = 1$$
$$z = 3.$$

Since elementary operations do not change the solution set, the solution set of the original system is the same as the solution set of the last system.

The following is an outline of our strategy for a system like (1) of three linear equations in three unknowns that has one and only one solution. We will carry out our strategy by using elementary row operations.

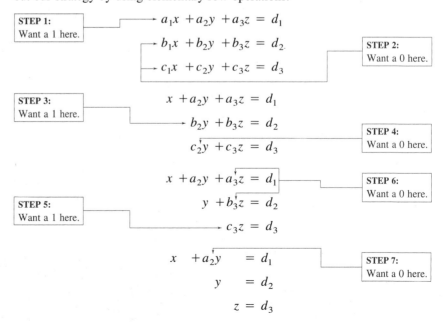

**STEP 1:** Want a 1 here.

$$a_1 x + a_2 y + a_3 z = d_1$$
$$b_1 x + b_2 y + b_3 z = d_2$$
$$c_1 x + c_2 y + c_3 z = d_3$$

**STEP 2:** Want a 0 here.

**STEP 3:** Want a 1 here.

$$x + a_2 y + a_3 z = d_1$$
$$b_2 y + b_3 z = d_2$$
$$c_2' y + c_3 z = d_3$$

**STEP 4:** Want a 0 here.

$$x + a_2 y + a_3 z = d_1$$
$$y + b_3' z = d_2$$
$$c_3 z = d_3$$

**STEP 5:** Want a 1 here.

**STEP 6:** Want a 0 here.

$$x + a_2 y = d_1$$
$$y = d_2$$
$$z = d_3$$

**STEP 7:** Want a 0 here.

The constants are *generic*, that is, they may change from step to step.

We continue with the following example.

**E X A M P L E   2**     **System With No Solution**

Solve

$$2x - 4y = 6$$
$$-3x + 6y = 1.$$

**Solution**

$$2x - 4y = 6 \quad \boxed{\begin{array}{l}\text{STEP 1:}\\ \text{Want a 1 here.}\end{array}} \qquad \tfrac{1}{2}E_1 \rightarrow E_1$$
$$-3x + 6y = 1$$

$$x - 2y = 3 \quad \boxed{\begin{array}{l}\text{STEP 2:}\\ \text{Want a 0 here.}\end{array}}$$
$$-3x + 6y = 1 \qquad\qquad\qquad E_2 + 3E_1 \rightarrow E_2$$

$$x - 2y = 3$$

$$0 = 10 \quad \text{contradiction}$$

Since the second equation yields a contradiction, the system has no solution.   ■

We shall see in the next section that some systems have infinite numbers of solutions. Thus there are three possibilities in solving a system of linear equations.

---

**The Three Possibilities in Solving a System of Linear Equations**

1. There can be precisely one solution.
2. There can be no solution.
3. There can be infinitely many solutions.

---

## Applications

The following is a typical (scaled-down) type of application in allocation theory that can be solved using the Gauss–Jordan method.

**E X A M P L E   3**     **Allocation of Machine Time**

A firm produces three products, a dress shirt, a casual shirt, and a sport shirt, and uses a cutting machine, a sewing machine, and a packaging machine in the process. To produce each gross of dress shirts requires 3 hours on the cutting machine and 2 hours on the sewing machine. To produce each gross of casual shirts requires 5 hours on the cutting machine and 1 hour on the sewing machine. To produce each gross of sport shirts requires 7 hours on the cutting machine and 3 hours on the sewing machine. It takes 2 hours to package each gross of each type of shirt. If the cutting machine is available for 480 hours, the sewing machine 170 hours, and the packaging machine 200 hours, determine the number of gross of each shirt that should be produced to use all of the available time on the three machines.

**Solution**   If $x$ is the number of gross of dress shirts, $y$ the number of gross of casual shirts, and $z$ the number of gross of sport shirts produced, then we saw at the beginning of this section that $x$, $y$, and $z$ must satisfy the following system of linear equations

$$2x + 2y + 2z = 200$$
$$3x + 5y + 7z = 480$$
$$2x + \phantom{5}y + 3z = 170.$$

We use the Gauss–Jordan method of solution. Thus

$$2x + 2y + 2z = 200$$
$$3x + 5y + 7z = 480$$
$$2x + \phantom{5}y + 3z = 170$$

**STEP 1:**
Want a 1 here.

$\frac{1}{2}E_1 \rightarrow E_1$

$$x + \phantom{5}y + \phantom{5}z = 100$$
$$3x + 5y + 7z = 480$$
$$2x + \phantom{5}y + 3z = 170$$

**STEP 2:**
Want a 0 here.

$E_2 - 3E_1 \rightarrow E_2$

$E_3 - 2E_1 \rightarrow E_3$

$$x + \phantom{5}y + \phantom{5}z = \phantom{-}100$$
$$2y + 4z = \phantom{-}180$$
$$-y + \phantom{5}z = -30$$

**STEP 3:**
Want a 1 here.

$\frac{1}{2}E_2 \rightarrow E_2$

$$x + y + \phantom{5}z = \phantom{-}100$$
$$y + 2z = \phantom{-}90$$
$$-y + \phantom{5}z = -30$$

**STEP 4:**
Want a 0 here.

$E_3 + E_2 \rightarrow E_3$

$$x + y + \phantom{5}z = 100$$
$$y + 2z = \phantom{5}90$$
$$3z = \phantom{5}60$$

**STEP 5:**
Want a 1 here.

$\frac{1}{3}E_3 \rightarrow E_3$

$$x + y + \phantom{5}z = 100$$
$$y + 2z = \phantom{5}90$$
$$z = \phantom{5}20$$

**STEP 6:**
Want a 0 here.

$E_1 - E_3 \rightarrow E_1$

$E_2 - 2E_3 \rightarrow E_2$

$$x + \phantom{5}y = 80$$
$$y = 50$$
$$z = 20$$

**STEP 7:**
Want a 0 here.

$E_1 - E_2 \rightarrow E_1$

$$x = 30$$
$$y = 50$$
$$z = 20.$$

Thus the firm should produce 30 gross of dress shirts, 50 gross of casual shirts, and 20 gross of sport shirts to use all the available time on the 3 machines. ∎

## SELF-HELP EXERCISE SET 2.1

1. Solve the following system of equations:

$$2x + 3y = 1$$
$$6x + 11y = 1.$$

2. Find the number $a$ such that the following system has no solution:

$$x + 2y = 3$$
$$2x + ay = 5.$$

3. A store sells 30 sweaters on a certain day. The sweaters come in three styles: A, B, and C. Style A costs $30, style B $40, and style C $50. If the store sold $1340 worth of these sweaters on that day and the number of C sold exceeded by 6 the sum of the other two styles sold, how many sweaters of each style were sold?

## EXERCISE SET 2.1

In Exercises 1 through 22 use the Gauss–Jordan method to find all solutions, if any exist, of the given systems.

1. $x + 2y = 12$
   $2x + 3y = 19$

2. $x + 3y = 2$
   $3x + 4y = 1$

3. $4x - 8y = 20$
   $-x + 3y = -7$

4. $3x - 12y = 3$
   $-2x + y = -9$

5. $-2x + 8y = -6$
   $-2x + 3y = -1$

6. $-3x + 12y = -21$
   $-10x + 2y = 6$

7. $3x + 6y = 0$
   $x - y = -3$

8. $2x + 4y = 8$
   $2x - 4y = 0$

9. $2x - 4y = 8$
   $-x + 2y = 4$

10. $-3x + 6y = 10$
    $x - 2y = 4$

11. $3x - 6y = 12$
    $-x + 2y = 5$

12. $-2x + 4y = 12$
    $x - 2y = 1$

13. $-x + 3y = 7$
    $2x - 6y = 14$

14. $x - 3y = -8$
    $-2x + 6y = 16$

15. $0.1x - 0.3y = 0.4$

$-2x + 6y = 4$

16. $-0.1x + 0.3y = 0.5$

$2x - 6y = 1$

17. $3x - 3y + 6z = -3$

$2x + y + 2z = 4$

$2x - 2y + 5z = -2$

18. $2x - 4y + 2z = -6$

$3x + 4y + 5z = 1$

$2x - y + z = -3$

19. $x + y + z = 10$

$x - y + z = 10$

$x + y - z = 0$

20. $x + y + z = 1$

$2x - y + z = 2$

$3x + 2y + 5z = 3$

21. $x + y + z = 1$

$x - y + z = 2$

$3x + y + 3z = 1$

22. $x + y + z = 1$

$x - y - z = 1$

$4x + 4y + 2z = 2$

23. A person has 25 coins, all of which are quarters and dimes. If the face value of the coins is $3.25, how many of each type of coin must this person have?

24. A person has three times as many dimes as quarters. If the total face value of these coins is $2.20, how many of each type of coin must this person have?

25. A person has 36 coins, all of which are nickels, dimes, and quarters. If there are twice as many dimes as nickels and if the face value of the coins is $4, how many of each type of coin must this person have?

26. A person has three times as many nickels as quarters and three more dimes than nickels. If the total face value of these coins is $2.40, how many of each type of coin must this person have?

## Applications

27. **Scheduling.** An insurance company has two types of documents to process, type A and type B. The type A document needs to be examined for 2 hours by the accountant and 3 hours by the attorney, while the type B needs to be examined for 3 hours by the accountant and 1 hour by the attorney. If the accountant has 38 hours and the attorney 36 hours each week to spend working on these documents, how many documents of each type can they process?

28. **Mixture.** A small store sells mint tea at $3.20 an ounce and peppermint tea at $4 an ounce. The store owner decides to make a batch of 100 ounces of tea that mixes both kinds and sell the mixture for $3.50 an ounce. How many ounces of each of the two varieties of tea should be mixed to obtain the same revenue as selling them unmixed?

29. **Investments.** An individual has a total of $1000 in two banks. The first bank pays 8 percent a year and the second pays 10 percent a year. If the person receives $86 of interest in the first year, how much money must have been deposited in each bank?

30. **Nutrition.** A dietitian must plan a meal for a patient using two fruits, oranges and strawberries. Each orange contains 1 gram of fiber and 75 mg of vitamin C, while each cup of strawberries contains 2 grams of fiber and 60 mg of vitamin

C. How much of each of these fruits needs to be eaten so that a total of 8 grams of fiber and 420 mg of vitamin C will be obtained?

31. **Production Scheduling.** A small plant with a cutting department and a sewing department produces two styles of sweaters, style A and style B. It takes 0.5 work-hours to cut a style A sweater and 0.6 work-hours to sew it. It takes 0.4 work-hours to cut a style B sweater and 0.3 work-hours to sew it. If the cutting department has 200 work-hours available each day and the sewing department has 186 work-hours available each day, how many sweaters of each style can be produced if both departments work at full capacity?

32. **Sales.** A store sells two types of sweaters: style A and style B. Style A costs $30 while style B costs $40. On a certain day the store sells twice as many sweaters of style A as style B. If the store sold $400 worth of these sweaters on that day, how many sweaters of each style were sold?

33. **Scheduling.** An insurance company has three types of documents to process, type A, type B, and type C. The type A document needs to be examined for 2 hours by the accountant and 3 hours by the attorney. The type B document needs to be examined for 4 hours by the accountant and 2 hours by the attorney. Finally, the last document needs to be examined for 2 hours by the accountant and 4 hours by the attorney. The secretary needs 3 hours to type each document. If the accountant has 34 hours, the attorney 35 hours, and the secretary 36 hours available to spend working on these documents, how many documents of each type can they process?

34. **Investments.** A pension fund manager has been given a total of $900,000 by a corporate client to invest in certain types of stocks and bonds over the next year. The client requires that twice as much money be in bonds as in stocks. Also stocks must be restricted to a certain class of low-risk stocks and another class of medium-risk stocks. Furthermore, the client demands an annual return of 8%. The pension fund manager assumes from historical data that the low risk stocks should return 9% annually, the medium risk stocks 11% annually, and bonds 7%. How should the pension fund manager allocate funds among the three groups to meet all the demands of the client?

35. **Investments.** Suppose in the previous problem the client demanded an annual return of 9%. What should the pension fund manager do?

36. **Production Scheduling.** A small plant with a cutting department, a sewing department, and a packaging department produces three styles of sweaters: styles A, B, and C. It takes 0.4 work-hours to cut a style A and 0.2 work-hours to sew it. It takes 0.3 work-hours to cut a style B sweater and the same to sew it. It takes 0.5 work-hours to cut a style C and 0.6 work-hours to sew it. It takes 0.1 work-hours to package each of the sweaters. If the cutting department has 110 work-hours available each day, the sewing department has 95 work-hours available each day, and the packaging department has 30 hours, how many sweaters of each style can be produced if all departments work at full capacity?

## Graphing Calculator Exercises

In Exercises 37 through 40 use the zoom feature of your graphing calculator to find the solution to two decimal places to each of the following systems.

**37.** $3x + 4y = 12$

$2x - 5y = 10$

**38.** $3x + 5y = 15$

$3x - 4y = 12$

**39.** $2.1x + 4.1y = 16$

$2.8x + 2.2y = 12$

**40.** $3.3x + 9.4y = 15.1$

$0.9x + 2.2y = 4.2$

In Exercises 41 through 42 use your graphing calculator to determine if there is one solution, an infinite number of solutions, or no solution to the given system.

**41.** $-2x + 4y = 12$

$x - 2y = 2$

**42.** $x - 4y = -8$

$-2x + 8y = 16$

## Solutions to Self-Help Exercise Set 2.1

**1.** Use the Gauss–Jordan method of solution and obtain

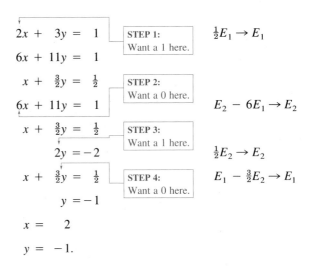

$$x = 2$$
$$y = -1.$$

**2.** Use the Gauss–Jordan method of solution and obtain

$$x + 2y = 3$$

$$2x + ay = 5$$

STEP 2:
Want a 0 here.    $E_2 - 2E_1 \rightarrow E_2$

$$x + 2y = 3$$

$$(a - 4)y = -1.$$

Now if $a = 4$, the last line reads $0 = -1$ and is a contradiction. Thus if $a = 4$, there is no solution to the system.

**3.** Let $x$ be the number of style A sweaters sold, $y$ the number of style B, and $z$ the number of style C. Since a total of 30 were sold, $x + y + z = 30$. The revenue in dollars from style A is $\$30x$, from style B is $\$40y$, and from style C is $\$50z$.

Since \$1340 worth of the sweaters were sold, $30x + 40y + 50z = 1340$. Finally, since the number of style C exceeded by 6 the sum of the other two styles, $z = x + y + 6$, or $-x - y + z = 6$. Thus we wish to solve the system

$$
\begin{aligned}
x + y + z &= 30 \\
30x + 40y + 50z &= 1340 \\
-x - y + z &= 6.
\end{aligned}
$$

Use the Gauss–Jordan method of solution and obtain

$$
\begin{aligned}
x + y + z &= 30 \\
30x + 40y + 50z &= 1340 \\
-x - y + z &= 6
\end{aligned}
$$
Want a 0 here.
$$
\begin{aligned}
E_2 - 30E_1 &\to E_2 \\
E_3 + E_1 &\to E_3
\end{aligned}
$$

$$
\begin{aligned}
x + y + z &= 30 \\
10y + 20z &= 440 \\
2z &= 36
\end{aligned}
$$
Want a 1 here.
$$
\begin{aligned}
\tfrac{1}{10}E_2 &\to E_2 \\
\tfrac{1}{2}E_3 &\to E_3
\end{aligned}
$$

$$
\begin{aligned}
x + y + z &= 30 \\
y + 2z &= 44 \\
z &= 18
\end{aligned}
$$
Want a 0 here.
$$
\begin{aligned}
E_1 - E_3 &\to E_1 \\
E_2 - 2E_3 &\to E_2
\end{aligned}
$$

$$
\begin{aligned}
x + y &= 12 \\
y &= 8 \\
z &= 18
\end{aligned}
$$
Want a 0 here.
$$
E_1 - E_2 \to E_1
$$

$$
\begin{aligned}
x &= 4 \\
y &= 8 \\
z &= 18.
\end{aligned}
$$

Thus the store sold 4 of style A, 8 of style B, and 18 of style C.

## **2.2**    GENERAL SYSTEMS OF LINEAR EQUATIONS

► *Augmented Matrix*
► *The Gauss–Jordan Method (Continued)*
► *Applications*

**The Gauss–Jordan Method in Ancient China**

Actually the Gauss–Jordan method was known in ancient China about 300 B.C. and was used as a practical method of solving concrete problems. The method was known as "the way of calculating by arrays." At this time in history, it had not occurred to anyone to set the unknowns in a problem equal to letters as we now do. Thus a system of linear equations such as

$$2x + 5y = 10$$
$$3x - 7y = 15$$

was written as the square array of numbers

$$
\begin{array}{rrr}
2 & 5 & 10 \\
3 & -7 & 15
\end{array}
$$

However, rods replaced the numbers in the array. The number 2 was replaced with a black rod, 2 units in length, while the number 5 was replaced with a black rod, 5 units in length. A negative number was represented by a red rod. The method of eliminating one unknown at a time then proceeded just as was done in the last section, where one imagined that the first column represented the first unknown, the second column the second unknown, and the third column the number to the right of the equality. We shall now follow this system.

**APPLICATION**
**ıputer Purchase**

A firm must purchase a total of 100 computers, some of small, some of medium, and some of large capacity. The small capacity computers cost $2000 each, the medium capacity cost $6000 each, and the large capacity cost $8000 each. If the firm plans to spend all of $400,000 on the total purchase, find the number of each type to be bought. For the answer see Example 5.

## Augmented Matrix

We now introduce the augmented matrix, which will save a little work. We then extend the Gauss–Jordan method that was introduced in the last section to linear systems with any number of equations and variables. We will then consider more applications.

In the last section we solved systems of linear equations such as

$$2x + 3y = 4$$
$$5y = -2.$$

It is convenient to write this system in the abbreviated form as

$$
\begin{bmatrix}
2 & 3 & 4 \\
0 & 5 & -2
\end{bmatrix}.
$$

Such a rectangular array of numbers is called a **matrix**. We will study matrices in detail in later sections. At this time we do need to know that the first line of the array is called the **first row** and the second line the **second row**, and so on if there are additional lines.

$$\begin{bmatrix} 2 & 3 & 4 \\ 0 & 5 & -2 \end{bmatrix} \begin{array}{l} \leftarrow \textbf{Row 1} \\ \leftarrow \textbf{Row 2} \end{array}$$

It is convenient to place a vertical line to separate the last column of the matrix from the rest of the matrix.

$$\begin{bmatrix} 2 & 3 & \big| & 4 \\ 0 & 5 & \big| & -2 \end{bmatrix}$$

In this form the matrix is called the **augmented matrix** for the system. The only difference between this augmented matrix and the original system is that in the augmented matrix the symbols $x$, $y$, and $=$ have been dropped. This spares us the work of always writing these symbols for each step of the solution. Now if we wish to interchange two equations in the system, we interchange the two corresponding rows in the augmented matrix. If we wish to multiply a constant times each side of one equation, we multiply a constant times each member of the corresponding row, etc. In general we have the following.

---

**Elementary Row Operations**

1.  Interchange the $i$th row with the $j$th row $(R_i \leftrightarrow R_j)$.
2.  Multiply each member of the $i$th row by a nonzero constant $k$ $(kR_i \rightarrow R_i)$.
3.  Replace each element in the $i$th row with the corresponding element in the $i$th row plus $k$ times the corresponding element in the $j$th row $(R_i + kR_j \rightarrow R_i)$.

---

## The Gauss–Jordan Method (Continued)

**E X A M P L E   1**      **System With Many Solutions**

Solve

$$-3x + 6y = 12$$
$$2x - 4y = -8.$$

**Solution**      We first replace this system with the augmented matrix

$$\begin{bmatrix} -3 & 6 & \big| & 12 \\ 2 & -4 & \big| & -8 \end{bmatrix}.$$

Using the Gauss–Jordan method, we have

$$\begin{bmatrix} -3 & 6 & | & 12 \\ 2 & -4 & | & -8 \end{bmatrix} \quad \boxed{\begin{array}{l}\textbf{STEP 1:} \\ \text{Want a 1 here.}\end{array}} \qquad -\tfrac{1}{3}R_1 \rightarrow R_1$$

$$\begin{bmatrix} 1 & -2 & | & -4 \\ 2 & -4 & | & -8 \end{bmatrix} \quad \boxed{\begin{array}{l}\textbf{STEP 2:} \\ \text{Want a 0 here.}\end{array}}$$

$$R_2 - 2R_1 \rightarrow R_2$$

$$\begin{bmatrix} 1 & -2 & | & -4 \\ 0 & 0 & | & 0 \end{bmatrix} \quad \text{or} \qquad \begin{aligned} x - 2y &= -4 \\ 0 &= 0. \end{aligned}$$

Since the second equation is automatically satisfied, the solutions of the system will be all solutions of $x - 2y = -4$. We can solve this last equation for $x$ or for $y$. We will solve for $x$ since this is easier. If we let $y = t$, where $t$ is any real number, and solve for $x$, we obtain $x = 2t - 4$. Then we have

$$x = 2t - 4$$
$$y = t,$$

and the pair $(2t - 4, t)$ is a solution for any real number $t$.   ■

The above variable $t$ is called a **parameter,** and replacing $t$ with any real number produces a **particular solution** to the system. For example, setting $t = 2$ yields the particular solution $(0, 2)$, while setting $t = 0$ yields the particular solution $(-4, 0)$. The solution $(2t - 4, t)$ in the previous problem is called the **general solution.**

Our goal is always to use elementary row operations to reduce the augmented matrix to a simple form where the solution can be read by inspection. When the augmented matrix is in this simple form, we called it the *reduced matrix*. More precisely, we have the following definition.

---

**Reduced Matrix**

A matrix is in **reduced form** if:

1. The leftmost nonzero element in each row is 1.
2. The column containing the leftmost 1 of any row has all other elements in that column equal to 0.
3. The leftmost 1 in any row is to the left of the leftmost 1 in a lower row.
4. Any row consisting of all zeros must be below any row with at least one nonzero element.

---

For example, all of the following matrices are in reduced form:

$$\begin{bmatrix} 1 & 0 & | & 2 \\ 0 & 1 & | & 0 \\ 0 & 0 & | & 0 \end{bmatrix}, \quad \begin{bmatrix} 1 & 4 & 0 & | & 3 \\ 0 & 0 & 1 & | & 2 \\ 0 & 0 & 0 & | & 0 \end{bmatrix}, \quad \begin{bmatrix} 1 & 0 & 3 & 0 & | & 3 \\ 0 & 1 & 2 & 0 & | & 2 \\ 0 & 0 & 0 & 1 & | & 3 \end{bmatrix}.$$

None of the following matrices are in reduced form

$$\begin{bmatrix} 1 & 0 & | & 2 \\ 0 & 2 & | & 2 \end{bmatrix}, \quad \begin{bmatrix} 1 & 2 & 3 & | & 4 \\ 0 & 0 & 1 & | & 2 \end{bmatrix}, \quad \begin{bmatrix} 0 & 1 & | & 1 \\ 1 & 0 & | & 2 \end{bmatrix}, \quad \begin{bmatrix} 1 & 0 & 0 & | & 1 \\ 0 & 0 & 0 & | & 0 \\ 0 & 1 & 0 & | & 2 \end{bmatrix}$$

since the first violates the first condition, the second violates the second condition, the third the third condition, and the fourth the fourth condition.

**E X A M P L E  2**     **Infinite Number of Solutions**

Finish the solution to the system that has the augmented matrix as follows:

$$\begin{bmatrix} 1 & 2 & 3 & | & 4 \\ 0 & 0 & 0 & | & 0 \\ 0 & 1 & 1 & | & 1 \end{bmatrix}.$$

**Solution**

$$\begin{bmatrix} 1 & 2 & 3 & | & 4 \\ 0 & 0 & 0 & | & 0 \\ 0 & 1 & 1 & | & 1 \end{bmatrix} \quad \boxed{\text{Need a 1 here.}} \quad R_2 \leftrightarrow R_3$$

$$\begin{bmatrix} 1 & 2 & 3 & | & 4 \\ 0 & 1 & 1 & | & 1 \\ 0 & 0 & 0 & | & 0 \end{bmatrix} \quad \boxed{\text{Need a 0 here.}} \quad R_1 - 2R_2 \rightarrow R_1$$

$$\begin{bmatrix} 1 & 0 & 1 & | & 2 \\ 0 & 1 & 1 & | & 1 \\ 0 & 0 & 0 & | & 0 \end{bmatrix} \quad \text{or} \quad \begin{aligned} x \quad\;\;\; + z &= 2 \\ y + z &= 1 \\ 0 &= 0 \end{aligned}$$

Now we have two linear equations in three unknowns. Thus any two of these unknowns can be solved in terms of the third. The following technique indicates just how this is to be done. Consider now any nonzero row in the reduced form. The leftmost nonzero element in this row must be a one, which we refer to as a **leading one**.

$$\textbf{Leading one} \longrightarrow \begin{bmatrix} 1 & 0 & 1 & | & 2 \\ 0 & 1 & 1 & | & 1 \\ 0 & 0 & 0 & | & 0 \end{bmatrix}$$

The variable in the column containing the leading one is called a **leading variable**.

Leading variable

$$\begin{array}{ccc} x & y & z \end{array}$$

$$\begin{bmatrix} 1 & 0 & 1 & | & 2 \\ 0 & 1 & 1 & | & 1 \\ 0 & 0 & 0 & | & 0 \end{bmatrix}$$

In the above case, $x$ and $y$ are leading variables. The remaining variables are called **free variables**. In the above case, $z$ is the free variable. The reduced form is such that it is always a simple matter to solve for the leading variables in terms of the free variables. In the above case, the free variable $z$ is "free" to be any real number. We then set $z = t$, where $t$ is any real number. Now we solve easily for the leading variables and obtain the general solution $(2 - t, 1 - t, t)$.  ■

**E X A M P L E   3**    **Infinite Number of Solutions**

Given the following augmented matrix, finish the problem of finding the solution to the corresponding system:

$$\begin{array}{ccc} x & y & z \end{array}$$

$$\begin{bmatrix} 1 & 2 & 3 & | & 4 \\ 0 & 0 & 0 & | & 0 \\ 0 & 0 & 0 & | & 0 \end{bmatrix}.$$

**Solution**    The augmented matrix is in reduced form. Now there is one leading variable $x$ and two free variables $y$ and $z$.

Leading variable

$$\begin{array}{ccc} x & y & z \end{array}$$

Leading one —$\begin{bmatrix} 1 & 2 & 3 & | & 4 \\ 0 & 0 & 0 & | & 0 \\ 0 & 0 & 0 & | & 0 \end{bmatrix}$

Thus both $y$ and $z$ are taken to be any real number. We indicate this by setting $y = s$ and $z = t$, where $s$ and $t$ are arbitrary numbers (that need not be equal). Thus the general solution has the form $(4 - 2s - 3t, s, t)$, where $s$ and $t$ are parameters. A particular solution can be obtained by setting each of the parameters equal to some number. For example, setting $s = 1$ and $t = 2$, gives $(-4, 1, 2)$ as a particular solution.  ■

E X A M P L E  4    **Infinite Number of Solutions**

Given the following augmented matrix, finish the problem of finding the solution to the corresponding system.

$$
\begin{array}{cccccc}
x & y & z & u & v & w \\
\end{array}
$$

$$
\left[\begin{array}{cccccc|c}
1 & 2 & 0 & 4 & 5 & 0 & 4 \\
0 & 0 & 1 & 8 & 9 & 0 & 3 \\
0 & 0 & 0 & 0 & 0 & 1 & 2
\end{array}\right]
$$

**Solution**    The augmented matrix is in reduced form with $x$, $z$, and $w$ the leading variables. Thus $y$, $u$, and $v$ are the free variables.

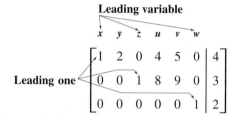

Setting $y = r$, $u = s$, and $v = t$, where $r$, $s$, and $t$ are the parameters, the solution can then immediately be written as $x = 4 - 2r - 4s - 5t$, $y = r$, $z = 3 - 8s - 9t$, $u = s$, $v = t$, $w = 2$, or as $(4 - 2r - 4s - 5t, r, 3 - 8s - 9t, s, t, 2)$. ∎

**Geometric Interpretations.**    It turns out that any equation of the form $ax + by + cz = d$ (with not all of the constants $a$, $b$, and $c$ being zero) is the equation of a plane in space. If there are 3 equations in 3 unknowns, Figure 2.4 indicates the possibilities. The three planes could intersect at a point indicating that the corresponding linear system of three equations in three unknowns has precisely one solution. The three planes could intersect in one line or in an entire plane, giving an infinite number of solutions to the corresponding system. Or there could

Figure 2.4

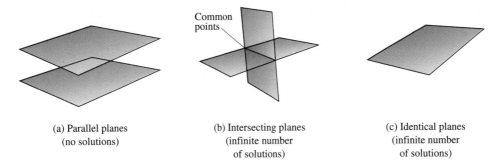

(a) Parallel planes
(no solutions)

(b) Intersecting planes
(infinite number
of solutions)

(c) Identical planes
(infinite number
of solutions)

**Figure 2.5**

be no point of intersection of the three planes, indicating that the corresponding system has no solution.

The geometry can give us further insights. For example, from Figure 2.5 we can see that two linear equations in three unknowns cannot have precisely one solution. If two linear equations in three unknowns had precisely one solution, this would mean that the corresponding two planes intersected in precisely one point, but this is impossible. Thus two linear equations in three unknowns has either no solution or an infinite number of solutions. This is true in general for any system with more variables than equations.

---

**More Variables Than Equations**

A system of equations that has more variables than equations has no solution or an infinite number of solutions.

---

By analyzing the reduced matrix one can see why this theorem must be true in general. Let us first assume the case in which there is no solution, that is, there is no row with all zeros except for the last entry. Then

$$\text{number of leading variables} = \text{number of leading ones}$$
$$\leq \text{number of equations}$$
$$< \text{number of variables}$$

Thus there is one variable left to be free, and, therefore, the system has an infinite number of solutions.

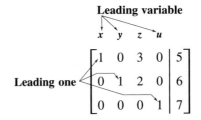

## Applications

**E X A M P L E   5**       **Computer Purchase**

A firm must purchase a total of 100 computers, some of small, some of medium, and some of large capacity. The small capacity computers cost $2000 each, the medium capacity cost $6000 each, and the large capacity cost $8000 each. If the firm plans to spend all of $400,000 on the total purchase, find the number of each type to be bought.

**Solution**       Let $x$, $y$, and $z$ be the respective number of small, medium, and large capacity computers. Then the first sentence indicates that $x + y + z = 100$.

The cost of purchasing $x$ small computers is $2000x$ dollars, of purchasing $y$ medium computers is $6000y$ dollars, and of purchasing $z$ large computers is $8000z$ dollars. Since the total cost is $400,000, we have $2000x + 6000y + 8000z = 400,000$ or, in terms of thousands of dollars, $2x + 6y + 8z = 400$.

We then have the system

$$x + \ y + \ \ z = 100$$
$$2x + 6y + 8z = 400.$$

Using the augmented matrix and the Gauss–Jordan method, we obtain

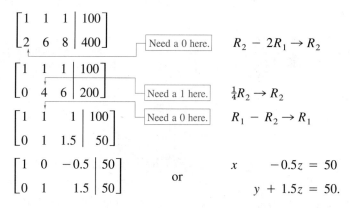

This gives

$$x = .50t + 50$$
$$y = 50 - 1.5t$$
$$z = t.$$

First notice that $t$ must be *even* in order to ensure that $x$ and $y$ will be *integers*. Of course, $x$, $y$, and $z$ must be nonnegative. Thus $t$ is nonnegative. Now $0 \le x = .50t + 50$ implies $-100 \le t$. However, this gives no new information. Also $0 \le y = 50 - 1.5t$. This implies that $t \le 100/3$. Thus the firm can purchase $.50t + 50$ small computers, $50 - 1.5t$ medium computers, and $t$ large computers, where $t$ is any even integer with $0 \le t \le 32$. For example, the firm can take $t = 20$

giving $x = .50(20) + 50 = 60$ small capacity computers, $y = 50 - 1.5(20) = 20$ medium capacity computers, and $t = 20$ large capacity computers, spending the entire \$400,000 and obtaining a total of 100 computers.   ■

## SELF-HELP EXERCISE SET 2.2

1. Given the following augmented matrix, finish the problem of finding the solution to the corresponding system:

$$\begin{array}{ccc} x & y & z \end{array}$$
$$\left[\begin{array}{ccc|c} 1 & 2 & 1 & 3 \\ 0 & 0 & 1 & 2 \\ 0 & 0 & 0 & 0 \end{array}\right].$$

2. Given the following augmented matrix, finish the problem of finding the solution to the corresponding system:

$$\begin{array}{ccc} x & y & z \end{array}$$
$$\left[\begin{array}{ccc|c} 0 & 1 & 0 & 1 \\ 0 & 0 & 1 & 2 \\ 0 & 0 & 0 & 0 \end{array}\right].$$

3. A contractor has 2000 hours of labor available for three projects. The cost per work-hour for each of the three projects is \$10, \$12, and \$14 respectively, and the total labor cost is \$25,000. Find the number of work-hours that should be allocated to each project if all the available work-hours are to be used and all \$25,000 spent on labor.

## EXERCISE SET 2.2

In Exercises 1 through 6 determine whether each matrix is in reduced form.

1. $\left[\begin{array}{ccc|c} 1 & 0 & 2 & 1 \\ 0 & 1 & 3 & 2 \end{array}\right]$

2. $\left[\begin{array}{ccc|c} 1 & 0 & 1 & 2 \\ 0 & 0 & 1 & 3 \end{array}\right]$

3. $\left[\begin{array}{ccccc|c} 1 & 2 & 4 & 0 & 3 & 1 \\ 0 & 0 & 0 & 1 & 2 & 2 \end{array}\right]$

4. $\left[\begin{array}{ccc|c} 0 & 1 & 0 & 1 \\ 0 & 0 & 1 & 1 \\ 0 & 0 & 0 & 0 \end{array}\right]$

5. $\left[\begin{array}{ccc|c} 1 & 0 & 0 & 3 \\ 0 & 0 & 0 & 0 \\ 0 & 0 & 1 & 2 \end{array}\right]$

6. $\left[\begin{array}{ccccccc|c} 0 & 0 & 1 & 2 & 0 & 4 & 2 & 1 \\ 0 & 0 & 0 & 0 & 1 & 2 & 3 & 2 \\ 0 & 0 & 0 & 0 & 0 & 0 & 0 & 0 \end{array}\right]$

In Exercises 7 through 16 each of the matrices is in reduced form. Write the corresponding linear system and solve.

7. $\begin{array}{cc} x & y \end{array}$ $\left[\begin{array}{cc|c} 1 & 0 & 2 \\ 0 & 1 & 3 \end{array}\right]$

8. $\begin{array}{cc} x & y \end{array}$ $\left[\begin{array}{cc|c} 1 & 3 & 0 \\ 0 & 0 & 0 \end{array}\right]$

9. $\begin{array}{cc} x & y \end{array}$ $\left[\begin{array}{cc|c} 1 & 2 & 4 \\ 0 & 0 & 0 \end{array}\right]$

10. $\begin{array}{ccc} x & y & z \end{array}$ $\left[\begin{array}{ccc|c} 1 & 0 & 2 & 1 \\ 0 & 1 & 0 & 0 \end{array}\right]$

11. $\begin{array}{cccc} x & y & z & u \end{array}$ $\left[\begin{array}{cccc|c} 1 & 0 & 2 & 3 & 4 \\ 0 & 1 & 2 & 3 & 5 \end{array}\right]$

12. $\begin{array}{cccc} x & y & z & u \end{array}$ $\left[\begin{array}{cccc|c} 1 & 0 & 0 & 0 & 3 \\ 0 & 1 & 2 & 3 & 4 \end{array}\right]$

13. $\begin{array}{cccc} x & y & z & u \end{array}$ $\left[\begin{array}{cccc|c} 1 & 0 & 2 & 4 & 6 \\ 0 & 1 & 2 & 3 & 1 \\ 0 & 0 & 0 & 0 & 0 \end{array}\right]$

14. $\begin{array}{ccc} x & y & z \end{array}$ $\left[\begin{array}{ccc|c} 0 & 1 & 0 & 1 \\ 0 & 0 & 1 & 2 \\ 0 & 0 & 0 & 0 \end{array}\right]$

15. $\begin{array}{cccccc} x & y & z & u & v & w \end{array}$ $\left[\begin{array}{cccccc|c} 0 & 1 & 2 & 0 & 2 & 0 & 1 \\ 0 & 0 & 0 & 1 & 1 & 0 & 2 \\ 0 & 0 & 0 & 0 & 0 & 1 & 3 \end{array}\right]$

16. $\begin{array}{cccc} x & y & z & u \end{array}$ $\left[\begin{array}{cccc|c} 0 & 0 & 1 & 0 & 2 \\ 0 & 0 & 0 & 1 & 0 \\ 0 & 0 & 0 & 0 & 0 \end{array}\right]$

In Exercises 17 through 44 solve each of the systems by replacing the system with the augmented matrix and using the Gauss–Jordan method.

17.  
$x + 2y = 5$  
$2x - 3y = -4$

18.  
$2x + 4y = 6$  
$4x - y = -6$

19.  
$x + y + z = 6$  
$2x + y + 2z = 10$  
$3x + 2y + z = 10$

20.  
$x + 2y + z = 6$  
$-x + 3y - 2z = -4$  
$2x - y - 3z = -8$

21.  
$x - y + 2z = -1$  
$2x + y - 3z = 6$  
$y - z = 2$

22.  
$x - 2y + z = 9$  
$3y - 2z = -11$  
$x + y + 4z = 3$

23.  
$2x - 4y + 8z = 2$  
$2x + 3y + 2z = 3$  
$2x - 3y + 5z = 0$

24.  
$2x - y - 3z = 1$  
$-x - 2y + z = -4$  
$3x + y - 2z = 9$

25.  
$-x + 2y + 3z = 14$  
$2x - y + 2z = 2$

26.  
$2x - 2y + 2z = -2$  
$3x + y - z = 5$

27.  
$-2x + 6y + 4z = 12$  
$3x - 9y - 6z = -18$

28.  
$3x - 3y + 6z = 15$  
$-2x + 2y - 4z = -10$

**29.**  $-x + 5y - 3z = 7$
  $2x - 10y + 6z = -14$

**30.**   $x - y + 3z = 4$
  $-2x + 2y - 6z = -8$

**31.**  $x + y + z = 1$
  $x - y - z = 2$
  $3x + y + z = 4$

**32.**   $x - y - z = 2$
  $2x + y + z = 1$
  $4x - y - z = 5$

**33.**  $2x + y - z = 0$
  $3x - y + 2z = 1$
  $x - 2y + 3z = 2$

**34.**   $x - y + z = 1$
  $2x + 3y - 2z = 1$
  $3x + 2y - z = 1$

**35.**   $x - 2y + 2z = 1$
  $2x + y - z = 2$
  $3x - y + z = 3$

**36.**   $x + y - 2z = 2$
  $-x + 2y + z = 3$
  $x + 4y - 3z = 7$

**37.**   $x + y + z + u = 6$
  $y - z + 2u = 4$
  $z + u = 3$
  $x + 2y + 3z - u = 5$

**38.**  $x - y - z + 2u = 1$
  $x - y + z + 3u = 9$
  $y + 2z - u = 5$
  $3y + z + 2u = 8$

**39.**  $x + 2y + z - u = -2$
  $x + 2y + 2z + 2u = 9$
  $y + z - u = -2$
  $y - 2z + 3u = 4$

**40.**   $x + y + z + u = 6$
  $x + 2y - z + u = 5$
  $x + y - z + 2u = 6$
  $2x + 2y + 2z + u = 10$

**41.**   $x + y = 4$
  $2x - 3y = -7$
  $3x - 4y = -9$

**42.**   $x - 3y = -7$
  $3x + 4y = 5$
  $2x - y = -4$

**43.**   $x + 2y = 4$
  $2x - 3y = 5$
  $x - 5y = 2$

**44.**   $x - 3y = 2$
  $3x + 2y = 1$
  $2x + 5y = 1$

**45.**  Show that the system of equations

$$x + 2y + az = b$$
$$3x + 4y + cz = d$$

has a solution no matter what $a$, $b$, $c$, and $d$ are.

**46.**  What condition must $a$, $b$, and $c$ satisfy so that the following system has a solution?

$$x + 2y = a$$
$$3x + 4y = b$$
$$2x + 3y = c$$

**47.** A person has 36 coins, all of which are nickels, dimes and quarters. If the total value of the coins is $4, how many of each type of coin must this person have?

**48.** A person has three more dimes than nickels. If the total value of these coins is $2.40, how many of each type of coin must this person have?

## Applications

**49. Scheduling.** An insurance company has three types of documents to process, type $A$, type $B$, and type $C$. The type $A$ document needs to be examined for 2 hours by the accountant and 3 hours by the attorney. The type $B$ document needs to be examined for 4 hours by the accountant and 2 hours by the attorney. The last document needs to be examined for 2 hours by the accountant and 4 hours by the attorney. If the accountant has 34 hours and the attorney 35 hours available to spend working on these documents, how many documents of each type can they process?

**50. Investments.** A pension fund manager has been given a total of $900,000 by a corporate client to invest in certain types of stocks and bonds over the next year. Stocks must be restricted to a certain class of low-risk stocks and another class of medium risk stocks. Furthermore, the client demands an annual return of 8%. The pension fund manager assumes from historical data that the low risk stocks should return 9% annually, the medium risk stocks 11% annually, and bonds 7%. How should the pension fund manager allocate funds among the three groups to meet all the demands of the client?

**51. Investments.** Suppose in the previous problem the client demanded an annual return of 9%. What should the pension fund manager do?

**52. Production Scheduling.** A small plant with a cutting department and a sewing department produces three styles of sweaters: styles A, B, and C. It takes 0.4 work-hours to cut a style A and 0.2 work-hours to sew it. It takes 0.3 work-hours to cut a style B sweater and the same to sew it. It takes 0.5 work-hours to cut a style C and 0.6 work-hours to sew it. If the cutting department has 110 work-hours available each day and the sewing department has 95 work-hours available each day, how many sweaters of each style can be produced if all departments work at full capacity?

**53. Sales.** A store sells a total of 300 sweaters in a certain month. The sweaters come in only three styles: styles A, B, and C. The sales of style B equaled the sum of the other two. Style A costs $30, style B costs $40, and style C $50. If the store sold $11,500 worth of these sweaters in that month, how many sweaters of each style were sold?

**54. Production.** Each day a firm produces 100 units of a perishable ingredient I and 200 units of another perishable ingredient II, both of which are used to manufacture 4 products, X, Y, Z, and U. The following table indicates how many units of each of the two perishable ingredients are used to manufacture each unit of the four products.

|   | I | II |
|---|-----|-----|
| X | 0.1 | 0.4 |
| Y | 0.3 | 0.2 |
| Z | 0.6 | 0.4 |
| U | 0.2 | 0.3 |

Let $x$, $y$, $z$, $u$ be respectively the number of units of X, Y, Z, U that are manufactured each day. If the manufacturer wishes to use all of the perishable ingredients that day, what are the options for the amounts of the four products to be manufactured?

55. **Production.** Refer to Exercise 54. Just before production for the day is to begin, an order for 500 units of product U is received. The production manager is told to fill this order this very same day (while still using all the perishable ingredients) or be fired. Will the production manager be fired? Why or why not?

56. **Production.** Answer the same question as in the previous problem if, instead of receiving an order for 500 units of product U, an order for 150 units of product Z is received.

57. **Nutrition.** A dietitian must plan a meal for a patient using three fruits: oranges, strawberries, and blackberries. Each orange contains 1 gram of fiber, 75 mg of vitamin C, and 50 mg of phosphorus. Each cup of strawberries contains 2 grams of fiber, 60 mg of vitamin C, and 50 mg of phosphorus. Each cup of blackberries contains 6 grams of fiber, 30 mg of vitamin C, and 40 mg of phosphorus. How much of each of these fruits needs to be eaten so that a total of 13 grams of fiber, 375 mg of vitamin C, and 290 mg of phosphorus is obtained?

58. **Chinese Farm Problem.** The ancient Chinese "way of calculating with arrays" can be found in Chapter 8 of the ancient text *Nine Chapters on the Mathematical Art*. The following is the first problem listed in Chapter 8. Solve it.

There are three grades of corn. After threshing, three bundles of top grade, two bundles of medium grade, and one bundle of low grade make 39 dou (a measure of volume). Two bundles of top grade, three bundles of medium grade, and one bundle of low grade make 34 dou. The yield of one bundle of top grade, two bundles of medium grade, and three bundles of low grade make 26 dou. How many dou are contained in each bundle of each grade?

59. **Traffic Flow.** The accompanying flow diagram indicates the traffic flow in numbers of vehicles per hour into and out of 4 intersections during rush hour. Traffic lights placed at each intersection can be timed to control the flow of traffic. Find all possible *meaningful* flow patterns. *Hint:* For each intersection write an equation that states that the total flow into the intersection must equal the total flow out. For example, the diagram indicates that

$$x_1 + x_4 = 300 + 400 = 700.$$

Now, using the diagram, find the other three equations. Solve by the Gauss–Jordan method.

*(© Stockphotos Inc./The Image Bank.)*

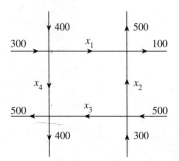

60.  **Traffic Flow.**   The accompanying flow diagram indicates the traffic flow into and out of 3 intersections during rush hour. Traffic lights placed at each intersection can be timed to control the flow of traffic. Find all possible *meaningful* flow patterns.

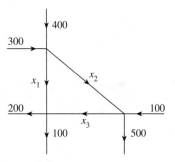

## Graphing Calculator Exercises

In Exercises 61 through 62 find the solution using the matrix menu on your graphing calculator to perform the elementary row operations for the Gauss–Jordan method.

**61.**  $1.2x + 2.1y + 3.4z = 54.2$          **62.**  $2.31x + 1.04y + 3.65z = 44.554$

$2.1x + 5.1y + 1.4z = 72.7$                    $1.32x + 2.87y + 5.21z = 59.923$

$3.7x + 1.4y + 3.8z = 62.9$                    $1.66x + 1.92y + 3.41z = 42.617$

In Exercises 63 through 64 graph the equations on your graphing calculator to see if there is a solution.

**63.**   $x + 2y = 4$                        **64.**   $x - 3y = 2$

$2x - 3y = 5$                                  $3x + 2y = 1$

$x - 5y = 3$                                   $2x + 5y = 2$

**65.**  **Demand and Supply.**   When the price $p$ of pork goes up, the demand $x_B$ for beef also goes up. In the same way, when the price $q$ of beef goes up, the demand $x_A$ of pork does also. Let the demand equations be given by

$$x_A = 6 - 2p + q \qquad x_B = 16 + p - 3q,$$

and let the supply equations be given as

$$p = 1 + 0.5x_A + x_B \qquad q = 1 + x_A + x_B.$$

Find the point of equilibrium, that is, the point at which all four equations are satisfied using the matrix menu to perform the elementary row operations required by the Gauss–Jordan method.

## Solutions to Self-Help Exercise Set 2.2

1. Use the Gauss–Jordan method and obtain

$$\begin{bmatrix} 1 & 2 & 1 & | & 3 \\ 0 & 0 & 1 & | & 2 \\ 0 & 0 & 0 & | & 0 \end{bmatrix} \quad \boxed{\text{Need a 0 here.}} \qquad R_1 - R_2 \rightarrow R_1$$

$$\begin{bmatrix} 1 & 2 & 0 & | & 1 \\ 0 & 0 & 1 & | & 2 \\ 0 & 0 & 0 & | & 0 \end{bmatrix} \qquad \text{or} \qquad \begin{aligned} x + 2y &= 1 \\ z &= 2 \\ 0 &= 0. \end{aligned}$$

The leading variables are $x$ and $z$, while $y$ is the free variable. Since $y$ is taken to be arbitrary, we set $y = t$, where $t$ is any real number. The general solution is $(1 - 2t, t, 2)$, where $t$ is any real number.

2. The matrix is already in reduced form. The leading variables are $y$ and $z$. The free variable is $x$. Setting $x$ equal to the parameter $t$ gives $z = 2$, $y = 1$, and $x = t$. The general solution is then $(t, 1, 2)$ for any $t$.

3. Let $x$, $y$, and $z$ be the amount of work-hours allocated to, respectively, the first, second, and third projects. Since the total number of work-hours is 2000, then $x + y + z = 2000$. The labor cost for the first project is $\$10x$, for the second is $\$12y$, and for the third is $\$14z$. Since the total labor funds is $\$25,000$, we then have $10x + 12y + 14z = 25,000$. These two equations are then written as

$$\begin{aligned} x + \phantom{0}y + \phantom{0}z &= \phantom{00}2000 \\ 10x + 12y + 14z &= 25,000. \end{aligned}$$

Using the augmented matrix and the Gauss–Jordan methods, we obtain

$$\begin{bmatrix} 1 & 1 & 1 & | & 2000 \\ 10 & 12 & 14 & | & 25,000 \end{bmatrix} \quad \boxed{\text{Need a 0 here.}} \qquad R_2 - 10R_1 \rightarrow R_2$$

$$\begin{bmatrix} 1 & 1 & 1 & | & 2000 \\ 0 & 2 & 4 & | & 5000 \end{bmatrix} \quad \boxed{\text{Need a 1 here.}} \qquad \tfrac{1}{2}R_2 \rightarrow R_2$$

$$\begin{bmatrix} 1 & 1 & 1 & | & 2000 \\ 0 & 1 & 2 & | & 2500 \end{bmatrix} \quad \boxed{\text{Need a 0 here.}} \qquad R_1 - R_2 \rightarrow R_1$$

$$\begin{bmatrix} 1 & 0 & -1 & | & -500 \\ 0 & 1 & 2 & | & 2500 \end{bmatrix} \qquad \text{or} \qquad \begin{aligned} x \phantom{0} - \phantom{0}z &= -500 \\ y + 2z &= \phantom{-}2500 \end{aligned}$$

This gives

$$x = t - 500$$

$$y = 2500 - 2t$$

$$z = t.$$

Since $0 \leq x = t - 500$, $t \geq 500$. Since $0 \leq y = 2500 - 2t$, $t \leq 1250$. Thus the contractor can allocate $t - 500$ work-hours to the first project, $2500 - 2t$ to the second, and $t$ to the third, where $500 \leq t \leq 1250$. For example, taking $t = 1000$, means the contractor can allocate 500 work-hours to the first project, 500 to the second, and 1000 to the third and use all of the available work-hours with all of the $25,000 spent on labor.

## 2.3 INTRODUCTION TO MATRICES

▶ *Basic Definitions*

▶ *Applications*

**James Joseph Sylvester,**
**1814–1897**

*(David Eugene Smith Collection, Columbia University.)*

The first mention of matrices occurred in 1850 in a paper by Sylvester. Sylvester was educated in England and became a professor at the University of Virginia in 1841. Unable to tolerate slavery, he left this position and the country after only two months. He returned to this country in 1871, accepting the chair of mathematics at Johns Hopkins University, where he was very productive, distinguishing himself as an outstanding mathematician. In 1878 he became the founding editor of the well regarded *American Journal of Mathematics*.

**A P P L I C A T I O N**

**Production**

A firm makes bulldozers (B), cranes (C), and tractors (T) at two locations, New York City and Los Angeles, and the number of each produced at each site for January and February is given in tabulated form as follows.

|     | *B* | *C* | *T* |
|-----|-----|-----|-----|
| NYC | 120 | 240 | 360 |
| LA  | 310 | 0   | 240 |

**January**

|     | *B* | *C* | *T* |
|-----|-----|-----|-----|
| NYC | 200 | 100 | 0   |
| LA  | 150 | 200 | 300 |

**February**

Find the total number of each product produced at each site for the January to February period in tabulated form. See Example 4 for the answer.

## Basic Definitions

Matrices were introduced in the last section. We will see that matrices are a natural way of presenting tabulated data. Certain calculations with this data correspond to certain matrix operations. These operations will be considered in this section, and some applications will be given.

As mentioned in the previous section, a matrix is a rectangular array of numbers. The matrix A

$$
\begin{array}{c}
\phantom{}\qquad\text{column 1}\quad\text{column 2}\quad\text{column 3} \\
\begin{array}{c}\text{row 1}\\ \text{row 2}\end{array}
\left[\begin{array}{ccc}
2 & 3 & 5 \\
0 & -2 & 4
\end{array}\right]
\end{array}
$$

is a matrix with 2 rows and 3 columns. The **order** of a matrix is $m \times n$ where $m$ is the number of rows and $n$ is the number of columns. Thus the order of the above matrix $A$ is $2 \times 3$.

Given a matrix such as $A$ above, the **entry** in the $i$th row and $j$th column is denoted by $a_{ij}$. For the above matrix $A$, $a_{11} = 2$, $a_{12} = 3$, $a_{13} = 5$, $a_{21} = 0$, $a_{22} = -2$, $a_{23} = 4$.

---

**Some Matrix Terminology**

A matrix $A$, also denoted by $(a_{ij})_{m \times n}$ or simply $(a_{ij})$, is a rectangular array of numbers.

The order of $A$ is denoted by $m \times n$ where $m$ is the number of rows and $n$ is the number of columns.

The element in the $i$th row and $j$th column is denoted by $a_{ij}$.

---

If $A = (a_{ij})_{m \times n}$, then we have

$$
A = 
\begin{array}{c}
\\
\end{array}
\left[\begin{array}{cccccc}
a_{11} & a_{12} & \cdots & a_{1j} & \cdots & a_{1n} \\
a_{21} & a_{22} & \cdots & a_{2j} & \cdots & a_{2n} \\
 & & \cdots & & & \\
a_{i1} & a_{i2} & \cdots & a_{ij} & \cdots & a_{in} \\
 & & \cdots & & & \\
a_{m1} & a_{m2} & \cdots & a_{mj} & \cdots & a_{mn}
\end{array}\right]
\begin{array}{l}
\text{first row} \\
\text{second row} \\
\cdots \\
\text{ith row} \\
\cdots \\
\text{mth row}
\end{array}
$$

first column   second column   jth column   nth column

E X A M P L E   1      **Determining the Order of Matrices**

Determine the order of

$$B = \begin{bmatrix} 1 & 2 \\ -1 & 0 \\ 4 & 8 \end{bmatrix}, \quad C = [1 \quad 2 \quad 5 \quad 0], \quad D = \begin{bmatrix} 1 \\ 3 \\ 4 \end{bmatrix}.$$

Also find $b_{22}$, $b_{32}$, $c_{12}$, $d_{31}$.

**Solution**      The order of $B$ is $3 \times 2$.
The order of $C$ is $1 \times 4$.
The order of $D$ is $3 \times 1$.
Also $b_{22} = 0$, $b_{32} = 8$, $c_{12} = 2$, $d_{31} = 4$.   ■

There are three special types of matrices that arise often: column matrices, row matrices, and square matrices.

---

**Row, Column, and Square Matrices**

A matrix of order $1 \times n$ (one row and $n$ columns) is called a **row matrix of dimension $n$**.

A matrix of order $m \times 1$ ($m$ rows and one column) is called a **column matrix of dimension $m$**.

A matrix is called a **square matrix** if the number of rows equals the number of columns.

---

The matrix $C$ above is a row matrix of dimension 4. The matrix $D$ above is a column matrix of dimension 3.

Naturally we need to know when two matrices are equal. Two matrices are equal when they are absolutely identical. The following definition says just that.

---

**Equality of Matrices**

Two matrices are equal if and only if they have the same order and all corresponding entries are equal.

---

For example, the two matrices

$$\begin{bmatrix} 1 & 2 \\ 3 & 4 \end{bmatrix} \quad \begin{bmatrix} 1 & 2 & 0 \\ 3 & 4 & 0 \end{bmatrix}$$

are not equal since they do not have the same order.

**E X A M P L E  2**    **Equal Matrices**

Find $x$, $y$, and $z$, such that

$$\begin{bmatrix} 1 & 2 & x \\ 3 & 0 & 5 \end{bmatrix} = \begin{bmatrix} 1 & 2 & 4 \\ y & 0 & z \end{bmatrix}.$$

**Solution**    Since corresponding entries must equal, $x = 4$, $y = 3$, and $z = 5$.  ∎

We now state three fundamental operations on matrices.

---

**Multiplication of a Matrix by Real Numbers**

If $c$ is a real number and $A$ is a matrix, then the matrix $cA$ is the matrix obtained by multiplying every entry in $A$ by $c$.

---

**Addition and Subtraction of Matrices**

Two matrices of the same order can be added (or subtracted) to obtain another matrix of the same order by adding (or subtracting) corresponding entries.

---

**E X A M P L E  3**    **Performing Elementary Matrix Operations**

Find the following.

(a)  $2 \begin{bmatrix} 3 & -1 & 0 \\ -2 & 3 & 10 \end{bmatrix} = \begin{bmatrix} 2(3) & 2(-1) & 2(0) \\ 2(-2) & 2(3) & 2(10) \end{bmatrix}$

$$= \begin{bmatrix} 6 & -2 & 0 \\ -4 & 6 & 20 \end{bmatrix}$$

(b)  $\begin{bmatrix} 1 & 2 \\ 3 & 5 \\ -1 & 2 \end{bmatrix} + \begin{bmatrix} 1 & -2 \\ 0 & 3 \\ 2 & -4 \end{bmatrix} = \begin{bmatrix} 1+1 & 2-2 \\ 3+0 & 5+3 \\ -1+2 & 2-4 \end{bmatrix} = \begin{bmatrix} 2 & 0 \\ 3 & 8 \\ 1 & -2 \end{bmatrix}$

(c)  $[1 \quad 2 \quad 3] + [2 \quad 5 \quad 6 \quad 0] = $ not defined

(d)  $\begin{bmatrix} 1 & 2 \\ 3 & 0 \end{bmatrix} - \begin{bmatrix} 2 & 1 \\ 1 & 3 \end{bmatrix} = \begin{bmatrix} 1-2 & 2-1 \\ 3-1 & 0-3 \end{bmatrix} = \begin{bmatrix} -1 & 1 \\ 2 & -3 \end{bmatrix}$

(e)  $2 \begin{bmatrix} 2 & 1 \\ 0 & 4 \end{bmatrix} - 3 \begin{bmatrix} -1 & 2 \\ 4 & 0 \end{bmatrix} = \begin{bmatrix} 4 & 2 \\ 0 & 8 \end{bmatrix} - \begin{bmatrix} -3 & 6 \\ 12 & 0 \end{bmatrix}$

$$= \begin{bmatrix} 7 & -4 \\ -12 & 8 \end{bmatrix} \quad ∎$$

The number 0 has the property that $x + 0 = 0 + x = x$ for any number $x$. There is a matrix, called the **zero matrix**, with an analogous property.

---

**The Zero Matrix**

The **zero matrix** of order $m \times n$ is the matrix with $m$ rows and $n$ columns, all of whose entries are zero.

---

Thus the zero matrix of order $2 \times 3$ is

$$\begin{bmatrix} 0 & 0 & 0 \\ 0 & 0 & 0 \end{bmatrix}$$

Notice that we have the following.

---

**Properties of the Zero Matrix**

If $A$ is any matrix of order $m \times n$ and $O$ is the zero matrix of the same order, then

$$A + O = O + A = A$$

$$A - A = O$$

---

The following properties of matrices can also be shown to be true.

---

**Further Properties of Matrices**

If $A$, $B$, and $C$ are three matrices all of the same order, then

$$A + B = B + A$$

$$A + (B + C) = (A + B) + C$$

---

## Applications

Data are often presented in matrix form. Suppose a firm makes bulldozers (B), cranes (C), and tractors (T) at two locations, New York City and Los Angeles, and the number of each produced at each location during January is given by the following table.

**January**

|      | B   | C   | T   |
|------|-----|-----|-----|
| NYC  | 120 | 240 | 360 |
| LA   | 310 | 0   | 240 |

A **production matrix** for January can be given as

$$J = \begin{bmatrix} 120 & 240 & 360 \\ 310 & 0 & 240 \end{bmatrix},$$

where it is understood that the first row refers to NYC, the second row to LA, the first column to the number of bulldozers, etc.

**E X A M P L E   4**   **Adding Production Matrices**

Now suppose that the same firm as above has a February production matrix given by

$$F = \begin{bmatrix} 200 & 100 & 0 \\ 150 & 200 & 300 \end{bmatrix}$$

Find (a) the production matrix for the January through February period and (b) the number of cranes produced in NYC during this period.

**Solution**   The matrix $J + F$ is the production matrix for the two-month period. Thus

(a)   $J + F = \begin{bmatrix} 120 & 240 & 360 \\ 310 & 0 & 240 \end{bmatrix} + \begin{bmatrix} 200 & 100 & 0 \\ 150 & 200 & 300 \end{bmatrix}$

$= \begin{bmatrix} 320 & 340 & 360 \\ 460 & 200 & 540 \end{bmatrix}$

(b)   This is the element in the first row and second column of $J + F$, which is 340.  ∎

**E X A M P L E   5**   **Multiplying a Number with a Production Matrix**

Suppose the production of the above firm in March increases by 10% from that of February for all products at all locations. Find (a) the production matrix $M$ for March and (b) how many tractors were produced in LA during March.

**Solution**   (a)   The production matrix $M$ for March is $1.1F$ or

$$M = 1.1F = \begin{bmatrix} 1.1(200) & 1.1(100) & 1.1(0) \\ 1.1(150) & 1.1(200) & 1.1(300) \end{bmatrix}$$

$= \begin{bmatrix} 220 & 110 & 0 \\ 165 & 220 & 330 \end{bmatrix}$

(b)   If $M = (m_{ij})$, then $m_{23}$ is the number of tractors produced in LA in March. We see that $m_{23} = 330$.  ∎

## SELF-HELP EXERCISE SET 2.3

1. Suppose $A = (a_{ij})$ is a matrix of order $2 \times 3$ and $a_{ij} = i + 2j$ for all $i$ and $j$. Write down the matrix.

2. Find $3\begin{bmatrix} 3 & 2 \\ -2 & 0 \\ 5 & 1 \end{bmatrix} - 2\begin{bmatrix} 1 & -3 \\ 0 & 2 \\ -3 & 2 \end{bmatrix}$

3. A person owns two furniture stores. The first store shows the following number of sales for the indicated two months.

**Store 1**

|       | Sofas | Love Seats | Chairs |
|-------|-------|------------|--------|
| April | 30    | 20         | 40     |
| May   | 30    | 30         | 50     |

The second store shows sales exactly 20% higher for all three types of furniture for each of the months. Use matrix operations to find the total sales in both stores.

## EXERCISE SET 2.3

In Exercises 1 through 14 let

$$A = \begin{bmatrix} 3 & 1 \\ -2 & 4 \\ 0 & 3 \\ 2 & 8 \end{bmatrix}, \quad B = \begin{bmatrix} -1 & 0 & 2 & 6 \\ 3 & 1 & -2 & 4 \\ 5 & 3 & 0 & -5 \end{bmatrix}, \quad C = [-1 \quad 2], \quad D = \begin{bmatrix} 0 \\ 1 \\ -3 \\ 4 \end{bmatrix}$$

Find the order of the following.

1. A      2. B      3. C      4. D

If A, B, C, and D are defined as above and $A = (a_{ij})$, $B = (b_{ij})$, $C = (c_{ij})$, $D = (d_{ij})$, find the following.

5. $a_{12}$      6. $a_{21}$      7. $a_{42}$      8. $a_{41}$      9. $b_{23}$

10. $b_{34}$      11. $b_{14}$      12. $b_{33}$      13. $c_{12}$      14. $d_{31}$

15. If $A = (a_{ij})$ is a matrix of order $2 \times 3$ and $a_{ij} = 5$ for all $i$ and $j$, write down $A$.

16.  If $A = (a_{ij})$ is a matrix of order $2 \times 3$ and $a_{ij} = i + j$ for all $i$ and $j$, write down $A$.

17.  If $A = (a_{ij})$ is a matrix of order $3 \times 3$ and $a_{ij} = 1$ if $i = j$, and $a_{ij} = 0$ if $i \neq j$, write down $A$.

18.  If $A = (a_{ij})$ is a matrix of order $3 \times 2$ and $a_{ij} = i$ for all $i$ and $j$, write down $A$.

19.  If $A = (a_{ij})$ is a matrix of order $3 \times 3$ and $a_{ij} = 0$ for all $i$ and $j$, write down $A$.

20.  If $A = (a_{ij})$ is a matrix of order $4 \times 3$ and $a_{ij} = 2i + j - 1$ for all $i$ and $j$, write down $A$.

21.  Find $x$, $y$, and $z$, such that

$$\begin{bmatrix} 3 & 2 \\ 4 & x \end{bmatrix} = \begin{bmatrix} 3 + y & z - 2 \\ 4 & 3 \end{bmatrix}$$

22.  Find $x$, $y$, and $z$, such that

$$\begin{bmatrix} x & 3 \\ -1 & 0 \\ 4 & 5 \end{bmatrix} = \begin{bmatrix} 1 & 3 \\ y - 1 & 0 \\ 3 & z + x \end{bmatrix}.$$

In Exercises 23 through 42 let

$$E = \begin{bmatrix} 1 & 4 \\ 3 & 6 \\ -3 & 7 \\ 3 & -2 \end{bmatrix}, \quad F = \begin{bmatrix} 4 & 3 & -1 & -6 \\ 5 & 0 & -2 & 7 \\ 8 & 1 & -3 & 0 \end{bmatrix}, \quad G = \begin{bmatrix} 1 & 4 \end{bmatrix}, \quad H = \begin{bmatrix} 3 \\ -1 \\ 5 \\ 1 \end{bmatrix}$$

and let A, B, C, D be the matrices given before Exercise 1. Find the indicated quantities.

| | | |
|---|---|---|
| 23. $A + E$ | 24. $B + F$ | 25. $C + G$ |
| 26. $D + H$ | 27. $A - E$ | 28. $B - F$ |
| 29. $C - G$ | 30. $D - H$ | 31. $2E$ |
| 32. $-3F$ | 33. $4G$ | 34. $\frac{1}{2}H$ |
| 35. $2A + E$ | 36. $B - 2F$ | 37. $2C + 3G$ |
| 38. $2D - \frac{1}{2}H$ | 39. $A - F$ | 40. $2C + D$ |
| 41. $E - E$ | 42. $H - H$ | |

In Exercises 43 through 47, let $A = (a_{ij})$, $B = (b_{ij})$, $C = (c_{ij})$, $D = (d_{ij})$, $X = (x_{ij})$ all be $2 \times 2$ matrices and let $O$ be the $2 \times 2$ zero matrix.

43.  Prove that $A + B = B + A$.

44. Prove that $A + (B + C) = (A + B) + C$.

45. Prove that $A + O = O + A = A$.

46. Prove that $(c + d)A = cA + dA$, where $c$ and $d$ are scalars.

47. Prove that $A - A = O$.

48. A matrix $D = (d_{ij})$ is a diagonal matrix if $d_{ij} = 0$ if $i \neq j$, that is, all entries off the "main diagonal" are zero. Suppose $A = (a_{ij})$ and $B = (b_{ij})$ are both diagonal matrices of order $(4 \times 4)$. Show that (a) $A + B$ is diagonal, (b) $cA$ is diagonal where $c$ is a real number.

49. Show that the matrix equation $X + A = B$ has the solution $X = B - A$.

50. A matrix $U = (u_{ij})$ is upper triangular if $u_{ij} = 0$ when $i > j$, that is, when all entries below the main diagonal are zero. Suppose $A = (a_{ij})$ and $B = (b_{ij})$ are both upper triangular matrices of order $(4 \times 4)$. Show that (a) $A + B$ is upper triangular and (b) $cA$ is upper triangular where $c$ is a real number.

## Applications

51. **Production.** Refer to Example 5 of the text. Suppose the April production matrix was

$$A = \begin{bmatrix} 300 & 200 & 100 \\ 200 & 200 & 400 \end{bmatrix}$$

Write a production matrix for the periods from March through April as a sum of two matrices and then carry out the calculation.

52. **Production.** In the previous problem find $A - M$, where $M$ is given in Example 5 of the text. What does $A - M$ stand for?

53. **Production.** Suppose that production of all products at all locations was reduced by 10% in May from what it was in April in the exercise before the last. Write the production matrix for May as a scalar times a matrix. Then carry out the calculation to obtain the production matrix for May.

54. **Production.** Refer to Example 5 of the text. Suppose the production for June of all products at all locations was the average of the January and February production. Write an expression using the matrices $J$ and $F$ for the June production matrix and then carry out the indicated calculation to obtain this matrix.

55. **Production.** A firm has a plant in NYC that has produced the following numbers of items of three products during the four quarters of the last year.

|  | First Quarter | Second Quarter | Third Quarter | Fourth Quarter |
| --- | --- | --- | --- | --- |
| Product A | 250 | 300 | 350 | 300 |
| Product B | 300 | 200 | 250 | 150 |
| Product C | 200 | 240 | 320 | 220 |

The following table gives the firm's production for the same period for its other production plant located in LA.

|  | First Quarter | Second Quarter | Third Quarter | Fourth Quarter |
|---|---|---|---|---|
| Product A | 300 | 350 | 450 | 400 |
| Product B | 320 | 240 | 280 | 250 |
| Product C | 250 | 260 | 420 | 280 |

**a.**   Write a production matrix $N$ for the NYC plant.
**b.**   Write a production matrix $L$ for the LA plant.
**c.**   Write an expression involving the matrices $N$ and $L$ that gives the production matrix for the firm for the last year. Carry out the indicated calculation.
**d.**   Calculate $L - N$ and indicate what this matrix stands for.
**e.**   Next year the firm is predicting production to increase by 20% at NYC. Write the production matrix for next year for NYC as a scalar times a matrix and then carry out the indicated calculation.
**f.**   Write a matrix expression that gives the production matrix that represents the average production of its two plants and then carry out the indicated calculation.

## Solutions to Self-Help Exercise Set 2.3

**1.**   $a_{11} = 1 + 2(1) = 3, a_{12} = 1 + 2(2) = 5, a_{13} = 1 + 2(3) = 7, a_{21} = 2 + 2(1) = 4, a_{22} = 2 + 2(2) = 6, a_{23} = 2 + 2(3) = 8.$ Thus $A = \begin{bmatrix} 3 & 5 & 7 \\ 4 & 6 & 8 \end{bmatrix}$

**2.**   $3\begin{bmatrix} 3 & 2 \\ -2 & 0 \\ 5 & 1 \end{bmatrix} - 2\begin{bmatrix} 1 & -3 \\ 0 & 2 \\ -3 & 2 \end{bmatrix} = \begin{bmatrix} 9 & 6 \\ -6 & 0 \\ 15 & 3 \end{bmatrix} - \begin{bmatrix} 2 & -6 \\ 0 & 4 \\ -6 & 4 \end{bmatrix}$

$$= \begin{bmatrix} 7 & 12 \\ -6 & -4 \\ 21 & -1 \end{bmatrix}$$

**3.**   The indicated matrix operations are

$$\begin{bmatrix} 30 & 20 & 40 \\ 30 & 30 & 50 \end{bmatrix} + 1.2\begin{bmatrix} 30 & 20 & 40 \\ 30 & 30 & 50 \end{bmatrix} = \begin{bmatrix} 30 & 20 & 40 \\ 30 & 30 & 50 \end{bmatrix} + \begin{bmatrix} 36 & 24 & 48 \\ 36 & 36 & 60 \end{bmatrix}$$

$$= \begin{bmatrix} 66 & 44 & 88 \\ 66 & 66 & 110 \end{bmatrix}$$

**2.4**   **MATRIX MULTIPLICATION**

___

▶ *A Row Matrix Times a Column Matrix*

▶ *Multiplication in General*

▶ *Applications*

▶ *Systems of Equations*

The definition of multiplication of two matrices may at first seem strange. But in doing applications the reader will see how this definition becomes natural. Furthermore, this definition of multiplication plays a central role in writing systems of equations in matrix form. In the next section we will see how systems of equations with the same number of equations as unknowns can be solved using matrix operations and this definition of multiplication.

### A Row Matrix Times a Column Matrix

To understand how to multiply two matrices, we first need to know how to multiply a row matrix with a column matrix of the same dimension. We therefore begin by defining such a multiplication.

---

**Row Matrix Times Column Matrix**

Let $R$ be the row matrix $R = [r_1 \quad r_2 \quad \cdots \quad r_n]$
and let $C$ be the column matrix

$$C = \begin{bmatrix} c_1 \\ c_2 \\ \vdots \\ c_n \end{bmatrix}$$

with the same dimension as $R$. Then the product $RC$ is defined to be

$$RC = [r_1 \quad r_2 \quad \cdots \quad r_n] \begin{bmatrix} c_1 \\ c_2 \\ \vdots \\ c_n \end{bmatrix} = r_1 c_1 + r_2 c_2 + \cdots + r_n c_n.$$

---

**R E M A R K .**     Notice that the answer is a real number.

**E X A M P L E   1**    **Multiplying Row Matrices with Column Matrices**

Perform the indicated operations.

**Solutions**

(a)  $[5 \quad \frac{1}{2}] \begin{bmatrix} 1 \\ 6 \end{bmatrix} = (5)(1) + (\frac{1}{2})(6) = 8.$

(b)  $[3 \quad 0 \quad -2] \begin{bmatrix} 3 \\ 2 \\ 1 \end{bmatrix} = (3)(3) + (0)(2) + (-2)(1) = 7.$

(c)  $[3 \quad 2 \quad 0] \begin{bmatrix} 3 \\ 5 \end{bmatrix} =$ undefined.  ∎

Notice from (c) that if the number of elements in the row is not equal to the number of elements in the column, then the product is not defined.

The following example illustrates an application and also indicates a motivation for the above definition. The item referred to could, for example, be a television set and the ingredients could be steel, plastic, copper, glass, etc.

**E X A M P L E   2**    **Application**

Suppose a firm makes an item with four ingredients $A$, $B$, $C$, and $D$. Each item uses 10 units of $A$, 5 of $B$, 7 of $C$, and 15 of $D$. The cost per item of each of these four ingredients is, respectively, $20, $10, $30, and $5. (a) Write a row vector that expresses the amount of each ingredient that goes into each item. (b) Write a column matrix that represents the cost of each unit of the ingredients. (c) Express the total cost of manufacturing one item using these four ingredients by a product of a row matrix with a column matrix.

**Solutions**

(a)  The appropriate row matrix is

$$R = [10 \quad 5 \quad 7 \quad 15].$$

(b)  The appropriate column matrix is

$$C = \begin{bmatrix} 20 \\ 10 \\ 30 \\ 5 \end{bmatrix}.$$

(c)  The cost of manufacturing one item using these four ingredients is

$$\text{cost of one item} = (\text{number of units of A}) \times (\text{cost of each unit of A})$$
$$+ (\text{number of units of B}) \times (\text{cost of each unit of B})$$
$$+ (\text{number of units of C}) \times (\text{cost of each unit of C})$$
$$+ (\text{number of units of D}) \times (\text{cost of each unit of D})$$

$$= (10)(20) + (5)(10) + (7)(30) + (15)(5)$$

$$= [10 \quad 5 \quad 7 \quad 15] \begin{bmatrix} 20 \\ 10 \\ 30 \\ 5 \end{bmatrix}.$$

The actual numerical answer is $535. ■

### Multiplication in General

We now turn to multiplying two general matrices and the condition that two matrices must satisfy before they can be multiplied.

We show how to multiply two general matrices by working out the specific example

$$AB = \begin{bmatrix} 3 & 1 \\ 2 & 4 \\ 5 & 0 \end{bmatrix} \begin{bmatrix} 2 & 3 \\ 4 & 6 \end{bmatrix}.$$

We obtain the product by the following rule. **Multiply the $i$th row of $A$ times the $j$th column of $B$ and place the resulting number in the $i$th row, $j$th column of the product.**

Notice for this process to work we must have the dimension of the $i$th row of $A$ equal to the dimension of the $j$th column of $B$. In other words **the number of columns of $A$ must equal the number of rows of $B$.**

We start by taking the product of the first row of $A$, $[3 \quad 1]$, and the first column of $B$, $\begin{bmatrix} 2 \\ 4 \end{bmatrix}$, and place the result in the first row and first column of the product

$$\begin{bmatrix} 3 & 1 \\ 2 & 4 \\ 5 & 0 \end{bmatrix} \begin{bmatrix} 2 & 3 \\ 4 & 6 \end{bmatrix} = \begin{bmatrix} 10 & \\ & \end{bmatrix}.$$

Now take the product of the first row of $A$, $[3 \quad 1]$, and the second column of $B$, $\begin{bmatrix} 3 \\ 6 \end{bmatrix}$, and place the result in the first row and second column of the product

$$\begin{bmatrix} 3 & 1 \\ 2 & 4 \\ 5 & 0 \end{bmatrix} \begin{bmatrix} 2 & 3 \\ 4 & 6 \end{bmatrix} = \begin{bmatrix} 10 & 15 \\ & \\ & \end{bmatrix}.$$

Since there are no more columns of $B$, we move on to the second row of $A$. Take the product of the second row of $A$, $[2 \quad 4]$, and the first column of $B$, $\begin{bmatrix} 2 \\ 4 \end{bmatrix}$, and place the result in the second row and first column of the product

$$\begin{bmatrix} 3 & 1 \\ 2 & 4 \\ 5 & 0 \end{bmatrix} \begin{bmatrix} 2 & 3 \\ 4 & 6 \end{bmatrix} = \begin{bmatrix} 10 & 15 \\ 20 & \\ & \end{bmatrix}.$$

Take the product of the second row of $A$, $[2 \quad 4]$, and the second column of $B$, $\begin{bmatrix} 3 \\ 6 \end{bmatrix}$, and place the result in the second row and second column of the product

$$\begin{bmatrix} 3 & 1 \\ 2 & 4 \\ 5 & 0 \end{bmatrix} \begin{bmatrix} 2 & 3 \\ 4 & 6 \end{bmatrix} = \begin{bmatrix} 10 & 15 \\ 20 & 30 \\ & \end{bmatrix}.$$

Since there are no more columns of $B$, we move on to the third row of $A$. Take the product of the third row of $A$, $[5 \quad 0]$, and the first column of $B$, $\begin{bmatrix} 2 \\ 4 \end{bmatrix}$, and place the result in the third row and first column of the product

$$\begin{bmatrix} 3 & 1 \\ 2 & 4 \\ 5 & 0 \end{bmatrix} \begin{bmatrix} 2 & 3 \\ 4 & 6 \end{bmatrix} = \begin{bmatrix} 10 & 15 \\ 20 & 30 \\ 10 & \end{bmatrix}.$$

Take the product of the third row of $A$, $[5 \quad 0]$, and the second column of $B$, $\begin{bmatrix} 3 \\ 6 \end{bmatrix}$, and place the result in the third row and second column of the product

$$\begin{bmatrix} 3 & 1 \\ 2 & 4 \\ 5 & 0 \end{bmatrix} \begin{bmatrix} 2 & 3 \\ 4 & 6 \end{bmatrix} = \begin{bmatrix} 10 & 15 \\ 20 & 30 \\ 10 & 15 \end{bmatrix}.$$

Since there are no more columns of $B$, we move on to the next row of $A$. Since there are no more rows of $A$, we are done.

**Multiplication of Matrices**

Given two matrices $A = (a_{ij})_{m \times p}$ and $B = (b_{ij})_{q \times n}$, the product $AB = C = (c_{ij})$ is defined if $p = q$, that is, if the number of columns of $A$ equals the number of rows of $B$. In such a case, the element in the $i$th row and $j$th column, $c_{ij}$, of the product matrix $C$ is obtained by taking the product of the $i$th row of $A$ and the $j$th column of $B$.

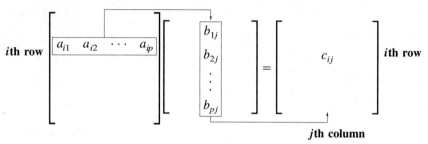

Notice that to obtain the first row of the product $C = AB$, we take the product of the first row of $A$ and the successive columns of $B$ until we exhaust the columns of $B$. This indicates that **the number of columns of the product $AB$ must equal the number of columns of $B$**. Also notice that a row of $B$ is obtained for each row of $A$. This indicates that **the number of rows of the product $AB$ must equal the number of rows of $A$**. Thus

**Order of AB**

Given the two matrices $A = (a_{ij})_{m \times p}$ and $B = (b_{ij})_{p \times n}$, the matrix $AB$ has order $m \times n$.

Applying this to the previous example we have

Order of A        Order of B

(3 × 2)            (2 × 2)

└─── 2 = 2 ───┘

└─── (3 × 2) ───┘

Order of AB

**E X A M P L E   3**     **Multiplying Matrices**

Perform the indicated operations.

(a)  $\begin{bmatrix} 1 & 2 \\ 3 & 4 \end{bmatrix} \begin{bmatrix} 5 & 6 \\ 7 & 8 \end{bmatrix} = \begin{bmatrix} (1)(5) + (2)(7) & (1)(6) + (2)(8) \\ (3)(5) + (4)(7) & (3)(6) + (4)(8) \end{bmatrix}$

$= \begin{bmatrix} 19 & 22 \\ 43 & 50 \end{bmatrix}$

(b)  $\begin{bmatrix} 5 & 6 \\ 7 & 8 \end{bmatrix} \begin{bmatrix} 1 & 2 \\ 3 & 4 \end{bmatrix} = \begin{bmatrix} (5)(1) + (6)(3) & (5)(2) + (6)(4) \\ (7)(1) + (8)(3) & (7)(2) + (8)(4) \end{bmatrix}$

$= \begin{bmatrix} 23 & 34 \\ 31 & 46 \end{bmatrix}$  ■

Notice that this example illustrates that in general $AB$ need not equal $BA$ even when they are both defined and of the same order.

Earlier we saw that

$$AB = \begin{bmatrix} 3 & 1 \\ 2 & 4 \\ 5 & 0 \end{bmatrix} \begin{bmatrix} 2 & 3 \\ 4 & 6 \end{bmatrix} = \begin{bmatrix} 10 & 15 \\ 20 & 30 \\ 10 & 15 \end{bmatrix}.$$

But notice that for this $A$ and $B$, the product $BA$ is not even defined, since the number of columns of $B$ does not equal the number of rows of $A$.

Order of B               Order of A

(2 × 2)                   (3 × 2)

2 ≠ 3

$BA$ undefined

$$BA = \begin{bmatrix} 2 & 3 \\ 4 & 6 \end{bmatrix} \begin{bmatrix} 3 & 1 \\ 2 & 4 \\ 5 & 0 \end{bmatrix} = \text{undefined}$$

At the beginning of this section we defined the product $RC$ where $R$ was a row matrix and $C$ a column matrix of the same order $n$. The answer we obtained was a number (a 1 × 1 matrix). What happens if we reverse the order of the multiplication? Checking the orders we have for $CR$,

Order of C               Order of R

($n$ × 1)                  (1 × $n$)

1 = 1

($n$ × $n$)

Order of CR

We obtain a square matrix of order $n \times n$.

E X A M P L E  4    **Multiplying Matrices**

Perform the indicated multiplication.

**Solution**

$$\begin{bmatrix} 1 \\ 2 \end{bmatrix}\begin{bmatrix} 3 & 4 \end{bmatrix} = \begin{bmatrix} (1)(3) & (1)(4) \\ (2)(3) & (2)(4) \end{bmatrix} = \begin{bmatrix} 3 & 4 \\ 6 & 8 \end{bmatrix} \quad \blacksquare$$

We now have several examples that indicate that, in general,

$$AB \neq BA,$$

even if both matrices $A$ and $B$ are square.

The following, however, are true.

---

**Properties of Matrix Multiplication**

For any matrices $A$, $B$, and $C$,

$$A(BC) = (AB)C$$

$$A(B + C) = AB + AC,$$

whenever the indicated products are defined.

---

If $a$ is any scalar, then the number 1 has the property that

$$a \cdot 1 = 1 \cdot a = a$$

There is a matrix analog to the number 1. This matrix, which must be a square matrix, is called the **identity matrix**.

---

**The Identity Matrix**

The $(n \times n)$ identity matrix, $I_n$, is the square matrix of order $(n \times n)$ with ones down the main diagonal and zeros elsewhere.

---

Thus

$$I_n = \begin{bmatrix} 1 & 0 & 0 & \cdots & 0 \\ 0 & 1 & 0 & \cdots & 0 \\ & & & \cdots & \\ 0 & 0 & 0 & \cdots & 1 \end{bmatrix} (n \text{ rows})$$

$(n \text{ columns})$

The identity matrix then has the following property.

---

**Multiplication Property of the Identity Matrix**

If $A$ is a square matrix of order $(n \times n)$, then

$$AI_n = I_nA = A.$$

---

For example,

$$AI_2 = \begin{bmatrix} a & b \\ c & d \end{bmatrix}\begin{bmatrix} 1 & 0 \\ 0 & 1 \end{bmatrix}$$

$$= \begin{bmatrix} a \cdot 1 + b \cdot 0 & a \cdot 0 + b \cdot 1 \\ c \cdot 1 + d \cdot 0 & c \cdot 0 + d \cdot 1 \end{bmatrix}$$

$$= \begin{bmatrix} a & b \\ c & d \end{bmatrix}$$

$$= A$$

## Applications

Matrix techniques are regularly used by firms in production planning.

We assume that a firm makes three products: $P_1$, $P_2$, and $P_3$. We assume that a production facility is located in St. Louis, where four ingredients, $I_1$, $I_2$, $I_3$, and $I_4$ are made that are then used to manufacture the three products. The following table and matrix gives the number of units of each ingredient needed for each product.

$$
\begin{array}{c}
\phantom{P_1} \\
P_1 \\
P_2 \\
P_3
\end{array}
\begin{array}{cccc}
I_1 & I_2 & I_3 & I_4 \\
\end{array}
$$

$$
\begin{array}{c}
P_1 \\
P_2 \\
P_3
\end{array}
\begin{bmatrix}
9 & 6 & 5 & 3 \\
5 & 5 & 8 & 4 \\
6 & 2 & 1 & 7
\end{bmatrix} = P
$$

Suppose the cost of each unit of the ingredients $I_1$, $I_2$, $I_3$, and $I_4$ is, respectively, \$10, \$30, \$20, \$40. We could summarize this in the following cost (column) matrix:

$$
C = \begin{bmatrix} 10 \\ 30 \\ 20 \\ 40 \end{bmatrix}
\begin{array}{l}
\textbf{cost of one unit of } I_1 \\
\textbf{cost of one unit of } I_2 \\
\textbf{cost of one unit of } I_3 \\
\textbf{cost of one unit of } I_4.
\end{array}
$$

The firm might naturally wish to know the total cost of the ingredients that go into each product. This is given by the (column) matrix

$$PC = \begin{matrix} P_1 \\ P_2 \\ P_3 \end{matrix} \begin{bmatrix} 9 & 6 & 5 & 3 \\ 5 & 5 & 8 & 4 \\ 6 & 2 & 1 & 7 \end{bmatrix} \begin{bmatrix} 10 \\ 30 \\ 20 \\ 40 \end{bmatrix} \begin{matrix} \text{cost of one unit of } I_1 \\ \text{cost of one unit of } I_2 \\ \text{cost of one unit of } I_3 \\ \text{cost of one unit of } I_4 \end{matrix}$$

$$= \begin{bmatrix} 490 \\ 520 \\ 420 \end{bmatrix} \begin{matrix} \text{total cost of } P_1 \\ \text{total cost of } P_2 \\ \text{total cost of } P_3 \end{matrix}$$

where the three numbers in the last column matrix represent the total cost of ingredients in the respective products $P_1$, $P_2$, and $P_3$.

Suppose the firm has an order for 9000 of product $P_1$, 6000 of $P_2$, and 7000 of $P_3$ and then wanted to know the total number of units of each of the ingredients needed. If we expressed the amounts ordered (in thousands) as a row matrix

$$R = [9 \quad 6 \quad 7],$$

then the matrix

$$RP = [9 \quad 6 \quad 7] \begin{bmatrix} 9 & 6 & 5 & 3 \\ 5 & 5 & 8 & 4 \\ 6 & 2 & 1 & 7 \end{bmatrix} = [153 \quad 98 \quad 100 \quad 100]$$

represents, respectively, the number of units of the ingredients $I_1$, $I_2$, $I_3$, and $I_4$ needed to fill the order.

Now suppose, after the ingredients are made in the St. Louis facility, they are shipped out to the two manufacturing plants, one in LA and one in NYC, where the three products are actually produced. The following table and matrix give the number in thousands of products manufactured at each of the two plants.

| | $P_1$ | $P_2$ | $P_3$ |
|---|---|---|---|
| LA | 2 | 1 | 1 |
| NYC | 3 | 1 | 0 |

$$A = \begin{bmatrix} 2 & 1 & 1 \\ 3 & 1 & 0 \end{bmatrix}$$

Suppose the firm wishes to keep track of how many units of each ingredient is going to LA and to NYC. The matrix

$$AP = \begin{bmatrix} 2 & 1 & 1 \\ 3 & 1 & 0 \end{bmatrix} \begin{bmatrix} 9 & 6 & 5 & 3 \\ 5 & 5 & 8 & 4 \\ 6 & 2 & 1 & 7 \end{bmatrix}$$

$$= \begin{matrix} \\ \text{LA} \\ \text{NYC} \end{matrix} \begin{matrix} I_1 & I_2 & I_3 & I_4 \\ \begin{bmatrix} 29 & 19 & 19 & 17 \\ 32 & 23 & 23 & 13 \end{bmatrix} \end{matrix}$$

summarizes this information. Thus the element in the first row and first column, 29, is the number of thousands of units of the ingredient $I_1$ that are used in the LA plant.

## Systems of Equations

We shall now see how to use matrices to represent systems of equations. Consider the following system of equations.

$$2x_1 + 3x_2 + 5x_3 = 2$$
$$5x_1 - 4x_2 + 3x_3 = 4$$
$$3x_1 + 2x_2 - 2x_3 = 1$$

Let

$$A = \begin{bmatrix} 2 & 3 & 5 \\ 5 & -4 & 3 \\ 3 & 2 & -2 \end{bmatrix}, \quad X = \begin{bmatrix} x_1 \\ x_2 \\ x_3 \end{bmatrix}, \quad B = \begin{bmatrix} 2 \\ 4 \\ 1 \end{bmatrix}$$

The matrix $A$ is called the **coefficient matrix**. Then notice that

$$AX = \begin{bmatrix} 2 & 3 & 5 \\ 5 & -4 & 3 \\ 3 & 2 & -2 \end{bmatrix} \begin{bmatrix} x_1 \\ x_2 \\ x_3 \end{bmatrix}$$

$$= \begin{bmatrix} 2x_1 + 3x_2 + 5x_3 \\ 5x_1 - 4x_2 + 3x_3 \\ 3x_1 + 2x_2 - 2x_3 \end{bmatrix}$$

$$= \begin{bmatrix} 2 \\ 4 \\ 1 \end{bmatrix} = B.$$

More generally, the system of equations

$$a_{11}x_1 + a_{12}x_2 + \cdots + a_{1n}x_n = b_1$$
$$a_{21}x_1 + a_{22}x_2 + \cdots + a_{2n}x_n = b_2$$
$$\cdots$$
$$a_{m1}x_1 + a_{m2}x_2 + \cdots + a_{mn}x_n = b_m$$

can be written as

$$AX = B,$$

if we define

$$A = \begin{bmatrix} a_{11} & a_{12} & \cdots & a_{1n} \\ a_{21} & a_{22} & \cdots & a_{2n} \\ & & \cdots & \\ a_{m1} & a_{m2} & \cdots & a_{mn} \end{bmatrix}, \quad X = \begin{bmatrix} x_1 \\ x_2 \\ \vdots \\ x_n \end{bmatrix}, \quad B = \begin{bmatrix} b_1 \\ b_2 \\ \vdots \\ b_m \end{bmatrix}.$$

## SELF-HELP EXERCISE SET 2.4

1. Perform the indicated operation.

$$\begin{bmatrix} 1 & 2 \\ -1 & 3 \end{bmatrix} \begin{bmatrix} 3 & 0 & 4 \\ 1 & -2 & 5 \end{bmatrix}$$

2. A firm makes two types of a product and uses three parts according to the following table.

|  | Part A | Part B | Part C |
|---|---|---|---|
| Type I | 3 | 2 | 1 |
| Type II | 1 | 2 | 4 |

$$= P$$

Both types of this product are manufactured at two different factories: Factory X and Factory Y. The time in work-hours that is required to install each part is given as follows:

|  | Factory X | Factory Y |
|---|---|---|
| Part A | 3 | 2 |
| Part B | 4 | 4 |
| Part C | 5 | 3 |

$$= Q$$

Compute $PQ$ and explain what each entry in this product matrix means.

3. Put the following system in matrix form $AX = B$, where $X$ is a column vector with the unknowns:

$$10x + 2y + 8z = 7$$
$$-3x + 5y - 7z = 5$$
$$-4x - 3y + 5z = 3.$$

## EXERCISE SET 2.4

In Exercises 1 through 6 perform the indicated multiplication if possible.

**1.**  $[2 \quad 3]\begin{bmatrix} 5 \\ 2 \end{bmatrix}$

**2.**  $[1 \quad 4 \quad -2]\begin{bmatrix} 3 \\ 2 \\ 5 \end{bmatrix}$

**3.**  $[-2 \quad 3 \quad 4 \quad 2]\begin{bmatrix} -2 \\ -1 \\ 3 \\ 5 \end{bmatrix}$

**4.**  $[2 \quad -3 \quad 0 \quad 4 \quad 2]\begin{bmatrix} 3 \\ 4 \\ 3 \\ -3 \\ 5 \end{bmatrix}$

**5.**  $[2 \quad 3]\begin{bmatrix} 5 \\ 2 \\ 0 \end{bmatrix}$

**6.**  $[0 \quad 0 \quad 0]\begin{bmatrix} 0 \\ 0 \end{bmatrix}$

In Exercises 7 through 16 find the order of $AB$ and $BA$ when either one exists.

**7.**  The order of $A$ is $(2 \times 3)$ and the order of $B$ is $(3 \times 4)$.

**8.**  The order of $A$ is $(3 \times 5)$ and the order of $B$ is $(5 \times 7)$.

**9.**  The order of $A$ is $(3 \times 4)$ and the order of $B$ is $(5 \times 3)$.

**10.**  The order of $A$ is $(5 \times 7)$ and the order of $B$ is $(8 \times 5)$.

**11.**  The order of $A$ is $(3 \times 3)$ and the order of $B$ is $(3 \times 5)$.

**12.**  The order of $A$ is $(5 \times 4)$ and the order of $B$ is $(4 \times 4)$.

**13.**  The order of $A$ is $(6 \times 6)$ and the order of $B$ is $(6 \times 6)$.

**14.**  The order of $A$ is $(8 \times 8)$ and the order of $B$ is $(8 \times 8)$.

**15.**  The order of $A$ is $(20 \times 2)$ and the order of $B$ is $(2 \times 30)$.

**16.**  The order of $A$ is $(3 \times 40)$ and the order of $B$ is $(50 \times 3)$.

In Exercises 17 through 42 perform the indicated multiplications if possible.

**17.**  $\begin{bmatrix} 2 & 5 \\ 4 & 3 \end{bmatrix}\begin{bmatrix} 2 \\ 5 \end{bmatrix}$

**18.**  $\begin{bmatrix} 5 & -3 \\ -2 & 4 \end{bmatrix}\begin{bmatrix} -3 \\ 4 \end{bmatrix}$

**19.**  $[3 \quad 7]\begin{bmatrix} -2 & 5 \\ 7 & 2 \end{bmatrix}$

**20.**  $[2 \quad 1]\begin{bmatrix} 3 & 20 \\ -3 & 5 \end{bmatrix}$

**21.**  $\begin{bmatrix} 0.2 & 0.1 \\ 0.4 & 0.5 \end{bmatrix}\begin{bmatrix} 2 & 3 \\ 5 & 2 \end{bmatrix}$

**22.**  $\begin{bmatrix} -1 & -3 \\ -2 & -4 \end{bmatrix}\begin{bmatrix} 0 & 2 \\ 2 & 5 \end{bmatrix}$

**23.** $\begin{bmatrix} 2 & 8 & 4 \\ 3 & 0 & 2 \end{bmatrix} \begin{bmatrix} 4 \\ 2 \\ 3 \end{bmatrix}$

**24.** $[4 \quad 2] \begin{bmatrix} 2 & -3 & 4 \\ -1 & 3 & 5 \end{bmatrix}$

**25.** $\begin{bmatrix} 0.1 & 0.4 & 0.2 \\ 0.5 & 0.2 & 0.1 \end{bmatrix} \begin{bmatrix} 2 & 4 \\ -1 & 5 \\ 5 & -2 \end{bmatrix}$

**26.** $\begin{bmatrix} 10 & 20 \\ 30 & 40 \end{bmatrix} \begin{bmatrix} 3 & 2 & 0 \\ 1 & 3 & 2 \end{bmatrix}$

**27.** $\begin{bmatrix} 3 & 5 \\ 8 & 2 \end{bmatrix} \begin{bmatrix} 1 & 5 & 2 & 3 \\ 0 & 3 & 5 & 1 \end{bmatrix}$

**28.** $\begin{bmatrix} 4 & 2 \\ -1 & 6 \\ 3 & -2 \end{bmatrix} \begin{bmatrix} 2 \\ 3 \end{bmatrix}$

**29.** $[0.01 \quad 0.03] \begin{bmatrix} 1 & 3 & 2 \\ 2 & 5 & 1 \end{bmatrix}$

**30.** $\begin{bmatrix} -1 & -2 & -5 & -2 \\ 3 & 5 & 1 & 0 \end{bmatrix} \begin{bmatrix} 0 & 3 \\ 2 & 0 \end{bmatrix}$

**31.** $\begin{bmatrix} 3 & 2 & 5 \\ 1 & 3 & 2 \\ 4 & 2 & 1 \end{bmatrix} \begin{bmatrix} 2 \\ 0 \\ 4 \end{bmatrix}$

**32.** $\begin{bmatrix} 4 & 0 & 1 \\ 2 & -2 & 2 \\ 1 & -1 & 0 \end{bmatrix} \begin{bmatrix} 2 \\ 4 \\ 1 \end{bmatrix}$

**33.** $[3 \quad -2 \quad 1] \begin{bmatrix} 2 & 1 & 5 \\ 5 & 0 & -2 \\ 0 & 1 & 2 \end{bmatrix}$

**34.** $[-1 \quad 0 \quad 1] \begin{bmatrix} 2 & 1 & 2 \\ 1 & -1 & -1 \\ 2 & 1 & 1 \end{bmatrix}$

**35.** $\begin{bmatrix} 1 & 4 & 2 \\ 0 & 2 & 0 \\ 5 & 2 & 1 \end{bmatrix} \begin{bmatrix} 3 & 0 & 1 \\ 0 & 2 & 4 \\ 1 & 1 & 2 \end{bmatrix}$

**36.** $\begin{bmatrix} 0 & -1 & 2 \\ -1 & 0 & 2 \\ 2 & -2 & 0 \end{bmatrix} \begin{bmatrix} 1 & 5 & 2 \\ 0 & 1 & 1 \\ 2 & 4 & 2 \end{bmatrix}$

**37.** $\begin{bmatrix} 1 & 2 \\ 0 & 0 \\ 1 & 0 \end{bmatrix} \begin{bmatrix} 2 & 3 \\ 1 & 1 \\ 0 & 0 \end{bmatrix}$

**38.** $\begin{bmatrix} 0 & 0 \\ 0 & 0 \end{bmatrix} \begin{bmatrix} 1 & 1 \\ 1 & 1 \\ 1 & 1 \end{bmatrix}$

**39.** $\begin{bmatrix} 2 \\ 5 \end{bmatrix} [3 \quad 5]$

**40.** $\begin{bmatrix} 2 \\ 0 \\ -1 \end{bmatrix} [2 \quad 0 \quad -1]$

**41.** $\begin{bmatrix} 0 & 0 & 0 \\ 0 & 0 & 0 \end{bmatrix} \begin{bmatrix} 0 & 0 \\ 0 & 0 \end{bmatrix}$

**42.** $\begin{bmatrix} 1 & 1 & 1 \\ 1 & 1 & 1 \end{bmatrix} \begin{bmatrix} 1 & 1 & 1 \\ 1 & 1 & 1 \end{bmatrix}$

**43.** Let

$$A = \begin{bmatrix} 1 & 1 \\ 1 & 1 \end{bmatrix} \quad \text{and} \quad B = \begin{bmatrix} 1 & 1 \\ -1 & -1 \end{bmatrix}$$

Show that $AB = O$. Thus a product of two matrices may be the zero matrix without either matrix being the zero matrix.

**44.** Consider the four matrices

$$I_2 = \begin{bmatrix} 1 & 0 \\ 0 & 1 \end{bmatrix}, \begin{bmatrix} 1 & 0 \\ 0 & -1 \end{bmatrix}, \begin{bmatrix} -1 & 0 \\ 0 & 1 \end{bmatrix}, \begin{bmatrix} -1 & 0 \\ 0 & -1 \end{bmatrix}$$

Show that the square of each of them is $I_2$. Thus the identity matrix has many square roots.

**45.** Let

$$A = \begin{bmatrix} 1 & 1 \\ -1 & -1 \end{bmatrix}.$$

Show that $A^2 = AA = O$.

**46.** Let

$$A = \begin{bmatrix} a_{11} & a_{12} & a_{13} \\ a_{21} & a_{22} & a_{23} \\ a_{31} & a_{32} & a_{33} \end{bmatrix}$$

Show that $AI_3 = I_3A = A$.

**47.** Let

$$A = \begin{bmatrix} 1 & 2 \\ 3 & 4 \end{bmatrix}, \quad B = \begin{bmatrix} 1 & -1 \\ 1 & 2 \end{bmatrix}.$$

Show that $(A + B)^2 \neq A^2 + 2AB + B^2$.

**48.** Let

$$A = \begin{bmatrix} 1 & 1 \\ 2 & 1 \end{bmatrix}, \quad B = \begin{bmatrix} 2 & 2 \\ 5 & 4 \end{bmatrix}, \quad C = \begin{bmatrix} 1 & 1 \\ -1 & -1 \end{bmatrix}.$$

Show $AC = BC$, but $A \neq B$.

**49.** Let $A, B, C$ be the matrices in Exercise 48. Show that

$$A(B + C) = AB + AC$$

$$A(BC) = (AB)C$$

**50.** A matrix $U = (u_{ij})$ is upper triangular if $u_{ij} = 0$ when $i > j$, that is, when all entries below the main diagonal are zero. Suppose $A = (a_{ij})$ and $B = (b_{ij})$ are both upper triangular matrices of order $(3 \times 3)$. Show that (a) $A^2$ is upper triangular, (b) $AB$ is upper triangular.

**51.** Let

$$A = \begin{bmatrix} 0 & 1 & 2 \\ 0 & 0 & 3 \\ 0 & 0 & 0 \end{bmatrix}.$$

Show $A^3 = AAA = O$.

52. A matrix $U = (u_{ij})$ is strictly upper triangular if $u_{ij} = 0$ when $i \geq j$, that is, when all entries below and on the main diagonal are zero. If $U = (u_{ij})$ is a strictly upper triangular matrix of order $(3 \times 3)$, show that $U^3 = O$.

In Exercises 53 through 60, find matrices $A$, $X$, and $B$ so that the given system of equations can be written as $AX = B$.

53. $2x_1 + 3x_2 = 5$

    $3x_1 - 5x_2 = 7$

54. $4x_1 - 5x_2 = 6$

    $2x_1 + 7x_2 = 9$

55. $2x_1 + 5x_2 + 3x_3 = 16$

    $4x_1 - 7x_2 - 2x_3 = 12$

    $5x_1 - 2x_2 + 6x_3 = 24$

56. $-3x_1 + 7x_2 + 2x_3 = 0$

    $- 7x_2 + 5x_3 = 2$

    $4x_1 + 3x_2 - 7x_3 = 4$

57. $2x_1 - 3x_2 + 3x_3 = 3$

    $5x_1 + 6x_2 - 2x_3 = 1$

58. $2x_1 - 7x_2 = 3$

    $x_1 + 8x_2 = 2$

    $3x_1 + 4x_2 = 5$

59. $2x_1 + 3x_2 = 7$

60. $3x_1 + 5x_2 - 6x_3 = 2$

## Applications

61. **Spare Parts Planning.** A car rental firm is planning a maintenance program involving certain parts of various makes of compact automobiles. The following matrix indicates the number of hundreds of automobiles available for rent in the three cities in which the firm operates.

|  | GM | Ford | Chrysler | Toyota |  |
|---|---|---|---|---|---|
| LA | 1 | 2 | 2 | 3 |  |
| NYC | 3 | 1 | 1 | 2 | $= N$ |
| SL | 2 | 2 | 1 | 1 |  |

Past repair records for the three cities indicate the number of each part needed per car per year. This is given in the following:

|  | Tires | Batteries | Plugs |  |
|---|---|---|---|---|
| GM | 4 | 0.5 | 6 |  |
| Ford | 3 | 0.5 | 7 | $= R.$ |
| Chrysler | 4 | 0.4 | 5 |  |
| Toyota | 3 | 1.0 | 4 |  |

a. Show that the quantity of each part needed in each city is given by $Q = NR$. Indicate for what each column and each row of $Q$ stands.

**b.** Let the elements in the column matrix

$$C = \begin{bmatrix} 100 \\ 40 \\ 1 \end{bmatrix}$$

be, respectively, the cost per each for tires, batteries, and plugs. Determine $QC$ and interpret what the elements stand for.

**62.** **Networks.** (The following application is a very scaled-down version of a technique used by the major airlines. The actual calculations are carried out by computers.) The diagram indicates the nonstop service between four cities, LA, Chicago, NYC, and Miami. For example, the arrows indicate that there is nonstop service in both directions between LA and NYC but only one-way service from Chicago to NYC. The network in the diagram can be summarized by an adjacency matrix $A = (a_{ij})$, as follows. We set $a_{ij} = 1$ if there is nonstop service from city $i$ to city $j$, otherwise $a_{ij}$ is zero. We then have

**To**

$$\text{From} \quad \begin{matrix} & \begin{matrix} \mathbf{L} & \mathbf{C} & \mathbf{N} & \mathbf{M} \end{matrix} \\ \begin{matrix} \mathbf{L} \\ \mathbf{C} \\ \mathbf{N} \\ \mathbf{M} \end{matrix} & \begin{bmatrix} 0 & 1 & 1 & 0 \\ 0 & 0 & 1 & 0 \\ 1 & 0 & 0 & 1 \\ 0 & 0 & 1 & 0 \end{bmatrix} \end{matrix} = A.$$

**a.** Compute $A^2$ and verify from the diagram that the element in the $i$th row and $j$th column of $A^2$ represents the number of one-stop routes between the city in the $i$th row and the city in the $j$th column.

**b.** Compute $A + A^2$ and interpret what this matrix means.

**c.** Compute $A^3$ and verify from the diagram that the element in the $i$th row and $j$th column of $A^3$ represents the number of two-stop routes between the city in the $i$th row and the city in the $j$th column.

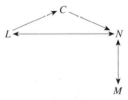

**63.** **Traffic Flow.** The figure shows the traffic flow through nine intersections. Traffic enters from two roads from the left and exits on four roads to the right. All roads are one-way. Given the number of cars $x_1$ and $x_2$ that enter the intersections $I_1$ and $I_3$, respectively, we wish to determine the number of cars exiting each of the four roads on the right. Notice, for example, according to the figure that 40% of the traffic that enters intersection $I_1$ exits due east. Thus $y_1 = 0.4x_1$. We also see that $y_2 = (0.6)(0.3)x_1 = 0.18x_1$ and $y_3 = (0.6)(0.7)x_1 + x_2$. Thus

$$y_1 = 0.4x_1$$

$$y_2 = 0.18x_1$$

$$y_3 = 0.42x_1 + x_2.$$

If

$$X = \begin{bmatrix} x_1 \\ x_2 \end{bmatrix}, \quad A = \begin{bmatrix} 0.40 & 0 \\ 0.18 & 0 \\ 0.42 & 1 \end{bmatrix}, \quad Y = \begin{bmatrix} y_1 \\ y_2 \\ y_3 \end{bmatrix},$$

then $Y = AX$.

**a.** Now by considering intersections $I_4, I_5, I_6$, find a $(3 \times 3)$ matrix $B$ such that $Z = BY$, where

$$Z = \begin{bmatrix} z_1 \\ z_2 \\ z_3 \end{bmatrix}.$$

**b.** Now by considering intersections $I_7, I_8, I_9$, find a $(4 \times 3)$ matrix $C$ such that $W = CZ$, where

$$W = \begin{bmatrix} w_1 \\ w_2 \\ w_3 \\ w_4 \end{bmatrix}.$$

**c.** Show that $W = (CBA)X$.
**d.** If $x_1 = 5000$ and $x_2 = 5000$, find $W$.

**64. Social Interaction.** The digraph indicates the influence in decision making that one individual has over another. For example, according to the digraph, $A$ has influence over $B$ and $D$ has influence over $A$, $B$, and $C$. The following matrix is then formed.

$$\begin{array}{c} \\ A \\ B \\ C \\ D \end{array} \begin{array}{cccc} A & B & C & D \\ \begin{bmatrix} 0 & 1 & 0 & 0 \\ 0 & 0 & 1 & 1 \\ 0 & 0 & 0 & 1 \\ 1 & 1 & 1 & 0 \end{bmatrix} \end{array} = T$$

A number one in the $i$th row and $j$th column indicates that the person in the $i$th row has direct influence over the person in the $j$th column, while a zero indicates no direct influence.

a. Calculate $T^2$ and show that a number 1 indicates that the individual in the row containing the 1 has influence over the individual in the column containing the 1 through one intermediary and that a number 2 indicates that the individual in the row containing the 2 has influence over the individual in the column containing the 2 through two different intermediaries.

b. Calculate $T + T^2$ and interpret what this matrix means.

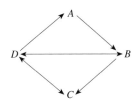

### Solutions to Self-Help Exercise Set 2.4

1.
$$\begin{bmatrix} 1 & 2 \\ -1 & 3 \end{bmatrix}\begin{bmatrix} 3 & 0 & 4 \\ 1 & -2 & 5 \end{bmatrix}$$

$$= \begin{bmatrix} (1)(3) + (2)(1) & (1)(0) + (2)(-2) & (1)(4) + (2)(5) \\ (-1)(3) + (3)(1) & (-1)(0) + (3)(-2) & (-1)(4) + (3)(5) \end{bmatrix}$$

$$= \begin{bmatrix} 5 & -4 & 14 \\ 0 & -6 & 11 \end{bmatrix}$$

2.
$$\begin{bmatrix} 3 & 2 & 1 \\ 1 & 2 & 4 \end{bmatrix}\begin{bmatrix} 3 & 2 \\ 4 & 4 \\ 5 & 3 \end{bmatrix}$$

$$= \begin{bmatrix} (3)(3) + (2)(4) + (1)(5) & (3)(2) + (2)(4) + (1)(3) \\ (1)(3) + (2)(4) + (4)(5) & (1)(2) + (2)(4) + (4)(3) \end{bmatrix}$$

$$\begin{array}{cc} \text{Factory X} & \text{Factory Y} \end{array}$$

$$= \begin{bmatrix} 22 & 17 \\ 31 & 22 \end{bmatrix}\begin{array}{l} \text{Type I} \\ \text{Type II} \end{array}$$

3. The system is in the form $AX = B$ if

$$A = \begin{bmatrix} 10 & 2 & 8 \\ -3 & 5 & -7 \\ -4 & -3 & 5 \end{bmatrix}, \quad X = \begin{bmatrix} x \\ y \\ z \end{bmatrix}, \quad B = \begin{bmatrix} 7 \\ 5 \\ 3 \end{bmatrix}$$

## 2.5    INVERSE OF A SQUARE MATRIX

▶ *Definition of the Inverse of a Matrix*

▶ *Finding the Inverse*

► *Systems of Equations*

► *Applications*

**Arthur Cayley,**
**1821–1895**

*(David Eugene Smith Collection,*
*Columbia University.)*

Arthur Cayley was the first to consider the idea of an inverse to a matrix in 1855. Cayley was educated in England and, unable to find a suitable teaching position, studied and practiced law for 14 years. During this time he published about 300 papers in mathematics. He did eventually become a professor at Cambridge in 1863. Cayley became acquainted with James Joseph Sylvester, mentioned in an earlier section, and together their collaborations resulted in many outstanding results in mathematics. Cayley was an enormously prolific researcher; his collected works contain 966 papers in thirteen volumes, each volume containing about 600 pages.

**A P P L I C A T I O N**

**Allocation of Funds**

An investment advisor has two mutual funds that she is recommending: a conservative bond fund with an assumed return of 7% a year and a stock fund with a projected return of 12% a year. One client has $100,000 to invest and wishes a return of 9% a year, and a second client has $50,000 and wants a 10% annual return. How should she allocate the money of each client between the bond fund and the stock fund? The answer is found in Example 5.

## Definition of the Inverse of a Matrix

In this section we will define the inverse of a square matrix and will see that the inverse matrix is an extremely powerful tool in solving systems of equations where the number of unknowns is the same as the number of equations. We will find the inverse of a square matrix using the technique of Gauss–Jordan. It is this technique that is commonly used by computers to find the inverse.

The scalar equation $ax = 1$ has the solution $x = a^{-1}$ if $a \neq 0$. The number $a^{-1}$ has the property that $aa^{-1} = a^{-1}a = 1$. Given a square $(n \times n)$ matrix $A$, we wish to find a matrix $A^{-1}$ called the **inverse of A** with the property that

$$AA^{-1} = A^{-1}A = I_n$$

---

**Inverse Matrix**

The $n \times n$ matrix $B$ is said to be the **inverse** to the $n \times n$ matrix $A$ if

$$AB = BA = I_n.$$

If the inverse matrix $B$ exists, we write $B = A^{-1}$.

---

**R E M A R K .** Do not think that the inverse matrix $A^{-1}$ can be written as $\dfrac{1}{A}$. *We never divide matrices.*

**E X A M P L E   1**    **Showing a Matrix is the Inverse Matrix**

Let

$$A = \begin{bmatrix} 3 & 1 \\ 5 & 2 \end{bmatrix}.$$

Then show that

$$A^{-1} = \begin{bmatrix} 2 & -1 \\ -5 & 3 \end{bmatrix}.$$

**Solution**    The indicated matrix $A^{-1}$ is the inverse of $A$ since

$$AA^{-1} = \begin{bmatrix} 3 & 1 \\ 5 & 2 \end{bmatrix}\begin{bmatrix} 2 & -1 \\ -5 & 3 \end{bmatrix} = \begin{bmatrix} 1 & 0 \\ 0 & 1 \end{bmatrix} = I_2$$

and

$$A^{-1}A = \begin{bmatrix} 2 & -1 \\ -5 & 3 \end{bmatrix}\begin{bmatrix} 3 & 1 \\ 5 & 2 \end{bmatrix} = \begin{bmatrix} 1 & 0 \\ 0 & 1 \end{bmatrix} = I_2. \quad \blacksquare$$

Not every matrix has an inverse. Consider

$$B = \begin{bmatrix} 2 & 1 \\ 4 & 2 \end{bmatrix}.$$

Let us try to find a matrix $X$ of order $(2 \times 2)$ such that

$$\begin{bmatrix} 1 & 0 \\ 0 & 1 \end{bmatrix} = I_2 = BX = \begin{bmatrix} 2 & 1 \\ 4 & 2 \end{bmatrix}\begin{bmatrix} a & b \\ c & d \end{bmatrix} = \begin{bmatrix} 2a + c & 2b + d \\ 4a + 2c & 4b + 2d \end{bmatrix}$$

Notice that equating elements in the first column requires that

$$2a + c = 1$$

$$4a + 2c = 0.$$

But 2 times the first of these two equations yields $4a + 2c = 2$. This is inconsistent with the second equation, and thus no solution can exist. This shows that $B^{-1}$ does not exist.

## Finding the Inverse

We shall now develop a technique that will find the inverse of a square matrix, if it exists, or indicate that the inverse does not exist.

Consider

$$A = \begin{bmatrix} 1 & 2 \\ 3 & 7 \end{bmatrix}.$$

To find

$$A^{-1} = X = \begin{bmatrix} x_1 & y_1 \\ x_2 & y_2 \end{bmatrix},$$

we must solve the equation $AX = I_2$ or

$$AX = \begin{bmatrix} 1 & 2 \\ 3 & 7 \end{bmatrix}\begin{bmatrix} x_1 & y_1 \\ x_2 & y_2 \end{bmatrix} = \begin{bmatrix} x_1 + 2x_2 & y_1 + 2y_2 \\ 3x_1 + 7x_2 & 3y_1 + 7y_2 \end{bmatrix} = \begin{bmatrix} 1 & 0 \\ 0 & 1 \end{bmatrix}.$$

This gives the four equations

$$x_1 + 2x_2 = 1 \qquad y_1 + 2y_2 = 0$$
$$3x_1 + 7x_2 = 0 \qquad 3y_1 + 7y_2 = 1,$$

which have been written as two systems. We can then solve each system by first forming the augmented matrices

$$\begin{bmatrix} 1 & 2 & | & 1 \\ 3 & 7 & | & 0 \end{bmatrix} \quad \text{and} \quad \begin{bmatrix} 1 & 2 & | & 0 \\ 3 & 7 & | & 1 \end{bmatrix}.$$

But since both of these systems have the same coefficient matrix, they both will be solved by the Gauss–Jordan method, using precisely the same row operations. Thus we may as well solve them together. We do this by forming the augmented matrix

$$\begin{bmatrix} 1 & 2 & | & 1 & 0 \\ 3 & 7 & | & 0 & 1 \end{bmatrix}.$$

Now perform the usual row operations and obtain

$$\begin{bmatrix} 1 & 2 & | & 1 & 0 \\ 3 & 7 & | & 0 & 1 \end{bmatrix} R_2 - 3R_1 \rightarrow R_2$$

$$\begin{bmatrix} 1 & 2 & | & 1 & 0 \\ 0 & 1 & | & -3 & 1 \end{bmatrix} R_1 - 2R_2 \rightarrow R_1$$

$$\begin{bmatrix} 1 & 0 & | & 7 & -2 \\ 0 & 1 & | & -3 & 1 \end{bmatrix}.$$

This means that the solutions to the two systems are

$$x_1 = 7 \qquad y_1 = -2$$
$$x_2 = -3 \qquad y_2 = 1.$$

Thus

$$A^{-1} = \begin{bmatrix} 7 & -2 \\ -3 & 1 \end{bmatrix}.$$

In general we have the following technique.

> **To Find $A^{-1}$**
>
> 1.  Form the augmented matrix
> $$[A|I].$$
> 2.  Use row operations to reduce $[A|I]$ to
> $$[I|B]$$
> if possible.
> 3.  $A^{-1} = B$.

**E X A M P L E   2**    **Finding the Inverse of a Matrix**

Find $A^{-1}$ if

$$A = \begin{bmatrix} 1 & 1 & 2 \\ 2 & 3 & 2 \\ 1 & 1 & 3 \end{bmatrix}.$$

**Solution**    Following the above technique, we have

$$\begin{bmatrix} 1 & 1 & 2 & | & 1 & 0 & 0 \\ 2 & 3 & 2 & | & 0 & 1 & 0 \\ 1 & 1 & 3 & | & 0 & 0 & 1 \end{bmatrix} \begin{matrix} \\ R_2 - 2R_1 \to R_2 \\ R_3 - R_1 \to R_3 \end{matrix}$$

$$\begin{bmatrix} 1 & 1 & 2 & | & 1 & 0 & 0 \\ 0 & 1 & -2 & | & -2 & 1 & 0 \\ 0 & 0 & 1 & | & -1 & 0 & 1 \end{bmatrix} \begin{matrix} R_1 - 2R_3 \to R_1 \\ R_2 + 2R_3 \to R_2 \\ \\ \end{matrix}$$

$$\begin{bmatrix} 1 & 1 & 0 & | & 3 & 0 & -2 \\ 0 & 1 & 0 & | & -4 & 1 & 2 \\ 0 & 0 & 1 & | & -1 & 0 & 1 \end{bmatrix} \begin{matrix} R_1 - R_2 \to R_1 \\ \\ \\ \end{matrix}$$

$$\begin{bmatrix} 1 & 0 & 0 & | & 7 & -1 & -4 \\ 0 & 1 & 0 & | & -4 & 1 & 2 \\ 0 & 0 & 1 & | & -1 & 0 & 1 \end{bmatrix}.$$

Thus

$$A^{-1} = \begin{bmatrix} 7 & -1 & -4 \\ -4 & 1 & 2 \\ -1 & 0 & 1 \end{bmatrix}. \quad \blacksquare$$

If a matrix does not have an inverse, the above technique will so indicate, as in the following.

**E X A M P L E   3**      **Showing That the Inverse Does Not Exist**

Find $A^{-1}$ if

$$A = \begin{bmatrix} 1 & 2 & 3 \\ 2 & 5 & 7 \\ 2 & 4 & 6 \end{bmatrix}.$$

**Solution**      Forming the augmented matrix and proceeding yields

$$\begin{bmatrix} 1 & 2 & 3 & | & 1 & 0 & 0 \\ 2 & 5 & 7 & | & 0 & 1 & 0 \\ 2 & 4 & 6 & | & 0 & 0 & 1 \end{bmatrix} \begin{matrix} \\ R_2 - 2R_1 \to R_2 \\ R_3 - 2R_1 \to R_3 \end{matrix}$$

$$\begin{bmatrix} 1 & 2 & 3 & | & 1 & 0 & 0 \\ 0 & 1 & 1 & | & -2 & 1 & 0 \\ 0 & 0 & 0 & | & -2 & 0 & 1 \end{bmatrix}.$$

The last row indicates that there can be no solution. Thus $A^{-1}$ does not exist.   ∎

## Systems of Equations

We shall now see how the inverse matrix can be used to solve systems of equations with the number of unknowns equal to the number of equations. Consider the system

$$a_{11}x_1 + a_{12}x_2 + \cdots + a_{1n}x_n = b_1$$

$$a_{21}x_1 + a_{22}x_2 + \cdots + a_{2n}x_n = b_2$$

$$\cdots$$

$$a_{n1}x_1 + a_{n2}x_2 + \cdots + a_{nn}x_n = b_n.$$

If

$$A = \begin{bmatrix} a_{11} & a_{12} & \cdots & a_{1n} \\ a_{21} & a_{22} & \cdots & a_{2n} \\ & & \cdots & \\ a_{n1} & a_{n2} & \cdots & a_{nn} \end{bmatrix}, X = \begin{bmatrix} x_1 \\ x_2 \\ \vdots \\ x_n \end{bmatrix}, B = \begin{bmatrix} b_1 \\ b_2 \\ \vdots \\ b_n \end{bmatrix},$$

then the system can be written as

$$AX = B.$$

Now if $A^{-1}$ exists, multiply both sides of $AX = B$ on the left by $A^{-1}$, obtaining

$$A^{-1}(AX) = A^{-1}B.$$

Now the left side is

$$A^{-1}(AX) = (A^{-1}A)X = I_nX = X.$$

Thus

$$X = A^{-1}B$$

is the solution of the system.

---

**Solution of $AX = B$**

Let $A$ be a square matrix of order $n$ and $B$ a column matrix of order $n$. If $A^{-1}$ exists, then the linear system of equations

$$AX = B$$

has the unique solution

$$X = A^{-1}B.$$

---

**E X A M P L E   4**     **Solving Systems of Equations Using Matrices**

Solve the two systems

$$
\begin{array}{cccc}
x + & y + 2z = 1 & \qquad x + & y + 2z = 2 \\
2x + & 3y + 2z = 2 & \qquad 2x + & 3y + 2z = 0 \\
x + & y + 3z = 0 & \qquad x + & y + 3z = 1
\end{array}
$$

**Solution**     Let

$$
A = \begin{bmatrix} 1 & 1 & 2 \\ 2 & 3 & 2 \\ 1 & 1 & 3 \end{bmatrix}, \quad
X = \begin{bmatrix} x \\ y \\ z \end{bmatrix}, \quad
B = \begin{bmatrix} 1 \\ 2 \\ 0 \end{bmatrix}, \quad
C = \begin{bmatrix} 2 \\ 0 \\ 1 \end{bmatrix}.
$$

Then notice that the two systems can be written as

$$AX = B \qquad AX = C.$$

The solutions are respectively

$$X = A^{-1}B \qquad X = A^{-1}C.$$

We already found $A^{-1}$ in Example 2. Thus the first system has the solution

$$
A^{-1}B = \begin{bmatrix} 7 & -1 & -4 \\ -4 & 1 & 2 \\ -1 & 0 & 1 \end{bmatrix} \begin{bmatrix} 1 \\ 2 \\ 0 \end{bmatrix} = \begin{bmatrix} 5 \\ -2 \\ -1 \end{bmatrix},
$$

or $x = 5$, $y = -2$, and $z = -1$.

The second system has the solution

$$A^{-1}C = \begin{bmatrix} 7 & -1 & -4 \\ -4 & 1 & 2 \\ -1 & 0 & 1 \end{bmatrix} \begin{bmatrix} 2 \\ 0 \\ 1 \end{bmatrix} = \begin{bmatrix} 10 \\ -6 \\ -1 \end{bmatrix},$$

or $x = 10$, $y = -6$, $z = -1$. ∎

## Applications

**E X A M P L E 5**     **Investments**
_____

An investment advisor has two mutual funds that she is recommending: a conservative bond fund with an assumed return of 7% a year and a stock fund with a projected return of 12% a year. One client has \$100,000 to invest and wishes a return of 9% a year, and a second client has \$50,000 and wants a 10% annual return. How should she allocate the money of each client between the bond fund and the stock fund?

**Solution**    Let $x$ be the amount invested in the bond fund and $y$ the amount invested in the stock fund. If $T$ is the total amount invested, then $x + y = T$. The return on the bond fund will be $0.07x$, and the return on the stock fund will be $0.12y$, for a total return of $0.07x + 0.12y$. If $R$ is the return desired, then $0.07x + 0.12y = R$. We then have

$$x + y = T$$
$$0.07x + 0.12y = R.$$

If

$$A = \begin{bmatrix} 1 & 1 \\ 0.07 & 0.12 \end{bmatrix}, \quad X = \begin{bmatrix} x \\ y \end{bmatrix}, \quad B = \begin{bmatrix} T \\ R \end{bmatrix},$$

then the system can be written as $AX = B$. The solution is $X = A^{-1}B$. To find $A^{-1}$ we use the Gauss–Jordan method.

$$[A|I] = \begin{bmatrix} 1 & 1 & | & 1 & 0 \\ 0.07 & 0.12 & | & 0 & 1 \end{bmatrix} \begin{matrix} \\ R_2 - 0.07R_1 \to R_2 \end{matrix}$$

$$\begin{bmatrix} 1 & 1 & | & 1 & 0 \\ 0 & 0.05 & | & -0.07 & 1 \end{bmatrix} 20R_2 \to R_2$$

$$\begin{bmatrix} 1 & 1 & | & 1 & 0 \\ 0 & 1 & | & -1.4 & 20 \end{bmatrix} R_1 - R_2 \to R_1$$

$$\begin{bmatrix} 1 & 0 & | & 2.4 & -20 \\ 0 & 1 & | & -1.4 & 20 \end{bmatrix}$$

Thus

$$A^{-1} = \begin{bmatrix} 2.4 & -20 \\ -1.4 & 20 \end{bmatrix}.$$

(a)    The first client has $100,000 to invest and wishes a return of (0.09)($100,000) = $9,000. Thus

$$X = A^{-1}B = \begin{bmatrix} 2.4 & -20 \\ -1.4 & 20 \end{bmatrix} \begin{bmatrix} 100,000 \\ 9,000 \end{bmatrix} = \begin{bmatrix} 60,000 \\ 40,000 \end{bmatrix}.$$

The first client should invest $60,000 in the bond fund and $40,000 in the stock fund.

(b)    The second client has $50,000 to invest and wishes a return of (0.10)($50,000) = $5,000. Thus

$$X = A^{-1}B = \begin{bmatrix} 2.4 & -20 \\ -1.4 & 20 \end{bmatrix} \begin{bmatrix} 50,000 \\ 5,000 \end{bmatrix} = \begin{bmatrix} 20,000 \\ 30,000 \end{bmatrix}.$$

The second client should invest $20,000 in the bond fund and $30,000 in the stock fund.    ∎

## SELF-HELP EXERCISE SET 2.5

1.    Find $A^{-1}$ if

$$A = \begin{bmatrix} 1 & 0 & 0 & 1 \\ 0 & 0 & 1 & 0 \\ 0 & 0 & 0 & 1 \\ 0 & 1 & 0 & 2 \end{bmatrix}.$$

2.    Solve the following system of equations using the inverse of the coefficient matrix:

$$
\begin{aligned}
x \quad\quad - 2z &= 1 \\
y + z &= 2 \\
x + y \quad\quad &= 4.
\end{aligned}
$$

3.    Solution $A$ is 3% alcohol, and solution $B$ is 7% alcohol. A lab technician needs to mix the two solutions to obtain 50 liters of a solution that is 4.2% alcohol. How many liters of each solution must be used? Use the inverse matrix to find the answer.

## EXERCISE SET 2.5

In Exercises 1 through 10 determine if the given pairs of matrices are inverses to each other by showing if their product is the identity matrix.

1. $\begin{bmatrix} 1 & 2 \\ 3 & 5 \end{bmatrix}$ and $\begin{bmatrix} -5 & 2 \\ 3 & -1 \end{bmatrix}$

2. $\begin{bmatrix} 10 & 3 \\ 3 & 1 \end{bmatrix}$ and $\begin{bmatrix} 1 & -3 \\ -3 & 10 \end{bmatrix}$

3. $\begin{bmatrix} 4 & 7 \\ 1 & 2 \end{bmatrix}$ and $\begin{bmatrix} 2 & -7 \\ -1 & 4 \end{bmatrix}$

4. $\begin{bmatrix} 13 & 4 \\ 3 & 1 \end{bmatrix}$ and $\begin{bmatrix} 1 & 4 \\ -3 & 13 \end{bmatrix}$

5. $\begin{bmatrix} 1 & 2 & 0 \\ 2 & 1 & -1 \\ 3 & 1 & 1 \end{bmatrix}$ and $\dfrac{1}{8}\begin{bmatrix} -2 & 2 & 2 \\ 5 & -1 & -1 \\ 1 & -5 & 3 \end{bmatrix}$

6. $\begin{bmatrix} 2 & 1 & 3 \\ 3 & 1 & 3 \\ 2 & 1 & 4 \end{bmatrix}$ and $\begin{bmatrix} -1 & 1 & 0 \\ 0 & -2 & -3 \\ 1 & 0 & 1 \end{bmatrix}$

7. $\begin{bmatrix} 1 & 1 & 0 \\ 0 & 1 & 1 \\ 0 & 0 & 1 \end{bmatrix}$ and $\begin{bmatrix} 1 & -1 & 1 \\ 0 & 1 & -1 \\ 0 & 0 & 1 \end{bmatrix}$

8. $\begin{bmatrix} 1 & 2 & 3 \\ 0 & 1 & 3 \\ 0 & 0 & 1 \end{bmatrix}$ and $\begin{bmatrix} 1 & -2 & 3 \\ 0 & 1 & -3 \\ 0 & 0 & 1 \end{bmatrix}$

9. $\begin{bmatrix} 2 & 0 & 1 & 2 \\ 1 & 1 & 0 & 2 \\ 2 & -1 & 3 & 1 \\ 3 & -1 & 4 & 3 \end{bmatrix}$ and $\dfrac{1}{3}\begin{bmatrix} 3 & 0 & 3 & -3 \\ -3 & 5 & 5 & -3 \\ -3 & 2 & 2 & 0 \\ 0 & -1 & -4 & 3 \end{bmatrix}$

10. $\begin{bmatrix} 1 & 0 & 0 & 0 \\ 0 & 1 & 0 & 0 \\ 0 & 0 & 5 & 2 \\ 0 & 0 & 1 & 1 \end{bmatrix}$ and $\begin{bmatrix} 1 & 0 & 0 & 0 \\ 0 & 1 & 0 & 0 \\ 0 & 0 & 1 & -2 \\ 0 & 0 & 1 & 5 \end{bmatrix}$

In Exercises 11 through 26 find the inverses of the given matrices, if they exist.

11. $\begin{bmatrix} 1 & 3 \\ 3 & 10 \end{bmatrix}$

12. $\begin{bmatrix} 1 & 3 \\ 1 & 4 \end{bmatrix}$

13. $\begin{bmatrix} 5 & -1 \\ -3 & 2 \end{bmatrix}$

14. $\begin{bmatrix} 5 & -1 \\ 4 & 2 \end{bmatrix}$

15. $\begin{bmatrix} 4 & 2 \\ 2 & 1 \end{bmatrix}$

16. $\begin{bmatrix} 1 & 1 \\ 1 & 1 \end{bmatrix}$

17. $\begin{bmatrix} 1 & 2 & 1 \\ 1 & 2 & 1 \\ 2 & 0 & 1 \end{bmatrix}$

18. $\begin{bmatrix} 1 & 0 & 1 \\ 0 & 2 & 1 \\ 1 & 0 & 2 \end{bmatrix}$

19. $\begin{bmatrix} 1 & 1 & 1 \\ 0 & 2 & 2 \\ -1 & 0 & 1 \end{bmatrix}$

**20.** $\begin{bmatrix} 1 & 0 & 1 \\ 0 & 1 & 1 \\ 1 & 1 & 1 \end{bmatrix}$

**21.** $\begin{bmatrix} 1 & 2 & 0 \\ 1 & 1 & 2 \\ 0 & -1 & -1 \end{bmatrix}$

**22.** $\begin{bmatrix} 2 & 4 & 4 \\ 1 & 2 & 1 \\ 1 & 0 & -1 \end{bmatrix}$

**23.** $\begin{bmatrix} 1 & 1 & 1 \\ 1 & 2 & 4 \\ 2 & 1 & 1 \end{bmatrix}$

**24.** $\begin{bmatrix} 0 & 1 & 1 \\ 1 & 1 & 0 \\ 1 & 0 & 1 \end{bmatrix}$

**25.** $\begin{bmatrix} 1 & 1 & 1 & 1 \\ 0 & 2 & 1 & 1 \\ 1 & 0 & 2 & 1 \\ 0 & 1 & 0 & 1 \end{bmatrix}$

**26.** $\begin{bmatrix} 0 & 1 & 1 & 0 \\ 1 & 0 & 0 & 2 \\ 2 & 0 & 1 & 0 \\ 2 & 0 & 0 & 1 \end{bmatrix}$

In Exercises 27 through 32 solve by using the inverse of the coefficient matrix.

**27.** $2x - 4y = 10$
$x - y = -2$

**28.** $2x - y = 7$
$x + 3y = 7$

**29.** $x + y = 6$
$2x + 3y = 9$

**30.** $x + y = 5$
$2x - 3y = -5$

**31.** $2x + 4y = 6$
$3x + 5y = 4$

**32.** $3x - 6y = 6$
$x - y = 9$

In Exercises 33 through 36 solve by using your answers to Exercises 11 through 14.

**33.** $x + 3y = 2$
$3x + 10y = 3$

**34.** $x + 3y = 1$
$x + 4y = 6$

**35.** $5x - y = 4$
$-3x + 2y = 3$

**36.** $5x - y = -3$
$4x + 2y = 5$

In Exercises 37 through 42 solve by using your answers to Exercises 21 through 26.

**37.** $x + 2y = 1$
$x + y + 2z = 0$
$-y - z = 2$

**38.** $2x + 4y + 4z = 1$
$x + 2y + z = 1$
$x - z = 3$

**39.** $x + y + z = 0$
$x + 2y + 4z = 1$
$2x + y + z = 2$

**40.** $y + z = -2$
$x + y = 2$
$x + z = 4$

**41.** $x_1 + x_2 + x_3 + x_4 = 1$
$2x_2 + x_3 + x_4 = 1$
$x_1 + 2x_3 + x_4 = 2$
$x_2 + x_4 = 2$

**42.** $x_2 + x_3 = 2$
$x_1 + 2x_4 = -3$
$2x_1 + x_3 = 4$
$2x_1 + x_4 = 6$

**43.** Let

$$A = \begin{bmatrix} a & b \\ c & d \end{bmatrix}$$

and assume that $D = ad - bc \neq 0$. Show that

$$A^{-1} = \frac{1}{D} \begin{bmatrix} d & -b \\ -c & a \end{bmatrix}.$$

**44.** If $A^{-1}$ and $B^{-1}$ exist, show that

$$(AB)^{-1} = B^{-1}A^{-1}.$$

## Applications

**45. Investment.** Redo Example 5 with a bond fund return of 8%.

**46. Investment.** Redo Example 5 with a bond fund return of 8% and a stock fund return of 13%.

**47. Scheduling.** The material and labor costs of manufacturing two styles of chair are given as follows:

|         | Material | Labor |
|---------|----------|-------|
| Style A | 40       | 50    |
| Style B | 30       | 40    |

If $10,000 is allocated for both material and labor, how many of each style of chair should be manufactured to exactly use the $10,000 if
**a.** $4400 is allocated for material and $5600 for labor;
**b.** $4300 is allocated for material and $5700 for labor.

**48. Mixture.** A small store sells mint tea at $3.20 an ounce and peppermint tea at $4 an ounce. The store owner decides to make a batch of 100 ounces of tea that mixes both kinds and to sell the mixture for $3.50 an ounce. How many ounces of each of the two varieties of tea should be mixed to obtain the same revenue as selling them unmixed?

## Graphing Calculator Exercises

In Exercises 49 through 56 use a calculator. Let

$$A = \begin{bmatrix} 1 & 4 & 5 & -3 & 6 \\ -11 & 6 & 4 & -2 & 3 \\ 4 & 16 & -1 & 8 & 1 \\ 2 & -8 & 1 & 4 & -1 \\ 1 & 9 & 10 & 4 & 8 \end{bmatrix} \qquad B = \begin{bmatrix} 1 & -4 & 3 & 2 & 1 \\ 6 & 1 & -8 & 5 & 1 \\ -2 & 8 & -5 & 7 & 3 \\ 4 & -5 & -16 & 2 & 4 \\ 1 & 3 & 1 & -5 & 1 \end{bmatrix}$$

**49.** Does $AB = BA$?

**50.** Find $A^{-1}$.

**51.** Find $B^{-1}$.

**52.** Find $(AB)^{-1}$.

**53.** Is $(AB)^{-1} = A^{-1}B^{-1}$?

**54.** Is $(AB)^{-1} = B^{-1}A^{-1}$?

**55.** Solve $AX = C$ where $\quad C = \begin{bmatrix} 10 \\ 8 \\ 20 \\ 22 \\ 37 \end{bmatrix}.$

**56.** Solve $BX = C$ where $C$ is given in the previous exercise.

## Solutions to Self-Help Exercise Set 2.5

**1.** We have

$$\left[\begin{array}{cccc|cccc} 1 & 0 & 0 & 1 & 1 & 0 & 0 & 0 \\ 0 & 0 & 1 & 0 & 0 & 1 & 0 & 0 \\ 0 & 0 & 0 & 1 & 0 & 0 & 1 & 0 \\ 0 & 1 & 0 & 2 & 0 & 0 & 0 & 1 \end{array}\right] \begin{array}{l} \\ R_2 \leftrightarrow R_4 \\ \\ \end{array}$$

$$\left[\begin{array}{cccc|cccc} 1 & 0 & 0 & 1 & 1 & 0 & 0 & 0 \\ 0 & 1 & 0 & 2 & 0 & 0 & 0 & 1 \\ 0 & 0 & 0 & 1 & 0 & 0 & 1 & 0 \\ 0 & 0 & 1 & 0 & 0 & 1 & 0 & 0 \end{array}\right] \begin{array}{l} \\ \\ R_3 \leftrightarrow R_4 \\ \end{array}$$

$$\left[\begin{array}{cccc|cccc} 1 & 0 & 0 & 1 & 1 & 0 & 0 & 0 \\ 0 & 1 & 0 & 2 & 0 & 0 & 0 & 1 \\ 0 & 0 & 1 & 0 & 0 & 1 & 0 & 0 \\ 0 & 0 & 0 & 1 & 0 & 0 & 1 & 0 \end{array}\right] \begin{array}{l} R_1 - R_4 \rightarrow R_1 \\ R_2 - 2R_4 \rightarrow R_2 \\ \\ \end{array}$$

$$\left[\begin{array}{cccc|cccc} 1 & 0 & 0 & 0 & 1 & 0 & -1 & 0 \\ 0 & 1 & 0 & 0 & 0 & 0 & -2 & 1 \\ 0 & 0 & 1 & 0 & 0 & 1 & 0 & 0 \\ 0 & 0 & 0 & 1 & 0 & 0 & 1 & 0 \end{array}\right].$$

Thus

$$A^{-1} = \begin{bmatrix} 1 & 0 & -1 & 0 \\ 0 & 0 & -2 & 1 \\ 0 & 1 & 0 & 0 \\ 0 & 0 & 1 & 0 \end{bmatrix}.$$

2.  Let

$$A = \begin{bmatrix} 1 & 0 & -2 \\ 0 & 1 & 1 \\ 1 & 1 & 0 \end{bmatrix}, X = \begin{bmatrix} x \\ y \\ z \end{bmatrix}, B = \begin{bmatrix} 1 \\ 2 \\ 4 \end{bmatrix}.$$

Then the system of equations can be written as $AX = B$. The solution is $X = A^{-1}B$. To find $A^{-1}$, we use the Gauss–Jordan method and obtain

$$\left[\begin{array}{ccc|ccc} 1 & 0 & -2 & 1 & 0 & 0 \\ 0 & 1 & 1 & 0 & 1 & 0 \\ 1 & 1 & 0 & 0 & 0 & 1 \end{array}\right] R_3 - R_1 \rightarrow R_3$$

$$\left[\begin{array}{ccc|ccc} 1 & 0 & -2 & 1 & 0 & 0 \\ 0 & 1 & 1 & 0 & 1 & 0 \\ 0 & 1 & 2 & -1 & 0 & 1 \end{array}\right] R_3 - R_2 \rightarrow R_3$$

$$\left[\begin{array}{ccc|ccc} 1 & 0 & -2 & 1 & 0 & 0 \\ 0 & 1 & 1 & 0 & 1 & 0 \\ 0 & 0 & 1 & -1 & -1 & 1 \end{array}\right] \begin{array}{l} R_1 + 2R_3 \rightarrow R_1 \\ R_2 - R_3 \rightarrow R_2 \end{array}$$

$$\left[\begin{array}{ccc|ccc} 1 & 0 & 0 & -1 & -2 & 2 \\ 0 & 1 & 0 & 1 & 2 & -1 \\ 0 & 0 & 1 & -1 & -1 & 1 \end{array}\right].$$

Thus

$$A^{-1} = \begin{bmatrix} -1 & -2 & 2 \\ 1 & 2 & -1 \\ -1 & -1 & 1 \end{bmatrix},$$

and

$$X = A^{-1}B$$

$$= \begin{bmatrix} -1 & -2 & 2 \\ 1 & 2 & -1 \\ -1 & -1 & 1 \end{bmatrix}\begin{bmatrix} 1 \\ 2 \\ 4 \end{bmatrix}$$

$$= \begin{bmatrix} (-1)(1) + (-2)(2) + (2)(4) \\ (1)(1) + (2)(2) + (-1)(4) \\ (-1)(1) + (-1)(2) + (1)(4) \end{bmatrix} = \begin{bmatrix} 3 \\ 1 \\ 1 \end{bmatrix}.$$

3.  Let $x$ and $y$ denote, respectively, the number of liters of solution A and B to be used. Then since there must be 50 liters, we have $x + y = 50$. The amount of alcohol in liters in solution A is $.03x$, in solution B is $.07y$, and in the mixture is $.042(50) = 2.1$. Thus $.03x + .07y = 2.1$. We then have the system of equations

$$x + \quad y = 50$$

$$.03x + .07y = 2.1.$$

This can be written in matrix form as $AX = B$ if

$$A = \begin{bmatrix} 1 & 1 \\ .03 & .07 \end{bmatrix}, \quad X = \begin{bmatrix} x \\ y \end{bmatrix}, \quad B = \begin{bmatrix} 50 \\ 2.1 \end{bmatrix}.$$

The solution is $X = A^{-1}B$. To find $A^{-1}$ we use the Gauss–Jordan method and obtain

$$[A|I] = \begin{bmatrix} 1 & 1 & | & 1 & 0 \\ 0.03 & 0.07 & | & 0 & 1 \end{bmatrix} \quad R_2 - 0.03R_1 \to R_2$$

$$\begin{bmatrix} 1 & 1 & | & 1 & 0 \\ 0 & 0.04 & | & -0.03 & 1 \end{bmatrix} \quad 25R_2 \to R_2$$

$$\begin{bmatrix} 1 & 1 & | & 1 & 0 \\ 0 & 1 & | & -.75 & 25 \end{bmatrix} \quad R_1 - R_2 \to R_1$$

$$\begin{bmatrix} 1 & 0 & | & 1.75 & -25 \\ 0 & 1 & | & -.75 & 25 \end{bmatrix}$$

Thus

$$A^{-1} = \begin{bmatrix} 1.75 & -25 \\ -.75 & 25 \end{bmatrix}.$$

Then

$$X = A^{-1}B = \begin{bmatrix} 1.75 & -25 \\ -.75 & 25 \end{bmatrix} \begin{bmatrix} 50 \\ 2.1 \end{bmatrix} = \begin{bmatrix} 35 \\ 15 \end{bmatrix}.$$

Thus the technician should use 35 liters of solution A and 15 liters of solution B.

## 2.6   LEONTIEF INPUT–OUTPUT MODELS

► *Introduction*

► *Open Leontief Input–Output Model*

► *Closed Leontief Input–Output Model*

# Introduction

There are three branches of economics that rely heavily on linear systems and matrix theory: game theory, input–output analysis, and linear programming. Game theory arrived on the scene first with the publication of a fundamental result by John von Neumann in 1928. The impact of game theory on economics, however, was considerably delayed. Game theory will be taken up in a later chapter in this text. Linear programming, which will be considered in the next chapter, appeared immediately after World War II. Input–output analysis was the second of these three branches of economics to appear. Input–output analysis was initiated by Wassily Leontief with the publication of a paper of his in 1936. In 1941 he published a full exposition of this method. For this work, he was awarded a Nobel prize in economics in 1973. Leontief divided the American economy into hundreds of sectors, such as automobiles, glass, copper, and so on, and then studied how these sectors interact with each other.

In order to effectively handle the hundreds of linear equations in the input–output analysis, one naturally needs to use computers. We will obtain a flavor for the method by considering an economy with only two or three sectors.

Consider an economy consisting of $n$ industries, each of which produces a single product. Demand for any one of the industry's products comes from demand from each of the industries (including possibly itself) and demand from outside these industries. For example, the energy industry will need energy for its own plants, will need energy to supply other industries, and will need to supply energy to the general public.

*(© Gary Benson/Comstock, Inc.)*

## Open Leontief Input–Output Model

In input–output analysis, we seek to determine the amount each industry must produce so that both the interindustry demand and the nonindustry demand are both met exactly, that is, so that supply and demand are exactly balanced.

The interindustry demand is summarized in an **input–output matrix** $A = (a_{ij})$. There may be as many as a hundred or even more industries in actual models. For illustrative purposes, we assume a very simple model of three industries: auto, energy, and transportation. Each of these industries acts as a supplier and a user of their products. We let $a_{ij}$ **represent the dollar amount of the output of industry $i$ required in producing one dollar of output in industry $j$.** The following matrix summarizes our simplified economy.

**User**

|  | Auto | Energy | Transportation |
|---|---|---|---|
| **Auto** | 0.2 | 0.4 | 0.1 |
| **Energy** | 0.1 | 0.2 | 0.2 |
| **Transportation** | 0.2 | 0.2 | 0.1 |

$= A$

Thus the element $a_{11} = 0.2$ indicates that of every dollar of output *from* the automobile industry, 20% of this value is contributed by the auto industry. The element $a_{12} = 0.4$ indicates that of every dollar of output *from* the energy industry, 40% of this value is contributed by the auto industry. The element $a_{13} = 0.1$ indicates that of every dollar of output *from* the transportation industry, 10% of this value is contributed by the auto industry. The element $a_{21} = 0.1$ indicates that of every dollar of output *from* the auto industry, 10% of this value is contributed by the energy industry. The element $a_{31} = 0.2$ indicates that of every dollar of output *from* the auto industry, 20% of this value is contributed by the transportation industry. The reader can now interpret the remaining elements.

Let $x_1, x_2,$ and $x_3$ represent respectively the output in dollars of auto, energy, and transportation produced. Then

$$0.2x_1 + 0.4x_2 + 0.1x_3$$

represents the interindustry demand for the first product, auto. Let $d_1, d_2,$ and $d_3$ be the nonindustry demand in dollars respectively for the three products. Then since

$$\text{auto output} = \left(\begin{matrix} \text{interindustry} \\ \text{demand for auto} \end{matrix}\right) + \left(\begin{matrix} \text{nonindustry} \\ \text{demand for auto} \end{matrix}\right)$$

$$x_1 \quad = \quad 0.2x_1 + 0.4x_2 + 0.1x_3 + \quad\quad d_1$$

Continuing in this manner, we obtain the following system of three equations:

$$x_1 = 0.2x_1 + 0.4x_2 + 0.1x_3 + d_1$$
$$x_2 = 0.1x_1 + 0.2x_2 + 0.2x_3 + d_2$$
$$x_3 = 0.2x_1 + 0.2x_2 + 0.1x_3 + d_3.$$

If we define the **production column matrix $X$** and the **demand column matrix $D$**, respectively, by

$$X = \begin{bmatrix} x_1 \\ x_2 \\ x_3 \end{bmatrix}, \quad D = \begin{bmatrix} d_1 \\ d_2 \\ d_3 \end{bmatrix},$$

then these last three equations can be written in matrix form as

$$X = AX + D,$$

where

$$A = \begin{bmatrix} .2 & .4 & .1 \\ .1 & .2 & .2 \\ .2 & .2 & .1 \end{bmatrix}.$$

This can be rewritten as

$$X - AX = D$$
$$IX - AX = D$$
$$(I - A)X = D.$$

If the matrix $(I - A)$ has an inverse, then the solution is

$$X = (I - A)^{-1}D.$$

E X A M P L E   1   **Finding the Production Matrix**

Find the production matrix $X$ in the previous problem if the demand matrix is

$$D = \begin{bmatrix} 474 \\ 948 \\ 474 \end{bmatrix}$$

**Solution**   The matrix $(I - A)$ is

$$I - A = \begin{bmatrix} 0.8 & -0.4 & -0.1 \\ -0.1 & 0.8 & -0.2 \\ -0.2 & -0.2 & 0.9 \end{bmatrix}.$$

Using the Gauss–Jordan method one obtains

$$(I - A)^{-1} = \frac{1}{474} \begin{bmatrix} 680 & 380 & 160 \\ 130 & 700 & 170 \\ 180 & 240 & 600 \end{bmatrix}.$$

Thus

$$X = (I - A)^{-1}D = \frac{1}{474} \begin{bmatrix} 680 & 380 & 160 \\ 130 & 700 & 170 \\ 180 & 240 & 600 \end{bmatrix} \begin{bmatrix} 474 \\ 948 \\ 474 \end{bmatrix} = \begin{bmatrix} 1600 \\ 1700 \\ 1260 \end{bmatrix}. \quad \blacksquare$$

If $A$ is an input–output matrix and $(I - A)^{-1}$ has only nonnegative elements, then for any demand matrix $D$ with nonnegative elements, $X = (I - A)^{-1}D$ will also have all nonnegative elements and thus a *meaningful* solution. This means that, in this situation, any outside (nonindustry) demand can be met. The following theorem can be proven.

---

**When Outside Demand can be Met**

If $A$ is an input–output matrix and the sum of the elements in each row of $A$ is less than one, then any outside demand can be met.

---

This was the case in Example 1.

## Closed Leontief Input–Output Model

In the closed input–output model, we assume that there is *no* demand from outside the $n$ industries under consideration. Let $f_{ij}$ be the fraction of the total output of the $j$th industry purchased by the $i$th industry. The problem then is to find suitable prices for each of the $n$ outputs so that the total expenditure of each industry equals its total income. Any such price structure represents an equilibrium for the economy.

Let $p_i$ be the price charged by the $i$th industry for its total output. For convenience, consider again an economy with the three industries: auto, energy, transportation. As usual, we allow for the possibility that each industry consumes (and therefore buys) some of its own product. The total income of the auto industry, $p_1$, is then the sum of its income from the three industries or

$$p_1 = f_{11}p_1 + f_{12}p_2 + f_{13}p_3.$$

In a similar fashion, we have

$$p_2 = f_{21}p_1 + f_{22}p_2 + f_{23}p_3$$

$$p_3 = f_{31}p_1 + f_{32}p_2 + f_{33}p_3.$$

If we set $F = (f_{ij})$ and let the **price matrix** be

$$P = \begin{bmatrix} p_1 \\ p_2 \\ p_3 \end{bmatrix},$$

then these last three equations can be written in matrix form as

$$P = FP.$$

This can be rewritten as

$$P - FP = O$$

$$IP - FP = O$$

$$(I - F)P = O$$

It turns out that $(I - F)$ does *not* have an inverse. The solutions can be found using the Gauss–Jordan method.

**E X A M P L E   2**    **Finding the Equilibrium Prices**

Let

$$F = \begin{bmatrix} \frac{1}{2} & \frac{1}{3} & \frac{1}{4} \\ \frac{1}{3} & \frac{1}{3} & \frac{1}{4} \\ \frac{1}{6} & \frac{1}{3} & \frac{1}{2} \end{bmatrix}$$

Find the equilibrium prices.

**Solution**

$$I - F = \begin{bmatrix} \frac{1}{2} & -\frac{1}{3} & -\frac{1}{4} \\ -\frac{1}{3} & \frac{2}{3} & -\frac{1}{4} \\ -\frac{1}{6} & -\frac{1}{3} & \frac{1}{2} \end{bmatrix}$$

Then

$$\begin{bmatrix} \frac{1}{2} & -\frac{1}{3} & -\frac{1}{4} & 0 \\ -\frac{1}{3} & \frac{2}{3} & -\frac{1}{4} & 0 \\ -\frac{1}{6} & -\frac{1}{3} & \frac{1}{2} & 0 \end{bmatrix} \begin{matrix} 2R_1 \rightarrow R_1 \\ \\ \\ \end{matrix}$$

$$= \begin{bmatrix} 1 & -\frac{2}{3} & -\frac{1}{2} & 0 \\ -\frac{1}{3} & \frac{2}{3} & -\frac{1}{4} & 0 \\ -\frac{1}{6} & -\frac{1}{3} & \frac{1}{2} & 0 \end{bmatrix} \begin{matrix} \\ R_2 + \frac{1}{3}R_1 \rightarrow R_2 \\ R_3 + \frac{1}{6}R_1 \rightarrow R_3 \end{matrix}$$

$$= \begin{bmatrix} 1 & -\frac{2}{3} & -\frac{1}{2} & 0 \\ 0 & \frac{4}{9} & -\frac{5}{12} & 0 \\ 0 & -\frac{4}{9} & \frac{5}{12} & 0 \end{bmatrix} \begin{matrix} \\ \frac{9}{4}R_2 \rightarrow R_2 \\ R_3 + R_2 \rightarrow R_3 \end{matrix}$$

$$= \begin{bmatrix} 1 & -\frac{2}{3} & -\frac{1}{2} & 0 \\ 0 & 1 & -\frac{15}{16} & 0 \\ 0 & 0 & 0 & 0 \end{bmatrix} \begin{matrix} R_1 + \frac{2}{3}R_2 \rightarrow R_1 \\ \\ \end{matrix}$$

$$= \begin{bmatrix} 1 & 0 & -\frac{9}{8} & 0 \\ 0 & 1 & -\frac{15}{16} & 0 \\ 0 & 0 & 0 & 0 \end{bmatrix}$$

Thus, letting $t$ be any real number, we have

$$p_1 = \frac{9}{8} t$$

$$p_2 = \frac{15}{16} t$$

$$p_3 = t. \quad \blacksquare$$

These are the suitable prices for each of the three outputs, that is, prices such that the total expenditure of each industry equals its total income.

Notice that the sum of the entries in each column of $F$ is one. This *must* be the case since each of the outputs of each industry is *completely* distributed among the three industries.

## SELF-HELP EXERCISE SET 2.6

1. Let the input–output matrix and the demand matrix be

$$A = \begin{bmatrix} 0.6 & 0.2 \\ 0.2 & 0.4 \end{bmatrix}, \quad D = \begin{bmatrix} 50,000,000 \\ 30,000,000 \end{bmatrix}$$

Find the production matrix.

## EXERCISE SET 2.6

In Exercises 1 through 10 you are given an input–output matrix $A$ and a demand matrix $D$. Assuming a Leontief open input–output model, find the production matrix.

1. $A = \begin{bmatrix} 0.3 & 0.2 \\ 0.4 & 0.3 \end{bmatrix}, \quad D = \begin{bmatrix} 10,000,000 \\ 20,000,000 \end{bmatrix}$

2. $A = \begin{bmatrix} 0.1 & 0.2 \\ 0.4 & 0.3 \end{bmatrix}, \quad D = \begin{bmatrix} 30,000,000 \\ 40,000,000 \end{bmatrix}$

3. $A = \begin{bmatrix} 0.5 & 0.2 \\ 0.4 & 0.2 \end{bmatrix}, \quad D = \begin{bmatrix} 100,000,000 \\ 200,000,000 \end{bmatrix}$

4. $A = \begin{bmatrix} 0.2 & 0.3 \\ 0.4 & 0.3 \end{bmatrix}, \quad D = \begin{bmatrix} 30,000,000 \\ 50,000,000 \end{bmatrix}$

5. $A = \begin{bmatrix} 0.5 & 0.1 \\ 0.4 & 0.1 \end{bmatrix}, \quad D = \begin{bmatrix} 200,000,000 \\ 300,000,000 \end{bmatrix}$

6. $A = \begin{bmatrix} 0.1 & 0.3 \\ 0.5 & 0.3 \end{bmatrix}, \quad D = \begin{bmatrix} 5,000,000 \\ 6,000,000 \end{bmatrix}$

7. $A = \begin{bmatrix} 0 & 0.1 & 0.2 \\ 0.4 & 0 & 0.1 \\ 0.2 & 0.3 & 0 \end{bmatrix}, \quad D = \begin{bmatrix} 20,000,000 \\ 40,000,000 \\ 10,000,000 \end{bmatrix}$

8. $A = \begin{bmatrix} 0.1 & 0.2 & 0.4 \\ 0.3 & 0.3 & 0.1 \\ 0.1 & 0.2 & 0.2 \end{bmatrix}, \quad D = \begin{bmatrix} 2,000,000 \\ 1,000,000 \\ 1,000,000 \end{bmatrix}$

9. $A = \begin{bmatrix} 0 & 0.1 & 0.1 \\ 0.2 & 0 & 0.3 \\ 0.1 & 0.3 & 0 \end{bmatrix}, \quad D = \begin{bmatrix} 30,000,000 \\ 20,000,000 \\ 20,000,000 \end{bmatrix}$

10. $A = \begin{bmatrix} 0.1 & 0.1 & 0.2 \\ 0.3 & 0.2 & 0.1 \\ 0.2 & 0.2 & 0.3 \end{bmatrix}$, $D = \begin{bmatrix} 1,000,000 \\ 2,000,000 \\ 1,000,000 \end{bmatrix}$

In Exercises 11 through 17 you are given an input–output matrix $F$ for a closed Leontief input–output model. Find the equilibrium price matrix.

11. $F = \begin{bmatrix} 0.3 & 0.2 \\ 0.7 & 0.8 \end{bmatrix}$
  12. $F = \begin{bmatrix} 0.5 & 0.4 \\ 0.5 & 0.6 \end{bmatrix}$

13. $F = \begin{bmatrix} 0.4 & 0.1 \\ 0.6 & 0.9 \end{bmatrix}$
  14. $F = \begin{bmatrix} 0.2 & 0.1 \\ 0.8 & 0.9 \end{bmatrix}$

15. $F = \begin{bmatrix} 0.3 & 0.1 & 0.2 \\ 0.3 & 0.4 & 0.2 \\ 0.4 & 0.5 & 0.6 \end{bmatrix}$
  16. $F = \begin{bmatrix} 0.2 & 0.1 & 0.3 \\ 0.4 & 0.3 & 0.4 \\ 0.4 & 0.6 & 0.3 \end{bmatrix}$

17. $F = \begin{bmatrix} 0.1 & 0.1 & 0.2 \\ 0.5 & 0 & 0.8 \\ 0.4 & 0.9 & 0 \end{bmatrix}$

18. **Closed Leontief Input–Output Model.**  Suppose in the closed Leontief input–output model for a three industry system, the input–output matrix is given by $A = (a_{ij})$. We then wish to solve $(I - A)X = O$. To solve we form the augmented matrix $[I - A|O]$. In the closed model we must have

$$a_{11} + a_{21} + a_{31} = 1$$
$$a_{12} + a_{22} + a_{32} = 1$$
$$a_{13} + a_{23} + a_{33} = 1$$

Using these three equations and the operations $R_3 + R_1 \rightarrow R_3$ and $R_3 + R_2 \rightarrow R_3$ on the above augmented matrix, show that the augmented matrix becomes

$$\begin{bmatrix} 1 - a_{11} & -a_{12} & a_{13} \\ -a_{21} & 1 - a_{22} & -a_{23} \\ 0 & 0 & 0 \end{bmatrix} \begin{bmatrix} 0 \\ 0 \\ 0 \end{bmatrix}.$$

 ## Graphing Calculator Exercise

19. Given the following input–output matrix $A$ and demand matrix $D$ and assuming a Leontief open input–output model, find the production matrix:

$$A = \begin{bmatrix} .20 & .20 & .15 & .10 \\ .40 & .10 & .20 & .15 \\ .02 & .03 & .01 & .12 \\ .25 & .04 & .01 & .10 \end{bmatrix}, \quad D = \begin{bmatrix} 2200 \\ 3400 \\ 2700 \\ 1600 \end{bmatrix}.$$

## Solutions to Self-Help Exercise Set 2.6

**1.** $I - A = \begin{bmatrix} 0.4 & -0.2 \\ -0.2 & 0.6 \end{bmatrix}$

Using the Gauss–Jordan method we have

$$\left[\begin{array}{cc|cc} 0.4 & -0.2 & 1 & 0 \\ -0.2 & 0.6 & 0 & 1 \end{array}\right] \frac{1}{0.4}R_1 \to R_1$$

$$\left[\begin{array}{cc|cc} 1 & -0.5 & 2.5 & 0 \\ -0.2 & 0.6 & 0 & 1 \end{array}\right] R_2 + 0.2R_1 \to R_2$$

$$\left[\begin{array}{cc|cc} 1 & -0.5 & 2.5 & 0 \\ 0 & 0.5 & 0.5 & 1 \end{array}\right] 2R_2 \to R_2$$

$$\left[\begin{array}{cc|cc} 1 & -0.5 & 2.5 & 0 \\ 0 & 1 & 1 & 2 \end{array}\right] R_1 + 0.5R_2 \to R_1$$

$$\left[\begin{array}{cc|cc} 1 & 0 & 3 & 1 \\ 0 & 1 & 1 & 2 \end{array}\right].$$

Thus

$$(I - A)^{-1} = \begin{bmatrix} 3 & 1 \\ 1 & 2 \end{bmatrix},$$

and

$$X = (I - A)^{-1}D = \begin{bmatrix} 3 & 1 \\ 1 & 2 \end{bmatrix}\begin{bmatrix} 50,000,000 \\ 30,000,000 \end{bmatrix} = \begin{bmatrix} 180,000,000 \\ 110,000,000 \end{bmatrix}.$$

## SUMMARY OUTLINE OF CHAPTER 2

- A **system of linear equations** is a set of linear equations.
- The set of all solutions to a system of linear equations is called the **solution set**.
- Two systems of equations with the same solution set are said to be **equivalent**.
- A **matrix** is a rectangular array of numbers.
- There are three **elementary operations** performed on a system of equations (matrix). Interchange two equations (rows). Replace an equation (row) with a constant times the equation (row). Replace an equation (row) with the equation (row) plus a constant times another equation (row).
- Elementary operations yield an equivalent system of equations.
- The **Gauss–Jordan** method is a procedure for systematically using elementary operations to reduce a system of linear equations to an equivalent system whose solution set can be immediately determined.

- There are three possibilities in solving a system of linear equations: There can be precisely one solution. There can be no solution. There can be infinitely many solutions.

- A matrix is in **reduced form** if:
  1. The leftmost nonzero element in each row is 1.
  2. The column containing the leftmost 1 of any row has all other elements in that column equal to 0.
  3. The leftmost 1 in any row is to the left of the leftmost 1 in a lower row.
  4. Any row consisting of all zeros must be below any row with at least one nonzero element.

- The leftmost nonzero element in a nonzero row of a reduced matrix is called a **leading one**.

- The variable in the column of a reduced matrix containing the leading one is called a **leading variable**. The remaining variables are called **free variables**.

- A **parameter** is any real number. A free variable is set equal to a parameter and is thus free to be any real number.

- The **order** of a matrix $A$ is denoted by $m \times n$ where $m$ is the number of rows and $n$ is the number of columns.

- The element in the $i$th row and $j$th column of a matrix $A$ is denoted by $a_{ij}$.

- A **row matrix** is a matrix with one row. A **column matrix** is a matrix with one column. A **square matrix** is a matrix with equal numbers of rows and columns.

- Two matrices are **equal** if, and only if, they have the same order and all corresponding entries are equal.

- **Multiplying a matrix by a real number.** If $c$ is a real number and $A$ is a matrix, then the matrix $cA$ is the matrix obtained by mutliplying every entry in $A$ by $c$.

- **Adding and subtracting matrices.** Two matrices of the same order can be added (or subtracted) to obtain another matrix of the same order by adding (or subtracting) corresponding entries.

- The **zero matrix** of order $m \times n$ is the matrix with $m$ rows and $n$ columns, all of whose entries are zero.

- **Some matrix properties.** Let A, B, C, and O be four matrices, all of the same order with O the zero matrix. Then

$$A + O = O + A = A$$

$$A - A = O$$

$$A + B = B + A$$

$$A + (B + C) = (A + B) + C.$$

- **Multiplying a row matrix times a column matrix.** Let $R$ be the row matrix $R = [r_1 \quad r_2 \quad \cdots \quad r_n]$ and let $C$ be the column matrix

$$C = \begin{bmatrix} c_1 \\ c_2 \\ \vdots \\ c_n \end{bmatrix}$$

with the same dimension as $R$. Then the product $RC$ is defined to be

$$RC = [r_1 \quad r_2 \quad \cdots \quad r_n] \begin{bmatrix} c_1 \\ c_2 \\ \cdot \\ \cdot \\ \cdot \\ c_n \end{bmatrix} = r_1 c_1 + r_2 c_2 + \cdots + r_n c_n.$$

- **Multiplying two matrices.** Given two matrices $A = (a_{ij})_{m \times p}$ and $B = (b_{ij})_{q \times n}$, the product $AB = C = (c_{ij})$ is defined if $p = q$, that is, if the number of columns of $A$ equals the number of rows of $B$. In such a case, the element in the $i$th row and $j$th column, $c_{ij}$, of the product matrix $C$ is obtained by multiplying the $i$th row of $A$ with the $j$th column of $B$.

- **The order of a product.** Given the two matrices $A = (a_{ij})_{m \times p}$ and $B = (b_{ij})_{p \times n}$, the matrix $AB$ has order $m \times n$.

- The $(n \times n)$ **identity matrix**, $I_n$, is the square matrix of order $(n \times n)$ with ones down the main diagonal and zeros elsewhere.

- **Properties of matrix multiplication.** For any matrices $A$, $B$, and $C$,

$$A(BC) = (AB)C$$

$$A(B + C) = AB + AC$$

$$AI_n = I_n A = A$$

whenever the indicated products are defined.

- Given a square $(n \times n)$ matrix $A$, the matrix $A^{-1}$ is called the **inverse** of $A$ if $AA^{-1} = A^{-1}A = I_n$.

- **To Find $A^{-1}$.** (1) Form the augmented matrix $[A|I]$. (2) Use row operations to reduce $[A|I]$ to $[I|B]$, if possible. (3) $A^{-1} = B$.

- The system of linear equations

$$a_{11}x_1 + a_{12}x_2 + \cdots + a_{1n}x_n = b_1$$

$$a_{21}x_1 + a_{22}x_2 + \cdots + a_{2n}x_n = b_2$$

$$\cdots$$

$$a_{n1}x_1 + a_{n2}x_2 + \cdots + a_{nn}x_n = b_n$$

can be put into the **matrix form** $AX = B$ by letting

$$A = \begin{bmatrix} a_{11} & a_{12} & \cdots & a_{1n} \\ a_{21} & a_{22} & \cdots & a_{2n} \\ & & \cdots & \\ a_{n1} & a_{n2} & \cdots & a_{nn} \end{bmatrix}, \quad X = \begin{bmatrix} x_1 \\ x_2 \\ \cdot \\ \cdot \\ \cdot \\ x_n \end{bmatrix}, \quad B = \begin{bmatrix} b_1 \\ b_2 \\ \cdot \\ \cdot \\ \cdot \\ b_n \end{bmatrix}.$$

The matrix $A$ is called the **coefficient matrix**.

- **Solution of $AX = B$.** Let $A$ be a square matrix of order $n$ and $B$ a column matrix of order $n$. If $A^{-1}$ exists, then the linear system of equations $AX = B$ has the unique solution $X = A^{-1}B$.

- The **input–output** matrix $A = a_{ij}$ represents the dollar amount of the output of industry $i$ required in producing one dollar of output in industry $j$.

- If $x_1$, $x_2$, and $x_3$ represent the respective outputs in dollars of a three-industry economy and $d_1$, $d_2$, and $d_3$ the respective nonindustry demand in dollars for the three products produced by the three industries, then we define the **production column matrix X** and the **demand column matrix D**, respectively, by

$$X = \begin{bmatrix} x_1 \\ x_2 \\ x_3 \end{bmatrix}, \quad D = \begin{bmatrix} d_1 \\ d_2 \\ d_3 \end{bmatrix}$$

- The **production matrix X** is given by $X = (I - A)^{-1}D$.
- In the **closed Leontief input–output model** no demand is assumed from outside the industries being considered. If $f_{ij}$ is the fraction of the total output of the $j$th industry purchased by the $i$th industry, and $p_i$ the price charged by the $i$th industry for its total output, then the matrix $P$ must satisfy the equation $(I - F)P = 0$, where $F = (f_{ij})$ is the **price matrix** and

$$P = \begin{bmatrix} p_1 \\ p_2 \\ p_3 \end{bmatrix}$$

is the equilibrium price matrix.

## Chapter 2 Review Exercises

In Exercises 1 through 8 find all solutions, if any exist, of the given systems using the Gauss–Jordan method.

**1.**  $x + 3y = 7$
   $3x + 4y = 11$

**2.**  $x + 3y = 1$
   $3x + 9y = 2$

**3.**  $x + 3y = 1$
   $3x + 9y = 3$

**4.**  $2x + 3y = 18$
   $3x - y = 5$

**5.**  $x + 3y - z = 4$
   $3x - y + z = 4$

**6.**  $x + y = 10$
   $2x + 2y + 2z = 20$

**7.**  $x + 2y - 3z = 4$
   $2x - y + z = 0$
   $5x - 3y + 2z = -2$

**8.**  $x + y + z = 5$
   $2x + y - z = 3$
   $y + 3z = 7$

In Exercises 9 through 20 let

$$A = \begin{bmatrix} 1 & 3 \\ 2 & 1 \\ -1 & 3 \end{bmatrix}, \quad B = \begin{bmatrix} 2 & 5 \\ 1 & 0 \\ 3 & -2 \end{bmatrix}, \quad C = [1 \quad 3], \quad D = \begin{bmatrix} 2 \\ 4 \end{bmatrix}$$

$$E = \begin{bmatrix} 2 & 0 & -1 \\ 3 & 1 & 2 \end{bmatrix}, \quad F = [2 \quad 4 \quad 1], \quad G = \begin{bmatrix} 1 \\ 3 \\ -2 \end{bmatrix}$$

and find the requested quantity if defined.

9.   $3A$                   10.   $-2B$               11.   $3A - 2B$

12.   $CD$                  13.   $DG$                14.   $AC$

15.   $EG$                  16.   $AF$                17.   $AE$

18.   $EA$                  19.   $AB$                20.   $FG$

21.   A matrix $D = (d_{ij})$ is a diagonal matrix if $d_{ij} = 0$ if $i \neq j$, that is, all entries off the main diagonal are zero. Suppose $A = (a_{ij})$ and $B = (b_{ij})$ are both diagonal matrices of order $(3 \times 3)$. Show that (a) $A^2$ is diagonal, (b) $AB$ is diagonal.

In Exercises 22 through 27 find the inverse of the given matrix, if it exists.

22.   $\begin{bmatrix} 5 & 1 \\ 2 & 1 \end{bmatrix}$           23.   $\begin{bmatrix} 4 & 2 \\ 2 & 1 \end{bmatrix}$           24.   $\begin{bmatrix} 2 & -3 \\ -3 & 1 \end{bmatrix}$

25.   $\begin{bmatrix} 1 & 3 & 0 \\ 2 & 5 & 1 \\ -1 & -2 & 2 \end{bmatrix}$      26.   $\begin{bmatrix} 1 & 4 & 0 \\ 2 & 9 & 1 \\ 5 & 22 & 2 \end{bmatrix}$      27.   $\begin{bmatrix} 1 & 1 & 1 \\ -1 & 2 & 2 \\ 2 & 1 & 2 \end{bmatrix}$

In Exercises 28 through 31 solve the systems using the inverses found in the previous exercises.

28.   $\begin{aligned} 5x + y &= 1 \\ 2x + y &= -2 \end{aligned}$

29.   $\begin{aligned} 2x - 3y &= 14 \\ -3x + y &= 35 \end{aligned}$

30.   $\begin{aligned} x + 3y &= -1 \\ 2x + 5y + z &= 3 \\ -x - 2y + 2z &= 2 \end{aligned}$

31.   $\begin{aligned} x + y + z &= 30 \\ -x + 2y + 2z &= -9 \\ 2x + y + 2z &= 0 \end{aligned}$

In Exercises 32 through 33 you are given an input–output matrix $A$ and a demand matrix $D$. Assuming a Leontief open input–output model, find the production matrix.

32.   $A = \begin{bmatrix} 0.2 & 0.2 \\ 0.7 & 0.3 \end{bmatrix}$, $D = \begin{bmatrix} 10{,}000{,}000 \\ 20{,}000{,}000 \end{bmatrix}$

33.   $A = \begin{bmatrix} 0 & 0.10 & 0.25 \\ 0.50 & 0 & 0.20 \\ 0.20 & 0.30 & 0 \end{bmatrix}$, $D = \begin{bmatrix} 10{,}000{,}000 \\ 20{,}000{,}000 \\ 30{,}000{,}000 \end{bmatrix}$

In Exercises 34 through 35 you are given an input–output matrix $F$ for a closed Leontief input–output model. Find the equilibrium price matrix.

34.   $F = \begin{bmatrix} 0.2 & 0.3 \\ 0.8 & 0.7 \end{bmatrix}$

35.   $F = \begin{bmatrix} 0.1 & 0.1 & 0.3 \\ 0.5 & 0 & 0.7 \\ 0.4 & 0.9 & 0 \end{bmatrix}$

**36.** A certain 9-hole golf course has the number of par 3 holes equal to one plus the number of par 5 holes. If par at this course for 9 holes is 35, how many par 3, 4, and 5 holes are there?

**37. Sales.**   A wholesaler receives an order for a total of a dozen standard and deluxe electric sweepers. The standard sweepers cost $200 and the deluxe $300. With the order is a check for $2900, but the order neglects to specify the number of each type of sweeper. Determine how to fill the order.

**38. Sales.**   A developer sells two sizes of condominiums. One sells for $75,000 and the other for $100,000. One year the developer sold 14 condominiums for a total of $1.2 million. How many of each size of condominiums did the developer sell?

**39. Sales.**   A small restaurant sold 50 specials one day. The ham special went for $10 and the beef special for $12. The specials brought in $560. The cook was too busy to keep track of how many of each were served. Find the answer for her.

**40. Mixture.**   Three acid solutions are available. The first has 20% acid, the second 30%, and the third 40%. The three solutions need to be mixed to form 300 liters of a solution with 32% acid. If the amount of the third solution must be equal to the sum of the amounts of the first two, find the amount of each solution.

**41. Distribution.**   A firm has 3 manufacturing plants located in different parts of the country. Three major wholesalers obtain the percentages of the product from the three plants according to the following table. Also included is the demand (in numbers) from each of the wholesalers. What should be the number produced by each of the plants to meet this demand with the given percentage distributions?

|              | Plant 1 | Plant 2 | Plant 3 | Demand |
| ------------ | ------- | ------- | ------- | ------ |
| Wholesaler A | 20%     | 30%     | 50%     | 105    |
| Wholesaler B | 40%     | 20%     | 40%     | 100    |
| Wholesaler C | 10%     | 30%     | 60%     | 100    |

Linear
programming is
used to determine
the most profitable
type of product to
be made by the
refinery. (Alan
Pitcairn/Grant
Heilman.)

# Linear Programming: The Graphical Method

3.1   Graphing Linear Inequalities

3.2   Graphical Solution of Linear Programming Problems

3.3   More Linear Programming

**T**his chapter introduces linear programming, one of the most important practical developments in economics and business and many other areas to appear since World War II. Linear programming was originally devised to help the U.S. Air Force optimize various goals in activities such as assignments, procurement, and maintenance. Linear programming is now used in a wide variety of practical situations. For example, AT&T uses linear programming to determine the most economical telephone links between cities. Oil refineries use linear programming to determine what crude oil to buy and what to produce with it. Airlines use linear programming to determine how to maximize profits associated with assigning aircrafts to their schedules. Many companies use linear programming to schedule transportation networks, to control inventories, to plan portfolios, and so on. According to some estimates, well over $100 million in human and computer time is spent yearly on the formulation and solution of linear programming problems.

This chapter will concentrate on translating the linear programming problems into mathematical models and also to finding solutions by geometric methods. Solving problems geometrically will give considerable insight into how linear programming works but is only practical for very small-scale problems. In the next chapter, a practical technique is introduced that can solve large systems.

## 3.1  GRAPHING LINEAR INEQUALITIES

▶ *Introduction*

▶ *Graphing One Linear Inequality*

▶ *Graphing Systems of Linear Inequalities*

George Dantzig, 1914–

*(Photo by Edward W. Souza News Service, Stanford University, Stanford, CA.)*

George Dantzig is considered the father of linear programming. His major accomplishments in linear programming were made during and immediately after World War II while working for the Air Force. He is often cited as a prime example of what can be accomplished with positive thinking. As a graduate student, Dantzig arrived late for a class one day, copied down what he thought was a homework assignment, and solved a famous unsolved problem in statistics.

APPLICATION

**Manufacture of Wooden Boats**

A small company manufactures two types of wooden boats: dinghies and skiffs. The manufacture of each boat must go through three operations: cutting (of the wood), assembly, and painting. The dinghies require 2 hours of cutting, 4 hours of assembling, and 2 hours of painting. The skiffs require 4 hours of cutting, 2 hours of assembly, and 2 hours of painting. The total work-hours available per week in the cutting section is 80, in the assembly section is 84, and in the painting section is 50. Write a system of inequalities that expresses this information.

### Introduction

The information given in the above problem can be summarized in the following table.

| Operation | Dinghies | Skiffs | Hours Available |
|-----------|----------|--------|-----------------|
| Cutting   | 2        | 4      | 80              |
| Assembly  | 4        | 2      | 84              |
| Painting  | 2        | 2      | 50              |

Let $x$ be the number of dinghies manufactured each week and $y$ be the number of skiffs manufactured each week. Each of the dinghies requires 2 hours of cutting and each of the skiffs requires 4 hours of cutting. Thus the cutting section will be occupied $2x$ hours with the dinghies and $4y$ hours with the skiffs, for a total of $2x + 4y$ hours. Since this must be at most 80, we have $2x + 4y \le 80$. Since each of the dinghies requires 4 hours to assemble and each of the skiffs requires 2 hours, the assembly section will be occupied $4x + 2y$ hours and therefore $4x + 2y \le 84$. Finally since each of the dinghies and skiffs requires 2 hours to paint, the painting section will be occupied $2x + 2y$ hours and therefore $2x + 2y \le 50$. Naturally we must have $x \ge 0$ and $y \ge 0$, which we refer to as the **nonnegativity constraint**. We then have the system of inequalities

$$2x + 4y \le 80 \qquad \text{cutting constraint}$$

$$4x + 2y \le 84 \qquad \text{assembly constraint}$$

$$2x + 2y \le 50 \qquad \text{painting constraint}$$

$$x, y \ge 0 \qquad \text{nonnegativity constraint.}$$

We would like to find the set of points $(x, y)$ that satisfies all of these inequalities simultaneously. This will tell us the possible numbers of each type of boat that can be made, given the labor constraints on the three operations. We will do this by looking at one inequality at a time.

## Graphing One Linear Inequality

We know that if $a$ and $b$ are not both zero, then the graph of the linear equation

$$ax + by = c$$

is a straight line in the $xy$-plane. We are now interested in the graph of linear inequalities such as

$$ax + by < c, \quad ax + by \le c, \quad ax + by > c, \quad ax + by \ge c.$$

**E X A M P L E   1**   **Graphing Linear Inequalities**

Graph (a) $x + y = 4$ (b) $x + y > 4$ (c) $x + y < 4$.

**Solutions**

(a) This can be written as $y = -x + 4$; the graph is shown in Figure 3.1a as the line $L$.

(b) This can be written as $y > -x + 4$. Consider now any point $(x_0, y_0)$ that lies on the straight line $y = -x + 4$ as shown in Figure 3.1b. Then $y_0 = -x_0 + 4$. Then, as shown in Figure 3.1b, if we take any point $(x_0, y)$ with $y > y_0 = -x_0 + 4$, the point $(x_0, y)$ must be *above* the line $L$. Since this must hold for any $x_0$, we conclude that the graph of $y > -x + 4$ or $x + y > 4$ is the half-plane *above* the line $L$. See Figure 3.2a, where the line $L$ has been drawn using a dotted line to indicate that the line is *not* on the graph.

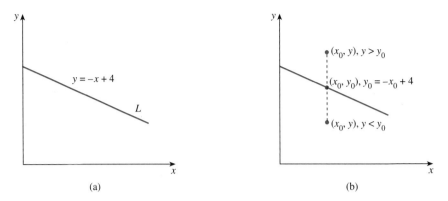

**Figure 3.1**

(c)   As Figure 3.1b indicates, if $y < y_0 = -x_0 + 4$, the point $(x_0, y)$ must lie *below* the line. We conclude that the graph of $x + y < 4$ is the half-plane *below* the line $L$. See Figure 3.2b.   ∎

There is a simpler way to graph the linear inequalities. Since we know that the graph of $x + y < 4$ must be a half-plane, we only need to decide *which* half-plane. A practical way of deciding is to choose a convenient test point $(x_0, y_0)$ and test to see if this point satisfies the inequality. If it does, then the half-plane containing this point must be the plane sought. If this point does not satisfy the inequality, then the half-plane containing this point is *not* the plane being sought. Thus the plane being sought is the other half-plane. Always pick the point $(0, 0)$ as a test point, if possible, since this is the easiest point to evaluate.

**Figure 3.2**

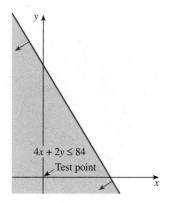

**Figure 3.3**

For example, in order to decide in which half-plane the graph of $x + y < 4$ lies, pick $(0, 0)$ as a test point. Since

$$0 + 0 = 0 < 4,$$

$(0, 0)$ satisfies the inequality, and thus the graph of the inequality is the half-plane containing $(0, 0)$. As Figure 3.2b indicates, $(0, 0)$ is below the line $L$, and thus the graph is the half-plane below the line $L$.

In order to decide in which half-plane the graph of $x + y > 4$ lies, pick $(0, 0)$ as a test point. Since

$$0 + 0 = 0 \not> 4,$$

$(0, 0)$ does not satisfy the inequality, and thus the graph of the inequality is the half-plane not containing $(0, 0)$. Since $(0, 0)$ is below the line $L$, the graph is the half-plane above the line $L$. See Figure 3.2a.

**E X A M P L E 2**  **Graphing a Linear Inequality**

Graph $4x + 2y \leq 84$, which is the assembly constraint given earlier.

**Solution**  We first graph $4x + 2y = 84$. This graph is the line $L$ shown in Figure 3.3. The graph of the linear inequality $4x + 2y \leq 84$ is either the half-plane above and including $L$, or the half-plane below and including $L$. We select the test point $(0, 0)$ and evaluate the inequality at this point and obtain

$$4(0) + 2(0) = 0 \leq 84.$$

Since $(0, 0)$ does satisfy the inequality and is *below* the line $L$, the graph we are seeking is the half-plane below $L$ and including $L$. This is indicated in Figure 3.3 using a solid line.  ∎

**E X A M P L E 3**  **Graphing a Linear Inequality**

Graph $x < -2$.

**Solution**  Any point $(x, y)$ that satisfies $x < -2$ must be to the left of the line $x = -2$, as shown in Figure 3.4. Alternately, one can pick the test point point $(0, 0)$ that is to the right of the line $x = -2$ and notice that $0 \not< -2$. This implies that $(0, 0)$ is *not* in the half-plane we are seeking. Thus the half-plane we are seeking is to the left of $x = -2$.  ∎

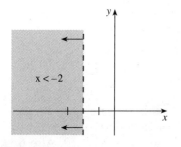

**Figure 3.4**

**Graphing a Linear Inequality**

Following are the steps required to graph the linear inequality

$$ax + by < c \text{ (or } \leq c \text{ or } > c \text{ or } \geq c).$$

1. First graph the straight line $L$ given by $ax + by = c$, using a dotted line if the inequalities "$<$" or "$>$" are given and a solid line if the inequalities "$\leq$" or "$\geq$" are given. The dotted line indicates the straight line $L$ is not part of the solution graph, while the solid line indicates that the line $L$ is part of the solution graph.

2. Pick any convenient test point not on $L$ and evaluate the inequality at that test point. If the inequality is true, then the solution graph is the half-plane on the same side of the line $L$ as the test point. If the inequality is not satisfied, then the solution graph is the half-plane on that side of $L$ that does not include the test point.

## Graphing Systems of Inequalities

If we are given *several* linear inequalities, as in the wooden boat example at the beginning of this section, then we find the solution set of each of the linear inequalities individually just as previously indicated. The solution set of the system is then the region common to all the solution sets of the individual inequalities. This common region is also called the **feasible region**. In other words, any point $(x, y)$ that satisfies all of the inequalities is in the feasible region. The feasible region is the set of all such points.

**E X A M P L E  4**    **Finding the Feasible Region**

A shop manufactures two styles of iron park benches. The first style requires 4 pounds of iron and 6 hours of labor while the second style requires 4 pounds of iron and 3 hours of labor. If there are 32 pounds of iron and 36 hours of labor available for the day, write a system of inequalities that expresses this information and find the feasible region.

**Solution**    The following table summarizes the information.

**Iron Park Benches**

|  | *Style 1* | *Style 2* | *Available* |
|---|---|---|---|
| Pounds of Iron | 4 | 4 | 32 |
| Hours of Labor | 6 | 3 | 36 |

Let $x$ be the number of style 1 benches made and $y$ the number of style 2 made. Then $4x + 4y$ is the number of pounds of iron needed to make these benches, while the total amount available is 32 pounds. Thus $4x + 4y \leq 32$. Also $6x + 3y$ is the number of hours needed to make these benches, while the total hours

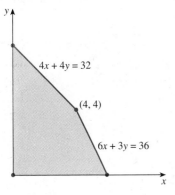

**Figure 3.5**

available is 36. Therefore, $6x + 3y \leq 36$. Naturally, $x$ and $y$ must be nonnegative. We then have

$$4x + 4y \leq 32$$

$$6x + 3y \leq 36$$

$$x, y \geq 0.$$

To determine the feasible region, first notice that since $x$ and $y$ are both nonnegative, the feasible region must be in the first quadrant. Now take $(0, 0)$ as a test point. Since $4(0) + 4(0) \leq 32$ and $6(0) + 3(0) \leq 36$, the point $(0, 0)$ satisfies both inequalities and thus lies below the line $4x + 4y = 32$ and also below the line $6x + 3y = 36$. Therefore the solution set for the first inequality is the half-plane below and including the line $4x + 4y = 32$, and the solution set for the second inequality is the half-plane below and including the line $6x + 3y = 36$. This is indicated in Figure 3.5. The solution set for the two inequalities is the shaded region, which is the set of points $(x, y)$ that satisfies *both* inequalities. The corner point $(4, 4)$ is determined by solving the system

$$4x + 4y = 32$$

$$6x + 3y = 36. \quad \blacksquare$$

**E X A M P L E   5**   **Finding the Feasible Region**

Find the feasible region for the system of inequalities that arose in the wooden boat manufacturing problem discussed at the beginning of this section.

**Solution**   We have the inequalities

$$2x + 4y \leq 80 \qquad \text{cutting constraint}$$

$$4x + 2y \leq 84 \qquad \text{assembly constraint}$$

$$2x + 2y \leq 50 \qquad \text{painting constraint}$$

$$x, y \geq 0 \qquad \text{nonnegativity constraint.}$$

Since the point $(0, 0)$ satisfies each of the first 3 inequalities and lies below the three lines $2x + 4y = 80$, $4x + 2y = 84$, and $2x + 2y = 50$, the solution set for each of these inequalities is the half-plane below and including each of the lines $2x + 4y = 80$, $4x + 2y = 84$, and $2x + 2y = 50$. This is indicated in Figure 3.6. The two inequalities $x \geq 0$ and $y \geq 0$ indicate that the solution set is in the first quadrant. The corner point $(10, 15)$ shown in Figure 3.6 can be found by solving the system

$$2x + 4y = 80$$

$$2x + 2y = 50,$$

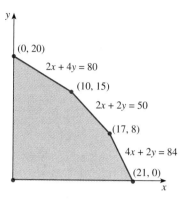

**Figure 3.6**

while the corner point (17, 8) shown in Figure 3.6 can be found by solving the system

$$4x + 2y = 84$$
$$2x + 2y = 50.$$

Finding the y-intercept for $2x + 4y = 80$ leads to the corner point (0, 20), while finding the x-intercept of $4x + 2y = 84$ leads to the corner (21, 0).    ■

**E X A M P L E   6**    **Finding the Feasible Region**

A farmer can buy two types of 100-pound bags of fertilizer, type A and type B. Each 100-pound bag of type A fertilizer contains 50 pounds of nitrogen, 30 pounds of phosphoric acid, and 20 pounds of potash, whereas each 100-pound bag of type B fertilizer contains 10 pounds of nitrogen, 30 pounds of phosphoric acid, and 60 pounds of potash. The farmer requires at least 1000 pounds of nitrogen, 1800 pounds of phosphoric acid, and 2800 pounds of potash. Write a system of inequalities that expresses the given information and graph the feasible region.

**Solution**    We organize the information in the following table.

| | *Pounds Per Bag* | | |
|---|---|---|---|
| | *Type A* | *Type B* | |
| *Number of Bags* | *x* | *y* | *Pounds Needed* |
| Nitrogen | 50 | 10 | 1000 |
| Phosphoric acid | 30 | 30 | 1800 |
| Potash | 20 | 60 | 2800 |

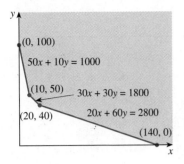

**Figure 3.7**

There are $x$ bags of type A each containing 50 pounds of nitrogen and $y$ bags of type B each containing 10 pounds of nitrogen for a total of $50x + 10y$ pounds. Since this must be at least 1000 pounds, we have $50x + 10y \geq 1000$. In a similar fashion we obtain the other two constraints together with the nonnegativity constraints $x \geq 0$ and $y \geq 0$ and have

$$50x + 10y \geq 1000 \qquad \text{nitrogen constraint}$$
$$30x + 30y \geq 1800 \qquad \text{phosphoric acid constraint}$$
$$20x + 60y \geq 2800 \qquad \text{potash constraint}$$
$$x, y \geq \quad 0 \qquad \text{nonnegativity constraint.}$$

These are then graphed in Figure 3.7. The corner point $(10, 50)$ is found by solving the system

$$50x + 10y = 1000$$
$$30x + 30y = 1800,$$

while the corner point $(20, 40)$ is found by solving the system

$$30x + 30y = 1800$$
$$20x + 60y = 2800.$$

The corner point $(0, 100)$ is obtained by finding the $y$-intercept of $50x + 10y = 1000$, while the corner point $(140, 0)$ is obtained by finding the $x$-intercept of $20x + 60y = 2800$. ∎

## SELF-HELP EXERCISE SET 3.1

1. Graph the feasible region.

$$3x + y \geq 5$$
$$2x + 3y \geq 8$$
$$x \quad \geq 0$$
$$y \quad \geq 0$$

2. Two types of metal connectors require for their manufacture spending a certain number of minutes on a stamping machine and also on a grinding machine. The first type of connector requires 4 minutes on the stamping machine and 3 minutes on the grinding machine, while the second connector requires 5 minutes on the stamping machine and 1 minute on the grinding machine. The stamping machine is available, at most, 50 minutes each hour, and the grinding machine is available, at most, 21 minutes each hour. Furthermore, each hour a total of at least two of the connectors must be produced. Write a system of inequalities that expresses this information and find the feasible region.

## EXERCISE SET 3.1

In Exercises 1 through 24 graph the solution set.

**1.** $x + y < 5$      **2.** $2x + y < 4$      **3.** $x + 2y \leq 1$

**4.** $3x + 2y \leq 6$      **5.** $2x + 3y \geq 12$      **6.** $4x + 5y \geq 20$

**7.** $-2x + 3y > 5$      **8.** $x - 2y > 8$      **9.** $3x - 2y < 6$

**10.** $-x + 2y < 8$      **11.** $-6x + 2y \leq 12$      **12.** $6x - 3y \leq 18$

**13.** $2x - 3y \geq -6$      **14.** $2x - 3y \geq -12$      **15.** $x > y$

**16.** $2x > y + 1$      **17.** $y \leq 3$      **18.** $y > 4$

**19.** $x > 2$      **20.** $x \geq 3$      **21.** $y \geq -2$

**22.** $x < -1$      **23.** $x \leq 0$      **24.** $y < 0$

In Exercises 25 through 42 graph the feasible region.

**25.** $x + y \leq 4$
$-x + y \leq 2$

**26.** $x + y \geq 4$
$-x + y \geq 2$

**27.** $x + y < 4$
$-x + y > 2$

**28.** $x + y > 4$
$-x + y < 2$

**29.** $x + y \leq 4$
$-x + y \geq 2$
$x, y \geq 0$

**30.** $x + y \leq 4$
$-x + y \geq 2$
$y \geq 0$

**31.** $x + y \leq 10$
$x \geq 2$
$y \geq 3$

**32.** $x + y \geq 10$
$y \leq 7$
$x \leq 8$

**33.** $x + y \geq 4$
$-x + y \geq 2$
$x \geq 0$

**34.** $x + y \leq 10$
$x + y \geq 4$
$x \geq 0$
$y \geq 0$

**35.** $-x + y \leq 4$
$x + y \geq 2$
$x \geq 0$
$y \geq 0$

**36.** $y \leq x$
$y \geq x - 2$
$y \leq 4$
$y \geq 0$

**37.** $-x + y \leq 1$
$x + y \geq 1$
$x + y \leq 4$
$y \geq 0$

**38.** $x + y \leq 4$
$x + y \geq 1$
$-x + y \geq -1$

**39.** $y \leq x$
$x + y \leq 2$

**40.** $x + y \leq 2$
$y \geq x$

**41.** $-x + y \geq 0$
$x \geq 1$
$y \leq 4$

**42.** $-x + y \leq 0$
$x \leq 4$
$y \geq 1$

## Applications

In Exercises 43 through 50 write a system of inequalities that expresses the given information. Graph the feasible region and label all corner points.

43. **Shipping.** A manufacturer supplies its product to two distributors. Distributor A needs at least 100 of the product next month while distributor B needs at least 50. At most 200 of the product can be manufactured and delivered to the distributors.

44. **Transportation.** A plane delivers two types of cargo between two destinations. Each crate of cargo I is 5 cubic feet in volume and 100 pounds in weight. Each crate of cargo II is 5 cubic feet in volume and 200 pounds in weight. The plane has available at most 400 cubic feet and 10,000 pounds for the crates. Finally, at least twice the number of crates of I than II must be shipped.

45. **Manufacturing.** Two products, A and B, require for their manufacture spending a certain amount of hours on a stamping machine, a grinding machine, and a polishing machine. Product A requires 1 hour on the stamping machine, 2 hours on the grinding machine, and 2 hours on the polishing machine. Product B requires 2 hours on the stamping machine, 2 hours on the grinding machine, and 1 hour on the polishing machine. The stamping machine can operate a maximum of 20 hours, the grinding machine a maximum of 22 hours, and the polishing machine a maximum of 15 hours each day.

46. **Diet.** An individual needs a daily supplement of at least 500 units of vitamin C and 200 units of vitamin E and agrees to obtain this supplement by eating two foods, I and II. Each ounce of food I contains 40 units of vitamin C and 10 units of vitamin E, while each ounce of food II contains 20 units of vitamin C and also 20 units of vitamin E. The total supplement of these two foods must be at most 30 ounces.

47. **Manufacturing.** A firm manufactures tables and desks. To produce each table requires 1 hour of labor, 10 square feet of wood, and 2 quarts of finish. To produce each desk requires 3 hours of labor, 20 square feet of wood, and 1 quart of finish. Available is at most 45 hours of labor, at most 350 square feet of wood, and at most 55 quarts of finish.

48. **Mixture.** A dealer has 7600 pounds of peanuts, 5800 pounds of almonds, and 3000 pounds of cashews to be used to make two mixtures. The first mixture consists of 60% peanuts, 30% almonds, and 10% cashews. The second mixture consists of 20% peanuts, 50% almonds, and 30% cashews.

49. **Fishery.** A certain lake has smallmouth and largemouth bass and also three types of food for these fish, I, II, and III. Each month the lake can supply 800 pounds of food I, 500 pounds of food II, and 700 pounds of food III. Each month each smallmouth bass requires 1 pound of food I, 1 pound of food II, and 2 pounds of food III, and each largemouth bass requires 4 pounds of food I, 2 pounds of food II, and 1 pound of food III.

50. **Scheduling.** A firm has two plants, A and B, that manufacture three products, $P_1$, $P_2$, and $P_3$. In one day Plant A can manufacture 30 of $P_1$, 30 of $P_2$, and 10 of $P_3$, whereas Plant B can manufacture 10 of $P_1$, 20 of $P_2$, and 30 of $P_3$. At least 240 of the first product, 390 of the second, and 410 of the third are needed.

*(© Alvis Upitis/The Image Bank.)*

*(© Grant Heilman Photography.)*

### Solutions to Self-Help Exercise Set 3.1

1. Since the point (0, 0) does not satisfy either of the first 2 inequalities and lies below the 2 lines $3x + y = 5$ and $2x + 3y = 8$, the solution set for the first two inequalities is the half-plane above and including the lines $3x + y = 5$ and $2x + 3y = 8$. This is indicated in the figure. The two inequalities $x \geq 0$ and $y \geq 0$ indicate that the solution set is in the first quadrant. The feasible region is then the shaded region in the following figure.

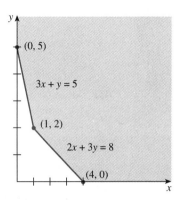

2. Most of the information can be summarized in the following table.

|  | First Connector | Second Connector | Minutes Available |
|---|---|---|---|
| Stamping Machine | 4 | 5 | 50 |
| Grinding Machine | 3 | 1 | 21 |

   Let $x$ be the number of the first type of connector manufactured each hour and $y$ be the number of the second type manufactured each hour. Each of the first type of connector requires 4 minutes on the stamping machine, and each of the second type of connector requires 5 minutes on this machine. Thus the stamping machine will be occupied $4x$ minutes with the first type of connector and $5y$ minutes with the second type of connector, for a total of $4x + 5y$ minutes. Since this must be at most 50, we have $4x + 5y \leq 50$.

   Each of the first type of connector requires 3 minutes on the grinding machine, and each of the second type of connector requires 1 minute on this machine. Thus the grinding machine will be occupied $3x + y$ minutes. Thus $3x + y \leq 21$. Since at least 2 connectors must be produced, we have $x + y \geq 2$. Naturally, we must have $x \geq 0$ and $y \geq 0$. Thus we have the five inequalities

$$4x + 5y \leq 50$$

$$3x + y \leq 21$$

$$x + y \geq 2$$

$$x, y \geq 0.$$

Since (0, 0) satisfies the first 2 inequalities and not the third and lies below each of the lines $4x + 5y = 50$, $3x + y = 21$, and $x + y = 2$, the solution set for the first two inequalities is the half-plane below and including the lines $4x + 5y = 50$ and $3x + y = 21$ and the half-plane above and including the line $x + y = 2$. This is indicated in the figure. The two inequalities $x \geq 0$ and $y \geq 0$ indicate that the solution set is in the first quadrant. The feasible region is then the shaded region in the following figure.

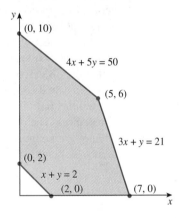

## 3.2   GRAPHICAL SOLUTION OF LINEAR PROGRAMMING PROBLEMS

► *Some Linear Programming Problems*
► *Problems with No Solution*
► *The Geometric Method*

**George Dantzig**   ❝*The final test of a theory is its capacity to solve the problems which originated it.*❞

**A P P L I C A T I O N**
**Maximizing Profit**

Suppose the small, wooden-boat manufacturer of the last section obtained a profit of $60 for every dinghy manufactured and $80 for every skiff. How many of each type of boat should be made in order to maximize the profits? The answer is found in Example 1.

### Some Linear Programming Problems

Business, economics, finance, and many other areas are replete with complex decision or allocation problems that involve the selection of specific values of many interrelated variables. Often one can focus attention on one specific objective that is to be maximized or minimized. Examples of such objectives are profit, revenue, cost, expected return on an investment, or perhaps some measure of social welfare for a government program. The variables are normally subject to constraints that take the form of a set of inequalities like the ones encountered in the previous section. These inequalities limit the possible selection of the values of the variables under consideration. In this section we will use linear programming to find the maximum or minimum of the objective function. We use the geometric approach on small-scale problems. The full power of linear programming will be taken up in the next section.

In this section we will wish to optimize (maximize or minimize) a *linear* function of two variables $z = Ax + By + C$. The function to be optimized is called the **objective function**. The variables will satisfy a system of *linear* inequalities called **constraints** that will give rise to a feasible region. Two of these inequalities will always be $x \geq 0$ and $y \geq 0$ and are called the **nonnegativity constraints**. The other constraints are called **problem constraints**. We will then attempt to maximize or minimize the function $z$ subject to the constraint that the points $(x, y)$ must lie in the feasible region. The following is a simple example.

**E X A M P L E   1**     **Geometric Method in Linear Programming**

Suppose the small wooden-boat manufacturer considered in the last section obtained a profit of $60 for every dinghy manufactured and $80 for every skiff. How many of each type of boat should be made in order to maximize the profits?

**Solution**     If $x$ is the number of dinghies manufactured each week and $y$ the number of skiffs, then the profit in dollars is $P = 60x + 80y$. Recalling the constraints that we found in the last section, we have the following linear programming problem:

| Maximize | $z = 60x + 80y$ | objective function |
|---|---|---|
| Subject to | $2x + 4y \leq 80$ | cutting constraint |
| | $2x + 2y \leq 50$ | painting constraint |
| | $4x + 2y \leq 84$ | assembly constraint |
| | $x, y \geq 0$ | nonnegativity constraint. |

In this section we take the geometric approach to solving such problems. We first need to graph the feasible region. This was done in the previous section and is shown in Figure 3.8, where we note that the five corners have coordinates $A = (10, 15)$, $B = (17, 8)$, $C = (21, 0)$, $D = (0, 0)$, and $E = (0, 20)$. Only points $(x, y)$ in the feasible region are feasible solutions.

First draw the straight lines $P = 60x + 80y = c$, called **isoprofit lines**, for a variety of values of $c$. For example, set $P$ equal to, say 1600. The equation $60x + 80y = 1600$ is a straight line with slope $-3/4$ and is shown in Figure 3.9. Since $P = 60x + 80y$, it is clear that we can take both $x$ and $y$ a little larger and *still remain in the feasible region* and obtain a larger profit. If we set $P = 60x$

**Figure 3.8**

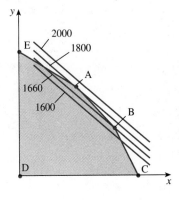

**Figure 3.9**

$+\ 80y\ =\ 1660$, we obtain the straight line shown in Figure 3.9. (Notice the slope is also $-3/4$.) Any point along this latter straight line gives a profit of 1660, with some points in the feasible region. But we see that we can do better. Take $P\ =\ 60x\ +\ 80y\ =\ 1800$. This line also has slope $-3/4$. Notice that this line goes through the corner $A$, since $A\ =\ (10,\ 15)$ and at this point $P\ =\ 60(10)\ +\ 80(15)\ =\ 1800$. Thus only at the point $A$ is the profit 1800 and also in the feasible region. If we take $P\ =\ 2000$, we see from Figure 3.9 that we obtain a straight line (parallel to the others) all of whose points are outside the feasible region. It thus appears that the profit is maximized at point $A$.

The lines for $P\ =\ 1600$, $P\ =\ 1660$, $P\ =\ 1800$, and $P\ =\ 2000$ are, as we noted, parallel. This can also be seen by setting $P\ =\ P_0$ and noting that

$$60x\ +\ 80y\ =\ P_0$$

$$y\ =\ -\frac{3}{4}x\ +\ \frac{1}{80}P_0.$$

Furthermore we notice that by increasing $P_0$ we increase the $y$-intercept, and thus the line moves up (and to the right). See Figure 3.10. We then see that our strategy is to increase $P_0$ as much as possible while still keeping the constant profit line $P\ =\ P_0$ in contact with the feasible region. We saw that $P_0\ =\ 1800$ meets this requirement and that the point in the feasible region at which the maximum occurs is the corner point $A$. (As we shall see, the optimum must always occur at a corner.)

Thus the solution to the problem is to manufacture 10 dinghies and 15 skiffs, and the weekly profit will be \$1800. ∎

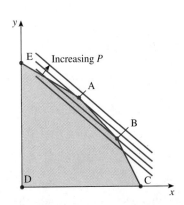

**Figure 3.10**

This problem has exactly one point at which the maximum occurs. Although this is typical, the maximum can occur at more than one point. The following is an example.

**E X A M P L E   2**    **Geometric Method in Linear Programming**

Redo the previous problem if the profit is given as $P = 70x + 70y$.

**Solution**    The feasible region is the same as in the previous example. If we set $P = 70x + 70y = P_0$, we obtain

$$y = -x + \frac{1}{70} P_0.$$

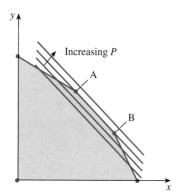

Figure 3.11

All of these lines have slope $-1$. Again, increasing $P_0$ increases the $y$-intercept, and thus the lines move up (and to the right), as indicated in Figure 3.11. But since the slopes of these lines are the same as the slope of the line from $A$ to $B$, the maximum occurs at all the points on the line segment from $A$ to $B$. But again the maximum does occur at a corner—in fact, at two adjacent corners. Thus two acceptable answers are $(10, 15)$ or $(17, 8)$, which correspond to manufacturing 10 dinghies and 15 skiffs or 17 dinghies and 8 skiffs with profits of

$$P = 70(10) + 70(15) = 1750$$

dollars per week. Other acceptable answers with the same profit are the points $(11, 14)$, $(12, 13)$, $(13, 12)$, $(14, 11)$, $(15, 10)$, and $(16, 9)$, all of which lie on the line segment between $A$ and $B$ and all of which will yield the same profit of $1750. ∎

We now give an example of a linear programming problem involving minimizing the objective function.

**E X A M P L E   3**    **A Minimizing Problem**

Solve the following linear programming problem.

Minimize        $z = 3x + 2y$        objective function

Subject to        $3x + y \geq 5$
                     $x + y \geq 3$        problem constraints
                     $x + 4y \geq 6$

                         $x, y \geq 0$        nonnegativity constraint

**Solution**    The feasible region is shown in Figure 3.12. If we set $z = z_0$, we obtain

$$3x + 2y = z_0$$

$$y = -\frac{3}{2}x + \frac{1}{2}z_0,$$

which is a straight line with slope $-3/2$. We see that decreasing $z_0$ brings the line $z = z_0$ closer to the origin. Furthermore, since the slope of the line segment from $B$ to $C$ is $-1$, while the slope of the line segment from $A$ to $B$ is $-3$, we

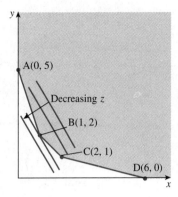

**Figure 3.12**

see, according to Figure 3.12, that the minimum occurs at the corner point $B$. At this point, $z = 3(1) + 2(2) = 7$. ∎

## Problems With No Solution

We have now seen two linear programming problems with exactly one solution and one with many solutions. We now encounter two linear programming problems with no solution.

**E X A M P L E   4**     **Problem With No Solution**

Solve the following linear programming problem.

$$\text{Maximize} \qquad P = x + y$$

$$\text{Subject to} \qquad x, y \geq 0$$

**Solution**     The feasible region is the first quadrant shown in Figure 3.13. It is clear that this problem has *no solution* since one can attain any profit $P = x + y$, no matter how large, simply by taking $x$ and $y$ sufficiently large. ∎

**E X A M P L E   5**     **Problem With No Solution**

Solve the following linear programming problem.

$$\text{Maximize} \qquad z = 3x + 2y \qquad \text{objective function}$$

$$\text{Subject to} \qquad x + y \leq 2 \qquad \text{problem constraint}$$

$$-x + y \geq 3 \qquad \text{problem constraint}$$

$$x, y \geq 0 \qquad \text{nonnegativity constraint}$$

**Figure 3.13**

**Figure 3.14**

**Solution**    We see in Figure 3.14 that the feasible region is empty. Thus there can be no optimal solution.    ■

## The Geometric Method

In all of the previous examples, the objective function was optimized at a corner point. Careful examination of other typical examples in Figure 3.15 indicates that this must be true in general. We then have the following theorem.

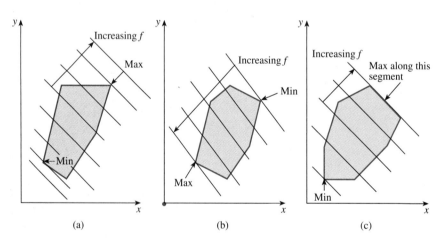

**Figure 3.15**

---

### Optimal Solution of a Linear Programming Problem

If a linear programming problem has an optimal solution, then the optimum must occur at a corner of the feasible region. If the objective function is optimized at two corners of the feasible region, then any point on the line segment connecting these two corners also optimizes the objective function.

We have already seen that an objective function need not have an optimal value. We now list two important cases for which an optimal value exists.

---

**Existence Theorem for Optimal Solutions**

1. If the feasible region is nonempty, bounded, and includes its boundary, then the objective function has both a minimum and a maximum in the feasible region.
2. If the feasible region $S$ includes its boundary, is unbounded, and the objective function $z = ax + by$ has the property that $a > 0$ and $b > 0$, then $z$ attains a minimum value in $S$ but has no maximum value in $S$.

---

The fact that an optimal solution, if it exists, must be at a corner gives rise to the following method of finding the optimal solution to a linear programming problem.

---

**Method of Corners**

Determine if a solution exists using the existence theorem for optimal solutions. If a solution exists then:

1. Graph the feasible region.
2. Locate all corners of the feasible region.
3. Evaluate the objective function at each of the corners.
4. The optimal solution will be the optimal value found in the previous step.

---

**E X A M P L E   6**     **The Method of Corners**

Find both the minimum and the maximum of $z = 5x + 4y$ subject to

$$-x + y \leq 4$$

$$x + y \leq 8$$

$$x + y \geq 4$$

$$x \leq 4$$

$$x, y \geq 0.$$

**Solution**     The feasible region is shown in Figure 3.16. Since the feasible region is nonempty, bounded, and includes its boundary, both the maximum and the minimum exist and can be found at a corner point. There are four corners, $A = (0, 4)$, $B = (2, 6)$, $C = (4, 4)$, and $D = (4, 0)$. Evaluating $z$ at these four corners yields

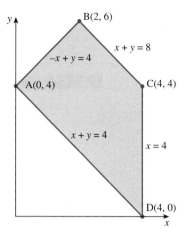

**Figure 3.16**

| Corner | $z = 5x + 4y$ |
|--------|---------------|
| $(0, 4)$ | $5(0) + 4(4) = 16 \leftarrow$ minimum |
| $(2, 6)$ | $5(2) + 4(6) = 34$ |
| $(4, 4)$ | $5(4) + 4(4) = 36 \leftarrow$ maximum |
| $(4, 0)$ | $5(4) + 4(0) = 20.$ |

Thus the maximum is $z = 36$ at the corner $C = (4, 4)$, while the minimum is $z = 16$ at the corner $A = (0, 4)$. ∎

**E X A M P L E   7**     **The Method of Corners**

A farmer can buy two types of 100-pound bags of fertilizer, type A and type B. Each 100-pound bag of type A fertilizer costs $20 and contains 50 pounds of nitrogen, 30 pounds of phosphoric acid, and 20 pounds of potash, whereas each 100-pound bag of type B fertilizer costs $30 and contains 10 pounds of nitrogen, 30 pounds of phosophoric acid, and 60 pounds of potash. The farmer requires at least 1000 pounds of nitrogen, 1800 pounds of phosphoric acid, and 2800 pounds of potash. How many bags of each type of fertilizer should be bought in order to minimize cost?

**Solution**     Let $x$ be the number of bags of type A fertilizer and $y$ the number of type B. Then we are to minimize the cost: $C = 20x + 30y$. The same constraints were given in Example 6 of the last section. Recalling our work there, we can write the linear programming problem as

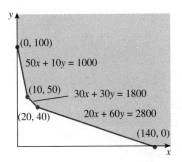

**Figure 3.17**

| Minimize | $z = 20x + 30y$ | objective function |
|---|---|---|
| Subject to | $50x + 10y \geq 1000$ | nitrogen constraint |
| | $30x + 30y \geq 1800$ | phosphoric acid constraint |
| | $20x + 60y \geq 2800$ | potash constraint |
| | $x, y \geq \quad 0$ | nonnegativity constraint. |

Figure 3.17 shows the feasible region found in the last section. The region is unbounded, and the objective function $z = 20x + 30y$ has positive coefficients of $x$ and $y$. Thus a minimum will exist that must be at a corner of the feasible region. We then have

| Corner | $z = 20x + 30y$ |
|---|---|
| (0, 100) | $20(0) + 30(100) = 3000$ |
| (10, 50) | $20(10) + 30(50) = 1700$ |
| (20, 40) | $20(20) + 30(40) = 1600 \leftarrow$ minimum |
| (140, 0) | $20(140) + 30(0) = 2800$ |

Thus the minimum cost is \$1600, requiring purchasing 20 bags of type A and 40 bags of type B. ∎

## SELF-HELP EXERCISE SET 3.2

**1.** Solve the following.

$$\text{Minimize} \qquad z = 4x + 6y$$
$$\text{Subject to} \qquad 7x + 2y \geq 20$$
$$x + 2y \geq \ 8$$
$$x, y \geq \ 0$$

**2.** Two types of metal connectors require for their manufacture spending a certain number of minutes on a stamping machine and on a grinding machine. The first type of connector requires 4 minutes on the stamping machine and 3 minutes on the grinding machine, while the second connector requires 5 minutes on the stamping machine and 1 minute on the grinding machine. The stamping machine is available at most 50 minutes each hour, and the grinding machine is available at most 21 minutes each hour. Furthermore, each hour a total of at least two of the connectors must be produced. If each of the first type of connector brings a profit of \$0.50 and each of the second type of connector a profit of \$0.60, find the number of each of these products that should be manufactured to maximize profits. (See Self-Help Exercise 2 in the last section.)

**3.** In the wooden-boat manufacturing problem of Example 1, suppose the profit is given by $P = ax + y$, with $a > 0$. Find all values of $a$ so that the profit is maximized at (a) (17, 8) (b) (21, 0).

## EXERCISE SET 3.2

In Exercises 1 through 6 you are given the feasible region. Find the maximum and minimum (if they exist) for the given objective function.

**1.**  $z = x + 2y$

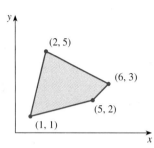

**2.**  $z = 2x - y$

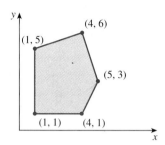

**3.**  $z = 2x + 3y$

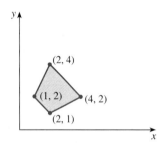

**4.**  $z = 10x + 20y$

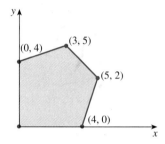

**5.**  $z = 3x + y$

**6.**  $z = x + y$

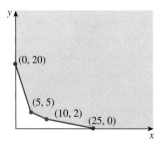

**7.**  Maximize $z = 20x + 30y$

Subject to   $x + 2y \le 90$

$x + y \le 50$

$x \le 30$

$x, y \ge 0$

**8.**  Maximize   $z = 4x + y$

Subject to   $x + y \le 100$

$3x + y \le 120$

$y \ge 50$

$x, y \ge 0$

9.  Maximize  $z = 2x + 3y$

Subject to $2x + y \le 100$

$y \le 40$

$x \ge 10$

$y \ge 20$

10.  Maximize $z = 2x - 3y$

Subject to             $y \le x$

$y \le 30$

$y \ge 10$

$x \le 40$

$x \ge 0$

11.  Minimize      $z = x + y$

Subject to $3x + y \ge 10$

$3x + 2y \ge 17$

$x + 2y \ge 11$

$x, y \ge 0$

12.  Minimize      $z = x + 3y$

Subject to $2x + y \ge 20$

$x + y \ge 18$

$x + 3y \ge 41$

$x, y \ge 0$

13.  Minimize $z = 2x + 3y$

Subject to  $x + y \le 10$

$x \ge 1$

$x \le 3$

$y \ge 0$

14.  Minimize      $z = 4x + 5y$

Subject to  $x + y \ge 2$

$-x + y \le 2$

$x + y \le 6$

$-x + y \ge -2$

15.  Maximize      $z = x + 10y$

Subject to $-x + y \le 0$

$-x + y \ge -4$

$y \ge 1$

$y \le 4$

16.  Maximize $z = 5x + 4y$

Subject to $-2x + y \le 0$

$-x + y \le 0$

$x + y \le 6$

17.  Minimize      $z = 4x + 5y$

Subject to      $x + y \ge 5$

$-x + y \le 1$

$-2x + y \le 2$

$y \ge 0$

18.  Minimize      $z = x + y$

Subject to $2x + y \ge 9$

$x + 2y \ge 9$

$x \le 7$

$y \le 7$

19.  Maximize $z = 2x + 4y$

Subject to   $x - y \le 0$

$x, y \ge 0$

20.  Maximize $z = 3x + 2y$

Subject to   $x - y \le 0$

$2x - y \ge 0$

$x, y \ge 0$

21.  Minimize      $z = x + 3y$

Subject to      $x + y \le 0$

$-2x + y \ge 1$

$x, y \ge 0$

22.  Minimize $z = x + y$

Subject to $x + y \le 2$

$y \ge 3$

$x \ge 0$

23.  Suppose that $z = ax + y$ is the objective function to be maximized over the

feasible region shown in the figure. Find all values of $a$ such that $z = ax + y$ will be maximized at (a) A, (b) B, (c) C, (d) D.

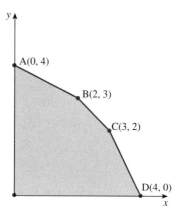

**24.** Suppose that $z = ax + y$ is the objective function to be minimized over the feasible region shown in the figure. Find all values of $a$ such that $z = ax + y$ will be minimized at (a) A, (b) B, (c) C, (d) D.

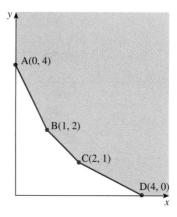

**25.** Suppose a linear programming problem has the feasible region shown in the figure. Determine the maximum point if the objective function is (a) $z = .20x + y$, (b) $z = .50x + y$, (c) $z = .70x + y$, (d) $z = 1.50x + y$, (e) $z = 4x + y$.

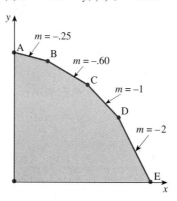

**26.** Suppose a linear programming problem has the feasible region shown in the figure. Determine the minimum point if the objective function is (a) $z = .20x + y$, (b) $z = .50x + y$, (c) $z = .70x + y$, (d) $z = 1.50x + y$, (e) $z = 4x + y$.

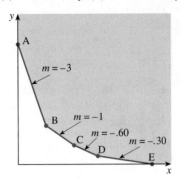

**27.** Suppose that $z = ax + by$ with $a > 0$ and $b > 0$ is the objective function to be maximized over the feasible region shown in the figure. Explain why the objective function must have a maximum at $A$.

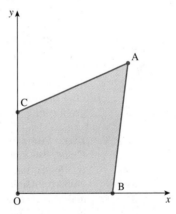

**28.** Suppose that $z = ax + by$ with $a > 0$ and $b > 0$ is the objective function to be minimized over the feasible region shown in the figure. Explain why the objective function must have a minimum at $B$.

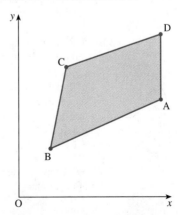

## Applications

29. **Shipping.** A manufacturer supplies its product to two distributors. Distributor A needs at least 100 of the product next month, while Distributor B needs at least 50. At most 200 of the product can be manufactured and delivered to the distributors. If it costs $30 to ship each product to Distributor A and $40 to B, find the number to be shipped to each distributor to minimize cost. Find the minimum cost. (Refer to Exercise 43 of the previous section.)

30. **Transportation.** A plane delivers two types of cargo between two destinations. Each crate of cargo I is 5 cubic feet in volume, 100 pounds in weight, and earns $20 in revenue. Each crate of cargo II is 5 cubic feet in volume, 200 pounds in weight, and earns $30 in revenue. The plane has available at most 400 cubic feet and 10,000 pounds for the crates. Finally, at least twice the number of crates of A than B must be shipped. Find the number of crates of each cargo to ship in order to maximize revenue. Find the maximum revenue. (Refer to Exercise 44 of the previous section.)

31. **Manufacturing.** Two products, A and B, require for their manufacture spending a certain number of hours on a stamping machine, a grinding machine, and a polishing machine. Product A requires 1 hour on the stamping machine, 2 hours on the grinding machine, and 2 hours on the polishing machine. Product B requires 2 hours on the stamping machine, 2 hours on the grinding machine, and 1 hour on the polishing machine. The stamping machine can operate 20 hours, the grinding machine 22 hours, and the polishing machine 15 hours each day. The profit on each of product A and B is, respectively, $40 and $50 each. Find the number of each product to be manufactured in order to maximize profits. Find the maximum profit. (Refer to Exercise 45 of the previous section.)

32. **Diet.** An individual needs a daily supplement of at least 500 units of vitamin C and 200 units of vitamin E and agrees to obtain this supplement by eating two foods, I and II. Each ounce of food I contains 40 units of vitamin C and 10 units of vitamin E, while each ounce of food II contains 20 units of vitamin C and also 20 units of vitamin E. The total supplement of these two foods must be at most 30 ounces. Unfortunately, food I contains 30 units of cholesterol per ounce, and food II contains 20 units of cholesterol per ounce. Find the appropriate amounts of the two food supplements so that cholesterol is minimized. Find the minimum amount of cholesterol. (Refer to Exercise 46 of the previous section.)

33. **Manufacturing.** A firm manufactures tables and desks. To produce each table requires 1 hour of labor, 10 square feet of wood, and 2 quarts of finish. To produce each desk requires 3 hours of labor, 20 square feet of wood, and 1 quart of finish. Available is at most 45 hours of labor, at most 350 square feet of wood, and at most 55 quarts of finish. The tables and desks yield profits of $4 and $3 respectively. Find the number of each product to be made in order to maximize profits. Find the maximum profit. (Refer to Exercise 47 of Section 3.1.)

34. **Mixture.** A dealer has 7600 pounds of peanuts, 5800 pounds of almonds, and 3000 pounds of cashews to be used to make two mixtures. The first mixture wholesales for $2 per pound and consists of 60% peanuts, 30% almonds, and

10% cashews. The second mixture wholesales for $3 per pound and consists of 20% peanuts, 50% almonds, and 30% cashews. How many pounds of each mixture should the dealer make in order to maximize revenue? Find the maximum revenue. (Refer to Exercise 48 of Section 3.1.)

35. **Fishery.** A certain lake has smallmouth and largemouth bass and also three types of food for these fish, I, II, and III. Each month the lake can supply 800 pounds of food I, 500 pounds of food II, and 700 pounds of food III. Each month each smallmouth bass requires 1 pound of food I, 1 pound of food II, and 2 pounds of food III, and each largemouth bass requires 4 pounds of food I, 2 pounds of food II, and 1 pound of food III. What is the maximum number of these fish that the lake can support? (Refer to Exercise 49 of Section 3.1.)

36. **Scheduling.** A firm has two plants, A and B, that manufacture three products, $P_1$, $P_2$, and $P_3$. In one day Plant A can manufacture 30 of $P_1$, 30 of $P_2$, and 10 of $P_3$ and costs $30,000 a day to operate, whereas Plant B can manufacture 10 of $P_1$, 20 of $P_2$, and 30 of $P_3$, and costs $40,000 a day to operate. If at least 240 of the first product, 390 of the second, and 410 of the third are needed, how many days must each plant operate in order to minimize cost? Find the minimum cost. (Refer to Exercise 50 of Section 3.1.)

37. **Scheduling.** Redo the previous problem if the cost of operating Plant A is $30,000 and of Plant B is $20,000.

38. **Diet.** Redo Exercise 32 if food II contains 25 units of cholesterol and everything else is the same.

39. **Fishery.** If in Exercise 35 $2 can be obtained from harvesting the smallmouth bass and $5 for the largemouth bass, how many of each type of fish should be harvested to maximize revenue? Find the maximum revenue.

40. **Manufacturing.** A firm is planning to manufacture and sell two products with the cost and profit (in thousands of dollars) given in the following table.

|        | Product 1 | Product 2 |
| ------ | --------- | --------- |
| Cost   | 2         | 1         |
| Profit | 2         | 3         |

The facilities are such that a total of at most 20 of the two products and at most 15 of product 2 can be manufactured in any one day, and at most $30,000 can be allocated for the costs in any one day. With these conditions the firm can sell all of the items manufactured. Find the number of each product to manufacture in order to maximize profit. Find the maximum profit.

41. **Manufacturing.** Redo the previous exercise if the profit is given as $P = x + y$.

## Graphing Calculator Exercises

In Exercises 42 through 44 solve by using the zoom feature of your graphing calculator to determine the corners of the feasible region (to two decimal places).

**42.**  Maximize      $z = 10.7x + 14.1y$

Subject to $1.22x + 3.15y \le 34.45$

$1.4x + 1.4y \le 19.6$

$1.31x + 2.46y \le 24.09$

$x, y \ge 0$

**43.**  Minimize      $z = 1.41x + 2.37y$

Subject to $1.26x + 4.17y \ge 44.31$

$2.17x + 2.17y \ge 41.23$

$1.68x + 1.03y \ge 24.12$

$x, y \ge 0$

**44.  Fishery.**   A lake has rainbow trout and brown trout. Three foods A, B, C are available for these trout in the lake. Each rainbow trout requires 3.42 units of food A, 2.87 units of food B, and 2.14 units of food C each day. Each brown trout requires 5.13 units of food A, 3 units of food B, and 6 units of food C each day. If at most 2701.8 units of food A, 1797.5 units of food B, and 2515 units of food C are available daily, what is the maximum number of rainbow and brown trout that this lake can support?

## Solutions to Self-Help Exercise Set 3.2

**1.**  The feasible region is shown in the figure. The feasible region includes its boundary, is unbounded, and the objective function $z = 4x + 6y$ has positive coefficients for $x$ and $y$. Thus a minimum exists and must occur at a corner point. By evaluating $z$ at the three corners we see that the minimum of $z = 26$ occurs when $x = 2$ and $y = 3$.

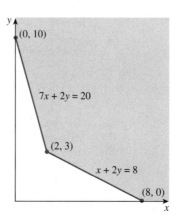

| Corner | z = 4x + 6y |
|--------|-------------|
| (0, 10) | 4(0) + 6(10) = 60 |
| (2, 3) | 4(2) + 6(3) = 26 ← minimum |
| (8, 0) | 4(8) + 6(0) = 32 |

2. With $x$ the number of the first type of connector manufactured and $y$ the number of the second type of connector, the profit is $P = .50x + .60y$. The feasible region and the corners were found in Self-Help Exercise 2 of the last section. Evaluating $P$ at the corners yields

| Corner | P = .50x + .60y |
|--------|-----------------|
| (2, 0) | .50(2) + .60(0) = 1.00 |
| (0, 2) | .50(0) + .60(2) = 1.20 |
| (0, 10) | .50(0) + .60(10) = 6.00 |
| (5, 6) | .50(5) + .60(6) = 6.10 ← maximum |
| (7, 0) | .50(7) + .60(0) = 3.50 |

Thus the maximum profit of $6.10 will occur if 5 of the first connectors and 6 of the second are made.

3. Setting $z = ax + y = P_0$ gives $y = -ax + P_0$, which has slope $m = -a$. Notice that the slope of line $AB$ is $-1$ and of $BD$ is $-2$. Thus if $1 \leq a \leq 2$, the constant profit line will be as shown in the figure and $B$ will yield the maximum. If $a \geq 2$, then the constant profit line will be steeper than the line $BD$ and therefore $D$ will yield the maximum.

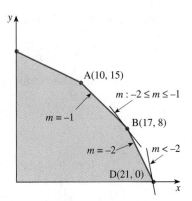

## 3.3    MORE LINEAR PROGRAMMING

▶ *Post-Optimal Analysis*
▶ *Portfolio Management*
▶ *A Transportation Problem*
▶ *Blending Problems*

**George Dantzig**    **"***Linear programming models have been successful because with them many large problems can be formulated so that they are acceptable to planners and solvable on a computer.***"**

### Post-Optimal Analysis

After an optimal solution to an applied linear programming problem has been found, further analysis is usually required. For example, analysis is needed to determine if an excess of some resource is being allocated. If this is the case, then management should consider reallocating or reducing this resource.

**E X A M P L E   1**    **Determining Excess Resources**

Consider the wooden-boat manufacturing problem considered in the last two sections. Recall that 10 dinghies and 15 skiffs were manufactured in order to maximize the profits subject to the three constraints on the available hours of labor for cutting, assembly, and painting. Identify any of the three operations that has an excess of needed labor.

**Solution**    Recall that $x$ was the number of dinghies manufactured each week and $y$ the number of skiffs, and the linear programming problem was

$$\text{Maximize} \quad z = 60x + 80y$$

$$\begin{aligned}
\text{Subject to} \quad & 2x + 4y \leq 80 && \text{hours available for cutting} \\
& 2x + 2y \leq 50 && \text{hours available for painting} \\
& 4x + 2y \leq 84 && \text{hours available for assembly} \\
& x, y \geq 0 && \text{nonnegativity constraint}
\end{aligned}$$

The feasible region is shown in Figure 3.18. Recall that the solution to the problem was (10, 15), that is, produce 10 dinghies and 15 skiffs. Notice from Figure 3.18 that the point (10, 15) is on the lines determined by the cutting and painting constraint. This means that all the labor hours allocated for cutting and painting are used. That is,

cutting hours used $= 2(10) + 4(15) = 80 =$ cutting hours available

painting hours used $= 2(10) + 2(15) = 50 =$ painting hours available.

Figure 3.18

However, if 10 dinghies and 15 skiffs are manufactured, the number of labor hours of assembly is

$$4(10) + 2(15) = 70,$$

which is *less* than the 84 hours allocated. Thus we see that management can reduce the number of hours allocated to the assembly section. ∎

## Portfolio Management

Applied linear programming problems routinely arise with dozens and even hundreds or thousands of decision variables and problem constraints. Such problems cannot be solved by the geometric method considered in the last section, nor can they be solved by hand using the simplex method in the next chapter. These problems are solved using computers. Of course the computer must be given the problem! Thus the steps that involve translating the problem into the linear programming format must be done by humans. Undoubtedly, this represents the most important skill to acquire. Once we identify the mathematical formulation of the linear programming problem, we can usually use standard software programs to determine the solution.

Despite the existence of such software programs, we should not underestimate the need to know the geometric method and the simplex method given in the next chapter. The more we know about how and why these methods work, the more likely we will be able to avoid catastrophic errors that can occur using the software and to make a sensible analysis of the answers given.

We now consider the formulation of several linear programming problems with three or more decision variables.

**E X A M P L E   2**   **Allocation of Funds to Maximize Return**

An individual wishes to distribute at most $100,000 among a money market mutual fund, 30-year U.S. Treasury bonds, and a stock index fund that mimics the Standard and Poor's 500. Being basically conservative, this individual wants at all times at least $30,000 in the money market fund, at least $30,000 in treasury bonds, and the amount of money in stocks less than or equal to the total in the money market and in treasury bonds. On the other hand, some exposure to equities is desired and at least $20,000 should at all times be in stocks. Based on historical patterns, this individual assumes that the money market, bonds, and stocks will earn annual returns of 6%, 8%, and 10% (including capital gains), respectively. How should this person allocate the funds among these three investments to maximize the return?

**Solution**    Let us denote by $x$, $y$, and $z$ the amount in thousands of dollars invested in, respectively, the money market, bonds, and stocks. Since the total must be at most $100,000, $x + y + z \leq 100$.

Since at least $30,000 must be in the money market and in bonds, we must have $x \geq 30$ and $y \geq 30$. Since at least $20,000 must be in stocks, $z \geq 20$. We also have the amount in stocks less than or equal to the totals in the money market and bonds. Thus $z \leq x + y$. In formulating linear programming problems, we always write the inequalities so that a constant is to the right of the inequality. We then rewrite this last inequality as $-x - y + z \leq 0$.

|  | *Money Market* | *Bonds* | *Stocks* |
|---|---|---|---|
| Variables | $x$ | $y$ | $z$ |
| Yield | .06 | .08 | .10 |

The return $R$ is $R = 0.06x + 0.08y + 0.10z$. This is to be maximized. The mathematical formulation in the linear programming format is then given as follows:

$$\text{Maximize} \quad R = 0.06x + 0.08y + .10z$$

$$
\begin{aligned}
\text{Subject to} \quad x \;+y \;+z &\leq 100 \\
-x \;-y \;+z &\leq\; 0 \\
x \qquad\qquad &\geq\; 30 \\
y \qquad &\geq\; 30 \\
z &\geq\; 20 \\
x, y, z &\geq\; 0. \quad \blacksquare
\end{aligned}
$$

## A Transportation Problem

Transportation is another area where linear programming is used.

In the remainder of this section the number of decision variables will be rather large, so we will find it convenient to change the decision variables from $x$ to $x_1$, $y$ to $x_2$, and so on.

**E X A M P L E   3**      **Minimizing the Cost of Transportation**

A firm has two stockpiles of an item stored in warehouses at two different locations, Denver and Santa Fe. The item must be shipped to two plants, located in Des Moines and Fort Wayne, for assembly into a final product. At the beginning of a week the stockpile at Denver has 150 items and the stockpile at Santa Fe has 180. The Des Moines plant needs at least 100 of these items, and the Fort Wayne plant needs at least 120 that week. The cost of sending each item from the stockpile at Denver to the plants in Des Moines and Fort Wayne is, respectively, $200 and $300, while the cost of sending each item from the stockpile at Santa Fe to the plants in Des Moines and Fort Wayne is, respectively, $300 and $500. Refer to Figure 3.19. Find the number of items that should be shipped from each stockpile to each plant in order to minimize cost.

**Solution**      We organize the information, as usual, in a table.

|  | *Des Moines Plant* | *Fort Wayne Plant* | *Supply* |
|---|---|---|---|
| Stockpile at Denver | $200 | $300 | 150 |
| Stockpile at Santa Fe | $300 | $500 | 180 |
| Demand | 100 | 120 |  |

Let $x_1$ and $x_2$ be the number shipped from the stockpile at Denver to the plants at Des Moines and Fort Wayne, respectively. Since the stockpile at Denver has 150 items, we then have the constraint $x_1 + x_2 \leq 150$. Now let $x_3$ and $x_4$ be the number shipped from the stockpile at Santa Fe to the plants at Des Moines and Fort Wayne, respectively. Since the stockpile at Santa Fe has 180 items, we then have the constraint $x_3 + x_4 \leq 180$. From the table we see that the Des Moines plant is demanding at least 100 items with $x_1$ coming from the stockpile at Denver and $x_3$ coming from the stockpile at Santa Fe. Thus $x_1 + x_3 \geq 100$. Also from the table we see that the Fort Wayne plant is demanding at least 120 items with $x_2$ coming from the stockpile at Denver and $x_4$ coming from the stockpile at Santa Fe. Thus $x_2 + x_4 \geq 120$.

The cost, according to the table, is $C = 200x_1 + 300x_2 + 300x_3 + 500x_4$. The linear programming problem can now be stated as follows:

$$\text{Minimize} \quad C = 200x_1 + 300x_2 + 300x_3 + 500x_4$$

$$\text{Subject to} \quad \begin{aligned} x_1 + x_2 & \leq 150 \\ x_3 + x_4 & \leq 180 \\ x_1 \quad + x_3 \quad & \geq 100 \\ x_2 \quad + x_4 & \geq 120 \\ x_1, x_2, x_3, x_4 & \geq 0. \quad \blacksquare \end{aligned}$$

**Figure 3.19**

## Blending Problems

Problems that involve the blending of various ingredients in some desired proportion to produce a final product are often amenable to linear programming. The following gives some examples of where blending occurs.

- Various types of metal alloys need to be blended to produce various types of steels.
- Various types of chemicals need to be blended to produce other chemicals.
- Various livestock feeds need to be blended to produce low-cost feed mixtures.
- Various types of crude oil must be blended to produce different types of gasoline or heating oil.

**E X A M P L E  4**    **Oil Refinery**

An oil refinery blends three types of crude oil, crude A, crude B, and crude C, to make regular and premium gasoline. Crude A has an octane rating of 90, a sulfur content of 1%, costs $20 a barrel, and there are 50,000 barrels on hand. Crude B has an octane rating of 96, a sulfur content of 2%, costs $24 a barrel, and there are 30,000 barrels on hand. Crude C has an octane rating of 98, a sulfur content of 3%, costs $25 a barrel, and there are 20,000 barrels on hand. The octane rating for regular gasoline must bc at least 92 and for premium 95. The sulfur content of regular must be at most 1.5% and for premium 2.1%. Regular gasoline sells for $26 a barrel and premium for $32. If the refinery has an order for 35,000 barrels of regular gasoline and 25,000 barrels of premium, how much of each type of crude should be used in each type of gasoline to maximize the profit?

**Solution**    The information is organized in the following tables.

### Available Crude Oil

| Crude | Octane | Sulfur | Cost/barrel | Supply (barrels) |
|-------|--------|--------|-------------|------------------|
| A | 90 | 1% | $20 | 50,000 |
| B | 96 | 2% | $24 | 30,000 |
| C | 98 | 3% | $25 | 20,000 |

### Needed Gasoline

| Gasoline | Octane | Sulfur | Price/barrel | Needed (barrels) |
|----------|--------|--------|--------------|------------------|
| Regular | 92 | 1.5% | $26 | 35,000 |
| Premium | 95 | 2.1% | $32 | 25,000 |

Let

$$x_1 = \text{barrels of crude A used to produce regular gasoline}$$

$$x_2 = \text{barrels of crude A used to produce premium gasoline}$$

$$x_3 = \text{barrels of crude B used to produce regular gasoline}$$

$$x_4 = \text{barrels of crude B used to produce premium gasoline}$$

$$x_5 = \text{barrels of crude C used to produce regular gasoline}$$

$$x_6 = \text{barrels of crude C used to produce premium gasoline.}$$

The total number of barrels of crude A used is $x_1 + x_2$, while the amount available is 50,000. Thus

$$x_1 + x_2 \leq 50{,}000.$$

Using the corresponding numbers for crude B and C gives

$$x_3 + x_4 \leq 30{,}000$$

$$x_5 + x_6 \leq 20{,}000.$$

The total number of barrels of regular gasoline produced is $x_1 + x_3 + x_5$, while the amount needed is 35,000. Thus

$$x_1 + x_3 + x_5 \geq 35{,}000.$$

Using the corresponding numbers for premium gasoline gives

$$x_2 + x_4 + x_6 \geq 25{,}000.$$

The sulfur constraint for regular gasoline can be given as

$$\frac{\text{total sulfur in regular}}{\text{number of barrels in regular mixture}} \leq 0.015$$

or

$$\frac{0.01x_1 + 0.02x_3 + .03x_5}{x_1 + x_3 + x_5} \leq 0.015.$$

This is not a linear inequality. But we can obtain such an inequality by multiplying both sides by $x_1 + x_3 + x_5$. Doing this gives

$$0.01x_1 + 0.02x_3 + 0.03x_5 \leq 0.015x_1 + 0.015x_3 + 0.015x_5$$

$$-0.005x_1 + 0.005x_3 + .015x_5 \leq 0.$$

For premium gasoline we have

$$\frac{0.01x_2 + 0.02x_4 + .03x_6}{x_2 + x_4 + x_6} \leq 0.021$$

$$0.01x_2 + 0.02x_4 + 0.03x_6 \leq 0.021x_2 + 0.021x_4 + 0.021x_6$$

$$-0.011x_2 - 0.001x_4 + .009x_6 \leq 0.$$

The octane levels of different crudes will blend linearly. Therefore, since $\dfrac{x_1}{x_1 + x_3 + x_5}$ is the fraction of crude A used in regular gasoline, $\dfrac{x_3}{x_1 + x_3 + x_5}$ is the fraction of crude B, and $\dfrac{x_5}{x_1 + x_3 + x_5}$ is the fraction of crude C, the octane rating for regular gasoline is

$$90\,\frac{x_1}{x_1 + x_3 + x_5} + 96\,\frac{x_3}{x_1 + x_3 + x_5} + 98\,\frac{x_5}{x_1 + x_3 + x_5}.$$

Since this must be at least 92, we have

$$90\,\frac{x_1}{x_1 + x_3 + x_5} + 96\,\frac{x_3}{x_1 + x_3 + x_5} + 98\,\frac{x_5}{x_1 + x_3 + x_5} \geq 92.$$

Again this is not a linear inequality. But multiplying both sides by $x_1 + x_3 + x_5$ gives

$$90x_1 + 96x_3 + 98x_5 \geq 92(x_1 + x_3 + x_5)$$

$$-2x_1 + 4x_3 \mid 6x_5 \geq 0.$$

For premium gasoline we then have

$$90\,\frac{x_2}{x_2 + x_4 + x_6} + 96\,\frac{x_4}{x_2 + x_4 + x_6} + 98\,\frac{x_6}{x_2 + x_4 + x_6} \geq 95$$

$$90x_2 + 96x_4 + 98x_6 \geq 95(x_2 + x_4 + x_6)$$

$$-5x_2 + x_4 + 3x_6 \geq 0.$$

Now we will determine the profit function. The cost of the crude is

$$C = 20(x_1 + x_2) + 24(x_3 + x_4) + 25(x_5 + x_6),$$

while the revenue is

$$R = 26(x_1 + x_3 + x_5) + 32(x_2 + x_4 + x_6).$$

Thus the profit is

$$
\begin{aligned}
P &= R - C \\
&= [26(x_1 + x_3 + x_5) + 32(x_2 + x_4 + x_6)] \\
&\quad - [20(x_1 + x_2) + 24(x_3 + x_4) + 25(x_5 + x_6)] \\
&= 6x_1 + 12x_2 + 2x_3 + 8x_4 + x_5 + 7x_6.
\end{aligned}
$$

We then have the following linear programming problem to solve.

Maximize         $P = 6x_1 + 12x_2 + 2x_3 + 8x_4 + x_5 + 7x_6$

$$
\begin{aligned}
\text{Subject to} \quad x_1 + \quad x_2 \quad &\leq 50{,}000 \\
x_3 + \quad x_4 \quad &\leq 30{,}000 \\
x_5 + \quad x_6 &\leq 20{,}000 \\
x_1 \quad + \quad x_3 \quad + \quad x_5 \quad &\geq 35{,}000 \\
x_2 \quad + \quad x_4 \quad + \quad x_6 &\geq 25{,}000 \\
-.005x_1 \quad + .005x_3 \quad + .015x_5 \quad &\leq 0 \\
-.011x_2 \quad - .001x_4 \quad + .009x_6 &\leq 0 \\
-2x_1 \quad + \quad 4x_3 \quad + 6x_5 \quad &\geq 0 \\
- \quad 5x_2 \quad + \quad x_4 \quad + \quad 3x_6 &\geq 0 \\
x_1, x_2, x_3, x_4, x_5, x_6 &\geq 0
\end{aligned}
$$

∎

## SELF-HELP EXERCISE SET 3.3

1. **Diet.**   A dietician has three foods, A, B, and C, to use to prepare a meal. The table indicates the constraints and the costs. If the total number of ounces must be at most 18, formulate a linear programming problem that will determine the number of ounces of each of the foods to use in order to minimize the cost.

|  | Units Per Ounce | | | Constraints in Units |
|---|---|---|---|---|
|  | A | B | C |  |
| Vitamin C | 100 | 0 | 120 | at least 500 |
| Vitamin E | 0 | 70 | 50 | at least 200 |
| Vitamin A | 1000 | 2000 | 0 | at most 25,000 |
| Calories | 200 | 150 | 75 | at most 3000 |
| Cost per ounce | $.60 | $.40 | $.70 |  |

## EXERCISE SET 3.3

In Exercises 1 through 20 translate into a mathematical formulation of a linear programming problem.

1. **Investments.**   An individual has at most $100,000 to invest in four vehicles: a money market fund, a bond fund, a conservative stock fund, and a speculative stock fund. These four investments are assumed to have annual returns (including capital gains) of 6%, 8%, 10%, and 13% respectively. The investor wants at least $10,000 in the conservative stock fund, at least $10,000 and at most $40,000 in the money market fund, the amount in speculative stocks less than or equal to the amount in the money market, and finally the amount in conser-

vative stocks less than or equal to the amount in bonds. Find the appropriate allocations of funds among these four investments to maximize the return.

2. **Investments.** Redo Example 2 of the text if the investor wants at least 40% invested in stocks with everything else remaining the same.

3. **Transportation.** Redo Example 3 of the text using the following table.

| | Plant A | Plant B | Supply |
|---|---|---|---|
| Stockpile I | $20 | $40 | 500 |
| Stockpile II | $50 | $20 | 100 |
| Demand | 150 | 250 | |

4. **Transportation.** A firm has an item on stockpile at two locations I and II. This item needs to be shipped in various quantities to three plants: A, B, and C. The following table gives the transportation cost of sending each item from the two stockpile locations to the three plants, the number of items at each stockpile, and the demanded number of this item for each of the three plants.

| | Plant A | Plant B | Plant C | Supply |
|---|---|---|---|---|
| Stockpile I | 2 | 1 | 3 | 600 |
| Stockpile II | 1 | 3 | 2 | 650 |
| Demand | 150 | 300 | 350 | |

Find the numbers of items to be shipped from each of the stockpiles to each of the plants in order to minimize transportation cost.

5. **Diet.** Three foods, A, B, and C, are being used by a dietician at a hospital to obtain certain minimum supplements of three vitamins, I, II, and III. The following table gives the number of units of each vitamin in each ounce of each food, the daily minimum requirements of each vitamin, and the cost in dollars per ounce of each of the foods.

| | Food A | Food B | Food C | Requirements |
|---|---|---|---|---|
| Vitamin I | 30 | 30 | 10 | 500 |
| Vitamin II | 20 | 20 | 10 | 329 |
| Vitamin III | 35 | 50 | 20 | 425 |
| Cost/oz | 4 | 3 | 1 | |

At most 20 ounces of the three foods are to be given. How much of each of the foods should be given if the cost is to be minimized?

6. **Investments.** An investment made for a retirement plan must pay particular attention to risk. Suppose for such an investment no more than $10,000 is placed in 3 investments with annual returns of 8%, 9%, and 10% respectively. The investor requires an annual return of at least $900 on the investment and wishes to minimize the risk. The second investment is twice as risky as the first, and the third investment is 4 times as risky as the first. How should the money be allocated to minimize the risk? *Hint:* Let risk $= x_1 + 2x_2 + 4x_3$.

7. **Plant Nutrition.** A farmer can buy three types of 100-pound bags of fertilizer, type A, type B, and type C. Each 100-pound bag of type A fertilizer costs $20 and contains 40 pounds of nitrogen, 30 pounds of phosphoric acid, and 10 pounds of potash. Each 100-pound bag of type B fertilizer costs $30 and contains 20 pounds of nitrogen, 20 pounds of phosphoric acid, and 60 pounds of potash. Each 100-pound bag of type C fertilizer costs $20 and contains no nitrogen, 30 pounds of phosphoric acid, and 40 pounds of potash. The farmer requires 4400 pounds of nitrogen, 1800 pounds of phosphoric acid, and 2800 pounds of potash. How many bags of each type of fertilizer should she buy in order to minimize cost?

8. **Cost.** An airline has three types of airplanes and has contracted with a tour group to provide accommodations for a minimum of each of 80 first class, 50 tourist, and 60 economy-class passengers. The first plane costs $4000 for the trip and can accommodate 40 first-class, 10 tourist, and 10 economy-class passengers; the second plane costs $5000 for the trip and can accommodate 10 first-class, 10 tourist, and 20 economy-class passengers; while the third plane costs $6000 for the trip and can accommodate 10 first-class, 12 tourist, and 24 economy-class passengers. How many of each type of airplane should be used to minimize the operating cost?

9. **Costs.** An oil company has three refineries and must refine and deliver at least 40,000 gallons of medium-grade gasoline and 24,000 gallons of high-grade gasoline to its stations. The first plant can refine 8000 gallons of medium-grade gasoline and 3000 gallons of high-grade gasoline and costs $40,000 a day to operate. The second plant can refine 4000 gallons of medium-grade gasoline and 6000 gallons of high-grade gasoline and costs $30,000 a day to operate. The third plant can refine 4500 gallons of medium-grade gasoline and 6500 gallons of high-grade gasoline and costs $45,000 a day to operate. How many days should each plant operate to minimize the cost?

10. **Production Scheduling.** A company has 3 plants that produce 3 different sizes of their product. The North plant can produce 2000 items in the small size, 1000 in the medium size, and 1000 in the large size per day. The Center plant can produce 1000 items in the small size, 1000 in the medium size, and 2000 in the large size per day. The South plant can produce 2000 items in the small size, 2000 in the medium size, and 1000 in the large size per day. The company needs to produce at least 9000 of the small items, 8000 of the medium items, and 9000 of the large items on a given day. The cost of producing the small item is $2, the medium item $3, and the large item $2. Find the number of items each plant should produce in order to minimize the cost.

11. **Production Scheduling.** A mining company has three mines that produce three grades of ore. The North mine can produce 2 tons of low-grade ore, 1 ton of medium-grade ore, and 1 ton of high-grade ore each day. The Center mine can produce 1 ton of low-grade, 1 ton of medium-grade, and 4 tons of high-

grade ore. The South mine can produce 1 ton of low-grade, 3 tons of medium-grade, and 3 tons of high-grade ore. At a particular time the company needs at least 21 tons of low-grade ore, 19 tons of medium-grade ore, and at least 25 tons of high-grade ore. It costs $2000 per day to operate the North mine, $3000 a day for the Center mine, and $4000 for the South mine. Find the number of days each mine should be operated to minimize cost.

12. **Purchasing.**   A store decides to have a sale on pink and red azaleas. Two nurseries, East and West, will supply these plants. East nursery charges $4 for a pink azalea and $6 for a red one. West nursery charges $7 for a pink azalea and $5 for a red one. East nursery can supply at most 200 azaleas of either color, while West nursery can supply at most 400 of these azaleas. The store needs at least 250 pink azaleas and at least 300 red azaleas. How many azaleas of each color should the store order from each nursery in order to minimize the cost?

13. **School Rezoning**[1].   A small city has two high schools, NW school and NE school, located in two predominately white districts and two high schools, SW school and SE school, located in two predominately minority districts. A decision is made to bus at least 400 minority students from SW and at least 300 minority students from SE to the two schools in the predominately white districts. NW can accommodate at most 200 additional students and NE at most 600. The weekly cost per student of busing from SW to NW is $4, from SE to NW $2, from SW to NE $2, and from SE to NE is $4. Determine the numbers of students that should be bused from each of the schools in the southern districts to each of the schools in the northern districts in order to minimize the cost.

14. **Oil Refinery.**   An oil refinery blends three types of crude oil, crude A, crude B, and crude C, to make regular, premium, and super gasoline. Crude A has an octane rating of 90, costs $20 a barrel, and there are 50,000 barrels on hand. Crude B has an octane rating of 96, costs $24 a barrel, and there are 30,000 barrels on hand. Crude C has an octane rating of 98, costs $26 a barrel, and there are 20,000 barrels on hand. The octane rating for regular gasoline must be at least 92, for premium at least 96, and for super at least 97. Regular gasoline sells for $26 a barrel, premium for $30, and super for $32. If the refinery has an order for 30,000 barrels of regular gasoline, 20,000 barrels of premium, and 28,000 barrels of super, how much of each type of crude should be used in each type of gasoline to maximize the profit?

15. **Mixture.**   You are going to produce two types of candy, both of which consist solely of sugar, nuts, and chocolate. You have in stock 200 oz of sugar, 30 oz of nuts, and 60 oz of chocolate. The mixture used to make the first candy must contain at least 25% nuts. The mixture used to make the second candy must contain at least 15% nuts and at least 10% chocolate. Each ounce of the first candy can be sold for 50 cents, and each ounce of the second candy for 65 cents. Determine the ounces of each ingredient needed to make each candy in order to maximize revenue.

16. **Farming.**   A farmer has two fields in which he grows wheat and corn. Because of different conditions, there are differences in the yield and costs of growing

---

[1]Hickman and Taylor, ''School Rezoning to Achieve Racial Balance: A Linear Programming Approach,'' *J. Socio-Econ. Planning Sci.* vol. 3 (1969–1970), pp. 127–134.

these crops in these two fields. The yields and cost are shown in the table. Each field has 100 acres available for cultivation; 15,000 bushels of wheat and 9000 bushels of corn must be grown. Determine a planting plan that will minimize the cost.

|                     | Field 1       | Field 2       |
| ------------------- | ------------- | ------------- |
| Corn yield/acre     | 150 bushels   | 200 bushels   |
| Cost/acre of corn   | $180          | $210          |
| Wheat yield/acre    | 40 bushels    | 50 bushels    |
| Cost/acre of wheat  | $140          | $120          |

17.   **Diet.**   A dietician has four foods, A, B, C, and D, to use to prepare a meal. The table indicates the constraints and the costs. If the total number of ounces must be at most 18, determine the number of ounces of each of the foods to use in order to minimize the cost.

|               | Units Per Ounce | | | | Constraints in Units |
|               | A | B | C | D | |
| ------------- | ----- | ----- | ----- | ----- | -------------------- |
| Iron          | 1     | 2     | 0     | 0     | at least 10          |
| Calcium       | 100   | 50    | 0     | 10    | at least 800         |
| Vitamin C     | 100   | 0     | 200   | 20    | at least 500         |
| Vitamin E     | 0     | 0     | 50    | 70    | at least 200         |
| Vitamin A     | 0     | 2000  | 0     | 5000  | at most 25,000       |
| Calories      | 200   | 150   | 75    | 250   | at most 3000         |
| Cholesterol   | 5     | 20    | 30    | 70    | at most 130          |
| Cost per ounce | $.50 | $.40  | $.60  | $.75  |                      |

18.   **Diet.**   Redo the previous exercise if the cholesterol is to be minimized and the cost is to be at most $10.

19.   **Work Schedule.**   A chemical plant must operate every day of the week and requires different numbers of employees on different days of the week. Each employee must work five consecutive days and then receive two days off. For example, an employee who works Tuesday to Saturday must be off on Sunday and Monday. The table indicates the required number of employees on each day. Formulate a linear programming problem that the plant managers can use to minimize the number of employees that must be used.

| Day | Employees Required |
|---|---|
| Monday | 82 |
| Tuesday | 87 |
| Wednesday | 77 |
| Thursday | 73 |
| Friday | 75 |
| Saturday | 42 |
| Sunday | 23 |

20. **Allocation.** A corporation owns three farms. The following table gives the size, usable land, and water allocation for each.

| Farm | Usable Land | Water Allocation (acre feet) |
|---|---|---|
| 1 | 1200 | 18000 |
| 2 | 1500 | 2000 |
| 3 | 1000 | 1300 |

The corporation grows corn, wheat, and sorghum. The following table gives the maximum acreage that can be planted in each crop, water consumption, and net return.

| Crop | Maximum Crop (acres) | Water Consumption (acre feet/acre) | Net Return (dollars/acre) |
|---|---|---|---|
| Corn | 1400 | 3 | 500 |
| Wheat | 1500 | 2 | 400 |
| Sorghum | 1700 | 1 | 100 |

Determine the number of acres of each crop to be planted on each farm to maximize net income.

21. **Manufacturing.** A firm manufactures tables and desks. To produce each table requires 1 hour of labor, 10 square feet of wood, and 2 quarts of finish. To produce each desk requires 3 hours of labor, 20 square feet of wood, and 1 quart of finish. Available is at most 45 hours of labor, at most 350 square feet of wood, and at most 55 quarts of finish. The tables and desks yield profits of $3 and $4 respectively. Find the number of each product to be made in order to maximize profits. Identify any excess resources available if the optimal number of each product is made. (Refer to Exercise 47 of Section 3.1 where the feasible region is the same.)

22. **Mixture.** A dealer has 7600 pounds of peanuts, 5800 pounds of almonds, and 3000 pounds of cashews to be used to make two mixtures. The first mixture

wholesales for $2 per pound and consists of 60% peanuts, 30% almonds, and 10% cashews. The second mixture wholesales for $4 per pound and consists of 20% peanuts, 50% almonds, and 30% cashews. How many pounds of each mixture should the dealer make in order to maximize revenue? At the optimal solution which type of nut will be in excess? (Refer to Exercise 48 of Section 3.1 where the feasible region is the same.)

23. **Fishery.** A certain lake has smallmouth and largemouth bass and also three types of food for these fish, I, II, and III. Each month the lake can supply 800 pounds of food I, 500 pounds of food II, and 700 pounds of food III. Each month each smallmouth bass requires 1 pound of food I, 1 pound of food II, and 2 pounds of food III; and each largemouth bass requires 4 pounds of food I, 2 pounds of food II, and 1 pound of food III. If $2 can be obtained from harvesting the smallmouth bass and $3 for the largemouth bass, how many of each type of fish should be harvested to maximize revenue? At the optimal solution which type of food will be in excess? (Refer to Exercise 49 of Section 3.1 where the feasible region is the same.)

24. **Scheduling.** A firm has two plants, A and B, in which are manufactured three products, $P_1$, $P_2$, and $P_3$. In one day Plant A can manufacture 30 of $P_1$, 30 of $P_2$, and 10 of $P_3$ and costs $30,000 a day to operate, whereas Plant B can manufacture 10 of $P_1$, 20 of $P_2$, and 30 of $P_3$, and costs $40,000 a day to operate. If at least 240 of the first product, 390 of the second, 410 of the third are needed, how many days must each plant operate in order to minimize cost? At the optimal solution, which product will be in excess of the stated minimum needed? (Refer to Exercise 36 of Section 3.2.)

25. **Fishery.** If in Exercise 23 $1 can be obtained from harvesting the smallmouth bass and $3 for the largemouth bass, how many of each type of fish should be harvested to maximize revenue? At the optimal solution which type of food will be in excess?

26. **Scheduling.** Redo Exercise 24 if the cost of operating Plant A is $30,000 and of Plant B is $20,000.

27. **Manufacturing.** In determining how many desks and chairs to manufacture in order to maximize revenue, the feasible region for the linear programming

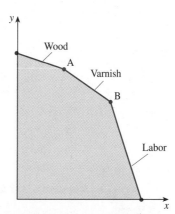

problem was determined to be the region shown in the figure where the first line comes from the constraint on the number of board-feet of wood available, the second line from the constraint on the number of gallons of varnish available, and the third line from the number of hours of labor available. Determine the excess resource if the maximum occurs at (a) A, (b) B.

28. **Mixture.**   In determining how many gallons of four fruit juices to mix in order to obtain a fruit drink to maximize revenue, the feasible region for the linear programming problem was determined to be the region shown in the figure where the indicated four lines come from the constraints on the number of gallons of orange juice, grape juice, grapefruit juice, and cranberry juice. Determine the excess resource(s) if the maximum occurs at (a) A, (b) B, (c) C.

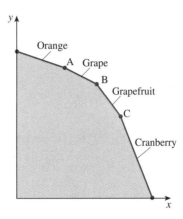

## Solutions to Self-Help Exercise Set 3.3

1. Let $x_1$, $x_2$, and $x_3$ be the respective number of ounces of foods A, B, and C. Then the linear programming problem is

Minimize     $C = .60x_1 + .40x_2 + .70x_3$

Subject to

$$
\begin{array}{llll}
x_1 + & x_2 + & x_3 \le & 18 & \text{total ounces} \\
100x_1 & + 120x_3 \ge & 500 & \text{vitamin C} \\
70x_2 + & 50x_3 \ge & 200 & \text{vitamin E} \\
1000x_1 + 2000x_2 & \le 25{,}000 & \text{vitamin A} \\
200x_1 + & 150x_2 + & 75x_3 \le & 3000 & \text{calories} \\
x_1, x_2, x_3 \ge & 0
\end{array}
$$

## SUMMARY OUTLINE OF CHAPTER 3

- A **linear inequality** is an inequality of the form

$$ax + by < c \text{ (or } \le c \text{ or } > c \text{ or } \ge c).$$

- A **system of linear inequalities** is a set of linear inequalities.

- The **feasible region** of a system of inequalities is the set of points that satisfies all the inequalities in the system.

- Graphing a Linear Inequality
  The following steps show how to graph the linear inequality
  $$ax + by < c \text{ (or } \le c \text{ or } > c \text{ or } \ge c).$$

  1. First graph the straight line $L$ given by $ax + by = c$, using a dotted line if the inequalities "$<$" or "$>$" are given and a solid line if the inequalities "$\le$" or "$\ge$" are given. The dotted line indicates the straight line $L$ is not part of the solution graph, while the solid line indicates that the line $L$ is part of the solution graph.
  2. Pick any convenient test point not on $L$ and evaluate the inequality at that test point. If the inequality is true, then the solution graph is the half-plane on the same side of the line $L$ as the test point. If the inequality is not satisfied, then the solution graph is the half-plane on that side of $L$ that does not include the test point.

- **Optimal Solution of a Linear Programming Problem**   If a linear programming problem has an optimal solution, then the optimum must occur at a corner of the feasible region. If the objective function is optimized at two corners of the feasible region, then any point on the line segment connecting these two corners also optimizes the objective function.

- **Existence Theorem for Optimal Solutions**

  1. If the feasible region is nonempty, bounded, and includes its boundary, then the objective function has both a minimum and a maximum in the feasible region.
  2. If the feasible region $S$ includes its boundary, is unbounded, and the objective function $z = ax + by$ has the property that $a > 0$ and $b > 0$, then $z$ attains a minimum value in $S$ but has no maximum value in $S$.

- **Method of Corners**   Determine if a solution exists using the existence theorem for optimal solutions. If a solution exists then:

  1. Graph the feasible region.
  2. Locate all corners of the feasible region.
  3. Evaluate the objective function at each of the corners.
  4. Note that the optimal solution will be the optimal value found in the previous step.

## Chapter 3 Review Exercises

1. A farmer wishes to raise corn and/or soybeans. The yield per acre and the price per bushel are the same for each. However, she must fertilize the fields with at least 100 pounds of nitrogen (N) per acre and at least 150 pounds of phosphorus (F) per acre. Two materials are available to supply these nutrients: sludge, which contains 6% N and 3% P, and bone meal, which contains 4% N and 12% P. If sludge costs $15 per 100-lb sack, and bone meal costs $20 per 100-pound sack, how many sacks of each should the farmer buy to minimize the cost of fertilizer?

2. An airline has two types of airplanes and has contracted with a tour group to provide accommodations for a minimum of each of 18 first-class, 12 tourist, and

6 economy-class passengers. The first plane costs $1 per 10 yards traveled to operate and can accommodate 4 first-class, 2 tourist, and 2 economy-class passengers, while the second plane costs $1 per 10 yards to operate and can accommodate 3 first-class, 3 tourist, and 3 economy-class passengers. The airline has available at most 6 of the first type of plane and 4 of the second type. How many of each type of airplane should be used to minimize the operating cost?

3. A furniture manufacturer produces chairs and sofas. The chairs require 5 feet of wood, 5 ounces of foam rubber, and 10 square yards of material. The sofas require 20 feet of wood, 10 ounces of foam rubber, and 30 square yards of material. The manufacturer has in stock 300 feet of wood, 210 ounces of foam rubber, and 500 square yards of material. If the chairs can be sold for $200 and the sofas for $500 each, how many of each should be produced to maximize the income?

4.  **a.**  For the optimal values in the previous exercise determine which of the three commodities used in the manufacturing process is in excess and by how much.
    **b.**  Redo the previous example if the chairs can be sold for $400 and everything else remains the same. For the optimal values found determine which of the three commodities used in the manufacturing process is now in excess and by how much.

5. A small shop manufactures two styles of park bench. To manufacture the first style requires one square foot of wood, 1 pound of iron, and 3 hours of labor. The second style requires 3 square feet of wood, 1 pound of iron, and 2 hours of labor. The shop has 21 square feet of wood, 9 pounds of iron, and 24 hours of available labor. If the shop makes a profit of $10 on the first style bench and $20 on the second style bench, find the number of each style of bench to manufacture that will maximize profits.

6.  **a.**  For the optimal values in the previous exercise determine whether iron, wood, or labor is in excess and by how much.
    **b.**  Redo the previous example if the first style of park bench yields a profit of $20 and the second $15 and everything else remains the same. For the optimal values found determine whether iron, wood, or labor is now in excess and by how much.

7. A mining company has two mines that produce three grades of ore. The North mine can produce 3 tons of low-grade ore, 1 ton of medium-grade ore, and 1 ton of high-grade ore each day. The South mine can produce 3 tons of low-grade ore, 2 tons of medium-grade ore, and 3 tons of high-grade ore each day. It costs $3000 per day to operate the North mine and $4000 a day to operate the South mine. At a particular time the company needs at least 15 tons of low-grade ore, 8 tons of medium-grade ore, and 9 tons of high-grade ore. Find the number of days each mine should be operated in order to minimize cost and find the minimum cost.

8. A farmer has 240 acres of land on which he can raise corn and/or soybeans. Raising corn yields 120 bushels per acre, which can be sold for $2.40 per bushel, and raising soybeans yields 40 bushels per acre and sells for $6.00 per bushel. To obtain this price, the farmer must be able to store his crop in bins with a total capacity of 12,000 bushels. The farmer must limit each crop to no more than 10,800 bushels. Assuming the costs of raising corn and soybeans are the same,

how much of each should the farmer plant in order to maximize his revenue? What is this maximum?

9.  A health food store manager is preparing two mixtures of breakfast cereal out of a supply of 10.5 pounds of oats, 0.9 pounds of almonds, 0.66 pounds of raisins, and 50 pounds of wheat. The first mixture contains 50% oats, 4% almonds, 2% raisins, and the rest wheat and sells for $5 a pound. The second mixture contains 10% oats, 2% almonds, 2% raisins, and the rest wheat and sells for $4 a pound. How much of each mixture should be made to maximize profits? What is the maximum profit?

Linear
programming is
used in the
manufacturing
process to determine
the most profitable
product mix and to
identify excess
resources.
(© Benjamin
Mendlowitz.)

# Linear Programming: The Simplex Method

4.1   Slack Variables and the Simplex Tableau

4.2   The Simplex Method: Standard Maximization Problems

4.3   The Dual

4.4   The Simplex Method: Nonstandard and Minimization Problems

4.5   Post-Optimal Analysis

In the last chapter we introduced linear programming using the geometric method. This method is practical only when the problem has a very small number of variables and constraints. For a typical problem encountered in the applications, the method is completely inadequate. Nevertheless, the geometric method gives substantial insights into the workings of the simplex method, the method considered in this chapter. The simplex method was discovered by George Dantzig in 1947.

Linear programming problems that arise in applications often involve thousands and even tens of thousands of unknowns and constraints. Thus a method is needed to solve such problems efficiently. The simplex method fills this need.

The simplex method can be readily programmed for a computer. Despite the considerable success of this method, important business and scientific problems abound that are too large in scope for this method to handle, even on the fastest known computers. Thus research continues in this area for better methods.

As we shall see, the simplex method follows the boundary of the feasible region, moving through a relatively small number of corners until the optimal corner is found. In 1979 L. G. Khachiyan discovered another way of solving linear programming problems when he constructed the "ellipsoid algorithm." The ellipsoid algorithm generates points from *outside* the feasible region that approach the optimal corner. Initially there was considerable excitement about this algorithm since it possesses certain theoretical properties that led some to conjecture that the method might be more efficient than the simplex method. Unfortunately, this turned out not to be the case in practice.

Then in 1984 Narendra Karmarkar of AT&T Bell Laboratories discovered what is now called Karmarkar's algorithm. This method generates a sequence of points in the *interior* of the feasible region that head to the optimal corner. This method is currently being evaluated to determine under what circumstances it is superior to the simplex method; it is now viewed as holding considerable promise.

## 4.1   SLACK VARIABLES AND THE SIMPLEX TABLEAU

▶ *The Need for an Efficient Method in Linear Programming*
▶ *Slack Variables*
▶ *Basic and Nonbasic Variables*
▶ *The Simplex Tableau and Pivoting*

### The Need for an Efficient Method in Linear Programming

An important problem in linear programming involves assigning individuals to jobs. Consider a modest assignment problem of 70 jobs with 70 individuals with the conditions that every one of the 70 individuals is assigned one of the 70 jobs and every one of the 70 jobs has an individual assigned to it. The number of corners in such a problem is large beyond comprehension. To obtain some idea of just how large, imagine that large mainframe computers of the type now used were available at the time of the generally assumed beginning of the universe, 15 billion years ago. Assume, furthermore, that the earth existed at that time and was filled with such computers, all working continuously to evaluate an objective function at the corners of the feasible region for this problem. By now these computers would not have had enough time to evaluate the objective function at all the corners! (The actual number of evaluations is approximately the number 1 followed by 100 zeros.)

It is clear that for practical problems an *efficient* method is needed to find the maximum or the minimum of the objective function. George Dantzig, the father of linear programming, discovered an efficient method in 1947 to find the solution by looking at a very modest number of corners. This method, called the **simplex method**, is considered in this chapter. The simplex method can solve important practical linear programming problems in reasonable time using computers available today.

## Slack Variables

We will now find it convenient to change the variables from $x$ to $x_1$, $y$ to $x_2$, and so on. In this section we consider linear programming problems in the following form:

$$
\begin{aligned}
\text{Maximize} \quad & z = c_1x_1 + c_2x_2 + \cdots + c_nx_n \\
\text{Subject to} \quad & a_{11}x_1 + a_{12}x_2 + \cdots + a_{1n}x_n \le b_1 \\
& a_{21}x_1 + a_{22}x_2 + \cdots + a_{2n}x_n \le b_2 \\
& \qquad\qquad \cdots \\
& a_{m1}x_1 + a_{m2}x_2 + \cdots + a_{mn}x_n \le b_m \\
& x_1, x_2, \ldots, x_n \ge 0
\end{aligned}
\tag{1}
$$

The variables $x_1, x_2, \ldots, x_n$ are called the **decision variables**.

Consider again the small manufacturer of wooden boats considered in the last chapter. If $x_1$ is the number of dinghies produced, and $x_2$ the number of skiffs, and $P$ is the profit, then recall that the linear programming problem was as follows:

$$
\begin{aligned}
\text{Maximize} \quad & P = 60x_1 + 80x_2 & & \text{objective function} \\
\text{Subject to} \quad & 2x_1 + 4x_2 \le 80 & & \text{cutting constraint} \\
& 2x_1 + 2x_2 \le 50 & & \text{painting constraint} \\
& 4x_1 + 2x_2 \le 84 & & \text{assembly constraint} \\
& x_1, x_2 \ge 0 & & \text{nonnegativity constraint.}
\end{aligned}
$$

The first constraint represents the constraint from the cutting operation, the second from the painting operation, and the third from the assembly operation. The feasible region, found in the last chapter, is shown in Figure 4.1.

Given the inequality $4 \le 7$, we can define $s = 7 - 4 = 3$ so that $4 + s = 7$. Thus the inequality has been converted to an *equality*, where $s$ takes up the slack and is referred to as a **slack variable**.

An important element in the simplex method is to move from a system of *inequalities* to a system of *equalities*. To accomplish this, we introduce a slack variable for each of the constraints. For example, for the first constraint define the first slack variable, $s_1$, so that

$$
2x_1 + 4x_2 + s_1 = 80.
$$

This can be done if $s_1 = 80 - 2x_1 - 4x_2$. Thus the inequality $2x_1 + 4x_2 \le 80$ becomes the equality $2x_1 + 4x_2 + s_1 = 80$, with $s_1$ taking up the slack. Notice that the original inequality $2x_1 + 4x_2 \le 80$ is true if and only if $s_1 \ge 0$.

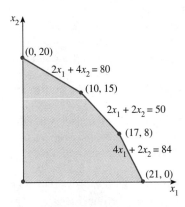

**Figure 4.1**

Doing the same for the other two constraints then leads to the three inequalities being converted to the following three equalities:

$$\left.\begin{array}{l} 2x_1 + 4x_2 + s_1 \phantom{+ s_2 + s_3} = 80 \\ 2x_1 + 2x_2 \phantom{+ s_1} + s_2 \phantom{+ s_3} = 50 \\ 4x_1 + 2x_2 \phantom{+ s_1 + s_2} + s_3 = 84 \end{array}\right\} \quad (2)$$

with the slack variables satisfying

$$\left.\begin{array}{l} s_1 = 80 - 2x_1 - 4x_2 \\ s_2 = 50 - 2x_1 - 2x_2 \\ s_3 = 84 - 4x_1 - 2x_2 \end{array}\right\}. \quad (3)$$

A solution of the system (2) consists of five numbers $(x_1, x_2, s_1, s_2, s_3)$.

In general we have the following.

---

**Slack Variables**

For each one of the constraints in problem (1) (not including the nonnegative constraints) one creates a slack variable. Using the slack variables $s_1$, $s_2$, $\ldots$, $s_m$, convert the constraints in problem (1) that involve inequalities to the following system of $m$ linear equations:

$$\left.\begin{array}{l} a_{11}x_1 + a_{12}x_2 + \cdots + a_{1n}x_n + s_1 \phantom{+ s_2} \phantom{+ s_m} = b_1 \\ a_{21}x_1 + a_{22}x_2 + \cdots + a_{2n}x_n \phantom{+ s_1} + s_2 \phantom{+ s_m} = b_2 \\ \phantom{xxxxxxxxxx} \cdots \\ a_{m1}x_1 + a_{m2}x_2 + \cdots + a_{mn}x_n \phantom{+ s_1 + s_2} + s_m = b_m \end{array}\right\}. \quad (4)$$

---

There are $m$ equations in system (4), and, therefore, the number of variables $x_1, \ldots, x_n, s_1, \ldots, s_m$ in system (4) exceeds the number of equations. Recall from the last chapter that this means that either there is no solution or an infinite number of solutions. In the system (2), for example, one readily sees that $x_1 = 0, x_2 = 0, s_1 = 80, s_2 = 50, s_3 = 84$ is a solution and since this system has at least one solution, the system (2) has an **infinite** number of solutions. We seek the solution that will maximize the objective function subject to the constraints on the decision variables.

We now consider the importance of determining whether the slack variables are negative or not. For example, notice that $s_1 = 0$ is the same as $2x_1 + 4x_2 = 80$. See Figure 4.2a. From our knowledge of linear inequalities, if $s_1 < 0$, and thus, $2x_1 + 4x_2 > 80$, then the point $(x_1, x_2)$ lies *above* the line $2x_1 + 4x_2 = 80$, and cannot be in the feasible region. On the other hand if $s_1 \geq 0$, then $2x_1 + 4x_2 \leq 80$ and the point $(x_1, x_2)$ lies *below or on* the line $2x_1 + 4x_2 = 80$ and has a chance of being in the feasible region of the system of inequalities.

(a)

(b)

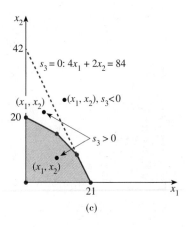

(c)

**Figure 4.2**

In the same way, if $s_2 < 0$, then the point $(x_1, x_2)$ lies above the line $2x_1 + 2x_2 = 50$ and cannot be in the feasible region, while if $s_2 \geq 0$, the point $(x_1, x_2)$ lies below or on the line $2x_1 + 2x_2 = 50$ and thus has a chance of lying in the feasible region. See Figure 4.2b.

Finally, if $s_3 < 0$, then the point $(x_1, x_2)$ lies above the line $4x_1 + 2x_2 = 84$ and cannot be in the feasible region, while if $s_3 \geq 0$, the point $(x_1, x_2)$ lies below or on the line $4x_1 + 2x_2 = 84$ and thus has a chance of lying in the feasible region. See Figure 4.2c.

This leads to the extremely important conclusion that, if any one of the slack variables (or, of course, any one of the decision variables $x_1, x_2$) is negative, then the point $(x_1, x_2)$ *cannot be in the feasible region*. And also if all of the slack variables (and of course all of the decision variables $x_1, x_2$) are nonnegative, then the point $(x_1, x_2)$ *will be in the feasible region*.

In general, we have the following.

---

**Algebraic Criteria for Determining the Feasible Region**

The point $(x_1, x_2, \ldots, x_n)$ will be in the feasible region of problem (1) if and only if all the slack variables $s_1, s_2, \ldots, s_m$ together with all the decision variables $x_1, x_2, \ldots, x_n$ are nonnegative.

---

For the wooden-boat manufacturing problem, this means that we have rewritten the original linear programming problem with inequality problem constraints to the following:

$$
\begin{aligned}
\text{Maximize} \qquad & P = 60x_1 + 80x_2 \\
\text{Subject to} \qquad & 2x_1 + 4x_2 + s_1 = 80 \\
& 2x_1 + 2x_2 + s_2 = 50 \\
& 4x_1 + 2x_2 + s_3 = 84 \\
& x_1, x_2, s_1, s_2, s_3 \geq 0.
\end{aligned}
$$

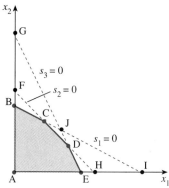

A = (0, 0): $s_1 = 80$, $s_2 = 50$, $s_3 = 84$

B = (0, 20): $s_1 = 0$, $s_2 = 10$, $s_3 = 44$

C = (10, 15): $s_1 = 0$, $s_2 = 0$, $s_3 = 14$

D = (17, 8): $s_1 = 14$, $s_2 = 0$, $s_3 = 0$

E = (21, 0): $s_1 = 38$, $s_2 = 8$, $s_3 = 0$

F = (0, 25): $s_1 = -20$, $s_2 = 0$, $s_3 = 34$

G = (0, 42): $s_1 = -88$, $s_2 = -34$, $s_3 = 0$

H = (25, 0): $s_1 = 30$, $s_2 = 0$, $s_3 = -16$

I = (40, 0): $s_1 = 0$, $s_2 = -30$, $s_3 = -76$

J = $(14\frac{2}{3}, 12\frac{2}{3})$: $s_1 = 0$, $s_2 = -4\frac{2}{3}$, $s_3 = 0$

**Figure 4.3**

## Basic and Nonbasic Variables

We saw in the last chapter that if the objective function in a linear programming problem attains a maximum or a minimum, then this maximum or minimum must occur at a corner point. Thus it is imperative that we be able to locate corner points. But we have abandoned the geometric method as inadequate for typical real applications. Thus we need an *algebraic* method that can be programmed on a computer for locating corner points. We have the first ingredient of such a method. Namely, the point $(x_1, x_2, \ldots, x_n)$ will at least be in the feasible region of problem (1) if, and only if, all the slack variables together with all the decision variables $x_1, x_2, \ldots, x_n$ are nonnegative.

Notice from Figure 4.3 that the feasible region in the wooden-boat manufacturing problem is bounded by the five straight lines $x_1 = 0$, $x_2 = 0$, $s_1 = 0$, $s_2 = 0$, $s_3 = 0$. Any corner point must be the point of intersection of two of these lines. For example, from Figure 4.3 we see that the corner point B is the intersection of $x_1 = 0$ and $s_1 = 0$, while the corner point C is the intersection of $s_1 = 0$ and $s_2 = 0$. Not all intersections of two of these five lines yield corner points. For example, the point of intersection, J, of the two lines $s_1 = 0$ and $s_3 = 0$ is *not* a corner point.

We then have the following procedure for finding the corner points. Determine the number of unknowns in the problem (1). For the wooden-boat manufacturer this is two. We then arbitrarily pick the same number, in this case two, of the variables $x_1, x_2, s_1, s_2, s_3$ and call them the **nonbasic variables**. Call the remaining three variables the **basic variables**. Now set the two nonbasic variables equal to zero and try to solve the three equations found in (2) for the three basic variables. Since we are dealing with a system of three equations in three unknowns, the system may have one solution, an infinite number of solutions, or no solution. If there is a solution and all five of the numbers $x_1, x_2, s_1, s_2, s_3$ are nonnegative, then the point $(x_1, x_2)$ is a corner point; otherwise it is not.

In general we have the following.

---

**Basic and Nonbasic Variables**

For problem (1) arbitrarily pick $n$ of the variables $x_1, \ldots, x_n, s_1, \ldots, s_m$, and call them the *nonbasic variables*. Call the remaining variables the *basic variables*. Set the $n$ nonbasic variables equal to zero in the system (4) and try to solve for the basic variables. If a solution exists and if all the numbers $x_1, \ldots, x_n, s_1, \ldots, s_m$, are nonnegative, then the point $(x_1, \ldots, x_n)$ is a corner point.

---

For example, in the wooden boat manufacturing problem pick $x_1$ and $s_1$ as the nonbasic variables. Setting $x_1 = 0$ and $s_1 = 0$ in the system (2), yields from the first equation in (2), $2(0) + 4x_2 + 0 = 80$, or $x_2 = 20$. This in turn yields from (3) that

$$s_2 = 50 - 2x_1 - 2x_2 = 50 - 2(0) - 2(20) = 10$$

$$s_3 = 84 - 4x_1 - 2x_2 = 84 - 4(0) - 2(20) = 44$$

Thus $(0, 20, 0, 10, 44)$ is a solution of (2). Since all of these numbers are nonnegative, the point $(0, 20)$ must be a corner point. From Figure 4.3 we see that this point is $B$ and is indeed a corner point.

For another example, pick $x_1$ and $s_2$ as the nonbasic variables. Then setting $x_1 = 0$ and $s_2 = 0$, yields, from the second equation in (2), $2(0) + 2x_2 + 0 = 50$, or $x_2 = 25$. This in turn yields from (3) that

$$s_1 = 80 - 2x_1 - 4x_2 = 80 - 2(0) - 4(25) = -20$$

$$s_3 = 84 - 4x_1 - 2x_2 = 84 - 4(0) - 2(25) = 34$$

Thus $(0, 25, -20, 0, 34)$ is a solution of (2). Since one of these numbers is negative, the point $(0, 25)$ must not be a corner point. From Figure 4.3 we see that this point is $F$ and is indeed not a corner point.

By systematically selecting two of the five variables to be the nonbasic variables and then setting them equal to zero in system (2) and solving for the three basic variables, we can obtain the results in Table 4.1.

**Table 4.1**

| $x_1$ | $x_2$ | $s_1$ | $s_2$ | $s_3$ | Point $(x_1, x_2)$ | | |
|---|---|---|---|---|---|---|---|
| 0 | 0 | 80 | 50 | 84 | A(0, 0) | corner | $s_1, s_2, s_3 \geq 0$ |
| 0 | 20 | 0 | 10 | 44 | B(0, 20) | corner | $s_1, s_2, s_3 \geq 0$ |
| 10 | 15 | 0 | 0 | 14 | C(10, 15) | corner | $s_1, s_2, s_3 \geq 0$ |
| 17 | 8 | 14 | 0 | 0 | D(17, 8) | corner | $s_1, s_2, s_3 \geq 0$ |
| 21 | 0 | 38 | 8 | 0 | E(21, 0) | corner | $s_1, s_2, s_3 \geq 0$ |
| 0 | 25 | −20 | 0 | 34 | F(0, 25) | not corner | $s_1 < 0$ |
| 0 | 42 | −88 | −34 | 0 | G(0, 42) | not corner | $s_1, s_2 < 0$ |
| 25 | 0 | 30 | 0 | −16 | H(25, 0) | not corner | $s_3 < 0$ |
| 40 | 0 | 0 | −30 | −76 | I(40, 0) | not corner | $s_2, s_3 < 0$ |
| $14\frac{2}{3}$ | $12\frac{2}{3}$ | 0 | $-4\frac{2}{3}$ | 0 | J($14\frac{2}{3}, 12\frac{2}{3}$) | not corner | $s_2 < 0$ |

Notice again from the table that the corner points have all variables nonnegative, while the points that are not corner points have at least one of the variables negative.

## The Simplex Tableau and Pivoting

We now rewrite the equation $P = z = 60x_1 + 80x_2$ as

$$-60x_1 - 80x_2 + z = 0$$

and add it to the system (2) and obtain

$$
\left.\begin{array}{rrrrr}
2x_1 + 4x_2 + s_1 & & & = 80 \\
2x_1 + 2x_2 & + s_2 & & = 50 \\
4x_1 + 2x_2 & & + s_3 & = 84 \\
-60x_1 - 80x_2 & & + z & = 0
\end{array}\right\}. \tag{5}
$$

For convenience we rewrite (5) as the following tableau.

| $x_1$ | $x_2$ | $s_1$ | $s_2$ | $s_3$ | $z$ | |
|---|---|---|---|---|---|---|
| 2 | 4 | 1 | 0 | 0 | 0 | 80 |
| 2 | 2 | 0 | 1 | 0 | 0 | 50 |
| 4 | 2 | 0 | 0 | 1 | 0 | 84 |
| $-60$ | $-80$ | 0 | 0 | 0 | 1 | 0 |

The simplex method will give a sequence of such tableaus. We now see how to associate a corner point with each such tableau. Recall that selecting 2 of the variables $x_1, x_2, s_1, s_2, s_3$ as the nonbasic variables and setting them equal to zero in the system (2) may or may not lead to a corner point. We want a procedure for selecting nonbasic variables that *will* lead to a corner point.

The key to doing this is to recognize unit and nonunit columns. A **unit column** is a column in the tableau with all entries zero except for one entry that is a one. In determining unit and nonunit columns we *ignore the last column of the tableau*. Thus in the above tableau the first and second columns are the nonunit columns and the third, fourth, fifth, and sixth columns are the unit columns.

| *Nonunit* | *Nonunit* | *Unit* | *Unit* | *Unit* | *Unit* | |
|---|---|---|---|---|---|---|
| $x_1$ | $x_2$ | $s_1$ | $s_2$ | $s_3$ | $z$ | |
| 2 | 4 | 1 | 0 | 0 | 0 | 80 |
| 2 | 2 | 0 | 1 | 0 | 0 | 50 |
| 4 | 2 | 0 | 0 | 1 | 0 | 84 |
| $-60$ | $-80$ | 0 | 0 | 0 | 1 | 0 |

*Pick the variables associated with the nonunit columns as the nonbasic variables.* This is $x_1$ and $x_2$ in the above tableau. Now set $x_1 = 0$ and $x_2 = 0$. Then we can immediately see the solution to the resulting system of equations.

$$
\begin{array}{l}
\boxed{2x_1} + \boxed{4x_2} + s_1 \phantom{+ s_2 + s_3 + z} = 80 \\
\boxed{2x_1} + \boxed{2x_2} \phantom{+ s_1} + s_2 \phantom{+ s_3 + z} = 50 \\
\boxed{4x_1} + \boxed{2x_2} \phantom{+ s_1 + s_2} + s_3 \phantom{+ z} = 84 \\
\boxed{-60x_1} - \boxed{80x_2} \phantom{+ s_1 + s_2 + s_3} + z = 0
\end{array}
$$

We then have $s_1 = 80$, $s_2 = 50$, $s_3 = 84$, and $z = 0$. This point corresponds to the origin $(0, 0)$ with a profit of zero. The origin is definitely a corner point. We can see this from the figures or from the fact that all five of the coordinates $x_1$, $x_2$, $s_1$, $s_2$, $s_3$ are nonnegative.

As we shall see in the next section, the simplex method will use elementary row operations to pivot about a certain element in the tableau. For example pivoting simply replaces the original equations (2) with an equivalent set of equations in which four of the columns are unit columns and two of the columns are nonunit columns. It is the variables associated with the latter two columns that we pick as the nonbasic variables and set equal to zero. The resulting equations will then immediately yield the remaining four basic variables. If the pivot element is chosen properly, the resulting point will actually be an adjacent corner point. Furthermore, the profit at the new corner point will be at least as great as the profit at the previous corner point. Thus the simplex method will move us from corner point to adjacent corner point, always improving the profit (or at least not decreasing the profit) until the maximum is reached.

In the next section we will determine the criteria needed to determine the column that the pivot element is in and the row that the pivot element is in. In this section we merely see how to pivot. In the process we will pivot about some less optimal elements (considering what we will learn in the next section), but we will then be able to compare what happens when pivoting about an optimal element with what happens when pivoting about a less optimal element.

We first pivot about the element in the first row and second column. We will see in the next section that this is the appropriate pivot. We want the pivot element to be a 1. This requires, in this case, dividing the first row by 4. Doing this we obtain the tableau:

| $x_1$ | $x_2$ | $s_1$ | $s_2$ | $s_3$ | $z$ | | |
|-------|-------|-------|-------|-------|-----|-----|-----|
| .50 | ① | .25 | 0 | 0 | 0 | 20 | |
| 2 | 2 | 0 | 1 | 0 | 0 | 50 | $R_2 - 2R_1 \rightarrow R_2$ |
| 4 | 2 | 0 | 0 | 1 | 0 | 84 | $R_3 - 2R_1 \rightarrow R_3$ |
| $-60$ | $-80$ | 0 | 0 | 0 | 1 | 0 | $R_4 + 80R_1 \rightarrow R_4$ |

Now our goal is to **obtain zeros in all the entries of the pivot column other than the pivot element**. We accomplish this by the row operations indicated to the right of the above tableau. This then leads to the new tableau:

| $x_1$ | $x_2$ | $s_1$ | $s_2$ | $s_3$ | $z$ | |
|---|---|---|---|---|---|---|
| .50 | 1 | .25 | 0 | 0 | 0 | 20 |
| 1 | 0 | −.50 | 1 | 0 | 0 | 10 |
| 3 | 0 | −.50 | 0 | 1 | 0 | 44 |
| −20 | 0 | 20 | 0 | 0 | 1 | 1600 |

Now notice from this last tableau that the second, fourth, fifth, and sixth columns are the unit columns and the first and third columns are the nonunit columns. These latter nonunit columns determine that the nonbasic variables are $x_1$ and $s_1$. If we now set the nonbasic variables equal to zero, then we can immediately solve for the remaining variables (which are associated with the unit columns and are the basic variables).

$$
\begin{aligned}
.50x_1 + x_2 + .25s_1 &= 20 \\
x_1 - .50s_1 + s_2 &= 10 \\
3x_1 - .50s_1 + s_3 &= 44 \\
-20x_1 + 20s_1 + z &= 1600
\end{aligned}
$$

We then obtain $x_1 = 0$, $x_2 = 20$, $s_1 = 0$, $s_2 = 10$, $s_3 = 44$. Since all of these variables are nonnegative, this point is a corner point. We recall that it is the point $B$ in the figures. Also notice from the last tableau that setting both nonbasic variables $x_1$ and $s_1$ equal to zero yields $z = 1600$. Thus we have moved from the origin where the profit is zero to an adjacent corner point $B$ where the profit is 1600.

We have the following rule for determining the solution from the tableaus.

---

**Determining the Solution from the Tableau**

1. For problem (1) pick $n$ of the variables $x_1, \ldots, x_n, s_1, \ldots, s_m$, to be the nonbasic variables. The remaining $m$ variables together with $z$ are the basic variables.
2. a. Let the nonbasic variables be variables that correspond to nonunit columns.
   b. If there are not $n$ nonunit columns, then we arbitrarily select some of the variables (but never $z$) associated with unit columns to be nonbasic variables so that there are $n$ nonbasic variables and so that no two basic variables have identical associated columns.
3. Set all of the nonbasic variables equal to zero.
4. We can then easily find the values of the basic variables from the tableau.

---

In the next section we shall see how to proceed on to the maximum. For now let us pivot about some other point and see what the consequences are.

**E X A M P L E   1**     **Pivoting**
_____

Pivot about the element in the third row and first column; determine the basic and nonbasic variables and the point that corresponds to the resulting tableau.

**Solution**     We first divide the third row by 4 and obtain a 1 in the pivot position.

| $x_1$ | $x_2$ | $s_1$ | $s_2$ | $s_3$ | $z$ | | |
|-------|-------|-------|-------|-------|-----|-----|---|
| 2 | 4 | 1 | 0 | 0 | 0 | 80 | |
| 2 | 2 | 0 | 1 | 0 | 0 | 50 | |
| ④ | 2 | 0 | 0 | 1 | 0 | 84 | $0.25R_3 \to R_3$ |
| $-60$ | $-80$ | 0 | 0 | 0 | 1 | 0 | |

| $x_1$ | $x_2$ | $s_1$ | $s_2$ | $s_3$ | $z$ | | |
|-------|-------|-------|-------|-------|-----|-----|---|
| 2 | 4 | 1 | 0 | 0 | 0 | 80 | $R_1 - 2R_3 \to R_1$ |
| 2 | 2 | 0 | 1 | 0 | 0 | 50 | $R_2 - 2R_3 \to R_2$ |
| ① | .50 | 0 | 0 | .25 | 0 | 21 | |
| $-60$ | $-80$ | 0 | 0 | 0 | 1 | 0 | $R_4 + 60R_3 \to R_4$ |

Now the indicated row operations will place zeros in all elements in the first column except for the pivot element.

| $x_1$ | $x_2$ | $s_1$ | $s_2$ | $s_3$ | $z$ | |
|-------|-------|-------|-------|-------|-----|------|
| 0 | 3 | 1 | 0 | $-.50$ | 0 | 38 |
| 0 | 1 | 0 | 1 | $-.50$ | 0 | 8 |
| 1 | .50 | 0 | 0 | .25 | 0 | 21 |
| 0 | $-50$ | 0 | 0 | 15 | 1 | 1260 |

This completes the pivoting.

We now see what solution this tableau corresponds to. The second and fifth columns are the nonunit columns. Thus the nonbasic variables are $x_2$ and $s_3$. The first, third, fourth, and sixth columns are the unit columns. Thus the basic variables are $x_1$, $s_1$, $s_2$, and $z$. Now we set both of the nonbasic variables $x_2$ and $s_3$ equal to zero. This immediately yields $x_1 = 21$, $x_2 = 0$, $s_1 = 38$, $s_2 = 8$, $s_3 = 0$. Since all of these variables are nonnegative, this is a corner point. In fact this is the point $E$ in Figure 4.3.  ∎

Notice that $E$ is the other corner adjacent to the origin. From the last tableau we see that the profit is only 1260. This is less than the profit of 1600 obtained at the point $B$. It is for this reason that this is the less optimal point, and the pivot element used here is the less optimal element. We will have much more to say about this point in the next section.

In the next section we will see that certain ratios play an important role in determining which is the pivot row. Let us examine one of these ratios and see what the ratio means. In the last example we pivoted about the element 4 in the third row and first column. The only operation that was done to the third row was division by 4. All the other elements in the first column became zero. Looking only at the third row and first column, we ended up with

| $x_1$ | $x_2$ | $s_1$ | $s_2$ | $s_3$ | $z$ | |
|---|---|---|---|---|---|---|
| 0 | | | | | | |
| 0 | | | | | | |
| 1 | .50 | 0 | 0 | .25 | 0 | $\frac{84}{4}$ |
| 0 | | | | | | |

It is important to notice that $x_1$ then will equal the ratio $\frac{84}{4}$.

**E X A M P L E   2**     **Pivoting and Finding Basic and Nonbasic Variables**

Pivot about the element in the second row and second column, find the basic and nonbasic variables and the point that corresponds to the tableau.

**Solution**     Start by dividing the second row by 2 and then perform row operations to obtain zeros in the second column except for the pivot element.

| $x_1$ | $x_2$ | $s_1$ | $s_2$ | $s_3$ | $z$ | | |
|---|---|---|---|---|---|---|---|
| 2 | 4 | 1 | 0 | 0 | 0 | 80 | |
| 2 | ② | 0 | 1 | 0 | 0 | 50 | $0.50R_2 \rightarrow R_2$ |
| 4 | 2 | 0 | 0 | 1 | 0 | 84 | |
| $-60$ | $-80$ | 0 | 0 | 0 | 1 | 0 | |

| $x_1$ | $x_2$ | $s_1$ | $s_2$ | $s_3$ | $z$ | | |
|---|---|---|---|---|---|---|---|
| 2 | 4 | 1 | 0 | 0 | 0 | 80 | $R_1 - 4R_2 \rightarrow R_1$ |
| 1 | ① | 0 | .50 | 0 | 0 | 25 | |
| 4 | 2 | 0 | 0 | 1 | 0 | 84 | $R_3 - 2R_2 \rightarrow R_3$ |
| $-60$ | $-80$ | 0 | 0 | 0 | 1 | 0 | $R_4 + 80R_2 \rightarrow R_4$ |

| $x_1$ | $x_2$ | $s_1$ | $s_2$ | $s_3$ | $z$ | |
|---|---|---|---|---|---|---|
| $-2$ | 0 | 1 | $-2$ | 0 | 0 | $-20$ |
| 1 | 1 | 0 | .50 | 0 | 0 | 25 |
| 2 | 0 | 0 | $-1$ | 1 | 0 | 34 |
| 20 | 0 | 0 | 40 | 0 | 1 | 2000 |

Taking the nonbasic variables to be $x_1$ and $s_2$, since the first and fourth columns are nonunit columns, and setting these two variables equal to zero then yields $x_1 = 0$, $x_2 = 25$, $s_1 = -20$, $s_2 = 0$, $s_3 = 34$. Notice that one of these variables is negative, and therefore the point $(x_1, x_2)$ cannot be a corner point. From the figures, we see that this is the point $F$.  ■

Once again let us consider a certain ratio. In the last example we pivoted about the element 2 in the second row and second column. The only operation performed on the second row was division by 2. The other elements in the second column became zero after the pivot. Thus the second row and second column became

| $x_1$ | $x_2$ | $s_1$ | $s_2$ | $s_3$ | $z$ | |
|---|---|---|---|---|---|---|
| | 0 | | | | | |
| 1 | 1 | 0 | .50 | 0 | 0 | $\frac{50}{2}$ |
| | 0 | | | | | |
| | 0 | | | | | |

Notice that the value of $x_2$ becomes the ratio $\frac{50}{2}$.

## SELF-HELP EXERCISE SET 4.1

1. Consider the following problem:

$$\text{Maximize} \quad z = 20x_1 + 35x_2$$
$$\text{Subject to} \quad 2x_1 + 3x_2 \le 20$$
$$2x_1 + 5x_2 \le 10$$
$$x_1, x_2 \ge 0$$

Introduce slack variables, converting the set of inequalities to a set of equalities. Also set up the corresponding tableau.

2. For the wooden-boat manufacturing problem considered in this section, take $x_1$ and $s_3$ as the nonbasic variables. Set the nonbasic variables equal to zero and solve equation (2) for the remaining variables.

3.  For the wooden-boat manufacturing problem considered in this section, pivot about the element in the third row and second column. Determine if the point corresponding to the resulting tableau is a corner point of the feasible region.

## EXERCISE SET 4.1

1.  Consider the following linear programming problem:

$$\text{Maximize} \quad z = 3x_1 + 2x_2$$

$$\text{Subject to} \quad x_1 + x_2 \leq 6$$

$$2x_1 + x_2 \leq 8$$

$$x_1, x_2 \geq 0.$$

   a.  Introduce two slack variables, $s_1$ and $s_2$ and convert the set of inequalities to a set of equalities.
   b.  Set up the corresponding tableau.
   c.  Graph the feasible region and locate all corner points.
   d.  Successively set two of the variables $x_1$, $x_2$, $s_1$, $s_2$ equal to zero and solve the equations found in part (a) for the remaining variables. Verify that a point is a corner point if, and only if, none of these variables are negative.

2.  Consider the following linear programming problem:

$$\text{Maximize} \quad z = 10x_1 + 20x_2$$

$$\text{Subject to} \quad x_1 + 2x_2 \leq 20$$

$$4x_1 + x_2 \leq 24$$

$$x_1, x_2 \geq 0$$

   Answer all the questions posed in the previous exercise.

3.  Consider the following linear programming problem:

$$\text{Maximize} \quad z = 20x_1 + 40x_2$$

$$\text{Subject to} \quad x_1 + 3x_2 \leq 30$$

$$x_1 + x_2 \leq 12$$

$$4x_1 + x_2 \leq 24$$

$$x_1, x_2 \geq 0$$

   Answer all the questions posed in the previous exercise. (There will be three slack variables.)

4.  Verify Table 4.1 in the text.

In Exercises 5 through 8 pivot about the element indicated for the tableau.

| $x_1$ | $x_2$ | $s_1$ | $s_2$ | $z$ | |
|---|---|---|---|---|---|
| 2 | 4 | 1 | 0 | 0 | 8 |
| 5 | 1 | 0 | 1 | 0 | 20 |
| −4 | −2 | 0 | 0 | 1 | 0 |

**5.** First row and first column     **6.** First row and second column

**7.** Second row and first column     **8.** Second row and second column

In Exercises 9 through 14 pivot about the element indicated for the tableau.

| $x_1$ | $x_2$ | $s_1$ | $s_2$ | $s_3$ | $z$ | |
|---|---|---|---|---|---|---|
| 2 | 4 | 1 | 0 | 0 | 0 | 8 |
| 5 | 1 | 0 | 1 | 0 | 0 | 25 |
| 3 | 6 | 0 | 0 | 1 | 0 | 15 |
| $-3$ | $-2$ | 0 | 0 | 0 | 1 | 0 |

**9.** First row and first column     **10.** First row and second column

**11.** Second row and first column     **12.** Second row and second column

**13.** Third row and first column     **14.** Third row and second column

In Exercises 15 through 24 find the basic and nonbasic variables according to the rules given in the text. Set the nonbasic variables equal to zero and solve. Determine if the point is a corner point.

**15.**

| $x_1$ | $x_2$ | $s_1$ | $s_2$ | $z$ | |
|---|---|---|---|---|---|
| 2 | 4 | 1 | 0 | 0 | 10 |
| 5 | 1 | 0 | 1 | 0 | 20 |
| $-4$ | $-2$ | 0 | 0 | 1 | 0 |

**16.**

| $x_1$ | $x_2$ | $s_1$ | $s_2$ | $z$ | |
|---|---|---|---|---|---|
| 3 | 5 | 1 | 0 | 0 | 15 |
| 2 | 2 | 0 | 1 | 0 | 8 |
| $-8$ | $-5$ | 0 | 0 | 1 | 0 |

**17.**

| $x_1$ | $x_2$ | $s_1$ | $s_2$ | $z$ | |
|---|---|---|---|---|---|
| 2 | 1 | 5 | 0 | 0 | 5 |
| 5 | 0 | 5 | 1 | 0 | 10 |
| $-7$ | 0 | $-3$ | 0 | 1 | 0 |

**18.**

| $x_1$ | $x_2$ | $s_1$ | $s_2$ | $z$ | |
|---|---|---|---|---|---|
| 2 | 0 | 1 | 2 | 0 | $-2$ |
| $-4$ | 1 | 0 | 5 | 0 | 4 |
| 9 | 0 | 0 | 0 | 1 | 0 |

**19.**

| $x_1$ | $x_2$ | $s_1$ | $s_2$ | $z$ | |
|---|---|---|---|---|---|
| 0 | 4 | 1 | 2 | 0 | $-2$ |
| 1 | 3 | 0 | 3 | 0 | 4 |
| 0 | $-5$ | 0 | $-2$ | 1 | 0 |

**20.**

| $x_1$ | $x_2$ | $s_1$ | $s_2$ | $z$ | |
|---|---|---|---|---|---|
| 1 | 10 | 1 | 0 | 0 | 50 |
| 0 | 10 | 30 | 1 | 0 | 20 |
| 0 | $-5$ | 0 | 0 | 1 | 0 |

**21.**

| $x_1$ | $x_2$ | $s_1$ | $s_2$ | $s_3$ | $z$ | |
|---|---|---|---|---|---|---|
| $-5$ | 0 | 1 | 0 | 2 | 0 | 21 |
| $-1$ | 0 | 0 | 1 | 1 | 0 | 30 |
| 2 | 1 | 0 | 0 | 8 | 0 | 2 |
| $-10$ | 0 | 0 | 0 | 40 | 1 | 30 |

**22.**

| $x_1$ | $x_2$ | $s_1$ | $s_2$ | $s_3$ | $z$ | |
|---|---|---|---|---|---|---|
| 1 | 0 | 0 | 1 | 1 | 0 | 5 |
| 3 | 0 | 1 | $-2$ | 0 | 0 | $-3$ |
| 7 | 1 | 0 | 1 | 0 | 0 | 6 |
| $-3$ | 0 | 0 | $-4$ | 0 | 1 | 10 |

**23.**

| $x_1$ | $x_2$ | $s_1$ | $s_2$ | $s_3$ | $z$ | |
|---|---|---|---|---|---|---|
| 0 | 1 | 1 | 3 | 0 | 0 | 0 |
| 1 | 8 | 0 | 6 | 0 | 0 | 2 |
| 0 | 1 | 0 | 5 | 1 | 0 | $-5$ |
| 0 | 0 | 0 | $-2$ | 0 | 1 | 8 |

**24.**

| $x_1$ | $x_2$ | $s_1$ | $s_2$ | $s_3$ | $z$ | |
|---|---|---|---|---|---|---|
| 0 | 1 | 0 | 1 | 5 | 0 | 10 |
| 0 | 5 | 1 | 0 | 2 | 0 | 4 |
| 1 | 2 | 0 | 0 | 1 | 0 | 1 |
| 0 | $-1$ | 0 | 0 | $-2$ | 1 | 10 |

## Solutions to Self-Help Exercise Set 4.1

**1.** Introduce two slack variables, rewrite the equation for the objective function, and obtain

$$2x_1 + 3x_2 + s_1 \qquad = 20$$
$$2x_1 + 5x_2 \qquad s_2 \quad = 10$$
$$-20x_1 - 35x_2 \qquad\qquad z = 0$$

The tableau is then

| $x_1$ | $x_2$ | $s_1$ | $s_2$ | $z$ | |
|---|---|---|---|---|---|
| 2 | 3 | 1 | 0 | 0 | 20 |
| 2 | 5 | 0 | 1 | 0 | 10 |
| $-20$ | $-35$ | 0 | 0 | 1 | 0 |

**2.** Setting $x_1 = 0$ and $s_3 = 0$ in the third equation in (2) yields $4(0) + 2x_2 = 84$, or $x_2 = 42$. Now use equation (3) to solve for $s_1$ and $s_2$ and obtain

$$s_1 = 80 - 2x_1 - 4x_2 = 80 - 2(0) - 4(42) = -88$$
$$s_2 = 50 - 2x_1 - 2x_2 = 50 - 2(0) - 2(42) = -34.$$

This yields the point $(0, 42, -88, -34, 0)$, which is the point $G$ in the figures. Since at least one of these coordinates is negative, the point $(0, 42)$ is not feasible and not a corner point.

**3.**

| $x_1$ | $x_2$ | $s_1$ | $s_2$ | $s_3$ | $z$ | | |
|---|---|---|---|---|---|---|---|
| 2 | 4 | 1 | 0 | 0 | 0 | 80 | |
| 2 | 2 | 0 | 1 | 0 | 0 | 50 | |
| 4 | ② | 0 | 0 | 1 | 0 | 84 | $0.50R_3 \to R_3$ |
| $-60$ | $-80$ | 0 | 0 | 0 | 1 | 0 | |

| $x_1$ | $x_2$ | $s_1$ | $s_2$ | $s_3$ | $z$ | | |
|---|---|---|---|---|---|---|---|
| 2 | 4 | 1 | 0 | 0 | 0 | 80 | $R_1 - 4R_3 \to R_1$ |
| 2 | 2 | 0 | 1 | 0 | 0 | 50 | $R_2 - 2R_3 \to R_2$ |
| 2 | ① | 0 | 0 | .50 | 0 | 42 | |
| $-60$ | $-80$ | 0 | 0 | 0 | 1 | 0 | $R_4 + 80R_3 \to R_4$ |

| $x_1$ | $x_2$ | $s_1$ | $s_2$ | $s_3$ | $z$ | |
|---|---|---|---|---|---|---|
| $-6$ | $0$ | $1$ | $0$ | $-2$ | $0$ | $-88$ |
| $-2$ | $0$ | $0$ | $1$ | $-1$ | $0$ | $-34$ |
| $2$ | $1$ | $0$ | $0$ | $.50$ | $0$ | $42$ |
| $100$ | $0$ | $0$ | $0$ | $40$ | $1$ | $3360$ |

The first and fifth columns are the nonunit columns. Thus take the nonbasic variables to be $x_1$ and $s_3$ and set these two variables equal to zero. This yields $x_1 = 0$, $x_2 = 42$, $s_1 = -88$, $s_2 = -34$, $s_3 = 0$. Notice that at least one of these variables is negative, and, therefore, the point $(x_1, x_2)$ cannot be a corner point. From the figures, we see that this point is the point $G$.

## 4.2 THE SIMPLEX METHOD: STANDARD MAXIMIZATION PROBLEMS

▶ *Standard Maximum Problems*
▶ *The Simplex Method*
▶ *Post-Optimal Analysis*

**Dantzig and the Simplex Method**

*(Photo by Edward W. Souza News Service, Stanford University, Stanford, CA.)*

As mentioned at the beginning of the last section, linear programming problems of even moderate size can have an unimaginably large number of corners to evaluate. In 1947 George Dantzig discovered the simplex algorithm for solving linear programming problems, which only requires evaluation of a relatively small number of corners to find the solution. Indeed Dantzig reported that most of the time the simplex method solved linear programming problems in $m$ equations in only $2m$ or $3m$ steps. The simplex method goes from corner to adjacent corner until the solution is found. In three dimensions the feasible region can be visualized as a cut and finished diamond with faces, edges, and corners. When Dantzig first formulated the method, he had little faith in it. He assumed that the method would spend an extremely large amount of time wandering from corner to corner and take forever to find the solution. In fact after formulating the method while with the Air Force, he left the lab he was at and spent some time visiting some top mathematicians and economists to find a more efficient method. When he returned to the lab in June 1948, his group informed him that the method was working with extreme efficiency on all the test problems. He was astounded. In fact, only in the last few years have people begun obtaining insight into why the simplex algorithm works so well.

### Standard Maximum Problems

We now develop the simplex method. This method is an algebraic method that can be readily programmed for use on computers. The method systematically and efficiently moves from one corner point of the feasible region to an adjacent corner point, always improving the objective function (or at least not making it worse) until the corner is found that maximizes the objective function.

In this section the simplex method is applied only to the so-called *standard maximum problems*. Subsequent sections will consider the simplex method for other problems. We have the following definition.

---

**Standard Maximum Problem**

A linear programming problem is said to be a **standard maximum problem** if the following are satisfied.

1. The decision variables are constrained to be nonnegative. This is called the *nonnegative condition*.
2. All of the problem constraint inequalities are $\leq$.
3. The constants in the problem constraints to the right of the inequality $\leq$ are never negative.
4. The objective function is to be maximized.

---

**E X A M P L E   1**    **Showing a Problem is a Standard Maximization Problem**

Show that the problem of the wooden-boat manufacturer of the last section is a standard maximization problem.

**Solution**    If $x_1$ is the number of dinghies produced, $x_2$ the number of skiffs, and $z$ is the profit, then recall that the linear programming problem was

$$\text{Maximize} \quad z = 60x_1 + 80x_2 \qquad \text{objective function}$$

$$\text{Subject to} \quad \left. \begin{array}{c} 2x_1 + 4x_2 \leq 80 \\ 2x_1 + 2x_2 \leq 50 \\ 4x_1 + 2x_2 \leq 84 \end{array} \right\} \qquad \text{problem constraints}$$

$$x_1, x_2 \geq 0 \qquad \text{nonnegativity constraint.}$$

It is then apparent that this problem satisfies all four conditions given above and is thus a standard maximization problem.  ∎

---

**The General Algebraic Form of the Standard Maximization Problem**

$$\text{Maximize} \qquad z = c_1x_1 + c_2x_2 + \cdots + c_nx_n$$

$$\text{Subject to} \quad a_{11}x_1 + a_{12}x_2 + \cdots + a_{1n}x_n \leq b_1$$

$$a_{21}x_1 + a_{22}x_2 + \cdots + a_{2n}x_n \leq b_2 \qquad (1)$$

$$\cdots$$

$$a_{m1}x_1 + a_{m2}x_2 + \cdots + a_{mn}x_n \leq b_m$$

$$x_1, x_2, \ldots, x_n \geq 0$$

where $b_1, b_2, \ldots, b_m$ are all nonnegative.

---

## The Simplex Method

The initial phases of the simplex method have already been completed in the previous section. First introduce a slack variable for every constraint and then form a simplex tableau.

For the problem of the wooden-boat manufacturer considered in the last section, this is

$$2x_1 + 4x_2 + s_1 \qquad\qquad = 80$$
$$2x_1 + 2x_2 \qquad + s_2 \qquad\qquad = 50$$
$$4x_1 + 2x_2 \qquad\qquad + s_3 \quad = 84$$
$$-60x_1 - 80x_2 \qquad\qquad\qquad + z = 0$$

and

| $x_1$ | $x_2$ | $s_1$ | $s_2$ | $s_3$ | $z$ | |
|---|---|---|---|---|---|---|
| 2 | 4 | 1 | 0 | 0 | 0 | 80 |
| 2 | 2 | 0 | 1 | 0 | 0 | 50 |
| 4 | 2 | 0 | 0 | 1 | 0 | 84 |
| $-60$ | $-80$ | 0 | 0 | 0 | 1 | 0 |

From what we learned in the last section, the nonbasic variables in the last tableau are taken to be $x_1$ and $x_2$. Setting $x_1 = 0$ and $x_2 = 0$ means we are starting at the point $(0, 0)$, which is a corner point. After setting $x_1 = 0$ and $x_2 = 0$, notice that $s_1$, $s_2$, and $s_3$ must be nonnegative since, in a standard maximization problem, the constants to the right of the constraint inequalities $\leq$ must be non-negative. In this case these constants are, respectively, 80, 50, and 84. Thus we do indeed begin with a corner point of the feasible region.

---

**Initial Step of Simplex Method**

Set up the initial tableau as indicated in the previous section. The initial tableau corresponds to setting all the variables $x_1, x_2, \ldots, x_n$ equal to zero and starting at the origin. The origin must be a feasible point in a standard maximization problem.

---

Now we must decide to which adjacent corner we will proceed. We must decide between setting $x_1 = 0$ and increasing $x_2$ or setting $x_2 = 0$ and increasing $x_1$. We wish to proceed to the adjacent corner that has the largest profit. The profit function is

$$z = 60x_1 + 80x_2.$$

Since 80 is larger than 60, it is clear that increasing $x_2$ by each unit will result in increasing the profit more than increasing $x_1$ by each unit. Thus we choose to increase $x_2$ and set $x_1 = 0$. In the tableau the profit equation appears as

$$-60x_1 - 80x_2 + z = 0.$$

Thus the correct choice is to *pick the variable with the largest negative value in the last row of the simplex tableau.*

| $x_1$ | $x_2$ | $s_1$ | $s_2$ | $s_3$ | $z$ | |
|-------|-------|-------|-------|-------|-----|-----|
| 2 | 4 | 1 | 0 | 0 | 0 | 80 |
| 2 | 2 | 0 | 1 | 0 | 0 | 50 |
| 4 | 2 | 0 | 0 | 1 | 0 | 84 |
| $-60$ | $-80$ | 0 | 0 | 0 | 1 | 0 |

$\uparrow$
pivot column

We have now established a general procedure for determining the pivot column.

> **Finding the Pivot Column**
>
> Select as the pivot column the column containing the largest negative value in the last row of the simplex tableau, ignoring the last entry. (If there is a tie, pick any one.)

Now that we have decided to increase $x_2$ while holding $x_1 = 0$, we must decide just how much to increase $x_2$. Naturally we wish to increase $x_2$ so that a corner point will be reached but do not wish to increase $x_2$ any further since the point will not be feasible. In algebraic terms, we wish to increase $x_2$ as much as possible without any of the slack variables becoming negative.

Recall that the slack variables are

$$s_1 = 80 - 2x_1 - 4x_2$$
$$s_2 = 50 - 2x_1 - 2x_2$$
$$s_3 = 84 - 4x_1 - 2x_2.$$

With $x_1 = 0$, this becomes

$$s_1 = 80 - 2(0) - 4x_2 = 80 - 4x_2$$
$$s_2 = 50 - 2(0) - 2x_2 = 50 - 2x_2$$
$$s_3 = 84 - 4(0) - 2x_2 = 84 - 2x_2.$$

Since the three slack variables must be nonnegative, the last set of equations becomes

$$80 - 4x_2 \geq 0$$

$$50 - 2x_2 \geq 0$$

$$84 - 2x_2 \geq 0.$$

Solving in terms of $x_2$, this becomes

$$\tfrac{80}{4} \geq x_2$$

$$\tfrac{50}{2} \geq x_2$$

$$\tfrac{84}{2} \geq x_2.$$

All three of these last equations will be satisfied if we use the smallest of the three ratios

$$\frac{80}{4} \quad \frac{50}{2} \quad \frac{84}{2}$$

This is $\tfrac{80}{4} = 20$.

If we then pivot about the first row, the row with the smallest of the ratios, we know from the last section that the new value for $x_2$ will be this ratio $\tfrac{80}{4} = 20$. Since we picked the smallest of the three ratios, we know that all the slack variables must be nonnegative. Thus we have a feasible point.

Recall from the last section that we pivoted this tableau about the first row and second column and obtained

| $x_1$ | $x_2$ | $s_1$ | $s_2$ | $s_3$ | $z$ |  |
|---|---|---|---|---|---|---|
| .50 | 1 | .25 | 0 | 0 | 0 | 20 |
| 1 | 0 | $-.50$ | 1 | 0 | 0 | 10 |
| 3 | 0 | $-.50$ | 0 | 1 | 0 | 44 |
| $-20$ | 0 | 20 | 0 | 0 | 1 | 1600 |

Notice that setting $x_1 = 0$ and $s_1 = 0$ immediately yields $x_2 = 20$, which is the first ratio. This gives the corner point we are seeking.

Had we pivoted about the second or third row of the second column as we did in Example 2 and Self-Help Exercise 3 of the last section, we would have increased $x_2$ too much and gone out of the feasible region. In fact $x_2$ becomes the second ratio in the first case and the third ratio in the second case.

In the problem under consideration, all the ratios are positive. Consider now the possibility that a ratio is negative. For example, suppose the third equation above is replaced with

$$4x_1 - 2x_2 + s_3 = 84.$$

(The only way the ratio can be negative is to have the coefficient of $x_2$ be negative, since in a standard maximization problem the number to the right of the inequality $\leq$, in this case the number 84, must be nonnegative.)

Setting $x_1 = 0$ as before then gives $s_3 = 84 + 2x_2$. Since $s_3$ must be nonnegative, this requires that $84 + 2x_2 \geq 0$, or $84 \geq -2x_2$. Dividing by $-2$ changes the sign of the inequality and yields

$$\frac{84}{-2} \leq x_2.$$

But this is true for any nonnegative $x_2$. Since this does not represent any restriction on $x_2$, *negative ratios can be ignored.*

Using similar arguments, it can be shown that a ratio with the denominator equal to zero does not impose any restriction on the variable. Thus *ratios obtained by dividing by zero can be ignored.* Also it can be shown that *zero ratios with division by a negative number can be ignored.*

Thus, when considering the ratios, we need only consider the nonnegative ratios, ignoring the negative ratios, the ratios that come from dividing by zero, and the zero ratios that are divided by a negative number.

We summarize these findings in the following rules.

---

**Finding the Pivot Row**

1.  For each row, except for the last, divide the entry in the last column by the corresponding entry in the pivot column.
2.  Choose as the pivot row the row with the smallest (nonnegative) ratio, ignoring zero ratios that are divided by a negative number.

---

Now that the pivot element has been found, perform the pivot operation in the manner shown in the last section. Now find the new pivot element and perform another pivot operation, and so on.

We now need a stopping procedure. If any element in the last row is negative (not including the last entry), we proceed since the value of $z$ can be increased some more. For example, from the last row of the last tableau we have that $-20x_1 + 20s_1 + z = 1600$ or $z = 20x_1 - 20s_1 + 1600$. Clearly $z$ can be increased some more by increasing $x_1$.

If there are no negative numbers in the last row (not including the last entry), we stop. In the last tableau there is one negative entry in the last row. Since there is precisely one negative entry, the pivot column is then just the column with this entry. Thus the pivot column is now the first column.

In order to find the pivot row, we proceed with finding the ratios.

| $x_1$ | $x_2$ | $s_1$ | $s_2$ | $s_3$ | $z$ | | ratios |
|---|---|---|---|---|---|---|---|
| .50 | 1 | .25 | 0 | 0 | 0 | 20 | $\frac{20}{.50} = 40$ |
| ① | 0 | $-.50$ | 1 | 0 | 0 | 10 | $\frac{10}{1} = 10 \leftarrow$ smallest ratio |
| 3 | 0 | $-.50$ | 0 | 1 | 0 | 44 | $\frac{44}{3} = 14\frac{2}{3}$ |
| $-20$ | 0 | 20 | 0 | 0 | 1 | 1600 | |

Since the second row yields the smallest of these ratios, the pivot row will be the second row. Thus the pivot element is the element in the second row and first column.

Then perform the elementary row operations $R_1 - .50R_2 \rightarrow R_1$, $R_3 - 3R_2 \rightarrow R_3$, $R_4 + 20R_2 \rightarrow R_4$, and obtain the new tableau

| $x_1$ | $x_2$ | $s_1$ | $s_2$ | $s_3$ | $z$ | |
|---|---|---|---|---|---|---|
| 0 | 1 | .50 | $-.50$ | 0 | 0 | 15 |
| 1 | 0 | $-.50$ | 1 | 0 | 0 | 10 |
| 0 | 0 | 1 | $-3$ | 1 | 0 | 14 |
| 0 | 0 | 10 | 20 | 0 | 1 | 1800 |

Since there are no more negative elements in the last row, we stop. Since the third and fourth columns are the nonunit columns, the nonbasic variables are $s_1$ and $s_2$. Setting the nonbasic variables $s_1$ and $s_2$ equal to zero then immediately yields $x_1 = 10$, $x_2 = 15$, and $s_3 = 14$. All the slack variables are nonnegative, and the maximum then occurs at the point $(x_1, x_2) = (10, 15)$. The actual maximum can be read from the last row and column of the tableau as 1800.

Notice how the simplex method started at the corner $(0, 0)$, then moved to $(0, 20)$, and then finally to $(10, 15)$, each time moving to an adjacent corner and each time improving the maximum value of the objective function. Refer to Figure 4.4.

We can now summarize the steps in the simplex method.

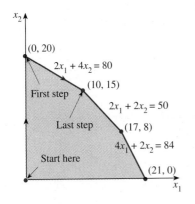

**Figure 4.4**

---

**Simplex Method for Problem (1)**

1. Introduce a slack variable for each of the constraints and create the initial tableau.

2. If none of the entries in the last row (ignoring the last entry) are negative, then we stop. The solution obtained from this tableau is the maximum. To obtain this solution:

   a. Pick $n$ of the variables to be the nonbasic variables. These variables correspond to the nonunit columns. Set all of these nonbasic variables equal to zero. If there are not $n$ nonunit columns, then pick variables (but never $z$) associated with unit columns to be nonbasic variables in such a manner that no two basic variables have identical associated columns.

    b.   The values of the basic variables can then be found immediately. (The basic variables always correspond to unit columns.)

3.   If one of the entries in the last row (except for the last entry) is negative, then construct a new tableau choosing the pivot element as follows:

    a.   Locate the most negative element (other than the last entry) in the last row of the tableau. The column with this entry will be the pivot column. If there is a tie, pick any one of these columns.

    b.   Divide each of the elements in the last column (except for the bottom most) by the corresponding elements in the pivot column. Choose as the pivot row the row that has the smallest of the nonnegative ratios, ignoring zero ratios that are divided by a negative number. If this is impossible to do, then stop. There is no solution.

4.   Obtain a new simplex tableau by doing the following.

    a.   Divide each entry in the pivot row by the pivot element, obtaining a 1 in the pivot position.

    b.   Use elementary row operations with the pivot row to obtain zeros in all elements in the pivot column other than the pivot position.

5.   Go to Step 2.

**E X A M P L E  2**    **Using the Simplex Method**

$$\text{Maximize} \quad z = 3x_1 + 2x_2$$
$$\text{Subject to} \quad -x_1 + x_2 \leq 2$$
$$2x_1 + x_2 \leq 8$$
$$x_1, x_2 \geq 0$$

**Solution**    Introducing two slack variables and rewriting the objective function yields

$$-x_1 + x_2 + s_1 \qquad\qquad = 2$$
$$2x_1 + x_2 \qquad + s_2 \qquad = 8$$
$$-3x_1 - 2x_2 \qquad\qquad + z = 0$$

The initial tableau then becomes

| $x_1$ | $x_2$ | $s_1$ | $s_2$ | $z$ | | ratio |
|---|---|---|---|---|---|---|
| $-1$ | 1 | 1 | 0 | 0 | 2 | |
| ② | 1 | 0 | 1 | 0 | 8 | $\frac{8}{2} \leftarrow$ smallest ratio |
| $-3$ | $-2$ | 0 | 0 | 1 | 0 | |

The most negative element in the last row is $-3$, which indicates that the pivot column is the first column. There is only one positive ratio, $\frac{8}{2}$, which is in the second row. Thus the second row is the pivot row. The pivot is the element in the second row and the first column. Dividing the second row by 2 yields

| $x_1$ | $x_2$ | $s_1$ | $s_2$ | $z$ | |
|---|---|---|---|---|---|
| $-1$ | 1 | 1 | 0 | 0 | 2 |
| ① | $\frac{1}{2}$ | 0 | $\frac{1}{2}$ | 0 | 4 |
| $-3$ | $-2$ | 0 | 0 | 1 | 0 |

Performing the row operations $R_1 + R_2 \rightarrow R_1$ and $R_3 + 3R_2 \rightarrow R_3$ yields the new tableau

| $x_1$ | $x_2$ | $s_1$ | $s_2$ | $z$ | | ratios |
|---|---|---|---|---|---|---|
| 0 | ③⁄② | 1 | $\frac{1}{2}$ | 0 | 6 | $\dfrac{6}{\frac{3}{2}} = 4 \leftarrow$ smallest ratio |
| 1 | $\frac{1}{2}$ | 0 | $\frac{1}{2}$ | 0 | 4 | $\dfrac{4}{\frac{1}{2}} = 8$ |
| 0 | $-\frac{1}{2}$ | 0 | $\frac{3}{2}$ | 1 | 12 | |

There is only one negative element in the last row, so the second column will be the pivot column. The two ratios indicate that the first row is the new pivot row. Dividing the first row by $\frac{3}{2}$ and performing the indicated row operations then yields

| $x_1$ | $x_2$ | $s_1$ | $s_2$ | $z$ | | |
|---|---|---|---|---|---|---|
| 0 | ① | $\frac{2}{3}$ | $\frac{1}{3}$ | 0 | 4 | |
| 1 | $\frac{1}{2}$ | 0 | $\frac{1}{2}$ | 0 | 4 | $R_2 - \frac{1}{2}R_1 \rightarrow R_2$ |
| 0 | $-\frac{1}{2}$ | 0 | $\frac{3}{2}$ | 1 | 12 | $R_3 + \frac{1}{2}R_1 \rightarrow R_3$ |

| $x_1$ | $x_2$ | $s_1$ | $s_2$ | $z$ | |
|---|---|---|---|---|---|
| 0 | 1 | $\frac{2}{3}$ | $\frac{1}{3}$ | 0 | 4 |
| 1 | 0 | $-\frac{1}{3}$ | $\frac{1}{3}$ | 0 | 2 |
| 0 | 0 | $\frac{1}{3}$ | $\frac{5}{3}$ | 1 | 14 |

Since the last row contains no negative elements, we stop. The solution is $x_1 = 2$ and $x_2 = 4$, with the maximum value of 14. Refer to Figure 4.5.   ∎

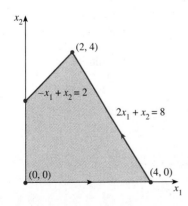

**Figure 4.5**

**E X A M P L E  3**    **Using the Simplex Method**

$$\text{Maximize} \qquad z = 10x_1 + 20x_2 + 10x_3$$

$$\text{Subject to} \qquad x_1 + x_2 + x_3 \le 16$$

$$-2x_1 + x_2 + 2x_3 \le 20$$

$$x_1, x_2, x_3 \ge 0$$

**Solution**    Introducing two slack variables yields

$$x_1 + x_2 + x_3 + s_1 = 16$$

$$-2x_1 + x_2 + 2x_3 + s_2 = 20$$

$$-10x_1 - 20x_2 - 10x_3 + z = 0$$

The initial tableau then becomes

| $x_1$ | $x_2$ | $x_3$ | $s_1$ | $s_2$ | $z$ | | ratios | |
|---|---|---|---|---|---|---|---|---|
| 1 | ① | 1 | 1 | 0 | 0 | 16 | $\frac{16}{1}$ ← smallest ratio | |
| $-2$ | 1 | 2 | 0 | 1 | 0 | 20 | $\frac{20}{1}$ | $R_2 - R_1 \to R_2$ |
| $-10$ | $-20$ | $-10$ | 0 | 0 | 1 | 0 | | $R_3 + 20R_1 \to R_3$ |

The most negative element in the last row is $-20$, which indicates that the pivot column is the second. The smallest ratio is 16, which indicates that the pivot row is the first. Following the indicated elementary row operations we obtain the tableau

| $x_1$ | $x_2$ | $x_3$ | $s_1$ | $s_2$ | $z$ | |
|---|---|---|---|---|---|---|
| 1 | 1 | 1 | 1 | 0 | 0 | 16 |
| $-3$ | 0 | 1 | $-1$ | 1 | 0 | 4 |
| 10 | 0 | 10 | 20 | 0 | 1 | 320 |

Since there are no negative elements in the last row, the solution is $x_1 = 0$, $x_2 = 16$, $x_3 = 0$, with maximum equal to 320.  ∎

**E X A M P L E  4**    **An Example with No Maximum**

$$\text{Maximize} \qquad z = 10x_1 + 20x_2$$

$$\text{Subject to} \qquad 4x_1 - 2x_2 \le 12$$

$$x_1 - x_2 \le 1$$

$$x_1, x_2 \ge 0$$

**Solution**     The initial tableau is

| $x_1$ | $x_2$ | $s_1$ | $s_2$ | $z$ | |
|---|---|---|---|---|---|
| 4 | $-2$ | 1 | 0 | 0 | 12 |
| 1 | $-1$ | 0 | 1 | 0 | 1 |
| $-10$ | $-20$ | 0 | 0 | 1 | 0 |

Since $-20$ is the most negative element in the last row, the pivot column is the second column. But all the ratios are negative! Since we cannot pick a pivot row, the simplex method stops. There is no solution.  ∎

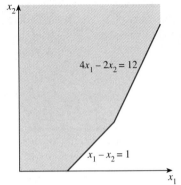

**Figure 4.6**

Recall from our previous analysis that having all ratios negative in the previous example indicates that $x_2$ can be increased without bound. Therefore, since $z = 10x_1 + 20x_2$, there can be no maximum.

The graph of the feasible region is shown in Figure 4.6. Notice the feasible region is unbounded. Since the objective function has positive numbers in front of the decision variables, no maximum can exist.

## Post-Optimal Analysis

After an optimal solution to an applied linear programming problem has been found, further analysis is usually required. For example, an analysis is needed to determine if an excess of some resource is being allocated. If this is the case, then management should consider reallocating or reducing this resource.

**E X A M P L E  5**     **Determining Excess Resources**

For the wooden-boat manufacturing problem considered earlier, identify any of the three operations that has an excess of needed labor and determine the amount.

**Solution**     To answer this question we need to examine the slack variables for the final tableau obtained when using the simplex method. Recall that $s_1 = 0$, $s_2 = 0$, and $s_3 = 14$. Recall also that $s_1$ and $s_2$ are associated with the cutting constraint and the painting constraint, respectively. Thus there is no slack in the cutting and painting departments. However, $s_3 = 14$, and since this slack variable is associated with the assembly constraint, there is a slack of 14 hours per week in the assembly department. Management should consider eliminating or reallocating these hours.  ∎

In general we can make the following statement.

---

**Economic Interpretation of Slack Variables**

The value of a slack variable in the final tableau indicates how much unused capacity is available for the associated constraint.

---

**REMARK.**      Recall that the assembly constraint was

$$4x_1 + 2x_2 \leq 84.$$

It would have simplified calculations if we had divided this inequality by two *first* and then added a slack variable obtaining

$$2x_1 + x_2 + s_3' = 42.$$

But now we must be very careful to note that the units of $s_3'$ are in *half* hours when we give an economic interpretation to this slack variable.

## SELF-HELP EXERCISE SET 4.2

1.  Two types of metal connectors require for their manufacture a certain amount of minutes on a stamping machine and on a grinding machine. The first type of connector requires 4 minutes on the stamping machine and 3 minutes on the grinding machine, while the second connector requires 5 minutes on the stamping machine and 1 minute on the grinding machine. The stamping machine is available at most 50 minutes each hour, and the grinding machine is available at most 21 minutes each hour. If each of the first type connector brings a profit of \$0.50 and each of the second type of connector a profit of \$0.60, use the simplex method to find the number of each of these products that should be manufactured to maximize profits. (See Self-Help Exercise 2 of Section 3.1 and Self-Help Exercise 2 of 3.2, where one of the constraints has been eliminated.)

## EXERCISE SET 4.2

In Exercises 1 through 10 use the simplex method to obtain the next tableau. If the next tableau should be the final tableau, then indicate the solution.

**1.**

| $x_1$ | $x_2$ | $s_1$ | $s_2$ | $z$ | |
|---|---|---|---|---|---|
| 2 | 4 | 1 | 0 | 0 | 8 |
| 5 | 1 | 0 | 1 | 0 | 25 |
| $-4$ | $-2$ | 0 | 0 | 1 | 0 |

**2.**

| $x_1$ | $x_2$ | $s_1$ | $s_2$ | $z$ | |
|---|---|---|---|---|---|
| 3 | 5 | 1 | 0 | 0 | 15 |
| 2 | 2 | 0 | 1 | 0 | 8 |
| $-8$ | $-5$ | 0 | 0 | 1 | 0 |

**3.**

| $x_1$ | $x_2$ | $s_1$ | $s_2$ | $z$ | |
|---|---|---|---|---|---|
| 2 | 1 | 5 | 0 | 0 | 5 |
| 5 | 0 | 5 | 1 | 0 | 10 |
| $-7$ | 0 | $-3$ | 0 | 1 | 0 |

**4.**

| $x_1$ | $x_2$ | $s_1$ | $s_2$ | $z$ | |
|---|---|---|---|---|---|
| 2 | 0 | 1 | 2 | 0 | 2 |
| $-4$ | 1 | 0 | 5 | 0 | 4 |
| 9 | 0 | 0 | 0 | 1 | 0 |

**5.**

| $x_1$ | $x_2$ | $s_1$ | $s_2$ | $z$ | |
|---|---|---|---|---|---|
| 0 | $-4$ | 1 | 2 | 0 | 2 |
| 1 | 3 | 0 | 3 | 0 | 3 |
| 0 | $-5$ | 0 | $-2$ | 1 | 0 |

**6.**

| $x_1$ | $x_2$ | $s_1$ | $s_2$ | $z$ | |
|---|---|---|---|---|---|
| 1 | 10 | 1 | 0 | 0 | 50 |
| 0 | 10 | 30 | 1 | 0 | 20 |
| 0 | $-5$ | 0 | 0 | 1 | 0 |

**7.**

| $x_1$ | $x_2$ | $s_1$ | $s_2$ | $s_3$ | $z$ | |
|---|---|---|---|---|---|---|
| $-5$ | 0 | 1 | 0 | 2 | 0 | 21 |
| $-1$ | 0 | 0 | 1 | 1 | 0 | 30 |
| 2 | 1 | 0 | 0 | 8 | 0 | 2 |
| $-10$ | 0 | 0 | 0 | 40 | 1 | 30 |

**8.**

| $x_1$ | $x_2$ | $s_1$ | $s_2$ | $s_3$ | $z$ | |
|---|---|---|---|---|---|---|
| 1 | 0 | 0 | 1 | 1 | 0 | 5 |
| 3 | 0 | 1 | $-2$ | 0 | 0 | 3 |
| 7 | 1 | 0 | 1 | 0 | 0 | 6 |
| $-3$ | 0 | 0 | $-4$ | 0 | 1 | 10 |

**9.**

| $x_1$ | $x_2$ | $s_1$ | $s_2$ | $s_3$ | $z$ | |
|---|---|---|---|---|---|---|
| 0 | 1 | 1 | $-3$ | 0 | 0 | 0 |
| 0 | 8 | 0 | 6 | 1 | 0 | 2 |
| 1 | 1 | 0 | 0 | 0 | 0 | 2 |
| 0 | 2 | 0 | $-2$ | 0 | 1 | 8 |

**10.**

| $x_1$ | $x_2$ | $s_1$ | $s_2$ | $s_3$ | $z$ | |
|---|---|---|---|---|---|---|
| 0 | 1 | 0 | 1 | 5 | 0 | 10 |
| 0 | 5 | 1 | 0 | 2 | 0 | 4 |
| 1 | 2 | 0 | 0 | 1 | 0 | 1 |
| 0 | $-1$ | 0 | 0 | $-2$ | 1 | 10 |

In Exercises 11 through 32 add the nonnegativity constraint and use the simplex method to solve. Where there are only 2 variables, graph the feasible region and note which corners the simplex method moves through.

**11.** Maximize $z = 2x_1 + 3x_2$

Subject to $x_1 + 2x_2 \le 24$
$5x_1 + x_2 \le 30$

**12.** Maximize $z = 10x_1 + 20x_2$

Subject to $x_1 + 3x_2 \le 36$
$5x_1 + x_2 \le 40$

**13.** Maximize $z = 3x_1 + 2x_2$

Subject to $-x_1 + 2x_2 \le 1$
$x_1 - x_2 \le 1$

**14.** Maximize $z = 2x_1 + x_2$

Subject to $x_1 + x_2 \le 8$
$2x_1 - x_2 \le 4$

15. Maximize $z = 2x_1 + x_2$

    Subject to  $x_1 + x_2 \leq 8$

    $x_1 \leq 2$

16. Maximize $z = x_1 + 2x_2$

    Subject to  $x_1 + x_2 \leq 8$

    $x_2 \leq 4$

17. Maximize  $z = x_1 + 2x_2$

    Subject to  $x_1 + x_2 \leq 6$

    $-x_1 + 2x_2 \leq 0$

    $x_2 \leq 4$

18. Maximize  $z = 2x_1 + x_2$

    Subject to $-2x_1 + x_2 \leq 0$

    $x_1 + x_2 \leq 6$

    $x_1 \leq 4$

19. Maximize  $z = x_1 + 2x_2$

    Subject to  $x_1 + 3x_2 \leq 36$

    $x_1 + x_2 \leq 16$

    $2x_1 + x_2 \leq 24$

20. Maximize  $z = 3x_1 + 2x_2$

    Subject to  $x_1 + 2x_2 \leq 28$

    $x_1 + x_2 \leq 16$

    $2x_1 + x_2 \leq 24$

21. Maximize  $z = x_1 + 2x_2 + x_3$

    Subject to  $x_1 + x_2 + x_3 \leq 12$

    $2x_1 + x_2 + 4x_3 \leq 24$

22. Maximize  $z = 2x_1 + x_2 + x_3$

    Subject to $x_1 + x_2 + x_3 \leq 4$

    $x_1 + 2x_2 + 3x_3 \leq 6$

23. Maximize  $z = x_1 + x_2 + 2x_3$

    Subject to  $x_1 + x_2 + x_3 \leq 4$

    $x_1 - 2x_2 + 2x_3 \leq 10$

    $2x_1 - x_2 + 3x_3 \leq 18$

24. Maximize  $z = x_1 + 3x_2 + x_3$

    Subject to  $x_1 + x_2 + 2x_3 \leq 2$

    $2x_1 - x_2 + 5x_3 \leq 3$

    $3x_1 + 2x_2 - 7x_3 \leq 6$

25. Maximize $z = 10x_1 + 20x_2 + 5x_3$

    Subject to    $x_1 + 3x_3 \leq 1$

    $2x_1 + x_2 + x_3 \leq 2$

    $x_1 - x_2 + 2x_3 \leq 0$

    $3x_1 - 2x_2 + 4x_3 \leq 5$

26. Maximize $z = x_1 + 2x_2 + 3x_3$

    Subject to $3x_1 - x_2 \leq 5$

    $2x_1 + x_2 - 4x_3 \leq 7$

    $x_1 + 2x_2 + x_3 \leq 1$

    $x_1 + 3x_2 + 2x_3 \leq 6$

**27.** Maximize $\quad z = 2x_1 + 4x_2 + x_3$

Subject to $\quad x_1 + x_2 + 2x_3 \le 200$

$$10x_1 + 8x_2 + 5x_3 \le 2000$$

$$2x_1 + x_2 \quad\quad \le 100$$

**28.** Maximize $\quad z = x_1 + 3x_2 + 4x_3$

Subject to $2x_1 + x_2 + 2x_3 \le 4$

$$x_1 + 2x_2 + 2x_3 \le 10$$

$$3x_1 + 2x_2 + 2x_3 \le 12$$

**29.** Maximize $z = x_1 + 2x_2 + 3x_3$

Subject to $2x_1 \quad\quad + x_3 \le 2$

$$x_1 + 2x_2 + x_3 \le 4$$

$$x_1 + x_2 + x_3 \le 8$$

**30.** Maximize $\quad z = x_1 + 2x_2 + x_3$

Subject to $x_1 - x_2 + 2x_3 \le 3$

$$x_1 + x_2 \quad\quad \le 1$$

$$2x_1 \quad\quad + x_3 \le 4$$

**31.** Maximize $\quad z = x_1 + 2x_2$

Subject to $\quad -x_1 + 2x_2 \le 4$

$$-3x_1 + x_2 \le 12$$

**32.** Maximize $\quad z = 3x_1 + 2x_2$

Subject to $-x_1 + 3x_2 \le 2$

$$x_1 - 4x_2 \le 2$$

## Applications

**33. Manufacturing.**  Three types of connectors, A, B, and C, require for their manufacture a certain number of minutes on each of three machines: a stamping machine, a grinding machine, and a polishing machine. Connector A requires 1 minute on the stamping machine, 2 minutes on the grinding machine, and 3 minutes on the polishing machine. Connector B requires 2 minutes on the stamping machine, 1 minute on the grinding machine, and 1 minute on the polishing machine. Connector C requires 2 minutes on the stamping machine, 2 minutes on the grinding machine, and 2 minutes on the polishing machine. The stamping machine can operate 12 minutes each hour, the grinding machine 8 minutes each hour, and the polishing machine 16 minutes each hour. The profit on each of the connectors A, B, and C is, respectively, $35, $40, and $50 each. Find the number of each of the connectors to be manufactured in order to maximize profits. Find the maximum profit.

**34. Manufacturing.**  A firm manufactures three products: tables, desks, and bookcases. To produce each table requires 1 hour of labor, 10 square feet of wood, and 2 quarts of finish. To produce each desk requires 3 hours of labor, 35 square feet of wood, and 1 quart of finish. To produce each bookcase requires 0.50 hours of labor, 15 square feet of wood, and 1 quart of finish. Available is at most 25 hours of labor, at most 350 square feet of wood, and at most 55 quarts of finish. The table yields a profit of $4, the desk $3, and the bookcase $3. Find the number of each product to be made in order to maximize profits. Find the maximum profit.

35. **Mixture.** A dealer has 8 pounds of cashews, 24 pounds of almonds, and 36 pounds of peanuts to be used to make three mixtures. Mixture A sells for $2 per pound and consists of 20% cashews, 20% almonds, and 60% peanuts. Mixture B sells for $4 per pound and contains 20% cashews, 40% almonds, and 40% peanuts. Mixture C sells for $3 per pound and contains 10% cashews, 30% almonds, and 60% peanuts. How many pounds of each mixture should the dealer make in order to maximize revenue? Find the maximum revenue.

36. **Fishery.** A certain lake has three species of trout, rainbow, brown, and brook, and also three types of food for these fish, I, II, and III. Each month the lake can supply 800 pounds of food I, 500 pounds of food II, and 1100 pounds of food III. Each month each rainbow trout requires 1 pound of food I, 1 pound of food II, and 2 pounds of food III. Each brown trout requires 4 pounds of food I, 2 pounds of food II, and 1 pound of food III. Each brook trout requires 0.50 pounds of each of the foods. What is the maximum number of these trout that the lake can support and the number of each type of trout?

37. **Manufacturing.** A furniture manufacturer produces chairs, sofas, and love seats. The chairs require 5 feet of wood, 1 pound of foam rubber, and 10 square yards of material. The sofas require 35 feet of wood, 2 pounds of foam rubber, and 20 square yards of material. The love seats require 9 feet of wood, 0.2 pounds of foam rubber, and 10 square yards of material. The manufacturer has in stock 405 feet of wood, 25 pounds of foam rubber, and 410 square yards of material. If the chairs yield a profit of $300, the sofas $200, and the love seats $220 each, how many of each should be produced to maximize the income? Find the maximum income.

38. **Fishery.** If in Exercise 36, $3 can be obtained from harvesting the rainbow trout, $2 for the brown trout, and $2 for the brook trout, how many of each trout should be harvested to maximize revenue? Find the maximum revenue.

39. **Profits.** A company makes 3 styles of sleds: Classic, Sport, and Deluxe. The Classic uses 1 board-foot of wood, 0.5 pounds of iron, and 1.25 hours of labor. The Sport uses 4 board-feet of wood, 2 pounds of iron, and 1 hour of labor. The Deluxe uses 1 board-foot of wood, 1.5 pounds of iron, and .25 hours of labor. There are 56 board-feet of wood, 44 pounds of iron, and 34 hours of labor available. The profit on the Classic is $3, on the Sport $8, and on the Deluxe $4. Find the number of each type sled to make in order to maximize the profits and find the maximum profits.

40. **Profits.** A firm manufactures 3 products, each of which uses 3 ingredients. The first product requires 2 units of the first ingredient, 1 unit of the second, and 1 unit of the third. The second product requires 1 unit of the first ingredient, 1 unit of the second, and 3 units of the third. The third product requires 1 unit of the first ingredient, 3 units of the second, and 3 units of the third. There are only 20 units of the first ingredient, 30 units of the second ingredient, and 40 units of the third ingredient available. The profit on each of the first product is $26, on the second is $24, and on the third is $30. Find the number of each product that should be made in order to maximize profits and find the maximum profits.

**41.** Give an economic interpretation of the slack variables associated with the optimal solutions found in Exercises 35, 37, and 39 and determine in each case which resources are in excess.

**42.** Give an economic interpretation of the slack variables associated with the optimal solutions found in Exercises 36, 38, and 40 and determine in each case which resources are in excess.

## Graphing Calculator Exercises

Do Exercises 43 through 44 using the row operations on your graphing calculator.

**43.** Maximize    $z = 90x_1 + 210x_2 + 52x_3$

Subject to $11x_1 + 12x_2 + 21x_3 \leq 2100$
$$12x_1 + 9x_2 + 6x_3 \leq 1700$$
$$8x_1 + 4x_2 \quad\quad\quad \leq 400$$
$$x_1, x_2, x_3 \geq 0$$

**44.** Maximize    $z = 81x_1 + 89x_2 + 153x_3$

Subject to $22x_1 + 9x_2 + 43x_3 \leq 3500$
$$19x_1 + 48x_2 + 102x_3 \leq 9000$$
$$42x_1 + 28x_2 + 8x_3 \leq 4900$$
$$x_1, x_2, x_3 \geq 0$$

## Solutions to Self-Help Exercise Set 4.2

**1.** If $x_1$ is the number of the first type connector and $x_2$ the number of the second type connector, then the problem is to maximize the objective function $z = 50x_1 + 60x_2$ subject to the constraints

$$4x_1 + 5x_2 \leq 50$$
$$3x_1 + x_2 \leq 21$$
$$x_1, x_2 \geq 0$$

The initial tableau is

| $x_1$ | $x_2$ | $s_1$ | $s_2$ | $z$ | | |
|---|---|---|---|---|---|---|
| 4 | ⑤ | 1 | 0 | 0 | 50 | $\dfrac{50}{5} = 10$ |
| 3 | 1 | 0 | 1 | 0 | 21 | $\dfrac{21}{1} = 21$ |
| $-50$ | $-60$ | 0 | 0 | 1 | 0 | |

Since $-60$ is the most negative element in the last row, the pivot column is the second column. The above ratios indicate that the pivot row is the first row. Dividing the first row by 5 yields

| $x_1$ | $x_2$ | $s_1$ | $s_2$ | $z$ | |
|---|---|---|---|---|---|
| $\frac{4}{5}$ | ① | $\frac{1}{5}$ | 0 | 0 | 10 |
| 3 | 1 | 0 | 1 | 0 | 21 |
| $-50$ | $-60$ | 0 | 0 | 1 | 0 |

$R_2 - R_1 \rightarrow R_2$

$R_3 + 60R_1 \rightarrow R_3$

Performing the indicated row operations then yields

| $x_1$ | $x_2$ | $s_1$ | $s_2$ | $z$ | |
|---|---|---|---|---|---|
| $\frac{4}{5}$ | 1 | $\frac{1}{5}$ | 0 | 0 | 10 |
| ⑪⁄₅ | 0 | $-\frac{1}{5}$ | 1 | 0 | 11 |
| $-2$ | 0 | 12 | 0 | 1 | 600 |

$\dfrac{10}{\frac{4}{5}} = 12.5$

$\dfrac{11}{\frac{11}{5}} = 5$

Notice that this tableau corresponds to the point $(0, 10)$.

The last row contains only one negative element, thus the pivot column is the first column. The two ratios indicate that the pivot row is the second row. Dividing the second row by $\frac{11}{5}$, then yields the next tableau

| $x_1$ | $x_2$ | $s_1$ | $s_2$ | $z$ | |
|---|---|---|---|---|---|
| $\frac{4}{5}$ | 1 | $\frac{1}{5}$ | 0 | 0 | 10 |
| ① | 0 | $-\frac{1}{11}$ | $\frac{5}{11}$ | 0 | 5 |
| $-2$ | 0 | 12 | 0 | 1 | 600 |

$R_1 - \frac{4}{5}R_2 \rightarrow R_1$

$R_3 + 2R_2 \rightarrow R_3$

Performing the indicated row operations then gives

| $x_1$ | $x_2$ | $s_1$ | $s_2$ | $z$ | |
|---|---|---|---|---|---|
| 0 | 1 | $\frac{3}{11}$ | $-\frac{4}{11}$ | 0 | 6 |
| 1 | 0 | $-\frac{1}{11}$ | $\frac{5}{11}$ | 0 | 5 |
| 0 | 0 | $11\frac{9}{11}$ | $\frac{10}{11}$ | 1 | 610 |

This yields $x_1 = 5$ and $x_2 = 6$. This solution was also found in the previous chapter by the geometric method. Notice the simplex method started at the corner $(0, 0)$, moved to $(0, 10)$, and then finally to $(5, 6)$. Since in the final tableau the two slack variables are zero, there is no slack in the operation of the two machines.

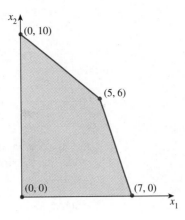

## 4.3 MINIMIZATION PROBLEMS AND THE DUAL

► *The Dual Problem*
► *Economic Interpretation of Dual Problems*
► *Matrix Formulation*

In the last two sections we considered maximization problems with the inequality $\leq$ in all the problem constraints. In this section we will consider minimization problems with the inequality $\geq$ in all the problem constraints. To solve these minimization problems we will need to consider an associated problem called the *dual*.

Associated with any linear program problem, referred to as the **primal problem**, is another linear programming problem, called the **dual**. The dual problem is important for several reasons. First, the solution of the dual problem can be used to find the solution of the primal problem. Secondly, the dual problem gives an interesting economic interpretation that adds additional insights into the primal problem.

### The Dual Problem

We first consider a typical minimization problem.

**EXAMPLE 1** **Dual of a Minimization Problem**

An animal rancher requires at least 2 units per day of nutrient A, 4 units of nutrient B, and 3 units of nutrient C for his animals. The rancher has two different mixtures of feeds available, feed 1 and 2. If each bag of feed 1 costs $12 and yields 1 unit of each of the nutrients and each bag of feed 2 costs $24 and yields 2 units of nutrient A, 1 unit of B, and 4 units of C, find the linear programming problem needed to find the number of bags of each feed that should be used to minimize the cost.

**Solution** The information can be summarized in the following table.

|            | Feed 1 | Feed 2 | Requirements |
|------------|--------|--------|--------------|
| Nutrient A | 1      | 2      | 2            |
| Nutrient B | 1      | 1      | 4            |
| Nutrient C | 1      | 4      | 3            |
| Costs      | 12     | 24     |              |

If $y_1$ is the number of bags of nutrient 1 and $y_2$ is the number of bags of nutrient 2, then the linear programming problem that we must solve is

$$\text{Minimize} \quad w = 12y_1 + 24y_2 \qquad \text{objective function}$$

$$\text{Subject to} \quad \left. \begin{array}{l} y_1 + 2y_2 \geq 2 \\ y_1 + \phantom{2}y_2 \geq 4 \\ y_1 + 4y_2 \geq 3 \end{array} \right\} \quad \text{problem constraints}$$

$$y_1, y_2 \geq 0 \qquad \text{nonnegativity constraint} \quad \blacksquare$$

We shall see how to solve this problem using the simplex method. However, to do this we must first formulate and solve an associated problem called the dual problem. We will refer to the original problem as the **primal problem**.

If the primal problem is a minimization problem with all problem constraints containing the inequality $\geq$, then the dual problem is a maximization problem with all problem constraints containing the inequality $\leq$. The number of decision variables in the dual problem equals the number of problem constraints in the primal problem, while the number of problem constraints in the dual problem equals the number of decision variables in the primal problem. Following are further details for determining the dual problem for Example 1.

**Primal Problem**                                **Dual Problem**

Minimize    $w = 12y_1 + 24y_2$      Maximize      $z = 2x_1 + 4x_2 + 3x_3$

Subject to    $y_1 + 2y_2 \geq 2$        Subject to      $x_1 + x_2 + \phantom{4}x_3 \leq 12$

$\phantom{Subject to}\quad y_1 + \phantom{2}y_2 \geq 4$ $\phantom{Subject to}\qquad 2x_1 + x_2 + 4x_3 \leq 24$

$\phantom{Subject to}\quad y_1 + 4y_2 \geq 3$ $\phantom{Subject to}\qquad\qquad x_1, x_2, x_3 \geq 0$

$\phantom{Subject to}\quad y_1, y_2 \geq 0$

First column becomes first row.

$$\text{coefficient matrix} \quad \begin{bmatrix} 1 & 2 \\ 1 & 1 \\ 1 & 4 \end{bmatrix} \quad \text{coefficient matrix} \quad \begin{bmatrix} 1 & 1 & 1 \\ 2 & 1 & 4 \end{bmatrix}$$

Second column becomes second row.

### Formation of the Dual Problem

Given a minimization problem with $\geq$ problem constraints, the following rules apply.

1. The dual problem is a maximization problem with $\leq$ problem constraints.
2. The coefficients of the decision variables in the primal problem become the constants to the right of the inequality $\leq$ in the problem constraints of the dual.
3. The constants to the right of the inequality $\geq$ in the problem constraints of the primal problem become the coefficients of the decision variables in the dual.
4. The coefficients of the $i$th decision variable in the problem constraints for the primal problem become the coefficients of the decision variables in the $i$th problem constraint of the dual.

More precisely, we have the following.

### The Dual Problem

If the primal problem has the form

$$
\left.
\begin{aligned}
\text{Maximize} \quad & z = c_1x_1 + c_2x_2 + \cdots + c_nx_n \\
\text{Subject to} \quad & a_{11}x_1 + a_{12}x_2 + \cdots + a_{1n}x_n \leq b_1 \\
& a_{21}x_1 + a_{22}x_2 + \cdots + a_{2n}x_n \leq b_2 \\
& \qquad\qquad \cdots \\
& a_{m1}x_1 + a_{m2}x_2 + \cdots + a_{mn}x_n \leq b_m \\
& \qquad\qquad x_1, x_2, \ldots, x_n \geq 0
\end{aligned}
\right\} \quad (1)
$$

then the dual problem is

$$
\left.
\begin{aligned}
\text{Minimize} \quad & w = b_1y_1 + b_2y_2 + \cdots + b_my_m \\
\text{Subject to} \quad & a_{11}y_1 + a_{21}y_2 + \cdots + a_{m1}y_m \geq c_1 \\
& a_{12}y_1 + a_{22}y_2 + \cdots + a_{m2}y_m \geq c_2 \\
& \qquad\qquad \cdots \\
& a_{1n}y_1 + a_{2n}y_2 + \cdots + a_{mn}y_m \geq c_n \\
& \qquad\qquad y_1, y_2, \ldots, y_m \geq 0
\end{aligned}
\right\} \quad (2)
$$

Also if the primal problem is (2), then the dual is (1).

**REMARK.** In this section we will reserve $x_1, x_2, \ldots$ for the decision variables in the maximization problems and $y_1, y_2, \ldots$ for the decision variables in the minimization problems.

E X A M P L E 2    **Dual of a Maximization Problem**

Find the dual problem for the wooden-boat manufacturing problem considered in this and the last chapter. Recall that the problem is

$$\text{Maximize} \quad z = 60x_1 + 80x_2$$
$$\text{Subject to} \quad 2x_1 + 4x_2 \leq 80$$
$$2x_1 + 2x_2 \leq 50$$
$$4x_1 + 2x_2 \leq 84$$
$$x_1, x_2 \geq 0$$

**Solution**    The dual must be a minimization problem with $\geq$ constraints and is

$$\text{Minimize} \quad w = 80y_1 + 50y_2 + 84y_3$$
$$\text{Subject to} \quad 2y_1 + 2y_2 + 4y_3 \geq 60$$
$$4y_1 + 2y_2 + 2y_3 \geq 80$$
$$y_1, y_2, y_3 \geq 0 \quad \blacksquare$$

To solve a minimization problem such as found in Example 1, first form the dual and then use the following theorem.

---

**Fundamental Theorem of Duality**

The objective function $z$ in the linear programming problem (1) assumes a maximum value if, and only if, the dual linear programming problem (2) assumes a minimum value. Furthermore,

1. The maximum value of $z$ in (1) equals the minimum value of $w$ in (2).
2. The optimal solution to the primal problem can be found under the slack variables in the last row of the final tableau associated with the dual problem.

---

We will now use this theorem to solve the problem in Example 1.

E X A M P L E 3    **Using the Fundamental Theorem of Duality**

Use the fundamental theorem of duality to solve the primal problem given in Example 1.

**Solution**    The dual problem was found in Example 1. The initial tableau for this dual problem is

| $x_1$ | $x_2$ | $x_3$ | $s_1$ | $s_2$ | $z$ | | ratios |
|---|---|---|---|---|---|---|---|
| 1 | ① | 1 | 1 | 0 | 0 | 12 | $\frac{12}{1}$ ← smallest ratio |
| 2 | 1 | 4 | 0 | 1 | 0 | 24 | $\frac{24}{1}$ |
| $-2$ | $-4$ | $-3$ | 0 | 0 | 1 | 0 | |

↑
pivot column

Pivoting about the first row and second column yields

| $x_1$ | $x_2$ | $x_3$ | $s_1$ | $s_2$ | $z$ | |
|---|---|---|---|---|---|---|
| 1 | 1 | 1 | 1 | 0 | 0 | 12 |
| 1 | 0 | 3 | $-1$ | 1 | 0 | 12 |
| 2 | 0 | 1 | 4 | 0 | 1 | 48 |

This is the final tableau. Now consider the last row.

| $x_1$ | $x_2$ | $x_3$ | $s_1$ | $s_2$ | $z$ | |
|---|---|---|---|---|---|---|
| 2 | 0 | 1 | 4 | 0 | 1 | 48 |

Looking under the slack variables in the last row, we see a 4 and a 0. Thus the solution to the primal problem is (4, 0). The minimum of the primal problem is the same as the maximum of the dual problem, which is found in the last entry of the last row to be 48.

Thus 4 bags of feed 1 and no bags of feed 2 should be purchased in order to obtain the minimum cost of $48 while meeting the nutritional requirement given in the constraints. ■

The following problem was formulated in Example 3 of Section 3.3.

**E X A M P L E   4**    **Using the Dual to Solve a Minimization Problem**

A firm has two stockpiles of an item stored in warehouses at two different locations, Denver and Santa Fe. The item must be shipped to two plants, located in Des Moines and Fort Wayne, for assembly into a final product. At the beginning of a week the stockpile at Denver has 150 items and the stockpile at Santa Fe has 180. The Des Moines plant needs at least 100 of these items, and the Fort Wayne plant needs at least 120 that week. The cost of sending each item from the stockpile at Denver to the plants in Des Moines and Fort Wayne is, respectively, $200 and $300, while the cost of sending each item from the stockpile at Santa Fe to

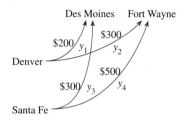

**Figure 4.7**

the plants in Des Moines and Fort Wayne is, respectively, $300 and $500. Refer to Figure 4.7. Find the number of items that should be shipped from each stockpile to each plant in order to minimize cost.

**Solution**    The following table summarizes the cost of shipping the items.

|  | *Des Moines Plant* | *Fort Wayne Plant* | *Supply* |
|---|---|---|---|
| Stockpile at Denver | $200 | $300 | 150 |
| Stockpile at Santa Fe | $300 | $500 | 180 |
| Demand | 100 | 120 | |

Let $y_1$ and $y_2$ be the number of items shipped from the stockpile in Denver to the plants in Des Moines and Fort Wayne, respectively. Let $y_3$ and $y_4$ be the number of items shipped from the stockpile in Santa Fe to the plants in Des Moines and Fort Wayne, respectively.

Then the problem is

$$\text{Minimize} \quad w = 200y_1 + 300y_2 + 300y_3 + 500y_4$$

$$\text{Subject to} \quad y_1 + y_2 \qquad\qquad \le 150$$
$$y_3 + y_4 \le 180$$
$$y_1 \qquad + y_3 \qquad \ge 100$$
$$y_2 \qquad + y_4 \ge 120$$
$$y_1, y_2, y_3, y_4 \ge \quad 0$$

The first two problem constraints have $\le$ inequalities. We correct this by multiplying both of these inequalities by $-1$ and obtain

$$\text{Minimize} \quad w = 200y_1 + 300y_2 + 300y_3 + 500y_4$$

$$\text{Subject to} \quad -y_1 - y_2 \qquad\qquad \ge -150$$
$$- y_3 - y_4 \ge -180$$
$$y_1 \qquad + y_3 \qquad \ge \quad 100$$
$$y_2 \qquad + y_4 \ge \quad 120$$
$$y_1, y_2, y_3, y_4 \ge \qquad 0$$

At first there seems to be a serious difficulty with this formulation of the problem since the constants to the right of the first two inequalities are *negative*. But we will solve this problem by first solving the dual problem, which does not have this difficulty. The dual problem is

$$\text{Maximize} \quad z = -150x_1 - 180x_2 + 100x_3 + 120x_4$$

$$\text{Subject to} \quad \begin{aligned} -x_1 \quad\quad + x_3 \quad\quad &\le 200 \\ -x_1 \quad\quad\quad\quad + x_4 &\le 300 \\ -x_2 + x_3 \quad\quad &\le 300 \\ -x_2 \quad\quad + x_4 &\le 500 \\ x_1, x_2, x_3, x_4 &\ge 0 \end{aligned}$$

As we see, this problem is a *standard maximization problem*. The initial tableau is

| $x_1$ | $x_2$ | $x_3$ | $x_4$ | $s_1$ | $s_2$ | $s_3$ | $s_4$ | $z$ | | ratios |
|---|---|---|---|---|---|---|---|---|---|---|
| $-1$ | $0$ | $1$ | $0$ | $1$ | $0$ | $0$ | $0$ | $0$ | $200$ | |
| $-1$ | $0$ | $0$ | ①  | $0$ | $1$ | $0$ | $0$ | $0$ | $300$ | $\frac{300}{1}$ ← smallest ratio |
| $0$ | $-1$ | $1$ | $0$ | $0$ | $0$ | $1$ | $0$ | $0$ | $300$ | |
| $0$ | $-1$ | $0$ | $1$ | $0$ | $0$ | $0$ | $1$ | $0$ | $500$ | $\frac{500}{1}$ |
| $150$ | $180$ | $-100$ | $-120$ | $0$ | $0$ | $0$ | $0$ | $1$ | $0$ | |

↑
pivot column

Pivoting about the second row and fourth column yields

| $x_1$ | $x_2$ | $x_3$ | $x_4$ | $s_1$ | $s_2$ | $s_3$ | $s_4$ | $z$ | | ratios |
|---|---|---|---|---|---|---|---|---|---|---|
| $-1$ | $0$ | ①  | $0$ | $1$ | $0$ | $0$ | $0$ | $0$ | $200$ | $\frac{200}{1}$ ← smallest ratio |
| $-1$ | $0$ | $0$ | $1$ | $0$ | $1$ | $0$ | $0$ | $0$ | $300$ | |
| $0$ | $-1$ | $1$ | $0$ | $0$ | $0$ | $1$ | $0$ | $0$ | $300$ | $\frac{300}{1}$ |
| $1$ | $-1$ | $0$ | $0$ | $0$ | $-1$ | $0$ | $1$ | $0$ | $200$ | |
| $30$ | $180$ | $-100$ | $0$ | $0$ | $120$ | $0$ | $0$ | $1$ | $36,000$ | |

↑
pivot column

Pivoting about the first row and third column yields

| $x_1$ | $x_2$ | $x_3$ | $x_4$ | $s_1$ | $s_2$ | $s_3$ | $s_4$ | $z$ | | ratios |
|---|---|---|---|---|---|---|---|---|---|---|
| $-1$ | 0 | 1 | 0 | 1 | 0 | 0 | 0 | 0 | 200 | |
| $-1$ | 0 | 0 | 1 | 0 | 1 | 0 | 0 | 0 | 300 | |
| ① | $-1$ | 0 | 0 | $-1$ | 0 | 1 | 0 | 0 | 100 | $\frac{100}{1}$ ← smallest ratio |
| 1 | $-1$ | 0 | 0 | 0 | $-1$ | 0 | 1 | 0 | 200 | $\frac{200}{1}$ |
| $-70$ | 180 | 0 | 0 | 100 | 120 | 0 | 0 | 1 | 56,000 | |

↑
pivot column

Pivoting about the third row and first column yields

| $x_1$ | $x_2$ | $x_3$ | $x_4$ | $s_1$ | $s_2$ | $s_3$ | $s_4$ | $z$ | |
|---|---|---|---|---|---|---|---|---|---|
| 0 | $-1$ | 1 | 0 | 0 | 0 | 1 | 0 | 0 | 300 |
| 0 | $-1$ | 0 | 1 | $-1$ | 1 | 1 | 0 | 0 | 400 |
| 1 | $-1$ | 0 | 0 | $-1$ | 0 | 1 | 0 | 0 | 100 |
| 0 | 0 | 0 | 0 | 1 | $-1$ | $-1$ | 1 | 0 | 100 |
| 0 | 110 | 0 | 0 | 30 | 120 | 70 | 0 | 1 | 63,000 |

This is the final tableau. Since the maximum of this problem is 63,000, the minimum of the primal problem is also 63,000. We find the solution to the primal problem under the slack variables in the last row to be (30, 120, 70, 0). ∎

**E X A M P L E   5**    **No Solution**

Solve

$$\text{Minimize} \qquad w = 2y_1 + 4y_2$$
$$\text{Subject to} \qquad y_1 - y_2 \geq 1$$
$$-2y_1 + y_2 \geq 3$$
$$y_1, y_2 \geq 0$$

**Solution**    The dual is

$$\text{Maximize} \qquad z = x_1 + 3x_2$$
$$\text{Subject to} \qquad x_1 - 2x_2 \leq 2$$
$$-x_1 + x_2 \leq 4$$
$$x_1, x_2 \geq 0$$

The initial tableau is

| $x_1$ | $x_2$ | $s_1$ | $s_2$ | $z$ | | ratio |
|---|---|---|---|---|---|---|
| 1 | −2 | 1 | 0 | 0 | 2 | |
| −1 | ① | 0 | 1 | 0 | 4 | $\frac{4}{1}$ ← smallest ratio |
| −1 | −3 | 0 | 0 | 1 | 0 | |

Pivoting about the second row, second column then gives

| $x_1$ | $x_2$ | $s_1$ | $s_2$ | $z$ | |
|---|---|---|---|---|---|
| −1 | 0 | 1 | 2 | 0 | 10 |
| −1 | 1 | 0 | 1 | 0 | 4 |
| −4 | 0 | 0 | 3 | 1 | 12 |

The pivot column is the first, however, there are only negative ratios. Thus, according to the simplex method, there is no solution. ■

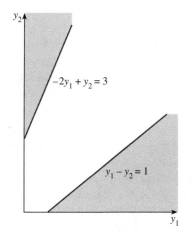

**Figure 4.8**

Figure 4.8 indicates that the feasible region for the primal problem in the last example is *empty*. Thus there can be no solution.

In the last section we noted that dividing a problem constraint by a constant will still lead to the correct answer, but that there will be a change in the economic interpretation of the corresponding slack variable. However, serious problems can arise when dividing a problem constraint by a constant in a dual problem! Suppose, for example, that we decide to solve a primal problem by solving the dual problem. *We must never divide any problem constraint in the dual problem by a constant since this will lead to an incorrect answer to the primal problem.* The difficulty arises because dividing a problem constraint by a constant in the dual problem will change the corresponding slack variable. Using the fundamental theorem of duality will then lead to an incorrect answer to the primal problem.

**W A R N I N G .**    When using the dual problem to solve a primal problem, never divide a problem constraint by a constant in the dual problem, since this will lead to an incorrect answer to the primal problem.

To illustrate this point, you are invited to divide the second problem constraint in the dual problem in Self-Help Exercise 2 by 2. The last row of the final tableau will yield slack variables of 2 and 6, which in turn would give the solution (2, 6) to the primal problem. This is not correct, the correct solution is (2, 3).

We can, however, divide a *primal* problem constraint by a constant without causing any difficulties.

### Economic Interpretation of Dual Problems

Let us now consider the economic interpretation of the dual problem to the problem of the wooden-boat builder of Example 2. First recall the table that gives the allocation of hours per week among the three operations.

| Operation | *Dinghies* | *Skiffs* | *Hours Available* |
|---|---|---|---|
| Cutting | 2 | 4 | 80 |
| Painting | 2 | 2 | 50 |
| Assembly | 4 | 2 | 84 |
| Profit | 60 | 80 | |

Suppose an entrepreneur wishes to purchase all of the resources of the wooden-boat company. Then the entrepreneur must determine the price to be paid for each of the resources. Let us define

$$y_1 = \text{price paid for 1 hour of labor in cutting}$$

$$y_2 = \text{price paid for 1 hour of labor in painting}$$

$$y_3 = \text{price paid for 1 hour of labor in assembly}.$$

Economists refer to these prices as **shadow prices**. (Shadow prices will be discussed in more detail in Section 4.5.) We will now show that the resource prices, $y_1$, $y_2$, $y_3$, should be determined by solving the dual problem. The total cost for all of these resources (per week) is $80y_1 + 50y_2 + 84y_3$. Since the entrepreneur wishes to minimize this cost, the problem is to

$$\text{minimize} \quad 80y_1 + 50y_2 + 84y_3.$$

This is precisely the objective function for the dual.

Now let us look at the constraints on the resource prices. Naturally the resource prices must be set high enough to induce the wooden-boat manufacturer to sell each one of the resources. For example, the boat manufacturer, in order to be induced to sell the production of dinghies, will demand that the cost of the labor, $2y_1 + 2y_2 + 4y_3$, be at least $60, or the boat manufacturer will simply continue to build the dinghies and make more profits. Thus for the entrepreneur to hope to buy the production of dinghies,

$$2y_1 + 2y_2 + 4y_3 \geq 60.$$

Similarly, to induce the boat manufacturer to sell the production of skiffs,

$$4y_1 + 2y_2 + 2y_3 \geq 80.$$

These are then the two constraints in the dual problem.

We now turn to the economic interpretation of a minimization problem.

**E X A M P L E   6**     **Economic Interpretation of the Dual**

Give an economic interpretation of the dual to the linear programming problem in Example 1.

**Solution**     The information can be summarized in the following table.

|  | Feed 1 | Feed 2 | Requirements |
|---|---|---|---|
| Nutrient A | 1 | 2 | 2 |
| Nutrient B | 1 | 1 | 4 |
| Nutrient C | 1 | 4 | 3 |
| Costs | 12 | 24 | |

The primal and dual problems are

| **Primal** | | **Dual** |
|---|---|---|

Minimize $\quad w = 12y_1 + 24y_2 \qquad$ Maximize $\quad z = 2x_1 + 4x_2 + 3x_3$

Subject to $\qquad y_1 + 2y_2 \geq 2 \qquad$ Subject to $\quad x_1 + x_2 + \quad x_3 \leq 12$

$$y_1 + \quad y_2 \geq 4 \qquad\qquad\qquad 2x_1 + x_2 + 4x_3 \leq 24$$

$$y_1 + 4y_2 \geq 3 \qquad\qquad\qquad\qquad x_1, x_2, x_3 \geq 0$$

$$y_1, y_2 \geq 0$$

To interpret the dual of this diet problem, suppose you are a salesman, selling nutrients A, B, and C. You want to ensure that the rancher will obtain the required amount of the three nutrients from you. You must determine

$$x_1 = \text{price per unit of nutrient A to charge rancher}$$

$$x_2 = \text{price per unit of nutrient B to charge rancher}$$

$$x_3 = \text{price per unit of nutrient C to charge rancher.}$$

You wish to maximize your revenue from selling the rancher the required nutrients for the animals. You will receive $2x_1 + 4x_2 + 3x_3$ dollars in revenue from the rancher. Thus you wish to

$$\text{maximize} \quad 2x_1 + 4x_2 + 3x_3.$$

This is precisely the objective function for the dual problem.

You must set the prices of the nutrients low enough to induce the rancher to buy all the needed nutrients from you. Your price for feed 1 is $x_1 + x_2 + x_3$. This must be at most \$12. Thus

$$x_1 + x_2 + x_3 \leq 12.$$

Similarly, considering your price for feed 2,

$$2x_1 + x_2 + 4x_3 \leq 24.$$

These are the two constraints in the dual problem.   ∎

## Matrix Formulations*

Problem (1) can be placed in a convenient matrix form no matter what the signs of the $a_{ij}$'s and the $b_i$'s. If we define

$$A = \begin{bmatrix} a_{11} & a_{12} & \cdots & a_{1n} \\ a_{21} & a_{22} & \cdots & a_{2n} \\ & & \cdots & \\ a_{m1} & a_{m2} & \cdots & a_{mn} \end{bmatrix}, \quad X = \begin{bmatrix} x_1 \\ x_2 \\ \vdots \\ x_n \end{bmatrix}, \quad Y = \begin{bmatrix} y_1 \\ y_2 \\ \vdots \\ y_m \end{bmatrix},$$

$$B = \begin{bmatrix} b_1 \\ b_2 \\ \vdots \\ b_m \end{bmatrix}, \quad C = [c_1 \quad c_2 \quad \cdots \quad c_n],$$

then (1) can be written as follows.

---

**Matrix Form of (1)**

$$\begin{aligned} \text{Maximize} \quad & z = CX \\ \text{Subject to} \quad & AX \leq B \\ & X \geq 0 \end{aligned}$$

---

We now introduce a convenient matrix operation. Given a matrix such as

$$\begin{bmatrix} 1 & 3 & 4 \\ 2 & 5 & 9 \end{bmatrix},$$

we can create another matrix whose rows are the columns of $A$. We call this matrix the **transpose** and denote the transpose by $A^T$. Thus if $A$ is given above,

$$A^T = \begin{bmatrix} 1 & 2 \\ 3 & 5 \\ 4 & 9 \end{bmatrix}.$$

---

*Optional

**Transpose of a Matrix**

The transpose of a matrix $A$, denoted by $A^T$, is the matrix obtained by interchanging the rows and columns of A, with the first row of $A$ becoming the first column of $A^T$, the second row of $A$ becoming the second column of $A^T$, and so on.

An examination of problems (1) and (2) indicates that the dual problem (2) can be written using the transpose notation as

**Matrix Form of (2)**

$$\text{Minimize} \quad w = B^T Y$$
$$\text{Subject to} \quad A^T Y \geq C^T$$
$$Y \geq 0$$

The definition of the dual can then be given in matrix form. We have the following.

**Matrix Form of a Dual Problem**

If the primal problem is

$$\left.\begin{array}{ll} \text{Maximize} & z = CX \\ \text{Subject to} & AX \leq B \\ & X \geq 0 \end{array}\right\} \quad (1)$$

Then the dual problem is

$$\left.\begin{array}{ll} \text{Minimize} & w = B^T Y \\ \text{Subject to} & A^T Y \geq C^T \\ & Y \geq 0 \end{array}\right\} \quad (2)$$

Also if (2) is the primal problem, then (1) is the dual problem.

**E X A M P L E   7**    **Finding the Matrix Formulation of a Linear Programming Problem**

Give the matrix formulation of the linear programming problem in the last chapter involving the maximization of profits in a wooden-boat builder.

$$\text{Maximize} \quad z = 60x_1 + 80x_2$$
$$\text{Subject to} \quad 2x_1 + 4x_2 \leq 80$$
$$2x_1 + 2x_2 \leq 50$$
$$4x_1 + 2x_2 \leq 84$$
$$x_1, x_2 \geq 0$$

**Solution**    If

$$A = \begin{bmatrix} 2 & 4 \\ 2 & 2 \\ 4 & 2 \end{bmatrix}, \quad X = \begin{bmatrix} x_1 \\ x_2 \end{bmatrix}, \quad B = \begin{bmatrix} 80 \\ 50 \\ 84 \end{bmatrix}, \quad C = [60 \quad 80],$$

then the problem can be written as

$$\begin{aligned} \text{Maximize} \quad & z = CX \\ \text{Subject to} \quad & AX \leq B \\ & X \geq 0 \quad \blacksquare \end{aligned}$$

**E X A M P L E  8**    **Finding the Dual Using the Matrix Formulation**

Give the matrix formulation of the dual problem to the linear programming problem in Example 7.

**Solution**    The matrix formulation of the dual problem is

$$\begin{aligned} \text{Minimize} \quad & w = B^T Y \\ \text{Subject to} \quad & A^T Y \geq C^T \\ & Y \geq 0, \end{aligned}$$

where

$$B^T = \begin{bmatrix} 80 \\ 50 \\ 84 \end{bmatrix}^T = [80 \quad 50 \quad 84], \quad A^T = \begin{bmatrix} 2 & 4 \\ 2 & 2 \\ 4 & 2 \end{bmatrix}^T = \begin{bmatrix} 2 & 2 & 4 \\ 4 & 2 & 2 \end{bmatrix},$$

$$C^T = [60 \quad 80]^T = \begin{bmatrix} 60 \\ 80 \end{bmatrix}. \quad \blacksquare$$

## SELF-HELP EXERCISE SET 4.3

1. Rambler Motors has two plants that produce two distinctive vehicles, the Rambler and the Convertible. The first plant can make three Ramblers and one convertible each month and costs $30,000 per month to operate. The second plant can make two Ramblers and two Convertibles each month and costs $40,000 per month to operate. If at least 12 Ramblers and at least 8 Convertibles are needed, how many months should each plant operate to minimize costs? For convenience let the cost function be in tens of thousands of dollars.

2. Give the economic interpretation of the dual in the previous exercise.

# EXERCISE SET 4.3

In Exercises 1 through 18 add the nonnegativity constraint and solve by using the dual. In some exercises you will need to multiply an inequality by $-1$ first.

**1.** Minimize $w = 3y_1 + 4y_2$

Subject to $\quad 2y_1 + y_2 \geq 3$

$\quad\quad\quad\quad y_1 + y_2 \leq 2$

**2.** Minimize $w = 4y_1 + 5y_2$

Subject to $3y_1 + y_2 \geq 4$

$\quad\quad\quad\quad y_1 + 2y_2 \geq 3$

**3.** Minimize $w = 3y_1 + 4y_2$

Subject to $3y_1 + 2y_2 \geq 11$

$\quad\quad\quad\quad y_1 \leq 3$

$\quad\quad\quad\quad y_2 \leq 4$

**4.** Minimize $w = y_1 + 2y_2$

Subject to $\quad y_1 + y_2 \geq 2$

$\quad\quad\quad\quad y_1 + y_2 \leq 4$

**5.** Minimize $w = 10y_1 + 20y_2$

Subject to $\quad y_1 + 2y_2 \geq 7$

$\quad\quad\quad\quad y_1 \leq 3$

$\quad\quad\quad\quad y_1 \geq 1$

$\quad\quad\quad\quad y_2 \leq 4$

**6.** Minimize $w = y_1 + 2y_2$

Subject to $\quad y_1 + y_2 \geq 2$

$\quad\quad\quad\quad y_1 \geq 1$

$\quad\quad\quad\quad y_2 \geq 1$

**7.** Minimize $w = 2y_1 + 3y_2$

Subject to $4y_1 + y_2 \geq 8$

$\quad\quad\quad\quad y_1 + y_2 \geq 5$

$\quad\quad\quad\quad y_1 + 2y_2 \geq 6$

**8.** Minimize $w = 20y_1 + 30y_2$

Subject to $\quad 3y_1 + y_2 \geq 8$

$\quad\quad\quad\quad y_1 + y_2 \geq 4$

$\quad\quad\quad\quad y_2 \geq 1$

**9.** Minimize $w = 3y_1 + 4y_2$

Subject to $3y_1 + 2y_2 \geq 14$

$\quad\quad\quad\quad y_1 \geq 2$

$\quad\quad\quad\quad y_2 \geq 1$

**10.** Minimize $w = y_1 + 2y_2$

Subject to $-y_1 + y_2 \geq 0$

$\quad\quad\quad\quad y_1 + y_2 \geq 4$

$\quad\quad\quad\quad y_2 \geq 1$

**11.** Minimize $w = 2y_1 + 2y_2 + y_3$

Subject to $\quad y_1 + y_2 + y_3 \leq 8$

$\quad\quad\quad\quad y_2 + y_3 \geq 1$

$\quad\quad\quad\quad y_1 + y_2 \quad\quad \geq 1$

**12.** Minimize $w = y_1 + 2y_2 + y_3$

Subject to $y_1 + 2y_2 + 2y_3 \geq 4$

$\quad\quad\quad\quad y_1 + 2y_2 + y_3 \leq 24$

$\quad\quad\quad\quad y_1 + y_2 + y_3 \geq 1$

**13.** Minimize $w = 16y_1 + 20y_2 + 9y_3$

Subject to $\quad y_1 + 4y_2 + y_3 \geq 66$

$\quad\quad\quad 4y_1 + 2y_2 + y_3 \geq 72$

$\quad\quad\quad 4y_1 + 10y_2 + 5y_3 \geq 200$

**14.** Minimize $w = y_1 + 2y_2 + 2y_3$

Subject to $y_1 + y_2 + y_3 \leq 4$

$$y_1 \geq 1$$

$$y_1 + y_2 + y_3 \geq 1$$

**15.** Minimize $w = 2y_1 + y_2 + y_3$

Subject to $y_1 + 2y_2 + y_3 \leq 4$

$$y_1 + y_2 \qquad \geq 1$$

$$y_3 \geq 2$$

**16.** Minimize $w = 7y_1 + 8y_2$

Subject to $y_1 + y_2 \geq 10$

$$4y_1 + 5y_2 \leq 80$$

$$3y_1 + 4y_2 \geq 36$$

**17.** Minimize $w = 2y_1 + 3y_2$

Subject to $y_1 - 2y_2 \geq 2$

$$-4y_1 + y_2 \geq 1$$

**18.** Minimize $w = 7y_1 + 8y_2$

Subject to $y_1 - y_2 \leq 1$

$$y_1 - 2y_2 \geq 2$$

**19.** Write the linear programming problems in Exercises 7 through 10 in matrix form. Then write the duals in matrix form.

**20.** Write the linear programming problems in Exercises 11 through 14 in matrix form by multiplying equations by $-1$ if necessary. Then write the duals in matrix form.

## Applications

**21. Nutrition.**   A dietitian must arrange a meal for a patient using two fruits— oranges and strawberries. Each orange costs 75 cents and contains 1 gram of fiber and 75 mg of vitamin C, while each cup of strawberries costs 75 cents and contains 2 grams of fiber and 60 mg of vitamin C. How many of each of these fruits needs to be eaten to minimize the cost and to ensure that a total of at least 8 grams of fiber and 420 mg of vitamin C will be consumed? What is the minimum cost?

**22. Nutrition.**   Suppose in the previous exercise the cost of each orange is 25 cents. Now answer the same questions.

**23. Costs.**   An oil company has two refineries and must refine and deliver at least 40,000 gallons of medium-grade gasoline and 24,000 gallons of high-grade gasoline to its stations. The first plant can refine 8000 gallons of medium-grade gasoline and 3000 gallons of high-grade gasoline and costs $40,000 a day to operate. The second plant can refine 4000 gallons of medium-grade gasoline and 6000 gallons of high-grade gasoline and costs $35,000 a day to operate. How many days should each plant operate to minimize the cost and what is the minimum cost? *Hint:* Minimize cost in thousands of dollars and divide each constraint by an appropriate constant.

**24. Cost.**   An airline has two types of airplanes and has contracted with a tour group to provide accommodations for a minimum of each of 80 first-class, 50 tourist, and 60 economy-class passengers. The first plane costs $4000 for the trip and can accommodate 40 first-class, 10 tourist, and 10 economy-class passengers, while the second plane costs $5000 for the trip and can accommodate

10 first-class, 10 tourist, and 20 economy-class passengers. How many of each type of airplane should be used to minimize the operating cost? *Hint:* Divide each of the constraints by 10 to simplify and minimize cost in thousands of dollars.

25. **Investments.** An investment made for a retirement plan must pay particular attention to risk. Suppose for such an investment no more than $10,250 is to be placed in three investments with estimated annual returns of 8%, 9%, and 10% respectively. The investor requires an annual return of at least $900 on the investment and wishes to minimize the risk. The second investment is twice as risky as the first, and the third investment is 4 times as risky as the first. How should the money be allocated to minimize the risk? *Hint:* Let risk = $y_1 + 2y_2 + 4y_3$.

26. **Diet.** An individual needs a daily supplement of at least 500 units of vitamin C and 200 units of vitamin E and agrees to obtain this supplement by eating two foods, I and II. Each ounce of food I contains 40 units of vitamin C and 10 units of vitamin E, while each ounce of food II contains 20 units of vitamin C and also 20 units of vitamin E. The total amount of these two foods must be at most 25 ounces. Unfortunately, food I contains 30 units of cholesterol per ounce, and food II contains 20 units of cholesterol per ounce. Find the appropriate amounts of the two food supplements so that cholesterol is minimized.

27. **Plant Nutrition.** A farmer can buy three types of 100-pound bags of fertilizer, type A, type B, and type C. Each 100-pound bag of type A fertilizer costs $20 and contains 40 pounds of nitrogen, 30 pounds of phosphoric acid, and 10 pounds of potash. Each 100-pound bag of type B fertilizer costs $30 and contains 20 pounds of nitrogen, 20 pounds of phosphoric acid, and 55 pounds of potash. Each 100-pound bag of type C fertilizer costs $20 and contains no nitrogen, 30 pounds of phosophoric acid, and 40 pounds of potash. The farmer requires 4000 pounds of nitrogen, 2000 pounds of phosphoric acid, and 2000 pounds of potash. How many bags of each type of fertilizer should be bought in order to minimize cost and what will be the minimum cost?

28. **Production Scheduling.** A company has 3 plants that produce 3 different sizes of their product. The North plant can produce 2000 items in the small size, 1000 in the medium size, and 1000 in the large size per day. The Center plant can produce 1000 items in the small size, 1000 in the medium size, and 2000 in the large size per day. The South plant can produce 2000 items in the small size, 2000 in the medium size, and 1000 in the large size per day. The company needs to produce at least 9000 of the small items, 8000 of the medium items, and 9000 of the large items on a given day. The cost per item of producing the small item is $2, the medium item $3, and the large item $2. Find the number of days each plant should operate in order to minimize the cost and find the minimum cost. *Hint:* Find the minimum cost in thousands of dollars and divide the constraints by an appropriate constant.

29. **Production Scheduling.** A mining company has three mines that produce three grades of ore. The North mine can produce 2 tons of low-grade ore, 1 ton of medium-grade ore, and 1 ton of high-grade ore each day. The Center mine can produce 1 ton of low-grade, 1 ton of medium-grade, and 4 tons of high-grade ore. The South mine can produce 1 ton of low-grade, 3 tons of medium-grade, and 3 tons of high-grade ore. At a particular time the company needs at

(© *Grant Heilman, Inc.*)

least 21 tons of low-grade ore, at least 19 tons of medium-grade ore, and at least 25 tons of high-grade ore. It costs $2000 per day to operate the North mine for a day, $3000 a day for the Center mine, and $4000 for the South mine. Find the number of days each mine should operate to minimize cost and find the minimum cost. *Hint:* Minimize cost in thousands of dollars.

30. **Transportation.**   Redo Example 4 of the text using the following table.

|              | Plant A | Plant B | Supply |
|--------------|---------|---------|--------|
| Stockpile I  | $20     | $40     | 500    |
| Stockpile II | $50     | $20     | 100    |
| Demand       | 150     | 250     |        |

*(© Mike Chuang/FPG International Corporation.)*

31. **Purchasing.**   A store decides to have a sale on pink and red azaleas. Two nurseries, East and West, will supply these plants. East nursery charges $4 for a pink azalea and $6 for a red one. West nursery charges $7 for a pink azalea and $5 for a red one. East nursery can supply at most 200 azaleas of either color while West nursery can supply at most 400 of these azaleas. The store needs at least 250 pink azaleas and at least 300 red azaleas. How many azaleas of each color should the store order from each nursery in order to minimize the cost? What is the minimum cost?

32. **Transportation.**   Redo Example 4 of the text using the following table. Notice there are now three plants.

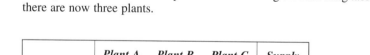

|              | Plant A | Plant B | Plant C | Supply |
|--------------|---------|---------|---------|--------|
| Stockpile I  | $20     | $30     | $40     | 500    |
| Stockpile II | $100    | $40     | $50     | 300    |
| Demand       | 150     | 200     | 100     |        |

33. **School Rezoning.**   A small city has two high schools, NW School and NE School, located in two predominately white districts and two high schools, SW School and SE School, located in two predominately minority districts. A decision is made to bus at least 400 minority students from SW and at least 300 minority students from SE to the two schools in the predominately white districts. NW can accommodate at most 200 additional students and NE at most 600. The weekly cost per student of busing from SW to NW is $4, from SW to NE $2, from SE to NW $2, and from SE to NE is $4. Determine the number of students that should be bused from each of the schools in the southern districts to each of the schools in the northern districts in order to minimize cost. What is the minimum cost per week?

34. **School Rezoning.**   Redo the previous exercise if the total miles bused needs to be minimized, where the distance from SW to NW is 4, from SE to NW is 2, from SW to NE is 3, and from SE to NE is 5.

35. **Profits.** Give an economic interpretation to the dual of the following problem. A firm manufactures 3 products, each of which uses 3 ingredients. The first product requires 2 units of the first ingredient, 1 unit of the second, and 1 unit of the third. The second product requires 1 unit of the first ingredient, 1 unit of the second, and 3 units of the third. The third product requires 1 unit of the first ingredient, 4 units of the second, and 3 units of the third. There are only 20 units of the first ingredient, 30 units of the second ingredient, and 40 units of the third ingredient available. The profit on each of the first product is $26, on the second is $24, and on the third is $30. Find the number of each product that should be made in order to maximize profits.

36. **Profits.** Give an economic interpretation to the dual of the following problem. A company makes 3 styles of sleds—Classic, Sport, and Deluxe. The Classic uses 2 board-feet of wood, 1 pound of iron, and 2 hours of labor. The Sport uses 1 board-foot of wood, 1 pound of iron, and 2 hours of labor. The Deluxe uses 1 board-foot of wood, 2 pounds of iron, and 1 hour of labor. There are 20 board-feet of wood, 30 pounds of iron, and 20 hours of labor available. The profit on the Classic is $12, on the Sport $11, and on the Deluxe $12. Find the number of each type of sled to make in order to maximize the profits.

37. **Nutrition.** State the dual problem to Exercise 21 and give an economic interpretation to the dual problem.

38. **Cost.** State the dual problem to Exercise 24 and give an economic interpretation to the dual problem.

39. **Plant Nutrition.** State the dual problem to Exercise 27 and give an economic interpretation to the dual problem.

40. **Diet.** State the dual problem to Exercise 26 and give an economic interpretation to the dual problem.

41. **Production Scheduling.** State the dual problem to Exercise 29 and give an economic interpretation to the dual problem.

42. **Production Scheduling.** State the dual problem to Exeircse 28 and give an economic interpretation to the dual problem.

## Graphing Calculator Exercises

Solve the next two exercises by solving the dual problem.

43. Minimize $\quad z = 79x_1 + 51x_2 + 118x_3$

$$\text{Subject to} \quad 9x_1 + 11x_2 + 12x_3 \geq 380$$
$$29x_1 + 11x_2 + 28x_3 \geq 800$$
$$27x_1 + 61x_2 + 82x_3 \geq 2200$$
$$x_1, x_2, x_3 \geq 0$$

44. Minimize $\quad z = 32x_1 + 16x_2 + 30x_3$

$$\text{Subject to } 48x_1 + 29x_2 + 41x_3 \geq 460$$
$$49x_1 + 58x_2 + 77x_3 \geq 1300$$
$$21x_1 + 6x_2 + 15x_3 \geq 310$$
$$x_1, x_2, x_3 \geq 0$$

## Solutions to Self-Help Exercise Set 4.3

**1.** Let $y_1$ and $y_2$ be the respective number of months that the first and second plants operate and let the cost $C$ be in tens of thousands of dollars. Then the problem is

$$\text{Minimize} \qquad C = 3y_1 + 4y_2$$
$$\text{Subject to} \qquad 3y_1 + 2y_2 \geq 12$$
$$y_1 + 2y_2 \geq 8$$
$$y_1, y_2 \geq 0$$

The dual of this problem is then

$$\text{Maximize} \qquad z = 12x_1 + 8x_2$$
$$\text{Subject to} \qquad 3x_1 + x_2 \leq 3$$
$$2x_1 + 2x_2 \leq 4$$
$$x_1, x_2 \geq 0$$

The initial tableau for the dual problem given above is

| $x_1$ | $x_2$ | $s_1$ | $s_2$ | $z$ | | |
|---|---|---|---|---|---|---|
| ③ | 1 | 1 | 0 | 0 | 3 | $\frac{3}{3} = 1$ |
| 2 | 2 | 0 | 1 | 0 | 4 | $\frac{4}{2} = 2$ |
| $-12$ | $-8$ | 0 | 0 | 1 | 0 | |

Pivoting about the first row and first column yields

| $x_1$ | $x_2$ | $s_1$ | $s_2$ | $z$ | | |
|---|---|---|---|---|---|---|
| 1 | $\frac{1}{3}$ | $\frac{1}{3}$ | 0 | 0 | 1 | $\dfrac{1}{\frac{1}{3}} = 3$ |
| 0 | $\left(\frac{4}{3}\right)$ | $-\frac{2}{3}$ | 1 | 0 | 2 | $\dfrac{2}{\frac{4}{3}} = 1.5$ |
| 0 | $-4$ | 4 | 0 | 1 | 12 | |

Pivoting about the second row and second column gives

| $x_1$ | $x_2$ | $s_1$ | $s_2$ | $z$ | |
|---|---|---|---|---|---|
| 1 | 0 | $\frac{1}{2}$ | $-\frac{1}{4}$ | 0 | $\frac{1}{2}$ |
| 0 | 1 | $-\frac{1}{2}$ | $\frac{3}{4}$ | 0 | $\frac{3}{2}$ |
| 0 | 0 | 2 | 3 | 1 | 18 |

This is the final tableau. Looking under the slack variables in the last row, we see a 2 and a 3. Thus the solution of the primal problem is (2, 3). This means that the first plant should operate for 2 months and the second for 3 months. The minimum of the primal problem is the same as the maximum of the dual problem, which is found in the last entry of the last row to be 18. Thus, considering the units, the cost is $18 \times 10{,}000 = 180{,}000$ dollars.

2. You have your own manufacturing plants that can produce these automobiles. You offer to sell to Rambling Motors a Rambler for $x_1$ tens of thousands of dollars and a Convertible for $x_2$ tens of thousands of dollars. Now 12 Ramblers and 8 Convertibles will yield revenue in tens of thousands of dollars equal to $12x_1 + 8x_2$. It is this revenue you wish to maximize. In the first plant Rambling Motors makes 3 Ramblers and 1 Convertible a month at a cost of $30,000. In the second plant Rambling Motors makes 2 Ramblers and 2 Convertibles a month at a cost of $40,000. Thus to get Rambling to buy the vehicles from you, you need to set prices $x_1$ and $x_2$ so that you can sell Rambling 3 Ramblers and 1 Convertible for *at most* $30,000 and also sell 2 Ramblers and 2 Convertibles for *at most* $40,000. That is,

$$3x_1 + x_2 \le 3$$
$$2x_1 + 2x_2 \le 4$$

Thus your problem is

$$\text{Maximize} \quad R = 12x_1 + 8x_2$$
$$\text{Subject to} \quad 3x_1 + x_2 \le 3$$
$$2x_1 + 2x_2 \le 4$$
$$x_1, x_2 \ge 0$$

## 4.4  THE SIMPLEX METHOD: NONSTANDARD AND MINIMIZATION PROBLEMS

► *Nonstandard Maximum Problem*

► *Minimization Problems*

**Tjalling Koopmans**

*(UPI/Bettmann.)*

Dantzig visited the well known mathematical economist Koopmans in June 1947 to talk to him about linear programming. Koopmans quickly realized that a good part of economics could be translated into the linear programming format. He then led a group of economists who developed the theory of allocation of resources and its relation to linear programming. For this work he was awarded the Nobel prize in economics in 1975. At that time he expressed regret that Dantzig was not sharing the honor with him.

In this section we will expand the simplex method to include maximization problems in which the constants to the right of the inequality $\le$ may be negative. We will also see how this technique can be used to solve minimization problems.

There are two methods for handling these problems: the big-M method, developed by Charnes and Cooper in 1961 and Crown's method, developed by Crown in 1982. We will use Crown's method since this method does not use the artificial variables of the other method and normally requires fewer tableaus.

## Nonstandard Maximum Problem

In a standard maximization problem the constraints are in the form using the inequality $\le$ and the constants to the right of the inequalities $\le$ are positive. In the last chapter we encountered maximization problems with inequalities of the form

$$x_1 + x_2 \ge 2.$$

We shall now multiply such an inequality by $-1$, yielding

$$-x_1 - x_2 \le -2.$$

This yields the inequality $\le$, but now a *negative* constant appears to the right of the inequality $\le$.

We shall see how to pivot about certain elements until the constants to the right of all the inequalities $\le$ become nonnegative. The problem is then in standard form. We then proceed as in the previous section.

Consider now the following example.

**E X A M P L E   1**   **Simplex Method on Nonstandard Problem**

Two types of metal connectors require for their manufacture spending a certain amount of minutes on a stamping machine and on a grinding machine. The first type of connector requires 4 minutes on the stamping machine and 3 minutes on the grinding machine, while the second connector requires 5 minutes on the stamping machine and 1 minute on the grinding machine. The stamping machine is available at most 50 minutes each hour and the grinding machine is available at most 21 minutes each hour. Furthermore, each hour a total of at least two of the connectors must be produced. If each of the first type connector brings a profit of $0.50 and each of the second type of connector a profit of $0.60, use the simplex method to find the number of each of these products that should be manufactured to maximize profits.

**Solution**   This problem was considered in the last chapter in Self-Help Exercise 3.2. If $x_1$ and $x_2$ are, respectively, the number of the first type connector and the second type connector manufactured, then the problem is

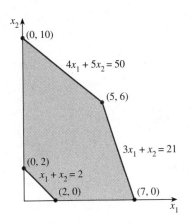

**Figure 4.9**

Maximize $\quad z = 50x_1 + 60x_2$

Subject to $\quad 4x_1 + 5x_2 \leq 50$

$$3x_1 + x_2 \leq 21$$

$$x_1 + x_2 \geq 2$$

$$x_1, x_2 \geq 0$$

For convenience $z$ is given in cents. Figure 4.9 shows the feasible region. The first step is to multiply the third constraint by $-1$. This then yields

$$4x_1 + 5x_2 \leq 50$$

$$3x_1 + x_2 \leq 21$$

$$-x_1 - x_2 \leq -2$$

$$x_1, x_2 \geq 0$$

The initial tableau is then

| $x_1$ | $x_2$ | $s_1$ | $s_2$ | $s_3$ | $z$ | | ratios |
|---|---|---|---|---|---|---|---|
| 4 | 5 | 1 | 0 | 0 | 0 | 50 | $\frac{50}{4} = 12.5$ |
| ③ | 1 | 0 | 1 | 0 | 0 | 21 | $\frac{21}{3} = 7 \leftarrow$ smallest of the $\frac{+}{+}$ ratios |
| $-1$ | $-1$ | 0 | 0 | 1 | 0 | $-2$ | |
| $-50$ | $-60$ | 0 | 0 | 0 | 1 | 0 | |

The $-2$ in the last column indicates that this tableau is not in standard form. From Figure 4.9 we see that $(0, 0)$ is not feasible.

We now consider the third row that contains the $-2$ in the last entry. We now **pick the most negative element in the third row**. Since in this case there is a tie, we will pick the first $-1$ and **use the column this element is in as the pivot column**.

Now form all the ratios as usual. First ignore all ratios that are of the form $\dfrac{\text{negative number}}{\text{negative number}}$, and pick the smallest nonnegative one according to the usual rules, picking the first to break any ties. The row this element is in then determines the pivot row. If this procedure should fail, that is, if there are no such ratios to be considered, then take the **largest** of the ratios of the form $\left(\dfrac{\text{negative}}{\text{negative}}\right)$ and use the row this element is in as the pivot row. (There must always be such a ratio.)

According to this rule, the second row is the pivot row. Dividing the second row by 3, we obtain

| $x_1$ | $x_2$ | $s_1$ | $s_2$ | $s_3$ | $z$ | | |
|---|---|---|---|---|---|---|---|
| 4 | 5 | 1 | 0 | 0 | 0 | 50 | $R_1 - 4R_2 \to R_1$ |
| ① | $\frac{1}{3}$ | 0 | $\frac{1}{3}$ | 0 | 0 | 7 | |
| $-1$ | $-1$ | 0 | 0 | 1 | 0 | $-2$ | $R_3 + R_2 \to R_3$ |
| $-50$ | $-60$ | 0 | 0 | 0 | 1 | 0 | $R_4 + 50R_2 \to R_4$ |

Performing the indicated row operations gives

| $x_1$ | $x_2$ | $s_1$ | $s_2$ | $s_3$ | $z$ | | ratios |
|---|---|---|---|---|---|---|---|
| 0 | $\textcircled{\frac{11}{3}}$ | 1 | $-\frac{4}{3}$ | 0 | 0 | 22 | $\frac{22}{\frac{11}{3}} = 6 \leftarrow$ smallest of the $\frac{+}{+}$ ratios |
| 1 | $\frac{1}{3}$ | 0 | $\frac{1}{3}$ | 0 | 0 | 7 | $\frac{7}{\frac{1}{3}} = 21$ |
| 0 | $-\frac{2}{3}$ | 0 | $\frac{1}{3}$ | 1 | 0 | 5 | |
| 0 | $-\frac{130}{3}$ | 0 | $\frac{50}{3}$ | 0 | 1 | 350 | |

Notice that the problem has now become a standard problem. The point corresponding to this last tableau is $(7, 0)$. The pivot column is the second and the above ratios indicate that the first row is the pivot row. Dividing the first row by $\frac{11}{3}$ yields

| $x_1$ | $x_2$ | $s_1$ | $s_2$ | $s_3$ | $z$ | | |
|---|---|---|---|---|---|---|---|
| 0 | ① | $\frac{3}{11}$ | $-\frac{4}{11}$ | 0 | 0 | 6 | |
| 1 | $\frac{1}{3}$ | 0 | $\frac{1}{3}$ | 0 | 0 | 7 | $R_2 - \frac{1}{3}R_1 \to R_2$ |
| 0 | $-\frac{2}{3}$ | 0 | $\frac{1}{3}$ | 1 | 0 | 5 | $R_3 + \frac{2}{3}R_1 \to R_3$ |
| 0 | $-\frac{130}{3}$ | 0 | $\frac{50}{3}$ | 0 | 1 | 350 | $R_4 + \frac{130}{3}R_1 \to R_4$ |

The above row operations yield

| $x_1$ | $x_2$ | $s_1$ | $s_2$ | $s_3$ | $z$ | |
|---|---|---|---|---|---|---|
| 0 | 1 | $\frac{3}{11}$ | $-\frac{4}{11}$ | 0 | 0 | 6 |
| 1 | 0 | $-\frac{1}{11}$ | $\frac{5}{11}$ | 0 | 0 | 5 |
| 0 | 0 | $\frac{2}{11}$ | $\frac{1}{11}$ | 1 | 0 | 9 |
| 0 | 0 | $\frac{130}{11}$ | $\frac{10}{11}$ | 0 | 1 | 610 |

This is the final tableau and gives the point (5, 6) as the point at which the objective function attains the maximum of 610.   ∎

We can think of the solution as a two-stage process. In the first stage we pivot about appropriate elements until the tableau is in standard form. Then in the second stage we proceed as in Section 4.2.

We now state the rules for applying the simplex method in the case when some constants to the right of the inequalities $\leq$ are negative.

---

**Extended Simplex Method**

If some constant to the right of an inequality $\leq$ is negative, then go to Stage I; otherwise, go to Stage II.

***Stage I***   Select the first negative element in the last column (ignoring the entry in the last row).

1.  Consider the row this element is in. Pick the most negative element in this row (ignoring the last entry). The column containing this element will be the pivot column. If there is no such negative entry, then stop. There is no solution.
2.  Form all the ratios as usual. First ignore all ratios that are of the form $\left(\dfrac{\text{negative}}{\text{negative}}\right)$ and pick the smallest nonnegative one according to the usual rules, picking the first to break any ties. The row this element is in then determines the pivot row. If this procedure should fail, that is, there are no such ratios to be considered, then take the **largest** ratio of the form $\left(\dfrac{\text{negative}}{\text{negative}}\right)$, and use the row this element is in as the pivot row. (There must always be such a ratio.)
3.  Pivot about the pivot element and repeat as necessary until the tableau is in standard form.

***Stage II***   Proceed as in Section 4.2.

---

## Minimization Problems

We can readily convert a minimization problem into a maximization problem by observing that the minimum of an objective function such as $z = 10 - 2x_1 + 3x_2$ occurs at the same point as the maximum of the negative of this function $-z = -10 + 2x_1 - 3x_2$. Furthermore, the minimum of the original objective function $z = 10 - 2x_1 + 3x_2$ is the maximum of the negative of this function $-z = -10 + 2x_1 - 3x_2$.

Thus, when asked to minimize an objective function, we instead maximize the negative of this function.

We now look at an important type of problem: the transportation problem. This problem was also done in the last section using the dual. We now see how to solve this problem without using the dual.

**E X A M P L E   2**    **Extended Simplex Method on Nonstandard Problem**

A firm has two stockpiles of an item stored in warehouses at two different locations, Denver and Santa Fe. The item must be shipped to two plants, located in Des Moines and Fort Wayne, for assembly into a final product. At the beginning of a week the stockpile at Denver has 150 items and the stockpile at Santa Fe has 180. The Des Moines plant needs at least 100 of these items and the Fort Wayne plant needs at least 120 that week. The cost of sending each item from the stockpile at Denver to the plants in Des Moines and Fort Wayne is, respectively, $200 and $300, while the cost of sending each item from the stockpile at Santa Fe to the plants in Des Moines and Fort Wayne is, respectively, $300 and $500. Find the number of items that should be shipped from each stockpile to each plant in order to minimize cost.

**Solution**    The following table summarizes the cost of shipping the items.

|  | Des Moines Plant | Fort Wayne Plant | Supply |
|---|---|---|---|
| Stockpile at Denver | $200 | $300 | 150 |
| Stockpile at Santa Fe | $300 | $500 | 180 |
| Demand | 100 | 120 | |

Let $x_1$ and $x_2$ be the number of items shipped from the stockpile in Denver to the plants in Des Moines and Fort Wayne, respectively. Let $x_3$ and $x_4$ be the number of items shipped from the stockpile in Santa Fe to the plants in Des Moines and Fort Wayne, respectively.

Then the problem is

$$\text{Minimize} \quad C = 200x_1 + 300x_2 + 300x_2 + 500x_4$$

$$
\begin{aligned}
\text{Subject to} \quad & x_1 + x_2 && && \le 150 \\
& && x_3 + x_4 && \le 180 \\
& x_1 && + x_3 && \ge 100 \\
& x_2 && + x_4 && \ge 120 \\
& x_1, x_2, x_3, x_4 && && \ge \quad 0
\end{aligned}
$$

This can then be considered as the following equivalent problem.

$$\text{Maximize} \quad z = -C = -200x_1 - 300x_2 - 300x_3 - 500x_4$$

$$
\begin{aligned}
\text{Subject to} \quad & x_1 + x_2 && && \le \quad 150 \\
& && x_3 + x_4 && \le \quad 180 \\
& -x_1 && - x_3 && \le -100 \\
& -x_2 && - x_4 && \le -120 \\
& x_1, x_2, x_3, x_4 && && \ge \quad 0
\end{aligned}
$$

The initial tableau is

| $x_1$ | $x_2$ | $x_3$ | $x_4$ | $s_1$ | $s_2$ | $s_3$ | $s_4$ | $z$ | | ratio |
|---|---|---|---|---|---|---|---|---|---|---|
| ① | 1 | 0 | 0 | 1 | 0 | 0 | 0 | 0 | 150 | $\frac{150}{1} \leftarrow$ smallest $\frac{+}{+}$ ratio |
| 0 | 0 | 1 | 1 | 0 | 1 | 0 | 0 | 0 | 180 | |
| $-1$ | 0 | $-1$ | 0 | 0 | 0 | 1 | 0 | 0 | $-100$ | |
| 0 | $-1$ | 0 | $-1$ | 0 | 0 | 0 | 1 | 0 | $-120$ | |
| 200 | 300 | 300 | 500 | 0 | 0 | 0 | 0 | 1 | 0 | |

Since the third row has a $-100$ as the last entry and is the first such negative encountered in the last column, we first consider the third row. There are two elements with a $-1$. We arbitrarily pick the first one and thus select the first column as the pivot column. There is only one ratio to consider. This indicates pivoting about the first row and first column. We then obtain

| $x_1$ | $x_2$ | $x_3$ | $x_4$ | $s_1$ | $s_2$ | $s_3$ | $s_4$ | $z$ | | ratios |
|---|---|---|---|---|---|---|---|---|---|---|
| 1 | 1 | 0 | 0 | 1 | 0 | 0 | 0 | 0 | 150 | $\frac{150}{1}$ |
| 0 | 0 | 1 | 1 | 0 | 1 | 0 | 0 | 0 | 180 | |
| 0 | ① | $-1$ | 0 | 1 | 0 | 1 | 0 | 0 | 50 | $\frac{50}{1} \leftarrow$ smallest $\frac{+}{+}$ ratio |
| 0 | $-1$ | 0 | $-1$ | 0 | 0 | 0 | 1 | 0 | $-120$ | |
| 0 | 100 | 300 | 500 | $-200$ | 0 | 0 | 0 | 1 | $-30,000$ | |

The $-120$ in the last entry of the fourth column results in selecting the fourth row for consideration. The first $-1$ is in the second column. Thus the second column becomes the pivot column. The two ratios indicate that the third row is the pivot row. Pivoting about the 1 in the third row and second column yields

| $x_1$ | $x_2$ | $x_3$ | $x_4$ | $s_1$ | $s_2$ | $s_3$ | $s_4$ | $z$ | | ratios |
|---|---|---|---|---|---|---|---|---|---|---|
| 1 | 0 | ① | 0 | 0 | 0 | $-1$ | 0 | 0 | 100 | $\frac{100}{1} \leftarrow$ smallest $\frac{+}{+}$ ratio |
| 0 | 0 | 1 | 1 | 0 | 1 | 0 | 0 | 0 | 180 | $\frac{180}{1}$ |
| 0 | 1 | $-1$ | 0 | 1 | 0 | 1 | 0 | 0 | 50 | |
| 0 | 0 | $-1$ | $-1$ | 1 | 0 | 1 | 1 | 0 | $-70$ | |
| 0 | 0 | 400 | 500 | $-300$ | 0 | $-100$ | 0 | 1 | $-35,000$ | |

The next pivot is in the first row and third column. Pivoting gives

| $x_1$ | $x_2$ | $x_3$ | $x_4$ | $s_1$ | $s_2$ | $s_3$ | $s_4$ | $z$ | | ratios |
|---|---|---|---|---|---|---|---|---|---|---|
| 1 | 0 | 1 | 0 | 0 | 0 | −1 | 0 | 0 | 100 | $\frac{100}{1}$ |
| −1 | 0 | 0 | 1 | 0 | 1 | 1 | 0 | 0 | 80 | |
| 1 | 1 | 0 | 0 | 1 | 0 | 0 | 0 | 0 | 150 | $\frac{150}{1}$ |
| ① | 0 | 0 | −1 | 1 | 0 | 0 | 1 | 0 | 30 | $\frac{30}{1}$ ← smallest $\frac{+}{+}$ ratio |
| −400 | 0 | 0 | 500 | −300 | 0 | 300 | 0 | 1 | −75,000 | |

The tableau is now in standard form. We continue as in Section 4.2 and pivot about the fourth row and first column, obtaining

| $x_1$ | $x_2$ | $x_3$ | $x_4$ | $s_1$ | $s_2$ | $s_3$ | $s_4$ | $z$ | |
|---|---|---|---|---|---|---|---|---|---|
| 0 | 0 | 1 | 1 | −1 | 0 | −1 | −1 | 0 | 70 |
| 0 | 0 | 0 | 0 | 1 | 1 | 1 | 1 | 0 | 110 |
| 0 | 1 | 0 | 1 | 0 | 0 | 0 | −1 | 0 | 120 |
| 1 | 0 | 0 | −1 | 1 | 0 | 0 | 1 | 0 | 30 |
| 0 | 0 | 0 | 100 | 100 | 0 | 300 | 400 | 1 | −63,000 |

This is the final tableau. The solution is the point (30, 120, 70, 0), and the minimum is the negative of the entry in the last row and last column, or $63,000. Thus ship 30 items from the stockpile in Denver to the plant in Des Moines, 120 items from the stockpile in Denver to the plant in Fort Wayne, 70 items from the stockpile in Santa Fe to the plant in Des Moines, and no items from the stockpile in Sante Fe to the plant in Fort Wayne. This will result in a minimum transportation cost of $63,000.  ∎

The same technique can readily be used to solve a transportation problem with any number of stockpiles and any number of plants.

**EXAMPLE 3**    **Simplex Method on Minimization Problem**

An animal rancher requires at least 12 units per day of nutrient A and at least 8 units per day of nutrient B for his animals. The rancher has two different mixtures of feeds available, feed 1 and 2. If each bag of feed 1 costs $3 and yields 3 units of nutrient A and 1 unit of B and each bag of feed 2 costs $4 and yields 2 units of nutrient A and 2 units of B, find the number of bags of each feed that should be used to minimize the cost.

**Solution**    If $x_1$ and $x_2$ are respectively the number of bags of feed 1 and feed 2, then the problem is

$$\text{Minimize} \qquad C = 3x_1 + 4x_2$$
$$\text{Subject to} \qquad 3x_1 + 2x_2 \geq 12$$
$$x_1 + 2x_2 \geq 8$$
$$x_1, x_2 \geq 0$$

We convert the problem to

$$\text{Maximize} \qquad w = -C = -3x_1 - 4x_2$$
$$\text{Subject to} \qquad -3x_1 - 2x_2 \leq -12$$
$$-x_1 - 2x_2 \leq -8$$
$$x_1, x_2 \geq 0$$

The initial tableau is

| $x_1$ | $x_2$ | $s_1$ | $s_2$ | $w$ | | ratios |
|---|---|---|---|---|---|---|
| $-3$ | $-2$ | $1$ | $0$ | $0$ | $-12$ | $\frac{-12}{-3} = 4$ |
| $\textcircled{-1}$ | $-2$ | $0$ | $1$ | $0$ | $-8$ | $\frac{-8}{-1} = 8 \leftarrow$ largest $\frac{(-)}{(-)}$ ratio. No $\frac{+}{+}$ ratio. |
| $3$ | $4$ | $0$ | $0$ | $1$ | $0$ | |

We begin with the first row. The most negative element in this row is in the first column. Thus the pivot column is the first column. Now there are only ratios of the form $\left(\dfrac{\text{negative}}{\text{negative}}\right)$. Picking the **largest** of these gives the second row as the pivot row. Pivoting gives

| $x_1$ | $x_2$ | $s_1$ | $s_2$ | $w$ | | ratio |
|---|---|---|---|---|---|---|
| $0$ | $\textcircled{4}$ | $1$ | $-3$ | $0$ | $12$ | $\frac{12}{4} = 3 \leftarrow$ smallest $\frac{+}{+}$ ratio |
| $1$ | $2$ | $0$ | $-1$ | $0$ | $8$ | $\frac{8}{2} = 4$ |
| $0$ | $-2$ | $0$ | $3$ | $1$ | $-24$ | |

This is in standard form. Pivoting about the first row and second column gives

| $x_1$ | $x_2$ | $s_1$ | $s_2$ | $w$ | |
|---|---|---|---|---|---|
| $0$ | $1$ | $\frac{1}{4}$ | $-\frac{3}{4}$ | $0$ | $3$ |
| $1$ | $0$ | $-\frac{1}{2}$ | $\frac{1}{2}$ | $0$ | $2$ |
| $0$ | $0$ | $\frac{1}{2}$ | $\frac{3}{2}$ | $1$ | $-18$ |

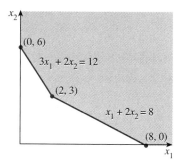

**Figure 4.10**

This is the final tableau. The minimum occurs at the point (2, 3) with minimum equal to $-(-18) = 18$. Thus the rancher should buy 2 bags of feed 1 and 3 bags of feed 2. The minimum cost is $18.

Figure 4.10 shows the feasible region. ∎

## SELF-HELP EXERCISE SET 4.4

1. Minimize $z = 2x_1 + 3x_2$ subject to the constraints

$$x_1 + x_2 \geq 3$$
$$x_1 \leq 2$$
$$x_2 \leq 2$$
$$x_1, x_2 \geq 0$$

## EXERCISE SET 4.4

In all of the following exercises use the extended simplex method to solve. In Exercises 1 through 26 assume the additional constraint that all the variables are nonnegative.

1. Maximize  $z = 2x_1 + x_2$
   Subject to  $x_1 + x_2 \leq 4$
   $-x_1 + 2x_2 \geq 2$

2. Maximize $z = x_1 + 2x_2$
   Subject to  $x_1 + x_2 \leq 3$
   $x_1 \geq 1$

3. Maximize  $z = x_1 + 2x_2$
   Subject to $x_1 + 3x_2 \geq 12$
   $x_1 + x_2 \leq 6$
   $x_2 \leq 4$

4. Maximize  $z = 3x_1 + 4x_2$
   Subject to $-x_1 + x_2 \leq -1$
   $x_1 + x_2 \leq 3$

5. Maximize $z = 2x_1 - 3x_2$
   Subject to $-x_1 + x_2 \leq 0$
   $x_1 + x_2 \leq 8$
   $x_1 \geq 2$

6. Maximize $z = 2x_1 + 3x_2$
   Subject to  $x_1 + x_2 \leq 3$
   $x_2 \leq 2$
   $x_2 \geq 1$

7. Maximize  $z = 2x_1 + 3x_2$
   Subject to $-x_1 + x_2 \leq 1$
   $-x_1 + x_2 \geq -1$
   $x_1 + x_2 \leq 5$
   $x_1 \geq 1$
   $x_2 \geq 1$

8. Maximize $z = 2x_1 + 3x_2$
   Subject to  $x_1 + x_2 \geq 2$
   $x_1 + x_2 \leq 4$
   $x_2 \leq 2$

9.  Maximize      $z = 2x_1 + 3x_2 + x_3$

    Subject to    $x_1 + x_2 + \phantom{3}x_3 \geq \phantom{0}660$
    $$4x_1 + 5x_2 + \phantom{1}3x_3 \leq 2600$$
    $$x_1 + 3x_2 + \phantom{1}2x_3 \leq 1360$$
    $$8x_1 + 5x_2 + 10x_3 \leq 7300$$

10. Maximize      $z = x_1 + \phantom{1}5x_2 + 3x_3$

    Subject to $3x_1 + \phantom{1}x_2 + \phantom{8}x_3 \geq 300$
    $$5x_1 + 16x_2 + 8x_3 \geq 120$$
    $$3x_1 + \phantom{1}6x_2 + 2x_3 \leq 450$$

11. Minimize  $z = 3x_1 + 4x_2$

    Subject to    $2x_1 + x_2 \geq 3$
    $$x_1 + x_2 \leq 2$$

12. Minimize $z = 4x_1 + 5x_2$

    Subject to $3x_1 + \phantom{2}x_2 \geq 4$
    $$x_1 + 2x_2 \geq 3$$

13. Minimize   $z = 3x_1 + 4x_2$

    Subject to $3x_1 + 2x_2 \geq 11$
    $$x_1 \leq \phantom{1}3$$
    $$x_2 \leq \phantom{1}4$$

14. Minimize $z = x_1 + 2x_2$

    Subject to   $x_1 + x_2 \geq 2$
    $$x_1 + x_2 \leq 4$$

15. Minimize $z = 10x_1 + 20x_2$

    Subject to      $x_1 + 2x_2 \geq 7$
    $$x_1 \leq 3$$
    $$x_1 \geq 1$$
    $$x_2 \leq 4$$

16. Minimize $z = x_1 + 2x_2$

    Subject to  $x_1 + x_2 \geq 2$
    $$x_1 \geq 1$$
    $$x_2 \geq 1$$

17. Minimize $z = 2x_1 + 3x_2$

    Subject to $4x_1 + \phantom{2}x_2 \geq 8$
    $$x_1 + \phantom{2}x_2 \geq 5$$
    $$x_1 + 2x_2 \geq 6$$

18. Minimize $z = 20x_1 + 30x_2$

    Subject to      $3x_1 + x_2 \geq 8$
    $$x_1 + x_2 \geq 4$$
    $$x_2 \geq 1$$

19. Minimize   $z = 3x_1 + 4x_2$

    Subject to $3x_1 + 2x_2 \geq 14$
    $$x_1 \geq \phantom{1}2$$
    $$x_2 \geq \phantom{1}1$$

20. Minimize   $z = x_1 + 2x_2$

    Subject to $-x_1 + x_2 \geq 0$
    $$x_1 + x_2 \geq 4$$
    $$x_2 \geq 1$$

21. Minimize $z = 2x_1 + 2x_2 + x_3$

    Subject to    $x_1 + x_2 + x_3 \leq 8$
    $$x_2 + x_3 \geq 1$$
    $$x_1 + x_2 \phantom{+ x_3} \geq 1$$

**22.** Minimize    $z = x_1 + 2x_2 + x_3$

Subject to $x_1 + 2x_2 + 2x_3 \geq 4$
$$x_1 + 2x_2 + x_3 \leq 24$$
$$x_1 + x_2 + x_3 \geq 1$$

**23.** Minimize    $z = 16x_1 + 20x_2 + 9x_3$

Subject to $x_1 + 4x_2 + x_3 \geq 66$
$$4x_1 + 2x_2 + x_3 \geq 72$$
$$4x_1 + 10x_2 + 5x_3 \geq 200$$

**24.** Minimize $z = x_1 + 2x_2 + 2x_3$

Subject to    $x_1 + x_2 + x_3 \leq 4$
$$x_1 \geq 1$$
$$x_1 + x_2 + x_3 \geq 1$$

**25.**  Minimize $z = 2x_1 + x_2 + x_3$        **26.**  Minimize  $z = 7x_1 + 8x_2$

Subject to $x_1 + 2x_2 + x_3 \leq 4$            Subject to  $x_1 + x_2 \geq 10$
$$x_1 + x_2 \geq 1 \qquad\qquad\qquad 4x_1 + 5x_2 \leq 80$$
$$x_3 \geq 2 \qquad\qquad\qquad 3x_1 + 4x_2 \geq 36$$

## Applications

**27.  Transportation.**  Redo Example 2 of the text using the following table.

|              | Plant A | Plant B | Supply |
|--------------|---------|---------|--------|
| Stockpile I  | $20     | $40     | 500    |
| Stockpile II | $50     | $20     | 100    |
| Demand       | 150     | 250     |        |

**28.  Transportation.**  Redo Example 2 of the text using the following table. Notice there are now three plants.

|              | Plant A | Plant B | Plant C | Supply |
|--------------|---------|---------|---------|--------|
| Stockpile I  | $20     | $30     | $40     | 500    |
| Stockpile II | $100    | $40     | $50     | 300    |
| Demand       | 150     | 200     | 100     |        |

**29.  Investments.**  In Example 2 of Section 3.3 the following linear programming problem was formulated (but not solved): Maximize $R = .06x + .08y + .10z$, subject to $x + y + z \leq 100$, $-x - y + z \leq 0$, $x \geq 30$, $y \geq 30$, $z \geq 20$. Solve this problem.

**30.  Investments.** An investment made for a retirement plan must pay particular attention to risk. Suppose for such an investment no more than $10,000 is placed in 3 investments with annual returns of 8%, 9%, and 10% respectively. The investor requires an annual return of at least $900 on the investment and wishes to minimize the risk. The second investment is twice as risky as the first, and the third investment is 4 times as risky as the first. How should the money be allocated to minimize the risk? *Hint:* Let risk $= x_1 + 2x_2 + 4x_3$.

**31.  Plant Nutrition.** A farmer can buy three types of 100-pound bags of fertilizer, type A, type B, and type C. Each 100-pound bag of type A fertilizer costs $20 and contains 40 pounds of nitrogen, 30 pounds of phosphoric acid, and 10 pounds of potash. Each 100-pound bag of type B fertilizer costs $30 and contains 20 pounds of nitrogen, 20 pounds of phosphoric acid, and 55 pounds of potash. Each 100-pound bag of type C fertilizer costs $20 and contains no nitrogen, 30 pounds of phosophoric acid, and 40 pounds of potash. The farmer requires 4000 pounds of nitrogen, 2000 pounds of phosphoric acid, and 2000 pounds of potash. How many bags of each type of fertilizer should the farmer buy in order to minimize cost? Find the minimum cost.

**32.  Cost.** An airline has two types of airplanes and has contracted with a tour group to provide accommodations for a minimum each of 80 first-class, 50 tourist, and 60 economy-class passengers. The first plane costs $4000 for the trip and can accommodate 40 first-class, 10 tourist, and 10 economy-class passengers, while the second plane costs $5000 for the trip and can accommodate 10 first-class, 10 tourist, and 20 economy-class passengers. How many of each type of airplane should be used to minimize the operating cost? Find the minimum cost. *Hint:* Divide each of the constraints by 10 to simplify and minimize cost in thousands of dollars.

*(© Guido A. Rossi/The Image Bank.)*

**33.  Costs.** An oil company has two refineries and must refine and deliver at least 40,000 gallons of medium-grade gasoline and 24,000 gallons of high-grade gasoline to its stations. The first plant can refine 8000 gallons of medium-grade gasoline and 3000 gallons of high-grade gasoline and costs $40,000 a day to operate. The second plant can refine 4000 gallons of medium-grade gasoline and 6000 gallons of high-grade gasoline and costs $30,000 a day to operate. How many days should each plant operate to minimize the cost? Find the minimum cost. *Hint:* Minimize cost in thousands of dollars and divide each constraint by an appropriate constant.

**34.  Production Scheduling.** A company has 3 plants that produce 3 different sizes of their product. The North plant can produce 2000 items in the small size, 1000 in the medium size, and 1000 in the large size per day. The Center plant can produce 1000 items in the small size, 1000 in the medium size, and 2000 in the large size per day. The South plant can produce 2000 items in the small size, 2000 in the medium size, and 1000 in the large size per day. The company needs to produce at least 9000 of the small items, 8000 of the medium items, and 9000 of the large items on a given day. The cost per item of producing the small item is $2, the medium item $3, and the large item $2. Find the number of days each plant should operate in order to minimize the cost; find the minimum cost. *Hint:* Find the minimum cost in thousands of dollars and divide the constraints by an appropriate constant.

**35.  Production Scheduling.** A mining company has three mines that produce three grades of ore. The North mine can produce 2 tons of low-grade ore, 1 ton of medium-grade ore, and 1 ton of high-grade ore each day. The Center mine can produce 1 ton of low-grade, 1 ton of medium-grade, and 4 tons of high-grade ore. The South mine can produce 1 ton of low-grade, 3 tons of medium-grade, and 3 tons of high-grade ore at least. At a particular time the company needs at least 21 tons of low-grade ore, at least 19 tons of medium-grade ore, and at least 25 tons of high-grade ore. It costs $2000 per day to operate the North mine for a day, $3000 a day for the Center mine, and $4000 for the South mine. Find the number of days each mine should operate to minimize cost; find the minimum cost. *Hint:* Take cost to be in thousands of dollars.

# Graphing Calculator Exercises

**36.** Minimize    $z = 16x_1 + 19x_2 + 10x_3$

Subject to $41x_1 + 19x_2 + 11x_3 \geq 680$

$\qquad 51x_1 + 59x_2 + 19x_3 \geq 1400$

$\qquad 42x_1 + 58x_2 + 27x_3 \geq 1380$

$\qquad\qquad\qquad x_1, x_2, x_3 \geq 0$

**37.** Maximize    $z = 51x_1 + 49x_2 + 87x_3$

Subject to $21x_1 + 9x_2 + 38x_3 \leq 870$

$\qquad 12x_1 + 31x_2 + 21x_3 \leq 1100$

$\qquad 28x_1 + 9x_2 + 29x_3 \leq 910$

$\qquad\qquad\qquad x_1, x_2, x_3 \geq 0$

# Solutions to Self-Help Exercise Set 4.4

**1.** Convert the problem to:

$$\text{Maximize} \qquad z = -2x_1 - 3x_2$$
$$\text{Subject to} \qquad -x_1 - x_2 \leq -3$$
$$x_1 \leq 2$$
$$x_2 \leq 2$$
$$x_1, x_2 \geq 0$$

The initial tableau is

| $x_1$ | $x_2$ | $s_1$ | $s_2$ | $s_3$ | $z$ | |
|---|---|---|---|---|---|---|
| $-1$ | $-1$ | 1 | 0 | 0 | 0 | $-3$ |
| ① | 0 | 0 | 1 | 0 | 0 | 2 |
| 0 | 1 | 0 | 0 | 1 | 0 | 2 |
| 2 | 3 | 0 | 0 | 0 | 1 | 0 |

Pivoting about the second row and first column gives

| $x_1$ | $x_2$ | $s_1$ | $s_2$ | $s_3$ | $z$ | | |
|---|---|---|---|---|---|---|---|
| 0 | $-1$ | 1 | 1 | 0 | 0 | $-1$ | |
| 1 | 0 | 0 | 1 | 0 | 0 | 2 | |
| 0 | ① | 0 | 0 | 1 | 0 | 2 | $\frac{2}{1}$ |
| 0 | 3 | 0 | $-2$ | 0 | 1 | $-4$ | |

Pivoting about the third row and second column gives

| $x_1$ | $x_2$ | $s_1$ | $s_2$ | $s_3$ | $z$ | | |
|---|---|---|---|---|---|---|---|
| 0 | 0 | 1 | 1 | ① | 0 | 1 | $\frac{1}{1}$ |
| 1 | 0 | 0 | 1 | 0 | 0 | 2 | |
| 0 | 1 | 0 | 0 | 1 | 0 | 2 | $\frac{2}{1}$ |
| 0 | 0 | 0 | $-2$ | $-3$ | 1 | $-10$ | |

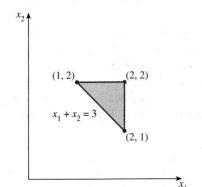

The tableau is now in standard form and corresponds to the point (2, 2). See the figure. Pivoting about the first row and fifth column yields

| $x_1$ | $x_2$ | $s_1$ | $s_2$ | $s_3$ | $z$ | |
|---|---|---|---|---|---|---|
| 0 | 0 | 1 | 1 | 1 | 0 | 1 |
| 1 | 0 | 0 | 1 | 0 | 0 | 2 |
| 0 | 1 | $-1$ | $-1$ | 0 | 0 | 1 |
| 0 | 0 | 3 | 1 | 0 | 1 | $-7$ |

This is the final tableau. The minimum occurs at the point (2, 1). The minimum is $-(-7) = 7$.

## 4.5   POST-OPTIMAL ANALYSIS*

► *Introduction*
► *Changes in the $b_i$'s*
► *Shadow Prices*
► *Complementary Slackness*

*Optional

APPLICATION

**Manufacturing**

Suppose the management of the manufacturer of wooden boats considered earlier in this chapter increased the cutting hours by 2, increased the painting hours by 4 and decreased the assembly hours by 1. Can you find the new optimal solution and value without doing the simplex from the beginning? For the answer see Example 1.

## Introduction

When an optimal solution has been found to a linear programming problem, usually the work has only begun. The assumption has been made that the constants in the linear programming problem are *known*. Usually these constants are only *estimates* based on what the research team believes conditions will be during the production process. Thus further analysis is normally needed when these "constants" are changed. For this reason the constants are usually referred to as *parameters* and can be changed within certain bounds. The constants, $b_i$, in particular, are very often set as a result of a company policy decision as to how much resource to allocate to the production process. A further analysis must be done to determine the consequences of changes in this allocation.

For all of these reasons it is important to perform a **sensitivity analysis** to determine the effect on the optimal solution and the maximum or minimum of any changes in the constants. Sensitivity analysis would be very expensive if each change in a constant required redoing the simplex method from the beginning. Fortunately, theoretical analysis has revealed certain relationships that can reduce the computations to a minimum and do not require the simplex method to be performed again.

## Changes in the $b_i$'s

In this section we shall restrict the discussion to the standard maximization problem

$$\begin{aligned} \text{Maximize} \quad & z = CX \\ \text{Subject to} \quad & AX \leq B \\ & X \geq 0 \end{aligned}$$

where all the components of the column matrix $B$ are nonnegative.

We will only look at the consequences of changing the $b_i$'s, that is, the components of the column matrix $B$. The analysis of the changes in the other constants involves more time than is available here.

We first look at the wooden-boat manufacturer again. Recall that the problem was

$$\text{Maximize} \qquad P = 60x_1 + 80x_2$$

$$\text{Subject to} \qquad 2x_1 + 4x_2 \leq 80$$

$$2x_1 + 2x_2 \leq 50$$

$$4x_1 + 2x_2 \leq 84$$

$$x_1, x_2 \geq 0.$$

Thus the column matrix $B$ is

$$B = \begin{bmatrix} 80 \\ 50 \\ 84 \end{bmatrix}$$

where the first component, 80, represents the constraint from the number of hours in the cutting operation, 50 represents the constraint from the number of hours in the painting operation, and 84 represents the constraint from the number of hours in the assembly operation.

The initial simplex tableau is

| $x_1$ | $x_2$ | $s_1$ | $s_2$ | $s_3$ | $z$ | |
|-------|-------|-------|-------|-------|-----|-----|
| 2 | 4 | 1 | 0 | 0 | 0 | 80 |
| 2 | 2 | 0 | 1 | 0 | 0 | 50 |
| 4 | 2 | 0 | 0 | 1 | 0 | 84 |
| $-60$ | $-80$ | 0 | 0 | 0 | 1 | 0 |

Suppose now we consider changing the first constraint to $80 + h_1$, changing the second constraint to $50 + h_2$, and the third to $84 + h_3$, where $h_1$, $h_2$, and $h_3$ could be positive or negative. We wish to analyze how changes in each of the components $h_1$, $h_2$, and $h_3$ will affect the optimal value of the objective function and the value of the optimal solution.

The simplex tableau remains the same except for the last column. The last column now becomes

$$\begin{bmatrix} 80 + h_1 \\ 50 + h_2 \\ 84 + h_3 \\ \hline 0 \end{bmatrix} = \begin{bmatrix} B + H \\ \hline 0 \end{bmatrix}$$

where

$$H = \begin{bmatrix} h_1 \\ h_2 \\ h_3 \end{bmatrix}.$$

If we now carry out the simplex method for this new last column, only the last column will change for any of the intermediate tableaus (and thus also for the last tableau). Therefore we simply follow the changes in the last column of the simplex method. Recalling these elementary row operations we obtain

| |
|---|
| $80 + h_1$ |
| $50 + h_2$ |
| $84 + h_3$ |
| $0$ |

$\frac{1}{4}R_1 \to R_1$

$\to$

| |
|---|
| $20 + \frac{1}{4}h_1$ |
| $50 + h_2$ |
| $84 + h_3$ |
| $0$ |

$R_2 - 2R_1 \to R_2$
$R_3 - 2R_1 \to R_3$
$R_4 + 80R_1 \to R_4$ $\to$

| |
|---|
| $20 + \frac{1}{4}h_1$ |
| $10 - \frac{1}{2}h_1 + h_2$ |
| $44 - \frac{1}{2}h_1 + h_3$ |
| $1600 + 20h_1$ |

$R_1 - \frac{1}{2}R_2 \to R_1$
$R_3 - 3R_2 \to R_3$
$R_4 + 20R_2 \to R_4$ $\to$

| |
|---|
| $15 + \frac{1}{2}h_1 - \frac{1}{2}h_2$ |
| $10 - \frac{1}{2}h_1 + h_2$ |
| $14 + h_1 - 3h_2 + h_3$ |
| $1800 + 10h_1 + 20h_2$ |

This is the last column of the final tableau. The optimal solution is $1800 + 10h_1 + 20h_2$ and the optimal point is

$$(10 - \tfrac{1}{2}h_1 + h_2,\ 15 + \tfrac{1}{2}h_1 - \tfrac{1}{2}h_2) \tag{1}$$

This latter statement is true *if all the elements in the last column (except for the last) remain nonnegative.* If values of $h_1$, $h_2$, and $h_3$ are chosen that make one of these elements negative, then the solution is not feasible. Thus in order for (1) to be the optimal solution, we must have

$$15 + \tfrac{1}{2}h_1 - \tfrac{1}{2}h_2 \geq 0$$
$$10 - \tfrac{1}{2}h_1 + h_2 \geq 0 \tag{2}$$
$$14 + h_1 - 3h_2 + h_3 \geq 0$$

It is of considerable interest to notice that when we increase the number of hours allocated to the assembly process, that is, for any positive $h_3$, *no change* occurs in the maximum. Thus more than enough hours have already been allocated to the assembly process. Perhaps management might consider changing this resource allocation.

Consider the following specific example.

**E X A M P L E   1**
_____

Suppose management increased the cutting hours by 2, increased the painting hours by 4, and decreased the assembly hours by 1. Find the new optimal solution and value.

**Solution**

We see that (2) is still satisfied and the new maximum profit becomes

$$1800 + 10h_1 + 20h_2 = 1800 + 10(2) + 20(4) = 1900$$

while the new optimal solution becomes

$$(10 - \tfrac{1}{2}(2) + (4), \ 15 + \tfrac{1}{2}(2) - \tfrac{1}{2}(4)) = (13, 14)$$

Thus the new maximum profit is $1900, and the new solution is to make 13 dinghies and 14 skiffs.  ∎

Naturally management would have to determine if the additional number of dinghies manufactured could actually be sold.

To repeat the above procedure for a large simplex tableau would represent an unacceptable cost. Fortunately, there is a simple way of seeing the result.

First recall that the final tableau is

| $x_1$ | $x_2$ | $s_1$ | $s_2$ | $s_3$ | $z$ | |
|---|---|---|---|---|---|---|
| 0 | 1 | .50 | $-.50$ | 0 | 0 | 15 |
| 1 | 0 | $-.50$ | 1 | 0 | 0 | 10 |
| 0 | 0 | 1 | $-3$ | 1 | 0 | 14 |
| 0 | 0 | 10 | 20 | 0 | 1 | 1800 |

If we let $D$ be the square matrix sitting under the slack variables and above the last line in the final tableau, and let $E$ be the row matrix under the slack variables in the last row of the final tableau, then it can be shown that the change in the last column of the final tableau can be written as $\boxed{\begin{array}{c} DH \\ \hline EH \end{array}}$

We notice above that

$$D = \begin{bmatrix} \tfrac{1}{2} & -\tfrac{1}{2} & 0 \\ -\tfrac{1}{2} & 1 & 0 \\ 1 & -3 & 1 \end{bmatrix}, \quad E = [10 \quad 20 \quad 0]$$

Then

$$DH = \begin{bmatrix} \tfrac{1}{2} & -\tfrac{1}{2} & 0 \\ -\tfrac{1}{2} & 1 & 0 \\ 1 & -3 & 1 \end{bmatrix} \begin{bmatrix} h_1 \\ h_2 \\ h_3 \end{bmatrix} = \begin{bmatrix} \tfrac{1}{2}h_1 - \tfrac{1}{2}h_2 \\ -\tfrac{1}{2}h_1 + h_2 \\ h_1 - 3h_2 + h_3 \end{bmatrix}$$

and

$$EH = \begin{bmatrix} 10 & 20 & 0 \end{bmatrix} \begin{bmatrix} h_1 \\ h_2 \\ h_3 \end{bmatrix} = 10h_1 + 20h_2$$

We now give the following theorem, which has been verified in the above case.

---

**Sensitivity Theorem**

If the linear programming problem is a standard maximization problem of the form

$$\text{Maximize} \qquad z = CX$$

$$\text{Subject to} \qquad AX \leq B$$

$$X \geq 0,$$

then changing the matrix $B$ to $B + H$ results in changing only the last column of the final tableau by adding the quantity $\dfrac{DH}{EH}$ to the last column of the old final tableau, where $D$ is the square matrix sitting under the slack variables and above the last line in the final tableau, and $E$ is the row matrix under the slack variables in the last row of the final tableau. This tableau is an acceptable final tableau as long as all the elements in the last column of the new tableau (except the last element) are nonnegative.

---

## Shadow Prices

It is no accident that in the wooden-boat manufacturing problem the row matrix $E$, which corresponds to the elements under the slack variables in the last row of the final tableau, is the **solution to the dual problem and also the shadow prices**. Thus if $Y$ is the column matrix that gives the solution to the dual problem, then $E = Y^T$. This is always the case in a standard maximization problem.

Shadow prices are defined in terms of marginal analysis. Thus we have the following definition.

---

**Shadow Prices**

The $i$th shadow price is the amount by which the optimal $z$-value is improved if the right-hand side of the $i$th constraint is increased by 1.

---

## E X A M P L E   2

Find the shadow prices in the wooden-boat manufacturing problem.

**Solution**    In the wooden-boat manufacturing problem, the optimal value is given by

$$z_{\text{opt}} = 1800 + EH = 1800 + 10h_1 + 20h_2$$

Thus increasing the cutting hours by 1 and leaving the other constraints unchanged is the same as setting $h_1 = 1$, $h_2 = 0$, and $h_3 = 0$, obtaining

$$z_{\text{opt}} = 1800 + 10(1) + 20(0) = 1800 + 10.$$

The increase, 10, is the first shadow price. Also, increasing the painting hours by 1 and leaving the other constraints unchanged is the same as setting $h_1 = 0$, $h_2 = 1$, and $h_3 = 0$, obtaining

$$z_{\text{opt}} = 10(1) = 1800 + 10(0) + 20(1) = 1800 + 20.$$

The increase, 20, is the second shadow price. Finally, increasing the assembly hours by 1 and leaving the other constraints unchanged is the same as setting $h_1 = 0$, $h_2 = 0$, and $h_3 = 1$, obtaining

$$z_{\text{opt}} = 10(1) = 1800 + 10(0) + 20(0) = 1800.$$

The increase, 0, is the third shadow price.    ∎

We now state the connection between shadow prices and the solution to the dual problem.

---

**Shadow Price Theorem**

Given a standard maximization problem

$$\begin{aligned} \text{Maximize} \quad & z = CX \\ \text{Subject to} \quad & AX \leq B \\ & X \geq 0 \end{aligned}$$

Let $Y$ be the solution of the dual problem

$$\begin{aligned} \text{Minimize} \quad & w = B^T Y \\ \text{Subject to} \quad & A^T Y \geq C^T \\ & Y \geq 0 \end{aligned}$$

Then the $i$th component of $Y$, $y_i$, is the $i$th shadow price.

---

## Complementary Slackness

We state one form of the complementary slackness theorem.

---

**Complementary Slackness Theorem**

1.  If the $i$th primal slack variable is $> 0$, then the $i$th dual variable equals zero.
2.  If the $i$th dual variable is $> 0$, then the $i$th primal slack variable is zero.

---

To interpret the first statement in the complementary slackness theorem in the wooden-boat manufacturing problem, recall that $s_3 > 0$. Thus according to (1) above, we must have $y_3 = 0$, that is, the third shadow price is zero. This makes sense. Since there is positive slack in the (third) assembly constraint, an extra hour of assembly labor is worthless.

To interpret the second part (2), recall that in the same problem, $y_1 > 0$ and $y_2 > 0$. Thus according to (2), we must have $s_1 = 0$ and $s_2 = 0$. This is indeed the case. This also is reasonable since a positive second and third shadow price means that an extra hour of cutting and an extra hour of painting have some value. This can only happen if all of the hours for cutting and painting are being used; that is, there is no slack in the cutting and painting constraints.

## SELF-HELP EXERCISE SET 4.5

In determining how many chairs ($x_1$), sofas ($x_2$), and love-seats ($x_3$) to manufacture to maximize revenue in a standard maximization problem, suppose the final tableau of the corresponding linear programming problem is

| $x_1$ | $x_2$ | $x_3$ | $s_1$ | $s_2$ | $s_3$ | $z$ | |
|---|---|---|---|---|---|---|---|
| 0 | 1 | 2 | 1 | 0 | 2 | 0 | 10 |
| 0 | 0 | 1 | 0 | 1 | $-1$ | 0 | 12 |
| 1 | 0 | 1 | 1 | 0 | 1 | 0 | 8 |
| 0 | 0 | 5 | 2 | 0 | 1 | 1 | 15 |

The first constraint refers to the number of yards of fabric available, the second constraint to the number of board-feet of wood available, and the third constraint to the number of hours of available labor.

1.  Determine the final tableau if the number of yards of fabric were changed by $h_1$, the number of board-feet of lumber by $h_2$, and the number of hours of labor by $h_3$.

2. Determine the constraints on $h_1$, $h_2$, and $h_3$ in order to assure that the tableau found in (1) is the final tableau.

3. Find the new optimal solution and the new maximum of the objective function if the number of yards of fabric is increased by 1, the number of board-feet of lumber is increased by 2, and the number of work-hours is increased by 3.

4. What are the shadow prices, and what do they mean?

## EXERCISE SET 4.5

1. To determine how many desks ($x_1$) and chairs ($x_2$) to manufacture to maximize revenue in a standard maximization problem, suppose the final tableau of the corresponding linear programming problem is

| $x_1$ | $x_2$ | $s_1$ | $s_2$ | $s_3$ | $z$ | |
|---|---|---|---|---|---|---|
| 1 | 0 | 0 | $-2$ | 0 | 0 | 20 |
| 0 | 0 | 1 | $-1$ | $-1$ | 0 | 30 |
| 0 | 1 | 0 | $-1$ | 3 | 0 | 25 |
| 0 | 0 | 0 | 4 | 3 | 1 | 40 |

The first constraint refers to the number of gallons of varnish available, the second constraint to the number of board-feet of wood available, and the third constraint to the number of hours of available labor.

   a. Determine the new tableau if the number of gallons of varnish available were changed by $h_1$, the number of board-feet of lumber available by $h_2$, and the number of hours of labor available by $h_3$.

   b. Determine the constraints on $h_1$, $h_2$, and $h_3$ to ensure that the new tableau found in (a) is the final tableau.

   c. Assuming that $h_1$, $h_2$, and $h_3$ satisfy the constraints in (b), find the new optimal solution.

   d. Assuming that $h_1$, $h_2$, and $h_3$ satisfy the constraints in (b), find the new maximum.

   e. How much can the second constraint be increased while keeping the first and third constraint unchanged without changing the optimal solution or the maximum value of the objective function?

   f. Find the new optimal solution and the new maximum of the objective function if the number of gallons of varnish is increased by 2, the number of board-feet of lumber is increased by 3, and the number of work-hours is decreased by 2.

   g. What are the shadow prices, and what do they mean?

2. In determining how many pounds of fudge 1 ($x_1$) and fudge 2 ($x_2$) to make to maximize revenue in a standard maximization problem, suppose the final tableau of the corresponding linear programming problem is

| $x_1$ | $x_2$ | $s_1$ | $s_2$ | $s_3$ | $z$ | |
|---|---|---|---|---|---|---|
| 0 | 1 | $-3$ | 4 | 0 | 0 | 30 |
| 1 | 0 | $-1$ | 2 | 0 | 0 | 20 |
| 0 | 0 | $-1$ | 1 | 1 | 0 | 25 |
| 0 | 0 | 5 | 3 | 0 | 1 | 50 |

The first constraint refers to the number of pounds of chocolate available, the second constraint to the number of pounds of marshmallow available, and the third constraint to the number of pounds of nuts available.

**a.** Determine the final tableau if the number of pounds of chocolate were changed by $h_1$, the number of pounds of marshmallow by $h_2$, and the number of pounds of nuts by $h_3$.

**b.** Determine the constraints on $h_1$, $h_2$, and $h_3$ to ensure that the new tableau found in (a) is the final tableau.

**c.** Assuming that $h_1$, $h_2$, and $h_3$ satisfy the constraints in (b), find the new optimal solution.

**d.** Assuming that $h_1$, $h_2$, and $h_3$ satisfy the constraints in (b), find the new maximum.

**e.** How much can the third constraint be decreased while keeping the first and second constraint unchanged, without changing the optimal solution or the maximum value of the objective function?

**f.** Find the new optimal solution and the new maximum of the objective function if the number of pounds of chocolate is increased by 3, the number of pounds of marshmallow is increased by 2, and the number of pounds of nuts is decreased by 1.

**g.** What are the shadow prices, and what do they mean?

**3.** In determining how many gallons of 4 fruit juices to mix to obtain fruit drink 1 ($x_1$), fruit drink 2 ($x_2$), and fruit drink 3 ($x_3$) in order to maximize revenue in a standard maximization problem, suppose the final tableau of the corresponding linear programming problem is

| $x_1$ | $x_2$ | $x_3$ | $s_1$ | $s_2$ | $s_3$ | $s_4$ | $z$ | |
|---|---|---|---|---|---|---|---|---|
| 0 | 0 | 0 | 1 | 4 | 1 | 5 | 0 | 20 |
| 1 | 0 | 0 | 3 | 2 | 0 | 1 | 0 | 30 |
| 0 | 1 | 0 | 2 | 1 | 0 | 2 | 0 | 25 |
| 0 | 0 | 1 | 1 | 3 | 0 | 6 | 0 | 40 |
| 0 | 0 | 0 | 2 | 4 | 0 | 5 | 1 | 90 |

The first constraint refers to the number of gallons of orange juice available, the second constraint to the number of gallons of grape juice available, the third constraint to the number of gallons of grapefruit juice available, and the fourth constraint to the number of gallons of cranberry juice available.

**a.** Determine the new tableau if the number of gallons of orange juice, grape juice, grapefruit juice, and cranberry juice available were changed respectively by $h_1$, $h_2$, $h_3$, and $h_4$.

**b.** Determine the constraints on $h_1$, $h_2$, $h_3$, and $h_4$ in order to assure that the new tableau found in (a) is the final tableau.

**c.** Assuming that $h_1$, $h_2$, $h_3$, and $h_4$ satisfy the constraints in (b), find the new optimal solution.

**d.** Assuming that $h_1$, $h_2$, $h_3$, and $h_4$ satisfy the constraints in (b), find the new maximum.

**e.** How much can the third constraint be decreased while keeping the other constraints unchanged, without changing the optimal solution or the maximum value of the objective function?

**f.** Find the new optimal solution and the new maximum of the objective function if the number of gallons of the 4 fruit juices is increased by 2, increased by 3, decreased by 4, and increased by 1, respectively.

**g.** What are the shadow prices, and what do they mean?

**4.** Redo Exercise 2 with the given tableau replaced with

| $x_1$ | $x_2$ | $s_1$ | $s_2$ | $s_3$ | $z$ | |
|---|---|---|---|---|---|---|
| 0 | 0 | 2 | 1 | 1 | 0 | 50 |
| 1 | 0 | 3 | 0 | 2 | 0 | 40 |
| 0 | 1 | 1 | 0 | 1 | 0 | 45 |
| 0 | 0 | 2 | 0 | 4 | 1 | 60 |

except that (e) is changed to the following.

**e.** How much can the second constraint be decreased while keeping the first and third constraint unchanged, without changing the optimal solution or the maximum value of the objective function?

If, when changing the column matrix from $B$ to $B + H$, one of the elements in the last column of the new tableau (not including the last entry) becomes negative, then one can use the new tableau as an intermediate tableau and proceed as in Section 4.4 to the final tableau. This, in general, will change the optimal solution and the optimal maximum of the objective function. Do this for the next four exercises to obtain the new solution and optimal value of $z$ for the indicated changes.

**5.** In the tableau in Exercise 1 decrease the third constraint by 9 without changing the other constraints.

**6.** In the tableau in Exercise 2 increase the first constraint by 7 without changing the other constraints.

**7.** In the tableau in Exercise 1 increase the second constraint by 11 without changing the other constraints.

**8.** In the tableau in Exercise 2 decrease the third constraint by 26 without changing the other constraints.

## Solutions to Self-Help Exercise Set 4.5

**1.** We notice above that

$$D = \begin{bmatrix} 1 & 0 & 2 \\ 0 & 1 & -1 \\ 1 & 0 & 1 \end{bmatrix}, \quad E = \begin{bmatrix} 2 & 0 & 1 \end{bmatrix}$$

Then

$$DH = \begin{bmatrix} 1 & 0 & 2 \\ 0 & 1 & -1 \\ 1 & 0 & 1 \end{bmatrix} \begin{bmatrix} h_1 \\ h_2 \\ h_3 \end{bmatrix} = \begin{bmatrix} h_1 + 2h_3 \\ h_2 - h_3 \\ h_1 + h_3 \end{bmatrix}$$

and

$$EH = \begin{bmatrix} 2 & 0 & 1 \end{bmatrix} \begin{bmatrix} h_1 \\ h_2 \\ h_3 \end{bmatrix} = 2h_1 + h_3$$

Thus the new final tableau is the same as the old one, except the last column is

$$\begin{array}{|c|}
\hline
10 + h_1 + 2h_3 \\
\hline
12 + h_2 - h_3 \\
\hline
8 + h_1 + h_3 \\
\hline
15 + 2h_1 + h_3 \\
\hline
\end{array}$$

**2.** The constraints are

$$10 + h_1 + 2h_3 \geq 0$$
$$12 + h_2 - h_3 \geq 0$$
$$8 + h_1 + h_3 \geq 0$$

**3.** Setting $h_1 = 1$, $h_2 = 2$, $h_3 = 3$, in the above constraints gives

$$10 + h_1 + 2h_3 = 10 + (1) + 2(3) = 17 \geq 0$$
$$12 + h_2 - h_3 = 12 + (2) - (3) = 11 \geq 0$$
$$8 + h_1 + h_3 = 8 + (1) + (3) = 12 \geq 0$$

Since all of these are satisfied, the new tableau will be the final tableau. The last entry in the last column is

$$15 + EH = 15 + 2h_1 + h_3 = 15 + 2(1) + (3) = 20,$$

so the final tableau is

| $x_1$ | $x_2$ | $x_3$ | $s_1$ | $s_2$ | $s_3$ | $z$ | |
|---|---|---|---|---|---|---|---|
| 0 | 1 | 2 | 1 | 0 | 2 | 0 | 17 |
| 0 | 0 | 1 | 0 | 1 | −1 | 0 | 11 |
| 1 | 0 | 1 | 1 | 0 | 1 | 0 | 12 |
| 0 | 0 | 5 | 2 | 0 | 1 | 1 | 20 |

The new optimal solution is (12, 17, 0) with maximum 20.

4.  The shadow prices are the elements of the row matrix $E$, and thus are 2, 0, 1. Thus increasing the amount of available fabric by 1 square yard will result in an increase of profit of $2; increasing the number of board-feet available by 1 will leave the profit unchanged; and increasing the number of work-hours by 1 will result in profits increasing by $1.

## SUMMARY OUTLINE OF CHAPTER 4

- Given a problem constraint in a linear programming problem such as $a_1x_1 + a_2x_2 + a_3x_3 \le b_1$, a **slack variable** $s_1$ is defined so that $a_1x_1 + a_2x_2 + a_3x_3 + s_1 = b_1$.

- **The General Algebraic Form of the Standard Maximization Problem**

  Maximize $\quad\quad z = c_1x_1 + c_2x_2 + \cdots + c_nx_n$

  Subject to $\quad\quad a_{11}x_1 + a_{12}x_2 + \cdots + a_{1n}x_n \le b_1$

  $$a_{21}x_1 + a_{22}x_2 + \cdots + a_{2n}x_n \le b_2$$

  $$\cdots$$

  $$a_{m1}x_1 + a_{m2}x_2 + \cdots + a_{mn}x_n \le b_m$$

  $$x_1, x_2, \ldots, x_n \ge 0$$

- **Algebraic Criteria for Determining the Feasible Region.** Given a linear programming problem with $n$ decision variables and $m$ problem constraints, the point $(x_1, x_2, \ldots, x_n)$ will be in the feasible region of a standard maximization problem if, and only if, all the slack variables $s_1, s_2, \ldots, s_m$, together with all the decision variables $x_1, x_2, \ldots, x_n$, are nonnegative.

- **Basic and Nonbasic Variables.** Arbitrarily pick $n$ of the variables $x_1, \ldots, x_n$, $s_1, \ldots, s_m$, and call them the *nonbasic variables*. Call the remaining variables the *basic variables*. Set the $n$ nonbasic variables equal to zero and try to solve for the basic variables. If a solution exists and if all the numbers $x_1, \ldots, x_n$, $s_1, \ldots, s_m$, are nonnegative, then the point $(x_1, \ldots, x_n)$ is a corner point.

- **Determining the Solution from the Tableau**

  1.  Pick $n$ of the variables $x_1, \ldots, x_n, s_1, \ldots, s_m$, to be the nonbasic variables. The remaining $m$ variables together with $z$ are the basic variables.

  2.  (a) Let the nonbasic variables be variables that correspond to nonunit columns. (b) If there are not $n$ nonunit columns, then we arbitrarily select some of the variables (but never $z$) associated with unit columns to be nonbasic

variables, so that there are $n$ nonbasic variables and so that no two basic variables have identical associated columns.

3.   Set all of the nonbasic variables equal to zero.
4.   We can then easily find the values of the basic variables from the tableau.

▪ **Simplex Method.**   See pages 207–208.

▪ **Dual Problem.**   See page 220.

▪ **Crown Method.**   See page 243.

## Chapter 4 Review Exercises

In Exercises 1 through 10 add the nonnegativity constraint and solve.

**1.**   Maximize        $z = x_1 + x_2$

Subject to   $x_1 + 4x_2 \le 800$
$x_1 + 2x_2 \le 500$
$2x_1 + x_2 \le 1100$

**2.**   Maximize        $z = 2x_1 + 4x_2$

Subject to   $x_1 + x_2 \le 200$
$10x_1 + 8x_2 \le 2000$
$2x_1 + x_2 \le 100$

**3.**   Maximize $z = x_1 + 2x_2$

Subject to   $x_1 - x_2 \le 3$
$x_1 + x_2 \le 1$
$x_1 \le 2$

**4.**   Maximize   $z = 2x_1 - 5x_2$

Subject to   $x_1 + 3x_2 \le 35$
$2x_1 - 4x_2 \ge 40$

**5.**   Maximize   $z = 3x_1 + 8x_2$

Subject to   $x_1 - x_2 \le 6$
$2x_1 + 6x_2 \le 21$
$x_1 - 2x_2 \ge 6$

**6.**   Maximize   $z = 4x_1 + 3x_2$

Subject to   $x_1 + 3x_2 \ge 12$
$2x_1 + x_2 \ge 8$
$5x_1 + 2x_2 \le 30$
$3x_1 + 4x_2 \le 28$

**7.**   Minimize   $z = 2x_1 + 3x_2$

Subject to $2x_1 + x_2 \ge 15$
$x_1 + x_2 \le 14$
$x_1 + 2x_2 \le 15$

**8.**   Minimize   $z = 2x_1 + 3x_2$

Subject to $2x_1 + x_2 \ge 31$
$x_1 + x_2 \ge 29$
$x_1 + 4x_2 \ge 35$

**9.**   Minimize   $z = 2x_1 + 3x_2$

Subject to $4x_1 + 2x_2 \ge 440$
$3x_1 + 2x_2 \le 180$
$x_1 + 6x_2 \le 280$

**10.**   Minimize $z = 4x_1 + 5x_2$

Subject to $4x_1 + x_2 \ge 8$
$x_1 + x_2 \ge 5$
$x_1 + 2x_2 \ge 6$

**11.   Farming.**   A farm has 2 fields to grow wheat and corn, the first field has 400 acres and the second 100 acres. In both fields the corn yield is 125 bushels per acre and the wheat yield is 40 bushels per acre. The cost per acre of growing corn is $90 for the first field and $120 for the second field. The cost per acre of growing wheat is $120 for the first field and $90 for the second field. If 13,750 bushels of corn and 8000 bushels of wheat must be grown, how many acres of each crop should be planted in each field in order to minimize cost? Find the minimum cost. Use the extended simplex method.

12. **Farming.**   Solve the previous problem by using the dual.

13. **Manufacturing.**   In determining how many leather purses $(x_1)$ and belts $(x_2)$ to manufacture to maximize revenue in a standard maximization problem, suppose the final tableau of the corresponding linear programming problem is

| $x_1$ | $x_2$ | $s_1$ | $s_2$ | $s_3$ | $z$ | |
|---|---|---|---|---|---|---|
| 0 | 1 | 1 | 0 | 1 | 0 | 50 |
| 1 | 0 | −1 | 0 | −2 | 0 | 60 |
| 0 | 0 | 1 | 1 | 3 | 0 | 40 |
| 0 | 0 | 2 | 0 | 4 | 1 | 70 |

The first constraint refers to the number of square feet of leather available, the second constraint to the number of ounces of brass available, and the third constraint to the number of hours of available labor.

a.   Find the new solution if the number of hours of labor available were changed by $+5$.

b.   What are the shadow prices and what do they mean?

How can we
determine the
monthly car
payment? (Chester
Higgins, Jr./Photo
Researchers, Inc.)

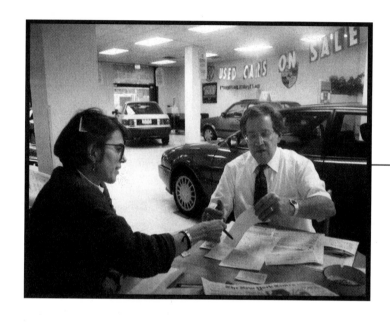

c h a p t e r **5** f i v e

# Mathematics of Finance

5.1 Simple Interest and Discount

5.2 Compound Interest

5.3 Annuities and Sinking Funds

5.4 Present Value of an Annuity and Amortization

**I**n the first two sections we consider what happens to an initial lump sum of money deposited into an account that is earning interest at some given rate and that is compounding in some given way. In the next two sections, we consider periodic payments into an account and payment schedules for loans.

## 5.1 SIMPLE INTEREST AND DISCOUNT

▶ *Simple Interest*
▶ *Present Value*
▶ *Discounts*

APPLICATION
**Simple Interest Rates**

A person deposits $1000 into an account earning simple interest. What simple interest rate is being charged if the amount at the end of 8 months is $1060? See Example 3 for the answer.

**A P P L I C A T I O N**
**Discounted Notes**

A borrower signs a note and agrees to pay $20,000 to a bank in 9 months at 10% simple discount. What is the discount and what does the borrower actually receive from the bank? See Example 5 for the answer.

## Simple Interest

If you lend someone a sum of money, the sum of money is called the **principal** or **present value**. You then charge for the use of this money. This charge is called **interest**. Thus the future amount $F$ of money at any time is given as the principal $P$ plus the interest $I$ or

$$\text{future amount} = \text{principal} + \text{interest}$$

$$F = P + I$$

Normally, the interest paid is given in terms of some **interest rate** expressed as a percent. To calculate the interest, this percent must be converted to a decimal. For example, if the account pays 5% interest per year, and the principal (present value) at the beginning of the year is $1000, then the interest rate is $r = 0.05$, and the interest is

$$I = Pr = \$1000(.05) = \$50$$

The **simple interest** after three years in the previous example is just 3 times $50 or $150. The simple interest for half a year is one half of $50 or $25. In general, we have the following.

---

**Simple Interest**

Suppose a sum of money $P$, called the principal or present value, is invested for $t$ years at an annual *simple* interest rate of $r$, where $r$ is given as a decimal. Then the interest $I$ at the end of $t$ years is given by

$$I = Prt.$$

The future value $F$ at the end of $t$ years is

$$F = P + I = P + Prt = P(1 + rt).$$

---

**E X A M P L E   1**     **Monthly Interest on Credit Card Account**

Suppose you have borrowed $1000 on a credit card that charges simple interest at an annual rate of 18%. What is your interest for the first month?

**Solution**     Here $r = 0.18$, $P = \$1000$, and $t = \frac{1}{12}$. Thus the interest for the first month is

$$I = Prt = \$1000(0.18)\frac{1}{12} = \$1000(0.015) = \$15. \quad \blacksquare$$

E X A M P L E   2

**The Future Value in an Account Paying Simple Interest**

An account with an initial amount of $1000 earns simple interest of 9% annually. How much is in the account after 4 years? After $t$ years?

**Solutions**

**Figure 5.1**

(a)   Here $P = \$1000$, $i = 0.09$, and $t = 4$. Thus the amount in this account after 4 years is

$$F = P(1 + rt) = \$1000[1 + (.09)(4)] = \$1360.$$

(b)   Here $P = \$1000$, $i = 0.09$, and $t = t$. Thus the dollar amount in this account after $t$ years is

$$F = P(1 + rt) = 1000[1 + (.09)(t)] = 1000 + 90t. \quad \blacksquare$$

**R E M A R K .**    We saw that the amount in the account in Example 2 after $t$ years was $1000 + 90t$. The equation $y = 1000 + 90t$ is the equation of a straight line with $y$-intercept 1000 and slope 90. See Figure 5.1.

E X A M P L E   3

**Determining the Annual Rate**

A person deposits $1000 into an account earning simple interest. What simple interest rate is being charged if the amount $F$ at the end of 8 months is $1060?

**Solution**

We substitute $P = 1000$, $F = 1060$, and $t = \frac{8}{12} = \frac{2}{3}$ into the equation $F = P + Prt$ and obtain

$$F = P + Prt$$

$$1060 = 1000 + 1000r\,\frac{2}{3}$$

$$60 = \frac{2000}{3}\,r$$

$$\frac{3}{2000}\,60 = r$$

$$r = .09,$$

which is an annual rate of 9%.   $\blacksquare$

## Present Value

Very often we find ourselves needing a certain amount of money at a specific time in the future. We will now consider how to determine the present amount or **present value** of money needed to deposit into an account earning simple interest in order to attain this future need. Suppose we need the future amount $F$ at the end of $t$ years where money is deposited at a simple annual rate of $r$, expressed as a decimal. Then to find the present amount needed to be deposited into this account, we solve for $P$ in the equation

$$F = P(1 + rt),$$

and obtain

$$P = \frac{F}{1 + rt}.$$

We then have the following.

---

**Present Value**

The **present value** needed to deposit into an account earning a simple annual rate of $r$, expressed as a decimal, in order to have a future amount $F$ after $t$ years is

$$P = \frac{F}{1 + rt}.$$

---

When asked how to amass $1 million the easy way, the banker quipped ''start with $950,000.''

**E X A M P L E　4**　　**Present Value**

How much should be placed into an account paying simple interest of 8% so that after 6 months the future value of the account will be $1,000,000?

**Solution**　　We have $F = \$1,000,000$, $r = .08$, and $t = \frac{6}{12} = \frac{1}{2}$. Thus

$$P = \frac{F}{1 + rt}$$

$$= \frac{\$1,000,000}{1 + (.08)\frac{1}{2}}$$

$$= \$961,538.46　\blacksquare$$

Thus you can have $1 million in this account in 6 months if you deposit $961,538.46 now.

## Discounts

Often when a bank loans money for some specific length of time, the bank will deduct interest when the loan is made. Such a loan or note is said to be **discounted**. The interest discounted from the loan is referred to as the **simple discount** or just **discount**. The amount that the borrower actually receives is called the **proceeds**, and the amount to be repaid is called the **maturity value**. Keep in mind that the bank gives the borrower the maturity value *less* the discount.

We can use the simple interest formula to obtain the discount for a simple discounted loan.

---

**Discount and Proceeds**

The discount $D$ on a discounted loan of $M$ dollars at a simple annual interest rate of $r$ for $t$ years is

$$D = Mrt,$$

where

$$D = \text{discount} \qquad \text{(interest paid at time of loan)}$$
$$M = \text{maturity value} \qquad \text{(amount borrowed)}$$
$$r = \text{discount rate} \qquad \text{(annual simple interest rate)}$$
$$t = \text{length of loan.}$$

The **proceeds** $P$ of the loan is the actual amount the borrower receives when the loan is made and is given by

$$P = M - D.$$

---

**E X A M P L E   5**      **Determining Discounts and Proceeds**

A borrower signs a note and agrees to pay $20,000 to a bank in 9 months at 10% simple discount. What is the discount and what does the borrower actually receive from the bank?

**Solution**      We have $M = \$20,000$, $r = .10$, and $t = \frac{9}{12} = \frac{3}{4}$. The discount is

$$D = Mrt = \$20,000(.10)\frac{3}{4} = \$1500.$$

The proceeds are

$$P = M - D = \$20,000 - \$1500 = \$18,500.$$

Thus the borrower will receive $18,500 at the time of the loan and pay the bank $20,000 in 9 months.  ■

**E X A M P L E   6**      **Discounted Loan with Given Proceeds**

Suppose the borrower in the previous example wanted to receive $20,000. What should the maturity value be?

**Solution**      We need to solve for the maturity value $M$ in the equation

$$P = M - D$$

$$= M - Mrt$$

$$= M(1 - rt).$$

We have the proceeds $P = \$20,000$, $r = .10$, and $t = \frac{3}{4} = .75$. Thus

$$M = \frac{P}{1 - rt}$$

$$= \frac{\$20,000}{1 - (.10)(.75)}$$

$$= \$21,621.62.$$

Thus the borrower will receive the proceeds of $20,000 at the time of the loan and must pay the bank $21,621.62 after 9 months. ∎

In Example 5 the bank quoted an annual interest rate of 10%. But what is the *actual* interest rate? The rate of interest actually paid is called the **effective rate**.

E X A M P L E   7     **Effective Rate of Interest**

What is the effective rate of interest for the borrower in Example 5?

**Solution**     The borrower must pay the bank $20,000 in 9 months after receiving $18,500. The discount or interest was $1500. We can calculate the effective interest rate $r_{\text{eff}}$ by using the formula $D = I = Pr_{\text{eff}}t$ and solving for $r_{\text{eff}}$, where $D = I = \$1500$, $P = \$18,500$, and $t = \frac{3}{4} = 0.75$. We have

$$D = Pr_{\text{eff}}t$$

$$r_{\text{eff}} = \frac{D}{Pt}$$

$$= \frac{1500}{18,500(.75)}$$

$$= .1081.$$

Thus the effective annual interest is 10.81%, whereas the advertised discount rate $r$ was 10%. ∎

We can calculate the effective interest rate using only the discount rate $r$ and the length $t$ of the loan. We know from Example 7 that

$$r_{\text{eff}} = \frac{D}{Pt}$$

$$= \frac{Mrt}{(M - Mrt)t}$$

$$= \frac{r}{1 - rt}.$$

> **Effective Rate of a Discounted Loan**
>
> The effective interest rate given on a discounted loan of length $t$ years with a discount rate of $r$ is
>
> $$r_{eff} = \frac{r}{1 - rt}.$$

**E X A M P L E  8**　**Effective Rate on a Discounted Loan**

Use the above formula to find the effective rate of the discounted loan in Example 5.

**Solution**　We have $r = .10$ and $t = .75$. Thus

$$r_{eff} = \frac{r}{1 - rt} = \frac{.10}{1 - .10(.75)} = .1081, \text{ or } 10.81\%. \quad \blacksquare$$

## SELF-HELP EXERCISE SET 5.1

1. A bank borrows $100,000 for 3 months at a simple interest rate of 6% per year. How much must be repaid at the end of the 3 months?

2. The United States government borrows substantial sums of money by issuing treasury bills (T-bills). T-bills do not specify an interest rate, but rather are sold at public auction. A bank wishes to purchase a 6-month $1 million T-bill. For such a T-bill the bank will receive $1 million at the end of 6 months. If the bank wishes to earn 5% simple discount interest on this T-bill, what should the bank bid?

3. If the bank obtains a 6-month T-bill at a discount rate of 5%, what is the effective yield?

## EXERCISE SET 5.1

In Exercises 1 through 4 an amount of $P$ dollars is borrowed for the given length of time at an annual interest rate of $r$. Find the simple interest that is owed.

1. $P = \$1000$,　$r = 8.0\%$,　4 months

2. $P = \$2000$,　$r = 6.0\%$,　3 months

3. $P = \$6000$,　$r = 4.0\%$,　2 years

4. $P = \$5000$,　$r = 7.0\%$,　3 years

In Exercises 5 through 8 an amount of $P$ dollars is borrowed for the given length of time at an annual simple interest rate of $r$. Find the amount due at the end of the given length of time.

*Future Value*

5. $P = \$2000$,   $r = 4.0\%$,   7 months

6. $P = \$5000$,   $r = 5.0\%$,   4 months

7. $P = \$1000$,   $r = 6.0\%$,   2 years

8. $P = \$8000$,   $r = 7.0\%$,   3 years

In Exercises 9 through 12 an amount of $P$ dollars is borrowed for the given length of time with the amount $F$ due at the end of the given length of time. Find the annual simple interest rate $r$.

*Future value* – $P(1 + It)$
*solve for r*

9. $P = \$2000$,   $F = \$2100$,   7 months

10. $P = \$5000$,   $F = \$5200$,   4 months

11. $P = \$1000$,   $F = \$1070$,   2 years

12. $P = \$8000$,   $F = \$9200$,   3 years

In Exercises 13 through 16 find the present amount needed to attain a future amount of $F$ dollars in the given time using an annual simple interest rate of $r$.

13. $F = \$2000$,   $r = 4.0\%$,   3 months

*make fraction?*

14. $F = \$5000$,   $r = 5.0\%$,   8 months

15. $F = \$6000$,   $r = 6.0\%$,   2 years

16. $F = \$8000$,   $r = 7.0\%$,   3 years

In Exercises 17 through 20 find the effective yield on a discount loan with the given discount rate $r$ and the time.

17. $r = 5\%$,   5 months

18. $r = 6\%$,   9 months

19. $r = 7\%$,   3 months

20. $r = 8\%$,   10 months

In Exercises 21 through 24 find the simple discount and the proceeds for the simple discounted loans.

21. $M = \$1000$,   5 months,   $r = .04$

22. $M = \$3000$,   8 months,   $r = .05$

23. $M = \$5000$,   20 months,   $r = .06$

24. $M = \$7000$,   2 years,   $r = .08$

## Applications

25. **Future Value.**   A principal of $5000 earns 8% per year simple interest. How long will it take for the future value to become $6000?

26. **Future Value.**   A principal of $2000 earns 6% per year simple interest. How long will it take for the future value to become $2300?

27. **Doubling Time.**   A principal earns 8% per year simple interest. How long will it take for the future value to double?

28. **Tripling Time.** A principal earns 6% per year simple interest. How long will it take for the future value to triple?

29. **Simple Discount.** A borrower signs a note and agrees to pay $50,000 to a bank in 5 months at 8% simple discount. What is the discount and what does the borrower actually receive from the bank?

30. **Simple Discount.** A borrower signs a note and agrees to pay $40,000 to a bank in 7 months at 6% simple discount. What is the discount and what does the borrower actually receive from the bank?

### Solutions to Self-Help Exercise Set 5.1

1. Since $P = \$100,000$, $r = .06$, and $t = \frac{3}{12} = .25$,

$$F = P(1 + rt) = \$100,000[1 + (.06)(.25)] = \$101,500.$$

2. Since $M = \$1,000,000$, $r = .05$, and $t = \frac{6}{12} = .50$,

$$P = M - D = M - Mrt = M(1 - rt)$$

$$= \$1,000,000[1 - (.05)(.50)] = \$975,000.$$

3. Since $r = .05$ and $t = .50$,

$$r_{eff} = \frac{r}{1 - rt} = \frac{.05}{1 - .05(.50)} = .0513,$$

or 5.13%.

## 5.2    COMPOUND INTEREST

▶ *Compound Interest*
▶ *Effective Yield*
▶ *Present Value*

APPLICATION

**Comparing Investments**

One bank advertises an annual rate of 9.1% compounded semiannually. A second bank advertises an annual rate of 9% compounded daily. In which bank would you deposit your money? The answer is given in Example 5.

APPLICATION

**Present Value of a Future Balance**

How much money must grandparents set aside at the birth of their grandchild if they wish to have $20,000 when the grandchild reaches eighteen? They can earn 9% compounded quarterly. The answer is given in Example 6.

## Compound Interest

The most common type of interest is **compound interest**. We now discuss this type of interest.

If the principal is invested for a certain fraction $t$ of a year at an annual *simple* interest rate of $r$, where $r$ is given as a decimal, then as we have seen in the last section, the amount at the end of the period is

$$\text{principal} + \text{interest} = P + Pi = P(1 + i), \qquad (1)$$

where $i = rt$. The interest $i = rt$ is called the **interest per period**.

If, for example, the annual simple interest on a credit card is 18% and the time period is one month, then $i = .18(\frac{1}{12}) = .015$ is the interest per month.

A useful way of looking at formula (1) is to think of it as the *new principal* equals the *old principal* times $(1 + i)$, or

$$\text{(new principal)} = \text{(old principal)}(1 + i). \qquad (2)$$

If the interest and principal are left in the account for more than one period and interest is calculated not only on the principal but also on the previous interest earned, we say that the interest is being **compounded**.

Consider now what happens during the second period of the compounding. At the start of the second period the *old principal* is $P(1 + i)$. Thus the amount of money in the account at the end of the second period of time (the *new principal*) is according to (2)

$$\text{(new principal)} = \text{(old principal)}(1 + i) = [P(1 + i)](1 + i) = P(1 + i)^2.$$

If this compounding continues for a third period of time, the amount of money in the account at the end of the third period of time (the *new principal*) is

$$\text{(new principal)} = \text{(old principal)}(1 + i) = [P(1 + i)^2](1 + i) = P(1 + i)^3.$$

Continuing in this manner, the future amount of money $F$ in the account at the end of $n$ periods will be

$$F = P(1 + i)^n.$$

**E X A M P L E   1**   **Finding Compound Interest**

Suppose $1000 is deposited into an account with an annual yield of 8% compounded quarterly. Find the amount in the account at the end of five years, ten years, twenty years, thirty years, and forty years.

**Solutions**     We have $P = \$1000$, while the interest per quarter is $i = .08(\frac{1}{4}) = 0.02$. Thus $F = \$1000(1 + .02)^n$. Now using either the $y^x$ key on a calculator or Table B.2 in Appendix B we can obtain

| Years | Periods ($n = 4t$) | Future Value |
|-------|--------------------|--------------|
| 5     | 20                 | $\$1000(1 + .02)^{20} = \$1485.95$ |
| 10    | 40                 | $\$1000(1 + .02)^{40} = \$2208.04$ |
| 20    | 80                 | $\$1000(1 + .02)^{80} = \$4875.44$ |
| 30    | 120                | $\$1000(1 + .02)^{120} = \$10,765.16$ |
| 40    | 160                | $\$1000(1 + .02)^{160} = \$23,769.91$ |

Also see Figure 5.2.  ■

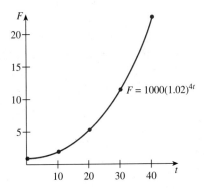

$F = 1000(1.02)^{4t}$

**Figure 5.2**

Notice that during the first ten years the account grows by about \$1200, but during the last ten years grows by about \$13,000. In fact, each year the account grows by more than in the previous year. Recall that for simple compounding, the account grows by the *same* amount every year.

**REMARK.**     We will use the letter $i$ to designate the interest rate for any period, whether annual or not. We will reserve the letter $r$ to always designate an annual rate.

Suppose $r$ is the annual interest rate expressed as a decimal, and interest is compounded $m$ times a year. Then the interest rate per time period is $i = r/m$. If the compounding goes on for $t$ years, then there are $n = mt$ time periods and the amount $F$ after $t$ years is

$$F = P(1 + i)^n = P\left(1 + \frac{r}{m}\right)^{mt}.$$

---

**Compound Interest**

Suppose a principal $P$ earns interest at the annual rate of $r$, expressed as a decimal, and interest is compounded $m$ times a year. Then the amount $F$ after $t$ years is

$$F = P(1 + i)^n = P\left(1 + \frac{r}{m}\right)^{mt},$$

where $n = mt$ is the number of time periods and $i = \dfrac{r}{m}$ is the interest per period.

---

**E X A M P L E   2**      **Finding Compounded Interest**

Suppose \$1000 is deposited into an account that yields 9% annually. Find the amount in the account at the end of the fifth year if the compounding is (a) quarterly, (b) monthly.

**Solutions**         (a)   Here $P = \$1000$, $r = .09$, $m = 4$, and $t = 5$. Using the $y^x$ key on your calculator gives

$$F = P\left(1 + \frac{r}{m}\right)^{mt} = \$1000\left(1 + \frac{.09}{4}\right)^{4(5)} = \$1560.51.$$

(b)   Here $P = \$1000$, $r = .09$, $m = 12$, and $t = 5$. Using the $y^x$ key on your calculator gives

$$F = P\left(1 + \frac{r}{m}\right)^{mt} = \$1000\left(1 + \frac{.09}{12}\right)^{12(5)} = \$1565.68. \quad\blacksquare$$

**R E M A R K .**     Notice that the amount at the end of the time period is larger if the compounding is done more often.

---

**E X A M P L E   3**      **Calculating Interest Over a Long Period of Time**

Suppose that when the Canarsie Indians sold Manhattan for \$24 in 1626, the money was deposited into a Dutch guilder account that yielded an annual rate of 6% compounded quarterly. How much would be in this account in 1994?

**Solution**      We have

$$F = P\left(1 + \frac{r}{m}\right)^{mt}$$

$$= \$24\left(1 + \frac{.06}{4}\right)^{368(4)}$$

$$\approx \$79 \text{ billion.} \quad\blacksquare$$

The following table indicates what the value of this account would have been at some intermediate times.

| Year | | Future Value |
|------|---|---|
| 1626 | | $24 |
| 1650 | $24 $(1 + \frac{.06}{4})^{4(24)} \approx$ | $100 |
| 1700 | $24 $(1 + \frac{.06}{4})^{4(74)} \approx$ | $2,000 |
| 1750 | $24 $(1 + \frac{.06}{4})^{4(124)} \approx$ | $39,000 |
| 1800 | $24 $(1 + \frac{.06}{4})^{4(174)} \approx$ | $760,000 |
| 1850 | $24 $(1 + \frac{.06}{4})^{4(224)} \approx$ | $15,000,000 |
| 1900 | $24 $(1 + \frac{.06}{4})^{4(274)} \approx$ | $293,000,000 |
| 1950 | $24 $(1 + \frac{.06}{4})^{4(324)} \approx$ | $6,000,000,000 |
| 1994 | $24 $(1 + \frac{.06}{4})^{4(368)} \approx$ | $79,000,000,000 |

**REMARK.**    It is, of course, very unlikely that any investment could have survived through the upheavals of wars and financial crises that occurred during this 368-year span of time. Nonetheless, it is examples like this one that inspire some to use the phrase "the wonders of compounding."

If an account earns interest at a rate of $i$, expressed as a decimal, per compounding period, how long will it take for the principal of $P$ to grow to some future value $F$? To answer this question we solve for the number $n$ of time periods in the equation

$$F = P(1 + i)^n,$$

by first taking common logarithms of each side.

$$\log F = \log P(1 + i)^n$$

$$= \log P + \log(1 + i)^n$$

$$= \log P + n \log(1 + i)$$

$$n \log(1 + i) = \log F - \log P$$

$$n \log(1 + i) = \log F/P$$

$$n = \frac{\log F/P}{\log(1 + i)}$$

An important case concerns the time for an account to double. For this case we must have $F = 2P$. This yields

$$n = \frac{\log F/P}{\log(1 + i)} = \frac{\log 2P/P}{\log(1 + i)} = \frac{\log 2}{\log(1 + i)}.$$

See Figure 5.3.

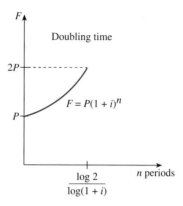

$F$

Doubling time

$2P$

$P$

$F = P(1 + i)^n$

$\dfrac{\log 2}{\log(1 + i)}$    $n$ periods

**Figure 5.3**

We have the following.

---

**The Time for Money to Grow from P to F**

The number of periods $n$ it takes for a principal of $P$ to grow to a future value of $F$ in an account that earns interest compounded at a rate of $i$ per period, expressed as a decimal, is

$$n = \frac{\log F/P}{\log(1 + i)}.$$

The number of periods for the principal to double is given by

$$n = \frac{\log 2}{\log(1 + i)}.$$

---

**E X A M P L E   4**    **Doubling Time**

Determine the time it takes for an account earning interest at an annual rate of 4% compounded annually to double.

**Solution**    Here $i = .04$. Since the time periods are measured in years, the number of years for this account to double is

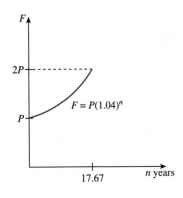

**Figure 5.4**

$$n = \frac{\log 2}{\log(1 + i)}$$

$$= \frac{\log 2}{\log(1 + .04)}$$

$$= 17.67 \qquad \text{using a calculator}$$

or almost exactly 17 years and 8 months. Refer to Figure 5.4.   ∎

The following table gives the doubling times for various annual interest rates for accounts that are compounding annually.

| Interest Rate | Doubling Time in Years |
|:---:|:---:|
| 4% | 17.67 |
| 5% | 14.21 |
| 6% | 11.90 |
| 7% | 10.24 |
| 8% | 9.01 |
| 9% | 8.04 |
| 10% | 7.27 |
| 15% | 4.96 |
| 20% | 3.80 |

## Effective Yield

If $1000 is invested at an annual rate of 9% compounded monthly, then at the end of a year there is

$$F = \$1000 \left(1 + \frac{.09}{12}\right)^{12} = \$1093.81$$

in the account. This is the same amount obtainable if the same principal of $1000 was invested for one year at an *annual* rate of 9.381% (or 0.09381, expressed as a decimal). We call the rate 9.381% the **effective annual yield**. The 9% annual rate is often referred to as the **nominal rate**.

Suppose $r$ is the annual interest rate expressed as a decimal and interest is compounded $m$ times a year. If the compounding goes on for one year, then the amount $F$ after one year is

$$F = P\left(1 + \frac{r}{m}\right)^{m}.$$

If we let $r_{eff}$ be the effective annual yield, then $r_{eff}$ must satisfy

$$P\left(1 + \frac{r}{m}\right)^m = P(1 + r_{eff}).$$

Solving for $r_{eff}$ we obtain

$$r_{eff} = \left(1 + \frac{r}{m}\right)^m - 1.$$

---

**Effective Yield**

Suppose a sum of money is invested at an annual rate of $r$ expressed as a decimal and is compounded $m$ times a year. The effective yield $r_{eff}$ is

$$r_{eff} = \left(1 + \frac{r}{m}\right)^m - 1.$$

---

**EXAMPLE 5** **Comparing Investments**

One bank advertises a nominal rate of 9.1% compounded semiannually. A second bank advertises a nominal rate of 9% compounded daily. What are the effective yields? In which bank would you deposit your money?

**Solutions** For the first bank $r = .091$ and $m = 2$. Then

$$r_{eff} = \left(1 + \frac{.091}{2}\right)^2 - 1 \approx .0931,$$

or as a percent, 9.31%.

For the second bank $r = .09$ and $m = 365$. Then

$$r_{eff} = \left(1 + \frac{.09}{365}\right)^{365} - 1 \approx .0942,$$

or as a percent, 9.42%.

Despite the higher nominal rate given by the first bank, the effective yield for the second bank is higher than the first. Thus money deposited in the second bank will grow faster than money deposited in the first bank. ■

## Present Value

If we have an account initially with $P$ and earning interest at an annual rate of $r$ expressed as a decimal and interest is compounded $m$ times a year, then the amount $F$ in the account after $t$ years is

$$F = P\left(1 + \frac{r}{m}\right)^{mt}.$$

If we wish to know how many dollars $P$ to set aside now in this account so that we will have a future amount of $F$ dollars after $t$ years, we simply solve the above expression for $P$. Thus

$$P = \frac{F}{\left(1 + \dfrac{r}{m}\right)^{mt}}.$$

This is called the **present value**.

---

**Present Value**

Suppose an account earns an annual rate of $r$ expressed as a decimal and is compounded $m$ times a year. Then the amount P, called the **present value**, needed presently in this account so that a future amount of F will be attained in $t$ years is given by

$$P = \frac{F}{\left(1 + \dfrac{r}{m}\right)^{mt}}.$$

---

E X A M P L E  6    **Finding the Present Value of a Future Balance**

How much money must grandparents set aside at the birth of their grandchild if they wish to have \$20,000 when the grandchild turns eighteen? They can earn 9% compounded quarterly.

**Solution**    Here $r = 0.09$, $m = 4$, $t = 18$, and $F = \$20,000$. Thus

$$P = \frac{F}{\left(1 + \dfrac{r}{m}\right)^{mt}}$$

$$= \frac{\$20,000}{\left(1 + \dfrac{.09}{4}\right)^{4(18)}} \approx \$4029.69 \quad \blacksquare$$

## SELF-HELP EXERCISE SET 5.2

1. An account with \$1000 earns interest at an annual rate of 8% compounded monthly. Find the amount in this account after 10 years.

2. Find the effective yield if the annual rate is 8% and the compounding is weekly.

3. How much money should be deposited in a bank account earning the annual interest rate of 8% compounded quarterly in order to have \$10,000 in the account at the end of 10 years?

# EXERCISE SET 5.2

In Exercises 1 through 10 find how much is in the accounts after the given years where $P$ is the initial principal and $r$ is the annual rate given as a percent with the indicated compounding.

1. $P = \$1000$, $r = 10\%$, compounded annually for 1 year

2. $P = \$2000$, $r = 9\%$, compounded annually for 1 year

3. After 1 year where $P = \$1000$, $r = 8\%$, compounded (a) annually, (b) quarterly, (c) monthly, (d) weekly, (e) daily

4. After 1 year where $P = \$1000$, $r = 10\%$, compounded (a) annually, (b) quarterly, (c) monthly, (d) weekly, (e) daily

5. After 40 years where $P = \$1000$, $r = 8\%$, compounded (a) annually, (b) quarterly, (c) monthly, (d) weekly, (e) daily

6. After 40 years where $P = \$1000$, $r = 10\%$, compounded (a) annually, (b) quarterly, (c) monthly, (d) weekly, (e) daily,

7. After 40 years where $P = \$1000$, compounded annually, $r$ equal to (a) 3%, (b) 5%, (c) 7%, (d) 9%, (e) 12%, (f) 15%

8. After 40 years where $P = \$1000$, compounded annually, $r$ equal to (a) 4%, (b) 6%, (c) 8%, (d) 10%, (e) 20%, (f) 25%

9. $P = \$1000$, $r = 9\%$, compounded annually, after (a) 5, (b) 10, (c) 15, (d) 30 years

10. $P = \$1000$, $r = 7\%$, compounded annually, after (a) 5, (b) 10, (c) 15, (d) 30 years

In Exercises 11 through 12 find the effective yield given the annual rate $r$ and the indicated compounding.

11. $r = 8\%$, compounded (a) semiannually, (b) quarterly, (c) monthly, (d) weekly, (e) daily

12. $r = 10\%$, compounded (a) semiannually, (b) quarterly, (c) monthly, (d) weekly, (e) daily

In Exercises 13 through 16 find the present value of the given amounts $F$ with the indicated annual rate of return $r$, the number of years $t$, and the indicated compounding

13. $F = \$10,000$, $r = 9\%$, $t = 20$, compounded (a) annually, (b) monthly, (c) weekly

14. $F = \$10,000$, $r = 10\%$, $t = 20$, compounded (a) annually, (b) quarterly, (c) daily

15. $F = \$100,000$, $r = 9\%$, $t = 40$, compounded (a) annually, (b) monthly, (c) weekly

16. $F = \$100,000$, $r = 10\%$, $t = 40$, compounded (a) annually, (b) quarterly, (c) daily

17. Your rich uncle has just given you your high school graduation present of $1,000,000. The present is in the form of a 40-year bond with an annual interest rate of 9% compounded annually. The bond says it will be worth $1,000,000 in 40 years. What is this $1,000,000 gift worth at the present time?

18. Redo Exercise 17 if the annual interest rate is 6%.

19. Your second rich uncle gives you a high school graduation present of $2,000,000. The present is in the form of a 50-year bond with an annual interest rate of 9% compounded annually. The bond says it will be worth $2,000,000 in 50 years. What is this $2,000,000 gift worth now? Compare your answer to Exercise 17.

20. Redo Exercise 19 with an annual interest rate of 6%.

21. In Example 3 find the amount in 1994 if the annual interest was 7% compounded quarterly. Compare your answer to that of Example 3 in the text.

22. In Example 3 find the amount in 1994 if the annual interest was 5% compounded quarterly. Compare your answer to that of Example 3 in the text.

23. An account grew from $1000 to $1100 in one year. The interest was compounded monthly. What was the nominal interest rate?

24. An account grew from $1000 to $1150 in one year. The interest was compounded quarterly. What was the nominal interest rate?

25. How long does it take an account earning interest at an annual rate of 6% and compounding monthly to grow from $1000 to $1500? Redo if the compounding is weekly.

26. How long does it take an account earning interest at an annual rate of 6% and compounding quarterly to grow from $1000 to $1600? Redo if the compounding is weekly.

27. How long does it take an account to double in value if the account earns interest at an annual rate of 6.5% and the compounding is (a) monthly, (b) daily?

28. How long does it take an account to double in value if the account earns interest at an annual rate of 5.5% and the compounding is (a) monthly, (b) daily?

## Applications

29. **Real Estate Appreciation.**   The United States paid about 4 cents an acre for the Louisiana Purchase in 1803. Suppose the value of this property grew at an annual rate of 5.5% compounded annually. What would an acre be worth in 1994? Does this seem realistic?

30. **Real Estate Appreciation.**   Redo the previous problem using a rate of 6% instead of 5.5%. Compare your answer with the answer to the previous problem.

31. **Coin Appreciation.**   What constant annual rate when compounded annually was necessary for the value of a United States half-penny to go from 0.50 cents in 1800 to $2000 in 1994?

*(Courtesy of* Coin *Magazine.)*

32. **Comparing Rates at Banks.**   One bank advertises a nominal rate of 6.5%

compounded quarterly. A second bank advertises a nominal rate of 6.6% compounded daily. What are the effective annual yields? In which bank would you deposit your money?

33. **Comparing Rates at Banks.** One bank advertises a nominal rate of 8.1% compounded semiannually. A second bank advertises a nominal rate of 8% compounded weekly. What are the effective yields? In which bank would you deposit your money?

34. **Saving for Machinery.** How much money should a company deposit in an account with a nominal rate of 8% compounded quarterly in order to have $100,000 for a certain piece of machinery in 5 years?

35. **Saving for Machinery.** Repeat the previous exercise with an annual rate of 7% and monthly compounding.

## Solutions to Self-Help Exercise Set 5.2

1. Here $P = \$1000$, $r = 0.08$, and $t = 10$, $m = 12$.
   Thus
   $$F = P\left(1 + \frac{r}{m}\right)^{mt} = \$1000\left(1 + \frac{.08}{12}\right)^{12(10)} = \$2219.64.$$

2. If the annual rate is 8%, the effective yield is given by
   $$r_{eff} = \left(1 + \frac{r}{m}\right)^{m} - 1 = \left(1 + \frac{.08}{52}\right)^{52} - 1 \approx .0832, \text{ or } 8.32\%.$$

3. The present value of $10,000 if the annual interest rate is 8% compounded quarterly for 10 years is
   $$P = \frac{F}{(1 + \frac{r}{m})^{mt}} = \frac{10,000}{(1 + \frac{.08}{4})^{4(10)}} = 4528.90.$$

   Thus a person must deposit $4528.90 in an account earning 8% compounded quarterly so that there will be $10,000 in the account after 10 years.

## 5.3 ANNUITIES AND SINKING FUNDS

► *Annuities*
► *Sinking Funds*

An individual is trying to save money for a down payment on a house to be purchased in 5 years. She can deposit $100 at the end of each month into an account that pays interest at an annual rate of 9% compounded monthly. How much is in this account after 5 years? (The answer can be found in Example 1.)

APPLICATION
**A Sinking Fund**

A corporation wishes to set up a fund in order to have the money necessary to replace a current machine. It is estimated that the machine will need to be replaced in 10 years and will cost $100,000. How much per quarter should be deposited into an account with an annual interest rate of 8% compounded quarterly to meet this future obligation? (The answer can be found in Example 2.)

## Annuities

We have previously studied lump sum payments. We now wish to study periodic payments.

An **annuity** is a sequence of equal payments made at equal time periods. An **ordinary annuity** is one in which the payments are made at the *end* of the time periods of compounding. The **term** of an annuity is the time from the beginning of the first period to the end of the last period. The total amount in the account, including interest, at the end of the term of an annuity is called the **future value of the annuity**. Examples of annuities are regular deposits into a savings account, monthly home mortgage payments, and monthly insurance payments.

We begin with an example of regular deposits into a savings account. Suppose we make deposits of $1000 at the end of each year into a savings account that earns 8% per year. We want to determine how much is in the account at the end of the sixth year. Notice that the first payment is made at the end of the first year and we want the total in the account immediately after the last deposit is made.

Figure 5.5 shows what happens to each deposit. For example, the first deposit of $1000 is in the account for 5 years and becomes $1000(1.08)^5$ at the end of the sixth year. The second deposit of $1000 is in the account for 4 years and becomes $1000(1.08)^4$ at the end of the sixth year, and so on. The amount in the account at the end of the sixth year is the sum of these six numbers or

$$S = 1000 + 1000(1.08) + 1000(1.08)^2 + 1000(1.08)^3 \qquad (1)$$
$$+ 1000(1.08)^4 + 1000(1.08)^5.$$

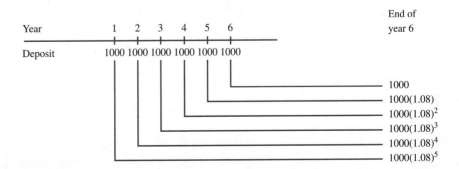

**Figure 5.5**

It is possible to greatly simplify this quantity. First notice that this quantity is in the form of

$$S = a + ab + ab^2 + \cdots + ab^{n-2} + ab^{n-1} \qquad (2)$$

where $a = 1000$, $b = 1.08$, and $n = 6$.

We now see how to find a simplified expression for any sum in the form of equation (2). If $b = 1$, then $S = na$. If $b \neq 1$, then multiply $S$ by $b$ and obtain

$$bS = ab + ab^2 + ab^3 + \cdots + ab^{n-1} + ab^n.$$

If we now subtract $S$ from $bS$, notice that all but two of the terms cancel and we have

$$bS - S = ab^n - a$$

$$(b - 1)S = a(b^n - 1)$$

$$S = a \frac{b^n - 1}{b - 1}.$$

Thus we have the following.

---

**Summation Formula**

For any positive integer $n$ and any real numbers $a$ and $b$, with $b \neq 1$,

$$a + ab + ab^2 + ab^3 + \cdots + ab^{n-1} = a \frac{b^n - 1}{b - 1}. \qquad (3)$$

---

Using formula (3) with $a = 1000$, $b = 1.08$, and $n = 6$, the sum of the six terms in (1) is

$$S = a \frac{b^n - 1}{b - 1} = 1000 \frac{(1.08)^6 - 1}{1.08 - 1} \approx 7335.93.$$

Thus there will be \$7335.93 in this account at the end of the sixth year.

Now more generally, suppose that $R$ dollars is deposited into an account at the end of each of $n$ periods with an interest rate of $i$ per period. We wish to know the amount at the end of the $n$th period. Figure 5.6 indicates what each of the deposits become by the end of the $n$th period. The value of the account at the end of the $n$th period is the sum of these $n$ values, denoted by $S$, and is given by

$$S = R + R(1 + i) + R(1 + i)^2 + \cdots + R(1 + i)^{n-1}.$$

We can now use the summation formula with $a = R$ and $b = (1 + i)$ and obtain

$$S = a \frac{b^n - 1}{b - 1} = R \frac{[(1 + i)^n - 1]}{1 + i - 1} = R \frac{[(1 + i)^n - 1]}{i}.$$

It is common to write

$$s_{n\rceil i} = \frac{[(1 + i)^n - 1]}{i}.$$

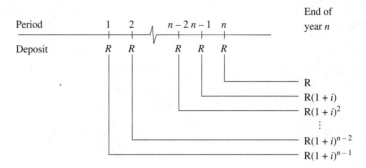

**Figure 5.6**

The symbol $s_{n\rceil i}$ is read "$s$ angle $n$ at $i$" and can be found in tables. It is also common to use the symbol $FV$ (future value) for $S$ and to use the symbol $PMT$ (payment) for $R$. We thus have the following.

---

**Future Value of an Ordinary Annuity**

The future value $FV$ of an ordinary annuity of $n$ payments of $PMT$ dollars paid at the end of each period into an account that earns interest at the rate of $i$ per period is

$$FV = PMT \frac{[(1 + i)^n - 1]}{i} = PMT s_{n\rceil i}. \tag{4}$$

---

**R E M A R K .**     When the periodic payment, PMT, is \$1, the future value $FT$ is just $s_{n\rceil i}$. Thus we have that $s_{n\rceil i}$ is the future value of an annuity after $n$ payments of \$1 have been made, where $i$ is the interest rate per period and the compounding is per the same period.

**E X A M P L E   1**   **Calculating the Future Value of an Ordinary Annuity**

An individual is trying to save money for a downpayment on a house to be purchased in 5 years. She can deposit \$100 at the end of each month into an account that pays interest at an annual rate of 9% compounded monthly. How much is in this account after 5 years? Also find the amount of interest that has been earned.

**Solution**   To use the above formula for $FV$, we need to find $PMT$, $i$ and $n$. The monthly payment is \$100, so $PMT = 100$. The interest rate per month is $0.09/12 = 0.0075$. The number of periods is $n = 12(5) = 60$. Thus

$$FV = PMT \frac{[(1 + i)^n - 1]}{i} \quad \text{or} \quad PMT\ s_{n\rceil i}$$

$$= \$100 \frac{[(1.0075)^{60} - 1]}{0.0075} \quad \text{or} \quad \$100\ s_{60\rceil 0.0075}$$

$$\approx \$100(75.4241) \qquad \text{Use calculator or}$$
$$\text{Table B.2 in Appendix B.}$$

$$= \$7542.41. \qquad \blacksquare$$

To find the interest, subtract the total deposits from $FV$. Total deposits are equal to the number of payments times the amount of each deposit or $60(\$100)$. Thus

$$\text{interest} = FV - nPMT = \$7542.41 - \$60(100) = \$1542.41.$$

## Sinking Funds

Often an individual or corporation knows that at some future date a certain amount $FV$ of money will be needed. Any account that is established for accumulating funds to meet a future need is called a **sinking fund**. The question is how much money $PMT$ should be put away periodically so that the deposits plus interest will be $FV$ at the needed point in time. In such a case, $FV$, $n$, and $i$ are known and one wishes to calculate the periodic payment $PMT$.

To find $PMT$ start with the formula

$$FV = PMT \frac{[(1 + i)^n - 1]}{i} = PMTs_{n\rceil i},$$

and solve for $PMT$. Thus

$$PMT = FV \frac{i}{[(1 + i)^n - 1]} = \frac{FV}{s_{n\rceil i}}.$$

We have the following.

---

**Sinking Fund Payment**

The periodic payment $PMT$ that is required to accumulate the sum $FV$ over $n$ periods of time with interest at the rate of $i$ per period is

$$PMT = FV \frac{i}{[(1 + i)^n - 1]} = \frac{FV}{s_{n\rceil i}}. \tag{5}$$

---

**E X A M P L E   2**    **Calculating a Sinking Fund Payment**

A corporation wishes to set up a sinking fund in order to have the funds necessary to replace a current machine. The machine will need to be replaced in 10 years and cost $100,000. How much should be deposited at the end of each quarter into an account with an annual interest rate of 8% compounded quarterly to meet this future obligation? What will be the total amount of the payments and what will be the interest earned?

**Solution**  We have $FV = 100,000$. Also $i = \frac{0.08}{4} = 0.02$. The number of periods is $n = 4(10) = 40$. Thus

$$PMT = FV \frac{i}{[(1 + i)^n - 1]} \qquad \text{or} \qquad \frac{FV}{s_{\overline{n}|i}}$$

$$= \$100,000 \frac{0.02}{[(1.02)^{40} - 1]} \qquad \text{or} \qquad \frac{\$100,000}{s_{\overline{40}|0.02}}$$

$$\approx \$100,000(0.0165557) \qquad \text{or} \qquad \frac{\$100,000}{60.40198} \quad \begin{array}{l} \text{Use a calculator or} \\ \text{Table B.2 in Appendix B.} \end{array}$$

$$= \$1655.57.$$

Thus if $1655.57 is placed into this sinking fund at the end of every quarter, there will be $100,000 in the fund at the end of 10 years.

Since there are 40 payments, the total amount of payments is $1655.57(40) = $66,222.80. The total interest earned is then $100,000 less the payments or $100,000 − $66,222.80 = $33,777.20. ∎

**E X A M P L E  3**  **Equity in a Sinking Fund**

Just 4 years after making the sinking fund payment in Example 2 the corporation decides to use the accumulated money (**equity**) for another purpose. Determine the equity.

**Solution**  We need to find the future value of an annuity with a quarterly payment of $1655.57 and an annual interest rate of 8% compounded quarterly at the end of 4 years. The number of periods is $n = 4(4) = 16$. We use formula (4).

$$FV = PMT \frac{[(1 + i)^n - 1]}{i} \qquad \text{or} \qquad PMT\, s_{\overline{n}|i}$$

$$= \$1655.57 \frac{[(1.02)^{16} - 1]}{0.02} \qquad \text{or} \qquad \$1655.57\, s_{\overline{16}|0.02}$$

$$\approx \$1655.57(18.6393) \qquad \begin{array}{l} \text{Use calculator or} \\ \text{Table B.2 in Appendix B.} \end{array}$$

$$= \$30,858.64$$

Thus after 4 years this fund has grown to $30,858.64. ∎

The following table indicates the equity in this sinking fund for selected times. The values can be found in the same way as in Example 3. The total interest is found as in Example 2, that is, by subtracting the total payments from the future value.

| End of Year | Number of $1655.57 Payments | Total Payments | Interest Earned (Equity less total payments) | Equity |
|---|---|---|---|---|
| 2 | 8 | $13,244.56 | $965.15 | $14,209.71 |
| 4 | 16 | 26,489.12 | 4369.52 | 30,858.64 |
| 6 | 24 | 39,733.68 | 10,631.84 | 50,365.52 |
| 8 | 32 | 52,978.24 | 20,242.70 | 73,220.94 |
| 10 | 40 | 66,222.80 | 33,777.20 | 100,000.00 |

The above corporation with a sinking fund will need to know the interest per quarter since this interest contributes to profits. To determine the interest for any period, we note that the value of the sinking fund increases in any one period due to one payment into the fund and the interest earned during that period. Thus we have

$$\text{Interest in } m\text{th period} = \begin{array}{l} \text{value at end of } m\text{th period} \\ \text{less value at end of previous period} \\ \text{less payment} \end{array}$$

or

$$\text{Interest in } m\text{th period} = PMT\, s_{\overline{m}|i} - PMT\, s_{\overline{m-1}|i} - PMT$$

$$= PMT(s_{\overline{m}|i} - s_{\overline{m-1}|i}) - PMT.$$

---

**Interest per Period in a Sinking Fund**

The interest earned during the $m$th period of a sinking fund with payments of PMT and earning interest at a rate of $i$ per period is

$$\text{interest in } m\text{th period} = PMT(s_{\overline{m}|i} - s_{\overline{m-1}|i}) - PMT.$$

$$= PMT(s_{\overline{m}|i} - s_{\overline{m-1}|i} - 1)$$

---

**E X A M P L E  4**    **Interest per Period in a Sinking Fund**

Find the interest earned by the sinking fund in Example 2 during the first quarter of the fifth year.

**Solution**    The first quarter of the fifth year is the 17th period. Taking $m = 17$ in the above formula and using Table B.2 in Appendix B gives

$$\begin{array}{l} \text{interest in} \\ \text{17th period} \end{array} = \$1655.57(s_{\overline{17}|0.02} - s_{\overline{16}|0.02}) - \$1655.57$$

$$= \$1655.57(20.012071 - 18.639285) - \$1655.57 = \$617.17$$

Thus $617.17 of interest was earned on this sinking fund during the first quarter of the fifth year.

## SELF-HELP EXERCISE SET 5.3

1. At the end of every six months an individual places $1000 into an account earning an annual rate of 10% compounded semiannually. Find the amount in the account at the end of 15 years.

2. A person will need $20,000 to start a small business in 5 years. How much should be deposited at the end of every 3 months into an account paying an annual rate of 9% compounded quarterly to meet this goal?

## EXERCISE SET 5.3

In Exercises 1 through 6 find the future values of each of the ordinary annuities at the given annual rate $r$ compounded as indicated. The payments are made to coincide with the periods of compounding.

1. $PMT = 1200$, $r = 0.08$, compounded annually for 10 years

2. $PMT = 600$, $r = 0.08$, compounded semiannually for 10 years

3. $PMT = 300$, $r = 0.08$, compounded quarterly for 10 years

4. $PMT = 100$, $r = 0.08$, compounded monthly for 10 years

5. $PMT = 100$, $r = 0.12$, compounded monthly for 20 years

6. $PMT = 100$, $r = 0.12$, compounded weekly for 40 years

In Exercise 7 through 10 find the future value of each of the annuities at the *end* of the given $n$th year where the payments are made at the *beginning* of each year and $r$ is the annual interest rate compounded annually. Such an annuity is called an **annuity due**.

7. $PMT = \$1000$, $r = 0.08$, 10 years

8. $PMT = \$1000$, $r = 0.10$, 20 years

9. $PMT = \$1000$, $r = 0.15$, 40 years

10. $PMT = \$500$, $r = 0.12$, 30 years

In Exercises 11 through 16 find the periodic payment for each sinking fund that is needed to accumulate the given sum under the given conditions.

11. $FV = \$20,000$, $r = 0.09$, compounded annually for 10 years

**12.** $FV = \$20,000$, $r = 0.09$, compounded monthly for 20 years

**13.** $FV = \$10,000$, $r = 0.10$, compounded quarterly for 10 years

**14.** $FV = \$1,000,000$, $r = 0.09$, compounded weekly for 40 years

**15.** $FV = \$1,000,000$, $r = 0.15$, compounded weekly for 40 years

**16.** $FV = \$1,000,000$, $r = 0.12$, compounded monthly for 40 years

**17.** In formula (4) in the text solve for $n$ and obtain

$$n = \frac{\log\left(\frac{iFV}{PMT} + 1\right)}{\log(1 + i)}.$$

**18.** Use the formula in Exercise 17 to find how many years it would take to accumulate $1 million with an annual payment of $500 and an interest rate of 15% per year compounded annually.

**19.** Use the formula in Exercise 17 to find how many years it would take to accumulate $1 million with an annual payment of $5000 and an interest rate of 9% per year compounded annually.

**20.** Show that the interest in $m$th period

$$PMT(s_{\overline{m}|i} - s_{\overline{m-1}|i} - 1) = PMT[(1 + i)^{m-1} - 1]$$

## Applications

**21. Retirement.**    An individual earns an extra $2000 each year and places this money at the end of each year into an Individual Retirement Account (IRA) in which both the original earnings and the interest in the account are not subject to taxation. If the account has an annual interest rate of 9% compounded annually, how much is in the account at the end of 40 years?

**22. Retirement.**    Repeat the previous problem if the annual interest rate is 12% per year.

**23. Retirement.**    Suppose the personal income tax is set at $33\frac{1}{3}\%$. An individual earns an extra $2000 each year and pays income taxes on these earnings and at the end of each year places the remaining funds $2000($\frac{2}{3}$)$ into a regular savings account in which the interest in the account is subject to the personal income tax mentioned before. If the account earns an annual interest rate of 9% compounded annually and the individual pays income taxes owed on the interest out of these funds, how much is in the account at the end of 40 years? Compare your answer to Exercise 21.

**24. Retirement.**    Repeat the previous problem if the annual interest rate is 12% per year. Compare your answer to Exercise 22.

**25. Education Fund.**    New parents wish to save for their newborn's education and wish to have $50,000 at the end of 18 years. How much should they place at the end of each year into a savings account that earns an annual rate of 9% compounded annually? How much interest would they earn over the life of this account? Determine the equity in this fund after 10 years. How much interest was earned during the tenth year?

26. **Education Fund.** Repeat the previous problem if the annual interest rate is 7%.

27. **House Down Payment.** A couple will need $20,000 at the end of 5 years for a down payment on a house. How much should they place at the end of each month into a savings account earning an annual rate of 9% compounded monthly to meet this goal? Determine the equity in this fund at the end of each year.

28. **House Down Payment.** Repeat the previous problem if the annual interest rate is 7%.

29. **Equipment.** A corporation creates a sinking fund in order to have $1 million to replace some machinery in 10 years. How much should be placed into this account at the end of each quarter if the annual interest rate is 10% compounded quarterly? Determine the equity in this fund at the end of every two years. How much interest was earned over the life of this fund? Determine the interest earned during the second quarter of the fifth year.

30. **Equipment.** Repeat the previous problem if the annual interest rate is 6%.

### Solutions to Self-Help Exercise Set 5.3

1. The amount in an account at the end of 15 years into which $1000 has been placed every 6 months and that pays 10% compounded semiannually is given by

$$FV = PMT \frac{[(1 + i)^n - 1]}{i},$$

where $PMT = \$1000$ and $i = \frac{.10}{2} = .05$ and $n = 2(15) = 30$. Thus

$$FV = \$1000 \frac{[(1.05)^{30} - 1]}{0.05} = \$66,438.85.$$

2. The periodic payment $PMT$ that is required to accumulate the sum of $FV = \$20,000$ over the $n = 4(5) = 20$ quarters of time with interest at the rate of $i = \frac{.09}{4} = .0225$ per quarter is

$$PMT = FV \frac{i}{[(1 + i)^n - 1]} = \$20,000 \frac{.0225}{[(1.0225)^{20} - 1]} = \$802.84.$$

## 5.4    PRESENT VALUE OF ANNUITIES AND AMORTIZATION

▶ *Present Value of Annuities*

▶ *Amortization*

▶ *Amortization Schedule*

**APPLICATION**

**Finding the Actual Value of $1,000,000 Lotto Prize**

Suppose you run a lottery for your state government and you have just had a "$1 million winner." The winner is immediately given $50,000 and you must make the arrangements that will ensure that this winner will receive

$50,000 at the end of each of the next 19 years. You notice that short, inter-
mediate, and long-term interest rates are all at 8%. How much money will it
presently cost your department to ensure that these 20 payments of $50,000
each will be made? In other words, how much has the "$1 million winner"
*really* won? (The answer to this question is given below.)

**A P P L I C A T I O N**

**Finding Car Payments**

You wish to borrow $12,000 from the bank to purchase a car. Interest is 1%
a month, and there are to be 48 equal monthly payments with the first to
begin in one month. What must the payments be so that the loan will be paid
off after 48 months? (The answer is given in Example 4.)

## Present Value of Annuities

We now answer the question posed in the first paragraph above.

You know that if an amount equal to $P$ earns interest at 8% per year, then
the amount after $n$ years is given by $F = P(1.08)^n$. Thus

$$P = \frac{F}{(1.08)^n} = F(1.08)^{-n}$$

is the present amount of money (present value) needed to attain the value $F$ after
$n$ years if compounded annually at 8%.

After making the initial payment of $50,000, we will now see how much
money must be placed into an account earning 8% per year so that a payment of
$50,000 can be made at the end of each of the next 19 years. If $P_1$ is the present
amount needed to be set aside now in order to make the $50,000 payment at the
end of the first year, then, according to the above formula,

$$P_1 = 50,000(1.08)^{-1}.$$

The present amount of money $P_2$ needed to be set aside now in order to make
the $50,000 payment at the end of the second year is

$$P_2 = 50,000(1.08)^{-2}.$$

And in general the present amount of money $P_n$ needed to be set aside now to
make the $50,000 payment at the end of the $n$th year is

$$P_n = 50,000(1.08)^{-n}.$$

See Figure 5.7. The sum of these 19 payments, denoted by $P$, is then the amount
needed.

$$P = P_1 + P_2 + \cdots + P_{19}$$

$$P = 50,000(1.08)^{-1} + 50,000(1.08)^{-2} + \cdots + 50,000(1.08)^{-19} \quad (1)$$

Recall the summation formula mentioned in the last section

**Figure 5.7**

$$a + ab + ab^2 + ab^3 + \cdots + ab^{n-1} = a\,\frac{b^n - 1}{b - 1}.$$

The sum of the 19 numbers in (1) can be found using this last formula with $a = 50{,}000(1.08)^{-1}$, $b = (1.08)^{-1}$, and $n = 19$, to obtain

$$P = a\,\frac{b^{19} - 1}{b - 1}$$

$$= 50{,}000(1.08)^{-1}\,\frac{(1.08)^{-19} - 1}{(1.08)^{-1} - 1}$$

$$= \frac{50{,}000}{(1.08)}\,\frac{1 - (1.08)^{-19}}{1 - (1.08)^{-1}}$$

$$= 50{,}000\,\frac{1 - (1.08)^{-19}}{1.08 - 1}$$

$$= 50{,}000\,\frac{1 - (1.08)^{-19}}{0.08}.$$

Thus we have

$$P = 50{,}000\,\frac{1 - (1.08)^{-19}}{0.08}. \tag{2}$$

If we carry out the calculation, we obtain

$$P \approx 50{,}000(9.6036) \approx 480{,}180.$$

So you need to put away into this 8% account only $480,180 to pay out the remaining $950,000 in 19 payments of $50,000 each. Thus the total cost to your department for this "$1 million" winner is just $480,180 + $50,000 = $530,180.

In general, if $R$ is the periodic payment, $i$ the rate of interest per period, and $n$ the number of periods, we can proceed as above and in analogy with formula (2) we obtain

$$P = R\,\frac{[1 - (1 + i)^{-n}]}{i}.$$

It is the custom to use *PV* for the present value, *PMT* for the periodic payment, and to use the notation

$$a_{\overline{n}|i} = \frac{[1 - (1 + i)^{-n}]}{i}.$$

We then have the following.

---

**Present Value of an Ordinary Annuity**

The present value *PV* of an ordinary annuity of *n* payments of *PMT* dollars each made at the end of the period with interest compounded at the rate of *i* per period is

$$PV = PMT \frac{[1 - (1 + i)^{-n}]}{i} = PMT a_{\overline{n}|i}. \tag{3}$$

---

**REMARK.** If the payment PMT is $1, then the present value is just $a_{\overline{n}|i}$. Thus $a_{\overline{n}|i}$ is the value of an account needed to provide for *n* payments (or withdrawals) of $1 each when interest is paid at a rate of *i* per period and the compounding is per the same period.

**E X A M P L E  1**     **Calculating the Present Value of an Ordinary Annuity**

You have announced to your company that you will retire in one year. Your pension plan requires the company to pay you $25,000 in a lump sum at the end of one year and every year thereafter until your demise. The company makes the assumption that you will live to receive 15 payments. Interest rates are 7% per year compounded annually. What amount of money should the company set aside now to ensure that they can meet their pension obligations to you?

**Solution**     We use formula (3) with *PMT* = 25,000, *i* = 0.07, and *n* = 15, and obtain

$$PV = PMT \frac{[1 - (1 + i)^{-n}]}{i} \qquad \text{or} \quad PMT \, a_{\overline{n}|i}$$

$$= \$25,000 \frac{[1 - (1.07)^{-15}]}{0.07} \quad \text{or} \quad \$25,000 \, a_{\overline{15}|0.07}$$

$$\approx \$25,000(9.107914) \qquad \qquad \text{Use a calculator or}$$
$$\qquad \qquad \qquad \qquad \qquad \qquad \text{Table B.2 in Appendix B.}$$

$$= \$227,698$$

Thus if the company now puts away $227,698 into this account, they can make payments of $25,000 each at the end of each of the next 15 years (with no money left in the account after the last payment). ■

EXAMPLE 2  **Comparing Investments**

You represent a bank that has $1 million in cash to invest. An oil company seeking money to hunt for oil approaches you wishing to borrow $1 million using one of their producing oil wells as collateral. They have the long-term oil contracts in hand and can guarantee you a payment of $100,000 at the end of each of the next 20 years. Interest rates available to the bank on the open market are at 8% per year compounded annually. Is this a good deal?

Solution  You ask yourself how much would be needed to invest at current interest rates to guarantee a payment of $100,000 at the end of each of the next 20 years. Use formula (3) with $PMT = 100,000$, $i = 0.08$, and $n = 20$, and obtain

$$PV = PMT \frac{[1 - (1 + i)^{-n}]}{i} \qquad \text{or} \quad PMT \, a_{n\rceil i}$$

$$= \$100,000 \frac{[1 - (1.08)^{-20}]}{0.08} \qquad \text{or} \quad \$100,000 \, a_{20\rceil 0.08}$$

$$\approx \$100,000(9.818147) \qquad \begin{array}{l}\text{Use a calculator or} \\ \text{Table B.2 in Appendix B.}\end{array}$$

$$= \$981,814.74$$

Thus only about $982,000 is needed at current interest rates to obtain the same cash payments as the oil company is offering for $1 million. This is not a good deal. ∎

EXAMPLE 3  **Capital Expenditure Analysis**

A corporation wishes to increase productivity and thus save money by purchasing new machinery. The corporation can buy a machine for $100,000 that will save $30,000 annually and has a useful life of 5 years or a second machine for $160,000 that will save $35,000 annually and has a useful life of 7 years. If the annual interest rate is 8% compounded annually, which machine should the corporation purchase? Assume that the annual savings occurs at the end of each year.

Solution  The first machine saves $30,000 annually for 5 years. The present value of an annuity with an annual payment of $30,000 for 5 years with an annual interest rate of 8% compounded annually is

$$PV = PMT \, a_{5\rceil 0.08}$$

$$= \$30,000(3.992710) \qquad \text{From Table B.2 in Appendix B}$$

$$= \$119,781.30$$

This means that an annual savings of $30,000 for 5 years is equivalent to a lump sum of $119,781.30 now for a savings of

$$\$119,781.30 - \$100,000 = \$19,781.30.$$

The second machine saves $35,000 annually for 7 years. The present value of an annuity with an annual payment of $35,000 for 7 years with an annual interest rate of 8% compounded annually is

$$PV = PMT \, a_{\overline{7}|0.08}$$

$$= \$35,000(5.206370) \qquad \text{From Table B.2 in Appendix B}$$

$$= \$182,222.95$$

This means that an annual savings of $35,000 for 7 years is equivalent to a lump sum of $182,222.95 now for a savings of

$$\$182,222.95 - \$160,000 = \$22,222.95.$$

The second machine offers the larger savings and should be bought.   ■

## Amortization

Now look again at Example 2 but this time from the perspective of the oil company, that is, from the perspective of the borrower. If the oil company wishes to borrow an amount $PV$ at an interest rate of $i$ with equal payments at the end of each year with interest charged on the unpaid balance so that the entire debt will be repaid after $n$ years, what will be the payments $PMT$? The very same formula used by the bank that relates all of these quantities must apply, namely,

$$PV = PMT \, \frac{[1 - (1 + i)^{-n}]}{i},$$

Now we are interested in solving for $PMT$. Solving for $PMT$, we obtain

$$PMT = PV \, \frac{i}{[1 - (1 + i)^{-n}]}.$$

This process of paying the debt is called **amortizing** (or killing) the debt. We then have

---

**Amortization Formula**

The periodic payment $PMT$ to be made at the end of each period on a loan of $PV$ dollars that is to be amortized over $n$ periods with interest at the rate of $i$ per period is

$$PMT = PV \, \frac{i}{[1 - (1 + i)^{-n}]} = \frac{PV}{a_{\overline{n}|i}}. \qquad (4)$$

---

**E X A M P L E   4**    **Calculating Payments on a Loan**

You wish to borrow $12,000 from the bank to purchase a car. The bank charges interest at an annual rate of 12%. There are to be 48 equal monthly payments with the first to begin in one month. What must the payments be so that the loan will be paid off after 48 months? Find the total interest paid on this loan.

**Solution**   The rate per month is the annual rate divided by 12 or $i = \frac{.12}{12} = .01$. Using formula (4) with $PV = 12,000$, $i = .01$, $n = 48$, we obtain

$$PMT = PV \frac{i}{[1 - (1 + i)^{-n}]} \quad \text{or} \quad \frac{PV}{a_{\overline{n}|i}}$$

$$= \$12,000 \frac{0.01}{[1 - (1.01)^{-48}]} \quad \text{or} \quad \frac{\$12,000}{a_{\overline{48}|0.01}}$$

$$\approx \$12,000(0.0263338) \quad \text{or} \quad \frac{\$12,000}{37.973959} \quad \begin{array}{l}\text{Use calculator} \\ \text{or Table B.2} \\ \text{in Appendix B.}\end{array}$$

$$= \$316.01$$

Thus you will have to make 48 monthly payments of $316.01 in order to pay off this loan at the given interest rate.

The total interest paid is the total payments less the loan, or

$$\$316.01(48) - \$12,000 = \$15,168.48 - \$12,000 = \$3168.48. \quad \blacksquare$$

## Amortization Schedule

We now look at what happens if you wish to make a lump-sum payment to pay off a loan. For example, it is routine for an individual to sell a house and buy another. This requires paying off the old mortgage. How much needs to be paid? To answer this we look closely at how much is owed at the end of each payment period. The next example indicates how to create such a schedule.

**E X A M P L E   5**   **Calculating an Amortization Schedule**

You agree to sell a small piece of property and grant a loan of $9000 to the buyer with annual interest at 10% compounded annually, with payments of equal amounts made at the end of each of the next 6 years. Construct a table that gives for each period the interest, payment toward principal, and the outstanding balance.

**Solution**   To find the payments we use formula (4) with $PV = 9000$, $i = 0.10$, $n = 6$ and obtain

$$PMT = PV \frac{i}{[1 - (1 + i)^{-n}]}$$

$$= \$9000 \frac{0.10}{[1 - (1.10)^{-6}]}$$

$$\approx \$9000(0.2296074) \qquad \text{Use calculator}$$

$$= \$2066.47$$

The interest paid for the first year is $9000(0.10) = $900. Thus the payment on the principal is

$$\$2066.47 - \$900 = \$1166.47.$$

The remaining principal is then

$$\$9000 - \$1166.47 = \$7833.53.$$

The interest paid for the second year is $7833.53(0.10) = $783.35. Thus the payment on the principal is

$$\$2066.47 - \$783.35 = \$1283.12.$$

The remaining principal is then

$$\$7833.53 - \$1283.12 = \$6550.41.$$

The following table summarizes the remaining calculations.

| End of Period | Repayment Made | Interest Charged | Payment Toward Principal | Outstanding Principal |
|---|---|---|---|---|
| 0 | | | | 9000 |
| 1 | 2066.47 | 900.00 | 1166.47 | 7833.53 |
| 2 | 2066.47 | 783.35 | 1283.12 | 6550.41 |
| 3 | 2066.47 | 655.04 | 1411.43 | 5138.98 |
| 4 | 2066.47 | 513.90 | 1552.57 | 3586.41 |
| 5 | 2066.47 | 358.64 | 1707.83 | 1878.58 |
| 6 | 2066.44 | 187.86 | 1878.61 | 0000.00 |

The last payment is the one that adjusts for any roundoff errors.  ■

E X A M P L E  6    **Calculating the Outstanding Principal**

In the loan in Example 4, suppose the individual wishes to pay off the loan after 20 payments. What is the outstanding principal or, alternatively, the **equity**?

**Solution**    One way of finding the outstanding principal or unpaid balance is to make a table as in the previous example, but there is a less tedious way. The bank views the loan as an annuity that they purchased from the borrower. Thus the unpaid balance of a loan with $(48 - 20) = 28$ remaining payments is the present value of the annuity that can be found using formula (3). We have the payments $PMT = \$316.01$, $i = 0.01$, and $n = 28$. Thus the present value is

$$PV = PMT \frac{[1 - (1 + i)^{-n}]}{i} \quad \text{or} \quad PMT \, a_{\overline{n}|i}$$

$$= \$316.01 \frac{[1 - (1.01)^{-28}]}{0.01} \quad \text{or} \quad \$316.01 \ a_{\overline{28}|0.01}$$

$$\approx \$316.01(24.316443) \qquad \text{Use calculator or Table B.2 in Appendix B.}$$

$$= \$7684.24$$

This is the amount the bank will require to pay off the loan at this point. ∎

In the previous example let $n = 48$ be the total number of payments and $m = 20$ be the number of payments made, then notice that

$$PV = PMT \ a_{\overline{28}|0.01} = PMT \ a_{\overline{n-m}|0.01}.$$

In general we have the following.

---

**Outstanding Principal**

Let $n$ be the total number of payments required to pay off an amortized loan and let $m$ be the number of payments of PMT made; then the outstanding principal $P$ is given by

$$P = PMT \ a_{\overline{n-m}|i}.$$

---

We can also use this formula to determine the amount of any payment that is applied to the principal. To see how to do this notice that

$$\text{the amount owed after } m - 1 \text{ payments} = PMT \ a_{\overline{n-m+1}|i}$$

$$\text{the amount owed after } m \text{ payments} = PMT \ a_{\overline{n-m}|i}$$

The difference of these two is the amount of the $m$th payment that is applied to the principal.

---

**Payment Toward Principal**

For an amortized loan requiring $n$ equal payments of PMT dollars each and interest at the rate of $i$ per period, the amount of the $m$th payment that is applied toward the principal is

$$\text{payment toward principal} = PMT(a_{\overline{n-m+1}|i} - a_{\overline{n-m}|i}).$$

---

**E X A M P L E  7**     **Finding the Payment Toward Principal**

For the amortized loan in Example 5 find the amount of the second payment $R$ made toward the principal.

**Solution**     Since each payment is $2066.47, $n = 6$, $m = 2$, and $i = .10$,

$$R = PMT(a_{n-m+1 \rceil i} - a_{n-m \rceil i})$$

$$= \$2066.47(a_{5 \rceil 0.10} - a_{4 \rceil 0.10})$$

$$= \$2066.47(3.790787 - 3.169865) \qquad \text{Use a calculator.}$$

$$= \$1283.12$$

This agrees with what we found in Example 5.   ■

## SELF-HELP EXERCISE SET 5.4

1.  Refer to Example 2 of the text. The very next day the Federal Reserve chairman announces a concerted attempt to control inflation. The financial markets respond immediately with a 1% drop in interest rates. Do you want the deal now?

2.  You borrow $100,000 to purchase a house. The bank charges interest of 9% per year (on the unpaid balance). Payments will be made every month for 30 years.
    a.  What are the monthly payments to ensure that the loan is paid in full by the end of 30 years?
    b.  What is the total interest paid on this mortgage?
    c.  What are the amounts of the 10th and 300th payments made toward the principal?

## EXERCISE SET 5.4

For Exercises 1 through 6 find the amount needed to deposit into an account today that will yield a typical pension payment of $25,000 at the end of each of the next 20 years for the given annual interest rates.

| 1. | 4% | 2. | 5% | 3. | 6% | 4. | 7% | 5. | 8% | 6. | 9% |

In Exercises 7 through 10 find the annual payment needed to amortize a $10,000 loan in 6 payments if the interest is compounded annually at the given rate. Then find the total interest paid on each loan.

| 7. | 10% | 8. | 12% | 9. | 15% | 10. | 18% |

In Exercises 11 through 14 find the monthly payment needed to amortize the following typical $100,000 mortgage loans amortized over 30 years at the given annual interest rate compounded monthly. Then find the total interest paid on each loan.

| 11. | 7% | 12. | 8% | 13. | 9% | 14. | 10% |

In Exercises 15 through 18 find the monthly payments needed to amortize the following typical $100,000 mortgage loans amortized over 15 years at the given annual interest rate compounded monthly. Then find the total interest paid on each loan and compare with your answers to Exercises 11 through 14.

| 15. | 7% | 16. | 8% | 17. | 9% | 18. | 10% |

In Exercises 19 through 22 find the monthly payment needed to amortize the following typical $10,000 automobile loans over 4 years at the given annual interest rate compounded monthly. Then find the total interest paid on each loan.

**19.** 10%  **20.** 11%  **21.** 12%  **22.** 14%

In Exercises 23 through 26 find the monthly payment needed to amortize the following typical $1,000 credit card loans over 3 years at the given annual interest rate compounded monthly. Then find the total interest paid on each loan.

**23.** 14%  **24.** 16%  **25.** 18%  **26.** 20%

**27.** An individual wishes to pay off the mortgage in Exercise 11 after 120 payments. What amount is owed? What are the amounts of the 15th and 340th payments paid toward the principal?  *example 6*

**28.** An individual wishes to pay off the mortgage in Exercise 12 after 120 payments. What amount is owed? What are the amounts of the 15th and 340th payments paid toward the principal?

**29.** An individual wishes to pay off the mortgage in Exercise 15 after 60 payments. What amount is owed? What are the amounts of the 10th and 150th payments paid toward the principal?

**30.** An individual wishes to pay off the mortgage in Exercise 16 after 60 payments. What amount is owed? What are the amounts of the 10th and 150th payments paid toward the principal?

In Exercises 31 through 34 prepare an amortization schedule as in the table in Example 5 for each of the following loans.  *TEST QUES.*

**31.** The loan in Exercise 7  **32.** The loan in Exercise 8

**33.** The loan in Exercise 9  **34.** The loan in Exercise 10

**35.** From either formula (3) or (4) in the text show that

$$n = \frac{\log \dfrac{PMT}{PMT - PV(i)}}{\log(1 + i)}$$

Suppose you wished to take out a $10,000 loan at 1% a month and could make a monthly payment of $200. Use this formula to determine the number of payments needed to amortize this loan.

**36.** Show that the payment toward principal

$$PMT\,(a_{n-m+1\rceil i} - a_{n-m\rceil i}) = PMT\,(1 + i)^{m-n-1}$$

## Applications

**37. Comparing Present Values.** Two oil wells are for sale. The first will yield payments of $10,000 at the end of each of the next 10 years, while the second will yield payments of $6500 at the end of each of the next 20 years. Interest rates are assumed to hold steady at 8% per year over the next 20 years. Which has the higher present value?

**38. Comparing Present Values.** Redo the previous exercise if interest rates were 6% over the next 20 years.

(© John Spragens, Jr./Photo Researchers, Inc.)

**39. Comparing Present Values.**   You are being offered a ''half a million dollar'' retirement package to be given in $50,000 payments at the end of each of the next 10 years. You are also given the option of accepting a $350,000 lump sum payment now. Interest rates are at 9% per year. Which looks better to you? Why?

**40. Comparing Present Values.**   You are being offered a ''million dollar'' retirement package to be given in $50,000 payments at the end of each of the next 20 years. You are also given the option of accepting a $500,000 lump-sum payment now. Interest rates are at 9% per year. Which looks better to you? Why?

**41. Capital Expenditure.**   A corporation wishes to increase productivity and thus save money by purchasing new machinery. The corporation can buy a machine for $95,000 that will save $30,000 annually and has a useful life of 5 years or a second machine for $160,000 that will save $35,000 annually and has a useful life of 7 years. Assuming that the annual savings occurs on the last day of each year, which machine should the corporation purchase if the annual interest rate is (a) 8% compounded annually, (b) 6% compounded annually?

**42. Leasing.**   A corporation can either lease a machine with a useful life of 6 years for $20,000 per year paid at the end of the year or buy it for $100,000. Which should the corporation do if annual interest rates are (a) 7% compounded annually (b) 5% compounded annually?

**43. Car Loan.**   You purchase a new car for $12,000 and make a $2000 down payment. You then owe $10,000 for the car which you intend to pay off in 3 years. Initially, you intended to finance your car loan through your credit union that charges 9% compounded monthly for new-car loans. The salesman tells you, however, that GM is having a promotion and you can finance your car through GMAC for 4% interest compounded monthly or you can receive a rebate of $800 that can be applied to the purchase price. Should you finance the full $10,000 with GMAC for the lower interest rate or take the rebate and finance $9200 with your credit union? Explain your answer.

## Solutions to Self-Help Exercise Set 5.4

1.  Do the same calculation as in Example 2 of the text but now use $i = 0.07$ and obtain

$$PV = PMT \frac{[1 - (1 + i)^{-n}]}{i} = 100,000 \frac{[1 - (1.07)^{-20}]}{0.07} \approx 1,059,400$$

Now it will take about $1,059,400 at current interest rates to obtain the same cash payments as the oil company is offering for less than $1 million. The deal is now a good one (without consideration of the riskiness of the deal).

2.  **a.**   The monthly payment of $PMT$ on a loan of $PV = \$100,000$ that is to be amortized over $n = 12(30) = 360$ periods with interest at the rate of $i = .09/12 = .0075$ per month is, using formula (4),

$$PMT = PV \frac{i}{[1 + (1 + i)^{-n}]} = 100,000 \frac{.0075}{[1 - (1.0075)^{-360}]} = 804.62$$

or $804.62 as the monthly payment.

**b.** The interest is the total payments less the principal or

$$804.62(360) - 100,000 = 189,664.14.$$

**c.** The amount of the 10th payment toward the principal is

$$R = \$804.62(a_{\overline{360-10+1}|0.0075} - a_{\overline{360-10}|0.0075})$$

$$= \$804.62(123.65224 - 123.57963)$$

$$= \$58.42$$

The amount of the 300th payment toward the principal is

$$R = \$804.62(a_{\overline{360-300+1}|0.0075} - a_{\overline{360-300}|0.0075})$$

$$= \$804.62(48.807319 - 48.173374)$$

$$= \$510.08$$

## SUMMARY OUTLINE OF CHAPTER 5

- **Simple Interest.** Suppose a sum of money P, called the principal or present value, is invested for $t$ years at an annual *simple* interest rate of $r$, where $r$ is given as a decimal. Then the interest $I$ at the end of $t$ years is given by

$$I = Prt.$$

The future value F at the end of $t$ years is

$$F = P + I = P + Prt = P(1 + rt).$$

- **Present Value.** The **present value** needed to deposit into an account earning a simple annual rate of $r$, expressed as a decimal, in order to have a future amount $F$ after $t$ years is

$$P = \frac{F}{1 + rt}.$$

- **Discount and Proceeds.** The discount $D$ on a discounted loan of $M$ dollars at a simple annual interest rate of $r$, expressed as a decimal, for $t$ years is

$$D = Mrt,$$

where $D$ is the discount (interest paid at time of loan), $M$ is the maturity value (amount borrowed), $r$ is the discount rate (annual simple interest rate), and $t$ is the length of the loan. The **proceeds** $P$ of the loan is the actual amount the borrower receives when the loan is made and is given by $P = M - D$.

- **Effective Rate of a Discounted Loan.** The effective interest rate given on a discounted loan of length $t$ with a discount rate of $r$, expressed as a decimal, is

$$r_{eff} = \frac{r}{1 - rt}.$$

- **Compound Interest.** Suppose a principal $P$ earns interest at the annual rate of $r$, expressed as a decimal, and interest is compounded $m$ times a year. Then the amount $F$ after $t$ years is

$$F = P(1 + i)^n = P\left(1 + \frac{r}{m}\right)^{mt},$$

where $n = mt$ is the number of time periods and $i = \frac{r}{m}$ is the interest per period.

- **The Time for Money to Grow from $P$ to $F$.**   The number of periods $n$ it takes for a principal of $P$ to grow to a future value of $F$ in an account that earns interest per period at a rate of $i$, expressed as a decimal, is

$$n = \frac{\log F/P}{\log(1 + i)}.$$

The number of periods for the principal to double is given by

$$n = \frac{\log 2}{\log(1 + i)}.$$

- **Effective Yield.**   Suppose a sum of money is invested at an annual rate of $r$ expressed as a decimal and is compounded $m$ times a year. The effective yield $r_{eff}$ is

$$r_{eff} = \left(1 + \frac{r}{m}\right)^m - 1.$$

- **Present Value.**   Suppose an account earns an annual rate of $r$ expressed as a decimal and is compounded $m$ times a year. Then the amount $P$, called the **present value**, needed presently in this account so that a future amount of $F$ will be attained in $t$ years is given by

$$P = \frac{F}{\left(1 + \frac{r}{m}\right)^{mt}}.$$

- An **annuity** is a sequence of equal payments made at equal time periods.

- An **ordinary annuity** is one in which the payments are made at the *end* of the time periods and the periods of compounding are the same time periods.

- The **term** of an annuity is the time from the beginning of the first period to the end of the last period.

- The total amount in the account, including interest, at the end of the term of an annuity is called the **future value of the annuity**.

- **Summation Formula.**   For any positive integer $n$ and any real numbers $a$ and $b$ with $b \neq 1$

$$a + ab + ab^2 + ab^3 + \cdots + ab^{n-1} = a\frac{b^n - 1}{b - 1}.$$

- **Future Value of an Ordinary Annuity.**   The future value $FV$ of an ordinary annuity of $n$ payments of $PMT$ dollars paid at the end of each period into an account that earns interest at the rate of $i$ per period is

$$FV = PMT\frac{[(1 + i)^n - 1]}{i} = PMT\, s_{n\rceil i}$$

- **Sinking Fund Payment.**   The periodic payment $PMT$ that is required to accumulate the sum $FV$ over $n$ periods of time with interest at the rate of $i$ per period is

$$PMT = FV\frac{i}{[(1 + i)^n - 1]} = \frac{FV}{s_{n\rceil i}}$$

- **Interest per Period in a Sinking Fund.** The interest earned during the $m$th period of a sinking fund with payments of $PMT$ and earning interest at a rate of $i$ per period is

$$\text{Interest in } m\text{th period} = PMT(s_{m\rceil i} - s_{m-1\rceil i}) - PMT.$$

- **Present Value of an Ordinary Annuity.** The present value $PV$ of an ordinary annuity of $n$ payments of $PMT$ dollars each made at the end of the period with interest compounded at the rate of $i$ per period is

$$PV = PMT \frac{[1 - (1 + i)^{-n}]}{i} = PMT\, a_{n\rceil i}$$

- **Amortization Formula.** The periodic payment $PMT$ to be made at the end of each period on a loan of $PV$ dollars that is to be amortized over $n$ periods with interest at the rate of $i$ per period is

$$PMT = PV \frac{i}{[1 - (1 + i)^{-n}]} = \frac{PV}{a_{n\rceil i}}$$

- **Outstanding Principal.** Let $n$ be the total number of payments required to pay off an amortized loan and let $m$ be the number of payments of $PMT$ made; then the outstanding principal $P$ is given by

$$P = PMT\, a_{n-m\rceil i}.$$

- **Payment Toward Principal.** For an amortized loan requiring $n$ equal payments of $PMT$ dollars each and interest at the rate of $i$ per period, the amount of the $m$th payment that is applied toward the principal is

$$\text{payment toward principal} = PMT(a_{n-m+1\rceil i} - a_{n-m\rceil i}).$$

## Chapter 5 Review Exercises

In Exercises 1 through 2 find how much is in the following accounts after the given $t$ years where $P$ is the initial principal, $r$ is the annual rate, and the compounding is as indicated.

1. $t = 5$, $P = \$1000$, $r = 8\%$, compounded (a) annually, (b) quarterly, (c) monthly, (d) weekly, (e) daily

2. $t = 5$, P = $\$1000$, $r = 10\%$, compounded (a) annually, (b) quarterly, (c) monthly, (d) weekly, (e) daily

In Exercises 3 through 4 find the effective yield given the annual rate $r$ and the indicated compounding.

3. $r = 9\%$, compounded (a) semiannually, (b) quarterly, (c) monthly, (d) weekly, (e) daily

4. $r = 11\%$, compounded (a) semiannually, (b) quarterly, (c) monthly, (d) weekly, (e) daily

In Exercises 5 through 6 find the present value of the given amounts $A$ with the indicated annual rate of return $r$, the number of years $t$, and the indicated compounding.

5. $A = \$10,000$, $r = 8\%$, $t = 10$, compounded (a) annually, (b) monthly, (c) weekly

6. $A = \$10,000, r = 9\%, t = 10$, compounded (a) annually, (b) quarterly, (c) daily

7. You have just won a $1,000,000 lottery. Your entire prize will be given to you in 20 years in a lump sum. If current interest rates are 8% compounded annually, what is this $1,000,000 prize worth at the present?

8. How much money should you deposit now in an account earning 9% annually in order to have $10,000 in 5 years?

9. What annual rate is required for an account that is compounded annually to double in 8 years?

10. If an account yields an annual return of $r$ where $r$ is expressed as a decimal and if $T$ is the time it takes for this account to triple, show that

$$T = \frac{\log 3}{\log(1 + r)}.$$

11. **Savings.**   A self-employed individual places $3000 a year into a Keogh account in which taxes are not paid on the interest. If interest rates remain at 9%, how much will be in this account after 30 years?

12. **Sinking Fund.**   A large corporation creates a sinking fund in order to have a $2 million cash bonus for the president of the company on his retirement in 10 years. How much should be placed into this account at the end of each quarter if the interest rate is 8% compounded quarterly? Over the life of this sinking fund how much interest accumulates? How much interest is earned in the third quarter of the fifth year?

13. **Lottery Prize.**   The state lottery has just had a $2 million winner and needs to pay the recipient $100,000 now and the same amount at the end of each of the next 19 years. The state does not wish to be involved in the administration of this disbursement and wishes to give this obligation together with a lump sum payment to an insurance company to handle the payments. You are an investment advisor for an insurance company. Interest rates are at 8%. What is the very least that you would accept for this obligation?

14. **Amortization.**   You plan to borrow $100,000 to buy a house. Interest rates are at 9% a year compounded monthly. What will be your monthly payments if the length of the mortgage is 25 years? 15 years? What will be the total interest paid in each case? In each case how much of the tenth payment will go toward the principal? The 120th?

How many different
poker hands are
there? (Dennis
Drenner.)

chapter **six**

# Sets and Counting

6.1   Sets

6.2   The Number of Elements in a Set

6.3   Multiplication Principal, Permutations

6.4   Combinations

6.5   Binomial Theorem, Partitions

The first section gives the fundamental ideas of set theory, which we will then use throughout the remainder of the text. In particular these ideas are used in the remaining sections of this chapter to develop means to count the number of elements in various sets. We shall see in the next chapter that this is fundamental to finding probability.

## **6.1**   **SETS**

► *Introduction*
► *Set Operations*
► *Additional Rules and Laws for Sets*
► *Applications*

**George Boole,**
**1815–1864**

*(The Bettmann Archive.)*

George Boole was born into a lower class family in Lincoln, England, and had only a common school education. He was largely self-taught and managed to become an elementary school teacher. Up to this time any rule of algebra such as $a(x + y) = ax + ay$ was understood to apply only to numbers and magnitudes. Boole developed an "algebra" of sets where the elements of the sets could be not just numbers but *anything*. This then laid down the foundations for a fundamental way of thinking. Bertrand Russell, a great mathematician and philosopher of the 20th century, said that the greatest discovery of the 19th century was the nature of pure mathematics, which he asserted was discovered by George Boole. Boole's pamphlet "The Mathematical Analysis of Logic" maintained that the essential character of mathematics lies in its form rather than in its content. Thus mathematics is not merely the science of measurement and number but any study consisting of symbols and precise rules of operation. Boole founded not only a new algebra of sets but also a formal logic.

## Introduction

Chapter 1 gave a brief introduction to sets. It would be beneficial to review that material before reading this section. This section discusses operations on sets and laws governing these set operations. These are fundamental notions that will be used throughout the remainder of this text. In the next two chapters we will see that probability and statistics are based on counting the elements in sets and manipulating set operations. Thus we first need to understand clearly the notion of sets and their operations.

---

**Subsets**

If every element of a set $A$ is also an element of another set $B$, we say that **$A$ is a subset of $B$** and write $A \subset B$. If $A$ is not a subset of $B$, we write $A \not\subset B$.

---

Thus $\{1, 2, 4\} \subset \{1, 2, 3, 4\}$, but $\{1, 2, 3, 4\} \not\subset \{1, 2, 4\}$.

---

**Empty Set**

The **empty set**, written as $\phi$, is the set with no elements.

---

The empty set can be used to conveniently indicate that an equation has no solution. For example

$$\{x \mid x \text{ is real and } x^2 = -1\} = \phi.$$

By the definition of subset, given any set $A$, we must have $\phi \subset A$ and $A \subset A$.

E X A M P L E   1     **Finding Subsets**

Find all the subsets of $\{a, b, c\}$.

**Solution**     The subsets are

$$\phi, \{a\}, \{b\}, \{c\}, \{a, b\}, \{a, c\}, \{b, c\}, \{a, b, c\}. \quad \blacksquare$$

The empty set is the set with no elements. At the other extreme is the **universal set**. This set is the set of all elements being considered and is denoted by $U$. If, for example, we are to take a national survey of voter satisfaction with the President, the universal set is the set of all voters in this country. If the survey is to determine the effects of smoking on pregnant women, the universal set is the set of all pregnant women. The context of the problem under discussion will determine the universal set for that problem. The universal set must contain every element under discussion.

A **Venn diagram** is a way of visualizing sets. For example, given a universal set $U$ and two subsets $A$ and $B$, Figure 6.1a is a Venn diagram that visualizes the concept that $A \subset U$, while Figure 6.1b is a Venn diagram that visualizes the concept $B \subset A$.

 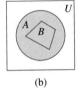

(a)          (b)

**Figure 6.1**

## Set Operations

The first set operation we consider is the complement.

$A^c$

**Figure 6.2**

---

**Complement**

Given a universal set $U$ and a set $A \subset U$, the **complement of $A$**, written $A^c$, is the set of all elements that are in $U$ but not in $A$, that is,

$$A^c = \{x \mid x \in U, x \notin A\}.$$

---

A Venn diagram visualizing $A^c$ is shown in Figure 6.2.

E X A M P L E   2     **Finding Complements of Sets**

Let $U = \{1, 2, 3, 4, 5, 6, 7, 8, 9\}$, $A = \{1, 3, 5, 7, 9\}$, $B = \{1, 2, 3, 4, 5\}$. Find $A^c$, $B^c$, $U^c$, $\phi^c$, $(A^c)^c$.

**Solution**    We have

$$A^c = \{2, 4, 6, 8\}$$

$$B^c = \{6, 7, 8, 9\}$$

$$U^c = \phi$$

$$\phi^c = U$$

$$(A^c)^c = \{2, 4, 6, 8\}^c$$

$$= \{1, 3, 5, 7, 9\} = A. \quad \blacksquare$$

If $U$ is a universal set, we must always have

$$U^c = \phi, \qquad \phi^c = U.$$

If $A$ is any subset of a universal set $U$, then

$$(A^c)^c = A.$$

This can be seen using the Venn diagram in Figure 6.2, since the complement of $A^c$ is all elements in $U$ but not in $A^c$. This can be seen in Figure 6.2 to be just the set $A$.

---

**Set Union**

The **union** of two sets $A$ and $B$, written $A \cup B$, is the set of all elements that belong to $A$, or to $B$, or to both. Thus

$$A \cup B = \{x | x \in A \text{ or } x \in B \text{ or both}\}.$$

---

**REMARK.**    Our convention will be to drop the phrase ''or both'' but still maintain the same meaning. Note very carefully that this gives a particular definition to the word ''or.'' Thus we will normally write

$$A \cup B = \{x | x \in A \text{ or } x \in B\}.$$

We can also say that $A \cup B$ is all elements in $A$ together with all elements in $B$. A Venn diagram is shown in Figure 6.3.

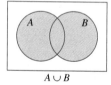

$A \cup B$

**Figure 6.3**

**E X A M P L E   3**    **Finding the Union of Two Sets**

Find $\{1, 2, 3, 4\} \cup \{1, 4, 5, 6\}$.

**Solution**    We add to the first set any elements in the second set that are not already there. Thus

$$\{1, 2, 3, 4\} \cup \{1, 4, 5, 6\} = \{1, 2, 3, 4, 5, 6\}. \quad \blacksquare$$

From Figure 6.2, we can see that if $U$ is a universal set and $A \subset U$, then

$$A \cup A^c = U.$$

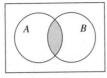

$A \cap B$

**Figure 6.4**

---

**Set Intersection**

The **intersection** of two sets $A$ and $B$, written $A \cap B$, is the set of all elements that belong to both the set $A$ and to the set $B$. Thus,

$$A \cap B = \{x \mid x \in A \text{ and } x \in B\}$$

---

A Venn diagram is shown in Figure 6.4.

**E X A M P L E   4**   **Find the Intersection of Two Sets**

Find $\{a, b, c, d\} \cap \{a, c, e\}$ and $\{a, b\} \cap \{c, d\}$.

**Solutions**   (a)   Only $a$ and $c$ are in both of the sets. Thus

$$\{a, b, c, d\} \cap \{a, c, e\} = \{a, c\}.$$

(b)   The two sets $\{a, b\}$ and $\{c, d\}$ have no elements in common. Thus

$$\{a, b\} \cap \{c, d\} = \phi. \quad \blacksquare$$

A and B disjoint

**Figure 6.5**

---

**Disjoint Sets**

Two sets $A$ and $B$ are **disjoint** if they have no elements in common, that is, if $A \cap B = \phi$. See Figure 6.5.

---

An examination of Figure 6.2 or referring to the definition of $A^c$ indicates that for any set $A$, $A$ and $A^c$ are disjoint. That is,

$$A \cap A^c = \phi.$$

## Additional Rules and Laws for Sets

The following rules can be established.

---

**Rules for Set Operations**

Let $U$ be a universal set and let $A$, $B$, and $C$ be any subsets of $U$. Then

| | |
|---|---|
| $A \cup B = B \cup A$ | Commutative law for union |
| $A \cap B = B \cap A$ | Commutative law for intersection |
| $A \cup (B \cup C) = (A \cup B) \cup C$ | Associative law for union |
| $A \cap (B \cap C) = (A \cap B) \cap C$ | Associative law for intersection |
| $A \cup (B \cap C) = (A \cup B) \cap (A \cup C)$ | Distributive law for union |
| $A \cap (B \cup C) = (A \cap B) \cup (A \cap C)$ | Distributive law for intersection |

$$(A \cup B)^c = A^c \cap B^c \qquad \text{De Morgan law}$$

$$(A \cap B)^c = A^c \cup B^c \qquad \text{De Morgan law}$$

**E X A M P L E   5**    **Establishing a De Morgan Law**

Use a Venn diagram to show that

$$(A \cup B)^c = A^c \cap B^c.$$

**Solution**    We first consider the right side of this equation. Figure 6.6 shows a Venn diagram of $A^c$ and $B^c$ and $A^c \cap B^c$. We then notice from Figure 6.3 that this is $(A \cup B)^c$. ∎

  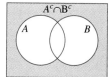

**Figure 6.6**

**E X A M P L E   6**    **Establishing the Distributive Law for Union**

Use a Venn diagram to show that

$$A \cup (B \cap C) = (A \cup B) \cap (A \cup C).$$

**Solution**    Consider first the left side of this equation. In Figure 6.7a the sets $A$, $B \cap C$, and the union of these two are shown. Now for the right side of the equation refer to Figure 6.7b, where the sets $A \cup B$, $A \cup C$, and the intersection of these two sets are shown. We see that we end up with the same set in both cases. ∎

**Figure 6.7**

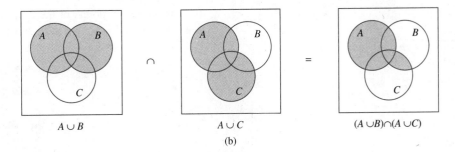

**Figure 6.7 (cont.)**

## Applications

E X A M P L E 7    **Using Set Operations to Write Expressions**

Let $U$ be the universal set consisting of the set of all students taking classes at the University of Hawaii and

$$B = \{x \in U | x \text{ is currently taking a business course}\}$$

$$E = \{x \in U | x \text{ is currently taking an English course}\}$$

$$M = \{x \in U | x \text{ is currently taking a math course}\}.$$

Write an expression using set operations for each of the following:

(a)   The set of students at the University of Hawaii taking a course in at least one of the above three fields.

(b)   The set of all students at the University of Hawaii taking both an English course and a math course but not a business course.

(c)   The set of all students at the University of Hawaii taking a course in exactly one of the three fields above.

**Solutions**    (a)   This is $B \cup E \cup M$. See Figure 6.8.

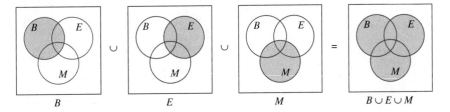

**Figure 6.8**

(b)   This can be described as the set of students taking an English course ($E$) and also (intersection) a math course ($M$) and also (intersection) not a business course ($B^c$) or

$$E \cap M \cap B^c.$$

This is the set of points in the universal set that are in both $E$ and $M$ but not in $B$ and is shown in Figure 6.9.

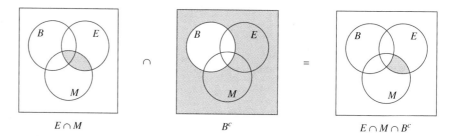

Figure 6.9

(c)   We describe this set as the set of students taking business but not taking English or math ($B \cap E^c \cap M^c$) together with (union) the set of students taking English but not business or math ($E \cap B^c \cap M^c$) together with (union) the set of students taking math but not business or English ($M \cap B^c \cap E^c$) or

$$(B \cap E^c \cap M^c) \cup (E \cap B^c \cap M^c) \cup (M \cap B^c \cap E^c).$$

This is the union of the three sets shown in Figure 6.10. The first, $B \cap E^c \cap M^c$, consists of those points in $B$ that are outside $E$ and also outside $M$. The second set $E \cap B^c \cap M^c$ consists of those points in $E$ that are outside $B$ and $M$. The third set $M \cap B^c \cap E^c$ is the set of points in $M$ that are outside $B$ and $E$. The union of these three sets is then shown on the right in Figure 6.10.  ∎

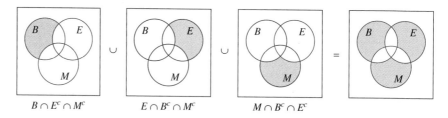

Figure 6.10

## SELF-HELP EXERCISE SET 6.1

1.   Let $U = \{1, 2, 3, 4, 5, 6, 7\}$, $A = \{1, 2, 3, 4\}$, $B = \{3, 4, 5\}$, $C = \{2, 3, 4, 5, 6\}$. Find the following.
     a.   $A \cup B$,
     b.   $A \cap B$,
     c.   $A^c$,
     d.   $(A \cup B) \cap C$,
     e.   $(A \cap B) \cup C$,
     f.   $A^c \cup B \cup C$.

2.  Let $U$ denote the set of all corporations in this country and $P$ those that made profits during the last year, $D$ those that paid a dividend during the last year, and $L$ those that increased their labor force during the last year. Describe the following using the three sets $P$, $D$, $L$, and set operations.
    a.  Corporations in this country that had profits and also paid a dividend last year
    b.  Corporations in this country that either had profits or paid a dividend last year
    c.  Corporations in this country that did not have profits last year
    d.  Corporations in this country that had profits, paid a dividend, and did not increase their labor force last year
    e.  Corporations in this country that had profits or paid a dividend, and did not increase their labor force last year

## EXERCISE SET 6.1

In Exercises 1 through 4 determine whether the statements are true or false.

1.  a.  $\phi \in A$,                                      b.  $A \in A$

2.  a.  $0 = \phi$,                                        b.  $\{x, y\} \in \{x, y, z\}$

3.  a.  $\{x \mid 0 < x < -1\} = \phi$,                    b.  $\{x \mid 0 < x < -1\} = 0$

4.  a.  $\{x \mid x(x - 1) = 0\} = \{0, 1\}$,              b.  $\{x \mid x^2 + 1 < 0\} = \phi$

5.  If $A = \{u, v, y, z\}$, determine whether the following statements are true or false.
    a.  $w \in A$,                                         b.  $x \notin A$,
    c.  $\{u, x\} \subset A$,                              d.  $\{y, z, v, u\} = A$

6.  If $A = \{u, v, y, z\}$, determine whether the following statements are true or false.
    a.  $x \notin A$                                       b.  $\{v, w\} \notin A$
    c.  $\{x, w\} \not\subset A$                           d.  $\phi \subset A$

7.  List all the subsets of (a) $\{3\}$, (b) $\{3, 4\}$.

8.  List all the subsets of (a) $\phi$, (b) $\{3, 4, 5\}$.

9.  Use Venn diagrams to indicate the following.
    a.  $A \subset U, B \subset U, A \subset B^c$          b.  $A \subset U, B \subset U, B \subset A^c$

10.  Use Venn diagrams to indicate the following.
    a.  $A \subset U, B \subset U, C \subset U, C \subset (A \cup B)^c$
    b.  $A \subset U, B \subset U, C \subset U, C \subset A \cap B$

11.  On the accompanying figure, indicate where the following sets are.
    a.  $A \cap B^c$                                       b.  $A \cap B$
    c.  $A^c \cap B$                                       d.  $A^c \cap B^c$

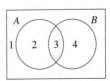

**12.** Can you find any two sets among the list of four sets given in Exercise 11 that are not disjoint?

On the accompanying figure indicate where the following sets are.

**13.** $A \cup B^c$

**14.** $A^c \cup B^c$

**15.** $(A \cup B)^c$

**16.** $(A \cap B)^c$

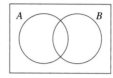

**17.** Match up the following sets to the appropriate region in the accompanying Venn diagram.

  **a.** $A \cap B \cap C$,    **b.** $A \cap B^c \cap C^c$    **c.** $A \cap B \cap C^c$,

  **d.** $B \cap A^c \cap C^c$,    **e.** $A^c \cap B^c \cap C^c$    **f.** $A \cap C \cap B^c$,

  **g.** $B \cap C \cap A^c$,    **h.** $C \cap A^c \cap B^c$

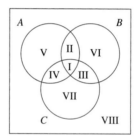

On the accompanying figure indicate where the following sets are.

**18.** $A \cup (B \cap C)$

**19.** $(A \cup B) \cap C^c$

**20.** $A \cap B \cap C^c$

**21.** $(A \cap B)^c \cap C$

**22.** $A^c \cap B^c \cap C^c$

**23.** $(A \cup B)^c \cap C$

**24.** $(A \cup B \cup C)^c \cap A$

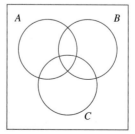

In Exercises 25 through 32 let $U = \{1, 2, 3, 4, 5, 6, 7, 8, 9, 10\}$, $A = \{1, 2, 3, 4, 5, 6\}$, $B = \{4, 5, 6, 7, 8\}$, $C = \{5, 6, 7, 8, 9, 10\}$ and find the indicated sets.

25.  **a.** $A \cap B$                                **b.** $A \cup B$

26.  **a.** $A^c$                                     **b.** $A^c \cap B$

27.  **a.** $A \cap B^c$                              **b.** $A^c \cap B^c$

28.  **a.** $A^c \cup B^c$                            **b.** $(A^c \cup B^c)^c$

29.  **a.** $A \cap B \cap C$                         **b.** $(A \cap B \cap C)^c$

30.  **a.** $A \cap (B \cup C)$                       **b.** $A \cap (B^c \cup C)$

31.  **a.** $A^c \cap B^c \cap C^c$                   **b.** $(A \cup B \cup C)^c$

32.  **a.** $A^c \cap B^c \cap C$                     **b.** $A^c \cap B \cap C^c$

Let $U$ be the set of all residents of your state and let

$A = \{x \in U | x \text{ owns an automobile}\}$
$H = \{x \in U | x \text{ owns a house}\}$.

In Exercises 33 through 36 describe each of the sets in words.

33.  **a.** $A^c$,              **b.** $A \cup H$,              **c.** $A \cup H^c$

34.  **a.** $H^c$,              **b.** $A \cap H$,              **c.** $A^c \cap H$

35.  **a.** $A \cap H^c$,       **b.** $A^c \cap H^c$,          **c.** $A^c \cup H^c$

36.  **a.** $(A \cap H)^c$,     **b.** $(A \cup H)^c$,          **c.** $(A^c \cap H^c)^c$

In Exercises 37 through 40 let $U$, $A$, and $H$ be as in the previous four problems, let $P = \{x \in U | x \text{ owns a piano}\}$, and describe each of the sets in words.

37.  **a.** $A \cap H \cap P$        **b.** $A \cup H \cup P$        **c.** $(A \cap H) \cup P$

38.  **a.** $(A \cup H) \cap P$      **b.** $(A \cup H) \cap P^c$    **c.** $A \cap H \cap P^c$

39.  **a.** $(A \cap H)^c \cap P$    **b.** $A^c \cap H^c \cap P^c$  **c.** $(A \cup H)^c \cap P$

40.  **a.** $(A \cup H \cup P)^c \cap A$   **b.** $(A \cup H \cup P)^c$    **c.** $(A \cap H \cap P)^c$

In Exercises 41 through 48 let $U$ be the set of major league baseball players and let

$N = \{x \in U | x \text{ plays for the New York Yankees}\}$

$S = \{x \in U | x \text{ plays for the San Francisco Giants}\}$

$F = \{x \in U | x \text{ is an outfielder}\}$

$H = \{x \in U | x \text{ has hit 20 homers in one season}\}$.

Write the set that represents the following descriptions.

41.  **a.** Outfielders for the New York Yankees
     **b.** New York Yankees who have never hit 20 homers in a season

42.  **a.**  San Francisco Giants who have hit 20 homers in a season
     **b.**  San Francisco Giants who do not play outfield

43.  **a.**  Major league ball players who play for the New York Yankees or the San Francisco Giants
     **b.**  Major league ball players who play for neither the New York Yankees nor the San Francisco Giants

44.  **a.**  San Francisco Giants who have never hit 20 homers in a season
     **b.**  Major league ball players who have never hit 20 homers in a season

45.  **a.**  New York Yankees or San Francisco Giants who have hit 20 homers in a season
     **b.**  Outfielders for the New York Yankees who have never hit 20 homers in a season

46.  **a.**  Outfielders for the New York Yankees or San Francisco Giants
     **b.**  Outfielders for the New York Yankees who have hit 20 homers in a season

47.  **a.**  Major league outfielders who have hit 20 homers in a season and do not play for the New York Yankees or the San Francisco Giants
     **b.**  Major league outfielders who have never hit 20 homers in a season and do not play for the New York Yankees or the San Francisco Giants

48.  **a.**  Major league players who do not play outfield, who have hit 20 homers in a season, and do not play for the New York Yankees or the San Francisco Giants
     **b.**  Major league players who play outfield, who have never hit 20 homers in a season, and do not play for the New York Yankees or the San Francisco Giants

In Exercises 49 through 54 let $U = \{1, 2, 3, 4, 5, 6, 7, 8, 9, 10\}$, $A = \{1, 2, 3, 4, 5\}$, $B = \{4, 5, 6, 7\}$, $C = \{5, 6, 7, 8, 9, 10\}$. Verify that the identities are true for these sets.

49.  $A \cup (B \cup C) = (A \cup B) \cup C$

50.  $A \cap (B \cap C) = (A \cap B) \cap C$

51.  $A \cup (B \cap C) = (A \cup B) \cap (A \cup C)$

52.  $A \cap (B \cup C) = (A \cap B) \cup (A \cap C)$

53.  $(A \cup B)^c = A^c \cap B^c$

54.  $(A \cap B)^c = A^c \cup B^c$

## Solutions to Self-Help Exercise Set 6.1

1.  **a.**  $A \cup B$ is the elements in A or B or both. Thus $A \cup B = \{1, 2, 3, 4, 5\}$.
    **b.**  $A \cap B$ is the elements in both $A$ and $B$. Thus $A \cap B = \{3, 4\}$.
    **c.**  $A^c$ is the elements not in $A$ (but in $U$). Thus $A^c = \{5, 6, 7\}$.
    **d.**  $(A \cup B) \cap C$ is those elements in $A \cup B$ and also in $C$. From (a) we have

    $$(A \cup B) \cap C = \{1, 2, 3, 4, 5\} \cap \{2, 3, 4, 5, 6\} = \{2, 3, 4, 5\}.$$

    **e.**  $(A \cap B) \cup C$ is those elements in $A \cap B$ or in $C$. Thus from (b)

    $$(A \cap B) \cup C = \{3, 4\} \cup \{2, 3, 4, 5, 6\} = \{2, 3, 4, 5, 6\}.$$

**f.**   $A^c \cup B \cup C$ is elements in $B$, or in $C$, or not in $A$. Thus

$$A^c \cup B \cup C = \{2, 3, 4, 5, 6, 7\}.$$

**2.   a.**   Corporations in this country that had profits and also paid a dividend last year is represented by $P \cap D$.

**b.**   Corporations in this country that either had profits or paid a dividend last year is represented by $P \cup D$.

**c.**   Corporations in this country that did not have profits last year is represented by $P^c$.

**d.**   Corporations in this country that had profits, paid a dividend, and did not increase their labor force last year is represented by $P \cap D \cap L^c$.

**e.**   Corporations in this country that had profits or paid a dividend, and did not increase their labor force last year is represented by $(P \cup D) \cap L^c$.

## 6.2   THE NUMBER OF ELEMENTS IN A SET

▶  *Counting*

▶  *Applications*

**Augustus De Morgan, 1806–1871**

*(David H. Smith Collection/Columbia University.)*

**A P P L I C A T I O N**

**Counting**

It was De Morgan who got George Boole interested in set theory and formal logic and then made significant advances upon Boole's epochal work. He discovered the De Morgan laws mentioned in the last section. Boole and De Morgan are together considered the founders of the algebra of sets and of mathematical logic. De Morgan was a champion of religious and intellectual toleration and on several occasions resigned his professorships in protest of the abridgements of academic freedom of others.

In a survey of 120 adults, 55 said they had an egg for breakfast that morning, 40 said they had juice for breakfast, and 70 said they had an egg or juice for breakfast. How many had both an egg and juice for breakfast? See Example 1 for the answer.

### Counting

This section shows the relationship between the number of elements in $A \cup B$ and the number of elements in $A$, $B$, and $A \cap B$. This is our first counting principle. The examples and exercises in this section give some applications. In other applications in Chapter 7 we will count the number of elements in various sets to find probability.

**Figure 6.11**

**Figure 6.12**

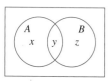

**Figure 6.13**

---

**The Notation $n(A)$**

If $A$ is a set with a finite number of elements, we denote the number of elements in $A$ by $n(A)$.

---

There are two results that are rather apparent. First, since the empty set $\phi$ has no elements, $n(\phi) = 0$. For the second refer to Figure 6.11 where the two sets $A$ and $B$ are disjoint.

---

**The Number in the Union of Disjoint Sets**

If the sets $A$ and $B$ are disjoint, then

$$n(A \cup B) = n(A) + n(B).$$

---

A consequence of the last result is the following. In Figure 6.12, we are given a universal set $U$ and a set $A \subset U$. Then since $A \cap A^c = \phi$ and $U = A \cup A^c$,

$$n(U) = n(A \cup A^c) = n(A) + n(A^c).$$

Now consider the more general case shown in Figure 6.13. We assume that $x$ is the number in the set $A$ that are not also in $B$, $z$ is the number in the set $B$ that are not also in $A$, and finally, $y$ is the number in both $A$ and $B$. Then

$$n(A \cup B) = x + y + z$$
$$= (x + y) + (y + z) - y$$
$$= n(A) + n(B) - n(A \cap B).$$

Alternatively, we can see that the total $n(A) + n(B)$ counts the number in the intersection $n(A \cap B)$ twice. Thus to obtain the number in the union $n(A \cup B)$, we must subtract $n(A \cap B)$ from $n(A) + n(B)$.

---

**The Number in the Union of Two Sets**

For any finite sets $A$ and $B$,

$$n(A \cup B) = n(A) + n(B) - n(A \cap B). \tag{1}$$

---

## Applications

**E X A M P L E   1**     **An Application of Counting**

---

In a survey of 120 adults, 55 said they had an egg for breakfast that morning, 40 said they had juice for breakfast, and 70 said they had an egg or juice for breakfast. How many had both an egg and juice for breakfast? How many had an egg but no juice for breakfast? How many had neither an egg nor juice for breakfast?

**Solutions**      Let $U$ be the set of adults surveyed, $E$ the subset that had an egg for breakfast, and $J$ the subset that had juice for breakfast. A Venn diagram is shown in Figure 6.14. From the survey, we have that

$$n(E) = 55, \qquad n(J) = 40, \qquad n(E \cup J) = 70.$$

                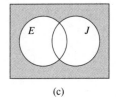

(a)                            (b)                            (c)

**Figure 6.14**

(a)   The number that had an egg and juice for breakfast is given by $n(E \cap J)$ and is shown as the shaded region in Figure 6.14a. Since $n(E \cup J) = n(E) + n(J) - n(E \cap J)$, we have that

$$n(E \cap J) = n(E) + n(J) - n(E \cup J)$$
$$= 55 + 40 - 70$$
$$= 25.$$

(b)   We first place the number 25, just found, in the $E \cap J$ area in the Venn diagram in Figure 6.14b. If we let $x$ be the number that had an egg but no juice, then according to Figure 6.14b,

$$x + 25 = n(E) = 55$$
$$x = 30.$$

(c)   We wish to find $n((E \cup J)^c)$. This is shown as the shaded region in Figure 6.14c. The unshaded region is $E \cup J$. We then have that

$$n(E \cup J) + n((E \cup J)^c) = n(U)$$
$$n((E \cup J)^c) = n(U) - n(E \cup J)$$
$$= 120 - 70$$
$$= 50. \quad \blacksquare$$

**E X A M P L E   2**      **An Application of Counting**

In a survey of 200 people that had just returned from a trip to Europe, the following information was gathered.

- 142 visited England
- 95 visited France

- 65 visited Germany
- 70 visited both England and France
- 50 visited both England and Germany
- 30 visited both France and Germany
- 20 visited all three of these countries

    (a)   How many went to England but not France or Germany?
    (b)   How many went to exactly one of these three countries?
    (c)   How many went to none of these three countries?

**Solutions**    Let $U$ be the set of 200 people that were surveyed and let

- $E = \{x \in U | x$ visited England$\}$
- $F = \{x \in U | x$ visited France$\}$
- $G = \{x \in U | x$ visited Germany$\}$.

We first note that the last piece of information from the survey indicates that

$$n(E \cap F \cap G) = 20.$$

Place this in the Venn diagram shown in Figure 6.15a. Recall that 70 visited both England and France, that is, $n(E \cap F) = 70$. If $a$ is number that visited England and France but not Germany, then, according to Figure 6.15a, $20 + a = n(E \cap F) = 70$. Thus $a = 50$. In the same way, if $b$ is the number that visited England and Germany but not France, then $20 + b = n(E \cap G) = 50$. Thus $b = 30$. Also if $c$ is the number that visited France and Germany but not England, then $20 + c = n(G \cap F) = 30$. Thus $c = 10$. All of this information is then shown in Figure 6.15b.

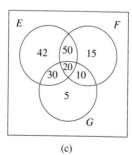

        (a)                            (b)                            (c)

**Figure 6.15**

    (a)   Let $x$ denote the number that visited England but not France or Germany. Then, according to Figure 6.15b, $20 + 30 + 50 + x = n(E) = 142$. Thus $x = 42$, that is, the number that visited England but not France or Germany is 42.

    (b)   Since $n(F) = 95$, the number that visited France but not England or Germany is given from Figure 6.15b by $95 - (50 + 20 + 10) = 15$.

Since $n(G) = 65$, the number that visited Germany but not England or France is, according to Figure 6.15b, given by $65 - (30 + 20 + 10) = 5$. Thus, according to Figure 6.15c, the number who visited just one of the three countries is

$$42 + 15 + 5 = 62.$$

(c)    According to Figure 6.15c, the number that visited none of these three countries is given by

$$200 - (42 + 15 + 5 + 50 + 30 + 10 + 20) = 200 - 172$$
$$= 28. \quad \blacksquare$$

## SELF-HELP EXERCISE SET 6.2

1. Given that $n(A \cup B) = 100$, $n(A \cap B^c) = 50$, and $n(A \cap B) = 20$, find $n(A^c \cap B)$.

2. The registrar reported that among 2000 students, 700 did not register for a math or English course, while 400 registered for both of these two courses. How many registered for exactly one of these courses?

## EXERCISE SET 6.2

1. If $n(A) = 100$, $n(B) = 75$, $n(A \cap B) = 40$, what is $n(A \cup B)$?

2. If $n(A) = 200$, $n(B) = 100$, $n(A \cup B) = 250$, what is $n(A \cap B)$?

3. If $n(A) = 100$, $n(A \cap B) = 20$, $n(A \cup B) = 150$, what is $n(B)$?

4. If $n(B) = 100$, $n(A \cup B) = 175$, $n(A \cap B) = 40$, what is $n(A)$?

5. If $n(A) = 100$ and $n(A \cap B) = 40$, what is $n(A \cap B^c)$?

6. If $n(U) = 200$ and $n(A \cup B) = 150$, what is $n(A^c \cap B^c)$?

7. If $n(A \cup B) = 500$, $n(A \cap B^c) = 200$, $n(A^c \cap B) = 150$, what is $n(A \cap B)$?

8. If $n(A \cap B) = 50$, $n(A \cap B^c) = 200$, $n(A^c \cap B) = 150$, what is $n(A \cup B)$?

9. In a survey of 1200 households, 950 said they had aspirin in the house, 350 said they had acetaminophen, and 200 said they had both aspirin and acetaminophen.
   a. How many in the survey had at least one of the two medications?
   b. How many in the survey had aspirin but not acetaminophen?
   c. How many in the survey had neither aspirin nor acetaminophen?

10. In a survey of 1000 households, 600 said they received the morning paper but not the evening paper, 300 said they received both papers, and 100 said they received neither paper.
    a. How many received the evening paper but not the morning paper?
    b. How many received at least one of the papers?

11. The registrar reported that among 1300 students, 700 students did not register for either a math or English course, 400 registered for an English course, and 300 registered for both types of courses.
    a. How many registered for an English course but not a math course?
    b. How many registered for a math course?

12. In a survey of 500 people, a pet food manufacturer found that 200 owned a dog but not a cat, 150 owned a cat but not a dog, and 100 owned neither a dog nor a cat.
    a. How many owned both a dog and a cat?
    b. How many owned a dog?

13. If $n(A \cap B) = 150$ and $n(A \cap B \cap C) = 40$, what is $n(A \cap B \cap C^c)$?

14. If $n(A \cap C) = 100$ and $n(A \cap B \cap C) = 60$, what is $n(A \cap B^c \cap C)$?

15. If $n(A) = 200$, $n(A \cap B \cap C) = 40$, $n(A \cap B \cap C^c) = 20$, $n(A \cap B^c \cap C) = 50$, what is $n(A \cap B^c \cap C^c)$?

16. If $n(B) = 200$, $n(A \cap B \cap C) = 40$, $n(A \cap B \cap C^c) = 20$, $n(A^c \cap B \cap C) = 50$, what is $n(A^c \cap B \cap C^c)$?

Given $n(U) = 100$, $n(A) = 40$, $n(B) = 37$, $n(C) = 35$, $n(A \cap B) = 25$, $n(A \cap C) = 22$, $n(B \cap C) = 24$, and $n(A \cap B \cap C^c) = 10$, find the following.

17. $n(A \cap B \cap C)$     18. $n(A^c \cap B \cap C)$     19. $n(A \cap B^c \cap C)$

20. $n(A \cap B^c \cap C^c)$     21. $n(A^c \cap B \cap C^c)$     22. $n(A^c \cap B^c \cap C)$

23. $n(A \cup B \cup C)$     24. $n((A \cup B \cup C)^c)$

25. Using a Venn diagram show that

$$n(A \cup B \cup C) = n(A) + n(B) + n(C) - n(A \cap B)$$
$$- n(A \cap C) - n(B \cap C) + n(A \cap B \cap C).$$

26. Give a proof of the formula in Exercise 25. *Hint:* set $B \cup C = D$ and use formula (1) of the text on $n(A \cup D)$. Now use formula (1) two more times, recalling from the last section that $A \cap (B \cup C) = (A \cap B) \cup (A \cap C)$.

## Applications

27. **Sales.** A survey by a fast-food chain of 1000 adults found that in the past month 500 had been to Burger King, 700 to McDonald's, 400 to Wendy's, 300 to Burger King and McDonald's, 250 to McDonald's and Wendy's, 220 to Burger King and Wendy's, and 100 to all three. How many went to
    a. Wendy's but not the other two?
    b. Only one of them?
    c. None of these three?

28. **Investments.** A survey of 600 adults over age 50 found that 200 owned stocks or real estate but no bonds, 220 owned real estate or bonds but no stock, 60 owned real estate but no stocks or bonds, and 130 owned both stocks and bonds. How many owned none of the three?

29. **Entertainment.** A survey of 500 adults found that 190 played golf, 200 skied, 95 played tennis, 100 played golf but did not ski or play tennis, 120 skied but did not play golf or tennis, 30 played golf and skied but did not play tennis, and 40 did all three.
    a. How many played golf and tennis but did not ski?
    b. How many played tennis but did not play golf or ski?
    c. How many participated in at least one of the three above sports?

30. **Transportation.** A survey of 600 adults found that during the last year, 100 traveled by plane but not by train, 150 traveled by train but not by plane, 120 traveled by bus but not by train or plane, 100 traveled by both bus and plane, 40 traveled by all three, and 360 traveled by plane or train. How many did not travel by any of these three modes of transportation?

31. **Education.** In a survey of 250 business executives, 40 said they did not read *Money*, *Fortune*, or *Business Week*, while 120 said they read exactly one of these three and 60 said they read exactly two of them. How many read all three?

32. **Sales.** A furniture store held a sale that attracted 100 people to the store. Of these, 57 did not buy anything, 9 bought both a sofa and loveseat, 8 bought both a sofa and chair, 7 bought both a loveseat and chair. There were 24 sofas, 18 loveseats, and 20 chairs sold. How many people bought all three items?

## Solutions to Self-Help Exercise Set 6.2

1. The accompanying Venn diagram indicates that $n(A \cap B^c) = 50$, $n(A \cap B) = 20$, and $x = n(A^c \cap B)$. Then, according to the diagram, $50 + 20 + x = n(A \cup B) = 100$. Thus $x = 30$.

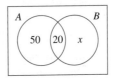

2. The number of students that registered for exactly one of the courses is the number that registered for math but not English, $n(M \cap E^c)$, plus the number that registered for English but not math, $n(M^c \cap E)$. Denote this sum by $x$. Then, according to the accompanying Venn diagram, $x + 400 + 700 = 2000$. Thus $x = 900$. That is, 900 students registered for exactly one of math or English.

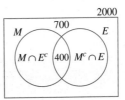

## 6.3 THE MULTIPLICATION PRINCIPLE AND PERMUTATIONS

▶ *Multiplication Principle*

▶ *Factorials*

▶ *Permutations*

### Multiplication Principle

Some of the basic counting techniques that will be used in the study of probability will now be considered. The "multiplication principle" given here is fundamental to all of the counting methods that will follow.

Before stating the multiplication principle, we consider the following example.

**E X A M P L E   1**     **A Simple Counting Problem**

A manufacturer makes four flavors of yogurt, and each flavor comes in two sizes. Blueberry comes in small and medium sizes, cherry comes in medium and large, strawberry comes in large and extra large, and vanilla comes in small and large. How many different possibilities are there?

**Solution**     We can think of the procedure as a sequence of two choices. The first is to pick a flavor and the second is to pick a size. We can denote the choice of picking cherry in the large size as the ordered pair (cherry, large). All possible outcomes can be visualized in the **tree diagram** in Figure 6.16. Counting all the possibilities gives 8.  ■

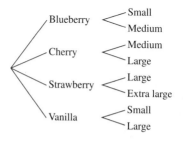

**Figure 6.16**

If we examine more closely why we obtained the answer 8, we can notice from Figure 6.16 that we had four groups of two each since, no matter what the flavor, there were the same number of choices for the size. This gives $4 \times 2 = 8$, where there were 4 choices for flavors, and for each flavor there were 2 choices for the size.

**E X A M P L E   2**     **A Simple Counting Problem**

A committee of three individuals, A, B, and C, must pick one of them to be chairman and a different one to be secretary. How many ways can this be done?

**Figure 6.17**

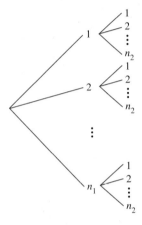

**Figure 6.18**

**Solution** The tree diagram in Figure 6.17 breaks this problem into an operation of first picking a chairman and then picking a secretary. Notice that there are three choices for chairman and, no matter who is picked for chairman, there are always two choices for secretary. Thus there are three groups of two each or $3 \times 2 = 6$ possible outcomes. ∎

Now suppose there is an operation that consists of making a sequence of two choices with $n_1$ possible outcomes for the first choice and no matter what the first choice there are always $n_2$ possible outcomes for the second choice. Then Figure 6.18 indicates a tree diagram. From this figure we see that there are $n_1$ groups, each with $n_2$ elements. Thus there is a total of $n_1 \cdot n_2$ possible outcomes. This is the **multiplication principle**.

---

**Multiplication Principle**

Suppose there is an operation that consists of making a sequence of two choices with $n_1$ possible outcomes for the first choice and, no matter what the first choice, there are always $n_2$ possible outcomes for the second choice. Then there are $n_1 \cdot n_2$ possible ways in which the sequence of two choices can be made.

---

**E X A M P L E   3**   **Using the Multiplication Principle**

Ten horses are running in the first race at Aqueduct. You wish to buy a Perfecta ticket, which requires picking the first and second finishers in order. How many possible tickets are there?

**Solution** You must make a choice: (first, second). Think of this as a sequence of two blanks (__, __) where we must fill in the blanks. There are 10 choices for first. No matter what horse you pick for first, there are always 9 left for second. This can be summarized as ($\underline{10}$, $\underline{9}$). Thus, by the multiplication principle, there are $10 \times 9 = 90$ possible tickets. ∎

We now can see in the same way the general multiplication principle.

---

**General Multiplication Principle**

Suppose there is an operation that consists of making a sequence of $k$ choices with $n_1$ possible outcomes for the first choice and, no matter what the first choice, there are always $n_2$ possible outcomes for the second choice, and, no matter what the first two choices, there are always $n_3$ possible outcomes for the third choice, and so on. Then there are $n_1 \cdot n_2 \cdots n_k$ possible ways in which the sequence of $k$ choices can be made.

---

For example, in Figure 6.19, there are $n_1 \cdot n_2$ elements in the second column, as we noted in Figure 6.18. Emanating from each one of these elements there are $n_3$ ''branches'' in the third column. Thus there are $n_1 \cdot n_2 \cdot n_3$ elements in the third column.

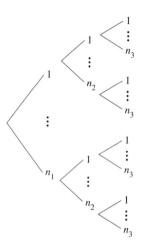

**Figure 6.19**

E X A M P L E 4    **Using the General Multiplication Principle**

Ten horses are running in the first race at Aqueduct. You wish to buy a Trifecta ticket that requires picking the first three finishers in order. How many possible tickets are there?

Solution    You must make a choice: (first, second, third). Thus we have a sequence of three blanks (__, __, __), and we must fill in the blanks. There are 10 choices for first. No matter what horse you pick for first there are always 9 left for second, and no matter which horses you pick for the first and second choices, there are always 8 left for third. This can be summarized as ($\underline{10}$, $\underline{9}$, $\underline{8}$). Thus by the general multiplication principle there are $10 \times 9 \times 8 = 720$ possible tickets.   ■

E X A M P L E 5    **Using the General Multiplication Principle**

In the seventh race at Aqueduct, which is running 10 horses, you have learned through impeccable sources that the overwhelming favorite will be held back from winning. You wish to be assured of having a winning Trifecta ticket. How many must you buy?

Solution    Again you must make a choice: (first, second, third). This time, from your inside knowledge, you know that there are only 9 choices for first, and, given any of these 9 choices, there remain 9 horses to pick for second. No matter what the first and second pick is, there are always 8 remaining choices for third. This can be summarized as (9, 9, 8). Thus, by the multiplication principle, there are $9 \times 9 \times 8 = 648$ possible choices. ■

E X A M P L E 6    **Using the General Multiplication Principle**

How many three letter words that all begin with consonants and have exactly one vowel can be made using the first seven letters of the alphabet where using a letter twice is permitted but having two consonants next to each other is not?

**Solution**    Since two consonants cannot be next to each other, one must make the choice (consonant, vowel, consonant). There are 5 choices for the first consonant (b, c, d, f, or g). No matter what the first choice there are 2 choices for vowels (a and e), and no matter what the first two choices, there remain the same 5 consonants to pick from. This can be summarized as (5, 2, 5). Thus the multiplication principle gives $5 \times 2 \times 5 = 50$ possible words. ■

## Factorials

We will be encountering expressions such as $6 \cdot 5 \cdot 4 \cdot 3 \cdot 2 \cdot 1$ and will need symbols to denote them. We have the following definition.

---

**Factorial**

For any natural number $n$

$$n! = n(n - 1)(n - 2) \cdots 3 \cdot 2 \cdot 1$$

$$0! = 1$$

---

E X A M P L E 7    **Calculating Some Factorials**

$$4! = 4 \cdot 3 \cdot 2 \cdot 1 = 24.$$

$$5! = 5 \cdot 4 \cdot 3 \cdot 2 \cdot 1 = 120.$$

$$6! = 6 \cdot 5 \cdot 4 \cdot 3 \cdot 2 \cdot 1 = 720. \quad ■$$

## Permutations

The set of elements $\{1, 2, 3, 4\}$ can be arranged in various orders. For example, they can be listed as $\{1, 3, 4, 2\}$ or perhaps $\{4, 3, 1, 2\}$. Each of these ordered arrangements is called a *permutation*. We have the following definition.

---

**Permutation**

A permutation of a set of elements is an ordered arrangement of all the elements.

---

By "arrangement" it is understood that elements can be used only once.

**E X A M P L E   8**    **Counting the Number of Permutations**

Find all permutations of the set $\{a, b, c\}$ and count the total number.

**Solution**    Given the set $\{a, b, c\}$, we can use a tree diagram to find all possible permutations. See Figure 6.20. There are 6 possibilities, and these can be seen also by the multiplication principle to be $3 \cdot 2 \cdot 1 = 3!$  ∎

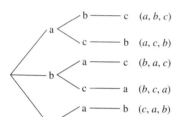

Suppose a set of $n$ distinct objects is given and we wish to find the number of permutations that are possible. We denote this number by $P(n, n)$. There are $n$ possibilities for the first one. No matter which one is chosen first, there always remain $(n - 1)$ for the second choice. No matter how the first two have been chosen, there remains $(n - 2)$ for the third choice, and so on. By the multiplication principle, we then have the total number of possibilities to be

$$P(n, n) = n(n - 1)(n - 2) \cdots 3 \cdot 2 \cdot 1 = n!$$

**Figure 6.20**    We thus have shown the following.

---

**Number of Permutations of $n$ Objects**

The number of permutations of $n$ distinct objects is given by
$$P(n, n) = n(n - 1)(n - 2) \cdots 3 \cdot 2 \cdot 1 = n!$$

---

**E X A M P L E   9**    **Counting the Number of Permutations**

In how many ways can a football team of eleven players arrange themselves to trot out onto the football field one at a time?

**Solution**    The answer is given by

$$P(11, 11) = 11! = 39{,}916{,}800. \quad ∎$$

There are situations in which we have a set of $n$ distinct objects and wish to select an ordered arrangement of $r$ of them. The total number of such arrangements is denoted by $P(n, r)$. We also refer to this as the **number of permutations of $r$ objects taken from a set of size $n$**. Example 4 was such a case where you selected a first, second, and third place from a set of 10 horses running in a race. Recall from the multiplication principle that we have

$$P(10, 3) = 10 \cdot 9 \cdot 8.$$

Given a set of $n$ distinct objects, the multiplication principle indicates that the number of ways of selecting an ordered arrangement of 3 of them is

$$P(n, 3) = n(n - 1)(n - 2),$$

of 4 of them is

$$P(n, 4) = n(n - 1)(n - 2)(n - 3),$$

and, in general, of $r$ of them is

$$P(n, r) = \underbrace{n(n - 1)(n - 2) \cdots (n - r + 1)}_{r \text{ factors}}.$$

It is sometimes convenient to write $P(n, r)$ in a different form as follows.

$$P(n, r) = n(n - 1)(n - 2) \cdots (n - r + 1)$$

$$= n(n - 1)(n - 2) \cdots (n - r + 1) \frac{(n - r)(n - r - 1) \cdots 2 \cdot 1}{(n - r)(n - r - 1) \cdots 2 \cdot 1}$$

$$= \frac{n(n - 1)(n - 2) \cdots (n - r + 1)(n - r)(n - r - 1) \cdots 2 \cdot 1}{(n - r)(n - r - 1) \cdots 2 \cdot 1}$$

$$= \frac{n!}{(n - r)!}$$

We have established the following.

---

**Number of Permutations of $r$ Objects Taken from a Set of Size $n$**

The number of permutations of $r$ distinct objects taken from a set of size $n$ is given by

$$P(n, r) = n(n - 1)(n - 2) \cdots (n - r + 1) = \frac{n!}{(n - r)!}.$$

---

Since $0! = 1$,

$$P(n, n) = \frac{n!}{(n - n)!} = \frac{n!}{1} = n!,$$

which is what we obtained earlier.

**EXAMPLE 10**  **Using the Multiplication Principle and Permutations**

A certain state uses license plates with three letters followed by three numbers with no repeats of letters or numbers. How many such license plates can be made?

**Solution**  Although this problem could be done directly from the multiplication principle with $k = 6$, we will instead use our knowledge of permutations. We can think of deciding on a license plate as a two-step process. First choose the three letters

*(Dennis Drenner.)*

and then choose the three numbers. The number of ways of arranging the three letters without repetitions and when order is important is the number of permutations of 26 letters taken 3 at a time or $P(26, 3)$. The number of ways of arranging the three numbers without repetitions and when order is important is the number of permutations of 10 digits taken 3 at a time or $P(10, 3)$. By the multiplication principle (when $k = 2$), the total number of different license plates will be

$$P(26, 3)P(10, 3) = (26 \cdot 25 \cdot 24)(10 \cdot 9 \cdot 8) = 11{,}232{,}000. \quad \blacksquare$$

## SELF-HELP EXERCISE SET 6.3

1. A restaurant serves 3 soups, 4 salads, 10 main dishes, and 6 desserts. How many different meals can be served if one of each category is chosen?

2. From a group of 6 junior executives, one is to be chosen to go to the Denver office, one to go to the St. Louis office, and one to the Atlanta office. In how many ways can this be done?

3. Six junior executives and 3 senior executives are to line up for a picture. The senior executives must all be lined up together on the left and the junior executives must be all together on the right. In how many ways can this be done?

## EXERCISE SET 6.3

In Exercises 1 through 8 evaluate the given expression.

1. $P(5, 3)$
2. $P(5, 2)$
3. $P(8, 5)$

4. $P(8, 3)$
5. $P(7, 7)$
6. $P(7, 1)$

7. $P(9, 8)$
8. $P(9, 1)$

9. A manufacturer offers 4 styles of sofas with 30 fabrics for each sofa. How many different sofas are there?

10. A manufacturer offers 3 grades of carpet with 10 colors for each grade. How many different carpets are there?

11. A restaurant offers 4 types of salads, 10 main dishes, and 5 desserts. How many different complete meals are there?

12. A card is picked from a standard deck of 52, and then a coin is flipped twice. How many possible outcomes are there?

13. A license plate has 6 digits with repetitions permitted. How many possible license plates of such type are there?

14. A license plate has 5 letters with repetitions not permitted. How many possible license plates of such type are there?

15. An automobile manufacturer offers a certain style car with two types of radios, 10 choices of exterior colors, 5 different interior colors, and 3 types of engines. How many different automobiles are offered?

16. A contractor has 4 styles of homes, each with 3 styles of garages, 4 styles of decks, and 5 styles of carpeting. How many possibilities are there?

17. A state makes license plates with three letters followed by three digits with repetitions permitted. How many of these license plates are there?

18. A state makes license plates with three letters followed by three numbers with no repetitions of letters permitted. How many of these license plates are there?

19. How many 3-letter words can be made from the first 8 letters of the alphabet if consonants cannot be next to each other and letters cannot be repeated?

20. How many 5-letter words can be made from the first 8 letters of the alphabet if consonants cannot be next to each other, vowels cannot be next to each other, and vowels cannot be repeated but consonants can be?

21. At an awards ceremony, 5 men and 4 women are to be called one at a time to receive an award. In how many ways can this be done if men and women must alternate?

22. At an awards ceremony, 5 women and 4 men are each to receive one award and are to be presented their award one at a time. Two of the awards are to be first given to two of the women, and then the remaining awards will alternate between men and women. How many ways can this be done?

23. In how many ways can the 5 members of a basketball team line up in a row for a picture?

24. In how many ways can the individuals in a foursome of golfers tee off in succession?

25. An executive is scheduling meetings with 12 people in succession. The first two meetings must be with two directors on the board, the second 4 with 4 vice-presidents, and the last 6 with 6 junior executives. How many ways can this schedule be made out?

26. An executive is scheduling trips to the company's European plants. First the 4 French plants will be visited, followed by the 3 Italian plants, and then the 5 German ones. How many ways can this schedule be made out?

27. The starting 9 players on the school baseball team and the starting 5 players on the basketball team are to line up for a picture with all members of the baseball team together on the left. How many ways can this be done?

28. The 7 starting offensive linemen and 4 starting offensive backs of the New York Giants are to line up for a picture with the 7 linemen in the middle. How many ways can this be done?

29. On a baseball team, the 3 outfielders can play any of the 3 outfield positions, and the 4 infielders can play any of the 4 infield positions. How many different arrangements of these 7 players can be made?

30. On a football team, the 7 linemen can play any of the 7 linemen positions and the 4 backs can play any of the backfield positions. How many different arrangements of these 11 players can be made?

31. A group of 12 must select a president, a vice-president, a treasurer, and a secretary. How many ways can this be done?

32. In the Superfecta, one must pick the first four finishers of a horse race in correct order. If there are 10 horses running in a race, how many different tickets are there?

33. A buyer for a furniture store selects 8 different style sofas from a group of 10 and has each style shipped on successive weeks. How many ways can this be done?

34. A chef can make 12 main courses. Every day a menu is formed by selecting 7 of the main courses and listing them in order. How many different such menus can be made?

35. Two groups are formed with 10 in the first group and 8 different people in the second. A president, vice-president, and a secretary/treasurer is to be chosen in each group. How many ways can this be done?

36. At a race track you have the opportunity to buy a ticket that requires you to pick the first and second place horses in the first two races. If the first race runs 8 horses and the second runs 10, how many different tickets are possible?

37. A picture is to be taken by lining up 4 of the 11 players from the football team on the left, then 3 of the 9 players from the baseball team in the center, and finally 2 of the 5 players from the basketball team on the right. How many ways can this arrangement be done?

38. A tourist has 8 cities in Great Britain, 6 in France, 5 in Italy, and 7 in Germany on a list she would like to visit. She decides that she will first go to Great Britain and visit 4 of the cities on her list, then on to France to visit 3 cities on the list, then on to Italy for 2 cities, and then on to Germany to visit 4 on the list. How many ways can her itinerary be made out?

## Solutions to Self-Help Exercise Set 6.3

1. A restaurant that serves 3 soups, 4 salads, 10 main dishes, and 6 desserts can by the multiplication principle serve $3 \times 4 \times 10 \times 6 = 720$ different meals.

2. The number of ways 6 junior executives can be sent to 3 different offices is the same as the number of permutations of 6 objects taken 3 at a time. This is

$$P(6, 3) = 6 \times 5 \times 4 = 120$$

ways.

3. Six junior executives can line up in 6! ways on the right. For each of these ways the 3 senior executives can line up in 3! ways on the left. Thus by the multiplication principle, the two groups can line up in

$$3! \times 6! = 4320$$

ways.

## 6.4 COMBINATIONS

▶ *Combinations*

▶ *Counting the Number of Sequences*

**A P P L I C A T I O N**

**Connecticut Lotto Game**

In the state of Connecticut's lotto game, 6 numbered ping pong balls are randomly selected without replacement from a set of 44 to determine a winning set of numbers (without regard to order). If no one picks these 6 numbers, the money wagered stays in the "pot" for the next drawing. After several consecutive games with no winners, the pot gets large and attracts a lot of attention and ticket sales. When the Connecticut lotto game started, the total number of ping pong balls was 36. After a number of years when there were very few long streaks with no winners, lotto officials changed the total number of ping pong balls from 36 to 44. Suppose a number of weeks has gone by without a winner and the pot has grown very large. You wish to organize a syndicate of investors that will purchase every possible ticket to ensure obtaining a winning ticket. How many tickets will the syndicate have to buy if there are 36 ping pong balls? If there are 44? The answer can be found in the discussion before Example 1.

### Combinations

When a permutation of $n$ distinct objects is taken $r$ at a time, one selects $r$ of the objects in a specific order. A **combination** of $r$ distinct objects taken from a set of size $n$ is merely a selection of $r$ of the objects (without concern for order). We consider combinations in this section.

Given the set $\{a, b, c\}$ we know from the last section that there are $P(3, 2) = 3 \times 2 = 6$ ways of selecting 2 of these at a time when order is important. The 6 ways are

$$(a, b), (b, a),$$
$$(a, c), (c, a),$$
$$(b, c), (c, b).$$

If now we wish to select 2 at a time when order is not important, then $(a, b)$ is the same as $(b, a)$, and $(a, c)$ is the same as $(c, a)$, and $(b, c)$ is the same as $(c, b)$. Thus there are only 3 ways of selecting 2 objects at a time when order is not important: $(a, b), (a, c), (b, c)$. These are referred to as *combinations*.

---

**Combinations**

A **combination** of $r$ distinct objects taken from a set of size $n$ is a selection of $r$ of the objects (without concern for order).

---

For example, if from a group of 4 people, we wish to select a president, a vice-president, and a secretary/treasurer, then order is important. If, on the other hand, we wished to select a committee of three people from a set of 4, then order is not important since the duties and title of each committee member is the same no matter in what order they are selected.

We know from the last section that the number of permutations of $r$ objects taken from a set of size $n$ is given by $P(n, r)$. We denote the number of combinations of $r$ objects taken from a set of size $n$ by $C(n, r)$. We wish to find a formula for $C(n, r)$.

We can see this by viewing the process of selecting all *permutations* of $r$ objects taken from a set of size $n$ as a sequence of two operations. In the first operation select $r$ objects. In the second operation place these $r$ objects in some order. That is, first select a *combination* of $r$ distinct objects and then order them. We know from the last section that any $r$ distinct objects can be ordered in $r!$ ways. Thus we can obtain all permutations of $r$ objects taken from a set of size $n$ by first selecting all combinations of $r$ objects taken from a set of size $n$ and then permuting each of the $r$ combinations in $(r!)$ ways. Notice that no matter what combination we take, we then always permute by the same number, namely $(r!)$. Thus by the multiplication principle we then must have

$$P(n, r) = C(n, r)r!$$

or

$$C(n, r) = \frac{1}{r!} P(n, r) = \frac{n!}{(r!)(n - r)!}.$$

See Figure 6.21 where the distinct combinations are listed as $C_1, C_2, \ldots, C(n, r)$. We have proven the following.

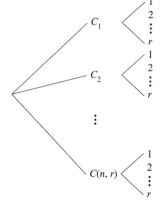

**Figure 6.21**

---

**Number of Combinations of $r$ Objects taken from a Set of Size $n$**

The number of combinations of $r$ distinct objects taken from a set of size $n$, denoted by $C(n, r)$, is given by

$$C(n, r) = \frac{n!}{(r!)(n - r)!}.$$

---

We now solve the problem posed at the beginning of this chapter. Since the order in which the numbers are selected is of no consequence, we are looking for $C(36, 6)$ in the first case and $C(44, 6)$ in the second case. These are

$$C(36, 6) = \frac{36!}{6!(36 - 6)!}$$

$$= \frac{36 \cdot 35 \cdot 34 \cdot 33 \cdot 32 \cdot 31 \cdot (30!)}{6!(30!)}$$

$$= \frac{36 \cdot 35 \cdot 34 \cdot 33 \cdot 32 \cdot 31}{6!}$$

$$= 1,947,792$$

and

$$C(44, 6) = \frac{44!}{6!(44 - 6)!}$$

$$= \frac{44 \cdot 43 \cdot 42 \cdot 41 \cdot 40 \cdot 39 \cdot (38!)}{6!(38!)}$$

$$= \frac{44 \cdot 43 \cdot 42 \cdot 41 \cdot 40 \cdot 39}{6!}$$

$$= 7,059,052.$$

Thus with 36 ping pong balls, the syndicate must buy about 2 million tickets. If there are 44 balls, then about 7 million tickets must be bought to guarantee a winning ticket.

*(Dennis Drenner.)*

### Standard Deck of 52 Playing Cards

A standard deck of 52 playing cards has four 13-card suits: clubs ♣, diamonds ◇, hearts ♡, and spades ♠. The diamonds and hearts are red, while the clubs and spades are black. Each 13-card suit contains cards numbered from 2 to 10, a jack, a queen, a king, and an ace. The jack, queen, king, and ace can be considered respectively as number 11, 12, 13, and 14. In poker the ace can be either a 14 or a 1.

**E X A M P L E   1**     **Calculating the Number of Different Poker Hands**

How many different 5-card poker hands can be dealt from a standard deck of 52 cards?

**Solution**     Since the order in which the cards in a poker hand are dealt is of no consequence, we are looking for $C(52, 5)$ which is

$$C(52, 5) = \frac{52!}{5!(52 - 5)!}$$

$$= \frac{52 \cdot 51 \cdot 50 \cdot 49 \cdot 48 \cdot (47!)}{5!(47!)}$$

$$= \frac{52 \cdot 51 \cdot 50 \cdot 49 \cdot 48}{5!}$$

$$= 2,598,960. \quad ■$$

**E X A M P L E   2**     **A Counting Problem in Poker**

Find the number of poker hands with 3 queens and 2 jacks.

**Solution**     View this as a sequence of two operations. In the first operation, select 3 queens from a deck with 4 queens. In the second operation, select 2 jacks from a deck with 4 jacks. Since order is not important in either case, the first selection can be made in $C(4, 3)$ ways and the second selection can be made in $C(4, 2)$ ways. Since no matter how the first selection is made, the second selection can always be made in $C(4, 2)$ ways, the multiplication principle then indicates that the number of hands with 3 queens and 2 jacks is given by

$$C(4, 3)C(4, 2) = \left(\frac{4!}{3!(4 - 3)!}\right)\left(\frac{4!}{2!(4 - 2)!}\right)$$

$$= \left(\frac{4(3!)}{3!}\right)\left(\frac{4 \cdot 3 \cdot 2}{2 \cdot 2}\right)$$

$$= (4)(6) = 24. \quad \blacksquare$$

**E X A M P L E  3**   **Counting Using Both Combinations and Permutations**

In how many ways can a committee be formed with a chairman, a vice chairman, a secretary/treasurer, and 4 additional people all chosen from a group of 10 people?

**Solution**   One way of doing this problem is to see the process as a sequence of two selections. The first is to pick the chairman, vice chairman, and secretary/treasurer from the set of 10. Since order is important, this can be done in $P(10, 3)$ ways. The second selection is to then pick the other 4 members of the committee from the remaining 7 people. This can be done in $C(7, 4)$ ways since order is not important. Furthermore, since the second selection can always be made in $C(7, 4)$ ways no matter how the first selection was made, the multiplication principle indicates that the total number of ways of selecting the committee is

$$P(10, 3)C(7, 4) = 10 \cdot 9 \cdot 8 \, \frac{7 \cdot 6 \cdot 5 \cdot 4}{4 \cdot 3 \cdot 2}. \quad \blacksquare$$

Another way of viewing this is to select 4 of the committee members from the group of 10 and then select the chairman, vice chairman, and secretary/treasurer from the remaining 6. This gives

$$C(10, 4)P(6, 3) = \frac{10 \cdot 9 \cdot 8 \cdot 7}{4 \cdot 3 \cdot 2} \, 6 \cdot 5 \cdot 4,$$

which is the same answer as before.

Yet another way of looking at this is to select the committee of 7 from the group of 10 and then select the chairman, vice chairman, and secretary/treasurer from the committee of 7. This gives

$$C(10, 7)P(7, 3) = \frac{10 \cdot 9 \cdot 8}{3 \cdot 2} \, 7 \cdot 6 \cdot 5,$$

which agrees with the other answers.

**E X A M P L E  4**

**A Counting Problem Using Combinations**

A committee of 15 people consists of 8 males and 7 females. In how many ways can a subcommittee of 5 be formed if the subcommittee consists of

(a)   Any 5 committee members?
(b)   All men?
(c)   At least 3 men?

**Solutions**

(a)   Since the order of selection is of no consequence, the answer is

$$C(15, 5) = \frac{15!}{5!(15-5)!}$$

$$= \frac{15 \cdot 14 \cdot 13 \cdot 12 \cdot 11 \cdot (10!)}{5!(10!)}$$

$$= \frac{15 \cdot 14 \cdot 13 \cdot 12 \cdot 11}{5!}$$

$$= 3003.$$

(b)   Here we must pick 5 men out of 8 possible men. This can be done in $C(8, 5)$ ways since again order is not important. Thus

$$C(8, 5) = \frac{8!}{5!(8-5)!}$$

$$= \frac{8 \cdot 7 \cdot 6 \cdot (5!)}{(5!)(3 \cdot 2)}$$

$$= 56.$$

(c)   At least 3 men means all subcommittees with 3 men and 2 women, plus all subcommittees with 4 men and 1 women, plus all with 5 men. This is

$$C(8, 3)C(7, 2) + C(8, 4)C(7, 1) + C(8, 5)C(7, 0)$$

$$= \frac{8 \cdot 7 \cdot 6}{3 \cdot 2}\left(\frac{7 \cdot 6}{2}\right) + \frac{8 \cdot 7 \cdot 6 \cdot 5}{4 \cdot 3 \cdot 2}(7)$$

$$+ \frac{8 \cdot 7 \cdot 6 \cdot 5 \cdot 4}{5 \cdot 4 \cdot 3 \cdot 2}(1)$$

$$= (56)(21) + (70)(7) + (56)(1)$$

$$= 1722. \quad \blacksquare$$

## Counting the Number of Sequences

**E X A M P L E  5**

**A Counting Problem Involving a Sequence**

An investor has selected a growth mutual fund from a large set of growth funds and will consider any of the next 10 years a success (S) if this mutual fund performs above average in the set of funds and a failure (F) otherwise.

(a)   How many different outcomes are possible?
(b)   How many different outcomes have exactly 6 successes?
(c)   How many different outcomes have at least 3 successes?

**Solutions**

(a)   An outcome consists of 10 operations in sequence. Each operation assigns a $S$ or $F$. One such example is

$$\{S, S, F, S, S, F, F, F, S, S\}.$$

No matter what the assignments of S's and F's in any of the prior years, there are always two possibilities for the current year: $S$ or $F$. Using the multiplication principle with $k = 10$ and $n_1 = n_2 = \ldots = n_{10} = 2$ yields

$$2^{10} = 1024$$

as the total number of possible outcomes.

(b)   An outcome with exactly 6 $S$'s was given in (a). Notice that this amounts to filling in 6 years with $S$'s and 4 years with $F$'s. A particular outcome will be determined once we fill in 6 $S$'s in 6 of the years. This can be done in $C(10, 6)$ ways. This is

$$C(10, 6) = \frac{10!}{6!(10 - 6)!} = \frac{10 \cdot 9 \cdot 8 \cdot 7}{4 \cdot 3 \cdot 2} = 210.$$

(c)   The answer to this is the number with exactly 3 $S$'s plus the number with exactly 4 $S$'s plus . . . the number with exactly 10 $S$'s, or

$$C(10, 3) + C(10, 4) + \cdots + C(10, 10).$$

A shorter way can be given by noticing that this is just the total number less the number with at most 2 $S$'s. This is

$$1024 - [C(10, 0) + C(10, 1) + C(10, 2)]$$
$$= 1024 - [1 + 10 + 45] = 968. \quad \blacksquare$$

## SELF-HELP EXERCISE SET 6.4

1.   A quinella ticket at a race track allows one to pick the first two finishers without regard to order. How many different tickets are possible in a race with 10 horses?

2.   A company wishes to select 4 junior executives from the San Francisco office, 5 from the Dallas office, and 5 from the Miami office to bring to the New York City headquarters. In how many ways can this be done if there are 10 junior executives in San Francisco, 12 in Dallas, and 15 in Miami?

## EXERCISE SET 6.4

In Exercises 1 through 13 calculate the indicated quantity.

1.   $C(8, 3)$            2.   $C(8, 4)$            3.   $C(8, 5)$

4.   $C(12, 12)$          5.   $C(12, 1)$           6.   $C(12, 0)$

**7.** $C(7, 4)$          **8.** $C(7, 3)$          **9.** $C(15, 2)$

**10.** $C(n, 1)$         **11.** $C(n, 0)$        **12.** $C(n, n - 1)$

**13.** $C(n, n)$

**14.** Show that $C(n, r) = C(n, (n - r))$.

**15.** Find all permutations of $\{a, b, c\}$ taken 2 at a time by first finding all combinations of the set taken 2 at a time and then permuting each combination. Construct a tree similar to Figure 6.21 in the text.

**16.** Find all permutations of $\{a, b, c, d\}$ taken 3 at a time by first finding all combinations of the set taken 3 at a time and then permuting each combination. Construct a tree similar to Figure 6.21 in the text.

**17.** If you have a penny, a nickel, a dime, a quarter, and a half-dollar in your pocket or purse, how many different tips can you leave using 3 coins?

**18.** From a list of 40 captains, 5 are to be promoted to major. In how many ways can this be done?

**19.** In a certain lotto game, 6 numbered ping pong balls are randomly selected without replacement from a set of balls numbered from 1 to 46 to determine a winning set of numbers (without regard to order). Find the number of possible outcomes.

**20.** A boxed Trifecta ticket at a horse track allows you to pick the first three finishers without regard to order. How many different tickets are possible in a race with 10 horses?

**21.** From a list of 20 recommended stocks from your brokerage firm, you wish to select 5 of them for purchase. In how many ways can you do this?

**22.** A restaurant offers 8 toppings on its pizza. In how many ways can you select 3 of them?

**23.** If you join a book club, you can purchase 4 books at a sharp discount from a list of 20. In how many ways can you do this?

**24.** A firm is considering expanding into 4 of 9 possible cities. In how many ways can this be done?

**25.** In how many ways can an inspector select 5 bolts from a batch of 40 for inspection?

**26.** A chef has 20 dinners that she can make. In how many ways can she select 6 of them for the menu for today?

**27.** In her last semester, a student must pick 3 mathematics courses and 2 computer science courses to graduate with a degree in mathematics with a minor in computer science. If there are 11 mathematics courses and 7 computer science courses available to take, how many different ways can this be done?

**28.** A committee of 12 U.S. Senators is to be formed with 7 Democrats and 5 Republicans. In how many ways can this be done if there are 53 Democratic senators and 47 Republican?

**29.** A chef has 20 main courses and 6 soups that he can prepare. How many different menus could he require if he always has 7 main courses and 3 soups on each menu?

**30.** A firm must select 4 out of a possible 10 sites on the East Coast and 3 out of a possible 8 sites on the West Coast for expansion. In how many ways can this be done?

**31.** A firm has 12 junior executives. Three are to be sent to Pittsburgh, one to Houston, one to Atlanta, and one to Boston. In how many ways can this be done?

**32.** Six prizes are to be given to six different people in a group of nine. In how many ways can a first prize, a second prize, a third prize, and three fourth prizes be given?

**33.** In a new group of 11 employees 4 are to be assigned to production, 1 to sales, and 1 to advertising. In how many ways can this be done?

**34.** A parent of 7 children wants 2 children to make dinner, 1 to dust, and 1 to vacuum. In how many ways can this be done?

**35.** In how many ways can the 9 member Supreme Court give a five-to-four decision upholding a lower court?

**36.** In how many ways can a committee of 5 reach a majority decision if there are no abstentions?

**37.** A coin is flipped 8 times in succession. In how many ways can exactly 5 heads occur?

**38.** A coin is flipped 8 times in succession. In how many ways can at least 6 heads occur?

**39.** A coin is flipped 8 times in succession. In how many ways can at least 2 heads occur?

**40.** A baseball team takes a road trip and plays 12 games. In how many ways could they win 7 and lose 5?

**41.** A banana split is made with 3 scoops of different flavors of ice cream, 3 different syrups, 2 different types of nuts, and with or without whipped cream. How many different banana splits can be made if there are 12 ice creams, 8 syrups, and 4 types of nuts to choose from?

**42.** A salesman has 10 customers in New York City, 8 in Dallas, and 6 in Denver. In how many ways can he see 4 customers in New York City, 3 in Dallas, and 4 in Denver?

**43.** Find the number of different poker hands that contain exactly 3 aces, while the remaining 2 cards do not form a pair.

**44.** Find the number of full houses in a poker hand, that is, the number of poker hands with 3 of a kind and 2 of a kind.

**45.** Find the number of poker hands with two pairs, that is, two different two of a kinds with the fifth card a third different kind.

**46.** In how many ways can a doubles game of tennis be arranged from 8 boys and 4 girls if each side must have 1 boy and 1 girl?

## Solutions to Self-Help Exercise Set 6.4

1. The number of tickets is the same as the number of ways of selecting 2 objects from 10 when order is not important. This is

$$C(10, 2) = \frac{10 \cdot 9}{2} = 45.$$

2. The number of ways that junior executives can be selected from San Francisco, Dallas, and Miami, is, respectively, $C(10, 4)$, $C(12, 5)$, and $C(15, 5)$. By the multiplication principle the total number of ways this can be done is

$$C(10, 4)C(12, 5)C(15, 5) = \frac{10!}{4!(10 - 4)!} \frac{12!}{5!(12 - 5)!} \frac{15!}{5!(15 - 5)!}$$

$$= \frac{10 \cdot 9 \cdot 8 \cdot 7}{4 \cdot 3 \cdot 2} \frac{12 \cdot 11 \cdot 10 \cdot 9 \cdot 8}{5 \cdot 4 \cdot 3 \cdot 2} \frac{15 \cdot 14 \cdot 13 \cdot 12 \cdot 11}{5 \cdot 4 \cdot 3 \cdot 2}$$

$$= (10 \cdot 3 \cdot 7)(11 \cdot 9 \cdot 8)(7 \cdot 13 \cdot 3 \cdot 11)$$

$$= 499,458,960.$$

## 6.5    BINOMIAL THEOREM, PARTITIONS*

> ► *Binomial Theorem*
> ► *Pascal's Triangle*
> ► *Ordered Partitions*

APPLICATION
**The Number of Options for an Executive**

A study committee reports to an executive that to solve a particular problem any number of 4 different actions can be taken, including doing nothing. How many options does the executive have? See Example 3 for the answer.

APPLICATION
**Counting the Ways of Dividing a Committee to Perform Tasks**

In how many ways can a group of 14 people be divided into 3 committees, each assigned a different task, the first committee consisting of 8 people, the second 4 people, and the third 2 people? See Example 4 for the answer.

### The Binomial Theorem

The expansion $(x + y)^2 = x^2 + 2xy + y^2$ is familiar. In the first part of this section formulas for the expansion of $(x + y)^n$ where $n$ is any integer will be given. In the last part of this section, ordered partitions are introduced. These are generalizations of the notion of combinations that have already been studied.

---

*This section is optional.

We wish to develop a systematic way of writing the expansion of expressions of the form $(x + y)^n$ where $n$ is a positive integer.

First calculating by direct multiplication we can obtain

$$(x + y)^0 = 1$$

$$(x + y)^1 = x + y$$

$$(x + y)^2 = x^2 + 2xy + y^2$$

$$(x + y)^3 = x^3 + 3x^2y + 3xy^2 + y^3$$

$$(x + y)^4 = x^4 + 4x^3y + 6x^2y^2 + 4xy^3 + y^4$$

$$(x + y)^5 = x^5 + 5x^4y + 10x^3y^2 + 10x^2y^3 + 5xy^4 + y^5$$

$$(x + y)^6 = x^6 + 6x^5y + 15x^4y^2 + 20x^3y^3 + 15x^2y^4 + 6xy^5 + y^6$$

Notice that in the expansions $(x + y)^n$ the powers of $x$ decrease by one and the powers of $y$ increase by one as we move to the next term. Also notice that for the expressions $x^a y^b$, we have $a + b = n$.

How can we predict the coefficients of such terms? Take $(x + y)^5$ as an example. Write

$$(x + y)^5 = (x + y)(x + y)(x + y)(x + y)(x + y) \qquad (1)$$

as the product of 5 factors. We can obtain an $x^2 y^3$ term in the product by selecting $y$ from exactly 3 of the factors on the right of (1). We can think of this as filling in the blanks of

$$\{\_\_, \_\_, \_\_, \_\_, \_\_\}$$

with exactly 3 $y$'s. (The other blanks then must be $x$.) For example, $\{y, x, y, y, x\}$ indicates that we have selected $y$ from only the first, third, and fourth factors on the right of (1). The number of ways we can select exactly 3 blanks to put a $y$ in from the 5 possible blanks is $C(5, 3)$. This is

$$C(5, 3) = \frac{5 \cdot 4 \cdot 3}{3 \cdot 2} = 10,$$

which agrees with the coefficient of $x^2 y^3$ in the above expansion of $(x + y)^5$.

In general, when looking at the expansion

$$(x + y)^n = (x + y)(x + y)(x + y) \cdots (x + y),$$

we will obtain an $x^{n-k}y^k$ term by selecting $y$ from exactly $k$ of the $n$ factors. This can be done in $C(n, k)$ ways. Thus the coefficient of the $x^{n-k}y^k$ term must be $C(n, k)$. We have proven the following theorem.

---

**The Binomial Theorem**

The coefficient of $x^{n-k}y^k$ in the expansion of $(x + y)^n$ is $C(n, k)$.

---

E X A M P L E   1   **Using the Binomial Theorem**

Find the coefficient of $x^7y^3$ in the expansion $(x + y)^{10}$.

**Solution**  According to the binomial theorem this must be

$$C(10, 3) = \frac{10 \cdot 9 \cdot 8}{3 \cdot 2} = 120. \quad \blacksquare$$

It is common to use the notation

$$\binom{n}{k} = C(n, k).$$

With this notation we then have

$$(x + y)^n = \binom{n}{0} x^n + \binom{n}{1} x^{n-1}y + \binom{n}{2} x^{n-2}y^2$$

$$+ \cdots +$$

$$\binom{n}{n - 2} x^2y^{n-2} + \binom{n}{n - 1} xy^{n-1} + \binom{n}{n} y^n.$$

E X A M P L E   2   **Using the Binomial Theorem**

Write out the expansion of $(a - 2b)^4$.

**Solution**  By setting $x = a$ and $y = -2b$ in the binomial theorem and using the above notation, we have

$$(a - 2b)^4 = (a + (-2b))^4$$

$$= \binom{4}{0} a^4 + \binom{4}{1} a^3(-2b) + \binom{4}{2} a^2(-2b)^2$$

$$+ \binom{4}{3} a(-2b)^3 + \binom{4}{4} (-2b)^4$$

$$= a^4 + 4a^3(-2b) + 6a^2(-2b)^2 + 4a(-2b)^3 + (-2b)^4$$

$$= a^4 - 8a^3b + 24a^2b^2 - 32ab^3 + 16b^4. \quad \blacksquare$$

A consequence of the binomial theorem is the following.

---

**The Number of Subsets of a Set**

A set with $n$ distinct elements has $2^n$ distinct subsets.

---

Before giving a proof, let us list all the subsets of $\{a, b, c\}$ by listing all subsets with 3 elements, all with 2 elements, all with 1 element, and all with no elements. We have

$$\{a, b, c\}$$

$$\{a, b\}, \{a, c\}, \{b, c\}$$

$$\{a\}, \{b\}, \{c\}$$

$$\phi.$$

There are 8, which is $8 = 2^3$.

To establish the theorem, the total number of subsets of a set with $n$ distinct elements is the number of subsets with $n$ elements, plus the number of subsets with $(n - 1)$ elements, plus the number with $(n - 2)$ elements, and so on. This is just

$$\binom{n}{n} + \binom{n}{n - 1} + \binom{n}{n - 2} + \cdots + \binom{n}{1} + \binom{n}{0}.$$

Now setting $x = y = 1$ in the binomial theorem gives

$$2^n = (1 + 1)^n$$

$$= \binom{n}{n} (1)^n + \binom{n}{n - 1} (1)^{n - 1}(1) + \binom{n}{n - 2} (1)^{n - 2}(1)^2$$

$$+ \cdots +$$

$$\binom{n}{1} (1)(1)^{n - 1} + \binom{n}{0} (1)^n$$

$$= \binom{n}{n} + \binom{n}{n - 1} + \binom{n}{n - 2} + \cdots + \binom{n}{1} + \binom{n}{0}.$$

which, as we have just seen, is the total number of subsets we are seeking.

**EXAMPLE 3**

**The Number of Options for an Executive**

A study committee reports to an executive that to solve a particular problem any number of 4 different actions can be taken, including doing nothing. How many options does the executive have?

**Solution**   The executive has the option of selecting any subset from a set with 4 elements in it. This can be done in $2^4 = 16$ ways.   ∎

## Pascal's Triangle

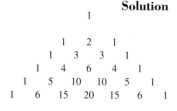

**Figure 6.22**

Figure 6.22 lists the coefficients from the expansions of $(x + y)^n$ given at the beginning of this section. This is called **Pascal's Triangle**, named after its discoverer, Blaise Pascal (1623–1662). Notice that there are always 1's at the two sides and that any coefficient inside the triangle can be obtained by adding the coefficient above and to the left with the coefficient above and to the right. This is another way the coefficients can be obtained.

## Ordered Partitions

We will now look at combinations from a slightly different perspective, which will permit us to generalize the notion of combinations to the notion of "ordered partitions."

If we have a set $S = \{a_1, a_2, \ldots, a_n\}$ with $n$ elements, that is, $n(S) = n$, and divide the set into two parts or subsets with the first part having $k$ elements and the second part having the remaining $n - k$ elements, we call this an **ordered partition of $S$ of type $(k, n - k)$**. We can think of this as a two-step process. In the first step we select a subset of $k$ elements. This can be done in $C(n, k)$ ways. In the second step we place the remaining $n - k$ elements in the second part, but this can be done in only one way. Thus the number of such ordered partitions is the same as the number of combinations of $n$ distinct elements taken $k$ at a time or

$$C(n, k) = \frac{n!}{k!(n - k)!}.$$

If we let $n_1 = k$ and $n_2 = n - k$, then this becomes

$$\frac{n!}{n_1! n_2!}$$

where

$$n_1 + n_2 = n.$$

We use the following notation

$$\binom{n}{n_1, n_2} = \frac{n!}{n_1! n_2!}.$$

Figure 6.23 gives a schematic of what we have done. We have divided the set $S$ into two parts, the first part is called $S_1$ and the second part is called $S_2$, with $n_1 = n(S_1)$, $n_2 = n(S_2)$, and $n_1 + n_2 = n = n(S)$. As we noted this can be done in

$$\binom{n}{n_1, n_2} = \frac{n!}{n_1! n_2!}$$

ways.

Thus if we have a group of 14 people and wish to divide the group into 2 committees, with the first committee consisting of 8 people to perform some task

$n_1 + n_2 = n = n(S)$

**Figure 6.23**

and the remaining 6 people into another committee to form another task, this can be done in

$$\binom{14}{8,\,6} = \frac{14!}{8!6!}$$

ways. This is referred to as an *ordered* partition since it is important which committee an individual is on.

**E X A M P L E   4**     **Counting the Ways of Dividing a Committee to Perform Tasks**

In how many ways can a group of 14 people be divided into 3 committees, each assigned a different task, the first committee consisting of 8 people, the second 4 people, and the third 2 people?

**Solution**     We can select 8 people for the first committee in $C(14, 8)$ ways and then can select 4 people for the second committee from among the remaining 6 people in $C(6, 4)$ ways. The remaining 2 people can then go into the third committee in only 1 way. By the multiplication principle, the total number of ways this can be done is

$$C(14,\,8)C(6,\,4) = \frac{14!}{8!6!}\,\frac{6!}{4!2!}$$

$$= \frac{14!}{8!4!2!}.$$

We leave the answer in this form in view of what we are about to do.   ■

We can think of the 3 committees in the previous problem as a division of the set of 14 people into three groups. We refer to this as an *ordered partition of type* (8, 4, 2). We use the term *ordered* since it is important which committee an individual will be on.

We now wish to generalize this notion. In Figure 6.24 we have a set $S$ of $n$ distinct elements and have partitioned the set into the 3 subsets shown. We call this an *ordered partition of type* $(n_1, n_2, n_3)$. Another way of looking at this is to think of the three rectangles in Figure 6.24 initially as empty boxes. Now select $n_1$ elements from $S$ and place them into $S_1$. Now select $n_2$ elements from the remaining $n - n_1$ elements in $S$ and place them into $S_2$. Now place the remaining $n_3$ elements of $S$ into $S_3$. In how many ways can this be done?

We can view this as a three-step process. First select $n_1$ elements from the set of $n$. This can be done in $C(n, n_1)$ ways. Then from among the remaining $n - n_1$ elements, select $n_2$ for $S_2$. This can be done in $C(n - n_1, n_2)$ ways. Now

$$n_1 + n_2 + n_3 = n = n(S)$$

**Figure 6.24**

for the third step place the remaining $n_3$ elements into $S_3$. This can be done in only one way. The total number of ways this can be done is

$$C(n, n_1)C(n - n_1, n_2) = \frac{n!}{n_1!(n - n_1)!} \frac{(n - n_1)!}{n_2!(n - n_1 - n_2)!}$$

$$= \frac{n!}{n_1!n_2!(n - n_1 - n_2)!}$$

$$= \frac{n!}{n_1!n_2!n_3!}.$$

We use the notation

$$\binom{n}{n_1, n_2, n_3} = \frac{n!}{n_1!n_2!n_3!}.$$

We have the more general definition.

---

**Ordered Partition of Type $(n_1, n_2, \ldots, n_k)$**

Given a set $S$ of $n$ elements, an ordered partition of $S$ of type $(n_1, n_2, \ldots, n_k)$ is a division of $S$ into $k$ subsets, $S_1, S_2, \ldots, S_k$, with order being important and where $n_1 = n(S_1)$, $n_2 = n(S_2)$, $\ldots$, $n_k = n(S_k)$. The number of such ordered partitions is denoted by

$$\binom{n}{n_1, n_2, \ldots, n_k}.$$

---

See Figure 6.25. Notice that since none of the sets $S_1, S_2, \ldots, S_k$, can overlap, we must have

$$n_1 + n_2 + \cdots + n_k = n.$$

$$n_1 + n_2 + \cdots + n_k = n(S) = n$$

**Figure 6.25**

The following is then true.

---

**Number of Ordered Partitions of Type $(n_1, n_2, \ldots, n_k)$**

$$\binom{n}{n_1, n_2, \ldots, n_k} = \frac{n!}{n_1! n_2! \cdots n_k!}$$

---

We have already established this for $k = 3$. The general proof is similar.

**E X A M P L E   5**     **The Number of Ordered Partitions of Stocks**

A brokerage firm follows 20 stocks for its clients. It divides the stocks into 4 groups: buy, hold, sell, short. Find the number of ways the firm can place 5 stocks into the buy group, 10 into the hold group, 3 into the sell group, and 2 into the short group.

**Solution**     From a set with 20 elements, we are to find the number of ordered partitions of type (5, 10, 3, 2). This is

$$\binom{20}{5, 10, 3, 2} = \frac{20!}{5! 10! 3! 2!} = 465,585,120. \quad \blacksquare$$

## SELF-HELP EXERCISES 6.5

1. Find the coefficient of $x^4 y^5$ in the expansion of $(x + y)^9$.

2. A company hires 10 junior executives. Five are to go to the Boston office, 3 to the Chicago office, and 2 to the San Diego office. In how many ways can this be done?

## EXERCISE SET 6.5

In Exercises 1 through 10 expand using the binomial theorem.

1. $(a - b)^5$
2. $(2a + b)^4$
3. $(2x + 3y)^5$

4. $(3x - 2y)^4$
5. $(1 - x)^6$
6. $(2 + x)^6$

7. $(2 - x^2)^4$
8. $(1 + 2x^2)^6$
9. $(s^2 + t^2)^6$

10. $(s^2 - 1)^5$

In Exercises 11 through 18 determine the first 3 and last 3 terms in the expansion of each of the terms.

11. $(a - b)^{10}$
12. $(a + b)^{12}$
13. $(x + y)^{11}$

**14.** $(x - y)^8$       **15.** $(1 - z)^{12}$       **16.** $(1 + x)^{10}$

**17.** $(1 - x^3)^{12}$       **18.** $(x^2 - 1)^{10}$

In Exercises 19 through 26 evaluate the given term.

**19.** $\dbinom{7}{2,\ 2,\ 3}$       **20.** $\dbinom{8}{2,\ 3,\ 3}$       **21.** $\dbinom{9}{2,\ 3,\ 4}$

**22.** $\dbinom{8}{2,\ 2,\ 2,\ 2}$       **23.** $\dbinom{10}{2,\ 2,\ 2,\ 2,\ 2}$       **24.** $\dbinom{12}{2,\ 3,\ 4,\ 3}$

**25.** $\dbinom{8}{2,\ 2,\ 2,\ 1,\ 1}$       **26.** $\dbinom{10}{2,\ 3,\ 2,\ 3}$

**27.** Determine the next row in Pascal's triangle shown in Figure 6.22.

**28.** Find the coefficient of $x^4 y^4$ in the expansion of $(x + y)^8$ using Figure 6.22.

**29.** A manufacturing company buys a certain component from 3 different vendors. In how many ways can the company order 8 components with 4 from the first vendor and 2 each from the other vendors?

**30.** A mutual fund has 20 stocks in its portfolio. On a given day 3 stocks move up, 15 stay the same, and 2 move down. In how many ways could this happen?

**31.** An advertising firm has 12 potential clients and 3 different salesmen. In how many ways can it divide the potential clients equally among the 3 salesmen?

**32.** In how many ways can a class of 10 students be assigned 1 A, 2 B's, 4 C's, 2 D's, and 1 F?

**33.** Two scholarships of $10,000 each, 3 of $5000 each, and 5 of $2000 each are to be awarded to 10 finalists. In how many ways can this be done?

**34.** The 12 directors of a company are to be divided equally into 3 separate committees to study sales, recent products, and labor relations. In how many ways can this be done?

**35.** Find the number of arrangements of each of the following words that can be distinguished. (a) $a_1 a_2 b$, (b) $aab$, (c) $a_1 a_2 b_1 b_2$, (d) $aab_1 b_2$, (e) $aabb$

**36.** Find the number of arrangements of each of the following words that can be distinguished. (a) $a_1 a_2 b_1 b_2 b_3$, (b) $aab_1 b_2 b_3$, (c) $aabbb$

**37.** Find the number of arrangements of each of the following words that can be distinguished. (a) Mississippi, (b) Tennessee

**38.** Suppose a word has $n$ symbols made from $k$ distinct elements with $n_1$ of the first element, $n_2$ of the second element, $\ldots$, $n_k$ of the $k$th element. If $n_1 + n_2 + \cdots + n_k = n$, show that the number of distinguishable arrangements of the $n$ symbol word is

$$\binom{n}{n_1,\ n_2,\ \ldots,\ n_k}.$$

Verify that this works for the previous exercise.

## Solutions to Self-Help Exercise Set 6.5

1. By the binomial theorem the coefficient of $x^4y^5$ in the expansion of $(x + y)^9$ is

$$C(9, 5) = \frac{9!}{5!(9 - 5)!} = \frac{9 \cdot 8 \cdot 7 \cdot 6 \cdot 5}{5 \cdot 4 \cdot 3 \cdot 2} = 126.$$

2. The 10 junior executives are partitioned into 3 groups of size 5, 3, and 2, where which group one is in is important. The number of ways this can be done is

$$\binom{10}{5, 3, 2} = \frac{10!}{5!3!2!} = \frac{10 \cdot 9 \cdot 8 \cdot 7 \cdot 6 \cdot 5!}{5!(3 \cdot 2)(2)} = 2520.$$

# SUMMARY OUTLINE OF CHAPTER 6

- If every element of a set $A$ is also an element of another set $B$, we say that **$A$ is a subset of $B$** and write $A \subset B$. If $A$ is not a subset of $B$, we write $A \not\subset B$.

- The **empty set**, written as $\phi$, is the set with no elements.

- Given a universal set $U$ and a set $A \subset U$, the **complement of $A$**, written $A^c$, is the set of all elements that are in $U$ but not in $A$, that is,

$$A^c = \{x | x \in U, x \notin A\}.$$

- The **union** of two sets $A$ and $B$, written $A \cup B$, is the set of all elements that belong to $A$, or to $B$, or to both. Thus

$$A \cup B = \{x | x \in A \text{ or } x \in B \text{ or both}\}.$$

- The **intersection** of two sets $A$ and $B$, written $A \cap B$, is the set of all elements that belong to both the set $A$ and to the set $B$.

- **Rules for Set Operations**

| | |
|---|---|
| $A \cup B = B \cup A$ | Commutative law for union |
| $A \cap B = B \cap A$ | Commutative law for intersection |
| $A \cup (B \cup C) = (A \cup B) \cup C$ | Associative law for union |
| $A \cap (B \cap C) = (A \cap B) \cap C$ | Associative law for intersection |
| $A \cup (B \cap C) = (A \cup B) \cap (A \cup C)$ | Distributive law for union |
| $A \cap (B \cup C) = (A \cap B) \cup (A \cap C)$ | Distributive law for intersection |
| $(A \cup B)^c = A^c \cap B^c$ | De Morgan law |
| $(A \cap B)^c = A^c \cup B^c$ | De Morgan law |

- If $A$ is a set with a finite number of elements, we denote the number of elements in $A$ by $n(A)$.

- If the sets $A$ and $B$ are disjoint, then

$$n(A \cup B) = n(A) + n(B)$$

- For any finite sets $A$ and $B$,

$$n(A \cup B) = n(A) + n(B) - n(A \cap B)$$

- **General Multiplication Principle.** Suppose there is an operation that consists of making a sequence of $k$ choices with $n_1$ possible outcomes for the first choice, and, no matter what the first choice, there are always $n_2$ possible outcomes for the second choice, and, no matter what the first two choices, there are always $n_3$ possible outcomes for the third choice, and so on. Then there are $n_1 \cdot n_2 \cdots n_k$ possible ways in which the sequence of $k$ choices can be made.

- For any natural number $n$

$$n! = n(n - 1)(n - 2) \cdots 3 \cdot 2 \cdot 1$$

$$0! = 1.$$

- A **permutation** of a set of elements is an ordered arrangement of all the elements.

- A **permutation of $r$ objects taken from a set of size $n$** is a selection of $r$ of the objects with order being important.

- The number of permutations of $n$ distinct objects is given by

$$P(n, n) = n(n - 1)(n - 2) \cdots 3 \cdot 2 \cdot 1 = n!$$

- The number of permutations of $r$ distinct objects taken from a set of size $n$ is given by

$$P(n, r) = n(n - 1)(n - 2) \cdots (n - r + 1) = \frac{n!}{(n - r)!}.$$

- A **combination** of $r$ distinct objects taken from a set of size $n$ is a selection of $r$ of the objects (without concern for order).

- The number of combinations of $r$ distinct objects taken from a set of size $n$, denoted by $C(n, r)$, is given by

$$C(n, r) = \frac{n!}{(r!)(n - r)!}.$$

- **The Binomial Theorem.**   The coefficient of $x^{n-k}y^k$ in the expansion of $(x + y)^n$ is $C(n, k)$.

- Given a set $S$ of $n$ elements, an ordered partition of $S$ of type $(n_1, n_2, \ldots, n_k)$ is a division of $S$ into $k$ subsets, $S_1, S_2, \ldots, S_k$, with order being important and where $n_1 = n(S_1)$, $n_2 = n(S_2), \ldots, n_k = n(S_k)$. The number of such ordered partitions is denoted by

$$\binom{n}{n_1, n_2, \ldots, n_k}.$$

- **Number of Ordered Partition of Type**   $(n_1, n_2, \ldots, n_k)$.

$$\binom{n}{n_1, n_2, \ldots, n_k} = \frac{n!}{n_1! n_2! \cdots n_k!}$$

## Chapter 6 Review Exercises

1. Determine which of the following are sets.
   a. Current members of the board of Chase Manhattan Bank.
   b. Past and present board members of Chase Manhattan Bank that have done an outstanding job.
   c. Current members of the board of Chase Manhattan Bank who are over 10 feet tall.

2.  Write in set-builder notation: {5, 10, 15, 20, 25, 30, 35, 40}.

3.  Write in roster notation: $\{x \mid x^3 - 2x = 0\}$.

4.  List all the subsets of $\{A, B, C\}$.

5.  On a Venn diagram indicate where the following sets are.

$$A \cap B \cap C \qquad A^c \cap B \cap C \qquad (A \cup B)^c \cap C$$

6.  Let $U = \{1, 2, 3, 4, 5, 6\}$, $A = \{1, 2, 3\}$, $B = \{2, 3, 4\}$, $C = \{4, 5\}$. Find the following sets.

$$A \cup B \qquad\qquad A \cap B \qquad\qquad B^c$$

$$A \cap B \cap C \qquad (A \cup B) \cap C \qquad A \cap B^c \cap C$$

7.  Let $U$ be the set of all your current instructors and let

$$H = \{x \in U \mid x \text{ is at least 6 feet tall}\}$$

$$M = \{x \in U \mid x \text{ is a male}\}$$

$$W = \{x \in U \mid x \text{ weighs more than 180 pounds}\}.$$

Describe each of the following sets in words.

(a) $H^c$           (b) $H \cup M$           (c) $M^c \cap W^c$

(d) $H \cap M \cap W$      (e) $H^c \cap M \cap W$      (f) $(H \cap M^c) \cup W$

8.  Using the set $H$, $M$, and $W$ in the previous exercise and set operations, write the set that represents the following statements.
    a.  My current female instructors
    b.  My current female instructors who weigh at most 180 pounds
    c.  My current male instructors who are at least 6 feet tall or else weigh more than 180 pounds

9.  For the sets given in Exercise 6, verify that

$$A \cup (B \cap C) = (A \cup B) \cap (A \cup C).$$

10. Use a Venn diagram to show that $(A \cap B)^c = A^c \cup B^c$.

11. If $n(A) = 100$, $n(B) = 40$, $n(A \cap B) = 20$, find $n(A \cup B)$.

12. If $n(A) = 40$ and $n(A \cap B^c) = 30$, find $n(A \cap B)$.

13. In a consumer advertising survey of 100 men, it was found that 20 watched the first game of the last World Series, 15 watched the first game of the last World Series and also watched the Super Bowl, while 30 did not watch either. How many watched the last Super Bowl but not the first game of the last World Series?

14. A consumer survey of 100 women with young daughters found that

57 had a Barbie doll
68 had a teddy bear
11 had a toy piano

45 had a Barbie doll and a teddy bear
8 had a teddy bear and a toy piano
7 had a Barbie doll and a toy piano
5 had all three.

    **a.**  How many had a Barbie doll and a teddy bear but not a toy piano?
    **b.**  How many had exactly 2 of these toys?
    **c.**  How many had none of these toys?

**15.**  Find $P(10, 4)$ and $C(10, 4)$.

**16.**  If instead of a social security *number*, we had a social security *word* where letters could be repeated, what would be the length of the words needed to have at least 250 million different words?

**17.**  Before 5 labor leaders and 6 management personnel begin negotiating a new labor contract, they decided to first take a picture with the 5 labor leaders together on the left. In how many different ways can such a picture be taken?

**18.**  In the previous exercise suppose a picture is to be taken of 3 of the labor leaders and 3 of the management personnel with the labor leaders grouped together on the right. In how many ways can this be done?

**19.**  Refer to Exercise 17. Suppose a committee of 3 labor leaders and 4 management personnel is formed to study the issue of pensions. In how many ways can this be done?

**20.**  Refer to Exercise 17. In a straw vote on a proposal from the labor leaders, the 6 management personnel each cast a vote with no abstentions. In how many ways can these 6 individuals come to a majority decision?

**21.**  An investor decides that her investment year is a success if her portfolio of stocks beats the S&P 500. In how many ways can her next 10 years have exactly 7 successes?

**22.**  Expand $(2 - x)^5$ using the binomial theorem.

**23.**  Find the last 3 terms in the expansion of $(1 - x)^9$.

**24.**  Find $\begin{pmatrix} 10 \\ 4, 4, 2 \end{pmatrix}$

**25.**  Find $\begin{pmatrix} 9 \\ 3, 2, 2, 2 \end{pmatrix}$

**26.**  In how many ways can a laboratory divide 12 scientists into 4 groups of equal size in order to perform 4 different experiments?

Probability problems in business are very similar to probability problems in gambling. (Michael Engler/The Image Bank.)

# Probability

7.1   The Basis of Probability

7.2   Empirical and Intuitive Probability

7.3   Mathematical Probability

7.4   Spaces with Equally Likely Outcomes

7.5   Conditional Probability

7.6   Bayes' Theorem

7.7   Bernoulli Trials

Solution of Game Show Problem

**P**robability is now used widely throughout business and the social and physical sciences. Probability deals with events that are random but yet display a certain statistical regularity. Thus, although it is not possible to predict what will happen next, it is possible to predict what will happen in the long run.

For example, we know that the outcomes of tossing a die or flipping a coin cannot be predicted, and thus the outcomes do not occur with a deterministic regularity. These are examples of *random phenomena*. After many repetitions or trials of either of these experiments, we know that we would observe a regularity of sorts. Namely, any one number on the die would show about as often as any other number, and heads would come up about as often as tails. We now wish to study this type of regularity in random phenomena.

There is a funny thing about probability. Everyone has an intuitive idea of what it is, but when it comes to actually determining probabilities, mistakes are very common. For a tricky subject such as probability we use a formal approach, carefully setting out the "spaces" on which we operate in the first section. This will help a great deal to keep the mistakes to a minimum. The second section continues with an intuitive definition of probability widely used in science, business, and everyday life. The third section presents the mathematical definition of probability, which is then shown to be consistent with the intuitive definition. Many useful properties of probability are then established using the mathematical definition of probability. The chapter continues with a number of important notions and applications in probability.

## 7.1   THE BASIS OF PROBABILITY

▶ *Preliminaries*
▶ *Events*

APPLICATION
**Popular Television Game
Show Problem**

A contestant is shown 3 closed doors and is correctly informed that behind one of the doors is a car and behind each of the other 2 doors is a goat. The contestant is then asked to select a door. After the contestant picks a door, the game show host, who knows where the goats are, then chooses one of the 2 doors not selected by the contestant that has a goat behind it and opens this door for the contestant to see that there is a goat behind this door. Now the host gives the contestant the opportunity to change the selection of the door. If the car is behind the door selected, the car is won. Should the contestant switch doors or not? Or does it not matter?

The correct solution to this problem was given in the "Ask Marilyn" column of *Parade* magazine. The solution caused enormous controversy with over 10,000 letters written to *Parade*, with the vast majority disagreeing with the solution given. Many who disagreed with the correct solution were well educated and well trained individuals. A discussion of the controversy appeared in *The New York Times*, July 21, 1991. At the end of the chapter is given a completely elementary solution.

### Preliminaries

We first take the time to define the basic notions that underlie probability theory. The reward will be a better understanding of the subject and the ability to obtain correct answers to probability problems. In the next section, the intuitive notion of probability is discussed.

We begin the preliminaries by stating the following definitions.

---

**Experiments and Outcomes**

An **experiment** is an activity that has observable results.
The results of the experiment are called **outcomes**.

---

The following are some examples of experiments. Flip a coin and observe whether it falls "heads" or "tails." Throw a die (a small cube marked on each face with from one to six dots) and observe the number of dots on the top face. Select a transistor from a bin and observe whether or not it is defective.

The following are some additional terms that are needed.

---

**Sample Spaces and Trials**

A **sample space** of an experiment is the set of all possible outcomes of the experiment. Each repetition of an experiment is called a **trial**.

---

For the experiment of throwing a die and observing the number of dots on the top face the sample space is the set

$$S = \{1, 2, 3, 4, 5, 6\}.$$

In the experiment of flipping a coin and observing whether it falls heads or tails, the sample space is $S = \{\text{heads, tails}\}$.

EXAMPLE 1    **Determining a Sample Space**

---

An experiment consists of noting whether the price of the stock of the Ford Corporation rose, fell, or remained unchanged on the most recent day of trading. What is the sample space for this experiment?

Solution    There are three possible outcomes depending on whether the price rose, fell, or remained unchanged. Thus the sample space $S$ is

$$S = \{\text{rose, fell, unchanged}\}. \quad \blacksquare$$

EXAMPLE 2    **Determining a Sample Space**

---

Two dice, identical except that one is white and the other is red, are tossed and the number of dots on the top face of each is observed. What is the sample space for this experiment?

Solution    The outcomes can be considered order pairs. For example (2, 3) will mean 2 dots on the top face of the white die and 3 dots on the top face of the red die. The sample space $S$ is

*(Dennis Drenner.)*

$$S = \{(1, 1), (1, 2), (1, 3), (1, 4), (1, 5), (1, 6),$$
$$(2, 1), (2, 2), (2, 3), (2, 4), (2, 5), (2, 6),$$
$$(3, 1), (3, 2), (3, 3), (3, 4), (3, 5), (3, 6),$$
$$(4, 1), (4, 2), (4, 3), (4, 4), (4, 5), (4, 6),$$
$$(5, 1), (5, 2), (5, 3), (5, 4), (5, 5), (5, 6),$$
$$(6, 1), (6, 2), (6, 3), (6, 4), (6, 5), (6, 6)\}. \blacksquare$$

If the experiment of tossing 2 dice consists of just observing the total number of dots on the top faces of the two dice, then the sample space would be

$$S = \{2, 3, 4, 5, 6, 7, 8, 9, 10, 11, 12\}.$$

In short, the sample space depends on the precise statement of the experiment.

**E X A M P L E   3**      **Determining the Sample Space**

A coin is flipped twice to observe whether heads or tails shows; order is important. What is the sample space for this experiment?

**Solution**      The sample space $S$ consists of the 4 outcomes

$$S = \{(H, H), (H, T), (T, H), (T, T)\}. \blacksquare$$

## Events

> **Events and Elementary Events**
>
> Given a sample space $S$ for an experiment, an **event** is any subset $E$ of $S$. An **elementary event** is an event with a single outcome.

**E X A M P L E   4**      **Determining an Event and an Elementary Event**

In Example 3 find the event: "At least one head comes up." Find the elementary event: "Two tails come up."

**Solutions**      (a)   "At least one head comes up."  $= \{(H, H), (H, T), (T, H)\}$
(b)   "Two tails come up."        $= \{(T, T)\} \blacksquare$

We can use union, intersection, and complement to describe events.

$E \cup F$

$E \cap F$

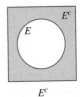

$E^c$

Figure 7.1

---

**Union of Two Events**

If $E$ and $F$ are two events, then $E \cup F$ is the union of the two events and consists of the set of outcomes that are in $E$ or $F$ or both.

---

Thus the event $E \cup F$ is the event that "E or F occurs." Refer to Figure 7.1.

---

**Intersection of Two Events**

If $E$ and $F$ are two events, then $E \cap F$ is the intersection of the two events and consists of the set of outcomes that are in both $E$ and $F$.

---

Thus the event $E \cap F$ is the event that "$E$ and $F$ both occur."

---

**Complement of an Event**

If $E$ is an event, then $E^c$ is the complement of $E$ and consists of the set of outcomes that are not in $E$.

---

Thus the event $E^c$ is the event that "$E$ does not occur."

**E X A M P L E   5**   **Determining Union, Intersection, and Complement**

Consider the sample space given in Example 2. Let $E$ consist of those outcomes for which the number of dots on the top faces of both dice is an even number. Let $F$ be the event that the sum of the number of dots on the top faces of the two dice is 6. Let $G$ be the event that the sum of the number of dots on the top faces of the two dice is less than 11. (a) List the elements of $E$ and $F$. (b) Find $E \cup F$. (c) Find $E \cap F$. (d) Find $G^c$.

**Solutions**

(a)  $E = \{(2, 2), (2, 4), (2, 6), (4, 2), (4, 4), (4, 6), (6, 2), (6, 4), (6, 6)\}$
$F = \{(1, 5), (2, 4), (3, 3), (4, 2), (5, 1)\}$

(b)  $E \cup F = \{(2, 2), (2, 4), (2, 6), (4, 2), (4, 4), (4, 6), (6, 2), (6, 4), (6, 6),$
$(1, 5), (3, 3), (5, 1)\}$

(c)   $E \cap F = \{(2, 4), (4, 2)\}$.

(d)   $G^c = \{(5, 6), (6, 5), (6, 6)\}$   ■

If $S$ is a sample space, $\phi \subset S$, and thus $\phi$ is an event. We call the event $\phi$ the **impossible event** since the event $\phi$ means that no outcome has occurred, whereas, in any experiment some outcome *must* occur.

---

**The Impossible Event**

The empty set, $\phi$, is called the **impossible event**.

---

For example, if $H$ is the event that a head shows on flipping a coin and $T$ is the event that a tail shows, then $H \cap T = \phi$. The event $H \cap T$ means that both heads and tails show, which is impossible.

Since $S \subset S$, $S$ is itself an event. We call $S$ the **certainty event** since any outcome of the experiment must be in $S$.

---

**The Certainty Event**

Let $S$ be a sample space. The event $S$ is called the **certainty event**.

---

We also have the following definition.

**Figure 7.2**

---

**Mutually Exclusive Events**

Two events $E$ and $F$ are said to be **mutually exclusive** or **disjoint** if

$$E \cap F = \phi.$$

---

See Figure 7.2.

**E X A M P L E   6**   **Determining if Sets are Mutually Exclusive**

Let a card be chosen from a standard deck of 52 cards. Let $E$ be the event consisting of drawing a 3. Let $F$ be the event of drawing a heart. Let $G$ be the event of drawing a jack. Are $E$ and $F$ mutually exclusive? Are $E$ and $G$?

**Solution**   Since
$$E \cap F = \{3\heartsuit\} \neq \phi$$

and
$$E \cap G = \phi,$$

$E$ and $F$ are not mutually exclusive while $E$ and $G$ are.   ■

## SELF-HELP EXERCISE SET 7.1

Tetrahedron

1. Two tetrahedrons (4 sided), each with equal sides numbered from 1 to 4, are identical except that one is red and the other white. If the two tetrahedrons are tossed and the number on the bottom face of each is observed, what is the sample space for this experiment?

2. Consider the sample space given in the previous exercise. Let $E$ consist of those outcomes for which both (tetrahedron) dice show an odd number. Let $F$ be the event that the sum of the two numbers on these dice is 5. Let $G$ be the event that the sum of the two numbers is less than 7. (a) List the elements of $E$ and $F$. (b) Find $E \cap F$. (c) Find $E \cup F$. (d) Find $G^c$.

## EXERCISE SET 7.1

1. Let $S = \{a, b, c\}$ be a sample space. Find all the events.

2. Let the sample space be $S = \{a, b, c, d\}$. How many events are there?

3. A coin is flipped 3 times, and heads or tails is observed after each flip. What is the sample space? Indicate the outcomes in the event "at least 2 heads are observed."

4. A coin is flipped, and it is noted whether heads or tails show. A die is tossed, and the number on the top face is noted. What is the sample space of this experiment?

5. A coin is flipped three times. If heads show, one is written down. If tails show, zero is written down. What is the sample space for this experiment? Indicate the outcomes if "one is observed at least twice."

6. Two tetrahedrons (4 sided), each with equal sides numbered from 1 to 4, are identical except that one is red and the other white. If the two tetrahedrons are tossed and the number on the bottom face of each is observed, indicate the outcomes in the event "the sum of the numbers is 4."

7. An urn holds 10 identical balls except that 1 is white, 4 are black, and 5 are red. An experiment consists of selecting a ball from the urn and observing its color. What is the sample space for this experiment? Indicate the outcomes in the event "the ball is not white."

8. For the urn in Exercise 7, an experiment consists of selecting 2 balls in succession without replacement and observing the color of each of the balls. What is the sample space of this experiment? Indicate the outcomes of the event "no ball is white."

9. An executive must select for promotion 3 people from a group of 5, $\{A, B, C, D, F\}$. What is the sample space? Indicate the outcomes of the event "A is selected."

10. An experiment consists of selecting a letter from the word CONNECTICUT

and observing it. What is the sample space? Indicate the outcomes of the event "a vowel is selected."

**11.** An inspector selects 10 transistors from the production line and notes how many are defective.

   **a.** Determine the sample space.

   **b.** Find the outcomes in the set corresponding to the event $E$ "at least 6 are defective."

   **c.** Find the outcomes in the set corresponding to the event $F$ "at most 4 are defective."

   **d.** Find and describe the sets $E \cup F, E \cap F, E^c, E \cap F^c, E^c \cap F^c$.

   **e.** Find all pairs of sets among the nonempty ones listed in (d) that are mutually exclusive.

**12.** A survey indicates first whether a person is in the lower income group (L), middle income group (M), or upper income group (U), and second which of these groups the father of the person is in.

   **a.** Determine the sample space using the letters L, M, and U.

   **b.** Find the outcomes in the set corresponding to the event $E$ "the person is in the lower income group."

   **c.** Find the outcomes in the set corresponding to the event $F$ "the person's father is in the higher income group."

   **d.** Find and describe the sets $E \cup F, E \cap F, E^c, E \cap F^c, E^c \cap F^c$.

   **e.** Find all pairs of sets listed in (d) that are mutually exclusive.

**13.** A corporate president decides that for each of the next 3 fiscal years success (S) will be declared if the earnings per share of the company go up at least 10% that year and failure (F) will occur if less than 10%.

   **a.** Determine the sample space using the letters S and F.

   **b.** Find the outcomes in the set corresponding to the event $E$ "at least 2 of the next 3 years is a success."

   **c.** Find the outcomes in the set corresponding to the event $G$ "the first year is a success."

   **d.** Find and describe the sets $E \cup G, E \cap G, G^c, E^c \cap G, (E \cup G)^c$.

   **e.** Find all pairs of sets listed in (d) that are mutually exclusive.

**14.** Let $E$ be the event that the life of a certain lightbulb is at least 100 hours and $F$ that the life is at most 200 hours. Describe the sets $E \cap F, F^c, E^c \cap F, (E \cup F)^c$.

In Exercises 15 through 20 $S$ is a sample space and $E$, $F$, and $G$ are 3 events. Use the symbols $\cup, \cap, {}^c$ to describe the given events.

**15.** $E$ but not $F$                    **16.** $F$ but not $E$

**17.** Not $F$ and not $E$           **18.** Not $F$ or not $E$

**19.** $E$ and $F$ but not $G$          **20.** Not $F$, nor $E$, nor $G$

**21.** Let $S$ be the 26 letters of the alphabet, $E$ be the 5 vowels $\{a, e, i, o, u\}$, $F$ the remaining 21 letters, and $G$ the first five letters of the alphabet. Find the events $E \cup F \cup G, E^c \cap F^c \cap G^c, E \cap F \cap G, E \cup F^c \cup G$.

**22.** Let $S$ be a sample space consisting of all the integers from 1 to 20 inclusive, $E$ the first 10 of these, and $F$ the last five of these. Find $E \cap F, E^c \cap F, (E \cup F)^c, E^c \cap F^c$.

## Solutions to Self-Help Exercise Set 7.1

1. Consider the outcomes as ordered pairs, with the number on the bottom of the red one the first number and the number on the bottom of the white one the second number. The sample space is

$$S = \{(1, 1), (1, 2), (1, 3), (1, 4),$$

$$(2, 1), (2, 2), (2, 3), (2, 4),$$

$$(3, 1), (3, 2), (3, 3), (3, 4),$$

$$(4, 1), (4, 2), (4, 3), (4, 4)\}.$$

2. **a.** $E = \{(1, 1), (1, 3), (3, 1), (3, 3)\}$
   $F = \{(1, 4), (2, 3), (3, 2), (4, 1)\}$
   **b.** $E \cap F = \phi$
   **c.** $E \cup F = \{(1, 1), (1, 3), (3, 1), (3, 3), (1, 4), (2, 3), (3, 2), (4, 1)\}$
   **d.** $G^c = \{(3, 4), (4, 3), (4, 4)\}$

## 7.2   EMPIRICAL AND INTUITIVE PROBABILITY

► *Empirical Probability*

► *Intuitive Probability*

► *Basic Properties of Probability*

**Blaise Pascal**
**1623–1662**

*(David H. Smith*
*Collection/Columbia*
*University.)*

**The Beginnings of Mathematical Probability**

In 1654 the famous mathematician Blaise Pascal had a friend, Chevalier de Méré, a member of the French nobility and a gambler, who wanted to adjust gambling stakes so that he would be assured of winning if he played long enough. This gambler raised questions with Pascal such as the following: In eight throws of a die a player attempts to throw a one, but after three unsuccessful trials the game is interrupted. How should he be indemnified? Pascal wrote to the leading mathematician of that day, Pierre de Fermat (1601–1665), about these problems, and their resulting correspondence represents the beginnings of the modern theory of mathematical probability.

### Empirical Probability

A very important type of problem that arises every day in business and science is to find a *practical* way to estimate the likelihood of certain events. For example, a food company may seek a practical method of estimating the likelihood that a new type of candy will be enjoyed by consumers. The most obvious procedure for the company to follow is to select a consumer, have the consumer taste the candy, and then record the result. This should be repeated many times and the final totals tabulated to give the fraction of tested consumers who enjoy the candy. This fraction is then a practical estimate of the likelihood that all consumers will enjoy this candy.

In mathematical terms, let $L$ be the event that the consumer enjoys the candy and $N$ the number of consumers that have been tested. If $f(L)$ is the number or frequency of times a consumer in this group enjoys the candy, then the fraction of consumers who enjoy the candy is given by $\dfrac{f(L)}{N}$. We call this fraction the **relative frequency**.

For a general event $E$ in a sample space $S$ we have the following.

---

**Relative Frequency**

If an experiment is repeated $N$ times and an event $E$ occurs $f(E)$ times, then the fraction

$$\frac{f(E)}{N}$$

is called the **relative frequency** of the event.

---

The food company will take $N$ as large as time and money will permit. The resulting relative frequency $\dfrac{f(L)}{N}$ is then a *practical* estimate of the likelihood that a randomly chosen consumer will enjoy this candy. We refer to such a relative frequency as the **empirical probability**.

In general we have the following.

---

**Empirical Probability**

Let $S$ be a sample space of an experiment and $E$ an event in $S$. If the experiment has actually been performed $N$ times, the relative frequency $\dfrac{f(E)}{N}$ is called the **empirical probability** of $E$ and is denoted by $p(E)$.

---

**The Beginnings of Empirical Probability**

Empirical probability began with the emergence of insurance companies. Insurance seems to have been originally used to protect merchant vessels and was in use even in Roman times. The first marine insurance companies began in Italy and Holland in the 14th century and spread to other countries by the 16th century. In fact, the famous Lloyd's of London was founded in the late 1600s. The first life insurance seems to have been written in the late 16th century in Europe. All of these insurance companies naturally needed to

> know with what relative frequencies certain events would occur. These relative frequencies were determined by collecting data over long periods of time.

The London merchant John Graunt (1620–1674) with the publication of *Natural and Political Observations Made upon the Bills of Mortality* in 1662 seems to have been the first person to have gathered data on mortality rates and determined relative frequencies from them. The data were extremely difficult to obtain. His then-famous London Life Table is reproduced below, showing the number of survivors through certain ages per 100 people.

**London Life Table**

| Age | 0 | 6 | 16 | 26 | 36 | 46 | 56 | 66 | 76 |
|-----|-----|-----|-----|-----|-----|-----|-----|-----|-----|
| Survivors | 100 | 64 | 40 | 25 | 16 | 10 | 6 | 3 | 1 |

**E X A M P L E   1**    **Finding Empirical Probability**

Using the London Life Table, find the empirical probability of a randomly chosen person living in London in the first half of the 17th century surviving until age 46.

**Solution**    In the London Life Table $N = 100$. If $E$ is the event "survive to age 46," then according to the table $f(E) = 10$. Thus the empirical probability of people living in London at that time surviving until age 46 was

$$p(E) = \frac{f(E)}{N} = \frac{10}{100} = 0.1. \quad \blacksquare$$

**E X A M P L E   2**    **Finding Empirical Probability**

In a survey of 1000 people selected at random throughout the country, 110 said that they planned to purchase an automobile, 30 said they planned to buy a house, and 200 planned to buy a major appliance during the next year. What is the empirical probability that a person during the next year in this country will (a) buy an automobile, (b) buy a house, or (c) buy a major appliance?

**Solutions**    The empirical probabilities are calculated from the given data using the relative frequencies. Thus the empirical probabilities are the following respective relative frequencies.

(a) $\dfrac{110}{1000} = .11$      (b) $\dfrac{30}{1000} = .03$      (c) $\dfrac{200}{1000} = .2$   $\blacksquare$

**E X A M P L E  3**  **Finding Empirical Probability**

An inspector finds 4 defective parts from among 100 removed from an assembly line. What is the empirical probability that a part on the assembly line is defective?

**Solution**  The relative frequency and thus the empirical probability is $4/100 = 0.04$. Thus based on this data alone, we expect about 4% of the parts to be defective. ∎

Consider now a poorly made die purchased at a discount store. Dice are made by drilling holes in the sides and then backfilling. Cheap dice are, of course, not carefully backfilled. So when a lot of holes are made in a face, such as for the side with 6, and they are not carefully backfilled, that side will not be quite as heavy as the others. Thus a 6 will tend to come up more often on the top. Even a die taken from a craps table in Las Vegas, where the dice are of very high quality, will have some tiny imbalance.

**E X A M P L E  4**  **Finding Empirical Probability**

A die with 6 sides numbered from 1 to 6, such as used in the game of craps, is suspected to be somewhat lopsided. A laboratory has tossed this die 1000 times and obtained the results shown in the table. Find the empirical probabilities of the elementary outcomes and also find the empirical probability that an even number will occur.

| Outcome | 1 | 2 | 3 | 4 | 5 | 6 |
|---|---|---|---|---|---|---|
| Number Observed | 161 | 179 | 148 | 177 | 210 | 125 |

**Solutions**  Dividing each of the numbers by 1000 gives the relative frequencies and thus the empirical probabilities. Thus

$$p(1) = .161, \qquad p(2) = .179, \qquad p(3) = .148,$$
$$p(4) = .177, \qquad p(5) = .210, \qquad p(6) = .125.$$

For convenience we write $p(1)$ instead of the more precise $p(\{1\})$.

To find the probability of an even number, denoted by $p(\{2, 4, 6\})$, we count the number of times a 2, 4, or 6 has occurred, which we denote by $f(\{2, 4, 6\})$ and divide by the number of trials $N = 1000$. Since the number of times an even number occurs is just the sum of the number of occurrences of each of the even numbers, or

$$f(\{2, 4, 6\}) = f(2) + f(4) + f(6),$$

we have

$$p(\{2, 4, 6\}) = \frac{f(\{2, 4, 6\})}{N}$$

$$= \frac{f(2) + f(4) + f(6)}{N}$$

$$= \frac{f(2)}{N} + \frac{f(4)}{N} + \frac{f(6)}{N}$$

$$= p(2) + p(4) + p(6)$$

$$= .179 + .177 + .125 = .481. \quad \blacksquare$$

It is very important to note that the empirical probability of an event $E = \{e_1, e_2, \ldots, e_r\}$ is simply obtained as the sum of the empirical probabilities of all of the individual outcomes in the event. This follows simply because, in repeated trials of an experiment, the total number of times an event $E$ occurs just equals the sum of the number of occurrences of each of the outcomes that comprise the event, that is,

$$f(\{e_1, e_2, \ldots, e_r\}) = f(e_1) + f(e_2) + \cdots + f(e_r).$$

This implies the following for relative frequencies:

$$\frac{f(\{e_1, e_2, \ldots, e_r\})}{N} = \frac{f(e_1)}{N} + \frac{f(e_2)}{N} + \cdots + \frac{f(e_r)}{N}.$$

Since the empirical probability is defined by

$$p(E) = \frac{f(E)}{N},$$

this last equation can also be written as

$$p(\{e_1, e_2, \ldots, e_r\}) = p(e_1) + p(e_2) + \cdots + p(e_r).$$

We thus have the following.

---

**Finding the Empirical Probability of an Event**

Suppose an experiment has been repeated $N$ times. Let $E = \{e_1, e_2, \ldots, e_r\}$ be any event, and $p(E)$ the empirical probability. Then

$$p(\{e_1, e_2, \ldots, e_r\}) = p(e_1) + p(e_2) + \cdots + p(e_r). \tag{1}$$

---

**Frederick Mosteller and the Dice Experiment**

Frederick Mosteller has been president of the American Association for the Advancement of Science, the Institute of Mathematical Statistics, and the American Statistical Association. He once decided that "It would be nice to see if the actual outcome of a real person tossing real dice would match up with the theory." He then engaged Willard H. Longcor to buy some dice, toss them, and keep careful records of the outcomes. Mr. Longcor then tossed the dice on his floor at home so that the dice would bounce on the floor and then up against the wall and then land back on the floor. After doing this several thousand times, his wife became troubled by the noise. He then placed a rug on the floor and on the wall, and then proceeded to quietly toss his dice *millions* of times, keeping careful records of the outcomes. In fact, he was so careful and responsible about his task, that he threw away his data on the first 100,000 tosses, since he had a nagging worry that he might have made some mistake keeping perfect track.

**E X A M P L E   5**          **Finding Empirical Probabilities**

In the previous example, find the empirical probability that the number on the die is at least 4.

**Solution**          We have

$$p(\{4, 5, 6\}) = p(4) + p(5) + p(6) = .177 + .210 + .125 = .512.$$  ■

## Intuitive Probability

We have just given a practical way to assign a number to an event that indicates the approximate likelihood that the event will occur. We now see how to assign a number to an event that gives the *exact* likelihood that the event will occur. This number is assigned on the basis of the percentage of times the event $E$ will occur over the long-term when the experiment is repeated over and over again. More specifically, the number is assigned on the basis of the long-term behavior of the relative frequency $\frac{f(E)}{N}$.

Consider a *theoretical* example. We flip a fair coin, that is, a coin that shows no preference for heads or tails. We naturally expect heads to show as often as tails. Thus we expect heads to show one-half the time, and if $H$ is the event that a head shows, we expect that as the number $N$ of times the coin is flipped becomes larger and larger, the relative frequencies $\frac{f(H)}{N}$ should become closer and closer to exactly .50. The number .50 is called the **intuitive probability** or just the **probability** of $H$, and is denoted by $p(H)$.

To obtain a specific example, the author flipped a fair coin 100,000 times. More precisely, the author used a random number generator on a computer to simulate flipping a fair coin.

Figure 7.3 shows in graphical form what happened to the relative frequencies. Figure 7.3(a) shows the relative frequencies for the first 10,000 "tosses" in increments of 100. Figure 7.3(b) shows the relative frequencies in increments of 5000 tosses. First notice that since we are dealing with a *random* phenomenon, we should not be surprised to see the relative frequencies bounce *randomly*. Very importantly, however, we notice that as $N$ becomes larger and larger, the relative frequency, although bouncing around, seems to be approaching the number $p(H)$ = .50, the probability of $H$. This is typical of random phenomena in general.

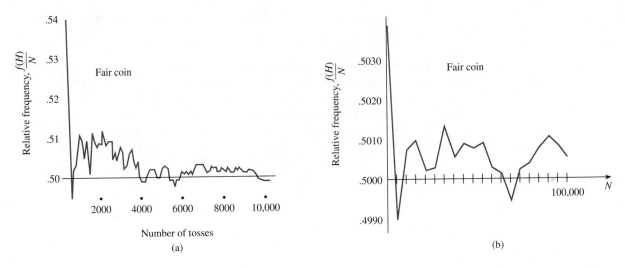

**Figure 7.3**

For a fair coin we know, on the basis of a theoretical analysis, that as $N$ becomes larger, the relative frequencies $\dfrac{f(H)}{N}$ must approach the number $p(H) = .50$. But if a coin were not known ahead of time to be fair, then how could we determine $p(H)$? Unfortunately, we cannot determine $p(H)$ *exactly*. However, if we flipped the coin a very large number $N$ of times, the empirical probability $\dfrac{f(H)}{N}$ should be approximately equal to $p(H)$. Thus in the situation in which we have no theoretical analysis that can give us the exact probability, we must repeat the experiment a large number of times and approximate the actual probability with the empirical probability.

For a general event $E$ in a sample space $S$, we have the following (intuitive) definition.

---

**The Probability of an Event E**

Let $S$ be the sample space of an experiment, let $E$ be an event in $S$, and let $N$ be the number of times the experiment has been repeated. The **probability of E**, denoted by $p(E)$, is the number that the relative frequency $\dfrac{f(E)}{N}$ approaches as $N$ becomes larger and larger.

---

**REMARK.** We also refer to probability defined in this way as **intuitive probability**.

In Section 8.5 we shall delve deeper into the long-term behavior of the relative frequency. In that section we shall see that the relative frequency $\dfrac{f(E)}{N}$ can be made as close to $p(E)$ as we like with any degree of certainty (short of absolute certainty) by taking $N$ sufficiently large, and we shall find formulas for $N$ toward this end.

Suppose we want to flip a fair coin enough times to be reasonably certain that the relative frequency is within .01 of $\frac{1}{2}$. In Section 8.5 we show that if a fair coin is flipped $N = 10{,}000$ times, then 95 times out of 100 the relative frequency of heads will be within 0.01 of $\frac{1}{2}$. We also show in the same section that if an experiment consists of flipping a fair coin $N = 1{,}000{,}000$ times, then 95 times out of 100 the relative frequency of heads will be within 0.001 of $\frac{1}{2}$.

If we wish even greater accuracy with even more certainty, then there will exist a number $N$ that will meet our increased demands. It is in this sense that we expect the relative frequency to approach the exact probability as $N$ becomes large.

## Some Basic Properties of Probability

We now consider some basic properties of probability. The definition immediately implies several things about the numbers $p(E)$. First, if $E$ is an event in a sample space $S$ and the experiment is repeated $N$ times, then the number of occurrences $f(E)$ of the event $E$ can be at most $N$ and, of course, at least 0. That is, $0 \le f(E) \le N$. This implies that

$$0 \le \frac{f(E)}{N} \le 1.$$

Since for large $N$, $\dfrac{f(E)}{N} \approx p(E)$, we then must have $0 \le p(E) \le 1$. Second, since some outcome must occur on every one of the $N$ trials, for every $N$ we have $f(S) = N$, and $\dfrac{f(S)}{N} = \dfrac{N}{N} = 1$. Thus we must have $p(S) = 1$. Finally, since some outcome must occur on every trial, no outcome is impossible and never happens. Therefore for every $N$ we have $f(\phi) = 0$, and $\dfrac{f(\phi)}{N} = \dfrac{0}{N} = 0$. Thus we must have $p(\phi) = 0$.

---

**Properties of Probability**

Let $E$ be an event in a sample space $S$. Then

**Property 1**  $\qquad\qquad\qquad\qquad 0 \le p(E) \le 1.$

**Property 2**  $\qquad\qquad\qquad\qquad p(S) = 1.$

**Property 3**  $\qquad\qquad\qquad\qquad p(\phi) = 0.$

---

We are now prepared to consider another important property of probability. We will now see that equation (1) is not only satisfied by empirical probability, but also by probability in general. We then have the following.

---

***Property 4*** Given any event $E = \{e_1, e_2, \ldots, e_r\}$, where $e_1, e_2, \ldots, e_r$, are all distinct, then

$$p(\{e_1, e_2, \ldots, e_r\}) = p(e_1) + p(e_2) + \cdots + p(e_r).$$

---

We have already indicated that Property 4 holds for empirical probability. To see why the fourth property is true in general, suppose the experiment is repeated $N$ times. Then in terms of relative frequencies

$$\frac{f(\{e_1, e_2, \ldots, e_r\})}{N} = \frac{f(e_1)}{N} + \frac{f(e_2)}{N} + \cdots + \frac{f(e_r)}{N}.$$

Now, for very large $N$, the left-hand side of this equation should be very close to $p(\{e_1, e_2, \ldots, e_r\})$ while the right-hand side should be very close to $p(e_1) + p(e_2) + \cdots + p(e_r)$. Thus $p(\{e_1, e_2, \ldots, e_r\})$ should equal $p(e_1) + p(e_2) + \cdots + p(e_r)$. This is then Property 4.

## DISCUSSION QUESTIONS

Answers to the following discussion questions can be found after the exercise set.

1. You buy a new die and toss it 1000 times. A 1 comes up 165 times. Is it true that the probability of a 1 showing on this die is .165?

2. A fair coin is to be tossed 100 times. Naturally you expect tails to come up 50 times. After 60 tosses, heads has come up 40 times. Is it true that now heads is likely to come up less often than tails during the next 40 tosses?

3. You are playing a game at a casino and have correctly calculated the probability of winning any one game to be 0.48. You have played for some time and have won 60% of the time. You are on a roll. Should you keep playing?

4. You are watching a roulette game at a casino and notice that red has been coming up unusually often. (Red should come up as often as black.) Is it true that, according to "the law of averages," black is likely to come up unusually often in the next number of games to "even things up"?

5. You buy a die and toss it 1000 times and notice that a 1 came up 165 times. You decide that the probability of a 1 on this die is .165. Your friend takes this die and tosses it 1000 times and notes that a 1 came up 170 times. He concludes that the probability of a one is .17. Who is correct?

6. People who frequent casinos and play lotteries are gamblers, but those who run the casinos and lotteries are not. Do you agree? Why or why not?

## SELF-HELP EXERCISE SET 7.2

A retail store that sells sweaters notes the following number of sweaters of each size that were sold last year.

| small | medium | large | extra large |
|-------|--------|-------|-------------|
| 10 | 40 | 30 | 20 |

1. What is the empirical probability that a customer that buys a sweater will buy one of size small? medium? large? extra large?

2. Of size medium or large?

3. Or size larger than small?

## EXERCISE SET 7.2

1. Suppose the probability of an event $E$ is $p(E) = .2317$. What is the expected frequency of occurrence $f(E)$ of the event $E$ if (a) $N = 100$, (b) $N = 1000$, (c) $N = 10,000$?

2. Suppose the probability of an event $E$ is $p(E) = \frac{1}{9}$. What is the expected frequency of occurrence $f(E)$ of the event $E$ if (a) $N = 100$, (b) $N = 900$, (c) $N = 1000$?

3. Using the London Life Table given in the text, find the empirical probability that a person living in London at that time would survive to age 16.

4. Repeat the previous exercise, replacing 16 with 76.

5. A somewhat lopsided die is tossed 1000 times with 1 showing on the top face 150 times. What is the empirical probability that a 1 will show?

6. A coin is flipped 10,000 times with heads showing 5050 times. What is the empirical probability that heads will show?

7. The speed of 500 vehicles on a highway with a speed limit of 55 mph was observed, with 400 going between 55 and 65 mph, 60 going less than 55 mph, and 40 going over 65 mph. What is the empirical probability that a vehicle chosen at random on this highway will be going (a) under 55 mph, (b) between 55 and 65 mph, (c) over 65 mph?

8. In a survey of 1000 randomly selected consumers, 50 said they bought brand A cereal, 60 said they bought brand B, and 80 said they bought brand C. What is the empirical probability that a consumer will purchase (a) brand A cereal, (b) brand B, (c) brand C?

9. Weather records indicate that during the past 10 years there has been a measurable snowfall in a certain city on a total of 62 days in January. What is the empirical probability that a day in January in this city will have a measurable snowfall?

10. A salesman has noticed that his records indicate that he sold at least one major appliance on 80 different days during the last year. If he worked 240 days last year, what is the empirical probability that he sells at least one major appliance on a working day?

11. A large dose of a suspected carcinogen has been given to 500 white rats in a laboratory experiment. During the next year, 280 rats get cancer. What is the empirical probability that a rat chosen randomly from this group of 500 will get cancer?

12. A new brand of sausage is tested on 200 randomly selected customers in grocery stores with 40 saying they like the product, the others saying they do not. What is the empirical probability that a consumer will like this brand of sausage?

13. Over a number of years the grade distribution in a mathematics course was observed to be

| A | B | C | D | F |
|---|---|---|---|---|
| 25 | 35 | 80 | 40 | 20 |

What is the empirical probability that a randomly selected student taking this course will receive a grade of A? B? C? D? F?

14. A store sells 4 different brands of VCRs. During the past year the following number of sales of each of the brands were found.

| Brand A | Brand B | Brand C | Brand D |
|---------|---------|---------|---------|
| 20 | 60 | 100 | 70 |

What is the empirical probability that a randomly selected customer who buys a VCR at this store will pick brand A? brand B? brand C? brand D?

15. A somewhat lopsided die is tossed 1000 times with the following results.

| 1 | 2 | 3 | 4 | 5 | 6 |
|---|---|---|---|---|---|
| 150 | 200 | 140 | 250 | 160 | 100 |

What is the empirical probability that each of the numbers will occur?

16. A retail store that sells sneakers notes the following number of sneakers of each size that were sold last year.

| 7 | 8 | 9 | 10 | 11 | 12 |
|---|---|---|----|----|----|
| 20 | 40 | 60 | 30 | 40 | 10 |

What is the empirical probability that a customer who buys a pair of sneakers will buy size 7? 8? 9? 10? 11? 12?

17. What is the empirical probability that one of the students in Exercise 13 will make a grade higher than a C?

18. What is the empirical probability that one of the students in Exercise 13 will make a grade of C or lower?

19. What is the empirical probability that one of the people in Exercise 14 will not select brand D?

20. What is the empirical probability that one of the people in Exercise 14 will not select brand A or B?

21. What is the empirical probability that an even number shows in Exercise 15?

22. What is the empirical probability that a number bigger than 4 shows in Exercise 15?

23. What is the empirical probability that a customer in Exercise 16 buys a pair of sneakers of size 7 or 12?

24. What is the empirical probability that a customer in Exercise 16 buys a pair of sneakers size 10 or larger?

25. If $E$ is any event in a sample space $S$, give an intuitive argument based on relative frequencies that indicates that $p(E^c) = 1 - p(E)$. *Hint:* Show $f(E) + f(E^c) = N$.

26. Let $E$ and $F$ be two mutually disjoint events in a sample space $S$. Give an intuitive argument based on relative frequencies that indicates that $p(E \cup F) = p(E) + p(F)$. *Hint:* Use the fact that $f(E) + f(F) = f(E \cup F)$.

27. Let the sample space $S = \{s_1, s_2, \ldots, s_n\}$ and let $p(s_1) = p_1$, $p(s_2) = p_2$, $\ldots, p(s_n) = p_n$. Then give an intuitive argument that indicates that $p_1 + p_2 + \cdots + p_n = 1$. *Hint:* Follow the discussion of the proof of Property 4 and use Property 2.

28. Let $E$ and $F$ be two events in a sample space $S$. Give an intuitive argument based on relative frequencies that indicates that $p(E \cup F) = p(E) + p(F) - p(E \cap F)$. *Hint:* Use the fact that $f(E \cup F) = f(E) + f(F) - f(E \cap F)$.

## Answers to Discussion Questions

1. You do not know what the actual probability is. You do know that the empirical probability is $\frac{f(H)}{N} = \frac{165}{1000} = .165$. This represents the best guess for the actual probability. But if you tossed the coin more times, the relative frequency and the new empirical probability would most likely change.

2. Since the coin is assumed to be fair, the probability of heads on any one toss is $\frac{1}{2}$. No matter how many heads or tails have occurred in the past, the coin remains a fair coin. Therefore, the probability must remain the same. Whether heads has come up often in the past or seldom in the past, the probability of heads on the

next toss must remain at $\frac{1}{2}$ and not change. Thus we still expect heads to come up as often as tails during the last 40 tosses.

3. The probabilities in the game are constant and do not change just because you are on a winning streak. Thus no matter what has happened to you in the past, the probability of winning any one game remains constant at .48. Thus if you continue to play, you should expect to win 48% of the time in the future. You have been lucky to have won 60% of the time up until now.

4. Since there are just as many red slots as black slots in a roulette game, the probability of a red must always equal the probability of a black. Just because an unusual number of reds has come up cannot affect this fundamental physical fact. The physical setup of the game and the physical laws of the universe do not change from one game to another. Thus no matter what has happened in the past, the probabilities for red and black remain the same.

5. After reading the first discussion problem above, we know that it is, in fact, impossible to determine with certainty the actual probability precisely. Since the die has been tossed a total of 2000 times and a one has come up 335 times, our best guess at the actual probability is $\frac{335}{2000} = .1675$.

6. Casinos and lotteries make money based on the fact that when the "experiment" of playing a casino game or picking a number is repeated a very large number of times, the relative frequency will very likely be very close to the actual probabilities. These probabilities have been carefully calculated so that the casino can be assured the customer will take back about 90 cents of every dollar bet, while lotteries generally give about 50 cents back for every dollar bet. Thus any individual player who plays for an extended number of times can count on the likelihood that the relative frequencies will be close to the actual probabilities, thus assuring losses for the player.

## Solutions to Self-Help Exercise Set 7.2

1. There were $10 + 40 + 30 + 20 = 100$ sold. Let the sample space be denoted by $\{s, m, l, x\}$ with the obvious designations. To find the empirical probability of each outcome, count the frequency of each outcome and divide by 100. The probabilities are as follows:

$$p(s) = \frac{f(s)}{N} = \frac{10}{100} = .10 \qquad p(m) = \frac{f(m)}{N} = \frac{40}{100} = .40$$

$$p(l) = \frac{f(l)}{N} = \frac{30}{100} = .30 \qquad p(x) = \frac{f(x)}{N} = \frac{20}{100} = .20$$

2. $p(\{m, l\}) = \dfrac{f(\{m, l\})}{N} = \dfrac{40 + 30}{100} = .70$

3. $p(\{m, l, x\}) = \dfrac{f(\{m, l, x\})}{N} = \dfrac{40 + 30 + 20}{100} = .90.$

## 7.3    MATHEMATICAL PROBABILITY

▶ *Mathematical Definition of Probability*

▶ *Additional Properties of Probability*

▶ *Odds*

---

**Further Historical Developments in Probability**

Neither Pascal nor Fermat published their initial findings on probability. Christian Huygens (1629–1695) became acquainted with the work of Pascal and Fermat and subsequently published in 1657 the first tract on probability: *On Reasoning in Games of Dice*. This little pamphlet remained the only published work on probability for the next 50 years. James Bernoulli (1654–1705) published the first substantial tract on probability when his *Art of Reasoning* appeared 7 years after his death. This expanded considerably on Huygens' work. The next major milestone in probability occurred with the publication in 1718 of Abraham De Moivre's work *Doctrine of Chance: A Method of Calculating the Probability of Events in Play*. Before 1770, probability was almost entirely restricted to the study of gambling and actuarial problems, although some applications in errors of observation, population, and certain political and social phenomena had been touched on. It was Pierre Simon Laplace (1749–1827) who broadened the mathematical treatment of probability beyond games of chance to many areas of scientific research. The theory of probability undoubtedly owes more to Laplace than to any other individual.

---

**A P P L I C A T I O N**

**Finding the Probability of Obtaining a Contract**

Three companies are competing for a contract. It is known that the first two companies have equal probability of obtaining the contract and that the third company has only one-half the probability of the first one. What is the probability that the first company will get the contract? See Example 4 for the answer.

## Mathematical Definition of Probability

In the last section we took the point of view that probability is part of the study of natural and human phenomena in the same way that physics or economics is. Probability was discussed only in the context of events in business or science. Subjects such as ''randomness'' and the long-term behavior of the relative frequency of an event were deemed intuitive and within the experience of the reader.

We take an alternate point of view of probability in this section. We will now regard probability as a part of *mathematics*. As such, we need to establish abstract definitions and construct a set of axioms that govern the theory, placing the theory on a firm foundation. Using the logic of mathematics we can then derive other properties that probability must satisfy. Such a mathematical theory is said to be a **model** of the phenomenon of probability. One important advantage of this approach is that all terms are clearly defined and "experience" is not a prerequisite for understanding the development. The axioms that describe the phenomenon of probability must be carefully chosen so that the resulting theory actually describes the phenomenon that we are studying. In other words, the theory must not contradict our intuitive notions of probability gained in the last section. How can we then choose the axioms that will govern the theory? Naturally we must look to the intuitive notions gained in the last section. The four properties of probability that we learned of in the last section play a fundamental role.

We have interpreted probability in terms of the long-term behavior of relative frequencies and found this very useful in applications. This concept is intuitive, however, and was not made mathematically precise in the last section. We could make this intuitive notion mathematically precise, but the process is not worth the effort since a more fruitful but still mathematically precise approach is available. Naturally any definition of probability must be consistent with the intuitive interpretation given in the last section so that the results can be used to model real applications.

We first need to have a set $S$ called the sample space. To define a probability, we must associate with each event $E$, that is with each subset of $S$, a real number $p(E)$ called the probability of $E$.

We choose as axioms the four properties of probability given in the last section. These four, in effect, capture the essence of what we intuitively think of as probability.

We then have the following definition.

---

**Definition of Probability**

Let $S$ be a sample space. Assign to each event $E$ in $S$ a number $p(E)$ so that the following four properties are satisfied.

***Property 1*** $\quad 0 \le p(E) \le 1$ for any event $E$.

***Property 2*** $\quad p(S) = 1$.

***Property 3*** $\quad p(\phi) = 0$.

***Property 4*** $\quad$ Given any event $E = \{e_1, e_2, \ldots, e_r\}$, where $e_1, e_2, \ldots, e_r,$ are all distinct, then

$$p(\{e_1, e_2, \ldots, e_r\}) = p(e_1) + p(e_2) + \cdots + p(e_r).$$

Then for any event $E$, we called $p(E)$ the **probability** of $E$ and say that $p$ is a probability on $S$.

---

Just how to assign the probabilities to each outcome must be determined by the information on hand. The following example indicates what is possible in a simple situation.

**E X A M P L E  1**    **Assigning Probabilities**

Let a sample space by $S = \{H, T\}$ and define

$$p(H) = p_1 \ge 0, \qquad p(T) = p_2 \ge 0, \qquad p(S) = 1, \qquad p(\phi) = 0,$$

and $p_1 + p_2 = 1$. Show that $p$ is a probability on $S$.

**Solution**    This $p$ certainly satisfies Properties 1, 2, and 3. For Property 4, notice that

$$p(\{H, T\}) = p(S) = 1,$$

while

$$p(H) + p(T) = p_1 + p_2 = 1.$$

This shows that $p$ is a probability on $S$.  ■

This probability could model a fair or unfair coin (depending on whether $p_1$ was .5 or was not) where $p(H) = p_1$. But this could also model any situation with 2 outcomes and where the probability of one outcome is $p_1$. In a 2 outcome experiment, we often refer to one outcome as "success" and the other as "failure." Then the sample space is {success, failure}. The probability of "success" is $p_1$ and the probability of "failure" is $p_2$, where $p_1 + p_2 = 1$. Also we are often interested in the probability of a specific event $E$ occurring or not occurring. In such a situation we have the two events: $E$ and $E^c$. As we will very shortly see, $p(E) + p(E^c)$ must be 1.

## Additional Properties of Probability

The four defining properties of probability have a number of consequences. We now give two.

---

**Property 5**

Let the sample space be $S = \{s_1, s_2, \ldots, s_n\}$ and let $p(s_1) = p_1, p(s_2) = p_2, \ldots, p(s_n) = p_n$. Then

$$p_1 + p_2 + \cdots + p_n = 1.$$

---

Property 5 follows by replacing $S$ with $E$ in Property 4 and using Property 2. Thus

$$1 = p(S) = p(s_1) + p(s_2) + \cdots + p(s_n) = p_1 + p_2 + \cdots + p_n.$$

---

**Property 6**

Let $G$ and $H$ be any two events. Then

$$p(G \cup H) = p(G) + p(H) \qquad \text{if} \qquad G \cap H = \phi.$$

---

To establish Property 6, let $G = \{g_1, \ldots, g_k\}$ and $H = \{h_1, \ldots, h_j\}$. Since $G \cap H = \phi$, all the elements in the set

$$\{g_1, \ldots, g_k, h_1, \ldots, h_j\}$$

are distinct. Thus from Property 4,

$$p(G \cup H) = p(\{g_1, \ldots, g_k, h_1, \ldots, h_j\})$$

$$= p(g_1) + \cdots + p(g_k) + p(h_1) + \cdots + p(h_j)$$

$$= [p(g_1) + \cdots + p(g_k)] + [p(h_1) + \cdots + p(h_j)]$$

$$= p(G) + p(H).$$

This establishes Property 6.

In the last section we had mentioned that probabilities can often be determined by theoretical means. We decided by an intuitive argument based on interpreting probabilities in terms of relative frequencies that if in a coin-flipping experiment heads is as likely as tails, then $p(H) = p(T) = .5$. We now give another reason for this, this time based on the above mathematical definition of probability.

**E X A M P L E  2**     **Finding the Probability of an Event**

Suppose in a coin-flipping experiment we assume that the probability of heads is the same as the probability of tails. What are $p(H)$ and $p(T)$?

**Solution**     We have $p(H) = p(T)$. By Property 6

$$1 = p(H) + p(T) = 2p(H).$$

Thus $p(H) = \frac{1}{2} = p(T)$.  ■

In the next two examples, the usefulness of Property 6 is shown.

**E X A M P L E  3**     **Finding the Probability of an Event**

The probability that any of the first 5 numbers of a loaded die will come up is the same while the probability that a 6 comes up is .25. What is the probability that a 1 will come up?

**Solution**     We are given that

$$p(1) = p(2) = p(3) = p(4) = p(5), \qquad p(6) = .25.$$

From Property 6

$$1 = p(1) + p(2) + p(3) + p(4) + p(5) + p(6)$$

$$= 5p(1) + .25$$

$$5p(1) = .75$$

$$p(1) = .15. \quad ■$$

**E X A M P L E   4**     **Finding the Probability of an Event**

Three companies are competing for a contract. It is known that the first two companies have equal probability of obtaining the contract and that the third company has only one-half the probability of the first one. What is the probability that the first company will get the contract?

**Solution**     Let $p(1)$, $p(2)$, $p(3)$, be respectively the probability of the first, second, and third company obtaining the contract. Then we have that $p(2) = p(1)$ and $p(3) = p(1)/2$. Thus from Property 6

$$1 = p(1) + p(2) + p(3) = p(1) + p(1) + \frac{1}{2}p(1) = \frac{5}{2}p(1).$$

Thus $p(1) = .40$  ∎

The first 6 properties have a number of further consequences. We now give some of them.

---

**Further Properties of Probability**

Let $p$ be a probability on a sample space $S$ and $E$ and $F$ two events in $S$. Then the following are true.

**Property 7**                    $p(E^c) = 1 - p(E)$.

**Property 8**     $p(E \cup F) = p(E) + p(F) - p(E \cap F)$.

**Property 9**            $p(F) \leq p(E)$    if    $F \subset E$.

**Property 10**   Suppose $E_1, E_2, \ldots, E_k$ are pairwise mutually disjoint, that is $E_i \cap E_j = \phi$ if $i \neq j$. Then

$$p(E_1 \cup E_2 \cup \cdots \cup E_k) = p(E_1) + p(E_2) + \cdots + p(E_k).$$

---

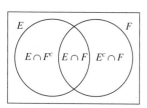

**Figure 7.4**

The proofs of Property 7, Property 9, and Property 10 are left as exercises. Property 10 represents an extension of Property 6 to any number of mutually disjoint sets.

To establish Property 8 notice from Figure 7.4 that since $E \cap F^c$ and $E \cap F$ are mutually exclusive and $E = (E \cap F^c) \cup (E \cap F)$, Property 6 implies that

$$p(E) = p(E \cap F^c) + p(E \cap F).$$

Thus

$$p(E \cap F^c) = p(E) - p(E \cap F).$$

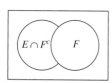

**Figure 7.5**

We can see from Figure 7.5 that we can write $E \cup F$ as the union of the two sets $E \cap F^c$ and $F$ and these two sets are mutually exclusive. Thus by Property 6 and the above

$$p(E \cup F) = p((E \cap F^c) \cup F)$$
$$= p(E \cap F^c) + p(F)$$
$$= p(E) - p(E \cap F) + p(F).$$

This gives Property 8.

E X A M P L E 5     **Finding the Probability of an Event**

A salesman makes two stops when in Pittsburgh. The first stop yields a sale 10% of the time, the second stop 15% of the time, and both stops yield a sale 4% of the time. What proportion of the time does a trip to Pittsburgh result in no sale?

Solution    Let $E$ be the event a sale is made at the first stop and $F$ the event that a sale is made at the second stop. Then we are given that

$$p(E) = .10, \quad p(F) = .15, \quad p(E \cap F) = .04.$$

Since $p(E \cap F) = .04$, we place .04 in the region $E \cap F$ in Figure 7.6. Now since $p(E) = .10$, we can see from Figure 7.6 that $p(E \cap F^c) = .10 - .04 = .06$. In a similar fashion we have $p(E^c \cap F) = .15 - .04 = .11$. Thus we readily see from Figure 7.6 that

$$p(E \cup F) = .06 + .04 + .11 = .21.$$

Then by Property 7

$$p((E \cup F)^c) = 1 - p(E \cup F) = 1 - .21 = .79.$$

Thus no sale is made in Pittsburgh 79% of the time.  ∎

R E M A R K .     We could have obtained $p(E \cup F)$ directly from Property 8 as follows.

$$P(E \cup F) = p(E) + p(F) - p(E \cap F) = .10 + .15 - .04 = .21.$$

Figure 7.6

## Odds*

One can interpret probabilities in terms of **odds** in a bet. Suppose in a sample space $S$ we are given an event $E$ with probability $p = p(E)$. Suppose we engage in a bet in which we agree to give $a$ dollars if $E$ does not occur and receive $b$ dollars if $E$ does occur. Let us decide what the relationship should be between $a$ and $b$ so that the bet will be fair. In a large number of such bets we expect to win $b$ dollars a fraction $p$ of the time and lose $a$ dollars a fraction $1 - p$ of the time. On average we expect to obtain $bp - a(1 - p)$ dollars. If the wager is to be fair, this should be zero. Thus $bp = a(1 - p)$ or

*Optional

$$\frac{a}{b} = \frac{p}{1 - p}.$$

This ratio is called the **odds** and is normally written as the ratio of two integers.

---

**Odds of an Event**

Let $p = p(E)$ be the probability of $E$, then the **odds for E** is defined to be

$$\text{Odds for } E = \frac{p}{1 - p} \qquad \text{if } p \neq 1.$$

---

If the odds are given as $\dfrac{a}{b}$, we say that **the odds are $a$ to $b$** and also write $a : b$.

**E X A M P L E   6**    **Determining the Odds of an Event**

You believe that a horse has a probability of $\frac{1}{4}$ of winning a race. What are the odds of this horse winning? What are the odds of this horse losing? What profit should a winning \$2 bet return for the bet to be fair?

**Solutions**    (a)    Since $p = \frac{1}{4}$, the odds for winning are

$$\frac{p}{1 - p} = \frac{\frac{1}{4}}{\frac{3}{4}} = \frac{1}{3},$$

that is, $1 : 3$.

(b)    Since the probability of winning is $\frac{1}{4}$, the probability of losing is $1 - \frac{1}{4} = \frac{3}{4}$. Then

$$\text{odds for losing} = \frac{\frac{3}{4}}{1 - \frac{3}{4}} = \frac{3}{1},$$

or $3 : 1$.

(c)    Since the fraction $\frac{1}{3}$ can also be written as $\frac{2}{6}$ with odds $2 : 6$, a fair \$2 bet should return a profit of \$6 for a winning bet.    ■

Notice that making this same bet many times, we expect to win \$6 one-fourth of the time and lose \$2 three-fourths of the time. Our average winnings would be

$$6(\tfrac{1}{4}) - 2(\tfrac{3}{4}) = 0$$

dollars.

The way in which probability can be interpreted in terms of odds is as follows. Suppose that the odds for an event $E$ occurring is given as $\dfrac{a}{b}$. If $p = p(E)$, then

$$\frac{a}{b} = \frac{p}{1 - p}$$

$$a(1 - p) = bp$$

$$a = bp + ap = p(a + b)$$

$$p = \frac{a}{a + b}.$$

---

**Obtaining Probability from Odds**

Suppose that the odds for an event $E$ occurring is given as $\frac{a}{b}$. If $p = p(E)$, then

$$p = \frac{a}{a + b}.$$

---

## SELF-HELP EXERCISE SET 7.3

1. If $S = \{a, b, c\}$ with $p(a) = p(b) = p(c)$, find $p(a)$.

2. A company has bids on two contracts. They believe that the probability of obtaining the first contract is .4 and of obtaining the second contract is .3, while the probability of obtaining both contracts is .1. (a) Find the probability that they will obtain exactly one of the contracts. (b) Find the probability that they will obtain neither of the contracts.

3. What are the odds that the company in the previous exercise will obtain both of the contracts?

## EXERCISE SET 7.3

In all the following, $S$ is assumed to be a sample space and $p$ a probability function on $S$.

1. Let $S = \{a, b, c\}$ with $p(a) = .1$, $p(b) = .4$, $p(c) = .5$. Let $E = \{a, b\}$ and $F = \{b, c\}$. Find $p(E)$ and $p(F)$.

2. Let $S = \{a, b, c, d, e, f\}$ with $p(a) = .1$, $p(b) = .2$, $p(c) = .25$, $p(d) = .15$, $p(e) = .12$, $p(f) = .18$. Let $E = \{a, b, c\}$ and $F = \{c, d, e, f\}$ and find $p(E)$ and $p(F)$.

3. Let $S = \{a, b, c, d, e, f\}$ with $p(b) = .2$, $p(c) = .25$, $p(d) = .15$, $p(e) = .12$, $p(f) = .1$. Let $E = \{a, b, c\}$ and $F = \{c, d, e, f\}$. Find $p(a)$, $p(E)$, and $p(F)$.

4. Let $S = \{a, b, c, d, e, f\}$ with $p(b) = .3$, $p(c) = .15$, $p(d) = .05$, $p(e) = .2$, $p(f) = .13$. Let $E = \{a, b, c\}$ and $F = \{c, d, e, f\}$. Find $p(a)$, $p(E)$, and $p(F)$.

5. If $S = \{a, b, c, d\}$ with $p(a) = p(b) = p(c) = p(d)$, find $p(a)$.

6. If $S = \{a, b, c\}$ with $p(a) = p(b)$ and $p(c) = .4$, find $p(a)$.

7. If $S = \{a, b, c, d, e, f\}$ with $p(a) = p(b) = p(c) = p(d) = p(e) = p(f)$, find $p(a)$.

8. If $S = \{a, b, c\}$ with $p(a) = 2p(b) = 3p(c)$, find $p(a)$.

9. If $S = \{a, b, c, d, e, f\}$ with $p(a) = p(b) = p(c), p(d) = p(e) = p(f) = .1$, find $p(a)$.

10. If $S = \{a, b, c, d, e\ f\}$ and if $p(a) = p(b) = p(c), p(d) = p(e) = p(f), p(d) = 2p(a)$, find $p(a)$.

11. If $E$ and $F$ are two mutually disjoint events in $S$ with $p(E) = .2$ and $p(F) = .4$, find $p(E \cup F), p(E^c),$ and $p(E \cap F)$.

12. Why is it not possible for $E$ and $F$ to be two mutually disjoint events in $S$ with $p(E) = .5$ and $p(F) = .7$?

13. If $E$ and $F$ are two mutually disjoint events in $S$ with $p(E) = .4$ and $p(F) = .3$, find $p(E \cup F), p(F^c), p(E \cap F), p((E \cup F)^c), p((E \cap F)^c)$.

14. Why is it not possible for $S = \{a, b, c\}$ with $p(a) = .3, p(b) = .4, p(c) = .5$?

15. Let $E$ and $F$ be two events in $S$ with $p(E) = .3, p(F) = .5, p(E \cap F) = .2$. Find $p(E \cup F)$ and $p(E \cap F^c)$.

16. Let $E$ and $F$ be two events in $S$ with $p(E) = .3, p(F) = .5, p(E \cap F) = .2$. Find $p(E^c \cap F)$ and $p(E^c \cap F^c)$.

17. Let $E$ and $F$ be two events in $S$ with $p(E) = .3, p(F) = .5, p(E \cup F) = .6$. Find $p(E \cap F)$ and $p(E \cap F^c)$.

18. Why is it not possible to have $E$ and $F$ two events in $S$ with $p(E) = .3$ and $p(E \cap F) = .5$?

19. Let $E$ and $F$ be two events in $S$ with $p(E) = .5, p(F) = .7$. Just how small could $p(E \cap F)$ possibly be?

20. Let $E$ and $F$ be two events in $S$ with $p(E) = .3, p(F) = .4$. Just how large could $p(E \cup F)$ possibly be?

In Exercises 21 through 24 let $E$, $F$, and $G$ be events in $S$ with $p(E) = .55, p(F) = .4, p(G) = .45, p(E \cap F) = .3, p(E \cap G) = .2, p(F \cap G) = .15, p(E \cap F \cap G) = .1$.

21. Find $p(E \cap F \cap G^c), p(E \cap F^c \cap G), p(E \cap F^c \cap G^c)$.

22. Using the results of the previous exercise, find $p(E^c \cap F \cap G), p(E^c \cap F \cap G^c), p(E^c \cap F^c \cap G)$.

23. Using the results of the previous two exercises, find $p(E \cup F \cup G)$.

24. Using the result of the previous exercise, find $p(E^c \cup F^c \cup G^c)$.

25. For the loaded die in Example 3 of the text, what are the odds that (a) a 2 will occur, (b) a 6 will occur?

26. For the loaded die in Example 3 of the text, what are the odds that (a) a 3 will occur, (b) a 1 will occur?

27. In Example 4 of the text, what are the odds that (a) the first company will receive the contract, (b) the third one will?

28. In Example 5 of the text, what are the odds that the salesman will make a sale on (a) the first stop, (b) on the second stop, (c) on both stops?

29. It is known that the odds that $E$ will occur are $1 : 3$ and that the odds that $F$ will occur are $1 : 2$, and that both $E$ and $F$ cannot occur simultaneously. What are the odds that $E$ or $F$ will occur?

30. Show that if the odds for $E$ are $a : b$, then the odds for $E$ not occurring are $b : a$.

31. If the odds for the Giants winning the World Series are $1 : 4$, what is the probability the Giants will win the Series?

32. If the odds for a successful marriage are $1 : 2$, what is the probability for a successful marriage?

33. Establish Property 7: For any event $E$, $p(E^c) = 1 - p(E)$. *Hint:* Notice that $E$ and $E^c$ are mutually exclusive and $E \cup E^c = S$. Now use Property 2 and Property 6.

34. Establish Property 9: If $E$ and $F$ are two events with $F \subset E$, then $p(F) \leq p(E)$. *Hint:* If $F \subset E$, the two sets $F$ and $E \cap F^c$ are mutually exclusive and have union equal to $E$. Use Property 6 and then Property 1.

35. Establish Property 10 for 3 pairwise mutually disjoint sets.

36. If $E$, $F$, and $G$ are events in a sample space $S$, show that

$$p(E \cup F \cup G) = p(E) + p(F) + p(G) - p(E \cap F) - p(E \cap G) - P(F \cap G) + p(E \cap F \cap G).$$

*Hint:* Let $H = F \cup G$ and use Property 8.

37. If $E$ and $F$ are two events in a sample space $S$, the event $(E \cap F^c) \cup (E^c \cap F)$ is the event that either $E$ or $F$ occur but not both, that is, that exactly one of the events $E$ or $F$ occurs. Show that

$$p((E \cap F^c) \cup (E^c \cap F)) = p(E) + p(F) - 2p(E \cap F).$$

38. For any two events $E$ and $F$ on a sample space $S$, show

$$p(E \cap F) \leq p(E) \leq p(E \cup F).$$

39. For any two events $E$ and $F$ on a sample space $S$, show

$$p(E^c \cap F) = p(F) - p(E \cap F).$$

## Applications

40. **Bidding on Contracts.** An aerospace firm has 3 bids on government contracts and knows that the contracts are most likely to be divided up among a number of companies. The firm decides that the probability of obtaining exactly one

contract is .6, of exactly two contracts is .15, and of exactly three contracts is .04. What is the probability that the firm will obtain at least one contract? No contracts?

41. **Quality Control.** An inspection of computers manufactured at a plant reveals that 2% of the monitors are defective, 3% of the keyboards are defective and 1% of the computers have both defects. (a) Find the probability that a computer at this plant has at least one of these defects. (b) Find the probability that a computer at this plant has none of these defects.

42. **Medicine.** A new medication produces headaches in 5% of the users, upset stomach in 15%, and both in 2%. (a) Find the probability that at least one of these side effects is produced. (b) Find the probability that neither of these side effects is produced.

43. **Manufacturing.** A manufactured item is guaranteed for one year and has three critical parts. It has been decided that during the first year the probability of failure of the first part is .03, of the second part .02, the third part .01, both the first and second .005, both the first and third .004, both the second and third .003, and all three parts .001.
    a. What is the probability that exactly one of these parts will fail in the first year?
    b. What is the probability that at least one of these parts will fail in the first year?
    c. What is the probability that none of these parts will fail in the first year?

44. **Marketing.** A survey of business executives found that 40% read *Business Week*, 50% read *Fortune*, 40% read *Money*, 17% read both *Business Week* and *Fortune*, 15% read both *Business Week* and *Money*, 14% read both *Fortune* and *Money*, and 8% read all three of these magazines.
    a. What is the probability that one of these executives reads exactly one of these three magazines?
    b. What is the probability that one of these executives reads at least one of these three magazines?
    c. What is the probability that one of these executives reads none of these three magazines?

45. **Advertising.** A firm advertises three different products, A, B, C, on television. From past experience, it expects 1.5% of listeners to buy exactly one of the products, 1% to buy exactly two of the products, 1.2% to buy A, .4% to buy both A and B, .3% to buy both A and C, and .6% to buy A but not the other two.
    a. Find the probability that a listener will buy only B or only C.
    b. Find the probability that a listener will buy all three.
    c. Find the probability that a listener will buy both B and C.
    d. Find the probability that a listener will buy none of the three.

46. **Sales.** A salesman always makes a sale at one of three stops in Atlanta and 30% of the time makes a sale at only the first stop, 15% at only the second stop, 20% at only the third stop, and 35% of the time at exactly two of the stops. Find the probability that the salesman makes a sale at all three stops in Atlanta.

### Solutions to Self-Help Exercise Set 7.3

1.  If $S = \{a, b, c\}$ and $p(a) = p(b) = p(c)$, then from Property 6

$$1 = p(a) + p(b) + p(c) = 3p(a).$$

Thus $p(a) = \frac{1}{3}$.

2.  **a.**  Let $E$ be the event that the company obtains the first contract and let $F$ be the event that the company obtains the second contract. The event that the company obtains the first contract but not the second is $E \cap F^c$, while the event that the company obtains the second contract but not the first is $E^c \cap F$. These two sets are mutually exclusive, so the probability that the company receives exactly one of the contracts is

$$p(E \cap F^c) + p(E^c \cap F).$$

Now $p(E) = .40$, $p(F) = .30$, and since $E \cap F$ is the event that the company receives both contracts, $p(E \cap F) = .10$.

Notice on the accompanying diagram that $E \cap F^c$ and $E \cap F$ are mutually disjoint and that $(E \cap F^c) \cup (E \cap F) = E$. Thus

$$p(E \cap F^c) + p(E \cap F) = p(E)$$

$$p(E \cap F^c) + .10 = .40$$

$$p(E \cap F^c) = .30.$$

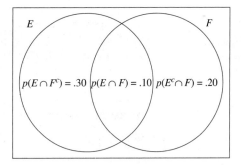

Also notice on the accompanying diagram that $E^c \cap F$ and $E \cap F$ are mutually disjoint and that $(E^c \cap F) \cup (E \cap F) = F$. Thus

$$p(E^c \cap F) + p(E \cap F) = p(F)$$

$$p(E^c \cap F) + .10 = .30$$

$$p(E^c \cap F) = .20.$$

Thus the probability that the company will receive exactly one of the contracts is

$$p(E \cap F^c) + p(E^c \cap F) = .30 + .20 = .50.$$

**2.  b.**  The event that the company obtains neither contracts is given by $(E \cup F)^c$. From the diagram

$$p(E \cup F) = .30 + .10 + .20 = .60.$$

Thus

$$p((E \cup F)^c) = 1 - p(E \cup F) = 1 - .60 = .40.$$

The probability that the company receives neither contract is .40.

**3.**  From the formula for odds, we have

$$\text{odds for } (E \cap F) = \frac{.10}{1 - .10} = \frac{1}{9}.$$

Thus the odds are 1 to 9 that the company will obtain both contracts.

## 7.4     SPACES WITH EQUALLY LIKELY OUTCOMES

▶  *Counting Techniques in Probability*

▶  *A Nonintuitive Example*

**A P P L I C A T I O N**

**Probability of Selecting Defective and Nondefective Transistors**

A bin contains 15 identical (to the eye) transistors except that 6 are defective and 9 are not. What is the probability of selecting 5 transistors from this bin with 2 defective and 3 not defective? See Example 3 for the answer.

### Counting Techniques in Probability

In this section we consider sample spaces in which the probability of any one elementary event is the same as any other. We shall see that for such spaces finding the probability reduces to counting the numbers in sets. To do this we will rely on the counting techniques that were given in the last chapter.

In this section we assume that the space $S$ is finite and has $n$ distinct elements with $S = \{s_1, s_2, \ldots, s_n\}$. We furthermore assume that the space $S$ is a **uniform sample space**, that is, the probability of each of the elementary events is the same. Thus

$$p(s_1) = p(s_2) = \cdots = p(s_n).$$

If we let $p$ be this common probability, then from Property 5 of the last section we have

$$1 = p(s_1) + p(s_2) + \cdots + p(s_n) = np.$$

Thus

$$p = \frac{1}{n}.$$

Now give any event $E = \{e_1, e_2, \ldots, e_r\}$, we have, using Property 4 of the last section, that

$$p(E) = p(\{e_1, e_2, \ldots, e_r\})$$
$$= p(e_1) + p(e_2) + \cdots + p(e_r)$$
$$= r\frac{1}{n}$$
$$= \frac{n(E)}{n(S)}$$

Thus we have the following.

---

**Probability of an Event in the Uniform Sample Space**

If $S$ is a finite uniform sample space and $E$ is any event, then

$$p(E) = \frac{\text{Number of elements in E}}{\text{Number of elements in S}} = \frac{n(E)}{n(S)}.$$

---

**E X A M P L E   1**  **Calculating Probability in a Uniform Sample Space**

A bin contains 15 identical (to the eye) transistors except that 6 are defective and 9 are not. What is the probability that a transistor selected at random is defective?

**Solution**  We can do an equivalent problem in which the 15 transistors are labeled from 1 to 15 in a manner that the human eye cannot detect, with the defective ones numbered from 1 to 6. Then the sample space is $S = \{1, 2, \ldots, 15\}$. Since the transistors are identical to the eye, the probability of selecting any one is the same as selecting any other. Since the event "the transistor is defective" is given by $E = \{1, 2, 3, 4, 5, 6\}$, we have

$$p(E) = \frac{n(E)}{n(S)} = \frac{6}{15} = .4. \quad \blacksquare$$

**E X A M P L E   2**  **Calculating Probability Using Counting Techniques**

Suppose one transistor is selected from the bin in Example 1 and then another is selected without replacing the first. What is the probability that both transistors are defective?

**Solution**    In order to solve this problem, we need to first know the sample space. The sample space $S$ consists of some subset of the ordered pairs $(x, y)$ with $x$ and $y$ integers from 1 to 15. We again label the transistors as before. Since order is not important, there are

$$C(15, 2) = \frac{15 \cdot 14}{2 \cdot 1} = 105$$

ways of selecting 2 transistors from the bin of 15. Thus there are 105 elementary events in the sample space.

Now we must decide what is the probability of each of these 105 elementary events. One might consider whether some particular selection, such as $\{2, 4\}$, is any more or less likely than another selection, say $\{3, 7\}$. Considering the physical set-up, there does not appear to be any reason to argue that one is any more or less likely than the other. Thus it would appear that the probability of each of the 105 elementary events is the same. In any such situation, we will always *assume* that this is the case, if not explicitly stated.

Then, since we are assuming the uniform probability, we count the number of ways of selecting 2 defective transistors from the 6 in the bin and divide by 105, which is the number of ways of selecting 2 transistors from the bin of 15. If $E$ is the event that both transistors are defective, we have

$$p(E) = \frac{n(E)}{n(S)} = \frac{C(6, 2)}{C(15, 2)} = \frac{\overset{6 \cdot 5}{\overline{2 \cdot 1}}}{105} = \frac{15}{105} = \frac{1}{7}. \quad \blacksquare$$

**E X A M P L E   3**    **Calculating Probability Using Counting Techniques**

What is the probability of selecting 5 transistors from the bin in Example 1 with 2 defective and 3 not defective?

**Solution**    The number of ways of selecting 5 from the 15 is

$$C(15, 5) = \frac{15 \cdot 14 \cdot 13 \cdot 12 \cdot 11}{5 \cdot 4 \cdot 3 \cdot 2 \cdot 1} = 7 \cdot 13 \cdot 3 \cdot 11 = 3003.$$

This is the number in the sample space $S$.

The number of ways of selecting 2 defectives from 6 is $C(6, 2)$, while the number of ways of selecting 3 nondefective ones from 9 is $C(9, 3)$. Thus by the multiplication principle, the number of ways of doing both is

$$C(6, 2) \cdot C(9, 3) = \frac{6 \cdot 5}{2 \cdot 1} \frac{9 \cdot 8 \cdot 7}{3 \cdot 2 \cdot 1} = 1260.$$

If $E$ is the probability of selecting 2 defective transistors and 3 nondefective ones, then

$$p(E) = \frac{n(E)}{n(S)} = \frac{1260}{3003} \approx .420. \quad \blacksquare$$

E X A M P L E  4    **Using Counting**

Two fair dice are tossed. Assuming that the probability of every elementary event shown in Example 2 of Section 7.1 is equally likely, find the probability that a sum shows that is equal to (a) 3, (b) 7.

Solutions    We first comment that the assumption that any elementary event is just as likely as any other seems reasonable since it does not seem possible to argue based on the physical set-up how one outcome, such as (2, 4), can be any more or less likely than any other outcome, such as (3, 6).

The total number of outcomes is $n(S) = 36$.

(a)    The event "the sum is 3" is given by $E = \{(1, 2), (2, 1)\}$. Thus

$$p(E) = \frac{n(E)}{n(S)} = \frac{2}{36} = \frac{1}{18}.$$

(b)    The event "the sum is 7" is given by

$$F = \{(1, 6), (2, 5), (3, 4), (4, 3), (5, 2), (6, 1)\}.$$

Thus

$$p(F) = \frac{n(F)}{n(S)} = \frac{6}{36} = \frac{1}{6}. \quad \blacksquare$$

E X A M P L E  5    **Using Counting Techniques**

A fair coin is tossed 6 times. Assuming that any outcome is as likely as any other, find the probability of obtaining exactly 3 heads.

Solution    Again the physical set-up implies the reasonableness of the assumption that each of the elementary events is equally likely.

By the multiplication principle of the last chapter, there are $2^6 = 64$ possible outcomes. The number of ways of obtaining exactly 3 heads is the number of ways of selecting 3 slots from among 6 to place the heads. This is

$$C(6, 3) = \frac{6 \cdot 5 \cdot 4}{3 \cdot 2 \cdot 1} = 20.$$

Thus if $E$ is the event that exactly 3 heads occur, then

$$p(E) = \frac{n(E)}{n(S)} = \frac{20}{64} = \frac{5}{16}. \quad \blacksquare$$

E X A M P L E  6    **The Probability of a Poker Hand**

Find the probability of drawing a flush, but not a straight flush, in a poker game, assuming that any 5-card hand is just as likely as any other.

Solution    A deck in poker has 52 cards. Thus there are

$$C(52, 5) = \frac{52 \cdot 51 \cdot 50 \cdot 49 \cdot 48}{5 \cdot 4 \cdot 3 \cdot 2 \cdot 1} = 2{,}598{,}960$$

possible hands.

A flush consists of 5 cards in a single suit. There are 4 suits, each of 13 cards. Thus the number of ways of obtaining a flush in a particular suit is

$$C(13, 5) = \frac{13 \cdot 12 \cdot 11 \cdot 10 \cdot 9}{5 \cdot 4 \cdot 3 \cdot 2 \cdot 1} = 1287.$$

A straight consists of 5 cards in sequence. There are 10 such straight flushes in each suit: $\{1, 2, 3, 4, 5\}, \ldots, \{10, J, Q, K, A\}$. Thus the number of flushes that are not straights is $4(1287 - 10) = 5108$. Thus if $E$ is the event of drawing a flush, but not a straight flush,

$$p(E) = \frac{n(E)}{n(S)} = \frac{5108}{2{,}598{,}960} \approx .0020. \quad \blacksquare$$

## A Nonintuitive Example

We now calculate the probability of a certain event and obtain a very surprising (nonintuitive) result.

**E X A M P L E  7**   **The Birthday Problem**

Suppose there are 50 people in a room with you. What is the probability that at least two of these people will have the same birthday?

**Solution**   Suppose more generally there are $n$ people in the room. Let $E$ be the event that at least two of these people will have the same birthday. We will ignore leap years and assume that every one of the 365 days of the year is just as likely to be a birthday as any other. It is much easier to first find $p(E^c)$, the probability that no two have the same birthday.

First notice that there are 365 possible birthdays for each individual. Thus, by the multiplication principle, there are $365^n$ possible birthdays for the $n$ individuals. This is the total number in the sample space. To find $p(E^c)$ notice that there are 365 possible birthdays for the first individual and since the second individual cannot have the same birthday as the first, there are 364 possible birthdays for the second, and then 363 for the third, and so on. We have

$$p(E) = 1 - p(E^c) = 1 - \frac{(365)(364) \cdots (365 - n + 1)}{365^n}.$$

The table gives this for several values of $n$. Notice the very surprising result that for $n$ equal to only 23 the probability is about .51.

| $n$ | 15 | 20 | 23 | 30 | 50 |
|------|-----|-----|-----|-----|-----|
| $p(E)$ | .25 | .41 | .51 | .71 | .97 |

With 50 people in the room the probability is approximately .97 that 2 or more of these people will have the same birthday. ■

## SELF-HELP EXERCISE SET 7.4

1. Two tetrahedrons (4 sided), each with equal sides, are identical except that one is red and the other white. Both have the four sides numbered from 1 to 4. The two tetrahedrons are tossed, and the number on the bottom face of each is observed. Let $E$ consist of those outcomes for which both tetrahedrons show an odd number. Let $F$ be the event that the sum of the two numbers on these tetrahedrons is 5. Let $G$ be the event that the sum of the two numbers is less than 7. (a) Find $p(E)$. (b) Find $p(F)$. (c) Find $p(E \cap F)$. (d) Find $p(E \cup F)$. (e) Find $p(G^c)$. (Refer to Self-Help Exercises 1 and 2 of Section 7.1.)

2. A lotto game consists of picking (in any order) the correct 6 numbers drawn from 1 to 42 without replacement. (a) What is the probability of any one pick winning? (b) What is the probability of a winning pick being all consecutive numbers?

## EXERCISE SET 7.4

In all the following exercises assume that all elementary events in the same sample space are equally likely.

1. A fair coin is flipped 3 times. What is the probability of obtaining exactly 2 heads? At least 1 head?

2. A family has 3 children. Assuming a boy is as likely as a girl to have been born, what is the probability that 2 are boys and 1 is a girl? That at least one is a boy?

3. A fair coin is flipped and a fair die is tossed. What is the probability of obtaining a head and a 3?

4. A fair coin is flipped twice and a fair die is tossed. What is the probability of obtaining 2 heads and a 3?

5. A pair of fair dice are tossed. What is the probability of obtaining a sum of 2? 4? 8?

6. A pair of fair dice are tossed. What is the probability of obtaining a sum of 5? 6? 11?

In Exercises 7 through 8, let an urn have 10 balls, identical except that 4 are white and 6 are red.

7. If 3 are selected randomly without replacement, what is the probability that 2 are white and 1 is red? At least 2 are white?

8. If 5 are selected randomly without replacement, what is the probability that 3 are white and 2 are red? At least 3 are white?

In Exercises 9 through 12, let an urn have 21 identical balls except that 6 are white, 7 are red, and 8 are blue.

9. What is the probability that there is one of each color if 3 are selected randomly without replacement?

10. What is the probability that all are white if 3 are selected randomly without replacement?

11. What is the probability that 3 are white, 2 are red, and 1 is blue if 6 are selected randomly without replacement?

12. What is the probability that at least 5 are white if 6 are selected randomly without replacement?

In Exercises 13 through 18 a 2-card hand is drawn from a standard deck of 52 cards. Find the probability that the hand contains the given cards.

13. Two kings

14. Two spades

15. A pair

16. Two of the same suit

17. Two consecutive cards

18. No face card

In Exercises 19 through 26 find the probability of obtaining each of the given in a 5-card poker hand. (*Hint:* the probabilities increase.)

19. Royal flush: ace, king, queen, jack, ten in the same suit

20. Straight flush: five cards in sequence in the same suit but not a royal flush

21. Four of a kind: four queens, four sevens, etc.

22. Full house: three of a kind together with a pair

23. Straight: five cards in sequence not all in the same suit

24. Three of a kind

25. Two pairs

26. One pair

27. Assume that the probability of an individual being born in any month is the same and that there are $n$ individuals in a room. Find the probability that at least two individuals have their birthdays in the same month when $n = 2, 3, 4, 5$.

28. Suppose $n$ different letters have been written with $n$ corresponding addressed envelopes, and the letters are inserted *randomly* into the envelopes. What is the probability that no letter gets into its correct envelope for $n = 2, 3, 4, 5$?

29. What is the probability that at least two members of the 434-member United States House of Representatives have their birthdays on the same day?

30. What is the probability that at least 2 of the 100 Senators of the U.S. Congress have the same birthday?

## Applications

31. **Quality Control.** A bin has 4 defective transistors and 6 nondefective ones. If 2 are picked randomly from the bin, what is the probability that both are defective?

32. **Stock Selection.** Among a group of 20 stocks, suppose that 10 stocks will perform above average and the other 10 below average. If you pick 3 stocks from this group, what is the probability that all 3 will be above average in performance?

33. **Mutual Funds.** Suppose in any year a certain mutual fund is just as likely to perform above average as not. Find the probability that this fund will perform above average in at least 8 of the next 10 years.

34. **Committees.** A committee of 3 is to be selected at random from a group of 3 senior and 4 junior executives. What is the probability that the committee will have more senior than junior executives?

35. **Testing.** A company places a dozen of the same product in one box. Before sealing, 3 of the product are tested. If any of the 3 is defective, the entire box will be rejected. Suppose a box has 2 defective products. What is the probability the box will be rejected?

36. **Awarding of Contracts.** Suppose that there are 3 corporations competing for 4 different government contracts. If the contracts are awarded randomly, what is the probability that each corporation will get a contract?

## Solutions to Self-Help Exercise Set 7.4

1. Referring to Solutions to Self-Help Exercise Set 7.1, Exercises 1 and 2, we see that $n(S) = 16$, $n(E) = 4$, $n(F) = 4$, $n(E \cap F) = 0$, $n(E \cup F) = 8$, $n(G^c) = 3$. Furthermore there is no reason to assume that any of the elementary events in $S$ are any more or less likely than any other. So we assume the uniform probability. Thus

$$p(E) = \frac{n(E)}{n(S)} = \frac{4}{16} = \frac{1}{4}$$

$$p(F) = \frac{n(F)}{n(S)} = \frac{4}{16} = \frac{1}{4}$$

$$p(E \cap F) = \frac{n(E \cap F)}{n(S)} = \frac{0}{16} = 0$$

$$p(E \cup F) = \frac{n(E \cup F)}{n(S)} = \frac{8}{16} = \frac{1}{2}$$

$$p(G^c) = \frac{n(G^c)}{n(S)} = \frac{3}{16}.$$

**2.    a.**    The number in the sample space $S$ is the number of ways of selecting 6 objects (without replacement) from a set of 42 where order is not important. This is

$$C(42, 6) = \frac{42 \cdot 41 \cdot 40 \cdot 39 \cdot 38 \cdot 37}{6 \cdot 5 \cdot 4 \cdot 3 \cdot 2 \cdot 1} = 5{,}245{,}786.$$

Thus the probability of any one pick is

$$\frac{1}{5{,}245{,}786}.$$

**2.    b.**    The picks in which the numbers are consecutive are

$$\{1, 2, 3, 4, 5, 6\}, \ldots, \{37, 38, 39, 40, 41, 42\}.$$

There are 37 such selections. Thus the probability of any one of these being the winning number is

$$\frac{37}{5{,}245{,}786}.$$

## 7.5    CONDITIONAL PROBABILITY

▶ *Conditional Probability*

▶ *Product Rule*

▶ *Probability Trees*

▶ *Independent Events*

A P P L I C A T I O N

**Locating Defective Parts**

A company makes the components for a product at a central location. These components are shipped to 3 plants, 1, 2, and 3, for assembly into a final product. The percentages of the product assembled by the 3 plants are, respectively, 50%, 20%, and 30%. The percentages of defective products coming from these 3 plants are, respectively, 1%, 2%, and 3%. What is the probability of randomly choosing a product made by this company that is defective from Plant 1? See Example 5 for the answer.

## Conditional Probability

The probability of an event is often affected by the occurrences of other events. For example, there is a certain probability that an individual will die of lung cancer. But if a person smokes heavily, then the probability that this person will die of lung cancer is higher. That is, the probability has changed with the additional information. In this section we study such *conditional probabilities*. This important idea is further considered in the next section.

Given two events $E$ and $F$, we call the probability that $E$ will occur given that $F$ has occurred the **conditional probability** and write $p(E|F)$.

Conditional probability normally arises in situations where the old probability is "updated" based on new information. Suppose, for example, a new family moves in next door. The real estate agent mentions that this new family has two children. Based on this information, you can calculate the probability that both children are boys. Now a neighbor mentions that they met the oldest child from the new family and this child is a boy. Now the probability that both children from the new family are boys has changed given this new information.

**E X A M P L E   1**   **Finding Conditional Probability**
_____

A new family has moved in next door and is known to have two children. (a) Find the probability that both children are boys. (b) Find the probability that both children are boys given that the oldest is a boy. Assume that a boy is as likely as a girl.

**Solutions**

(a)  With no information at all about either child, the sample space is

$$S = \{BB,\ BG,\ GB,\ GG\}$$

where the first element in each ordered pair refers to the youngest child and the second element to the oldest. We know from the last section that it is reasonable to assign uniform probability to this sample space of 4 outcomes and if $E = \{BB\}$, $p(E) = \frac{1}{4}$.

(b)  Given the information that the oldest child is a boy, the new sample space becomes $S_2 = \{BB,\ GB\}$. There is no reason to assign anything other than the uniform probability to this new sample space with 2 outcomes. Doing this we then obtain the probability of the event $\{BB\}$ in the new sample space is $\frac{1}{2}$. That is, if $F = \{BB,\ GB\}$, $p(E|F) = \frac{1}{2}$.  ∎

**E X A M P L E   2**   **Finding Conditional Probability**
_____

A card is drawn randomly from a deck of 52 cards. (a) What is the probability that this card is an ace? (b) What is the probability that this card is an ace given that the card is known to be a 10 or higher?

**Solutions**

(a)  The sample space consists of 52 cards and has the uniform probability. Thus, from the last section, if $E = \{\text{ace}\}$, $p(E) = \frac{4}{52}$.

(b)  With the new information, the new sample space $S_2$ is now the space consisting of the 20 cards that are 10 or higher. Since there is no reason to assume that any of these is any more or less likely than any other, we assume the uniform probability on $S_2$. Letting $F$ be the event in $S$ consisting of all cards 10 or higher (which of course is also $S_2$), then the probability of obtaining an ace, given that the card is a 10 or higher, is

$$p(E|F) = \frac{4}{20}. \quad \blacksquare$$

We have seen in the previous two examples that finding the conditional probability requires identifying the *new* sample space and determining the *new* probability. The procedure used illustrated these important points but is not adequate in general. We now see how to define conditional probability in a more precise and useful manner.

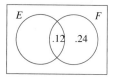

**Figure 7.7**

To do this we return to our intuitive notions of probability. Suppose an experiment yields a sample space $S$, and $E$ and $F$ are two events in $S$ with $p(E \cap F) = .12$ and $p(F) = .36$ as shown in Figure 7.7. If $N = 100$ is the number of times the experiment has been repeated, then we expect the frequency of occurrence of $E \cap F$ and $F$ to be $f(E \cap F) = 12$ and $f(F) = 36$. The conditional probability $p(E|F)$ is the expected fraction of those experiments in which $F$ occurred that $E$ also occurred. Therefore we expect

$$p(E|F) = \frac{f(E \cap F)}{f(F)} = \frac{12}{36} = \frac{1}{3}.$$

Another instructive way of viewing this is to divide the numerator and the denominator of the last fraction by $N = 100$ and obtain

$$p(E|F) = \frac{f(E \cap F)/100}{f(F)/ 100} = \frac{.12}{.36} = \frac{p(E \cap F)}{p(F)}.$$

More generally, suppose we are given a sample space $S$ and any two events $E$ and $F$ with $p(F) > 0$. If $N$ is the number of times the experiment is repeated and $N$ is very large, then we expect that

$$\frac{f(F)}{N} \approx p(F) \quad \text{and} \quad \frac{f(E \cap F)}{N} \approx p(E \cap F).$$

On the other hand, the conditional probability $p(E|F)$ is the expected fraction of those experiments in which $F$ occurred that $E$ also occurred. We therefore expect

$$p(E|F) \approx \frac{f(E \cap F)}{f(F)}.$$

We now divide the numerator and denominator of the last fraction by $N$ and obtain

$$p(E|F) \approx \frac{f(E \cap F)}{f(F)} = \frac{f(E \cap F)/N}{f(F)/N} \approx \frac{p(E \cap F)}{p(F)}.$$

This motivates the following definition.

---

**Conditional Probability**

Let $E$ and $F$ be two events in a sample space $S$. The conditional probability that $E$ occurs given that $F$ has occurred is defined to be

$$p(E|F) = \frac{p(E \cap F)}{p(F)} \quad \text{if} \quad p(F) > 0.$$

---

**REMARK.** It is worthwhile to notice that the events $E$, $F$, and $E \cap F$ are all in the original sample space $S$ and $p(E)$, $p(F)$, and $p(E \cap F)$ are all probabilities defined on $S$. However $p(E|F)$ is a probability defined on the new sample space $S' = F$. See Figure 7.8.

Old sample space     New sample space

**Figure 7.8**

**EXAMPLE 3**     **Calculating a Conditional Probability**

A new family has moved in next door and is known to have two children. Find the probability that both children are boys given that at least one is a boy. Assume that a boy is as likely as a girl.

**Solution**     Let $E$ be the event that both children are a boy and $F$ the event that at least one is a boy. Then

$$S = \{BB, BG, GB, GG\},$$

$E = \{BB\}$, $F = \{BB, BG, GB\}$, $E \cap F = \{BB\}$, and

$$p(E|F) = \frac{p(E \cap F)}{p(F)} = \frac{\frac{1}{4}}{\frac{3}{4}} = \frac{1}{3}. \quad \blacksquare$$

## Product Rule

We will now see how to write $p(E \cap F)$ in terms of a product of two probabilities. From the definition of conditional probability, we have

$$p(E|F) = \frac{p(E \cap F)}{p(F)}, \qquad p(F|E) = \frac{p(F \cap E)}{p(E)},$$

if $p(E) > 0$ and $p(F) > 0$. Solving for $p(E \cap F)$ and $p(F \cap E)$, we obtain

$$p(E \cap F) = p(F)p(E|F), \qquad p(F \cap E) = p(E)p(F|E).$$

Since $E \cap F = F \cap E$, it follows that

$$p(E \cap F) = p(F)p(E|F) = p(E)p(F|E).$$

This is called the product rule.

---

**Product Rule**

If $E$ and $F$ are two events in a sample space $S$ with $p(E) > 0$ and $p(F) > 0$, then

$$p(E \cap F) = p(F)p(E|F) = p(E)p(F|E).$$

---

**E X A M P L E   4**    **Using the Product Rule**

Two bins contain transistors. The first has 5 defective and 15 nondefective while the second has 3 defective and 17 nondefective ones. If the probability of picking either bin is the same, what is the probability of picking the first bin and a good transistor?

**Solution**    The sample space is $S = \{1D, 1N, 2D, 2N\}$ where the number refers to picking the first or second bin and the letter refers to picking a defective ($D$) or nondefective ($N$) transistor.

   If $E$ is the event "pick the first bin" and $F$ is the event "pick a nondefective transistor," then $E = \{1D, 1N\}$ and $F = \{1N, 2N\}$. The probability of picking a nondefective transistor given that the first bin has been picked is the conditional probability $p(F|E) = \frac{15}{20}$. The event "picking the first bin and a nondefective transistor" is $E \cap F$. From the product rule

$$p(E \cap F) = p(E)p(F|E) = \frac{1}{2}\frac{15}{20} = \frac{3}{8}. \qquad \blacksquare$$

## Probability Trees

We shall now consider a finite sequence of experiments in which the outcomes and associated probabilities of each experiment depend on the outcomes of the preceding experiments. Such a finite sequence of experiments is called a **finite stochastic process**. Stochastic processes can be effectively described by probability trees that we now consider. The following example should be studied carefully since we will return to it at the beginning of the next section.

**E X A M P L E   5**    **Using a Probability Tree**

A company makes the components for a product at a central location. These components are shipped to 3 plants, 1, 2, and 3, for assembly into a final product. The percentages of the product assembled by the 3 plants are, respectively, 50%,

20%, and 30%. The percentages of defective products coming from these 3 plants are, respectively, 1%, 2%, and 3%. (a) What is the probability of randomly choosing a product made by this company that is defective from Plant 1? 2? 3? (b) What is the probability of randomly choosing a product made by this company that is defective?

Before proceeding, some notational conventions will be discussed that will be used in this problem and others. The sample space is

$$S = \{1N, 1D, 2N, 2D, 3N, 3D\},$$

where for example "$1D$" means selecting Plant 1 and a defective product from this plant. Thus the event "Plant 1 has been selected" is given by $\{1N, 1D\}$. However, this is rather cumbersome and we follow the usual convention and just call this event "1." The event "a defective product has been selected" is given by $\{1D, 2D, 3D\}$. Again this is rather inconvenient, and we just refer to this event as "$D$." Then the event "select a defective product from Plant 1" is the intersection of the previous two events, expressed in our shorthand notation as

$$1 \cap D = \{1N, 1D\} \cap \{1D, 2D, 3D\} = \{1D\}.$$

**Solutions**    Using these conventions, the probability of the part being assembled in Plant 1 is $p(1) = .5$, in Plant 2 is $p(2) = .2$, and in Plant 3 is $p(3) = .3$. We are also given the conditional probabilities, $p(D|1) = .01$, $p(D|2) = .02$, $p(D|3) = .03$. This is all shown in the tree diagram in Figure 7.9.

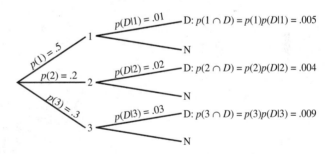

**Figure 7.9**

(a)   Use the product rule and obtain

$$p(1D) = p(1 \cap D) = p(1)p(D|1) = (.5)(.01) = .005$$

$$p(2D) = p(2 \cap D) = p(2)p(D|2) = (.2)(.02) = .004$$

$$p(3D) = p(3 \cap D) = p(3)p(D|3) = (.3)(.03) = .009.$$

(b)   Using (a) gives

$$p(D) = p(\{1D, 2D, 3D\}) = p(1D) + p(2D) + p(3D)$$

$$= .005 + .004 + .009 = .018. \quad \blacksquare$$

EXAMPLE 6    **Using a Probability Tree**

A box contains 3 defective ($D$) parts and 4 nondefective ($N$) ones. One randomly selects a part (without replacement) until a nondefective one is obtained. What is the probability that the number of parts selected is (a) one, (b) two, (c) three?

Solutions

(a) There are 4 nondefective parts in a box of 7. So the probability of selecting a nondefective part is $\frac{4}{7}$.

We note for further reference that the probability of selecting a defective part is $\frac{3}{7}$.

(b) The only way it can take 2 selections to obtain a nondefective part is for the first selection to be defective. In this case there are 2 defective and 4 nondefective parts left in the box. Thus the probability of selecting a nondefective part given that a defective part was chosen first is $\frac{4}{6}$. From the tree in Figure 7.10, the probability of the branch $DN$ is $\frac{3}{7}\frac{4}{6} = \frac{2}{7}$.

We note for further reference that the probability of selecting a defective part on the second selection given that a defective part was selected first is $\frac{2}{6}$. See Figure 7.10.

(c) The only way it can take 3 selections to obtain a nondefective part is for the first two selections to be defective. In this case there are 1 defective and 4 nondefective parts left in the box. Thus the probability of selecting a nondefective part given that a defective part was chosen the first and second time is $\frac{4}{5}$. From the tree in Figure 7.10, the probability of the branch $DDN$ is $\frac{3}{7}\frac{2}{6}\frac{4}{5} = \frac{4}{35}$. ∎

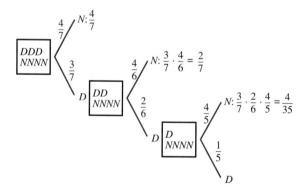

**Figure 7.10**

## Independent Events

Intuitively, two events $E$ and $F$ are **independent** if the outcome of one does not affect the outcome of the other. For example, the probability of obtaining a head on a second flip of a coin is independent from what happened on the first flip. This is intuitively clear since the coin cannot have any memory of what happened on the first flip. Indeed the laws of physics determine the probability of heads occurring. Thus for the probability of heads to be different on the second flip the laws of physics must be different on the second flip. On the other hand, the probability of selecting an ace on the second draw (without replacement) clearly depends on what happens on the first draw. Thus drawing an ace on the second draw without replacement is not independent from drawing an ace on the first draw.

Intuitively, two events $E$ and $F$ are independent if

$$p(E|F) = p(E) \quad \text{and} \quad p(F|E) = p(F),$$

that is, the probability of $E$ given that $F$ has occurred is just the probability of $E$, and, similarly, the probability of $F$ given that $E$ has occurred is just the probability of $F$.

---

**Independent Events**

Two events $E$ and $F$ are said to be independent if

$$p(E|F) = p(E) \quad \text{and} \quad p(F|E) = p(F)$$

---

We shall now obtain a result that is more convenient to apply when attempting to determine if two events are independent.

If then two events $E$ and $F$ are independent, the previous comments together with the product rule indicate that

$$p(E \cap F) = p(E|F)p(F) = p(E)p(F).$$

Now consider the case that $p(E) > 0$ and $p(F) > 0$. Assume that $p(E \cap F) = p(E)p(F)$, then

$$p(E|F) = \frac{p(E \cap F)}{p(F)} = \frac{p(E)p(F)}{p(F)} = p(E)$$

$$p(F|E) = \frac{p(E \cap F)}{p(E)} = \frac{p(E)p(F)}{p(E)} = p(F).$$

This discussion then yields the following theorem.

---

**Independent Events Theorem**

Let $E$ and $F$ be two events with $p(E) > 0$ and $p(F) > 0$. Then $E$ and $F$ are independent if, and only if,

$$p(E \cap F) = p(E)p(F).$$

---

Although at times one is certain whether or not two events are independent, often one can only tell by doing the calculations.

**E X A M P L E   7**     **Determining if Two Events are Independent**

A study of 1000 men over 65 indicated that 250 smoked and 50 of these smokers had some signs of heart disease, while 100 of the nonsmokers showed some signs of heart disease. Let $E$ be the event ''smokes'' and $H$ be the event ''has signs of heart disease.'' Are these two events independent?

**Solution**     The Venn diagram is given in Figure 7.11. From this diagram $p(H) = .15$, $p(E) = .25$, and $p(H \cap E) = .05$. Thus

$$p(H)p(E) = (.15)(.25)$$
$$= .0375$$
$$\neq .05$$
$$= p(H \cap E),$$

**Figure 7.11**

and the two events are not independent.   ∎

**E X A M P L E   8**     **Determining if Two Events are Independent**

Answer the previous question if a study of 1000 men over 65 indicated that 500 smoked and 100 of these smokers have some sign of heart disease, while 100 of the nonsmokers showed some signs of heart disease.

**Solution**     From the Venn diagram in Figure 7.12, we have

$$p(H)p(E) = (.20)(.50)$$
$$= .10$$
$$= p(H \cap E).$$

**Figure 7.12**

Thus these events are independent.   ∎

**W A R N I N G .**     Notice that saying two events are independent is not the same as saying that they are mutually exclusive. The sets $E$ and $H$ in both previous examples were not mutually exclusive, but in one case the sets were independent and in the other case they were not.

The notion of independence can be extended to any number of finite sets.

**Independent Set of Events**

A set of events $\{E_1, E_2, \ldots, E_n\}$ is said to be independent if, for any $k$ of these events, the probability of the intersection of these $k$ events is the product of the probabilities of each of the $k$ events. This must hold for any $k = 2, 3, \ldots, n$.

For example, for $\{E, F, G\}$ to be independent all of the following must be true:

$$p(E \cap F) = p(E)p(F), \quad p(E \cap G) = p(E)p(G), \quad p(F \cap G) = p(F)p(G),$$

$$p(E \cap F \cap G) = p(E)p(F)p(G).$$

It is intuitively clear that if two events $E$ and $F$ are independent, then so also are $E$ and $F^c$, $E^c$ and $F$, $E^c$ and $F^c$. (See Exercises 38 and 39.) Similar statements are true about a set of events.

**E X A M P L E   9**   **Independent Events and Safety**

An aircraft has a system of three computers, each independently able to exercise control of the flight. The computers are considered 99.9% reliable during a routine flight. What is the probability of having a failure of the control system during a routine flight?

**Solution**   Let the events $E_i$, $i = 1, 2, 3$ be the three events given by the reliable performance of respectively the first, second, and third computer. Since the set of events $\{E_1, E_2, E_3\}$ is independent, so is the set of events $\{E_1^c, E_2^c, E_3^c\}$. The system will fail only if all three computers fail. Thus the probability of failure of the system is given by

$$p(E_1^c \cap E_2^c \cap E_3^c) = p(E_1^c)p(E_2^c)p(E_3^c) = (.001)^3,$$

which, of course, is an extremely small number.   ∎

## SELF-HELP EXERCISE SET 7.5

1.  Three companies A, B, and C are competing for a contract. The probabilities that they receive the contract are, respectively, $p(A) = \frac{1}{6}$, $p(B) = \frac{1}{3}$, and $p(C) = \frac{1}{2}$. What is the probability that A will receive the contract if C pulls out of the bidding?

2.  Two bins contain transistors. The first has 4 defective and 15 nondefective, while the second has 3 defective and 22 nondefective ones. If the probability of picking either bin is the same, what is the probability of picking the second bin and a defective transistor?

3.  Success is said to breed success. Suppose you are in a best-of-3-game tennis match with an evenly matched opponent. However, if you win a game, your probability of winning the next increases from $\frac{1}{2}$ to $\frac{2}{3}$. Suppose, however, that if you lose, the probability of winning the next match remains the same. (Success does not breed success for your opponent.) What is your probability of winning the match? *Hint:* Draw a tree.

4.  A family has 3 children. Let $E$ be the event "at most one boy" and $F$ the event "at least one boy and at least one girl." Are $E$ and $F$ independent if a boy is as likely as a girl? *Hint:* Write down every element in the sample space $S$ and the events $E$, $F$, $E \cap F$, and find the appropriate probabilities by counting.

## EXERCISE SET 7.5

In Exercises 1 through 6 refer to the accompanying Venn diagram to find the conditional probabilities.

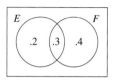

1.  **a.** $p(E|F)$,                    **b.**  $p(F|E)$

2.  **a.** $p(E^c|F)$,                  **b.**  $p(F^c|E)$

3.  **a.** $p(E|F^c)$,                  **b.**  $p(F|E^c)$

4.  **a.** $p(E^c|F^c)$,                **b.**  $p(E \cup F|F)$

5.  **a.** $p(F|E \cap F)$,             **b.**  $p(F^c|F)$

6.  **a.** $p(E^c \cap F|F)$,           **b.**  $p(E \cap F^c|F)$

In Exercises 7 through 12, let $p(E) = .4$, $p(F) = .6$, and $p(E \cap F) = .2$. Draw a Venn diagram and find the conditional probabilities.

7.  **a.** $p(E^c|F)$,                  **b.**  $p(F^c|E)$

8.  **a.** $p(E|F)$,                    **b.**  $p(F|E)$

9.  **a.** $p(F^c|E^c)$,                **b.**  $p(E \cup F|E)$

10.  **a.** $p(E|F^c)$,                 **b.**  $p(F|E^c)$

11.  **a.** $p(E^c \cap F|E)$,          **b.**  $p(E \cap F^c|E)$

12.  **a.** $p(F|E \cap F)$,            **b.**  $p(E^c|E)$

In Exercises 13 through 20, determine if the given events $E$ and $F$ are independent.

13.  $p(E) = .3, p(F) = .5, p(E \cap F) = .2$

14.  $p(E) = .5, p(F) = .7, p(E \cap F) = .3$

15.  $p(E) = .2, p(F) = .5, p(E \cap F) = .1$

16.  $p(E) = .4, p(F) = .5, p(E \cap F) = .2$

17.  $p(E) = .4, p(F) = .3, p(E \cup F) = .6$

18.  $p(E \cap F^c) = .3, p(E \cap F) = .2, p(E^c \cap F) = .2$

19.  $p(E \cap F^c) = .3, p(E \cap F) = .3, p(E^c \cap F) = .2$

20.  $p(E) = .2, p(F) = .5, p(E \cup F) = .6$

21.  A pair of fair dice is tossed. What is the probability that a sum of seven has been tossed if it is known that at least one of the numbers is a 3?

22.  A single fair die is tossed. What is the probability that a 3 occurs on the top if it is known that the number is a prime?

23.  A fair coin is flipped 3 times. What is the probability that heads occurs 3 times if it is known that heads occurs at least once?

24.  A fair coin is flipped 4 times. What is the probability that heads occurs 3 times if it is known that heads occurs at least twice?

25.  Three cards are randomly drawn without replacement from a standard deck of 52 cards.
     **a.**  What is the probability of drawing an ace on the third draw?
     **b.**  What is the probability of drawing an ace on the third draw given that at least one ace was drawn on the first 2 draws?

26.  Three balls are randomly drawn from an urn that contains 4 white and 6 red balls.
     **a.**  What is the probability of drawing a red ball on the third draw?
     **b.**  What is the probability of drawing a red ball on the third draw given that at least one red ball was drawn on the first 2 draws?

27.  From the tree diagram find (a) $p(A \cap E)$, (b) $p(A)$, (c) $p(A|E)$.

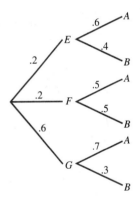

28.  From the tree diagram find (a) $p(A \cap E)$, (b) $p(A)$, (c) $p(A|E)$.

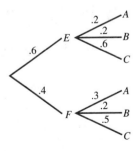

29.  An urn contains 5 white, 3 red, and 2 blue balls. Two balls are randomly drawn. What is the probability that one is white and one is red if the balls are drawn (a) without replacement? (b) with replacement after each draw?

**30.** An urn contains 4 white and 6 red balls. Two balls are randomly drawn. If the first one is white, the ball is replaced. If the first one is red, the ball is not replaced. What is the probability of drawing at least one white ball?

**31.** In a family of 4 children, let $E$ be the event "at most one boy" and $F$ the event "at least one girl and at least one boy." If a boy is as likely as a girl, are these two events independent?

**32.** A fair coin is flipped three times. Let $E$ be the event "at most one head" and $F$ the event "at least one head and at least one tail." Are these two events independent?

**33.** The two events $E$ and $F$ are independent with $p(E) = .3$, and $p(F) = .5$. Find $p(E \cup F)$.

**34.** The two events $E$ and $F$ are independent with $p(E) = .4$, and $p(F) = .6$. Find $p(E \cup F)$.

**35.** The three events $E$, $F$, and $G$ are independent with $p(E) = .2$, $p(F) = .3$, $p(G) = .5$. What is $p(E \cup F \cup G)$?

**36.** The three events $E$, $F$, and $G$ are independent with $p(E) = .3$, $p(F) = .4$, $p(G) = .6$. What is $p(E^c \cup F^c \cup G^c)$?

**37.** Given the probabilities shown in the accompanying Venn diagram, show that the events $E$ and $F$ are independent if, and only if,

$$p_1 p_3 = p_2 p_4.$$

What must the Venn diagram look like if the sets are mutually disjoint?

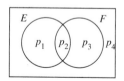

**38.** Show that if $E$ and $F$ are independent, then so are $E$ and $F^c$.

**39.** Show that if $E$ and $F$ are independent, then so are $E^c$ and $F^c$.

**40.** Show that two events are indepenent if they are mutually exclusive and the probability of one of them is zero.

**41.** Show that two events $E$ and $F$ are not independent if they are mutually exclusive and both have nonzero probability.

**42.** Show that if $E$ and $F$ are independent events, then

$$p(E \cup F) = 1 - p(E^c)p(F^c).$$

**43.** If $p(F) > 0$, then show that

$$p(E^c|F) = 1 - p(E|F).$$

**44.** If $E$, $F$, $G$ are three events and $p(G) > 0$, show that

$$p(E \cup F|G) = p(E|G) + p(F|G) - p(E \cap F|G).$$

**45.** If $E$ and $F$ are two events with $F \subset E$, then show that $p(E|F) = 1$.

**46.** If $E$ and $F$ are two events with $E \cap F = \phi$, then show that $p(E|F) = 0$.

**47.** If $E$ and $F$ are two events, show that

$$p(E|F) + p(E^c|F) = 1.$$

## Applications

**48. Manufacturing.** Two machines turn out all the products in a factory, with the first machine producing 40% of the product and the second 60%. The first machine produces defective products 2% of the time and the second machine 4% of the time.
   **a.** What is the probability that a defective product is produced at this factory given that it was made on the first machine?
   **b.** What is the probability that a defective product is produced at this factory?
   **c.** Given a defective product, what is the probability it was produced on the first machine?

**49. Manufacturing.** A plant has 3 assembly lines with the first line producing 50% of the product and the second 30%. The first line produces defective products 1% of the time, the second line 2% of the time, and the third 3% of the time.
   **a.** What is the probability that a defective product is produced at this plant given that it was made on the second assembly line?
   **b.** What is the probability that a defective product is produced at this plant?
   **c.** Given a defective product, what is the probability it was produced on the second assembly line?

**50. Advertising.** A television ad for a company's product has been seen by 20% of the population. Of those who see the ad, 10% then buy the product. Of those who do not see the ad, 2% buy the product. Find the probability that a person buys the product.

**51. Suppliers.** A manufacturer buys 40% of a certain part from one supplier and the rest from a second supplier. It notes that 2% of the parts from the first supplier are defective, and 3% are defective from the second supplier. Given that a part is defective, what is the probability it came from the first supplier?

**52. Psychology and Sales.** A door-to-door salesman expects to makes a sale 10% of the time when starting the day. But making a sale increases his enthusiasm so much that the probability of a sale to the next customer is .2. If he makes no sale, the probability for a sale stays at .1. What is the probability that he will make at least 2 sales with his first 3 visits?

**53. Quality Control.** A box contains 2 defective (D) parts and 5 nondefective (N) ones. You randomly select a part (without replacement) until you get a nondefective part. What is the probability that the number of parts selected is (a) one, (b) two, (c) three?

**54. Medicine.** The probability of residents of a certain town contracting cancer is .01. Let $x$ be the percent of residents that work for a certain chemical plant and suppose that the probability of both working for this plant and of contracting cancer is .001. What must $x$ be for the two events "gets cancer" and "works for the chemical plant" to be independent?

55. **Medicine.** In a study of 250 men over 65, 100 smoked, 60 of the smokers had some signs of heart disease, and 90 of the nonsmokers showed some signs of heart disease. Let $E$ be the event "smokes" and $H$ be the event "has signs of heart disease." Are these two events independent?

56. **Sales.** A company sells machine tools to two firms in a certain city. In 40% of the years it makes a sale to the first firm, 30% of the years to the second firm, and 10% to both. Are the two events "a sale to the first firm" and "a sale to the second firm" independent?

57. **Stocks.** A firm checks the last 200 days on which its stock has traded. On 100 of these occasions the stock has risen in price with a broad-based market index also rising on 70 of these particular days. The same market index has risen on 90 of the 200 trading days. Are the movement of the firm's stock and the movement of the market index independent?

58. **Contracts.** A firm has bids on two contracts. It is known that the awarding of these two contracts are independent events. If the probability of receiving these contracts are .3 and .4 respectively, what is the probability of not receiving either?

59. **Reliability.** A firm is making a very expensive optical lens to be used in an earth satellite. To be assured that the lens has been ground correctly, three independent tests using entirely different techniques are used. The probability is .99 that any of one of these tests will detect a defect in the lens. What is the probability that the lens has a defect even though none of the three tests so indicates?

*(Photo courtesy of NASA.)*

## Solutions to Self-Help Exercise Set 7.5

1. If $E$ is the event that $A$ obtains the contract and $F$ the event that either A or B obtain the contract, then $E = \{A\}$, $F = \{A, B\}$, and $E \cap F = \{A\}$, and the conditional probability that $A$ will receive the contract if C pulls out of the bidding is

$$p(E|F) = \frac{p(E \cap F)}{p(F)} = \frac{\frac{1}{6}}{\frac{1}{2}} = \frac{1}{3}.$$

2. The sample space is $S = \{1D, 1N, 2D, 2N\}$ where the number refers to picking the first or second bin and the letter refers to picking a defective ($D$) or nondefective ($N$) transistor. If $E$ is picking the first bin and $F$ is picking a defective transistor, then

$$p(E \cap F) = p(E)p(F|E) = \frac{1}{2}\frac{3}{25} = \frac{3}{50}.$$

3. The appropriate tree is given. The probability of winning the match is then

$$\frac{1}{3} + \frac{1}{12} + \frac{1}{6} = \frac{7}{12}.$$

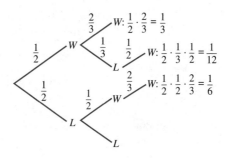

4. The elements in the spaces $S$, $E$, $F$, and $E \cap F$ are

$$S = \{BBB, BBG, BGB, BGG, GBB, GBG, GGB, GGG\}$$

$$E = \{BGG, GBG, GGB, GGG\}$$

$$F = \{BBG, BGB, BGG, GBB, GBG, GGB\}$$

$$E \cap F = \{BGG, GBG, GGB\}.$$

Thus counting elements gives

$$p(E \cap F) = \frac{3}{8} \qquad p(E)p(F) = \frac{4}{8}\frac{6}{8} = \frac{3}{8}.$$

Since these two numbers are the same, the two events are independent.

## 7.6  BAYES' THEOREM

▶ *Bayes' Theorem*

▶ *Applications*

**APPLICATION**

**Finding a Probability of a Defective Part**

Recall Example 5 of the last section. A company makes the components for a product at a central location. These components are shipped to 3 plants, 1, 2, and 3, for assembly into a final product. The percentages of the product assembled by the 3 plants are, respectively, 50%, 20% and 30%. The percentages of defective products coming from these 3 plants are, respectively, 1%, 2%, and 3%. Given a defective product, what is the probability it was assembled at Plant 1? At Plant 2? At Plant 3? See the discussion immediately below for the answers to these questions.

### Bayes' Theorem

We have been concerned with finding the probability of an event that will occur in the future. We now look at calculating probabilities after the events have occurred.

We begin this section by answering the question posed above. Refer to Figure 7.13. Recall from the last section that we had

$$p(1D) = p(1 \cap D) = p(1)p(D|1) = (.5)(.01) = .005$$

$$p(2D) = p(2 \cap D) = p(2)p(D|2) = (.2)(.02) = .004$$

$$p(3D) = p(3 \cap D) = p(3)p(D|3) = (.3)(.03) = .009,$$

and thus

$$p(D) = p(\{1D, 2D, 3D\}) = p(1D) + p(2D) + p(3D)$$

$$= .005 + .004 + .009$$

$$= .018.$$

Using this information and the definition of conditional probability, we then have

$$p(1|D) = \frac{p(1 \cap D)}{p(D)} = \frac{.005}{.005 + .004 + .009} = \frac{5}{18}.$$

We now notice that the numerator of this fraction is the probability of the branch $1D$, while the denominator is the sum of all the probabilities of all branches that end in $D$. We also have

$$p(2|D) = \frac{p(2 \cap D)}{p(D)} = \frac{.004}{.018} = \frac{4}{18}.$$

Now notice that the numerator of this fraction is the probability of the branch $2D$, while the denominator is the sum of all the probabilities of all branches that end in $D$. A similar statement can be made for

$$p(3|D) = \frac{p(3 \cap D)}{p(D)} = \frac{.009}{.018} = \frac{9}{18}.$$

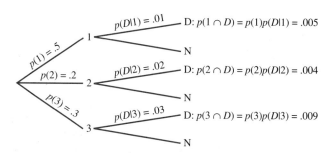

**Figure 7.13**

We will see that this is an example of **Bayes' theorem**. This theorem was discovered by the Presbyterian minister Thomas Bayes (1702–1763). We will now indicate how to establish this result and the exact conditions needed.

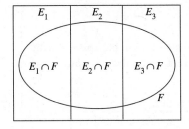

**Figure 7.14**

We suppose that we are given a sample space $S$ and three mutually exclusive events $E_1$, $E_2$, $E_3$, with $E_1 \cup E_2 \cup E_3 = S$ as indicated in Figure 7.14. Notice that the three events divide the space $S$ into 3 partitions. Given another event $F$, the tree diagram of possibilities is shown in Figure 7.15. We now wish to show that the probability $p(F|E_1)$ is the fraction with the numerator given by the probability of the branch $E_1F$ while the denominator is the sum of all the probabilities of all branches that end in $F$.

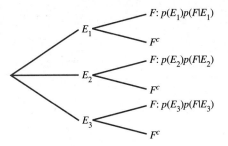

**Figure 7.15**

To see this first notice from Figure 7.14 that

$$F = (E_1 \cap F) \cup (E_2 \cap F) \cup (E_3 \cap F)$$

and

$$p(F) = p(E_1 \cap F) + p(E_2 \cap F) + p(E_3 \cap F).$$

Using the product rule, this can be written as

$$p(F) = p(E_1)p(F|E_1) + p(E_2)p(F|E_2) + p(E_3)p(F|E_3).$$

On the other hand notice that

$$p(F|E_1) = \frac{p(E_1 \cap F)}{p(E_1)} \quad \text{implies} \quad p(E_1 \cap F) = p(E_1)p(F|E_1).$$

Using this then gives

$$p(E_1|F) = \frac{p(E_1 \cap F)}{p(F)} = \frac{p(E_1)p(F|E_1)}{p(F)}$$

$$= \frac{p(E_1)p(F|E_1)}{p(E_1)p(F|E_1) + p(E_2)p(F|E_2) + p(E_3)p(F|E_3)}.$$

Similar statements can be made about the other probabilities. For example, $p(F|E_2)$ is the fraction with the numerator given by the probability of the branch $E_2F$, while the denominator is the sum of all the probabilities of all branches that end in $F$. This gives Bayes' theorem, which is true for any number of events $E_i$.

**Bayes' Theorem**

Let $E_1, E_2, \ldots, E_n$, be mutually exclusive events in a sample space $S$ with $E_1 \cup E_2 \cup \cdots \cup E_n = S$. If $F$ is any event in $S$, then for any $i = 1, 2, \ldots, n$,

$$p(E_i|F) = \frac{\text{probability of branch } E_i F}{\text{sum of all probabilities of all branches that end in } F}$$

$$= \frac{p(E_i)p(F|E_i)}{p(E_1)p(F|E_1) + p(E_2)p(F|E_2) + \cdots + p(E_n)p(F|E_n)}$$

## Applications

A very interesting example occurs in medical tests for disease. We begin with a test that gives every appearance of being excellent, but an important consequence may be disappointing.

**E X A M P L E   1**   **A Medical Application of Bayes' Theorem**

The standard tine test for tuberculosis attempts to identify carriers, that is, people who have been infected by the tuberculin bacteria. The probability of a false negative is .08, that is, the probability of the tine test giving a negative reading to a carrier is $P(-|C) = .08$. The probability of a false positive is .04, that is, the probability of the tine test giving a positive indication when a person is a noncarrier is $P(+|N) = .04$. The probability of a random person in the United States having tuberculosis is .0075. Find the probability that a person is a carrier given that the tine test gives a positive indication.

**Solution**   The probability we are seeking is $P(C|+)$. Figure 7.16 shows the appropriate tree diagram where $C$ is the event "is a carrier," $N$ the event "is a noncarrier," $+$ the event "test yields positive result." Then Bayes' theorem can be used and $P(C|+)$ is the probability of branch $C+$ divided by the sum of all probabilities that end in $+$. Then

$$p(C|+) = \frac{p(C)p(+|C)}{p(C)p(+|C) + p(N)p(+|N)}$$

$$= \frac{(.0075)(.92)}{(.0075)(.92) + (.9925)(.04)}$$

$$\approx .15.$$

This says that only 15% of people with positive tine test results actually carry TB.   ∎

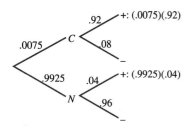

**Figure 7.16**

This number is surprisingly low. Does this indicate that the test is of little value? As Self-Help Exercise 1 will show, a person whose tine test is negative has a probability of .999 of not having tuberculosis. Such an individual can feel safe. The individuals whose tine test is positive are probably okay also but will need to undergo further tests, such as a chest x-ray.

In some areas of the United States the probability of being a carrier can be as high as .10. The following example examines the tine test under these conditions.

E X A M P L E  2    **A Medical Application of Bayes' Theorem**

Find $P(C|+)$ again when $P(C) = .10$.

**Solution**    See Figure 7.17. Using Bayes' Theorem exactly as before, we obtain

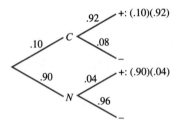

$$p(C|+) = \frac{p(C)p(+|C)}{p(C)p(+|C) + p(N)p(+|N)}$$

$$= \frac{(.1)(.92)}{(.1)(.92) + (.9)(.04)}$$

$$\approx .72.$$

Thus 72% of these individuals who have a positive tine test result will be carriers.  ■

**Figure 7.17**

Thus in the first example, when the probability of being a carrier is low, the tine test is useful for determining those who do not have TB. In the second example, when the probability of being a carrier is much higher, the tine test is useful for determining those who are carriers, although, naturally, these latter individuals will undergo further testing.

E X A M P L E  3    **An Application of Bayes' Theorem**

Suppose there are only 4 economic theories that can be used to predict expansions and contractions in the economy. By polling economists on their beliefs on which theory is correct, the probability that each of the theories is correct has been determined as follows:

$$p(E_1) = .40, \qquad p(E_2) = .25, \qquad p(E_3) = .30, \qquad p(E_4) = .05.$$

The economists who support each theory then use the theory to predict the likelihood of a recession ($R$) in the next year. These are as follows:

$$p(R|E_1) = .01, \qquad p(R|E_2) = .02, \qquad p(R|E_3) = .03, \qquad p(R|E_4) = .90.$$

Now suppose a recession actually occurs in the next year. How would the probabilities of the correctness of the fourth and first theories be changed?

**Solution**    We first note that the fourth theory $E_4$ has initially a low probability of being correct. Also notice that this theory is in sharp disagreement with the other 3 on whether there will be a recession in the next year.

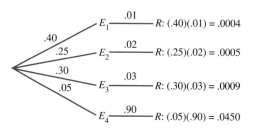

**Figure 7.18**

Bayes' theorem gives $p(E_4|R)$ as the probability of the branch $E_4R$ divided by the sum of all the probabilities of all the branches that end in $R$. See Figure 7.18. Similarly for the others. Thus

$$p(E_4|R) = \frac{p(E_4)p(R|E_4)}{p(E_1)p(R|E_1) + p(E_2)p(R|E_2) + p(E_3)p(R|E_3) + p(E_4)p(R|E_4)}$$

$$= \frac{(.05)(.9)}{(.4)(.01) + (.25)(.02) + (.3)(.03) + (.05)(.9)}$$

$$= \frac{.045}{.063} \approx .71,$$

and

$$p(E_1|R) = \frac{p(E_1)p(R|E_1)}{p(E_1)p(R|E_1) + p(E_2)p(R|E_2) + p(E_3)p(R|E_3) + p(E_4)p(R|E_4)}$$

$$= \frac{(.4)(.01)}{(.4)(.01) + (.25)(.02) + (.3)(.03) + (.05)(.9)}$$

$$= \frac{.004}{.063} \approx .06.$$

Thus, given that the recession did occur in the next year, the probability that $E_4$ is correct has jumped dramatically, while the probability that $E_1$ is true has plunged. ∎

Although this is an artificial example, probabilities are indeed reevaluated in this way based on new information.

## SELF-HELP EXERCISE SET 7.6

1. Referring to Example 1 of the text, find the probability that an individual in the United States is not a carrier given that the tine test is negative.

2. Suppose a new tine test had the remarkable property that $p(+|C) = 1$ but $P(+|N)$ is still the same .04. Now find the probability that an individual in the United States is a carrier given a positive reading on this new tine test.

## EXERCISE SET 7.6

1. Given the tree diagram, find $p(E_1|F)$ and $p(E_1|F^c)$.

2. Given the same tree diagram as in the previous exercise, find $p(E_2|F)$ and $p(E_2|F^c)$.

3. Given the tree diagram, find $p(E_1|F)$ and $p(E_1|F^c)$.

4. Given the same tree diagram as in the previous exercise, find $p(E_3|F)$ and $p(E_3|F^c)$.

5. Two urns each contain 10 balls. The first contains 2 white and 8 red, the second 7 white and 3 red. An urn is selected, and a ball is randomly drawn from this urn. If this ball is white and if the probability of selecting the first urn is twice the probability of selecting the second, find the probability that the first urn was selected.

6. In the previous problem, find the probability that the second urn was selected.

7. Three urns each contain 10 balls. The first contains 2 white and 8 red balls, the second 5 white and 5 red, and the third all 10 white. An urn is selected, and a ball is randomly drawn from this urn. If this ball is white and if the probability of selecting any urn is the same, find the probability that the first urn was selected.

8. In the previous problem, find the probability that the third urn was selected.

9. Using the urns in Exercise 5, an urn is selected with the given probabilities and two balls are drawn from this urn (without replacement); both are white. What is the probability that the urn selected was the (a) first one, (b) the second one?

10. Using the urns in Exercise 5, an urn is selected with the given probabilities and one ball is drawn, replaced, and then another drawn. Suppose both are white. What is the probability that the urn selected was the (a) first one, (b) the second one?

11. Using the urns in Exercise 7, an urn is selected with the given probabilities and two balls are drawn from this urn (without replacement); both are white. What is the probability that the urn selected was the (a) first one, (b) the second one?

12. Using the urns in Exercise 7, an urn is selected with the given probabilities and one ball is drawn, replaced, and then another drawn. Suppose both are white. What is the probability that the urn selected was the (a) first one, (b) the second one?

## Applications

13. **Medicine.**  Do Example 2 of the text if all the information remains the same except that the tine test has the remarkable property that $p(+|C) = 1$. Compare your answer to the one in Example 2.

14. **Medicine.**  Using the information in Example 2 of the text, find $p(N|-)$, where $-$ is the event "test shows negative."

15. **Economics.**  Using the information in Example 3 of the text, find $p(E_1|R^c)$ and $p(E_4|R^c)$. Compare your answer with $p(E_1)$ and $p(E_4)$.

16. **Manufacturing.**  For the example discussed in the beginning of this section (p. 421), find $p(E_1|F^c)$.

17. **Quality Control.**  One of two bins is selected at random, one as likely to be selected as the other, and from the bin selected a transistor is chosen at random. The transistor is tested and found to be defective. It is known that the first bin contains 2 defective and 4 nondefective transistors, while the second bin contains 5 defective and 1 nondefective transistors. Find the probability that the second bin was selected.

18. **Quality Control.**  Suppose in the previous exercise there is a third bin with 5 transistors, all of which are defective. Now one of the 3 bins is selected at random, one as likely to be selected as any other, and from this bin a transistor is chosen at random. If the transistor is defective, find the probability it came from the third bin.

19. **Quality Control.** A typical box of 100 transistors contains only 1 defective one. It is realized that among the last 10 boxes, one box has 10 defective transistors. An inspector picks one box at random, and the first transistor selected is found to be defective. What is the probability that this box is the bad one?

20. **Quality Control.** A typical box of 100 transistors contains only 1 defective one. It is realized that among the last 10 boxes, one box has 10 defective transistors. An inspector picks one box at random, inspects two transistors from this box on a machine and discovers that one of them is defective and one is not. What is the probability that this box is the bad one?

21. **Manufacturing.** A manufacturing firm has 4 machines that produce the same component. Using the table, given that a component is defective find the probability that the defective component was produced by (a) machine 1, (b) machine 2.

| Machine | Percentage of Components Produced | Percentage of Defective Components |
|---------|-----------------------------------|------------------------------------|
| 1 | 20 | 1 |
| 2 | 30 | 2 |
| 3 | 40 | 3 |
| 4 | 10 | 4 |

22. **Social Sciences.** A man claimed not to be the father of a certain child. On the basis of evidence presented, the court felt that this man was twice as likely to be the father as not, and, hardly satisfied with these odds, required the man to take a blood test. The mother of the child had a different blood type than the child; therefore the blood type of the child was completely determined by the father. If the man's blood type was different than the child, then he could not be the father. The blood type of the child occurred in only 10% of the population. The blood tests indicated that the man had the same blood type as the child. What is the probability that the man is the father?

23. **Medical Diagnosis.** A physician examines a patient and, on the basis of the symptoms, determines that he may have one of four diseases; the probability of each is given in the table. She orders a blood test, which indicates that the blood is perfectly normal. Data are available on the percentage of patients with each disease whose blood tests are normal. On the basis of the normal blood test, find all probabilities that the patient has each disease.

| Diseases | Probability of Disease | Percentage of Normal Blood With this Disease |
|----------|------------------------|----------------------------------------------|
| 1 | .1 | 60 |
| 2 | .2 | 20 |
| 3 | .3 | 20 |
| 4 | .4 | 10 |

**Solutions to Self-Help Exercise Set 7.6**

1. Bayes' theorem can be used if $n = 2$, $E_1 = N$, $E_2 = C$, and $F = -$. Then

$$p(N|-) = \frac{p(N)p(-|N)}{p(N)p(-|N) + p(C)p(-|C)}$$

$$= \frac{(.9925)(.96)}{(.9925)(.96) + (.0075)(.08)}$$

$$\approx .999.$$

2. Using Bayes' theorem we obtain

$$p(C|+) = \frac{p(C)p(+|C)}{p(C)p(+|C) + p(N)p(+|N)}$$

$$= \frac{(.0075)(1)}{(.0075)(1) + (.9925)(.04)}$$

$$\approx .16.$$

This is about the same as before.

## 7.7    BERNOULLI TRIALS

▶ *Bernoulli Trials*

▶ *Applications*

A computer manufacturer uses 8 microchips in each of its computers. It knows that 1% of these chips are defective. What is the probability that at least 6 are good? See Example 3 for the answer.

### Bernoulli Trials

In this section we consider the simplest possible experiments: those with just two outcomes.

We refer to experiments in which there are just 2 outcomes as **Bernoulli trials**. Some examples are as follows. Flip a coin and see if heads or tails turns up. Test a transistor to see if it is defective or not. Examine a patient to see if a particular disease is present or not. Take a free throw in basketball and make the basket or not.

We commonly refer to the two outcomes of a Bernoulli trial as "success" (S) or "failure" (F). We agree always to write $p$ for the probability of "success" and $q$ for the probability of "failure." Naturally $q = 1 - p$.

In this section we are actually not so much interested in performing an experiment with two outcomes once, but rather many times. We refer to this as a **repeated Bernoulli trial**. We make a very fundamental assumption: The successive Bernoulli trials are independent of one another. Thus, for example, flipping a coin 10 times is a repeated Bernoulli trial. Tossing a die 20 times and seeing if an even number or an odd number occurs each time is another example.

Given a Bernoulli trial repeated $n$ times, we are interested in determining the probability that a specific number of successes occurs. If $k$ is the number of successes, then we denote $b(k : n, p)$ the probability of exactly $k$ successes in $n$ repeated Bernoulli trials where the probability of success is $p$.

**E X A M P L E   1**   **Finding the Probability of an Outcome in a Repeated Bernoulli Trial**

Suppose a basketball player makes on average two free throws of every three attempted and that success and failure on any one free throw does not depend on the outcomes of the other shots. If the player shoots 10 free throws, find the probability of making exactly 6 of them, that is, find $b(6 : 10, \frac{2}{3})$.

**Solution**   We let $S$ designate "making the basket" and $F$ "not making the basket." A typical sequence of exactly 6 successes looks like

$$SSFSFSSFFS.$$

Now the product rule indicates that the probability of this occurring is

$$\left(\frac{2}{3}\right)\left(\frac{2}{3}\right)\left(\frac{1}{3}\right)\left(\frac{2}{3}\right)\left(\frac{1}{3}\right)\left(\frac{2}{3}\right)\left(\frac{2}{3}\right)\left(\frac{1}{3}\right)\left(\frac{1}{3}\right)\left(\frac{2}{3}\right)$$

or

$$\left(\frac{2}{3}\right)^6\left(\frac{1}{3}\right)^4.$$

*(Frank A. Cezus/FPG International Corporation.)*

We might think of the process of obtaining exactly 6 successes as lining up 10 boxes in a row and picking exactly 6 of them in which to place an $S$. Thus the previous sequence of successes and failures would be

$$\boxed{S}\boxed{S}\boxed{\phantom{S}}\boxed{S}\boxed{\phantom{S}}\boxed{S}\boxed{S}\boxed{\phantom{S}}\boxed{\phantom{S}}\boxed{S}.$$

Since every such sequence must contain exactly 6 $S$'s and 4 $F$'s, the probability of any one of these occurring is always $(\frac{2}{3})^6(\frac{1}{3})^4$. To find $b(6 : 10, \frac{2}{3})$, we then must count the number of ways there can be exactly 6 successes. But this is the number of ways of selecting 6 of the boxes to place an $S$ inside. This can be done in $C(10, 6)$ ways. Thus

$$b\left(6 : 10, \frac{2}{3}\right) = C(10, 6)\left(\frac{2}{3}\right)^6\left(\frac{1}{3}\right)^4$$

$$= \frac{10 \cdot 9 \cdot 8 \cdot 7 \cdot 6 \cdot 5}{6 \cdot 5 \cdot 4 \cdot 3 \cdot 2 \cdot 1}\left(\frac{2}{3}\right)^6\left(\frac{1}{3}\right)^4$$

$$= 210\left(\frac{2}{3}\right)^6\left(\frac{1}{3}\right)^4$$

$$\approx .228. \quad \blacksquare$$

In general, to find the probability of exactly $k$ successes in $n$ repeated Bernoulli trials, first notice that any particular sequence with exactly $k$ successes must have exactly $n - k$ failures. Thus by the product rule, the probability of any one of these occurring is

$$p^k q^{n-k}.$$

The number of ways of obtaining exactly $k$ successes in $n$ trials is the number of ways of selecting $k$ objects (the boxes with $S$'s inside in the above discussion) from a total of $n$. This is $C(n, k)$. Thus

$$\boxed{b(k : n, p) = C(n, k) p^k q^{n-k}.}$$

## Applications

Ken Griffey, Jr. played baseball for the Seattle Mariners during the 1990 season and was the youngest big league player that year although 1990 was Griffey, Jr.'s second season. His batting average that year was .300, that is, the probability of his getting a hit on any official time at bat[1] was .300.

**E X A M P L E  2**     **Finding the Probability of an Outcome in a Repeated Bernoulli Trial**

What was the probability of Griffey, Jr. getting at least 2 hits in a game in 1990 if we assume that he came to bat 4 official times in that game and if we assume coming to bat each time is an independent trial.

**Solution**     If we designate success as a hit, then $p = .300$ and $q = 1 - p = .700$. We then are looking for

$$b(2 : 4, .300) + b(3 : 4, .300) + b(4 : 4, .300).$$

We can save some calculations if we notice that this is one less the probability of at most 1 hit, which is

$$1 - [b(0 : 4, .300) + b(1 : 4, .300)],$$

or

$$1 - [C(4, 0)(.300)^0(.700)^4 + C(4, 1)(.300)^1(.700)^3] = 1 - [(.7)^4 + 4(.3)(.7)^3]$$
$$= .3483.$$

So about 35% of the time Griffey, Jr. should have obtained at least 2 hits out of 4 times-at-bat.  ∎

---

[1]An official time at bat does not include a time when the batter walked, hit a sacrifice fly or bunt, or was hit by a pitched ball.

E X A M P L E  3    **Finding Probabilities of Outcomes in a Repeated Bernoulli Trial**

A computer manufacturer uses 8 microchips in each of its computers. It knows that 1% of these chips is defective. What is the probability that (a) all 8 chips are good? (b) at least 6 are good?

**Solutions**    If we let $S$ be the event "not defective," then $p = .99$ and $q = .01$.

(a)  The answer is

$$b(8:8, .99) = C(8, 8)(.99)^8(.01)^0$$
$$= (.99)^8 \approx .92274.$$

(b)  The answer is the sum of the three probabilities

$$b(8:8, .99) + b(7:8, .99) + b(6:8, .99).$$

This is

$$.92274 + C(8, 7)(.99)^7(.01)^1 + C(8, 6)(.99)^6(.01)^2$$
$$= .92274 + 8(.99)^7(.01) + 28(.99)^6(.01)^2$$
$$= .92274 + .07457 + .00264 = .99995. \quad \blacksquare$$

## SELF-HELP EXERCISE SET 7.7

1.  If a repeated Bernoulli trial is performed 6 times, find the probability of obtaining 2 successes and 4 failures if $p = .20$.

2.  A retail store sells 2 brands of TVs, with the first brand comprising 60% of these sales. What is the probability that the next 5 sales of TVs will consist of at most 1 of the first brand?

## EXERCISE SET 7.7

For Exercises 1 through 6, a repeated Bernoulli trial is performed. Find the probability of obtaining the indicated number of successes and failures for the indicated value of $p$.

1.  4 S's, 1 F, $p = .2$            2.  4 S's, 3 F's, $p = .3$

3.  3 S's, 4 F's, $p = .5$          4.  4 S's, 4 F's, $p = .5$

5.  4 S's, 4 F's, $p = .25$         6.  2 S's, 2F's, $p = .1$

Evaluate the term in Exercises 7 through 12.

7.  $b(3:6, .5)$                    8.  $b(4:6, .5)$

9.  $b(4:7, .1)$                    10.  $b(3:4, .2)$

11.  $b(3:5, .1)$                   12.  $b(3:8, .2)$

In Exercises 13 through 18 flip a fair coin 10 times. Find the probability of getting the following outcomes.

**13.**   Exactly 8 heads

**14.**   Exactly 3 heads

**15.**   At least 8 heads

**16.**   At least 7 heads

**17.**   At most 1 head

**18.**   At most 2 heads

In Exercises 19 through 23 an event $E$ has probability $p = p(E) = .6$ in some sample space. Suppose the experiment that yields this sample space is repeated 7 times and the outcomes are independent. Find the probability of getting the following outcomes.

**19.**   $E$ exactly 6 times

**20.**   $E$ exactly 3 times

**21.**   $E$ at least 6 times

**22.**   $E$ at least 5 times

**23.**   $E$ at most 2 times

**24.**   Show

$$b(k : n, p) = b(n - k : n, 1 - p).$$

Ty Cobb has the highest lifetime batting average of any big league baseball player with a remarkable average of .367. Assume in Exercises 25 through 28 that Cobb came to bat officially 4 times in every game played.

**25.**   What would be Cobb's probability of getting at least one hit in a game?

**26.**   What would be Cobb's probability of getting at least 3 hits in a game?

**27.**   What would be Cobb's probability of getting at least one hit in 10 successive games? (Use the result in Exercise 25.)

**28.**   What would be Cobb's probability of getting at least one hit in 20 successive games? (Use the result in Exercise 25.)

Babe Ruth holds the record for the highest lifetime percent (8.5%) of home runs per time-at-bat. (Ralph Kiner comes in a distant second at 7.1%.) Assume in Exercise 29 through 32 that Ruth came to bat officially 4 times in every game played.

**29.**   What would be Ruth's probability of getting at least two home runs in a game?

**30.**   What would be Ruth's probability of getting at least one home run in a game?

**31.**   What would be Ruth's probability of getting at least two home runs in 3 successive games? (Use the result in Exercise 29.)

**32.**   What would be Ruth's probability of getting 4 home runs in a game?

## Applications

**Oil Drilling.**   An oil company estimates that only 1 oil well in 20 will yield commercial quantities of oil. Assume that successively drilled wells represent independent events. If 12 wells are drilled, find the probability of obtaining a commercially successful well for the following number of times.

**33.**   Exactly 1

**34.**   None

**35.**   At most 2

**36.**   Exactly 4

**Personnel.**   A company finds that one out of five workers it hires turns out to be unsatisfactory. Assume that the satisfactory performance of any hired worker is independent of that of any other hired workers. If the company hires 20 people, what is the probability that the following number of people will turn out satisfactory?

**37.**  Exactly 10                                **38.**  At most 2

**39.**  At least 18                               **40.**  Exactly 20

**Medicine.**   A certain type of heart surgery in a certain hospital results in mortality in 5% of the cases. Assume that the death of a person undergoing this surgery is independent from the death of any others who have undergone this same surgery. If 20 people have this heart surgery at this hospital, find the probability that the following number of people will not survive the operation.

**41.**  Exactly 2                                 **42.**  At most 2

**43.**  At most 3                                 **44.**  Exactly 10

## Solutions to Self-Help Exercise Set 7.7

**1.**   In a repeated Bernoulli trial the probability of obtaining 2 successes and 4 failures if $p = .20$ is

$$b(k : n, p) = C(n, k)p^k q^{n-k} = C(6, 2)(.20)^2(.80)^4 = \frac{6 \cdot 5}{2 \cdot 1}(.20)^2(.80)^4 \approx .246.$$

**2.**   If a retail store sells 2 brands of TVs, with the first brand comprising 60% of these sales, then the probability that the next 5 sales of TVs will consist of at most 1 of the first brand is the probability that the next 5 sales will consist of exactly 0 of the first brand plus exactly 1. This is

$$C(5, 0)(.60)^0(.40)^5 + C(5, 1)(.60)^1(.40)^4 = (.40)^5 + 5(.60)(.40)^4$$

$$= .01024 + .0768 = .08704.$$

## SUMMARY OUTLINE OF CHAPTER 7

- An **experiment** is an activity that has observable results. The results of the experiment are called **outcomes**.

- A **sample space** of an experiment is the set of all possible outcomes of the experiment. Each repetition of an experiment is called a **trial**.

- Given a sample space $S$ for an experiment, an **event** is any subset $E$ of $S$. An **elementary event** is an event with a single outcome.

- If $E$ and $F$ are two events, then $E \cup F$ is the union of the two events and consists of the set of outcomes that are in $E$ or $F$ or both.

- If $E$ and $F$ are two events, then $E \cap F$ is the intersection of the two events and consists of the set of outcomes that are in both $E$ and $F$.

- If $E$ is an event, then $E^c$ is the **complement** of $E$ and consists of the set of outcomes that are not in $E$.

- The empty set, $\phi$, is called the **impossible event**.

- Let $S$ be a sample space. The event $S$ is called the **certainty event**.

- Two events $E$ and $F$ are said to be **mutually exclusive** or **disjoint** if $E \cap F = \phi$.

- If an experiment is repeated $N$ times and an event $E$ occurs $f(E)$ times, then the fraction $\dfrac{f(E)}{N}$ is called the **relative frequency** of the event.

- Let $S$ be the sample space of an experiment and $E$ an event in $S$. If the experiment has actually been performed $N$ times, the relative frequency $\dfrac{f(E)}{N}$ is called the **empirical probability** of $E$.

- Let $S$ be the sample space of an experiment, let $E$ be an event in $S$, and let $N$ be the number of times the experiment has been repeated. The **intuitive probability of $E$**, denoted by $p(E)$, is the number that the relative frequency $\dfrac{f(E)}{N}$ approaches as $N$ becomes larger and larger.

- **Definition of Probability.**   Let $S$ be a sample space. Assign to each event $E$ in $S$ a number $p(E)$ so that the following four properties are satisfied.

   ***Property 1.***   $0 \le p(E) \le 1$ for any event $E$.

   ***Property 2.***   $p(S) = 1$.

   ***Property 3.***   $p(\phi) = 0$.

   ***Property 4.***   Given any event $E = \{e_1, e_2, \ldots, e_r\}$, where $e_1, e_2, \ldots, e_r$, are all distinct, then

   $$p(\{e_1, e_2, \ldots, e_r\}) = p(e_1) + p(e_2) + \cdots + p(e_r).$$

   Then for any event $E$, we called $p(E)$ the **probability** of $E$ and say that $p$ is a probability on $S$.

- **Further Properties of Probability.**   Let $p$ be a probability on a sample space $S = \{s_1, s_2, \ldots, s_n\}$ and $E$ and $F$ two events in $S$. Then the following are true.

   ***Property 5.***   $p(s_1) + p(s_2) + \cdots + p(s_n) = 1$.

   ***Property 6.***   $p(G \cup H) = p(G) + p(H)$      if      $G \cap H = \phi$.

   ***Property 7.***   $p(E^c) = 1 - p(E)$.

   ***Property 8.***   $p(E \cup F) = p(E) + p(F) - p(E \cap F)$.

   ***Property 9.***   $p(F) \le p(E)$      if      $F \subset E$.

   ***Property 10.***   Suppose $E_1, E_2, \ldots, E_k$ are pairwise mutually disjoint, that is, $E_i \cap E_j = \phi$ if $i \ne j$. Then

   $$p(E_1 \cup E_2 \cup \cdots \cup E_k) = p(E_1) + p(E_2) + \cdots + p(E_k).$$

- Let $p = p(E)$ be the probability of $E$, then the **odds for $E$** is defined to be

   $$\text{Odds for } E = \frac{p}{1 - p} \qquad \text{if } p \ne 1.$$

- Suppose that the odds for an event $E$ occurring is given as $\dfrac{a}{b}$. If $p = p(E)$, then

   $$p = \frac{a}{a + b}.$$

- If $S$ is a finite uniform sample space and $E$ is any event, then

$$p(E) = \frac{\text{Number of elements in } E}{\text{Number of elements in } S} = \frac{n(E)}{n(S)}.$$

- Let $E$ and $F$ be two events in a sample space $S$. The conditional probability that $E$ occurs given that $F$ has occurred is defined to be

$$p(E|F) = \frac{p(E \cap F)}{p(F)} \quad \text{if} \quad p(F) > 0.$$

- If $E$ and $F$ are two events in a sample space $S$ with $p(E) > 0$ and $p(F) > 0$, then

$$p(E \cap F) = p(F)p(E|F) = p(E)p(F|E).$$

- Two events $E$ and $F$ are said to be independent if

$$p(E|F) = p(E) \quad \text{and} \quad p(F|E) = p(F).$$

- **Let $E$ and $F$ be two events with $p(E) > 0$ and $p(F) > 0$. Then $E$ and $F$ are independent if, and only if, $p(E \cap F) = p(E)p(F)$.**
  A set of events $\{E_1, E_2, \ldots, E_n\}$ is said to be independent if, for any $k$ of these events, the probability of the intersection of these $k$ events is the product of the probabilities of each of the $k$ events. This must hold for any $k = 2, 3, \ldots, n$.

- **Bayes' Theorem**   Let $E_1, E_2, \ldots, E_n$, be mutually exclusive events in a sample space $S$ with $E_1 \cup E_2 \cup \cdots \cup E_n = S$. If $F$ is any event in $S$, then for any $i = 1, 2, \ldots, n$,

$$p(E_i|F) = \frac{\text{probability of branch } E_iF}{\text{sum of all probabilities of all branches that end in } F}$$

$$= \frac{p(E_i)p(F|E_i)}{p(E_1)p(F|E_1) + p(E_2)p(F|E_2) + \cdots + p(E_n)p(F|E_n)}.$$

- The probability of exactly $k$ successes in $n$ repeated Bernoulli trials where the probability of success is $p$ and failure is $q$ is given by $b(k : n, p) = C(n, k)p^k q^{n-k}$.

## Chapter 7 Review Exercises

1.  During a recent four-round golf tournament, the following frequencies of scores were recorded on a par 5 hole.

| Score     | 3 | 4  | 5   | 6  | 7 | 8 |
|-----------|---|----|-----|----|---|---|
| Frequency | 4 | 62 | 157 | 22 | 4 | 1 |

   **a.**   Find the probability that each of the scores was made by a random player.
   **b.**   Find the probability that a score of par or lower was recorded.
   **c.**   Find the probability that a score of less than par was recorded.

2.  An urn has 10 white, 5 red, and 15 blue balls. A ball is drawn at random. What is the probability that the ball will be (a) red? (b) red or white? (c) not white?

3.  If $E$ and $F$ are mutually disjoint sets in a sample space $S$ with $p(E) = .25$ and $p(F) = .35$, find (a) $p(E \cup F)$, (b) $p(E \cap F)$, and (c) $p(E^c)$.

4.  If $E$ and $F$ are two events in the sample space $S$ with $p(E) = .20$, $p(F) = .40$, and $p(E \cap F) = .05$, find (a) $p(E \cup F)$, (b) $p(E^c \cap F)$, and (c) $p((E \cup F)^c)$.

5.  Consider the sample space $S = \{a, b, c, d\}$ and suppose that $p(a) = p(b)$, $p(c) = p(d)$, and $p(d) = 2p(a)$. Find $p(b)$.

6.  If the odds for a company obtaining a certain contract are 3 to 1, what is the probability that the company will receive the contract?

7.  A furniture manufacturer notes that 6% of its reclining chairs have a defect in the upholstery, 4% a defect in the reclining mechanism, and 1% have both defects. (a) Find the probability that a recliner has at least one of these defects. (b) Find the probability that a recliner has none of these defects.

8.  A survey of homeowners indicated that during the last year: 22% had planted some vegetables, 30% some flowers, 10% some trees, 9% vegetables and flowers, 7% vegetables and trees, 5% flowers and trees, and 4% all three of these. (a) Find the probability that a homeowner planted vegetables but not flowers. (b) Find the probability that exactly 2 of the items were planted. (c) Find the probability that none of these three items were planted.

9.  For the urn in Exercise 2 what is the probability that there is one of each color if 3 are selected randomly without replacement? With replacement?

10. For the urn in Exercise 2, what is the probability that 3 are white, 4 are red, and 2 are blue, if 9 are selected randomly without replacement? With replacement?

11. Let $p(E) = .3$, $p(F) = .5$, and $p(E \cap F) = .2$. Draw a Venn diagram and find the indicated conditional probabilities: (a) $p(E|F)$, (b) $p(E^c|F)$, and (c) $p(F^c|E^c)$.

12. If $p(E) = .5$, $p(F) = .6$, and $p(E \cap F) = .4$, determine if $E$ and $F$ are independent events.

13. **Reliability.**   A spacecraft has 3 batteries that can operate all systems independently. If the probability that any battery will fail is .05, what is the probability that all three will fail?

14. **Basketball.**   A basketball player sinks a free-throw 80% of the time. If she sinks one, the probability of sinking the next goes to .90. If she misses, the probability of sinking the next goes to .70. Find the probability that she will sink exactly 2 out of 3 free-throws.

15. **Manufacturing.**   A manufacturing firm has 5 machines that produce the same component. Using the table, find the probability that a defective component was produced by (a) machine 1, (b) machine 4.

| Machine | Percentage of Components Produced | Percentage of Defective Components |
|---------|-----------------------------------|-------------------------------------|
| 1 | 20 | 1 |
| 2 | 30 | 2 |
| 3 | 30 | 3 |
| 4 | 10 | 4 |
| 5 | 10 | 10 |

16. **Assembly Line.**   A machine on an assembly line is malfunctioning randomly and produces defective parts 30% of the time. What is the probability that this machine will produce exactly 3 defective parts among the next 6?

17. **Teaching Methods.**   An instructor finds that only 55% of her college algebra students pass the course. She then tries a new teaching method and finds that 85% of the 20 students in the first class with the new method pass. Assuming that the probability at this school of passing college algebra is .55 and that any one student passing is independent of any other student passing, what is the probability that at least 85% of a college algebra class of 20 will pass this course? Is the instructor justified in claiming that the new method is superior to the old method?

18. **Drug Testing.**   A company tests its employees for drug usage with a test that gives a positive reading 95% of the time when administered to a drug user and gives a negative reading 95% of the time when administered to a nondrug user. If 5% of the employees are drug users, find the probability that an employee is a nondrug user given that this person had a positive reading on the test. (The answer is shocking and illustrates the care that must be exercised in using such tests in determining guilt.)

## Solution to Game Show Problem

The solution given here was selected since it relies on the commonly held intuitive notions of probability.

A contestant is shown 3 closed doors and is correctly informed that behind one of the doors is a car and behind each of the other 2 doors is a goat. The contestant is then asked to select a door. After the contestant picks a door, the game show host then chooses one of the 2 doors not selected that has a goat behind it and opens this door for the contestant to see the goat. Now the host gives the contestant the opportunity to change the selection of the door. If the car is behind the door selected, the contestant wins the car. Should the contestant switch doors or not, or does it matter?

**Solution**   Our approach to the solution of this problem is to first assume that the contestant is allowed to play the game a large number of times, say 300. For the sake of definiteness, assume that *the third door is always initially selected.* (We will see that this is unimportant.)

Naturally the game show will place the car behind a door in a random fashion. Thus, from our intuitive understanding of probability, the car should be behind the first door about 100 times, behind the second door about 100 times, and behind the third door about 100 times.

Let us now assume that *the car is behind the first door exactly 100 times, behind the second door exactly 100 times, and behind the third door exactly 100 times.* Then if the contestant does not switch doors, she will win 100 out of 300 times, or one-third of the time.

Now let us see what happens if the contestant always switches. First consider the 100 times that the car is behind the first door: (car, goat, goat). In every one of these 100 cases, the third door has been selected and the car is behind the first door, and thus in every one of these 100 cases, the game show has no choice but to reveal the goat behind the second door. The contestant then switches the selection to the first door and wins every time in these 100 cases.

Consider now the 100 cases when the car is behind the second door: (goat, car, goat). This is very similar to the previous situation. Since the game show is forced to reveal the goat behind the first door, the contestant switches to the second door and wins every one of these 100 cases.

Finally consider the 100 cases when the car is behind the third door: (goat, goat, car). Since in every one of these 100 cases the contestant switches to either the first or second door, every one of these 100 cases will be lost.

If the contestant always initially selects another door besides the third one, the analysis is the same and leads to the same conclusion. The reader is invited to do this analysis.

Thus for the 300 cases, switching results in winning 200 times and losing 100 times. That is, the probability of winning by switching is $\frac{2}{3}$, and of losing by switching is $\frac{1}{3}$.

The importance of this example is to realize that, when confronted with a confusing problem, it never hurts to return to the basic intuitive notation of probability.

The cost of disaster insurance is based on the likelihood of the disaster occurring. (National Center for Atmospheric Research/National Science Foundation.)

# Statistics and Probability Distributions

8.1   Probability Distributions

8.2   Measures of Central Tendency

8.3   Measures of Dispersion

8.4   The Normal Distribution

8.5   Normal Approximation to Binomial Distribution

**T**he application of statistics to business and the sciences can be divided into two broad areas. In the first area, large masses of data need to be described by graphs or by certain descriptive numbers such as the "average," "standard deviation," and so on. This is referred to as *descriptive statistics*. The second area is involved in making inferences about the data based on sampling. Since this chapter is a short introduction to the area of statistics, we will just touch on the second area.

Fundamental to a statistical analysis is knowledge of how the underlying probabilities are distributed. We will discuss a number of important probability distributions and their applications. These include the binomial and the normal distributions.

## 8.1    PROBABILITY DISTRIBUTIONS

> ▶ *Models and the Need for Statistics*
> ▶ *Random Variables*
> ▶ *Probability Distributions*

A P P L I C A T I O N
**Picking a Mutual Fund**

Suppose you wish to select a mutual fund from a list of 64 funds, all of which involve similar risks. You notice that one single fund has performed above the average of these 64 funds in every one of the last 6 years. Is this a clear indication that the managers of this fund have the ability to keep their fund above average in performance? See the discussion immediately below for the answer.

### Models and the Need for Statistics

In this section we wish to see how to present numerical data in a manner that will permit making interpretations and comparisons. We begin with an example from finance mentioned above.

A very important problem in the theory of finance is to determine if it is possible for investment advisors or theories to predict, with some degree of accuracy, the movements in the various financial markets. One very highly touted way of investing is to pick an investment advisor on the basis of past performance. This certainly appears reasonable.

Unfortunately, a great deal of research has tended to indicate that the future performance of an advisor is *independent* of the past performance. Certainly the research in this area indicates that if there is a connection between past and future performance, it is a weak and subtle one.

The data in Table 8.1 refer to the past 6-year annual performance records of 64 mutual funds considered to be growth funds. Each of the 64 funds were ranked according to whether their percent increase in asset value (including capital gains and dividends) for any one year was above the average for that year of the group of 64 or not. The number of years a fund was above average was then counted. The table gives the number of funds that had an above-average performance for a given number of years during the 6-year time period. For example, the table indicates that 14 funds had an above-average performance in 4 of the 6 years studied and 22 had an above-average performance in 3 of the 6 years. It is interesting to notice that one fund had an above-average performance in every single year of the 6-year period and one fund did not perform above average for any of these years. Would it make sense *based solely on this data* to invest in the one

Table 8.1

| No. of Yrs. Above Avg. | 0 | 1 | 2 | 3 | 4 | 5 | 6 |
|---|---|---|---|---|---|---|---|
| Frequency | 1 | 4 | 17 | 22 | 14 | 5 | 1 |
| Empirical Probability | $\frac{1}{64}$ | $\frac{4}{64}$ | $\frac{17}{64}$ | $\frac{22}{64}$ | $\frac{14}{64}$ | $\frac{5}{64}$ | $\frac{1}{64}$ |

fund with above-average performance all 6 years and to avoid the one that never had an above-average performance during this period?

For later use, the last line of Table 8.1 gives the empirical probabilities (the relative frequencies).

Table 8.1 contains fictitious data. However, it does mirror actual data.[1] In the actual studies performance was adjusted to take into account the risk each fund was assuming. Studies indicate that higher rates of return can, in general, be obtained by taking higher risks. Precisely how to define and measure risk will be taken up in a subsequent section. For the 64 funds in Table 8.1 we suppose they all assumed the same risk. This enormously simplifies the analysis and is the reason the actual published data is not used here.

Let us assume that Table 8.1 contains actual data. Let us now consider a model that we think might explain why the data looks as it does. In our model suppose we have 64 funds and let us assume that the performance of a fund in this group in any year is *independent* from its performance in the previous years and furthermore assume that the probability of the performance of any fund being above average is .50. Our model is saying that a fund is determined to be above average or not in any year the same way we determine heads or tails by flipping a fair coin. Then we can view the performance of the funds in any year as a Bernoulli trial with $p = .50$ and the performance over the 6-year period of time as a repeated Bernoulli trial with $n = 6$. Using the formulas in the last section, the probability that a fund (under the stated assumptions in the model) has an above average performance exactly $k$ of the 6 years is given by

$$b(k : 6, .50) = C(6, k)(.5)^k(.5)^{6-k} = C(6, k)(.5)^6.$$

By recalling Pascal's triangle, the binomial coefficients are 1, 6, 15, 20, 15, 6, 1. From this and the fact that $(.5)^6 = \frac{1}{64}$ we can create Table 8.2, which gives the probability for every value of $k$ in the model.

What is striking is the similarity of Table 8.2 to Table 8.1. This can be further demonstrated by presenting the data in the two tables in graphical form. The numbers 0, 1, 2, 3, 4, 5, 6, represent the possible outcomes. These numbers are located on the horizontal axis as in Figure 8.1. Above each number is drawn a rectangle with base equal to one unit and height equal to the probability of that number. For example, above the number 5 is a rectangle with height $\frac{5}{64}$ in the first case and $\frac{6}{64}$ in the second case. Such graphs are called **histograms**; they give a vivid description of how the probability is distributed. As a consequence, comparison can be made more easily between the two histograms than between the two tables of probability.

**Table 8.2**

| No. of Yrs. Above Avg. | 0 | 1 | 2 | 3 | 4 | 5 | 6 |
|---|---|---|---|---|---|---|---|
| Theoretical Prob. in Model | $\frac{1}{64}$ | $\frac{6}{64}$ | $\frac{15}{64}$ | $\frac{20}{64}$ | $\frac{15}{64}$ | $\frac{6}{64}$ | $\frac{1}{64}$ |

[1]Cf. Michael C. Jensen, "The Performance of Mutual Funds in the Period 1945–64," *Journal of Finance*, vol. 23 no. 2 (May 1968), pp. 389–416.

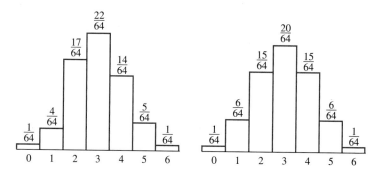

**Figure 8.1**

Given the remarkable similarity between the *theoretical* model we created and the actual data in Table 8.1, we might be very inclined to believe that the assumptions underlying the model are correct. That is, if the assumptions of the model are correct, then the data shown in Table 8.2 can be viewed as a prediction of what should be happening. Since the predictions indicated in Table 8.2 are so close to the actual data given in Table 8.1, we would tend to think that the assumptions of the model are correct.

These assumptions state that determining whether a fund is above average in any year is the same as flipping a fair coin and interpreting the occurrence of (say) ''heads'' with ''above average performance.'' The model is then the same as having 64 people each flip a fair coin 6 times in succession and declaring their performance to be above average if a head occurs. In such a model we naturally expect, for example, one person to obtain heads all 6 times. Would you conclude that this person is a talented ''head flipper'' and then wager large sums of money on this person's alleged ability to flip heads?

Thus in our model we fully expect one fund to be above average in performance every single one of the 6 years. Now we see the possibility that the single fund in Table 8.1 that had the above-average performance in every one of the 6 years may not be a result at all of the brilliance of the management but an inevitable result of chance.

It is apparent that some measurements of the data in the two tables are needed that would measure the similarity of the two sets of data. This would be especially helpful in cases where the similarity is not so obvious. We will do this in subsequent sections.

Critical to the above analysis was the need to find the probability of every outcome in each of the cases given in the two tables. We then needed to see how the probability was *distributed* in each case. Furthermore, we needed to be able to compare the manner in which the probability was distributed in Table 8.1 with the manner it was distributed in Table 8.2. In the next two sections we will develop ways of making such comparisons. This will enable us to state in what ways two sets of data are similar and in what ways they are not. This will be especially helpful in cases where the similarity is not so obvious as it is when comparing Table 8.1 with Table 8.2.

## Random Variables

As we already noted, the outcomes in the above experiments were any of the numbers 0, 1, 2, 3, 4, 5, and 6. Of course, outcomes of experiments are not always real numbers. The outcomes of experiments can be "heads," "above average," or "defective"—none of which are real numbers. It is often useful to assign a real number to each outcome of an experiment. For example, the outcome of the experiment "your grade in math" is a *letter*. But for purposes that you are well aware of, each letter is assigned a real number. Usually 4 is assigned to an A, 3 to a B, 2 to a C, 1 to a D, and 0 to an F. Usually there is some rational basis on which the assignment of a real number to an outcome is made. For example, since a grade of *A* is "better" than a grade of *B*, it makes sense to assign a *higher* number to the grade A than to B. Also the assignment of the numbers to each letter is done in such a manner that the numerical difference between two successive letters will be equal. Assigning zero to F makes sense since no credit is given for a grade of F. Finally, using the numbers 0, 1, 2, 3, and 4 makes it easy to calculate the "average" grade and for this number to have a readily understood meaning.

When numbers are assigned to the outcomes of experiments according to some rule, the rule is referred to as a **random variable**. As we have seen in assigning numbers to letter grades, the assignments of numbers to the outcomes of experiments are normally done in a manner that is reasonable and, most important, in a manner that permits these numbers to be used for interpretation and comparison.

---

**Random Variables**

A **random variable** is a rule that assigns precisely one real number to each outcome of an experiment.

---

**R E M A R K .**    Unless otherwise specified, when the outcomes of an experiment are themselves numbers, the random variable is the rule that simply assigns each number to itself.

There are three types of random variables, as we now indicate.

---

**Finite Discrete, Infinite Discrete, and Continuous Random Variables**

1.  A random variable is **finite discrete** if it assumes only a finite number of values.
2.  A random variable is **infinite discrete** if it takes on an infinite number of values that can be listed in a sequence, so that there is a first one, a second one, a third one, and so on.
3.  A random variable is said to be **continuous** if it can take any of the infinite number of values in some interval of real numbers.

If a random variable denotes the number of years one of the funds mentioned at the beginning of this section has an above-average performance in a 6-year period, then the random value can take the finite number of values in the set $\{0, 1, 2, \ldots, 6\}$, and thus is finite discrete. If the random variable denotes the number of flips it takes to obtain a head, then the random value can take any of the infinitely many values in the set $\{1, 2, \ldots\}$ and thus is infinite discrete. If a random variable denotes the heights in feet of adult men in this country, then the random variable can take on any value in the interval $[2, 8]$ and thus is continuous.

## Probability Distributions

Suppose now that the outcomes of an experiment are the finite set of real numbers $\{x_1, x_2, \ldots, x_n\}$ and that these numbers are all the different real numbers assigned to the outcomes of an experiment. We have usually denoted the probability of the outcome $x_k$ by $p(x_k)$. We now denote this same probability by

$$P(X = x_k).$$

We will use this latter notation since it is widely used in statistics and has some advantages, as we shall soon see. The **probability distribution of the random variable** $X$ is a listing of all the probabilities associated with all possible values of the random variable. Such a listing is often given in a table.

---

**Probability Distribution of the Random Variable $X$**

Suppose the random variable $X$ can take the values $x_1, \ldots, x_n$. The probability distribution of the random variable $X$ is a listing of all the probabilities associated with all possible values of the random variable, that is, $p_1, \ldots, p_n$, where $p_1 = p(x_1), \ldots, p_n = p(x_n)$.

---

Using the new notation, we can rewrite Table 8.2 as follows.

| $x$ | 0 | 1 | 2 | 3 | 4 | 5 | 6 |
|---|---|---|---|---|---|---|---|
| $P(X = x)$ | $\frac{1}{64}$ | $\frac{6}{64}$ | $\frac{15}{64}$ | $\frac{20}{64}$ | $\frac{15}{64}$ | $\frac{6}{64}$ | $\frac{1}{64}$ |

This particular distribution is referred to as a **binomial distribution** since the listed probabilities are the terms in the binomial expansion of $(p + q)^6 = (\frac{1}{2} + \frac{1}{2})^6$. In general, if $p + q = 1$,

$$1 = (p + q)^n = C(n, n)p^n q^0 + C(n, n - 1)p^{n-1}q + \cdots + C(n, 0)p^0 q^n.$$

Recalling the material on Bernoulli trials this can be written as

$$b(n : n, p) + b(n - 1 : n, p) + \cdots + b(0 : n, p).$$

---

**Binomial Distribution**

Given a sequence of $n$ Bernoulli trials with the probability of success $p$ and the probability of failure $q$, the **binomial distribution** is given by

$$P(X = k) = b(k : n, p) = C(n, k)p^k q^{n-k}.$$

---

**E X A M P L E  1**  **Finding a Probability Distribution and Corresponding Histogram**

Suppose a pair of fair dice are tossed. Let $X$ denote the random variable that gives the sum of the top faces. Find the probability distribution. Draw a histogram.

**Solution**  The random variable can take any of the values 2, 3, ..., 12. Using the new notation and recalling the work done in Example 2 of Section 7.1, we have

$$P(X = 2) = p(\{(1, 1)\}) = \frac{1}{36}$$

$$P(X = 3) = p(\{(1, 2), (2, 1)\}) = \frac{2}{36},$$

and so forth. This then yields the probability distribution given in the following table.

| $x$ | 2 | 3 | 4 | 5 | 6 | 7 | 8 | 9 | 10 | 11 | 12 |
|---|---|---|---|---|---|---|---|---|---|---|---|
| $P(X = x)$ | $\frac{1}{36}$ | $\frac{2}{36}$ | $\frac{3}{36}$ | $\frac{4}{36}$ | $\frac{5}{36}$ | $\frac{6}{36}$ | $\frac{5}{36}$ | $\frac{4}{36}$ | $\frac{3}{36}$ | $\frac{2}{36}$ | $\frac{1}{36}$ |

The histogram is drawn in Figure 8.2.  ■

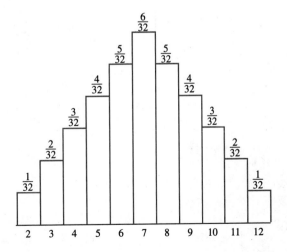

**Figure 8.2**

E X A M P L E   2    **Finding a Probability Distribution**

Suppose a pair of fair dice is tossed. Let $X$ be the random variable given by 1 if the sum of the top two faces is a prime number and 0 otherwise. Find the probability distribution.

**Solution**    Since the primes are 2, 3, 5, 7, and 11, the table in the previous example indicates that

$$P(X = 1) = \frac{1 + 2 + 4 + 6 + 2}{36} = \frac{15}{36}.$$

Thus

| Event | *Prime* | *Not a Prime* |
|---|---|---|
| Random Variable $x$ | 1 | 0 |
| $P(X = x)$ | $\frac{15}{36}$ | $\frac{21}{36}$ |

E X A M P L E   3    **Finding a Probability Distribution and Corresponding Histogram**

Ted Williams had a batting average of .406 in 1941. During that year he had 112 singles, 33 doubles, 3 triples, and 37 home runs in 456 (official) times-at-bat. Let the random variable $X$ be 0 for no hit, 1 for a single, 2 for a double, 3 for a triple, and 4 for a home run. Find a probability distribution for Ted Williams obtaining a 0, 1, 2, 3, or 4 for each time-at-bat during 1941. Draw a histogram.

**Solution**    The probability of hitting a home run is $37/456 = .081$, for a triple is $3/456 = .007$, for a double is $33/456 = .072$, for a single is $112/456 = .246$. For no hit, we have $[456 - (37 + 3 + 33 + 112)]/456 = 271/456 = .594$.

| Event | *No Hit* | *Single* | *Double* | *Triple* | *Home Run* |
|---|---|---|---|---|---|
| Random Variable $x$ | 0 | 1 | 2 | 3 | 4 |
| Frequency | 271 | 112 | 33 | 3 | 37 |
| $P(X = x)$ | .594 | .246 | .072 | .007 | .081 |

See Figure 8.3 for the histogram.  ■

We will now use histograms to find the probability of events by measuring the areas under appropriate rectangles.

**Figure 8.3**

E X A M P L E  4

### Finding Probability Using Histograms

Use the histogram in Figure 8.2 to find the probability that the sum of the top faces of two fair die when tossed will be at least 9 and less than 12.

**Solution**    We are seeking

$$P(X = 9) + P(X = 10) + P(X = 11).$$

The probability that $X = 9$, $P(X = 9)$ is the area of the rectangle above 9, $P(X = 10)$ is the area of the rectangle above 10, and so on. Thus, we are seeking the area of the shaded region in Figure 8.4. This is

$$\frac{4}{36} + \frac{3}{36} + \frac{2}{36} = \frac{1}{4}. \quad \blacksquare$$

A shorthand notation for the probability found in this example is $P(9 \leq X < 12)$.

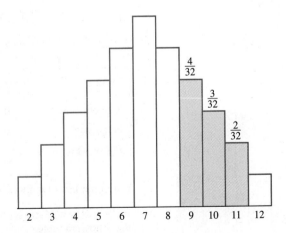

**Figure 8.4**

E X A M P L E   5     **Finding Probability Using Histograms**

Using Figure 8.1a, find the (empirical) probability that a fund had at most 2 years of above-average returns.

**Solution**     We are seeking $P(X \leq 2)$. This is the area of the shaded region under the first 3 rectangles shown in Figure 8.5. This is

$$P(X \leq 2) = \frac{1}{64} + \frac{4}{64} + \frac{17}{64} = \frac{11}{32}. \quad \blacksquare$$

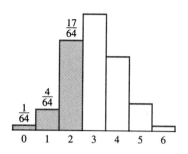

**Figure 8.5**

## SELF-HELP EXERCISE SET 8.1

1.  Use the histogram in Figure 8.3 to find the probability that Ted Williams obtained an extra-base hit (double, triple, or home run) during 1941.

2.  Find the probability distribution of the binomial distribution $b(k; 6, .75)$ and draw a histogram.

## EXERCISE SET 8.1

In Exercises 1 through 4, determine the possible values of the given random variable and indicate whether the random variable is finite discrete, infinite discrete, or continuous.

1.  The number of times a coin must be flipped before two heads appear in succession

2.  The number of times two heads appear in succession when a coin is flipped 1,000,000 times

3.  The number of hours spent studying math each week

4.  The number of rotten eggs in a dozen

5.  The grade distribution in a certain math class is given in the following table.

| Event | A | B | C | D | F |
|---|---|---|---|---|---|
| Random Variable $x$ | 0 | 1 | 2 | 3 | 4 |
| Frequency | 6 | 10 | 19 | 11 | 4 |
| $P(X = x)$ | | | | | |

Fill in the last row. Draw a histogram.

6.  In 1961 Roger Maris set the record for most home runs in a season. Complete the following table.

| Event | No Hit | Single | Double | Triple | Home Run |
|---|---|---|---|---|---|
| Random Variable $x$ | 0 | 1 | 2 | 3 | 4 |
| Frequency | 431 | 78 | 16 | 4 | 61 |
| $P(X = x)$ | | | | | |

7. The student ratings of a particular mathematics professor are given in the following table. Complete the last line and draw a histogram.

| Event | 1 | 2 | 3 | 4 | 5 | 6 | 7 | 8 | 9 | 10 |
|---|---|---|---|---|---|---|---|---|---|---|
| Frequency | 5 | 2 | 3 | 0 | 4 | 10 | 10 | 11 | 3 | 2 |
| $P(X = x)$ | | | | | | | | | | |

8. A basketball player has a probability of .8 of sinking a foul shot. Assume that sinking a shot is independent of what occurred before. Let $X$ be the random variable given by the number of foul shots sunk by this player in 5 successive shots. Find the probability distribution. Draw a histogram.

9. A pair of fair dice is tossed. Let $X$ be the random variable given by the absolute value of the difference of the numbers on the top faces of the dice. Find the probability distribution. Draw a histogram.

10. On a true–false test with 5 questions, let $X$ denote the random variable given by the total number of questions correctly answered by guessing. Find the probability distribution. Draw a histogram.

11. In the previous problem let $X$ denote the random variable given by twice the total number of correct answers minus the number of incorrect answers. Find the probability distribution. Draw a histogram.

12. Let $X$ denote the random variable given by the number of girls in a family of 4 children. Find the probability distribution if a girl is as likely as a boy. Draw a histogram.

13. Two balls are selected at random from an urn that contains 3 white and 7 red balls. Let the random variable $X$ denote the number of white balls drawn. Find the probability distribution. Draw a histogram.

14. In the previous problem let the random variable $X$ denote the number of white balls times the number of red balls drawn. Find the probability distribution.

15. The probability distribution of the random variable $X$ is given in the following table. What must $z$ be?

| Random Variable $x$ | 0 | 1 | 2 | 3 | 4 |
|---|---|---|---|---|---|
| $P(X = x)$ | .2 | .1 | .1 | $z$ | .3 |

**16.** The probability distribution of the random variable $X$ is given in the following table.

| Random Variable $x$ | $-2$ | $-1$ | 0 | 1 | 2 | 3 |
|---|---|---|---|---|---|---|
| $P(X = x)$ | .20 | .15 | .05 | .35 | .15 | .10 |

Find (a) $P(X = 0)$, (b) $P(X \le 0)$, (c) $P(-1 < X \le 4)$, (d) $P(X \ge 1)$. Draw a histogram and identify each of the probabilities as an area.

**17.** The probability distribution of the random variable $X$ is given in the following table.

| Random Variable $x$ | 4 | 5 | 6 | 7 | 8 | 9 |
|---|---|---|---|---|---|---|
| $P(X = x)$ | .15 | .26 | .14 | .22 | .18 | .05 |

Find (a) $P(X = 4)$, (b) $P(X \le 2)$, (c) $P(5 \le X \le 6)$, (d) $P(X \ge 8)$. Draw a histogram and identify each of the probabilities as an area.

**18.** Let $Y$ be the random variable given by $X^2$ where $X$ is given in Exercise 16. Find the probability distribution of $Y$ by completing the following table.

| Random Variable $y$ | 0 | 1 | 4 | 9 |
|---|---|---|---|---|
| $P(X^2 = y)$ | | | | |

**19.** Let $Y$ be the random variable given by $X + 2$ where $X$ is given in Exercise 16. Find the probability distribution of $Y$.

## Applications

In Exercises 20 through 23 use a histogram to represent the requested data.

| Height | Percent Distribution of Population by Height | | | |
|---|---|---|---|---|
| | Males 25–34 yrs | Females 35–44 yrs | Males 25–34 yrs | Females 35–44 yrs |
| <5'2" | .42 | .87 | 19.11 | 18.24 |
| 5'2"–5'4" | 1.13 | 2.81 | 28.32 | 31.66 |
| 5'4"–5'6" | 7.96 | 8.79 | 27.49 | 28.17 |
| 5'6"–5'8" | 17.18 | 14.01 | 18.56 | 15.36 |
| 5'8"–5'10" | 28.66 | 30.61 | 6.21 | 5.76 |
| 5'10"–6' | 26.23 | 24.14 | .31 | .81 |
| >6' | 18.42 | 18.77 | .00 | .00 |

20. **Height Distribution.** Males ages 25–34

21. **Height Distribution.** Males ages 35–44

22. **Height Distribution.** Females ages 25–34

23. **Height Distribution.** Females ages 35–44

24. **Sales.** A car dealership tracks the number of cars sold each week. During the past year they sold from zero to six cars per week with the frequencies indicated in the following table. Find the probability distribution.

| Number Sold in One Week | 0 | 1 | 2 | 3 | 4 | 5 | 6 |
|---|---|---|---|---|---|---|---|
| Frequency | 4 | 11 | 8 | 10 | 10 | 5 | 2 |

25. **Quality Control.** Suppose 2 light bulbs are chosen at random from an assembly line that has 5% defective light bulbs. Let $X$ denote the random variable that gives the number of defective light bulbs chosen. Find the probability distribution of $X$. Draw a histogram.

## Solutions to Self-Help Exercise Set 8.1

1. The appropriate area in the accompanying histogram has been shaded. The area and thus the probability that Ted Williams obtained an extra-base hit in 1941 is then

$$P(X > 1) = .072 + .007 + .081 = .160.$$

2. From Section 7.7,

$$b(k : 6, .75) = C(6, k)\left(\frac{3}{4}\right)^k \left(\frac{1}{4}\right)^{6-k} = C(6, k)3^k \left(\frac{1}{4}\right)^6.$$

Let $h = (.25)^6 \approx .000244$. Then for $k = 0, 1, \ldots, 6$, $b(k:6, .75)$ are successively $1 \cdot 3^0 h$, $6 \cdot 3^1 h$, $15 \cdot 3^2 h$, $20 \cdot 3^3 h$, $15 \cdot 3^4 h$, $6 \cdot 3^5 h$, $1 \cdot 3^6 h$, or $h$, $18h$, $135h$, $540h$, $1215h$, $1458h$, $729h$. The histogram is shown in the figure. Notice how the probability distribution is skewed to the right with $p > .5$.

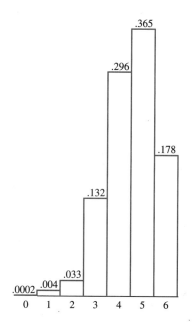

## 8.2   MEASURES OF CENTRAL TENDENCY

▶ *Expected Value*

▶ *Applications*

▶ *Mean of the Binomial Distribution*

▶ *Median and Mode*

APPLICATION

**Expected Profit of an Insurance Policy**

An insurance company sells a 65-year-old man a 1-year life insurance policy. The policy, which costs $100, pays $5000 in the event the man dies during the next year. Suppose there is a 1% chance that the man will die in the next year. If the insurance company were to sell a large number of the same insurance policies under the same conditions, what should they expect for an average profit per such policy? See Example 5 for the answer.

## Expected Value

In this section we introduce several statistics that will be used to measure the "center" of the data.

In the last section we studied probability distributions, that is, how probability was distributed over the various outcomes. We also saw the need for comparing two probability distributions. The *expected value* will now be discussed. We will see that this number is a generalization of the average and conveys certain very useful information about the data. The expected value will be seen to be one way of locating the "center" of the data. In another typical application, we will use the expected value to compare two investments in order to determine which will be more profitable.

We begin with a notion of *average* which is familiar to most. If you have three test grades of 95, 85, and 93, the average is the sum of all the numbers divided by the total number or

$$\frac{95 + 85 + 93}{3} = \frac{273}{3} = 91.$$

In general we have the following definition of *average* or *mean*.

---

**Average or Mean**

The **average** or **mean** of the $n$ numbers $x_1, x_2, \ldots, x_n$, denoted by $\bar{x}$ or the Greek letter $\mu$, is given by

$$\mu = \frac{x_1 + x_2 + \cdots + x_n}{n}.$$

---

**E X A M P L E   1**    **Calculating the Mean**

You are thinking of buying a small manufacturer of quality yachts and have obtained the following data from the owners that give how often various numbers of yacht sales occur per quarter.

| Number Sold in One Quarter | 1 | 2 | 3 | 4 | 5 |
|---|---|---|---|---|---|
| Frequency of Occurrence | 10 | 15 | 16 | 7 | 2 |

Find the average or mean number sold per quarter.

**Solution**    We first need to find the total number of quarters. This is

$$10 + 15 + 16 + 7 + 2 = 50.$$

Now we need to find the total sold and divide by 50. To find the total sold, notice that on 10 occasions exactly 1 was sold, on 15 occasions 2 were sold, and so on. Thus we must sum up the number one 10 times plus sum up the number two 15 times plus . . . and then divide by 50, that is, we have

$$\frac{\overbrace{1 + \cdots + 1}^{10 \text{ times}} + \overbrace{2 + \cdots + 2}^{15 \text{ times}} + \overbrace{3 + \cdots + 3}^{16 \text{ times}} + \overbrace{4 + \cdots + 4}^{7 \text{ times}} + \overbrace{5 + 5}^{2 \text{ times}}}{50}$$

or

$$\mu = \frac{1 \cdot 10 + 2 \cdot 15 + 3 \cdot 16 + 4 \cdot 7 + 5 \cdot 2}{50} = 2.52. \quad \blacksquare$$

We can gain considerable insight into this mean if we write the above fraction in a different form. We can write

$$\mu = \frac{1 \cdot 10 + 2 \cdot 15 + 3 \cdot 16 + 4 \cdot 7 + 5 \cdot 2}{50}$$

$$= 1\frac{10}{50} + 2\frac{15}{50} + 3\frac{16}{50} + 4\frac{7}{50} + 5\frac{2}{50}$$

$$= 1 \cdot p_1 + 2 \cdot p_2 + 3 \cdot p_3 + 4 \cdot p_4 + 5 \cdot p_5$$

where $p_1, p_2, p_3, p_4, p_5$, are, respectively, the (empirical) probabilities of the 5 possible outcomes 1, 2, 3, 4, 5. If we now let $x_1 = 1, x_2 = 2, x_3 = 3, x_4 = 4, x_5 = 5$, this can be written as

$$x_1 p_1 + x_2 p_2 + x_3 p_3 + x_4 p_4 + x_5 p_5.$$

This is the form we are seeking. Notice that each term is the product of an outcome with the probability of that outcome.

In general suppose there are the $n$ outcomes $x_1, x_2, \ldots, x_n$, with respective probabilities $p_1, p_2, \ldots, p_n$. Now suppose the experiment has been repeated a large number of times $N$ so that $x_1$ is observed to occur a frequency of $f_1$ times, $x_2$ is observed $f_2$ times, and so on. (We must have $f_1 + f_2 + \cdots + f_n = N$). Then the mean is

$$\text{mean} = \frac{\overbrace{x_1 + \cdots + x_1}^{f_1 \text{ times}} + \overbrace{x_2 + \cdots + x_2}^{f_2 \text{ times}} + \cdots + \overbrace{x_n + \cdots + x_n}^{f_n \text{ times}}}{N}$$

$$= \frac{x_1 f_1 + x_2 f_2 + \cdots + x_n f_n}{N}$$

$$= x_1 \frac{f_1}{N} + x_2 \frac{f_2}{N} + \cdots + x_n \frac{f_n}{N}$$

But as $N$ gets larger we expect the relative frequency $\dfrac{f_1}{N}$ to approach $p_1$. Similarly for the other relative frequencies. Thus the right-hand side of the last displayed line approaches

$$x_1 p_1 + x_2 p_2 + \cdots + x_n p_n.$$

We call this the **expected value**. In the language of random variables we then have the following.

---

**Expected Value or Mean**

Let $X$ denote the random variable that takes on the values $x_1, x_2, \ldots, x_n$, and let the associated probabilities be $p_1, p_2, \ldots, p_n$, then the **expected value** or **mean** of the random variable $X$, denoted by $E(X)$ or by $\mu$, is

$$E(X) = x_1 p_1 + x_2 p_2 + \cdots + x_n p_n.$$

---

Often we do not know the probabilities, but an experiment has been performed $N$ times. Then the probabilities $p_1, p_2, \ldots, p_n$, are the relative frequencies (empirical probabilities). This was illustrated in Example 1.

Notice that the expected value is an average in some sense. As we saw above, if the random phenomenon is repeated a large number of times, the mean or average of the observed outcomes will approach the expected value.

$E(X)$ is what we "expect" over the long term. But we must realize that $E(X)$ need not be an actual outcome. In Example 1, the expected value of sales per quarter was 2.52. We cannot sell 2.52 yachts. Nonetheless, it is extremely useful to think as though we could. (Actually, if the yacht manufacturer were incorporated with 100 shares outstanding and you owned one of these shares, it would make a great deal of sense to say you sold 2.52 yachts.)

In the case that all the $n$ outcomes are equally likely, the probabilities are just $\dfrac{1}{n}$ and then

$$E(X) = x_1 \frac{1}{n} + x_2 \frac{1}{n} + \cdots + x_n \frac{1}{n}$$

$$= \frac{x_1 + x_2 + \cdots + x_n}{n},$$

which is the average.

## Applications

**E X A M P L E   2**   **Calculating the Expected Value**

Redo Example 1 using the formula for $E(X)$.

**Solution**   Using the relative frequencies calculated earlier as the (empirical) probability yields Table 8.3.

**Table 8.3**

| Number Sold in One Quarter | 1 | 2 | 3 | 4 | 5 |
|---|---|---|---|---|---|
| Probability of Occurrence | $\frac{10}{50} = .20$ | $\frac{15}{50} = .30$ | $\frac{16}{50} = .32$ | $\frac{7}{50} = .14$ | $\frac{2}{50} = .04$ |

Then

$$E(X) = 1(.20) + 2(.30) + 3(.32) + 4(.14) + 5(.04) = 2.52. \quad \blacksquare$$

Suppose in considering buying the yacht manufacturer in Example 1, you wish to take into account a principal competitor who also manufactures yachts. Knowledge of the competitor's profits would be useful in determining how competitive the two manufacturers are and would heavily influence whether you will actually make the purchase. It would be convenient to know the details of the competitor's operations, but this is privileged information you are not likely to obtain. You then hire some analysts who have considerable knowledge of the yacht business and ask them to prepare a table like the one in Table 8.3 that you can use as a comparison. Naturally they are unable to give you exact figures, but they give you the data in Table 8.4.

**Table 8.4**

| Number Sold in One Quarter | 0 | 1 | 2 | 3 | 4 | 5 |
|---|---|---|---|---|---|---|
| Probability of Occurrence | .10 | .20 | .30 | .25 | .10 | .05 |

Notice that they have assessed from their collective knowledge a *probability* that the various outcomes will occur.

**E X A M P L E  3**   **Calculating the Expected Value**

Suppose in Example 1 you are told that the average profit per yacht is $20,000 and for the competitor is $21,000. Can you decide which company will have the higher profits per quarter?

**Solution**   For Example 1, the average quarterly sales are 2.52 and the average profit per yacht is $20,000, thus the average quarterly profit is

$$(2.52)(\$20,000) = \$50,400.$$

Looking now at the competitor, the "expected" quarterly sales is

$$E(X) = 0(.10) + 1(.20) + 2(.30) + 3(.25) + 4(.10) + 5(.05) = 2.2.$$

(William J. Kennedy/The
Image Bank.)

Thus you expect the competitor to sell 2.2 yachts per quarter. Their quarterly profit is then expected to be

$$(2.2)(\$21,000) = \$46,200.$$

The profits of the competitor are less, so you feel more comfortable about the purchase.   ∎

There are a number of versions of roulette. In the next two examples, we will find the expected losses of two types of bets for one version. In this version the wheel has the numbers from 0 to 36 located at 37 equally spaced slots. Half of the numbers from 1 to 36 are red and the other half are black. You place a bet. The wheel spins, and a ball falls randomly into one of the 37 slots.

**E X A M P L E   4**

### Calculating the Expected Value

In this bet select any of the 37 numbers. Making a $1 bet results in winning back $36 if your number occurs and, of course, you lose your $1 if your number does not occur. Find the expected return.

**Solution**

A tree diagram is shown in Figure 8.6. There is a probability of $\frac{1}{37}$ of winning and of $\frac{36}{37}$ of losing. Let the random variable $X$ be $35 if you win and $-\$1$ if you lose. Then the expected return is

$$E(X) = 35\,\frac{1}{37} - 1\,\frac{36}{37} \approx -.027.$$

**Figure 8.6**

On this bet you expect to lose about $.027 or 2.7% per bet in the long run.   ∎

**E X A M P L E   5**

### Calculating the Expected Value

An insurance company sells a 65-year-old man a 1-year life insurance policy. The policy, which costs $100, pays $5000 in the event the man dies during the next year. If there is a 1% chance that the man will die in the next year, find the probability distribution for this financial transaction for the insurance company and find their expected return. What meaning does this expected value have to the insurance company?

**Solution**

If the man lives for the next year, $100 is made. If the man dies within the next year, $5000 − $100 = $4900 is lost. The random variable and the associated probability is described in the following table.

| Event | *Lives for One Year* | *Dies within the Next Year* |
|---|---|---|
| Random Variable $x$ | 100 | −4900 |
| $P(X = x)$ | .99 | .01 |

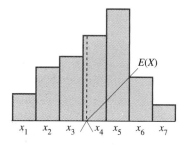

**Figure 8.7**

Thus

$$E(X) = 100(.99) - 4900(.01) = 50$$

dollars. If the insurance company were to sell a large number of the same insurance policies under the same conditions, they would expect a $50 return per such policy.  ∎

There is a physical interpretation of expected value that is sometimes useful. Draw a histogram of the probability distribution as indicated in Figure 8.7. Turn the rectangle above $x_1$, the area of which is the probability $p_1$ of $x_1$ occurring, into a thin rectangle with uniform thickness with *weight* equal to $p_1$. Do the same for each of the points $x_i$. Then the point $E(X)$ is the point at which the histogram will be balanced.

## Mean of the Binomial Distribution

Figure 8.1b of the last section gives the histogram for the binomial distribution $b(k : 6, .50)$. Notice the perfect symmetry about the point $x = 3$. By the physical interpretation of expected value, we should have $E(X) = 3$. This is indeed the case.

The histogram for the binomial distribution $b(k : 6, .75)$ is shown in the Solution to Self-Help Exercise 2 in the last section. One expects from the physical interpretation of $E(X)$ that $E(X)$ should be greater than 3; it is. Indeed the following is true.

---

**Expected Value of Binomial Distribution**

The expected value of the binomial distribution

$$p_k = b(k : n, p) = C(n, k)p^k q^{n-k}    p + q = 1$$

is

$$E(X) = np.$$

---

**E X A M P L E   6**    **Finding the Expected Value of a Binomial Distribution**

Find the expected value of the binomial distribution $b(k: 6, .75)$ mentioned above. Is it to the right of 3?

**Solution**    From (1) we obtain

$$E(X) = np = 6(.75) = 4.5,$$

which is greater than 3.  ∎

E X A M P L E 7 **Finding the Expected Value of a Binomial Distribution**

A manufacturer of light bulbs produces 1% of them with defects. In a case of 100 bulbs, what will be the average number of defective bulbs?

Solution If $p = .01$, then choosing a defective light bulb is a binomial distribution where "success" is picking a defective light bulb. Using (1) yields

$$E(X) = np = 100(.01) = 1. \quad \blacksquare$$

## Median and Mode

Suppose 10 families live on Millionaire Avenue. One family has an annual income of $10 million, while the other 9 have annual incomes of $20,000 each. The mean annual income is over $1 million. Does this number convey what we would like the "average" to convey? Hardly. This simple example illustrates what happens to the mean when one or a few extreme values exist in a set of numbers.

Two other measures of the center of a set of numbers are the **median** and the **mode**. Roughly speaking, the median of a collection of numbers is the middle number when arranged in increasing (or decreasing) order, while the mode is the number that occurs most often. A more precise definition, covering all possible cases, is given in the following.

---

**Median and Mode**

1. The **median** of a set of numbers is the middle number when the numbers are arranged in order of size and there are an odd number in the set. In the case that the number in the set is even, the median is the mean of the two middle numbers.
2. The **mode** of a set of numbers is the number that occurs more frequently than the others. If the frequency of occurrence of two numbers is the same and also greater than the frequency of occurrence of all the other numbers, then we say the set is **bimodal** and has two modes. If no one or two numbers occurs more frequently than the others, we say that the set has no mode.

---

E X A M P L E 8 **Finding the Mean, Median, and the Mode**

Find the mean, median, and the mode of the set of data $\{1, 1, 1, 2, 3, 6, 6, 6\}$.

Solution The mean is $(1 + 1 + 1 + 2 + 3 + 6 + 6 + 6)/8 = 3.25$. Since there is an even number in the collection, the median is the mean of the middle two numbers 2 and 3. This is 2.5. The set is bimodal with modes 1 and 4. $\quad \blacksquare$

The following example illustrates that the mean is very sensitive to extreme values whereas the median and mode are not.

**E X A M P L E  9**    **Finding the Mean, Median, and the Mode**

Find the mean, median, and the mode of the set of data {1, 1, 1, 2, 3, 6, 6, 6, 8974} and compare your answers to the answers in the previous example.

**Solution**    The mean is $(1 + 1 + 1 + 2 + 3 + 6 + 6 + 6 + 8974)/9 = 1000$, which is considerably different from the mean in the previous example. Since there is an odd number in the collection, the median is the middle number 3. This has changed only slightly. The set is bimodal with modes 1 and 6.    ■

## SELF-HELP EXERCISE SET 8.2

1.  Find the expected value of the random variable having the probability distribution given in the following table.

| Random Variable $x$ | $-2$ | $-1$ | 0 | 1 | 3 | 5 |
|---|---|---|---|---|---|---|
| $P(X = x)$ | .20 | .15 | .30 | .05 | .20 | .10 |

2.  In this version of roulette you bet on a color, say red. If red occurs, you win back twice your bet. If black occurs, you lose your stake. If 0 occurs, the wheel is spun again until 0 does not occur. If red occurs, then you win your original stake back. If black comes up, you lose your original stake. Find the expected return.

## EXERCISE SET 8.2

1.  If the random variable $X$ denotes the real valued outcomes given in the following table with the given frequency of occurrence of each of the outcomes, fill in the last line of the table and find $E(X)$.

| Outcome | 0 | 2 | 4 | 6 | 8 | 10 |
|---|---|---|---|---|---|---|
| Frequency | 50 | 5 | 10 | 5 | 20 | 10 |
| $P(X = x)$ | | | | | | |

2.  If the random variable $X$ denotes the real valued outcomes given in the following table with the given frequency of occurrence of each of the outcomes, fill in the last line of the table and find $E(X)$.

| Outcome | $-2$ | $-1$ | 0 | 1 | 2 | 10 |
|---|---|---|---|---|---|---|
| Frequency | 10 | 11 | 9 | 9 | 5 | 6 |
| $P(X = x)$ | | | | | | |

3.  Find the expected value of the random variable having the probability distribution given in the following table.

| Random Variable $x$ | $-30$ | $-10$ | 0 | 5 | 10 | 20 |
|---|---|---|---|---|---|---|
| $P(X = x)$ | .15 | .25 | .05 | .10 | .25 | .20 |

4.  Find the expected value of the random variable having the probability distribution given in the following table.

| Random Variable $x$ | $-4$ | $-1$ | 0 | 1 | 2 | 3 |
|---|---|---|---|---|---|---|
| $P(X = x)$ | .10 | .10 | .40 | .10 | .10 | .20 |

5.  Babe Ruth established the all-time highest "slugging average" for a single season in 1920. With the random variable $X$ assigned as indicated in the following table with the associated probability, find $E(X)$ rounded to 3 decimal places. This is the "slugging average."

| Outcome | *No Hit* | *Single* | *Double* | *Triple* | *Home Run* |
|---|---|---|---|---|---|
| Random Variable $x$ | 0 | 1 | 2 | 3 | 4 |
| $P(X = x)$ | .6245 | .1594 | .0786 | .0197 | .1179 |

6.  A student takes a single course during summer school and decides that the probability of obtaining each grade is as shown in the following table. Find the expected grade-point-average for this summer; that is, find the expected value of the random variable $X$.

| Outcome | A | B | C | D | F |
|---|---|---|---|---|---|
| Random Variable $x$ | 4 | 3 | 2 | 1 | 0 |
| $P(X = x)$ | .6 | .2 | .1 | .05 | .05 |

7.  The annual returns for an investment over the last 3 years are 50%, 0%, $-50\%$. Find the average annual return. Find the actual return over the 3-year period.

8.  A 2-year investment breaks even at the end of two years but yielded a 50% loss in the first year. Find the average annual rate of return for this investment.

9.  For an advertising promotion a company is giving away $10,000. All you have to do is submit a letter with your address on the entry form. Your entry will then be randomly selected with your chance of winning the same as anyone else's. It turns out that 100,000 people enter this contest. If your only cost is a 29 cent stamp, what is your expected return?

**10.** In the daily numbers game, a person picks any of the 1000 numbers from 0 to 999. A $1 wager on one number will return $600 if that number hits; otherwise nothing is returned. If the probability of any one number is the same as any other, find the expected return of this wager.

**11.** One version of roulette has a wheel with 38 numbers (1 through 36 plus 0 and 00) in 38 equally spaced slots. Half of the numbers from 1 to 36 are red, and the other half are black. The numbers 0 and 00 are green. Consider a $1 bet on red. If red comes up, you are returned $2, otherwise you lose the $1 wagered. Find the expected return for this bet.

**12.** In the version of roulette in the previous exercise, consider a $1 bet on a number from 1 to 36. If your number hits, you have $36 returned to you; otherwise you lose your $1 bet. Find the expected return for this bet.

**13.** In the version of roulette in the previous exercise, where the 0 and 00 are adjacent, consider a $1 bet that can be made on two adjacent numbers. If one of the numbers comes up, the player receives $18; otherwise nothing is received. What is the expected value of this bet?

**14.** A fair die is tossed. If $X$ denotes the random variable giving the number of the top face of the die, find $E(X)$.

**15.** A lottery has a grand prize of $500,000, 2 runner-up prizes of $100,000 each, and 100 consolation prizes of $1000 each. If 1 million tickets are sold for $1 each and the probability of any ticket winning is the same as any other, find the expected return on a $1 ticket.

**16.** A lottery has a grand prize of $50,000, 5 runner-up prizes of $5,000 each, 10 third-place prizes of $1000, and 100 consolation prizes of $10 each. If 100,000 tickets are sold for $1 each and the probability of any ticket winning is the same as any other, find the expected return on a $1 ticket.

**17.** Two fair dice are tossed. If you roll a total of 7, you win $6; otherwise you lose $1. What is the expected return of this game?

**18.** Two coins are taken at random (without replacement) from a bag containing 5 nickels, 4 dimes, and 1 quarter. Let $X$ denote the random variable given by the total value of the two coins. Find $E(X)$.

**19.** Two people are asked to find the average of 200 numbers. They decide that one will find the average of the first 100 and the other will find the average of the second 100. They will then take the average of these two averages. Is this the average of the 200 numbers? Why or why not?

**20.** A person is asked to find the average of the 100 numbers 1234(1), 1234(2), ..., 1234(100). They decide to factor out the number 1234, then find the average of the integers from 1 to 100, and then multiply this last average by 1234. Are they obtaining the correct average? Why or why not?

**21.** A person is asked to find the average of the 100 integers from 1234 to 1333. They decide to ''move each data point 1234 units to the left, find the average of the resulting data, then move the average over 1234.'' In other words, they find the average of all the integers from 0 to 99, and then add 1234 to this average. Is this the correct average? Why or why not?

22. Let $X$ be a discrete random variable and $c$ a constant. Define $Y$ to be the random variable given by $Y = cX$.
    a. Show $E(Y) = cE(X)$, that is, show $E(cX) = cE(X)$.
    b. Given the random variable $X$ with the probability distribution shown in the following table, find $E(X)$.

| Random Variable $x$ | 12 | 24 | 36 |
|---------------------|----|----|----|
| $P(X = x)$          | .1 | .6 | .3 |

    c. Now using the fact that the probability distribution for $\frac{1}{12}X$ must be the same as that for $X$, find $E(\frac{1}{12}X)$. Verify that $E(X) = 12E(\frac{1}{12}X)$.

23. Let $X$ be a discrete random variable and $c$ a constant. Define $Y$ to be the random variable given by $Y = X - c$.
    a. Show $E(Y) = E(X) - c$, that is, show $E(X - c) = E(X) - c$, or $E(X) = E(X - c) + c$
    b. Given the random variable $X$ with the probability distribution shown in the following table, find $E(X)$.

| Random Variable $x$ | 12 | 13 | 14 |
|---------------------|----|----|----|
| $P(X = x)$          | .2 | .5 | .3 |

    c. Now find the probability distribution for $X - 13$ and then find $E(X - 13)$. Verify that $E(X) = E(X - 13) + 13$.

24. Let $X$ and $Y$ be two discrete random variables associated with the outcomes of the same experiment. Define the random variable $Z$ by $Z = X + Y$. Show $E(Z) = E(X) + E(Y)$, that is, show $E(X + Y) = E(X) + E(Y)$.

25. If the mean for a binomial distribution $b(k; 10, p)$ is .9, find $p$.

## Applications

26. **Insurance.** An insurance company sells a $10,000, 5-year term life insurance policy to an individual for $700. Find the expected return for the company if the probability that the individual will live for the next 5 years is .95.

27. **Sales.** The number of sales per week and the associated probabilities of a car salesman are given in the following table. Find the expected number of sales per week.

| Number of Sales per Week | 0 | 1 | 2 | 3 | 4 |
|--------------------------|-----|-----|-----|-----|-----|
| Probability              | .50 | .30 | .10 | .07 | .03 |

28. **Employee Attendance.** A bank has 4 tellers. The following table gives the probabilities that a given number will be at work on any given day.

| Number at Work | 4 | 3 | 2 | 1 | 0 |
|---|---|---|---|---|---|
| Probability | .80 | .10 | .07 | .02 | .01 |

Find the expected number that show up for work on a given day.

29. **Investment Returns.** The following tables give all the possible returns and the associated probabilities of two investments, A and B. Find the expected value of each investment and compare the two.

| Outcome of A | $1000 | $2000 | $1000 |
|---|---|---|---|
| Probability | .1 | .8 | .1 |

| Outcome of B | −$1000 | $0 | $9000 |
|---|---|---|---|
| Probability | .8 | .1 | .1 |

30. **Investment Advisor.** An investment advisor informs you that his average annual return for the last 3 years is 100%. Furthermore, he says his annual return during each of the first 2 years of this 3-year period was 200%. Find how the people did who followed his advice by finding the return for the third year of this 3-year period.

31. **Quality Control.** Electrical switches are manufactured with the probability of .05 that any one is defective. If 50 are chosen at random, what is the expected number of defective switches in this batch?

32. **Medicine.** For a certain heart operation, the probability of survival is .95. If 10 of these operations are performed every week, what is the expected number of deaths due to this operation?

33. **Comparing Investments.** Two car dealerships are up for sale. The following two tables give the number of cars sold per day together with the associated probabilities.

| | *First* | | | | *Second* | | | | |
|---|---|---|---|---|---|---|---|---|---|
| Number Sold in One Day | 0 | 1 | 2 | 3 | 0 | 1 | 2 | 3 | 4 |
| Probability of Occurrence | .50 | .30 | .15 | .05 | .60 | .20 | .05 | .05 | .10 |

The average profit per car at the first dealership is $400 and at the second is $300. Which dealership will yield the highest daily profit?

34. **Comparing Investments.** Two motels are up for sale. The following two tables give the number of rooms rented per day together with the associated probabilities.

| Number Rented in One Day | 5 | 6 | 7 | 8 | 9 | 10 |
|---|---|---|---|---|---|---|
| Probability of Occurrence | .10 | .30 | .40 | .10 | .05 | .05 |

| Number Rented in One Day | 3 | 4 | 5 | 6 | 7 |
|---|---|---|---|---|---|
| Probability of Occurrence | .05 | .05 | .10 | .20 | .60 |

The average profit per room rented at the first motel is $20 and at the second is $21. Which motel will yield the highest daily profit?

35. **Cases Argued Before Supreme Court.** Given below is the number of cases argued before the U.S. Supreme Court for 5 selected years. Find the mean of these 5 numbers.

| Year | 1970 | 1975 | 1980 | 1985 | 1990 |
|---|---|---|---|---|---|
| Cases argued | 151 | 179 | 154 | 171 | 125 |

36. **Voter Turnout.** The following table gives the voter turnout as a percentage of the voting-age population in the 5 presidential elections from 1972 to 1988. Find the mean percentage turnout for these five presidential elections.

| Year | 1972 | 1976 | 1980 | 1984 | 1988 |
|---|---|---|---|---|---|
| % of VAP voting | 55 | 54 | 53 | 53 | 50 |

37. **Number of Federal Employees.** The following table gives the number of federal employees as a percentage of the population. Find the mean of these percentages.

| Year | 1960 | 1965 | 1970 | 1975 | 1980 | 1985 | 1990 |
|---|---|---|---|---|---|---|---|
| Percentage | 1.32 | 1.28 | 1.44 | 1.32 | 1.23 | 1.24 | 1.20 |

38. **Oil Pollution.** The number of oil polluting incidents in and around U.S. waters for recent years is given in the following table. Find the average number per year of such incidents.

| Year | 1986 | 1987 | 1988 | 1989 | 1990 |
|---|---|---|---|---|---|
| Incidents | 6539 | 6352 | 6791 | 8225 | 7114 |

**39. Oil Pollution.**   The number of millions of gallons of oil polluted in and around U.S. waters for recent years is given in the following table. Find the average amount per year.

| Year | 1986 | 1987 | 1988 | 1989 | 1990 |
|------|------|------|------|------|------|
| Millions of gallons | 4.6 | 3.9 | 6.6 | 13.6 | 4.3 |

**40. Health Care Unit Location.**   A state government is trying to locate a new health care unit to serve the needs of the city of Adams and the four smaller cities indicated in the figure. The table gives the population and coordinates of the centers of the five cities. For computational purposes we assume that *all* residents of a city are located precisely at the center of each city and that there are no residents between cities. The health care unit should be located at the population ''center'' of these five cities. The probability that a randomly selected individual in this community is in North Adams is just the population of North Adams divided by the total population of this community of five cities. Similarly for the other four cities. If $X$ and $Y$ are the random variables that take on the values of $x_i$ and $y_i$, respectively, then the population ''center'' and therefore the location of the health unit should be located at the coordinate $(E(X), E(Y))$. Find this point.

| City | Population | Population Coordinates | |
|------|-----------|------|------|
|  |  | $x_i$ | $y_i$ |
| North Adams | 11,000 | 3 | 7 |
| West Adams | 12,000 | 1 | 4 |
| Adams | 42,000 | 4 | 3 |
| East Adams | 17,000 | 6 | 5 |
| South Adams | 18,000 | 5 | 1 |

## Graphing Calculator Exercises

Solve Exercises 41 through 42 by using the statistical routines on your graphing calculator.

**41. Public Expenditure on Education.** Following is a table of the public expenditure on education as a percentage of GNP in the United States. Find the mean and also find the mean by using the results of Exercise 23 with $c = 6.7$.

| Year | 1980 | 1981 | 1982 | 1983 | 1984 | 1985 | 1986 | 1987 | 1988 | 1989 | 1990 |
|------|------|------|------|------|------|------|------|------|------|------|------|
| % of GNP | 6.6 | 6.7 | 6.5 | 6.7 | 6.7 | 6.6 | 6.7 | 6.9 | 6.9 | 6.9 | 6.9 |

**42. Lives Lost in Tornadoes.** The following table gives the number of lives lost from tornadoes in the United States. Find the mean number of lives lost for these 10 years.

| Year | 1981 | 1982 | 1983 | 1984 | 1985 | 1986 | 1987 | 1988 | 1989 | 1990 |
|------|------|------|------|------|------|------|------|------|------|------|
| Deaths | 24 | 64 | 34 | 122 | 94 | 15 | 59 | 32 | 50 | 53 |

## Solutions to Self-Help Exercise Set 8.2

1. $E(X) = -2(.20) - 1(.15) + 0(.30) + 1(.05) + 3(.20) + 5(.10) = 0.6$.

2. A tree diagram is shown in the figure. On a $1 bet there is a probability on the first spin of $\frac{18}{37}$ of red occurring and of winning $1 ($2 less $1 bet). There is a probability of $\frac{18}{37}$ of black occurring and of losing $1. Then there is a probability of $\frac{1}{2}\frac{1}{37}$ of 0 and then red, which results in breaking even and a probability of $\frac{1}{2}\frac{1}{37}$ of 0 and then black with a subsequent loss of $1. The expected value is

$$1\frac{18}{37} - 1\frac{18}{37} + 0\frac{1}{2}\frac{18}{37} - 1\frac{1}{2}\frac{18}{37} \approx -.0135.$$

On this bet you expect to lose about 1.35% per bet in the long run.

Clearly then, in general, this bet will permit one to play twice as long as the bet in Example 2 with the same stake. Many people will play the first bet because of the excitement of occasionally obtaining a substantially larger winning amount. But notice that the casino makes the customer pay for this excitement by taking a bigger cut of the money bet.

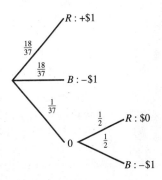

### 8.3 MEASURES OF DISPERSION

▶ *Variance and Standard Deviation*
▶ *An Alternate Form of the Variance*
▶ *Variance as a Measure of Risk*
▶ *Chebyshev's Inequality*
▶ *Variance of the Binomial Distribution*

**A P P L I C A T I O N**
**Quality Control**

A manufacturer of precision ball bearings has just started up two assembly lines. Industry standards require these bearings to be 1 cm in diameter with at most a 1% error. In other words, the diameter of a ball bearing must lie in the interval [.99, 1.01] or be rejected. The first 4 ball bearings coming off each assembly line are very carefully measured in the lab. Initial word comes back that the average diameter of the 4 off the first assembly line have an average right at 1 cm, which looks very good, but that the average diameter of the 4 off the second assembly line is somewhat in error at 1.005 cm. At this point would you accept the products from the first assembly line? The second? Why or why not? For the answers see below.

## Variance and Standard Deviation

In the last section we saw that the mean or the expected value is one measure of the "center" of the data. In this section, we will use the variance and the standard deviation to measure the extent of the dispersion of the data from the mean. We will then use Chebyshev's inequality to give crude estimates of any probability distribution, knowing only the mean and variance.

Further word comes from the lab of the manufacturer of ball bearings mentioned above indicating that none of the first 4 ball bearings are acceptable, having diameters of 1.020, .980, 1.030, and .970. Apparently the first assembly line produces an average right on target, but the actual diameters are so widely dispersed that the first assembly line is apparently useless for production of acceptable ball bearings.

Word also arrives that all the ball bearings from the second assembly line meet the standards, having diameters 1.006, 1.004, 1.007, 1.003. The average 1.005 is a little off being exactly 1, but the actual diameters are very close to this average and are dispersed only a very small amount.

The above example indicates how the *dispersion* of the data can play as important a role as the average. In this section we will define numbers called the **variance** and the **standard deviation** that measure the dispersion of data about the expected value. If the data is clustered close to the expected value, then the variance and the standard deviation will be small. If the data is widely dispersed away from the expected value, then the variance and standard deviation will be large.

*(Barry L. Runk/Grant Heilman Photography, Inc.)*

Suppose an experiment has the $n$ outcomes $x_1, x_2, \ldots, x_n$, which we assume for the present are all equally likely and $\mu$ is the mean. Then it is tempting to measure the dispersion by taking the average of all the differences

$$x_1 - \mu, x_2 - \mu, \ldots, x_n - \mu.$$

However, the average of these differences must be zero. (Why?) To avoid this we take the average of the *square* of the above differences and call this the **variance**. Thus

$$\text{Var} = \frac{(x_1 - \mu)^2 + (x_2 - \mu)^2 + \cdots + (x_n - \mu)^2}{n}.$$

**E X A M P L E   1**    **Finding the Variance**

Find the variance of each of the sets of 4 ball bearings that were the first to come off the assembly line mentioned above.

**Solutions**    (a)    Here the data are 1.020, .980, 1.030, .970. The mean is $\mu = 1.000$. Thus

$$\text{Var} = \frac{(x_1 - \mu)^2 + (x_2 - \mu)^2 + (x_3 - \mu)^2 + (x_4 - \mu)^2}{4}$$

$$= \frac{(1.020 - 1)^2 + (.980 - 1)^2 + (1.030 - 1)^2 + (.970 - 1)^2}{4}$$

$$= \frac{.0004 + .0004 + .0009 + .0009}{4}$$

$$= .00065.$$

(b)    Here the data are 1.006, 1.004, 1.007, 1.003. The mean is $\mu = 1.005$. Thus

$$\text{Var} = \frac{(x_1 - \mu)^2 + (x_2 - \mu)^2 + (x_3 - \mu)^2 + (x_4 - \mu)^2}{4}$$

$$= \frac{(1.006 - 1.005)^2 + (1.004 - 1.005)^2 + (1.007 - 1.005)^2 + (1.003 - 1.005)^2}{4}$$

$$= \frac{.000\ 001 + .000\ 001 + .000\ 004 + .000\ 004}{4}$$

$$= .000\ 0025. \quad \blacksquare$$

As one can see, the variance in the second case is substantially smaller than in the first case.

There is one unsettling aspect of the variance we just found. The units are *squared*. Thus in the first case above, the variance is .00065 and is in cm² and thus is difficult to relate back to the original problem, which is in cm. To remedy this we define the **standard deviation** to be the square root of the variance and denote it by $\sigma$. Thus $\sigma = \sqrt{\text{Var}}$.

**E X A M P L E   2**    **Finding the Standard Deviation**

Find the standard deviation of the two sets of data in the previous example.

**Solutions**
(a)   $\sigma = \sqrt{.00065} \approx .025$

Now this number .025 cm can be compared to the acceptable error of the ball bearings, which was .01 cm, and thus viewed as large.

(b)   $\sigma = \sqrt{.000\ 0025} \approx .0016$

Now this number .0016 cm can be compared to the acceptable error of the ball bearings, which was .01 cm, and thus viewed as small.   ■

In general the outcomes of the experiment are not equally likely. We now consider this general situation.

In general suppose there are the $n$ outcomes $x_1, x_2, \ldots, x_n$, with respective probabilities $p_1, p_2, \ldots, p_n$. Now suppose the experiment has been repeated a large number of times $N$ so that $x_1$ is observed to occur a frequency of $f_1$ times, $x_2$ is observed $f_2$ times, and so on. (We must have $f_1 + f_2 + \cdots + f_n = N$.) Then if we set $\mu = E(X)$, the variance is

$$\text{Var} = \frac{\overbrace{(x_1 - \mu)^2 + \cdots + (x_1 - \mu)^2}^{f_1 \text{ times}} + \cdots + \overbrace{(x_n - \mu)^2 + \cdots + (x_n - \mu)^2}^{f_n \text{ times}}}{N}$$

$$= \frac{(x_1 - \mu)^2 f_1 + (x_2 - \mu)^2 f_2 + \cdots + (x_n - \mu)^2 f_n}{N}$$

$$= (x_1 - \mu)^2 \frac{f_1}{N} + (x_2 - \mu)^2 \frac{f_2}{N} + \cdots + (x_n - \mu)^2 \frac{f_n}{N}.$$

But as $N$ gets larger, we expect the relative frequency $\dfrac{f_1}{N}$ to approach $p_1$, $\dfrac{f_2}{N}$ to approach $p_2$, and so forth. Thus the right-hand side of the last displayed line approaches

$$(x_1 - \mu)^2 p_1 + (x_2 - \mu)^2 p_2 + \cdots + (x_n - \mu)^2 p_n.$$

We call this term the **variance**. In the language of random variables, we then have the following.

---

**Variance and Standard Deviation**

Let $X$ denote the random variable that takes on the values $x_1, x_2, \ldots, x_n$, and let the associated probabilities be $p_1, p_2, \ldots, p_n$. Then if $\mu = E(X)$, the **variance** of the random variable $X$, denoted by $\text{Var}(X)$, is

$$\text{Var}(X) = (x_1 - \mu)^2 p_1 + (x_2 - \mu)^2 p_2 + \cdots + (x_n - \mu)^2 p_n.$$

The **standard deviation**, denoted by $\sigma(X)$ is

$$\sigma(X) = \sqrt{\text{Var}(X)}.$$

EXAMPLE 3    **Comparing Variances of Two Probability Distributions**

Figure 8.8 shows the histograms of 2 probability distributions both with means equal to 1. Find the variance of each and compare the two.

Solutions    The variance of the first probability distribution shown in Figure 8.8 is

$$(0 - 1)^2(0.10) + (1 - 1)^2(.80) + (2 - 1)^2(0.10) = 0.20.$$

The variance of the second probability distribution shown in Figure 8.8 is

$$(0 - 1)^2\frac{1}{3} + (1 - 1)^2\frac{1}{3} + (2 - 1)^2\frac{1}{3} = \frac{2}{3}.$$

The variance of the first probability distribution is much smaller than the variance of the second. This just reflects the fact, seen from Figure 8.8, that the first probability distribution is less dispersed from the mean than the second one. ∎

Often the probabilities are not known, but an experiment has been performed $N$ times. Then the probabilities $p_1, p_2, \ldots, p_n$, are the relative frequencies (empirical probabilities). This is illustrated in the next problem.

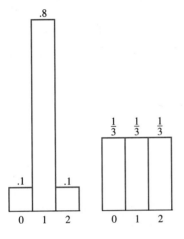

Figure 8.8

EXAMPLE 4    **Finding the Standard Deviation**

You have completed 40 courses, each with the same number of credits, with the frequency of each grade given in the following table.

| Outcome | A | B | C | D | F |
|---|---|---|---|---|---|
| Random Variable $X$ | 4 | 3 | 2 | 1 | 0 |
| Frequency | 25 | 11 | 3 | 1 | 0 |

Find the expected value of the random variable $X$; that is, find the grade point average, and then find $\text{Var}(X)$.

Solution    We first notice that the total number is given by

$$f_1 + f_2 + f_3 + f_4 + f_5 = 25 + 11 + 3 + 1 + 0 = 40.$$

Dividing each of the frequencies by 40 gives the relative frequencies or the empirical probability. Doing this gives the following table.

| Outcome | A | B | C | D | F |
|---|---|---|---|---|---|
| Random Variable $x$ | 4 | 3 | 2 | 1 | 0 |
| Frequency | 25 | 11 | 3 | 1 | 0 |
| Relative Frequency | $\frac{25}{40}$ | $\frac{11}{40}$ | $\frac{3}{40}$ | $\frac{1}{40}$ | 0 |

The expected value is then

$$\mu = 4\frac{25}{40} + 3\frac{11}{40} + 2\frac{3}{40} + 1\frac{1}{40} + 0\frac{0}{40} = 3.5.$$

Then

$$\text{Var}(X) = (4 - 3.5)^2\frac{25}{40} + (3 - 3.5)^2\frac{11}{40} + (2 - 3.5)^2\frac{3}{40}$$
$$+ (1 - 3.5)^2\frac{1}{40} + 0 = .55. \quad \blacksquare$$

## An Alternate Form of the Variance

The computation of

$$\text{Var}(X) = (x_1 - \mu)^2 p_1 + (x_2 - \mu)^2 p_2 + \cdots + (x_n - \mu)^2 p_n$$

can be simplified by using an alternative form. To see how, write the terms on the right-hand side as

$$(x_1 - \mu)^2 p_1 = x_1^2 p_1 - 2x_1\mu p_1 + \mu^2 p_1$$

$$\cdots$$

$$(x_n - \mu)^2 p_n = x_n^2 p_n - 2x_n\mu p_n + \mu^2 p_n.$$

Summing up the left-hand sides of these equations gives $\text{Var}(x)$. Doing this and rearranging the sum of the right-hand sides gives

$$\text{Var}(X) = (x_1^2 p_1 + \cdots + x_n^2 p_n) - 2\mu(x_1 p_1 + \cdots + x_n p_n)$$
$$+ \mu^2(p_1 + \cdots + p_n).$$

But recall that

$$\mu = x_1 p_1 + \cdots + x_n p_n$$

while

$$p_1 + \cdots + p_n = 1.$$

Using these last two facts then yields

$$\text{Var}(X) = (x_1^2 p_1 + \cdots + x_n^2 p_n) - 2\mu\mu + \mu^2 =$$
$$(x_1^2 p_1 + \cdots + x_n^2 p_n) - \mu^2.$$

This will save a substantial amount of work. We have the following.

---

**Alternate Form of Variance**

An alternate form of the variance is given by

$$\text{Var}(X) = (x_1^2 p_1 + x_2^2 p_2 + \cdots + x_n^2 p_n) - \mu^2.$$

---

EXAMPLE 5     **Using the Alternate Form of the Variance**

Use this last form to find $\text{Var}(X)$ in the last example.

Solution     Form the following table.

| $x_1$ | $p_i$ | $x_i p_i$ | $x_i^2 p_i$ |
|-------|-------|-----------|-------------|
| 4 | $\frac{25}{40}$ | 2.500 | 10.000 |
| 3 | $\frac{11}{40}$ | 0.825 | 2.475 |
| 2 | $\frac{3}{40}$ | 0.150 | 0.300 |
| 1 | $\frac{1}{40}$ | 0.025 | 0.025 |
| 0 | $\frac{0}{40}$ | 0.000 | 0.000 |
| | | Sum | Sum |
| | | = | = |
| | | 3.5 | 12.8 |

Then

$$\text{Var}(X) = (x_1^2 p_1 + x_2^2 p_2 + \cdots + x_n^2 p_n) - \mu^2 = 12.8 - (3.5)^2 = 0.55. \quad \blacksquare$$

## Variance as a Measure of Risk

It is a basic tenet in the modern theory of finance that the amount that the value of an asset varies over time is a measure of the risk of the asset. The standard measure of this variability is the standard deviation.

If everyone has a great deal of confidence what the future value of an investment is, then the value of the investment should vary little. For example, everyone has essentially complete confidence what the future performance of money in the bank will be. Thus the value of this asset (the principal) does not vary in time. This reflects the fact that this investment is (essentially) risk-free. Everyone has very high confidence in the future performance of a high quality bond. However, there is some variability in price as investors change somewhat from week to week their assessment of the economy, inflation, and other factors that may affect the price of the bond. This reflects the fact that a bond does have some risk.

Moving on to stocks, the future earnings and dividend performance of a very large and regulated company such as AT&T is estimated by many analysts with reasonable confidence and so the price of this security does not vary significantly. However, the future earnings of a small bio-tech firm is likely to be very unpredictable. Will it be able to come up with new products? Will it be able to fend off competitors? Will some key personnel leave? Thus the weekly assessment of such a firm is likely to change substantially more than that of AT&T. As a consequence, the weekly price changes of this company will vary significantly more, reflecting the fact that an investment in the small company is more risky.

Therefore the risk of a security is defined as the standard deviation of the daily (or perhaps weekly) prices of the security.

Table 8.5 gives the weekly price changes (during 5 consecutive weeks in 1991) in percent of 4 assets: money in the bank, a high quality bond (IBM), AT&T, and Scan Optics.

**Table 8.5**

| Principal in Bank | IBM 7$\frac{7}{8}$04 Bond | AT&T Stock | Scan Optics Stock |
|---|---|---|---|
| 0.0 | 0.0 | 0.3 | 4.6 |
| 0.0 | −0.2 | 0.0 | 4.7 |
| 0.0 | −0.1 | 3.1 | 0.0 |
| 0.0 | 0.0 | −2.0 | 8.7 |
| 0.0 | −0.5 | 2.1 | −4.2 |

According to Exercises 17 and 18 at the end of this section, the standard deviation of the four assets over this period of time was 0, 0.19, 1.77, 4.44, respectively. This reflects the fact that these assets are increasingly risky in the order listed.

## Chebyshev's Inequality

Knowing only the mean and variance of a probability law is not sufficient in general to determine the probability law. However, crude estimates of the probability law can be made, which suffice for many purposes, from a knowledge of just the mean and variance. One such estimate is Chebyshev's inequality.

---

**Chebyshev's Inequality**

Let $X$ be a random variable with expected value $\mu$ and standard deviation $\sigma$. Then the probability that the random variable associated with a random outcome lies between $\mu - h\sigma$ and $\mu + h\sigma$ is at least $1 - 1/h^2$, that is,

$$P(\mu - h\sigma \leq X \leq \mu + h\sigma) \geq 1 - \frac{1}{h^2}.$$

---

A proof is not given here.

For Chebyshev's inequality to give new information, clearly we must have $h \geq 1$.

E X A M P L E  6     **Using Chebyshev's Inequality**

A probability distribution of a random variable $X$ has mean $\mu = 14$ and standard deviation $\sigma = 4$. Use Chebyshev's inequality to give a lower bound for the probability that $X$ is within 12 units of $\mu$.

**Solution**     Setting $12 = h\sigma = h \cdot 4$, yields $h = 3$. Then $\mu - h\sigma = 2$ and $\mu + h\sigma = 26$ and

$$P(2 \le X \le 26) \ge 1 - \frac{1}{3^2} \approx .89.$$

Thus there is at least an 89% chance that the random variable lies on the interval $[2, 26]$.  ∎

E X A M P L E  7     **Using Chebyshev's Inequality**

When cereal is packaged, the actual amount of cereal varies from box to box. Suppose a sample indicates that the average in a 16-ounce box is 16.0 ounces with a standard deviation of 0.5. Using Chebyshev's inequality, find the minimum percent of boxes that will have 15 to 17 ounces of cereal.

**Solution**     Let $X$ be the random variable equal to the number of ounces in a box of cereal. We are given $\mu = 16$ and $\sigma = 0.5$. We are seeking an estimate of $P(15 \le X \le 17)$ or in terms of $\mu = 16$, $P(\mu - 1 \le X \le \mu + 1)$. In the terminology of Chebyshev's inequality, we then must have $1 = h\sigma = h(0.5)$. This implies $h = 2$. Thus

$$P(15 \le X \le 17) \ge 1 - \frac{1}{h^2} = 1 - \frac{1}{4} = .75.$$

Thus at least 75% of these boxes will have between 15 and 17 ounces of cereal.  ∎

## Variance of the Binomial Distribution

We give without proof the following additional useful fact about binomial distributions.

---

**Variance of the Binomial Distribution**

The variance of the binomial distribution with $n$ trials and the probability of "success" equal to $p$ and of "failure" equal to $q$ is

$$\mathrm{Var}(X) = npq.$$

---

## SELF-HELP EXERCISE SET 8.3

1. The following table gives two probability distributions associated with the same outcomes.

| Outcome | 1 | 2 | 3 | 4 | 5 |
|---|---|---|---|---|---|
| (a) Probability | .1 | .2 | .4 | .2 | .1 |
| (b) Probability | .4 | .1 | 0 | .1 | .4 |

Draw a histogram of each probability distribution. By inspecting these histograms (making no calculations), determine the means and determine which one has the largest variance. Now calculate the means, variances, and standard deviations.

2. Use Chebyshev's inequality to estimate the minimum probability that the outcome is between 2 and 4 for a probability distribution with $\mu = 3$ and $\sigma = 0.25$.

## EXERCISE SET 8.3

In Exercises 1 through 2 you are given a table with two probability distributions associated with the same outcomes. For each exercise draw a histogram of each probability distribution. Using only these histograms and making no calculations, determine the means and determine which one has the largest variance. Now calculate the means, variances, and standard deviations.

1.

| Outcome | 1 | 2 | 3 | 4 | 5 |
|---|---|---|---|---|---|
| (a) Probability | .2 | .2 | .2 | .2 | .2 |
| (b) Probability | .1 | .2 | .4 | .2 | .1 |

2.

| Outcome | 5 | 6 | 7 | 8 | 9 |
|---|---|---|---|---|---|
| (a) Probability | .1 | .1 | .6 | .1 | .1 |
| (b) Probability | .3 | .1 | .2 | .1 | .3 |

3. In the June 1991 Atlantic City Golf Classic, the top 3 finishers had the following scores for the three rounds of the tournament.

| Jane Geddes | 71 | 68 | 69 |
|---|---|---|---|
| Cindy Schreger | 70 | 70 | 69 |
| Amy Alcott | 69 | 68 | 72 |

Find the mean, variance, and standard deviation for the three rounds for each of the players. Which had the lowest average? Which was the most consistent?

4. In the June 1991 McDonald's Golf Championship, the top 5 finishers had the following scores for the four rounds of the tournament.

| Beth Daniel | 67 | 71 | 67 | 68 |
|---|---|---|---|---|
| Pat Bradley | 69 | 67 | 70 | 71 |
| Sally Little | 67 | 69 | 67 | 74 |
| M. McGann | 70 | 66 | 72 | 70 |
| A. Okamoto | 70 | 65 | 73 | 70 |

Find the mean, variance, and standard deviation for the four rounds for each of the players. Who had the lowest average? Who was the most consistent? Least consistent?

5. Find the mean, variance, and standard deviation of the random variable with the probability distribution given in the following table.

| Random Variable $x$ | $-3$ | 0 | 1 |
|---|---|---|---|
| $P(X = x)$ | .2 | .5 | .3 |

6. Find the mean, variance, and standard deviation of the random variable with the probability distribution given in the following table.

| Random Variable $x$ | $-2$ | 1 | 3 |
|---|---|---|---|
| $P(X = x)$ | .1 | .3 | .6 |

7. Find the mean, variance, and standard deviation of the random variable with the probability distribution given in the following table.

| Random Variable $x$ | $-2$ | 0 | 1 | 2 | 4 |
|---|---|---|---|---|---|
| $P(X = x)$ | .1 | .3 | .1 | .2 | .3 |

8. Find the mean, variance, and standard deviation of the random variable with the probability distribution given in the following table.

| Random Variable $x$ | $-2$ | $-1$ | 1 | 2 | 5 |
|---|---|---|---|---|---|
| $P(X = x)$ | .2 | .1 | .25 | .25 | .2 |

9. By inspection of the following two histograms (and making no calculations), determine the means and determine which one has the largest variance.

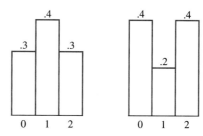

10. By inspection of the following two histograms (and making no calculations), determine the means and determine which one has the largest variance.

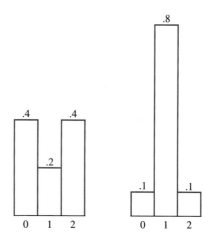

11. Let $X$ be a discrete random variable and $c$ a constant. Define $Y$ to be the random variable given by $Y = cX$.

   a. Show $\text{Var}(Y) = c^2\text{Var}(X)$, that is, show $\text{Var}(cX) = c^2\text{Var}(X)$. *Hint:* Use the fact given in Exercise 22 of the last section that $E(cX) = cE(X)$, or $\mu(cX) = c\mu(X)$, and the second form of the variance.

   b. Given the random variable $X$ with the probability distribution shown in the following table, find $\text{Var}(X)$.

   | Random Variable $x$ | 3 | 6 | 9 |
   |---|---|---|---|
   | $P(X = x)$ | .1 | .6 | .3 |

   c. Now using the fact that the probability distribution for $\frac{1}{3}X$ must be the same as for $X$, find $\text{Var}(\frac{1}{3}X)$. Verify that $\text{Var}(X) = 9\,\text{Var}(\frac{1}{3}X)$.

**12.** Let $X$ be a discrete random variable and $c$ a constant. Define $Y$ to be the random variable given by $Y = X - c$.

   **a.** Show $\text{Var}(Y) = \text{Var}(X)$, that is, show $\text{Var}(X - c) = \text{Var}(X)$. *Hint:* Use the fact given in Exercise 23 of the last section that $E(X) - c = E(X - c)$, or $\mu(X - c) = \mu(X) - c$, and the first form of the variance.

   **b.** Given the random variable $X$ with the probability distribution shown in the following table, find $\text{Var}(X)$.

| Random Variable $x$ | 12 | 13 | 14 |
|---|---|---|---|
| $P(X = x)$ | .2 | .5 | .3 |

   **c.** Now find the probability distribution for $X - 13$ and then find $\text{Var}(X - 13)$. Verify that $\text{Var}(X) = \text{Var}(X - 13)$.

**13.** A probability distribution has a mean of 30 and a standard deviation of 2. Use Chebyshev's inequality to find the minimum probability that an outcome is between 20 and 40.

**14.** A probability distribution has a mean of 90 and a standard deviation of 3. Use Chebyshev's inequality to find the minimum probability that an outcome is between 60 and 120.

**15.** A probability distribution has a mean of 60 and a standard deviation of 1. Use Chebyshev's inequality to find the number $c$ so that the probability that the numerical outcome is between $60 - c$ and $60 + c$ is at least .99.

**16.** A probability distribution has a mean of $\mu$ and a standard deviation of $\sigma$. Use Chebyshev's inequality to find a lower bound for the percentage of outcomes between $\mu - 5\sigma$ and $\mu + 5\sigma$.

## Applications

**17. Finance.** Find the mean, variance, and standard deviation of the weekly percentage change in the value of the principal in the bank account and for the bond in Table 8.5 of the text. Which is varying the most? How do you correlate your answer with the risk of each investment?

**18. Finance.** Find the mean, variance, and standard deviation of the weekly percentage change in the values of the 2 stocks given in Table 8.5 of the text. Which is varying the most? How do you correlate your answer with the risk of each investment?

**19. Finance.** During a 5-week period in 1991, the stock of a utility company and the stock of a small company showed the following weekly percentage changes.

| Company | Weekly Price Change in % | | | | |
|---|---|---|---|---|---|
| United Illuminating | −1.1 | 0.4 | 1.5 | −2.5 | −2.2 |
| U. S. Surgical | 5.4 | −1.2 | 5.9 | 1.0 | −3.2 |

Find the variance of the weekly price changes of each. Relate the two variances found to the riskiness of the two stocks.

20. **Finance.**   During a 5-week period in 1991, the stock of a large insurance company and the stock of a small hi-tech company showed the following weekly percentage change.

| Company | Weekly Price Change in % | | | | |
|---------|------|------|-----|------|------|
| Aetna Life | −0.3 | −2.8 | 1.2 | 1.8 | 0.0 |
| Data Switch | 2.9 | −5.9 | 6.3 | −11.8 | −3.2 |

Find the variance of the weekly price changes of each. Relate the two variances found to the riskiness of the two stocks.

21. **Sales.**   The number of sales per week and the associated probabilities of 2 car salesmen, A and B, are given in the following table. Find the mean, variance, and standard deviation for the sales per week of each. Which one will sell the most cars? Which one is the most consistent?

| Number of Sales per Week | 0 | 1 | 2 | 3 | 4 |
|--------------------------|-----|-----|-----|-----|-----|
| Probability A | .50 | .30 | .10 | .07 | .03 |
| Probability B | .40 | .20 | .20 | .10 | .10 |

22. **Employee Absences.**   The following table gives the probabilities that two employees, A and B, will have the given number of absences from work per month. Find the mean, variance, and standard deviation for each. Which attendance is the best? Which attendance is the most consistent?

| Number of Absences per Month | 0 | 1 | 2 | 3 | 4 |
|------------------------------|-----|-----|-----|-----|-----|
| Probability A | .90 | .04 | .03 | .02 | .01 |
| Probability B | .85 | .05 | .05 | .05 | .00 |

23. **Venereal Disease.**   The following table gives the number of thousands of reported cases of gonorrhea and syphilis in the United States. Find the average, variance, and standard deviation of each. Which is varying the most?

| Year | 1985 | 1986 | 1987 | 1988 | 1989 | 1990 |
|------|------|------|------|------|------|------|
| Gonorrhea | 911 | 901 | 781 | 720 | 733 | 690 |
| Syphilis | 68 | 68 | 87 | 103 | 111 | 134 |

24. **Mortality Rates from Legal Abortions and Childbirth.**   The following table gives the mortality rates in numbers per 100,000 from legal abortions and child-

birth in the United States. Find the average, variance, and standard deviation of each. Which is varying the most?

| Year | 1981 | 1982 | 1983 | 1984 | 1985 |
|------|------|------|------|------|------|
| Abortion | 0.4 | 0.8 | 0.7 | 0.7 | 0.4 |
| Childbirth | 7.2 | 7.9 | 8.0 | 7.9 | 7.8 |

25. **Temperatures.** The following table gives the average monthly temperatures for two cities for four selected months. Find the average, variance, and standard deviation of each. Which is varying the most?

| Month | Jan | Apr | Jul | Oct |
|-------|-----|-----|-----|-----|
| San Diego | 65 | 68 | 76 | 75 |
| Chicago | 29 | 59 | 83 | 64 |

26. **Health.** The following table gives the number of reported cases of plague and polio in the United States. Find the average, variance, and standard deviation of each. Which is varying the most?

| Year | 1985 | 1986 | 1987 | 1988 | 1989 | 1990 |
|------|------|------|------|------|------|------|
| Plague | 17 | 10 | 12 | 15 | 4 | 2 |
| Polio, acute | 7 | 8 | 6 | 9 | 5 | 7 |

27. **Drug Use.** The following table gives the percentage of current users of marihuana and cocaine in the 18 to 25-year-old age group in the United States. Find the average, variance, and standard deviation of each. Which is varying the most?

| Year | 1974 | 1979 | 1982 | 1985 | 1988 | 1991 |
|------|------|------|------|------|------|------|
| Marihuana | 25 | 35 | 27 | 22 | 16 | 13 |
| Cocaine | 3 | 9 | 7 | 8 | 5 | 2 |

28. **Drug Use.** The following table gives the percentage of current users of alcohol and cigarettes in the 18 to 25-year-old age group in the United States. Find the average, variance, and standard deviation of each. Which is varying the most?

| Year | 1974 | 1979 | 1982 | 1985 | 1988 | 1991 |
|------|------|------|------|------|------|------|
| Alcohol | 69 | 76 | 71 | 71 | 65 | 64 |
| Cigarettes | 49 | 43 | 40 | 37 | 35 | 32 |

29. **Scholastic Aptitude Test Scores.** The following table gives the verbal SAT scores for recent years. Find the average, variance, and standard deviation of each. Which is varying the most?

| Year | 1987 | 1988 | 1989 | 1990 | 1991 |
|------|------|------|------|------|------|
| Males | 435 | 435 | 434 | 429 | 426 |
| Females | 425 | 422 | 421 | 419 | 418 |

30. **Scholastic Aptitude Test Scores.** The following table gives the math SAT scores for recent years. Find the average, variance, and standard deviation of each. Which is varying the most?

| Year | 1987 | 1988 | 1989 | 1990 | 1991 |
|------|------|------|------|------|------|
| Males | 500 | 498 | 500 | 499 | 497 |
| Females | 453 | 455 | 454 | 455 | 453 |

31. **Air Pollutant Concentration.** The following table gives the concentration of two pollutants, in appropriate units, for recent years. Find the average, variance, and standard deviation of each. Which is varying the most?

| Year | 1986 | 1987 | 1988 | 1989 | 1990 |
|------|------|------|------|------|------|
| Carbon Monoxide | 7.1 | 6.7 | 6.4 | 6.3 | 5.9 |
| Ozone | .12 | .13 | .14 | .12 | .11 |

32. **Air Pollutant Concentration.** The following table gives the concentration of two pollutants, in appropriate units, for recent years. Find the average, variance, and standard deviation of each. Which is varying the most?

| Year | 1986 | 1987 | 1988 | 1989 | 1990 |
|------|------|------|------|------|------|
| Sulfur Dioxide | 9 | 9 | 9 | 8 | 8 |
| Nitrogen Dioxide | 24 | 24 | 24 | 23 | 22 |

33. **Death Rates by Causes.** The death rate per 100,000 in the United States for viral hepatitis and meningitis for selected years is given in the following table. Find the average, variance, and standard deviation of each. Which is varying the most?

| Year | 1970 | 1980 | 1985 | 1990 |
|------|------|------|------|------|
| Viral Hepatitis | .5 | .4 | .4 | .7 |
| Meningitis | .8 | .6 | .5 | .5 |

34. **Death Rates by Causes.** The death rate per 100,000 in the United States for accidents from automobiles and for all other accidents for selected years is given in the following table. Find the average, variance, and standard deviation of each. Which is varying the most?

| Year | 1970 | 1980 | 1985 | 1990 |
|------|------|------|------|------|
| Automobiles | 27 | 24 | 19 | 19 |
| All Others | 30 | 23 | 20 | 18 |

35. **Sales.** The expected amount of sales for each person on a sales force of a company is $100,000 per month with a standard deviation of $20,000. Find the minimum probability using Chebyshev's inequality that a salesman will sell between $70,000 and $130,000 in a particular month.

36. **Quality Control.** The expected number of defective parts produced on an assembly line per shift is 40 with a standard deviation of 10. Find the minimum probability using Chebyshev's inequality that the number of defective parts on a particular shift will be between 20 and 60.

37. **Useful Life of a Product.** The expected useful life of a certain brand of VCR is 5 years with standard deviation equal to 1 year. Find the minimum probability using Chebyshev's inequality that the useful life of one of these VCRs will be between 3 and 7 years.

38. **Weights.** If the mean weight of a delivered ton of top soil is 2000 pounds with a standard deviation 200 pounds, find the minimum probability using Chebyshev's inequality that the weight of a delivered "ton" is between 1700 and 2300 pounds.

## Graphing Calculator Exercises

In Exercises 39 through 40 solve by using the statistical routines on your graphing calculator.

39. **Birth and Death Rates.** The following table gives the birth and death rates in the United States per 1000 total population for each of the 10 years from 1980 to 1989. Find the mean, variance, and standard deviation of each. Which is varying the most? Also use the results of Exercise 23 of the last section and Exercise 12 above with $c = 15.5$ for the birthrate and $c = 8.7$ for the death rate.

| Year | 1980 | 1981 | 1982 | 1983 | 1984 | 1985 | 1986 | 1987 | 1988 | 1989 |
|------|------|------|------|------|------|------|------|------|------|------|
| Birth rate | 15.9 | 15.8 | 15.9 | 15.5 | 15.5 | 15.8 | 15.6 | 15.7 | 15.9 | 16.2 |
| Death rate | 8.8 | 8.6 | 8.5 | 8.6 | 8.6 | 8.7 | 8.7 | 8.7 | 8.8 | 8.7 |

40. **Marriage and Divorce Rates.** The following table gives the marriage and divorce rates in the United States per 1000 total population for each of the 10 years from 1980 to 1989. Find the mean, variance, and standard deviation of

each. Which is varying the most? Use the results of Exercise 23 of the last section and Exercise 12 above with $c = 10.6$ for the marriage rate and $c = 5.0$ for the divorce rate.

| Year | 1980 | 1981 | 1982 | 1983 | 1984 | 1985 | 1986 | 1987 | 1988 | 1989 |
|------|------|------|------|------|------|------|------|------|------|------|
| Marriage rate | 10.6 | 10.6 | 10.6 | 10.5 | 10.5 | 10.1 | 10.0 | 9.9 | 9.7 | 9.7 |
| Divorce rate | 5.2 | 5.3 | 5.0 | 4.9 | 5.0 | 5.0 | 4.8 | 4.8 | 4.8 | 4.7 |

## Solutions to Self-Help Exercise Set 8.3

**1.** The following figure shows the histograms.

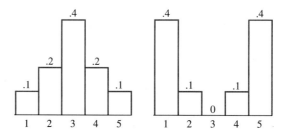

From the histograms one sees that the probability is evenly distributed about 3 in each case. Thus the mean should be 3. In the first case the probability is clustered about the mean, whereas in the second case the probability is more dispersed away from the mean. Thus the variance in the second case should be larger.

To calculate the mean and variance, make the following table.

| | | **Probability (a)** | | **Probability (b)** | | |
|---|---|---|---|---|---|---|
| $x_i$ | $p_i$ | $x_i p_i$ | $x_i^2 p_i$ | $p_i$ | $x_i p_i$ | $x_i^2 p_i$ |
| 1 | .1 | 0.1 | 0.1 | .4 | 0.4 | 0.4 |
| 2 | .2 | 0.4 | 0.8 | .1 | 0.2 | 0.4 |
| 3 | .4 | 1.2 | 3.6 | .0 | 0.0 | 0.0 |
| 4 | .2 | 0.8 | 3.2 | .1 | 0.4 | 1.6 |
| 5 | .1 | 0.5 | 2.5 | .4 | 2.0 | 10.0 |
| | | Sum = 3.0 | Sum = 10.2 | | Sum = 3.0 | Sum = 12.4 |

**a.**   Thus from the table $\mu = 3$ and using the second form for variance gives

$$\text{Var} = 10.2 - 3^2 = 1.2$$

$$\sigma = \sqrt{1.2} \approx 1.095.$$

**b.** From the table $\mu = 3$ and using the second form for variance gives

$$\text{Var} = 12.4 - 3^2 = 3.4$$

$$\sigma = \sqrt{3.4} \approx 1.84.$$

As we see, the means are the same, and the variance of the second is larger.

2. If $X$ is the random variable equal to the outcome, then we are seeking an estimate of $P(2 \leq X \leq 4)$. Since $\mu = 3$, this can be written as $P(\mu - 1 \leq X \leq \mu + 1)$. This implies that $1 = h\sigma = h(0.25)$. Thus $h = 4$ and, from Chebyshev's inequality,

$$P(\mu - h\sigma \leq X \leq \mu + h\sigma) \geq 1 - \frac{1}{h^2} = 1 - \frac{1}{(4)^2} = .9375.$$

Thus there is a probability of at least .9375 that the outcome will lie between 2 and 4.

## 8.4   THE NORMAL DISTRIBUTION

▶ *Introduction*

▶ *Standard Normal Distribution*

▶ *Normal Distribution*

APPLICATION

**Determination of Warranty Periods**

A manufacturer of washing machines has collected data indicating that the time before one of its washers needs its first repair is normally distributed with mean 2 years and standard deviation 0.608 years. What length of time should the manufacturer set for the warranty period so that at least 95% of the machines will get through the warranty period without need of a repair? See Example 5 for the answer.

### Introduction

In this section we will study the normal distribution, arguably the most important probability distribution. One reason for its importance is that many natural phenomena obey a normal distribution or at least have a probability distribution that closely approximates a normal distribution. Examples of such random variables are: the weights of house cats, the heights of women in Spain, the waiting time in a line at a bank, the diameter of a ball bearing coming off an assembly line, the pounds of fertilizer in a 100-pound bag.

A normally distributed random variable has the property that it can take any value in the interval $(-\infty, +\infty)$ and thus is a continuous random variable. We are familiar with viewing probabilities associated with a discrete random variable

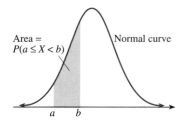

Area =
$P(a \leq X < b)$
Normal curve

$a$  $b$

**Figure 8.9**

as areas in a histogram. In a similar fashion we can view probabilities associated with a continuous random variable that is normally distributed as areas under a **normal curve**. Figure 8.9 shows a typical normal curve. The probability $P(a \leq X \leq b)$ that the random variable $X$ is between $a$ and $b$ is the area under the normal curve on the interval $[a, b]$. The area under the entire curve is 1. This is merely due to the fact that the probability that the random variable $X$ takes some value is 1.

Another reason for the importance of the normal distribution, is the "central limit theorem of probability theory." This theorem can be stated approximately as follows. Suppose $X$ is a random variable with an unknown probability distribution. Given a random sample of $n$ observations with $n$ large, the means

$$\bar{x} = \frac{x_1 + x_2 + \cdots + x_n}{n}.$$

have a distribution that is approximately normal. An important consequence of this theorem is that, under various conditions, the normal probability distribution closely approximates many other probability distributions. For example, we shall see in the next section how to approximate a binomial distribution with a normal distribution.

## Standard Normal Distribution

We now consider the *standard normal distribution*. As we shall see later in this section, to find the probability for a normally distributed random variable we find the probability of a certain standard normally distributed random variable. We thus need to work first with the standard normal distribution. We state the following definition.

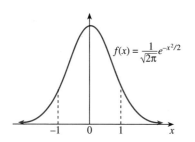

$f(x) = \frac{1}{\sqrt{2\pi}} e^{-x^2/2}$

−1  0  1  $x$

**Figure 8.10**  Standard normal distribution

---

**Standard Normal Distribution**

The random variable $X$ has a standard normal distribution on the interval $(-\infty, +\infty)$ if the probability $P(a \leq X \leq b)$ that $X$ is between $a$ and $b$ is the area under the standard normal curve given by

$$y = \frac{1}{\sqrt{2\pi}} e^{-0.5x^2}$$

on the interval $[a, b]$, where $\pi \approx 3.14159$ and $e \approx 2.71828$.

---

A graph of this curve is shown in Figure 8.10.
The curve has a number of important characteristics.

### Characteristics of the Standard Normal Curve

The graph of the standard normal curve shown in Figure 8.10 has the following characteristics.

1. It is bell-shaped.
2. It is symmetric about $x = 0$.
3. It lies above the $x$-axis.
4. It is "asymptotic" to both the positive and negative $x$-axis.
5. The curve "bends downward" on the interval $(-1, 1)$ and "bends upward" outside this interval.
6. The area under the entire curve is exactly 1.

From the symmetry, it appears that the mean is zero. In fact, it can be shown that for the standard normal distribution, $\mu = 0$. It can also be shown that the standard deviation $\sigma = 1$.

Since the normal distribution is so extraordinarily important, a table has been constructed that gives the areas under the standard normal curve. Given any value $b$, the table gives the area under the curve to the left of $b$, shown as the shaded region in Figure 8.11. We designate this area as $\mathcal{A}(b)$. Table B.1 in Appendix B can be used to find this number.

**CONVENTION.** The random variable associated with the standard normal distribution is designated by $Z$.

If we are asked for $P(Z \leq b)$, this is just $\mathcal{A}(b)$. Thus

$$P(Z \leq b) = \mathcal{A}(b),$$

where we can find $\mathcal{A}(b)$ in Table B.1 in Appendix B.

The probability $P(b \leq Z)$ is the area under the curve to the right of $b$, which is also the unshaded area under the curve in Figure 8.11. This area is also the entire area under the curve (which is 1) less the area under the curve to the left of $b$. Thus

$$P(b \leq Z) = 1 - \mathcal{A}(b).$$

Finally, $P(a \leq Z \leq b)$ is the area under the curve from $a$ to $b$ shown in Figure 8.12. This can be obtained as the area to the left of $b$, $\mathcal{A}(b)$, less the area to the left of $a$, $\mathcal{A}(a)$. Thus

$$P(a \leq Z \leq b) = \mathcal{A}(b) - \mathcal{A}(a),$$

where both the numbers $\mathcal{A}(b)$ and $\mathcal{A}(a)$ are obtained in Table B.1 in the Appendix.

**Figure 8.11**

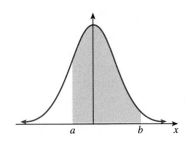

**Figure 8.12**

**E X A M P L E   1**     **Finding Probabilities for the Standard Normal Distribution**

Let $Z$ be a random variable with a standard normal distribution. Use Table B.1 in Appendix B for the standard normal distribution to find (a) $P(Z \leq 1.23)$ (b) $P(Z \geq 1.23)$ (c) $P(-1 \leq Z \leq 1)$ (d) $P(-2 \leq Z \leq 2)$.

**Solutions**    (a)   Here

$$P(Z \le 1.23) = \mathcal{A}(1.23),$$

which according to Table B.1 in Appendix B is .8907.

(b)   Here

$$P(Z \ge 1.23) = 1 - \mathcal{A}(1.23) = 1 - .8907 = .1093.$$

(c)   Here

$$P(-1 \le Z \le 1) = \mathcal{A}(1) - \mathcal{A}(-1),$$

which according to Table B.1 in Appendix B is

$$.8413 - .1587 = .6826.$$

(d)   Here

$$P(-2 \le Z \le 2) = \mathcal{A}(2) - \mathcal{A}(-2),$$

which according to Table B.1 in Appendix B is

$$.9772 - .0028 = .9544. \quad \blacksquare$$

It is instructive to notice that the answer to (c) is approximately two thirds and to (d) is approximately .95. That is, if the random variable is normally distributed with $\mu = 0$ and $\sigma = 1$, the probability that the random variable is within one standard deviation of the mean is about two thirds, and the probability that the random variable is within two standard deviations of the mean is about .95.

**E X A M P L E   2**    **Finding a Symmetric Interval About the Mean with Given Probability**

Let $Z$ be a random variable with a standard normal distribution. Using Table B.1 of the standard normal distribution found in Appendix B, find the interval $[-c, c]$ so that the probability that the random variable is in this interval is .99.

**Solution**    Given the symmetry of the normal distribution, we need only find $c$ such that the probability that the random variable is less than $c$ is .995, that is, $P(Z \le c) = .995$. The area of the "tail" in Figure 8.13 is then .005. Symmetry then implies that the area of the "tail" to the left of $-c$ is also .005. Thus since the area under the entire curve is 1, the area under the curve from $-c$ to $+c$ is then $1 - .005 - .005 = .99$.

If we look up the value of $c$ in Table B.1 for which $\mathcal{A}(c) = .995$, we obtain 2.575 after extrapolation. Thus $c = 2.575$, that is, $P(-2.575 \le Z \le 2.575) = .99$.

This also says that there is a 99% chance that the random variable $Z$ is within about 2.5 standard deviations of the mean.    $\blacksquare$

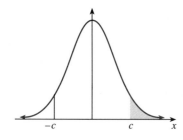

**Figure 8.13**

This gives a good indication of just how fast the standard normal curve is approaching the $x$-axis. The area under the curve on the interval $[-2.575, 2.575]$ is .99, so that the remaining area under all the rest of the curve is just .01.

## Normal Distribution

We now consider the general normal distribution.

---

### The Normal Probability Distribution

The random variable $X$ has a **normal probability distribution with mean** $\boldsymbol{\mu}$ **and standard deviation** $\boldsymbol{\sigma}$ on $(-\infty, +\infty)$ if the probability $P(a \leq X \leq b)$ that $X$ is between $a$ and $b$ is the area under the normal curve given by

$$y = \frac{1}{\sigma\sqrt{2\pi}}\, e^{-0.5[(x-\mu)/\sigma]^2}$$

on the interval $[a, b]$.

---

Figure 8.14 indicates a typical such curve. The curve has a number of important characteristics.

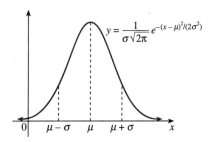

**Figure 8.14**   Normal distribution

---

### Characteristics of the Normal Curve

The graph of the normal curve shown in Figure 8.14 has the following characteristics.

1. It is bell shaped.
2. It is symmetric about $x = \mu$.
3. It lies above the $x$-axis.
4. It is "asymptotic" to both the positive and negative $x$-axis.
5. The curve "bends downward" on the interval $(\mu - \sigma, \mu + \sigma)$ and "bends upward" outside this interval.
6. The area under the entire curve is exactly one.

---

Figure 8.15 indicates 3 such curves with the same standard deviation $\sigma = 1$ but different means. The curves with $\mu = -2$ and 2 are identical to the one with $\mu = 0$, except that they have been shifted over by an amount equal to $\mu$.

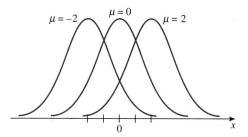

**Figure 8.15**

Figure 8.16 indicates 3 normal curves all with $\mu = 0$ and different standard deviations. Notice that the larger $\sigma$ is the more spread out the curve is.

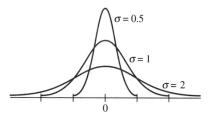

**Figure 8.16**

One might think that finding the area under a normal curve with mean $\mu$ and standard deviation $\sigma$ would require a different table for different values of $\mu$ and $\sigma$. Fortunately this is not the case. We can readily find $P(a \le X \le b)$ in terms of certain probabilities of the standard normal distribution.

It turns out that if $X$ is normally distributed with mean $\mu$ and standard deviation $\sigma$, and

$$Z = \frac{X - \mu}{\sigma},$$

then $Z$ is normally distributed with mean $\mu = 0$ and $\sigma = 1$. The situation is summarized as follows.

---

**Finding the Probability of a Normal Distribution**

Let $X$ be a random variable with mean $\mu$ and standard deviation $\sigma$. Then

$$P(a \le X \le b) = P\left(\frac{a - \mu}{\sigma} \le Z \le \frac{b - \mu}{\sigma}\right)$$

$$= \mathcal{A}\left(\frac{b - \mu}{\sigma}\right) - \mathcal{A}\left(\frac{a - \mu}{\sigma}\right).$$

---

**E X A M P L E  3**

**Finding the Probability for a Normal Distribution**

Suppose that the number of pounds in 100-pound bags of fertilizer is normally distributed with a mean of 100 and a standard deviation of 4. What is the probability that the bag will contain less than 95 pounds?

**Solution**

We are being asked for $P(X \le 95)$ for a random variable that is normally distributed with mean $\mu = 100$ and standard deviation $\sigma = 4$. We form the $Z$ variable as

$$Z = \frac{95 - \mu}{\sigma} = \frac{95 - 100}{4} = -1.25.$$

Then

$$P(X \le 95) = P(Z \le -1.25) = .1056$$

from Table B.1 in Appendix B. Thus we expect about 10.56% of the bags to have less than 95 pounds of fertilizer in them.  ■

---

**E X A M P L E  4**

**Quality Control Using the Normal Distribution**

The diameters of ball bearings produced at a plant are normally distributed with a mean of 1 cm and standard deviation of .004 cm. A ball bearing is acceptable if the diameter lies in the interval [.99, 1.01]. What percentage of ball bearings will be rejected?

**Solution**

We will first find the probability that one will be accepted. If $X$ is the random variable that denotes the diameters of the ball bearings, then this probability is $P(.99 \le X \le 1.01)$. Then

$$P(.99 \le X \le 1.01) = P\left(\frac{a - \mu}{\sigma} \le Z \le \frac{b - \mu}{\sigma}\right)$$

$$= \mathcal{A}\left(\frac{b - \mu}{\sigma}\right) - \mathcal{A}\left(\frac{a - \mu}{\sigma}\right)$$

$$= \mathcal{A}\left(\frac{1.01 - 1.00}{.004}\right) - \mathcal{A}\left(\frac{.99 - 1.00}{.004}\right)$$

$$= \mathcal{A}(2.5) - \mathcal{A}(-2.5)$$

$$= .9938 - .0062 = .9876. \text{ From Table B.1}$$

To find the probability that one will be rejected is $1 - .9876 = .0124$. So about 1.24% of these ball bearings will be rejected. ∎

E X A M P L E  5    **Determining a Warranty Period**

A manufacturer of washing machines has collected data indicating that the time before one of its washers needs its first repair is normally distributed with mean 2 years and standard deviation 0.608 years. What length of time should the manufacturer set for the warranty period so that at least 95% of the machines will get through the warranty period without need of a repair?

**Solution**    If $X$ denotes the life of a washer and $L$ denotes the length of the warranty period, we are asking that the probability that $X \le L$ should be .95; that is, $P(X \le L) = .95$. Since $X$ is normally distributed with mean $\mu = 2$ and standard deviation $\sigma = 0.608$, using the $Z$ statistic, this is the same as

$$P\left(Z \le \frac{L - \mu}{\sigma}\right).$$

We wish this probability to be .95. Using Table B.1 in Appendix B, we see that this will occur if

$$\frac{L - \mu}{\sigma} = 1.645.$$

Solving for $L$, we obtain

$$L = 1.645\sigma + \mu,$$

or

$$L = 1.645(.608) + 2 \approx 3.$$

Thus the manufacturer should set the warranty period at 3 years, since 95% of the washers will go through a 3-year period without any need for repairs. ∎

## SELF-HELP EXERCISE SET 8.4

1.  Let $X$ be a normally distributed random variable with mean $\mu = 4$ and standard deviation $\sigma = 2$. Find (a) $P(X \le 3)$. (b) $P(X \ge 3)$. (c) $P(1 \le X \le 3)$.

2.  Let $Z$ be a random variable with a standard normal distribution. Find the value $Z_0$ such that $P(Z \ge Z_0) = .10$.

3.  Suppose the daily production $X$, in tons, of a steel manufacturer is normally distributed with $\mu = 100$ tons and $\sigma = 10$ tons. Management decides to offer the employees a bonus whenever the daily production level is in the top 10% of the daily production distribution. What is the level of production that will produce the bonus?

## EXERCISE SET 8.4

In Exercises 1 through 12 find the probabilities given that $Z$ is a random variable with a standard normal distribution.

**1.** $P(Z \leq 0.7)$

**2.** $P(Z \leq 0.9)$

**3.** $P(Z \leq -0.5)$

**4.** $P(Z \leq -1.2)$

**5.** $P(Z \geq 1.2)$

**6.** $P(Z \geq 0.7)$

**7.** $P(Z \geq -0.6)$

**8.** $P(Z \geq -1.22)$

**9.** $P(1.0 \leq Z \leq 2)$

**10.** $P(0 \leq Z \leq 1.25)$

**11.** $P(-1.5 \leq Z \leq -1.0)$

**12.** $P(-0.5 \leq Z \leq 0.75)$

In Exercises 13 through 20 find the probabilities assuming that $X$ is a random variable with a normal distribution with the indicated mean and standard deviation.

**13.** $P(X \leq 26)$, $\mu = 10$, $\sigma = 8$

**14.** $P(X \leq 150)$, $\mu = 100$, $\sigma = 50$

**15.** $P(X \geq 0.01)$, $\mu = 0.006$, $\sigma = 0.002$

**16.** $P(X \geq -10)$, $\mu = 20$, $\sigma = 30$

**17.** $P(100 \leq X \leq 200)$, $\mu = 150$, $\sigma = 100$

**18.** $P(5 \leq X \leq 15)$, $\mu = 5$, $\sigma = 10$

**19.** $P(0.01 \leq X \leq 0.02)$, $\mu = 0.005$, $\sigma = 0.01$

**20.** $P(-2 \leq X \leq 8)$, $\mu = 4$, $\sigma = 10$

## Applications

**21. Sales.** Suppose the sales in thousands of dollars of a certain company is a normal random variable $X$ with mean $\mu = 20$ and $\sigma = 4$. Find the probability that sales on a certain day are
**a.** between \$16,000 and \$24,000, **b.** over \$30,000.

**22. Temperature.** Suppose the thermostat in a room is set at 70 degrees and that the actual temperature in the room is given by a normal random variable with $\mu = 70$ and $\sigma = 0.5$. What is the probability that the temperature in the room is
**a.** less than 69 degrees? **b.** between 69 degrees and 71 degrees?

**23. Temperature.** Suppose if the thermostat in a room is set at $t$ degrees, then the actual temperature in the room is given by a normal random variable with $\mu = t$ and $\sigma = 0.5$. What is the lowest setting of the thermostat that will maintain a temperature of at least 70 degrees with a probability of .99?

**24. Packaging.** A certain package of wrapping tape comes in 10-foot lengths. However, the actual length in feet of any tape in any package is a normal random variable with $\mu = 12$ and $\sigma = 0.25$. Find the probability that the length of tape in a package is

a.   at least 9.5 feet,   b.   at most 9.5 feet,   c.   between 9.5 and 10.5 feet.

25. **Construction.**   The finished inside diameter of a piston ring is normally distributed with a mean of 3.50 inches and $\sigma = 0.005$ inches. What is the probability of obtaining a piston ring with a diameter exceeding 3.51 inches?

26. **Quality Control.**   The inside diameter of a certain cast-iron pipe is normally distributed with $\mu = 4.02$ inches and $\sigma = 0.04$ inches. The pipe will be rejected unless the actual diameter is within 0.05 inches of 4.00. What percentage of pipes will be rejected?

27. **IQ Scores.**   Suppose the IQ scores of a new group of employees is normally distributed with $\mu = 115$ and $\sigma = 10$. Find the percentage that have IQs
a.   above 125,   b.   below 100,   c.   between 110 and 125.

28. **Life of a Product.**   Suppose the life of a pair of stockings worn once a week is normally distributed with $\mu = 12$ months and $\sigma = 3$ months. What should the manufacturer set as the warranty period in order to have at most 5% wear out during this period?

29. **Management.**   Refer to Self-Help Exercise 3. Suppose management gives a bonus when daily production reaches the 95% level. What should the production level be then?

30. **Management.**   Redo the previous exercise changing 95% to 99%.

31. **Grades.**   Suppose the scores on a mathematics exam are assumed to be normally distributed. The instructor calculates the mean to be 75 and the standard deviation to be 15. If the instructor wishes to have 10% of the class obtain an A, 20% a B, 40% a C, 20% a D, and 10% a F, find the numerical grade corresponding to each letter grade.

## Solutions to Self-Help Exercise Set 8.4

1. If $X$ is a normally distributed random variable with mean $\mu = 4$ and standard deviation $\sigma = 2$, then if $Z$ is the random variable for the standard normal distribution,

a.   $P(X \le 3) = P\left(Z \le \dfrac{3 - \mu}{\sigma}\right) = P\left(Z \le \dfrac{3 - 4}{2}\right) = \mathcal{A}(-0.5) = .3085,$

b.   $P(X \ge 3) = 1 - P(X \le 3) = 1 - .3085 = .6915,$

c.   $P(1 \le X \le 3) = \mathcal{A}\left(\dfrac{3 - 4}{2}\right) - \mathcal{A}\left(\dfrac{1 - 4}{2}\right) = .3085 - .0668 = .2417.$

2. Let $Z$ be a random variable with a standard normal distribution. The value $Z_0$, such that $P(Z \ge Z_0) = .10$, is given by the value $Z_0$ such that $\mathcal{A}(Z_0) = .90$. Looking through the Table B.1 in Appendix B for this $Z_0$, we find that $Z_0 = 1.28$.

3. With the daily production, $X$, in tons of a steel manufacturer normally distributed with $\mu = 100$ tons and $\sigma = 10$ tons, the $Z$ statistic is given by

$$Z = \frac{X - \mu}{\sigma}.$$

We know from the previous example that $\mathcal{A}(1.28) = .90$. Thus we have

$$X = \mu + Z\sigma = 100 + 1.28(10) = 112.8.$$

Thus when the daily production is 112.8 tons, 90% of the time the daily production is less than this number.

## 8.5   NORMAL APPROXIMATION TO THE BINOMIAL DISTRIBUTION

▶ *Normal Approximation to the Binomial Distribution*
▶ *Law of Large Numbers*

A P P L I C A T I O N
**Lot Acceptance Sampling**

An electronics manufacturer randomly selects 200 stamped circuits from the production for one day where a 5% rate of defectives is acceptable. Determine approximately the probability that 18 or more defectives are discovered in this sample. Is this alarming? The answer is given in Example 4.

### Normal Approximation to the Binomial Distribution

Finding probabilities that arise using the binomial distribution can be extremely tedious. In this section we will see how, under certain conditions, the normal distribution can be used as an effective approximation to the binomial distribution.

According to the National Cancer Institute, the 5-year survival rate for cancer of the cervix diagnosed in the 1980–85 period was approximately 66%. Suppose at a particular hospital, 15 patients were diagnosed for cancer of the cervix. What is the probability that at least 10 of these patients will survive for 5 years using the above probability of survival?

If success for a patient diagnosed as having cervical cancer is survival for 5 years, then the situation above can be described as a repeated Bernoulli trial with $n = 15$ and $p = .66$, $q = .34$. The answer can readily be written down in terms of the binomial distribution as

$$C(15,\ 10)(.66)^{10}(.34)^5 + \mathbf{C}(15,\ 11)(.66)^{11}(.34)^4$$
$$+ \cdots + \mathbf{C}(15,\ 15)(.66)^{15}(.34)^0.$$

Unfortunately, this will be a lengthy calculation. Figure 8.17 shows a histogram of the probability of the number of successes for 15 repeated Bernoulli trials with the probability of success equal to .66. As we have noted in a previous section, the answer to the question posed above is the area of the shaded rectangles shown in the histogram of Figure 8.17. In this section we will see that this area is very close to a certain area under a certain normal curve, which will be easy to find using tables. We will now see how to do this.

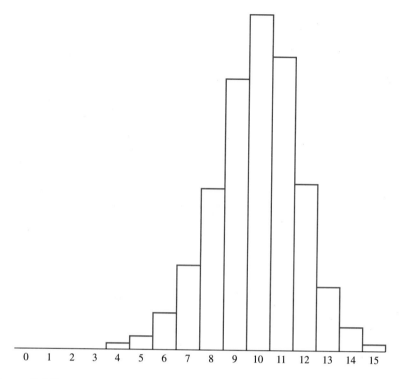

**Figure 8.17**

Figure 8.18, Figure 8.19, and Figure 8.20 show histograms of the binomial distributions for $p = .66$ and $n = 3$, 5, and 10, respectively. Notice how as $n$ increases, the histograms are more bell shaped and appear more similar to a normal curve. This is indeed the case. Figure 8.21 shows the binomial distribution

**Figure 8.18**

**Figure 8.19**

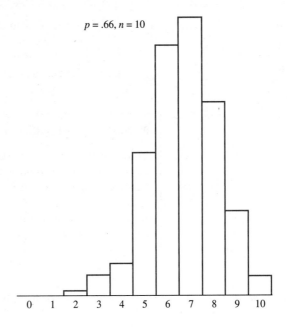

$p = .66, n = 10$

**Figure 8.20**

**Figure 8.21**

of Figure 8.17 overlaid with the normal distribution with the same mean $\mu = np = 15(.66) = 9.9$ and the same standard deviation $\sigma = \sqrt{npq} = \sqrt{15(.66)(.34)} \approx 1.835$ as the binomial distribution shown. Notice how good an approximation the normal curve is to the histogram for the binomial distribution. This is true under a wide set of conditions, as the following theorem indicates.

---

**Normal Approximation to the Binomial Distribution**

In a sequence of $n$ repeated (independent) Bernoulli trials with probability of success equal to $p$ and probability of failure equal to $q$, let the random variable $X$ denote the number of successes. Then the histogram for the probability distribution of $X$ is approximated by the normal curve with mean $\mu = np$ and standard deviation $\sigma = \sqrt{npq}$. This is a particularly close approximation when $np \geq 5$ and $nq \geq 5$.

---

**E X A M P L E   1**    **Using the Normal Approximation**

Using the normal approximation, find the probability that at least 10 of the patients mentioned above will survive for 5 years.

**Solution**    First notice that $np = 9.9 \geq 5$ and $nq = 5.1 \geq 5$. Thus the normal approximation should be excellent.

To find the area of the shaded rectangles, we use the normal approximation with the normal curve with mean $\mu = np = 15(.66) = 9.9$ and standard deviation $\sigma = \sqrt{npq} = \sqrt{15(.66)(.34)} \approx 1.835$. Refer to Figure 8.21. Notice that since we want to include all of the histogram to the right of 10 and including 10, we must include the entire rectangle corresponding to 10. Thus to use the approximating normal curve we must take the area to the right of $10 - .50 = 9.5$. That is, using the $Z$ substitution,

$$P(X \geq 9.5) = 1 - \mathscr{A}\left(\frac{9.5 - \mu}{\sigma}\right)$$

$$= 1 - \mathscr{A}\left(\frac{9.5 - 9.9}{1.835}\right)$$

$$= 1 - \mathscr{A}(-.22)$$

$$= 1 - .4129 = .5871. \text{ From Table B.1}$$

This compares with the actual value of .5968.  ∎

---

**E X A M P L E   2**    **Using the Normal Approximation**

A coin has been flipped 900 times and heads has been observed 496 times. Is there some justification in claiming that this is an unfair coin?

**Solution**   First notice that $np = nq = 450 \geq 5$ so the normal approximation can be used.

If we assume the coin is fair, then we have 900 repeated Bernoulli trials with $p = .5 = q$. The mean is $\mu = np = 450$. We will calculate the probability that the number of heads of a fair coin could deviate by more than 45 from the mean. This is one less the probability that it could deviate by as much as 46 from the mean. If $X$ is the random variable that denotes the number of heads, then this latter probability is denoted by $P(405 \leq X \leq 495)$ and is

$$C(900, 405)(.5)^{405}(.5)^{495} + C(900, 406)(.5)^{406}(.5)^{494}$$
$$+ \cdots + C(900, 495)(.5)^{495}(.5)^{405}.$$

The need for approximating this by the area under a normal curve is apparent. We use the normal curve with $\mu = 450$ and $\sigma = \sqrt{npq} = \sqrt{900(.5)(.5)} = 15$ and let the random variable $Y$ denote the continuous random value associated with this. From Figure 8.22 we see that we must include the rectangles corresponding to 405 and 495. Then $P(405 \leq X \leq 495)$ is approximately the area under this normal curve from $405 - .50 = 404.5$ to $495 + .50 = 495.5$ and is denoted by $P(404.5 \leq Y \leq 495.5)$. Then

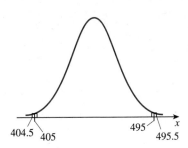

$$P(404.5 \leq Y \leq 495.5) = \mathscr{A}\left(\frac{495.5 - 450}{15}\right) - \mathscr{A}\left(\frac{404.5 - 450}{15}\right)$$

$$= \mathscr{A}(3.03) - \mathscr{A}(-3.03)$$

$$\approx .9987 - .0013 = .9974. \text{ From Table B.1}$$

**Figure 8.22**

One less this is .0026 and is the probability that the number of heads can deviate by more than 45 from the mean when a fair coin is flipped 900 times. Since this is so low, we can conclude that the coin is not a fair coin. ∎

We saw in the last example that a fair coin has a very small probability of deviating very far from the mean. We now examine how likely it is for the number of heads to be very close to the mean.

**E X A M P L E   3**   **Using the Normal Approximation**

For the experiment in the previous example, find the probability that the number of heads varies at most by 1 from the mean.

**Solution**   With $X$ denoting the number of heads in 900 flips of the fair coin, the mean is 450, and we are seeking the probability $P(449 \leq X \leq 451)$, which is the area under the three rectangles shown in Figure 8.23. If $Y$ is the same as in the previous example, we see from Figure 8.23 that we approximate this by $P(448.5 \leq Y \leq 451.5)$, the area under the approximating normal curve from 448.5 to 451.5. We then obtain

$$P(448.5 \leq Y \leq 451.5) = \mathscr{A}\left(\frac{451.5 - 450}{15}\right) - \mathscr{A}\left(\frac{448.5 - 450}{15}\right)$$

$$= \mathscr{A}(0.10) - \mathscr{A}(-0.10)$$

$$\approx .5398 - .4602 = .0796. \text{ From Table B.1}$$

**Figure 8.23**

So there is about an 8% chance of the number of heads being within 1 of the mean. Thus we see that just as there is a small probability of the number of heads deviating far from the mean, so too is the probability that the number of heads will be very close to the mean.  ■

Many electronic items can be mass produced cheaply because the solid-state circuitry used can be stamped by a machine. Controlling the quality of the items is a serious problem and is often handled by using **lot acceptance sampling**. In this method a sample is taken from a large lot of items and inspected. The entire lot will be accepted or rejected on the basis of the number of defective items in the sample.

**E X A M P L E   4**          **Lot Acceptance Sampling**

An electronics manufacturer randomly selects 200 stamped circuits from the production for one day where a 5% rate of defectives is acceptable. Determine approximately the probability that 18 or more defectives are discovered in this sample. Should this discovery be cause for concern?

**Solution**          First, suppose that the defective rate is 5%. Then the random variable that gives the number of defective circuits is binomially distributed with $n = 200$, $p = .05$, and $q = .95$. Since $np = 10 \geq 5$ and $nq = 190 \geq 5$, we expect the normal approximation with mean $\mu = np = 10$ and standard deviation $\sigma = \sqrt{npq} = \sqrt{200(.05)(.95)} \approx 3.1$ to be a good approximation for the binomial distribution.

To find the approximating area corresponding to $X \geq 18$, refer to Figure 8.24. Since we must include the rectangle that corresponds to 18, we take the area under the approximating curve to start at $18 - .50 = 17.5$. Then

$$P(X \geq 17.5) = 1 - \mathcal{A}\left(\frac{17.5 - 10}{3.1}\right)$$

$$\approx 1 - .9922 = .0078. \text{ From Table B.1}$$

Thus the approximate probability that at least 18 defective circuits are in the sample of 200 is .0078 *if in fact the true defective rate is 5%*. Since the probability of .0078 is extremely small, the manufacturer should be concerned that the defective rate of the sample is, in fact, much higher than 5%.  ■

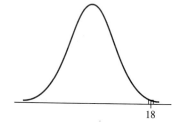

18

**Figure 8.24**

## The Law of Large Numbers

Assume an experiment with $n$ repeated independent Bernoulli trials with the probability of success $p$ equal to the probability of failure $q$. Then $p = q = \frac{1}{2}$. Let $X$ denote the random variable given by the number of successes. We seek an estimate of the probability that $X$ differs from the mean $\mu = np = \frac{1}{2}n$ by at most $h$ standard deviations.

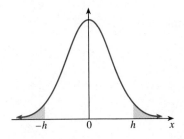

**Figure 8.25**

Since the standard deviation is $\sigma = \sqrt{npq} = \frac{1}{2}\sqrt{n}$, we are seeking a good estimate of

$$P\left(\left|X - \frac{1}{2}n\right| \le \frac{h}{2}\sqrt{n}\right) = P(|X - \mu| \le h\sigma) = P\left(-h \le \frac{X - \mu}{\sigma} \le h\right).$$

If $Y$ is the random variable given by the normal distribution with mean $\mu = \frac{1}{2}n$ and standard deviation $\sigma = \frac{1}{2}\sqrt{n}$, then by the normal approximation, this probability is approximated by $P\left(-h \le \frac{Y - \mu}{\sigma} \le h\right)$. But using the $Z$ statistic for the standard normal distribution, this is $P(-h \le Z \le h)$. From Figure 8.25 we note that the area under the entire standard normal curve is 1 and that by symmetry the area under each "tail" is the same. Thus

$$P(-h \le Z \le h) = 1 - 2P(Z \le -h).$$

Thus we have established the important estimate that

$$P\left(\left|X - \frac{1}{2}n\right| \le \frac{h}{2}\sqrt{n}\right) \approx 1 - 2P(Z \le -h) = 1 - 2\mathscr{A}(-h). \quad (1)$$

Now if we are given a specific value of $h$, we can look up $P(Z \le -h) = \mathscr{A}(-h)$ in Table B.1 in Appendix B. For example, if $h = 2$, then looking in this table, we have

$$P(-2 \le Z \le 2) = 1 - 2P(Z \le -2) = 1 - 2(.0228) = .9544.$$

Thus we can say that there is about a 95% chance that the number of successes will be within $2\sigma = \sqrt{n}$ of $\mu = \frac{1}{2}n$.

E X A M P L E   5    **The Law of Large Numbers**

In the experiment of flipping a fair coin let $H$ be the event that a head occurs. Estimate how large $n$ must be so that there is a probability of .95 that the relative frequency $\dfrac{f(H)}{n}$ is within .01 of $\frac{1}{2}$. Within .001.

**Solution**    If we let the random variable $X = f(H)$, then we can use the estimate given by equation (1). We already noticed above that if $h = 2$, then $1 - 2\mathscr{A}(-h) \approx .95$. Thus with $h = 2$ equation (1) becomes

$$P\left(\left|X - \frac{1}{2}n\right| \le \sqrt{n}\right) \approx .95.$$

Setting $X = f(H)$ and dividing by $n$, this can be rewritten as

$$P\left(\left|\frac{f(H)}{n} - \frac{1}{2}\right| \le \frac{1}{\sqrt{n}}\right) \approx .95. \quad (2)$$

If we want the relative frequency $\dfrac{f(H)}{n}$ to be within $\epsilon$ of $\frac{1}{2}$, we want

$$\left| \frac{f(H)}{n} - \frac{1}{2} \right| \le \epsilon.$$

Looking at equation (2), we then want

$$\epsilon = \frac{1}{\sqrt{n}}$$

or

$$n = \frac{1}{\epsilon^2}. \tag{3}$$

Now using equation (3) if $\epsilon = .01$, then $n = 10,000$. If $\epsilon = .001$, then $n = 1,000,000$. This means, for example, that if $n = 10,000$, the relative frequency $\dfrac{f(H)}{n}$ will be within the interval $(.49, .51)$ with a probability of .95. More precisely, this means that 95% of the time that a fair coin is flipped 10,000 times, the relative frequency will be within the interval $(.49, .51)$. The second answer means that 95% of the time that a fair coin is flipped 1,000,000 times, the relative frequency will be within the interval $(.499, .501)$.

## SELF-HELP EXERCISE SET 8.5

1.  About 3% of the flashlights produced by a manufacturer are defective. Using the normal curve approximation, find the probability that in a shipment of 1000 to a distributor, at least 35 will be defective.

2.  Assume an experiment with $n$ repeated independent Bernoulli trials with the probability of success equal to $p$ and the probability of failure equal to $q = 1 - p$. Let $X$ denote the random variable given by the number of successes. Find the value of $h$ so that the probability is about .99 that $X$ differs from the mean $\mu = np$ by at most $h\sigma = h\sqrt{npq}$.

## EXERCISE SET 8.5

1.  A fair coin has been tossed 1600 times. Use the normal approximation to find the probability that (a) the number of heads is at least 825, (b) the number of heads is between 775 and 825.

2.  A fair die has been tossed 180 times. Use the normal approximation to find the probability that (a) the number 1 appears at most 35 times, (b) the number 3 appears between 25 and 35 times.

3.  A baseball player has a batting average of .300. What is the probability that he will have a batting average of at least .400 over the next 100 official times at bat?

4. A basketball player hits 80% of her free-throws. What is the probability that she will hit 90% of her next 50 free-throws?

## Applications

5. **Bladder Cancer.** According to the National Cancer Institute, the 5-year survival rate for cancer of the bladder diagnosed in 1980–85 was 78%. Among 1000 such diagnosed cases, find the probability that (a) 750 or more survived 5 years, (b) between 750 and 800 survived 5 years.

6. **Thyroid Cancer.** Repeat the previous exercise for cancer of the thyroid with a 5-year survival rate of 93%.

7. **Heart Disease.** Aproximately 43% of deaths in this country are attributed to heart disease. Among 1000 randomly chosen deaths, what is the probability that heart disease was the cause in at least 400 instances? Between 400 and 450?

8. **Accidental Deaths.** Approximately 4.5% of deaths in this country are attributed to accidents. Among 1000 randomly chosen deaths, what is the probability that an accident was the cause in at most 50 instances? Between 30 and 60?

9. **Use of Marihuana.** Recently, 37% of college students reported using marihuana during the previous year. What is the probability that among 300 randomly chosen college students in 1987 that (a) at least 120 used marihuana during the previous year? (b) between 100 and 120?

10. **Use of Cocaine.** Recently, 8% of high school seniors reported using cocaine during the last year. What is the probability that among 300 randomly chosen high school students in 1988 that (a) at least 30 used cocaine during the previous year? (b) between 20 and 40?

11. **Immigration.** In recent years about 41% of immigrants admitted to this country were Asian. Among 100 randomly selected immigrants to this country, what is the probability that (a) at least 30 are Asian? (b) between 30 and 50 are Asian?

12. **Arrests.** Approximately 82% of those arrested each year in this country are males. Given 100 randomly chosen arrests, what is the probability that at least 10 were female? Between 10 and 20 were female?

13. **Heart Disease.** In Exercise 7 find the probability that the number of deaths due to heart disease differs from the mean by at most 2.5 standard deviations.

14. **Accidental Deaths.** In Exercise 8 find the probability that the number of deaths due to accidents differs from the mean by at most 3 standard deviations.

15. **Quality Control.** About 3% of a certain product of a manufacturer has been found to be defective. A new system of quality control has been introduced with 18 defective products found in the first 1000 after the new system went into effect. Is the new system effective? *Hint:* Find the probability of at most 18 defectives under the old system.

16. **Safety.** In a certain plant about 5% of the workers were suffering some accident each year. New safety procedures were instituted that resulted in 30 accidents among 1000 workers during the next year. Are the safety procedures working?

17. **Advertising.** A company received a 15% response to its mail advertisements. After changing to a new advertisement, it noticed a 20% response among the next 1000 mailings. Is the new advertisement effective?

18. **Drug Effectiveness.** For a certain disease, 20% of individuals with the disease will recover without any treatment. A new experimental drug is administered to 40 individuals with the disease and 16 recover. Is this drug promising? *Hint:* Find the probability that at least 16 patients would recover without the drug.

19. **The Law of Large Numbers.** In the experiment of flipping a fair coin let $H$ be the event that a head occurs. Estimate how large $n$ must be so that there is a probability of .50 that the relative frequency $\dfrac{f(H)}{n}$ is within .01 of $\frac{1}{2}$. Within .001?

20. **The Law of Large Numbers.** In the experiment of flipping a fair coin, let $H$ be the event that a head occurs. Estimate how large $n$ must be so that there is a probability of .50 that the relative frequency $\dfrac{f(H)}{n}$ is within .0001 of $\frac{1}{2}$. Within .00001?

21. **The Law of Large Numbers.** In the experiment of flipping a fair coin, let $H$ be the event that a head occurs. Estimate how large $n$ must be so that there is a probability of .90 that the relative frequency $\dfrac{f(H)}{n}$ is within .01 of $\frac{1}{2}$. Within .001?

22. **The Law of Large Numbers.** In the experiment of flipping an unfair coin, let $H$ be the event that a head occurs and $p = p(H) = .60$. Estimate how large $n$ must be so that there is a probability of .95 that the relative frequency $\dfrac{f(H)}{n}$ is within .01 of $p = .60$. Within .001?

## Solutions to Self-Help Exercise Set 8.5

1. The situation is a repeated Bernoulli trial with $n = 1000$ and $p = .03$. If $X$ denotes the number of defective flashlights, then we are seeking $P(X \geq 35)$. If $Y$ denotes the random variable associated with the normal distribution with mean $\mu = np = 1000(.03) = 30$ and standard deviation $\sigma = \sqrt{npq} = \sqrt{1000(.03)(.97)} = 5.3944$, then

$$P(X \geq 35) \approx P(Y \geq 34.5)$$

$$= P\left(Z \geq \frac{34.5 - 30}{5.3944}\right)$$

$$= P(Z \geq .83)$$

$$= 1 - \mathcal{A}(.83)$$

$$= 1 - .7967 = .2033.$$

Thus there is about a 20% chance that at least 35 flashlights will be defective in a shipment of 1000.

2.  Equation (1) says that

$$P(|X - np| \leq h\sqrt{npq}) \approx 1 - 2P(Z \leq -h) = 1 - 2\mathscr{A}(-h).$$

Thus we are seeking the value of $h$ for which

$$.99 = 1 - 2\mathscr{A}(-h).$$

This can be written as

$$\mathscr{A}(-h) = .005.$$

Looking for an $h$ in Table B.2 in Appendix B for which this is true, we find that $h = 2.575$. Thus we can say that $X$ differs from the mean $\mu = np$ by $2.575\sigma = 2.575\sqrt{npq}$ with probability of .99.

## SUMMARY OUTLINE OF CHAPTER 8

- A **random variable** is a rule that assigns precisely one real number to each outcome of an experiment.

- A random variable is **finite discrete** if it assumes only a finite number of values. A random variable is **infinite discrete** if it takes on an infinite number of values that can be listed in a sequence, so that, there is a first one, a second one, a third one, and so on. A random variable is said to be **continuous** if it can take any of the infinite number of values in some interval of real numbers.

- Suppose the random variable $X$ can take the values $x_1, \ldots, x_n$. The **probability distribution** of the random variable $X$ is a listing of all the probabilities associated with all possible values of the random variable, that is, $p_1, \ldots, p_n$, where $p_1 = p(x_1), \ldots, p_n = p(x_n)$.

- Given a sequence of $n$ Bernoulli trials with the probability of success $p$ and the probability of failure $q$, the **binomial distribution** is given by

$$P(X = k) = b(k : n, p) = C(n, k)p^k q^{n-k}.$$

- The **average** or **mean** of the $n$ numbers $x_1, x_2, \ldots, x_n$, denoted by $\bar{x}$ or the Greek letter $\mu$, is given by

$$\mu = \frac{x_1 + x_2 + \cdots + x_n}{n}.$$

- Let $X$ denote the random variable that takes on the values $x_1, x_2, \ldots, x_n$, and let the associated probabilities be $p_1, p_2, \ldots, p_n$, then the **expected value** or **mean** of the random variable $X$, denoted by $E(X)$ or by $\mu$, is

$$E(X) = x_1 p_1 + x_2 p_2 + \cdots + x_n p_n.$$

- The expected value of the binomial distribution

$$p_k = b(k : n, p) = C(n, k)p^k q^{n-k} \qquad p + q = 1$$

is $E(X) = np$.

- The **median** of a set of numbers is the middle number when the numbers are arranged in order of size and there are an odd number in the set. If the number in the set is even, the median is the mean of the two middle numbers.

- The **mode** of a set of numbers is the number that occurs more frequently than the others. If the frequency of occurrence of two numbers is the same and also greater than the frequency of occurrence of all the other numbers, then we say the set is **bimodal** and has two modes. If no one or two numbers occur more frequently than the others, we say that the set has no mode.

- Let $X$ denote the random variable that takes on the values $x_1, x_2, \ldots, x_n$, and let the associated probabilities be $p_1, p_2, \ldots, p_n$. Then if $\mu = E(X)$, the **variance** of the random variable $X$, denoted by $\text{Var}(X)$, is

$$\text{Var}(X) = (x_1 - \mu)^2 p_1 + (x_2 - \mu)^2 p_2 + \cdots + (x_n - \mu)^2 p_n.$$

  The **standard deviation**, denoted by $\sigma(X)$ is

$$\sigma(X) = \sqrt{\text{Var}(X)}.$$

- An alternate form of the variance is given by

$$\text{Var}(X) = (x_1^2 p_1 + x_2^2 p_2 + \cdots + x_n^2 p_n) - \mu^2.$$

- **Chebyshev's Inequality.**  Let $X$ be a random variable with expected value $\mu$ and standard deviation $\sigma$. Then the probability that the random variable associated with a random outcome lies between $\mu - h\sigma$ and $\mu + h\sigma$ is at least $1 - 1/h^2$, that is,

$$P(\mu - h\sigma \leq X \leq \mu + h\sigma) \geq 1 - \frac{1}{h^2}.$$

- The variance of the binomial distribution with $n$ trials and the probability of "success" equal to $p$ and of "failure" equal to $q$ is $\text{Var}(X) = npq$.

- The random variable $X$ has a **standard normal distribution** on the interval $(-\infty, +\infty)$ if the probability $P(a \leq X \leq b)$ that $X$ is between $a$ and $b$ is the area under the normal curve given by

$$y = \frac{1}{\sqrt{2\pi}} e^{-0.5x^2}$$

  on the interval $[a, b]$, where $\pi \approx 3.14159$ and $e \approx 2.71828$.

- The random variable $X$ has a **normal probability distribution with mean $\mu$ and standard deviation $\sigma$** on $(-\infty, +\infty)$ if the probability $P(a \leq X \leq b)$ that $X$ is between $a$ and $b$ is the area under the normal curve given by

$$y = \frac{1}{\sigma\sqrt{2\pi}} e^{-0.5[(x-\mu)/\sigma]^2}$$

  on the interval $[a, b]$.

- Let $X$ be a normal random variable with mean $\mu$ and standard deviation $\sigma$. Then

$$P(a \leq X \leq b) = P\left(\frac{a - \mu}{\sigma} \leq Z \leq \frac{b - \mu}{\sigma}\right) = \mathscr{A}\left(\frac{b - \mu}{\sigma}\right) - \mathscr{A}\left(\frac{a - \mu}{\sigma}\right)$$

  where $\mathscr{A}(b)$ is the area under the standard normal curve to the left of $b$, which can be found in Table B.1 in Appendix B.

- In a sequence of $n$ repeated (independent) Bernoulli trials with probability of success equal to $p$ and probability of failure equal to $q$, let the random variable $X$ denote the number of successes. Then the histogram for the probability distribution of $X$ is approximated by the normal curve with mean $\mu = np$ and standard deviation $\sigma = \sqrt{npq}$. This is a particularly close approximation when $np \geq 5$ and $nq \geq 5$.

## Chapter 8 Review Exercises

1.  A baseball player has a batting average of .300. Let $X$ be the number of hits in the next 5 official times at bat. Find the probability distribution. Draw a histogram.

2.  The probability distribution of the random variable $X$ is given in the following table.

    | Random Variable $x$ | 0 | 1 | 2 | 3 | 4 | 5 |
    |---|---|---|---|---|---|---|
    | $P(X = x)$ | .20 | .10 | .05 | .15 | .18 | .32 |

    a.  Draw a histogram. Find **(b)** $P(X = 0)$, **(c)** $P(X \leq 2)$, **(d)** $P(0 < X \leq 3)$, (e) $P(X \geq 2)$. Identify each of the probabilities as an area.

3.  Find the expected value of the random variable given in the previous exercise.

4.  **Lottery.** A lottery has a grand prize of $1,000,000, a second prize of $100,000, and 10 consolation prizes of $2000 each. If 1 million tickets are sold and the probability of any ticket winning is the same as any other, find the expected return on a $1 ticket.

5.  **Life Insurance.** An insurance company sells a $10,000, 5-year term life insurance policy to an individual for $800. Find the expected return for the company if the probability that the individual will live for the next 5 years is .96.

6.  The pitcher Cy Young holds the all-time record for the most wins in a lifetime with 511, a record that is unlikely to be matched. The following table gives his win and loss record for the 5-year period beginning with 1900.

    | Year | *1900* | *1901* | *1902* | *1903* | *1904* |
    |---|---|---|---|---|---|
    | No. of Wins | 20 | 33 | 32 | 28 | 27 |
    | No. of Losses | 18 | 10 | 10 | 9 | 16 |

    Find the mean, variance, and standard deviation for his wins and also for his losses for this 5-year period. Which is varying the most?

7.  **AIDS.** The following table gives the number of deaths due to acquired immunodeficiency syndrome (AIDS) for recent years in the U.S. by two age groups. Find the average, variance, and standard deviation of each. Which is varying the most?

    | Year | *1987* | *1988* | *1989* | *1990* | *1991* |
    |---|---|---|---|---|---|
    | 13–29 | 2864 | 3531 | 4598 | 4745 | 3574 |
    | 30–39 | 6535 | 8091 | 11308 | 11927 | 8830 |

8.  Find the variance and the standard deviation of the random variable given in Exercise 2 above.

9. A probability distribution has a mean of 20 and a standard deviation of 2. Use Chebyshev's inequality to find the minimum probability that an outcome is between 10 and 30.

10. Let $Z$ be a random variable with a standard normal distribution. Find (a) $P(Z \leq .87)$, (b) $P(Z \geq .87)$, (c) $P(-.50 \leq Z \leq .87)$.

11. Let $Z$ be a random variable with a standard normal distribution. Find the interval $[-c, c]$ so that the probability that the random variable is in this interval is .98.

12. **Manufacturing.** Suppose that in a production run of 100 twelve-ounce sodas, the mean number of ounces in these cans was 12 and the standard deviation was 0.25 ounces. What is the probability that a can will have less than 11.5 ounces if the number of ounces in the cans is normally distributed?

13. The heights of professional soccer players are normally distributed with a mean of 71 inches and a standard deviation of 2 inches. What percentage of players are within 1 inch of the mean?

14. A coin has been flipped 900 times and heads has been observed 485 times. What is the probability of obtaining at least 485 heads if the coin were fair?

15. **Quality Control.** About 2% of a certain product is found to be defective. What is the probability that in a shipment of 1000 of these products at least 25 will be defective?

Which restaurant will you go to this week? (Dennis Drenner.)

# Markov Processes

9.1   Markov Processes

9.2   Regular Markov Processes

9.3   Absorbing Markov Processes

**I**n this chapter we consider a sequence of finite experiments, all with the same possible outcomes, in which the outcomes and associated probabilities of each experiment depend only on the outcome of the immediately preceding one. Such a sequence of experiments is called a Markov process. The basic results are presented together with a number of applications including scheduling and charge accounts.

## 9.1   MARKOV PROCESSES

► *Two-State Markov Processes*

► *Three-State Markov Processes*

► *m-State Markov Processes*

APPLICATION

**Percentage of Sons that Go to College**

A survey of a certain ethnic group indicated that among the sons of college-educated fathers, 85% went to college, while among the sons of fathers that did not attend college, 35% went to college. What percentage of the third generation went to college if initially 30% of the male population was college educated? For the answer see Example 3.

## Two-State Markov Processes

In an earlier chapter, we considered finite stochastic processes. Recall that a finite stochastic process is a finite sequence of experiments in which the outcomes and associated probabilities of each experiment depend on the outcomes of all the preceding experiments. Recall that probability trees are effective in describing such processes.

We now wish to consider finite stochastic processes in which the outcomes and associated probabilities depend only on the preceding experiment. Such a process is called a **Markov process**. The outcome of any experiment is called the **state** of the experiment.

---

**Markov Processes**

A finite stochastic process in which the outcomes and associated probabilities depend only on the preceding experiment is called a **Markov process**.

---

We first consider Markov processes with two states.

EXAMPLE 1     **A Markov Process with Two States**

A survey of a certain ethnic group indicated that 85% of the sons of fathers who attended college also attended college, and 65% of the sons of fathers who did not attend college also did not attend college. Show that this is a Markov process and use a probability tree to describe the transition from one state to another.

Solution     This is a two-state Markov process. Each stage represents a generation. There are 2 states: ''attended college,'' ''did not attend college.'' The probabilities of each depend only on those of the previous stage.

Since 85% of the sons of college-educated fathers go to college and 15% do not, the probability that a randomly selected son of a college-educated father in this group goes to college is .85 and the probability that he does not is .15. This is shown in the tree diagram in Figure 9.1a. Notice that these probabilities must sum to one.

Since 35% of the sons of non–college-educated fathers go to college and 65% do not, the probability that a randomly selected son of a non–college-educated father in this group goes to college is .35 and the probability that he

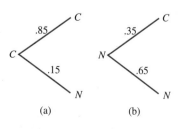

**Figure 9.1**

does not is .65. This is shown in the tree diagram in Figure 9.1b. Notice that these probabilities must sum to one. ∎

**E X A M P L E   2**   **A Markov Process with Two States**

Use the results of Example 1 and a tree diagram to find the probability that a son in this group is college educated and the probability that a son is not college educated if 30% of the fathers were college educated.

**Solution**   The initial probability that a father from this group is college educated is given as .30 and thus of not being college educated is .70. We then draw the tree diagram shown in Figure 9.2. The probability that a son will be college educated is then seen to be

$$(.30)(.85) + (.70)(.35) = .50,$$

while the probability that a son will not be college educated is then seen to be

$$(.30)(.15) + (.70)(.65) = .50. \quad ∎$$

As will be seen shortly, it is very useful to describe a Markov process by a **transition matrix**. The transition matrix gives the probabilities of going from one state to another. For example, if we create a matrix

$$T = \begin{bmatrix} p_{11} & p_{12} \\ p_{21} & p_{22} \end{bmatrix}$$

with the 4 probabilities in Figure 9.1 taken in the order that they appear there, we have

**Current state**

State 1   State 2

$$\begin{bmatrix} .85 & .35 \\ .15 & .65 \end{bmatrix} \begin{matrix} \text{State 1} \\ \text{State 2} \end{matrix} \quad \text{Next state}$$

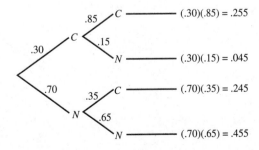

**Figure 9.2**

where State 1 is college educated and State 2 is not college educated. Thus the probability of going from State 1 to State 1, that is from college-educated father to college-educated son, is $p_{11} = .85$. As a conditional probability this is

$$p_{11} = p(\text{next in state } 1 \mid \text{currently in state } 1).$$

The probability of going from State 1 to State 2, or from college-educated father to non-college-educated son, is $p_{21} = .15$. As a conditional probability this is

$$p_{21} = p(\text{next in state } 2 \mid \text{currently in state } 1).$$

Similarly

$$.35 = p_{12} = p(\text{next in state } 1 \mid \text{currently in state } 2),$$

$$.65 = p_{22} = p(\text{next in state } 2 \mid \text{currently in state } 2).$$

For notational convenience, the fathers are described as generation 0; the current or initial generation. The sons are then described as the first generation.

Now let us see why the transition matrix is so useful. Suppose the percent of the initial generation (the fathers) that was college educated was $100x_0$ and the percent that was not was $100y_0$. If $x_1$ is the fraction of the first generation (the sons) that is college educated, then, according to Figure 9.3,

$$x_1 = .85x_0 + .35y_0.$$

If $y_1$ is the fraction of the first generation that are not college educated, then according to Figure 9.3,

$$y_1 = .15x_0 + .65y_0.$$

These last 2 equations can be written in matrix terms as

$$\begin{bmatrix} x_1 \\ y_1 \end{bmatrix} = \begin{bmatrix} .85 & .35 \\ .15 & .65 \end{bmatrix} \begin{bmatrix} x_0 \\ y_0 \end{bmatrix} = \begin{bmatrix} .85x_0 + .35y_0 \\ .15x_0 + .65y_0 \end{bmatrix}$$

If we let

$$X_0 = \begin{bmatrix} x_0 \\ y_0 \end{bmatrix} \quad \text{and} \quad X_1 = \begin{bmatrix} x_1 \\ y_1 \end{bmatrix},$$

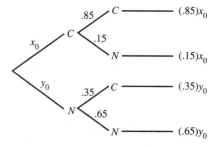

**Figure 9.3**

then the last matrix equation can be written as

$$X_1 = TX_0. \tag{1}$$

Now suppose we wish the distribution

$$X_2 = \begin{bmatrix} x_2 \\ y_2 \end{bmatrix}$$

of college- and non-college-educated males in the second generation. We can view the distribution in the first generation as the "initial" distribution and use formula (1) and obtain

$$X_2 = TX_1$$
$$= T(TX_0)$$
$$= T^2X_0$$

Similarly, if $X_3$ is the distribution in the third generation, then $X_2$ can be viewed as the "initial" distribution in the second generation and we can use formula (1) again. This gives

$$X_3 = TX_2$$
$$= T(T^2X_0)$$
$$= T^3X_0$$

We now see that if $X_n$ is the distribution in the $n$th generation, then

$$X_n = T^nX_0. \tag{2}$$

We also saw above that

$$X_n = TX_{n-1}, \tag{3}$$

where $X_{n-1}$ can be viewed as the "initial" distribution in generation $(n - 1)$. We shall need equation (3) in the next section.

E X A M P L E  3    **A Markov Process with Two States**

Determine the population distribution in the second and third generations if the initial distribution was the same as in Example 2.

**Solution**    Since

$$T^2 = \begin{bmatrix} .85 & .35 \\ .15 & .65 \end{bmatrix}\begin{bmatrix} .85 & .35 \\ .15 & .65 \end{bmatrix} = \begin{bmatrix} .775 & .525 \\ .225 & .475 \end{bmatrix},$$

$$X_2 = T^2X_0 = \begin{bmatrix} .775 & .525 \\ .225 & .475 \end{bmatrix}\begin{bmatrix} .30 \\ .70 \end{bmatrix} = \begin{bmatrix} .60 \\ .40 \end{bmatrix}.$$

Thus, in the second generation 60% of the males are college educated. Also since

$$T^3 = \begin{bmatrix} .775 & .525 \\ .225 & .475 \end{bmatrix} \begin{bmatrix} .85 & .35 \\ .15 & .65 \end{bmatrix} = \begin{bmatrix} .7375 & .6125 \\ .2625 & .3875 \end{bmatrix},$$

$$X_3 = T^3 X_0 = \begin{bmatrix} .7375 & .6125 \\ .2625 & .3875 \end{bmatrix} \begin{bmatrix} .30 \\ .70 \end{bmatrix} = \begin{bmatrix} .65 \\ .35 \end{bmatrix}.$$

Thus, in the third generation, 65% of the males are college educated and 35% are not. ■

## Three-State Markov Processes

We now look at Markov processes with 3 states.

**E X A M P L E   4**     **A Markov Process with Three States**

*(Dennis Drenner.)*

There are 3 restaurants in town: Abby's, Barney's, and Cathy's. The 3 restaurants start a vigorous advertising campaign for Sunday business. A survey indicates that among those who eat their Sunday meal at Abby's, 50% return next Sunday, whereas 30% go to Barney's and 20% go to Cathy's. Among those who eat their Sunday meal at Barney's, 60% return next Sunday, whereas 30% go to Abby's and 10% go to Cathy's. Finally, among those who eat their Sunday meal at Cathy's, 80% return next Sunday, whereas 10% go to Abby's and 10% go to Barney's. Show that this is a Markov process. Draw a tree diagram describing the movements from one state to the other and give a transition matrix for this Markov process.

**Solution**     This is a 3-state Markov process. Each stage is a Sunday. The 3 states are "ate at Abby's," "ate at Barney's," "ate at Cathy's." The probabilities depend only on those of the previous stage.

The tree diagram is shown in Figure 9.4. Notice that the probabilities of each of the 3 branches must sum to one.

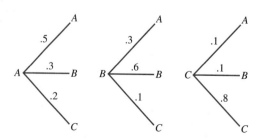

**Figure 9.4**

The transition matrix is read from Figure 9.4 to be

$$T = \begin{bmatrix} .50 & .30 & .10 \\ .30 & .60 & .10 \\ .20 & .10 & .80 \end{bmatrix}. \quad \blacksquare$$

Notice that each of the columns in the transition matrix sums to 1.

**E X A M P L E  5**    **A Markov Process with Three States**

Suppose initially Abby's has $100x_0\%$ of the Sunday business, Barney's had $100y_0\%$, and Cathy's had $100z_0\%$. (a) Draw a tree diagram indicating the distribution of business for the next Sunday. (b) Find the equations that give the distribution for next Sunday in terms of $x_0$, $y_0$, and $z_0$. (c) Verify that these equations can be written in the form of formula (1).

**Solution**    (a)  Figure 9.5 gives the tree diagram.

(b)  Let $x_1$, $y_1$, and $z_1$ be the distribution of business for the next Sunday for the respective 3 restaurants. Then, from Figure 9.5,

$$x_1 = .50x_0 + .30y_0 + .10z_0$$

$$y_1 = .30x_0 + .60y_0 + .10z_0$$

$$z_1 = .20x_0 + .10y_0 + .80z_0.$$

(c)  If

$$X_0 = \begin{bmatrix} x_0 \\ y_0 \\ z_0 \end{bmatrix} \quad \text{and} \quad X_1 = \begin{bmatrix} x_1 \\ y_1 \\ z_1 \end{bmatrix}$$

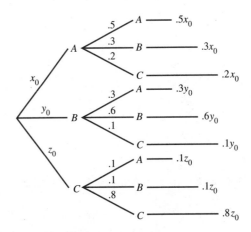

**Figure 9.5**

then

$$X_1 = \begin{bmatrix} x_1 \\ y_1 \\ z_1 \end{bmatrix} = \begin{bmatrix} .50x_0 + .30y_0 + .10z_0 \\ .30x_0 + .60y_0 + .10z_0 \\ .20x_0 + .10y_0 + .80z_0 \end{bmatrix}$$

$$= \begin{bmatrix} .50 & .30 & .10 \\ .30 & .60 & .10 \\ .20 & .10 & .80 \end{bmatrix} \begin{bmatrix} x_0 \\ y_0 \\ z_0 \end{bmatrix} = TX_0.$$

Since $X_1 = TX_0$, it then follows as before that

$$X_n = T^n X_0. \quad \blacksquare$$

**E X A M P L E  6**    **A Markov Process with Three States**

Assume initially that Abby's and Barney's each have half of the Sunday business. What is the distribution of business after 1 week? After 2 weeks?

**Solution**    After 1 week the distribution is given by

$$X_1 = TX_0 = \begin{bmatrix} .50 & .30 & .10 \\ .30 & .60 & .10 \\ .20 & .10 & .80 \end{bmatrix} \begin{bmatrix} .50 \\ .50 \\ 0 \end{bmatrix} = \begin{bmatrix} .40 \\ .45 \\ .15 \end{bmatrix}$$

To find the distribution for the second Sunday, first notice that

$$T^2 = \begin{bmatrix} .50 & .30 & .10 \\ .30 & .60 & .10 \\ .20 & .10 & .80 \end{bmatrix} \begin{bmatrix} .50 & .30 & .10 \\ .30 & .60 & .10 \\ .20 & .10 & .80 \end{bmatrix} = \begin{bmatrix} .36 & .34 & .16 \\ .35 & .46 & .17 \\ .29 & .20 & .67 \end{bmatrix}.$$

Then

$$X_2 = T^2 X_0 = \begin{bmatrix} .36 & .34 & .16 \\ .35 & .46 & .17 \\ .29 & .20 & .67 \end{bmatrix} \begin{bmatrix} .50 \\ .50 \\ 0 \end{bmatrix} = \begin{bmatrix} .350 \\ .405 \\ .245 \end{bmatrix}. \quad \blacksquare$$

**E X A M P L E  7**    **A Markov Process with Three States**

What are the probabilities that an individual will go to each of the 3 restaurants two Sundays after going to Barney's?

**Solution**    In this case the initial distribution is

$$X_0 = \begin{bmatrix} 0 \\ 1 \\ 0 \end{bmatrix}$$

Then

$$X_2 = T^2 X_0 = \begin{bmatrix} .36 & .34 & .16 \\ .35 & .46 & .17 \\ .29 & .20 & .67 \end{bmatrix} \begin{bmatrix} 0 \\ 1 \\ 0 \end{bmatrix} = \begin{bmatrix} .34 \\ .46 \\ .20 \end{bmatrix}$$

Thus there is a 34%, 46%, 20% chance he will go to respectively Abby's, Barney's, and Cathy's in two Sundays.

Another way of viewing this is to realize that $T^2$ represents the transition matrix from the initial stage to the stage after next. Thus

**Current state**

| State 1 | State 2 | State 3 | |
|---------|---------|---------|---|
| .36 | .34 | .16 | State 1 |
| .35 | .46 | .17 | State 2   State after next |
| .29 | .20 | .67 | State 3 |

Then, for example, the entry .34 in the first row and second column represents the probability of going initially from state 2 (Barney's) to state 1 (Abby's) for the stage after next.  ∎

## *m*-State Markov Processes

The above analysis works for a Markov process with any number of (finite) states. This is summarized in the following.

---

### *m*-State Markov Process

If a Markov process has $m$ states and the probabilities of moving from one state to another are given by the tree diagrams in Figure 9.6, then the **transition matrix** is given by

$$T = \begin{bmatrix} p_{11} & p_{12} & \cdots & p_{1m} \\ p_{21} & p_{22} & \cdots & p_{2m} \\ & & \cdots & \\ p_{m1} & p_{m2} & \cdots & p_{mm} \end{bmatrix},$$

where the element in the $i$th row and $j$th column, $p_{ij}$, is the conditional probability

$$p_{ij} = p(\text{next in state } i \mid \text{currently in state } j).$$

If the column matrix $X_n$ of dimension $m$

$$X_n = \begin{bmatrix} x_1 \\ x_2 \\ \dots \\ x_m \end{bmatrix}$$

gives the state after the $n$th experiment, then

$$X_n = T^n X_0,$$

where the column matrix $X_0$ of dimension $m$ is the initial state.

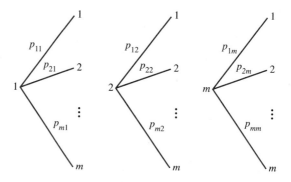

**Figure 9.6**

We also have the following definition.

---

**Stochastic Matrix**

We say that a matrix $T$ is a **stochastic matrix** if

1. The matrix $T$ is square.
2. All elements in the matrix $T$ are nonnegative.
3. The sum of all the elements in any column is 1.

---

## SELF-HELP EXERCISE SET 9.1

1. Consider migration between the North and the South of the United States. Suppose every year 6% of Northerners migrate south and 2% of Southerners migrate

north. For simplicity, assume the populations of the two regions change only due to migration between the regions and that the population is constant. Show that this is a Markov process. Use a probability tree to describe the transition from one state to another and find the transition matrix.

2. Determine the population distribution after 1 year and after 2 years if initially 70% of the population was in the North.

## EXERCISE SET 9.1

In Exercises 1 through 12 determine whether or not the given matrix is stochastic.

1. $\begin{bmatrix} .1 & .4 \\ .2 & .4 \\ .7 & .2 \end{bmatrix}$

2. $\begin{bmatrix} .5 & .4 & .3 \\ .5 & .6 & .7 \end{bmatrix}$

3. $\begin{bmatrix} .5 & .2 & .7 \\ .8 & .5 & -.5 \\ -.3 & .3 & .8 \end{bmatrix}$

4. $\begin{bmatrix} 2 & .5 \\ -1 & .5 \end{bmatrix}$

5. $\begin{bmatrix} .87 & .7 \\ .2 & .2 \end{bmatrix}$

6. $\begin{bmatrix} .9 & .5 \\ .1 & .6 \end{bmatrix}$

7. $\begin{bmatrix} .6 & .4 \\ .3 & .7 \end{bmatrix}$

8. $\begin{bmatrix} .4 & .6 \\ .1 & .9 \end{bmatrix}$

9. $\begin{bmatrix} .4 & .7 \\ .5 & .3 \end{bmatrix}$

10. $\begin{bmatrix} .5 & .1 \\ .5 & .9 \end{bmatrix}$

11. $\begin{bmatrix} 0 & 1 \\ 1 & 0 \end{bmatrix}$

12. $\begin{bmatrix} .5 & 0 \\ .5 & 1 \end{bmatrix}$

In Exercises 13 through 18 is given a transition matrix for a Markov process and an initial distribution. Find $X_1$, $X_2$, and $X_4$.

13. $T = \begin{bmatrix} .5 & .4 \\ .5 & .6 \end{bmatrix}$, $X_0 = \begin{bmatrix} 1 \\ 0 \end{bmatrix}$

14. $T = \begin{bmatrix} .5 & 1 \\ .5 & 0 \end{bmatrix}$, $X_0 = \begin{bmatrix} 0 \\ 1 \end{bmatrix}$

15. $T = \begin{bmatrix} .3 & .2 \\ .7 & .8 \end{bmatrix}$, $X_0 = \begin{bmatrix} .5 \\ .5 \end{bmatrix}$

16. $T = \begin{bmatrix} .1 & .4 \\ .9 & .6 \end{bmatrix}$, $X_0 = \begin{bmatrix} .5 \\ .5 \end{bmatrix}$

17. $T = \begin{bmatrix} .5 & 0 & .5 \\ .5 & .5 & 0 \\ 0 & .5 & .5 \end{bmatrix}$, $X_0 = \begin{bmatrix} 1 \\ 0 \\ 0 \end{bmatrix}$

18. $T = \begin{bmatrix} .4 & .3 & 0 \\ .6 & 0 & .2 \\ 0 & .7 & .8 \end{bmatrix}$, $X_0 = \begin{bmatrix} 0 \\ 1 \\ 0 \end{bmatrix}$

19. A town has only two electricians. Of those who call the first electrician, 70% will call again next time. Of those who call the second electrician, 50% will call again next time. Write the transition matrix. If a person calls the first electrician, what is the probability he or she will call the first electrician again the call after next?

20. It is known that a certain businessman always wears a white shirt or a blue shirt to work and that 60% of the time when wearing a white shirt he changes the color the next business day, and 70% of the time when wearing a blue shirt he changes the color the next business day. Write the transition matrix. He was

observed wearing a white shirt on Monday. What is the probability that he will be wearing a white shirt again 4 business days later?

21. Every Mother's Day, a certain individual sends her mother either roses or carnations. If she sends roses one year, 30% of the time she will send roses again the next year. If she sends carnations one year, 80% of the time she will send roses the next year. Write the transition matrix. If she gave roses this year, what is the probability that she will give roses again 2 years from now?

22. A certain person always buys either vanilla or chocolate ice cream. If he buys vanilla, 30% of the time he will buy vanilla again the next time. If he buys chocolate, 10% of the time he will buy chocolate again the next time. Write the transition matrix. If vanilla was bought last time, what is the probability of buying vanilla again 4 times from now?

23. Suppose a person moves into the town mentioned in Exercise 19 and that she is just as likely to call one electrician as the other. What is the probability that she will call the first electrician on her third call?

24. On the first day back to work after vacation, the businessman mentioned in Exercise 20 is just as likely to wear a white shirt as a blue one. What is the probability that he will be wearing a white shirt 4 business days later?

25. The individual in Exercise 21 forgot to send flowers to her Mother last year. The probability that she will send roses this year is .90. What is the probability that she will send roses 2 Mother's Days after this year?

26. The person mentioned in Exercise 22 was given a free strawberry ice cream the last time he had ice cream and now has the same chance of buying vanilla as chocolate the next time he buys ice cream. What is the probability he will buy chocolate 4 times from now?

## Applications

27. **Marketing.**   A person always buys either a Ford, GM, or Chrysler product. If she buys a Ford, next time she will buy a Ford again 60% of the time and will be just as likely to buy a GM as a Chrysler. If she buys a GM, next time she will buy a GM again 80% of the time and will be just as likely to buy a Ford as a Chrysler. If she buys a Chrysler, next time she will buy a Chrysler again 70% of the time and will be twice as likely to buy a GM as a Ford. Find the transition matrix. The car that she now owns is a Nissan that she won in a contest. For her next car, she is equally likely to buy a Ford, GM, or Chrysler. What is the probability that she will buy a GM 3 purchases from now?

28. **Politics.**   In a certain state, when a Democrat is governor 30% of the time a Democrat will be governor next time, 10% of the time an Independent will be, and 60% of the time a Republican will be. When an Independent is governor, 20% of the time an Independent will be governor next time, 50% of the time a Democrat, and 30% of the time a Republican. When a Republican is governor, 40% of the time a Republican will be governor next time, 50% of the time a Democrat, and 10% of the time an Independent. Write the transition matrix. For the next election, the experts are giving the Democrat the same chance as the Republican of being elected governor, with the Independent given a 20%

chance. What is the probability that an Independent will be elected 2 elections from now?

29. **Biology.** A house cat tries to catch a mouse every night in one of 3 neighboring fields. If some night she visits the first field, then she has an 80% chance of returning to the first field the next night and an equal chance of going to one of the other 2 fields. If she visits the second field, she has a 90% chance of returning to the second field the next night and a 10% chance of going to the first. If she visits the third field, she has a 70% chance of returning to the third field the next night and a 30% chance of going to the second. If initially she has the same chance of visiting one field as any other, what is the chance she will be in each of the fields two nights later?

30. **Marketing.** A survey indicates that people in a certain area take their summer vacations either at the beach, at the lake, or in the mountains. The survey finds that among people who have gone to the beach, 20% go to the beach next summer, 30% go to the lake, and 50% go to the mountains. The survey finds also that among people who have gone to the lake, 20% go to the beach next summer, 20% go to the lake, and 60% go to thc mountains. Finally the survey finds that among people who have gone to the mountains, 50% go to the beach next summer and 50% go to the lake. Last year the beach suffered an oil spill and nobody went there. Half went to the lake and half went to the mountains. What is the probability that an individual will go to each of the 3 places for vacation two summers from now? Four summers from now?

### Solutions to Self-Help Exercise Set 9.1

1. This is a 2-state Markov process. The two states are: "live in the North," "live in the South." The probabilities of each state depend only on those of the previous stage. Each stage is one year.

   Since 94% of Northerners stay in the North and 6% go to the South, the probability that a randomly selected Northerner will stay in the North is .94 and will go to the South is .06. This is shown in the accompanying tree diagram.

   Since 2% of Southerners go to the North and 98% stay in the South, the probability that a randomly selected Southerner will head to the North is .02 and will stay in the South is .98. This is shown in the accompanying tree diagram.

   The transition matrix is then seen from the tree diagrams to be

   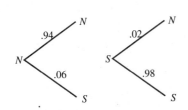

   $$\begin{bmatrix} .94 & .02 \\ .06 & .98 \end{bmatrix}.$$

2. The distribution after the first generation is

   $$X_1 = TX_0 = \begin{bmatrix} .94 & .02 \\ .06 & .98 \end{bmatrix}\begin{bmatrix} .70 \\ .30 \end{bmatrix} = \begin{bmatrix} .664 \\ .336 \end{bmatrix}$$

   Since

   $$T^2 = \begin{bmatrix} .94 & .02 \\ .06 & .98 \end{bmatrix}\begin{bmatrix} .94 & .02 \\ .06 & .98 \end{bmatrix} = \begin{bmatrix} .8848 & .0384 \\ .1152 & .9616 \end{bmatrix}$$

   the distribution after the second generation is

$$X_2 = T^2 X_0 = \begin{bmatrix} .8848 & .0384 \\ .1152 & .9616 \end{bmatrix} \begin{bmatrix} .70 \\ .30 \end{bmatrix} = \begin{bmatrix} .6309 \\ .3691 \end{bmatrix}$$

After the second generation about 63% of the population is in the North and about 37% in the South.

## 9.2    REGULAR MARKOV PROCESSES

- ▶ *The Long Run*
- ▶ *Finding the Steady-State Distribution*

**APPLICATION**

**What Happens in the Long-Term**

In the last section we studied a Markov process involving three restaurants: Abby's, Barney's, and Cathy's. What happens in the long-term? For the answer see Example 3.

### The Long Run

In this section we will consider a certain type of Markov process that has the remarkable property that no matter what the initial distribution, the distribution at the $n$th stage will tend toward the very same limiting distribution as $n$ becomes large.

Consider again the first Markov process mentioned in the last section. Recall that initially 30% of the fathers of this group attended college. The subsequent distributions over the next 3 generations were calculated to be

$$X_1 = \begin{bmatrix} .50 \\ .50 \end{bmatrix}, \quad X_2 = \begin{bmatrix} .60 \\ .40 \end{bmatrix}, \quad X_3 = \begin{bmatrix} .65 \\ .35 \end{bmatrix}$$

Let us continue the calculations further and obtain

$$X_4 = \begin{bmatrix} .675 \\ .325 \end{bmatrix}, \quad X_5 = \begin{bmatrix} .6875 \\ .3125 \end{bmatrix}, \quad X_6 = \begin{bmatrix} .69275 \\ .30625 \end{bmatrix}, \quad X_7 = \begin{bmatrix} .6969 \\ .3031 \end{bmatrix}$$

$$X_8 = \begin{bmatrix} .6984 \\ .3016 \end{bmatrix}, \quad X_9 = \begin{bmatrix} .6992 \\ .3008 \end{bmatrix}, \quad X_{10} = \begin{bmatrix} .6996 \\ .3004 \end{bmatrix}, \quad X_{11} = \begin{bmatrix} .6998 \\ .3002 \end{bmatrix}$$

We see that $X_n$ is approaching

$$\overline{X} = \begin{bmatrix} .7 \\ .3 \end{bmatrix}$$

That is, in the long run, 70% of the males will attend college and 30% of the males will not attend college. The limiting column matrix $\overline{X}$ is referred to as the **steady-state distribution** for the Markov process.

In order to understand what is happening, we need to look at what is happening to the powers of the transition matrix $T$, that is, to $T^n$. Recall from the last section that

$$X_n = T^n X_0 \tag{1}$$

Since $X_0$ is fixed and $X_n$ approaches the specific matrix $\overline{X}$, one suspects from equation (1) that $T^n$ must approach some particular matrix. Calculating the powers of $T$, we obtain the following

$$T = \begin{bmatrix} .85 & .35 \\ .15 & .65 \end{bmatrix}, \quad T^2 = \begin{bmatrix} .775 & .525 \\ .225 & .475 \end{bmatrix}, \quad T^3 = \begin{bmatrix} .7375 & .6125 \\ .2625 & .3875 \end{bmatrix}$$

$$T^4 = \begin{bmatrix} .71875 & .65625 \\ .28125 & .34375 \end{bmatrix}, \quad T^5 = \begin{bmatrix} .7094 & .6781 \\ .2906 & .3219 \end{bmatrix}, \quad T^6 = \begin{bmatrix} .7047 & .6891 \\ .2953 & .3109 \end{bmatrix}$$

$$T^7 = \begin{bmatrix} .7023 & .6945 \\ .2977 & .3055 \end{bmatrix}, \quad T^8 = \begin{bmatrix} .7012 & .6973 \\ .2988 & .3027 \end{bmatrix}, \quad T^9 = \begin{bmatrix} .7006 & .6986 \\ .2994 & .3014 \end{bmatrix}$$

$$T^{10} = \begin{bmatrix} .7003 & .6993 \\ .2997 & .3007 \end{bmatrix}, \quad T^{11} = \begin{bmatrix} .7001 & .6997 \\ .2999 & .3003 \end{bmatrix}.$$

Thus the matrix $T^n$ is heading for the limiting matrix

$$L = \begin{bmatrix} .7 & .7 \\ .3 & .3 \end{bmatrix}.$$

The matrix $L$ is referred to as the **steady-state matrix** for the Markov process. Notice that all the columns of $L$ are just $\overline{X}$.

Since $X_n = T^n X_0$ and $T^n \to L$, this shows why $X_n$ is also heading for some limiting column matrix. In fact as $n$ becomes large without bound,

$$T^n X_0 \to \begin{bmatrix} .7 & .7 \\ .3 & .3 \end{bmatrix} \begin{bmatrix} x_0 \\ y_0 \end{bmatrix} = \begin{bmatrix} 0.7 (x_0 + y_0) \\ 0.3 (x_0 + y_0) \end{bmatrix} = \begin{bmatrix} 0.7 \\ 0.3 \end{bmatrix}$$

since $x_0 + y_0$ must equal one. We have thus discovered that *no matter what the initial probability distribution, the limiting distribution is always the same.* Thus no matter how the system starts initially, the system will head for the limiting distribution $\overline{X}$. Markov processes that have this property are called **regular** and the associated transition matrix $T$ is called a **regular stochastic matrix**.

Since not all Markov processes are regular, it would be convenient if there were a way of easily recognizing when a Markov process was regular. We first give the following definition.

---

**Regular Stochastic Matrix**

A stochastic matrix is said to be **regular** if some power has all positive entries.

The following is then true.

> **Theorem 1**
>
> A Markov process is regular if the associated transition matrix is regular.

**E X A M P L E   1**     **Identifying Regular Stochastic Matrices**

Determine which of the following stochastic matrices are regular.

$$A = \begin{bmatrix} .2 & 1 \\ .8 & 0 \end{bmatrix} \qquad B = \begin{bmatrix} 0 & 1 \\ 1 & 0 \end{bmatrix}$$

**Solution**     The matrix $A$ does not have all entries positive, but

$$A^2 = \begin{bmatrix} .2 & 1 \\ .8 & 0 \end{bmatrix}\begin{bmatrix} .2 & 1 \\ .8 & 0 \end{bmatrix} = \begin{bmatrix} .84 & .20 \\ .16 & .80 \end{bmatrix}$$

does. Thus $A$ is regular.

The matrix $B$ does not have all entries positive. Furthermore,

$$B^2 = \begin{bmatrix} 0 & 1 \\ 1 & 0 \end{bmatrix}\begin{bmatrix} 0 & 1 \\ 1 & 0 \end{bmatrix} = \begin{bmatrix} 1 & 0 \\ 0 & 1 \end{bmatrix} = I$$

also does not have all entries positive. But $B^3 = B$ and $B^4 = B^2 = I$ and $B^5 = B$, and so on. Thus no power of $B$ will have all entries positive, and $B$ is not regular.   ■

## Finding the Steady-State Solution

Suppose now we are dealing with a regular Markov process and thus that the associated transition matrix $T$ is a regular stochastic matrix. No matter what the initial state $X_0$ we know that there is a limiting state, $\overline{X}$. There is an easier way to find $\overline{X}$ then by calculating many powers of $T$.

Recall that

$$X_n = TX_{n-1} \tag{2}$$

Since as $n$ becomes large without bound, $X_n \to \overline{X}$, then also $X_{n-1} \to \overline{X}$. Thus the left-hand side of equation (2) approaches $\overline{X}$, while the right-hand side approaches $T\overline{X}$. We conclude that

$$\overline{X} = T\overline{X} \tag{3}$$

Equation (3) together with the fact that the sum of the entries in the column matrix $\overline{X}$ must be one will then determine $\overline{X}$.

> **Finding the Steady-State Distribution**
>
> If $T$ is a regular stochastic matrix, one can find the steady-state distribution $\overline{X}$, by solving the equations
>
> $$\overline{X} = T\overline{X}$$
>
> together with the fact that the sum of the entries in $\overline{X}$ must be one.

Let us now apply this procedure to the example considered earlier.

**E X A M P L E  2**    **Finding the Steady-State Distribution**

Find the steady-state distribution $\overline{X}$ for the Markov process with the regular stochastic transition matrix

$$\begin{bmatrix} .8 & .4 \\ .2 & .6 \end{bmatrix}.$$

**Solution**    Let

$$\overline{X} = \begin{bmatrix} x \\ y \end{bmatrix},$$

then we wish to solve

$$T\overline{X} = \overline{X}$$
$$x + y = 1$$

or

$$x + y = 1$$
$$.80x + .40y = x$$
$$.20x + .60y = y$$

This can be written as

$$x + y = 1$$
$$-.20x + .40y = 0$$
$$.20x - .40y = 0$$

Since the last equation is just the negative of the second equation this is just

$$x + y = 1$$
$$-.20x + .40y = 0$$

Now use the Gauss–Jordan method. Replace the second equation with .20 times the first plus the second and obtain

$$x + y = 1$$

$$.60y = .20.$$

Divide the last equation by .60 and obtain

$$x + y = 1$$

$$y = \frac{1}{3}.$$

Now replace the first equation with the first minus the second and obtain

$$x = \frac{2}{3}$$

$$y = \frac{1}{3}.$$

This agrees with what we found earlier.   ∎

The following is a summary of what has been done.

---

**Regular Markov Processes**

Let $T$ be a regular stochastic matrix associated with a regular Markov process. Then

1.  For any initial distribution $X_0$, $X_n = T^n X_0$ approaches the same steady-state distribution $\overline{X}$, as $n$ becomes large without bound.
2.  As $n$ becomes large without bound, $T^n$ approaches the steady-state matrix $L$.
3.  Every column of $L$ is just $\overline{X}$.
4.  The steady-state distribution can be found by solving

    $$\overline{X} = T\overline{X}$$

    and using the fact that the entries in $\overline{X}$ sum to one.
5.  A stochastic matrix $T$ will be regular if some power of $T$ has all positive entries.

---

**E X A M P L E   3**    **Determining the Steady-State**

Find the steady-state distribution $\overline{X}$ and the steady-state matrix $L$ for the problem in Example 5 in the last section concerning the 3 restaurants in town where the transition matrix was

$$T = \begin{bmatrix} .50 & .30 & .10 \\ .30 & .60 & .10 \\ .20 & .10 & .80 \end{bmatrix}$$

**Solution**     If

$$\overline{X} = \begin{bmatrix} x \\ y \\ z \end{bmatrix},$$

then we must solve

$$x + y + z = 1$$
$$T\overline{X} = \overline{X}$$

or

$$x + y + z = 1$$
$$.50x + .30y + .10z = x$$
$$.30x + .60y + .10z = y$$
$$.20x + .10y + .80z = z$$

The following summarizes the Gauss–Jordan method.

$$\begin{vmatrix} 1 & 1 & 1 & 1 \\ -.5 & .3 & .1 & 0 \\ .3 & -.4 & .1 & 0 \\ .2 & .1 & -.2 & 0 \end{vmatrix} \rightarrow \begin{vmatrix} 1 & 1 & 1 & 1 \\ 0 & .8 & .6 & .5 \\ 0 & -.7 & -.2 & -.3 \\ 0 & -.1 & -.4 & -.2 \end{vmatrix} \rightarrow$$

$$\begin{vmatrix} 1 & 1 & 1 & 1 \\ 0 & 1 & \frac{3}{4} & \frac{5}{8} \\ 0 & 0 & \frac{13}{40} & \frac{11}{80} \\ 0 & 0 & -\frac{13}{40} & -\frac{11}{80} \end{vmatrix} \rightarrow \begin{vmatrix} 1 & 1 & 1 & 1 \\ 0 & 1 & \frac{3}{4} & \frac{5}{8} \\ 0 & 0 & 1 & \frac{11}{26} \\ 0 & 0 & 0 & 0 \end{vmatrix} \rightarrow$$

$$\begin{vmatrix} 1 & 1 & 0 & \frac{15}{26} \\ 0 & 1 & 0 & \frac{8}{26} \\ 0 & 0 & 1 & \frac{11}{26} \\ 0 & 0 & 0 & 0 \end{vmatrix} \rightarrow \begin{vmatrix} 1 & 0 & 0 & \frac{7}{26} \\ 0 & 1 & 0 & \frac{8}{26} \\ 0 & 0 & 1 & \frac{11}{26} \\ 0 & 0 & 0 & 0 \end{vmatrix}$$

Thus after a large number of Sundays, approximately $\frac{7}{26}$ will eat at Abby's, approximately $\frac{8}{26}$ will eat at Barney's, and approximately $\frac{11}{26}$ at Cathy's. The limiting matrix $L$ is

$$L = \begin{bmatrix} \frac{7}{26} & \frac{7}{26} & \frac{7}{26} \\ \frac{8}{26} & \frac{8}{26} & \frac{8}{26} \\ \frac{11}{26} & \frac{11}{26} & \frac{11}{26} \end{bmatrix}. \quad \blacksquare$$

## SELF-HELP EXERCISE SET 9.2

1.  Determine if the stochastic matrix

$$T = \begin{bmatrix} 0 & .50 \\ 1 & .50 \end{bmatrix}$$

is regular.

2.  Find the steady-state solution for the Markov process found in Self-Help Exercise Set 9.1.

## EXERCISE SET 9.2

In Exercises 1 through 10, determine if the given stochastic matrix is regular.

1.  $\begin{bmatrix} 0 & .4 \\ 1 & .6 \end{bmatrix}$

2.  $\begin{bmatrix} .7 & 1 \\ .3 & 0 \end{bmatrix}$

3.  $\begin{bmatrix} 1 & .5 \\ 0 & .5 \end{bmatrix}$

4.  $\begin{bmatrix} .25 & 0 \\ .75 & 1 \end{bmatrix}$

5.  $\begin{bmatrix} .5 & .2 & 0 \\ .5 & .5 & .4 \\ 0 & .3 & .6 \end{bmatrix}$

6.  $\begin{bmatrix} 0 & .5 & .1 \\ .4 & .5 & .4 \\ .6 & 0 & .5 \end{bmatrix}$

7.  $\begin{bmatrix} 0 & .5 & 0 \\ 1 & 0 & 1 \\ 0 & .5 & 0 \end{bmatrix}$

8.  $\begin{bmatrix} 0 & .3 & 0 \\ 1 & .4 & 1 \\ 0 & .3 & 0 \end{bmatrix}$

9.  $\begin{bmatrix} 0 & 0 & .25 \\ 1 & 0 & 0 \\ 0 & 1 & .75 \end{bmatrix}$

10.  $\begin{bmatrix} 0 & .1 & 1 \\ 1 & .8 & 0 \\ 0 & .1 & 0 \end{bmatrix}$

Exercises 11 through 20 give regular stochastic matrices. Find the steady-state distribution for each.

11.  $\begin{bmatrix} .8 & .5 \\ .2 & .5 \end{bmatrix}$

12.  $\begin{bmatrix} .5 & .4 \\ .5 & .6 \end{bmatrix}$

13.  $\begin{bmatrix} .7 & .2 \\ .3 & .8 \end{bmatrix}$

14.  $\begin{bmatrix} .6 & .7 \\ .4 & .3 \end{bmatrix}$

15.  $\begin{bmatrix} .7 & .4 \\ .3 & .6 \end{bmatrix}$

16.  $\begin{bmatrix} .3 & 1 \\ .7 & 0 \end{bmatrix}$

**17.** $\begin{bmatrix} .6 & .3 & 0 \\ .4 & .4 & .6 \\ 0 & .3 & .4 \end{bmatrix}$ **18.** $\begin{bmatrix} .4 & .5 & .2 \\ .3 & .4 & .6 \\ .3 & .1 & .2 \end{bmatrix}$ **19.** $\begin{bmatrix} 0 & 0 & 1 \\ 1 & .4 & 0 \\ 0 & .6 & 0 \end{bmatrix}$

**20.** $\begin{bmatrix} .1 & .4 & .1 \\ .1 & .4 & .2 \\ .8 & .2 & .7 \end{bmatrix}$

**21.** Find the steady-state distribution for Exercise 19 of the last section.

**22.** Find the steady-state distribution for Exercise 20 of the last section.

**23.** Find the steady-state distribution for Exercise 21 of the last section.

**24.** Find the steady-state distribution for Exercise 22 of the last section.

## Applications

**25. Marketing.** Find the steady-state distribution for Exercise 27 of the last section.

**26. Politics.** Find the steady-state distribution for Exercise 28 of the last section.

**27. Biology.** Find the steady-state distribution for Exercise 29 of the last section.

**28. Marketing.** Find the steady-state distribution for Exercise 30 of the last section.

## Graphing Calculator Exercises

**29.** Find the steady-state distribution for the stochastic matrix $A$ given in Exercise 17 by finding on your calculator $A^2$, $A^4$, $A^8$, $A^{16}$, ..., until your answers are repeated.

**30.** Repeat the previous exercise for the stochastic matrix given in Exercise 18.

## Solutions to Self-Help Exercise Set 9.2

**1.** The matrix $T$ does not have all positive entries, but

$$T^2 = \begin{bmatrix} 0 & .50 \\ 1 & .50 \end{bmatrix}\begin{bmatrix} 0 & .50 \\ 1 & .50 \end{bmatrix} = \begin{bmatrix} .50 & .25 \\ .50 & .75 \end{bmatrix}$$

has all entries positive. Thus $T$ is regular.

**2.** Since

$$T = \begin{bmatrix} .94 & .02 \\ .06 & .98 \end{bmatrix},$$

the steady-state solution will be the solution to the equations

$$x + y = 1$$
$$.94x + .02y = x$$
$$.06x + .98y = y$$

or

$$x + y = 1$$
$$-.06x + .02y = 0$$
$$.06x - .02y = 0$$

The last equation is just the negative of the second; therefore, it can be eliminated. Using the Gauss–Jordan method, replace the second equation with the second equation plus .06 times the first and obtain

$$x + y = 1$$
$$.08y = .06$$

or

$$x + y = 1$$
$$y = .75$$

Subtracting gives

$$x = .25$$
$$y = .75$$

Thus, in the long run, 25% of the population will be in the North and 75% will be in the South.

## 9.3    ABSORBING MARKOV PROCESSES

▶ *Absorbing Markov Processes*
▶ *Long-Term Behavior*

APPLICATION
**Long-Term Behavior**

Suppose in the restaurant example of the last section that Cathy's responds to the competition by making the food, service, ambiance, and price so extraordinary that once anyone eats at Cathy's, they always eat at Cathy's. Assume furthermore that eating at Barney's results in eating next at Cathy's, Abby's, or Barney's, with probability .20, .70, and .10, respectively. Eating at Barney's results in eating at one of the 3 restaurants next time with probability .10, .10, .80, respectively. What happens in the long-term? For the answer see Example 3.

## Absorbing Markov Processes

We now consider Markov processes with at least one state that has the property that when entered it is impossible to leave. Such a state is called **absorbing**. As we shall shortly see, Markov processes with at least one absorbing state are very different from the regular Markov processes studied in the last section. Some applications are given including an important application to charge accounts and production lines in the Exercises.

We will study Markov processes with at least one **absorbing state**, that is, a state that is impossible to leave.

We begin with the following definition.

---

**Absorbing Markov Process**

A Markov process is called **absorbing** if the following two conditions are satisfied.

1. There is at least one absorbing state.
2. It is possible to move from any nonabsorbing state to one of the absorbing states in a finite number of stages.

The transition matrix for an absorbing Markov process is said to be an **absorbing stochastic matrix**.

---

We begin with the following example of an absorbing Markov process.

**EXAMPLE 1**  **An Absorbing Markov Process**

---

Consider a particle in Figure 9.7 that can be at 0, 1, 2, or 3. If at 1 or 2, the particle moves left one unit with probability .50 and right one unit with probability .50. If the particle is at 0 or 3, it can never leave. Assume each stage takes one second. Draw a tree diagram that describes the situation. Indicate why this is an absorbing Markov process. Write the transition matrix.

**Figure 9.7**

**Solution**  Figure 9.8 is a tree diagram that describes the situation. Since the particle can never leave states 0 and 3, these are absorbing states. From Figure 9.8 one can see that the transition matrix is

**State**

$$
\begin{array}{cccc}
0 & 1 & 2 & 3
\end{array}
$$

$$
\begin{bmatrix}
1 & .50 & 0 & 0 \\
0 & 0 & .50 & 0 \\
0 & .50 & 0 & 0 \\
0 & 0 & .50 & 1
\end{bmatrix}
\begin{array}{l}
0 \\
1 \\
2 \\
3
\end{array}
\quad \textbf{State}
$$

One can also see that, starting at any of the 2 nonabsorbing states, the system can go to at least one of the absorbing states in a finite number of stages. For example, starting at 1, one can go $1 \to 0$. (One can also go $1 \to 2 \to 3$). This is then an absorbing Markov process. ∎

This is an example of a *random walk*. Random walk not only plays an important role in physics but also in many other areas, including finance. Random walk is sometimes used to describe the behavior of movements of individual stocks or other financial securities. In these applications the probability that the particle (or price) moves one unit up or down need not be equal.

Example 1 can also be considered as "gambler's ruin." In this interpretation, two gamblers are playing each other with the states representing the number of dollars that the first gambler has. When the number is zero, the first gambler is broke and the game cannot continue. When the number is 3, the second gambler is broke and again the game cannot continue.

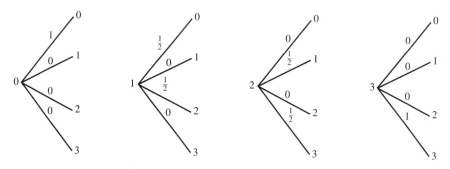

**Figure 9.8**

E X A M P L E   2     **Transition Matrix**

Determine if the stochastic matrix

$$
\begin{bmatrix}
.10 & 0 & .30 \\
.40 & 1 & .20 \\
.50 & 0 & .50
\end{bmatrix}
$$

is a transition matrix for an absorbing Markov process.

**Solution**    The first state is clearly not absorbing, since the first column indicates that it is possible to leave state 1. In the same way, the third column indicates that the third state is not absorbing. The second column, however, indicates that the second state is absorbing, since the probability of going from state 2 to state 2 is one. Thus, once in state 2, always in state 2. Finally, notice that the system can go from either of the nonabsorbing states 1 and 3 to the absorbing state 2 in just one stage. Thus this Markov process is absorbing.   ■

The above analysis indicates that the $i$th state will be absorbing if the $i$th column has all zeros except for the $i$th, which is 1.

We continue with another example of an absorbing Markov process.

**E X A M P L E   3**    **An Absorbing Markov Process**

Suppose in the restaurant example of the last section that Cathy's responds to the competition by making the food, service, ambiance, and price so extraordinary that once anyone eats at Cathy's, they always eat at Cathy's. Assume furthermore that eating at Barney's results in eating next at Cathy's, Abby's, or Barney's, with probability .20, .70, and .10, respectively. Eating at Barney's results in eating at one of the 3 restaurants next time with probability .10, .10, .80, respectively. Write the transition matrix and show why this is an absorbing Markov process.

**Solution**    Since Cathy's is absorbing, we place Cathy's first. The transition matrix is then

$$
\begin{array}{ccc}
\textbf{Cathy's} & \textbf{Abby's} & \textbf{Barney's}
\end{array}
$$

$$
\begin{bmatrix}
1 & .20 & .10 \\
0 & .70 & .10 \\
0 & .10 & .80
\end{bmatrix}
\begin{array}{l}
\textbf{Cathy's} \\
\textbf{Abby's} \\
\textbf{Barney's}
\end{array}
$$

The first column indicates that the first state is an absorbing state. The other 2 columns indicate that the other two states are not absorbing and one can go from any of the 2 nonabsorbing states to the absorbing state in just one stage. Thus the Markov process is absorbing.   ■

## Long-Term Behavior

Consider now the long-term behavior of the Markov process given above in Example 1. If a particle starts in state 1 or 2, then the only way for the particle to avoid being absorbed is to bounce back and forth between 1 and 2. The probability that this can occur will naturally decrease steadily toward zero in time. Thus in the long-term we expect that **the particle will be absorbed** by either 0 or 3. This is typical of absorbing Markov processes. In fact we have the following theorem.

> **Theorem 1**
>
> In an absorbing Markov process, the probability of going from any state to some absorbing state is one.

Another important observation is that if the particle is initially at 1, the probability that the particle will be absorbed by 0 is certainly larger than the probability it will be absorbed by 3. On the other hand, if the particle is initially at 2, then the probability it will be absorbed by 3 is greater than the probability it will be absorbed by 0. Thus **the long-term behavior may be different for different initial states**. This is in sharp contrast with regular Markov processes.

In order to see analytically exactly what is happening in the long-term, reorganize the states into two groups: the first is the absorbing states, the second is the nonabsorbing states. Doing this we have

$$
\begin{array}{cc}
\textbf{Absorbing} & \textbf{Nonabsorbing}
\end{array}
$$

$$
\begin{array}{cccc}
\textbf{State 0} & \textbf{State 3} & \textbf{State 1} & \textbf{State 2}
\end{array}
$$

$$
\begin{bmatrix}
1 & 0 & .50 & 0 \\
0 & 1 & 0 & .50 \\
0 & 0 & 0 & .50 \\
0 & 0 & .50 & 0
\end{bmatrix}
\begin{array}{l}
\textbf{State 0} \\
\textbf{State 3} \\
\textbf{State 1} \\
\textbf{State 2}
\end{array}
$$

Now to see the long-term behavior we need to find $T^n$ for large values of $n$. Before doing just that, let us anticipate at least part of the answer. Consider the element in the third row and third column of $T^n$. This is the probability that after $n$ stages the particle will have moved from state 1 to state 1. This can only happen if the particle bounces back and forth between 1 and 2. We have already indicated that the probability that this occurs should be steadily decreasing toward zero as $n$ becomes larger. Similarly, the element in the fourth row and third column of $T^n$ represents the probability that the particle has moved from state 1 to state 2. Again, this can happen only if the particle bounces back and forth between 1 and 2. Thus we expect the element in the third row and fourth column of $T^n$ to steadily approach zero. The same for the third and fourth elements in the fourth column.

Also the first column of $T^n$ should always be

$$
\begin{bmatrix}
1 \\
0 \\
0 \\
0
\end{bmatrix}
$$

since the probability that the particle moves from state 0 to states 0, 3, 1, and 2 must always be 1, 0, 0, 0, respectively, since 0 is an absorbing state. In the same way the second column of the matrices $T^n$ must be

$$
\begin{bmatrix} 0 \\ 1 \\ 0 \\ 0 \end{bmatrix}
$$

Now actually calculating some of the $T^n$ gives

$$
T^2 = \begin{bmatrix} 1 & 0 & .50 & .25 \\ 0 & 1 & .25 & .50 \\ 0 & 0 & .25 & 0 \\ 0 & 0 & 0 & .25 \end{bmatrix}, \quad
T^4 = \begin{bmatrix} 1 & 0 & .625 & .3125 \\ 0 & 1 & .3125 & .625 \\ 0 & 0 & .0625 & 0 \\ 0 & 0 & 0 & .0625 \end{bmatrix}
$$

$$
T^8 = \begin{bmatrix} 1 & 0 & .6641 & .3320 \\ 0 & 1 & .3320 & .6641 \\ 0 & 0 & .0039 & 0 \\ 0 & 0 & 0 & .0039 \end{bmatrix}, \quad
T^{16} = \begin{bmatrix} 1 & 0 & .6667 & .3333 \\ 0 & 1 & .3333 & .6667 \\ 0 & 0 & .00002 & 0 \\ 0 & 0 & 0 & .00002 \end{bmatrix}
$$

It then appears that this has the limiting value

$$
L = \begin{bmatrix} 1 & 0 & \frac{2}{3} & \frac{1}{3} \\ 0 & 1 & \frac{1}{3} & \frac{2}{3} \\ 0 & 0 & 0 & 0 \\ 0 & 0 & 0 & 0 \end{bmatrix}
$$

The number $\frac{2}{3}$ in the first row and third column can be interpreted to mean that the probability (in the long-term) that a particle will go from state 1 to state 0 is $\frac{2}{3}$. The number $\frac{1}{3}$ in the first row and fourth column can be interpreted to mean that the probability (in the long-term) that a particle will go from state 2 to state 0 is $\frac{1}{3}$. To take another example, the zero in the fourth row and third column can be interpreted to mean that the probability (in the long-term) of the particle going from state 1 to state 2 is zero.

Later in this section we give a sketch of the proof of how to obtain the other elements in the limiting matrix. The main result is given in the following general theorem.

**Theorem 2**

Let $T$ be the transition matrix of an absorbing Markov process with $a$ absorbing states and $b$ nonabsorbing states. Reorder the states so that the first $a$ are absorbing and the remaining $b$ are nonabsorbing. Partition the matrix as follows

$$\begin{array}{c} \quad\quad\quad \text{Absorbing} \quad \text{Nonabsorbing} \\ \left[ \begin{array}{c|c} I_{a \times a} & A_{a \times b} \\ \hline O_{b \times a} & B_{b \times b} \end{array} \right] \begin{array}{l} \text{Absorbing} \\ \\ \text{Nonabsorbing} \end{array} \end{array}$$

Then,

1. As $n$ becomes large without bound, the matrix $T^n$, which is the transition matrix from the initial stage to the $n$th stage, heads for the limiting matrix

$$L = \left[ \begin{array}{c|c} I_{a \times a} & A(I - B)^{-1} \\ \hline O_{b \times a} & O_{b \times b} \end{array} \right]$$

where the identity matrix, $I$, in the expression $(I - B)^{-1}$ is the same dimension as $B$, that is, $b \times b$.

2. The entry in the $i$th row and $j$th column of the matrix $A(I - B)^{-1}$ gives the probability that the system will end up in the $i$th absorbing state when initially in the $j$th nonabsorbing state.

3. All column sums of $A(I - B)^{-1}$ are 1, and thus everything is expected to be absorbed.

**E X A M P L E   4**    **Determining the Long-Term Behavior**

Use the above formulas to find the long-term behavior in Example 3.

**Solution**    There is 1 absorbing state and 2 nonabsorbing states. Thus $a = 1$ and $b = 2$. We then write the transition matrix in partitioned form as

$$\left[ \begin{array}{c|c} I_{1 \times 1} & A_{1 \times 2} \\ \hline O_{2 \times 1} & B_{2 \times 2} \end{array} \right] = \left[ \begin{array}{c|cc} 1 & .20 & .10 \\ \hline 0 & .70 & .10 \\ 0 & .10 & .80 \end{array} \right]$$

Thus

$$A = [.20 \quad .10], \quad B = \left[ \begin{array}{cc} .70 & .10 \\ .10 & .80 \end{array} \right]$$

Then

$$I - B = \left[ \begin{array}{cc} .30 & -.10 \\ -.10 & .20 \end{array} \right]$$

and

$$(I - B)^{-1} = \left[ \begin{array}{cc} 4 & 2 \\ 2 & 6 \end{array} \right]$$

Thus

$$A(I - B)^{-1} = [.20 \quad .10]\begin{bmatrix} 4 & 2 \\ 2 & 6 \end{bmatrix} = [1 \quad 1]$$

Then the limiting matrix is

$$L = \begin{bmatrix} 1 & 1 & 1 \\ 0 & 0 & 0 \\ 0 & 0 & 0 \end{bmatrix}. \quad \blacksquare$$

Notice that this limiting matrix indicates that no matter which state is the initial state, the limiting state will be state 1. That is, eventually everyone will be eating at Abby's and nobody will eat at the other two restaurants.

## SELF-HELP EXERCISE SET 9.3

1. Determine if the stochastic matrix

$$\begin{bmatrix} 1 & 0 & 0 \\ 0 & 0 & 1 \\ 0 & 1 & 0 \end{bmatrix}$$

is a transition matrix for an absorbing Markov process.

2. Find the fundamental matrix for the absorbing Markov process of Example 1 using the formula $F = (I - B)^{-1}$.

## EXERCISE SET 9.3

In Exercises 1 through 4 determine if the given matrix is an absorbing stochastic matrix.

1. $\begin{bmatrix} 0 & 0 & .10 \\ .30 & 1 & .20 \\ .70 & 0 & .70 \end{bmatrix}$
2. $\begin{bmatrix} 1 & .20 & 0 \\ 0 & .20 & 0 \\ 0 & .60 & 1 \end{bmatrix}$

3. $\begin{bmatrix} 0 & 1 & 0 \\ 1 & 0 & 0 \\ 0 & 0 & 1 \end{bmatrix}$
4. $\begin{bmatrix} 0 & 0 & 1 \\ 0 & 1 & 0 \\ 1 & 0 & 0 \end{bmatrix}$

Exercises 5 through 12 give a transition matrix for an absorbing Markov process. Find the limiting matrix.

**5.**
$$\begin{bmatrix} 1 & 0 & .10 \\ 0 & 1 & .20 \\ 0 & 0 & .70 \end{bmatrix}$$

**6.**
$$\begin{bmatrix} 1 & 0 & .30 \\ 0 & 1 & .20 \\ 0 & 0 & .50 \end{bmatrix}$$

**7.**
$$\begin{bmatrix} 1 & 0 & 0 & .10 \\ 0 & 1 & 0 & .20 \\ 0 & 0 & 1 & .30 \\ 0 & 0 & 0 & .40 \end{bmatrix}$$

**8.**
$$\begin{bmatrix} 1 & 0 & 0 & .20 \\ 0 & 1 & 0 & .30 \\ 0 & 0 & 1 & .40 \\ 0 & 0 & 0 & .10 \end{bmatrix}$$

**9.**
$$\begin{bmatrix} 1 & .20 & .10 \\ 0 & .50 & .50 \\ 0 & .30 & .40 \end{bmatrix}$$

**10.**
$$\begin{bmatrix} 1 & .10 & .30 \\ 0 & .20 & .20 \\ 0 & .70 & .50 \end{bmatrix}$$

**11.**
$$\begin{bmatrix} 1 & 0 & .10 & .20 \\ 0 & 1 & .30 & .50 \\ 0 & 0 & .20 & .20 \\ 0 & 0 & .40 & .10 \end{bmatrix}$$

**12.**
$$\begin{bmatrix} 1 & 0 & .20 & .20 \\ 0 & 1 & .20 & .30 \\ 0 & 0 & .30 & .10 \\ 0 & 0 & .30 & .40 \end{bmatrix}$$

## Applications

**13. Estes Learning Model.** Assume an individual is in state 1 if he has learned a particular task and in state 2 if he has not. Once the individual learns a task, he never forgets it. If a person has not learned the task, there is a probability of .25 that he will learn it during the next time period. Write the transition matrix for this Markov process. Show that this Markov process is absorbing. Determine the long-term behavior.

**14. Estes Learning Model.** Repeat the previous problem if for the individual who has not learned the task, the probability that he will learn the task in the next time period is $p$, with $0 < p < 1$.

**15. Gambler's Ruin.** A person plays a game in which the probability of winning $1 is .40 and the probability of losing $1 is .60. If he goes broke or reaches $3, he quits. Find the long-term behavior if he starts with $1 or $2.

**16. Gambler's Ruin.** A person plays a game in which the probability of winning $1 is .50 and the probability of losing $1 is .50. If she goes broke or reaches $4, she quits. Find the long-term behavior if she starts with $1, $2, or $3. *Hint:*

$$\begin{bmatrix} 1 & -\frac{1}{2} & 0 \\ -\frac{1}{2} & 1 & -\frac{1}{2} \\ 0 & -\frac{1}{2} & 1 \end{bmatrix}^{-1} = \begin{bmatrix} \frac{3}{2} & 1 & \frac{1}{2} \\ 1 & 2 & 1 \\ \frac{1}{2} & 1 & \frac{3}{2} \end{bmatrix}$$

*(Courtesy of TropWorld Casino and Entertainment Resort.)*

**17. Production Line.** A certain manufacturing process consists of 2 manufacturing states and a completion state together with a fourth state in which the item is scrapped if improperly manufactured. At the end of each of the 2 manufacturing states each item is inspected. At each inspection there is a probability of $\frac{2}{3}$ that the item will be passed on to the next state (the first manufacturing state to the second or the second manufacturing state to the completion state), probability $\frac{1}{6}$ that it will be sent back to the same state for reworking, and probability

$\frac{1}{6}$ that the item will be scrapped. Naturally, an item that is complete stays complete, and an item that is scrapped stays scrapped. Write a transition matrix for this Markov process. Show that this is an absorbing Markov process with 2 absorbing states and 2 nonabsorbing states. Determine the long-term behavior.

18. **Production Line.** Repeat the previous problem with the following changes. At each inspection there is a probability of .70 that the item will be passed on to the next stage, probability .20 that it will be sent back to the same stage for reworking, and probability .10 that the item will be scrapped.

19. **Charge Accounts.** A charge account is classified at the end of each month as "paid," "overdue less than 30 days," "overdue less than 60 days," or "overdue 60 days or more." In the latter case, the debt is judged "bad" and the charge account is canceled. Each month the debt can age by at most one month, keep the same or lesser age by means of a payment, or be paid off. If paid off, the account remains paid off. Over a period of time, the company has obtained figures that give the following transition matrix

$$
\begin{array}{cccc}
\textbf{Paid} & \textbf{<30} & \textbf{<60} & \textbf{Bad} \\
\end{array}
$$

$$
\begin{bmatrix}
1 & \frac{1}{3} & \frac{1}{6} & 0 \\
0 & \frac{1}{3} & \frac{1}{3} & 0 \\
0 & \frac{1}{3} & \frac{1}{3} & 0 \\
0 & 0 & \frac{1}{6} & 1
\end{bmatrix}
\begin{array}{l}
\textbf{Paid} \\
\textbf{<30} \\
\textbf{<60} \\
\textbf{Bad}
\end{array}
$$

By looking at the long-term behavior, determine the probability that the debt will be paid off and the probability that the debt will default in each of the cases: <30 and <60.

20. **Charge Accounts.** The charge card company in the previous exercise includes a threat in the "less than 60" category debt that going to 60 days or more will result in "serious consequences" for the customer. As a result the transition matrix changes to

$$
\begin{array}{cccc}
\textbf{Paid} & \textbf{<30} & \textbf{<60} & \textbf{Bad} \\
\end{array}
$$

$$
\begin{bmatrix}
1 & \frac{1}{3} & \frac{1}{6} & 0 \\
0 & \frac{1}{3} & \frac{1}{3} & 0 \\
0 & \frac{1}{3} & \frac{5}{12} & 0 \\
0 & 0 & \frac{1}{12} & 1
\end{bmatrix}
\begin{array}{l}
\textbf{Paid} \\
\textbf{<30} \\
\textbf{<60} \\
\textbf{Bad}
\end{array}
$$

Now answer the same questions as in the previous exercise. Carefully compare how the probability of default has now gone down dramatically with this small change.

## Graphing Calculator Exercises

21. Using the stochastic matrix $A$ in Example 3 of the text, determine the long-term behavior by calculating $A^2$, $A^4$, $A^8$, $A^{16}$, ..., until your answers repeat themselves.

22. Repeat the previous exercise using the 4 × 4 stochastic matrix discussed in the text after Theorem 1.

23. Repeat Exercise 21 using the stochastic matrix found in Exercise 9.

24. Repeat Exercise 21 using the stochastic matrix found in Exercise 10. Determine the fraction that each decimal in each entry represents.

25. **Charge Accounts.**   A charge account is classified at the end of each month as "paid," "overdue less than 30 days," "overdue less than 60 days," "overdue less than 90 days," or "overdue 90 days or more." In the latter case, the debt is judged "bad" and the charge account is canceled. Each month the debt can age by at most one month, keep the same or lesser age by means of a payment, or be paid off. If paid off, the account remains paid off. Over a period of time, the company has obtained figures that give the following transition matrix.

| | Paid | <30 | <60 | <90 | Bad | |
|---|---|---|---|---|---|---|
| | 1 | .6 | .1 | 0 | 0 | Paid |
| | 0 | .3 | .2 | .1 | 0 | <30 |
| | 0 | .1 | .4 | .4 | 0 | <60 |
| | 0 | 0 | .3 | .2 | 0 | <90 |
| | 0 | 0 | 0 | .3 | 1 | Bad |

By looking at the long-term behavior, determine the probability that the debt will be paid off and the probability that the debt will default in each of the cases: <30, <60, and <90.

26. **Charge Accounts.**   The charge card company in the previous exercise includes a threat in the "less than 60" category debt that going to 60 days or more will result in "serious consequences" for the customer. As a result the transition matrix changes to

| | Paid | <30 | <60 | <90 | Bad | |
|---|---|---|---|---|---|---|
| | 1 | .7 | .25 | 0 | 0 | Paid |
| | 0 | .2 | .25 | .1 | 0 | <30 |
| | 0 | .1 | .25 | .4 | 0 | <60 |
| | 0 | 0 | .25 | .2 | 0 | <90 |
| | 0 | 0 | 0 | .3 | 1 | Bad |

Now answer the same questions as in the previous exercise. Carefully compare how the probability of default has now gone down dramatically with this small change.

## Solutions to Self-Help Exercise Set 9.3

1. The first state is absorbing. Thus this Markov process has at least one absorbing state. But notice from the transition matrix that if the system is in state 2, it must go to state 3, while if the system is in state 3, it must go to state 2. Thus, if the

system starts in either of the nonabsorbing states, state 2 or state 3, the system cannot enter the absorbing state 1. Thus the system fails to satisfy condition (2) above and is not an absorbing Markov process

2.  We have

$$I - B = \begin{bmatrix} 1 & 0 \\ 0 & 1 \end{bmatrix} - \begin{bmatrix} 0 & \frac{1}{2} \\ \frac{1}{2} & 0 \end{bmatrix} = \begin{bmatrix} 1 & -\frac{1}{2} \\ -\frac{1}{2} & 0 \end{bmatrix}$$

and

$$(I - B)^{-1} = \begin{bmatrix} \frac{4}{3} & \frac{2}{3} \\ \frac{2}{3} & \frac{4}{3} \end{bmatrix}$$

# SUMMARY OUTLINE OF CHAPTER 9

- A finite stochastic process in which the outcomes and associated probabilities depend only on the preceding experiment is called a **Markov process**.
- The outcome of any experiment is called the **state** of the experiment.
- A **transition matrix** for a Markov process gives the probabilities of going from one state to another.
- If the column matrix $X_n$ gives the state after the $n$th experiment, then $X_n = T^n X_0$, where $T$ is the transition matrix and $X_0$ is the initial state.
- A Markov process is said to be **regular** if no matter what the initial state, the limiting distribution is always the same. This limiting distribution is called the **steady-state distribution**.
- A stochastic matrix is said to be **regular** if some power has all positive entries.
- A Markov process is regular if the associated transition matrix is regular.
- If $T$ is a regular stochastic matrix, one can find the steady-state distribution $\overline{X}$, by solving the equations $\overline{X} = T\overline{X}$ together with the fact that the sum of the entries in $\overline{X}$ must be one.
- **Regular Markov Processes.**   Let $T$ be a regular stochastic matrix associated with a Markov process. Then
  1.  For any initial distribution $X_0$, $X_n = T^n X_0$ approaches the same steady-state distribution $\overline{X}$, as $n$ becomes large without bound.
  2.  As $n$ becomes large without bound, $T^n$ approaches the steady-state matrix $L$.
  3.  Every column of $L$ is just $\overline{X}$.
  4.  The steady-state distribution can be found solving

     $$\overline{X} = T\overline{X}$$

     and using the fact that the entries in $\overline{X}$ sum to one.
  5.  A stochastic matrix $T$ will be regular if some power of $T$ has all positive entries.
- A Markov process is called **absorbing** if there is at least one absorbing state and it is possible to move from any nonabsorbing state to one of the absorbing states in a finite number of stages.

- The transition matrix for an absorbing Markov process is said to be an **absorbing stochastic matrix**.

- In an absorbing Markov process, the probability of going from any state to some absorbing state is one.

- Let $T$ be the transition matrix of an absorbing Markov process with $a$ absorbing states and $b$ nonabsorbing states. Reorder the states so that the first $a$ are absorbing and the remaining $b$ are nonabsorbing. Partition the matrix as follows

**Absorbing  Nonabsorbing**

$$\left[\begin{array}{c|c} I_{a\times a} & A_{a\times b} \\ \hline O_{b\times a} & B_{b\times b} \end{array}\right] \begin{array}{l} \textbf{Absorbing} \\ \textbf{Nonabsorbing} \end{array}$$

Then,

1. As $n$ becomes large without bound, the matrix $T^n$, which is the transition matrix from the initial stage to the $n$th stage, heads for the limiting matrix

$$L = \left[\begin{array}{c|c} I_{a\times a} & A(I - B)^{-1} \\ \hline O_{b\times a} & O_{b\times b} \end{array}\right]$$

where the identity matrix, $I$, in the expression $(I - B)^{-1}$ is the same dimension as $B$, that is, $b \times b$.

2. The entry in the $i$th row and $j$th column of the matrix $A(I - B)^{-1}$ gives the probability that the system will end up in the $i$th absorbing state when initially in the $j$th nonabsorbing state.

3. All column sums of $A(I - B)^{-1}$ are 1, and thus, everything is expected to be absorbed.

## Chapter 9 Review Exercises

In Exercises 1 through 6 determine if the given matrix is stochastic, regular stochastic, absorbing stochastic, or none of these.

**1.** $\begin{bmatrix} .10 & .90 \\ .50 & .50 \end{bmatrix}$

**2.** $\begin{bmatrix} .20 & .40 \\ .80 & .60 \end{bmatrix}$

**3.** $\begin{bmatrix} 0 & .50 & .30 \\ .40 & 0 & .70 \\ .60 & .50 & 0 \end{bmatrix}$

**4.** $\begin{bmatrix} 1 & 0 & 0 \\ 0 & .40 & .50 \\ 0 & .60 & .50 \end{bmatrix}$

**5.** $\begin{bmatrix} 0 & .5 & 0 \\ 1 & .5 & 0 \\ 0 & 0 & 1 \end{bmatrix}$

**6.** $\begin{bmatrix} 1 & .10 & 0 \\ 0 & .30 & .50 \\ 0 & .60 & .50 \end{bmatrix}$

Exercises 7 through 8 give a regular stochastic matrix, $T$. Find $T^2$ and $T^4$ and interpret what these matrices represent.

**7.** $\begin{bmatrix} .10 & .60 \\ .90 & .40 \end{bmatrix}$

**8.** $\begin{bmatrix} .30 & .20 & .10 \\ .30 & .20 & .40 \\ .40 & .60 & .50 \end{bmatrix}$

9. Find the steady-state distribution for the regular stochastic matrix found in Exercise 7.

10. Find the steady-state distribution for the regular stochasitc matrix found in Exercise 8.

11. Find the limiting matrix for the absorbing stochastic matrix

$$\begin{bmatrix} 1 & 0 & .20 \\ 0 & 1 & .30 \\ 0 & 0 & .50 \end{bmatrix}.$$

12. Find the limiting matrix for the absorbing stochastic matrix

$$\begin{bmatrix} 1 & .10 & .20 \\ 0 & .50 & .50 \\ 0 & .40 & .30 \end{bmatrix}.$$

13. **Car Rental.**   A car rental company has two locations in a certain city. A car rented at the first location will be returned to this location 70% of the time and 30% of the time to the second location. A car rented at the second location will be returned to this location 80% of the time and 20% of the time to the first location. If initially the cars are evenly distributed between the two locations, what will the distribution of cars eventually be?

14. **Social Mobility.**   Suppose that the son of a father in the upper class has a probability of .50 of being in the upper class, .40 of being in the middle class, and .10 of being in the lower class. For a son of a father in the middle class, suppose the probabilities are .20, .70, and .10, respectively, while for a father of a son in the lower class, the probabilities are .10, .30, and .60, respectively. Determine what happens in the long-term.

15. **Personnel.**   A firm has 2 types of mechanics: mechanic and master mechanic. Each year a mechanic has a 20% chance of leaving the firm and a 20% chance of being promoted to a master mechanic. A master mechanic has a 10% chance of leaving. If a mechanic or master mechanic leave, they never return. A master mechanic is never demoted to a mechanic. Consider the 4 states: leave as master mechanic, leave as mechanic, mechanic, master mechanic. Determine the long-term behavior.

Game theory can help a firm develop the best competitive strategy. (Gregory Heisler/The Image Bank.)

# Game Theory

10.1   Strictly Determined Games

10.2   Mixed Strategy Games

10.3   Linear Programming and Game Theory

66 **T**he theory of games rests on the notion that there is a close analogy between parlor games of skill, on the one hand, and conflict situations in economics, political, and military life, on the other. In any of these situations there are a number of participants with incompatible objectives, and the extent to which each participant attains his objectives depends upon what all the participants do. The problem faced by each participant is to lay his plans so as to take account of the actions of his opponents, each of whom, of course, is laying his own plans so as to take account of the first participant's actions. Thus each participant must surmise what each of his opponents will expect him to do and how these opponents will react to these expectations." Dorfman, Samuelson, Solow in *Linear Programming and Economic Analysis.*

## 10.1    STRICTLY DETERMINED GAMES

▶ *Strictly Determined Games*
▶ *A Nonstrictly Determined Game*

**John von Neuman,**
**1903–1957**

*(The Bettmann Archive.)*

Born in Hungary, John von Neuman began his teaching career in Germany. In 1933 he became a permanent member of the Institute for Advanced Study at Princeton. Von Neuman was one of the most creative and versatile mathematicians of the 20th century. He made deep contributions to logic and set theory and in 1928 developed the central theorem of game theory. The impact of game theory on economics was delayed until the publication of *The Theory of Games and Economic Behavior* by von Neumann and Morgenstern in 1944. Von Neuman was also deeply involved in quantum theory and in 1955 was appointed to the Atomic Energy Commission. He was most responsible for initiating the first fully electronic calculator and for the concept of a stored digital computer.

## Strictly Determined Games

We use the term **game** to refer to situations of conflict between players. The players can be individuals, coalitions, companies or even the weather. We first consider the simplest games: strictly determined games. In these games the strategies of both players and the outcome are all precisely determined.

To obtain an idea of what is involved in game theory, consider the following very simplified decision problem.

**E X A M P L E    1**    **Analysis of a Strictly Determined Game**

Two car dealers, R and C, in a certain town need to clear out their inventory of new cars in August to make room for the new models that will arrive in September. They decide to have either a $300 or $500 over-the-cost sale. From past price wars they know that if both set prices at $500 above cost, dealer R will obtain 70% of the total sales. If both set prices at $300 above cost, dealer R will obtain 50% of sales. Also if dealer R sets prices $300 above cost while dealer C sets prices $500 above cost, dealer R will obtain 75% of sales. Finally if dealer R sets prices $500 above cost while dealer C sets price $300 above cost, dealer R will obtain 35% of sales. What is the best strategy for both dealers to clear out their inventory if we make the important assumption that *both dealers are rational and intelligent?*

*(Dennis Drenner.)*

**Solution**    We summarize the situation as follows:

$$
\begin{array}{c}
\textbf{Dealer C} \\[4pt]
\begin{array}{cc}
\$500 & \$300
\end{array}
\end{array}
$$

$$
\textbf{Dealer R}
\begin{array}{c}
\$500 \\
\$300
\end{array}
\begin{bmatrix}
70\% & 35\% \\
75\% & 50\%
\end{bmatrix}
\tag{1}
$$

Consider now the strategy for dealer R. Suppose dealer R sets his prices at $500 above cost (pick row 1 in matrix (1)). As soon as dealer C got wind of this, dealer C would set his prices at $300 above cost, resulting in dealer R obtaining only 35% of sales. This would be a disaster for dealer R. On the other hand, if dealer R sets his prices $300 above cost (pick row 2), then he knows that dealer C will respond in kind, and dealer R will obtain 50% of the sales. The best strategy for dealer R is to set his prices at $300 above cost (pick row 2) and obtain 50% of the sales.

Now consider the point of view of dealer C. If he sets his prices at $500 above cost (picks column 1), then dealer R will set prices at $300 above cost, resulting in dealer R obtaining 75% of sales. This is a disaster for dealer C. If dealer C sets his prices at $300 above cost (picks column 2 in matrix (1)), then dealer R will be expected to set his prices at $300 above cost, resulting in dealer R obtaining 50% of the sales. Clearly then, the best strategy for dealer C is to set prices at $300 above cost (pick column 2).

The final result is that both dealers will set their prices at $300 above cost and each will obtain 50% of the sales. ∎

The above game is an example of a **two-person zero-sum game**.

---

**Two-Person Zero-Sum Game**

A game is called a **two-person zero-sum game** if the following are satisfied.

1. There are two players (called the row player and the column player).
2. The row player must choose one of $m$ strategies and simultaneously, the column player must choose one of $n$ strategies.
3. If the row player chooses her $i$th strategy and the column player chooses his $j$th strategy, the row player receives a reward of $a_{ij}$ and the column player loses an amount $a_{ij}$.

---

The game is called zero-sum since the row player's gain is the column player's loss, and vice versa. Thus the 2 players are in total conflict, and no cooperation between the two is possible.

The following indicates the general situation.

**Column's strategy**

|  | Column 1 | Column 2 |  | Column $n$ |
|---|---|---|---|---|

$$
\begin{array}{c}
\text{Row's strategy}
\end{array}
\begin{array}{c}
\text{Row 1} \\
\text{Row 2} \\
\vdots \\
\text{Row } m
\end{array}
\begin{bmatrix}
a_{11} & a_{12} & \cdots & a_{1n} \\
a_{21} & a_{22} & \cdots & a_{2n} \\
\vdots & \vdots & \cdots & \vdots \\
a_{m1} & a_{m2} & \cdots & a_{mn}
\end{bmatrix}
\qquad (2)
$$

---

**Payoff Matrix**

The $m \times n$ matrix $A = (a_{ij})$ in (2) is called the **payoff matrix**.

---

In Example 1 each of the elements $a_{ij}$ in the payoff matrix represented a percent. We now consider a game in which each of the elements $a_{ij}$ in the payoff matrix represents a dollar amount.

**E X A M P L E   2**      **Determining the Best Strategies**

Consider the game with the following payoff matrix

$$
\begin{bmatrix}
1 & -2 & -5 & 1 \\
6 & 1 & 2 & 3 \\
0 & -3 & 3 & -6
\end{bmatrix}
$$

where the element in the $i$th row and $j$th column represents the amount in dollars that the row player wins from the column player. (A negative win means the row player loses this amount.) Thus if the row player chooses row 1 and the column player chooses column 1, then the row player wins \$1 from the column player. If the row player chooses row 1 and the column player chooses column 2, then the row player loses \$2 to the column player. Find the best strategies for each player.

**Solution**      We first look at the row player's strategy. If the row player chooses the first row, that is, row strategy 1, then she can count on the column player picking column 3, which contains the minimum of the numbers in row 1. This results in a loss of \$5 to the row player. We place a small "m" after the $-5$ in the first row. We refer to the $-5$ as the minimum of row 1. See matrix (3). If the row player chooses the second row, that is, row strategy 2, then she can count on the column player picking the column that contains the smallest number in the second row. This is column 2. This results in a gain of \$1 to the row player. We place a small "m" after the 1 in the second row. If the row player chooses the third row, then this row minimum is $-6$ and is in column 4. Thus the column player will pick column 4. This results in a loss of \$6 to the row player. We place a small "m" after the $-6$ in the third row. The best strategy for the row player is then to pick the row containing the *maximum* of the row minimums. The row minimums are the numbers with a small "m" after them. Since the maximum of $\{-5, 1, -6\}$ is 1 and is in row 2, the best strategy for the row player is to pick row 2. This results in a certain gain of at least \$1.

$$\begin{bmatrix} +1 & -2 & -5m & +1 \\ +6 & +1m & +2 & +3 \\ 0 & -3 & +3 & -6m \end{bmatrix} \tag{3}$$

If the column player chooses column 1, then the row player will pick row 2 since 6 is the column maximum for the first column. This results in a gain of \$6 for the row player. This is indicated in matrix (4) by a capital "M" after the 6. If the column player chooses column 2, then the row player will pick row 2, since the 1 in column 2 is that column's maximum. This is indicated in matrix (2) by a capital "M" after the 1. Selecting column 3 results in the row player selecting row 3 since 3 is the column maximum for that column. Selecting column 4 results in the row player selecting row 2, since 3 is the fourth column's maximum. The best strategy for the column player is to then pick the column that contains the minimum of the column maximums. This is column 2. This results in a gain of \$1 for the row player.

$$\begin{bmatrix} +1 & -2 & -5 & +1 \\ +6M & +1M & +2 & +3M \\ 0 & -3 & +3M & -6 \end{bmatrix} \tag{4}$$

Thus the only rational outcome of this game is for the row player to pick the second row and the column player to pick the second column, resulting in row winning \$1.  ∎

The value \$1 in the previous game is called the **payoff value**. If the payoff value is zero, the game is called **fair**.

Let us now see what happens when we overlay matrix (3) with matrix (4). This is shown in matrix (5).

$$\begin{bmatrix} +1 & -2 & -5m & +1 \\ +6M & +1mM & +2 & +3M \\ 0 & -3 & +3M & -6m \end{bmatrix} \tag{5}$$

Notice the payoff value 1 in the second row and second column has both a small "m" and a capital "M" next to it. This indicates that this element is both the row minimum and the column maximum for the row and column that it is in. Such an element is called a **saddle point**. One can think of riding a horse in the direction $\rightarrow$. Directly in front of you and behind you, the saddle rises. To each side, where your legs are, the saddle drops. In the same way, as we move horizontally away (left or right) from the saddle point 1 in the previous example, the values increase, whereas, as we move vertically (up or down) away from 1, the values decrease. This is in analogy with the saddle. As may be suspected from the previous example, saddle points play an important role in game theory.

Let the payoff matrix of a game have $a_{ij}$ in the $i$th row and $j$th column.

---

**Saddle Point**

If there is an entry, $a_{hk}$, that is the minimum in its row (row $h$) and also the maximum in its column (column $k$), we call the entry $a_{hk}$ a **saddle point**.

---

**Strictly Determined Game**

A game with a saddle point is said to be **strictly determined**.

---

If a game has a saddle point $a_{hk}$, then the number $a_{hk}$ is called the **value** of the game. If there is more than one saddle point, they will both yield the same value for the game. (This is not proven here.) In a game with a saddle, the optimal strategy for the row player is to chose row $R_h$ (the row containing a saddle). The optimal strategy for the column player is to chose column $C_k$ (the column containing a saddle).

---

**Locating a Saddle Point (The m & M method)**

1. For each row, find the minimum value and place a small ''m'' next to it.
2. For each column, find the maximum value and place a capital ''M'' next to it.
3. Any entry with both a small ''m'' and a capital ''M'' next to it is a saddle point.

---

A saddle point can also be thought of as an **equilibrium point**. The word equilibrium is used because the row player's payoff will decrease if the row player unilaterally changes to another row other than the optimal row. Thus the row player will move back to the optimal row. If the column player changes the column from the optimal column, the result is an increased gain for the row player. Thus the column player would return to the optimal choice. The saddle point is then in ''equilibrium'' since if either player moves away from it, he or she will quickly return.

---

**E X A M P L E   3**     **Determining Optimal Strategies and the Value of a Game**

Determine if the game with the following payoff matrix is strictly determined. If it is, find the value of the game and the optimal strategies of each of the players.

$$A = \begin{bmatrix} -3 & 0 & -2 \\ 1 & 2 & 0 \\ 3 & 1 & -1 \end{bmatrix}$$

**Solution**   We place a small "m" next to the $-3$ in the first row of $A$, the 0 in the second row, and the $-1$ in the third row. Then place a capital "M" next to the 3 in the first column, the 2 in the second column, and the 0 in the third column. Notice that there is one entry with both a small m and a capital M next to it. This is the saddle point, which indicates that the game is strictly determined with value 0. This game is a fair game. The saddle point is in the second row and third column. Thus the optimal strategy for the row player is to select row 2 and the optimal strategy for the column player is to select column 3.

$$\begin{bmatrix} -3m & 0 & -2 \\ +1 & +2M & 0mM \\ +3M & +1 & -1m \end{bmatrix} \quad ■$$

The next example has more than 2 saddle points.

**E X A M P L E   4**   **Finding Saddle Points**

Find all saddle points for the game with the following payoff matrix

$$\begin{bmatrix} 0 & 1 & 1 & 0 \\ 1 & 1 & 2 & 1 \\ 1 & 2 & 3 & 1 \end{bmatrix}.$$

**Solution**   Placing the m's and the M's yields

$$\begin{bmatrix} 0m & +1 & +1 & 0m \\ +1mM & +1m & +2 & +1mM \\ +1mM & +2M & +3M & +1mM \end{bmatrix}.$$

There are 4 saddle points. The value of this game is 1.   ■

## A Nonstrictly Determined Game

Consider the game of Odds and Evens.

**E X A M P L E   5**   **A Game that is not Strictly Determined**

Two players, called Odd and Even, simultaneously put out either 1 or 2 fingers. If the sum of all the fingers from both players is odd, then Odd wins $1 from Even. If the sum of all the fingers from both players is even, then Odd loses $1 to Even. Let Odd be the row player and Even the column player. Find the payoff matrix and determine if there is a saddle point.

**Solution**    The result of the various outcomes are summarized in the following.

**Even**

1 finger    2 fingers

$$\begin{array}{c} \text{Odd} \end{array} \begin{array}{c} \text{1 finger} \\ \text{2 fingers} \end{array} \begin{bmatrix} -1 & +1 \\ +1 & -1 \end{bmatrix}$$

Now adding the m's and M's gives

$$\begin{bmatrix} -1m & +1M \\ +1M & -1m \end{bmatrix}.$$

Since no entry has both a m and a M next to it, there are no saddle points and the game is not strictly determined. ■

We will study games that are not strictly determined in the next section.

## SELF-HELP EXERCISE SET 10.1

Determine if any of the following matrices are payoff matrices for strictly determined games. If one is, then find the value of the game and the optimal strategies for each player.

1. $\begin{bmatrix} -2 & 2 \\ 1 & -1 \\ -1 & 3 \end{bmatrix}$

2. $\begin{bmatrix} -3 & -1 & 0 \\ -2 & 0 & 2 \\ -4 & 0 & -3 \end{bmatrix}$

## EXERCISE SET 10.1

In Exercises 1 through 12 determine if any of the matrices are payoff matrices for strictly determined games. If one is, then find the value of the game and the optimal strategies for each player.

1. $\begin{bmatrix} 1 & 0 \\ 2 & 3 \end{bmatrix}$

2. $\begin{bmatrix} -1 & 0 \\ -2 & -3 \end{bmatrix}$

3. $\begin{bmatrix} -3 & -2 & -4 \\ 0 & -2 & -1 \end{bmatrix}$

4. $\begin{bmatrix} 3 & 2 & 4 \\ 0 & 1 & 1 \end{bmatrix}$

**5.** $\begin{bmatrix} 1 & 2 \\ 3 & 4 \\ 5 & 6 \end{bmatrix}$
    **6.** $\begin{bmatrix} -1 & -2 \\ -3 & -4 \\ -5 & -6 \end{bmatrix}$

**7.** $\begin{bmatrix} -3 & -2 & -4 \\ 0 & -1 & -1 \end{bmatrix}$
    **8.** $\begin{bmatrix} 1 & 1 & 3 \\ 1 & 1 & 2 \end{bmatrix}$

**9.** $\begin{bmatrix} 1 & 0 \\ 0 & 1 \end{bmatrix}$
    **10.** $\begin{bmatrix} -1 & 0 \\ 0 & -2 \end{bmatrix}$

**11.** $\begin{bmatrix} 2 & -1 & -5 & -2 \\ 4 & -1 & 8 & 7 \\ 3 & 0 & 1 & 4 \end{bmatrix}$
    **12.** $\begin{bmatrix} 4 & -4 & -2 & 0 \\ 0 & 0 & -1 & 0 \\ 9 & 3 & -3 & -2 \end{bmatrix}$

**13.** Show that the game with the payoff matrix

$$\begin{bmatrix} 0 & a \\ 1 & 2 \end{bmatrix}$$

is strictly determined no matter what $a$ is.

**14.** Show that the game with the payoff matrix

$$\begin{bmatrix} 2 & 4 \\ 1 & b \end{bmatrix}$$

is strictly determined no matter what $b$ is.

In Exercises 15 through 18, write the payoff matrix for the given game and determine if the game is strictly determined. If it is, find the optimal strategies for each player.

**15.** Each of two players R and C has, in their respective pockets, 3 coins: a penny, a nickel, and a dime. Each one selects a coin, and simultaneously each lays a coin on the table. If the coins are the same, no payment is made. If the coins are different, the one who played the coin with the smallest denomination wins both coins.

**16.** In this version of three-finger Morra, each player simultaneously shows either 1, 2, or 3 fingers. Player C agrees to pay player R an amount of dollars equal to the number of fingers shown by player R less the number shown by player C.

**17.** A person secretly places a penny, nickel, or dime in his fist. If you guess the correct coin, you win the coin. If you guess incorrectly, you give him the difference between your guess and the coin held.

**18.** In this version of four-finger Morra, each player simultaneously shows either 1, 2, 3, or 4 fingers. If the sum of fingers shown is even, you win an amount in dollars equal to the sum. If the sum of fingers is odd, you lose an amount equal to the sum.

### Solutions to Self-Help Exercise Set 10.1

1. For this matrix we have

$$\begin{bmatrix} -2m & 2 \\ 1M & -1m \\ -1m & 3M \end{bmatrix}.$$

Since there is no entry with both a m and a M next to it, there is no saddle. Thus the game is not strictly determined.

2. For this matrix we have

$$\begin{bmatrix} -3m & -1 & 0 \\ -2mM & 0M & 2M \\ -4m & 0M & -3 \end{bmatrix}.$$

The entry $-2$ has both a m and a M next to it, indicating that this entry is a saddle point. The game is strictly determined with value $-2$. The optimal strategy is for the row player to chose row 2 and the column player to chose column 1.

## 10.2   MIXED STRATEGY GAMES

▶ *Mixed Strategies*

▶ *Expectation*

▶ *A Geometric Method*

▶ *Domination*

### Mixed Strategies

Games that are not strictly determined are considered in this section. We will find the value and the optimal strategies for nonstrictly determined two-person, zero-sum games.

Consider again the game of Odds and Evens. This game is simple enough that a strategy can easily be developed and the value of the game determined.

Two players, called Odd and Even, simultaneously put out either 1 or 2 fingers. If the sum of all the fingers from both players is odd, then Odd wins $1 from Even. If the sum of all the fingers from both players is even, then Odd loses $1 to Even. Let Odd be the row player and Even the column player. Find the payoff matrix and determine if there is a saddle point.

The results of the various outcomes are summarized in the following.

**Even**

1 finger    2 fingers

$$\text{Odd} \begin{array}{c} \text{1 finger} \\ \text{2 fingers} \end{array} \begin{bmatrix} -1 & +1 \\ +1 & -1 \end{bmatrix}$$

We saw in the last section that this was not a strictly determined game. Consider how such a game might be played. If the row player always plays 1 finger, then the column player can counter by always playing 2 fingers and win all the time. If the row player always plays 2 fingers, then the column player can counter with always playing 1 finger, and win all the time.

Consider now the possibility of the row player selecting either 1 finger or 2 fingers in some **random** fashion. For example, in playing the game a number of times, the row player might select 1 finger one-third of the time and 2 fingers two-thirds of the time with the choice being made in a random fashion. Once the column player realizes that this particular strategy has been selected, one that favors 2 fingers, then the column player will counter by always putting out 2 fingers. Such a strategy for the row player will then result in losses. If the row player makes a random selection by favoring 1 finger, then the column player can counter by always putting out 1 finger, again resulting in losses for the row player.

Thus for this simple game, it is clear that if the row player is to avoid losses, she must not favor either of the two strategies. In other words, she will need to select 1 finger or 2 fingers equally often and in a random manner. This should lead to her breaking even. Similar reasoning implies a similar strategy for the column player. Thus we suspect that this game has a value of zero. Each player can assure that they are selecting each of the two strategies randomly and with equal probability by determining their selections by, say, flipping a fair coin. This way, the column player, for example, cannot possibly guess how many fingers will be put out by the row player since the row player will not know herself until the coin is flipped.

A **mixed strategy**

$$p = \begin{bmatrix} p_1 \\ p_2 \end{bmatrix},$$

or written in the more convenient transposed form, $p^T = [p_1 \quad p_2]$, is a strategy in which the first and second strategy is selected randomly with probability $p_1$ and $p_2$, respectively. Thus a strategy of $p^T = [\frac{1}{3} \quad \frac{2}{3}]$, mentioned earlier, means selecting randomly the first strategy with probability $\frac{1}{3}$ and the second with probability $\frac{2}{3}$.

We will always use the letter $p$ to denote the strategy of the row player and the letter $q$ for the column player. Thus a strategy of $q^T = [\frac{1}{4} \quad \frac{3}{4}]$ means that the column player will randomly select his first strategy with probability .25 and the second with probability .75.

## Expectation

In the discussion of the Odds and Evens game, we decided that if the row player used a strategy that favored selecting 2 fingers, then the column player could counter with a strategy that favored selecting 2 fingers and cause the row player to lose money. Let us in fact now consider the expected gain or loss for the row player if the row player used the strategy $p^T = [\frac{1}{3} \quad \frac{2}{3}]$ in the game of Odds and Evens and the column player used the strategy $q^T = [\frac{1}{4} \quad \frac{3}{4}]$.

Figure 10.1 is a probability tree showing the probability of each of the 4 possible outcomes using these two strategies of the row and column player. The expected value for these two selections of strategies is then

$$E = \frac{1}{3} \cdot \frac{1}{4}(-1) + \frac{1}{3} \cdot \frac{3}{4}(+1) + \frac{2}{3} \cdot \frac{1}{4}(+1) + \frac{2}{3} \cdot \frac{3}{4}(-1) = -\frac{1}{6}.$$

As we strongly suspected, this favoring of 2 fingers by both players will result in losses for the row player.

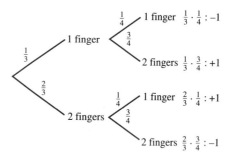

**Figure 10.1**

To generalize, consider a game with a payoff matrix

$$A = \begin{bmatrix} a_{11} & a_{12} \\ a_{21} & a_{22} \end{bmatrix}$$

and assume the row player selects the mixed strategy $p^T = [p_1 \quad p_2]$ and the column player selects the mixed strategy $q^T = [q_1 \quad q_2]$. Then since the probabilities of the 4 possible outcomes are given by the tree diagram in Figure 10.2, the expectation is

$$E(p, q) = p_1 q_1 a_{11} + p_1 q_2 a_{12} + p_2 q_1 a_{21} + p_2 q_2 a_{22}.$$

This can be written in the more convenient form

$$E(p, q) = p^T A q,$$

as the reader can easily verify.

Similar calculations work for a general $m \times n$ payoff matrix $A$. We then have

### Expected Value

Suppose a game has a payoff matrix

$$A = \begin{bmatrix} a_{11} & a_{12} & \cdots & a_{1n} \\ a_{21} & a_{22} & \cdots & a_{2n} \\ & & \cdots & \\ a_{m1} & a_{m2} & \cdots & a_{mn} \end{bmatrix}$$

and row plays the strategy $p^T = [p_1 \quad p_2 \quad \cdots \quad p_m]$, and column selects $q^T = [q_1 \quad q_2 \quad \cdots \quad q_n]$, then the expected value for these two strategies is

$$E(p, q) = p^T A q = [p_1 \quad p_2 \quad \cdots \quad p_m] \begin{bmatrix} a_{11} & a_{12} & \cdots & a_{1n} \\ a_{21} & a_{22} & \cdots & a_{2n} \\ & & \cdots & \\ a_{m1} & a_{m2} & \cdots & a_{mn} \end{bmatrix} \begin{bmatrix} q_1 \\ q_2 \\ \vdots \\ q_n \end{bmatrix}.$$

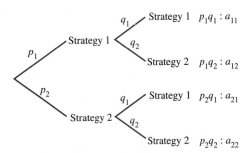

**Figure 10.2**

We say that the strategy $[p_1 \quad p_2 \quad \cdots \quad p_m]$ is a **pure strategy** if any one of the $p_i$ equals 1. (In such a case all the other $p_i$'s must be zero.) Thus in the Odds and Evens game the strategy $[0 \quad 1]$ means always selecting 2 fingers. Notice that pure strategies were the only kind considered in the last section and that a pure strategy is just a special case of a mixed strategy.

**E X A M P L E  1**     **Finding Expected Values**

Consider the game with the payoff matrix

$$A = \begin{bmatrix} -1 & 1 \\ 2 & -4 \end{bmatrix}.$$

Find the expected value if (a) $p^T = [.2 \quad .8]$, $q^T = [1 \quad 0]$. (b) $p^T = [.7 \quad .3]$, $q^T = [1 \quad 0]$.

**Solutions**

(a) In this case

$$E(p, q) = p^T A q = [.2 \quad .8]\begin{bmatrix} -1 & 1 \\ 2 & -4 \end{bmatrix}\begin{bmatrix} 1 \\ 0 \end{bmatrix} = 1.4.$$

(b) In this case

$$E(p, q) = p^T A q = [.7 \quad .3]\begin{bmatrix} -1 & 1 \\ 2 & -4 \end{bmatrix}\begin{bmatrix} 1 \\ 0 \end{bmatrix} = -0.1. \quad \blacksquare$$

We will see in the next section, using our knowledge of linear programming, that for any game with a $m \times n$ payoff matrix $A$, there exists an optimal row-strategy $\bar{p}$ and a value $v_r$, such that

$$E(\bar{p}, q) = \bar{p}^T A q \geq v_r$$

for any $q$. Thus $\bar{p}$ is optimal in the sense that by selecting the mixed strategy $\bar{p}$, row can assure herself a profit of at least $v_r$, if $v_r \geq 0$ (or a loss of at most $v_r$ if $v_r \leq 0$), no matter what strategy column selects.

Also we shall see in the next section that there exists an optimal column strategy $\bar{q}$ and a value $v_c$, such that

$$E(p, \bar{q}) = p^T A \bar{q} \leq v_c$$

for any $p$. Thus $\bar{q}$ is optimal in the sense that by selecting the mixed strategy $\bar{q}$, the column player can assure himself that the row player will obtain a profit of at most $v_c$, if $v_c \geq 0$ (or a loss of at least $v_c$ if $v_c \leq 0$), no matter what strategy the row player selects.

The remarkable fact is that $v_r = v_c$, as will be seen in the next section. This common value, denoted by $v$, is called the **value** of the game. This is all summarized in the following.

---

**Fundamental Theorem of Game Theory**

Given any game with a $m \times n$ payoff matrix $A$, there exist strategies, $\bar{p}$ and $\bar{q}$, for row and column respectively, such that

1. $$E(\bar{p}, q) = \bar{p}^T A q \geq v$$

 for all $q$ and

2. $$E(p, \bar{q}) = p^T A \bar{q} \leq v$$

 for all $p$.

---

Since $E(\bar{p}, q) = \bar{p}^T A q \geq v$ for any $q$, then

$$E(\bar{p}, \bar{q}) = \bar{p}^T A \bar{q} \geq v. \tag{1}$$

Also since $E(p, \bar{q}) = p^T A \bar{q} \leq v$ for any $p$, then

$$E(\bar{p}, \bar{q}) = \bar{p}^T A \bar{q} \leq v. \qquad (2)$$

Equations (1) and (2) then imply that

$$E(\bar{p}, \bar{q}) = \bar{p}^T A \bar{q} = v. \qquad (3)$$

**Thus if both players use their optimal strategies, the expected outcome is $v$.** Furthermore, from parts 1 and 2 of the fundamental theorem, this is the best each player can hope to do. This is why we call $v$ the value of the game.

## A Geometric Method

We shall now give a geometric method of finding optimal strategies and the value of the game. This technique works readily if the payoff matrix has at most 2 rows, or at most 2 columns, or if it can be reduced to such a case. The more general situation will be solved in the next section using linear programming. The advantage of considering the geometric method given here is that it gives considerable insight into an important result that will be needed in the next section. (This result says that adding the value $k$ to every entry in the payoff matrix $A$ results in a new game that has the *same* optimal strategies as the original game, with the value of the new game $v + k$ where $v$ is the value of the original game.)

E X A M P L E  2  **Finding Optimal Strategies**

Find the optimal strategies and the value of the game given in Example 1.

**Solution**  We will find Row's optimal strategy and reserve finding Column's for a Self-Help Exercise. Since $p_1 + p_2 = 1$, $p_2 = 1 - p_1$. Thus any mixed strategy $[p_1 \quad p_2]$ can be written as $[p_1 \quad 1 - p_1]$. Suppose then that Row chooses a particular mixed strategy $[p_1 \quad 1 - p_1]$. Let us find the expected value that will result for each of the possible 2 selections that Column can make.

If Column selects strategy 1, that is, $q^T = [1 \quad 0]$, then

$$E(p, q) = [p_1 \quad 1 - p_1] \begin{bmatrix} -1 & 1 \\ 2 & -4 \end{bmatrix} \begin{bmatrix} 1 \\ 0 \end{bmatrix} = 2 - 3p_1.$$

If Column selects strategy 2, that is, $q^T = [0 \quad 1]$, then

$$E(p, q) = [p_1 \quad 1 - p_1] \begin{bmatrix} -1 & 1 \\ 2 & -4 \end{bmatrix} \begin{bmatrix} 0 \\ 1 \end{bmatrix} = -4 + 5p_1.$$

Both of these are straight lines, as shown in Figure 10.3.

Column will naturally always select the strategy that yields the smallest expected return for Row. This is shown as the two line segments in color in Figure 10.3. Row will naturally maximize her expected return and choose the point $p_1$ that yields the highest value on the line segments marked in color. This

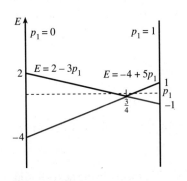

**Figure 10.3**

can be determined as the intersection of the two lines. Setting

$$2 - 3p_1 = -4 + 5p_1$$

gives $p_1 = \frac{3}{4}$. Thus the optimal strategy for Row is $p^T = [.75 \quad .25]$. The value can be computed using either one of the lines. Using the first gives

$$v = 2 - 3(.75) = -.25. \quad \blacksquare$$

**E X A M P L E  3**    **Using the Geometric Technique to Find Optimal Strategies**

Use the geometric technique to find the value of the Odds and Evens game and the optimal strategies for Column.

**Solution**    Recall that the payoff matrix is given by

$$\begin{bmatrix} -1 & +1 \\ +1 & -1 \end{bmatrix}.$$

Since $q_1 + q_2 = 1$, $q_2 = 1 - q_1$. Thus any mixed strategy $[q_1 \quad q_2]$ can be written as $[q_1 \quad 1 - q_1]$. Suppose then that Column chooses a particular mixed strategy $[q_1 \quad 1 - q_1]$. Let us find the expected value that will result for each of the possible 2 selections that Row can make.

If Row selects strategy 1, that is, $p^T = [1 \quad 0]$, then

$$E(p, q) = [1 \quad 0] \begin{bmatrix} -1 & +1 \\ +1 & -1 \end{bmatrix} \begin{bmatrix} q_1 \\ 1 - q_1 \end{bmatrix} = 1 - 2q_1.$$

If Row selects strategy 2, that is, $p^T = [0 \quad 1]$, then

$$E(p, q) = [0 \quad 1] \begin{bmatrix} -1 & +1 \\ +1 & -1 \end{bmatrix} \begin{bmatrix} q_1 \\ 1 - q_1 \end{bmatrix} = -1 + 2q_1.$$

These lines are shown in Figure 10.4.

Since Row will select the strategy that maximizes her return, she will select the strategy shown as the colored line segments. Column will then minimize Row's expected return by choosing the point $q_1$ that gives the lowest value on the colored line segments in Figure 10.4. The value of $q_1$ can be found as the intersection of the two lines. Thus setting

$$1 - 2q_1 = -1 + 2q_1$$

yields $q_1 = .50$ and a value of

$$v = 1 - 2q_1 = 1 - 2(.50) = 0.$$

Thus the optimal strategy for Column is $[.50 \quad .50]$, and the value of the game is 0.  $\blacksquare$

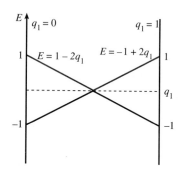

**Figure 10.4**

We will establish the fundamental theorem of game theory in the next section, using linear programming. The proof requires that all entries in the payoff matrix be positive. Consider now a game with a payoff matrix with some nonpositive elements. One can always select a number $k$ so that adding this number to every entry in the payoff matrix will result in a new payoff matrix that has all entries positive. The following theorem indicates that the game with the new payoff matrix has the same optimal strategies and a value that differs by $k$.

---

**Translated Games**

Consider a game with the $m \times n$ payoff matrix $A$ and value $v$. Let $E$ be the $m \times n$ matrix with every entry equal to 1 and let $k$ be any constant. Then the game with payoff matrix $A + kE$ has value $v + k$ and the same optimal strategies as the game with payoff matrix $A$.

---

The proof of this theorem is clear from the geometric methods used in this section. Adding $k$ to every entry in the matrix $A$ will merely move all the line segments in the previous discussion up by $k$ units if $k$ is positive and down by $k$ units if $k$ is negative. Thus it is apparent that the optimal strategies will remain the same and the value will be changed by $k$.

## Domination

There are certain rows (columns) for some payoff matrices that the row (column) player should never choose. These rows and column can then be deleted from consideration and the payoff matrix then reduced to a lower order.

Consider the game with the following payoff matrix.

$$\begin{bmatrix} 2 & -1 & 3 \\ 1 & -2 & 1 \\ -1 & 1 & 2 \end{bmatrix}$$

Let us compare the first and second rows. No matter what choice Column makes, Row will do better by selecting the first row over the second row. We say from Row's point of view that the first row **dominates** the second row. The second row can be eliminated obtaining

$$\begin{bmatrix} 2 & -1 & 3 \\ -1 & 1 & 2 \end{bmatrix}.$$

Now from Column's point of view, the first or second column is always a better choice than the third column. Thus the third column can be eliminated, leaving the game

$$\begin{bmatrix} 2 & -1 \\ -1 & 1 \end{bmatrix}.$$

We have the following definition.

---

**Domination**

Row $i$ is said to **dominate** row $j$ if every element in row $i$ is greater than or equal to the corresponding element in row $j$. Also column $i$ is said to dominate column $j$ if every element in column $i$ is less than or equal to the corresponding element in column $j$.

---

**E X A M P L E  4**    **Finding the Value of a Game and the Optimal Strategy**

Find the value and the optimal strategy for Row for the game with the payoff matrix

$$B = \begin{bmatrix} -2 & -1 & 0 \\ -1 & 0 & 1 \\ 2 & 3 & -4 \end{bmatrix}.$$

**Solution**    The second row dominates the first row since every element in the second row is greater than or equal to the corresponding element in the first row. Thus the first row can be eliminated to obtain

$$\begin{bmatrix} -1 & 0 & 1 \\ 2 & 3 & -4 \end{bmatrix}.$$

Now the first column dominates the second column since every element in the first column is smaller or equal to every corresponding element in the second column. Thus the second column can be eliminated to obtain the payoff matrix

$$A = \begin{bmatrix} -1 & 1 \\ 2 & -4 \end{bmatrix}.$$

We recognize this payoff matrix as the same one in as in Example 2. Since the value of the game with payoff matrix $A$ is $-.25$, the value of the game with payoff matrix $B$ is also $-.25$. Recall that the optimal strategy for Row in the game with payoff matrix $A$ is $\bar{p}^T = [.75 \quad .25]$. Thus, since the first row is never chosen, the optimal strategy for Row in the game with payoff matrix $B$ is $\bar{p}^T = [0 \quad .75 \quad .25]$. ∎

## SELF-HELP EXERCISE SET 10.2

1. For the game of Odds and Evens, determine the optimal strategy for Row using the geometric methods presented in this section.

2. For the game in Example 2, determine the optimal strategy for Column using the geometric methods presented in this section.

3. Given the game with the payoff matrix

$$\begin{bmatrix} 0 & -1 & 1 & 1 \\ 1 & 0 & -1 & -2 \\ -1 & 0 & -2 & -2 \end{bmatrix},$$

reduce it to the smallest order possible by eliminating any row or column that is dominated.

## EXERCISE SET 10.2

In Exercises 1 through 4 find the expected return for the given mixed strategies, using the game with the indicated payoff matrix

$$\begin{bmatrix} 1 & -1 \\ -2 & 3 \end{bmatrix}.$$

1. $p^T = [.50 \quad .50]$, $q^T = [.25 \quad .75]$

2. $p^T = [.50 \quad .50]$, $q^T = [.50 \quad .50]$

3. $p^T = [.30 \quad .70]$, $q^T = [.80 \quad .20]$

4. $p^T = [.60 \quad .40]$, $q^T = [.30 \quad .70]$

In Exercises 5 through 8 find the expected return for the given mixed strategies, using the game with the indicated payoff matrix

$$\begin{bmatrix} 2 & -1 & 1 \\ -1 & 3 & 4 \\ 2 & -1 & 3 \end{bmatrix}$$

5. $p^T = [.50 \quad .50 \quad 0]$, $q^T = [.20 \quad 0 \quad .80]$

6. $p^T = [.10 \quad .20 \quad .70]$, $q^T = [0 \quad .50 \quad .50]$

7. $p^T = [.30 \quad .30 \quad .40]$, $q^T = [.40 \quad .20 \quad .40]$

8. $p^T = [.30 \quad .60 \quad .10]$, $q^T = [0 \quad .70 \quad .30]$

Exercises 9 through 18 are payoff matrices. Find the value and the optimal strategies using the geometric methods of this section. Use dominance to eliminate any rows or columns.

9. $\begin{bmatrix} 1 & -2 \\ -1 & 2 \end{bmatrix}$

10. $\begin{bmatrix} 0 & -1 \\ -2 & 1 \end{bmatrix}$

11. $\begin{bmatrix} 0 & 2 \\ 1 & -1 \end{bmatrix}$

12. $\begin{bmatrix} -1 & 0 \\ 1 & -2 \end{bmatrix}$

13. $\begin{bmatrix} 1 & 1 & 0 \\ -1 & 0 & 2 \end{bmatrix}$

14. $\begin{bmatrix} 1 & 1 & -1 \\ -1 & -2 & 1 \end{bmatrix}$

**15.** $\begin{bmatrix} -1 & 0 \\ 2 & -1 \\ 2 & -1 \end{bmatrix}$     **16.** $\begin{bmatrix} -2 & 0 \\ 1 & 0 \\ 1 & -1 \end{bmatrix}$

**17.** $\begin{bmatrix} 2 & 0 & 0 \\ 2 & 1 & 0 \\ 0 & -1 & 2 \end{bmatrix}$     **18.** $\begin{bmatrix} 0 & -1 & 1 \\ 4 & 3 & -2 \\ 3 & 3 & 0 \end{bmatrix}$

The geometric theory given in the text can be used to find the optimal strategy for Row for any game with a payoff matrix with two rows. Find the optimal strategy for Row in Exercises 19 through 20.

**19.** $\begin{bmatrix} 3 & 0 & -1 \\ -5 & -4 & 2 \end{bmatrix}$     **20.** $\begin{bmatrix} 3 & 2 & -1 \\ -2 & 0 & 1 \end{bmatrix}$

The geometric theory given in the text can be used to find the optimal strategy for Column in any game with a payoff matrix with two columns. Find the optimal strategy for Column in Exercises 21 through 22.

**21.** $\begin{bmatrix} 4 & -2 \\ 1 & 2 \\ 3 & 1 \end{bmatrix}$     **22.** $\begin{bmatrix} 3 & -5 \\ -2 & 0 \\ -1 & -3 \end{bmatrix}$

## Applications

*(Dennis Drenner.)*

*(Dennis Drenner.)*

**23. Setting Prices.** Every August and January two carpet retailers in town run their sales. They must decide whether to have a 20% or 30% sale. From many previous years of experience with these sales, they know that if both set prices at 20% off the first retailer will get 80% of the business, and if both set prices at 30% off the first retailer will get 70% of the business. If the first retailer takes 20% off and the second 30% off, then the first retailer gets 30% of the business, while if the first takes 30% off and the second 20% off, the first retailer gets 60% of the business. What should their strategies be and what is the value of this game?

**24. Price Wars.** Every day two movie theaters in town must decide on the price to set on their movies for the next day so the advertisement can go into the morning paper for the next day. They know from long experience that if they both set their prices at $5 or both at $6, then the first theater will get 70% of the business. If the first theater sets prices at $5 and the second at $6, then the first theater gets 60% of the business, while if the first theater charges $6 and the second $5, then the first theater gets 40% of the business. What should their strategies be and what is the value of this game?

**25. Advertising.** Two competing electronics stores must decide each week whether to advertise in one and only one of the three media: TV, radio, or newspaper. From past experience they know that the payoff matrix in terms of percent of the market gained or lost to the other is

|        | **TV** | **Radio** | **Paper** |
|--------|--------|-----------|-----------|
| **TV**    | 1  | 1  | 0  |
| **Radio** | −1 | 0  | −1 |
| **Paper** | 2  | −1 | 2  |

Find the value of this game and the optimal strategies.

26. **Marketing.** Early in each year two local growers compete for the freshly cut gladiola market, committing themselves entirely to one and only one color of gladiola. Past experience indicates the following payoff matrix given in terms of the percentage of the market the first grower gains or loses to the second.

|        | **Pink** | **Yellow** | **Red** |
|--------|----------|------------|---------|
| **Pink**   | 20  | 10  | −20 |
| **Orange** | −10 | −20 | −10 |
| **Red**    | 30  | 20  | −20 |

Find the value of this game and the optimal strategies.

## Solutions to Self-Help Exercise Set 10.2

1. As usual any mixed strategy $[p_1 \quad p_2]$ can be written as $[p_1 \quad 1 - p_1]$. Suppose then that Row chooses a particular mixed strategy $[p_1 \quad 1 - p_1]$. Let us find the expected value that will result for each of the possible 2 selections that Column can make.

   If Column selects strategy 1, that is, $q^T = [1 \quad 0]$, then

$$E(p, q) = [p_1 \quad 1 - p_1]\begin{bmatrix} -1 & +1 \\ +1 & -1 \end{bmatrix}\begin{bmatrix} 1 \\ 0 \end{bmatrix} = 1 - 2p_1.$$

   If Column selects strategy 2, that is, $q^T = [0 \quad 1]$, then

$$E(p, q) = [p_1 \quad 1 - p_1]\begin{bmatrix} -1 & +1 \\ +1 & -1 \end{bmatrix}\begin{bmatrix} 1 \\ 0 \end{bmatrix} = -1 + 2p_1.$$

   Both of these are straight lines and are shown in the figure.

   Column will always select the strategy that yields the smallest expected return for Row. This is shown in color in the figure. Row will naturally maximize her expected return and choose the point $p_1$ that yields the highest value on the line segments marked in color. This can be determined as the intersection of the two lines. Setting

$$1 - 2p_1 = -1 + 2p_1$$

   gives $p_1 = .50$. Thus the optimal strategy for Row is $p = [.50 \quad .50]$.

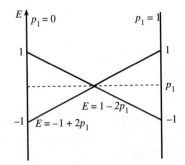

2. As usual, the mixed strategy $[q_1 \quad q_2]$ can be written as $[q_1 \quad 1 - q_1]$. Suppose then that Column chooses a particular mixed strategy $[q_1 \quad 1 - q_1]$. Let us find

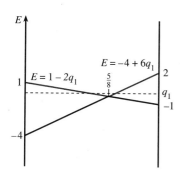

the expected value that will result for each of the possible 2 selections that Row can make.

If Row selects strategy 1, that is, $p^T = [1 \quad 0]$, then

$$E(p, q) = [1 \quad 0] \begin{bmatrix} -1 & +1 \\ +2 & -4 \end{bmatrix} \begin{bmatrix} q_1 \\ 1 - q_1 \end{bmatrix} = 1 - 2q_1.$$

If Row selects strategy 2, that is, $p^T = [0 \quad 1]$, then

$$E(p, q) = [0 \quad 1] \begin{bmatrix} -1 & +1 \\ +2 & -4 \end{bmatrix} \begin{bmatrix} q_1 \\ 1 - q_1 \end{bmatrix} = -4 + 6q_1.$$

These lines are shown in the figure.

Since Row will select the strategy that maximizes her return, she will select the strategy shown as the colored line segments. Column will then minimize Row's expected return by choosing the point $q_1$ that gives the lowest value on the colored line segments in the figure. The value of $q_1$ can be found as the intersection of the two lines. Thus setting

$$1 - 2q_1 = -4 + 6q_1$$

yields $q_1 = \frac{5}{8}$.

Thus the optimal strategy for Column is $[\frac{5}{8} \quad \frac{3}{8}]$.

3. The second row dominates the third row. Removing the third row gives

$$\begin{bmatrix} 0 & -1 & 1 & 1 \\ 1 & 0 & -1 & -2 \end{bmatrix}.$$

Now notice that the second column dominates the first column. Thus the first column can be eliminated. Also notice that the fourth column dominates the third column. Thus the third column can be eliminated. This leaves

$$\begin{bmatrix} -1 & 1 \\ 0 & -2 \end{bmatrix}$$

# 10.3   LINEAR PROGRAMMING AND GAME THEORY

▶ *Matrices with Positive Entries*

▶ *Linear Programming and Game Theory*

In the last section we saw how to find the optimal strategies and the value of nonstrictly determined games using mixed strategies. A geometric method was given to find the optimal strategy for Row in a game with a payoff matrix with at most 2 rows, or the optimal strategy for Column for any game with a payoff matrix with at most 2 columns. We need to use linear programming, however, to solve games with payoff matrices of larger order. The theory of linear programming is also used to establish the basic theorems of game theory.

## Matrices with Positive Entries

To solve a game using the techniques of linear programming presented here, we will see that the payoff matrix must have all entries that are positive. This turns out not to be a restriction. To see why, consider the game given in the last section with payoff matrix

$$A = \begin{bmatrix} -1 & 1 \\ 2 & -4 \end{bmatrix}.$$

We know from the last section that adding, for example 5, to every entry in the payoff matrix gives the payoff matrix

$$B = \begin{bmatrix} 4 & 6 \\ 7 & 1 \end{bmatrix}$$

which will have the very same optimal strategies as $A$ and value equal to the value of $A$ plus 5.

In the same way we can always add some positive constant $k$ to all the entries of any payoff matrix $A$ with value $v_A$ that have some nonnegative entries and obtain a new payoff matrix $B$ with the same optimal strategies and value $v_B = v_A + k$. If we then solve the game with payoff matrix $B$, the very same optimal strategies for this game will also be optimal for the original game and the value of the original game will be the value of the game with payoff matrix $B$ *minus k*. This is, $v_A = v_B - k$.

## Linear Programming and Game Theory

We now consider a game with payoff matrix

$$A = \begin{bmatrix} a_{11} & a_{12} \\ a_{21} & a_{22} \end{bmatrix}$$

for which all the entries are positive. We will show how to use linear programming to solve this game.

Consider first the point of view of Row. Row is interested in finding a strategy $\bar{p}$ and the largest possible number $v_R$ such that

$$E(\bar{p}, q) = \bar{p}^T A q \geq v_R \tag{1}$$

for any choice of mixed strategy $q$.

Since (1) must hold for all mixed strategies $q$, it certainly must hold for the two particular strategies $q_1^T = [1 \quad 0]$ and $q_2^T = [0 \quad 1]$. Replacing $q_1$ and $q_2$ successively with $q$ in equation (1) then yields

$$[p_1 \quad p_2] \begin{bmatrix} a_{11} & a_{12} \\ a_{21} & a_{22} \end{bmatrix} \begin{bmatrix} 1 \\ 0 \end{bmatrix} \geq v_R, \qquad [p_1 \quad p_2] \begin{bmatrix} a_{11} & a_{12} \\ a_{21} & a_{22} \end{bmatrix} \begin{bmatrix} 0 \\ 1 \end{bmatrix} \geq v_R$$

or

$$a_{11}p_1 + a_{21}p_2 \geq v_R$$

$$a_{12}p_1 + a_{22}p_2 \geq v_R.$$

Although it is tempting to use the substitution $p_2 = 1 - p_1$, we take a more useful tactic and divide each inequality by $v_R$. Since all entries in the matrix $A$ are positive, $v_R > 0$. We then have

$$a_{11}\frac{p_1}{v_R} + a_{21}\frac{p_2}{v_R} \geq 1 \tag{2}$$

$$a_{12}\frac{p_1}{v_R} + a_{22}\frac{p_2}{v_R} \geq 1.$$

It is more convenient to let

$$y_1 = \frac{p_1}{v_R}, \qquad \text{and} \qquad y_2 = \frac{p_2}{v_R}.$$

Notice that $y_1 \geq 0$ and $y_2 \geq 0$. Equation (2) then becomes

$$a_{11}y_1 + a_{21}y_2 \geq 1$$

$$a_{12}y_1 + a_{22}y_2 \geq 1 \tag{3}$$

$$y_1, y_2 \geq 0.$$

Row then wishes to maximize $v_R$ subject to the constraints given in equation (3). But for a linear programming problem, the function to be maximized or minimized must be a linear function of $y_1$ and $y_2$. To see how to obtain this function, notice that

$$y_1 + y_2 = \frac{p_1}{v_R} + \frac{p_2}{v_R} = \frac{p_1 + (1 - p_1)}{v_R} = \frac{1}{v_R}.$$

We will then maximize $v_R$ by minimizing $1/v_R = y_1 + y_2$. We then have the following linear programming problem:

$$\text{Minimize} \qquad w = \frac{1}{v_R} = y_1 + y_2$$

$$\text{Subject to} \qquad a_{11}y_1 + a_{21}y_2 \geq 1 \tag{4}$$

$$a_{12}y_1 + a_{22}y_2 \geq 1$$

$$y_1, y_2 \geq 0.$$

Now consider the point of view of the column player. He wishes to find a strategy $\overline{q}$ and the smallest number $v_C$ such that

$$E(p, \overline{q}) = p^T A \overline{q} \leq v_C \tag{5}$$

for all strategies $p$.

Since equation (5) must hold for all $p$, it must hold for the two particular values of $p$ given by $p_1^T = [1 \quad 0]$ and $p_2^T = [0 \quad 1]$. Substituting these two values for $p$ in equation (5) then yields

$$a_{11}q_1 + a_{12}q_2 \leq v_C \tag{6}$$

$$a_{21}q_1 + a_{22}q_2 \leq v_C.$$

In a similar fashion as before, we divide both inequalities in equation (6) by $v_C$ and set

$$x_1 = \frac{q_1}{v_C} \quad \text{and} \quad x_2 = \frac{q_2}{v_C}.$$

This gives

$$a_{11}x_1 + a_{12}x_2 \leq 1$$

$$a_{21}x_1 + a_{22}x_2 \leq 1 \tag{7}$$

$$x_1, x_2 \geq 0.$$

Now notice that

$$x_1 + x_2 = \frac{q_1}{v_C} + \frac{q_2}{v_C} = \frac{q_1 + (1 - q_1)}{v_C} = \frac{1}{v_C}.$$

We then minimize $v_C$ by maximizing $1/v_C$. We then have the following linear programming problem

$$\text{Maximize} \quad z = \frac{1}{v_C} = x_1 + x_2$$

$$\text{Subject to} \quad a_{11}x_1 + a_{12}x_2 \leq 1 \tag{8}$$

$$a_{21}x_1 + a_{22}x_2 \leq 1$$

$$x_1, x_2 \geq 0.$$

We notice that this last linear programming problem is a standard linear programming problem and that the linear programming problems in (4) and (8) are the duals of each other. From our knowledge of the duals, we know that the maximum $v_C$ found in (8) must equal the minimum $v_R$ found in (4). This common value, denoted by $v$, is then the value of the game.

The general case of a $m \times n$ payoff matrix $A$ will follow in a similar manner. We then have the following.

---

**Optimal Strategy for Column**

Let a game have the $m \times n$ payoff matrix

$$A = \begin{bmatrix} a_{11} & a_{12} & \cdots & a_{1n} \\ a_{21} & a_{22} & \cdots & a_{2n} \\ & & \cdots & \\ a_{m1} & a_{m2} & \cdots & a_{mn} \end{bmatrix}.$$

Let $x_1, x_2, \ldots, x_n$, be the solution to the standard linear programming problem

Maximize $\qquad\qquad z = x_1 + x_2 + \cdots + x_n$

Subject to $\qquad a_{11}x_1 + a_{12}x_2 + \cdots + a_{1n}x_n \leq 1$
$$a_{21}x_1 + a_{22}x_2 + \cdots + a_{2n}x_n \leq 1$$
$$\vdots$$
$$a_{m1}x_1 + a_{m2}x_2 + \cdots + a_{mn}x_n \leq 1$$
$$x_1, x_2, \ldots, x_n \geq 0.$$

Then the value $v$ of the game is the reciprocal of this maximum, and the optimal strategy for Column is

$$\bar{q} = \begin{bmatrix} q_1 \\ q_2 \\ \vdots \\ q_n \end{bmatrix} = v \begin{bmatrix} x_1 \\ x_2 \\ \vdots \\ x_n \end{bmatrix} = \begin{bmatrix} vx_1 \\ vx_2 \\ \vdots \\ vx_n \end{bmatrix}.$$

**Optimal Strategy for Row**

Let a game have the $m \times n$ payoff matrix

$$A = \begin{bmatrix} a_{11} & a_{12} & \cdots & a_{1n} \\ a_{21} & a_{22} & \cdots & a_{2n} \\ & & \cdots & \\ a_{m1} & a_{m2} & \cdots & a_{mn} \end{bmatrix}.$$

Let $y_1, y_2, \ldots, y_m$, be the solution to the linear programming problem

Minimize $\qquad\qquad w = y_1 + y_2 + \cdots + y_m$

Subject to $\qquad a_{11}y_1 + a_{21}y_2 + \cdots + a_{m1}y_m \geq 1$
$$a_{12}y_1 + a_{22}y_2 + \cdots + a_{m2}y_n \geq 1$$
$$\vdots$$
$$a_{1n}y_1 + a_{2n}y_2 + \cdots + a_{mn}y_m \geq 1$$
$$y_1, y_2, \ldots, y_m \geq 0.$$

Then the value $v$ of the game is the reciprocal of this minimum, and the optimal strategy for Row is

$$\overline{p} = \begin{bmatrix} p_1 \\ p_2 \\ \vdots \\ p_m \end{bmatrix} = v \begin{bmatrix} y_1 \\ y_2 \\ \vdots \\ y_m \end{bmatrix} = \begin{bmatrix} vy_1 \\ vy_2 \\ \vdots \\ vy_m \end{bmatrix}.$$

The minimum value of $w$ is the same as the maximum value of $z$ found in solving the linear programming problem for Column. In fact the two linear programming problems are the duals of each other. If Row and Column play their optimal strategies $\overline{p}$ and $\overline{q}$, respectively, then the expected return for the game is

$$E(\overline{p}, \overline{q}) = v.$$

For any $p$, $E(p, \overline{q}) \le v$. While for any $q$, $E(\overline{p}, q) \ge v$.

**E X A M P L E   1**   **Using Linear Programming in Game Theory**

Use linear programming to find the value of the game and the optimal strategy for Row in the game with payoff matrix

$$A = \begin{bmatrix} -1 & 1 \\ 2 & -4 \end{bmatrix}.$$

**Solution**   Since some entries are nonnegative, the first step is to add 5 to every entry in the matrix to obtain the payoff matrix

$$B = \begin{bmatrix} 4 & 6 \\ 7 & 1 \end{bmatrix}.$$

Row then must solve the following linear programming problem:

Minimize    $w = \dfrac{1}{v_R} = y_1 + y_2$

Subject to    $4y_1 + 7y_2 \ge 1$
$6y_1 + y_2 \ge 1$
$y_1, y_2 \ge 0.$

Since the number of constraints and the number of variables is only 2, we will solve by using the geometric method of linear programming rather than the simplex method.

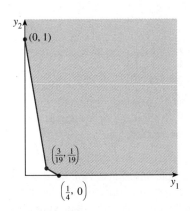

Figure 10.5 shows the admissible region and the 3 corners $(0, 1)$, $(\frac{3}{19}, \frac{1}{19})$, $(\frac{1}{4}, 0)$. The function $w = y_1 + y_2$ takes the values of 1, $\frac{4}{19}$, and $\frac{1}{4}$, respectively, at these corners. Thus the minimum of $w$ is $\frac{4}{19}$ at the point $(\frac{3}{19}, \frac{1}{19})$. The value of the game with payoff matrix $B$ is the reciprocal of this minimum, that is $v = \frac{19}{4}$.

**Figure 10.5**

The optimal strategy for Row for the game with the payoff matrix $B$ or $A$ is then

$$\bar{p} = v \begin{bmatrix} y_1 \\ y_2 \end{bmatrix} = \frac{19}{4} \begin{bmatrix} \frac{3}{19} \\ \frac{1}{19} \end{bmatrix} = \begin{bmatrix} \frac{3}{4} \\ \frac{1}{4} \end{bmatrix}.$$

The value of the original game with payoff matrix $A$ is

$$v_A = v - 5 = -\frac{1}{4}.$$

This all agrees with the solution to this problem found in the last section, using the method of that section.  ∎

**E X A M P L E  2**     **Using Linear Programming in Game Theory**

Use linear programming to find the value of the game and the optimal strategy for Column in the game in the previous example.

**Solution**     Using the matrix $B$ in Example 1, Column must solve the following linear programming problem:

$$\begin{array}{ll} \text{Maximize} & z = x_1 + x_2 \\ \text{Subject to} & 4x_1 + 6x_2 \le 1 \\ & 7x_1 + \phantom{6}x_2 \le 1 \\ & x_1, x_2 \ge 0. \end{array}$$

The admissible region is shown in Figure 10.6. The 4 corner points are $(0, 0)$, $(0, \frac{1}{6})$, $(\frac{5}{38}, \frac{3}{38})$, $(\frac{1}{7}, 0)$. The value of $z$ at these 4 corners is, respectively, $0$, $\frac{1}{6}$, $\frac{4}{19}$, and $\frac{1}{7}$. The largest of these is $\frac{4}{19}$, occurring at the corner point $(\frac{5}{38}, \frac{3}{38})$. The value of the game with payoff matrix $B$ is then the reciprocal of $\frac{4}{19}$, or $\frac{19}{4}$. Thus the optimal strategy for Column is

$$\bar{q} = v \begin{bmatrix} x_1 \\ x_2 \end{bmatrix} = \frac{19}{4} \begin{bmatrix} \frac{5}{38} \\ \frac{3}{38} \end{bmatrix} = \begin{bmatrix} \frac{5}{8} \\ \frac{3}{8} \end{bmatrix}.$$

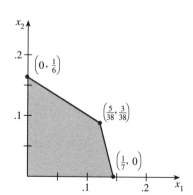

**Figure 10.6**

The value of the original game with payoff matrix $A$ is $v_A = v - 5 = -\frac{1}{4}$, which agrees with Example 1. This all agrees with the solution to this problem found in the last section, using the method of that section.  ∎

Using the simplex method has one considerable advantage. One need solve only one of the linear programming problems. Since the other one is the dual, the solution can be read off as the numbers in the last row under the slack variables of the final tableau.

**E X A M P L E   3**   **Using the Simplex Method in Game Theory**

Use the simplex method to find the value of the game and the optimal strategy for Column in the game with the payoff matrix in the previous example. Use the theory for the dual problem to find the optimal stragegy for Row by using the information in the final tableau.

**Solution**   The initial tableau for the game with the payoff matrix $B$ is

| $x_1$ | $x_2$ | $s_1$ | $s_2$ | $z$ | | |
|-------|-------|-------|-------|-----|---|---|
| 4 | 6 | 1 | 0 | 0 | 1 | $\frac{1}{4}$ |
| ⑦ | 1 | 0 | 1 | 0 | 1 | $\frac{1}{7}$ |
| $-1$ | $-1$ | 0 | 0 | 1 | 0 | |

Since there is a tie, we arbitrarily pick the first column as the pivot column. The fractions then indicate that the pivot row is the second row. Pivoting on the element in the second row and first column gives

| $x_1$ | $x_2$ | $s_1$ | $s_2$ | $z$ | | |
|-------|-------|-------|-------|-----|---|---|
| 0 | ㉞⁄₇ | 1 | $-\frac{4}{7}$ | 0 | $\frac{3}{7}$ | $\frac{3/7}{38/7} = \frac{3}{38}$ |
| 1 | $\frac{1}{7}$ | 0 | $\frac{1}{7}$ | 0 | $\frac{1}{7}$ | 1 |
| 0 | $-\frac{6}{7}$ | 0 | $\frac{1}{7}$ | 1 | $\frac{1}{7}$ | |

The pivot column is the second, and the fractions to the right of the tableau indicate that the pivot row is the first. Pivoting about the first row and second column gives the final tableau.

| $x_1$ | $x_2$ | $s_1$ | $s_2$ | $z$ | |
|-------|-------|-------|-------|-----|---|
| 0 | 1 | $\frac{7}{38}$ | $-\frac{2}{19}$ | 0 | $\frac{3}{38}$ |
| 1 | 0 | $-\frac{1}{38}$ | $\frac{3}{19}$ | 0 | $\frac{5}{38}$ |
| 0 | 0 | $\frac{3}{19}$ | $\frac{1}{19}$ | 1 | $\frac{4}{19}$ |

The solution is $(\frac{5}{38}, \frac{3}{38})$ with maximum $\frac{4}{19}$. This all agrees with the answer in the previous exercise. The remaining part of the problem then proceeds as in the previous problem.

Now reading off the numbers in the last row under the slack variables gives $(\frac{3}{19}, \frac{1}{19})$. This is the solution found in Exercise 1.   ■

## SELF-HELP EXERCISE SET 10.3

1. Consider the game with payoff matrix

$$A = \begin{bmatrix} 5 & -1 & 2 \\ -2 & 5 & 1 \end{bmatrix}.$$

   Find the value of the game and the optimal strategy for Row using linear programming (without using the simplex method).

2. For the game with the payoff matrix in the previous exercise, find the value of the game and the optimal strategy for Column using the simplex method. Using the final tableau, find the solution to the dual problem and check this with the solution to the linear programming problem solved in the previous exercise.

## EXERCISE SET 10.3

In Exercises 1 through 8 find the value and the optimal strategies for both Row and Column for the game with the indicated payoff matrix. Use geometric linear programming.

1. $\begin{bmatrix} 1 & -2 \\ -3 & 2 \end{bmatrix}$
   
2. $\begin{bmatrix} 0 & -1 \\ -2 & 2 \end{bmatrix}$

3. $\begin{bmatrix} 0 & 3 \\ 1 & -1 \end{bmatrix}$

4. $\begin{bmatrix} -1 & 0 \\ 1 & -3 \end{bmatrix}$

5. $\begin{bmatrix} 1 & -1 \\ -1 & 0 \end{bmatrix}$

6. $\begin{bmatrix} 1 & 2 \\ 3 & -2 \end{bmatrix}$

7. $\begin{bmatrix} -1 & 0 \\ 2 & -1 \end{bmatrix}$

8. $\begin{bmatrix} -2 & 4 \\ 1 & 0 \end{bmatrix}$

In Exercises 9 through 10 find the value and the optimal strategy for Row for the game with the indicated payoff matrix. Use geometric linear programming.

9. $\begin{bmatrix} 3 & 0 & -1 \\ -5 & -4 & 2 \end{bmatrix}$

10. $\begin{bmatrix} 3 & 2 & -1 \\ -2 & 0 & 1 \end{bmatrix}$

In Exercises 11 through 12 find the value and the optimal strategy for Column for the game with the indicated payoff matrix. Use geometric linear programming.

11. $\begin{bmatrix} 4 & -2 \\ 1 & 2 \\ 3 & 1 \end{bmatrix}$

12. $\begin{bmatrix} 3 & -5 \\ -2 & 0 \\ -1 & -3 \end{bmatrix}$

In Exercises 13 through 16 use the simplex method and dual theory in the indicated problems to find the optimal strategies for both row and column.

**13.** Exercise 1 above.                    **14.** Exercise 2 above.

**15.** Exercise 3 above.                    **16.** Exercise 4 above.

**17.** Find the optimal strategy for Column by using the simplex method in Exercise 9. Use the final tableau and dual theory to find the optimal strategy for Row. Compare with Exercise 9.

**18.** Find the optimal strategy for Column by using the simplex method in Exercise 10. Use the final tableau and dual theory to find the optimal strategy for Row. Compare with Exercise 10.

**19.** Find the optimal strategy for Row by using the simplex method in Exercise 11. Use the final tableau and dual theory to find the optimal strategy for Column. Compare with Exercise 11.

**20.** Find the optimal strategy for Row by using the simplex method in Exercise 12. Use the final tableau and dual theory to find the optimal strategy for Column. Compare with Exercise 12.

**21.** Solve the Odds and Evens game given in the last section by linear programming.

**22.** In the stone, paper, scissors game, two players each simultaneously say one of the words: stone, paper, scissors. The game is a draw if both players say the same word. A player will win $1 from the other player according to the following rules: scissors defeats (cuts) paper, paper defeats (covers) stone, stone defeats (breaks) scissors. Determine the payoff matrix and find the value and optimal strategies for both players.

## Applications

Solve Exercises 23 through 26 using linear programming. Note that these are Exercises 23 through 26 of the last section.

**23. Setting Prices.** Every August and January two carpet retailers in town run their sales. They must decide whether to have a 20% or 30% sale. From many previous years of experience with these sales, they know that if they both set prices at 20% off, the first retailer will get 80% of the business and if they both set prices at 30% off, the first retailer will get 70% of the business. If the first retailer takes 20% off and the second 30% off, then the first retailer gets 30% of the business, while if the first takes 30% off and the second 20% off, the first retailer gets 60% of the business. What should their strategies be, and what is the value of this game?

**24. Price Wars.** Every day two movie theaters in town must decide on the price to set on their movies for the next day so the advertisement can go into the morning paper for the next day. They know from long experience that if both set their prices at $5 or both at $6, then the first theater will get 70% of the business. If the first theater sets prices at $5 and the second at $6, then the first theater gets 60% of the business, while if the first theater charges $6 and the second $5, then the first theater gets 40% of the business. What should their strategies be and what is the value of this game?

25. **Advertising.**   Two competing electronics stores must decide each week whether to advertise in one and only one of the three media: TV, radio, or newspaper. From past experience they know that the payoff matrix in terms of percent of the market gained or lost to the other is

$$
\begin{array}{cc}
 & \begin{array}{ccc} \textbf{TV} & \textbf{Radio} & \textbf{Paper} \end{array} \\
\begin{array}{c} \textbf{TV} \\ \textbf{Radio} \\ \textbf{Paper} \end{array} &
\left[ \begin{array}{ccc}
1 & 1 & 0 \\
-1 & 0 & -1 \\
2 & -1 & 2
\end{array} \right].
\end{array}
$$

Find the value of this game and the optimal strategies.

26. **Marketing.**   Early in each year two local growers compete for the freshly cut gladiola market, committing themselves entirely to one and only one color of gladiola. Past experience indicates the following payoff matrix given in terms of the percentage of the market the first grower gains or loses to the second.

$$
\begin{array}{cc}
 & \begin{array}{ccc} \textbf{Pink} & \textbf{Yellow} & \textbf{Red} \end{array} \\
\begin{array}{c} \textbf{Pink} \\ \textbf{Orange} \\ \textbf{Red} \end{array} &
\left[ \begin{array}{ccc}
20 & 10 & -20 \\
-10 & -20 & -10 \\
30 & 20 & -20
\end{array} \right]
\end{array}
$$

Find the value of this game and the optimal strategies.

## Solutions to Self-Help Exercise Set 10.3

1. First add 3 to each entry in the payoff matrix $A$ and obtain the new payoff matrix

$$
B = \begin{bmatrix} 8 & 2 & 5 \\ 1 & 8 & 4 \end{bmatrix}.
$$

Row must then solve the following linear programming problem:

$$
\begin{aligned}
\text{Minimize} \quad & w = y_1 + y_2 \\
\text{Subject to} \quad & 8y_1 + y_2 \geq 1 \\
& 2y_1 + 8y_2 \geq 1 \\
& 5y_1 + 4y_2 \geq 1 \\
& y_1, y_2 \geq 0.
\end{aligned}
$$

The feasible region is shown in the figure. The 4 corners are $(0, 1)$, $(\frac{1}{9}, \frac{1}{9})$, $(\frac{1}{8}, \frac{3}{32})$, $(\frac{1}{2}, 0)$. The values of $w = y_1 + y_2$ at these 4 corners are, respectively, $1$, $\frac{2}{9} = .22222$, $\frac{7}{32} = .21875$, $\frac{1}{2}$. The minimum is then $\frac{7}{32}$ at the point $(\frac{1}{8}, \frac{3}{32})$. The value of the game with payoff matrix $B$ is the reciprocal of the minimum value of $w$ or $v_B = \frac{32}{7}$. The optimal strategy for Row is then

$$
\overline{p} = \frac{32}{7} \begin{bmatrix} \frac{1}{8} \\ \frac{3}{32} \end{bmatrix} = \begin{bmatrix} \frac{4}{7} \\ \frac{3}{7} \end{bmatrix}.
$$

Since we added 3 to each entry in $A$, the value of $A$ is

$$v_A = v_B - 3 = \frac{32}{7} - 3 = \frac{11}{7}.$$

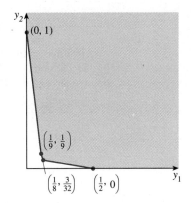

2.  Referring to the payoff matrix $B$ in the previous exercise, Column must first solve the linear programming problem:

$$\text{Maximize} \qquad\qquad z = x_1 + x_2$$
$$\text{Subject to} \qquad 8x_1 + 2x_2 + 5x_3 \le 1$$
$$x_1 + 8x_2 + 4x_3 \le 1$$
$$x_1, x_2 \ge 0.$$

The initial tableau is

| $x_1$ | $x_2$ | $x_3$ | $s_1$ | $s_2$ | $z$ | |
|---|---|---|---|---|---|---|
| 8 | 2 | ⑤ | 1 | 0 | 0 | 1 | $\frac{1}{5}$ |
| 1 | 8 | 4 | 0 | 1 | 0 | 1 | $\frac{1}{4}$ |
| $-1$ | $-1$ | $-1$ | 0 | 0 | 1 | 0 |

Arbitrarily pick the pivot column to be the third. Then the pivot row must be the first. Pivoting yields

| $x_1$ | $x_2$ | $x_3$ | $s_1$ | $s_2$ | $z$ | |
|---|---|---|---|---|---|---|
| $\frac{8}{5}$ | $\frac{2}{5}$ | 1 | $\frac{1}{5}$ | 0 | 0 | $\frac{1}{5}$ | $\frac{1}{2}$ |
| $-\frac{27}{5}$ | $\frac{32}{5}$ | 0 | $-\frac{4}{5}$ | 1 | 0 | $\frac{1}{5}$ | $\frac{1}{32}$ |
| $\frac{3}{5}$ | $-\frac{3}{5}$ | 0 | $\frac{1}{5}$ | 0 | 1 | $\frac{1}{5}$ |

Pivoting about the second row and second column yields the final tableau.

| $x_1$ | $x_2$ | $x_3$ | $s_1$ | $s_2$ | $z$ | |
|---|---|---|---|---|---|---|
| $\frac{31}{16}$ | 0 | 1 | $\frac{1}{4}$ | $-\frac{1}{16}$ | 0 | $\frac{3}{16}$ |
| $-\frac{27}{32}$ | 1 | 0 | $-\frac{1}{8}$ | $\frac{5}{32}$ | 0 | $\frac{1}{32}$ |
| $\frac{3}{32}$ | 0 | 0 | $\frac{1}{8}$ | $\frac{3}{32}$ | 1 | $\frac{7}{32}$ |

The maximum of $\frac{7}{32}$ occurs at the point $(0, \frac{1}{32}, \frac{3}{16})$. Then $v_B$ is the reciprocal of this maximum or $\frac{32}{7}$. This agrees with that found in the previous exercise. The optimal strategy for Column is then

$$\overline{q} = \frac{32}{7} \begin{bmatrix} 0 \\ \frac{1}{32} \\ \frac{3}{16} \end{bmatrix} = \begin{bmatrix} 0 \\ \frac{1}{7} \\ \frac{6}{7} \end{bmatrix}.$$

The solution to the dual problem can be found under the slack variables on the last line in the final tableau. This is $(\frac{1}{8}, \frac{3}{32})$. This agrees with the solution of the linear programming problem in the previous exercise, since this linear programming problem is the dual.

# SUMMARY OUTLINE OF CHAPTER 10

- **Two-Person Zero-Sum Game.** A game is called a **two-person zero-sum game** if the following are satisfied.
  1. There are two players (called the row player and the column player).
  2. The row player must choose one of $m$ strategies, and, simultaneously, the column player must choose one of $n$ strategies.
  3. If the row player chooses her $i$th strategy and the column player chooses his $j$th strategy, the row player receives a reward of $a_{ij}$ and the column player loses an amount $a_{ij}$.

- The matrix $A = (a_{ij})$ given in the prior item is called the **payoff matrix**.

- If there is an entry, $a_{hk}$, that is the minimum in its row (row h) and also the maximum in its column (column k), we call the entry $a_{hk}$ a **saddle point**.

- A game with a saddle point is said to be **strictly determined**.

- **Locating a Saddle Point (The m & M method)**
  1. For each row, find the minimum value and place a small ''m'' next to it.
  2. For each column, find the maximum value and place a capital ''M'' next to it.
  3. Any entry with both a small ''m'' and a capital ''M'' next to it is a saddle point.

- A **mixed strategy**

$$p = \begin{bmatrix} p_1 \\ p_2 \end{bmatrix}$$

is a strategy in which the first and second strategies are selected randomly with probability $p_1$ and $p_2$, respectively.

- Suppose a game has a payoff matrix $A$ and Row plays the strategy $p^T = [p_1 \quad p_2 \quad \cdots \quad p_m]$ and Column selects $q^T = [q_1 \quad q_2 \quad \cdots \quad q_n]$, then the **expected value** for these two strategies is $E(p, q) = p^T A q$.

- **Fundamental Theorem of Game Theory.**   Given any game with a $m \times n$ payoff matrix $A$, there exist strategies $\bar{p}$ and $\bar{q}$ for row and column, respectively, such that
$$E(\bar{p}, q) = \bar{p}^T A q \geq v \text{ for all } q \qquad \text{and} \qquad E(p, \bar{q}) = p^T A \bar{q} \leq v \text{ for all } p.$$

- **Translated Games.**   Consider a game with the $m \times n$ payoff matrix $A$ and value $v$. Let $E$ be the $m \times n$ matrix with every entry equal to 1 and let $k$ be any constant. Then the game with payoff matrix $A + kE$ has value $v + k$ and the same optimal strategies as the game with payoff matrix $A$.

- Row $i$ is said to **dominate** row $j$ if every element in row $i$ is greater than or equal to the corresponding element in row $j$. Also column $i$ is said to dominate column $j$ if every element in column $i$ is less than or equal to the corresponding element in column $j$.

- **Optimal Strategies for Column and Row.**   Let a game have the $m \times n$ payoff matrix $A = (a_{ij})$ and let $x_1, x_2, \ldots, x_n$, be the solution to the standard linear programming problem

$$\text{Maximize} \qquad\qquad z = x_1 + x_2 + \cdots + x_n$$

$$\text{Subject to} \qquad a_{11}x_1 + a_{12}x_2 + \cdots + a_{1n}x_n \leq 1$$
$$a_{21}x_1 + a_{22}x_2 + \cdots + a_{2n}x_n \leq 1$$
$$\vdots$$
$$a_{m1}x_1 + a_{m2}x_2 + \cdots + a_{mn}x_n \leq 1$$
$$x_1, x_2, \ldots, x_n \geq 0$$

Then the value $v$ of the game is also reciprocal of this maximum.

Let $y_1, y_2, \ldots, y_m$, be the solution to the dual of the above linear programming problem. Then the value $v$ of the game is also the reciprocal of the minimum to the dual problem. The optimal strategy for Column is $\bar{q} = vX$ while the optimal strategy for Row is $\bar{p} = vY$.

## Chapter 10 Review Exercises

In Exercises 1 through 8 find the value of the game and the optimal strategies for both players for the games with the given payoff matrix.

**1.** $\begin{bmatrix} 3 & 2 & 0 \\ 2 & -2 & -1 \end{bmatrix}$

**2.** $\begin{bmatrix} 4 & -1 & 2 & 1 \\ 3 & -2 & 1 & 2 \end{bmatrix}$

**3.** $\begin{bmatrix} 2 & 1 & 0 \\ 2 & 2 & 1 \\ 1 & 3 & 3 \end{bmatrix}$

**4.** $\begin{bmatrix} 3 & 2 & 1 \\ 1 & 1 & 0 \\ 2 & 1 & 2 \end{bmatrix}$

5. $\begin{bmatrix} 2 & -1 \\ -1 & 1 \end{bmatrix}$

6. $\begin{bmatrix} 0 & 2 \\ 3 & -3 \end{bmatrix}$

7. $\begin{bmatrix} -1 & 0 & 1 \\ 2 & -1 & 0 \end{bmatrix}$

8. $\begin{bmatrix} -2 & 3 & 5 \\ 1 & 0 & 1 \end{bmatrix}$

9. Determine the value of the game and the optimal strategy for Row for the game with the payoff matrix given by

$$\begin{bmatrix} 1 & 0 & -1 \\ -3 & -1 & 2 \end{bmatrix}.$$

10. Determine the value of the game and the optimal strategy for Column for the game with the payoff matrix given by

$$\begin{bmatrix} 6 & 1 \\ 5 & 4 \\ 0 & 5 \end{bmatrix}.$$

11. Determine the value of the game and the optimal strategy for Column for the game with the payoff matrix in Exercise 9.

12. Determine the value of the game and the optimal strategy for Row for the game with the payoff matrix in Exercise 10.

13. Determine the value of the game and the optimal strategy for Row and for Column for the game with the payoff matrix

$$\begin{bmatrix} 1 & 0 & 0 \\ 0 & 1 & 0 \\ 0 & 0 & 1 \end{bmatrix}.$$

14. Show that every game with a payoff matrix

$$\begin{bmatrix} a & a \\ b & c \end{bmatrix}$$

must be strictly determined.

Firms have both fixed and variable costs. (© Dennis Drenner.)

# Functions and Graphs

11.1   Functions

11.2   Some Special Functions

11.3   Combinations of Functions

## Introduction

**This chapter presents some basic material that will be used throughout the rest of the text. The first section gives an elementary presentation of the fundamentally important notion of function. The second section explains some special functions that will be used often and gives some techniques of graphing. The third section shows how functions can be broken up into combinations of simpler functions to provide insight into the original function.**

## 11.1   FUNCTIONS

- ► *Definition of Function*
- ► *Graphs of Functions*
- ► *Applications*

Lejeune Dirichlet
1805–1859

*(Bettmann/Hulton.)*

Dirichlet was one of the mathematical giants of the 19th century. He formulated the notion of function that is currently used. He is also known for the Dirichlet series, the Dirichlet function, the Dirichlet principle, and the Dirichlet problem. The Dirichlet problem is fundamental to thermodynamics and electrodynamics. Although described as noble, sincere, humane, and possessing a modest disposition, he was known as a dreadful teacher. Dirichlet was also a failure as a family correspondent. When his first child was born, he neglected to inform his father-in-law. When the father-in-law found out about the event, he commented that Dirichlet might have at least written a note saying "2 + 1 = 3."

APPLICATION
**Volume of an Open Box**

From all 4 corners of a rectangular piece of cardboard of dimensions 10 inches by 20 inches squares are cut with dimensions $x$ by $x$. The sides are turned up to form an open box. What is the volume $V$ of the box as a function of $x$? See Example 11 for the answer.

## Definition of Function

*(© Dennis Drenner.)*

We are all familiar with corresponding an element in one set with an element in another set. For example, to each house there corresponds a house number, to each automobile there corresponds a license number, to each individual there corresponds a name, to each size candy bar there corresponds a price. We call any rule that assigns or corresponds to each element in one set precisely one element in another set a **function**.

Typical functions that we encounter will arise in equations. For example, suppose you are going a steady 40 miles per hour in a car. Then in 1 hour you will travel 40 miles. In 2 hours you will travel 80 miles, and so on. The distance you travel depends on (corresponds to) the time. Indeed the equation relating distance ($d$), velocity ($v$), and time ($t$) is $d = v \cdot t$. In our example, we have $d = 40 \cdot t$. We can view this as a correspondence or rule: given the time $t$ in hours, the rule gives a distance $d$ in miles according to $d = 40 \cdot t$. Thus given $t = 3$, $d = 40 \cdot 3 = 120$. Notice carefully how this rule is *unambiguous*. That is, given any time $t$, the rule specifies one and only one distance $d$. Therefore, this rule is a function. The correspondence is between time and distance.

Often the letter $f$ is used to denote a function. Thus in the previous example, we can write $d = f(t) = 40 \cdot t$. The symbol $f(t)$ is read "f of t." One can think of $t$ as the "input" and the value of $d = f(t)$ as the "output." For example an input of $t = 4$ results in an output of $d = f(4) = 40 \cdot 4 = 160$.

The previous example was an example of a linear function. We have already studied such functions earlier. However, functions need not be linear. The following gives a general definition of function.

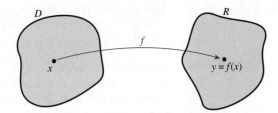

**Figure 11.1**

---

**Definition of Function**

Let $D$ and $R$ be two nonempty sets. A **function** $f$ from $D$ to $R$ is a rule that assigns to each element $x$ in $D$ one and only one element $y = f(x)$ in $R$. See Figure 11.1.

---

The set $D$ in the definition is called the **domain** of $f$. We might think of the domain as the set of inputs. We then can think of the values $f(x)$ as outputs.

Another helpful way to think of a function is shown in Figure 11.2. Here the function $f$ accepts the input $x$ from the conveyor belt, operates on $x$, and outputs (assigns) the new value $f(x)$.

**Figure 11.2**

The set of all possible outputs is called the **range** of $f$. The letter representing elements in the domain is called the **independent variable**, and the letter representing the elements in the range is called the **dependent variable**. Thus if $y = f(x)$, $x$ is the independent variable, and $y$ is the dependent variable since $y$ *depends* on $x$. In the equation $d = 40t$, we can write $d = f(t) = 40t$ with $t$ as the independent variable. We are free to set the independent variable $t$ equal to any number of values. The dependent variable is $d$. Notice that $d$ will *depend* on the particular value of $t$ that is used.

EXAMPLE 1        **Determining When Rules are Functions**

Which of the rules in Figure 11.3 are functions? Find the range of each function.

Solution        The rule indicated in Figure 11.3a is a function from $A = \{a, b, c\}$ to $B = \{p, q, r, s\}$ since each element in $A$ is assigned a unique element in $B$. The domain of this function is $A$, and the range is $\{p, r\}$. The rule indicated in Figure 11.3b is not a function since the element $a$ has been assigned more than one element in $B$. The rule indicated in Figure 11.3c is a function with domain $A$ and with range the single point $\{q\}$. ∎

**REMARK.**    Notice that it is possible for the same output to result from different inputs. For example, inputting $a$ and $b$ into the function in (a) results in the same output $p$. In (c), the output is always the same regardless of the input.

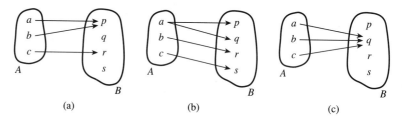

   (a)      (b)      (c)

**Figure 11.3**

EXAMPLE 2        **Determining if a Rule is a Function**

Let $A = [0, +\infty)$. For each $x \in A$, assign $y$ to be any value such that $y^2 = x$. Is this rule a function?

Solution        If we take some value in $A$ such as 4, then there are *two* numbers whose square is 4. These two values are 2 and $-2$. Since the rule assigns more than one value to each positive input, this rule cannot be a function. ∎

EXAMPLE 3        **Determining if a Rule is a Function**

Let $A = [0, +\infty)$. For any $x \in A$, assign $y$ to be any nonnegative value such that $y^2 = x$. Is this rule a function?

Solution        Given any nonnegative $x$ there is precisely one *nonnegative* value $y$ such that $y^2 = x$. Thus this rule is a function. ∎

  Functions are often given by equations, as we saw for the function $d = f(t) = 40 \cdot t$ at the beginning of this section. As another example, let both the sets $D$ and $R$ be the set of real numbers, and let $f$ be the function from $D$ to $R$ defined by $f(x) = 3x^2 + 1$. In words the output is obtained by taking three times the square of the input and adding one to the result. Thus the formula

$$f(x) = 3x^2 + 1$$

can also be viewed as

$$f(\text{input}) = 3(\text{input})^2 + 1.$$

If we want $f(2)$ then we replace $x$ (or the input) with 2 and obtain

$$f(2) = 3(2)^2 + 1 = 13.$$

If we want $f(-4)$ then we replace $x$ (or the input) with $-4$ and obtain

$$f(-4) = 3(-4)^2 + 1 = 49.$$

If we want $f(s + 1)$ then we replace $x$ (or the input) with $s + 1$ and obtain

$$f(s + 1) = 3(s + 1)^2 + 1 = 3(s^2 + 2s + 1) + 1 = 3s^2 + 6s + 4.$$

If we want $f(x + 2)$ then we replace $x$ (or the input) with $x + 2$ and obtain

$$f(x + 2) = 3(x + 2)^2 + 1 = 3(x^2 + 4x + 4) + 1 = 3x^2 + 12x + 13$$

C O N V E N T I O N .    When the domain of a function $f$ is not stated explicitly, we shall agree to let the domain be all values $x$ for which $f(x)$ makes sense and is real. For example, if $f(x) = \dfrac{1}{x}$, the domain is assumed to be all real numbers other than zero.

EXAMPLE 4

**Finding the Domain of a Function**

Let $f(x) = \sqrt{2x - 4}$. Find the domain of $f$. Evaluate $f(2)$, $f(4)$, $f(2t + 2)$.

**Solutions**

Since $f(x)$ must be real, we must have $2x - 4 \geq 0$ or $x \geq 2$. Thus the domain is $[2, +\infty)$.

$$f(2) = \sqrt{2(2) - 4} = \sqrt{0} = 0$$

$$f(4) = \sqrt{2(4) - 4} = \sqrt{4} = 2$$

$$f(2t + 2) = \sqrt{2(2t + 2) - 4} = \sqrt{4t} = 2\sqrt{t}, \, t \geq 0 \quad \blacksquare$$

The next example is a problem that will come up shortly in the calculus.

EXAMPLE 5

**Evaluating Functions**

Let $f(x) = 3x^2 + 2x$. For $h \neq 0$, compute

$$\frac{f(x + h) - f(x)}{h}.$$

**Solution**

$$\frac{f(x + h) - f(x)}{h} = \frac{[3(x + h)^2 + 2(x + h)] - [3x^2 + 2x]}{h}$$

$$= \frac{3x^2 + 6hx + 3h^2 + 2x + 2h - 3x^2 - 2x}{h}$$

$$= \frac{6xh + 2h + 3h^2}{h}$$

$$= 6x + 2 + 3h \quad \blacksquare$$

## Graphs of Functions

When the domain and range of a function are sets of real numbers, the function can be graphed in the same way that equations were graphed in Section 1.1.

---

**Graph of a Function**

The **graph** of a function $f$ consists of all points $(x, y)$ such that $x$ is in the domain of $f$ and $y = f(x)$.

---

**E X A M P L E   6**    **Graphing a Function**

Graph the functions $f$ and $g$ defined as follows. (a) $y = f(x) = x^2$, (b) $y = g(x) = |x|$.

**Solutions**    In both cases the domain is $(-\infty, +\infty)$.

(a)   In order to graph $y = f(x) = x^2$, we evaluate the function at a number of points as indicated by Table 11.1. The graph is then shown in Figure 11.4.

(b)   If $x \geq 0$, then $g(x) = |x| = x$. This is a line through the origin with slope 1. This line is graphed in the first quadrant in Figure 11.5. Also if $x \leq 0$, then $g(x) = |x| = -x$, which is a line through the origin with slope $-1$. This is shown in Figure 11.5.    $\blacksquare$

In applications we often encounter functions with domains divided into two or more parts with a different rule applied to each part. We call such functions **piecewise-defined functions**. The following is such an example.

**Table 11.1**

| $x$ | $-3$ | $-2$ | $-1$ | 0 | 1 | 2 | 3 |
|------|------|------|------|---|---|---|---|
| $f(x)$ | 9 | 4 | 1 | 0 | 1 | 4 | 9 |

**Figure 11.4**

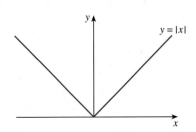

**Figure 11.5**

**E X A M P L E   7**   **A Piecewise-Defined Function**

Graph the function

$$f(x) = \begin{cases} 2x + 1 & \text{if } x \le 0 \\ 2 & \text{if } 0 < x < 2 \\ -x + 2 & \text{if } x \ge 2 \end{cases}$$

and evaluate $f(-1)$, $f(0)$, $f(1)$, $f(4)$.

**Solution**   When $x \le 0$, the function is given by $2x + 1$, which is a line with slope 2 and $y$-intercept 1. When $0 < x < 2$, the function is the constant 2. When $x \ge 0$, the function is $-x + 2$ which is a line with slope $-1$ and $x$-intercept 2. Putting these pieces together gives the graph shown in Figure 11.6.

Since $-1 \le 0$, $f(-1) = 2(-1) + 1 = -1$.
Since $0 \le 0$, $f(0) = 2(0) + 1 = 1$.
Since $0 < 1 < 2$, $f(1) = 2$.
Since $4 \ge 2$, $f(4) = -(4) + 2 = -2$.   ■

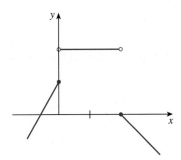

**Figure 11.6**

**E X A M P L E   8**   **Determining if a Graph is the Graph of a Function**

A graph is shown in Figure 11.7. Does this represent the graph of a function of $x$?

**Solution**   Figure 11.7 indicates that given any number $x$ there is one and only one value of $y$. You can see that starting at any value of $x$ and moving vertically until you strike the graph, then moving horizontally until you strike the $y$-axis will never result in more than one value of $y$. Thus this process assigns unambiguously one single value $y$ to any value $x$ on [0, 5]. Thus this graph describes a function.  ■

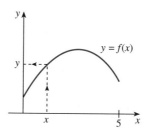

**Figure 11.7**

**E X A M P L E   9**   **Determining if a Graph is the Graph of a Function**

A graph is shown in Figure 11.8. Does this represent the graph of a function of $x$?

**Solution**   Notice that for the value $a$ indicated in Figure 11.8, drawing a vertical line from $a$ results in striking the graph in *two* places resulting in assigning two values to $x$ when $x = a$. This cannot then be the graph of a function of $x$.  ■

This works in general.

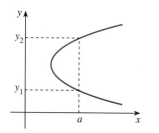

**Figure 11.8**

---

**Vertical Line Test**

A graph in the $xy$-plane represents a function of $x$, if and only if every vertical line intersects the graph in at most one place.

---

**E X A M P L E   1 0**   **A Graph That Fails to be the Graph of a Function**

Since $\sqrt{x^2 + y^2}$ is the distance from the point $(x, y)$ to the origin, the graph of the equation $x^2 + y^2 = 1$ is a circle centered at the origin with radius 1. See Figure 11.9a. Does this represent a function of $x$?

**Solution**   As Figure 11.9b indicates, the vertical line test fails, so this cannot be the graph of a function of $x$.  ■

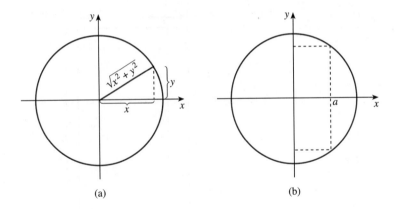

(a)  (b)

**Figure 11.9**

**REMARK.** In the previous example, if we try to solve for $y$ we obtain $y = \pm\sqrt{1 - x^2}$. Thus for any value of $x$ in the interval $(-1, 1)$, there are *two* corresponding values of $y$. We are not told which of these two values to take. Thus the rule is ambiguous and not a function.

## Applications

In order to solve any applied problem, one must first take the problem and translate it into mathematics. Doing this will require creating equations and functions. We already did this at the beginning of the section when we related distance traveled with velocity and time. We also have created many other *linear* equations in earlier chapters of this text. In the next example, we will find a *nonlinear* function that gives the volume of a certain package. Later in the text we will then use this function to find important information such as the dimensions that yield the maximum volume. For now, we will just find the function.

**EXAMPLE 11**   **Finding a Function**

From all 4 corners of a rectangular piece of cardboard of dimensions 10 inches by 20 inches squares are cut with dimensions $x$ by $x$. The sides are turned up to form an open box. See Figure 11.10. Find the volume $V$ of the box as a function of $x$.

**Solution**   The volume of a box is given by the product of the width (w), the length (l), and the height (h), that is, $V = w \cdot l \cdot h$. The width of the box is $w = 10 - 2x$, the length is $20 - 2x$, and the height is $x$. Thus the volume is

$$V = V(x) = (10 - 2x)(20 - 2x)x. \quad \blacksquare$$

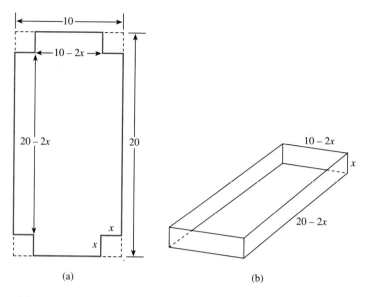

(a)

(b)

**Figure 11.10**

## SELF-HELP EXERCISE SET 11.1

**1.** Which of the graphs in the figures represents the graph of a function?

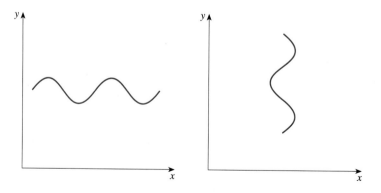

**2.** Find the domain of $f(x) = \dfrac{\sqrt{x+1}}{x}$.

**3.** For the function in the previous exercise evaluate (a) $f(-1)$, (b) $f(3)$, (c) $f(a+2)$, (d) $f(x+2)$.

4. A telephone company charges $1.25 for the first minute for a certain long-distance call and $.25 for each additional minute or fraction thereof. If $C$ is the cost of a call and $x$ is the length of the call in minutes, find the cost as a function of $x$ and sketch a graph. Assume $x \leq 3$.

## EXERCISE SET 11.1

For Exercises 1 through 6, determine which of the rules in the figures represent functions.

1.

2.

3.

4.

5.

6.

7. Specify the domain and range for each rule that represents a function in Exercises 1, 2, and 3.

8. Specify the domain and range for each rule that represents a function in Exercises 4, 5, and 6.

For Exercises 9 through 24, find the domain of each function.

9. $\dfrac{x}{x - 1}$

10. $\dfrac{x - 1}{x + 2}$

11. $\sqrt{x}$

12. $\sqrt{x - 1}$

13. $\left| \dfrac{1}{\sqrt{x - 1}} \right|$

14. $\sqrt{2 - x}$

15. $\dfrac{x - 1}{x^2 - 5x + 6}$

16. $\dfrac{x}{x^2 - 1}$

17. $\sqrt{|x - 1|}$

18. $|\sqrt{x - 1}|$

19. $\dfrac{\sqrt{x + 2}}{x}$

20. $\dfrac{x}{\sqrt{x - 1}}$

**21.**  $\sqrt{x^2 + 1}$

**22.**  $\dfrac{1}{x^2 + 1}$

**23.**  $\dfrac{1}{\sqrt{x^2 + 1}}$

**24.**  $\sqrt{x^2 - 1}$

For Exercises 25 through 30, find the domain and range of each function.

**25.**

**26.**

**27.**

**28.**

**29.**

**30.**

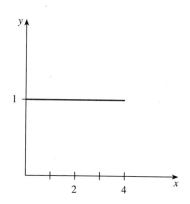

In Exercises 31 through 38 evaluate the given function at the given values.

**31.**  $f(x) = 2x^3 + 3$, $f(1)$, $f(-1)$, $f(0)$, $f(\sqrt{3})$, $f(x^2)$, $f(\frac{1}{x})$.

**32.**  $f(x) = 2x^4 - 1$, $f(0)$, $f(2)$, $f(-\frac{1}{2})$, $f(-\sqrt{2})$, $f(x^2)$, $f(\sqrt{x})$.

**33.** $f(x) = \dfrac{1}{x+1}$, $f(1)$, $f(\frac{1}{2})$, $f(-2)$, $f(\sqrt{2}-1)$, $f(x+2)$, $f\left(\frac{1}{x+1}\right)$.

**34.** $f(x) = \dfrac{1}{x^2+1}$, $f(2)$, $f(\frac{1}{2})$, $f(-3)$, $f(2\sqrt{2})$, $f(\sqrt{x})$, $f(x-1)$.

**35.** $f(x) = \sqrt{2x-1}$, $f(5)$, $f(\frac{1}{2})$, $f(x^2 - \frac{1}{2})$, $f(\frac{1}{x})$, $f(-x)$, $-f(x)$.

**36.** $f(x) = \sqrt{2 - 3x}$, $f(-1)$, $f(0)$, $f(x+1)$, $f(\frac{1}{x})$, $f(-x)$, $-f(x)$.

**37.** $f(x) = |x+1|$, $f(-2)$, $-f(-2)$, $f(x-1)$, $f(x^2-1)$.

**38.** $f(x) = \dfrac{x-1}{x+1}$, $f(1)$, $f(-1)$, $f(x+1)$, $f(x^2)$.

**39.** Let $f(x)$ be the function given by the graph in Exercise 27. Find $f(0)$, $f(1)$, $f(1.5)$, $f(2)$.

**40.** Let $f(x)$ be the function given by the graph in Exercise 28. Find $f(-1)$, $f(1)$, $f(2)$.

**41.** Let $f(x)$ be the function given by the graph in Exercise 29. Find $f(0)$, $f(2)$, $f(4)$, $f(6)$.

**42.** Let $f(x)$ be the function given by the graph in Exercise 30. Find $f(0)$, $f(1)$, $f(2)$.

For Exercises 43 through 50, determine whether each graph represents a function of $x$.

**43.**

**44.**

**45.**

**46.**

**47.**

**48.**

**49.**

**50.**

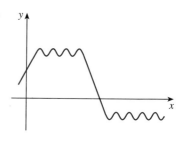

In Exercises 51 through 58 an equation involving $x$ and $y$ is given. Determine if $y$ can be written as a function of $x$. If it can, find it.

**51.** $2x + 5y = 15$        **52.** $x - y = 2$        **53.** $x^2 + y^2 = 4$

**54.** $(x - 1)^2 + (y - 2)^2 = 1$        **55.** $y^3 = x$

**56.** $y^4 = x$        **57.** $x + y^2 - 1 = 0$        **58.** $x^2 + y - 1 = 0$

For Exercise 59 through 66 graph the indicated function.

**59.** $f(x) = \begin{cases} 3x - 1 & \text{if } x \le -1 \\ x & \text{if } x > -1 \end{cases}$        **60.** $f(x) = \begin{cases} x & \text{if } x \le 1 \\ -x & \text{if } x > 1 \end{cases}$

**61.** $f(x) = \begin{cases} x^2 & \text{if } x \le 0 \\ x & \text{if } x > 0 \end{cases}$        **62.** $f(x) = \begin{cases} -1 & \text{if } x < 0 \\ 1 & \text{if } x \ge 0 \end{cases}$

**63.** $f(x) = \begin{cases} x^2 & \text{if } x \le 0 \\ x & \text{if } 0 < x \le 1 \\ 2x & \text{if } x > 1 \end{cases}$        **64.** $f(x) = \begin{cases} 1 & \text{if } x < 0 \\ 0 & \text{if } 0 \le x \le 1 \\ -1 & \text{if } x > 1 \end{cases}$

**65.** $f(x) = \begin{cases} 1 & \text{if } 0 \le x < 1 \\ 2 & \text{if } 1 \le x < 2 \\ 3 & \text{if } 2 \le x \le 3 \end{cases}$

66. $f(x) = \begin{cases} -x & \text{if } x \le -1 \\ 1 & \text{if } -1 < x < 1 \\ x & \text{if } x \ge 1 \end{cases}$

In Exercises 67 through 72 find $\dfrac{f(x + h) - f(x)}{h}$ for the indicated functions.

67. $f(x) = 3x + 1$       68. $f(x) = x^2 - 1$       69. $f(x) = 3x^2 + 1$

70. $f(x) = |x|$       71. $\dfrac{1}{x}$       72. $\sqrt{x}$

73. Let $f(x) = mx$. Show $f(a + b) = f(a) + f(b)$.

74. Let $f(x) = x^2$. Find $x$ such that $f(x + 1) = f(x + 2)$.

75. Let $f(x) = 3^x$. Show that $3f(x) = f(x + 1)$ and that $f(a + b) = f(a) \cdot f(b)$.

76. Let $f(x) = x^2 + 1$. Find $f[f(1)]$ and $f[f(x)]$.

77. Find the area $A$ of a circle as a function of the radius $r$.

78. Find the area $A$ of a square as a function of the length $s$ of a side.

## Applications

79. **Packaging.**  A box has a square base with each side of length $x$ and height equal to $3x$. Find the volume $V$ as a function of $x$.

80. **Packaging.**  Find the surface area $S$ of the box in the previous exercise as a function of $x$.

81. **Velocity.**  A car travels at a steady 60 miles per hour. Write the distance $d$ in miles that the car travels as a function of time $t$ in hours.

82. **Navigation.**  Two ships leave port at the same time. The first ship heads due north at 5 miles per hour while the second heads due west at 3 miles per hour. Let $d$ be the distance between the ships in miles and $t$ the time in hours since they left port. Find $d$ as a function of $t$.

83. **Revenue.**  A company sells a certain style of shoe for $60. If $x$ is the number of shoes sold, what is the revenue $R$ as a function of $x$?

84. **Delivery Fees.**  A store will deliver a sofa for free if you are within 10 miles of the store and charges $2 per mile (including fractions thereof) beyond 10 miles. Let $x$ be the miles from the store and $C$ the cost of delivery. Write $C$ as a function of $x$ and sketch a graph of this function.

85. **Sales Commission.**  A salesman receives a commission of $1 per square yard for the first 500 yards of carpeting sold in a month and $2 per square yard for any additional carpet sold during the same month. If $x$ is the number of yards of carpet sold and $C$ is the commission, find $C$ as a function of $x$ and graph this function.

86. **Taxes.**   A certain state has a tax on electricity of 1% of the monthly electricity bill with the first $50 of the bill exempt from tax. Let $P(x)$ be the percent one pays in taxes and $x$ the amount of the bill in dollars. Graph this function.

87. **Taxes.**   In the previous problem, let $T(x)$ be the total amount in dollars paid on the tax where $x$ is the amount of the monthly bill. Graph $T(x)$.

## Graphing Calculator Exercises

Graph the remaining functions on a graphing calculator and find the domain and range of each function.

88. $f(x) = x^2 - 4$

89. $f(x) = -x^2 + 8$

90. $f(x) = |-x^2 + 8|$

91. $f(x) = \sqrt{x^2 + 4}$

92. $f(x) = \sqrt{9 - x^2}$

93. $f(x) = \sqrt{x^2 - 9}$

94. $f(x) = \dfrac{\sqrt{x}}{x - 5}$

95. $f(x) = \dfrac{\sqrt{1 - x}}{x + 4}$

96. The function given in (a) Exercise 59, (b) Exercise 63.

97. The function given in (a) Exercise 60, (b) Exercise 66.

## Solutions to Self-Help Exercise Set 11.1

1. The vertical line test indicates that the first figure represents a function but the second does not.

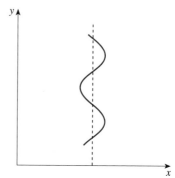

2. The function $f(x) = \dfrac{\sqrt{x + 1}}{x}$ is defined if $x + 1 \geq 0$, or $x \geq -1$, and if $x \neq 0$. Thus the domain is all points on $[-1, 0)$ together with all points on $(0, +\infty)$.

3. For the function in the previous exercise

   **a.** $f(-1) = \dfrac{\sqrt{-1 + 1}}{-1} = 0$,   **b.** $f(3) = \dfrac{\sqrt{3 + 1}}{3} = \dfrac{2}{3}$,

c.  $f(a + 2) = \dfrac{\sqrt{a + 2} + 1}{a + 2} = \dfrac{\sqrt{a} + 3}{a + 2}$,

d.  $f(x + 2) = \dfrac{\sqrt{x + 2} + 1}{x + 2} = \dfrac{\sqrt{x} + 3}{x + 2}$

4.  For the first minute the cost is a constant \$1.25. During the second minute the cost is a constant \$1.25 + \$0.25 = \$1.50. During the third minute the cost is \$1.75. We can write the function as

$$C(x) = \begin{cases} 1.25 & \text{if } 0 < x \le 1 \\ 1.50 & \text{if } 1 < x \le 2 \\ 1.75 & \text{if } 2 < x \le 3 \end{cases}$$

The graph is shown in the figure.

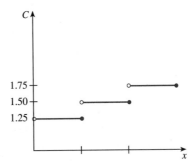

## 11.2   SOME SPECIAL FUNCTIONS

► *Quadratic Functions*
► *Models Using Quadratics*
► *Other Special Functions*

A P P L I C A T I O N

**Maximizing Revenue**

Suppose the demand equation for a certain commodity is given by $p = -2x + 12$, where $x$ is given in thousands per week. For what value of $x$ will the revenue be maximum? The answer can be found in Example 5.

### Quadratic Functions

In Section 1.3 we assumed a *linear* model for both the cost function and the revenue function. As a consequence the profit function was linear. This is adequate if the number sold is not too large. If the profit function is linear, as in Figure 11.11, the graph implies that profits are becoming large without bound.

**Figure 11.11**

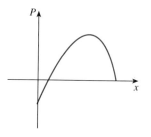

**Figure 11.12**

This we know is not what actually happens. If, for example, we try to flood the market with pens, we will most likely have to set the price per pen so low as to result in losses, not profits. Thus a more reasonable profit curve is one that rises at first and then falls, such as shown in Figure 11.12. Some quadratic functions behave like this. Quadratics can also be used in other applications and are often more realistic than a linear model. Thus we need to study quadratic functions.

The function

$$f(x) = ax^2 + bx + c, \qquad a \neq 0.$$

is a **quadratic function**. In this section we will graph quadratic functions, find where each one takes either a maximum or minimum, and consider models based on quadratic functions.

We shall see that all quadratics have graphs that look basically like the graphs of $ax^2$. Thus we begin with the following example.

**E X A M P L E   1**    **Graphs of Some Simple Quadratics**

Graph the quadratics $y = \frac{1}{2}x^2$, $y = x^2$, $y = 2x^2$, $y = -x^2$.

**Solution**    Using the values found in Table 11.2, the graphs are then given in Figure 11.13.

As Figure 11.13 indicates, the graphs of $f(x) = ax^2$ when $a > 0$ all look similar; they open upward and are narrower when $a$ is larger. The graphs of $ax^2$ when $a < 0$ are similar except that they open downward. ■

**Figure 11.13**

**Table 11.2**

| $x$ | $-3$ | $-2$ | $-1$ | $0$ | $1$ | $2$ | $3$ |
|-----|------|------|------|-----|-----|-----|-----|
| $0.5x^2$ | 4.5 | 2 | 0.5 | 0 | 0.5 | 2 | 4.5 |
| $x^2$ | 9 | 4 | 1 | 0 | 1 | 4 | 9 |
| $2x^2$ | 18 | 8 | 2 | 0 | 2 | 8 | 18 |
| $-x^2$ | $-9$ | $-4$ | $-1$ | 0 | $-1$ | $-4$ | $-9$ |

E X A M P L E  2 **Shifting a Graph Up or Down**

Graph $y = x^2 + 3$ and $y = x^2 - 1$.

**Solution** In Figure 11.14a we give the graph of $y = x^2$ found in Figure 11.13. To graph $y = x^2 + 3$ in Figure 11.14b, we simply add 3 to every $y$ value obtained in Figure 11.14a. This results in the graph of $y = x^2$ shifted upward 3 units in order to obtain the graph of $y = x^2 + 3$. To obtain the graph of $y = x^2 - 1$ in Figure 11.14c, we shift the graph of $y = x^2$ down 1 unit. ■

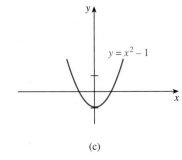

(a)  (b)  (c)

**Figure 11.14**

E X A M P L E  3 **Shifting a Graph Left or Right**

Graph $y = (x - 2)^2$ and $y = (x + 1)^2$.

**Solution** First notice that these are indeed quadratics since

$$(x - 2)^2 = x^2 - 4x + 4, \qquad (x + 1)^2 = x^2 + 2x + 1.$$

Table 11.3 compares the values of $x^2$ with $(x - 2)^2$. Notice that the $y$ values of $y = (x - 2)^2$ are the same as the $y$ values of $y = x^2$ except that they occur 2 units to the right on the $x$-axis. This is seen in Figure 11.15a.

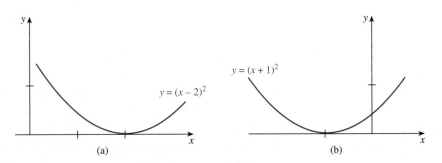

(a)  (b)

**Figure 11.15**

**Table 11.3**

| $x$ | $x^2$ | $(x-2)^2$ |
|---|---|---|
| −5 | 25 | 49 |
| −4 | 16 | 36 |
| −3 | 9 | 25 |
| −2 | 4 | 16 |
| −1 | 1 | 9 |
| 0 | 0 | 4 |
| 1 | 1 | 1 |
| 2 | 4 | 0 |
| 3 | 9 | 1 |
| 4 | 16 | 4 |
| 5 | 25 | 9 |

**Table 11.4**

| $x$ | $x^2$ | $(x+1)^2$ |
|---|---|---|
| −5 | 25 | 16 |
| −4 | 16 | 9 |
| −3 | 9 | 4 |
| −2 | 4 | 1 |
| −1 | 1 | 0 |
| 0 | 0 | 1 |
| 1 | 1 | 4 |
| 2 | 4 | 9 |
| 3 | 9 | 16 |
| 4 | 16 | 25 |
| 5 | 25 | 36 |

Table 11.4 compares the values of $x^2$ with $(x+1)^2$. Notice that the $y$ values of $y = (x+1)^2$ are the same as the $y$ values of $y = x^2$ except that they occur 1 unit to the left on the $x$-axis. This is seen in Figure 11.15b.  ∎

**E X A M P L E  4**     **Graphing a Quadratic**

Sketch the graph of $y = 2x^2 - 16x + 14$.

**Solution**     We complete the square.

STEP 1.   Factor out the coefficient of $x^2$.

$$2x^2 - 16x + 14 = 2(x^2 - 8x + 7)$$

STEP 2.   Complete the square inside the bracket. Do this by adding and subtracting the square of one half of the coefficient of $x$.

$$\left(\frac{-8}{2}\right)^2 = (-4)^2 = 16$$

| | |
|---|---|
| $2(x^2 - 8x + 7) = 2(x^2 - 8x + 16 - 16 + 7)$ | Add and subtract 16. |
| $= 2([x^2 - 8x + 16] - 9)$ | Collect terms. |
| $= 2([x - 4]^2 - 9)$ | Complete the square. |
| $= 2(x - 4)^2 - 18$ | Multiply by 2. |

In view of the previous examples, the graph of $2(x - 4)^2 - 18$ is obtained by shifting the graph of $2x^2$ four units to the right and 18 units down, as shown in Figure 11.16.  ∎

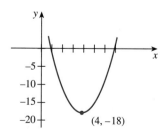

**Figure 11.16**

The same procedure can be used for any quadratic $ax^2 + bx + c$, with $a \neq 0$.

$$f(x) = ax^2 + bx + c$$

$$= a\left(x^2 + \frac{b}{a}x + \frac{c}{a}\right) \qquad \text{Factor out } a.$$

$$= a\left(\left[x^2 + \frac{b}{a}x + \frac{b^2}{4a^2}\right] - \frac{b^2}{4a^2} + \frac{c}{a}\right) \qquad \text{Add and subtract } \left(\frac{b}{2a}\right)^2.$$

$$= a\left(\left[x + \frac{b}{2a}\right]^2 + \frac{c}{a} - \frac{b^2}{4a^2}\right) \qquad \text{Complete the square.}$$

$$= a\left(x + \frac{b}{2a}\right)^2 + \left(c - \frac{b^2}{4a}\right) \qquad \text{Multiply by } a.$$

If we set

$$h = -\frac{b}{2a} \qquad \text{and} \qquad k = c - \frac{b^2}{4a},$$

the quadratic can be written in **standard form**

$$f(x) = a(x - h)^2 + k.$$

The point $(h, k)$ is called the **vertex**. With the quadratic in the standard form $y = a(x - h)^2 + k$, and $a > 0$, notice that $y$ is smallest when the term $a(x - h)^2 = 0$. This can only happen if $x = h$. Thus, when $a > 0$, $y = a(x - h)^2 + k$ takes a *minimum* when $x = h$. The minimum is $y = k$.

With the quadratic in the standard form $y = a(x - h)^2 + k$, and $a < 0$, notice that $y$ is largest when the term $a(x - h)^2 = 0$. This can only happen if $x = h$. Thus, when $a < 0$, $y = a(x - h)^2 + k$ takes a *maximum* when $x = h$. The maximum is $y = k$.

The graphs, which are called **parabolas**, are given in Figure 11.17 in the two cases $a > 0$ and $a < 0$. Notice that when $a > 0$, the parabola opens upward and $y = k$ is the *minimum* value of $y$. When $a < 0$, the parabola opens downward and $y = k$ is the *maximum* value of $y$.

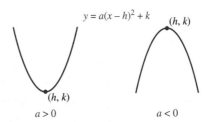

$y = a(x - h)^2 + k$

$(h, k)$

$(h, k)$

$a > 0$                    $a < 0$

**Figure 11.17**

---

**The Quadratic $ax^2 + bx + c$**

Every quadratic with $a \neq 0$ can be put into the standard form

$$ax^2 + bx + c = a(x - h)^2 + k,$$

where

$$h = -\frac{b}{2a} \quad \text{and} \quad k = c - \frac{b^2}{4a}.$$

The graphs of the quadratics are shown in Figure 11.17. When $a > 0$, the parabola opens upward and $y = k$ is the *minimum* value of $y$. When $a < 0$, the parabola opens downward and $y = k$ is the *maximum* value of $y$.

---

## Models Using Quadratics

We will now use quadratic functions to model the revenue equation. The exercise set includes quadratics that model cost and profit equations.

**EXAMPLE 5**   **A Quadratic Revenue Function**

Suppose the demand equation for a certain commodity is given by $p = -2x + 12$, where $x$ is given in thousands per week and $p$ is the price of one item. Find the value of $x$ for which the revenue will be maximum. Find the value(s) of $x$ for which the revenue is zero.

**Solution**   If $R(x)$ is the revenue in thousands of dollars per week, then

$$R(x) = xp = x(-2x + 12) = -2x^2 + 12x.$$

Completing the square, we obtain

$$
\begin{aligned}
R(x) &= -2x^2 + 12x \\
&= -2(x^2 - 6x) && \text{Factor out } -2. \\
&= -2([x^2 - 6x + 9] - 9) && \text{Add and subtract } \left(\frac{-6}{2}\right)^2 = 9. \\
&= -2([x - 3]^2 - 9) && \text{Complete the square.} \\
&= -2(x - 3)^2 + 18 && \text{Multiply by } -2.
\end{aligned}
$$

The quadratic is now in standard form. Thus we see that the maximum revenue occurs when $x = 3$ or when 3000 of the items are sold. See Figure 11.18.
Naturally we could have used the formula

$$h = -\frac{b}{2a} = -\frac{12}{2(-2)} = 3.$$

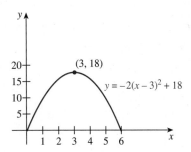

**Figure 11.18**

To answer the second question we factor $R(x)$ to obtain

$$R(x) = 2x(6 - x).$$

This shows that the zeros of $R$ occur at $x = 0$ and $x = 6$. Thus revenue is zero when no items are sold and also when 6000 items are sold. ■

The graph of the revenue function shown in Figure 11.18 is typical. In order to sell an unusually large number of an item to consumers, the price will need to be lowered to such an extent as to begin lowering the (total) revenue.

## Other Special Functions

In later sections we will often refer to the functions $y = x^2$ and $y = |x|$, graphed in Figures 11.14 and 11.15 of the last section, together with quadratic functions. Three additional functions that we will often encounter are $y = x^3$, $y = \sqrt{x}$, and $y = \sqrt[3]{x}$. We will now determine the domains and graphs of these functions.

The domain of $y = x^3$ and $y = \sqrt[3]{x}$ is the entire real line while the domain of $y = \sqrt{x}$ is $[0, +\infty)$. From Tables 11.5, 11.6, and 11.7, which show a number of values for these functions, the graphs are constructed in Figures 11.19, 11.20, and 11.21.

**Table 11.5**

| $x$ | $-3$ | $-2$ | $-1$ | 0 | 1 | 2 | 3 |
|---|---|---|---|---|---|---|---|
| $x^3$ | $-27$ | $-8$ | $-1$ | 0 | 1 | 8 | 27 |

**Table 11.6**

| $x$ | 0 | 1 | 4 | 9 | 16 | 25 | 36 |
|---|---|---|---|---|---|---|---|
| $\sqrt{x}$ | 0 | 1 | 2 | 3 | 4 | 5 | 6 |

**Table 11.7**

| $x$ | $-27$ | $-8$ | $-1$ | 0 | 1 | 8 | 27 |
|---|---|---|---|---|---|---|---|
| $\sqrt[3]{x}$ | $-3$ | $-2$ | $-1$ | 0 | 1 | 2 | 3 |

**Figure 11.19**

**Figure 11.20**

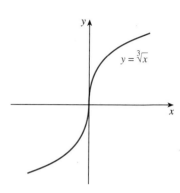

**Figure 11.21**

---

**Definition of Polynomial Function**

A **polynomial function** of degree $n$ is a function of the form

$$f(x) = a_n x^n + a_{n-1} x^{n-1} + \cdots + a_1 x + a_0 \qquad (a_n \neq 0)$$

where $a_n, a_{n-1}, \ldots, a_1, a_0$ are constants and $n$ is a nonnegative integer. The domain is all real numbers.

---

A polynomial function of degree 1 ($n = 1$), $f(x) = a_1 x + a_0$, ($a_1 \neq 0$) is a linear function and has already been encountered earlier. A polynomial function of degree 2 ($n = 2$), $f(x) = a_2 x^2 + a_1 x + a_0$, ($a_2 \neq 0$) is a quadratic function. The functions

$$f(x) = 5x^3 - x^2 + 3x - 5 \qquad \text{and} \qquad g(x) = 0.02x^6 - x^3 + x - 7$$

are polynomials of degree 3 and 6, respectively.

We continue with the definition of *rational function.*

---

**Rational Function**

A **rational function** $R(x)$ is the quotient of two polynomial functions and thus is of the form

$$R(x) = \frac{f(x)}{g(x)}$$

where $f(x)$ and $g(x)$ are polynomial functions. The domain of $R(x)$ is all real numbers for which $g(x) \neq 0$.

---

Thus

$$R(x) = \frac{x^5 + x - 2}{x - 3} \qquad \text{and} \qquad Q(x) = \frac{x + 1}{x^2 - 1}$$

are rational functions with the domain of $R(x)$ all real numbers other than $x = 3$ and the domain of $Q(x)$ is all real numbers other than $x = \pm 1$.

We have considered earlier in this section the functions $f(x) = \sqrt{x} = x^{1/2}$ and $g(x) = \sqrt[3]{x} = x^{1/3}$. These are examples of **power functions**.

---

**Power Functions**

A **power function** is a function of the form

$$f(x) = x^r$$

where $r$ is any real number.

---

If $r > 0$, the domains of power functions are either all real numbers or else all nonnegative numbers, depending on $r$. For example, the two rational functions $f(x) = x^{1/2}$ and $g(x) = x^{5/4}$ have $[0, +\infty)$ as their domains whereas $f(x) = x^{1/3}$ and $g(x) = x^{4/5}$ have all real numbers as their domains. If $r < 0$, then $x = 0$ cannot be in the domain. For example, if $f(x) = x^{-1/3} = 1/x^{1/3}$, then the domain is the real line except for $x = 0$. If $x = x^{-1/2} = 1/x^{1/2}$, then the domain is $(0, +\infty)$.

## SELF-HELP EXERCISE SET 11.2

1. Graph $y = (x + 3)^2 - 8$ and locate the vertex.

2. Suppose in Example 5 the cost function, in thousands of dollars per week, is given by $C(x) = 4x + 6$. Find the $x$ for which the profits will be maximized. Compare this value of $x$ to the value that maximizes revenue. Find any break-even quantities.

3. A bus company charges $70 per person for a sightseeing trip if 40 people travel in a group. If for each person above 40 the company reduces the charge per person by $1, how many people will maximize the total revenue for the bus company? What is the maximum revenue?

## EXERCISE SET 11.2

In Exercises 1 through 24 graph the given quadratic function and locate the vertex.

1. $y = -0.5x^2$
2. $y = -x^2$
3. $y = -2x^2$

4. $y = -4x^2$
5. $y = (x - 1)^2$
6. $y = (x - 3)^2$

7. $y = (x + 2)^2$
8. $y = (x + 4)^2$
9. $y = (x - 1)^2 + 2$

10. $y = 2(x + 2)^2 + 3$
11. $y = (x - 1)^2 - 2$

12. $y = 3(x - 2)^2 - 3$
13. $y = 2(x + 2)^2 - 3$

14. $y = (x + 1)^2 - 4$
15. $y = 3(x + 2)^2 + 4$

**16.** $y = (x + 3)^2 + 1$

**17.** $y = -(x - 1)^2 + 3$

**18.** $y = -2(x - 2)^2 + 1$

**19.** $y = -2(x - 2)^2 - 4$

**20.** $y = -(x - 1)^2 - 2$

**21.** $y = -(x + 1)^2 - 2$

**22.** $y = -3(x + 2)^2 - 1$

**23.** $y = -3(x + 3)^2 + 5$

**24.** $y = -(x + 1)^2 + 2$

In Exercises 25 through 40 complete the square and put in standard form. Locate the vertex and graph.

**25.** $y = x^2 - 4x + 4$

**26.** $y = x^2 - 8x + 16$

**27.** $y = x^2 + 2x + 1$

**28.** $y = x^2 + 6x + 9$

**29.** $y = x^2 - 2x + 4$

**30.** $y = 2x^2 - 12x + 20$

**31.** $y = x^2 - 2x$

**32.** $y = 2x^2 - 8x + 7$

**33.** $y = -x^2 + 4x - 2$

**34.** $y = -2x^2 + 4x + 1$

**35.** $y = -2x^2 + 12x - 20$

**36.** $y = -x^2 + 2x - 4$

**37.** $y = -x^2 - 2x - 2$

**38.** $y = -3x^2 - 6x - 4$

**39.** $y = -4x^2 - 8x - 3$

**40.** $y = -5x^2 - 10x - 4$

In Exercises 41 through 46 find where the revenue function is maximized using the given linear demand function.

**41.** $p = -x + 10$          **42.** $p = -2x + 3$          **43.** $p = -5x + 15$

**44.** $p = -0.5x + 20$       **45.** $p = -0.01x + 10$       **46.** $p = -0.2x + 12$

In Exercises 47 through 52 find the profit function using the given revenue and cost functions. Find the value of $x$ that maximizes the profit. Find the break-even quantities (if they exist), that is, the value of $x$ for which the profits are zero. Graph the solution.

**47.** $R(x) = -2x^2 + 30x,\ C(x) = 10x + 42$

**48.** $R(x) = -3x^2 + 20x,\ C(x) = 2x + 15$

**49.** $R(x) = -x^2 + 30x,\ C(x) = 15x + 36$

**50.** $R(x) = -2x^2 + 15x,\ C(x) = 5x + 8$

**51.** $R(x) = -2x^2 + 8x,\ C(x) = x^2 + 4x + 3$

**52.** $R(x) = -4x^2 + 10x,\ C(x) = 3x^2 + 3x + 10$

In Exercises 53 through 56 determine the degree of each of the given polynomial functions and their domains.

**53.** $p(x) = x^8 - x + 2$

**54.** $p(x) = 0.001x^6 + x^5 + x - 3$

**55.** $p(x) = 1000x^9 - 10x^7 + x^2$

**56.** $p(x) = 100x^{101} + x^{50} - x^9 + 2$

In Exercises 57 through 62 determine the domains of each of the given functions.

**57.** $\dfrac{x + 1}{x - 2}$

**58.** $\dfrac{x^3 - 1}{x^2 - 4}$

**59.** $\dfrac{x^4 + 1}{x^2 + 1}$

**60.** $x^{1/4}$

**61.** $x^{-1/5}$

**62.** $x^{-3/4}$

**63.** The sum of two numbers is 9. What is the largest possible value for their product?

**64.** The difference between two numbers is 2. Find the smallest possible value for the sum of their squares.

**65.** If 200 feet of fence is used to fence in a rectangular field and $x$ is the width of the field, the enclosed area is given by $A = x(100 - x)$. Determine the width and length that give the maximum area.

## Applications

*(John Colwell/Grant Heilman Photography, Inc.)*

**66. Cost.** A farmer wishes to enclose a rectangular field of area 200 square feet using an existing wall as one of the sides. The cost of the fence to go on the other three sides is $1 per foot. Find the dimensions of the rectangular field that minimizes the cost of the fence.

**67. Revenue.** An apple orchard produces annual revenue of $60 per tree when planted with 100 trees. Due to overcrowding, the annual revenue per tree is reduced by $0.50 for each additional tree planted. How many trees should be planted to maximize the revenue from the orchard?

**68. Profit.** If in the previous problem the cost of maintaining each tree is $5 per year, how many trees should be planted to maximize the profit from the orchard?

**69. Spread of Rumors.** It has been estimated that a rumor spreads at a rate $R$ that is proportional both to the ratio $r$ of individuals who have heard the rumor and to the ratio $(1 - r)$ who have not. Thus

$$R = kr(1 - r),$$

where $k$ is a positive constant. For what value of $r$ does the rumor spread fastest?

*(© Grant Heilman Photography, Inc.)*

**70. Biology.** When predators are not stimulated to search very hard for prey, the rate of change $R$ of the number $N$ of prey attacked per unit of area with respect to the number $n$ of prey is given by

$$R(x) = kx(ax - N),$$

where $x$ is the number of predators, $k$ is the per capita effectiveness of the predators, and $a$ is the maximum per capita effectiveness of the predators over an extended period of time. Find the number of predators that minimizes $R$.

**71. Physics.** If an object is initially at a height above the ground of 17 feet and is thrown straight upward with an initial velocity of 64 feet per second, then from physics it can be shown that the height in feet above the ground is given by

$$s(t) = -16t^2 + 64t + 17,$$

where $t$ is in feet. Find how long it takes for the object to reach maximum height. Find when the object hits the ground.

**72.   Physics.**   If an object is initially at a height above the ground of $s_0$ feet and is thrown straight upward with an initial velocity of $v_0$ feet per second, then from physics it can be shown that the height in feet above the ground is given by

$$s(t) = -16t^2 + v_0t + s_0,$$

where $t$ is in seconds. Find how long it takes for the object to reach maximum height. Find when the object hits the ground.

## Graphing Calculator Exercises

In Exercises 73 through 80 use a graphing calculator to graph the given functions on the same screen.

**73.   a.**   $f(x) = \frac{1}{3}x^2$,   **b.**   $f(x) = x^2$,   **c.**   $f(x) = 3x^2$

**74.   a.**   $f(x) = x^2$,   **b.**   $f(x) = x^2 + 5$,   **c.**   $f(x) = x^2 - 5$

**75.   a.**   $f(x) = (x - 6)^2$,   **b.**   $f(x) = (x + 6)^2$

**76.   a.**   $f(x) = (x - 6)^2 + 4$,   **b.**   $f(x) = (x + 6)^2 - 4$

**77.   a.**   $f(x) = 2(x - 5)^2 + 4$,   **b.**   $f(x) = 2(x + 5)^2 - 4$

**78.   a.**   $f(x) = \frac{1}{2}x^3$,   **b.**   $f(x) = x^3$,   **c.**   $f(x) = 2x^3$

**79.   a.**   $f(x) = |3x|$,   **b.**   $f(x) = |3x| + 5$,   **c.**   $f(x) = |3x| - 5$

**80.   a.**   $f(x) = |x|$,   **b.**   $f(x) = |x - 4|$,   **c.**   $f(x) = |x + 4|$

For each of the next two linear demand equations, use a graphing calculator to graph the revenue function and find the point where the revenue attains a maximum.

**81.**   $p = -x + 5$                    **82.**   $p = -2x + 7$

**83.**   On your graphing calculator, graph on the same screen the equations $y = \sqrt[2]{x}$, $y = \sqrt[3]{x}$, $y = \sqrt[4]{x}$, $y = \sqrt[5]{x}$. What can you now say about the graphs of $y = \sqrt[n]{x}$ when $n$ is even and when $n$ is odd?

## Solutions to Self-Help Exercise Set 11.2

**1.**   The quadratic $y = (x + 3)^2 - 8$ is in standard form $y = a(x - h)^2 + k$ with $a = 1, h = -3$, and $k = -8$. Since $a > 0$, the parabola opens upward and the vertex $(h, k) = (-3, -8)$ is the point on the graph where the function attains its minimum.

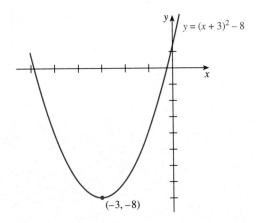

$$y = (x + 3)^2 - 8$$

$(-3, -8)$

**2.**   First complete the square.

$$P(x) = R(x) - C(x)$$
$$= [-2x^2 + 12x] - [4x + 6]$$
$$= -2x^2 + 8x - 6$$
$$= -2(x^2 - 4x + 3)$$
$$= -2([x^2 - 4x + 4] - 4 + 3) \qquad \left(\frac{-4}{2}\right)^2 = 4$$
$$= -2([x - 2]^2 - 1)$$
$$= -2(x - 2)^2 + 2$$

The quadratic is now in standard form. Thus we see that profits are maximized at $x = 2$ or 2000 units.

We could have used the formula

$$h = -\frac{b}{2a} = -\frac{8}{2(-2)} = 2.$$

To find the break-even quantities, we must set $P = 0$. This gives

$$0 = P(x) = -2x^2 + 8x - 6 = -2(x^2 - 4x + 3) = -2(x - 1)(x - 3).$$

From this factorization it is apparent that $x = 1$ and $x = 3$ are the break-even quantities. See the figure.

Notice the point where profit is maximized is not the same point for which revenue is maximized. This figure is typical of profit functions. As more items are produced and sold, initial losses turn into profits. But if too many items are made and sold, this flooding of the market will result in a downturn in profits.

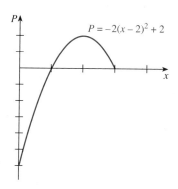

$$P = -2(x - 2)^2 + 2$$

3. The revenue $R$ for the bus company is the number $n$ of people who take the trip times the price $p$ each person pays, that is, $R = n \cdot p$. Let $x$ be the number of people in excess of 40 that take the trip. Then the total number that take the trip is $n = 40 + x$. For each person in excess of 40 who takes the trip, the company lowers the price by \$1. Thus the price per person is $p = 70 - x$. Revenue is then

$$R = n \cdot p = (40 + x)(70 - x) = 2800 + 30x - x^2.$$

This is a quadratic. Completing the square yields

$$2800 + 30x - x^2 = -(x^2 - 30x - 2800)$$

$$= -(x^2 - 30x + 225 - 225 - 2800) \qquad \left(\frac{-30}{2}\right)^2 = 225$$

$$= -(x^2 - 30x + 225 - 3025)$$

$$= -(x - 15)^2 + 3025$$

The quadratic is now in standard form. The maximum occurs at $x = 15$, and the maximum is 3025. Thus the bus company will have maximum revenue if the number that take the trip is $40 + x = 40 + 15 = 55$. The maximum revenue will be \$3025.

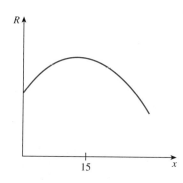

## 11.3 COMBINATIONS OF FUNCTIONS

► *The Algebra of Functions*
► *The Composition of Functions*

**A P P L I C A T I O N**
**Area of Forest Fire**

Suppose a forest fire is spreading in a circular manner with the radius of the burnt area increasing at a steady rate of 3 feet per minute. What is the area of the circular burnt region as a function of the time? See Example 6 for the answer.

*(© Kent and Donna Dannen/Photo Researchers, Inc.)*

### The Algebra of Functions

A function such as $h(x) = x^4 + \sqrt{x + 1}$ can be viewed as the sum of two simpler functions: $f(x) = x^4$ and $g(x) = \sqrt{x + 1}$. We have always written profit as revenue less costs, that is, $P(x) = R(x) - C(x)$. A function such as

$$r(x) = \frac{1 + 10x^3}{1 + x^2},$$

can be viewed as the quotient of the two polynomials $p(x) = 1 + 10x^3$ and $q(x) = 1 + x^2$. The two polynomials are much simpler than the original rational function. Breaking a complicated function into less complicated parts can yield insight into the behavior of the original function.

As we noted above, functions can be added, subtracted, multiplied, and divided. We state now the formal definitions of these concepts.

---

**The Algebra of Functions**

Let $f$ and $g$ be two functions and define

$$D = \{x \mid x \in \text{(domain of } f), x \in \text{(domain of } g)\}$$

Then for all $x \in D$, we define

The sum $f + g$ by $\qquad (f + g)(x) = f(x) + g(x)$
The difference $f - g$ by $\qquad (f - g)(x) = f(x) - g(x)$
The product $f \cdot g$ by $\qquad (f \cdot g)(x) = f(x) \cdot g(x)$

The quotient $f/g$ by $\qquad \left(\dfrac{f}{g}\right)(x) = \dfrac{f(x)}{g(x)} \qquad$ for $x$ with $g(x) \neq 0$.

---

**E X A M P L E   1**   **Examples of Combining Functions**

Let $f(x) = \sqrt{x}$ and $g(x) = 1 - x$. Find $(f + g)(x)$, $(f - g)(x)$, $(f \cdot g)(x)$, $(f/g)(x)$, and find the domains of each of them.

**Solution**     The domain of $f$ is $[0, +\infty)$ and the domain of $g$ is all real numbers. Thus the common domain of the first three functions is $D = [0, +\infty)$.

$$(f + g)(x) = f(x) + g(x) = \sqrt{x} + (1 - x) = \sqrt{x} + 1 - x$$

$$(f - g)(x) = f(x) - g(x) = \sqrt{x} - (1 - x) = \sqrt{x} - 1 + x$$

$$(f \cdot g)(x) = f(x) \cdot g(x) = \sqrt{x}(1 - x)$$

$$\left(\frac{f}{g}\right)(x) = \frac{f(x)}{g(x)} = \frac{\sqrt{x}}{1 - x}$$

The domain of $f/g$ is all nonnegative numbers other than 1.   ∎

## The Composition of Functions

Consider the function $h$ given by the equation

$$h(x) = \sqrt{25 - x^2}.$$

In order to evaluate $h(3)$, for example, we would first take $25 - (3)^2 = 16$ and then take the square root of 16 to obtain $h(3) = 4$. Thus, we can view this function as the combination of two simpler functions as follows. Let

$$y = f(u) = \sqrt{u}$$

$$u = g(x) = 25 - x^2,$$

then

$$y = \sqrt{u} = \sqrt{25 - x^2}.$$

In terms of the two functions $f$ and $g$,

$$h(x) = \sqrt{25 - x^2} = \sqrt{g(x)} = f[g(x)].$$

The function $h(x)$ is called the **composition** of the two functions $f$ and $g$. We use the notation $h(x) = (f \circ g)(x)$.

The domain of $h$ consists of those values in the domain of $g$ for which $g(x)$ is in the domain of $f$. In the above case the domain of $g$ is $(-\infty, +\infty)$, while the domain of $f$ is the set of nonnegative numbers. Thus the domain of $h$ consists of those numbers $x$ for which $g(x) = 25 - x^2$ is nonnegative. Thus the domain of $h$ consists of all numbers for which $x^2 \leq 25$ or the set $[-5, 5]$. See Figure 11.22.

We now define in general terms the composition of two functions. See Figure 11.23.

---

**Definition of Composite Function**

Let $f$ and $g$ be two functions. The composite function $f \circ g$ is defined by

$$(f \circ g)(x) = f[g(x)].$$

The domain of $f \circ g$ is the set of all $x$ in the domain of $g$ for which $g(x)$ is in the domain of $f$.

---

Figure 11.22

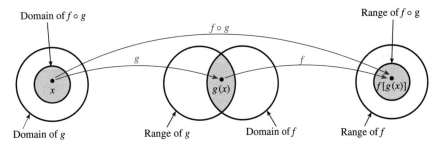

Figure 11.23

Sometimes all of the range of $g$ is in the domain of $f$, as shown in Figure 11.24. In this special case the domain of $f \circ g$ is just the domain of $g$.

Another way of looking at the composition is shown in Figure 11.25. Here the conveyor belt inputs $x$ into the function $g$, which outputs $g(x)$. The conveyor belt then inputs the value $g(x)$ into the function $f$, which then outputs $f[g(x)]$.

Figure 11.24

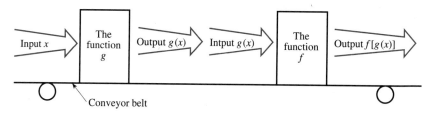

**Figure 11.25**

EXAMPLE 2    **Finding a Composite Function**

Let $f(x) = x + 1$ and $g(x) = x^3$. Find the domains of $f \circ g$ and $g \circ f$. Find $(f \circ g)(2)$ and $(g \circ f)(2)$. Find $(f \circ g)(x)$ and $(g \circ f)(x)$.

Solutions    Since the domains of $f$ and $g$ are all the real numbers, the domains of $f \circ g$ and $g \circ f$ are also all the real numbers.

$$(f \circ g)(2) = f[g(2)] = f[2^3] = f[8] = 8 + 1 = 9$$

$$(g \circ f)(2) = g[f(2)] = g[2 + 1] = g[3] = 3^3 = 27$$

$$(f \circ g)(x) = f[g(x)] = f[x^3] = x^3 + 1$$

$$(g \circ f)(x) = g[f(x)] = g[x + 1] = (x + 1)^3 \quad \blacksquare$$

REMARK.    Notice that $f \circ g$ and $g \circ f$ are two very different functions.
   In the next example we need to be more careful about the domain of the composite function.

EXAMPLE 3    **Finding a Composite Function and its Domain**

Let $g(x) = \sqrt{x}$ and let $f(x) = x^2 + 5$. Find $(f \circ g)(x)$ and its domain.

Solution    $$(f \circ g)(x) = f[g(x)] = f[\sqrt{x}] = [\sqrt{x}]^2 + 5 = x + 5$$

One might naturally be inclined to think (incorrectly) that since one can take $x + 5$ for any $x$, that the domain of $f \circ g$ is all real numbers. But notice that the domain of $f \circ g$ is all $x$ in the *domain of g* for which $g(x)$ is in the domain of $f$. But the domain of $g$ is $[0, +\infty)$, and the domain of $f$ is all real numbers. Thus the domain of $f \circ g$ is $[0, +\infty)$.    $\blacksquare$

REMARK.    To avoid possible mistakes keep in mind that the domain of $f \circ g$ is always a subset of the domain of $g$ and that the range of $f \circ g$ is always a subset of the range of $f$.

E X A M P L E   4

### Writing a Function as the Composition of Simpler Functions

Let $h(x) = (3x + 1)^9$. Find two functions $f$ and $g$ such that $(f \circ g)(x) = h(x)$.

**Solution**     There are many possibilities. But perhaps the simplest is to notice that we are taking the ninth power of something. The something is $3x + 1$. Thus we might set $g(x) = 3x + 1$ and $f(x) = x^9$. Then

$$(f \circ g)(x) = f[g(x)] = f[3x + 1] = (3x + 1)^9 = h(x). \quad \blacksquare$$

E X A M P L E   5

### Measurement Conversions

Let $x$ be the number of bushels, $q$ the number of quarts, and $c$ the number of cups. Given that there are 32 quarts to a bushel, find the function $h$ such that $q = h(x)$. Given that there are 4 cups to a quart, find the function $f$ such that $c = f(q)$. Now determine $c$ as a function of $x$ and relate this to the composition of two functions. Explain your formula in words.

**Solution**     We have $q = 32x = h(x)$ and $c = 4q = f(q)$. Then

$$c = 4q = 4(32x) = 128x.$$

In terms of the functions $f$ and $h$, this is the same as

$$c = (f \circ h)(x) = f[h(x)] = f[32x] = 4[32x] = 128x.$$

This says that there are $128x$ cups in $x$ bushels.   $\blacksquare$

E X A M P L E   6

### Area of a Forest Fire

Suppose a forest fire is spreading in a circular manner with the radius of the burnt area increasing at a steady rate of 3 feet per minute. Write the area of the circular burnt region as a function of the time.

**Solution**     The area of a circle of radius $r$ is given by $A(r) = \pi r^2$. Refer to Figure 11.26. If $r$ is the radius of the burnt region, then $r(t) = 3t$, where $t$ is measured in minutes and $r$ in feet. Thus

$$A = A[r(t)] = A[3t] = \pi(3t)^2 = 9\pi t^2. \quad \blacksquare$$

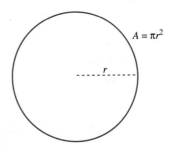

$A = \pi r^2$

$r$

**Figure 11.26**

## SELF-HELP EXERCISE SET 11.3

1. If $f(x) = x^2 + 1$ and $g(x) = x - 3$, find (a) $(f + g)(x)$, (b) $(f - g)(x)$, (c) $(f \cdot g)(x)$, and (d) $\left(\dfrac{f}{g}\right)(x)$, and the domains of each.

2. If $f(x) = x^2 + 1$ and $g(x) = \sqrt{x - 3}$, find $(f \circ g)(x)$ and the domain.

## EXERCISE SET 11.3

In Exercises 1 through 2 let $f(x) = 3x + 1$ and $g(x) = x^2$ and find the indicated quantity.

1.  **a.** $(f + g)(2)$, **b.** $(f - g)(2)$, **c.** $(f \cdot g)(2)$, **d.** $\left(\frac{f}{g}\right)(2)$

2.  **a.** $(g - f)(-3)$, **b.** $(g \cdot f)(-3)$, **c.** $\left(\frac{g}{f}\right)(-3)$, **d.** $(g + f)(-3)$

In Exercises 3 through 10 you are given a pair of functions, $f$ and $g$. In each case find $(f + g)(x)$, $(f - g)(x)$, $(f \cdot g)(x)$, and $\left(\dfrac{f}{g}\right)(x)$, and the domains of each.

3.  $f(x) = 2x + 3, g(x) = x^2 + 1$
4.  $f(x) = x^3, g(x) = 3$
5.  $f(x) = 2x + 3, g(x) = x + 1$
6.  $f(x) = x^3, g(x) = x - 1$
7.  $f(x) = \sqrt{x + 1}, g(x) = x + 2$
8.  $f(x) = \sqrt{x + 1}, g(x) = x - 3$
9.  $f(x) = 2x + 1, g(x) = 1/x$
10. $f(x) = \sqrt{x + 2}, g(x) = 1/x$

Let $f(x) = 2x + 3$ and $g(x) = x^3$. Find the indicated quantity.

11. **a.** $(g \circ f)(1)$, **b.** $(g \circ f)(-2)$, **c.** $(f \circ g)(1)$, **d.** $(f \circ g)(-2)$

In Exercises 12 through 17 you are given a pair of functions, $f$ and $g$. In each case find $(f \circ g)(x)$ and $(g \circ f)(x)$ and the domains of each.

12. $f(x) = 2x + 1, g(x) = 3x - 2$
13. $f(x) = 2x + 3, g(x) = x^3$
14. $f(x) = x^2, g(x) = \sqrt{x}$
15. $f(x) = x^3, g(x) = \sqrt[3]{x}$
16. $f(x) = |x|, g(x) = \sqrt{x}$
17. $f(x) = \sqrt{x}, g(x) = x^4 + 1$

In Exercises 18 through 19 let $f(x) = 2x + 1$ and $g(x) = 2x^3$. Find the indicated quantity.

18. **a.** $(f \circ g)(1)$, **b.** $(g \circ f)(1)$, **c.** $(f \circ f)(1)$ **d.** $(g \circ g)(1)$

19. **a.** $(f \circ g)(x)$, **b.** $(g \circ f)(x)$ **c.** $(f \circ f)(x)$, **d.** $(g \circ g)(x)$

In Exercises 20 through 28 express each of the given functions as the composition of two functions. Find the two functions that seem the simplest.

20. $(3x - 4)^7$      21. $(x + 5)^5$      22. $\sqrt{x^3 + 1}$

**23.** $\sqrt[3]{x + 1}$      **24.** $|2x - 3|$      **25.** $|x^2 - 1|$

**26.** $1/(3x + 2)$      **27.** $1/(x^2 + 1)$      **28.** $\sqrt{x^3 + 1}$

## Applications

**29. Cost.** A manufacturing firm has a daily cost function of $C(x) = 3x + 10$, where $x$ is the number of thousands of an item produced and $C$ is in thousands of dollars. Suppose the number of items that can be manufactured is given by $x = n(t) = 3t$, where $t$ is measured in hours. Find $(C \circ n)(t)$ and state what this means.

**30. Revenue.** A firm has a revenue function given by $R(p) = 10p - p^2$, where $p$ is the price of a chocolate bar sold in dollars and $R$ is measured in thousands of dollars per day. Suppose the firm is able to increase the price of each bar by 5 cents each year (without affecting demand). If $t$ is time measured in years, write an equation for the revenue as a function of $t$ if the price of a candy bar starts out at 25 cents.

**31. Revenue and Cost.** If $R(x)$ is the revenue function and $C(x)$ is the cost function, what does the function $(R - C)(x)$ stand for?

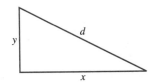

**32. Distance.** Two ships leave the same port at the same time. The first travels due north at 4 miles per hour and the second travels due east at 5 miles per hour. Find the distance $d$ between the ships as a function of time $t$ measured in hours. *Hint:* $d = \sqrt{x^2 + y^2}$. Now find both $x$ and $y$ as functions of $t$.

**33. Medicine.** A cancerous spherical tumor originally 30 millimeters in radius is decreasing at the rate of 2 millimeters per month after treatment. Write an equation for the volume of the tumor as a function of time $t$ measured in months. Note that the volume $V$ of a sphere of radius $r$ is given by

$$V(r) = \frac{4}{3} \pi r^3.$$

**34. Revenue.** Suppose a firm's annual revenue function is given by $R(x) = 20x + .01x^2$, where $x$ is the number of items sold and $R$ is in dollars. The firm sells 1000 items now and anticipates its sales to increase by 100 each of the next number of years. If $t$ is the number of years from now, write the number of sales as a function of $t$ and also write the revenue as a function of $t$.

**35. Measurement Conversion.** Let $x$ be the length in furlongs, $r$ in rods, and $y$ in yards. Given that there are 40 rods to a furlong, find the function $g$ such that $r = g(x)$. Given that there are 5.5 yards to a rod, find the function $f$ such that $y = f(r)$. Now determine $y$ as a function of $x$ and relate this to the composition of two functions. Explain your formula in words.

**36. Measurement Conversion.** Let $x$ be the size in hectares, $a$ in acres, and $y$ in square yards. Given that there are 100 acres to a hectare, find the function $g$ such that $r = g(x)$. Given that there are 4840 square yards to an acre, find the function $f$ such that $y = f(a)$. Now determine $y$ as a function of $x$ and relate this to the composition of two functions. Explain your formula in words.

### Solutions to Self-Help Exercise Set 11.3

1.  If $f(x) = x^2 + 1$ and $g(x) = x - 3$, then
    a.  $(f + g)(x) = f(x) + g(x) = (x^2 + 1) + (x - 3) = x^2 + x - 2$,
    b.  $(f - g)(x) = f(x) - g(x) = (x^2 + 1) - (x - 3) = x^2 - x + 4$,
    c.  $(f \cdot g)(x) = f(x) \cdot g(x) = (x^2 + 1)(x - 3) = x^3 - 3x^2 + x - 3$,
    d.  $\left(\dfrac{f}{g}\right)(x) = \dfrac{f(x)}{g(x)} = \dfrac{x^2 + 1}{x - 3}$

    The domains of both $f$ and $g$ are the set of all real numbers. The domains of the first three functions are also the set of all real numbers. The domain of the quotient is the set of all real numbers except 3.

2.  If $f(x) = x^2 + 1$ and $g(x) = \sqrt{x - 3}$, then

    $$(f \circ g)(x) = f[g(x)] = (\sqrt{x - 3})^2 + 1 = (x - 3) + 1 = x - 2.$$

    The domain of $f$ is $(-\infty, +\infty)$ and the domain of $g$ is $[3, +\infty)$. The domain of $(f \circ g)$ is then $[3, +\infty)$.

## SUMMARY OUTLINE OF CHAPTER 11

- A **function** $f$ from the set D to the set R is a rule that assigns to each element $x$ in D one and only one element $y = f(x)$ in R.

- The set D above is called the **domain**.

- One thinks of the domain as the set of inputs and the values $y = f(x)$ as the outputs.

- The set of all possible outputs is called the **range**.

- The letter representing the elements in the domain is called the **independent variable**.

- The letter representing the elements in the range is called the **dependent variable**.

- The **graph** of the function $f$ consists of all points $(x, y)$ such that $x$ is in the domain of $f$ and $y = f(x)$.

- **Vertical line test.**   A graph in the $xy$-plane represents a function if, and only if, every vertical line intersects the graph in at most one place.

- The function $f(x) = ax^2 + bx + c, a \neq 0$, is called a **quadratic function**.

- The graph of a quadratic function is called a **parabola**.

- Every quadratic function $f(x) = ax^2 + bx + c$ with $a \neq 0$ can be put in **standard form** $f(x) = a(x - h)^2 + k$ where $h = -\dfrac{b}{2a}$ and $k = c - \dfrac{b^2}{4a}$ by **completing the square**.

- The point $(h, k)$ is called the **vertex**.

- When $a < 0$, the graph of the quadratic $y = ax^2 + bx + c$ looks like $\frown$ and assumes the maximum value $k$ when $x = h$.

- When $a > 0$, the graph of the quadratic $y = ax^2 + bx + c$ looks like $\smile$ and assumes the minimum value $k$ when $x = h$.

- A **polynomial function** of degree $n$ is a function of the form $f(x) = a_n x^n + a_{n-1}x^{n-1} + \cdots + a_1 x + a_0, a_n \neq 0$. The domain is the set of all real numbers.

- A **rational function** is the quotient of two polynomials. The domain is the set of all real numbers for which the denominator is not zero.

- A **power function** is a function of the form $f(x) = x^r$, where $r$ is any real number.

- The **sum** $f + g$ is defined by $(f + g)(x) = f(x) + g(x)$.

- The **difference** $f - g$ is defined by $(f - g)(x) = f(x) - g(x)$.

- The **product** $f \cdot g$ is defined by $(f \cdot g)(x) = f(x) \cdot g(x)$.

- The **quotient** $\dfrac{f}{g}$ is defined by $\left(\dfrac{f}{g}\right)(x) = \dfrac{f(x)}{g(x)}$, if $g(x) \neq 0$.

- The **composition** $f \circ g$ is defined by $(f \circ g)(x) = f[g(x)]$ for all $x$ in the domain of $g$ for which $g(x)$ is in the domain of $f$.

## Chapter 11 Review Exercises

In Exercises 1 through 6 determine which graphs are graphs of functions.

**1.**

**2.**

**3.**

**4.**

**5.**     **6.**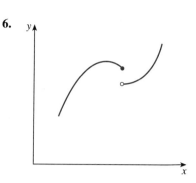

In Exercises 7 through 8 find the domain of each of the functions.

**7.** $\dfrac{x + 1}{x - 3}$    **8.** $\dfrac{x + 1}{\sqrt{x - 5}}$

**9.** Let $f(x) = \dfrac{1}{x^2 + 1}$. Evaluate $f(0)$, $f(\sqrt{2})$, $f(a - 1)$, $f(x + 1)$.

**10.** Graph

$$f(x) = \begin{cases} x^2 & \text{if } x \le 0 \\ x + 1 & \text{if } 0 < x \le 1 \\ -x + 2 & \text{if } x > 1 \end{cases}$$

**11.** Graph $y = -(x + 3)^2 + 5$.

**12.** Graph $y = |x - 5| - 4$.

In Exercises 13 through 16 let $f(x) = x^3 + 4$ and $g(x) = 2x - 1$. Find the indicated quantity and the domains for each.

**13.** $(f + g)(x)$    **14.** $(f - g)(x)$

**15.** $(f \cdot g)(x)$    **16.** $\left(\dfrac{f}{g}\right)(x)$

In Exercises 17 through 18 find $(f \circ g)(x)$ and $(g \circ f)(x)$ for the given $f$ and $g$ and find the domains for each.

**17.** $f(x) = x^2 + 2$, $g(x) = 3x + 1$    **18.** $f(x) = \sqrt{x - 1}$, $g(x) = x + 1$

**19. Packaging.** A box with top has length twice the width and height equal to the width. If $x$ denotes the width and $S$ the surface area, find $S$ as a function of $x$.

**20. Postage.** The postage on a first class letter is $0.29 for the first ounce, and $.23 for each additional ounce or any fraction thereof. If $x$ denotes the number of ounces that a letter weighs and $C(x)$ the first class postage, graph the function $C(x)$ for $0 < x \le 3$.

**21. Spread of Infestation.** A certain pine disease is spreading in a circular fashion through a pine forest with the radius of the infested region increasing at a rate of 2 miles per year. Write the area of the infested circular region as a function of time, $t$, measured in years.

**The velocity of the eagle changes from instant to instant. (© Mark Wilson/ Photo Researchers, Inc.)**

# The Derivative

12.0   Introduction to Calculus

12.1   Limits

12.2   Continuity

12.3   Rates of Change

12.4   The Derivative

12.5   Derivatives of Constants, Powers, and Sums

12.6   Derivatives of Products and Quotients

12.7   The Chain Rule

## Introduction

This chapter begins the study of calculus. The notions of limits and continuity are discussed first. Then the derivative is introduced with applications. The second half of the chapter develops various formulas used to find derivatives, and further applications are given.

# 12.0   INTRODUCTION TO CALCULUS

**Isaac Newton,
1642–1727**

*(The Bettmann
Archive)*

Newton is considered the inventor of the calculus. He also made very significant discoveries in the sciences and is often considered the most accomplished scientific thinker of recent centuries. He attended public schools in England, where he was undistinguished. Indeed he was reported to be idle and inattentive. He blossomed quickly however, and, while confining himself to his room for two years from 1665–1667 to avoid the plague that was raging at the time, he laid the foundations for his future accomplishments. During these golden years he laid the foundations for calculus, discovered that white light can be split into different colors, and discovered the law of universal gravitation. Any one of these discoveries would have made him an outstanding figure in the history of science.

**Gottfried Leibniz,
1646–1716**

*(The Bettmann
Archive)*

Leibniz is considered the coinventor of the calculus. He was a precocious child and mentioned once that at age 14 he would go for walks in the woods, ''comparing and contrasting the principles of Aristotle with those of Democritus.'' He invented a calculating machine that could multiply and divide. He commented on the machine that ''It is unworthy of excellent men to lose hours like slaves in the labor of calculation.'' He received a degree in law and served in the diplomatic service, only later taking up mathematics. He developed much of the currently used notation in the calculus and a number of the elementary formulas.

*(© Elisabeth Weiland/Photo
Researchers, Inc.)*

Except possibly for the creation of Euclidean geometry, the calculus represents the greatest creation in all of mathematics. Armed with the calculus, major scientific problems of the 17th century were solved. Even today, the use of calculus continues to spread into new areas such as business, economics, and biology.

There were four major problems in the seventeenth century solved by the calculus. The first problem was to find a means of calculating the (instantaneous) velocity of a moving body. To everyone who walks, runs, rides a horse, or flies in a plane (Figure 12.1), it is physically apparent that a moving body has an instantaneous velocity. If the functions $s = s(t)$ gives the distance a body has moved in the time $t$, how can one calculate the (instantaneous) velocity if the velocity is different at different times? One can calculate the average velocity as distance divided by time. But since no distance or time elapses in an instant, this formula will give the meaningless expression $\frac{0}{0}$ if used to calculate the instantaneous velocity.

The second problem was to find the slope of the line that is tangent to a curve at a given point. Refer to Figure 12.2. This problem was not just a problem in pure geometry, but was also of considerable scientific interest in the 17th century.

**Figure 12.1**

Figure 12.2

Figure 12.3

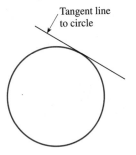

Figure 12.4

At that time men such as Fermat and Huygens were interested in the construction of lenses. This required a knowledge of the angle between the ray of light and the normal line to the lens shown in Figure 12.3. Since the normal line is perpendicular to the tangent line, knowledge of the tangent line will immediately give the normal line.

Actually there was a serious problem as to what tangent actually meant. For example, looking at Figure 12.4, it is easy to agree that a tangent to a figure, such as a circle, is that line that touches the figure at only one point and lies entirely on one side of the figure. This was known by the Greeks. But more complex curves such as shown in Figure 12.5 were being considered in the seventeenth century. Is the line shown in this figure a tangent? (We will see later that it is.)

The third problem was to find the maximum or minimum value of a function on some given interval. For us, we might be interested in finding the point at which maximum profits occur in Figure 12.6.

The fourth type of problem was to find the area under a curve. Refer to Figure 12.7. We will see in a later chapter why this is so important in the applications.

Figure 12.5

Figure 12.6

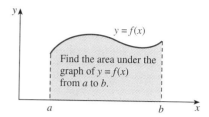

**Figure 12.7**

In the first problem considered above, velocity measures the rate at which distance is changing with respect to a change in time. In exact analogy, the calculus will then enable us to determine, among other things, the rate at which profits are changing with respect to the changes in the number of items produced and sold, the rate at which a population is growing with respect to changes in the time, and the rate at which demand for an item drops with respect to increases in the price of the item. Thus we will see that the calculus can be used not only to solve the great scientific problems of the past centuries but also to solve many important problems in modern business, economics, and biology.

## 12.1   LIMITS

▶ *Introduction to Limits*

▶ *Two Fundamentally Important Applications*

▶ *Some Limits that Do Not Exist*

▶ *Rules for Limits*

**Augustin Louis Cauchy,
1789–1857**

*(The Bettmann
Archive)*

It is interesting to note that Newton and Leibniz, who are given credit for the discovery of the calculus, did *not* develop the theory on the basis of a clear and logical foundation. Indeed they, and the many individuals that expanded the calculus in the subsequent 150 years, used arguments that were intuitively plausible rather than mathematically exact. From the beginning, the calculus, as presented, evoked considerable controversy, with many critics pointing out (correctly) that many fundamental concepts were vague. The success of the calculus in solving many physical problems is what gave individuals the confidence that the mathematics must somehow be correct. It was Cauchy who finally placed the calculus on a sound footing. He developed an acceptable theory of limits and used this theory as the logical underpinnings of the calculus. Thus Cauchy has become known as the creator of the calculus in the modern sense.

APPLICATION

**Velocity Problem**

If a baseball is dropped from the side of a building as in Figure 12.8, it can be shown using basic principles of physics that the distance $s$ in feet that the ball falls is given by $s = s(t) = 16t^2$, where $t$ is measured in seconds. What is the velocity at the instant $t = 1$? For the answer see Example 4.

APPLICATION

**Tangent Line Problem**

What is the slope of the line tangent to the curve $y = f(x) = x^2$ at the point $x = 1$? Refer to Figure 12.9. For the answer see Example 5.

## Introduction to Limits

Before attacking the two major problems given above, we first need to intuitively establish the notion of a limit and to develop some elementary properties of limits.

As our first example consider the graph given in Figure 12.10 of some function $y = f(x)$. Notice that the function is not defined for the value $x = 1$. Thus the point $(1, 2)$ is not on the graph. This is denoted in Figure 12.10 by a circle surrounding the point $(1, 2)$. We seek to understand what is happening to $f(x)$ as $x$ approaches (gets closer and closer to) 1 without $x$ actually attaining the value 1.

First notice that it is meaningless to have $x$ actually attain the value 1 since $f(1)$ is not even defined. In Figure 12.10 we can see that as $x$ approaches 1 from the right, the point $(x, f(x))$ slides along the graph and approaches the point $(1, 2)$ and $f(x)$ approaches the number 2. Also we can see that as $x$ approaches 1 from the left, the point $(x, f(x))$ slides along the graph and approaches the point $(1, 2)$ and $f(x)$ again approaches the number 2. We then see that no matter how $x$ approaches 1, $f(x)$ approaches the same number $L = 2$.

**Figure 12.8**

**Figure 12.9**

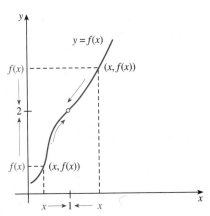

**Figure 12.10**

In such a case we say that $L = 2$ is the **limit** of $f(x)$ as $x$ approaches 1 and write

$$\lim_{x \to 1} f(x) = 2.$$

We now give the following intuitive definition of limit of a function.

---

**Definition of Limit of a Function**

Suppose that the function $f(x)$ is defined for all values of $x$ near $a$, but not necessarily at $a$. If as $x$ approaches $a$ (without actually attaining the value $a$), $f(x)$ approaches the number $L$, then we say that $L$ is that limit of $f(x)$ as $x$ approaches $a$, and write

$$\lim_{x \to a} f(x) = L.$$

---

We say that this definition is intuitive since the definition does not define the term "approaches" in a strictly mathematical sense.

**REMARK.**    Notice that in this definition, $f(a)$ need not be defined. Furthermore, even if $f(a)$ is defined, the limit process never requires knowledge of $f(a)$, since this process only requires knowledge of $f(x)$ for $x$ near $a$ but not *equal* to $a$. Thus even if $f(a)$ is defined, this value has nothing to do with the limit as $x$ approaches $a$.

In determining the limit of $f(x)$ as $x$ approaches 1, we found it convenient to consider the limit of $f(x)$ as $x$ approaches 1 from the left and also the limit of $f(x)$ as $x$ approaches 1 from the right. These two limits are written as

$$\lim_{x \to 1-} f(x), \qquad \lim_{x \to 1+} f(x),$$

respectively. In the above example, we discovered that both of these limits exist and equal the very same number $L = 2$. We then concluded that the limit of $f(x)$ as $x$ approaches 1 existed and equaled $L = 2$. This reasoning is true in general. We then have the following theorem.

---

**Existence of a Limit**

If the function $f(x)$ is defined near $x = a$ but not necessarily at $x = a$, then

$$\lim_{x \to a} f(x) = L$$

if, and only if, both the limits

$$\lim_{x \to a-} f(x) \text{ and } \lim_{x \to a+} f(x)$$

exist and are equal to the same number $L$.

---

The following example should further clarify the idea of the limit.

E X A M P L E   1     **Finding the Limit Geometrically**

Define the function $g(x)$ to be the very same function as $f(x)$, except that $g(1) = 1.5$. See Figure 12.11. Find $\lim\limits_{x \to 1} g(x)$ if it exists.

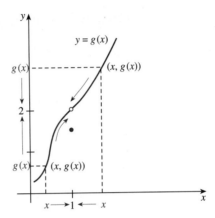

**Figure 12.11**

**Solution**     Notice that from Figure 12.11 that we still have

$$\lim_{x \to 1-} g(x) = 2 \quad \text{and} \quad \lim_{x \to 1+} g(x) = 2.$$

Thus                         $\lim\limits_{x \to 1} g(x) = 2.$ ■

**R E M A R K .**     Notice that this last statement is true despite the fact that $g(1) = 1.5$, since the limit process does not involve the value of the function at $x = 1$.

The two functions $f(x)$ and $g(x)$ given above are not actually the same function since they differ for the single value $x = 1$. However, they are equal for all other values, and, therefore, the limits as $x$ approaches 1 of these two functions must be equal.

This reasoning applies in general. That is, if two functions are equal everywhere except possibly at the limiting value, then the limits are equal. In practice this means that if $f(x) = g(x)$ except possibly at $x = a$, then we can evaluate $\lim\limits_{x \to a} f(x)$ by replacing $f(x)$ with $g(x)$. We have the following.

---

**The Replacement Theorem**

Suppose $f(x) = g(x)$ for all $x \neq a$. If $\lim\limits_{x \to a} g(x)$ exists, then $\lim\limits_{x \to a} f(x)$ exists and

$$\lim_{x \to a} f(x) = \lim_{x \to a} g(x).$$

---

**E X A M P L E  2**   **Finding the Limit Using Numerical Calculations**

Find $\lim\limits_{x \to 2} (3x + 5)$ if it exists.

**Solution**    It seems intuitive that as $x$ approaches 2, $3x$ approaches 6, and $3x + 5$ approaches $6 + 5 = 11$. See Figure 12.12. The calculations summarized in Table 12.1 indicate this.

Thus                                $\lim\limits_{x \to 2} (3x + 5) = 11.$ ∎

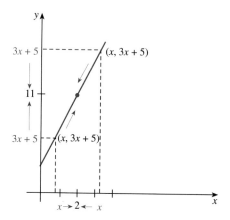

**Figure 12.12**

**Table 12.1**

| $x$ | 1.9 | 1.99 | 1.999 | → | 2 | ← | 2.001 | 2.01 | 2.1 |
|---|---|---|---|---|---|---|---|---|---|
| $3x + 5$ | 10.7 | 10.97 | 10.997 | → | 11 | ← | 11.003 | 11.03 | 11.3 |

E X A M P L E 3    **Finding the Limit Using Cancellation**

Find $\lim\limits_{x \to 2} \left( \dfrac{3x^2 - x - 10}{x - 2} \right)$ if it exists.

**Solution**    First notice that the function $f(x)$ is not defined at $x = 2$, since setting $x = 2$ in the fraction yields $\frac{0}{0}$. Intuitively, if we look at the numerator, we see that as $x$ approaches 2, $3x^2$ approaches $3(2)^2 = 12$, and $3x^2 - x - 10$ approaches $12 - 2 - 10 = 0$. On the other hand, the denominator, $x - 2$, also approaches 0 as $x$ approaches 2. So the fraction approaches $\frac{0}{0}$. At this point it is not clear if the limit exists.

To get some idea of what is happening as $x$ approaches 2, we should evaluate the function for various values of $x$. The results are shown in Table 12.2.

**Table 12.2**

| $x$ | 1.9 | 1.99 | 1.999 | $\to$ | 2 | $\leftarrow$ | 2.001 | 2.01 | 2.1 |
|---|---|---|---|---|---|---|---|---|---|
| $f(x)$ | 10.7 | 10.97 | 10.997 | $\to$ | 11 | $\leftarrow$ | 11.003 | 11.03 | 11.3 |

As Table 12.2 indicates, $f(x)$ seems to approach the number 11 as $x$ approaches $x = 2$ from the right and also seems to approach the number 11 as $x$ approaches $x = 2$ from the left. Thus we have every reason to suspect that the limit is equal to 11.

We can clarify the situation with some algebra. Notice that the numerator can be factored as

$$3x^2 - x - 10 = (3x + 5)(x - 2).$$

Therefore

$$f(x) = \frac{(3x + 5)(x - 2)}{(x - 2)} = 3x + 5 \text{ for } x \neq 2.$$

Since $f(x) = 3x + 5$ except at $x = 2$, the replacement theorem tells us that the two functions, $f(x)$ and $3x + 5$, must have the same limits as $x$ approaches 2. Thus

$$\lim\limits_{x \to 2} \frac{3x^2 - x - 10}{x - 2} = \lim\limits_{x \to 2} \frac{(3x + 5)(x - 2)}{(x - 2)} \qquad \text{factor}$$

$$= \lim\limits_{x \to 2} (3x + 5) \qquad \text{replacement theorem}$$

$$= 11 \qquad \text{Example 2} \quad \blacksquare$$

## Two Fundamentally Important Applications

We are now in a position to solve the two problems mentioned at the beginning of this section. We first consider the problem of finding the instantaneous velocity of an object.

**E X A M P L E   4**    **Calculating the Instantaneous Velocity**

**Solution**

If a baseball is dropped from the side of a building as in Figure 12.8, it can be shown using basic principles of physics that the distance $s$ in feet that the ball falls is given by $s = s(t) = 16t^2$, where $t$ is measured in seconds. Find the velocity at the instant $t = 1$.

The formula

$$\text{velocity} = \frac{\text{distance}}{\text{time}}$$

is familiar. But no time elapses in the instant $t = 1$ and no distance is traveled in this instant of time. The formula for velocity then becomes $\frac{0}{0}$ and is meaningless.

Let us consider another approach. We can evaluate the (average) velocity on any interval of time $[1, 1 + h]$, $h > 0$. Refer to Figure 12.13. Here, $h$ represents the change in time on this interval and is often denoted by $\Delta t$ (pronounced "delta t"). The change in the distance during this time interval is $s(1 + h) - s(1)$, which is often denoted as $\Delta s$. Thus the average velocity on the interval $[1, 1 + h]$ is given by

$$\text{average velocity} = \frac{\Delta s}{\Delta t} = \frac{\text{change in distance}}{\text{change in time}} = \frac{s(1 + h) - s(1)}{h}.$$

The smaller we take $h$, the smaller the interval of time $[1, 1 + h]$ will be for which we are measuring average velocity and the closer we will come to the instantaneous velocity at the instant $t = 1$. We then find the limit as $h$ goes to zero of the average velocity. This limit will be the instantaneous velocity.

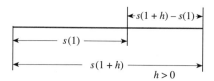

**Figure 12.13**

We then have

$$\text{average velocity} = \frac{\Delta s}{\Delta t} \qquad \frac{\text{change in distance}}{\text{change in time}}$$

$$= \frac{s(1 + h) - s(1)}{h} \qquad \text{substitute}$$

$$= \frac{16(1 + h)^2 - 16}{h} \qquad (1 + h)^2 = 1 + 2h + h^2$$

$$= \frac{16 + 32h + 16h^2 - 16}{h} \qquad \text{simplify}$$

$$= \frac{(32 + 16h)(h)}{h} \qquad \text{cancel } h$$

$$= 32 + 16h$$

where the $h$ can be canceled, since $h$ will never be zero.

Now let us take the decreasing positive values of $h$ to be 0.1, 0.01, 0.001, .0001. This yields the intervals [1, 1.1], [1, 1.01], [1, 1.001], [1, 1.0001] of decreasing size. The average velocity on these intervals is respectively 33.6, 32.16, 32.016, 32.0016. These latter numbers seem to be approaching the number 32. That is, as $h$ approaches 0 from the right, the average velocity on the interval [1, $1 + h$] approaches 32 or

$$\lim_{h \to 0+} (\text{average velocity}) = \lim_{h \to 0+} (32 + 16h) = 32. \qquad (1)$$

To complete this analysis, we also need to consider $h < 0$ and the average velocity on the interval [$1 + h$, 1]. See Figure 12.14. First notice that since $h$ is *negative*, the change in time is $\Delta t = -h$. Then

$$\text{average velocity} = \frac{\Delta s}{\Delta t} \qquad \frac{\text{change in distance}}{\text{change in time}}$$

$$= \frac{s(1) - s(1 + h)}{-h}$$

$$= \frac{s(1 + h) - s(1)}{h}$$

which is the very same expression as found before. Now taking the negative

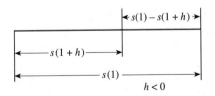

**Figure 12.14**

**Table 12.3**

| $h$ | $-.1$ | $-.01$ | $-.001$ | $-.0001$ | $\rightarrow$ | $0$ | $\leftarrow$ | $.0001$ | $.001$ | $.01$ | $.1$ |
|---|---|---|---|---|---|---|---|---|---|---|---|
| $v_{\text{ave}}$ | 30.4 | 31.84 | 31.984 | 31.9984 | $\rightarrow$ | 32 | $\leftarrow$ | 32.0016 | 32.016 | 32.16 | 33.6 |

values of $h$ to be $-0.1$, $-0.01$, $-0.001$, $-.0001$ then yields the intervals [.9, 1], [.99, 1], [.999, 1], [.9999, 1] of decreasing size with average velocity respectively 30.4, 31.84, 31.984, 31.9984. See Table 12.3. These numbers seem to be approaching the number 32. That is, as $h$ approaches 0 from the left, the average velocity on the interval $[1 + h, 1]$ approaches 32 or

$$\lim_{h \to 0-} (\text{average velocity}) = \lim_{h \to 0-} (32 + 16h) = 32. \tag{2}$$

Thus no matter how $h$ approaches 0, the average velocity approaches 32 and therefore the instantaneous velocity at $t = 1$ is 32 feet per second. ∎

**REMARK.**  In terms of the limit notation, this says that

$$\lim_{h \to 0} (\text{average velocity}) = \lim_{h \to 0} \frac{s(1 + h) - s(1)}{h} = 32.$$

We now consider a second fundamental problem in the calculus.

**E X A M P L E  5**   **Finding the Slope of the Tangent Line**

Find the slope of the line tangent to the curve $y = f(x) = x^2$ at the point $(1, 1)$.

**Solution**   In order to find the slope of the tangent line, we first need to find the slope $m_{\text{sec}}$ of the secant line drawn from $P(1, 1)$ to $Q(1 + h, f(1 + h))$ in Figure 12.15, where $h > 0$. To find this slope we must take the ratio of the rise and the run. The rise is $f(1 + h) - f(1)$, the change in $y$, which is also denoted by $\Delta y$. The run is $h$, the change in $x$, which is also denoted by $\Delta x$. We then have

$$m_{\text{sec}} = \frac{\Delta y}{\Delta x} \qquad \frac{\text{change in } y}{\text{change in } x}$$

$$= \frac{f(1 + h) - f(1)}{h}$$

$$= \frac{(1 + h)^2 - 1}{h}$$

$$= \frac{1 + 2h + h^2 - 1}{h}$$

$$= \frac{(2 + h)(h)}{h}$$

$$= 2 + h$$

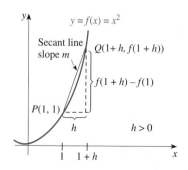

**Figure 12.15**

where $h$ can be canceled, since $h$ will never be zero.

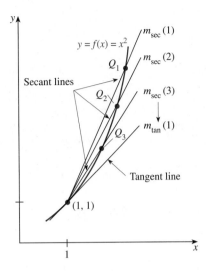

**Figure 12.16**

Figure 12.16 indicates geometrically what happens to the slope of the secant line when $h$ becomes smaller and smaller and heads for 0. From a careful examination of Figure 12.16, we see that as $h$ approaches the value 0, $m_{sec}$ approaches $m_{tan}(1)$, where $m_{tan}(1)$ is the slope of the tangent line to the curve at (1, 1). On the other hand, setting $h = 0.1, 0.01, 0.001$, yields values for $m_{sec}$ of 2.1, 2.01, and 2.001, respectively. Refer to Table 12.4. These latter values are approaching the number 2. Thus we conclude that

$$\lim_{h \to 0+} m_{sec} = 2.$$

**Table 12.4**

| $h$ | $-.1$ | $-.01$ | $-.001$ | $\to$ | 0 | $\leftarrow$ | .001 | .01 | .1 |
|---|---|---|---|---|---|---|---|---|---|
| $m_{sec}$ | 1.9 | 1.99 | 1.999 | $\to$ | 2 | $\leftarrow$ | 2.001 | 2.01 | 2.1 |

Now consider the case that $h < 0$. Then from Figure 12.17 the slope of the secant line from $(1 + h, 1)$ to (1, 1) is

$$\frac{f(1) - f(1 + h)}{-h} = \frac{f(1 + h) - f(1)}{h}$$

$$= 2 + h.$$

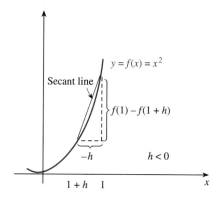

**Figure 12.17**

This is the very same expression as before. Now setting $h = -0.1, -0.01, -0.001$, yields values for $m_{sec}$ of 1.9, 1.99, and 1.999, respectively. These values for $m_{sec}$ are also approaching the number 2. Thus we conclude that

$$\lim_{h \to 0^-} m_{sec} = 2.$$

Therefore we conclude that

$$\lim_{h \to 0} \frac{f(1 + h) - f(1)}{h} = \lim_{h \to 0} m_{sec} = 2$$

and that the slope of the tangent line to the graph of the curve $y = x^2$ at the point $(1, 1)$ is $m_{tan}(1) = 2$.  ∎

## Some Limits That Do Not Exist

We now look at two cases where limits do not exist.

**E X A M P L E   6**    **Different Limits From Each Side**

$$\text{Let } f(x) = \begin{cases} x + 1 & \text{if } x < 0 \\ 0 & \text{if } x = 0 \\ 2 & \text{if } x > 0 \end{cases} \quad \text{Does } \lim_{x \to 0} f(x) \text{ exist?}$$

**Solution**    This function is graphed in Figure 12.18. Notice that

$$\lim_{x \to 0^-} f(x) = 1, \quad \text{while} \quad \lim_{x \to 0^+} f(x) = 2.$$

Since the limits from each side are not the same, $\lim_{x \to 0} f(x)$ does not exist.  ∎

**Figure 12.18**

E X A M P L E 7 **Limit Does Not Exist If Function Unbounded**

Let $f(x) = \dfrac{1}{x}$ for $x \neq 0$. Does $\lim\limits_{x \to 0} f(x)$ exist?

**Solution** In Table 12.5, we see the result of evaluating $f(x)$ at values of $x$ that approach 0. The values of the function $f(x)$ are getting large without bound if $x > 0$ and are getting small without bound if $x < 0$ and therefore are not approaching any particular number $L$. See Figure 12.19. Thus $\lim\limits_{x \to 0} f(x)$ does not exist.

**Table 12.5**

| $x$ | $-.1$ | $-.01$ | $-.001$ | $\to$ | 0 | $\leftarrow$ | .001 | .01 | .1 |
|-----|-------|--------|---------|-------|---|--------------|------|-----|----|
| $1/x$ | $-10$ | $-100$ | $-1000$ | $\to$ | ? | $\leftarrow$ | 1000 | 100 | 10 |

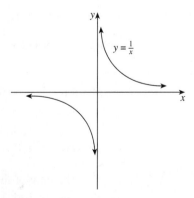

**Figure 12.19**

Whenever the graph of a curve, such as in Figure 12.19, approaches a vertical line $x = x_0$, we say that $x = x_0$ is a **vertical asymptote**. ∎

**REMARK .** We have now seen two basic ways in which a limit may not exist. If a function "blows up" near some value, such as in Figure 12.19, then the limit does not exist at that value. If the limits from each side of $x = a$ exist but are unequal, as in Figure 12.18, then the limit does not exist at the value $x = a$.

## Rules For Limits

We have just seen that graphs and tables can be used to find limits or to determine that limits do not exist. However, there are rules concerning limits that can be used to evaluate limits efficiently. We now give a number of such rules. For convenience let us use the notation $x \to a$ to represent the phrase "$x$ approaches $a$" and $f(x) \to L$ to represent the phrase "$f(x)$ approaches $L$."

Consider two functions, $f(x)$ and $g(x)$. Suppose that as $x \to a$, $f(x) \to L$ and $g(x) \to M$. Then it is reasonable that $cf(x) \to cL$, $f(x) \pm g(x) \to L \pm M$, $f(x)g(x) \to LM$, $\dfrac{f(x)}{g(x)} \to \dfrac{L}{M}$ (if $M \neq 0$), and $[f(x)]^n \to L^n$ when $L^n$ makes sense. This then gives the following five rules.

---

Assume that $\lim\limits_{x \to a} f(x) = L$    and    $\lim\limits_{x \to a} g(x) = M$. Then

**Rule 1.** $\lim\limits_{x \to a} cf(x) = c \lim\limits_{x \to a} f(x) = cL$.

**Rule 2.** $\lim\limits_{x \to a} (f(x) \pm g(x)) = \lim\limits_{x \to a} f(x) \pm \lim\limits_{x \to a} g(x) = L \pm M$.

**Rule 3.** $\lim\limits_{x \to a} (f(x) \cdot g(x)) = \left( \lim\limits_{x \to a} f(x) \right) \cdot \left( \lim\limits_{x \to a} g(x) \right) = L \cdot M$.

**Rule 4.** $\lim\limits_{x \to a} \dfrac{f(x)}{g(x)} = \dfrac{\lim\limits_{x \to a} f(x)}{\lim\limits_{x \to a} g(x)} = \dfrac{L}{M}$ if $\lim\limits_{x \to a} g(x) = M \neq 0$.

**Rule 5.** $\lim\limits_{x \to a} (f(x))^n = L^n$, $n$ any real number, $L^n$ defined.

---

**REMARK .** The following is a verbal interpretation of the above rules.

1. The limit of a constant times a function is the constant times the limit of the function.
2. The limit of the sum or difference equals the sum or difference of the limits.
3. The limit of the product is the product of the limits.
4. The limit of a quotient is the quotient of the limits if the limit of the denominator is not zero.

5. The limit of a function raised to a power is the limit of the function raised to the power provided the terms make sense.

**REMARK.** In Rule 5, $L^n$ will not be defined if for example $L < 0$ and $n = \frac{1}{2}, \frac{1}{4}$, and so forth, or if $L = 0$ and $n \leq 0$.

Consider also the following. If $f(x) = c$, then since $f(x)$ is always the constant $c$, the limit as $x \to a$ must also be $c$. Thus

$$\lim_{x \to a} c = c.$$

Furthermore, it is intuitively clear that

$$\lim_{x \to a} x = a.$$

**E X A M P L E  8** **The Limit of Some Quadratic Polynomials**

Find (a) $\lim_{x \to 3} (2x^2)$, (b) $\lim_{x \to 3} (2x^2 + 4)$.

**Solutions**

(a) $\lim_{x \to 3} (2x^2) = 2\lim_{x \to 3} (x^2)$       Rule 1

$$= 2 \left( \lim_{x \to 3} x \right) \cdot \left( \lim_{x \to 3} x \right) \quad \text{Rule 3}$$

$$= 2 \cdot 3 \cdot 3 = 2(3)^2 = 18$$

(b) Using this result, we then have

$$\lim_{x \to 3} (2x^2 + 4) = \lim_{x \to 3} (2x^2) + \lim_{x \to 3} (4) \quad \text{Rule 2}$$

$$= 2(3)^2 + 4 = 22. \quad \blacksquare$$

**REMARK.** It is extremely important to notice that if we let $p(x) = 2x^2 + 4$, then we showed that

$$\lim_{x \to 3} p(x) = p(3).$$

This is intuitive, since as $x \to 3$, it is reasonable that $2x^2 + 4 \to 2(3)^2 + 4$.

This reasoning applies in general and applies to any polynomial function $p(x) = a_n x^n + a_{n-1} x^{n-1} + \cdots + a_1 x + a_0$. We then have the following result.

---

**The Limit of a Polynomial Function**

If $p(x)$ is any polynomial, then for any number $a$

$$\lim_{x \to a} p(x) = p(a).$$

---

Thus, in the case of a polynomial function, the limit can be obtained by *substitution*.

**E X A M P L E   9**     **The Limit of a Polynomial Function**

Find $\lim\limits_{x \to 2} (3x^3 + 2x - 3)$.

**Solution**              $$\lim\limits_{x \to 2} (3x^3 + 2x - 3) = 3(2)^3 + 2(2) - 3 = 25 \quad \blacksquare$$

**E X A M P L E   1 0**   **Using the Rules of Limits**

Evaluate (a) $\lim\limits_{x \to 3} \dfrac{x^2 + 11}{x^2 - 4}$,  (b) $\lim\limits_{x \to 1} (x^2 - 9)^{1/3}$,  (c) $\lim\limits_{x \to 4} \dfrac{x^2 - 16}{x^2 - 2x - 8}$.

**Solutions**

(a) $\lim\limits_{x \to 3} \dfrac{x^2 + 11}{x^2 - 4} = \dfrac{\lim\limits_{x \to 3} (x^2 + 11)}{\lim\limits_{x \to 3} (x^2 - 4)}$        Rule 4

$\qquad\qquad = \dfrac{20}{5} = 4$                  limit of polynomial functions

(b) $\lim\limits_{x \to 1} (x^2 - 9)^{1/3} = \left(\lim\limits_{x \to 1} (x^2 - 9)\right)^{1/3}$     Rule 5

$\qquad\qquad = (-8)^{1/3} = -2$                limit of polynomial functions

(c) Using Rule 4 and substitution gives $\frac{0}{0}$. Therefore, we must try to factor the numerator and denominator to find a common factor.

$\lim\limits_{x \to 4} \dfrac{x^2 - 16}{x^2 - 2x - 8} = \lim\limits_{x \to 4} \dfrac{(x + 4)(x - 4)}{(x + 2)(x - 4)}$        factor

$\qquad\qquad = \lim\limits_{x \to 4} \dfrac{(x + 4)}{(x + 2)}$          replacement theorem

$\qquad\qquad = \dfrac{\lim\limits_{x \to 4} (x + 4)}{\lim\limits_{x \to 4} (x + 2)}$          Rule 4

$\qquad\qquad = \dfrac{8}{6} = \dfrac{4}{3}$                limit of polynomial functions   $\blacksquare$

Whenever you encounter an algebraic fraction involving a radical in the denominator, you may have learned to simplify the fraction by rationalizing the denominator, that is, to eliminate the radical from the denominator. However, sometimes in calculus we must rationalize the *numerator*, that is, eliminate the radical from the numerator. The following example illustrates such a case.

**E X A M P L E   1 1**  **Finding a Limit by Rationalizing the Numerator**

Find $\lim\limits_{x \to 1} \dfrac{\sqrt{x} - 1}{x - 1}$.

**Solution**

$$\lim_{x \to 1} \frac{\sqrt{x} - 1}{x - 1} = \lim_{x \to 1} \frac{\sqrt{x} - 1}{x - 1} \frac{\sqrt{x} + 1}{\sqrt{x} + 1} \qquad \text{Rationalize the numerator.}$$

$$= \lim_{x \to 1} \frac{(x - 1)}{(\sqrt{x} + 1)(x - 1)}$$

$$= \lim_{x \to 1} \frac{1}{\sqrt{x} + 1} \qquad \text{Replacement theorem}$$

$$= \frac{\lim\limits_{x \to 1} (1)}{\lim\limits_{x \to 1} \sqrt{x} + \lim\limits_{x \to 1} (1)} \qquad \text{Rules 4, 2}$$

$$= \frac{1}{2} \qquad \text{Rule 5} \quad \blacksquare$$

**R E M A R K .**    Notice in Example 10c that the limit as $x$ approaches 4 exists even though the function $\dfrac{x^2 - 16}{x^2 - 2x - 8}$ is not defined at $x = 4$. Also, in Example 11 the limit as $x$ approaches 1 exists even though the function $\dfrac{\sqrt{x} - 1}{x - 1}$ is not defined at $x = 1$. These two examples further illustrate that the limit of a function $f(x)$ as $x$ approaches $a$ never depends on what happens at $x = a$.

## SELF-HELP EXERCISE SET 12.1

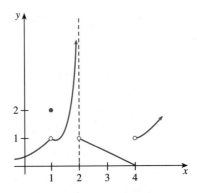

**1.** For the function in the figure, find $\lim\limits_{x \to a^-} f(x)$, $\lim\limits_{x \to a^+} f(x)$, $\lim\limits_{x \to a} f(x)$ for $a = 1, 2,$ 3, 4.

2. For $s(t) = 16t^2$, find the velocity $v(2)$ at $t = 2$ by finding

$$\lim_{h \to 0} \frac{s(2 + h) - s(2)}{h}$$

3. Find the slope of the line tangent to the curve $y = f(x) = x^2$ at $x = -1$ by finding

$$\lim_{h \to 0} \frac{f(-1 + h) - f(-1)}{h}$$

4. Find $\displaystyle\lim_{x \to 2} \sqrt{\frac{x^2 - 8}{x^3 - 9}}$

## EXERCISE SET 12.1

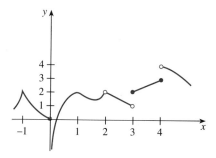

1. Refer to the figure. Find $\displaystyle\lim_{x \to a-} f(x)$, $\displaystyle\lim_{x \to a+} f(x)$, $\displaystyle\lim_{x \to a} f(x)$ at the indicated value for the function given in the above figure.
   a. $a = -1$,          b. $a = 0$,          c. $a = 1$,
   d. $a = 2$,          e. $a = 3$,          f. $a = 4$

In Exercises 2 through 7, fill in the table, using a calculator if necessary, and then estimate the limit of the given function at the given value.

2. $a = 2$ and $f(x) = 2x + 3$.

| $x$ | 1.9 | 1.99 | 1.999 | $\rightarrow$ | 2 | $\leftarrow$ | 2.001 | 2.01 | 2.1 |
|-----|-----|------|-------|---------------|---|--------------|-------|------|-----|
| $f(x)$ | | | | $\rightarrow$ | ? | $\leftarrow$ | | | |

3. $a = 2$ and $f(x) = x^3 - 4$.

| $x$ | 1.9 | 1.99 | 1.999 | $\rightarrow$ | 2 | $\leftarrow$ | 2.001 | 2.01 | 2.1 |
|-----|-----|------|-------|---------------|---|--------------|-------|------|-----|
| $f(x)$ | | | | $\rightarrow$ | ? | $\leftarrow$ | | | |

**4.** $a = -1$ and $f(x) = \dfrac{|x + 1|}{x + 1}$

| $x$ | $-1.1$ | $-1.01$ | $-1.001$ | $\rightarrow$ | $-1$ | $\leftarrow$ | $-0.999$ | $-0.99$ | $-0.9$ |
|---|---|---|---|---|---|---|---|---|---|
| $f(x)$ | | | | $\rightarrow$ | ? | $\leftarrow$ | | | |

**5.** $a = 2$ and $f(x) = \dfrac{1}{x - 2}$.

| $x$ | 1.9 | 1.99 | 1.999 | $\rightarrow$ | 2 | $\leftarrow$ | 2.001 | 2.01 | 2.1 |
|---|---|---|---|---|---|---|---|---|---|
| $f(x)$ | | | | $\rightarrow$ | ? | $\leftarrow$ | | | |

**6.** $a = 4$ and $f(x) = \dfrac{\sqrt{x} - 2}{x - 2}$.

| $x$ | 3.9 | 3.99 | 3.999 | $\rightarrow$ | 4 | $\leftarrow$ | 4.001 | 4.01 | 4.1 |
|---|---|---|---|---|---|---|---|---|---|
| $f(x)$ | | | | $\rightarrow$ | ? | $\leftarrow$ | | | |

**7.** $a = 4$ and $f(x) = \dfrac{x - 4}{\sqrt{x} - 2}$.

| $x$ | 3.9 | 3.99 | 3.999 | $\rightarrow$ | 4 | $\leftarrow$ | 4.001 | 4.01 | 4.1 |
|---|---|---|---|---|---|---|---|---|---|
| $f(x)$ | | | | $\rightarrow$ | ? | $\leftarrow$ | | | |

In Exercises 8 through 20 find $\lim\limits_{x \to a^-} f(x)$, $\lim\limits_{x \to a^+} f(x)$, and $\lim\limits_{x \to a} f(x)$ at the indicated value for the indicated function by graphing the function.

**8.** $a = 0$, $f(x) = |x|$

**9.** $a = 2$, $f(x) = |x - 2|$

**10.** $a = 1$, $f(x) = \begin{cases} x & \text{if } x < 1 \\ 2 & \text{if } x = 1 \\ x - 1 & \text{if } x > 1 \end{cases}$

**11.** $a = 0$, $f(x) = \begin{cases} -x + 1 & \text{if } x < 0 \\ x^2 & \text{if } x \geq 0 \end{cases}$

**12.** $a = 1$, $f(x) = \begin{cases} -x + 1 & \text{if } x < 1 \\ x^2 - 1 & \text{if } x > 1 \end{cases}$

**13.** $a = 1$, $f(x) = \begin{cases} x^2 & \text{if } x < 1 \\ 2 & \text{if } x = 1 \\ x & \text{if } x > 1 \end{cases}$

**14.** $a = 1$, $f(x) = \begin{cases} -x + 2 & \text{if } x < 1 \\ 0 & \text{if } x = 1 \\ x^2 & \text{if } x > 1 \end{cases}$

**15.** $a = 0$, $f(x) = \begin{cases} x & \text{if } x < 0 \\ \frac{1}{x} & \text{if } x > 0 \end{cases}$

**16.** $a = 2$, $f(x) = \dfrac{x - 2}{|x - 2|}$

**17.** $a = -1$, $f(x) = \dfrac{x + 1}{|x + 1|}$

**18.** $a = 0$, $f(x) = \dfrac{1}{x^2}$

**19.** $a = 1$, $f(x) = \dfrac{1}{x - 1}$

**20.** $a = 1$, $f(x) = \begin{cases} -x + 1 & \text{if } x < 1 \\ \dfrac{1}{1 - x} & \text{if } x > 1 \end{cases}$

In Exercises 21 through 26 find the velocity $v(t_0)$ by evaluating

$$\lim_{h \to 0} \frac{s(t_0 + h) - s(t_0)}{h}$$

at the indicated time $t_0$ for the indicated function $s = s(t)$.

**21.** $t_0 = 4$, $s(t) = 16t^2$

**22.** $t_0 = 3$, $s(t) = 16t^2$

**23.** $t_0 = 1$, $s(t) = t^2 + 1$

**24.** $t_0 = 1$, $s(t) = 2t^2 + 1$

**25.** $t_0 = 1$, $s(t) = -t^2 + 2$

**26.** $t_0 = 1$, $s(t) = t^2 + t$

In Exercises 27 through 32 find the slope $m_{\tan}(x_0)$ of the line tangent to the given curve at the point $(x_0, f(x_0))$ at the indicated value $x_0$ by evaluating

$$\lim_{h \to 0} \frac{f(x_0 + h) - f(x_0)}{h}$$

**27.** $x_0 = -1$, $f(x) = 3x^2$

**28.** $x_0 = 0$, $f(x) = x^2$

**29.** $x_0 = 0$, $f(x) = 2x^2 + 1$

**30.** $x_0 = 0$, $f(x) = 2x^2 + 1$

**31.** $x_0 = 0$, $f(x) = x^3$

**32.** $x_0 = 1$, $f(x) = x^3$

In Exercises 33 through 61 use the rules of limits to find the indicated limits if they exist.

**33.** $\lim\limits_{x \to 0} 13$

**34.** $\lim\limits_{x \to 1} 3\pi$

**35.** $\lim\limits_{x \to 2} (2x + 1)$

**36.** $\lim\limits_{x \to -1} (3x - 2)$

**37.** $\lim\limits_{x\to 2} (3x^3 - 2x^2 - 2x + 1)$     **38.** $\lim\limits_{x\to -1} (3x^3 + x - 2)$

**39.** $\lim\limits_{x\to 0} \dfrac{x}{x^2 + x + 1}$     **40.** $\lim\limits_{x\to 1} \dfrac{x^2 + 2}{x^2 - 2}$

**41.** $\lim\limits_{x\to 2} \dfrac{x + 1}{x - 3}$     **42.** $\lim\limits_{x\to 1} \dfrac{3x + 1}{x + 2}$

**43.** $\lim\limits_{x\to 2} \dfrac{x + 3}{x - 3}(2x^2 - 1)$     **44.** $\lim\limits_{x\to 1} \dfrac{\sqrt{x}}{x - 3}\dfrac{x + 2}{x^2 + 1}$

**45.** $\lim\limits_{x\to -1} \dfrac{x^3 + 1}{x^2 + 1}$     **46.** $\lim\limits_{x\to 2} \dfrac{x^2 + 3}{x - 1}$

**47.** $\lim\limits_{x\to 2} \sqrt{x^2 - 3}$     **48.** $\lim\limits_{x\to 2} \sqrt{x^3 - x - 2}$

**49.** $\lim\limits_{x\to 2} \sqrt[3]{2x + 4}$     **50.** $\lim\limits_{x\to 0} \sqrt[3]{x - 8}$

**51.** $\lim\limits_{x\to 2} \dfrac{x^2 - 4}{x - 2}$     **52.** $\lim\limits_{x\to 2} \dfrac{x^2 - x - 2}{x - 2}$

**53.** $\lim\limits_{x\to 1} \dfrac{x^2 - 1}{x - 1}$     **54.** $\lim\limits_{x\to -1} \dfrac{x^2 + 3x + 2}{x + 1}$

**55.** $\lim\limits_{x\to -1} \dfrac{x^2 - 1}{x + 1}$     **56.** $\lim\limits_{x\to -2} \dfrac{x^2 + 4}{x + 2}$

**57.** $\lim\limits_{x\to 0} \dfrac{x + 1}{x^2 + x}$     **58.** $\lim\limits_{x\to 0} \dfrac{x^3 + x}{x}$

**59.** $\lim\limits_{x\to 1} \dfrac{\sqrt{x} - 1}{x - 1}$     **60.** $\lim\limits_{x\to 4} \dfrac{x - 4}{\sqrt{x} - 2}$

**61.** $\lim\limits_{h\to 0} \dfrac{(2 + h)^3 - 8}{h}$

## Applications

**62.** **Average Cost.**  An electric utility company estimates that the cost, $C(x)$, of producing $x$ kilowatts of electricity is given by $C(x) = a + bx$, where $a$ represents the fixed costs and $b$ represents the unit costs. The *average* cost, $\overline{C}(x)$, is defined as the total cost of producing $x$ kilowatts divided by $x$, that is, $\overline{C}(x) = \dfrac{C(x)}{x}$. What happens to the average cost if the amount of kilowatts produced, $x$, heads for zero? That is, find $\lim\limits_{x\to 0+} \overline{C}(x)$. How do you interpret your answer in economic terms?

**63.** **Postal Rates.**  First-class postage in 1994 was $0.29 for the first ounce (or any fraction thereof) and $0.23 for each additional ounce (or fraction thereof). if $C(x)$ is the cost of postage for a letter weighing $x$ ounces, then for $x \le 3$,

$$C(x) = \begin{cases} 0.29 & \text{if } 0 < x \le 1 \\ 0.52 & \text{if } 1 < x \le 2 \\ 0.75 & \text{if } 2 < x \le 3 \end{cases}$$

Graph and find

**a.**  $\lim\limits_{x \to 1-} C(x)$, $\lim\limits_{x \to 1+} C(x)$, and $\lim\limits_{x \to 1} C(x)$,

**b.**  $\lim\limits_{x \to 1.5-} C(x)$, $\lim\limits_{x \to 1.5+} C(x)$, and $\lim\limits_{x \to 1.5} C(x)$.

(NASA)

**64.    Cost.**    It is estimated that the cost, $C(x)$, in billions of dollars, of maintaining the atmosphere in the United States at an average $x$ percent free of a chemical toxin is given by $C(x) = \dfrac{10}{100 - x}$. Find the cost of maintaining the atmosphere at levels 90, 99, 99.9, and 99.99 percent free of this toxin. What is happening as $x \to 100-$? Does your answer make sense?

**65.    Cost.**    A large corporation already has cornered 30 percent of the toothpaste market and now wishes to undertake an extensive advertising campaign to increase its share of this market. It is estimated that the advertizing cost, $C(x)$, in billions of dollars, of attaining an $x$ percent share of this market is given by $C(x) = \dfrac{1}{100 - x}$. What would it cost to attain a 90 percent share? 99 percent? 99.9 percent? What is happening to $C(x)$ as $x \to 100-$? Does this make sense?

**66.    Medicine.**    Two rules have been suggested by Cowling and also by Young to adjust adult drug dosage levels for young children. Let $a$ denote the adult dosage and let $t$ be the age (in years) of the child. The two rules are given, respectively, by

$$C = \frac{t + 1}{24} a, \quad \text{and} \quad Y = \frac{t}{t + 12} a.$$

Find the limit as $t \to 0+$ for both of these. Which seems most realistic for a newborn baby?

## Graphing Calculator Exercises

The program limit found in C.2 of Appendix C finds the values of a function at the points $a \pm 0.1$, $\pm 0.01$, ... and stores them in the data section. Use this program on your graphing calculator to find the limit, the limit from the left, and the limit from the right (if any exist) of each of the remaining functions at the indicated value.

**67.**  $f(x) = x^2 + 2x - 3$, $a = 1$

**68.**  $f(x) = \dfrac{|x - 3|}{x - 3}$, $a = 3$

**69.**  $f(x) = \dfrac{1}{x - 5}$, $a = 5$

**70.**  $f(x) = \dfrac{x^2 - 2x - 8}{x - 4}$, $a = 5$

**71.**  $a = 1$, $f(x) = \begin{cases} x - 1 & \text{if } x < 1 \\ x^3 + x - 2 & \text{if } x > 1 \end{cases}$

72.  $a = 2, f(x) = \begin{cases} x - 1 & \text{if } x < 2 \\ 5 + x^2 - x^3 & \text{if } x > 2 \end{cases}$

73.  $a = 1, f(x) = \begin{cases} x - 1 & \text{if } x < 1 \\ x^3 + 2x - 2 & \text{if } x > 1 \end{cases}$

74.  $a = 2, f(x) = \begin{cases} x - 1 & \text{if } x < 2 \\ 4 + x^2 - x^3 & \text{if } x > 2 \end{cases}$

## Solutions to Self-Help Exercise Set 12.1

1.  From the graph

$$\lim_{x \to 1-} f(x) = 1, \qquad \lim_{x \to 1+} f(x) = 1, \qquad \lim_{x \to 1} f(x) = 1$$

$$\lim_{x \to 2-} f(x) \text{ does not exist}, \qquad \lim_{x \to 2+} f(x) = 1, \qquad \lim_{x \to 2} f(x) \text{ does not exist}$$

$$\lim_{x \to 3-} f(x) = \tfrac{1}{2}, \qquad \lim_{x \to 3+} f(x) = \tfrac{1}{2}, \qquad \lim_{x \to 3} f(x) = \tfrac{1}{2}$$

$$\lim_{x \to 4-} f(x) = 0, \qquad \lim_{x \to 4+} f(x) = 1, \qquad \lim_{x \to 4} f(x) \text{ does not exist}$$

2.  First find the average velocity

$$\text{average velocity} = \frac{s(2 + h) - s(2)}{h}$$

$$= \frac{16(2 + h)^2 - 64}{h}$$

$$= \frac{64 + 64h + 16h^2 - 64}{h}$$

$$= 64 + 16h$$

Now take the limit of this expression and obtain

$$v(2) = \lim_{h \to 0} (\text{average velocity}) = \lim_{h \to 0} (64 + 16h) = 64$$

3.  First find the slope of the secant line

$$m_{\text{sec}} = \frac{f(-1 + h) - f(-1)}{h}$$

$$= \frac{(-1 + h)^2 - 1}{h}$$

$$= \frac{1 - 2h + h^2 - 1}{h}$$

$$= -2 + h$$

Now take the limit of this expression and obtain

$$m_{\tan}(-1) = \lim_{h \to 0} m_{\sec} = \lim_{h \to 0} (-2 + h) = -2$$

4.    $$\lim_{x \to 2} \sqrt{\frac{x^2 - 8}{x^3 - 9}} = \sqrt{\lim_{x \to 2} \frac{x^2 - 8}{x^3 - 9}}$$    Rule 5

$$= \sqrt{\frac{\displaystyle\lim_{x \to 2} (x^2 - 8)}{\displaystyle\lim_{x \to 2} (x^3 - 9)}}$$    Rule 4

$$= \sqrt{\frac{-4}{-1}} = 2$$    limit of polynomial functions

## 12.2    CONTINUITY

▶ *Continuous Functions*
▶ *Rules for Continuous Functions*
▶ *Applications*
▶ *Sign Charts of Continuous Functions*

APPLICATION

**Interest on a Savings Certificate**

A \$100,000 savings certificate has annual interest of 8% compounded quarterly credited to the account only at the end of each quarter. Determine the amount $A(t)$ in the account during the first year and find points at which the graph of the function has a "gap." Of what significance are these points? The answer is found in Example 3.

APPLICATION

**State Income Tax Cliffs**

A certain state levies an income tax of 1% on income over \$20,000 but less than or equal to \$25,000 and 2% on entire income if income is over \$25,000. At what income levels will an individual's tax take a sudden jump? Might a person's behavior be affected by the presence of these jumps? (In the parlance of the tax experts, these are referred to as "cliffs.") The answer is found in Example 4.

## Continuous Functions

Geometrically, a function is said to be **continuous** on some interval if the graph of the function has no "holes," "gaps," or "tears." According to this loose definition, one can draw the graph of a continuous function without lifting a pen from the paper.

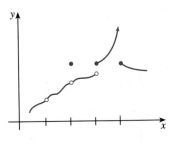

**Figure 12.20**

For example, in Figure 12.20, the graph of the function has a "hole" or "gap" at $x = 1$ as a result of the function being undefined at $x = 1$. So this function is not continuous (*is discontinuous*) at $x = 1$. The function is defined at $x = 2$, but the graph has a "tear" at this value, so is discontinuous at $x = 2$. The graph of the function has a "tear" or "jump" at $x = 3$ so the function is not continuous at $x = 3$. The function "blows up" to the left of $x = 4$, so the graph has a "substantial tear" at $x = 4$, and is not continuous at $x = 4$.

Analytically, what happens at a discontinuity such as $x = 3$ in Figure 12.20 is simply that $x$ can be very close to 3 but $f(x)$ will *not* necessarily be very close to $f(3)$. The following definition, using limits, captures this notion.

---

**Definition of a Function Continuous at a Point**

The function $f(x)$ is *continuous* at $x = a$ if the following three conditions are met:

1. $\lim\limits_{x \to a} f(x)$ exists.
2. $f(a)$ is defined.
3. $\lim\limits_{x \to a} f(x) = f(a)$.

---

We can summarize these three conditions by the one equation

$$\lim_{x \to a} f(x) = f(a).$$

Condition 1 says that the limit must exist. This ensures that the graph of the function has no jumps or "blowups." So if the limit does *not* exist, the function is discontinuous. For example the function in Figure 12.20 is discontinuous at $x = 3$ and $x = 4$ since the limits do not exist for these values. What differentiates continuity from limits is the last two conditions above. Condition 2 ensures that a "hole" in the graph such as at $x = 1$ does not occur. Condition 3 ensures that $f(a)$ is not removed from $\lim\limits_{x \to a} f(x)$ such as at $x = 2$ in Figure 12.20.

When determining $\lim\limits_{x \to a} f(x)$, we need only be concerned with the values of $f(x)$ for $x$ *near a*. The value of $f(a)$ plays no role in the limit process. Thus for the example shown in Figure 12.20, $\lim\limits_{x \to 1} f(x)$ and $\lim\limits_{x \to 2} f(x)$ both exist. In the notion of *continuity*, however, the value of $f(a)$ plays a vital role. In other words, in the notion of continuity we are concerned about what happens "when we get there."

---

**Definition of Continuous Function on an Open Interval**

We say that a function $f(x)$ is **continuous on an open interval** $(a, b)$ if $f(x)$ is continuous at every point of the interval.

---

In the last section we already noticed that if $p(x)$ is any polynomial function and $a$ is any real number, then

$$\lim_{x \to a} p(x) = p(a).$$

This says that a polynomial function is continuous everywhere.

---

**Continuity of Polynomial Functions**

A polynomial function is continuous everywhere.

---

## Rules for Continuous Functions

Continuous functions have rules very similar to the rules for limits.

---

*Rule 1.*   A constant times a continuous function is continuous.

*Rule 2.*   The sum or difference of continuous functions is continuous.

*Rule 3.*   The product of two continuous functions is continuous.

*Rule 4.*   The quotient of two continuous functions is continuous at points for which the denominator is not zero.

*Rule 5.*   A continuous function $f(x)$ raised to any real power $r$ is continuous at points where $[f(x)]^r$ is defined.

---

These rules follow directly from the rules of limits.

Since a rational function is a quotient of polynomial functions and since a polynomial function is continuous everywhere, Rule 4 indicates that a rational function will be continuous at every point at which the denominator is not zero.

---

**Continuity of Rational Functions**

A rational function is continuous at every point at which the denominator is not zero.

---

**E X A M P L E   1**    **Determining Where Functions Are Continuous**

Determine where the following functions are continuous.

(a)   $f(x) = 2x^2 - 3x + 4$,   (b)   $f(x) = \dfrac{x + 1}{(x - 1)(x + 3)}$,

(c)   $f(x) = \sqrt[3]{x^2 - 1}$

**Solutions**    (a)   Since $f(x)$ is a polynomial, $f(x)$ is continuous for all $x$.

(b)   Since $f(x)$ is a rational function, $f(x)$ is continuous for all $x$ except the points $x = -3$ and $x = 1$, where the denominator is zero.

(c)   Since $x^2 - 1$ is a polynomial, it is continuous for all $x$. Then $f(x)$ is continuous for all $x$ by Rule 5.   ■

For a function defined on a closed interval $[a, b]$ we cannot take limits at $a$ and $b$ since the function $f(x)$ is only defined to one side of these values. But we can take $\lim\limits_{x \to a+} f(x)$ and $\lim\limits_{x \to b-} f(x)$. We then use these one-sided limits to define continuity at an endpoint.

---

**Continuity at an Endpoint of an Interval**

Suppose $f(x)$ is defined on a closed interval $[a, b]$.

a.   We say that $f$ is continuous at $x = a$ if $\lim\limits_{x \to a+} f(x) = f(a)$.

b.   We say that $f$ is continuous at $x = b$ if $\lim\limits_{x \to b-} f(x) = f(b)$.

---

Corresponding definitions can be made for the intervals $[a, +\infty)$ and $(-\infty, b]$.

**E X A M P L E   2**

**Continuity on a Closed Interval**

Determine where the function $f(x) = \sqrt{x - 1}$ is continuous.

**Solution**   Since $x - 1$ is a polynomial, it is continuous for all $x$. Also $x - 1$ is positive for all $x > 1$. Rule 5 then assures us that $f(x)$ is continuous for all $x > 1$. Also $f(x)$ is continuous at $x = 1$ since $\lim\limits_{x \to 1+} \sqrt{x - 1} = 0 = f(1)$. Thus $\sqrt{x - 1}$ is continuous on the closed interval $[1, +\infty)$.   ■

## Applications

Functions encountered in real world applications can be discontinuous, but the discontinuities usually occur at isolated points, as the following examples indicate.

**E X A M P L E   3**

**Interest on a Savings Certificate**

A \$100,000 savings certificate has annual interest of 8% compounded quarterly credited to the account only at the end of each quarter. Determine the amount $A(t)$ in the account during the first year and find points at which the function is discontinuous. What is the significance of these points?

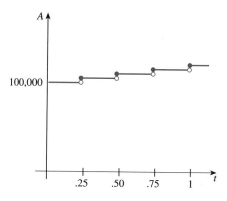

**Figure 12.21**

**Solution**    The graph of the function is shown in Figure 12.21. The discontinuities occur at the quarter-year marks when the interest is credited. During any fraction of the last second of the first quarter, the certificate is worth $100,000. During any fraction of the first second of the second quarter, the certificate is worth $102,000. Thus if you wished to cash this certificate, you would wait, if possible, until the beginning of a quarter to do so.   ■

**E X A M P L E   4**    **State Income Tax Cliffs**

A certain state levies an income tax of 1% on incomes over $20,000 but less than or equal to $25,000 and 2% on entire income if income is over $25,000. At what income levels will the tax be discontinuous? Of what significance are these points? (In the parlance of the tax experts, these are referred to as "cliffs.")

**Solution**    If $x$ represents an individual's income and $t(x)$ the tax, then

$$t(x) = \begin{cases} 0 & \text{if } x \le 20{,}000 \\ (.01)x & \text{if } 20{,}000 < x \le 25{,}000 \\ (.02)x & \text{if } x > 25{,}000 \end{cases}$$

The graph is shown in Figure 12.22, where it can be seen that there are

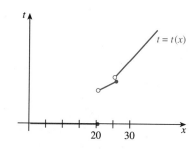

**Figure 12.22**

discontinuities at $x = 20,000$ and $x = 25,000$. If you earn $20,000 in this state, you pay no taxes, but if you earn $20,000.01 you pay taxes of $200. In the parlance of tax experts, your tax bill falls off a "cliff." Naturally you will try your best not to earn that extra one cent beyond the discontinuity at $20,000. Thus such discontinuities will affect a person's behavior. ∎

## Sign Chart of Continuous Functions

We now consider a particularly important property of continuous functions that will be used extensively in the next chapter.

If we look at Figure 12.23, we see that the function is known to be positive at $x = a$ and negative at $x = b$, where $a < b$. Is it possible to draw a graph of a *continuous* function connecting the two points $(a, f(a))$ and $(b, f(b))$ with*out* touching the $x$-axis? That is, place your pencil at the point $(a, f(a))$ in Figure 12.23 and draw a curve, without lifting your pencil, that connects this point with the point $(b, f(b))$ *without* ever touching the $x$-axis. This is impossible!

This means that if $f(x)$ is continuous, $a < b$, and $f(x)$ is of opposite signs at $a$ and $b$, then there must be a point on the interval $(a, b)$ for which $f(x)$ is zero. As a consequence, we have the following theorem.

---

**Constant Sign Theorem**

If $f(x)$ is continuous on $(a, b)$ and $f(x) \neq 0$ for any $x$ in $(a, b)$, then either $f(x) > 0$ for all $x$ in $(a, b)$ or $f(x) < 0$ for all $x$ in $(a, b)$.

---

If the conclusion were not true, then $f(x)$ would be positive at one point and negative at another. From the preceding discussion, we know that $f(x)$ would have to be zero at some intermediate point. But this would contradict the hypothesis that $f(x)$ is never zero on $(a, b)$.

This theorem gives us a very practical way (called a sign chart) to determine the sign of a continuous function over an interval. We will use sign charts in the next chapter.

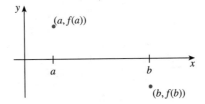

**Figure 12.23**

## SELF-HELP EXERCISE SET 12.2

1.  Find all points at which the function shown in the graph is not continuous.

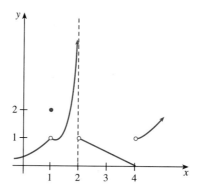

2.  A furniture store gives free delivery to any location less than or equal to 10 miles from the store and charges a delivery fee of $2 times the mileage if the mileage is greater than 10 miles. Find the function $C(x)$, where $x$ is the distance from the store and $C(x)$ the cost in dollars of the delivery. Find the discontinuities of this function and determine their significance. In particular, determine the cost of delivery if you are one inch less than 10 miles and one inch more than 10 miles from the store.

## EXERCISE SET 12.2

1.  Find where the function given in the following graph is not continuous.

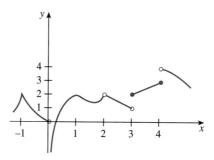

In Exercises 2 through 13 draw a graph to find where each of the given functions is not continuous. (Refer to Exercises 8 through 19 of the last section.)

2.  $f(x) = |x|$

3.  $f(x) = |x - 2|$

4.  $f(x) = \begin{cases} x & \text{if } x < 1 \\ 2 & \text{if } x = 1 \\ x - 1 & \text{if } x > 1 \end{cases}$

5.  $f(x) = \begin{cases} -x + 1 & \text{if } x < 0 \\ x^2 & \text{if } x \geq 0 \end{cases}$

**6.** $f(x) = \begin{cases} -x + 1 & \text{if } x < 1 \\ x^2 - 1 & \text{if } x > 1 \end{cases}$

**7.** $f(x) = \begin{cases} x^2 & \text{if } x < 1 \\ 2 & \text{if } x = 1 \\ x & \text{if } x > 1 \end{cases}$

**8.** $f(x) = \begin{cases} -x + 2 & \text{if } x < 1 \\ 0 & \text{if } x = 1 \\ x^2 & \text{if } x > 1 \end{cases}$

**9.** $f(x) = \begin{cases} x & \text{if } x < 0 \\ \dfrac{1}{x} & \text{if } x > 0 \end{cases}$

**10.** $f(x) = \dfrac{x - 2}{|x - 2|}$

**11.** $f(x) = \dfrac{x + 1}{|x + 1|}$

**12.** $f(x) = \dfrac{1}{x^2}$

**13.** $f(x) = \begin{cases} -x + 1 & \text{if } x < 1 \\ \dfrac{1}{1 - x} & \text{if } x > 1 \end{cases}$

In Exercises 14 through 22 use the rules of continuity to determine where the given functions are continuous.

**14.** $5x^7 - 3x^2 + 4$

**15.** $\dfrac{x}{x^2 + x}$

**16.** $\dfrac{x^2 + x}{x}$

**17.** $\dfrac{x}{x^2 - 1}$

**18.** $\dfrac{x - 1}{x^2 - 1}$

**19.** $\sqrt[5]{1 - x^2}$

**20.** $\sqrt[3]{\dfrac{x - 1}{x^2 + 3}}$

**21.** $\sqrt[4]{1 - x^2}$

**22.** $\sqrt{\dfrac{x - 1}{x^2 + 3}}$

**23.** Suppose $f(x)$ is continuous on $(0, 2)$ and never zero there and $f(1) > 0$. What can you say about the sign of $f$ on $(0, 2)$?

**24.** Find a function $f(x)$ defined on $[0, 1]$ that changes its sign on $[0, 1]$ but is never zero on $[0, 1]$. Can this function be continuous on $[0, 1]$?

**25.** Suppose $f(x)$ is continuous on $(0, 2)$ and never zero there and $f(1) < 0$. What can you say about the sign of $f$ on $(0, 2)$?

**26.** Suppose $f(x)$ is continuous on $(0, 4)$, $f(2) = 0$, and $f$ has no other zeros on $(0, 4)$. If $f(1) = 2$ and $f(3) = -4$, what can you say about the sign of $f$ on $(0, 4)$?

**27.** Suppose $f(x)$ is continuous on $(0, 4)$, $f(2) = 0$, and $f$ has no other zeros on $(0, 4)$. If $f(1) = -1$ and $f(3) = 2$, what can you say about the sign of $f$ on $(0, 4)$?

## Applications

**28.** **Taxes.** A certain state has a tax on electricity of 1% of the monthly electricity bill with the first \$50 of the bill exempt from tax. Let $P(x)$ be the percent one pays in taxes and $x$ the amount of the bill in dollars. Graph this function and find where it is continuous. What is the significance of the points of discontinuity?

**29.** **Taxes.** In the previous problem, let $T(x)$ be the total amount in dollars paid on the tax where $x$ is the amount of the monthly bill. Graph $T(x)$ and find where it is continuous. What is the significance of the points of discontinuity? Will some people be affected by the point(s) of discontinuity?

30. **Costs.**   A telephone company charges $1.50 for the first minute for a certain long-distance call and $.25 for each additional minute. Graph this cost function and determine where the function is continuous. What is the significance of the points of discontinuity? Will some people be affected by these points?

31. **Salary.**   A certain computer salesman earns a base salary of $20,000 plus 5% of the total sales if sales are above $100,000 and 10% of total sales if sales are above $1,000,000. Let $S(x)$ be his salary and $x$ the amount in dollars of computers he sells. Find the discontinuities of $S(x)$. What is the significance of these discontinuities? Do you think that the salesman will be affected if his sales are almost at one of the discontinuities?

## Graphing Calculator Exercises

In the remaining exercises, draw graphs of the given functions on your graphing calculator and use your graphs to determine where the function is not continuous.

32. $f(x) = |x^2 - 9x + 18|$

33. $f(x) = \dfrac{x - 1}{x^2 + 1}$

34. $f(x) = \dfrac{x^3 - 2}{x - 5}$

35. $f(x) = \dfrac{x - 3}{x^2 - 4}$

36. $f(x) = \sqrt{x^2 - 4}$

37. $f(x) = \sqrt{9 - x^2}$

38. $f(x) = \begin{cases} -x + 1 & \text{if } x < 1 \\ x^2 - 1 & \text{if } 1 \le x < 2 \\ -x + 4 & \text{if } x \ge 2 \end{cases}$

39. $f(x) = \begin{cases} -x^3 + x + 1 & \text{if } x < 0 \\ x^4 - 2x^2 + 1 & \text{if } 0 \le x < 2 \\ -x + 10 & \text{if } x \ge 2 \end{cases}$

40. $f(x) = \begin{cases} -x - 1 & \text{if } x \le -1 \\ \dfrac{x}{x + 1} & \text{if } -1 < x \le 0 \\ x + 1 & \text{if } x > 0 \end{cases}$

41. $f(x) = \begin{cases} -x^2 + x & \text{if } x \le 0 \\ \dfrac{x}{x - 1} & \text{if } 0 < x \le 1 \\ x + 1 & \text{if } x > 1 \end{cases}$

## Solutions to Self-Help Exercise Set 12.2

1. Since $\lim\limits_{x \to 2} f(x)$ and $\lim\limits_{x \to 4} f(x)$ do not exist, the function is not continuous at

$x = 2$ and $x = 4$. Although $\lim\limits_{x \to 1} f(x) = 1$, $f(1) = 2 \neq 1$. Thus $f(x)$ is not continuous at $x = 1$.

2.   The cost function for delivery is

$$C(x) = \begin{cases} 0 & \text{if } 0 \le x \le 10 \\ 2x & \text{if } x > 10 \end{cases}$$

The graph is shown in the figure. A discontinuity is at $x = 10$. At this point, the cost of delivery suddenly jumps. If the distance is one inch less than 10 miles, the delivery is free. If the distance is one inch greater than 10 miles, the cost of the delivery is $20.00, rounded to the nearest cent.

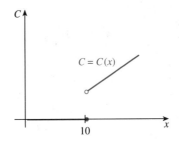

# 12.3   RATES OF CHANGE

▶ *Velocity*

▶ *Other Rates of Change*

▶ *Slope of the Tangent Line*

APPLICATION

**Velocity**

It is known from physics that if a baseball is dropped from the top of a building, the distance $s$ in feet that it will fall is given by $s = s(t) = 16t^2$, where $t$ is in seconds. What is the velocity at any time $t$? See Example 1 for the answer.

APPLICATION

**Instantaneous Rate of Change**

A small fertilizer plant sells all the fertilizer it produces. Suppose that the profit, $P(x)$, in dollars of producing $x$ tons of fertilizer is given by

$$P(x) = 100(-x^2 + 12x - 20).$$

What is the instantaneous rate of change of profits at any $x$? See Example 6 for the answer.

APPLICATION

**Slope of the Tangent Line**

What is the slope of the tangent line to the curve given in the previous example at $x = 5$? See Example 8 for the answer.

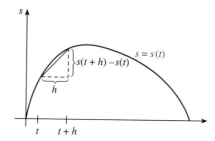

**Figure 12.24**

## Velocity

In the first section of this chapter we determined a method of finding the velocity of an object at an instant of time. Let $s(t)$ be the distance the object has gone in time $t$. Assume now that the graph of the function $s = s(t)$ is given in Figure 12.24. On the interval $[t, t + h]$, $h > 0$, $h$ represents the change in time, $\Delta t$, and $s(t + h) - s(t)$ is the change in distance, $\Delta s$. The ratio

$$v_{ave} = \frac{\Delta s}{\Delta t} = \frac{\text{change in distance}}{\text{change in time}} = \frac{s(t + h) - s(t)}{h}$$

is the average velocity over the interval $t$ to $t + h$. If $h < 0$, then the average velocity on the interval $[t + h, h]$ is

$$v_{ave} = \frac{\Delta s}{\Delta t} = \frac{s(t) - s(t + h)}{-h} = \frac{s(t + h) - s(t)}{h}.$$

which is the same expression as before. Thus, for this expression of $v_{ave}$, we need not distinguish between $h$ positive and $h$ negative. We also refer to the quantity $v_{ave}$ as the **average rate of change** of $s$ with respect to $t$ from $t$ to $t + h$, or simply the average rate of change.

---

**Average Velocity**

Suppose $s = s(t)$ describes the position of an object at time $t$. Then the average velocity from $t$ to $t + h$ is

$$\text{average velocity} = v_{ave} = \frac{\Delta s}{\Delta t} = \frac{s(t + h) - s(t)}{h}.$$

---

**E X A M P L E   1**    **Finding the Average Velocity**

It is known from physics that if a baseball is dropped from the top of a building, the distance $s$ in feet that it has fallen is given by $s = s(t) = 16t^2$, where $t$ is in seconds. Find the average velocity from $t$ to $t + h$ for any $t$ and any $h \neq 0$. Simplify the expression.

**Solution**    We have

$$v_{\text{ave}} = \frac{\Delta s}{\Delta t} \qquad\qquad \frac{\text{change in } s}{\text{change in } t}$$

$$= \frac{s(t + h) - s(t)}{h}$$

$$= \frac{16(t + h)^2 - 16t^2}{h}$$

$$= \frac{16t^2 + 32th + 16h^2 - 16t^2}{h}$$

$$= \frac{(32t + 16h)h}{h}$$

$$= 32t + 16h \quad\blacksquare$$

**E X A M P L E   2**   **Finding the Average Velocity**

Suppose a ball is thrown straight upward with an initial velocity (that is, velocity at the time of release) of 64 ft/sec. Suppose, furthermore, that the point at which the ball is released is considered to be at zero height. Then from physics it is known that the height $s(t)$ in feet of the ball at time $t$ in seconds is given by $s(t) = -16t^2 + 64t$. Find the average velocity from $t$ to $t + h$ for any $t$ and $h \neq 0$.

**Solution**    Since $s(t) = 16t(4 - t)$, the function is easily graphed, as shown in Figure 12.25. (Do not confuse the *graph* in the figure with the actual *path* of the ball. The ball goes straight up and then straight down.)

The average velocity from $t$ to $t + h$ is

$$\frac{s(t + h) - s(t)}{h} = \frac{\Delta s}{\Delta t} \qquad\qquad \frac{\text{change in } s}{\text{change in } t}$$

$$= \frac{[-16(t + h)^2 + 64(t + h)] - [-16t^2 + 64t]}{h}$$

$$= \frac{-16t^2 - 32th - 16h^2 + 64t + 64h + 16t^2 - 64t}{h}$$

$$= \frac{h(-32t - 16h + 64)}{h}$$

$$= -32t - 16h + 64 \quad\blacksquare$$

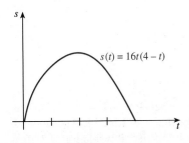

$$s(t) = 16t(4 - t)$$

**Figure 12.25**

To find the instantaneous velocity or the **instantaneous rate of change** at time $t$, we must take the limit as $h$ goes to zero of the average velocity from $t$ to $t + h$. We then have the following.

**Instantaneous Velocity**

Suppose $s = s(t)$ describes the position of an object at time $t$. Then the *instantaneous velocity* (or simply velocity) $v(t)$ at time $t$ is

$$v(t) = \lim_{h \to 0} v_{\text{ave}} = \lim_{h \to 0} \frac{s(t + h) - s(t)}{h}$$

if this limit exists.

**REMARK.** Velocity is referred to as the *instantaneous rate of change* of distance with respect to time, since velocity is a measurement of how fast distance is changing with respect to time.

We already found the velocity of a baseball dropped from the top of a building at the time $t = 1$ in the first section of this chapter. We now find the velocity at any time.

**E X A M P L E   3**    **Finding the Velocity**

For the ball in Example 1, find the velocity at any time $t$.

**Solution**    We first find the average velocity on the interval from $t$ to $t + h$ and simplify the expression. This was done in Example 1. Then to find the instantaneous velocity, or the instantaneous rate of change, we must take the limit of this expression as $h \to 0$. Doing this and using the rules of limits gives

$$v(t) = \lim_{h \to 0} \frac{s(t + h) - s(t)}{h} = \lim_{h \to 0} (32t + 16h) = 32t. \quad \blacksquare$$

**E X A M P L E   4**    **Finding the Velocity**

For the ball in Example 2, find the velocity at any time $t$. Also find $v(1)$, $v(2)$, and $v(3)$, and interpret the meaning of these numbers.

**Solutions**    (a)    To find the instantaneous velocity we must take the limit as $h \to 0$ of the expression for average velocity found in Example 2. Doing this and using the rules of limits gives

$$v(t) = \lim_{h \to 0} \frac{s(t + h) - s(t)}{h}$$
$$= \lim_{h \to 0} (-32t + 64 - 16h) = -32t + 64.$$

(b)    Since $v(t) = -32t + 64$, $v(1) = -32(1) + 64 = 32$. Thus the instantaneous velocity of the ball at the time $t = 1$ is 32 feet per second. This means that if the ball continued at this rate, it would travel 32 feet upward in one second. Notice from Figure 12.25 that the ball is rising at this point.

(c) Since $v(t) = -32t + 64$, $v(2) = -32(2) + 64 = 0$. Thus the instantaneous velocity at time $t = 2$ is zero. The ball goes up and at some point in time, turns and falls downward. At the "turning point," which in this case occurs at $t = 2$, the instantaneous velocity is zero.

(d) Using the same formula gives $v(3) = -32(3) + 64 = -32$. Thus the ball is falling at the instantaneous rate of 32 feet per second. This means that if the ball continued at this rate, it would travel 32 feet downward in one second. Notice from Figure 12.25 that the ball is falling at this point. ∎

## Other Rates of Change

We can take the average rate of change and the instantaneous rate of change for any function $y = f(x)$. We will now do this and take care to note any geometric interpretations.

Consider the graph of the function $y = f(x)$ shown in Figure 12.26. The term $h = \Delta x$ represents the change in $x$ on the interval $[x, x + h]$ and $f(x + h) - f(x) = \Delta y$ represents the corresponding change in $y$ on this interval. In analogy with average velocity, we have the following.

---

**Average Rate of Change**

The average rate of change of $y$ with respect to $x$ from $x$ to $x + h$ is the ratio

$$\frac{\text{change in } y}{\text{change in } x} = \frac{\Delta y}{\Delta x} = \frac{f(x + h) - f(x)}{h}.$$

---

Taking the limit of this last expression as $h \to 0$ will yield the **instantaneous rate of change**. We have the following.

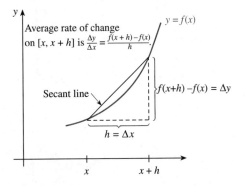

**Figure 12.26**

---

### Instantaneous Rate of Change

Given a function $y = f(x)$, the instantaneous rate of change of $y$ with respect to $x$ is given by

$$\lim_{h \to 0} \frac{f(x + h) - f(x)}{h}$$

if this limit exists.

---

**E X A M P L E  5**    **Finding the Average Rate of Change**

A small fertilizer plant sells all the fertilizer it produces. Suppose that the profit, $P(x)$, in dollars of producing $x$ tons of fertilizer is given by

$$P(x) = 100(-x^2 + 12x - 20).$$

Find the average rate of change of profits on the interval (a) [2, 4], (b) [5, 7] (c) [8, 10].

**Solutions**    Since $P(x) = -100(x - 2)(x - 10)$, the function is easily graphed as in Figure 12.27.

(a)  Here $x = 2$ and $h = 2$. Thus

$$\frac{\Delta P}{\Delta x} = \frac{P(x + h) - P(x)}{h} = \frac{P(4) - P(2)}{2} = \frac{1200 - 0}{2} = 600.$$

(b)  Here $x = 5$ and $h = 2$. Thus

$$\frac{\Delta P}{\Delta x} = \frac{P(x + h) - P(x)}{h} = \frac{P(7) - P(5)}{2} = \frac{1500 - 1500}{2} = 0.$$

(c)  Here $x = 8$ and $h = 2$. Thus

$$\frac{\Delta P}{\Delta x} = \frac{P(x + h) - P(x)}{h} = \frac{P(10) - P(8)}{2} = \frac{0 - 1200}{2} = -600.$$

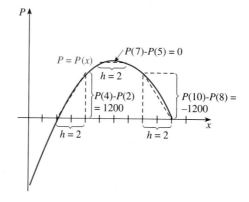

**Figure 12.27**

Thus on the first interval profits are increasing at the average rate of $600 per ton, on the second interval are not increasing or decreasing on average, and on the third interval are decreasing on average by $600 per ton.   ∎

E X A M P L E   6        **Finding the Instantaneous Rate of Change**

Find the instantaneous rate of change of profits at any point $x > 0$ in Example 5.

**Solution**      The average rate of change of $P(x) = 100(-x^2 + 12x - 20)$ as $x$ changes from $x$ to $x + h$ is

$$\frac{P(x + h) - P(x)}{h} = 100 \, \frac{[-(x + h)^2 + 12(x + h) - 20] - [-x^2 + 12x - 20]}{h}$$

$$= 100 \, \frac{-x^2 - 2xh - h^2 + 12x + 12h - 20 + x^2 - 12x + 20}{h}$$

$$= 100 \, \frac{(-2x + 12 - h)h}{h}$$

$$= 100(-2x + 12 - h)$$

Taking the limit as $h \to 0$ of the expression for the average rate of change gives the instantaneous rate of change

$$\lim_{h \to 0} \frac{P(x + h) - P(x)}{h} = \lim_{h \to 0} \, [100(-2x + 12 - h)] = 100(-2x + 12)   ∎$$

E X A M P L E   7        **Finding the Sign of the Instantaneous Rate of Change**

Find where the instantaneous rate of change of profits is positive, zero, and negative in Example 6.

**Solution**      The instantaneous rate of change of profits from the last example is $100(-2x + 12) = 200(6 - x)$. This is positive on the interval $(0, 6)$, zero at $x = 6$, and negative on the interval $(6, +\infty)$.   ∎

For example, at $x = 5$ tons, the instantaneous rate of change of profits with respect to tons is $100(-2(5) + 12) = 200$. This means that at $x = 5$ tons, profits are increasing at the rate of $200 per ton. Since at $x = 6$ tons the instantaneous rate of change of profits is 0, profits are not increasing or decreasing. Since at $x = 7$ tons the instantaneous rate of change of profits is $-200$, profits are decreasing at the rate of $200 per ton.

## Slope of the Tangent Line

The average rate of change and the instantaneous rate of change have an important geometric interpretation which we now consider.

Notice from Figure 12.26 that the average rate of change from $x_0$ to $x_0 + h$ is also the slope $m_{sec}$ of the secant line from the point $P(x_0, f(x_0))$ to the point $Q(x_0 + h, f(x_0 + h))$.

---

**Average Rate of Change is Slope of Secant Line**

The average rate of change from $x_0$ to $x_0 + h$ is the slope of the secant line from $P(x_0, f(x_0))$ to the point $Q(x_0 + h, f(x_0 + h))$:

average rate of change from $x_0$ to $x_0 + h$

$$= \frac{\Delta y}{\Delta x} = \frac{f(x_0 + h) - f(x_0)}{h} = m_{sec}.$$

---

Figure 12.28 shows what happens to the slopes of the secant lines as $h$ approaches 0. Close examination of this figure indicates that $m_{sec}$ approaches $m_{tan}(x_0)$, the slope of the tangent line to the curve at the point $x_0$. That is

$$\lim_{h \to 0} \frac{f(x_0 + h) - f(x_0)}{h} = \lim_{h \to 0} m_{sec} = m_{tan}(x_0),$$

provided the limit exists.

The notion of a tangent line to a curve has until now been presented as a geometric notion. But the graphs of some functions can be extremely complicated, requiring an *analytic* definition of tangent line. Given the observations made in the previous paragraph, we then make the following definition.

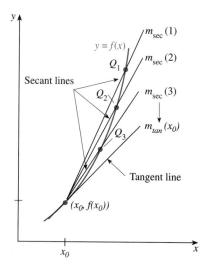

**Figure 12.28**

**Definition of Tangent Line**

The **tangent line** to the graph of $y = f(x)$ at $x = x_0$ is the line through the point $(x_0, f(x_0))$ with slope

$$m_{\tan}(x_0) = \lim_{h \to 0} \frac{f(x_0 + h) - f(x_0)}{h},$$

provided this limit exists.

Notice now that the above limit (if it exists) that defines the slope $m_{\tan}(x_0)$ of the tangent line is the very same limit that defines the instantaneous rate of change at $x_0$. We then have the following.

**Instantaneous Rate of Change and the Slope of the Tangent Line**

If the instantaneous rate of change of $f(x)$ with respect to $x$ exists at a point $x_0$, then it is the slope of the tangent line at that point.

instantaneous rate of change at $x_0 = \lim_{h \to 0} \dfrac{f(x_0 + h) - f(x_0)}{h} = m_{\tan}(x_0)$

**E X A M P L E   8**     **Slope of Tangent Line**

Find the slope of the line tangent to the curve in Example 6 at (a) $x = 5$, (b) $x = 6$, (c) $x = 7$.

**Solution**     We saw above that for the profit function $P(x) = 100(-x^2 + 12x - 20)$ in Example 6, the instantaneous rates of change at $x = 5, 6$, and 7 were respectively 200, 0, and $-200$. Thus

(a)   $m_{\tan}(5) = 200$,   (b)   $m_{\tan}(6) = 0$,   (c)   $m_{\tan}(7) = -200$

This is indicated in Figure 12.29.   ■

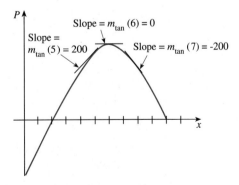

**Figure 12.29**

## SELF-HELP EXERCISE SET 12.3

1. Suppose a ball is thrown straight upward with an initial velocity (that is, velocity at the time of release) of 96 ft/sec. Suppose, furthermore, that the point at which the ball is released is considered to be at zero height. Then from physics it is known that the height $s(t)$ in feet of the ball at time $t$ in seconds is given by $s(t) = -16t^2 + 96t$.
   a. Find the average velocity from $t$ to $t + h$ for any $t$ and any $h \neq 0$.
   b. Using the rules of limits, take the limit as $h \to 0$ of the expression for average velocity found in (a) and find the instantaneous velocity $v(t)$.

2. For the ball in the previous example, find $v(1)$, $v(3)$, and $v(4)$ and interpret these numbers both physically and geometrically.

## EXERCISE SET 12.3

1. Let $s(t) = -t^2 + 8t$ be the height in feet of an object at time $t$ in seconds. (a) Find the average velocity on $[t, t + h]$ for any $t$ and for any $h > 0$, and the average velocity on $[0, 3]$, $[3, 5]$, $[5, 7]$. (b) Find the instantaneous velocity $v(t)$. (c) Find $v(3)$, $v(4)$, $v(5)$. Interpret physically and geometrically what these numbers mean.

2. Repeat the previous exercise if $s(t) = -2t^2 + 16t - 12$.

3. Let $y = f(x) = 2x + 5$. (a) Find the average rate of change on the interval $[x, x + h]$ for any $x$ and for any $h > 0$. (b) Find the instantaneous rate of change at any point $x$.

4. Let $y = f(x) = -2x + 3$. (a) Find the average rate of change on the interval $[x, x + h]$ for any $x$ and for any $h > 0$. (b) Find the instantaneous rate of change at any point $x$.

5. Let $y = f(x) = x^2 - 4x$. (a) Find the average rate of change on the interval $[x, x + h]$ for any $x$ and for any $h > 0$. (b) Find the instantaneous rate of change at any point $x$.

6. Repeat the previous exercise if $y = f(x) = -x^2 + 8x + 1$.

7. Let $y = f(x) = 3x^2 - 4$. (a) Find the slope of the secant line from $(x, f(x))$ to $(x + h, f(x + h))$ for any $x$ and for any $h > 0$. (b) Find the slope of the tangent line at any point $x$.

8. Repeat the previous exercise if $y = f(x) = -x^2 + 8$.

9. Find the slope of the line tangent to the graph of the curve in Exercise 5 at (a) $x_0 = 1$, (b) $x_0 = 2$, (c) $x_0 = 3$.

10. Find the slope of the line tangent to the graph of the curve in Exercise 6 at (a) $x_0 = 3$, (b) $x_0 = 4$, (c) $x_0 = 5$.

In Exercises 11 through 24 find the instantaneous rates of change of the given functions.

**11.** $f(x) = 2x + 3$

**12.** $f(x) = -3x + 4$

**13.** $f(x) = x^2 - 1$

**14.** $f(x) = -x^2 + 3$

**15.** $f(x) = -2x^2 + 3$

**16.** $f(x) = -3x^2 - 1$

**17.** $f(x) = x^2 + 2x + 3$

**18.** $f(x) = x^2 - 3x + 4$

**19.** $f(t) = -2t^2 - 4t - 2$

**20.** $f(r) = 3r^2 - 4r + 2$

**21.** $f(u) = u^3$

**22.** $f(v) = -v^3 + 1$

**23.** $f(t) = t^3 + 4$

**24.** $f(z) = z^3 + z - 2$

## Applications

**25.** The following table gives the number of law enforcement officers killed in the line of duty in the United States.

| Year | 1979 | 1980 | 1981 | 1982 | 1983 | 1984 | 1985 | 1986 | 1987 | 1988 |
|---|---|---|---|---|---|---|---|---|---|---|
| Killed | 164 | 165 | 157 | 164 | 152 | 147 | 148 | 131 | 147 | 155 |

Find the average rate of change of those killed on the interval (a) [1979, 1982], (b) [1982, 1984], (c) [1985, 1988], (d) [1986, 1988].

**26.** The following table gives the total U.S. consumption of petroleum in millions of barrels per day.

| Year | 1950 | 1955 | 1960 | 1965 | 1970 | 1975 | 1980 | 1985 |
|---|---|---|---|---|---|---|---|---|
| Amount | 6.5 | 8.5 | 9.8 | 11.5 | 14.7 | 16.3 | 17.1 | 15.7 |

Find the average rate of change of the consumption of petroleum on the interval (a) [1950, 1980], (b) [1975, 1985], (c) [1980, 1985].

**27.** The following table gives the number of work stoppages in the United States.

| Year | 1960 | 1965 | 1970 | 1975 | 1980 | 1985 |
|---|---|---|---|---|---|---|
| Number | 222 | 268 | 381 | 235 | 187 | 54 |

Find the average rate of change of the number of work stoppages on the intervals (a) [1960, 1970], (b) [1960, 1975], (c) [1970, 1985].

**28.** **Population.** The following table gives the population in hundreds of thousands of a city over a period of years. Find the average rate of change of the population with respect to time measured in years during (a) 1985 to 1987, (b) 1985 to 1988, (c) 1987 to 1989, (d) 1987 to 1991, (e) 1985 to 1991.

| Year | 1985 | 1986 | 1987 | 1988 | 1989 | 1990 | 1991 |
|---|---|---|---|---|---|---|---|
| Number | 3.0 | 2.8 | 2.6 | 2.6 | 2.9 | 3.1 | 3.2 |

29. **Profits.**  The profits in millions of dollars of a truck manufacturer is given by $P = -x^2 + 10x - 16$ where $x$ is the number in thousands of trucks sold. Find the instantaneous rate of change when $x$ is (a) 3, (b) 5, (c) 7. Interpret your answer.

30. **Profits.**  Find where the instantaneous rate of change of profits in the previous exercise is (a) positive, (b) zero, (c) negative.

31. **Revenue.**  The revenue in hundreds of dollars for a kitchen sink manufacturer is given by $R = 10x - 0.01x^2$ where $x$ is the number of sinks sold. Find the instantaneous rate of change when $x$ is (a) 100, (b) 500, (c) 600. Interpret your answer.

32. **Revenue.**  Find where the instantaneous rate of change of revenue in the previous exercise is (a) positive, (b) zero, (c) negative.

33. **Sales.**  Suppose that $t$ weeks after the end of an advertising campaign the weekly sales are $100(t^2 - 8t - 20)$. At what instantaneous rate are sales changing when $t$ equals (a) 2, (b) 4, (c) 6. Interpret your answer.

34. **Sales.**  Find where the instantaneous rate of change of sales in the previous exercise is (a) positive, (b) zero, (c) negative.

35. **Cost.**  Suppose that the cost, in thousands of dollars, for a jacket manufacturer is given by $C = 6 + 2x^2$, where $x$ is the number of thousands of jackets produced. Find the instantaneous rate of change of costs when $x$ is (a) 0, (b) 1, (c) 4.

36. **Cost.**  Find where the instantaneous rate of change of cost in the previous exercise is (a) positive, (b) zero, (c) negative.

37. **Medicine.**  Suppose that the weight in mg of a cancerous tumor at time $t$ measured in weeks is given by $W(t) = 0.1t^2$. What is the instantaneous rate of growth of the tumor in mg per week when $t = 4$?

## Graphing Calculator Exercises

The program SecantN found in C.2 of Appendix C calculates the slope of the secant line for $h = \pm 0.1$, $h = \pm 0.01$, ... and stores the values in the data section. In Exercises 38 through 45 find the instantaneous rate of change of the given function at the indicated point by using this program on your graphing calculator.

38.  $f(x) = x^4 - 10$, $a = 2$

39.  $f(x) = x^3 - 5$, $a = 5$

40.  $f(x) = \sqrt{x}$, $a = 25$

41.  $f(x) = \dfrac{1}{x}$, $a = 1$

**42.**  $f(x) = \dfrac{1}{x + 2}, a = 3$ 　　　　　　　 **43.**  $f(x) = \sqrt{x + 1}, a = 3$

**44.**  $f(x) = \dfrac{1 + x}{2 + x}, a = 1$ 　　　　　　 **45.**  $f(x) = \dfrac{1 + x}{1 + x^2}, a = 1$

The program SecantG in C.2 of Appendix C graphs $y = f(x)$, the tangent line at the indicated point, and a sequence of secant lines for values of $h$ approaching zero and thus demonstrates Figure 12.28 of the text. In Exercises 46 through 49 use this program on the given functions at the indicated points. As usual you will need to adjust the RANGE variables on your calculator to obtain the best viewing rectangle.

**46.**  $f(x) = x^3, a = 1$ 　　　　　　　　　 **47.**  $f(x) = x^2, a = 2$

**48.**  $f(x) = 1/x, a = 0.5$ 　　　　　　　 **49.**  $f(x) = \sqrt{x}, a = 1$

## Solutions to Self-Help Exercise Set 12.3

**1.**  Since $s(t) = 16t(6 - t)$, the function is easily graphed.

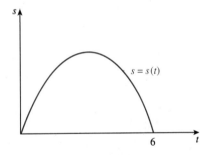

**a.**  The average velocity on the interval $[t, t + h]$ is

$$\frac{s(t + h) - s(t)}{h} = \frac{[-16(t + h)^2 + 96(t + h)] - [-16t^2 + 96t]}{h}$$

$$= 16 \frac{[-(t + h)^2 + 6(t + h)] - [-t^2 + 6t]}{h}$$

$$= 16 \frac{-t^2 - 2th - h^2 + 6t + 6h + t^2 - 6t}{h}$$

$$= 16 \frac{h(-2t + 6 - h)}{h}$$

$$= -32t + 96 - 16h.$$

**b.**  Taking the limit as $h \to 0$ of the expression for the average velocity found in (a) gives

$$\lim_{h \to 0} \frac{s(t + h) - s(t)}{h} = \lim_{h \to 0} (-32t + 96 - 16h) = -32t + 96.$$

**2.**  **a.**  Since $v(t) = -32t + 96$, $v(1) = -32(1) + 96 = 64$. Thus the instantaneous velocity of the ball at the time $t = 1$ is 64 feet per second. This means that if the ball continued at this rate, it would travel 64 feet upward in one second. The fact that the instantaneous velocity is positive at $t = 1$ indicates that the ball is rising at this time. Notice from the accompanying figure that the slope of the tangent line at $t = 1$ is positive and equal also to 64.

    **b.**  Since $v(t) = -32t + 96$, $v(3) = -32(3) + 96 = 0$. Thus the instantaneous velocity at time $t = 3$ is zero. The ball goes up and at some point in time, turns and falls downward. At the "turning point," which occurs at $t = 3$, the instantaneous velocity is zero. The figure indicates that the slope $m_{\tan}(3)$ of the tangent line at $t = 3$ is zero.

    **c.**  Using the same formula, gives $v(4) = -32(4) + 96 = -32$. Thus the ball is falling at the instantaneous rate of 32 feet per second. From the figure, the slope of the tangent line to the curve at $t = 4$ is $-32$.

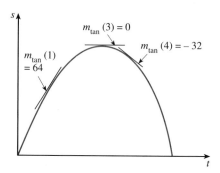

## 12.4   THE DERIVATIVE

▶ *The Derivative*

▶ *Equation of Tangent Line*

▶ *Nonexistence of the Derivative*

▶ *The Derivative and Continuity*

**A P P L I C A T I O N**

**Instantaneous Rate of Change**

Suppose that $p(x) = \frac{1}{x}$ is the price in dollars of a pen where $x$ is the number of these pens sold per year (in millions). Find the instantaneous rate of change of the price with respect to the number of pens sold per year (in millions) when 2 million pens are sold. See Example 2 for the answer.

## The Derivative

Given a function $y = f(x)$ for which

$$\lim_{h \to 0} \frac{f(x + h) - f(x)}{h}$$

exists, we have seen that this limit represents the instantaneous rate of change of $y$ with respect to $x$ and also the slope of the line tangent to the graph of the function at the point $(x, f(x))$. This limit (when it exists) is called the **derivative** and is denoted by $f'(x)$.

We then have the following definition.

---

**Definition of Derivative**

If $y = f(x)$, the derivative of $f(x)$, denoted by $f'(x)$, is defined to be

$$f'(x) = \lim_{h \to 0} \frac{f(x + h) - f(x)}{h},$$

if this limit exists.

---

**REMARK.**   Other notations for this derivative are $y'$, $\dfrac{dy}{dx}$, and $\dfrac{d}{dx} f(x)$.

**E X A M P L E   1**   **Finding the Derivative of a Function**

Let $y = f(x) = \dfrac{1}{x}$, $x \neq 0$. Find $f'(x)$.

**Solution**   First find an expression for the average rate of change of $y$ with respect to $x$ from $x$ to $x + h$ and simplify. This gives

$$\frac{f(x + h) - f(x)}{h} = \frac{\Delta y}{\Delta x} \qquad \frac{\text{change in } y}{\text{change in } x}$$

$$= \frac{\dfrac{1}{x + h} - \dfrac{1}{x}}{h}$$

$$= \frac{1}{h} \left[ \frac{1}{x + h} - \frac{1}{x} \right]$$

$$= \frac{1}{h} \left[ \frac{x - (x + h)}{x(x + h)} \right]$$

$$= \frac{1}{h} \left[ \frac{-h}{x(x + h)} \right] \qquad \text{Cancel the } h.$$

$$= -\frac{1}{x(x + h)}$$

Now to find the derivative, take the limit of this expression as $h \to 0$, obtaining

$$f'(x) = \lim_{h \to 0} \frac{f(x + h) - f(x)}{h} = \lim_{h \to 0} \frac{-1}{x(x + h)} = -\frac{1}{x^2},$$

provided $x \neq 0$. ∎

**REMARK.** Notice that the derivative $f'(x)$ is itself a *function*. In the previous example the original function was $f(x) = \frac{1}{x}$, whereas the derivative of this function is the function $f'(x) = -\frac{1}{x^2}$, which for any $x$ gives the slope of the tangent line at $x$ to the graph of $y = f(x)$ (or the instantaneous rate of change at $x$).

**EXAMPLE 2**  **Finding an Instantaneous Rate of Change**

Suppose that $p(x) = \frac{1}{x}$ is the price in dollars of a pen where $x$ is the number of these pens sold per year (in millions). Find the instantaneous rate of change of the price with respect to the number of pens sold per year (in millions) when 2 million pens are sold.

**Solution**  Since the instantaneous rate of change is the same as the derivative, the instantaneous rate of change of price with respect to the number of millions of pens sold per year when 2 million pens are sold is $p'(2)$, where $p(x) = \frac{1}{x}$. From the previous example, we have $p'(2) = -\frac{1}{(2)^2} = -0.25$. Thus when 2 million pens are being sold, the price is dropping at a rate of $0.25 per million sold per year. ∎

The general procedure for finding the derivative is summarized as the following three-step process.

---

**To Find the Derivative of $f(x)$.**

1.  Find $\dfrac{f(x + h) - f(x)}{h}$.

2.  Simplify.

3.  Take the limit as $h \to 0$ of the simplified expression. That is,

$$f'(x) = \lim_{h \to 0} \frac{f(x + h) - f(x)}{h}.$$

E X A M P L E  3     **Finding the Derivative of a Function**

Find the derivative of $y = f(x) = \sqrt{x}$.

**Solution**     Doing the first 2 steps gives

$$\frac{f(x + h) - f(x)}{h} = \frac{\sqrt{x + h} - \sqrt{x}}{h}$$

$$= \frac{\sqrt{x + h} - \sqrt{x}}{h} \cdot \frac{\sqrt{x + h} + \sqrt{x}}{\sqrt{x + h} + \sqrt{x}}$$

Rationalize the numerator.

$$= \frac{x + h - x}{h(\sqrt{x + h} + \sqrt{x})}$$

$$= \frac{h}{h(\sqrt{x + h} + \sqrt{x})} \quad \text{Cancel the } h.$$

$$= \frac{1}{\sqrt{x + h} + \sqrt{x}}.$$

Now take the limit as $h \to 0$ of the last expression, using the rules of limits. This gives

$$\frac{dy}{dx} = \lim_{h \to 0} \frac{f(x + h) - f(x)}{h} = \lim_{h \to 0} \frac{1}{\sqrt{x + h} + \sqrt{x}} = \frac{1}{2\sqrt{x}}, \quad \text{if } x > 0. \quad \blacksquare$$

## Equation of Tangent Line

If $(x_0, y_0)$ is on a line that has slope $m$, then the equation of this line is

$$y - y_0 = m(x - x_0).$$

Thus if $m_{tan}(x_0)$ is the slope of the tangent line at $(x_0, f(x_0))$ to the curve $y = f(x)$, then the equation of the tangent line is given by

$$y - f(x_0) = m_{tan}(x_0)(x - x_0).$$

But since $m_{tan}(x_0) = f'(x_0)$, the equation of the tangent line can be written as

$$y - f(x_0) = f'(x_0)(x - x_0).$$

See Figure 12.30.

**Equation of the Tangent Line**

Given $y = f(x)$ and given that $f'(x_0)$ exists, the equation of the line tangent to the curve $y = f(x)$ at $x_0$ is

$$y - f(x_0) = f'(x_0)(x - x_0).$$

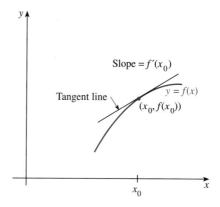

**Figure 12.30**

E X A M P L E  4    **Finding the Equation of the Tangent Line**

Find the equation of the tangent line to $y = \frac{1}{x}$ at $(2, \frac{1}{2})$. See Figure 12.31.

**Solution**    Here $x_0 = 2$ and $f(x_0) = f(2) = 1/2$. Also from Example 1, since $f'(x) = -1/x^2$, $f'(2) = -1/4$. Thus the equation of the tangent line is

$$y - \tfrac{1}{2} = -\tfrac{1}{4}(x - 2). \quad \blacksquare$$

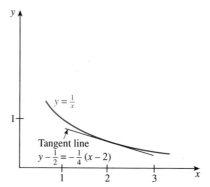

**Figure 12.31**

## Nonexistence of the Derivative

We know that $f'(x_0) = m_{\tan}(x_0)$, the slope of the line tangent to the graph of the curve $y = f(x)$ at the point $x_0$. Thus, if $m_{\tan}(x_0)$ does not exist, then $f'(x_0)$ does not exist.

**E X A M P L E   5**     **A Function Whose Derivative Does Not Exist at One Point**

Let $f(x) = |x|$. Does $f'(0)$ exist?

**Solution**     In view of the previous remarks, we shall try to determine if $m_{tan}(0)$ exists. From the graph of this function in Figure 12.32 we readily see that the secant line from $P = (0, 0)$ to any point $Q$ on the curve where $Q$ is to the right of $P$ is given by $y = x$. Thus $m_{sec} = 1$ if $Q$ is to the right of $P$. This implies that

$$\lim_{h \to 0+} \frac{f(0 + h) - f(0)}{h} = \lim_{h \to 0+} \frac{h - 0}{h} = \lim_{h \to 0+} 1 = 1.$$

On the other hand, from this figure the secant line from $P = (0, 0)$ to any $Q$ on the curve where $Q$ is to the left of $P$ is given by $y = -x$. Thus $m_{sec} = -1$ if $Q$ is to the left of $P$. This then implies that

$$\lim_{h \to 0-} \frac{f(0 + h) - f(0)}{h} = \lim_{h \to 0-} \frac{-h - 0}{h} = \lim_{h \to 0-} -1 = -1.$$

Since the limit from the left does not equal the limit from the right,

$$\lim_{h \to 0} \frac{f(0 + h) - f(0)}{h}$$

does not exist, and thus the tangent line does not exist. Therefore $m_{tan}(0)$ and hence $f'(0)$ does not exist.   ∎

**R E M A R K .**     The problem here is that an infinite number of lines seem to qualify as possible "tangent lines." Figure 12.33 indicates two of them.

**Figure 12.32**

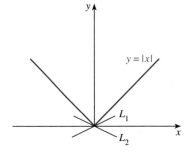

**Figure 12.33**

**R E M A R K .**     The very same situation occurs whenever a function has a corner. Thus when the graph of a function has a corner, such as in Figure 12.34, the function has no (unique) tangent line and thus no derivative at the corner.

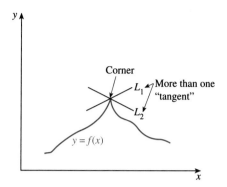

**Figure 12.34**

E X A M P L E   6     **Drawing a Graph of $y = f'(x)$**

Find $f'(x)$ and graph $y = f'(x)$ if $f(x) = |x|$.

**Solution**     From Figure 12.32 we notice that for $x > 0$, $f(x) = |x| = x$ is a line with slope equal to 1. Thus for $x > 0$, $f'(x) = m_{\tan}(x) = 1$. In a similar fashion we have for $x < 0$, $f'(x) = m_{\tan}(x) = -1$. The graph is shown in Figure 12.35, where we notice that no value is given for $x = 0$, since we saw in the previous example that $f'(0)$ does not exist.   ■

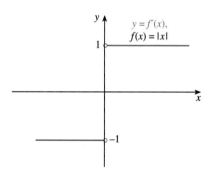

**Figure 12.35**

E X A M P L E   7     **A Function With a Vertical Tangent**

If $f(x) = \sqrt[3]{x}$, determine if $f'(0)$ exists.

**Solution**     A very carefully drawn graph of this function, such as the one shown in Figure 12.36, indicates that the curve *does* have a tangent line at $x = 0$. However, since the tangent line is *vertical*, the slope does not exist. Therefore $f'(0)$ cannot exist.

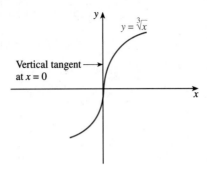

**Figure 12.36**

We can also verify this by the following work.

$$f'(0) = \frac{\Delta y}{\Delta x} \qquad \frac{\text{change in } y}{\text{change in } x}$$

$$= \lim_{h \to 0} \frac{f(0 + h) - f(0)}{h}$$

$$= \lim_{h \to 0} \frac{h^{\frac{1}{3}}}{h}$$

$$= \lim_{h \to 0} \frac{1}{h^{\frac{2}{3}}},$$

which does not exist since $\dfrac{1}{h^{\frac{2}{3}}}$ becomes large without bound as $h \to 0$. ■

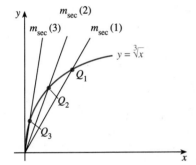

**Figure 12.37**

The last equation indicates that the slope of the secant line is becoming large without bound. See Figure 12.37.

### The Derivative and Continuity

There is an important connection between the existence of the derivative and continuity. We have the following theorem.

---

**A Differentiable Function is Continuous**

If $y = f(x)$ has a derivative at $x = x_0$, then $f(x)$ is continuous at $x = x_0$.

---

We give the following proof.*

$$\lim_{x \to x_0} f(x) - f(x_0) = \lim_{x \to x_0} (f(x) - f(x_0))$$

$$= \lim_{x \to x_0} (x - x_0) \cdot \frac{f(x) - f(x_0)}{x - x_0}$$

$$= \left( \lim_{x \to x_0} (x - x_0) \right) \left( \lim_{x \to x_0} \frac{f(x) - f(x_0)}{x - x_0} \right)$$

$$= 0 \cdot f'(x_0) = 0$$

$$\lim_{x \to x_0} f(x) = f(x_0)$$

This is exactly the definition of continuity at $x = x_0$.

**REMARK.**     The converse of this theorem is not true. That is, a function may be continuous at points where it is not differentiable. For example, the functions graphed in Figures 12.32, 12.34, and 12.36 are continuous at all points in their domains, but each has one value for which the function is not differentiable.

We now summarize the circumstances for which the derivative does not exist.

---

**When the Derivative Fails to Exist**

The derivative fails to exist in the following three circumstances.

1.  The graph of the function has a corner.
2.  The graph of the function has a vertical tangent.
3.  The graph of the function has a break (discontinuity).

---

**E X A M P L E   8**        **Determining Graphically When the Derivative Exists**

Find the points where the derivative of the function whose graph is shown in Figure 12.38 fails to exist.

**Solution**        The derivative does not exist at $x = 2$, 4, and 6 since at $x = 2$ the tangent is vertical, at $x = 4$ the graph has a corner, and at $x = 6$ the function is not continuous.   ∎

---

*Optional

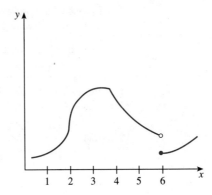

**Figure 12.38**

## SELF-HELP EXERCISE SET 12.4

1.  If $y = f(x) = \dfrac{1}{x + 1}$, find $f'(x)$ when $x \neq -1$.

2.  Find the equation of the line tangent to the curve $y = \sqrt{x}$ at the point $(4, 2)$.

## EXERCISE SET 12.4

In Exercises 1 through 6 find $f'(x)$ using the limit definition of $f'(x)$.

1.  $f(x) = 5x - 3$

2.  $f(x) = -2x + 3$

3.  $f(x) = x^2 + 4$

4.  $f(x) = 2x^2 - x$

5.  $f(x) = 3x^2 + 3x - 1$

6.  $f(x) = -3x^2 - x + 1$

In Exercises 7 through 12 find $\dfrac{dy}{dx}$.

7.  $y = -2x^3 + x - 3$

8.  $y = 3x^3 - 4x + 1$

9.  $y = \dfrac{1}{x + 2},\ x \neq -2$

10. $y = \dfrac{2}{3x + 4},\ x \neq -\dfrac{4}{3}$

11. $y = \dfrac{1}{2x - 1},\ x \neq \dfrac{1}{2}$

12. $y = \dfrac{3}{2 - 3x},\ x \neq \dfrac{2}{3}$

In Exercises 13 through 16 find $\dfrac{dy}{dt}$.

13. $y = \sqrt{t + 1},\ t > -1$

14. $y = \sqrt{t - 3},\ t > 3$

15. $y = \sqrt{2t + 5},\ t > -\dfrac{5}{2}$

16. $y = \sqrt{1 - t},\ t < -1$

**17.** In each of the problems in Exercises 1, 3, and 5, find the equation of the tangent line to the curve at the point $(1, f(1))$.

**18.** In each of the problems in Exercises 2, 4, and 6, find the equation of the tangent line to the curve at the point $(-1, f(-1))$.

In Exercises 19 through 28 sketch a graph of $y = f(x)$ to find where $m_{\tan}(x)$ and thus $f'(x)$ exists.

**19.** $f(x) = \begin{cases} 0 & \text{if } x \le 0 \\ x & \text{if } x > 0 \end{cases}$

**20.** $f(x) = -|x|$

**21.** $f(x) = -|x - 1|$

**22.** $f(x) = \begin{cases} x & \text{if } x \le 1 \\ -x + 2 & \text{if } x > 1 \end{cases}$

**23.** $f(x) = \begin{cases} x & \text{if } x \le 1 \\ 1 & \text{if } x > 1 \end{cases}$

**24.** $f(x) = \begin{cases} -x & \text{if } x \le 0 \\ 0 & \text{if } x > 0 \end{cases}$

**25.** $f(x) = \begin{cases} -\sqrt{1 - (x + 1)^2} & \text{if } -1 \le x < 0 \\ \sqrt{1 - (x - 1)^2} & \text{if } 0 \le x \le 1 \end{cases}$

**26.** $f(x) = \begin{cases} \sqrt{1 - (x + 1)^2} & \text{if } -1 \le x < 0 \\ -\sqrt{1 - (x - 1)^2} & \text{if } 0 \le x \le 1 \end{cases}$

**27.** $f(x) = \begin{cases} \sqrt{1 - (x - 1)^2} & \text{if } 1 \le x \le 2 \\ -\sqrt{1 - (x - 3)^2} & \text{if } 2 \le x < 3 \end{cases}$

**28.** $f(x) = \begin{cases} -\sqrt{1 - (x + 1)^2} & \text{if } -1 \le x \le 0 \\ \sqrt{1 - (x - 1)^2} & \text{if } 0 \le x \le 2 \\ \sqrt{1 - (x - 3)^2} & \text{if } 2 \le x \le 3 \end{cases}$

In Exercises 29 through 32 use the limit definition of $f'(x)$ to determine if $f'(x)$ exists at the indicated point.

**29.** $\sqrt[3]{x}, x_0 = 0$

**30.** $\sqrt[3]{x - 1}, x_0 = 1$

**31.** $\sqrt[3]{x^2}, x_0 = 0$

**32.** $\sqrt[3]{(x - 4)^2}, x_0 = 4$

In Exercises 33 through 38, draw a graph of $y = f'(x)$ for the given graph.

**33.**

**34.**

**35.**

**36.**

**37.**

**38.**

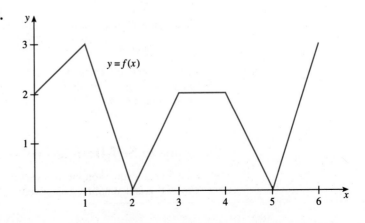

## Applications

39. **Profits.**   The profits in millions of dollars of a firm is given by $P = \sqrt{4x + 1} - 2$ where $x$ is the number in thousands of items sold. Find the instantaneous rate of change when $x$ is (a) 2, (b) 6. Interpret your answer.

40. **Revenue.**   The revenue function for a firm is given by $R = \sqrt{2x + 5}$. Find the instantaneous rate of change when $x$ is (a) 2, (b) 10. Interpret your answer.

41. **Sales.**   Suppose that $t$ weeks after the end of an advertising campaign the weekly sales are $s = \frac{100}{t + 1}$. At what rate are sales changing when $t$ equals (a) 1, (b) 4. Interpret your answer.

42. **Cost.**   Suppose that the cost function for a firm is given by $C = x + \sqrt{x}$. Find the instantaneous rate of change of costs when $x$ is (a) 1, (b) 4, (c) 9.

43. **Price.**   Suppose the price of an item as a function of the number sold is given by $p(x) = \frac{5}{3x + 1}$. Find the instantaneous rate of change of the price when $x$ is (a) 3, (b) 8.

## Graphing Calculator Exercises

In Exercises 44 through 51 determine if the derivative of the given function exists at the indicated point by using the program SecantN in C.2 of Appendix C.

44.  $f(x) = |x^5|, a = 0$

45.  $f(x) = |x^3|, a = 0$

46.  $f(x) = \sqrt[5]{x}, a = 0$

47.  $f(x) = \sqrt[3]{x}, a = 0$

48.  $f(x) = \begin{cases} x^3 & \text{if } x \le 0 \\ x & \text{if } x > 0 \end{cases}$

49.  $f(x) = \begin{cases} x & \text{if } x \le 0 \\ x^2 & \text{if } x > 0 \end{cases}$

50.  $f(x) = \begin{cases} 2x + 1 & \text{if } x \le 0 \\ x^2 + 2x + 1 & \text{if } x > 0 \end{cases}$

51.  $f(x) = \begin{cases} x^3 & \text{if } x \le 0 \\ x^2 & \text{if } x > 0 \end{cases}$

52.  Your graphing calculator estimates the derivative of a function $f(x)$ at a point $a$ by calculating $\dfrac{f(a + h) - f(a - h)}{2h}$ for small values of $h$. Use this operation on your graphing calculator to estimate $f'(0)$ when $f(x) = |x|$. We know from Example 5 that $f'(0)$ does not exist. Explain why your calculator is giving a misleading answer.

## Solutions to Self-Help Exercise Set 12.4

1.  First find an expression for the average rate of change of $y$ with respect to $x$ from $x$ to $x + h$ and simplify. This gives

$$\frac{f(x + h) - f(x)}{h} = \frac{\dfrac{1}{x + 1 + h} - \dfrac{1}{x + 1}}{h}$$

$$= \frac{1}{h}\left[\frac{1}{x + 1 + h} - \frac{1}{x + 1}\right]$$

$$= \frac{1}{h}\left[\frac{x + 1 - (x + 1 + h)}{(x + 1)(x + 1 + h)}\right]$$

$$= \frac{1}{h}\left[\frac{-h}{(x + 1)(x + 1 + h)}\right]$$

$$= -\frac{1}{(x + 1)(x + 1 + h)}.$$

Now to find the derivative, take the limit of this expression as $h \to 0$, obtaining

$$f'(x) = \lim_{h \to 0} \frac{f(x + h) - f(x)}{h} = \lim_{h \to 0} \frac{-1}{(x + 1)(x + 1 + h)} = -\frac{1}{(x + 1)^2},$$

provided $x \neq -1$.

2. From Example 3 of the text, we have that $\dfrac{d}{dx}(\sqrt{x}) = \dfrac{1}{2\sqrt{x}}$. Thus

$$m_{\text{tan}}(4) = f'(4) = \frac{1}{2\sqrt{4}} = \frac{1}{4}.$$

Thus the equation of the tangent line at $x = 4$ is

$$y - 2 = \frac{1}{4}(x - 4).$$

## 12.5   DERIVATIVES OF CONSTANTS, POWERS, AND SUMS

► *Derivatives of Constants*
► *Derivatives of Powers*
► *Derivative of a Constant Times a Function*
► *Derivatives of Sums and Differences*
► *Marginal Analysis*
► *Other Rates of Change*

APPLICATION

**Cost of Next Item**

Suppose the total cost to produce $x$ barrels of a beverage is given by $C(x) = 50 + x^2$. Use the derivative to find the approximate cost of the next barrel when $x = 100$. See Example 6 for the answer.

**A P P L I C A T I O N**

**Instantaneous Rate of Change of Population**

Assume that a population of mice with an initial size of 1000 grows according to the law

$$p(t) = 1000 + 2t^3,$$

where $t$ is measured in months. What is the instantaneous rate of change of the population with respect to time at the end of the fourth month? The answer is found in Example 10.

## Derivatives of Constants

In the last section, we learned how to find the derivatives of functions by taking the limit, that is, by using the definition of derivative. In this section we will develop some basic rules for derivatives that can then be used to find the derivatives of a wide variety of functions without having to take limits.

Suppose $f(x)$ is a constant function, that is, $f(x) = C$, where $C$ is a constant. The graph of this function is given in Figure 12.39 and is just a horizontal straight line. The slope of any such horizontal straight line is zero. Thus we expect that the derivative should also be zero. Using the definition of derivative, we have

$$f'(x) = \lim_{h \to 0} \frac{f(x + h) - f(x)}{h}$$

$$= \lim_{h \to 0} \frac{C - C}{h}$$

$$= \lim_{h \to 0} \frac{0}{h}$$

$$= \lim_{h \to 0} 0$$

$$= 0.$$

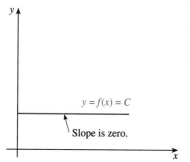

**Figure 12.39**

We can write this as a basic rule.

---

**Derivative of a Constant**

For any constant $C$

$$\frac{d}{dx}(C) = 0.$$

---

**E X A M P L E   1**    **Finding the Derivative of Constants**

Find (a) $\dfrac{d}{dx}(3.4)$, (b) $f'(x)$ if $f(x) = -6$, (c) $\dfrac{dy}{dx}$ if $y = \dfrac{1}{3}$.

**Solutions**

(a) $\dfrac{d}{dx}(3.4) = 0$, (b) $f'(x) = \dfrac{d}{dx}(-6) = 0$,

(c) $\dfrac{dy}{dx} = \dfrac{d}{dx}\left(\dfrac{1}{3}\right) = 0$ ∎

## Derivatives of Powers

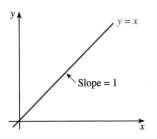

Figure 12.40

In the last section we found the derivatives of some powers. For example, we found that $\dfrac{d}{dx}(x^{1/2}) = \dfrac{1}{2}x^{-1/2}$ and $\dfrac{d}{dx}(x^{-1}) = -x^{-2}$.

Furthermore, since the function $f(x) = x$ represents a straight line with constant slope equal to one, see Figure 12.40, we suspect that $\dfrac{d}{dx}(x) = 1$. (This is in fact true.)

The rule for finding the derivative of $f(x) = x^n$ for any $n$ follows.

---

### Derivative of a Power

If $n$ is any real number ($n$ may or may not be an integer),

$$\frac{d}{dx}(x^n) = nx^{n-1}$$

---

We shall give the proof only in the case that $n$ is a positive integer, the general case being somewhat difficult.

If $n$ is any positive integer, then

$$f(x + h) = (x + h)^n = x^n + nx^{n-1}h$$

$$+ \frac{n(n-1)}{2}x^{n-2}h^2 + \cdots + nxh^{n-1} + h^n.$$

Thus

$$f(x + h) - f(x) = (x + h)^n - x^n = nx^{n-1}h$$

$$+ \frac{n(n-1)}{2}x^{n-2}h^2 + \cdots + nxh^{n-1} + h^n.$$

Then we see that

$$\frac{f(x + h) - f(x)}{h} = \frac{(x + h)^n - x^n}{h} = nx^{n-1}$$

$$+ \frac{n(n-1)}{2}x^{n-2}h + \cdots + nxh^{n-2} + h^{n-1}.$$

Now to find $f'(x)$ we let $h \to 0$ in this last expression to obtain

$$f'(x) = \lim_{h \to 0} \frac{f(x + h) - f(x)}{h}$$

$$= \lim_{h \to 0} \frac{(x + h)^n - x^n}{h}$$

$$= nx^{n-1} + \frac{n(n - 1)}{2} x^{n-2}(0) + \cdots + nx(0)^{n-2} + (0)^{n-1}$$

$$= nx^{n-1}$$

This completes the proof for the derivative of $x^n$ in the case that $n$ is a positive integer.

**E X A M P L E   2**     **Finding the Derivatives of Some Powers**

Find (a) $\dfrac{d}{dx} (x^7)$, (b) $f'(t)$ if $f(t) = t^{-4}$, (c) $\dfrac{dy}{dv}$ if $y = v^{4/3}$.

**Solutions**

(a)   $\dfrac{d}{dx} (x^7) = 7x^{7-1} = 7x^6$,

(b)   $f'(t) = \dfrac{d}{dt} (t^{-4}) = -4t^{-4-1} = -4t^{-5}$,

(c)   $\dfrac{dy}{dv} = \dfrac{d}{dv} (v^{4/3}) = \dfrac{4}{3} v^{(4/3)-1} = \dfrac{4}{3} v^{1/3}$   ∎

Sometimes a function $f(x)$ is actually $x$ to some power, without this being apparent at first.

**E X A M P L E   3**     **Finding the Derivatives of Some Powers**

Find (a) $\dfrac{dy}{dx}$ if $y = \sqrt[3]{x^2}$, (b) $\dfrac{d}{dx} f(x)$ if $f(x) = 1/x^2$.

**Solutions**

(a)   $\dfrac{dy}{dx} = \dfrac{d}{dx} (\sqrt[3]{x^2}) = \dfrac{d}{dx} (x^{2/3}) = \dfrac{2}{3} x^{-1/3}$,

(b)   $\dfrac{d}{dx} f(x) = \dfrac{d}{dx} (1/x^2) = \dfrac{d}{dx} (x^{-2}) = -2x^{-3}$   ∎

## Derivative of a Constant Times a Function

We now establish a rule for finding the derivative of a constant times a function.

---

**Derivative of a Constant Times a Function**

If $f'(x)$ exists, then

$$\frac{d}{dx}(Cf(x)) = C\frac{d}{dx}f(x)$$

---

We use the definition of derivative to give a proof of this rule.

$$[Cf(x)]' = \lim_{h \to 0} \frac{Cf(x+h) - Cf(x)}{h}$$

$$= \lim_{h \to 0} C\frac{f(x+h) - f(x)}{h}$$

$$= C \lim_{h \to 0} \frac{f(x+h) - f(x)}{h}$$

$$= Cf'(x).$$

**E X A M P L E   4**   **Finding the Derivative of a Constant Times a Function**

Find the derivatives of (a) $-5x^3$, (b) $\sqrt{2}x^{5/2}$, (c) $\dfrac{3}{x^8}$.

**Solutions**

(a)  $\dfrac{d}{dx}(-5x^3) = -5\dfrac{d}{dx}(x^3) = -5(3x^{3-1}) = -15x^2,$

(b)  $\dfrac{d}{dx}(\sqrt{2}x^{5/2}) = \sqrt{2}\dfrac{d}{dx}(x^{5/2}) = \sqrt{2}\dfrac{5}{2}x^{3/2},$

(c)  $\dfrac{d}{dx}\left(\dfrac{3}{x^8}\right) = \dfrac{d}{dx}(3x^{-8}) = 3\dfrac{d}{dx}(x^{-8}) = 3(-8x^{-9}) = -24x^{-9}$  ∎

## Derivatives of Sums and Differences

The last rule is for derivatives of sums and differences.

---

**Derivative of Sum or Difference**

If $f'(x)$ and $g'(x)$ exist, then $\dfrac{d}{dx}[f(x) \pm g(x)]$ exists and

$$\frac{d}{dx}[f(x) \pm g(x)] = \frac{d}{dx}f(x) \pm \frac{d}{dx}g(x)$$

---

We give the proof for the derivative of a sum, the proof for the derivative of a difference being similar. If $p(x) = f(x) + g(x)$, then

$$p'(x) = \lim_{h \to 0} \frac{p(x + h) - p(x)}{h}$$

$$= \lim_{h \to 0} \frac{[f(x + h) + g(x + h)] - [f(x) + g(x)]}{h}$$

$$= \lim_{h \to 0} \frac{[f(x + h) - f(x)] + [g(x + h) - g(x)]}{h}$$

$$= \lim_{h \to 0} \frac{f(x + h) - f(x)}{h} + \lim_{h \to 0} \frac{g(x + h) - g(x)}{h}$$

$$= f'(x) + g'(x)$$

**E X A M P L E   5**    **Derivatives of Sums and Differences**

Find the derivatives of (a) $x^4 + x^3$, (b) $\dfrac{2}{x^3} - 3\sqrt{x^3} + \pi$.

**Solutions**

(a) $\dfrac{d}{dx}(x^4 + x^3) = \dfrac{d}{dx}(x^4) + \dfrac{d}{dx}(x^3) = 4x^3 + 3x^2,$

(b) $\dfrac{d}{dx}\left(\dfrac{2}{x^3} - 3\sqrt{x^3} + \pi\right) = 2\dfrac{d}{dx}(x^{-3}) - 3\dfrac{d}{dx}(x^{3/2}) + \dfrac{d}{dx}(\pi) =$

$-6x^{-4} - \dfrac{9}{2}x^{1/2}$ ∎

## Marginal Analysis

In order to make intelligent business and economic decisions, one needs to know the impact of changes in production and sales on costs, revenues and profits. That is, one needs to know the rate of change of cost, revenue, and profits. For example, knowing if the rate of change of profits with respect to an increase in production is positive or negative would be very useful information. In economics, the word *marginal* refers to an instantaneous rate of change, that is, to the derivative. Thus if $C(x)$ is the cost function, we define the marginal cost to be $C'(x)$. If the revenue and profit functions are given by $R(x)$ and $P(x)$, we also define $R'(x)$ and $P'(x)$ to be the marginal revenue and marginal profit, respectively.

It would be convenient if a manager knew what the additional cost would be of producing one additional unit, or what the change in revenue would be on selling one single additional unit. We shall see that under reasonable conditions, the derivative represents an approximation to the actual change in cost or revenue on production or sale of one additional unit.

To be more specific, suppose a manager knows that the cost of producing $x$ units is given by some function $C(x)$. Suppose the plant has produced $x$ units and the manager wishes to know to some reasonable approximation what the cost of the next item will be. Naturally the exact cost is given by

$$C(x + 1) - C(x).$$

We shall now see that under reasonable conditions $C'(x)$ is an approximation to this number.

The actual definition of $C'(x)$ is given by

$$C'(x) = \lim_{h \to 0} \frac{C(x + h) - C(x)}{h}.$$

By definition of the limit, we can say that if $h$ is "small," then

$$C'(x) \approx \frac{C(x + h) - C(x)}{h}.$$

If we assume that $h = 1$ is small, then we have that

$$C'(x) \approx \frac{C(x + h) - C(x)}{h}$$

$$= \frac{C(x + 1) - C(x)}{1}$$

$$= C(x + 1) - C(x).$$

This shows that $C'(x)$ is the **approximate cost of the next unit**.

Figure 12.41 gives a geometric interpretation.

**Figure 12.41**

In the same way one can show for the revenue and profit functions that

$$R(x + 1) - R(x) \approx R'(x),$$

and

$$P(x + 1) - P(x) \approx P'(x).$$

This says that the revenue obtained from the sale of the next item is approximately equal to $R'(x)$, and the profit from the sale of the next item is approximately equal to $P'(x)$.

---

**Economic Interpretation of Derivative**

If $C(x)$, $R(x)$, and $P(x)$ gives the cost, revenue, and profit functions respectively, then under reasonable circumstances $C'(x)$, $R'(x)$, and $P'(x)$ are, respectively, the approximate cost, revenue, and profit, associated with the $(x + 1)$st item. The terms $C'(x)$, $R'(x)$, and $P'(x)$ are referred to, respectively, as the **marginal cost**, **marginal revenue**, and **marginal profit**.

---

**E X A M P L E   6**     **Marginal Cost**

Suppose the total cost in dollars to produce $x$ barrels of a beverage is given by $C(x) = 50 + x^2$. Find the marginal cost and the cost of the next barrel when (a) $x = 100$, (b) $x = 1$.

**Solutions**

(a)   Since $C'(x) = 2x$, $C'(100) = 200$, which is the marginal cost when $x = 100$. The actual cost of the next barrel is given by

$$C(101) - C(100) = 201,$$

which is remarkably close to the marginal cost. We expect this since $x = 100$ is large relative to $h = 1$.

(b)   The marginal cost is $C'(1) = 2$. The actual cost of the next barrel is

$$C(2) - C(1) = 54 - 51 = 3,$$

which is not at all close (on a percentage basis) to the marginal cost. But we do not expect the marginal cost to be close to the actual cost in this case since $h = 1$ is not small relative to $x = 1$.  ∎

---

**E X A M P L E   7**     **Marginal Cost**

Suppose the cost function is given by $C(x) = 0.001x^3 - 0.3x^2 + 50x + 100$. Find the marginal cost.

**Solution**

The marginal cost is given by

$$C'(x) = 0.003x^2 - 0.6x + 50.$$

This is graphed in Figure 12.42.  ∎

Notice in Example 7 that the marginal cost decreases initially and then at some value the marginal cost begins to increase. This is typical of cost functions. Because of economies of scale, unit costs normally will decrease initially. But if

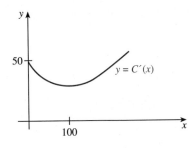

**Figure 12.42**

production is increased too much, unit costs will increase due to the use of less efficient equipment and payment of overtime, and so on.

If a manufacturer sells $x$ units of an item and each item is sold at a price of $\$p$, then the total revenue, $R$, is given by

$$R = (\text{Number sold})(\text{Price of each}) = xp.$$

Sometimes the price $p$ is a function $p = p(x)$ of the number sold $x$. This function is called the **demand equation** and was introduced in Section 1.3.

**E X A M P L E  8**     **Marginal Revenue**

Suppose that a manufacturer knows that the demand equation is given by $p(x) = \frac{9}{2} - \frac{1}{2}x$. Find the marginal revenue.

**Solution**     Since $R = xp$ and $p = \frac{9}{2} - \frac{1}{2}x$,

$$R = x\left(\frac{9}{2} - \frac{1}{2}x\right) = \frac{9}{2}x - \frac{1}{2}x^2,$$

and therefore marginal revenue is given by

$$R'(x) = \frac{9}{2} - x.$$

Notice that $R'(2) = 2.5$, which is positive. Thus revenue is increasing at this point. Notice that $R'(8) = -3.5$ indicating that revenue is decreasing at this point. ■

**E X A M P L E  9**     **Marginal Profit and Break-Even Quantities**

Suppose that a manufacturer knows that the cost in thousands of dollars of producing $x$ items is given by $C(x) = \frac{1}{2}x + 6$ and that the demand equation is known to be $p(x) = \frac{9}{2} - \frac{1}{2}x$. Find the marginal profit and the break-even quantities.

**Solution**     If $P$ is profit, then since $R(x) = xp(x)$

$$P(x) = R(x) - C(x)$$
$$= x\left(\frac{9}{2} - \frac{1}{2}x\right) - \left(\frac{1}{2}x + 6\right)$$
$$= -\frac{1}{2}x^2 + 4x - 6$$

Thus marginal profit is given by

$$P'(x) = -x + 4.$$

Notice that $P'(2) = 2$, implying that profits are increasing at this point. But $P'(6) = -2$, indicating that profits are decreasing at this point.

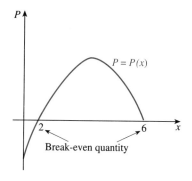

$P = P(x)$

2

6

Break-even quantity

**Figure 12.43**

To find the break-even quantities, we must set profits equal to zero or, equivalently, set revenue equal to cost. Thus setting $P = 0$ yields

$$0 = -\tfrac{1}{2}(x^2 - 8x + 12)$$
$$= -\tfrac{1}{2}(x - 2)(x - 6).$$

Thus the break-even quantities are $x = 2$ and $x = 6$. See Figure 12.43. ∎

## Other Rates of Change

Instantaneous rates of change can be found using the formulas for derivatives developed in this section as in the following example.

**E X A M P L E   1 0**

### Instantaneous Rate of Change of Population

Assume that a population of mice with an initial size of 1000 grows according to the law

$$p(t) = 1000 + 2t^3,$$

where $t$ is measured in months. Find the instantaneous rate of change of the population with respect to time at the end of the fourth month.

**Solution**    Since

$$\frac{d}{dt} p(t) = p'(t) = 6t^2,$$

$p'(4) = 6(4)^2 = 96$. Thus the population is increasing at the rate of 96 per month at the end of the fourth month. ∎

## SELF-HELP EXERCISE SET 12.5

1.  Find the derivative of each of the following.
    a. $x^7 + 4x^{2.5}$,  b. $\sqrt{t} + \dfrac{1}{\sqrt{t}}$

2.  Let $f(x) = 0.125x^2 + 0.50$.
    a.  Find $f'(x)$,
    b.  Find the equation of the line tangent to the curve $y = f(x)$ at the point $x = 2$.

3.  Suppose the cost function is given by $C(x) = 10 + .02x^2$. Find $C'(30)$. Use this to estimate the cost of the 31st item. Compare this estimate with the actual cost of the 31st item.

4. Lightning strikes a tree in a forest and ignites a fire that spreads in a circular manner. What is the instantaneous rate of change of the burnt area with respect to the radius $r$ of the burnt area when $r = 50$ feet?

## EXERCISE SET 12.5

In Exercises 1 through 27 find $\dfrac{d}{dx} f(x)$.

1. $f(x) = 13$

2. $f(x) = \sqrt{2}$

3. $f(x) = 3 + \pi$

4. $f(x) = 4\pi^2$

5. $f(x) = x^{23}$

6. $f(x) = x^7$

7. $f(x) = x^{1.4}$

8. $f(x) = x^{3.1}$

9. $f(x) = x^{0.4}$

10. $f(x) = x^{0.03}$

11. $f(x) = x^{4/3}$

12. $f(x) = x^{2/5}$

13. $f(x) = 3x^{-1.7}$

14. $f(x) = \frac{1}{3}x^{-2.4}$

15. $f(x) = .24\sqrt{x}$

16. $f(x) = .24\sqrt[3]{x}$

17. $f(x) = 2x - 1$

18. $f(x) = 4 - 3x$

19. $f(x) = \pi x + 2$

20. $f(x) = 3 - \pi^2 x$

21. $f(x) = 4 - x - x^2$

22. $f(x) = 3x^2 + x - 1$

23. $f(x) = \frac{1}{\pi}x^2 + \pi x - 1$

24. $f(x) = \pi^2 x^2 + \dfrac{1}{\sqrt{2}} x + \sqrt{3}$

25. $f(x) = 1 - x - x^2 - x^3$

26. $f(x) = 2x^3 - 2x^2 + 3x - 10$

27. $f(x) = x^5 - x + 1$

In Exercises 28 through 39 find $f'(x)$.

28. $f(x) = 1 - x^4 - x^8$

29. $f(x) = .35x^2 - .3x + .45$

30. $f(x) = 3 - .001x^2$

31. $f(x) = 4 - .01x^2 - .03x^3$

32. $f(x) = -.003x^4 + .01x^3$

33. $f(x) = \dfrac{x^2 - 2x + 4}{6}$

34. $f(x) = \dfrac{x^2 - 3x - 6}{3x}$

35. $f(x) = 3x^3 + \dfrac{3}{x^3}$

36. $f(x) = \dfrac{1}{x} + x$

37. $f(x) = x^2 - \dfrac{1}{x}$

38. $f(x) = x^{-2} + x^{-3} + 3$

39. $f(x) = 4 - 3x^{-3} + 4x^2$

In Exercises 40 through 45 find $\dfrac{d}{dt} f(t)$.

40. $f(t) = (2t)^5$

41. $f(t) = (-3t)^3$

**42.** $f(t) = \dfrac{1}{t^2} + \dfrac{2}{t^3}$

**43.** $f(t) = \dfrac{1}{t} + \dfrac{1}{t^2} + \dfrac{1}{t^3}$

**44.** $f(t) = \sqrt{t} - \dfrac{1}{\sqrt{t}}$

**45.** $f(t) = \sqrt[3]{t} - \dfrac{1}{\sqrt[3]{t}}$

In Exercises 46 through 51 find $\dfrac{d}{du} f(u)$.

**46.** $f(u) = 2u^{3/2} + 4u^{3/4}$

**47.** $f(u) = 3u^{5/3} - 3u^{2/3}$

**48.** $f(u) = (u - 1)^3$

**49.** $f(u) = (u + 1)(2u + 1)$

**50.** $f(u) = \dfrac{(u + 1)^2}{u}$

**51.** $f(u) = \dfrac{u + 3/u}{\sqrt{u}}$

In Exercises 52 through 64 find $\dfrac{dy}{dx}$.

**52.** $y = \dfrac{x^2 + x - 2}{x}$

**53.** $y = \sqrt{4x} - \dfrac{1}{\sqrt{9x}}$

**54.** $y = x^{1.5} - x^{-1.5}$

**55.** $y = \pi x^{\pi}$

**56.** $y = \dfrac{1}{\pi} x^{1/\pi}$

**57.** $y = x^{\sqrt{2}} - \dfrac{1}{x^{\sqrt{2}}}$

**58.** $y = 3x^{-3/5} - 6x^{-5/2}$

**59.** $y = x^{-3/2} + x^{-4/5}$

**60.** $y = x^{-2} + 3x^{-4}$

**61.** $y = \sqrt[3]{x^2}$

**62.** $y = \sqrt{x^3}$

**63.** $y = \dfrac{1}{\sqrt[5]{x^2}}$

**64.** $y = \dfrac{1}{\sqrt[3]{x^4}}$

In Exercises 65 through 70 find the equation of the tangent line of the given function at the indicated point.

**65.** $y = f(x) = x^2 + x + 1, x_0 = 2$

**66.** $y = f(x) = x^4 + 3x^2 - 1, x_0 = -1$

**67.** $y = f(x) = x^{-1} + x^{-2}, x_0 = 1$     **68.** $y = f(x) = \sqrt{x} + x, x_0 = 4$

**69.** $y = f(x) = \dfrac{1}{x} + x^2, x_0 = -1$     **70.** $y = f(x) = x^{-3/2}, x_0 = 4$

In Exercises 71 through 76 find the points where the tangent to the graph of $y = f(x)$ is horizontal.

**71.** $y = f(x) = x^2 - 1$

**72.** $y = f(x) = x^3 + 3$

**73.** $y = f(x) = x^3 - 9x + 3$

**74.** $y = f(x) = 7 + 12x - x^3$

**75.** $y = f(x) = 2x^3 + 3x^2 - 12x + 4$

**76.** $y = f(x) = x^3 - x^2 - x - 1$

In Exercises 77 through 88 estimate the cost of the next item or the revenue or profits from the sale of the next item for the indicated functions and the indicated $x$.

**77.** $C(x) = 20 + .01x^2, x = 30$

**78.** $C(x) = 5 + 2x + .03x^2, x = 30$

**79.** $C(x) = \sqrt{x}(x + 12) + 50, x = 36$

**80.** $C(x) = \sqrt{x^3} + 100, x = 100$

**81.** $R(x) = x^2 - 10x, x = 20$          **82.** $R(x) = 0.1x^3 - 8x, x = 20$

**83.** $R(x) = \sqrt{x}(x - 24), x = 16$          **84.** $R(x) = \sqrt{x + 10}, x = 90$

**85.** $P(x) = x^2 - 17x - 60, x = 50$          **86.** $P(x) = x^3 - 13x - 30, x = 10$

**87.** $P(x) = \sqrt[3]{x}(x - 54) - 50, x = 27$

**88.** $P(x) = \sqrt{2x - 1}, x = 41$

## Applications

**89. Revenue.** If the demand equation for a product is given by $p = -2x + 24$, find the marginal revenue.

**90. Costs.** If the cost equation for a product is given by $C = 2x^2 + 32$, find the marginal cost.

**91. Profits.** If the demand equation and cost equation for a product are as given in the previous two problems, find the marginal profit and the break-even quantities.

**92. Demand.** If the demand equation for a product is given by

$$p = \frac{1000}{\sqrt{x}},$$

find the marginal revenue.

**93. Velocity.** A particle moves according to the law $s(t) = 2t^3 - 4t + 10$. Find $v(t)$.

**94. Velocity.** A particle moves according to the law

$$s(t) = -t^3 + 10 + \frac{2}{t}.$$

Find $v(t)$.

**95. Biology.** A population of bacteria is growing according to the law $N(t) = 0.03t^4 + 10t + 1000$ where $t$ is measured in hours. What is the rate of change of the population with respect to time when $t = 10$?

**96. Medicine.** A spherical tumor of radius $r$ cm is growing. What is the rate of change of the volume of the tumor with respect to the radius when $r = 4$?

*(Gena Husman/FPG International Corporation)*

**97.** **Medicine.**   Poiseuille's law states that the total resistance $R$ to blood flow in a blood vessel of constant length $l$ and radius $r$ is given by

$$R = \frac{al}{r^4},$$

where $a$ is a positive constant. Find the rate of change of resistance with respect to the radius when $r = 3$.

**98.** **Biology.**   The free water vapor diffusion coefficient $D$ in soft woods is given by

$$D(T) = \frac{167.2}{P}\left(\frac{T}{273}\right)^{1.75},$$

where $T$ is the temperature and $P$ is the constant atmospheric pressure measured in appropriate units. Find the rate of change of $D$ with respect to $T$.

## SOLUTIONS TO SELF-HELP EXERCISE SET 12.5

**1. a.**   $\dfrac{d}{dx}(x^7 + 4x^{2.5}) = \dfrac{d}{dx}(x^7) + 4\dfrac{d}{dx}(x^{2.5}) = 7x^6 + 10x^{1.5}$

   **b.**   $\dfrac{d}{dx}\left(\sqrt{t} + \dfrac{1}{\sqrt{t}}\right) = \dfrac{d}{dt}(t^{1/2} + t^{-1/2}) = \dfrac{1}{2}t^{-1/2} - \dfrac{1}{2}t^{-3/2}.$

**2. a.**   $f'(x) = \dfrac{d}{dx}(0.125x^2 + 0.50) = 0.125\dfrac{d}{dx}(x^2) + \dfrac{d}{dx}(0.50) = 0.25x$

   **b.**   Since $x_0 = 2$, $y_0 = f(x_0) = f(2) = 1$. The slope of the tangent line at $x = 2$ is $f'(2) = 0.25(2) = 0.50$. Thus the equation of the line tangent to the curve at $x = 2$ is

$$y - 1 = 0.50(x - 2)$$

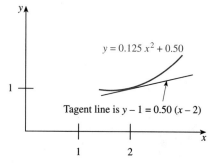

**3.**   $C'(x) = \dfrac{d}{dx}(10 + .02x^2) = \dfrac{d}{dx}(10) + .02\dfrac{d}{dx}(x^2) = .04x.$ Then $C'(30) = .04(30) = 1.20.$ Thus the approximate cost of the 31st item is \$1.20. The actual cost of the 31st item is $C(31) - C(30) = \$1.22.$

4.  The area of a circle of radius $r$ is given by $A = A(r) = \pi r^2$. Then

$$\frac{dA}{dr} = A'(r) = 2\pi r,$$

and $A'(50) = 2\pi(50) \approx 314$. Thus the burnt area is increasing at the rate of approximately 314 square feet per one foot increase in $r$ when $r = 50$.

## 12.6    DERIVATIVES OF PRODUCTS AND QUOTIENTS

► *Product Rule*
► *Quotient Rule*
► *Proofs*

APPLICATION
**Derivative of a Product**

What is $h'(x)$ if $h(x) = (\sqrt{x} + 1)(x^6 + 3x^3 - 5)$? See Example 2 for the answers.

APPLICATION
**Marginal Revenue**

If the demand equation for a certain product is given by

$$p = \frac{1}{x^4 + x^2 + 4},$$

what is the marginal revenue? See Example 5 for the answer.

### Product Rule

We continue to develop formulas that permit us to take derivatives without having to resort to using the definition of the derivative. In this section we will develop the rules for differentiating the product of two functions and the quotient of two functions.

Since the derivative of the sum is the sum of the derivatives, one might suspect that the derivative of the product is the product of the derivatives. This is not true in general. The correct rule for taking the derivative of the product of two functions follows.

---

**Product Rule**

If both $f'(x)$ and $g'(x)$ exist, then $\dfrac{d}{dx}(f(x) \cdot g(x))$ exists and

$$\frac{d}{dx}(f(x) \cdot g(x)) = f(x) \cdot \frac{d}{dx}g(x) + g(x) \cdot \frac{d}{dx}f(x)$$

---

**E X A M P L E  1**

### Using the Product Rule

Find $h'(x)$ if $h(x) = (x^2 + 3)(2x^3 - 4)$.

**Solution**    Think of $h(x)$ as a product of the two boxed functions:

$$h(x) = \boxed{(x^2 + 3)} \cdot \boxed{(2x^3 - 4)}.$$

Let $f(x)$ equal the first boxed function and $g(x)$ equal the second boxed function. That is,

$$f(x) = (x^2 + 3), \qquad g(x) = (2x^3 - 4).$$

Then

$$h'(x) = f(x) \cdot \frac{d}{dx} g(x) + g(x) \cdot \frac{d}{dx} f(x)$$

$$= (x^2 + 3)\frac{d}{dx}(2x^3 - 4) + (2x^3 - 4)\frac{d}{dx}(x^2 + 3)$$

$$= (x^2 + 3)(6x^2) + (2x^3 - 4)(2x)$$

$$= 10x^4 + 18x^2 - 8x \quad \blacksquare$$

**E X A M P L E  2**

### Using the Product Rule

Find $h'(x)$ if $h(x) = (\sqrt{x} + 1)(x^6 + 3x^3 - 5)$.

**Solution**    We think of $h(x)$ as $f(x) \cdot g(x)$ where

$$f(x) = (\sqrt{x} + 1), \qquad g(x) = (x^6 + 3x^3 - 5).$$

Then

$$h'(x) = f(x) \cdot \frac{d}{dx} g(x) + g(x) \cdot \frac{d}{dx} f(x)$$

$$= (x^{1/2} + 1)\frac{d}{dx}(x^6 + 3x^3 - 5) + (x^6 + 3x^3 - 5)\frac{d}{dx}(x^{1/2} + 1)$$

$$= (x^{1/2} + 1)(6x^5 + 9x^2) + (x^6 + 3x^3 - 5)\left(\frac{1}{2}x^{-1/2}\right) \quad \blacksquare$$

**Revenue and Demand Functions.**   In the last section we discussed the formula

$$R(x) = xp(x),$$

where $x$ is the number of units sold, $p$ is the price of each unit, and $R$ is the revenue from selling $x$ units at a price of $p$ dollars each. We can use the product rule of differentiation to determine an alternative way of finding the marginal revenue. To use the product rule to find the marginal revenue, we first need to think of $R$ as the product of two functions:

$$R(x) = \boxed{x} \cdot \boxed{p(x)}$$

$$= \boxed{f(x)} \cdot \boxed{g(x)}$$

Set

$$f(x) = x, \qquad g(x) = p(x)$$

in the formula for the product rule. Since $f'(x) = 1$, the product rule yields

$$R'(x) = f(x) \cdot \frac{d}{dx} g(x) + g(x) \cdot \frac{d}{dx} f(x)$$

$$= x \frac{d}{dx} p(x) + p(x)(1)$$

$$= xp'(x) + p(x)$$

Thus we have that

$$R'(x) = xp'(x) + p(x). \tag{1}$$

**E X A M P L E  3**  **Finding Marginal Revenue Using the Alternate Formula (1)**

Use the above formula (1) to obtain the marginal revenue for the demand equation $p(x) = \frac{9}{2} - \frac{1}{2}x$. Compare with solution to Example 8 in Section 12.5.

**Solution**  Recall that $p(x) = \frac{9}{2} - \frac{1}{2}x$ and thus $p'(x) = -\frac{1}{2}$. Then

$$R'(x) = xp'(x) + p(x)$$

$$= x\left(-\frac{1}{2}\right) + \left(\frac{9}{2} - \frac{1}{2}x\right)$$

$$= \frac{9}{2} - x$$

which, of course, agrees with the answer we obtained in Example 8 of the previous section. ■

## Quotient Rule

Now that we know that the derivative of the product is not the product of the derivatives, we suspect that the derivative of the quotient is not the quotient of the derivatives. This is indeed the case. The correct formula for the derivative of the quotient is now given.

> **Quotient Rule**
>
> If $f'(x)$ and $g'(x)$ exist, then $\dfrac{d}{dx}\left(\dfrac{f(x)}{g(x)}\right)$ exists and
>
> $$\frac{d}{dx}\left(\frac{f(x)}{g(x)}\right) = \frac{g(x) \cdot \dfrac{d}{dx} f(x) - f(x) \cdot \dfrac{d}{dx} g(x)}{g^2(x)}$$
>
> at points where $g(x) \neq 0$.

**E X A M P L E   4**     **Using the Quotient Rule**

Find $h'(x)$ if $h(x) = \dfrac{x^3 - 1}{x^2 + 1}$.

**Solution**     Here we think of $h(x)$ as the quotient of two functions written as

$$h(x) = \frac{\boxed{x^3 - 1}}{\boxed{x^2 + 1}}$$

Thus in the quotient rule set

$$f(x) = x^3 - 1, \qquad g(x) = x^2 + 1,$$

and then

$$h'(x) = \frac{g(x) \cdot \dfrac{d}{dx} f(x) - f(x) \cdot \dfrac{d}{dx} g(x)}{[g(x)]^2}$$

$$= \frac{(x^2 + 1) \dfrac{d}{dx} (x^3 - 1) - (x^3 - 1) \dfrac{d}{dx} (x^2 + 1)}{(x^2 + 1)^2}$$

$$= \frac{(x^2 + 1)(3x^2) - (x^3 - 1)(2x)}{(x^2 + 1)^2}$$

$$= \frac{x^4 + 3x^2 + 2x}{(x^2 + 1)^2} \quad \blacksquare$$

**E X A M P L E   5**     **Marginal Revenue**

If the demand equation for a certain product is given by $p = \dfrac{1}{x^4 + x^2 + 4}$, what is the marginal revenue?

**Solution**     Since $R(x) = xp(x) = \dfrac{x}{x^4 + x^2 + 4}$, we think of $R(x)$ as the quotient of two functions, $\dfrac{f(x)}{g(x)}$, where $f(x) = x$ and $g(x) = x^4 + x^2 + 4$. Then

$$R'(x) = \frac{g(x) \cdot \dfrac{d}{dx} f(x) - f(x) \cdot \dfrac{d}{dx} g(x)}{[g(x)]^2}$$

$$= \frac{(x^4 + x^2 + 4) \dfrac{d}{dx}(x) - (x) \dfrac{d}{dx}(x^4 + x^2 + 4)}{(x^4 + x^2 + 4)^2}$$

$$= \frac{(x^4 + x^2 + 4)(1) - (x)(4x^3 + 2x)}{(x^4 + x^2 + 4)^2}$$

$$= \frac{-3x^4 - x^2 + 4}{(x^4 + x^2 + 4)^2} \quad \blacksquare$$

---

**Average Cost**

If $C(x)$ is the cost of producing $x$ items, then the **average cost**, denoted by $\overline{C}(x)$, is the cost of all the items divided by the number of items, that is,

$$\overline{C}(x) = \frac{C(x)}{x}.$$

The marginal average cost is given by $\overline{C}'(x)$.

---

**E X A M P L E  6**  **Application of the Quotient Rule**

Use the quotient rule to find a formula for the marginal average cost in terms of the cost and the marginal cost.

**Solution**  Using the quotient rule we have

$$\overline{C}'(x) = \frac{d}{dx}\left(\frac{C(x)}{x}\right)$$

$$= \frac{x \cdot \dfrac{d}{dx} C(x) - C(x) \cdot \dfrac{d}{dx}(x)}{x^2}$$

$$= \frac{xC'(x) - C(x)}{x^2}$$

$$= \frac{1}{x}\left(C'(x) - \frac{C(x)}{x}\right)$$

$$= \frac{1}{x}(C'(x) - \overline{C}(x)) \quad \blacksquare$$

In words, this formula says that to find marginal average cost take marginal cost less average cost and divide the result by the number of items sold.

**E X A M P L E   7**    **Application of the Quotient Rule**

If the cost function is given by $C(x) = 2x^3 - x^2 + 10$, find (a) marginal cost, (b) average cost, (c) marginal average cost.

**Solution**    Marginal cost is given by

$$C'(x) = 6x^2 - 2x.$$

Average cost is given by

$$\overline{C}(x) = 2x^2 - x + \frac{10}{x}.$$

Marginal average cost is given by

$$\overline{C}'(x) = \frac{d}{dx}\left(2x^2 - x + \frac{10}{x}\right) = 4x - 1 - \frac{10}{x^2}.$$

To obtain marginal average cost we could also use the formula given in Example 6. This gives

$$\overline{C}'(x) = \frac{1}{x}\left((6x^2 - 2x) - \left(2x^2 - x + \frac{10}{x}\right)\right)$$

$$= \frac{1}{x}\left(4x^2 - x - \frac{10}{x}\right)$$

$$= 4x - 1 - \frac{10}{x^2} \quad \blacksquare$$

## Proofs

**Proof of Product Rule.**   Let $y = p(x) = f(x)g(x)$, then

$$\frac{p(x + h) - p(x)}{h} = \frac{f(x + h)g(x + h) - f(x)g(x)}{h}$$

$$= \frac{f(x + h)g(x + h) - f(x)g(x + h) + f(x)g(x + h) - f(x)g(x)}{h}$$

$$= \frac{f(x + h)g(x + h) - f(x)g(x + h)}{h} + \frac{f(x)g(x + h) - f(x)g(x)}{h}$$

$$= \frac{f(x + h) - f(x)}{h}g(x + h) + f(x)\frac{g(x + h) - g(x)}{h}$$

Now taking the limit of this expression and using the rules of limits, we obtain

$$p'(x) = \lim_{h \to 0} \frac{p(x+h) - p(x)}{h}$$

$$= \lim_{h \to 0} \frac{f(x+h) - f(x)}{h} \lim_{h \to 0} g(x+h) + f(x) \lim_{h \to 0} \frac{g(x+h) - g(x)}{h}$$

$$= f'(x)g(x) + f(x)g'(x)$$

where $\lim_{h \to 0} g(x+h) = g(x)$ since $g(x)$ is continuous (since $g'(x)$ exists).

**Proof of Quotient Rule.** Let

$$y = q(x) = \frac{f(x)}{g(x)},$$

Then

$$\frac{q(x+h) - q(x)}{h} = \frac{1}{h} \left[ \frac{f(x+h)}{g(x+h)} - \frac{f(x)}{g(x)} \right]$$

$$= \frac{1}{h} \left[ \frac{f(x+h)g(x) - f(x)g(x+h)}{g(x+h)g(x)} \right]$$

$$= \frac{1}{g(x+h)g(x)} \left[ \frac{f(x+h)g(x) - f(x)g(x+h)}{h} \right]$$

$$= \frac{1}{g(x+h)g(x)} \left[ \frac{f(x+h)g(x) - g(x)f(x) - f(x)g(x+h) + g(x)f(x)}{h} \right]$$

$$= \frac{1}{g(x+h)g(x)} \left[ g(x) \frac{f(x+h) - f(x)}{h} - f(x) \frac{g(x+h) - g(x)}{h} \right]$$

We now let $h \to 0$ and obtain

$$q'(x) = \lim_{h \to 0} \frac{q(x+h) - q(x)}{h}$$

$$= \left[ \lim_{h \to 0} \frac{1}{g(x+h)g(x)} \right] \left[ g(x) \lim_{h \to 0} \frac{f(x+h) - f(x)}{h} - f(x) \lim_{h \to 0} \frac{g(x+h) - g(x)}{h} \right]$$

$$= \frac{1}{g(x)g(x)} [g(x)f'(x) - f(x)g'(x)]$$

$$= \frac{g(x)f'(x) - f(x)g'(x)}{[g(x)]^2}$$

where $\lim_{h \to 0} g(x+h) = g(x)$

since $g(x)$ is continuous (since $g'(x)$ exists).

## SELF-HELP EXERCISE SET 12.6

Find the derivative of each of the following. Do not simplify.

**1.** $(x^2 - 1)(x^3 + x + 1)$

**2.** $\dfrac{x^{3/2} + 1}{x^4 + 1}$

## EXERCISE SET 12.6

In Exercises 1 through 9 use the product rule to find the derivative.

**1.** $(2x + 1)(3x - 2)$     **2.** $(5x - 3)(2x + 5)$     **3.** $(4 - x)(3 + x)$

**4.** $(3 - 2x)(2 - 5x)$     **5.** $(x + 1)(x^2 - 3)$     **6.** $(1 - x)(2x^3 + 3)$

**7.** $(4 - x^5)(2 - 3x)$     **8.** $(x^3 - 3)(2x + 3)$     **9.** $(x^7 - 3)(x^5 + 5)$

In Exercises 10 through 16 use the product rule to find $f'(x)$.

**10.** $f(x) = (2x^8 + 1)(3x^2 - 1)$     **11.** $f(x) = (x^2 + x + 1)(x^2 - x - 3)$

**12.** $f(x) = (x^2 + 1)(x^3 - 1)$     **13.** $f(x) = (x + 1)(\sqrt{x} + 1)$

**14.** $f(x) = (x^2 - 3)(\sqrt[3]{x} - 1)$     **15.** $f(x) = (2x - 3)\left(x + \dfrac{1}{x}\right)$

**16.** $f(x) = \left(x + \dfrac{1}{x}\right)\left(1 + \dfrac{1}{x^2}\right)$

In Exercises 17 through 28 use the quotient rule to find the derivative.

**17.** $\dfrac{1}{x + 2}$     **18.** $\dfrac{1}{x^2 + 3}$     **19.** $\dfrac{1}{x^3 + 1}$

**20.** $\dfrac{x}{x + 3}$     **21.** $\dfrac{x}{x + 2}$     **22.** $\dfrac{x + 1}{x - 2}$

**23.** $\dfrac{x - 3}{x + 5}$     **24.** $\dfrac{3}{x + 3}$     **25.** $\dfrac{-4}{x - 2}$

**26.** $\dfrac{2x - 3}{4x - 1}$     **27.** $\dfrac{2 - 3x}{2x - 1}$     **28.** $\dfrac{3 - 2x}{1 - 2x}$

In Exercises 29 through 36 use the quotient rule to find $\dfrac{dy}{du}$.

**29.** $y = \dfrac{3}{u^2 + 1}$     **30.** $y = \dfrac{-4}{u^4 + 3}$     **31.** $y = \dfrac{u + 1}{u^2 + 2}$

**32.** $y = \dfrac{2u - 1}{u^3 - 1}$     **33.** $y = \dfrac{u^3 - 1}{u^2 + 3}$     **34.** $y = \dfrac{\sqrt{u}}{u^2 + u + 1}$

**35.** $y = \dfrac{\sqrt[3]{u}}{u^3 - u - 1}$     **36.** $y = \dfrac{\sqrt[4]{u}}{u^2 + 1}$

## Applications

**37. Revenue.** If the demand equation for a certain product is given by

$$p = \frac{1}{1 + x^2},$$

find the marginal revenue.

**38. Cost.** It is estimated that the cost, $C(x)$, in millions of dollars, of maintaining the toxic emisions of a certain chemical plant $x$ percent free of the toxins is given by

$$C(x) = \frac{4}{100 - x}.$$

Find the rate of change of $C$ with respect to $x$.

**39. Cost.** A large corporation already has cornered 20 percent of the disposable diaper market and now wishes to undertake a very extensive advertising campaign to increase its share of this market. It is estimated that the advertising cost, $C(x)$, in billions of dollars, of attaining an $x$ percent share of this market is given by

$$C(x) = \frac{1}{10 - 0.1x}.$$

What is the rate of change of $C$ with respect to $x$?

**40. Productivity.** The number of items, $n(t)$, produced per employee per day over the last ten years in a certain plant is given by

$$n(t) = \frac{30t}{t + 1},$$

where $t$ is the number of years from 10 years ago. What is the rate of change of $n$ with respect to $t$?

**41. Propensity to Save.** The amount of money a family saves is a function $S(I)$ of its income $I$. The *marginal propensity to save* is $\frac{dS}{dI}$. If

$$S(I) = \frac{2I^2}{5(I + 50,000)},$$

find the marginal propensity to save.

**42. Spread of Technology.** If $t$ is the number of years after the introduction of a certain new technology, the percentage of the firms in this industry using the new technology is given by

$$p(t) = \frac{100t^2}{t^2 + 1}.$$

What is the rate of change of $p$ with respect to $t$?

**43.   Medicine.**   Suppose that the concentration $c$ of a drug in the blood $t$ hours after it is taken orally is given by

$$c = \frac{3t}{2t^2 + 5}.$$

What is the rate of change of $c$ with respect to $t$?

**44.   Chemistry.**   Salt water with a concentration 0.1 pounds of salt per gallon flows into a large tank that initially holds 100 gallons of pure water. If 5 gallons of salt water per minute flows into the tank, show that the concentration of salt in the tank is given by

$$c(t) = \frac{t}{200 + 5t},$$

where $t$ is measured in minutes. What is the rate of change of $c$ with respect to $t$?

**45.   Biology.**   A population $N$ of deer in a state forest follows the law

$$N(t) = \frac{1000t^4}{1 + t^4}$$

where $t$ is in years. Find the rate of change of $N$ with respect to $t$.

### Solutions to Self-Help Exercise Set 12.6

**1.**

$$\frac{d}{dx}[(x^2 - 1)(x^3 + x + 1)] = (x^2 - 1)\frac{d}{dx}(x^3 + x + 1) + (x^3 + x + 1)\frac{d}{dx}(x^2 - 1)$$

$$= (x^2 - 1)(3x^2 + 1) + (x^3 + x + 1)(2x)$$

**2.**  $\dfrac{d}{dx}\left(\dfrac{x^{3/2} + 1}{x^4 + 1}\right) = \dfrac{(x^4 + 1)\dfrac{d}{dx}(x^{3/2} + 1) - (x^{3/2} + 1)\dfrac{d}{dx}(x^4 + 1)}{(x^4 + 1)^2}$

$$= \frac{(x^4 + 1)(\tfrac{3}{2}x^{1/2}) - (x^{3/2} + 1)(4x^3)}{(x^4 + 1)^2}$$

## 12.7    THE CHAIN RULE

▶ *The Chain Rule*

▶ *The General Power Rule*

▶ *Derivatives of Complex Expressions*

▶ *Justification of the Chain Rule*

**APPLICATION**

**Finding the Derivative of Complex Expressions**

What is the derivative of $\dfrac{1}{(x^3 + x - 5)^4}$? See Example 3 for the answer.

**APPLICATION**

**Finding Marginal Cost**

What is the marginal cost if $C(x) = \dfrac{x^3}{\sqrt{x^2 + 1}}$? See Example 5 for the answer.

## The Chain Rule

Consider the situation in Figure 12.44 where $y = (f \circ g)(x) = f[g(x)]$ is the composition of the two functions $f(x)$ and $g(x)$. We wish to find a formula for the derivative of $(f \circ g)(x)$ in terms of the derivatives of $f(x)$ and $g(x)$. The formula is perhaps the most important rule of differentiation.

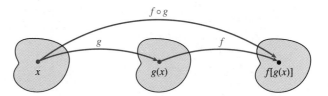

**Figure 12.44**

To gain some insight into how this formula can be obtained, let us set $u = g(x)$. Then we have $y = f(u)$, where $u = g(x)$. Since we are looking for the derivative of $y$ with respect to $x$, we need to determine the relationship between the change in $y$ and the change in $x$. We can write

$$\frac{\Delta y}{\Delta x} = \frac{\text{change in } y}{\text{change in } x}$$

$$= \frac{\text{change in } y}{\text{change in } u} \cdot \frac{\text{change in } u}{\text{change in } x}$$

$$= \frac{\Delta y}{\Delta u} \cdot \frac{\Delta u}{\Delta x}$$

This formula suggests (correctly) that

$$\frac{dy}{dx} = \frac{dy}{du} \cdot \frac{du}{dx}.$$

We now state two forms of the chain rule.

---

**Chain Rule**

If $y = f(u)$ and $u = g(x)$ and both of these functions are differentiable, then the composite function $y = f[g(x)]$ is differentiable and

$$\frac{dy}{dx} = \frac{dy}{du} \cdot \frac{du}{dx}$$

or

$$\frac{d}{dx} f[g(x)] = f'[g(x)] \cdot g'(x).$$

---

**E X A M P L E   1**    **Verifying the Chain Rule**

The function $y = 6x - 21 = 3(2x - 7)$ can be written as $y = f[g(x)]$ where $y = f(u) = 3u$ and $u = g(x) = 2x - 7$. Verify the chain rule in this case.

**Solution**    We have

$$\frac{dy}{dx} = 6, \qquad \frac{dy}{du} = 3, \qquad \frac{du}{dx} = 2.$$

Since $6 = 3 \cdot 2$,

$$\frac{dy}{dx} = \frac{dy}{du} \cdot \frac{du}{dx}. \quad \blacksquare$$

**E X A M P L E   2**    **Using the Chain Rule**

Find the derivative of $y = (x^2 + 3)^{78}$.

**Solution**    We could expand this function using the binomial theorem and obtain a polynomial of degree 156 and then differentiate, a very time-consuming process. An alternate way of finding the derivative is to use the chain rule. First recognize that this function is *something* raised to the 78th power, where the something is $(x^2 + 3)$. This suggests that $y$ can be viewed as the composition of two functions, the first function takes $x$ into $x^2 + 3$ and then the second function raises this quantity to the 78th power. In terms of the chain rule, set $y = f(u) = u^{78}$ and $u = g(x) = x^2 + 3$. Then

$$y = f(u) = f[g(x)] = [g(x)]^{78} = (x^2 + 3)^{78}.$$

Since $\dfrac{dy}{du} = \dfrac{d}{du}(u^{78}) = 78u^{77}$ and $\dfrac{du}{dx} = \dfrac{d}{dx}(x^2 + 3) = 2x$, the chain rule yields

$$\frac{dy}{dx} = \frac{dy}{du} \cdot \frac{du}{dx}$$

$$= 78u^{77} \cdot (2x)$$

$$= 78(x^2 + 3)^{77}2x$$

$$= 156x(x^2 + 3)^{77} \quad \blacksquare$$

## The General Power Rule

Example 2 suggests how to take the derivative of any *function* raised to any power. We will now see how to find the derivative of

$$[g(x)]^n,$$

where $n$ is any real number. Here we recognize $y = [g(x)]^n$ as the composition of two functions, the first function takes $x$ to $g(x)$ and the second function raises $g(x)$ to the $n$th power. If we set $y = f(u) = u^n$ and $u = g(x)$, then

$$y = f[g(x)] = [g(x)]^n.$$

Then the chain rule yields

$$\frac{dy}{dx} = \frac{dy}{du} \cdot \frac{du}{dx}$$

$$= nu^{n-1} \cdot \frac{d}{dx}g(x)$$

$$= n[g(x)]^{n-1} \cdot \frac{d}{dx}g(x)$$

We state this result as the general power rule.

---

**General Power Rule**

If $n$ is any real number and $g'(x)$ exists, then

$$\frac{d}{dx}[g(x)]^n = n[g(x)]^{n-1} \cdot \frac{d}{dx}g(x).$$

---

W A R N I N G .    A common error is to neglect the $\dfrac{d}{dx}g(x)$ term in the general power formula.

E X A M P L E   3    **Using the General Power Rule**

Find the derivatives of (a) $(x^2 + 3)^{78}$, (b) $\dfrac{1}{(x^3 + x - 5)^4}$ using the general power rule.

**Solutions**

(a)  Using the general power rule when $g(x) = x^2 + 3$ and $n = 78$ yields

$$\frac{d}{dx}[(x^2 + 3)^{78}] = 78(x^2 + 3)^{77}\frac{d}{dx}(x^2 + 3)$$

$$= 78(x^2 + 3)^{77}(2x)$$

$$= 156x\,(x^2 + 3)^{77}$$

(b)  We must recognize this as some function raised to a power. The power is *minus* four. Then we can use the general power formula to obtain

$$\frac{d}{dx}\left[\frac{1}{(x^3 + x - 5)^4}\right] = \frac{d}{dx}[(x^3 + x - 5)^{-4}]$$

$$= -4(x^3 + x - 5)^{-5}\frac{d}{dx}(x^3 + x - 5)$$

$$= -4(x^3 + x - 5)^{-5}(3x^2 + 1)$$

$$= -4(3x^2 + 1)(x^3 + x - 5)^{-5} \quad \blacksquare$$

Notice that the general power rule reduces to the ordinary power rule discussed in a previous section. Thus using the general power rule yields

$$\frac{d}{dx}(x^n) = nx^{n-1}\frac{d}{dx}(x)$$

$$= nx^{n-1}(1)$$

$$= nx^{n-1}$$

## Derivatives of Complex Expressions

We now consider some examples that use the general power rule together with other differentiation rules learned earlier.

E X A M P L E   4    **Finding the Derivatives of Complex Expressions**

Find the derivative of $x^2(x^3 + 5)^{10}$.

**Solution** We first think of this expression as the product of two functions $f(x)g(x)$, where $f(x) = x^2$ and $g(x) = (x^3 + 5)^{10}$, and then use the product rule of differentiating. We then use the general power rule where necessary. Thus we obtain

$$\frac{d}{dx}[x^2(x^3 + 5)^{10}] = x^2 \frac{d}{dx}[(x^3 + 5)^{10}] + (x^3 + 5)^{10} \frac{d}{dx}(x^2) \qquad \text{product rule}$$

$$= x^2 10(x^3 + 5)^9 \frac{d}{dx}(x^3 + 5) + (x^3 + 5)^{10}(2x) \qquad \text{general power rule}$$

$$= x^2 10(x^3 + 5)^9(3x^2) + 2x(x^3 + 5)^{10}$$

$$= 2x(x^3 + 5)^9[15x^3 + (x^3 + 5)]$$

$$= 2x(x^3 + 5)^9(16x^3 + 5) \quad \blacksquare$$

**E X A M P L E  5** **Finding the Derivative of Complex Expressions**

Find the marginal cost if $C(x) = \dfrac{x^3}{\sqrt{x^2 + 1}}$.

**Solution**
$$C'(x) = \frac{d}{dx}\left(\frac{x^3}{(x^2 + 1)^{1/2}}\right)$$

$$= \frac{(x^2 + 1)^{1/2} \dfrac{d}{dx}(x^3) - x^3 \dfrac{d}{dx}[(x^2 + 1)^{1/2}]}{[(x^2 + 1)^{1/2}]^2} \qquad \text{quotient rule}$$

$$= \frac{(x^2 + 1)^{1/2} 3x^2 - x^3 \dfrac{1}{2}(x^2 + 1)^{-1/2} \dfrac{d}{dx}(x^2 + 1)}{(x^2 + 1)} \qquad \text{general power rule}$$

$$= \frac{(x^2 + 1)^{1/2} 3x^2 - x^3 \dfrac{1}{2}(x^2 + 1)^{-1/2}(2x)}{(x^2 + 1)}$$

$$= \frac{3x^2(x^2 + 1)^{1/2} - x^4(x^2 + 1)^{-1/2}}{x^2 + 1}$$

$$= x^2(x^2 + 1)^{-1/2} \frac{3(x^2 + 1) - x^2}{x^2 + 1}$$

$$= x^2 \frac{2x^2 + 3}{(x^2 + 1)^{3/2}} \quad \blacksquare$$

## Justification of the Chain Rule*

We present here a more detailed explanation of why the chain rule works. Assume that $f(x)$ and $g(x)$ are differentiable. Since we are interested in $\dfrac{d}{dx} f[g(x)]$, naturally we wish to consider

$$\frac{f[g(x + h)] - f[g(x)]}{h}.$$

So long as

$$k = g(x + h) - g(x) \neq 0,$$

we have

$$\frac{f[g(x + h)] - f[g(x)]}{k} \frac{k}{h} = \frac{f[g(x + h)] - f[g(x)]}{k} \frac{g(x + h) - g(x)}{h}.$$

In order to find $\dfrac{d}{dx} f[g(x)]$, we need to take the limit of this last expression as $h \to 0$.

Since $g'(x)$ exists, $g(x)$ is continuous and

$$\lim_{h \to 0} g(x + h) = g(x).$$

Thus as $h \to 0$,

$$k = g(x + h) - g(x) \to g(x) - g(x) = 0.$$

Thus $h \to 0$ implies that $k \to 0$. Then we see that

$$\frac{d}{dx} f[g(x)] = \lim_{h \to 0} \left( \frac{f[g(x + h)] - f[g(x)]}{k} \frac{g(x + h) - g(x)}{h} \right)$$

$$= \left( \lim_{h \to 0} \frac{f[g(x + h)] - f[g(x)]}{k} \right) \cdot \left( \lim_{h \to 0} \frac{g(x + h) - g(x)}{h} \right)$$

$$= \left( \lim_{k \to 0} \frac{f[g(x) + k] - f[g(x)]}{k} \right) \cdot \left( \lim_{h \to 0} \frac{g(x + h) - g(x)}{h} \right)$$

$$= f'[g(x)] \cdot g'(x).$$

## SELF-HELP EXERCISE SET 12.7

1.  Find the derivative of $\sqrt{3x^2 + 7}$.

2.  Use the general power rule and the product rule to derive the quotient rule. *Hint:* Write

$$\frac{f(x)}{g(x)} = f(x) \cdot [g(x)]^{-1}$$

---

*Optional

## EXERCISE SET 12.7

In Exercises 1 through 20 find the derivative.

**1.** $(2x + 1)^7$      **2.** $(3x - 2)^{11}$      **3.** $(x^2 + 1)^4$

**4.** $(1 - x^2)^5$      **5.** $(2x - 3)^{3/2}$      **6.** $(3 - x)^{2/3}$

**7.** $(x^3 + 1)^{1.1}$      **8.** $(x^2 + x + 1)^{0.9}$      **9.** $\sqrt{2x + 1}$

**10.** $\sqrt{1 - 3x}$      **11.** $6\sqrt[3]{x^2 + 1}$      **12.** $-15\sqrt[5]{2x^3 - 3}$

**13.** $\dfrac{1}{x + 1}$      **14.** $\dfrac{4}{x^2 + 1}$      **15.** $-\dfrac{2}{x^3 + 1}$

**16.** $\dfrac{1}{\sqrt{x + 1}}$      **17.** $\dfrac{1}{\sqrt{x^2 + 1}}$      **18.** $\dfrac{3}{\sqrt{x^4 + 1}}$

**19.** $(2x + 1)(3x - 2)^5$      **20.** $(2x - 3)^5(4x + 7)$

In Exercises 21 through 36 find $\dfrac{dy}{dx}$.

**21.** $y = (1 - 4x)(x^2 + 1)^8$      **22.** $y = (1 - x^3)^4(1 - x)$

**23.** $y = (1 - x)^4(2x - 1)^5$      **24.** $y = (7x + 3)^3(x^2 - 4)^6$

**25.** $y = \sqrt{x}(4x + 3)^3$      **26.** $y = -3\sqrt{x}(1 - x^3)^3$

**27.** $y = \dfrac{3}{(x + 1)^4}$      **28.** $y = \dfrac{-3}{(3x + 1)^3}$

**29.** $y = \dfrac{x + 1}{(x + 3)^3}$      **30.** $y = \dfrac{x^3 + x^2 - 1}{(x + 2)^5}$

**31.** $y = \sqrt{\sqrt{x} + 1}$      **32.** $y = \sqrt[3]{\sqrt{x} + 1}$

**33.** $y = (\sqrt{x} - 1)(\sqrt{x} + 3)^{1/2}$      **34.** $y = (2x - 1)\sqrt{\sqrt{x} + 1}$

**35.** $y = \sqrt{\dfrac{x - 1}{x + 1}}$      **36.** $y = \sqrt{\dfrac{x + 1}{x^2 + 1}}$

## Applications

**37. Demand.** The demand equation for a certain product is given by

$$p = \frac{1}{\sqrt{1 + x^2}}.$$

Find the rate of change of $p$ with respect to $x$.

**38. Revenue.** Find the marginal revenue if the demand equation is given in the previous problem.

39. **Compound Interest.** If $1000 is invested at an interest rate of $i$ compounded monthly, the amount in the account at the end of 4 years is given by

$$A = 1000 \left( 1 + \frac{1}{12}i \right)^{48}.$$

Find the rate of change of $A$ with respect to $i$.

40. **Forestry.** The volume, $V$, in board feet of certain trees was found to be given by $V = 10 + .008(d - 4)^3$ for $d > 10$ where $d$ is the diameter. Find the rate of change of $V$ with respect to $d$ when $d = 12$.

41. **Fishery.** Let $n$ denote the number of prey in a school of fish, $r$ the detection range of the school by a predator, and $c$ a constant that defines the average spacing of fish within the school. The volume of the region within which a predator can detect the school, called the *visual volume*, has been given by biologists as

$$V = \frac{4\pi r^3}{3} \left( 1 + \frac{c}{r} n^{1/3} \right)^3.$$

With $c$ and $r$ considered as constants, one then has $V = V(n)$. Find the rate of change of V with respect to $n$.

42. **Fishery.** In the previous problem suppose $c$ and $n$ are constants. Thus $V = V(r)$. Find the rate of change of V with respect to $r$.

43. **Psychology.** The psychologist L. L. Thurstone suggested that the relationship between the learning time $T$ of memorizing a list of length $n$ is a function of the form

$$T = an\sqrt{n - b},$$

where $a$ and $b$ are positive constants. Find the rate of change of $T$ with respect to $n$.

## Solutions to Self-Help Exercise Set 12.7

1. We must recognize this expression as some function raised to a power. In this case the power is $n = \frac{1}{2}$. Using the general power rule we obtain

$$\frac{d}{dx} (\sqrt{3x^2 + 7}) = \frac{d}{dx} [(3x^2 + 7)^{1/2}]$$

$$= \frac{1}{2} (3x^2 + 7)^{-1/2} \frac{d}{dx} (3x^2 + 7)$$

$$= \frac{1}{2} (3x^2 + 7)^{-1/2}(6x)$$

$$= 3x(3x^2 + 7)^{-1/2}$$

2. First apply the product rule and then use the general power rule when appropriate to obtain

$$\frac{d}{dx}\left(\frac{f(x)}{g(x)}\right) = \frac{d}{dx}\left(f(x) \cdot g^{-1}(x)\right)$$

$$= g^{-1}(x) \cdot \frac{d}{dx} f(x) + f(x) \cdot \frac{d}{dx}\left(g^{-1}(x)\right) \qquad \text{product rule}$$

$$= g^{-1}(x) \cdot \frac{d}{dx} f(x) + f(x)\left(-g^{-2}(x) \frac{d}{dx} g(x)\right)$$

$$\qquad\qquad\qquad\qquad\qquad\qquad\qquad\qquad \text{general power rule}$$

$$= \frac{\dfrac{d}{dx} f(x)}{g(x)} \cdot \frac{g(x)}{g(x)} - \frac{f(x) \dfrac{d}{dx} g(x)}{g^2(x)}$$

$$= \frac{g(x) \cdot \dfrac{d}{dx} f(x) - f(x) \cdot \dfrac{d}{dx} g(x)}{g^2(x)}$$

## SUMMARY OUTLINE OF CHAPTER 12

- **Limit of a function.** If as $x$ approaches $a$, $f(x)$ approaches the number $L$, then we say that $L$ is the limit of $f(x)$ as $x$ approaches $a$, and write $\lim_{x \to a} f(x) = L$.

- **Limits from left and right.** If as $x$ approaches $a$ from the left, $f(x)$ approaches the number $M$, then we write $\lim_{x \to a-} f(x) = M$. If as $x$ approaches $a$ from the right, $f(x)$ approaches the number $N$, then we write $\lim_{x \to a+} f(x) = N$.

- **Existence of a limit.** $\lim_{x \to a} f(x) = L$ if, and only if, both of the limits $\lim_{x \to a-} f(x)$ and $\lim_{x \to a+} f(x)$ exist and equal $L$.

- **Replacement theorem.** Suppose $f(x) = g(x)$ for all $x \neq a$. If $\lim_{x \to a} g(x)$ exists, then $\lim_{x \to a} f(x)$ exists and $\lim_{x \to a} f(x) = \lim_{x \to a} g(x)$.

- **Rules for limits.** Assume that $\lim_{x \to a} f(x) = L$ and $\lim_{x \to a} g(x) = M$. Then

  $$\lim_{x \to a} cf(x) = cL \qquad\qquad \lim_{x \to a} (f(x) \pm g(x)) = L \pm M$$

  $$\lim_{x \to a} (f(x) \cdot g(x)) = L \cdot M. \qquad \lim_{x \to a} \frac{f(x)}{g(x)} = \frac{L}{M} \text{ if } M \neq 0.$$

  $$\lim_{x \to a} (f(x))^n = L^n, \text{ if } L^n \text{ is defined.}$$

- **Limit of polynomial function.** If $p(x)$ is a polynomial function, then $\lim_{x \to a} p(x) = p(a)$

- **Continuity.** The function $f(x)$ is continuous at $x = a$ if $\lim_{x \to a} f(x) = f(a)$.

- A polynomial function is continuous everywhere.

- **Rules for continuity.**   A constant times a continuous function, the sum or difference of continuous functions, and the product of continuous functions are all continuous. The quotient of continuous functions is continuous at all points where the denominator is not zero. A continuous function $f(x)$ raised to any real power $r$ is continuous at points where $(f(x))^r$ is defined.

- A rational function is continuous at every point at which the denominator is not zero.

- **Constant sign theorem.**   If $f(x)$ is continuous on $(a, b)$ and $f(x) \neq 0$ on $(a, b)$, then $f(x)$ is either always positive or always negative on $(a, b)$.

- **Average and instantaneous velocity.**   These are, respectively,
$$\frac{s(t + h) - s(t)}{h} \quad \text{and} \quad \lim_{h \to 0} \frac{s(t + h) - s(t)}{h}.$$

- **Average and instantaneous rate of change** for $f(x)$ are, respectively,
$$\frac{f(x + h) - f(x)}{h} \quad \text{and} \quad \lim_{h \to 0} \frac{f(x + h) - f(x)}{h}.$$

- **The slope of the tangent line** to the graph of $y = f(x)$ at $x = x_0$ is $\lim_{h \to 0} \dfrac{f(x_0 + h) - f(x_0)}{h}$, if this limit exists.

- The instantaneous rate of change of $f(x)$ at $x_0$ equals the slope of the tangent line to the graph of $f(x)$ at $x_0$.

- **Definition of derivative.**   The derivative of a function $y = f(x)$, denoted by $f'(x)$ or $\dfrac{dy}{dx}$, is defined to be $f'(x) = \lim_{h \to 0} \dfrac{f(x + h) - f(x)}{h}$ if this limit exists.

- **Rules for derivatives.**   If $f(x)$ and $g(x)$ are continuous, $C$ a constant, and $n$ a real number, then

$$\frac{d}{dx}(C) = 0 \qquad\qquad \frac{d}{dx}(x^n) = nx^{n-1}$$

$$\frac{d}{dx}(Cf(x)) = C\frac{d}{dx}f(x) \qquad\qquad \frac{d}{dx}(f(x) \pm g(x)) = \frac{d}{dx}f(x) \pm \frac{d}{dx}g(x)$$

$$\frac{d}{dx}(f(x) \cdot g(x)) = f(x) \cdot \frac{d}{dx}g(x) + g(x) \cdot \frac{d}{dx}f(x)$$

$$\frac{d}{dx}\left(\frac{f(x)}{g(x)}\right) = \frac{g(x) \cdot \dfrac{d}{dx}f(x) - f(x) \cdot \dfrac{d}{dx}g(x)}{g^2(x)} \qquad \text{if } g(x) \neq 0.$$

$$\frac{d}{dx}(f[g(x)]) = f'[g(x)] \cdot \frac{d}{dx}g(x) \qquad \frac{d}{dx}([g(x)]^n) = n[g(x)]^{n-1} \cdot \frac{d}{dx}g(x)$$

- The **equation of the tangent line** to the graph of the curve $y = f(x)$ at $x = x_0$ is $y - f(x_0) = f'(x_0)(x - x_0)$.

- A differentiable function is continuous.

- If $C(x)$, $R(x)$, and $P(x)$ are the cost, revenue, and profit functions, respectively, then $C'(x)$, $R'(x)$, and $P'(x)$ are the **marginal cost, marginal revenue**, and **marginal profit**, respectively.

- The terms $C'(x)$, $R'(x)$, $P'(x)$, respectively, approximate the cost, revenue, and profit associated with the next item.

## Chapter 12 Review Exercises

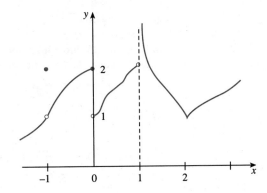

1.  For the function in the figure, find

$$\lim_{x \to a-} f(x), \qquad \lim_{x \to a+} f(x), \qquad \lim_{x \to a} f(x)$$

for $x = -1, 0, 1, 2$.

2.  For the function in the above figure, find all points where $f(x)$ is not continuous.

3.  For the function in the above figure, find all points where $f(x)$ does not have a derivative.

In Exercises 4 through 18 find the limits if they exist.

4.  $\lim\limits_{x \to 1} 12.7$

5.  $\lim\limits_{x \to 1} \pi^3$

6.  $\lim\limits_{x \to 1} (x^2 + 2x - 3)$

7.  $\lim\limits_{x \to -1} (x^3 - 3x - 5)$

8.  $\lim\limits_{x \to 0} (3x^5 + 2x^2 - 4)$

9.  $\lim\limits_{x \to 1} \dfrac{2x^2 + 1}{x + 2}$

10. $\lim\limits_{x \to 2} \dfrac{x + 5}{x - 3}$

11. $\lim\limits_{x \to 1} \dfrac{x^2 + x - 2}{x - 1}$

12. $\lim\limits_{x \to 3} \dfrac{x^2 + x - 12}{x - 3}$

13. $\lim\limits_{x \to 2} \dfrac{x^2 + 1}{x - 2}$

14. $\lim\limits_{x \to -2} \dfrac{x + 3}{x + 2}$

15. $\lim\limits_{x \to 2} \sqrt{x^3 - 4}$

16. $\lim\limits_{x \to -1} \sqrt{10 - x^2}$

17. $\lim\limits_{x \to 4} \left( \sqrt{x} - \dfrac{1}{\sqrt{x}} \right)$

18. $\lim\limits_{x \to -2} \sqrt{1 - x^2}$

In Exercises 19 through 24 use the definition of limit to find the derivative.

19. $f(x) = 2x^2 + x - 3$

20. $f(x) = x^2 + 5x + 1$

**21.** $f(x) = \dfrac{1}{x^2}$

**22.** $f(x) = \dfrac{1}{2x + 5}$

**23.** $f(x) = \sqrt{x + 7}$

**24.** $f(x) = \sqrt{2x - 1}$

In Exercises 25 through 28 find the equation of the tangent line to the graph of each of the functions at the indicated point.

**25.** $y = x^3 + 1$, $x_0 = 2$

**26.** $y = x^{3/2} + 2$, $x_0 = 4$

**27.** $y = x^5 + 3x^2 + 2$, $x_0 = 1$

**28.** $y = \dfrac{1}{\sqrt{x}}$, $x_0 = 9$

In Exercises 29 through 56 find the derivative.

**29.** $f(x) = x^6 - 3x^2 + 1$

**30.** $f(x) = 4x^{6/5} + 5x^{2/5} + 3$

**31.** $f(x) = 2x^{-3} + 3x^{-2/3}$

**32.** $f(x) = 2x^{1.3} + 3x^{-2.4}$

**33.** $f(x) = \pi^2 x^2 + 2x - \pi$

**34.** $f(x) = \pi^3 x^3 + e^2 x^2 - \pi^2$

**35.** $f(x) = \sqrt[3]{x^2}$

**36.** $f(x) = \sqrt[3]{x^{-5}}$

**37.** $f(x) = (x + 3)(x^3 + x + 1)$

**38.** $f(x) = (x^2 + 5)(x^5 - 3x + 2)$

**39.** $f(x) = (\sqrt{x})(x^4 - 2x^2 + 3)$

**40.** $f(x) = (\sqrt[3]{x})(x^3 - 3x + 5)$

**41.** $f(x) = (x^{3/2})(x^2 + 4x + 7)$

**42.** $f(x) = (x^{-7/2})(x^3 + 2x + 1)$

**43.** $f(x) = \dfrac{x - 1}{x + 1}$

**44.** $f(x) = \dfrac{\sqrt{x}}{2x + 1}$

**45.** $f(x) = \dfrac{x^3}{x^2 + 1}$

**46.** $f(x) = \dfrac{x^2 + x + 1}{3x - 2}$

**47.** $f(x) = \dfrac{x^2 - 1}{x^4 + 1}$

**48.** $f(x) = (3x + 1)^{10}$

**49.** $f(x) = (x^2 + 4)^{20}$

**50.** $f(x) = (\sqrt{x} - 5)^{5/2}$

**51.** $f(x) = \sqrt{x^2 + x + 1}$

**52.** $f(x) = x^3(x^2 + x + 2)^5$

**53.** $f(x) = (x^2 + 3)(x^4 + x + 1)^5$

**54.** $f(x) = \dfrac{x^2 + 1}{(x^4 + x + 1)^4}$

**55.** $f(x) = \sqrt{x^{3/2} + 1}$

**56.** $f(x) = \sqrt[3]{\sqrt{x} + 1}$

**57.** If $f(2) = 3$ and $f(x) = \dfrac{x^2 - x - 2}{x - 2}$ if $x \neq 2$, is $f(x)$ continuous at $x = 2$?

**58.** If $f(1) = 3$ and $f(x) = \dfrac{x^2 + 3x - 4}{x - 1}$ if $x \neq 1$, is $f(x)$ continuous at $x = 1$?

**59.** **Average Cost.** A person making a long-distance telephone call to a neighboring town is charged $1.00 for the first minute or fraction thereof and $.20 for each additional minute or fraction thereof. If $t$ is the length of the telephone call and $C(t)$ the cost of the call, then the average cost, $\overline{C}(t)$, is defined to be

$$\overline{C}(t) = \frac{C(t)}{t}.$$

Thus if $t = 2$, the average cost is $\$1.20/2 = \$0.60$ per minute. If $t = 1$, the average cost is $\$1.00$ per minute. If $t = 1/2$, then the average cost is $\$2.00$ per minute. What happens to the average cost as the length of the call, $t$, approaches zero, that is, what is

$$\lim_{t \to 0+} \overline{C}(t)?$$

How do you interpret your answer in economic terms?

**60.   Biology.**   According to Wiess's law, the intensity of an electric current required to excite a living tissue (threshold strength) is given by $i = \dfrac{A}{t} + B$, where $t$ is the duration of the current and $A$ and $B$ are positive constants. What happens as $t \to 0$? Does your answer seem reasonable?

**61.   Medicine.**   Suppose that the concentration $c$ of a drug in the blood stream $t$ hours after it is taken orally is given by

$$c = \frac{5t}{t^2 + 2}.$$

What happens to the drug concentration as $t \to 0+$? Does this make sense?

**62.   Costs.**   A telephone company charges $\$1.00$ for the first minute or fraction thereof for a certain long-distance call and $\$.50$ for each additional minute or fraction thereof. Determine where this cost function is continuous.

**63.   Profits.**   The daily profits of a firm in dollars is given by $P = 100(-x^2 + 8x - 12)$ where $x$ is the number of items sold. Find the instantaneous rate of change when $x$ is (a) 3, (b) 4, (c) 5.

**64.   Velocity.**   A particle moves according to the law $s(t) = t^2 - 8t + 15$ where $s$ is in feet and $t$ in seconds. Find the instantaneous velocity at $t = 3, 4, 5$. Interpret your answers.

**65.   Costs.**   The weekly total cost function for a company in dollars is $C(x)$. If $C'(100) = 25$, what is the approximate cost of the 101st item?

**66.   Demand.**   If the demand for a product is given by

$$p = \frac{300}{3x + 1},$$

find the marginal demand.

**What should be the price of a ticket in order to maximize revenue? (Dennis Drenner.)**

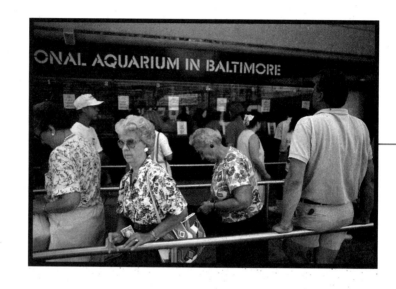

# 13

# Applications of the Derivative

13.1   Increasing and Decreasing

13.2   Relative Extrema

13.3   The Second Derivative

13.4   Limits at Infinity; More Curve Sketching

13.5   Absolute Extrema

13.6   Optimization Problems

I n this chapter we learn how to sketch the graph of a function and how to find the maximum and the minimum of a function on a given interval. To mention just a few applications we will look at the Laffer curve that was used to justify decreasing the personal federal income tax rates in the Reagan administration; we will consider the management of renewable natural resources, considering the whaling industry, specifically; and we will see how to find the quantity of sales that will produce the most profit.

## 13.1    INCREASING AND DECREASING

► *Increasing and Decreasing Functions*
► *Federal Tax Revenue*

Let $x$ be the percent of the sum of total income of individuals and total profits of corporations in the United States collected as federal taxes and $R(x)$ the total tax revenue. If $x$ is small, then increasing $x$ should result in an increase in the revenue $R(x)$. If $x$ is large, is it possible increasing $x$ further will decrease revenue? If this is so, could decreasing the rate $x$ of taxation result in an *increase* in the revenue $R(x)$ collected? For the answers given by some of the Reagan administration economists, see the discussion preceding Example 5.

Suppose in the above discussion that $R(x) = \sqrt{x}(100 - x)$. Where is the function increasing and where is it decreasing? See Example 5 for the answer.

## Increasing and Decreasing Functions

All functions $y = f(x)$ considered in this text are assumed to be "reasonable." That is, $f$ and $f'$ are continuous, except possibly at a few isolated values.

In Figure 13.1 we see that the population of the United States has steadily *increased* over the time period shown, whereas in Figure 13.2, we see that the

**Figure 13.1**

**Figure 13.2**

percentage of the population that are farmers has steadily *decreased.* In Figure 13.3 we see that the median age of the U.S. population increased from 1860 to 1950, decreased from 1950 to 1970, and then increased again from 1970 to 1990.

In this section the sign of the first derivative will be seen to determine whether the function is increasing or decreasing. First we give two definitions.

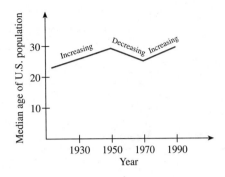

**Figure 13.3**

---

**Definition of Increasing and Decreasing Function**

A function $y = f(x)$ is said to be **increasing** ($\nearrow$) on the interval $(a, b)$ if the graph of the function rises while moving left to right or, equivalently, if

$$f(x_1) < f(x_2) \qquad \text{when} \qquad a < x_1 < x_2 < b.$$

A function $y = f(x)$ is said to be **decreasing** ($\searrow$) on the interval $(a, b)$ if the graph of the function falls while moving left to right or, equivalently, if

$$f(x_1) > f(x_2) \qquad \text{when} \qquad a < x_1 < x_2 < b.$$

---

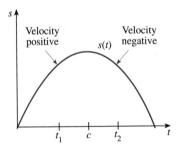

**Figure 13.4**

We already know that for straight lines, a positive slope indicates the line is rising and therefore the function is increasing, while if the slope is negative the line is falling and therefore the function is decreasing. A similar situation exists for curves that are not lines.

Suppose a ball is thrown upward and the function $s(t)$ gives the height of the ball in feet where $t$ is measured in seconds. See Figure 13.4. We already know that the derivative $s'(t)$ gives the velocity $v(t)$ at time $t$ and also the slope of the tangent line to the graph of the curve at $t$. If the slope or velocity $s'(t_1)$ is positive, then we know that the ball is rising at $t_1$. If the slope or velocity $s'(t_2)$ is negative, then we know that the ball is falling at $t_2$. We can see in Figure 13.4 that for values of $t$ less than $c$, the slope or derivative is positive and the function is increasing, while for $t > c$, the slope or derivative is negative and the function is decreasing.

Indeed it can be shown that if $f'(x)$ is positive on an entire interval $(a, b)$, then $f(x)$ is increasing on this interval and if $f'(x)$ is negative on an entire interval $(a, b)$, then $f(x)$ is decreasing on the interval. Examination of the graph in Figure 13.4 indicates that the function is increasing where the slope is positive and decreasing where the slope is negative. We now state the following test without proof.

**Test for Increasing or Decreasing Functions**

If for all $x \in (a, b)$, $f'(x) > 0$, then $f(x)$ is increasing ($\nearrow$) on $(a, b)$.
If for all $x \in (a, b)$, $f'(x) < 0$, then $f(x)$ is decreasing ($\searrow$) on $(a, b)$.

**E X A M P L E   1**    **Using the Test for Increasing or Decreasing Functions**

Find where $y = f(x) = x^3 + x$ is increasing and also find where $f$ is decreasing.

**Solution**    Since $f'(x) = 3x^2 + 1$ is always positive, the function $y = f(x) = x^3 + x$ must always be increasing. A graph is drawn in Figure 13.5.   ∎

For a function like the one graphed in Figure 13.6, to find the intervals on which $f(x)$ is either always increasing or always decreasing, we need to find the intervals on which $f'(x)$ is either always positive or always negative. The simplest way to accomplish this is to first find all values where the derivative is neither positive nor negative since there will normally only be a small number of such values. Obviously, all values where the derivative is *zero* fall into this category, but also in this category are all values where the derivative does *not exist*. These two types of values are so important that they are given a special name, **critical values**.

Figure 13.5

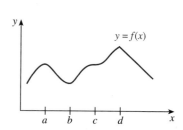

Figure 13.6

**Definition of Critical Value**

A value $x = c$ is a **critical value** if

1. $c$ is in the domain of the function $f(x)$ and
2. $f'(c) = 0$ or $f'(c)$ does not exist.

E X A M P L E   2     **Finding Critical Values Graphically**

Find the critical values for the function shown in Figure 13.6.

**Solution**     Since the derivative of the function is zero at the values $a$, $b$, and $c$, these 3 values are critical values. Since the derivative does not exist at the value $d$, this is also a critical value. ∎

In this text we will always restrict ourselves to functions that satisfy the following assumptions.

---

**Basic Assumptions of this Text**

1.  $f(x)$ is continuous on its domain of definition, except possibly for a finite number of values.
2.  $f'(x)$ is continuous at all values that are not critical values or discontinuities of $f$.
3.  $f$ has only a finite number of critical values on any finite interval, unless $f$ is constant.

---

In this and the next two sections, we will only encounter functions that are continuous throughout their domains of definition.

Condition (3) is a very important assumption, since it eliminates certain "pathological" functions that can oscillate infinitely often near a point. Since these pathological functions are not found in business, finance, and economics, we lose little. Furthermore, their elimination will greatly simplify the analysis that would otherwise be required in this and subsequent sections. Notice that since the constant functions have $f'(x) = 0$ at every value, every value is a critical value. The constant functions are the only functions that we will consider in this text that have an infinite number of critical values on a finite interval.

Given a function $y = f(x)$ that satisfies the above basic assumption, we first locate an interval $(a, b)$ on which both $f$ and $f'$ are continuous and which contains no critical values of $f$. Since there are no critical values of $f$ on $(a, b)$, $f'(x)$ is never zero on $(a, b)$. Since $f'(x)$ is continuous and never zero on the interval $(a, b)$, the constant sign theorem of Section 12.2 assures us that $f'(x)$ must maintain the same sign throughout this interval. Thus $f'(x)$ is either always positive on $(a, b)$ or always negative there. Therefore $f(x)$ is either always increasing throughout $(a, b)$ or always decreasing throughout $(a, b)$.

If $f'(x)$ is either always positive on $(a, b)$ or always negative there, then we can select a test value on the interval, say $x_1$, and evaluate $f'(x_1)$. If $f'(x_1)$ is positive, then $f'(x)$ is positive on the entire interval. If $f'(x_1)$ is negative, then $f'(x)$ is negative on the entire interval.

Let us now see how these techniques can be used to solve the following problems.

**E X A M P L E  3**       **Finding Where $f$ is Increasing or Decreasing**

Find the critical values and the intervals where $f(x) = x^3 - 3x^2 + 1$ is increasing or decreasing. Use this information to sketch a graph.

**Solution**       **Step 1:**   Find the critical values.

   (a)   Notice that $f'(x) = 3x^2 - 6x$ exists and is continuous everywhere. Thus the only critical values are values at which $f'(x) = 0$.
   (b)   Setting $f'(x) = 3x^2 - 6x = 3x(x - 2) = 0$ gives $x = 0$ and $x = 2$. Thus the critical values are $x = 0$ and $x = 2$.

**Step 2:**   Find the subintervals on which $f'(x)$ has constant sign.

The two critical values then divide the real line into the three subintervals: $(-\infty, 0)$, $(0, 2)$, and $(2, +\infty)$. The derivative is continuous and never zero on each of these subintervals. Therefore the sign of $f'(x)$ is constant on each of these subintervals.

**Step 3:**   Determine the sign of $f'$ on each subinterval.

Take the three test values to be $x_1 = -1$, $x_2 = 1$ and $x_3 = 3$. These three values fall respectively in the three subintervals given in the preceding step. See Figure 13.7. Evaluating $f'$ at these three values gives

$$f'(-1) = +9, \qquad f'(1) = -3, \qquad f'(3) = +9.$$

Thus $f'(x) > 0$ on the first interval, $f'(x) < 0$ on the second interval, and $f'(x) > 0$ on the third interval.

**Step 4:**   Use the test for increasing or decreasing functions.

We conclude that $f(x)$ is increasing ($\nearrow$) on the first interval, decreasing ($\searrow$) on the second, and increasing ($\nearrow$) on the third.

Now using this information and the fact that $f(0) = 1$ and $f(2) = -3$, we obtain the graph in Figure 13.8.   ■

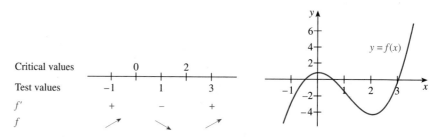

Figure 13.7                                        Figure 13.8

**W A R N I N G .**       In Step 1 it is easy to overlook the solution $x = 0$. A common mistake is to set $f'(x) = 3x^2 - 6x = 0$, cancel $x$, and obtain $3x - 6 = 0$ or $x = 2$, as the *only* solution.

> **Procedure for Finding Where a Function is Increasing or Decreasing**
>
> **Step 1:**  Find the critical values. To do this:
>
>    (a)  Locate all values of $c$ for which $f'(c)$ does not exist but $f(c)$ does.
>    (b)  Locate all values of $c$ for which $f'(c) = 0$.
>
> **Step 2:**  Find the subintervals on which $f'(x)$ has constant sign.
>
> **Step 3:**  Select a convenient test value on each subinterval found in the previous step and evaluate $f'$ at this value. The sign of $f'$ at this test value will be the sign of $f'$ on the whole subinterval.
>
> **Step 4:**  Use the test for increasing or decreasing functions to determine whether $f$ is increasing or decreasing on each subinterval.

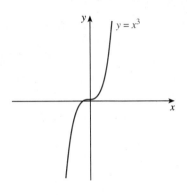

**Figure 13.9**

**REMARK.**     We have seen that if the derivative is always positive on an interval, then the function increases there, and, if the derivative is always negative on an interval, the function always decreases there. However, a function may increase on an interval without the derivative always being positive there. The graph of the function $f(x) = x^3$ in Figure 13.9 illustrates that this function is increasing throughout the interval $(-\infty, +\infty)$. However $f'(x) = 3x^2$ and is zero at $x = 0$. A similar remark applies to decreasing functions.

**E X A M P L E   4**      **Finding Where $f$ is Increasing or Decreasing**

Find the critical values and the intervals on which $f(x) = \sqrt[3]{x^2}$ is increasing or decreasing.

**Solution**      **Step 1:**  Find the critical values.

(a)  Since
$$f(x) = x^{2/3},$$

$$f'(x) = \frac{2}{3}x^{-1/3} = \frac{2}{3\sqrt[3]{x}}.$$

Note that $f'(x)$ exists and is continuous everywhere except at $x = 0$. Since $f'(0)$ does not exist, while $f(0)$ does exist, $x = 0$ is a critical value.

(b)  There are no values for which $f'(x) = 0$, thus $x = 0$ is the only critical value.

**Step 2:**  We divide the real line into the two intervals: $(-\infty, 0)$ and $(0, +\infty)$. On each of these intervals $f'(x)$ is never zero and $f'(x)$ is continuous. Thus the sign of $f'(x)$ is constant on each of these intervals.

**Step 3:**  Take the convenient test values $x_1 = -1$ and $x_2 = 1$ and notice that

$$f'(-1) = \frac{2}{3\sqrt[3]{-1}} = -\frac{2}{3} < 0, \qquad f'(1) = \frac{2}{3\sqrt[3]{1}} = \frac{2}{3} > 0.$$

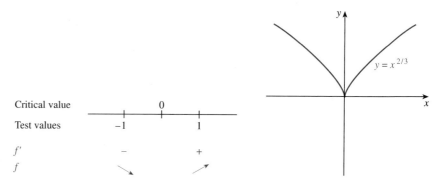

Critical value

Test values

$f'$

$f$

**Figure 13.10**

**Figure 13.11**

Thus $f'(x) < 0$ on the first interval and $f'(x) > 0$ on the second. See Figure 13.10.

**Step 4:**   Thus $f(x)$ is decreasing ($\searrow$) on $(-\infty, 0)$ and increasing ($\nearrow$) on $(0, +\infty)$. See Figure 13.11.   ∎

**Figure 13.12**

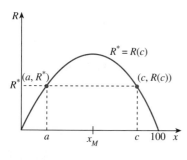

**Figure 13.13**

## Federal Tax Revenue

In the early 1980s the Reagan administration was struggling with the consequences of lowering the tax rates. Some had the view that if you lowered the tax rate you must lower the total income to the government. But others argued that lowering the tax rates would *increase* total tax revenue.

One of these economists was Arthur Laffer of the University of Southern California. He argued as follows. Let $R = R(x)$ where $x$ is the percent of income taken by the federal government in the form of taxes and $R(x)$ the total revenue from taxes from a rate of $x$ percent. Obviously if the rate is set at zero, the government will have no revenue. That is, $R(0) = 0$. If the rate, $x$, is increased somewhat, surely the revenue $R(x)$ will be increased somewhat. So $R(x)$ must be an increasing function at least for small values of $x$. On the other hand, it is reasonable to expect that if $x = 100$, that is, the government takes everything, then probably nobody will wish to earn any money and federal revenue will be zero. Thus $R(100) = 0$. If $x$ is a little less than 100, then there will probably be some people who will earn some money, resulting in some income for the federal government. So one can assume that for $x$ near 100, $R(x)$ will be decreasing. At this point the graph of the curve $R = R(x)$ looks like the solid curve in Figure 13.12. Now Laffer made the assumption that the rest of the curve will probably look like the dotted curve in Figure 13.12 at the intermediate values. (This curve is called a Laffer curve.) Label as $x_M$ the value at which this curve changes direction from increasing to decreasing.

Suppose currently $x = c$. The argument was then made that $x_M$ was *less* than $c$, as in Figure 13.13. But $R(x)$ is *decreasing* near the value $c$. Thus increasing $x$

will result in a *decrease* of revenue. Then decreasing $x$ will result in an increase of revenue.

The Reagan administration wished only to lower the tax rate—not to increase federal tax revenue. Drawing a horizontal line at height $R^* = R(c)$ in Figure 13.13 results in this line intersecting the graph of $y = R(x)$ at *two* points, $(c, R^*)$ of course, but also $(a, R^*)$ shown in the figure. Thus according to this model, the income tax rate could be decreased all the way down to $x = a$ and the government would still have the same revenue.

**E X A M P L E   5**   **Finding Where Tax Revenue Increases or Decreases**

Suppose in the above discussion that $R(x) = \sqrt{x}(100 - x)$. Find where the function is increasing and where it is decreasing.

**Solution**   We are only interested in $0 < x < 100$.

**Step 1:**

(a)   Since $R(x) = 100x^{1/2} - x^{3/2}$,

$$R'(x) = 50x^{-1/2} - \frac{3}{2}x^{1/2}$$

and $R'$ exists and is continuous everywhere on $(0, 100)$. Thus the only critical values on $(0, 100)$ are values where the derivative is zero.

(b)   Setting the derivative equal to zero yields

$$50x^{-1/2} - \frac{3}{2}x^{1/2} = 0 \qquad \text{Multiply by } x^{1/2}.$$

$$50 = \frac{3}{2}x$$

$$x = \frac{100}{3}$$

Thus the only critical value on $(0, 100)$ is at $\dfrac{100}{3}$.

**Step 2:**   This single critical value then divides $(0, 100)$ into the two intervals $\left(0, \dfrac{100}{3}\right)$ and $\left(\dfrac{100}{3}, 100\right)$.

**Step 3:**   Pick the convenient test values $x_1 = 25$ and $x_2 = 64$ and note that

$$R'(25) = \frac{50}{\sqrt{25}} - \frac{3}{2}\sqrt{25} = \frac{5}{2}, \qquad R'(64) = \frac{50}{\sqrt{64}} - \frac{3}{2}\sqrt{64} = -\frac{23}{4}.$$

See Figure 13.14.

**Figure 13.14**

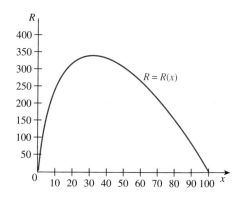

**Figure 13.15**

**Step 4.**   Thus $R'(x) > 0$ on $\left(0, \dfrac{100}{3}\right)$ and thus $R$ is increasing ($\nearrow$) there and $R'(x) < 0$ on $\left(\dfrac{100}{3}, 100\right)$ and thus $R$ is decreasing ($\searrow$) on this second interval. The graph is shown in Figure 13.15.  ∎

It has been pointed out that when the Kennedy administration lowered the tax rate, federal tax revenue *increased.* The Reagan administration, however, had inconclusive results, perhaps due to other factors operating in the economy at that time.

Of course, it is very unrealistic to think that one could have knowledge of an *explicit* formula for $R(x)$. But knowledge merely of the general *shape* of the graph is very important.

### SELF-HELP EXERCISE SET 13.1

1.  Suppose a firm has a monopoly on VCRs and $x$ is the number of millions of VCRs sold in a year and $P(x) = -0.1x^2 + 1.1x - 1.0$ is the profit in tens of millions of dollars for selling $x$ million VCRs. (a) Find where $P(x)$ is increasing and find where it is decreasing. Of what significance is this to the firm? (b) Find the break-even values. (c) Draw a graph of $P = P(x)$. (d) Is this model realistic?

2.  Find the critical values and the intervals on which $f(x) = 3x^4 + 8x^3 + 5$ is increasing or decreasing. Draw a graph. Must $f'$ change sign at a critical value?

### EXERCISE SET 13.1

In Exercises 1 through 8 find all critical values and the largest open intervals on which the given functions in the following graphs are increasing and the largest open intervals on which the given functions are decreasing.

**1.**

**2.**

**3.**

**4.**

**5.**

**6.**

**7.**

**8.**

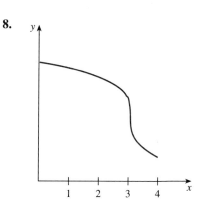

In Exercises 9 through 20 $f'(x)$ is given in factored form and $p(x)$ is a function that is always positive. Find all critical values and the largest open intervals on which the given functions are increasing and the largest open intervals on which the given functions are decreasing. Using the other information, sketch a rough graph of $f$.

**9.**  $f'(x) = x - 2,\ f(2) = 1$

**10.**  $f'(x) = 4 - x,\ f(4) = 1$

**11.**  $f'(x) = p(x)(x - 1)(x - 3),\ f(1) = 5,\ f(3) = 1$

**12.**  $f'(x) = -p(x)(x + 3)(x - 1),\ f(-3) = -17,\ f(1) = 15$

**13.**  $f'(x) = p(x)(x - 1)(x - 3)(x - 5),\ f(1) = 1,\ f(3) = 2,\ f(5) = 1$

**14.**  $f'(x) = p(x)(x + 2)(x + 1)(x - 3)$, where $f(-2) = 1,\ f(-1) = 5,$
$f(3) = -5$

**15.**  $f'(x) = -p(x)(x + 2)(x - 1)(x - 3)$, where $f(-2) = 6,\ f(1) = 1,$
$f(3) = 8$

**16.**  $f'(x) = p(x)(x - 1)(x - 3)^2(x - 5)$, where $f(1) = 2,\ f(3) = 0,\ f(5) = -2$

**17.**  $f'(x) = p(x)(x + 1)^2(x - 2)^3(x - 4)$, where $f(-1) = 1,\ f(2) = 2,$
$f(4) = 0$

**18.**  $f'(x) = p(x)(x - 1)^2(x - 3)^3(x - 5)$, where $f(1) = 1,\ f(3) = 5,\ f(5) = 2$

**19.**  $f'(x) = p(x)x^2(x - 2)^3(x - 4)^4$, where $f(0) = 3,\ f(2) = 2,\ f(4) = 3$

**20.**  $f'(x) = p(x)x(x - 2)^3(x - 4)^5$, where $f(0) = 1,\ f(2) = 2,\ f(4) = 0$

In Exercises 21 through 48 find all critical values and all open intervals on which the given functions are increasing and all open intervals on which the given functions are decreasing. Sketch a rough graph of $f$.

**21.**  $f(x) = 4x - 5$

**22.**  $f(x) = -2x + 1$

**23.**  $f(x) = 10 - x - x^3$

**24.**  $f(x) = 2x^3 + x$

**25.** $f(x) = x^2 + 3x - 1$

**26.** $f(x) = 2x^2 - 2x + 4$

**27.** $f(x) = 3 + 2x - x^2$

**28.** $f(x) = 5 - 12x - 3x^2$

**29.** $f(x) = 2x^3 - 9x^2 + 12x + 3$

**30.** $f(x) = 2x^3 - 6x - 5$

**31.** $f(x) = -4x^3 + 3x - 2$

**32.** $f(x) = -2x^3 + 9x^2 + 24x - 3$

**33.** $f(x) = 3x^4 - 4x^3 + 1$

**34.** $f(x) = -x^4 + 4x^3 - 2$

**35.** $f(x) = x^5 - 5x^4 + 1$

**36.** $f(x) = -4x^5 + 9x^4 - 3$

**37.** $f(x) = x^4 - 2x^2 + 1$

**38.** $f(x) = 3x^4 + 8x^3 - 18x^2 + 2$

**39.** $f(x) = -3x^4 - 14x^3 + 12x^2 + 5$

**40.** $f(x) = -3x^4 + 6x^3 + 6x^2 - 4$

**41.** $f(x) = \sqrt[3]{x^4}$

**42.** $f(x) = \sqrt[5]{x^2}$

**43.** $f(x) = \sqrt{x^2 + 1}$

**44.** $f(x) = \sqrt{x^4 + 1}$

**45.** $f(x) = (1 + x)^{1/3}$

**46.** $f(x) = (1 - x)^{1/3}$

**47.** $f(x) = |x - 1|$

**48.** $f(x) = |x + 2|$

**49.** For the special cubic equation

$$f(x) = ax^3 + bx + c,$$

show that $f(x)$ is always increasing if both $a$ and $b$ are positive and that $f(x)$ is always decreasing if both $a$ and $b$ are negative.

**50.** For the function

$$\frac{x + a}{x + b},$$

show that, except for $x = -b$, this function is always increasing if $b > a$. Also show that except for $x = -b$, this function is always decreasing if $b < a$.

## Applications

**51. Revenue.** The demand function is given by $p = -0.1x + 10$. Find where the revenue is increasing and where the revenue is decreasing.

**52. Cost.** The cost function for a firm is given by $C(x) = -x^4 + 16x^3 - 5$. Find where the cost is increasing and where the cost is decreasing.

**53. Profits.** The profit function for a firm is given by $P(x) = -x^2 + 12x + 10$. Find where the profit is increasing and where the profit is decreasing.

**54. Appreciation.** If a deposit of $P$ dollars is invested at a yearly rate of interest $i$ compounded $m$ times per year for $n$ years, the compounded amount is given by

$$A = P\left(1 + \frac{i}{m}\right)^{mn}.$$

Fix $m$ and $n$ and let $A$ be a function of $i$, that is, $A = A(i)$. Where is $A(i)$ increasing and where is it decreasing? Does your answer make sense?

**55. Federal Tax Revenue.**   Let $x$ be the percent of income taken by the federal government in the form of taxes and $R(x) = x^2(100 - x)$ be the total revenue obtained. Find where $R(x)$ is increasing and find where $R(x)$ is decreasing.

**56. Federal Tax Revenue.**   Answer the same questions as in the previous problem if $R(x) = \sqrt[3]{x}(100 - x)$.

**57. Biology.**   According to Wiess's law, the intensity of an electric current required to excite a living tissue (threshold strength) is given by $i = \frac{A}{t} + B$, where $t$ is the duration of the current and $A$ and $B$ are positive constants. Where is the intensity increasing and where is the intensity decreasing?

**58. Biology.**   According to Poiseuille's law, the total resistance, $R$ to blood flow in a blood vessel of constant length $L$ and radius $r$ is given by

$$R = \frac{kL}{r^4}$$

where $k$ is a constant determined by the viscosity of the blood. Where is $R(r)$ increasing and where is it decreasing?

**59. Biology.**   Theoretical studies of photosynthesis assume that the gross rate $R$ of photosynthesis per unit of leaf area is given by

$$R(E) = \frac{aE}{1 + bE},$$

where $E$ is the incident radiation per unit leaf area, and $a$ and $b$ are positive constants. Show that $R$ is an increasing function of $E$.

## Graphing Calculator Exercises

The program Tangent located in C.2 of Appendix C plots tangent lines along a curve. In Exercises 60 through 65 use this program and notice that a positive slope corresponds to an increasing function while a negative slope corresponds to a decreasing function.

**60.** $y = x^2$        **61.** $y = 5 - x^2$        **62.** $y = x^2 - 10x + 21$

**63.** $y = 30 + x - x^2$        **64.** $y = x^3$        **65.** $y = -x^3$

## Solutions to Self-Help Exercise Set 13.1

**1. a.**   Taking the derivative gives

$$P'(x) = -0.2x + 1.1.$$

Notice that $P'(x)$ is continuous everywhere. Thus the only critical values are values at which $P'(x) = 0$. Setting $P'(x) = -0.2x + 1.1 = 0$ gives $x = 5.5$ as the only critical value. This divides the interval $(0, +\infty)$ into the 2 intervals $(0, 5.5)$ and $(5.5, +\infty)$. Take the test values to be, say, $x_1 = 1$ and $x_2 = 10$. Then

$$P'(1) = -0.2(1) + 1.1 = 0.9 > 0,$$

$$P'(10) = -0.2(10) + 1.1 = -0.9 < 0.$$

Thus $P(x)$ is increasing on $(0, 5.5)$ and decreasing on $(5.5, +\infty)$.

Thus if $0 < x < 5.5$, $P(x)$ is increasing. The firm would be wise to produce and sell more items since profits would increase. If $5.5 < x$, $P'(x) < 0$ and $P(x)$ is decreasing. Producing and selling more items means less profits.

**b.** Since

$$P(x) = -0.1(x^2 - 11x + 10) = -0.1(x - 10)(x - 1),$$

$P(1) = 0$ and $P(10) = 0$. This means that profits will be zero if the firm sells 1 million or 10 million VCRs in a year. (Apparently, a very substantial price reduction will be necessary to sell such a large number of VCRs.)

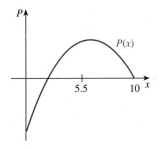

**c.** The graph is given in the figure.

**d.** Is this model realistic? It is clearly very unrealistic to think that any firm would know the explicit form of the function $P(x)$. However, experience has shown that the general shape of the graph of the profit function is realistic. Firms are well aware of the fact that flooding the market with their products will lead to *decreased* profits and even losses.

**2.** Taking the derivative gives

$$f'(x) = 12x^3 + 24x^2 = 12x^2(x + 2),$$

The critical values are $x = -2, 0$ where $f'$ is zero. Since the derivative exists everywhere, there are no other critical values.

We divide the real line into three intervals: $(-\infty, -2), (-2, 0), (0, +\infty)$.

Pick the three test values $-3, -1$, and $1$ that lie in the intervals given in the preceding step. See the figure. Notice

$$f'(-3) = -108, \qquad f'(-1) = +12, \qquad f'(1) = +36.$$

Thus $f'(x)$ is negative on the first interval, positive on the second and third intervals. Thus $f$ is decreasing on the first interval, increasing on the second and third intervals. Notice that $f(-2) = -11$ and $f(0) = 5$. The graph is given in the figure. Note that $f'(0) = 0$, but $f'$ does not change sign at this value.

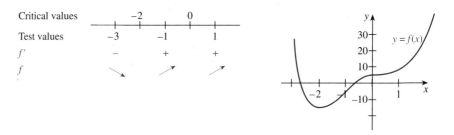

## 13.2    RELATIVE EXTREMA

▶ *Relative Extrema*

▶ *The First Derivative Test*

▶ *Management of Renewable Natural Resources*

APPLICATION
**Maximum Sustainable Yield**

Suppose $x$ is the number of thousands of pounds of a certain species of fish in a small lake and $y = f(x) = x - 0.1x^2$ is the increase in fish biomass over the next year measured in thousands of pounds. What is the maximum number of pounds of fish that can be harvested each year indefinitely? See Example 3 for the answer.

### Relative Extrema

In this section we will see how the first derivative can be used to find the maxima and minima of functions. We only consider functions $f$ that are continuous on their domains of definition and for which $f'$ is continuous everywhere except possibly at the critical values of which there are at most a finite number on any finite interval.

We shall first be interested in studying relative maxima and relative minima, that is, peaks and troughs on the graph of a continuous function. Figure 13.16 is a graph typical of functions we will be studying. The graph has peaks at $(x_1, f(x_1))$ and $(x_5, f(x_5))$ and troughs at $(x_3, f(x_3))$ and $(x_6, f(x_6))$. We call the peaks **relative maxima** and the troughs **relative minima**.

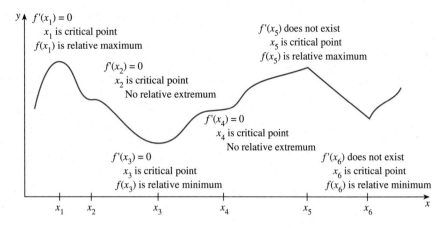

$f'(x_1) = 0$
$x_1$ is critical point
$f(x_1)$ is relative maximum

$f'(x_5)$ does not exist
$x_5$ is critical point
$f(x_5)$ is relative maximum

$f'(x_2) = 0$
$x_2$ is critical point
No relative extremum

$f'(x_4) = 0$
$x_4$ is critical point
No relative extremum

$f'(x_3) = 0$
$x_3$ is critical point
$f(x_3)$ is relative minimum

$f'(x_6)$ does not exist
$x_6$ is critical point
$f(x_6)$ is relative minimum

**Figure 13.16**

---

### Definition of Relative Maximum and Relative Minimum

We say that the quantity $f(c)$ is a relative maximum if $f(x) \leq f(c)$ for all $x$ in some open interval $(a, b)$ that contains $c$.

We say that the quantity $f(c)$ is a relative minimum if $f(x) \geq f(c)$ for all $x$ in some open interval $(a, b)$ that contains $c$.

---

**REMARK.**    Notice that for the function whose graph is given in Figure 13.16 there are values of $x$ for which $f(x)$ is *larger* than $f(x_5)$ and also values of $x$ for which $f(x)$ is *smaller* than $f(x_6)$. But $f(x_5)$ is a relative maximum since $f(x_5)$ is the maximum value of $f(x)$ for all $x$ that are near $x_5$. And $f(x_6)$ is a relative minimum since $f(x_6)$ is the minimum value of $f(x)$ for all $x$ that are near $x_6$.

For convenience we also make the following definition.

---

### Definition of Relative Extremum

We say that $f(c)$ is a **relative extremum** if $f(c)$ is a relative maximum or a relative minimum.

---

**REMARK.**    Notice that according to our definitions *an endpoint cannot be a relative extremum.*

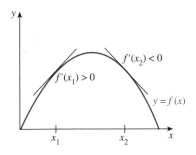

**Figure 13.17**

### The First Derivative Test

If we are hunting for a value where a function $f(x)$ attains a relative extremum on an open interval $(a, b)$, we can easily discard any value where the derivative is positive or negative. If $f'(x_1) > 0$, then the tangent line has positive slope. And as we see in Figure 13.17, the graph must be rising near this point. In the same way if $f'(x_2) < 0$, then the tangent line has negative slope, so that the graph must be falling near this point. In each of these two cases $f(x_1)$ and $f(x_2)$ could be neither a relative minimum nor a relative maximum.

Thus if $f(c)$ is a relative extremum of $f$, then $f'(c) = 0$ or $f'(c)$ does not exist. This means that $c$ is a critical value. Thus we have given the proof of the following theorem.

---

**Suspects for Relative Extrema**

If $f$ is defined on the interval $(a, b)$, $c \in (a, b)$, and $f(c)$ is a relative extremum, then $c$ is a critical value, that is, a value where $f'(c) = 0$ or $f'(c)$ does not exist.

---

In other words a relative extremum must occur at a critical value. Thus when searching for values at which the function attains relative extrema, we need only look at the critical values. Thus all critical values are "suspects." But just as not all suspects are guilty, so also not all critical values are values at which the function attains relative extrema. For example, for the function in Figure 13.16, there are 6 "suspects": $x_1$ through $x_6$. But only 4 of them are "guilty." The 2 values $x_2$ and $x_4$ are critical values, but the function does not attain relative extrema at these values. Thus one must heed the following warning.

W A R N I N G .    If $c$ is a critical value of the function $f(x)$, the function need NOT attain a relative extremum there.

Often the function $f'$ exists everywhere in the domain of definition of the function $f$. For this important case we state the following corollary.

---

**Suspects for Relative Extrema when $f'$ Exists Everywhere**

If $f'$ exists everywhere on the interval $(a, b)$, $c \in (a, b)$, and $f(c)$ is a relative extremum, then $f'(c) = 0$.

---

In other words if the derivative exists everywhere, then the suspect values for a relative extremum are values at which the derivative is zero.

Let us now return to the case where $f'$ may not exist at some values of the interval $(a, b)$ under consideration (but $f$ still satisfies the basic assumptions). As in the previous section, the critical values divide the original interval into subintervals on which the sign of $f'$ is constant. There are four possibilities for the sign chart of $f'$ as $x$ goes from less than $c$ to greater than $c$, where $c$ is a critical value. These four possible sign charts are.

$$+\,-\,,\qquad -\,+\,,\qquad +\,+\,,\qquad -\,-\,.$$

Figure 13.18 summarizes what happens in each instance.

In the first case, Figure 13.18a, $f'(x)$ is positive $(+)$ to the left of $c$ and negative $(-)$ to the right of $c$. Then $f(x)$ is increasing $(\nearrow)$ to the left of $c$ and decreasing $(\searrow)$ to the right of $c$. Thus we have the situation $\nearrow\searrow$, and $f(c)$ is then a relative maximum.

In the second case, Figure 13.18b, $f'(x)$ is negative $(-)$ to the left of the critical value $c$ and positive $(+)$ to the right of $c$. Then $f(x)$ is decreasing $(\searrow)$ to the left of $c$ and increasing $(\nearrow)$ to the right of $c$. Thus we have the situation $\searrow\nearrow$, and $f(c)$ is then a relative minimum.

It is important to notice that *whenever f changes sign at a critical value, f attains a relative extremum there.*

In the third case, Figure 13.18c, $f'$ is positive on both sides of $c$. Thus $f$ is increasing $(\nearrow)$ to the left of $c$ and also increasing $(\nearrow)$ to the right of $c$. Thus $f$ cannot attain a relative extremum at $c$.

In the fourth case, Figure 13.18d, $f'$ is negative on both sides of $c$. Thus $f$ is decreasing $(\searrow)$ to the left of $c$ and also decreasing $(\searrow)$ to the right of $c$. Thus $f$ cannot attain a relative extremum at $c$.

The last two cases indicate that *if the derivative is of the same sign on both sides of the critical value, then the function does not attain a relative extremum there.*

We summarize this discussion as the first derivative test for finding relative extrema.

---

**First Derivative Test**

Suppose $f$ is defined on $(a, b)$ and $c$ is a critical value in the interval $(a, b)$.

1.  If $f'(x) > 0$ for $x$ near and to the left of $c$ and $f'(x) < 0$ for $x$ near and to the right of $c$, then we have $\nearrow\searrow$ and $f(c)$ is a relative maximum.
2.  If $f'(x) < 0$ for $x$ near and to the left of $c$ and $f'(x) > 0$ for $x$ near and to the right of $c$, then we have $\searrow\nearrow$ and $f(c)$ is a relative minimum.
3.  If the sign of $f'(x)$ is the same on both sides of $c$, then $f(c)$ is not a relative extremum.

---

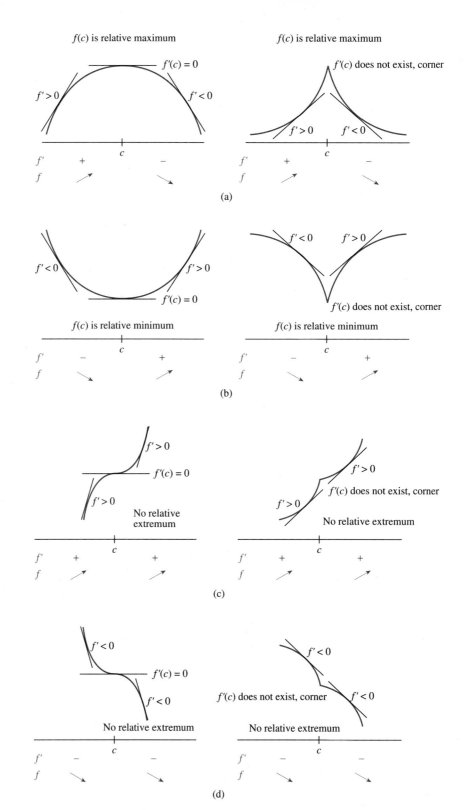

**Figure 13.18**

E X A M P L E   1      **Finding Relative Extrema**

Find the relative extrema of $f(x) = 3x^5 - 20x^3 + 1$ and graph the function.

**Solution**      **Step 1:**   Find the critical values.

(a)   The derivative $f'(x) = 15x^4 - 60x^2$ exists and is continuous everywhere. Thus the only critical values are values at which $f'(x) = 0$.

(b)   Setting $f'(x) = 15x^4 - 60x^2 = 15x^2(x^2 - 4) = 0$ yields $x = 0$ and
$x = \pm 2$. Thus the critical values are at $-2$, $0$, and $+2$.

**Step 2:**   These three critical values divide the real line into the four subintervals
$(-\infty, -2)$, $(-2, 0)$, $(0, 2)$, and $(2, +\infty)$.

**Step 3:**   Use the test values $-3$, $-1$, $1$, and $3$. See Figure 13.19. Then notice
that

$$f'(-3) > 0, \qquad f'(-1) < 0, \qquad f'(1) < 0, \qquad f'(3) > 0.$$

This leads to the sign chart in Figure 13.19.

**Step 4:**   (a)   Since $f'(x) > 0$ to the left of $c = -2$ and $f'(x) < 0$ to the right
of $-2$, $f(x)$ is increasing ($\nearrow$) on $(-\infty, -2)$ and decreasing ($\searrow$) on
$(-2, 0)$. Thus we have $\nearrow\searrow$, and $f$ has a relative maximum at $x = -2$.

(b)   Since $f'(x) < 0$ on either side of $c = 0$, $f(x)$ is decreasing ($\searrow$) on
$(-2, 0)$ and $(0, 2)$ and thus $f$ is actually decreasing on the entire interval
$(-2, 2)$ and has no relative extremum at $c = 0$.

(c)   Since $f'(x) < 0$ to the left of $c = 2$ and $f'(x) > 0$ to the right of $2$, $f$
is decreasing ($\searrow$) on $(0, 2)$ and increasing ($\nearrow$) on $(2, +\infty)$. Thus we
have $\searrow\nearrow$, and $f$ has a relative minimum at $x = 2$.

Notice $f(-2) = 65$, $f(0) = 1$, and $f(2) = -63$. The graph is then given
in Figure 13.20.   ■

**R E M A R K .**      Sometimes calculating the *numerical* value of the derivative at
the test values can be tedious. But observe that we are only interested in the *sign*
of the derivative. Calculating the *sign* can often be less tedious if one has $f'$
factored. Thus, in the previous example, one factor of $f'$ was $15x^2$. But this factor

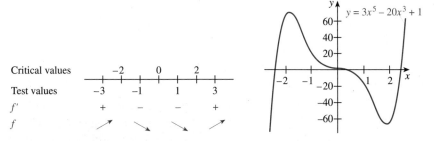

**Figure 13.19**                                              **Figure 13.20**

is clearly positive at all of the test values, and thus the sign of $f'$ at the test values is determined by the sign of the remaining factors $(x + 2)(x - 2)$. Thus one need only evaluate $(x + 2)(x - 2)$ at the test values. In more complicated situations this can be very useful.

**E X A M P L E  2**    **Finding the Relative Extrema**

Find the relative extrema of $f(x) = 6\sqrt[3]{x^5} + 15\sqrt[3]{x^2} = 6x^{5/3} + 15x^{2/3}$.

**Solution**    **Step 1:**    Find the critical values.

(a)    The derivative $f'(x) = 10x^{2/3} + 10x^{-1/3}$ exists and is continuous everywhere except at $x = 0$, where $f'$ does not exist but $f$ does. Thus $x = 0$ is a critical value.

(b)    Now setting $f'(x) = 10x^{2/3} + 10x^{-1/3} = 0$ yields

$$10x^{2/3} + 10x^{-1/3} = 0$$

$$x^{2/3} = -\frac{1}{x^{1/3}}$$

$$x^{2/3}(x^{1/3}) = -\frac{1}{x^{1/3}}(x^{1/3})$$

$$x = -1$$

The critical values are then $x = -1$ and $x = 0$.

**Step 2:**    These two values divide the real line into the three intervals $(-\infty, -1)$, $(-1, 0)$, and $(0, +\infty)$, on each of which $f'$ is continuous and not zero.

**Step 3:**    Use test values $-8$, $-\frac{1}{8}$, and 1. See Figure 13.21. Then

$$f'(-8) = +35, \qquad f'\left(-\frac{1}{8}\right) = -\frac{35}{2}, \qquad f'(1) = +20.$$

This leads to the rest of the sign chart given in Figure 13.21.

**Figure 13.21**

**Step 4:**    The first derivative test indicates that $f(-1)$ is a relative maximum and $f(0)$ a relative minimum. Note also that $f$ is increasing on $(-\infty, -1)$ and $(0, +\infty)$ and that $f$ is decreasing on $(-1, 0)$. Noting further that $f(-1) = 9$ and $f(0) = 0$, one can put all this together to give the graph shown in Figure 13.22.    ∎

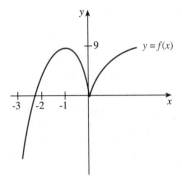

**Figure 13.22**

## Management of Renewable Natural Resources

As an important application we now consider a mathematical model of a fishery. We can also apply the ideas presented here to other renewable natural resources. To avoid confusion we emphasize that we are interested in the maximization of productive resources rather than the preservation of natural environments.

A fishery can consist of all or part of an ocean, a lake, a river, a natural or artificial pond, or even an area of water that has been fenced in. To fix a particular situation in our minds, we assume that a corporation has the sole rights to the fish in a fairly small lake. Of course the corporation wishes to harvest the maximum number of fish with the minimum effort and cost. We assume that the corporation does no active management of the fish other than simply harvesting.

Let $y = f(x)$, where $x$ represents the total pounds (**biomass**) of fish at the beginning of a year and $y$ represents the total increase in the biomass of fish over the next year. Thus the term $x + y$ represents the total biomass one year later.

For convenience we always harvest at the end of the year. We are interested in the *increase* in population since it is this increase that we wish to harvest. We would like to predict how much of the resource we can harvest and still allow the resource to replenish itself. The largest possible increase in biomass is called the **maximum sustainable yield**. We want to know what the biomass of fish should be to obtain the maximum sustainable yield.

We know that if the biomass is zero, then the increase in the biomass will also be zero. Mathematically, this says that $f(0) = 0$. Some fish in the lake will result in an increase in the population and biomass, and even more fish will result in even more of an increase in the biomass. Thus $f$ should be increasing, at least initially.

Clearly our small lake has only a finite amount of resources and can only hold so much biomass. This number is referred to as the **carrying capacity**, which we label $K$. Since $K$ is the saturation level of the species, there can be *no* increase in biomass beyond this level. Thus $f(K) = 0$. If the biomass, $x$, is a little less than $K$, then we do expect some increase in biomass. Thus moving from right to

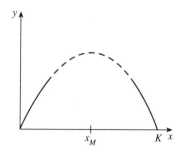

**Figure 13.23**

left starting at $x = K$, $y$ should increase. Thus moving from left to right and just to the left of $K$, the function $f$ should be decreasing.

At this point in the discussion the function looks like the solid curve in Figure 13.23. Biologists now normally make the assumption that the actual curve at the intermediate values is described by the dotted curve in Figure 13.23.

Thus it is to be expected that the increase in the biomass, $y = f(x)$, will become larger and larger (increase) until some value is reached, say $x = x_M$, after which the increase in the biomass begins to get smaller and smaller (decrease). The number $M = f(x_M)$ is then the maximum sustainable yield. For the corporation interested in the long haul, the harvesting should be managed so that $x = x_M$. Thus, given these assumptions, if the biomass is at $x = x_M$ at the beginning of the year, the corporation can harvest $f(x_M)$ pounds of fish at the end of the year and see the population renew itself over the subsequent year so that the biomass returns once again a year later to $x_M$. Furthermore, this can go on indefinitely.

**E X A M P L E   3**    **Finding the Maximum Sustainable Yield**

Suppose $x$ is the number of thousands of pounds of a certain species of fish in a small lake described above and $y = f(x) = x - 0.1x^2$ is the increase in fish biomass over the next year measured in thousands of pounds. What is the carrying capacity and the maximum sustainable yield? What should management do if the fish biomass is at 1000 pounds? At 9000 pounds?

**Solution**    **Step 1:**  Find the critical values.

(a)  The derivative

$$f'(x) = 1 - 0.2x = 0.2(5 - x)$$

exists and is continuous everywhere. Thus the only critical values are values for which $f'(x) = 0$.

(b)  From the factorization of $f'$ it is clear that $f'(x) = 0$ implies that $x = 5$. Thus $x = 5$ is the only critical value.

**Step 2:**  This single critical value then divides the nonnegative numbers into two subintervals: $[0, 5)$ and $(5, +\infty)$.

**Step 3:**  It is clear from the factorization of $f'(x)$ that if $0 < x < 5$, $f'(x) > 0$ and if $5 < x$, $f'(x) < 0$.

**Step 4:**  Thus $f$ is increasing on $[0, 5)$ and decreasing thereafter and therefore $f$ attains a relative maximum at $x = 5$. In fact, since $f$ always increases to the left of $x = 5$ and always decreases to the right of $x = 5$, it is clear that the function $f$ attains its largest value at $x = 5$.

Furthermore, since

$$f(x) = x - 0.1x^2$$
$$= x(1 - 0.1x),$$

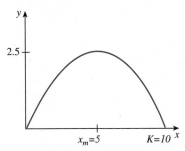

**Figure 13.24**

$f(0) = 0$ and $f(10) = 0$. Thus the carrying capacity is $K = 10$, or 10,000 pounds of fish. The maximum sustainable yield is $f(5) = 2.5$ or 2500 pounds. See Figure 13.24, where $x_M = 5$ corresponds to 5000 pounds.

If the fish biomass is 1000 pounds, then $x = 1$ and $y = f(1) = 0.9$ or 900 pounds. Since $x + y = 1.9$, there will be 1900 pounds of fish at the end of the year. Since this is well below 5000 pounds, no harvesting should be done. If the biomass is 9000 pounds, management can immediately harvest 4000 pounds of fish, resulting in a biomass right at 5000 pounds. ∎

Sometimes, an argument is made that the amount of fish (or some other animal species) should be kept at the carrying capacity. One might speculate on the condition of the fish in the fishery if the biomass were at the carrying capacity. Since the resources are such that no additional pound of fish can survive an additional year, the fish must be obtaining the absolute minimal amount of food for survival. This clearly does not represent pleasant living conditions for the fish.

**Fish Catches by Region, 1988**

| Region | Total Catch (in metric tons, live weight catches) |
|---|---|
| Africa | 5,310,300 |
| Asia | 43,601,200 |
| Europe | 12,877,400 |
| North America | 9,567,900 |
| Oceania | 887,300 |
| South America | 14,412,200 |
| (Former) U.S.S.R. | 11,332,100 |
| **World Total** | **97,985,400** |

**The Pacific Halibut Fishery**

Halibut is a highly regarded table fish, and the halibut fishery is second only to the Pacific salmon fishery. Halibut was heavily fished until the population dropped dramatically below its maximum sustainable yield. Under the supervision of the Pacific Halibut Commission, founded in 1924, the overfished halibut stocks were steadily built to near maximal sustainable yields by the 1950s. However, recent incursions by Japanese and Russians have seriously depleted the halibut stocks again.

**The Coastal Shark Fishery**

Rising demand for shark fins in Asia and for shark meat in the United States has caused a sharp rise in the killing of sharks over the last decade, with the result that large coastal sharks are being killed in numbers that outstrip their capacity to reproduce. As a result the National Marine Fisheries Service of the Federal Government has imposed controls on shark fishing within 200 miles of the Atlantic and Gulf coasts by issuing regulations that will place a total quota of 2436 metric tons, dressed weight, on the catch of large coastal sharks by commercial fishermen in 1993 and 2570 metric tons in 1994. Twenty-two shark species are in this management category, including the tiger, lemon, hammerhead, bull, and great white. The total catch of large coastal sharks in the United States waters has substantially exceeded 4000 metric tons in each of the last five years, according to the fisheries service. The service has calculated that the maximum sustainable yield of these sharks is 3800 metric tons a year, and that the 1993 and 1994 quotas will enable that level to be achieved in 1996.

## SELF-HELP EXERCISE SET 13.2

1. Suppose a ball-point pen manufacturer believes that the demand function is given by $p = -0.5x + 10$, where $p$ is the price of each pen in dollars and $x$ is the number sold in millions. Find the number to be sold if revenue is to be maximized. Draw a graph of $R = R(x)$.

2. Suppose the pen manufacturer in the previous exercise has a cost function given by $C(x) = 8 + 5x$, where $C$ is given in millions of dollars. Find the number to be sold if profit is to be maximized, and find the maximum profit. Draw a graph of $P = P(x)$. Compare the value at which profit is maximum with the value where revenue is maximum.

## EXERCISE SET 13.2

In Exercises 1 through 8 find all critical values and values where the function attains a relative maximum or minimum.

**1.**

**2.**

**3.**

**4.**

**5.**

**6.**

**7.**

**8.**
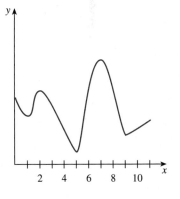

In Exercises 9 through 20 $f'(x)$ is given in factored form and $p(x)$ is a function that is always positive. Find all critical values of $f$, find the largest open intervals on which $f$ is increasing, find the largest open intervals on which $f$ is decreasing, find all relative maxima and all relative minima. Using the additional information given for each function, draw a rough graph of $f$.

**9.**  $f'(x) = x - 3$, $f(3) = 1$          **10.**  $f'(x) = 4 - x$, $f(4) = 1$

**11.**  $f'(x) = p(x)(x - 2)(x - 4)$, $f(2) = 2$, $f(4) = -2$

**12.**  $f'(x) = -p(x)(x + 2)(x - 1)$, $f(-2) = -15$, $f(1) = 12$

13. $f'(x) = p(x)(x - 1)(x - 3)(x - 5)$, $f(1) = 1$, $f(3) = 4$, $f(5) = 0$

14. $f'(x) = p(x)(x + 1)(x - 2)(x - 5)$, $f(-1) = 3$, $f(2) = 10$, $f(5) = 4$

15. $f'(x) = p(x)(x - 1)(x - 3)(x - 5)^2$, $f(1) = 1$, $f(3) = -2$, $f(5) = 5$

16. $f'(x) = p(x)(x + 2)^2(x - 5)^3(x - 10)$, $f(-2) = -2$, $f(5) = 3$, $f(10) = 0$

17. $f'(x) = p(x)(x + 1)^2(x - 1)^3(x - 2)$, $f(-1) = -2$, $f(1) = 3$, $f(2) = 0$

18. $f'(x) = p(x)(x + 1)^2(x - 1)^4(x - 2)$, $f(-1) = 5$, $f(1) = 3$, $f(2) = 0$

19. $f'(x) = p(x)(x + 1)(x - 1)^3(x - 2)^5$, $f(-1) = 1$, $f(1) = 3$, $f(2) = 0$

20. $f'(x) = p(x)(x + 1)^2(x - 1)^4(x - 2)^4$, $f(-1) = -2$, $f(1) = 3$, $f(2) = 4$

In Exercises 21 through 42 find all critical values, find the largest open intervals on which $f$ is increasing, find the largest open intervals on which $f$ is decreasing, find all relative maxima and all relative minima. Draw a rough graph of $f$.

21. $f(x) = x^2 - 4x + 1$       22. $f(x) = -2x^2 + 8x - 1$

23. $f(x) = x^3 + 3x + 1$       24. $f(x) = 1 - 5x - x^5$

25. $f(x) = x^4 + 2x^2 + 1$       26. $f(x) = -4x^6 - 6x^4 + 5$

27. $f(x) = 1 - (1 - x)^{2/3}$       28. $f(x) = (x + 3)^{2/3}$

29. $f(x) = x^3 - 3x + 1$       30. $f(x) = x^3 + 3x^2 - 9x + 3$

31. $f(x) = 8x^3 - 24x^2 + 18x + 6$       32. $f(x) = -8x^3 + 6x - 1$

33. $f(x) = x^4 - 8x^2 + 3$       34. $f(x) = 2x^6 - 3x^4 + 1$

35. $f(x) = -x^5 + 5x - 1$       36. $f(x) = 5 + 20x - x^5$

37. $f(x) = 3x^5 - 20x^3 + 5$       38. $f(x) = 3x^5 - 5x^3 + 1$

39. $f(x) = 2\sqrt[3]{x^5} - 5\sqrt[3]{x^2}$       40. $f(x) = 6\sqrt[3]{x^2} - 4x$

41. $f(x) = 5\sqrt[5]{x^4} - 4x$       42. $f(x) = 7\sqrt[7]{x^6} - 6x$

## Applications

43. **Revenue.**   If the revenue function for a firm is given by $R(x) = 10x - 0.01x^2$, find the value of $x$ for which the revenue is maximum.

44. **Revenue.**   If the demand equation for a firm is given by $p = -0.2x + 16$, find the value where the revenue is maximum.

45. **Profit.**   If the cost function for a firm is given by $C(x) = 5 + 8x$ and the demand equation is the same as in the previous problem, find the value where the profit is maximum.

46. **Revenue.**   If the revenue function for a firm is given by $R(x) = -x^3 + 36x$, find the value where revenue is maximized.

47. **Profit.**   If the cost function for a firm is given by $C(x) = 9x + 30$ and the revenue function is the same as in the previous problem, find the value where the profit is maximized.

48. **Average Revenue.** The total revenue function for producing $x$ items is given by $R(x) = 100x^2 - 20x^3$. How many items should be produced to maximize $\overline{R}(x) = R(x)/x$, the average revenue per item?

49. **Average Revenue.** In the previous problem show that average revenue is maximized at the value at which $\overline{R} = R'$.

50. **Average Revenue.** Suppose the total revenue function for producing $x$ items is given by $R(x)$ and the average revenue per item is given by $\overline{R}(x) = R(x)/x$. If the average revenue is at a maximum at some value on $(0, +\infty)$; then show that at this value $\overline{R} = R'$.

51. **Fishery.** If in Example 3 of the text, $f(x) = 16x^3 - x^4$, find the carrying capacity and the maximum sustainable yield.

52. **Fishery.** If in Example 3 of the text, $f(x) = 20\sqrt[3]{x} - \sqrt[3]{x^4}$, find the carrying capacity and the maximum sustainable yield.

53. **Auto-catalytic Reaction.** The rate $V$ of a certain auto-catalytic reaction is proportional to the amount $x$ of the product and also proportional to the remaining substance being catalyzed. Thus if $I$ is the initial amount of the substance, then $V = kx(I - x)$, where $k$ is a positive constant. At what concentration $x$ will the rate be maximal?

54. **Medicine.** It is well known that when a person coughs, the diameter of the trachea and bronchi shrinks. It has been shown that the flow of air $F$ through the windpipe is given by

$$F(r) = k(R - r)r^4,$$

where $r$ is the radius of the windpipe under the pressure of the air being released from the lungs, $R$ is the radius with no pressure, and $k$ is a positive constant. Find the radius at which the flow will be maximum.

55. **Biology.** Biologists have proposed that the rate of production $P$ of photosynthesis is related to the light intensity $I$ by the formula

$$P(I) = \frac{aI}{b + I^2},$$

where $a$ and $b$ are positive constants. Find the light intensity that produces the maximum rate of production of photosynthesis.

## Graphing Calculator Exercises

In Exercises 56 through 61 use the program Tangent found in the Appendix C.2 to graph tangent lines on either side of the critical values of the given functions in order to demonstrate the cases shown in Figure 13.18.

56. $y = x^2$

57. $y = 8 - x^2$

58. $y = x^3$

59. $y = -x^3$

60. $y = x^{2/3}$

61. $y = 4 - x^{2/3}$

## Solutions to Self-Help Exercise Set 13.2

1. **Step 1:**  Since $R(x) = xp(x)$,

$$R(x) = x(-0.5x + 10) = -0.5x^2 + 10x,$$

and
$$R'(x) = -x + 10,$$

and $x = 10$ is the only critical value.

**Step 2:**  This single critical value divides the nonnegative numbers into 2 sub-intervals: $[0, 10)$ and $(10, +\infty)$.

**Step 3:**  If $x < 10$, $R'(x) > 0$, while if $x > 10$, $R'(x) < 0$.

**Step 4:**  Thus by the first derivative test, $R$ reaches a relative maximum at $x = 10$, or 10 million pens. Since the function is always increasing before $x = 10$ and always decreasing after $x = 10$, it is clear that revenues will be maximum at $x = 10$.

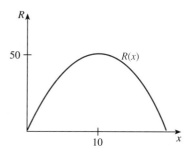

2. **Step 1:**  Since $C'(x) = 5$ and $P = R - C$,

$$P'(x) = R'(x) - C'(x) = (-x + 10) - (5) = -x + 5,$$

and the only critical value is at $x = 5$.

**Step 2:**  This single critical value then divides the nonnegative numbers into two subintervals: $[0, 5)$ and $(5, +\infty)$.

**Step 3:**  If $x < 5$, then $P'(x) > 0$, while if $x > 5$, then $P'(x) < 0$.

**Step 4:**  Thus by the first derivative test, profits attain a relative maximum when $x = 5$, or when 5 million pens are sold.

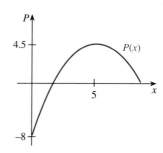

The graph indicates that maximum profits are attained when $x = 5$.
Maximum profits are then

$$P(5) = R(5) - C(5) = -0.5(5)^2 + 10(5) - 8 - 5(5) = 4.5,$$

or 4.5 million dollars.

Notice from the figure that profit is maximized at a different value than revenue is.

## 13.3   THE SECOND DERIVATIVE

▶ *Concavity*

▶ *The S-curve*

▶ *Acceleration*

▶ *The Second Derivative Test*

**A P P L I C A T I O N**
**A Population of Mice**

A population of mice on a small, isolated island follows the population law given by

$$p(t) = 5 \, \frac{3 + 5.01t^2}{3 + .01t^2},$$

where $t$ represents the number of years starting from now. It can be readily shown that $p(t)$ is increasing. Where is the rate of increase of the population itself increasing and where is this rate decreasing? Interpret your results. See Example 4 for the answer.

### Concavity

In this section we will see that the second derivative measures the rate at which the first derivative is increasing or decreasing. This can then be used in graphing and also used to give another test for relative extrema.

As in the previous two sections we assume that any function $f$ under consideration is *continuous* on its domain of definition and that $f'$ is continuous everywhere except possibly at the critical values of which there are at most a finite number on any finite interval.

Given a function $y = f(x)$, the second derivative, denoted by $f''(x)$, is defined to be the derivative of the first derivative.

---

**Definition of Second Derivative**

Given a function $y = f(x)$, the second derivative, denoted by $f''(x)$, is defined to be the derivative of the first derivative. Thus

$$f''(x) = \frac{d}{dx} (f'(x)).$$

---

**R E M A R K .**     The second derivative is also denoted by $\dfrac{d^2}{dx^2}\, f(x)$ or $\dfrac{d^2y}{dx^2}$.

For example, given $y = f(x) = x^4$,

$$\frac{d^2y}{dx^2} = f''(x) = \frac{d}{dx}\,[f'(x)] = \frac{d}{dx}\,(4x^3) = 12x^2.$$

**R E M A R K .**     Notice that the second derivative is a function, just like the first derivative is a function.

In Figures 13.25a and 13.25b, we see two functions that have one characteristic in common. They both have derivatives that are *increasing*. Notice that

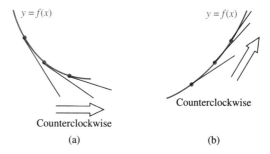

**Figure 13.25**

this means that the slopes are rotating *counterclockwise* as $x$ increases. When this happens we say that the curve is **concave up** (denoted by $\smile$). Also notice that in such a case the curve always lies *above* its tangent.

We shall now see that a curve will be concave up if $f''(x) > 0$. Recall that if the derivative of a function is positive, then the function must be increasing. But the second derivative is the derivative of the first derivative. Thus if

$$f''(x) = \frac{d}{dx}\,(f'(x)) > 0,$$

then the function $f'(x)$ is increasing. Therefore the curve is concave up $\smile$.

Similarly in Figure 13.26a and b, we see two functions that have both derivatives that are *decreasing*, that is the slopes are rotating *clockwise* as $x$ increases. When this happens we say that the curve is **concave down** (denoted by $\frown$). In such a case the curve always lies *below* its tangent.

**R E M A R K .**     The reader should carefully note that concavity has nothing to do with the function itself increasing or decreasing. In Figures 13.25a and 13.25b, the graphs of both functions are concave up, but the first function is decreasing while the second function is increasing. In Figure 13.26a and b, the graphs of both functions are concave down, but the first function is decreasing while the second function is increasing.

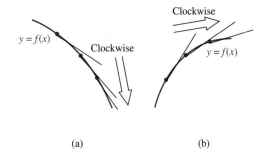

(a)                                                (b)

**Figure 13.26**

We summarize all of this in the following definition and theorem where we assume that $f'(x)$ exists on $(a, b)$.

---

**Definition of Concave Up and Down**

1. We say that the graph of $f$ is concave up ($\smile$) on $(a, b)$ of $f'(x)$ is increasing on $(a, b)$.
2. We say that the graph of $f$ is concave down ($\frown$) on $(a, b)$ if $f'(x)$ is decreasing on $(a, b)$.

---

**Test for Concavity**

1. If $f''(x) > 0$ on $(a, b)$, then the graph of $f$ is concave up ($\smile$) on $(a, b)$.
2. If $f''(x) < 0$ on $(a, b)$, then the graph of $f$ is concave down ($\frown$) on $(a, b)$.

---

A point $(c, f(c))$ on the graph of $f$ where the concavity changes is called an **inflection point** and $c$ an **inflection value**. More precisely we have the following definition.

---

**Definition of Inflection Point**

A point $(c, f(c))$ on the graph of $f$ is an **inflection point** and $c$ an **inflection value** if $f(c)$ is defined and the concavity of the graph of $f$ changes at $(c, f(c))$.

---

Figure 13.27 shows four inflection points for four functions. Notice that in each case the graph of the function must cross the tangent line.

In order to find where the graph of $f$ is concave up and where the graph of $f$ is concave down, we need to find the values where $f''(x)$ changes sign. This will occur at values $c$ where $f''(c) = 0$ or where $f''(c)$ does not exist.

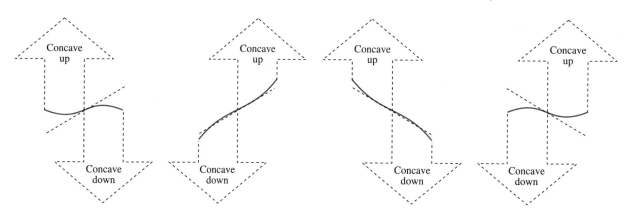

**Figure 13.27**

**E X A M P L E  1**   **Determining Concavity**

Discuss the concavity of $f$ if $f(x) = x^2$.

**Solution**   Since $f''(x) = 2$, the second derivative is always positive and therefore the graph of $f$ is concave up for all values of $x$. There are no inflection points. See Figure 13.28.  ■

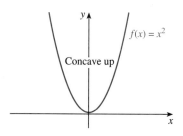

**Figure 13.28**

**E X A M P L E  2**   **Determining Concavity**

Discuss the concavity of $f$ if $f(x) = x^3$.

**Solution**   Since $f''(x) = 6x$, $f''(x) < 0$ if $x < 0$, and $f''(x) > 0$ if $x > 0$. Thus the graph of $f$ is concave down for all values of $x$ on $(-\infty, 0)$ and concave up for all values of $x$ on $(0, +\infty)$, and $(0, f(0)) = (0, 0)$ is an inflection point. See Figure 13.29.  ■

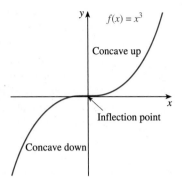

**Figure 13.29**

E X A M P L E  3

**Determining Concavity**

Discuss the concavity of $f$ if $f(x) = 9x^{1/3}$.

**Solution**

Since $f'(x) = 3x^{-2/3}$, $f''(x) = -2x^{-5/3}$. Thus $f''(x) > 0$ if $x < 0$, and $f''(x) < 0$ if $x > 0$. So the graph of $f$ is concave up for all values of $x$ on $(-\infty, 0)$ and concave down for all values of $x$ on $(0, +\infty)$. Also, $(0, f(0)) = (0, 0)$ is a point of inflection. If we notice that $f'(x)$ is always positive (except at $x = 0$), we can obtain the graph shown in Figure 13.30. ∎

## The S-Curve

In 1837 the Dutch mathematician–biologist Verhulst developed a theory of the growth of populations that predicts the manner of growth of populations that are subject to limited available resources. If $p(t)$ represents the population at time $t$, then according to this theory, the graph of the population will look similar to the graph in Figure 13.31, which looks like an "S."

**Figure 13.30**

**Figure 13.31**

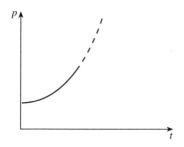

**Figure 13.32**

Biologists have data on many species that indicate that this model does have considerable predictive ability. Furthermore, economists in the early 1960s used a similar model to predict the spread of technological innovation throughout an industry. In this case $p$ stands for the percent of the products that use the innovation in their manufacture. This will be taken up in the exercise set.

If we examine the S-curve in Figure 13.31 and think of $p(t)$ as population, then we notice that there is an initial period when the curve is concave up. This means that the rate at which the population is increasing is itself increasing. Geometrically, the slope of the tangent line is increasing. This is a period of time when the population is often characterized as "exploding." If one were to do simple extrapolation, one might think that the curve would continue as the dotted curve in Figure 13.32.

In fact the concavity of the curve changes at some time and from that time on the curve is concave down. This means that the slope of the tangent line, and thus the rate of increase of $p(t)$, is decreasing. What typically happens is that some population, say of mice, initially "explodes" but then the number of mice becomes so large that the growth then becomes inhibited by the finite amount of food and natural resources available to the mice. At this point the growth in the population steadily decreases. As we shall see in the next section the growth eventually almost comes to a halt.

**E X A M P L E   4**    **Determining the Concavity of a Population Curve**

A population of mice on a small, isolated island follows the population law given by

$$p(t) = 5\,\frac{3 + 5.01t^2}{3 + .01t^2},$$

where $t$ represents the number of years starting from now. Find where $p$ is increasing and find where the graph of $p$ is concave up and where the graph is concave down. Interpret your results.

**Solution**    Using the quotient rule for differentiation

$$p'(t) = 5\,\frac{(3 + .01t^2)(10.02t) - (3 + 5.01t^2)(.02t)}{(3 + .01t^2)^2}$$

$$= 150\,\frac{t}{(3 + .01t^2)^2}.$$

Since this is positive for positive time, the population, $p$, is always increasing.

The reader can verify that

$$p''(t) = 450\,\frac{1 - .01t^2}{(3 + .01t^2)^3}.$$

Then $p''(t) > 0$ for $0 \le t < 10$, $p''(10) = 0$, and $p''(t) < 0$ for $t > 10$. Thus the graph of $p$ is concave up for all values of $t$ on $(0, 10)$, concave down for all values of $t$ on $(10, +\infty)$, and $t = 10$ is an inflection value.

**Figure 13.33**                                                        **Figure 13.34**

Since $p(0) = 5$, the initial population is 5. Since $p(10) = 630$, the population "explodes" to 630 mice in 10 years, after which the population begins to level off. See Figure 13.33   ∎

Biologists have observed that the actual long-term behavior is not what is shown in Figure 13.31. When the mouse population moves up to some large value corresponding to some high point on the S-curve, the general health of the population becomes rather poor. This is due to the lack of resources available per individual. As a consequence, plagues strike the population, spreading rapidly and depleting the population quickly. The population then resumes, from a lower level, its climb of the S-curve, as indicated in Figure 13.34.

## Acceleration

If $s(t)$ is the distance traveled in time $t$, then the velocity, $v(t)$, is given by $v(t) = s'(t)$. The **acceleration**, denoted by $a(t)$, is defined as the rate of change of velocity. Thus $a(t) = v'(t) = s''(t)$.

---

**Definition of Acceleration**

Acceleration, denoted by $a(t)$, is defined to be the derivative of the velocity, that is,

$$a(t) = v'(t) = s''(t).$$

---

There are physical situations when one can actually "feel" the second derivative. When a person steps heavily on the accelerator of a car, the velocity of the car increases rapidly and the driver is pressed back against the seat. In this situation the rate of change of velocity is large. That is, $a(t) = s''(t)$ is large. If the driver now slams on the brakes, the velocity decreases rapidly. In this case the acceleration $a(t) = s''(t)$ is negative and the driver lurches forward in the seat.

### The Second Derivative Test

We shall now see that the second derivative can sometimes be used to find relative extrema. Suppose $(a, b)$ is in the domain of definition of $f$ and $c \in (a, b)$. If the second derivative exists and is not zero at a point, $c$, where $f'(c) = 0$, then we shall see that the sign of $f''(c)$ will determine if the value $f(c)$ is a relative minimum or a relative maximum.

Consider, for example, the situation shown in Figure 13.35 in which $f'(c) = 0$ and $f''(c) > 0$. We will now show that in this case $f(c)$ is a relative minimum. To see this, notice that since $f''(c) > 0$, so $f'(x)$ is increasing for $x$ near $c$ and so $f'(x)$ must be smaller than $f'(c) = 0$ for $x$ just to the left of $c$ and $f'(x)$ must be larger than $f'(c) = 0$ for $x$ just to the right of $c$. Thus $f'(x) < 0$ for $x$ to the left of $c$ and $f'(x) > 0$ for $x$ to the right of $c$. The sign chart of $f'$ is then $-0+$ and we therefore conclude from the first derivative test that $f(c)$ is a relative minimum.

For a second example, consider the situation shown in Figure 13.36 in which we also have $f'(c) = 0$ but this time $f''(c) < 0$. Then, using a similar argument, we see that the sign chart of $f'$ must be $+0-$, and $f(c)$ must be a relative maximum.

We summarize this as *the second derivative test*.

---

**Second Derivative Test**

Suppose $f$ is defined on $(a, b)$, $f'(c) = 0$, and $c \in (a, b)$.

1.  If $f''(c) > 0$, then $f(c)$ is a relative minimum.
2.  If $f''(c) < 0$, then $f(c)$ is a relative maximum.

---

Figure 13.35

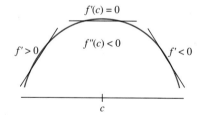

Figure 13.36

**E X A M P L E   5**   **Using the Second Derivative Test**

Use the second derivative test to find the relative extrema of $f$ if $f(x) = x^2$.

**Solution**   Since $f'(x) = 2x$, $f'(x) = 0$ at $x = 0$. Furthermore, since $f''(x) = 2$, $f''(0) = 2 > 0$, and thus by the second derivative test, $f(0)$ is a relative minimum. See Figure 13.37. ∎

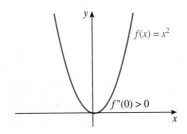

$f(x) = x^2$

$f''(0) > 0$

**Figure 13.37**

E X A M P L E  6   **Using the Second Derivative Test**

Consider again the function $f(x) = x^3 - 3x^2 + 1$ considered in Example 3 of the Section 13.1. Find where the graph of $f$ is concave up and where it is concave down. Find any inflection points. Use the second derivative test to find the relative extrema. Graph.

**Solution**   We have

$$f'(x) = 3x^2 - 6x = 3x(x - 2).$$

Thus $f'$ is continuous everywhere and is zero at $x = 0$ and $x = 2$. Also

$$f''(x) = 6x - 6 = 6(x - 1).$$

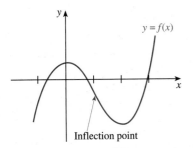

$y = f(x)$

Inflection point

**Figure 13.38**

Since $f''(x) < 0$ on $(-\infty, 1)$ and $f''(x) > 0$ on $(1, +\infty)$, the graph of $f$ is concave down for all values of $x$ on $(-\infty, 1)$ and concave up for all values of $x$ on $(1, +\infty)$. Since $f(1)$ is defined and the second derivative changes sign at $x = 1$, $(1, f(1)) = (1, -1)$ is an inflection point.

Since $f''(0) = -6 < 0$, the second derivative test indicates that $f(0)$ is a relative maximum. Since $f''(2) = 6 > 0$, the second derivative test indicates that $f(2)$ is a relative minimum.

Recalling that $f$ is increasing on $(-\infty, 0)$ and $(2, +\infty)$ and decreasing on $(0, 2)$, we can sketch a graph as shown in Figure 13.38. ∎

R E M A R K .   The second derivative test can be useful if the second derivative can be easily calculated. A serious problem with the second derivative test is that it can fail to work. For example, if the second derivative is zero or does not exist at a point where the first derivative is zero, then the second derivative fails to supply any information.

E X A M P L E  7   **Failure of the Second Derivative Test**

Try to use the second derivative test on the following functions. (a) $x^4$, (b) $-x^4$, (c) $x^3$, (d) $9x^{4/3}$.

**Solutions**   (a)   Here $f'(x) = 4x^3$ and $f''(x) = 12x^2$. Thus the only critical value is at $x = 0$ and $f''(0) = 0$. Thus the second derivative test fails, but from

**Figure 13.39**

**Figure 13.40**

**Figure 13.41**

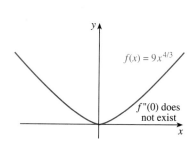

**Figure 13.42**

the *first* derivative test it is easily seen that $f(0)$ is a relative minimum. See Figure 13.39.

(b) Here $f'(x) = -4x^3$ and $f''(x) = -12x^2$. Thus the only critical value is at $x = 0$ and $f''(0) = 0$. Thus the second derivative test fails, but from the *first* derivative test it is easily seen that $f(0)$ is a relative maximum. See Figure 13.40.

(c) Here $f'(x) = 3x^2$ and $f''(x) = 6x$. Thus the only critical value is $x = 0$ at which $f''(0) = 0$. Thus the second derivative test fails, but using the first derivative one can see that $f(0)$ is not a relative minimum or maximum. See Figure 13.41.

(d) Here $f'(x) = 12x^{1/3}$ and $f''(x) = 4x^{-2/3}$. Thus the only critical value is $x = 0$, since $f'(0) = 0$ and $f(0)$ exists. But $f''(0)$ does not exist, and so the second derivative test cannot be applied. One can see by the first derivative test that $f(0)$ is a relative minimum. See Figure 13.42.  ∎

## SELF-HELP EXERCISE SET 13.3

1. Suppose a firm believes that the percent, $p(t)$, of the market of its product that it can attain by spending a certain amount on advertising each year for the next six years is given by $p(t) = -\frac{1}{3}t^3 + 3t^2 + 14$ where $t$ is given in years. Find where

the graph of the function is concave up and where it is concave down. Discuss the graph of the curve and interpret.

**2.** Suppose the second derivative of some function $y = f(x)$ is given as

$$f''(x) = (x - 1)(x - 2)^2(x - 3).$$

Find the intervals where the graph of the function is concave up and where it is concave down. Find all inflection values.

## EXERCISE SET 13.3

In Exercises 1 through 6 find $f''(x)$.

**1.** $f(x) = 4x^5 + x^2 + x + 3$

**2.** $f(x) = x^3 + x$

**3.** $f(x) = \dfrac{1}{(x + 1)}$

**4.** $f(x) = \dfrac{1}{(x^2 + 1)}$

**5.** $f(x) = \sqrt{2x + 1}$

**6.** $f(x) = \sqrt{x^2 + 1}$

Define $f'''(x) = \dfrac{d}{dx}f''(x)$ and $f^{iv}(x) = \dfrac{d}{dx}f'''(x)$. In Exercises 7 through 12 find $f'''(x)$ and $f^{iv}(x)$ for the given functions.

**7.** $f(x) = 4x^5 + x^2 + x + 3$

**8.** $f(x) = x^3 + x$

**9.** $f(x) = \dfrac{1}{x}$

**10.** $f(x) = \sqrt{x}$

**11.** $f(x) = \sqrt[3]{x^2}$

**12.** $f(x) = \dfrac{1}{\sqrt{x}} + \sqrt{x}$

In Exercises 13 through 20 find all inflections values, find the largest open intervals on which the graph of $f$ is concave up, and find the largest open intervals on which the graph of $f$ is concave down for the given functions.

**13.**

**14.**

**15.**

**16.**

**17.**

**18.**

**19.**

**20.**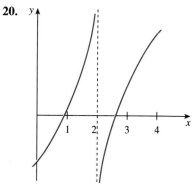

In the following, $f''(x)$ is given in factored form. Find all inflections values, find the largest open intervals on which the graph of $f$ is concave up, and find the largest open intervals on which the graph of $f$ is concave down.

**21.** $f''(x) = x - 2$

**22.** $f''(x) = 3 - 2x$

**23.** $f''(x) = (x - 1)(x - 3)$

**24.** $f''(x) = (x + 1)(x - 4)$

**25.** $f''(x) = (x - 1)(x - 3)(x - 5)$

**26.** $f''(x) = (x - )(x - 3)^3(x - 5)$

**27.** $f''(x) = x(x - 2)^4(x - 4)$

**28.** $f''(x) = x(x - 2)^2(x - 4)^3$

**29.** $f''(x) = (x + 2)x^2(x - 2)^2$

**30.** $f''(x) = (x - 1)^4(x - 3)^2(x - 5)^2$

**31.** $f''(x) = (x + 2)x^3(x - 2)^5$

**32.** $f''(x) = (x + 2)^2x^3(x - 2)^4$

In Exercises 33 through 62 find all critical values and inflection values; find the largest open intervals on which the given function is increasing, decreasing, concave up, and concave down. Find the x-coordinate of each relative minimum and the x-coordinate of each relative maximum. Graph.

**33.**  $f(x) = x^5$

**34.**  $f(x) = x^7$

**35.**  $f(x) = -x^4$

**36.**  $f(x) = x^6$

**37.**  $f(x) = x^2 - 8x + 1$

**38.**  $f(x) = 1 + 4x - x^2$

**39.**  $f(x) = 1 + x - x^2$

**40.**  $f(x) = 1 + x + x^2$

**41.**  $f(x) = x^3 - 3x^2 + 2$

**42.**  $f(x) = 2 - x - x^3$

**43.**  $f(x) = 1 - 9x + 6x^2 - x^3$

**44.**  $f(x) = 1 + 15x + 6x^2 - x^3$

**45.**  $f(x) = 4x^5 - 5x^4 + 1$

**46.**  $f(x) = 6x^7 - 7x^6 + 1$

**47.**  $f(x) = \sqrt{x}, x > 0$

**48.**  $f(x) = \dfrac{1}{\sqrt{x + 1}}, x > -1$

**49.**  $f(x) = \sqrt{1 - x}, x < 1$

**50.**  $f(x) = \dfrac{1}{\sqrt{1 - x}}, x < 1$

**51.**  $f(x) = (x - 2)^3$

**52.**  $f(x) = (4 - x)^3$

**53.**  $f(x) = 3 - x + x^2$

**54.**  $f(x) = 4 - 2x - x^2$

**55.**  $f(x) = x^4 + 6x^2 - 2$

**56.**  $f(x) = 3 - x^2 - x^4$

**57.**  $f(x) = \sqrt{x^2 + 1}$

**58.**  $f(x) = \dfrac{1}{\sqrt{x^2 + 1}}$

**59.**  $f(x) = (x - 3)^4$

**60.**  $f(x) = -(x - 2)^6$

**61.**  $f(x) = 5 + 8x^3 - 3x^4$

**62.**  $f(x) = 3x^4 - 16x^3 + 3$

**63.**  Show that the graph of the quadratic function

$$f(x) = ax^2 + bx + c \qquad (a \neq 0)$$

is always concave up if $a > 0$ and always concave down if $a < 0$.

**64.**  Show that the quadratic function

$$f(x) = ax^2 + bx + c \qquad (a \neq 0)$$

has an absolute minimum at $x = -\dfrac{b}{2a}$ if $a > 0$ and an absolute maximum at $x = -\dfrac{b}{2a}$ if $a < 0$.

**65.**  Show that the graph of the special fourth order polynomial function

$$f(x) = ax^4 + bx^2 + cx + d$$

is always concave up if both $a$ and $b$ are positive and always concave down if both $a$ and $b$ are negative.

**66.** Show that the graph of the cubic function

$$f(x) = ax^3 + bx^2 + cx + d \qquad (a \neq 0)$$

has one and only one inflection point. Find this point.

**67.** Show that the special cubic function

$$f(x) = x^3 - 3bx + c \qquad (b > 0)$$

has exactly one relative minimum and exactly one relative maximum. Locate and identify them.

## Applications

**68. Revenue.** If the revenue function for a firm is given by $R(x) = -x^3 + 144x$, find where the revenue is maximized.

**69. Profit.** If the cost function for a firm is given by $C(x) = 36x + 5$ and the revenue function is the same as in the previous problem, find where the profit is maximized.

**70. Population Dynamics.** A population model that includes the finite resource restriction that is very similar to the one used by the Dutch mathematician Verhultst is given by

$$p(t) = \frac{ap_0(1 + t^2)}{a + bp_0t^2}$$

where $p_0$ is the initial population, that is, $p(0) = p_0$, and $a$ and $b$ are positive constants. We assume $p_0 < a/b$.
  **a.** Show that $p(t)$ is always increasing for $t > 0$.
  **b.** Show that the graph of $p(t)$ is concave up at first, then turns concave down, and that the inflection point has $t$-coordinate $\sqrt{a/3bp_0}$.
  **c.** Draw a graph.

**71. Spread of Technology.** When a new, useful technology is introduced into a particular industry, it naturally spreads through this industry. (See E. Mansfield, "Technical Change and the Rate of Imitation," *Econometrica*, vol. 29, no. 4, [Oct. 1961]). A certain model (very similar to that of Mansfield) of this effect indicates that if $t$ is the number of years after the introduction of the new technology, the percentage of the firms in this industry using the new technology is given by

$$p(t) = \frac{100t^2}{t^2 + 300}.$$

  **a.** Find where $p(t)$ is increasing and where it is decreasing.
  **b.** Find where the graph of $p(t)$ is concave up and where it is concave down. Find all inflection values.
  **c.** Draw a graph.

**72. Diminishing Returns.** Suppose a firm believes that the percent, $p(t)$, of the market that it can attain of its product by spending a certain amount on advertising each year for the next three years is given by $p(t) = -2t^3 + 12t^2$

+ 20 where $t$ is given in years. Find where the graph of the function is concave up and where it is concave down. Find the inflection point. Where is the point of diminishing returns?

**73. Diminishing Returns.** Suppose a firm believes that the percent, $p(t)$, of the market that it can attain of its product by spending a certain amount on advertising each year for the next ten years is given by

$$p(t) = 0.1 \frac{t^4 + 125}{t^4 + 135}$$

where $t$ is given in years. Find where the graph of the function is concave up and where it is concave down. Find the inflection point. Where is the point of diminishing returns?

**74. Average Cost.** An electric utility company estimates that the cost, $C(x)$, of producing $x$ kilowatts of electricity is given by $C(x) = a + bx^2$, where $a$ represents the fixed costs and $b$ is a positive constant. The *average* cost, $\overline{C}(x)$, is defined to be the total cost of producing $x$ kilowatts divided by $x$, that is,

$$\overline{C}(x) = \frac{C(x)}{x}.$$

Find where the average cost attains a minimum.

**75. Acceleration.** A body moves according to the law $s(t) = t^4 - t^2 + 1$. Find the acceleration and find where it is positive and where it is negative.

**76. Acceleration.** A body moves according to the law $s(t) = t^3 - 3t^2 + 1$. Find the acceleration and find where it is positive and where it is negative.

**77. Spread of Rumors.** It has been estimated that a rumor spreads at a rate $R$ that is proportional both to the ratio $r$ of individuals who have heard the rumor and to the ratio $(1 - r)$ who have not. Thus

$$R = kr(1 - r),$$

where $k$ is a positive constant. For what value of $r$ does the rumor spread fastest?

**78. Biology.** When predators are not stimulated to search very hard for prey, the rate of change $R$ of the number $N$ of prey attacked per unit of area with respect to the number $n$ of prey is given by

$$R(x) = kx(ax - N),$$

where $x$ is the number of predators, $k$ is the per capita effectiveness of the predators, and $a$ is the maximum per capita effectiveness of the predators over an extended period of time. Find the number of predators that minimizes $R$.

## Graphing Calculator Exercises

In Exercises 79 through 84 use the program Tangent located in the Appendix C.2 to demonstrate that the tangent line to the graph of a function moves in a counterclockwise direction when the curve is concave up and in a clockwise direction when the curve is concave down.

**79.** $y = x^2$        **80.** $y = 8 - x^2$        **81.** $y = x^3$

**82.** $y = -x^3$        **83.** $y = x^3 - 4x$        **84.** $y = 6x - x^3$

In Exercises 85 through 92 you will see that you cannot solve the equations $f'(x) = 0$ exactly to find the critical points. Use the zoom feature of your graphing calculator to find the approximate solution to the equations $f'(x) = 0$, and then use the second derivative test to determine the relative extrema. Now sketch a graph of $y = f(x)$, and then check your graph by using the graphing calculator.

**85.** $f(x) = x^4 + x^2 + x + 1$        **86.** $f(x) = x^4 + x^3 + x + .50$

**87.** $f(x) = x^5 + x^4 - 3x + 6$        **88.** $f(x) = .20x^5 + .50x^2 - x$

**89.** $f(x) = x^4 - 8x^2 + 2x + 1$        **90.** $f(x) = x^4 - 4x^3 + 2x + 1$

**91.** $f(x) = x^4 - 10x^2 - 2x + 1$        **92.** $f(x) = x^4 - 2x^2 + x + 3$

## Solutions to Self-Help Exercise Set 13.3

**1.** Since

$$p'(t) = -t^2 + 6t = t(6 - t),$$

$p$ is always increasing on the interval $(0, 6)$.
    Since

$$p''(t) = -2t + 6 = 2(3 - t),$$

$p''(t) > 0$, and thus the graph of $p$ is concave up on $(0, 3)$ and $p''(t) < 0$, and thus the graph of $p$ is concave down on $(3, 6)$. On the interval $(0, 3)$, $p$ is increasing ever faster, while on $(3, 6)$, $p$ is increasing ever slower. We might characterize the interval $(3, 6)$ as the period of diminishing returns.

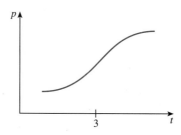

**2.** According to the given factorization of $f''(x)$, $f''(x) > 0$ on the intervals $(-\infty, 1)$ and $(3, +\infty)$, while $f''(x) < 0$ on $(1, 2)$ and $(2, 3)$. Thus the graph of $f$ is concave up on the intervals $(-\infty, 1)$ and $(3, +\infty)$, and concave down on $(1, 2)$ and $(2, 3)$.
    Since the second derivative changes sign at $x = 1$ and $x = 3$, these values are inflection values.

## 13.4    LIMITS AT INFINITY; MORE CURVE SKETCHING

▶ *Limits at Infinity*
▶ *More Curve Sketching*

APPLICATION

**A Mouse Population in the Long Term**

Let

$$p(t) = 5 \frac{3 + 5.01t^2}{3 + 0.01t^2}$$

be the mouse population given in Example 4 of the last section. What happens to the population in the long term? See Example 1 for the answer.

APPLICATION

**Fence Construction**

Three sides of a rectangular fence are to be constructed with a straight river making up the fourth side. See Figure 13.43. The area enclosed must be 2 square miles. What is the perimeter as a function of $x$ if $x$ is in miles, and what is the graph of the resulting function? See Example 3 for the answer.

**Figure 13.43**

### Limits at Infinity

Suppose we are given a demand equation

$$f(p) = \frac{1}{p^2}$$

where $p$ is the price of an item and $f(p)$ is the number of items sold at a price equal to $p$. Let us see what happens to $f$ as $p$ becomes very large.

From Table 13.1, as $p$ becomes large, $f(p)$ approaches zero. In such a case we say that *the limit of $f(p)$ as $p$ goes to $+\infty$ is zero* and write

$$\lim_{p \to +\infty} \frac{1}{p^2} = 0.$$

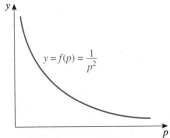

**Figure 13.44**

**Table 13.1**

| $p$ | 1 | 10 | 100 | $\to$ | $+\infty$ |
|---|---|---|---|---|---|
| $\frac{1}{p^2}$ | 1 | .01 | .0001 | $\to$ | 0 |

This makes sense. If we set the price very high, then the number of items sold should be very small. See Figure 13.44.

We have the following definitions.

> **Definition of Limits at Infinity**
>
> 1.  If $f(x)$ approaches the number $L$ as $x$ becomes large without bound, then we say that $L$ is the limit of $f(x)$ as $x$ approaches $+\infty$, and we write
>
> $$\lim_{x \to +\infty} f(x) = L.$$
>
> 2.  If $f(x)$ approaches the number $K$ as $x$ becomes a large negative number without bound, then we say that $K$ is the limit of $f(x)$ as $x$ approaches $-\infty$, and we write
>
> $$\lim_{x \to -\infty} f(x) = K.$$
>
> We refer to the lines $y = L$ and $y = K$ as **horizontal asymptotes**.

Thus for $y = f(p) = 1/p^2$, $y = 0$ is a horizontal asymptote.

**E X A M P L E   1**   **Mouse Population and Carrying Capacity**

Let

$$p(t) = 5\,\frac{3 + 5.01t^2}{3 + 0.01t^2}$$

be the mouse population given in Example 4 of the last section. Find $\lim\limits_{t \to +\infty} p(t)$ and interpret the result.

**Solution**   To determine what is happening for large $t$ to a rational expression such as this, *divide the numerator and the denominator by $t$ to the highest power of $t$ present.* This is $t^2$. Thus

$$p(t) = 5\,\frac{\dfrac{3}{t^2} + 5.01}{\dfrac{3}{t^2} + .01}.$$

Limits at infinity satisfy the same rules as limits studied in Section 12.1. Thus, for example, the limit of the quotient is the quotient of the limits, provided the limit in the denominator is not zero. Since we just saw that $\lim\limits_{t \to +\infty} \dfrac{1}{t^2} = 0$, we then have

$$\lim_{t \to +\infty} p(t) = 5 \frac{0 + 5.01}{0 + .01}$$

$$= 2505.$$

Recall from the last section that we found that the mouse population always increases. Thus the population steadily increases and heads for the limiting population of 2505. ∎

The limiting population 2505 is the carrying capacity. It is typical that populations grow rapidly until the numbers become so large that the population growth becomes inhibited by the finite amount of food and natural resources available. Eventually the population levels out since any habitat can only hold so many individuals. We can see this geometrically in Figure 13.45 where for large $t$ the function is almost a constant. This means there is little change in the slope and thus in the second derivative.

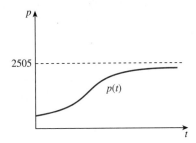

**Figure 13.45**

## More Curve Sketching

In the last three sections we have seen how to use both the first and the second derivative to sketch the graph of a function. We now consider a systematic approach to sketching the graph of a function.

The following checklist is a useful procedure for graphing a function.

---

**Checklist for Graphing a Function**

1. Determine the domain of the function and the intervals on which the function is continuous.
2. Determine if the function is symmetric about the $y$-axis or the origin.
3. Find the critical values.
4. Find intervals where the function is increasing and decreasing.
5. Find all relative extrema.

6. Find intervals where the graph of the function is concave up and concave down.
7. Find all inflection values.
8. Find all vertical asymptotes.
9. Find all horizontal asymptotes.
10. Find where the function crosses the axes.
11. Find the values of $f$ at the critical values and inflection values and graph these points.

We will see how to apply this checklist by working through the following examples.

**E X A M P L E    2**    **Sketching a Graph**

Sketch a graph of the function $f(x) = x^4 - 6x^2 + 4$.

**Solution**

1. The domain of this polynomial is the entire real line. The function is continuous everywhere.
2. Since $f(-x) = f(x)$, the function is symmetric about the $y$-axis. Thus we need only consider the set of points $x \geq 0$. (However for practice and purposes of illustration, negative values of $x$ will be considered.)
3. Since

$$f'(x) = 4x^3 - 12x = 4x(x^2 - 3) = 4x(x + \sqrt{3})(x - \sqrt{3}),$$

there are three critical values, $-\sqrt{3}, 0, +\sqrt{3}$.

4. The three critical values divide the real line into the four intervals $(-\infty, -\sqrt{3}), (-\sqrt{3}, 0), (0, \sqrt{3})$, and $(\sqrt{3}, +\infty)$. Figure 13.46 gives the critical values and the test values. At the test values

$$f'(-2) = -8, \quad f'(-1) = +8, \quad f'(1) = -8, \quad f'(2) = +8.$$

This then implies the rest of Figure 13.46. Figure 13.46 indicates that on the successive intervals $(-\infty, -\sqrt{3}), (-\sqrt{3}, 0), (0, \sqrt{3})$, and $(\sqrt{3}, +\infty)$, $f'(x)$ is negative, positive, negative, positive. Thus on the four intervals, $f(x)$ is successively decreasing ($\searrow$), increasing ($\nearrow$), decreasing ($\searrow$), and increasing ($\nearrow$).

5. By the *first* derivative test, $f(-\sqrt{3})$ and $f(\sqrt{3})$ are relative minima and $f(0)$ is a relative maximum.

**Figure 13.46**

**Figure 13.47**

6. Now notice that

$$f''(x) = 12x^2 - 12 = 12(x + 1)(x - 1).$$

The second derivative exists and is continuous everywhere and is zero at the two values $-1$ and $1$. These two values divide the real line into the three subintervals: $(-\infty, -1)$, $(-1, 1)$, and $(1, +\infty)$. Figure 13.47 lists the values where the second derivative is zero and the test values for the three intervals. At the test values,

$$f''(-2) = +36, \qquad f''(0) = -12, \qquad f''(2) = +36.$$

From this is constructed a sign chart for $f''$. From the sign chart of $f''$ we see that $f''(x)$ is positive on the intervals $(-\infty, -1)$ and $(1, +\infty)$, and thus the graph of $f$ is concave up $(\smile)$ on these two intervals. From the sign chart of $f''$ we see that $f''(x)$ is negative on the interval $(-1, 1)$, and thus the graph of $f$ is concave down $(\frown)$ on this interval.

7. Since $f''(x)$ changes sign at $x = -1$ and $x = +1$, the inflection values are $\pm 1$.

8. There are no vertical asymptotes.

9. We notice that for very large values of $|x|$,

$$f(x) = x^4 \left(1 - \frac{1}{x^2} + \frac{4}{x^4}\right) \approx x^4.$$

Thus as $x \to +\infty$ or as $x \to -\infty$, $f(x)$ becomes a large positive number without bound. The following notation is used to describe this situation.

$$\lim_{x \to +\infty} f(x) = +\infty, \qquad \lim_{x \to -\infty} f(x) = +\infty.$$

This notation does not say that the limits exist (they do not), nor does the notation imply that $+\infty$ or $-\infty$ are numbers (they are not). Rather the notation is a shorthand way of conveying the information that the function increases without bound as $x \to +\infty$ or $x \to -\infty$.

10. The curve crosses the $y$-axis when $x = 0$ at the point $(0, 4)$. Although normally one cannot hope to find the zeros of a fourth degree polynomial, in this case it is possible. Setting $z = x^2$ and then setting $f = 0$ gives $z^2 - 6z + 4 = 0$. Thus by the quadratic formula, $z = 3 \pm \sqrt{5}$ and $x = \pm \sqrt{(3 \pm \sqrt{5})}$. These latter numbers are approximately $\pm 2.3$ and $\pm .9$.

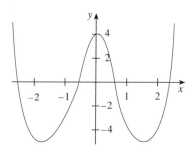

**Figure 13.48**

11. The values of $f$ at the critical values $-\sqrt{3}$, $0$, and $\sqrt{3}$ are $-5$, $4$, and $-5$. The values of $f$ at the inflection values $-1$ and $+1$ are respectively $-1$ and $-1$.

All of this information is then put together to give the graph shown in Figure 13.48. ∎

In the last example, the *second* derivative test could have been used to find the relative extrema. Since $f''(-\sqrt{3}) > 0$ and $f''(\sqrt{3}) > 0$, the second derivative test indicates that $f(-\sqrt{3})$ and $f(\sqrt{3})$ are relative minima. Since $f''(0) < 0$, the second derivative test indicates that $f(0)$ is a relative maximum.

**E X A M P L E   3**    **Fence Construction**

Three sides of a rectangular fence are to be construced with a straight river making up the fourth side. See Figure 13.43. The area enclosed must be 2 square miles. Write the perimeter as a function of $x$ for $x$ measured in miles; graph.

**Solution**    According to Figure 13.43, the area is $A = xy$. Since we are given that $A = 2$, we then have that $xy = 2$ and $y = 2/x$. The perimeter is then given by

$$f = x + 2y = x + 2\frac{2}{x} = x + \frac{4}{x}.$$

Since $x$ is the length of one side of the fence, naturally we must have $x > 0$ to make physical sense. But for the sake of practice *we will graph the function for all $x$ in the domain of definition of the function.*

1. The domain of the function is all real numbers except for $x = 0$. Since $f$ is continuous everywhere except for $x = 0$, divide the real line into the two intervals $(-\infty, 0)$ and $(0, +\infty)$, on each of which $f$ is continuous.
2. Since $f(-x) = -f(x)$, we only need to sketch the graph for $x > 0$, the graph being symmetric about the origin. However, we will ignore this so that we may better illustrate the remaining steps.
3. Setting $f'(x) = 1 - 4x^{-2} = 0$ yields $x^2 = 4$, which gives the two critical values $x = \pm 2$.
4. These 2 critical values, together with the point of discontinuity $x = 0$, then divide the real line into the four subintervals $(-\infty, -2)$, $(-2, 0)$, $(0, 2)$, and $(2, +\infty)$. Test values are then chosen for these 4 subintervals, as indicated in Figure 13.49. We have at the test values

$$f'(-4) = \frac{3}{4} > 0, \qquad f'(-1) = -3 < 0,$$

$$f'(1) = -3 < 0, \qquad f'(4) = \frac{3}{4} > 0.$$

Thus $f$ is increasing on the first and fourth subintervals and decreasing on the second and third intervals.

| Critical values | | | | −2 | | | | | | 2 | | | |
|---|---|---|---|---|---|---|---|---|---|---|---|---|---|

Discontinuity               0

Test values     −4       −1     1        4

$f'$             +        −     −        +

$f$             ↗        ↘     ↘        ↗

**Figure 13.49**

5. The first derivative test indicates that $f(-2)$ is a relative maximum and $f(2)$ is a relative minimum.
6. Since $f''(x) = 8x^{-3}$, $f''$ is negative on $(-\infty, 0)$ and positive on $(0, +\infty)$. Thus the graph of $f$ is concave down on the first interval, while the graph of $f$ is concave up on the second interval.
7. There are no inflection points since $f(0)$ is not defined.
8. As $x \to 0+$, $f(x)$ increases without bound. We use the notation

$$\lim_{x \to 0+} f(x) = +\infty$$

to describe this situation. As $x \to 0-$, $f(x)$ decreases without bound. We use the notation

$$\lim_{x \to 0-} f(x) = -\infty$$

to describe this situation. All of this certainly implies that $x = 0$ is a vertical asymptote.
9. For very large values of $|x|$,

$$f(x) = x + \frac{4}{x} \approx x.$$

10. The curve does not cross the $x$- or $y$-axes.
11. $f(-2) = -4$ and $f(2) = 4$.

Putting all of this information together then gives the sketch of the graph in Figure 13.50. ∎

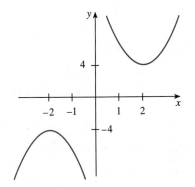

**Figure 13.50**

## SELF-HELP EXERCISE SET 13.4

1. Find $\displaystyle\lim_{x \to +\infty} \frac{2x^2 + x - 1}{x^3 + 1}$

2. Sketch a graph of $f(x) = \dfrac{x^2}{x^2 - 1}$.

## EXERCISE SET 13.4

In Exercises 1 through 18 find the limits.

**1.** $\displaystyle \lim_{x \to +\infty} \frac{2x^2 + 1}{3x^2 - 1}$

**2.** $\displaystyle \lim_{x \to +\infty} \frac{2x^3 - 1}{3x^3 + 1}$

**3.** $\displaystyle \lim_{x \to -\infty} \frac{4x^5 - x - 1}{3x^5 + 1}$

**4.** $\displaystyle \lim_{x \to -\infty} \frac{4x^2 - x - 1}{3x^2 + 1}$

**5.** $\displaystyle \lim_{x \to +\infty} \frac{3x^2 + x + 1}{2x^4 - 1}$

**6.** $\displaystyle \lim_{x \to +\infty} \frac{x^3 - 1}{2x^4 + 1}$

**7.** $\displaystyle \lim_{x \to -\infty} (x^2 + 1)$

**8.** $\displaystyle \lim_{x \to -\infty} (x^3)$

**9.** $\displaystyle \lim_{x \to -\infty} \frac{x^4 - 1}{x^2 + 2}$

**10.** $\displaystyle \lim_{x \to -\infty} \frac{x^4 - x^2 + x - 1}{x^3 + 1}$

**11.** $\displaystyle \lim_{x \to +\infty} \frac{\sqrt{x} + 1}{x + 2}$

**12.** $\displaystyle \lim_{x \to +\infty} \frac{x^2 - 1}{\sqrt{x} - 3}$

**13.** $\displaystyle \lim_{x \to +\infty} \sqrt{4 - \frac{1}{x}}$

**14.** $\displaystyle \lim_{x \to -\infty} \sqrt{1 - \frac{1}{x^2}}$

**15.** $\displaystyle \lim_{x \to +\infty} \sqrt[3]{8 - \frac{1}{x}}$

**16.** $\displaystyle \lim_{x \to -\infty} \sqrt[3]{1 - \frac{1}{x^3}}$

**17.** $\displaystyle \lim_{x \to +\infty} (\sqrt{x^2 + 1} - x)$

**18.** $\displaystyle \lim_{x \to +\infty} (x - \sqrt{x^2 - 4})$

In Exercises 19 through 42 go through the checklist and obtain as much information as possible and then sketch a graph of the given function.

**19.** $f(x) = x^4 + 4x^3 + 1$

**20.** $f(x) = 3x^4 - 16x^3 + 24x^2 - 1$

**21.** $f(x) = x^4 - 32x^2 + 10$

**22.** $f(x) = x^4 - 2x^3 + x^2$

**23.** $f(x) = 3x^5 - 40x^3 + 20$

**24.** $f(x) = -3x^5 + 10x^3 - 5$

**25.** $f(x) = \dfrac{1}{x - 1}$

**26.** $f(x) = \dfrac{x}{x - 1}$

**27.** $f(x) = \dfrac{x + 1}{x + 2}$

**28.** $f(x) = \dfrac{x + 1}{x - 1}$

**29.** $f(x) = \dfrac{1}{x^2 + 1}$

**30.** $f(x) = \dfrac{1}{x^4 + 1}$

**31.** $f(x) = x + \dfrac{9}{x}$

**32.** $f(x) = x + \dfrac{25}{x}$

**33.** $f(x) = x + \dfrac{1}{3x^3}$

**34.** $f(x) = x + \dfrac{16}{3x^3}$

**35.** $f(x) = \dfrac{x^2}{x - 1}$

**36.** $f(x) = -\dfrac{x^2}{x + 2}$

**37.** $f(x) = \dfrac{x}{x^2 + 1}$

**38.** $f(x) = 2 - \dfrac{1}{x^2 + 4}$

**39.** $f(x) = \dfrac{x - 1}{x^2}$

**40.** $f(x) = \dfrac{x^2}{1 + x^2}$

**41.** $f(x) = \dfrac{x^2 + 2x - 4}{x^2}$

**42.** $f(x) = \dfrac{1}{x^2 - x - 6}$

## Applications

**43. Cost.** It is estimated that the cost, $C(x)$, in billions of dollars of maintaining the atmosphere in the United States at an average $x$ percent free of a chemical toxin is given by

$$C(x) = \frac{10}{100 - x}.$$

Sketch a graph of the function on $(0, 100)$.

**44. Cost.** A large corporation already has cornered 30 percent of the toothpaste market and now wishes to undertake a very extensive advertising campaign to increase its share of this market. It is estimated that the advertising cost, $C(x)$, in billions of dollars, of attaining an $x$ percent share of this market is given by

$$C(x) = \frac{1}{100 - x}.$$

Sketch a graph of the function on $(0, 100)$.

**45. Productivity.** A plant manager has worked hard over the past years using various innovative means to increase the number of items produced per employee in his plant and has noticed that the number of items, $n(t)$, produced per employee per day over the last ten years is given by

$$n(t) = 9 \cdot \frac{t}{t + 1},$$

where $t$ is the number of years from 10 years ago. Sketch a graph of the function on $(0, +\infty)$.

**46. Biology.** The rate of production $P$ of photosynthesis is related to the light intensity $I$ by the formula

$$P(I) = \frac{aI}{b + I^2}$$

where $a$ and $b$ are positive constants. Sketch a graph of this function.

**47. Average Cost.** Suppose the cost function is given by $C(x) = a + bx^2$ where $a$ and $b$ are positive constants. Sketch a graph of the average costs.

**48.  Biology.**  According to Wiess's Law, the intensity of an electric current required to excite a living tissue (threshold strength) is given by $i = \frac{A}{t} + B$, where $t$ is the duration of the current and $A$ and $B$ are positive constants. What happens at $t \to +\infty$. Does your answer seem reasonable?

**49.  Medicine.**  Suppose that the concentration $c$ of a drug in the blood $t$ hours after it is taken orally is given by

$$c = \frac{5t}{t^2 + 2}.$$

What happens to the drug concentration as $t \to +\infty$? Does your answer make sense?

**50.  Biology.**  A certain model of the human population of the entire earth is given by

$$p(t) = \frac{20t^2}{t^2 + 4}$$

where $t$ is measured in years from 1990 and $p(t)$ is measured in billions. What happens as $t \to +\infty$?

**51.  Spread of Technology.**  When a new useful technology is introduced into a particular industry, it naturally spreads through this industry. A certain model of this effect indicates that if $t$ is the number of years after the introduction of the new technology, the percentage of the firms in this industry using the new technology is given by

$$p(t) = \frac{100t^4}{t^4 + 1}.$$

What happens to $p(t)$ as $t \to +\infty$?

**52.  Chemistry.**  Salt water with a concentration 0.1 pounds of salt per gallon flows into a large tank that initially holds 100 gallons of pure water. If 5 gallons of salt water per minute flows into the tank, show that the concentration of salt in the tank is given by

$$c(t) = \frac{1}{200 + 20t},$$

where $t$ is measured in minutes. What happens to $c(t)$ as $t \to +\infty$?

## Graphing Calculator Exericses

**53.**  Use your graphing calculator to graph the functions found in Exercises 19 through 42.

## Solutions to Self-Help Exercise Set 13.4

**1.**  Divide numerator and denominator by $x$ to the highest power, which is $x^3$, and obtain

$$\lim_{x \to +\infty} \frac{2x^2 + x - 1}{x^3 + 1} = \lim_{x \to +\infty} \frac{\frac{2}{x} + \frac{1}{x^2} - \frac{1}{x^3}}{1 + \frac{1}{x^3}}$$

$$= \frac{0 + 0 - 0}{1 + 0} = 0.$$

2.  Use the checklist for graphing a function.
    1.  The domain of $f$ is all real values except for $x = \pm 1$. The function is continuous at all values except for $x = \pm 1$.
    2.  Since $f(-x) = f(x)$, the curve is symmetric about the $y$-axis.
    3.  Since

    $$f'(x) = \frac{2x(x^2 - 1) - x^2(2x)}{(x^2 - 1)^2} = \frac{-2x}{(x^2 - 1)^2},$$

    the only critical value is $x = 0$.
    4.  Since the denominator in the above expression for $f'(x)$ is always positive (except for $x = \pm 1$ where $f'$ is not defined), the sign of $f'$, where it exists, is the same as the sign of $-x$. Thus on the two intervals $(-\infty, -1)$ and $(-1, 0)$, $f'(x) > 0$ and $f$ is increasing. On the two intervals $(0, 1)$ and $(1, +\infty)$, $f'(x) < 0$ and $f$ is decreasing.

    5.  Since $f'$ is positive to the left of 0 and negative to the right, the first derivative test indicates that $f(0) = 0$ is a relative maximum.
    6.  We have

    $$f''(x) = \frac{-2(x^2 - 1)^2 + (2x)2(x^2 - 1)2x}{(x^2 - 1)^4} = \frac{6x^2 + 2}{(x^2 - 1)^3}.$$

    The numerator of this expression is always positive. Thus the sign of $f''$ is the same as the sign of $(x^2 - 1)$. Thus $f''(x)$ is positive on the two intervals $(-\infty, -1)$ and $(1, +\infty)$ and negative on the interval $(-1, 1)$. Thus the graph of $f$ is concave up on the first two of these intervals and concave down on the last one.
    7.  There are no points of inflection since $x = +1$ and $-1$ are not in the domain of definition of $f$.
    8.  We have

    $$\lim_{x \to -1-} f(x) = +\infty, \qquad \lim_{x \to -1+} f(x) = -\infty$$

    while

    $$\lim_{x \to 1-} f(x) = -\infty, \qquad \lim_{x \to 1+} f(x) = +\infty.$$

    Certainly $x = \pm 1$ are two vertical asymptotes.

9.  Notice that

$$\lim_{x \to \pm\infty} \frac{x^2}{x^2 - 1} = \lim_{x \to \pm\infty} \frac{1}{1 - \frac{1}{x^2}} = \frac{1}{1} = 1.$$

Thus $y = 1$ is a horizontal asymptote.

10. Setting $x = 0$ gives $y = 0$. Setting $y = 0$ gives $x = 0$. Thus the curve crosses the axes at the point $(0, 0)$.

11. $f(0) = 0$.

Putting all this information together gives the following figure.

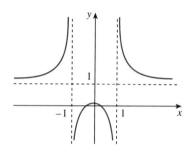

## 13.5    ABSOLUTE EXTREMA

▶  *Absolute Extrema*

▶  *Finding Extrema on a Closed and Bounded Interval*

▶  *Management of Renewable Natural Resources*

**APPLICATION**

**The Antarctic Blue Whale**

Let $x$ be the number of thousands of Antarctic blue whales and $f(x)$ the increase in thousands in the number over the next year, where

$$f(x) = 0.000002(-2x^3 + 303x^2 - 600x).$$

What is the maximum sustainable yield? Also what is an estimate of the minimum population level below which the whales will inevitably become extinct? What is the carrying capacity? See Example 6 for the answers.

### Absolute Extrema

In previous sections we studied relative extrema and gave the first and second derivative tests to determine them. In this section we study absolute extrema and find conditions that ensure that a continuous function attains its absolute extrema on a given interval.

As usual, we assume that all our functions $f$ are continuous on some interval

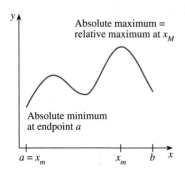

Absolute maximum = relative maximum at $x_M$

Absolute minimum at endpoint $a$

$a = x_m$  $x_M$  $b$  $x$

**Figure 13.51**

and have a finite number of critical values on any finite subinterval and that $f'$ is continuous, except possibly at the critical values.

Notice that the function in Figure 13.51 attains its smallest value at the left end-point and its largest vaue at $x_M$.

We refer to the smallest value as the **absolute minimum** and to the largest value as the **absolute maximum**. An **absolute extremum** is either an absolute maximum or an absolute minimum. We have the following definitions.

---

**Definition of Absolute Maximum and Minimum**

Suppose $f(x)$ is defined on some interval $I$.

1. We say that $f(x_M)$ is an **absolute maximum** on the interval $I$ if $x_M \in I$ and

$$f(x) \leq f(x_M) \qquad \text{for all } x \in I.$$

2. We say that $f(x_m)$ is an **absolute minimum** on the interval $I$ if $x_m \in I$ and

$$f(x_m) \leq f(x) \qquad \text{for all } x \in I.$$

---

**E X A M P L E   1**   **Finding the Absolute Extrema on an Infinite Interval**

Find the absolute extrema of $f$ (if they exist) where $f(x) = \dfrac{2}{1 + x^2}$.

**Solution**   The derivative is

$$f'(x) = \frac{-4x}{(1 + x^2)^2}.$$

Thus the derivative is positive on $(-\infty, 0)$ and $f$ is increasing on this interval. Also $f'(x) < 0$ on $(0, +\infty)$, and $f$ is decreasing on this latter interval. Since $f$ is always increasing to the left of $x = 0$ and decreasing to the right of $x = 0$, $f$ must attain its maximum at $x = 0$. This absolute maximum is $f(0) = 2$.

Notice furthermore that $f(x)$ is always strictly positive and that

$$\lim_{x \to -\infty} \frac{2}{1 + x^2} = 0 \qquad \lim_{x \to +\infty} \frac{2}{1 + x^2} = 0.$$

The graph is shown in Figure 13.52.

Notice that the function does not have an absolute minimum on the interval $(-\infty, +\infty)$.  ∎

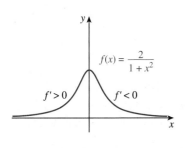

$f(x) = \dfrac{2}{1 + x^2}$

$f' > 0$   $f' < 0$

**Figure 13.52**

**R E M A R K .**   The function in the previous example had one single critical value on the interval. This is typical of functions in the applications. For example, if the derivative is always positive to the left of the critical value and always negative to the right of the critical value, the function is always increasing to the left of the critical value and always decreasing to the right. Thus the critical value is an *absolute* maximum.

**E X A M P L E   2**     **Finding the Absolute Extrema on an Infinite Interval**

Find the absolute minimum of $f$ on $(0, +\infty)$ if $f(x) = x + \dfrac{1}{x}$.

**Solution**     The first derivative is

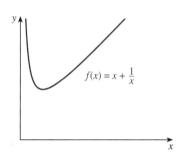

$f(x) = x + \dfrac{1}{x}$

**Figure 13.53**

$$f'(x) = 1 - \frac{1}{x^2} = \frac{1}{x^2}(x^2 - 1),$$

and is continuous on $(0, +\infty)$. Setting $f'(x) = 0$, then yields $x^2 = 1$ or $x = +1$ as the only critical value on the given interval. The first derivative is negative on $(0, 1)$ and positive on $(1, +\infty)$ and thus $f$ is decreasing on $(0, 1)$ and increasing on $(1, +\infty)$. From this it is clear that $f$ assumes an absolute minimum at $x = 1$.  ∎

The function in Example 2 has no absolute maximum on $(0, +\infty)$ since $f(x) = x + \dfrac{1}{x}$ is large without bound for $x$ large or for $x$ small. A graph is shown in Figure 13.53.

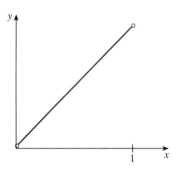

**Figure 13.54**

## Finding Extrema on a Closed and Bounded Interval

Continuous functions on a closed and bounded interval have the following very important property.

> **Extreme Value Theorem**
>
> If $f(x)$ is continuous on the closed and bounded interval $[a, b]$, then $f$ attains both its absolute maximum and its absolute minimum on $[a, b]$.

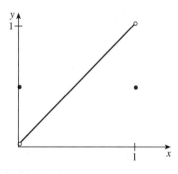

**Figure 13.55**

The conclusion of this theorem need not be true if the interval is not bounded, as we have already seen in Example 1 and Example 2. In Example 1 there was no absolute minimum on $(-\infty, +\infty)$ and in Example 2 there was no absolute maximum on $(0, +\infty)$. But the conclusion also need not be true if the interval is not closed or if the function is not continuous on all of $[a, b]$.

In Figure 13.54, the function $f(x) = x$ is continuous on $(0, 1)$ but attains neither an absolute maximum nor an absolute minimum on $(0, 1)$.

In Figure 13.55, the function $f(x)$ equals $x$ on $(0, 1)$, $f(0) = 1/2 = f(1)$, and $f(x)$ is not continuous on $[0, 1]$ and attains neither an absolute maximum nor an absolute minimum on $[0, 1]$.

From Figure 13.51 we can see that an absolute extremum for a function can be a relative extremum or *one of the end points*. There can be no other possibilities. If $f'(x_0) > 0$ or $f'(x_0) < 0$ and $x_0 \in (a, b)$, then we know from the previous sections that $f(x)$ will be increasing in the first case and decreasing in the second. Thus in neither case can $f(x)$ attain an absolute extremum. The only values that are left are the critical values and the end points.

The procedure for finding the absolute extrema on a closed and bounded interval can then be given as follows.

---

**To Find the Absolute Extrema of $f(x)$ on $[a, b]$.**

1.  Locate all critical values.
2.  Evaluate $f(x)$ at all the critical values and also at the two values $a$ and $b$.
3.  The absolute maximum of $f(x)$ on $[a, b]$ will be the largest number found in step 2, while the absolute minimum of $f(x)$ on $[a, b]$ will be the smallest number found in step 2.

---

**E X A M P L E   3**    **Finding the Absolute Extrema**

Find the absolute extrema of $f(x) = x + \dfrac{1}{x}$ on the interval $[\frac{1}{2}, 3]$, if they exist.

**Solution**    We notice that on the interval $[\frac{1}{2}, 3]$, $f$ is continuous, so by the extreme value theorem $f$ will have absolute extrema.

This is the same function considered in Example 2. Recall that $x = 1$ is the only critical value on the positive axis. Thus $x = 1$ is the only critical value in the given interval $[\frac{1}{2}, 3]$.

Thus the only "suspects" for the relative extrema are the critical value $x = 1$ together with the end points of the interval $\frac{1}{2}$ and 3. We then make a list of the numbers $f(\frac{1}{2})$, $f(1)$, and $f(3)$. The largest of these three numbers will be the absolute maximum while the smallest of the three will be the absolute minimum.

$$f(\tfrac{1}{2}) = \frac{5}{2}$$

$$f(1) = 2 \quad \leftarrow \text{absolute minimum}$$

$$f(3) = \frac{10}{3} \quad \leftarrow \text{absolute maximum} \quad \blacksquare$$

Although the previous work is sufficient to obtain the answers, it is generally wise to draw a rough sketch of the graph as a check on your work. A graph is shown in Figure 13.56.

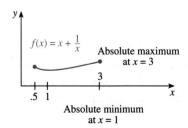

Figure 13.56

**E X A M P L E   4**    **Finding the Absolute Extrema**

Find the absolute extrema of $f(x) = x + \dfrac{1}{x}$ on $[-2, 3]$ if they exist.

**Solution**    It would be extremely easy for the reader to do this problem incorrectly. The temptation is to charge right ahead and find the critical values, which are at $-1$ and $1$, both of which are on the given interval $[-2, 3]$ and then claim that the absolute extrema must be among the four numbers $f(-2)$, $f(-1)$, $f(1)$, and $f(3)$. But the problem with this is that the function is *not* continuous on the given interval $[-2, 3]$ and thus the extreme value theorem does not guarantee the existence of absolute extrema. We therefore must proceed with caution. As Figure 13.57 indicates, there is *no* absolute maximum and *no* absolute minimum on $[-2, 3]$.  ∎

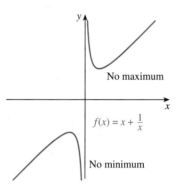

**Figure 13.57**

**E X A M P L E   5**    **Finding the Absolute Extrema**

Find the absolute extrema of $f(x) = 3x^4 + 4x^3 - 12x^2 + 1$ on $[-1, 2]$ if they exist.

**Solution**    Since the function is continuous on $[-1, 2]$, $f$ will attain both an absolute minimum and an absolute maximum.

Since

$$f'(x) = 12x^3 + 12x^2 - 24x = 12x(x^2 + x - 2) = 12x(x - 1)(x + 2),$$

the function is differentiable everywhere and the critical values are $-2$, $0$, and $1$, with only the latter two on the given interval $[-1, 2]$. Thus the absolute maximum and the absolute minimum must be among the list of numbers, $f(-1)$, $f(0)$, $f(1)$, $f(2)$.

$$f(-1) = -12 \leftarrow \text{absolute minimum}$$

$$f(0) = 1$$

$$f(1) = -4$$

$$f(2) = 33 \leftarrow \text{absolute maximum} \quad \blacksquare$$

Although this is sufficient to obtain the answer, a graph can be constructed using the factorization of $f'(x)$ to find where the function is increasing and decreasing. The function is increasing on $(-1, 0)$, decreasing on $(0, 1)$, and increasing on $(1, 2)$. Notice that by itself this information is not enough to determine where the absolute extrema are. We have to evaluate the function at the four points to determine this. A graph is shown in Figure 13.58.

## Management of Renewable Natural Resources

In a previous section we studied a fishery in a relatively small lake. We now look at a fishery in a relatively large lake.

In a relatively small lake, we decided that if there were a few fish in the lake then they would breed and produce more fish. But if the lake is relatively large and there are only a small number of fish in the lake, the fish may not be able to find each other in order to breed. (For some species of whales the feeding grounds amount to as much as 10 million square miles.) Thus instead of the population and biomass *increasing*, the population and biomass would actually *decrease* due to other natural causes.

Assume as before that the biomass in pounds of fish is $x$ and the increase in the biomass over the next year is given by $y = f(x)$. Thus the biomass one year later is $x + y$ and we wish to harvest $y$.

Figure 13.59 indicates that in a relatively large lake the increase in the biomass is *negative*, that is, the biomass decreases for small numbers of fish for the reasons indicated earlier. For some larger amount of fish, indicated by $K_0$, the increase in biomass will move to zero, after which the biomass will then increase.

The number $K_0$ is very important and is called the **minimum viable biomass level**. If the biomass of fish in the lake ever gets *below* $K_0$, then the population will head eventually to zero. Thus overfishing to the point where the biomass goes below the minimum viable biomass level results in an irrevocable catastrophe (assuming no intervention).

In the following problem we take for convenience $x$ to be the *population* of whales.

**Figure 13.58**          **Figure 13.59**

**E X A M P L E   6**    **Management of a Fishery**

Suppose the fishery consisting of the Antarctic blue whale has the function $f(x)$ in the above discussion given by

$$f(x) = 0.000\,002(-2x^3 + 303x^2 - 600x),$$

where $x$ is in thousands of whales. Find the maximum sustainable yield. Also estimate the minimum viable population level and the carrying capacity. Draw a graph of $f(x)$.

**Solution**    We see that

$$f'(x) = 0.000002(-6x^2 + 606x - 600)$$

$$= -0.000012(x^2 - 101x + 100)$$

$$= -0.000012(x - 1)(x - 100).$$

Thus the only critical values are $x = 1$ and $x = 100$. Furthermore, from the above factorization, $f'$ is negative on $(0, 1)$, positive on $(1, 100)$, and negative again on $(100, +\infty)$. Thus $f(1)$ is a relative minimum and $f(100)$ a relative maximum. Furthermore, we can readily see that $f(100)$ is an absolute maximum on $(0, +\infty)$, since $f(0) = 0$, whereas $f(100) = 1.940$.

The maximum sustainable yield is $f(100) = 1.940$, or 1940 whales.

If we now set $f(x) = 0$, we obtain

$$2x^2 - 303x + 600 = 0.$$

Using the quadratic formula and a calculator yields $x \approx 2$ and $x \approx 150$. Thus the minimum viable population level is about 2000 whales, and the carrying capacity is 150,000 whales. See Figure 13.60.    ∎

**Figure 13.60**

Whaling scientists have estimated the above numbers for the carrying capacity and the maximum sustainable yield of the Antarctic blue whale. The minimum viable population level is unknown.

---

**The Antarctic Blue Whale Fishery**

In the 1960s the population of the Antarctic blue whale was reduced to levels thought to lead to extinction. Under the protection of the International Whaling Commission the population increased to almost 10,000 in about 10 years and to about 50,000 in 1994.

---

## SELF-HELP EXERCISE SET 13.5

**1.**  Find the absolute extrema of $f(x) = \dfrac{x}{1 + x^2}$ on $[0, +\infty)$.

**2.**  Find the absolute extrema of $f(x) = x^3 - 12x + 1$ on $[-3, 5]$.

## EXERCISE SET 13.5

In Exercises 1 through 8, locate the value(s) of $x$ where each function attains an absolute maximum and the value(s) where each function attains an absolute minimum.

**1.**

**2.**

**3.**

**4.**

**5.**

**6.**

**7.**

**8.**

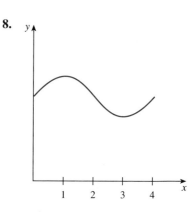

In Exercises 9 through 14 let $f(x) = x^3 - 3x + 1$. Locate the value(s) where $f$ attains an absolute maximum and the value(s) where $f$ attains an absolute minimum on each interval.

**9.** $[2, 4]$        **10.** $[-1, 1]$        **11.** $[0, 2]$

**12.** $[-2, 0]$        **13.** $[-2, 2]$        **14.** $[-3, 2]$

In Exercises 15 through 20 let $f(x) = -2x^3 + 3x^2$. Locate the value(s) where $f$ attains an absolute maximum and the value(s) where $f$ attains an absolute minimum on each interval.

**15.** $[-2, 0]$        **16.** $[0, 1]$        **17.** $[0, 2]$

**18.** $[-1, 1]$        **19.** $[-2, 2]$        **20.** $[-2, 1]$

In Exercises 21 through 38 locate the value(s) where each function attains an absolute maximum and the value(s) where the function attains an absolute minimum, if they exist, of the given function on the given interval.

**21.** $f(x) = x^2 - 4x + 1$ on $[-1, 1]$      **22.** $f(x) = x^2 - 2x + 1$ on $[2, 4]$

**23.** $f(x) = -x^2 + 4x - 2$ on $[0, 3]$      **24.** $f(x) = x^2 + 2x + 1$ on $[-3, 0]$

**25.** $f(x) = x^3 - 3x + 1$ on $[-1, 3]$      **26.** $f(x) = x^3 - 3x^2 + 2$ on $[0, 4]$

**27.** $f(x) = x^3 + 3x^2 - 2$ on $[-3, 2]$

**28.** $f(x) = -8x^3 + 6x - 1$ on $[-1, +1]$

**29.** $f(x) = x^4 - 8x^2 + 3$ on $[-3, 3]$

**30.** $f(x) = -x^4 + 8x^2 + 1$ on $[-1, 3]$

**31.** $f(x) = x^2 + 16x^{-2}$ on $[1, 8]$      **32.** $f(x) = x^2 - 16x^{-2}$ on $[1, 8]$

**33.** $f(x) = 3 - x + x^2$ on $(-\infty, +\infty)$

**34.** $f(x) = 4 - 2x - x^2$ on $(-\infty, +\infty)$

**35.** $f(x) = x^4 + 6x^2 - 2$ on $(-\infty, +\infty)$

**36.** $f(x) = 3 - x^2 - x^4$ on $(-\infty, +\infty)$

**37.** $f(x) = \sqrt{x^2 + 1}$ on $(-\infty, +\infty)$      **38.** $f(x) = (x - 3)^4$ on $(-\infty, +\infty)$

In Exercises 39 through 46 let $f(x) = 3x + x^{-3}$. On each interval locate the value(s) (if they exist) where $f$ attains an absolute extremum.

**39.**  [0.5, 2]                    **40.**  $[-2, -0.5]$                   **41.**  $[-1, 3]$

**42.**  $[-2, 3]$                   **43.**  (0, 3]                         **44.**  $[-3, 0)$

**45.**  $(-\infty, 0)$              **46.**  $(0, +\infty)$


## Applications

**47.  Fishery.**  Suppose the function $f(x)$ in the text under the heading "Management of Renewable Natural Resources" is given by

$$f(x) = -x^3 + 15x^2 - 27x.$$

Find the maximum sustainable yield. Also find the minimum viable population level and the carrying capacity.

**48.  Fishery.**  Redo the previous problem if

$$f(x) = -x^5 + 5x^4 - 4x^3.$$

In Exercises 49 through 51 draw a rough graph *before* evaluating the appropriate function at the "suspect" values. The graph will clearly indicate that one of the suspect values is not the extremum being sought and will eliminate the need to use a calculator.

**49.**  A wire 12 inches long is cut into two pieces with one piece used to construct a square and the other piece used to construct a circle. Where should the wire be cut to maximize the sum of the areas enclosed by the two figures?

**50.**  A wire 12 inches long is cut into two pieces with one piece used to construct a square and the other piece used to construct an equilateral traingle. Where should the wire be cut to maximize the sum of the areas enclosed by the two figures? (The area of an equilateral triangle is $\dfrac{\sqrt{3}}{4} r^2$, where $r$ is the length of a side.)

**51.**  An individual is planning to construct two fenced enclosures, one square and one circular. The cost per yard of constructing the circular enclosure is twice that of the square enclosure. If the total cost is fixed at $T$ dollars, what should be the dimensions of the two figures in order to maximize the area enclosed by the two figures?

## SOLUTIONS TO SELF-HELP EXERCISE SET 13.5

**1.** We have

$$f'(x) = \frac{x^2 + 1 - x(2x)}{(x^2 + 1)^2} = \frac{1 - x^2}{(x^2 + 1)^2}.$$

Since $f'$ is positive on $[0, 1)$ and negative on $(1, +\infty)$, $f$ is increasing to the left of $x = 1$ and decreasing to the right of $x = 1$. Thus $f(1) = \frac{1}{2}$ is the absolute maximum. The graph is shown in the figure. The absolute minimum is $f(0) = 0$.

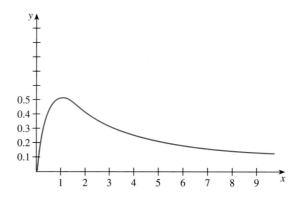

**2.** We have

$$f'(x) = 3x^2 - 12 = 3(x^2 - 4) = 3(x + 2)(x - 2).$$

Thus $f'$ is continuous everywhere on the interval $[-3, 5]$ and the extreme value theorem assures that the function attains an absolute minimum and an absolute maximum on this interval. Thus the extrema must be attained at the end points, $-3$ or $5$, or the critical values $-2$ or $+2$. Evaluating $f$ at these 4 values yields

$$f(-3) = 10$$

$$f(-2) = 17$$

$$f(2) = -15 \leftarrow \text{absolute miniumum}$$

$$f(5) = 66 \quad \leftarrow \text{absolute maximum}$$

One can do some further work and construct the following graph.

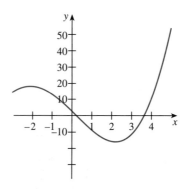

**13.6** **OPTIMIZATION PROBLEMS**

▶ *Miscellaneous Examples*
▶ *Inventory Cost Model*

In this section we put together all that has been learned in the earlier sections to solve various applied problems.

## Miscellaneous Examples

**E X A M P L E   1** **Construction of Fencing**

A farmer has 500 yards of fencing with which to fence in three sides of a rectangular pasture. A straight river will form the fourth side. Find the dimensions of the pasture of largest area that the farmer can fence.

**Solution** The situation is shown in Figure 13.61. Since we do not know either of the dimensions of the rectangle, we label them $x$ and $y$ as shown in the figure. It is these two dimensions that we must determine, where we take the dimensions in yards.

**Figure 13.61**

The area $A$ of the pasture is then just $A = xy$. It is $A$ that we wish to maximize. We cannot maximize $A$ in this form. We need to have $A$ as a function of *one* variable not two.

But the two variables $x$ and $y$ are *not* both independent. Rather these two variables are connected by the fact that, on the one hand, we are given that the total amount of fencing is 500 yards, while, on the other hand, Figure 13.61 indicates that the total amount of fencing is $x + 2y$. Thus

$$500 = x + 2y.$$

We can now solve this equation for one of the variables, say $y$, obtaining

$$y = 250 - \frac{1}{2}x.$$

Now that we have $y$ in terms of $x$ we can substitute this value of $y$ in the formula for area, $A = xy$, and obtain

$$A = xy = x\left(250 - \frac{1}{2}x\right) = 250x - \frac{1}{2}x^2.$$

Finally, this gives us $A$ as a function of one variable $x$.

Before proceeding as in the previous sections, we should first determine the domain of the function $A(x)$. Naturally we must have $x \geq 0$. On the other hand, we must have $x \leq 500$, since $y \geq 0$ and $500 = x + 2y$.

Let us now summarize what we have. We wish to find the maximum value of the *continuously differentiable* function $A = A(x)$ given above on the interval $[0, 500]$. This is now just like problems we have worked in previous sections.

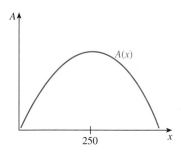

**Figure 13.62**

We then differentiate, obtaining

$$A'(x) = 250 - x.$$

It follows that $A'(x) > 0$ and $A(x)$ is increasing on $(0, 250)$, while $A'(x) < 0$ and $A(x)$ is decreasing on $(250, 500)$. Thus $A(x)$ will attain its absolute maximum at $x = 250$. See Figure 13.62.

Since $y = 250 - \frac{1}{2}x$, setting $x = 250$ will yield

$$y = 250 - \frac{1}{2}(250) = 125.$$

Thus $x = 250$ and $y = 125$.

The maximum area is

$$A = xy = 250 \times 125 = 31{,}250. \quad \blacksquare$$

It is not possible to give an explicit procedure for solving the problems we encounter in this section so that the solution will be automatic. However, the following six-step procedure should be helpful.

---

**Six-Step Procedure for Solving Word Problems**

1.  Identify the unknowns and denote each with a symbol.
2.  Identify the variable that is to be maximized or minimized and express it as a function of the other variables.
3.  Find any relationships between the variables.
4.  Express the quantity to be maximized or minimized as a function of only one variable. To do this you may need to solve for one of the variables in one of the equations in Step 3 and substitute this into the formula found in Step 2.
5.  Determine the interval over which the function is to be minimized or maximized.
6.  Steps 1 through 5 leave you with a function of a single variable on a certain interval to be maximized or minimized. Proceed now as in the previous sections.

---

The six-step procedure applies to Exercise 1 as follows.

1.  We identified the two sides of the rectangle and the area as unknowns.
2.  We identified $A$ as the variable to be maximized and wrote $A = xy$.
3.  We noted that $500 = x + 2y$.
4.  We solved for $y$ in the formula $500 = x + 2y$ and substituted the result into the formula for $A = xy$ to obtain $A$ as a function of $x$ only.
5.  We noted that $x$ must lie between 0 and 500.
6.  We noted that $A(x) = 250x - \frac{1}{2}x^2$ was to be maximized on the interval $[0, 500]$.

E X A M P L E  2

**Maximizing Revenue**

A theater owner charges $5 per ticket and sells 250 tickets. By checking other theaters, the owner decides that for every one dollar she raises the ticket price, she will lose 10 customers. What should she charge to maximize revenue?

**Solution**  If $p$ is the price of a ticket, $n$ is the number of tickets sold, and $R$ is the revenue, the formula for revenue is given by

$$\text{revenue} = (\text{price per ticket}) \times (\text{number of tickets sold}),$$

or  $$R = pn.$$

Thus her current revenue is

$$(5) \times (250) = 1250.$$

If she raises the price by one dollar, the price of a ticket will be $p = 5 + 1$, the number of tickets sold will be $n = 250 - 10$, and the revenue will be

$$R = pn = (5 + 1)(250 - 10) = 1440.$$

If she raises the price by two dollars, the price of a ticket will be $p = 5 + 2$, the number of tickets sold will be $n = 250 - 20$, and the revenue will be

$$R = pn = (5 + 2)(250 - 20) = 1610.$$

If she raises the price by $x$ dollars, the price of a ticket will be $p = 5 + x$, the number of tickets sold will be $n = 250 - 10x$, and the revenue will be

$$R(x) = pn = (5 + x)(250 - 10x) = 1250 + 200x - 10x^2.$$

This is an everywhere continuously differentiable function. Since the owner clearly wishes to have the price of each ticket *positive*, we must have $5 + x = p > 0$ or $x > -5$. Also the owner clearly wants to sell at least one ticket, that is, $250 - 10x = n > 0$ or $x < 25$.

We thus wish to maximize the function $R$ given above on the interval $(-5, 25)$.

Since

$$R'(x) = 200 - 20x = 20(10 - x),$$

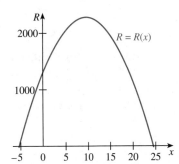

**Figure 13.63**

there is one critical value at $x = 10$. Also $R'(x) > 0$ on $(-5, 10)$ and $R'(x) < 0$ on $(10, 25)$. Thus $R$ attains an absolute maximum at $x = 10$. See Figure 13.63. It is thus clear that the revenue, $R$, will be maximum when $x = 10$. This means that the price of each ticket should be $5 + 10 = 15$ dollars.

The maximum revenue will be

$$R(10) = (5 + 10)(250 - 10 \times 10) = 2250. \quad\blacksquare$$

E X A M P L E  3

**Maximizing the Volume of a Package**

A package mailed in this country must have length plus girth of no more than 108 inches. Find the dimensions of a rectangular package with square base of greatest volume that can be mailed. (The girth is the circumference of a rectangular cross section.)

**Solution**

**Figure 13.64**

See Figure 13.64 where the length of the package is given as $l$ and each side of the square base is given by $s$. We wish to maximize the volume, $V$, which is given by

$$V = s^2 l.$$

Here $V$ is a function of *two* variables. We need to make $V$ a function of only *one* variable.

But the two variables $s$ and $l$ are connected by the fact that the length plus the girth can be at most 108 inches. It is clear that, if we are to maximize the volume, then we should take the length plus the girth to be as large as possible, which is 108. Thus take

$$108 = \text{length} + \text{girth} = l + 4s.$$

From this equation we can solve for $l$ and obtain

$$l = 108 - 4s.$$

Now substitute this into the formula for $V$ and obtain

$$V = s^2 l = s^2(108 - 4s) = 108s^2 - 4s^3.$$

We now have $V$ as a function of one variable.

This function is continuous and differentiable everywhere. Naturally we must have $s \geq 0$ and also $s \leq 108/4 = 27$ inches.

In summary, we wish to find the absolute maximum of the continuously differentiable function $V = V(s) = 108s^2 - 4s^3$ on the interval $[0, 27]$.

To do this we first differentiate $V(s)$ obtaining

$$V'(s) = 216s - 12s^2 = 12s(18 - s).$$

Since $V'(s) = 0$ implies that $s = 0$ or $s = 18$ inches, the only critical value of $V(s)$ on $(0, 27)$ is at $s = 18$ inches. Furthermore, from the factorization of $V'(s)$, we see that $V'(s) > 0$ on $(0, 18)$ and $V'(s) < 0$ on $(18, 27)$. Thus $V$ is increasing on $(0, 18)$ and decreasing on $(18, 27)$. From this it is clear that $V(18)$ is an absolute maximum on the interval $[0, 27]$. See Figure 13.65. Then

$$l = 108 - 4s = 108 - 4(18) = 36.$$

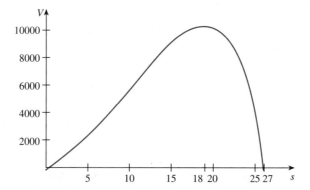

**Figure 13.65**

The maximum volume is

$$V = s^2 l = 18^2 \times 36 = 11,664 \text{ in}^3. \quad \blacksquare$$

**E X A M P L E  4**  **Minimizing the Cost of Materials**

A manufacturer must produce a sturdy rectangular container with a square base and a volume of 128 cubic feet. The cost of materials making up the top and four sides is $2 per square foot while the cost of the materials making up the bottom (which must be reinforced) is $6 per square foot. Find the dimensions of the box that minimizes the cost of the materials.

**Solution**    In Figure 13.66, each side of the square base is labeled $x$ and the height is labeled $h$. We wish to minimize the total cost. The total cost is the cost of the bottom, plus the cost of the top, plus the cost of the four sides.

The area of the bottom is $x^2$, so the cost of the bottom is $6(x^2) = 6x^2$.

The area of the top is $x^2$, so the cost of the top is $2(x^2) = 2x^2$.

The area of the four sides is $4xh$, so the cost of the four sides is $2(4xh) = 8xh$.

Thus the total cost, $C(x)$, is given by

$$C(x) = 6x^2 + 2x^2 + 8xh = 8x^2 + 8xh.$$

The volume must be 128, and, by Figure 13.66, the volume is also $x^2 h$. Thus

$$128 = V = x^2 h.$$

Solving this for $h$ gives

$$h = \frac{128}{x^2}.$$

We substitute this into the formula for $C(x)$ and obtain

$$C(x) = 8x^2 + 8x \left( \frac{128}{x^2} \right) = 8x^2 + 1024x^{-1}.$$

Now we have $C$ as a function of one variable $x$ and we wish to find the absolute minimum of $C$ on the interval $(0, +\infty)$. Differentiating, we obtain

$$C'(x) = 16x - 1024 \frac{1}{x^2} = \frac{16}{x^2}(x^3 - 64).$$

Setting this equal to zero yields

$$x^3 = 64.$$

This gives $x = 4$ as the only critical value. From the above factorization of $C'(x)$ we see that $C'(x) < 0$ on $(0, 4)$ and that $C'(x) > 0$ on $(4, +\infty)$. Thus $C$ is decreasing on $(0, 4)$ and increasing on $(4, +\infty)$. See Figure 13.67. It is then clear that $C(4)$ is the absolute minimum.

**Figure 13.66**

**Figure 13.67**

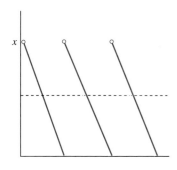

**Figure 13.68**

When $x = 4$ feet

$$h = \frac{128}{4^2} = 8 \text{ feet}.$$

The minimum cost is

$$
\begin{aligned}
C &= 8(x^2 + xh) \\
&= 8(4^2 + 4 \times 8) \\
&= \$384. \quad \blacksquare
\end{aligned}
$$

## Inventory Cost Model

Suppose a manufacturer intended to produce and sell a large number of items during the next year. All of the items could be manufactured at the start of the year. Because of the economies of mass production, this would certainly hold down the manufacturing costs. However, this would result in a large cost of storing the large number of items until they can be sold. If, on the other hand, a large number of production runs were made and spread out over the year, there would be substantially fewer unsold items that would require storage. Thus storage costs would be much smaller. However, because of the lack of economies inherent in small production runs, the manufacturing costs will be large. Presumably there will be some middle ground that will minimize the total costs.

We make the simplifying assumption that sales are made at a steady rate throughout the year and the production run occurs when the number of unsold items has been reduced to zero. We assume that each run produces the same number of items, and we designate this number by $x$. Figure 13.68 shows the number of items in storage. From this it is apparent that on average there are $\frac{1}{2}x$ items in storage at any time. If $s$ is the cost of storing an item for one year, then the cost of storage will be given by the cost of storing one item for one year times the average number in storage or

$$\text{cost of storage} = s\left(\frac{1}{2}x\right).$$

**E X A M P L E   5**   **Minimizing Total Cost**

Suppose the start-up costs of each production run is $50, it costs $5 to manufacture each item, and it costs $.50 to store each item for one year. Determine the number of items in each run and the number of runs in order to minimize the total cost if the total number of items to be produced and sold is 5000.

**Solution**   As we have seen above,

$$\text{cost of storage} = s\left(\frac{1}{2}x\right) = .50\left(\frac{1}{2}x\right) = \frac{1}{4}x,$$

where $x$ is the number of items produced in each run.

Let the number of runs be given by $n$. Then the total number of items produced must be given by the number of runs times the number of items in each run or

$$5000 = nx,$$

or

$$n = \frac{5000}{x}.$$

The total cost of preparing the plant for $n$ runs is the cost of preparing the plant for one run times the number of runs or

$$\text{total preparation cost} = 50n = 50\,\frac{5000}{x} = 250000\,\frac{1}{x}.$$

The total cost of manufacturing the 5000 items is given by 5000 times the cost of each item or

$$\text{total manufacturing cost} = 5000(5) = 25000.$$

The total cost, $C(x)$, is the sum of the previous three costs or

$$C(x) = 25000 + \frac{1}{4}x + 250000\,\frac{1}{x}.$$

We then wish to find the absolute minimum of this function on the interval [1, 5000]. The function is continuous on this interval.

Differentiating we obtain

$$C'(x) = \frac{1}{4} - 250000\,\frac{1}{x^2} = \frac{1}{4x^2}(x^2 - 1{,}000{,}000).$$

This derivative is continuous on [1, 5000]. Setting $C'(x) = 0$, we obtain

$$x^2 = 1{,}000{,}000.$$

This gives $x = 1000$ as the only critical value on [1, 5000].

From the factorization of $C'(x)$, we see that $C'(x) < 0$ on (0, 1000) and $C'(x) > 0$ on (1000, $+\infty$). Thus $C$ is decreasing on (1, 1000) and increasing on (1000, 5000). See Figure 13.69. From this it is clear that $C(1000)$ is the absolute minimum of $C$ on [1, 5000].

The number of runs is

$$n = \frac{5000}{x} = \frac{5000}{1000} = 5.$$

The minimum cost is

$$C(1000) = 25{,}000 + \frac{1}{4}(1000) + \frac{250{,}000}{1000}$$

$$= 25{,}500. \quad \blacksquare$$

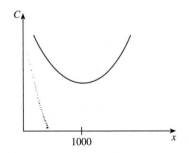

**Figure 13.69**

## SELF-HELP EXERCISE SET 13.6

1.  The figure shows a lake. A medical team is located at point A and must get to an injured person at point D as quickly as possible. The team can row a boat directly to point D, or can row the boat directly across the lake to point B and then run along the shore to D, or can row to some intermediate point C and then run down the shore to D. If the team can row at 6 mph and run at 10 mph, where on the opposite shore should they land the boat?

2.  A farmer wishes to enclose a rectangular field of area 200 square feet using an existing wall as one of the sides. The cost of the fence for the other three sides is $1 per foot. Find the dimensions of the rectangular field that minimizes the cost of the fence.

## EXERCISE SET 13.6

1.  Find two numbers whose sum is 16 and whose product is maximum.

2.  Find two numbers whose sum is 24 and whose product is maximum.

3.  Find two numbers whose sum is 20 and for which the sum of the squares is a minimum.

4.  Find two numbers whose sum is 32 and for which the sum of the squares is a minimum.

5.  Find two nonnegative numbers $x$ and $y$ with $2x + y = 12$ for which the term $xy^2$ is maximized.

6.  Find two nonnegative numbers $x$ and $y$ with $x + y = 60$ for which the term $x^2y$ is maximized.

7.  Find two numbers $x$ and $y$ with $x - y = 20$ for which the term $xy$ is minimized.

8.  Find two numbers $x$ and $y$ with $x - y = 36$ for which the term $xy$ is minimized.

9.  What is the area of the largest rectangle that can be enclosed by a circle of radius $a$?

10. What is the area of the largest rectangle that can be enclosed by a semicircle of radius $a$?

11. Find the coordinates of the point $P$ that maximizes the area of the rectangle shown in the figure.

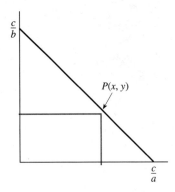

12. A triangle has base on the $x$-axis, one side on the $y$-axis, and the third side goes through the point $(1, 2)$ as shown in the figure. Find the slope of the line through $(1, 2)$ if the area of the triangle is to be a minimum.

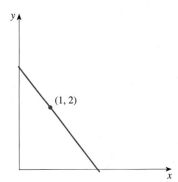

## Applications

13. **Fencing.** Where a river makes a right-angle turn, a farmer with 400 feet of fencing wishes to construct a rectangular fenced-in pasture that uses the river for two sides. What should be the dimensions of the rectangle in order to maximize the area enclosed?

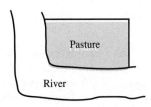

14. **Fencing.**   A farmer with 400 feet of fencing wishes to construct a rectangular pasture with maximum area. What should the dimensions be?

15. **Packaging.**   From a 9 inch by 9 inch piece of cardboard, square corners are cut out so that the sides can be folded up to form a box with no top. What should $x$ be to maximize the volume?

16. **Packaging.**   Find the dimensions of a cylindrical package with circular base of largest volume that can be mailed in this country.

17. **Cost.**   A fence is to be built around a 200-square-foot rectangular field. Three sides are to be made of wood costing $10 per foot while the other side is made of stone costing $30 per foot. Find the dimensions of the enclosure that minimizes total cost.

18. **Cost.**   A farmer wishes to enclose a rectangular field of area 450 square feet using an existing wall as one of the sides. The cost of the fence for the other three sides is $3 per foot. Find the dimensions of the rectangular field that minimizes the cost of the fence.

19. **Cost.**   What are the dimensions of the rectangular field of 20,000 square feet that will minimize the cost of fencing if one side costs three times as much per unit length as the other three?

20. **Fencing.**   A rectangular corral of 32 square yards is to be fenced off and then divided by a stone fence into two sections. The outer fence costs $10 per yard, while the special dividing stone fence costs $20 per yard. Find the dimensions of the corral that minimizes the cost.

21. **Revenue.**   A bus company charges $10 per person for a sightseeing trip if 30 people travel in a group. If for each person above 30 the company reduces the charge per person by $0.20, how many people will maximize the total revenue for the bus company?

22. **Revenue.** A hotel has 10 luxury units and during the peak season will rent all out at $300 per day. From experience management knows that one unit will become vacant for each $50 increase in charge per day. What rent should be charged to maximize revenue?

23. **Profit.** Suppose in the previous problem, the daily cost of maintaining and cleaning a rented luxury room was $100 per day. What rent should be charged to maximize profit?

24. **Profit.** An apple orchard produces annual revenue of $50 per tree when planted with 1000 trees. Due to overcrowding, the annual revenue per tree is reduced by 2 cents for each additional tree planted. If the cost of maintaining each tree is $10 per year, how many trees should be planted in order to maximize total profit from the orchard?

25. **Packaging.** A company wishes to design a rectangular box with square base and no top that will have a volume of 32 cubic inches. What should the dimensions be to yield a minimum surface area? What is the minimum surface area?

26. **Packaging.** A company wishes to design a rectangular box with square base and no top that will have a volume of 27 cubic inches. The cost of the bottom is twice the cost of the sides. What should the dimensions be to minimize the cost? What is the minimum cost?

27. **Cost.** A closed rectangular box of volume 324 cubic inches is to be made with a square base. If the material for the bottom costs twice as much as the material for the sides and top, find the dimensions of the box that minimize the cost of materials.

28. **Construction.** Find the most economical proportions for a closed cylindrical can that will hold 16 cubic inches.

29. **Construction.** Pipe is to be laid connecting $A$ with $B$ in the figure. The cost along the level stretch from $A$ to $C$ is $10 per foot whereas the cost along the difficult (away from the road) stretch from $C$ to $B$ is $20 per foot. The distance from $A$ to $D$ is 40 feet and the distance from $B$ to $D$ is 30 feet. At what point $C$ will the cost of laying pipe from $A$ to $B$ to $C$ be minimum?

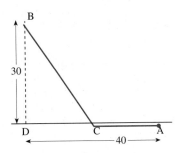

30. **Construction.** A wire 12 inches long is cut into two pieces with one piece used to construct a square and the other piece used to construct a circle. Where should the wire be cut to minimize the sum of the areas enclosed by the two figures?

31. **Construction.** A wire 12 inches long is cut into two pieces with one piece used to construct a square and the other piece used to construct an equilateral triangle. Where should the wire be cut to minimize the sum of the areas enclosed by the two figures? (The area of an equilateral triangle is $\dfrac{\sqrt{3}}{4}\,r^2$, where $r$ is the length of a side.)

32. **Construction.** An individual is planning to construct two fenced enclosures, one square and one circular. The cost per yard of constructing the circular enclosure is $8 and of the square enclosure is $4. If the total cost is fixed at $1600 dollars, what should be the dimensions of the two figures in order to minimize the area enclosed by the two figures?

33. **Construction.** An individual is planning to construct two fenced enclosures, one square and one an equilateral triangle. The cost per yard of constructing the triangle is $6 and of the square enclosure is $3. If the total cost is fixed at $1000 dollars, what should be the length of each side of the triangular enclosure in order to minimize the area enclosed by the two figures? (The area of an equilateral triangle is $\dfrac{\sqrt{3}}{4}\,r^2$, where $r$ is the length of a side.)

34. **Inventory Control.** Suppose the start-up costs of each production run is $1000 and that it costs $20 to manufacture each item and it costs $2 to store each item for one year. Determine the number of items in each run and the number of runs in order to minimize total cost if the total number of items to be produced and sold is 16,000.

35. **Inventory Control.** Suppose the start-up costs of each production run is $2500 and that it costs $20 to manufacture each item and it costs $2 to store each item for one year. Determine the number of items in each run and the number of runs in order to minimize total cost if the total number of items to be produced and sold is 10,000.

36. **Inventory Control.** In the discussion on inventory control in the text, let $T$ be the total number of items to be produced and sold, $k$ is the amount of money needed to prepare each production run, and $s$ the cost of storing one item for an entire year. Show that the number $x$ of items in each run needed in order to minimize the total cost is

$$x = \sqrt{\frac{2kT}{s}}.$$

Notice that the answer does not depend on the cost $p$ of producing each item. (If the total cost is $C(x)$, show that $C''(x) = 2kTx^{-3}$.)

## Solutions to Self-Help Exercise Set 13.6

1.  Let $x$ be the distance from B to C. Since

$$\text{time} = \frac{\text{distance}}{\text{rate}},$$

the time spent in the boat is $\dfrac{\sqrt{x^2 + 16}}{6}$, and the time spent running along the

shore is $\dfrac{10 - x}{10}$. Thus the total time of the trip is given by

$$T(x) = \frac{\sqrt{x^2 + 16}}{6} + \frac{10 - x}{10}.$$

We wish to find the minimum of $T$ on the interval $[0, 10]$. We have

$$T'(x) = \frac{x}{6\sqrt{x^2 + 16}} - \frac{1}{10}.$$

Then $T'(x) = 0$ if, and only if, $5x = 3\sqrt{x^2 + 16}$, or $25x^2 = 9x^2 + 144$, or $16x^2 = 144$ or $x = 3$. Evaluating $T(x)$ at the critical value $x = 3$ and the end points $x = 0$ and $x = 10$ gives

$$T(0) = \frac{5}{3} \approx 1.67, \quad T(3) = \frac{5}{6} + \frac{7}{10} \approx 1.53, \quad T(10) = \frac{\sqrt{116}}{6} \approx 1.80.$$

Thus the minimum occurs at $x = 3$. This means that the medical team should row to a point C located 3 miles below point B and run the remaining 7 miles to the injured person at point D.

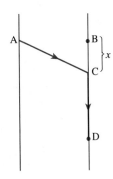

2.  Let $x$ and $y$ denote the sides of the rectangular enclosure shown in the figure. Then the area is given by $A = xy = 200$. The cost in dollars is the same as the number of feet of fence. Thus

$$C(x) = x + 2y = x + 2\frac{200}{x} = x + \frac{400}{x},$$

and

$$C'(x) = 1 - \frac{400}{x^2} = \frac{1}{x^2}(x^2 - 400).$$

Then $C'(x) = 0$ if and only if $x^2 = 400$ or $x = 20$. From the factorization of $C'(x)$ we see that $C'(x) < 0$ on $(0, 20)$ and $C'(x) > 0$ on $(20, +\infty)$. Thus $C$ is decreasing on $(0, 20)$ and increasing on $(20, +\infty)$. From this it is clear that $C(20)$ is an absolute minimum.

Solving for $y$ in the equation $xy = 200$ then gives $y = 10$. Thus the area should be 20 by 10.

# SUMMARY OUTLINE OF CHAPTER 13

- A function is said to be **increasing** on $(a, b)$ if $f(x_1) < f(x_2)$ when $x_1 < x_2$ and **decreasing** if $f(x_1) > f(x_2)$.

- The number $c$ is a **critical value** for the function $f(x)$ if $f'(c) = 0$ or if $f'(c)$ does not exist but $f(c)$ is defined.

- $f(x_M)$ is a **relative maximum** for the function $f(x)$ if $f(x) \le f(x_M)$ for all $x$ on some interval $(a, b)$ that contains $x_M$, and $f(x_m)$ is a **relative minimum** if $f(x) \ge f(x_m)$ for all $x$ on some interval $(a, b)$ that contains $x_m$.

- **First Derivative Test.**   The sign charts for $f'$ on either side of a critical value $c$ have the following consequences: $+ -$ implies a relative maximum, $- +$ implies a relative minimum, and $+ +$ and $- -$ both indicate no extremum.

- The **second derivative** $f''(x)$ of $y = f(x)$ is the derivative of the first derivative, that is, $f''(x) = \dfrac{d}{dx}[f'(x)]$. We also use the notation $f''(x) = \dfrac{d^2y}{dx^2}$.

- The graph of a function $f(x)$ is **concave up** ($\smile$) if the derivative $f'(x)$ is an increasing function. This will happen if $f''(x) > 0$. The graph of a function $f(x)$ is **concave down** ($\frown$) if the derivative $f'(x)$ is an decreasing function. This will happen if $f''(x) < 0$.

- An **inflection point** is a point $(c, f(c))$ on the graph of $f$ where the concavity changes sign. The value $c$ is called an **inflection value**.

- Acceleration $a(t) = v'(t)$, where $v(t)$ is the velocity.

- **Second Derivative Test.**   Let $f'(c) = 0$. If $f''(c) > 0$, then $f$ assumes a relative minimum at $c$. If $f''(c) < 0$, then $f$ assumes a relative maximum at $c$.

- If the graph of a function $y = f(x)$ approaches the horizontal line $y = a$, then the line $y = a$ is a **horizontal asymptote**. If the graph approaches the vertical line $x = b$, then the line $x = b$ is a **vertical asymptote**.

- $f(x_M)$ is the **absolute maximum** of the function $f(x)$ on the interval $I$ if $f(x)$ $\leq f(x_M)$ for all $x \in I$, and an **absolute minimum** on the interval $I$ if $f(x)$ $\geq f(x_M)$ for all $x \in I$.

- **Extreme Value Theorem.**   A continuous function on a closed and bounded interval attains its maximum and its minimum.

## Chapter 13 Review Exercises

1.  For the function whose graph is shown in the accompanying figure, find all critical values and the largest open intervals on which the function is increasing and the largest open intervals on which the function is decreasing.

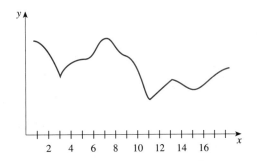

2.  Suppose the derivative of a function $y = f(x)$ is given by

$$f'(x) = (x + 2)(x - 1).$$

Find all the largest open intervals on which the function is increasing, the largest open intervals on which the function is decreasing, all critical values, and the $x$-coordinate of all relative extrema.

3.  Repeat the previous exercise if

$$f'(x) = (x + 4)(x - 2)^2(x - 5).$$

In Exercises 4 through 21 find the intervals on which the function is increasing, decreasing, is concave up and concave down, the critical values, the $x$-coordinate of each relative maximum and minimum, the inflection values, the vertical asymptotes, and the horizontal asymptotes. Then sketch a graph.

4.  $y = f(x) = 2x^2 + 3x - 4$       5.  $y = f(x) = -x^2 + 2x + 5$

6.  $y = f(x) = -3x^2 - 6x + 1$       7.  $y = f(x) = 3x^2 + 12x - 10$

8.  $y = f(x) = \dfrac{1}{3}x^3 - x + 1$       9.  $y = f(x) = -x^3 + 3x + 2$

10.  $y = f(x) = x^4 - 4x^3 + 4$       11.  $y = f(x) = x^5 - 15x^4 + 5$

12.  $y = f(x) = x^4 - 18x^2 + 15$       13.  $y = f(x) = -x^5 + 15x^3 - 10$

**14.** $y = f(x) = x + \dfrac{4}{x^2}$

**15.** $y = f(x) = x^2 + \dfrac{1}{x^2}$

**16.** $y = f(x) = x - \dfrac{1}{x^2}$

**17.** $y = f(x) = x^2 - \dfrac{27}{x^2}$

**18.** $y = f(x) = \dfrac{x}{x - 5}$

**19.** $y = f(x) = \dfrac{x - 1}{x - 5}$

**20.** $y = f(x) = \sqrt{x^6 + 1}$

**21.** $y = f(x) = (9 - x)^{1/5}$

In Exercises 22 through 31 find the location of all absolute maxima and absolute minima on the given intervals.

**22.** $f(x) = \dfrac{x}{3 + x^4}, \ (-\infty, +\infty)$

**23.** $f(x) = 1 - \dfrac{x}{48 + x^4}, \ (-\infty, +\infty)$

**24.** $f(x) = x - \dfrac{1}{x}, \ [1, +\infty)$

**25.** $f(x) = \dfrac{1}{x^3} - x, \ [1, +\infty)$

**26.** $f(x) = 2x^3 - 3x^2 + 1, \ [-1, 2]$

**27.** $f(x) = 2x^3 - 9x^2 + 12x + 1, \ [0, 3]$

**28.** $f(x) = (x - 2)^2, \ [0, 3]$

**29.** $f(x) = x + \dfrac{4}{x^2}, \ [1, 3]$

**30.** $f(x) = x^2 + \dfrac{16}{x^2}, \ [1, 3]$

**31.** $f(x) = x - \dfrac{1}{x^2}, \ [1, 4]$

**32.** Show that the graph of the special fourth order polynomial function

$$f(x) = ax^4 + bx^2 + cx + d$$

is always concave up if both $a$ and $b$ are positive and always concave down if both $a$ and $b$ are negative.

**33.** If $b > a$ and $c$ is any constant, show that the polynomial function

$$f(x) = -2x^3 + 3(a + b)x^2 - 6abx + c$$

has exactly one relative minimum and exactly one relative maximum. Locate and identify them.

**34.** **Average Cost.**   An electric utility company estimates that the cost, $C(x)$, of producing $x$ kilowatts of electricity is given by $C(x) = a + bx^2$, where $a$ represents the fixed costs and $b$ is the unit cost. The *average* cost, $\overline{C}(x)$, is defined to be the total cost of producing $x$ kilowatts divided by $x$, that is,

$$\overline{C}(x) = \frac{C(x)}{x}.$$

Find where the average cost increases and where it decreases.

**35.** **Average Cost.**   The total cost function for producing $x$ items is given by

$C(x) = 1000 + 10x^2$. How many items should be produced to minimize $\overline{C}(x) = C(x)/x$, the average cost per item?

36. **Average Cost.** In the previous problem show that average cost is minimized at the value at which $\overline{C} = C'$.

37. **Average Cost.** Suppose the total cost function for producing $x$ items is given by $C(x)$ and the average cost per item is given by $\overline{C}(x) = C(x)/x$. If average cost is at a minimum at some value on $(0, +\infty)$, then show that at this value $\overline{C} = C'$.

38. **Biology.** In the attack equation the rate of change $R$ of the number attacked with respect to the number $N$ vulnerable to attack is given by

$$R(x) = x^{1-a} (kx - N),$$

where $x$ is the number of attackers, $a$ and $k$ are positive constants with $a < 1$. Find the number of attackers that minimizes $R$.

39. **Average Cost.** An electric utility company estimates that the cost, $C(x)$, of producing $x$ kilowatts of electricity is given by $C(x) = a + bx^3$, where $a$ represents the fixed costs and $b$ is a positive constant. The *average* cost, $\overline{C}(x)$, is defined to be the total cost of producing $x$ kilowatts divided by $x$, that is,

$$\overline{C}(x) = \frac{C(x)}{x}.$$

Find where the average cost attains a minimum.

40. **Cost.** An individual is planning to construct two fenced enclosures, one square and one an equilateral triangle. The cost per yard of constructing the triangle is $20 and of the square enclosure is $10. If the total cost is fixed at $1000, what should be the length of each side of the square in order to maximize the area enclosed by the two figures? (The area of an equilateral triangle is $\frac{\sqrt{3}}{4} r^2$, where $r$ is the length of a side.)

41. **Revenue.** A bus company charged $60 per person for a sightseeing trip and obtained 40 people for the trip. The company has data that indicate that for the same trip, each $2 increase in the price above $60 results in the loss of one customer. What should the company charge to maximize revenue?

42. **Profits.** The bus company in the previous exercise has fixed costs of $3000 for each trip, and, in addition, there are costs of $4 per customer. What should the company charge to maximize profits?

43. **Packaging.** A manufacturer wishes to make a cylindrical can of 108 cubic inches. The top and bottom will cost twice as much per square inch as the sides. What should the dimensions of the can be to minimize the cost?

44. **Inventory Control.** A company produces 1000 cases of perfume a year. The start-up cost of each production run is $125, it costs $50 to manufacture each case, and $4 to store each case for a year. Determine the number of cases to manufacture in each run and the number of runs in order to minimize the total cost.

**At what rate is the distance between two boats changing?** (© Frank Gordon/ FPG International Corporation.)

# More on the Derivative

14.1   The Linear Approximation and Differentials

14.2   Implicit Differentiation

14.3   Related Rates and Elasticity of Demand

I n this chapter we continue to see how we can use the derivative to our advantage. In the first section we will see that near a point where a function is differentiable, the graph of the function is approximately the same as the graph of the line tangent to the curve at the point. In the second section we will see how, when given an equation involving $x$ and $y$ for which we cannot solve explicitly for $y$, we can still find the derivative. In the third section we will see how, in many important instances, one rate of change can be found if we know some other rate of change. Also in this section we will see how economists measure the sensitivity of price to changes in demand.

## 14.1   THE LINEAR APPROXIMATION AND DIFFERENTIALS

▶ *The Tangent Line Approximation*

▶ *Differentials*

One month after using an experimental drug, the radius of a spherical cancer was measured to have decreased from 3 centimeters to 2.8 centimeters. Using the derivative, what linear function can be used to estimate the change in the volume of the spherical cancer? See Example 1 for the answer.

What is an easy-to-calculate approximation to $\sqrt{9.3}$ without using a calculator or computer? See Example 2 for the answer.

## The Tangent Line Approximation

Given a function $y = f(x)$ differentiable at $x_0$, we know that

$$f'(x_0) = \lim_{h \to 0} \frac{f(x_0 + h) - f(x_0)}{h}.$$

From our knowledge of limits we know that if $h$ is small, then

$$f'(x_0) \approx \frac{f(x_0 + h) - f(x_0)}{h}.$$

Multiplying both sides of this expression by $h$ then yields

$$f(x_0 + h) - f(x_0) \approx f'(x_0)h.$$

The term $f(x_0 + h) - f(x_0)$ is the actual change in $y$ and is shown in Figure 14.1. The term $f'(x_0)h$ gives the change in $y$ that occurs when moving along the tangent line. According to Figure 14.1, we expect the change in $y$ when moving along the tangent line, which is given by $f'(x_0)h$, to be an increasingly better approximation to the actual change in $y$, which is given by $f(x_0 + h) - f(x_0)$, as $h$ becomes smaller. The term $E = [f(x_0 + h) - f(x_0)] - f'(x_0)h$ represents

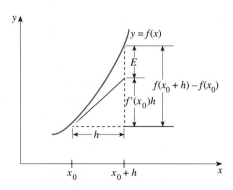

**Figure 14.1**

the error in using the term $f'(x_0)h$ to approximate the actual change in $y$. We see that as $h$ approaches zero, this error $E$ will also approach zero.

---

**Approximating the Change in $y$**

If $y = f(x)$ is a differentiable function at $x_0$, then the change in $y$, given by $f(x_0 + h) - f(x_0)$, can be approximated by the linear function $f'(x_0)h$. That is,

$$f(x_0 + h) - f(x_0) \approx f'(x_0)h \tag{1}$$

if $h$ is small.

---

The following example illustrates a way in which this approximation can be used.

**E X A M P L E   1**     **Approximating the Change in a Cancerous Growth**

One month after using an experimental drug, the radius of a spherical cancer was measured to have decreased from 3 centimeters to 2.8 centimeters. Use Equation (1) to approximate the change in volume.

**Solution**     If $r$ is the radius, then the volume $V$ is given by $V = \frac{4}{3}\pi r^3$. We note that $V'(r) = 4\pi r^2$. Since $r$ goes from 3 to 2.8, we take $r_0 = 3$ and $h = -0.2$. The actual change in volume is given by

$$V(r_0 + h) - V(r_0) = V(2.8) - V(3).$$

Using Equation (1) to approximate this change, we have

$$V(2.8) - V(3) = V(r_0 + h) - V(r_0)$$
$$\approx V'(r_0)h$$
$$= 4\pi(3)^2(-0.2)$$
$$\approx -22.6$$

Compare this number to the actual change in volume

$$\Delta V = V(2.8) - V(3) = -21.1. \quad \blacksquare$$

We have already used the tangent line approximation earlier when considering marginal profits, costs, and revenues. For example, suppose a profit function $P = P(x)$ is given. Then according to Equation (1)

$$P(x_0 + h) - P(x_0) \approx P'(x_0)h.$$

If we take $h = 1$, then we obtain

$$P(x_0 + 1) - P(x_0) \approx P'(x_0).$$

This says that $P'(x_0)$ is approximately equal to the profit of the $(x_0 + 1)$st item. We first discovered this in Section 12.5.

We will now turn our attention to establishing a fundamentally important geometrical fact concerning the behavior of a function near a point where the function is differentiable. Briefly stated, if $f'(x_0)$ exists, then for values of $x$ near to $x_0$, the graph of the function $y = f(x)$ is approximately the same as the graph of the tangent line through $(x_0, f(x_0))$.

We now see why this is true. In Equation (1) move the $f(x_0)$ term to the right-hand side, and obtain

$$f(x_0 + h) \approx f(x_0) + f'(x_0)h.$$

Think of $x_0$ as fixed and $h$ as the variable and let $x = x_0 + h$. Then the last equation becomes

$$f(x) \approx f(x_0) + f'(x_0)(x - x_0).$$

Recall that the right-hand side of this last equation is the equation of the line tangent to the curve at $x_0$. Thus we are saying that the graph of $y = f(x)$ is approximately the same as the graph of the line tangent to the curve at $x = x_0$. See Figure 14.2.

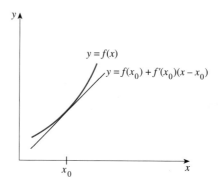

**Figure 14.2**

---

**Tangent Line Approximation**

If $y = f(x)$ is differentiable at $x_0$, then for values of $x$ near $x_0$

$$f(x) \approx f(x_0) + f'(x_0)(x - x_0) \tag{2}$$

Thus for values of $x$ near $x_0$ the graph of the curve $y = f(x)$ is approximately the same as the graph of the tangent line through the point $(x_0, f(x_0))$.

E X A M P L E   2          **The Tangent Line Approximation**

Use the tangent line approximation (2) to find an approximation to $\sqrt{9.3}$ without resorting to a calculator.

**Solution**          Let $f(x) = \sqrt{x}$, $x = 9.3$, and $x_0 = 9$. Then $f'(x) = \dfrac{1}{2\sqrt{x}}$ and using the approximation (2) we have

$$\sqrt{9.3} = f(9.3)$$

$$\approx f(9) + f'(9)(9.3 - 9)$$

$$= \sqrt{9} + \frac{1}{2\sqrt{9}}(0.3)$$

$$= 3 + \frac{0.3}{6}$$

$$= 3.05 \quad \blacksquare$$

One should compare this to the actual $\sqrt{9.3}$ which to four decimal places is 3.0496. This approximation is off by only .0004. In Figure 14.1, this means that $E \cong -.0004$.

## Differentials

Using the notation $\Delta x = h$ and $\Delta y = f(x_0 + \Delta x) - f(x_0)$, the approximation formula (1) can be written as

$$\Delta y \approx f'(x_0)\Delta x.$$

Replacing $x_0$ with any (fixed) $x$, we can also write

$$\Delta y \approx f'(x)\Delta x.$$

Continuing to think of $x$ as *fixed*, we now define the **differential** $dy$ as the *linear* function of $dx$ given by

$$dy = f'(x)\, dx,$$

where $dx$ is the independent variable and $dy$ the dependent variable. Since $x$ is *fixed*, we do not think of $dy$ as a function of $x$, rather, we think of $dy$ as a function of $dx$.

---

**Definition of the Differential**

If $y = f(x)$ is differentiable at $x$, then the **differential**, denoted by $dy$, is defined to be the linear function of $dx$ given by

$$dy = f'(x)\, dx.$$

---

Figure 14.3 is the same as Figure 14.1 but using the new notation. In Figure 14.3, $\Delta y$ gives the actual change in $y$ on the interval $[x, x + dx]$, whereas the differential $dy = f'(x) \, dx$ gives the change that occurs when moving along the tangent line. With this notation we then have the following.

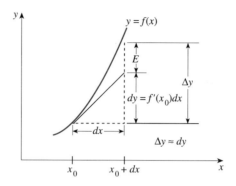

**Figure 14.3**

---

**Linear Approximation Using Differentials**

If $y = f(x)$ is differentiable at $x$ and $\Delta x = dx$, then for $\Delta x$ close to 0,

$$\Delta y \approx dy,$$

where

$$\Delta y = f(x + \Delta x) - f(x)$$

$$dy = f'(x) \, dx.$$

---

**REMARK.**    Notice that the formula for the differential $dy = f'(x) \, dx$ could also be obtained by first writing

$$\frac{dy}{dx} = f'(x),$$

and multiplying by $dx$. Thus we may regard the derivative $\dfrac{dy}{dx}$ as a quotient of differentials.

**E X A M P L E   4**    **Differentials and Linear Approximations**

Use the differential to approximate the volume of a coating of material 0.1 inch thick covering a cube with side of 4 inches.

**Solution**    The volume of a cube of length $s$ is given by $V = f(s) = s^3$. The coating increases each dimension by *twice* 0.1. Taking $s_0 = 4$ and $ds = 0.2$ gives

$$dV = f'(4)\ ds$$

$$= 3(4)^2\ (0.2)$$

$$= 9.6 \quad \blacksquare$$

Compare this to the actual volume of $f(4.2) - f(4) = 10.088$.

## SELF-HELP EXERCISE SET 14.1

1. Use the tangent line approximation $f(x) \approx f(x_0) + f'(x_0)(x - x_0)$ to find an approximation to $\sqrt[3]{7.7}$ without resorting to a calculator.

2. Evaluate the differential $dy$ for $y = f(x) = \sqrt[3]{x}$ at the point $x_0 = 8$ in terms of $dx$. Then find $dy$ for $dx = 0.1$.

## EXERCISE SET 14.1

In Exercises 1 through 10 use the linear approximation $f(x) \approx f(x_0) + f'(x_0)(x - x_0)$ to approximate the given quantity. Use a calculator to find the actual answer and compare your approximation to this. (Of course the answer on your calculator is also an approximation.)

1. $\sqrt{4.1}$     2. $\sqrt{16.2}$     3. $\sqrt{3.9}$     4. $\sqrt{15.7}$

5. $\sqrt[3]{8.12}$     6. $\sqrt[3]{27.54}$     7. $\sqrt[3]{7.76}$     8. $\sqrt[3]{26.19}$

9. $\sqrt[4]{16.64}$     10. $\sqrt[4]{80}$

In Exercises 11 through 16 estimate the cost of the next item or the revenue or profits from the sale of the next item for the indicated functions and the indicated $x$.

11. $C(x) = 30 + .02x^2, x = 30$     12. $C(x) = 10 + 3x + .04x^2, x = 30$

13. $R(x) = x^2 - 2x, x = 20$     14. $R(x) = 0.2x^3 - 5x, x = 20$

15. $P(x) = x^2 - 5x - 60, x = 50$     16. $P(x) = x^3 - 10x - 30, x = 10$

In Exercises 17 through 24 find (a) an expression for $dy$ in terms of $x$ and $dx$, (b) an expression for $dy$ in terms of $dx$ at the indicated point $x_0$, (c) $dy$ for the indicated values of $x_0$ and $dx$.

17. $y = f(x) = x^2 + x + 1, x_0 = 3, dx = 0.1$

18. $y = f(x) = x^3 - 2x + 10, x_0 = 2, dx = 0.2$

19. $y = f(x) = 2x^3 + x - 3, x_0 = -2, dx = -0.1$

**20.** $y = f(x) = x^3 - 2x + 10$, $x_0 = 1$, $dx = -0.2$

**21.** $y = f(x) = \sqrt{x^3}$, $x_0 = 4$, $dx = 0.01$

**22.** $y = f(x) = \sqrt{x}$, $x_0 = 9$, $dx = -.15$

**23.** $y = f(x) = \dfrac{x}{1 + x^2}$, $x_0 = -1$, $dx = 0.1$

**24.** $y = f(x) = \dfrac{1 + x}{3 + x^2}$, $x_0 = 2$, $dx = -0.1$

**25.** If the radius of a circle increases its size from 4 inches to 4.1 inches, use the linear approximation formula to estimate the change in area.

**26.** If the side of a square decreases from 4 inches to 3.8 inches, use the linear approximation formula to estimate the change in area.

**27.** If the side of a cube decreases from 10 feet to 9.6 feet, use the linear approximation formula to estimate the change in volume.

**28.** A cone has a height of 6 inches. The radius has been measured as one inch with a possible error of 0.01 in. Estimate the maximum error in the volume ($\pi r^2 h/3$) of the cone.

## Applications

**29. Demand.** The demand equation for a certain product is given by

$$p(x) = \frac{10}{\sqrt{x + 1}}.$$

Use the linear approximation formula to estimate the change in price $p$ as $x$ changes from 99 to 102.

**30. Revenue.** The revenue equation for a certain product is given by

$$R(x) = 10\sqrt{x}(100 - x).$$

Use the linear approximation formula to estimate the change in revenue as $x$ changes from 400 to 402.

**31. Cost.** The cost equation for a certain product is given by

$$C(x) = \sqrt{x^3}(x + 100) + 1000.$$

Use the linear approximation formula to estimate the change in costs as $x$ changes from 900 to 904.

**32. Profit.** The profit equation for a certain product is given by

$$P(x) = \sqrt{x}(1000 - x) - 5000.$$

Use the linear approximation formula to estimate the change in profits as $x$ changes from 25 to 27.

*(© Omikron Collection/Photo Researchers, Inc.)*

**33.   Biology.**  Use the linear approximation formula to estimate the change in volume of a spherical bacterium as the radius increases from 2 microns to 2.1 microns.

**34.   Biology.**  An anatomist wishes to measure the surface area of a bone assumed to be a cylinder. The length $l$ can be measured with great accuracy and can be assumed to be known exactly. However, the radius of the bone varies slightly throughout its length, making it difficult to decide what the radius should be. Discuss how slight changes in the radius $r$ will approximately affect the surface area.

**35.   Learning.**  The time $T$ it takes to memorize a list of $n$ items is given by

$$T(n) = 3n\sqrt{n} - 3.$$

Find the approximate change in time required from memorizing a list of 12 items to memorizing a list of 14 items.

## Graphing Calculator Exercises

In Exercises 36 through 41 enter the function $y = f(x)$ and the equation of the tangent line at the indicated point $x_0$ into your graphing calculator and draw the graphs one at a time. Then use the zoom feature to see a magnified view of the graph near the point $(x_0, f(x_0))$. Repeat this process a number of times and show that the graph of the curve approximates the graph of the tangent line.

**36.**  $y = x^2, x_0 = -1$                    **37.**  $y = x^2, x_0 = 1$

**38.**  $y = 5 - x^2, x_0 = 0$                  **39.**  $y = 6x - x^2, x_0 = 2$

**40.**  $y = 6x - x^2, x_0 = 4$                 **41.**  $y = 6x - x^2, x_0 = 2$

## Solutions to Self-Help Exercise Set 14.1

**1.**   Let $y = f(x) = \sqrt[3]{x}$, $x = 7.7$, and $x_0 = 8$. Then since $f'(x) = \dfrac{1}{3\sqrt[3]{x^2}}$,

$$\sqrt[3]{7.7} = f(x) \approx f(x_0) + f'(x_0)(x - x_0)$$

$$= \sqrt[3]{8} + \frac{1}{3\sqrt[3]{8^2}}(-0.3) = 1.975.$$

On a calculator, one obtains $\sqrt[3]{7.7} \approx 1.9747$ to four decimal places.

**2.**   Since $f(x) = \sqrt[3]{x}$, $f'(x) = \dfrac{1}{3}x^{-2/3}$ and at $x_0$

$$dy = \frac{1}{3\sqrt[3]{8^2}}\,dx = \frac{1}{12}\,dx.$$

If $dx = 0.1$, then

$$dy = \frac{1}{12}(0.1) = 0.00833.$$

## 14.2     IMPLICIT DIFFERENTIATION

▶ *Implicit Differentiation*

▶ *Applications*

A P P L I C A T I O N

**Rate of Change for
a Demand Function**

The demand function for a certain commodity is given by

$$p = \frac{13}{1 + x^2 + x^3}.$$

What is $\frac{dx}{dp}$ when $x = 2$? See Example 2 for the answer.

## Implicit Differentiation

We often have an expression such as $F(x, y) = 0$ in which we are unable to solve for $y$ as an explicit function of $x$. Nevertheless, we may still need to find $\frac{dy}{dx}$

at some specific point $x = x_0$. In this section we shall see how to find $\frac{dy}{dx}$ at a

point without knowing $y$ as an explicit function of $x$.

We usually have encountered functions of the form $y = f(x)$, which express $y$ explicitly as a function of $x$. But there have been equations that we have had to deal with such as

$$x^2 + y^2 = 1,$$

that do not give $y$ as an explicit function of $x$. Still, this last equation can define $y$ as a function of $x$. If we set $x$ equal to some specific value, then the resulting equation can be solved for one or more values of $y$. If we specify which value of $y$ to take, we have a function. We then say that $y$ is an **implicit** function of $x$.

For example, if $x^2 + y^2 = 1$, and we set $x = 0$, then there results the equation $y^2 = 1$, which has solutions $y = -1$ and $y = +1$.

If we were interested in finding the slope to the curve $x^2 + y^2 = 1$ at the point $(1/\sqrt{2}, 1/\sqrt{2})$, we would have to solve for $y$ as an *explicit* function of $x$ and then find the derivative. In this case we would have

$$y = f(x) = \sqrt{1 - x^2},$$

and

$$f'(x) = -\frac{x}{\sqrt{1 - x^2}}.$$

Then

$$f'(1/\sqrt{2}) = -1.$$

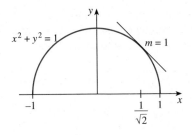

**Figure 14.4**

See Figure 14.4.

However, there are equations such as

$$x^4 - 2xy^3 + y^5 = 32$$

for which it is not possible to solve for $y$ explicitly in terms of $x$. In such cases we still need to know the slope of the curve at a given point. It may seem impossible to ever be able to find $\dfrac{dy}{dx}$ at a point if we do not know $y$ as a function of $x$. But the technique of *implicit differentiation* will enable us to do precisely this.

Let us now see how we can find the slope of the tangent line to the curve $x^2 + y^2 = 1$ given above at the point $(1/\sqrt{2}, 1/\sqrt{2})$ without finding $y$ as an explicit function of $x$. We first *assume* that it is possible at least theoretically to solve for $y$ explicitly as, say, $y = f(x)$. Then

$$x^2 + [f(x)]^2 = x^2 + y^2 = 1.$$

Thus

$$x^2 + [f(x)]^2 = 1.$$

We now differentiate this last equation, using the rules we have already learned and obtain

$$\frac{d}{dx}x^2 + \frac{d}{dx}[f(x)]^2 = \frac{d}{dx}(1).$$

This gives

$$2x + 2f(x)f'(x) = 0.$$

We then solve for $f'(x)$ and obtain

$$\frac{dy}{dx} = f'(x) = -\frac{x}{f(x)} = -\frac{x}{y}.$$

At the point under consideration $x = 1/\sqrt{2}$ and also $y = 1/\sqrt{2}$. Using this in the previous equation yields

$$f'(1/\sqrt{2}) = -\frac{x}{y} = -\frac{1/\sqrt{2}}{1/\sqrt{2}} = -1.$$

Notice that we were able to find $f'(1/\sqrt{2})$ without ever knowing what $f(x)$ was explicitly!

**E X A M P L E   1**   **Finding the Derivative of an Implicitly Defined Function**

Let $x^4 - 2xy^3 + y^5 = 32$.

(a)   Find $\dfrac{dy}{dx}$.

(b)   Find the slope of the tangent line to the curve at $x_0 = 0$.

**Solutions**

(a)   It would be extremely difficult, if not impossible, to solve for $y$ *explicitly* as a function of $x$. We then need to differentiate *implicitly*. We assume that $y = f(x)$ for some $f(x)$ and substitute this into the last equation giving

$$x^4 - 2x[f(x)]^3 + [f(x)]^5 = 32.$$

Differentiating we obtain

$$\frac{d}{dx}x^4 - \frac{d}{dx}2x[f(x)]^3 + \frac{d}{dx}[f(x)]^5 = \frac{d}{dx}(32).$$

Using familiar rules we obtain

$$4x^3 - 2[f(x)]^3 - (2x)3[f(x)]^2 f'(x) + 5[f(x)]^4 f'(x) = 0.$$

We now solve this equation for $f'(x)$ and obtain

$$(-6x[f(x)]^2 + 5[f(x)]^4)f'(x) = 2[f(x)]^3 - 4x^3,$$

and finally

$$f'(x) = \frac{2[f(x)]^3 - 4x^3}{-6x[f(x)]^2 + 5[f(x)]^4}.$$

or

$$\frac{dy}{dx} = f'(x) = \frac{2y^3 - 4x^3}{-6xy^2 + 5y^4}.$$

(b)   We wish to find $f'(0)$. But when $x = 0$, the original equation becomes

$$0^4 - 2(0)y^3 + y^5 = 32,$$

or $y = 2$. Thus we substitute $x = 0$ and $y = 2$ in the answer to part (a) and obtain

$$f'(0) = \frac{2(2)^3 - 0}{0 + 5(2)^4} = \frac{1}{5}. \quad \blacksquare$$

**REMARK.**    In the above analysis we *assumed* that we could at least theoretically solve for $y = f(x)$ for some $f(x)$. There is a theorem called the *implicit function theorem* that gives specific conditions under which this assumption is guaranteed. We do not go into these conditions in this text.

**REMARK.**    Notice in the previous work that the equation involving $f'(x)$ was *linear* in $f'$ and therefore it was easy to solve for $f'$. *This is always the case.*

## Applications

**E X A M P L E  2**    **Finding the Derivative of an Implicitly Defined Function**

The demand function for a certain commodity is given by

$$p = \frac{13}{1 + x^2 + x^3}.$$

Find the instantaneous rate of change of the number sold with respect to the price when $x = 2$.

**Solution**    It would be extremely difficult, if not impossible, to solve for $x$ explicitly as a function of $p$ in order to find $\dfrac{dx}{dp}$. Thus we try implicit differentiation.

We assume that $x = f(p)$ for some function $f(p)$. Then

$$p = \frac{13}{1 + [f(p)]^2 + [f(p)]^3}.$$

Clearing the fraction, we obtain

$$p + p[f(p)]^2 + p[f(p)]^3 = 13.$$

Now differentiate this with respect to $p$ and obtain

$$\frac{d}{dp}p + \frac{d}{dp}p[f(p)]^2 + \frac{d}{dp}p[f(p)]^3 = \frac{d}{dp}(13).$$

Now use the differentiation techniques from previous sections and obtain

$$1 + [f(p)]^2 + 2pf(p)f'(p) + [f(p)]^3 + 3p[f(p)]^2 f'(p) = 0.$$

This is a *linear* equation if $f'$. Solve for $f'$ and obtain

$$f'(p) = -\frac{1 + [f(p)]^2 + [f(p)]^3}{2pf(p) + 3p[f(p)]^2},$$

or

$$\frac{dx}{dp} = f'(p) = -\frac{1 + x^2 + x^3}{2px + 3px^2}.$$

Now when $x = 2$,

$$p = \frac{13}{1 + 2^2 + 2^3} = 1.$$

Then

$$f'(1) = -\frac{1 + 2^2 + 2^3}{2 \cdot 1 \cdot 2 + 3 \cdot 1 \cdot 2^2} = -\frac{13}{16}. \quad \blacksquare$$

## SELF-HELP EXERCISE SET 14.2

1.   If $2x^2 \sqrt{y} + y^3 + x^4 = 27$, find $\dfrac{dy}{dx}$.

2.   Find the slope of the tangent line to the curve in the previous example at the point $x_0 = 0$.

## EXERCISE SET 14.2

In Exercises 1 through 14 differentiate implicitly and find the slope of the curve at the indicated point.

1. $xy + 2x + y = 6, (1, 2)$

2. $xy + x + y = -1, (-2, -1)$

3. $xy + x + y^2 = 7, (1, 2)$

4. $x^2y + x - y^3 = -1, (2, -1)$

5. $3x^2 - y^2 = 23, (3, 2)$

6. $x^3 - 2y^3 = -15, (1, 2)$

7. $x^2y - 2x + y^3 = 1, (2, 1)$

8. $\sqrt{xy} + x^2 - y^4 = 17, (4, 1)$

9. $x^2 + y^2 + xy^3 = 4, (0, 2)$

10. $\sqrt{x} - \sqrt{y} = 1, (9, 4)$

11. $x^2y^2 + xy^4 = 2, (1, 1)$

12. $xy^2 + xy + y^3 = 5, (2, 1)$

13. $x\sqrt{y} - y^2 + x^2 = -13, (1, 4)$

14. $\sqrt{xy} + x^3y^2 = 84, (1, 9)$

## Applications

In Exercises 15 through 22 the price of a commodity is given as a function of the demand $x$. Use implicit differentiation to find $\dfrac{dx}{dp}$ for the indicated $x$.

15. $p = -2x + 15, x = 3$

16. $p = -3x + 20, x = 5$

17. $p = \dfrac{3}{1 + x}, x = 2$

18. $p = \dfrac{5}{1 + x^2}, x = 2$

19. $p = \dfrac{15}{3 + x + x^2}, x = 3$

20. $p = \dfrac{5}{3 + x + x^3}, x = 1$

21. $p = \sqrt{25 - x^2}, x = 4$

22. $p = \sqrt{25 - x^4}, x = 2$

23. **Demand.** Suppose the demand $x$ and the price $p$ are related by the equation

$$16x^2 + 100p^2 = 500.$$

Find $\dfrac{dx}{dp}$ at the point where $p = 1$.

24. **Demand.** Suppose the demand $x$ and the price $p$ are related by the equation

$$x^2 + 100p^2 = 20{,}000.$$

Find $\dfrac{dp}{dx}$ at the point where $x = 100$.

25. **Medicine.** Poiseuille's law states that the total resistance $R$ to blood flow in a blood vessel of constant length $l$ and radius $r$ is given by

$$R = \frac{al}{r^4},$$

where $a$ is a positive constant. Use implicit differentiation to find $\dfrac{dr}{dR}$ when $R = 1$.

**26.** **Medicine.**  It is well known that when a person coughs, the diameter of the trachea and bronchi shrinks. It has been shown that the flow of air $F$ through the windpipe is given by

$$F = k(R - r)r^4,$$

where $r$ is the radius of the windpipe under the pressure of the air being released from the lungs, $R$ is the radius with no pressure, and $k$ is a positive constant. Suppose $k = 1$ and $R = 2$. Use implicit differentiation to find $\dfrac{dr}{dF}$ when $F = 1$.

**27.** **Biology.**  Biologists have proposed that the rate of production $P$ of photosynthesis is related to the light intensity $I$ by the formula

$$P = \frac{aI}{b + I^2},$$

where $a$ and $b$ are positive constants. Suppose $a = 6$ and $b = 8$. Use implicit differentiation to find $\dfrac{dI}{dP}$ when $P = 1$ and $I = 2$.

**28.** **Biology.**  The free water vapor diffusion coefficient $D$ in soft woods is given by

$$D = \frac{167.2}{P}\left(\frac{T}{273}\right)^{1.75},$$

where $T$ is the temperature and $P$ is the constant atmospheric pressure measured in appropriate units. Suppose $P = 1$. Use implicit differentiation to find $\dfrac{dT}{dD}$ when $D = 1$.

## Solutions to Self-Help Exercise Set 14.2

**1.**  Assume that we have $y = f(x)$, then

$$2x^2\sqrt{f(x)} + [f(x)]^3 + x^4 = 27.$$

Now differentiate the above expression and obtain

$$4x[f(x)]^{1/2} + x^2[f(x)]^{-1/2}f'(x) + 3[f(x)]^2f'(x) + 4x^3 = 0.$$

Solving for $f'(x)$ gives

$$f'(x) = -\frac{4x[f(x)]^{1/2} + 4x^3}{x^2[f(x)]^{-1/2} + 3[f(x)]^2}$$

$$= -4\,\frac{x\sqrt{y} + x^3}{\dfrac{x^2}{\sqrt{y}} + 3y^2}$$

$$= -4\,\frac{xy + x^3\sqrt{y}}{x^2 + 3y^{5/2}}$$

2. Setting $x_0 = 0$ in the original equation gives

$$0 + y^3 + 0 = 27,$$

or $y = 3$. Substituting $x = 0$ and $y = 3$ in the answer to the previous exercise yields

$$f'(0) = -4 \frac{(0)(3) + (0)(\sqrt{3})}{0 + (3)3^{5/2}} = 0.$$

## 14.3   RELATED RATES AND ELASTICITY OF DEMAND

▶ *Related Rates*

▶ *Elasticity of Demand*

APPLICATION
**Spread of an Oil Slick**

An oil slick is spreading in a circular fashion. The radius of the circular slick is observed to be increasing at 0.1 miles per day. What is the rate of increase of the area of the slick when $r = 2$ miles? See Example 2 for the answer.

APPLICATION
**Relating Changes in Demand to Changes in Price**

Suppose a demand equation is

$$x = f(p) = 100 - 2p.$$

When $p = 40$ how are changes in demand related to changes in price? See Example 4 for the answer.

## Related Rates

In the first part of this section the rate of change of one quantity is related to the rate of change of another quantity by the chain rule. In the second part of this section, we consider a way in which economists analyze the demand function.

It is often the case when given a function, such as a cost function, $C = C(x)$, that the number of items sold is a function of time. For example, a company may estimate that their sales may increase on a yearly basis at the same rate as the increase in population. Thus the number sold, $x$, is actually a function of the time $t$ or $x = f(t)$.

If now we wish to find the rate of change of costs with respect to time, then we are interested in the quantity $\frac{dC}{dt}$. But since $C = C(f(t))$, the chain rule then gives

$$\frac{dC}{dt} = \frac{dC}{dx} \frac{dx}{dt}.$$

*(© Larry Leferer/Grant Heilman Photography, Inc.)*

EXAMPLE 1      **Finding a Related Rate**

Suppose the cost function in thousands of dollars is given by $C(x) = 100 + x^2$ where $x$ is in thousands of units. The company estimates that its rate of change of sales with respect to time will be .04. What will be the rate of increase of costs with respect to time if the company produces and sells 10,000 items?

Solution       Since the rate of change of sales with respect to time is .04,

$$\frac{dx}{dt} = .04.$$

Then

$$\frac{dC}{dt} = \frac{dC}{dx}\frac{dx}{dt} = (2x)(.04).$$

When $x = 10$, this yields

$$\frac{dC}{dt} = 2(10)(.04) = .8$$

That is, costs are increasing at an annual rate of $800 when 10,000 items are being produced.   ∎

EXAMPLE 2      **Finding a Related Rate of the Spread of an Oil Slick**

An oil slick is spreading in a circular fashion. The radius of the circular slick is observed to be increasing at 0.1 mile per day. What is the rate of increase of the area of the slick when $r = 2$ miles?

Solution       If $r$ is the radius of the circular slick in miles, then the area as shown in Figure 14.5 is given by $A = \pi r^2$. Also $\frac{dr}{dt} = 0.1$. Thus

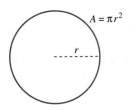

$A = \pi r^2$

$$\frac{dA}{dt} = \frac{dA}{dr}\frac{dr}{dt} = 2\pi r(0.1).$$

When $r = 2$, this becomes

$$\frac{dA}{dt} = 2\pi 2(0.1) \approx 1.26.$$

**Figure 14.5**

Thus when the radius of the slick is two miles, the area of the slick is increasing at a rate of 1.26 square miles per day.   ∎

EXAMPLE 3      **Change of Distance Between Two Ships**

Two ships leave the same port at 1:00 P.M. The first ship heads due north and at 2:00 P.M. (one hour later) is observed to be 3 miles due north of port and going at 6 miles per hour at that instant. The second ship leaves port at the same time as the first ship and heads due east and one hour later is observed to be 4 miles due east of port and going at 7 miles per hour. At what rate is the distance between the two ships changing at 2:00 P.M.?

**Solution**    Let $a(t)$ and $b(t)$ be the respective distances that the first and second ships are from port at time $t$ after leaving. See Figure 14.6.

If $d(t)$ is the distance between the ships then

$$d^2(t) = a^2(t) + b^2(t).$$

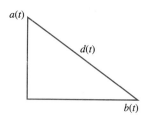

**Figure 14.6**

Differentiating this expression gives

$$2d(t)d'(t) = 2a(t)a'(t) + 2b(t)b'(t).$$

Thus

$$d'(t) = \frac{a(t)a'(t) + b(t)b'(t)}{d(t)}.$$

Now we are given that $a(1) = 3$, $b(1) = 4$, $a'(1) = 6$, and $b'(1) = 7$. It follows that

$$d(1) = \sqrt{3^2 + 4^2} = 5.$$

Therefore

$$d'(1) = \frac{3 \cdot 6 + 4 \cdot 7}{5} = 9.2.$$

The ships are moving apart at 9.2 miles per hour at 2:00 P.M.    ■

Notice how we were able to do this problem without knowing explicitly the two functions $a(t)$ and $b(t)$.

## Elasticity of Demand

Let us now consider gauging the sensitivity of the demand for a product to changes in the price of the product. Any reasonable demand curve given by $x = f(p)$ will be a *decreasing* function. See Figure 14.7. That is, an increase in price will lead to a decrease in demand.

Economists measure this sensitivity of demand to changes in price by the ratio of the percentage change in demand to the percentage change in price. As usual, we let $\Delta p$ represent the change in price and $\Delta x$ represent the change in demand. This is shown in Figure 14.7. Notice that if $\Delta p$ is positive then $\Delta x$ must be *negative* since the function is decreasing. Also if $\Delta p$ is negative then $\Delta x$ must be positive. (That is, a decrease in price will lead to greater demand.) We conclude that $\Delta p$ and $\Delta x$ must be of opposite signs.

The percentage change in demand is given by the expression

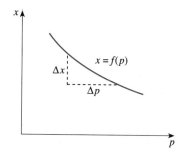

**Figure 14.7**

$$\frac{\Delta x}{x} \, 100,$$

while the percentage change in price is given by the expression

$$\frac{\Delta p}{p} \, 100.$$

The ratio is then

$$\frac{\frac{\Delta x}{x} \, 100}{\frac{\Delta p}{p} \, 100} = \frac{p}{x} \frac{\Delta x}{\Delta p}.$$

As noted above, $\Delta p$ and $\Delta x$ must be of opposite signs, and, therefore, this ratio is *negative*. Since economists perfer to work with positive numbers, they actually work with

$$-\frac{p}{x} \frac{\Delta x}{\Delta p}.$$

If this ratio is large, this means that a small change in price leads to a relatively large change in demand. If this ratio is small (say 0.1), then it will take a large change in price to effect a relatively small change in demand.

Now if $x = f(p)$ is differentiable, then

$$\lim_{\Delta p \to 0} \frac{\Delta x}{\Delta p} = \frac{dx}{dp}.$$

Then we define the term $E$ by

$$E = \lim_{\Delta p \to 0} -\frac{p}{x} \frac{\Delta x}{\Delta p} = -\frac{p}{x} \frac{dx}{dp}.$$

This term is called the *elasticity of demand* and measures the instantaneous sensitivity of demand to price.

We say that demand is *inelastic* if $E < 1$, *elastic* if $E > 1$ and unit elasticity if $E = 1$.

Thus if the percentage change in demand is less than the percentage change in price, demand is inelastic and $E < 1$. If the percentage change in demand is greater than the percentage change in price, demand is elastic and $E > 1$. Finally, if the percentage change in demand is equal to the percentage change in price, demand is of unit elasticity and $E = 1$.

---

**Elasticity of Demand**

If the demand equation is given by $x = f(p)$, then

$$E = -\frac{p}{x} \frac{dx}{dp}$$

is called the **elasticity of demand**.

1.  If $E > 1$, then demand is **elastic**.
2.  If $E < 1$, then demand is **inelastic**.
3.  If $E = 1$, then demand has **unit elasticity**.

---

E X A M P L E  4    **Finding Elasticity of Demand**

Suppose

$$x = f(p) = 100 - 2p.$$

Find the elasticity of demand when $p$ equals (a) 10, (b) 40, (c) 25.

**Solutions**    Since $\dfrac{dx}{dp} = -2,$

$$E = -\frac{p}{x}\frac{dx}{dp} = -\frac{p}{100 - 2p}(-2) = \frac{2p}{100 - 2p}.$$

(a)  For $p = 10$,

$$E = \frac{2p}{100 - 2p} = \frac{20}{80} = 0.25.$$

Demand is inelastic. The percentage change in demand is one fourth the percentage change in price.

(b)  For $p = 40$,

$$E = \frac{2p}{100 - 2p} = \frac{80}{20} = 4.$$

Demand is elastic. The percentage change in demand is four times the percentage change in price.

(c)  For $p = 25$,

$$E = \frac{2p}{100 - 2p} = \frac{50}{50} = 1.$$

Demand is of unit elasticity. The percentage change in demand is the same as the percentage change in price.  ∎

There is an interesting connection between the revenue R and elasticity E. We suppose we have $x = f(p)$. Then since $R = px$,

$$\frac{dR}{dp} = x + p\frac{dx}{dp}$$
$$= x + x\frac{p}{x}\frac{dx}{dp}$$
$$= x - xE$$
$$= x(1 - E)$$

If demand is inelastic, $E < 1$, and then $R'(p) > 0$ and therefore $R(p)$ is increasing. If demand is elastic, $E > 1$, and then $R'(p) < 0$, and, therefore, $R(p)$ is decreasing.

Therefore, if $E < 1$, increasing price increases revenue. If $E > 1$, increasing price decreases revenue. Revenue is optimized when $E = 1$. See Figure 14.8.

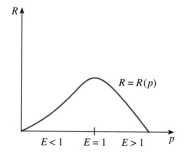

**Figure 14.8**

## SELF-HELP EXERCISE SET 14.3

1. After using an experimental drug, the radius of a spherical tumor is observed to be decreasing at the rate of 0.1 cm per week when the radius is 2 cm. At what rate is the volume of the tumor decreasing?

2. Let $x = 82 - p^2$. Find the elasticity of demand if (a) $p = 1$, (b) $p = 9$.

## EXERCISE SET 14.3

In Exercises 1 through 12 find $\dfrac{dy}{dt}$ given the indicated information.

1. $y = 3 + 2x, \dfrac{dx}{dt} = 3$

2. $y = 3x - 4, \dfrac{dx}{dt} = 2$

3. $y = 2 + x^2, \dfrac{dx}{dt} = 4, x = 2$

4. $y = 3 - 2x^2, \dfrac{dx}{dt} = 2, x = 3$

5. $y = 3 - 2x^3, \dfrac{dx}{dt} = -2, x = 1$

6. $y = 1 - x^4, \dfrac{dx}{dt} = -1, x = 2$

7. $y = \dfrac{1 - x}{1 + x}, \dfrac{dx}{dt} = 2, x = 1$

8. $y = \dfrac{1 - x}{1 + x^2}, \dfrac{dx}{dt} = -2, x = -1$

9. $x^2 + y^2 = 5, \dfrac{dx}{dt} = 2, x = 1, y = 2$

10. $x^2 - y^2 = 3, \dfrac{dx}{dt} = -2, x = 2, y = -1$

11. $x^3 + xy + y^2 = 7, \dfrac{dx}{dt} = 2, x = 1, y = 2$

12. $x + x^2y - x^2 = -1, \dfrac{dx}{dt} = 3, x = -1, y = 1$

## Applications

13. **Demand.**  If the demand equation is given by $x = 10 - 0.1p$ and the number of items manufactured (and sold) is increasing at the rate of 20 per week, find the rate of change of $p$ with respect to time.

14. **Cost.**  If the cost equation is given by $C(x) = 8 + 0.2x^2$ and the number of items manufactured is increasing at the rate of 20 per week, find the rate of change of $C$ with respect to time when $x = 4$.

15. **Revenue.**  If the demand equation is given by $x = 10 - 0.1p$ and the number of items manufactured (and sold) is increasing at the rate of 20 per week, find the rate of change of revenue with respect to time when $x = 4$.

16. **Profits.** If the demand and cost equations are the same as in the two previous problems, find the rate of change of profits with respect to time.

17. **Distance.** A ship is observed to be 4 miles due north of port and traveling due north at 5 miles per hour. At the same time another ship is observed to be 3 miles due west of port and traveling due *east* on its way back to port at 4 miles per hour. What is the rate at which the distance between the ships is changing?

18. **Distance.** Answer the same question as in the previous problem if the first ship is traveling due *south* at 5 miles per hour, everything else remaining the same.

19. **Physics.** A 10-foot ladder leans against a wall and slides down with the foot of the ladder observed to be moving away from the wall at 4 ft/sec when 6 feet from the wall. At what speed is the top of the ladder moving downward?

20. **Medicine.** After using an experimental drug, the volume of a spherical tumor is observed to be decreasing at the rate of 2 cubic cm per month when the radius is 5 cm. At what rate is the radius of the tumor decreasing?

21. **Physics.** A water tank in the shape of a circular cone (see figure) has a radius of 4 yards and a height of 10 yards. If water is being poured into the tank at the rate of 5 cubic yards per minute, find the rate at which the water level is rising when the water level is at 2 yards. *Hint:* $V = \frac{1}{3}\pi r^2 h$. Use similar triangles to find $r$ as a function of $h$ and substitute this into the last equation. Then differentiate the latter expression.

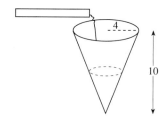

22. **Geometry.** If the volume of a cube is increasing at 6 cubic feet per sec, what is the rate at which the sides are increasing when the sides are 10 ft long?

23. **Geometry.** If the sides of a square are increasing at 2 ft/sec, what is the rate at which the area is changing when the sides are 10 ft long?

In Exercises 24 through 32 find the elasticity $E$ at the given points and determine if demand is inelastic, elastic, or unit elastic.

24. $x = 20 - 4p$, (a) $p = 1$, (b) $p = 2.5$, (c) $p = 4$

25. $x = 24 - 3p$, (a) $p = 1$, (b) $p = 4$, (c) $p = 6$

26. $x = 20 - 5p$, (a) $p = 2$, (b) $p = 3$, (c) $p = 1$

27. $x = \dfrac{1}{10 + p}$, (a) $p = 5$, (b) $p = 10$, (c) $p = 100$

28. $x = 10 + \dfrac{1}{p}$, (a) $p = 0.1$, (b) $p = 1$

29. $x = 10 + \dfrac{1}{p^2}$, (a) $p = 0.01$, (b) $p = 1$

30. $x = \dfrac{10}{p}$, (a) $p = 2$, (b) $p = 3$

31. $x = \dfrac{10}{p^2}$, (a) $p = 1$, (b) $p = 3$

32. $x = \dfrac{10}{\sqrt{p}}$, (a) $p = 1$, (b) $p = 3$

In Exercises 33 through 36 find (a) the value (if any) of $x$ when $E = 1$, (b) the value(s), if any, of $x$ for which the revenue is maximized.

33. $x = 30 - 10p$

34. $x = 10 - 0.1p$

35. $x = \dfrac{1}{p^3}$

36. $x = \dfrac{1}{1 + p}$

## Solutions to Self-Help Exercise Set 14.3

1. The relationship between the volume $V$ in cubic cm and the radius $r$ in cm of the tumor is given by $V = \frac{4}{3}\pi r^3$. Thus

$$\frac{dV}{dt} = 4\pi r^2 \frac{dr}{dt},$$

where $t$ is measured in weeks. We are given that $\dfrac{dr}{dt} = -0.1$ when $r = 2$. Putting this information into the last displayed line gives

$$\frac{dV}{dt} = 4\pi(2)^2(-0.1) \approx -5.03$$

cubic cm per week.

2. We have

$$E = -\frac{p}{x}\frac{dx}{dp} = \frac{-p}{82 - p^2}(-2p) = \frac{2p^2}{82 - p^2}.$$

   a. When $p = 1$, $E = \frac{2}{81} < 1$. Thus demand is inelastic.
   b. When $p = 9$, $E = 162 > 1$. Thus demand is elastic.

## SUMMARY OUTLINE OF CHAPTER 14

- **Tangent Line Approximation.** If $y = f(x)$ is a differentiable function at $x_0$, then $f(x_0 + h) - f(x_0) \approx f'(x_0)h$ if $h$ is small.
- As $x$ goes from $x_0$ to $x_1$, the increment of $x$, denoted by $\Delta x$, and the increment of $y$, denoted by $\Delta y$, are defined as follows: $\Delta x = x_1 - x_0$, $\Delta y = f(x_1) - f(x_0) = f(x_0 + \Delta x) - f(x_0)$.

- If $y = f(x)$ is differentiable at $x$, then the **differential**, denoted by $dy$, is defined to be the linear function of $dx$ given by $dy = f'(x)\, dx$.

- **Elasticity of Demand.**   If the demand equation is given by $x = f(p)$, then
$E = -\dfrac{p}{x}\dfrac{dx}{dp}$ is called the **elasticity of demand**. If $E > 1$, then demand is **elastic**.
If $E < 1$, then demand is **inelastic**. If $E = 1$, then demand has **unit elasticity**.

## Chapter 14 Review Exercises

In Exercises 1 through 4 find $\dfrac{dy}{dx}$ at any point $(x, y)$ and then find the slope of the tangent line at the given point.

**1.** $x^3 y^4 - xy = 6$, $(2, 1)$

**2.** $\sqrt{x} + x\sqrt{y} + y^2 = 7$, $(4, 1)$

**3.** $\dfrac{x + y}{x - y} = 3y^2$, $(2, 1)$

**4.** $\sqrt[3]{y^2 + 7} - x^2 y^3 = 2$, $(0, 1)$

In Exercises 5 through 8 find $dy$ in terms of $x$ and $dx$.

**5.** $y = \sqrt[5]{x}$

**6.** $y = \sqrt[3]{x^2}$

**7.** $y = \dfrac{1}{x^3}$

**8.** $y = \sqrt{x} + \dfrac{1}{\sqrt{x}}$

In Exercises 9 through 12 use the differential found in the previous four exercises and the linear approximation to find approximations of the following quantities.

**9.** $\sqrt[5]{1.05}$

**10.** $\sqrt[3]{(7.7)^2}$

**11.** $\dfrac{1}{(2.08)^3}$

**12.** $\sqrt{4.016} + \dfrac{1}{\sqrt{4.016}}$

**13.** Given the demand curve $x = 10 - 2p$, determine if the demand is elastic, inelastic, or unit elastic if (a) $p = 2$, (b) $p = 2.5$, (c) $p = 3$.

**14.** **Biology.**   Use the differential and the linear approximation formula to estimate the change in volume of a cylindrical bacterium as the radius increases from 2 microns to 2.1 microns while the length stays at 10 microns.

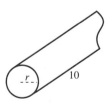

**15.** **Physics.**   A man 6 feet tall walks along a level sidewalk away from a lampost. On the post is a light 18 feet above the ground. At what rate is the top of his shadow moving if he is walking at 3 feet per second?

Exponents and
logarithms are used
to date fossils.
(© Tom McHugh/
Photo Researchers,
Inc.)

# Exponential and Logarithmic Functions

15.1   Exponential Functions

15.2   Natural Logarithms

15.3   Derivatives of Exponential and Logarithmic Functions

15.4   Growth and Decay

**T**his chapter introduces exponential functions and their basic properties and defines the exponential function to the base $e$, showing its use in continuous compounding. Logarithmic functions are introduced as the inverse of exponential functions. The examples and exercises provide applications of exponential and logarithmic functions in many diverse areas. For example, we will see the manner in which an infectious disease such as AIDS may spread and will estimate the limiting human population of the earth.

## 15.1   EXPONENTIAL FUNCTIONS

▶ *Irrational Powers*

▶ *Graphs of Exponential Functions*

▶ *The Base* e *and Continuous Compounding*

▶ *Effective Yield*

▶ *Present Value*

A P P L I C A T I O N

**Continuous Compounding**

Suppose $2000 is invested at an annual rate of 9% compounded continuously. How much is in the account after 1 year? See Example 4 for the answer.

A P P L I C A T I O N

**Effective Yield**

What must be the annual yield that would give the same amount of interest as a rate of 9% compounded continuously? See Example 5 for the answer.

## Irrational Powers

If $a$ is a positive number and $r$ is a rational number, that is, $r = m/n$, where $m$ and $n$ are integers, then we know that

$$a^r = a^{m/n} = \sqrt[n]{a^m}.$$

This simply means multiply the number $a$ by itself $m$ times obtaining a number, say $y$, and then find the $n$th root of $y$. This should be clear. But what happens if we wish to take $a^x$ where $x$ is an irrational number? In this case, it is impossible to write $x$ as the quotient of integers. How can one give a meaning, then, to the expression $a^x$? For example, it is known that $\pi$ is an irrational number. What does the expression $2^\pi$ mean?

In order to answer this question, we first note that every irrational number has a decimal expansion. For example $\pi = 3.1415926 \ldots$ and the decimal is infinite.

The sequence of numbers, 3, 3.1, 3.14, 3.141, 3.1415, 3.14159, 3.141592, 3.1415926, ... are all *rational* and have $\pi$ as a limit. Thus, in the first place, it makes perfectly good sense to raise 2 to any of these powers. Doing this we obtain the sequence $2^3$, $2^{3.1}$, $2^{3.14}$, $2^{3.141}$, $2^{3.1415}$, $2^{3.14159}$, $2^{3.141592}$, $2^{3.1415926}$, ..., or, to six decimal places, 8, 8.574188, 8.815241, 8.821353, 8.824411, 8.824962, 8.824974, 8.824978, .... But it can be shown that this sequence has a limit. We then *define* $2^\pi$ to be this limit. It is in this manner that we can define what is meant by $a^x$ when $x$ is a real and irrational number. Furthermore, this definition assures us that the function $a^x$, so defined, will be **continuous**.

With $a^x$ defined now for all real numbers, it can be shown that all the usual rules for exponents (see Appendix A.1) are still satisfied.

With these preliminaries dispensed with, we now can deal with $a^x$ where $x$ is any real number.

## Graphs of Exponential Functions

We shall see that the graphs of functions of the form $a^x$ for $a$ a positive number fall into three categories: $a > 1$, $a = 1$, $0 < a < 1$. The case when $a = 1$ is not very interesting since $1^x = 1$ for all real numbers $x$. We now consider the other cases.

**E X A M P L E   1**     **Finding the Graphs of Exponential Functions**

Graph the functions (a) $y = 2^x$, (b) $y = (\frac{1}{2})^x = 2^{-x}$.

**Solutions**

**Figure 15.1**

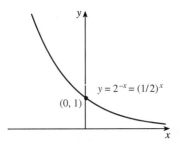

**Figure 15.2**

(a)  To graph $y = 2^x$ set $x$ equal to various values and check the corresponding values of $2^x$ as in Table 15.1. See Figure 15.1.
        Notice that the graph of $2^x$ goes through the point $(0, 1)$, is increasing, and is always above the $x$-axis. This latter point means that even if $x$ is *negative*, $2^x$ is still *positive*. From Table 15.1, it appears that $\lim\limits_{x \to -\infty} a^x = 0$ if $a > 1$. This can be shown to be true. Thus $y = 0$ is a horizontal asymptote.

(b)  To graph $y = (\frac{1}{2})^x = 2^{-x}$ set $x$ equal to various values and check the corresponding values of $y = (\frac{1}{2})^x = 2^{-x}$ as in Table 15.1. See Figure 15.2.
        Notice that the graph goes through the point $(0, 1)$, is decreasing, and is always above the $x$-axis. Notice that as $x$ becomes larger and larger, $y = (\frac{1}{2})^x = 2^{-x}$ becomes closer and closer to zero but never attains zero, that is, the graph never touches the $x$-axis. From Table 15.1, it appears that $\lim\limits_{x \to +\infty} a^x = 0$ if $0 < a < 1$. This can be shown to be true. Thus $y = 0$ is a horizontal asymptote.   ■

**Table 15.1**

| $x$ | $-4$ | $-3$ | $-2$ | $-1$ | $0$ | $1$ | $2$ | $3$ | $4$ |
|---|---|---|---|---|---|---|---|---|---|
| $2^x$ | $\frac{1}{16}$ | $\frac{1}{8}$ | $\frac{1}{4}$ | $\frac{1}{2}$ | $1$ | $2$ | $4$ | $8$ | $16$ |
| $(\frac{1}{2})^x$ | $16$ | $8$ | $4$ | $2$ | $1$ | $\frac{1}{2}$ | $\frac{1}{4}$ | $\frac{1}{8}$ | $\frac{1}{16}$ |

The graph of $y = a^x$ looks like Figure 15.3 if $a > 1$ and like Figure 15.4 if $0 < a < 1$.
        Notice that any exponential function $a^x$, with $a > 0$ and $a \neq 1$, has domain $(-\infty, +\infty)$, range $(0, +\infty)$, intercept $(0, 1)$, and is continuous everywhere. We now give some further general properties of $a^x$ that are very useful.

**Figure 15.3**

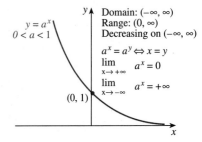

**Figure 15.4**

> **Properties of $a^x$**
>
> **Property 1**   If $a > 1$, the function $a^x$ is increasing.
>
> **Property 2**   If $0 < a < 1$, the function $a^x$ is decreasing.
>
> **Property 3**   If $a \neq 1$, then $a^x = a^y$ if and only if $x = y$.
>
> **Property 4**   If $a > 1$, then $\lim\limits_{x \to -\infty} a^x = 0$ and $\lim\limits_{x \to +\infty} a^x = +\infty$.
>
> If $0 < a < 1$, then $\lim\limits_{x \to -\infty} a^x = +\infty$ and $\lim\limits_{x \to +\infty} a^x = 0$.

**E X A M P L E   2**     **Solving Equations with Exponents**

Solve (a) $2^x = 8$, (b) $9^x = 27^{4x-10}$.

**Solutions**
(a)   First rewrite $2^x = 8$ as $2^x = 2^3$. Then by Property 3, $x = 3$.

(b)   We first have $(3^2)^x = (3^3)^{4x-10}$. Then

$$3^{2x} = 3^{12x-30}$$

$$2x = 12x - 30 \qquad \text{Property 3}$$

$$x = 3 \qquad\qquad \blacksquare$$

## The Base $e$ and Continuous Compounding

We will now consider the exponential function $e^x$ where $e$ is a certain irrational number equal to 2.71828 (to five decimal places). This number comes up in various situations as naturally as the irrational number $\pi$ does. For example, as we shall see, $e$ is used in continuous compounding, which is the most advantageous way of compounding to an individual earning interest in a bank account, and is a common way of compounding that is used throughout the banking industry. There are many other applications of the exponential function in many diverse areas. These will be taken up later in this chapter.

We first consider how the exponential function with base $e$ arises in continuous compounding. What is continuous compounding? Suppose you had some money in the bank. Consider what happens if the money is compounded annually, monthly, weekly, daily, hourly, by the minute, and so on. We know from Chapter 5 that the amount will increase accordingly as the number of compoundings increases. But it turns out that the amount in the account will head for a definite number as the number of compoundings increases. The number so obtained will be the amount attained by **continuous compounding**. The following example illustrates this. Recall that if an account with an initial principal $P$ and an annual rate of $r$, where $r$ is expressed as a decimal, is compounded $m$ times a year, then the future amount in the account after $t$ years is

$$F(t) = P \left( 1 + \frac{r}{m} \right)^{mt}.$$

E X A M P L E  3   **Compounding Over Smaller and Smaller Time Intervals**

Suppose $1000 is deposited into an account that yields 6% annually. Find the principal at the end of the first year if compounded
(a) annually, (b) monthly, (c) weekly, (d) daily, (e) hourly, (f) by the minute.

**Solutions**   In all cases $P = \$1000$, $r = .06$, and $t = 1$.

(a)  Here $m = 1$. Thus, using a calculator, we have

$$F = P(1 + r) = \$1000(1.06) = \$1060.00.$$

(b)  Here $m = 12$. Thus

$$F = P\left(1 + \frac{r}{m}\right)^{m} = \$1000\left(1 + \frac{.06}{12}\right)^{12} = \$1061.68.$$

(c)  Here $m = 52$. Thus

$$F = P\left(1 + \frac{r}{m}\right)^{m} = \$1000\left(1 + \frac{.06}{52}\right)^{52} = \$1061.80.$$

(d)  Here $m = 365$. Thus

$$F = P\left(1 + \frac{r}{m}\right)^{m} = \$1000\left(1 + \frac{.06}{365}\right)^{365} = \$1061.83.$$

(e)  Here $m = 8760$. Thus

$$F = P\left(1 + \frac{r}{m}\right)^{m} = \$1000\left(1 + \frac{.06}{8760}\right)^{8760} = \$1061.84.$$

(f)  Here $m = 525600$. Thus

$$F = P\left(1 + \frac{r}{m}\right)^{m}$$

$$= \$1000\left(1 + \frac{.06}{525600}\right)^{525600} = \$1061.84. \quad \blacksquare$$

Thus as the number of compoundings per year increases, the amount of money in the account seems to be heading for a definite limit (see Table 15.2).

**Table 15.2**

| Compounded | Amount at end of 1 year |
|---|---|
| yearly | $1060.00 |
| monthly | 1061.68 |
| weekly | 1061.80 |
| daily | 1061.83 |
| hourly | 1061.84 |
| by the minute | 1061.84 |

We then need to analyze in general what happens as the number of compoundings increases without bound, that is, as $m \to +\infty$.

To do this set $n = m/r$ in the above formula for $F$. This gives $m = rn$ and $r/m = 1/n$. Then

$$F(t) = P\left(1 + \frac{r}{m}\right)^{mt}$$

$$= P\left(1 + \frac{1}{n}\right)^{rnt}$$

$$= P\left[\left(1 + \frac{1}{n}\right)^{n}\right]^{rt}$$

Since $n = m/r$, as the number $m$ of compoundings increases without bound, that is, $m \to +\infty$, then also $n \to +\infty$. Thus we wish to find

$$\lim_{n \to +\infty} \left(1 + \frac{1}{n}\right)^{n}$$

Table 15.3 indicates that this limit is 2.71828 (to five decimal places).

**Table 15.3**

| $n$ | 10 | 100 | 1000 | 10,000 | 100,000 | 1,000,000 |
|---|---|---|---|---|---|---|
| $\left(1 + \dfrac{1}{n}\right)^{n}$ | 2.593742 | 2.704814 | 2.716924 | 2.718146 | 2.718268 | 2.718280 |

This number is of such importance that a special letter is set aside for it. This letter is $e$.

---

**The Number $e$**

The number $e$ is defined to be

$$e = \lim_{n \to +\infty} \left(1 + \frac{1}{n}\right)^{n},$$

and

$$e = 2.71828 \ldots$$

---

Using this in the above formula for $F(t)$ then yields

$$\lim_{n \to +\infty} F(t) = \lim_{n \to +\infty} P\left[\left(1 + \frac{1}{n}\right)^{n}\right]^{rt}$$

$$= P\left[\lim_{n \to +\infty} \left(1 + \frac{1}{n}\right)^{n}\right]^{rt}$$

$$= P[e]^{rt}$$

$$= Pe^{rt}$$

We refer to this as *continuous* compounding. The compounding is not done just every day, or every hour, or every minute, or every second, but *continuously*.

**EXAMPLE 4** **Finding Amounts in a Continuously Compounded Account**

Suppose $2000 is invested at an annual rate of 9% compounded continuously. How much is in the account after (a) 1 year, (b) 3 years?

**Solutions**

(a) Here $P = \$2000$, $r = .09$, and $n = 1$. Thus, using a calculator, we have

$$F = Pe^{rt} = \$2000e^{.09} = \$2000(1.094174) = \$2188.35.$$

(b) Here $P = \$2000$, $r = .09$, and $n = 3$. Thus

$$F = Pe^{rt} = \$2000e^{(.09)(3)} = \$2000(1.309964) = \$2619.93. \quad \blacksquare$$

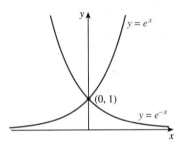

**Figure 15.5**

Since $e > 1$, $e^x$ is an increasing function and thus satisfies Properties 1 and 3 and the first part of Property 4. Furthermore, since $2 < e < 3$, the graph of $e^x$ lies between the graphs of $2^x$ and $3^x$. The graphs of $y = e^x$ and $y = e^{-x}$ are given in Figure 15.5.

## Effective Yield

Suppose $r$ is the annual interest rate expressed as a decimal and that interest is compounded continuously. If the compounding goes on for one year, then the amount F after one year is

$$F = Pe^{r(1)} = Pe^r.$$

If we let $r_{\text{eff}}$ be the effective annual yield, then $r_{\text{eff}}$ must satisfy

$$Pe^r = P(1 + r_{\text{eff}}).$$

Solving for $r_{\text{eff}}$ we obtain

$$r_{\text{eff}} = e^r - 1.$$

---

**Effective Yield**

Suppose a sum of money is invested at an annual rate of $r$ expressed as a decimal and is compounded continuously. The effective yield $r_{\text{eff}}$ is

$$r_{\text{eff}} = e^r - 1,$$

and is the annual rate that gives the same amount at the end of one year as the continuous compounding does.

EXAMPLE 5    **Finding the Effective Yield**

Find the effective annual yield if 9% is compounded continuously.

Solution    Here $r = .09$. Then

$$r_{\text{eff}} = e^{0.09} - 1 = 0.09417$$

or, as a percent, 9.417%.   ■

## Present Value

If we now have an account initially at P and earning an annual rate of $r$ expressed as a decimal, and interest is compounded continuously, then the amount $F$ in the account after $t$ years is $F = Pe^{rt}$.

If we wish to know how many dollars $P$ to set aside now in this account so that we will have $F$ dollars after $t$ years, we simply solve the above expression for $P$. Thus

$$P = Fe^{-rt}.$$

This is called the **present value**.

---

**Present Value**

Suppose an account earns an annual rate of $r$ expressed as a decimal and is compounded continuously. Then the amount P, called the **present value**, needed presently in this account so that a future balance of $F$ will be attained in $t$ years is given by

$$P = Fe^{-rt}.$$

---

EXAMPLE 6    **Calculating Present Value**

How much money must grandparents set aside at the birth of their grandchild if they wish to have $20,000 by the grandchild's 18th birthday? They can earn 9% compounded continuously.

Solution    Here $r = 0.09$, $t = 18$, and $F = \$20,000$. Thus

$$P = Fe^{-rt} = \$20,000e^{-0.09(18)} = \$3957.97. \quad ■$$

## SELF-HELP EXERCISE SET 15.1

1. Solve $e^{2x+1} = \dfrac{1}{e^x}$.

2. Suppose $1000 is invested at an annual rate of 8% compounded continuously. Find the amount after 5 years.

**3.**   What is the effective yield for an annual rate of 8% compounded continuously?

**4.**   How much should be deposited in an account that will earn 8% compounded continuously so that there will be $10,000 in the account after 10 years?

## EXERCISE SET 15.1

In Exercises 1 through 8 graph each of the functions.

**1.**  $y = 5^x$      **2.**  $y = 7^x$      **3.**  $y = 3^{x^2}$      **4.**  $y = 3^{-x^2}$

**5.**  $y = (\frac{1}{2})^x$      **6.**  $y = (0.1)^x$      **7.**  $y = 10^x$      **8.**  $y = 5^{-x}$

In Exercises 9 through 16 solve for $x$.

**9.**   $3^x = 9$                     **10.**   $4^x = 8$                     **11.**   $16^x = 8$

**12.**   $(\frac{1}{2})^x = \frac{1}{16}$                **13.**   $(\frac{1}{16})^x = \frac{1}{8}$               **14.**   $(\frac{1}{4})^x = 8$

**15.**   $4^{2x} = 8^{9x+15}$              **16.**   $5^{3x} = 125^{4x-4}$

In Exercises 17 through 24 graph each of the functions.

**17.**   $y = e^{3x}$                   **18.**   $y = e^{-2x}$                  **19.**   $y = e^{x^2}$

**20.**   $y = e^{-x^2}$                  **21.**   $y = 2 - e^{-x}$               **22.**   $y = 2 + e^{-x}$

**23.**   $y = \frac{1}{2}(e^x + e^{-x})$          **24.**   $y = \frac{1}{2}(e^x - e^{-x})$

In Exercises 25 through 38 solve for $x$.

**25.**   $e^x = 1/e$                   **26.**   $e^x = 1/e^2$                 **27.**   $e^x = 1/e^3$

**28.**   $e^{5x} = e^{x+8}$              **29.**   $(e^{x+1})^2 = e$             **30.**   $e^{2x-1}e^x = 1/e^x$

**31.**   $e^{x+1} = 1/e^x$              **32.**   $e^{x^2+2x} = e^{-x}$          **33.**   $e^{-x^2+2x} = e$

**34.**   $xe^x = e^x$                  **35.**   $e^{-x}(x + 1) = 0$           **36.**   $x^2 e^{2x} = e^{2x}$

**37.**   $xe^{-x} = x^2 e^{-x}$           **38.**   $(x^2 + x - 2)e^x = 0$

In Exercises 39 through 46 find the limits.

**39.**   $\lim\limits_{x \to +\infty} (1 + 2e^{-x})$                          **40.**   $\lim\limits_{x \to -\infty} (3 - 2e^x)$

**41.**   $\lim\limits_{x \to +\infty} \dfrac{10}{2 + e^{-x}}$                    **42.**   $\lim\limits_{x \to +\infty} \dfrac{2}{1 + e^x}$

**43.**   $\lim\limits_{x \to +\infty} \dfrac{e^x - 1}{e^x + 1}$                   **44.**   $\lim\limits_{x \to +\infty} \dfrac{e^x - e^{-x}}{e^x + e^{-x}}$

**45.**   $\lim\limits_{x \to +\infty} \dfrac{4e^x}{e^x + 1}$                     **46.**   $\lim\limits_{x \to +\infty} \dfrac{3e^x}{2e^x - 1}$

## Applications

In Exercises 47 through 54 find how much is in the following accounts after the given $n$ years where $P$ is the initial principal, $r$ is the annual rate given as a percent, and the compounding is continuous.

**47.** $n = 1, P = \$1000, r = 8\%$

**48.** $n = 1, P = \$1000, r = 10\%$

**49.** $n = 40, P = \$1000, r = 8\%$

**50.** $n = 40, P = \$1000, r = 10\%$

**51.** $n = 20, P = \$1000$, compounded continuously, $r$ equal to (a) 3%, (b) 5%, (c) 7%, (d) 9%, (e) 12%, (f) 15%

**52.** $n = 40, P = \$1000$, compounded continuously, $r$ equal to (a) 3%, (b) 5%, (c) 7%, (d) 9%, (e) 12%, (f) 15%

**53.** $P = \$1000, r = 9\%$, compounded continuously, $n$ equal to (a) 5, (b) 10, (c) 15, (d) 30

**54.** $P = \$1000, r = 6\%$, compounded continuously, $n$ equal to (a) 5, (b) 10, (c) 15, (d) 30

In Exercises 55 through 56 find the effective yield given the annual rate is $r$ and the compounding is continuous.

**55.** $r = 8\%$

**56.** $r = 10\%$

In Exercises 57 through 58 find the present value of the given future amounts $F$ with the indicated annual rate of return $r$, the number of years $n$, and the compounding is continuous.

**57.** $F = \$10,000, r = 9\%, n = 20$

**58.** $F = \$10,000, r = 10\%, n = 20$

## Graphing Calculator Exercises

**59.** Graph the three functions $y = 2^x, y = 3^x, y = 5^x$, together in the same viewing rectangle.

**60.** Graph the three functions $y = 2^{-x}, y = 3^{-x}, y = 5^{-x}$, together in the same viewing rectangle.

## Solutions to Self-Help Exercise Set 15.1

**1.** Write the equation as $e^{2x+1} = e^{-x}$. Then from property 3, we must have $2x + 1 = -x$ and $x = -\frac{1}{3}$.

**2.** If \$1000 is invested at an annual rate of 8% compounded continuously, then after 5 years the amount in the account will be

$$F = Pe^{rt} = \$1000e^{(.08)(5)} = \$1491.82.$$

3.  The effective yield for an annual rate of 8% compounded continuously is given by

$$r_{\text{eff}} = e^r - 1 = e^{0.08} - 1 \approx .083287, \text{ or } 8.3287\%.$$

4.  The amount that should be deposited in an account that will earn 8% compounded continuously so that there will be $10,000 in the account after 10 years is

$$P = Fe^{-rt} = \$10{,}000e^{-(.08)(10)} = \$4493.29.$$

## 15.2   NATURAL LOGARITHMS

- ▶  *Basic Properties of Natural Logarithms*
- ▶  *Solving Equations with Exponents and Logarithms*
- ▶  *Writing Exponential Functions in Terms of the Natural Exponent*
- ▶  *Applications*

**John Napier, 1550–1617**

Napier invented logarithms around 1594 in order to ease the burden of numerical calculations needed in astronomy. Most of his time was spent engaged in the political and religious controversies of his day. He published an extremely widely read attack on the Roman Catholic Church, which he sincerely believed was his primary achievement. For relaxation he amused himself with the study of mathematics and science.

**A P P L I C A T I O N**
**Time Needed to**
**Double an Account**

An account pays an annual rate of 9% compounded continuously. How long will it take for the money in this account to double? See Example 7 for the answer.

### Basic Properties of Natural Logarithms

In the applications we will see the need to solve equations such as $2 = e^x$ for $x$. In order to accomplish this, we will need to invent a function, which we temporarily call $l(x)$, that "undoes" the exponential. That is, we want $l(x)$ to have the property that for all $x$, $l(e^x) = x$. Thus, in Figure 15.6, the conveyor belt inputs $x$ into the exponential function, which outputs $e^x$. The term $e^x$ is then inputed into the $l$ function, which outputs $x$. An equation such as $2 = e^x$ can then be solved by applying $l$ to each side, obtaining

$$l(2) = l(e^x) = x.$$

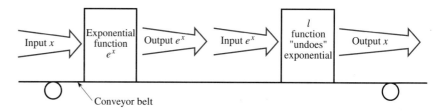

**Figure 15.6**

The values of the function $l(x)$ will be tabulated in a table or can be obtained on a calculator. In this way we will solve for $x$.

You are already familiar with a similar process. For example, in order to solve the equation $2 = x^3$, we apply the cube root function $g(x) = \sqrt[3]{x}$ to each side. The cube root function $g(x) = \sqrt[3]{x}$ "undoes" the cube function $f(x) = x^3$, that is, $g(x^3) = \sqrt[3]{x^3} = x$. Thus we can solve the equation $2 = x^3$ for $x$ by applying $g$ to each side, obtaining

$$\sqrt[3]{2} = \sqrt[3]{x^3} = x.$$

A table or calculator gives the value of $\sqrt[3]{2}$. In this way we solve for $x$.

When given any $x$, Figure 15.7 indicates geometrically how one finds $y = e^x$. Since the function $l$ is to "undo" the exponential function $y = e^x$, then, given any $y > 0$, we see in Figure 15.7 how to obtain $l(y)$ by "reversing the arrows" in Figure 15.8. That is, given any $y > 0$,

$$l(y) = x \quad \text{if, and only if,} \quad y = e^x.$$

Thus $l(y)$ is that exponent to which $e$ must be raised in order to obtain $y$. We normally write our functions with $x$ as the independent variable. So interchanging $x$ and $y$ above gives for any $x > 0$

$$y = l(x) \quad \text{if, and only if,} \quad x = e^y.$$

The function $l(x)$ is usually written as $l(x) = \ln x$ and is called the **natural logarithm of** $x$. In terms of this notation we have the following.

**Figure 15.7**

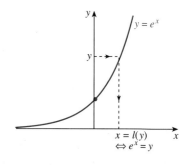

**Figure 15.8**

> ## Natural Logarithm
>
> If $x > 0$, the natural logarithm of $x$, denoted by $\ln x$, is defined as follows:
>
> $$y = \ln x \quad \text{if, and only if,} \quad x = e^y.$$

**E X A M P L E   1**

## Using the Definition of Natural Logarithm

Solve for $x$ in (a) $\ln x = -\dfrac{3}{2}$, (b) $\ln e^2 = x$.

**Solutions**

(a) $\ln x = -\dfrac{3}{2}$ if, and only if, $x = e^{-3/2}$.

(b) $\ln e^2 = x$ if, and only if, $e^x = e^2$. This is true if and only if $x = 2$. ∎

**R E M A R K .** Notice that one can only have $x > 0$ in the definition of the natural logarithm since $x = e^y$ is always positive. Thus the domain of the function $y = \ln x$ is $(0, +\infty)$.

**R E M A R K .** Notice that $y = \ln x$ means that $x = e^y = e^{\ln x}$. Thus $\ln x$ is an *exponent*. In fact **ln x is that exponent to which e must be raised in order to obtain x.**

According to the definition of natural logarithm, for any $x > 0$, $y = \ln e^x$ if, and only if, $e^x = e^y$. But this implies that $x = y = \ln e^x$. This shows that the function $\ln x$ "undoes" the exponential $e^x$.

> ## Basic Properties of the Natural Logarithmic Function
>
> 1. $e^{\ln x} = x$     if     $x > 0$.
> 2. $\ln e^x = x$     for all     $x$.

**R E M A R K .** It has already been mentioned that Property 2 means that the logarithmic function $\ln x$ "undoes" the exponential function $e^x$. Property 1 says that the exponential function $e^x$ "undoes" the logarithmic function $\ln x$.

**R E M A R K .** To find any intercepts of the graph of $y = \ln x$, set $0 = \ln x$. Then one must have $x = e^0 = 1$. Thus $\ln 1 = 0$ and $(1, 0)$ is the only intercept of the graph of $y = \ln x$.

**E X A M P L E   2**

## Using the Basic Properties of Natural Logarithm

Find (a) $\ln e^{-2}$, (b) $e^{\ln \pi}$

**Solutions**

(a) Use property 2 to obtain $\ln e^{-2} = -2$.

(b) Use property 1 to obtain $e^{\ln \pi} = \pi$. ∎

**Figure 15.9**

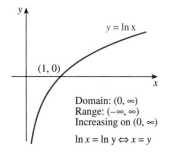

**Figure 15.10**

We now consider the graph of the function $y = \ln x$. First notice that a point $(a, b)$ is on the graph of $y = e^x$ if, and only if, $b = e^a$. However, according to the definition of the natural logarithm, $b = e^a$ if, and only if, $a = \ln b$. This says that $(b, a)$ is on the graph of $y = \ln x$. Thus $(a, b)$ lies on the graph of $y = e^x$ if, and only if, $(b, a)$ lies on the graph of $y = \ln x$. This means that the graphs of $y = e^x$ and $y = \ln x$ are symmetrical about the line $y = x$, as shown in Figure 15.9. In this way we are able to construct the graph of $y = \ln x$ shown in Figures 15.9 and 15.10 where the properties of $y = \ln x$ are also given.

From Figure 15.10 we see that $\ln x$ is increasing. We now see how fast it is increasing. We use the fact that $\ln e^t = t$ for all $t$ to construct Table 15.4. For example, rounding off to 2 decimal places, we have $1 = \ln e^1 = \ln e \approx \ln 2.72$, $2 = \ln e^2 \approx \ln 7.39$.

**Table 15.4**

| $x$ | 2.71 | 7.39 | 54.60 | 2981 | 8,886,111 |
|------|------|------|-------|------|-----------|
| $\ln x$ | 1 | 2 | 4 | 8 | 16 |

Notice how slowly $y = \ln x$ is increasing!
The natural logarithm $\ln x$ obeys the following rules.

---

**Rules for the Natural Logarithm**

1. $\ln xy = \ln x + \ln y$.

2. $\ln \dfrac{x}{y} = \ln x - \ln y$.

3. $\ln x^a = a \ln x$.

---

To establish these rules, we need to use $x = e^{\ln x}$, $y = e^{\ln y}$, $xy = e^{\ln xy}$, $\dfrac{x}{y} = e^{\ln(x/y)}$.

To establish Rule 1, notice that

$$e^{\ln xy} = xy = e^{\ln x}e^{\ln y} = e^{\ln x + \ln y}.$$

Then equating exponents gives Rule 1.

We can establish Rule 2 in a similar manner. Notice that

$$e^{\ln(x/y)} = \frac{x}{y} = \frac{e^{\ln x}}{e^{\ln y}} = e^{\ln x - \ln y}.$$

Then equating exponents gives Rule 2.

To establish Rule 3, notice that

$$e^{\ln x^a} = x^a = (e^{\ln x})^a = e^{a \ln x}.$$

Equating exponents gives Rule 3.

John Napier invented logarithms so that the numerical calculations demanded by astronomy, navigation, engineering, and other areas could be done more quickly. This invention was enthusiastically and quickly adopted throughout Europe. The great mathematician Laplace commented that the invention of logarithms "by shortening the labors, doubled the life of the astronomer."

Let us now see how logarithms can aid in the calculation of a complex expression.

**E X A M P L E  3**    **Using Logarithms to Simplify Calculations**

Express

$$\ln \frac{x^2 \sqrt{y}}{z^3}$$

in terms of the natural logarithms of $x$, $y$, and $z$.

**Solution**    Using the properties of logarithms

$$\ln \frac{x^2 \sqrt{y}}{z^3} = \ln \frac{x^2 y^{1/2}}{z^3}$$

$$= \ln x^2 y^{1/2} - \ln z^3$$

$$= \ln x^2 + \ln y^{1/2} - \ln z^3$$

$$= 2 \ln x + \frac{1}{2} \ln y - 3 \ln z. \quad \blacksquare$$

If the natural logarithms of $x$, $y$, and $z$ are known (from a table, for example), then this last expression $v = 2 \ln x + \frac{1}{2} \ln y - 3 \ln z$ can be easily evaluated. One then looks in the table to find the number $u$ for which $\ln u = v$. The number $u$ is then $\frac{x^2 \sqrt{y}}{z^3}$. This was how complex calculations were done in Napier's time and up until the advent of the modern computer. (Napier expended enormous amounts of time creating a table of logarithms that then circulated throughout the scientific community.)

Although tables of logarithms are no longer used, the logarithmic function will be seen in this text to play a vital role in business and nature and to solve equations involving exponents.

## Solving Equations with Exponents and Logarithms

If we have an equation with an exponent, then at an appropriate point we must take the logarithm to "undo" the exponent as in the following example.

**E X A M P L E   4**     **Solving an Equation with an Exponent**

Solve $15 = 5e^{4x-1}$.

**Solution**
$$15 = 5e^{4x-1}$$
$$3 = e^{4x-1}$$
$$\ln 3 = \ln e^{4x-1} = 4x - 1 \qquad \text{Property 2}$$
$$x = \frac{1}{4}(1 + \ln 3) \quad \blacksquare$$

If we have an equation with a logarithm, then at an appropriate point we can use the definition of natural logarithm, as in the following example.

**E X A M P L E   5**     **Solving an Equation with a Logarithm**

Solve $2\ln(2x + 5) + 6 = 0$.

**Solution**
$$2\ln(2x + 5) + 6 = 0$$
$$\ln(2x + 5) = -3$$
$$2x + 5 = e^{-3} \qquad \text{definition of logarithm}$$
$$x = \frac{1}{2}(e^{-3} - 5) \quad \blacksquare$$

**R E M A R K .**     In Example 5, we had $\ln(2x + 5) = -3$ at one point. We could have applied the exponential function to each side of this equation to "undo" the logarithm. Thus

$$e^{-3} = e^{\ln(2x+5)} = 2x + 5$$

by Property 1.

## Writing Exponential Functions in Terms of the Natural Exponent

We shall now see that any exponential function $y = a^x$, $a > 0$, can be written in terms of the natural exponential function. To see this, just notice that

$$a^x = (e^{\ln a})^x = e^{x \ln a}.$$

Thus we have the following.

> **Writing Exponential Functions in Terms of the Natural Exponent**
>
> For any $a > 0$
>
> $$a^x = e^{kx},$$
>
> where $k = \ln a$.

**E X A M P L E   6**   **Writing an Exponential Function in Terms of the Natural Exponent**

Write $2^x$ as $e^{kx}$.

**Solution**   Since to 5 decimal places $\ln 2 \approx 0.69315$,

$$2^x \approx e^{0.69315x}. \quad\blacksquare$$

## Applications

There will be many applications of the logarithmic function in this text. One is given here.

**E X A M P L E   7**   **Time to Double an Account**

Suppose an account yields 9% per year compounded continuously. How long until this account doubles in value?

**Solution**   The amount in this account after $t$ years is given by

$$F(t) = P_0 e^{0.09t},$$

where $P_0$ is the initial amount. Let $T$ be the time it takes for this account to double. Then $F(T) = 2P_0$. Thus

$$2P_0 = F(T) = P_0 e^{0.09T},$$

and canceling $P_0$ yields

$$2 = e^{0.09T}$$

$$\ln 2 = \ln e^{0.09T} = .09T$$

$$T = \frac{\ln 2}{.09} \approx \frac{.693}{.09} \approx 7.7,$$

or about eight years. Refer to Figure 15.11.   $\blacksquare$

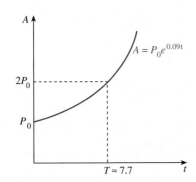

**Figure 15.11**

## SELF-HELP EXERCISE SET 15.2

1. Solve for $x$ in the equation $e^{x^3} = 5$.

2. An account earns an annual rate of $r$, expresssed as a decimal, and is compounded continuously. The account has $1000 initially and 5 years later had $1500. What is $r$?

## EXERCISE SET 15.2

In Exercises 1 through 6 solve for $x$.

**1.** $\ln x = 2$          **2.** $\ln x = -3$          **3.** $\ln x = -\pi$

**4.** $\ln x = -\frac{3}{4}$          **5.** $3 \ln x = 5$          **6.** $2 \ln x = 9$

In Exercises 7 through 16 simplify.

**7.** $\ln e^4$          **8.** $\ln \sqrt{e}$          **9.** $\ln \dfrac{1}{e}$          **10.** $\ln \dfrac{1}{\sqrt{e}}$

**11.** $e^{\ln 2\pi}$          **12.** $e^{\ln \sqrt{2}}$          **13.** $e^{3 \ln 2}$          **14.** $e^{0.5 \ln 9}$

**15.** $(e^{\ln 3})^4$          **16.** $\sqrt{e^{\ln 2}}$

In Exercises 17 through 22 write the given quantity in terms of $\ln x$, $\ln y$, and $\ln z$.

**17.** $\ln x^2\sqrt{yz}$          **18.** $\ln\sqrt{xyz}$          **19.** $\ln \dfrac{\sqrt{xy}}{z}$

**20.** $\ln \dfrac{xy^2}{z^3}$          **21.** $\ln \sqrt{\dfrac{xy}{z}}$          **22.** $\ln \dfrac{x^2y^3}{\sqrt{z}}$

In Exercises 23 through 28 write the given quantity as one natural logarithm.

**23.** $2 \ln x + \ln y$          **24.** $2 \ln x - \ln y$

**25.** $\frac{1}{2} \ln x - \frac{1}{3} \ln y$          **26.** $2 \ln x - \frac{1}{2} \ln y + \ln z$

**27.** $3 \ln x + \ln y - \frac{1}{3} \ln z$          **28.** $\sqrt{2} \ln x - \ln y$

In Exercises 29 through 40 solve the equation for $x$.

**29.** $10e^{5x} = 3$          **30.** $2e^{3x} = 5$          **31.** $6 = 2e^{-2x}$

**32.** $3 = 4e^{-0.5x}$          **33.** $2e^{3x-1} = 1$          **34.** $3e^{2-5x} = 4$

**35.** $e^{x^2} = 4$          **36.** $e^{\sqrt{x}} = 4$

**37.** $2 \ln(x + 7) + 3 = 0$          **38.** $\ln x^2 = 9$

**39.** $\ln(\ln 4x) = 0$          **40.** $\ln 3x = \ln 6$

**41.** Find $k$ such that $10^x = e^{kx}$.

42.  Find $k$ such that $3^x = e^{kx}$.

43.  Find $k$ such that $2^x = e^{kx}$.

44.  Establish the change of base theorem:

$$\log_a x = \frac{\log_b x}{\log_b a}.$$

*Hint:* Start with $x = a^{\log_a x}$ and take a logarithm to the base $b$ of each side.

45.  Use the change of base theorem to find $\log_{10} x$ in terms of natural logarithms.

46.  Use the change of base theorem to find $\log_2 x$ in terms of natural logarithms.

## Applications

47.  **Doubling Time.**   If an account has an annual yield of 8% compounded continuously, how long before the account doubles?

48.  **Tripling Time.**   If an account has an annual yield of 8% compounded continuously, how long before the account triples?

49.  **Doubling Time.**   If an account has an annual yield of 6% compounded quarterly, how long before the account doubles?

50.  **Doubling Time.**   If an account has an annual yield of 9% compounded monthly, how long before the account doubles?

51.  **Continuous Compounding.**   If an account pays an annual rate of $r$ compounded continuously and goes from $100 to $180 in 10 years, what is $r$?

52.  **Continuous Compounding.**   If an account pays an annual rate of $r$ compounded continuously and triples in 12 years, what is $r$?

53.  **Continuous Compounding.**   If an account pays an annual rate of $r$ compounded continuously and goes from $100 to $150 in 10 years, what is $r$?

54.  **Continuous Compounding.**   An individual deposits $2000 into an account with an annual yield of 6% compounded continuously and $1000 into an account with an annual yield of 9% compounded continuously. How long will it take for the amount in the second account to equal the amount in the first account?

55.  **Continuous Compounding.**   Repeat the previous problem if the second account yields 10%.

## Graphing Calculator Exercises

56.  On the same viewing rectangle graph $y = e^x$, $y = x$, and $y = \ln x$, using the zoom feature if necessary, to reproduce Figure 15.9 of the text.

57.  Show that neither of the following equations are true by graphing each side of each equation in the same viewing rectangle.
     **a.** $\ln(x + 4) = \ln x + \ln 2$,   **b.** $\ln x^2 = (\ln x)^2$

**58.** Graph $y = e^x$ and $y = 2$ in the same viewing rectangle. Using the zoom feature find the solution of $e^x = 2$ to four decimal places. (The exact solution is $x = \ln 2$.)

**59.** An account with an initial amount of $1 grows at 25% a year compounded continuously. By graphing two appropriate functions in the same viewing rectangle and using the zoom feature, find the time (to two decimal places) for this account to triple in value.

**60.** One account with an initial amount of $1 grows at 100% a year compounded continuously, and a second account with an initial amount of $2 grows at 50% a year compounded continuously. Graph the two appropriate functions in the same viewing rectangle and, using the zoom feature, find the time (to two decimal places) for the two accounts to have equal amounts.

### Solutions to Self-Help Exercise Set 15.2

**1.** Taking the natural logarithm of each side gives

$$\ln 5 = \ln e^{x^3} = x^3.$$

Thus $x = \sqrt[3]{\ln 5} \approx \sqrt[3]{1.6094} \approx 1.1719$.

**2.** If $F(t)$ is the amount in the account after $t$ years, then $F(t) = 1000e^{rt}$. Then

$$1500 = F(5) = 1000e^{5r}$$

$$1.5 = e^{5r}$$

$$\ln 1.5 = \ln e^{5r} = 5r$$

$$r = \frac{1}{5} \ln 1.5 \approx 0.081$$

or 8.1%.

## 15.3   DERIVATIVES OF EXPONENTIAL AND LOGARITHMIC FUNCTIONS

▶ *Derivatives of Exponential Functions*
▶ *Derivatives of Logarithmic Functions*

**APPLICATION**

**Graphing a Revenue Function**

Let the demand equation be given by $p(x) = e^{-x}$. Where is the revenue $R(x)$ increasing and where is it decreasing? Where are all relative and absolute maxima and minima? Where is the graph of $R(x)$ concave up and where is it concave down? What is the graph? See Example 3 for the answer.

## Derivatives of Exponential Functions

We will show that the derivative of $e^x$ is again $e^x$, i.e.,

$$\frac{d}{dx}(e^x) = e^x.$$

In order to see this, let $f(x) = e^x$. Use the limit definition of the derivative and obtain

$$f'(x) = \lim_{h \to 0} \frac{f(x + h) - f(x)}{h}$$

$$= \lim_{h \to 0} \frac{e^{x+h} - e^x}{h}$$

$$= \lim_{h \to 0} \frac{e^x e^h - e^x}{h}$$

$$= \lim_{h \to 0} e^x \frac{e^h - 1}{h}$$

$$= e^x \lim_{h \to 0} \frac{e^h - 1}{h}$$

To complete this analysis, we must evaluate

$$\lim_{h \to 0} \frac{e^h - 1}{h}.$$

We do this by evaluating $\dfrac{e^h - 1}{h}$ for ever smaller values of $h$ in the following table

| $h$ | $-0.1$ | $-0.01$ | $-0.001$ | $\to$ | 0 | $\leftarrow$ | 0.001 | 0.01 | 0.1 |
|---|---|---|---|---|---|---|---|---|---|
| $\frac{e^h - 1}{h}$ | .9516 | .9950 | .9995 | $\to$ | ? | $\leftarrow$ | 1.0005 | 1.0050 | 1.0517 |

The table then indicates that

$$\lim_{h \to 0} \frac{e^h - 1}{h} = 1.$$

In fact, this can be proven to be the case, but we do not do so here.

Using this information, we then have

$$f'(x) = e^x \cdot \lim_{h \to 0} \frac{e^h - 1}{h} = e^x \cdot 1 = e^x.$$

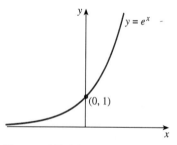

**Figure 15.12**

The Derivative of $e^x$

The derivative of $e^x$ is again $e^x$, that is,

$$\frac{d}{dx}(e^x) = e^x.$$

Since $\dfrac{d}{dx}(e^x) = e^x$, we also have $\dfrac{d^2}{dx^2}(e^x) = e^x$. But since $e^x$ is positive for all $x$, both the first and second derivatives of $e^x$ are positive. This shows that $e^x$ is an everywhere increasing function that is always concave up. See Figure 15.12.

**E X A M P L E   1**    **Calculating the Derivatives of Expressions with Exponentials**

Find the derivatives of (a) $f(x) = (x^2 + x + 1)e^x$, (b) $g(x) = \sqrt{e^x + 1}$.

**Solutions**    (a)   Use the product rule of differentiating and obtain

$$f'(x) = e^x \frac{d}{dx}(x^2 + x + 1) + (x^2 + x + 1)\frac{d}{dx}(e^x)$$

$$= e^x(2x + 1) + (x^2 + x + 1)(e^x)$$

$$= (x^2 + 3x + 2)e^x$$

(b)   Use the rule for differentiating a function raised to a power and obtain

$$g'(x) = \frac{d}{dx}((e^x + 1)^{1/2})$$

$$= \frac{1}{2}(e^x + 1)^{-1/2}\frac{d}{dx}(e^x + 1)$$

$$= \frac{1}{2}(e^x + 1)^{-1/2}e^x \quad \blacksquare$$

We need to be able to differentiate functions of the form $y = e^{f(x)}$. To see how to do this set $y = e^u$ where $u = f(x)$, then

$$\frac{dy}{du} = \frac{d}{du}e^u = e^u.$$

Now use the chain rule and obtain

$$\frac{d}{dx}(e^{f(x)}) = \frac{dy}{dx} = \frac{dy}{du} \cdot \frac{du}{dx} = e^u \frac{du}{dx} = e^{f(x)}f'(x).$$

The Derivative of $e^{f(x)}$

$$\frac{d}{dx}(e^{f(x)}) = e^{f(x)}f'(x)$$

**E X A M P L E  2**  **The Derivatives of *e* Raised to a Function**

Find the derivatives of (a) $e^{3x}$, (b) $xe^{x^3+x}$, (c) $\dfrac{10}{1 + e^{-2x}}$.

**Solutions**  (a)  Using the above formula with $f(x) = 3x$ yields

$$\frac{d}{dx}(e^{3x}) = e^{3x}\frac{d}{dx}(3x) = 3e^{3x}.$$

(b)  Using the product rule yields

$$\frac{d}{dx}(xe^{x^3+x}) = e^{x^3+x}\frac{d}{dx}(x) + x\frac{d}{dx}(e^{x^3+x})$$

$$= e^{x^3+x} + x\,(e^{x^3+x})(3x^2 + 1)$$

$$= (3x^3 + x + 1)e^{x^3+x}$$

(c)  Using the quotient rule for differentiating yields

$$\frac{d}{dx}\left(\frac{10}{1 + e^{-2x}}\right) = 10\frac{d}{dx}\left(\frac{1}{1 + e^{-2x}}\right)$$

$$= 10\,\frac{(1 + e^{-2x})\dfrac{d}{dx}(1) - (1)\dfrac{d}{dx}(1 + e^{-2x})}{(1 + e^{-2x})^2}$$

$$= 10\,\frac{(1 + e^{-2x})(0) - (1)(-2e^{-2x})}{(1 + e^{-2x})^2}$$

$$= 20\,\frac{e^{-2x}}{(1 + e^{-2x})^2} \quad\blacksquare$$

**E X A M P L E  3**  **Graphing a Revenue Function**

Let the demand equation be given $p(x) = e^{-x}$. Find where the revenue $R(x)$ is increasing and where it is decreasing. Find all relative and absolute maxima and minima. Find where the graph of $R(x)$ is concave up and where it is concave down. Graph the revenue function.

**Solution**  We have

$$R(x) = xp(x) = xe^{-x}.$$

We first need to find the derivative and to factor the result. Thus

$$R'(x) = \frac{d}{dx}(xe^{-x})$$

$$= e^{-x}\frac{d}{dx}(x) + x\frac{d}{dx}(e^{-x})$$

$$= e^{-x}(1) + x(-e^{-x})$$

$$= e^{-x}(1 - x)$$

First notice that the derivative is continuous everywhere. The term $e^{-x}$ is always positive. Thus from the factorization of $R'(x)$ we see that if $x < 1$, $R'(x) > 0$, while if $x > 1$, $R'(x) < 0$. Also $R'(1) = 0$. So $R(x)$ is increasing on $(-\infty, 1)$ and decreasing on $(1, +\infty)$. This implies that $R$ has an absolute maximum at $x = 1$ and no relative or absolute minima.

To find the concavity we must find $R''(x)$ and factor. Thus

$$R''(x) = \frac{d}{dx}(e^{-x}(1 - x))$$

$$= e^{-x}\frac{d}{dx}(1 - x) + (1 - x)\frac{d}{dx}(e^{-x})$$

$$= e^{-x}(-1) - (1 - x)e^{-x}$$

$$= e^{-x}(x - 2)$$

Since the factor $e^{-x}$ is always positive, we see that if $x < 2$, $R''(x) < 0$, while if $x > 2$, $R''(x) > 0$.

Thus the graph of $R(x)$ is concave down on $(-\infty, 2)$ and concave up on $(2, +\infty)$.

Finally, using a calculator one can obtain $R(5) = 0.033\ 690$, $R(10) = 0.000\ 454$, and $R(15) = 0.000\ 005$. Thus we see that $\lim\limits_{x \to +\infty} R(x) = 0$. The graph is given in Figure 15.13.  ∎

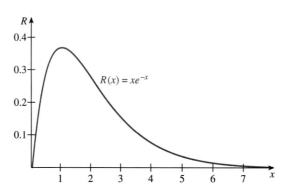

**Figure 15.13**

We have already seen how to differentiate the expression $e^{f(x)}$. We now see how to differentiate the expression $a^{f(x)}$ for any positive value of $a$. First recall that for any $a > 0$, $a^x = e^{x \ln a}$. Differentiating this yields

$$\frac{d}{dx}(a^x) = \frac{d}{dx}(e^{x \ln a}) = e^{x \ln a}\frac{d}{dx}(x \ln a) = e^{x \ln a}\ln a = a^x \ln a.$$

Using the chain rule we can write the more general form of this as follows.

---

**The Derivative of $a^{f(x)}$**

$$\frac{d}{dx}\left(a^{f(x)}\right) = a^{f(x)}f'(x)\ln a$$

---

E X A M P L E   4      **Finding the Derivative of $a^{f(x)}$**

Find $\dfrac{d}{dx}\left(2^{x^2+x+1}\right)$. We have from the previous formula

$$\frac{d}{dx}\left(2^{x^2+x+1}\right) = 2^{x^2+x+1}\frac{d}{dx}(x^2 + x + 1) \cdot \ln 2$$

$$= 2^{x^2+x+1}(2x + 1)\ln 2. \quad \blacksquare$$

## Derivatives of Logarithmic Functions

We will show the following.

---

**Derivative of $\ln x$**

$$\frac{d}{dx}(\ln x) = \frac{1}{x}$$

---

To show this suppose $x > 0$ and let $f(x) = \ln x$. This is equivalent to saying that

$$x = e^{f(x)}.$$

Now we differentiate this last expression and obtain

$$\frac{d}{dx}(x) = \frac{d}{dx}e^{f(x)},$$

which is just

$$1 = e^{f(x)}f'(x).$$

Then

$$\frac{d}{dx}(\ln x) = f'(x) = \frac{1}{e^{f(x)}} = \frac{1}{x}.$$

**E X A M P L E   5**    **Finding the Derivative of Expressions with Logarithms**

For $x > 0$, find the derivative of $f(x) = e^x \ln x$.

**Solution**

$$f'(x) = \frac{d}{dx}(e^x \ln x)$$

$$= e^x \frac{d}{dx}(\ln x) + (\ln x)\frac{d}{dx}(e^x)$$

$$= (e^x)\left(\frac{1}{x}\right) + (\ln x)(e^x)$$

$$= e^x \left(\frac{1}{x} + \ln x\right) \quad \blacksquare$$

When $x > 0$, $\dfrac{d}{dx}(\ln x) = \dfrac{1}{x} > 0$. Thus $\ln x$ is a strictly increasing function for $x > 0$.

To investigate the concavity of the graph of $\ln x$ we have

$$\frac{d^2}{dx^2}(\ln x) = \frac{d}{dx}\left(\frac{d}{dx}(\ln x)\right)$$

$$= \frac{d}{dx}\left(\frac{1}{x}\right)$$

$$= -\frac{1}{x^2}$$

But for $x > 0$ this is always negative. Thus for $x > 0$, the graph of $\ln x$ is concave down. See Figure 15.14.

We shall now find the derivative of an expression such as $y = \ln f(x)$. We first let $u = f(x)$ and notice that

$$\frac{dy}{du} = \frac{d}{du}\ln u = \frac{1}{u}.$$

Now the chain rule shows that

$$\frac{d}{dx}(\ln f(x)) = \frac{dy}{dx} = \frac{dy}{du} \cdot \frac{du}{dx} = \frac{1}{u}\frac{du}{dx} = \frac{1}{f(x)}f'(x).$$

Thus

$$\frac{d}{dx}(\ln f(x)) = \frac{f'(x)}{f(x)}.$$

This formula can be used to show that

$$\frac{d}{dx}\ln |x| = \frac{1}{x}.$$

To see this consider $y = \ln(-x)$ where $x < 0$ and set $f(x) = -x$. Then

$$\frac{d}{dx}(\ln(-x)) = \frac{f'(x)}{f(x)} = \frac{-1}{-x} = \frac{1}{x}.$$

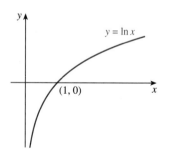

**Figure 15.14**

This in turn can be used to establish the following more general formula.

---

**Derivative of Logarithmic Function**

$$\frac{d}{dx}\left(\ln |f(x)|\right) = \frac{f'(x)}{f(x)}$$

---

**REMARK.** Notice that when taking the derivative of an expression such as $\ln |f(x)|$, there is *no* logarithm in the answer.

**EXAMPLE 6** **Derivative of a Logarithmic Function**

Find the derivative of $\ln |x^2 + x + 1|$.

**Solution** Use the above formula with $f(x) = x^2 + x + 1$. Then

$$\frac{d}{dx}\left(\ln |x^2 + x + 1|\right) = \frac{\dfrac{d}{dx}(x^2 + x + 1)}{x^2 + x + 1}$$

$$= \frac{2x + 1}{x^2 + x + 1} \quad \blacksquare$$

The properties of logarithms can be very useful when differentiating expressions involving the logarithm, as the following example illustrates.

**EXAMPLE 7** **Differentiating Using Properties of Logarithm**

Find the derivative of $\ln \dfrac{x + 1}{x^2 + 1}$ for $x > -1$.

**Solution** Use the fact that

$$\ln \frac{x + 1}{x^2 + 1} = \ln(x + 1) - \ln(x^2 + 1).$$

Then

$$\frac{d}{dx}\left(\ln \frac{x + 1}{x^2 + 1}\right) = \frac{d}{dx}\left(\ln(x + 1)\right) - \frac{d}{dx}\left(\ln(x^2 + 1)\right)$$

$$= \frac{\dfrac{d}{dx}(x + 1)}{x + 1} - \frac{\dfrac{d}{dx}(x^2 + 1)}{x^2 + 1}$$

$$= \frac{1}{x + 1} - \frac{2x}{x^2 + 1} \quad \blacksquare$$

**E X A M P L E   8**     **Sketching a Graph**

Consider $f(x) = \dfrac{\ln x}{x}$, for $x > 0$. Find where $f$ is increasing and where $f$ is decreasing. Find any relative or absolute maxima or minima. Find where the graph of $f$ is concave up and where the graph of $f$ is concave down.

**Solution**     First notice that the sign of $f(x)$ on $(0, +\infty)$ is the same as the sign of $\ln x$. Thus $f(x)$ is negative on $(0, 1)$, zero at $x = 1$, and positive on $(1, +\infty)$.

$$f'(x) = \frac{d}{dx}(x^{-1} \ln x)$$

$$= (x^{-1})\frac{d}{dx}(\ln x) + (\ln x)\frac{d}{dx}(x^{-1})$$

$$= (x^{-1})\left(\frac{1}{x}\right) + (\ln x)(-x^{-2})$$

$$= \frac{1}{x^2}(1 - \ln x)$$

The first factor $1/x^2$ is always positive on $(0, +\infty)$. To analyze the factor $(1 - \ln x)$, we recall that $\ln x$ is increasing and $1 - \ln x = 0$ when $x = e$. Thus $1 - \ln x > 0$ on $(0, e)$ and therefore $f'(x) > 0$ on $(0, e)$. Also $1 - \ln x < 0$ on $(e, +\infty)$ and therefore $f'(x) < 0$ on $(e, +\infty)$. Since $1 - \ln e = 0$, $f'(e) = 0$. Thus $f(x)$ is increasing on $(0, e)$ and decreasing on $(e, +\infty)$. This implies that $f(x)$ has an absolute maximum at $x = e$ and no relative or absolute minima on $(0, +\infty)$.

Calculating the second derivative gives

$$f''(x) = (1 - \ln x)\frac{d}{dx}\left(\frac{1}{x^2}\right) + \frac{1}{x^2}\frac{d}{dx}(1 - \ln x)$$

$$= (1 - \ln x)\left(-\frac{2}{x^3}\right) + \frac{1}{x^2}\left(-\frac{1}{x}\right)$$

$$= \frac{2}{x^3}\left(\ln x - \frac{3}{2}\right)$$

The first factor $\dfrac{2}{x^3}$ is positive for $x > 0$. Notice that $\ln x - \dfrac{3}{2} = 0$ when $x = e^{3/2}$ and that $\ln x$ is an increasing function. Therefore if $0 < x < e^{3/2}$, $\ln x - \dfrac{3}{2}$ and $f''(x)$ are negative, while if $x > e^{3/2}$, $\ln x - \dfrac{3}{2}$ and $f''(x)$ are positive. Thus the graph of $f$ is concave down on $(0, e^{3/2})$ and concave up on $(e^{3/2}, +\infty)$.

Finally since $f(100) = 0.046\ 052$, $f(1000) = 0.006\ 908$, and $f(10,000) = 0.000\ 921$, $\lim\limits_{x \to +\infty} f(x) = 0$.

The graph is given in Figure 15.15.   ∎

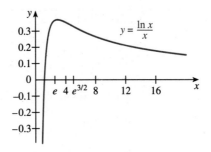

**Figure 15.15**

## SELF-HELP EXERCISE SET 15.3

1.  Sketch a graph of $y = f(x) = e^{-0.50x^2}$.

2.  Sketch a graph of $y = g(x) = x \ln x$ for $x > 0$. *Hint:* By using a table of small values of $x$, show $\lim_{x \to 0+} g(x) = 0$.

## EXERCISE SET 15.3

In Exercises 1 through 42 find the derivative.

1.  $e^{4x}$
2.  $e^{-0.1x}$
3.  $e^{-3x}$

4.  $4e^{\pi x}$
5.  $e^{2x^2 + 1}$
6.  $e^{4x^3 - x - 1}$

7.  $e^{\sqrt{x}}$
8.  $e^{\sqrt[3]{x}}$
9.  $x^2 e^x$

10.  $\sqrt{x} e^x$
11.  $x^3 e^x$
12.  $(x^3 + 1)e^x$

13.  $\dfrac{x}{e^x}$
14.  $(x^3 - 2)e^x$
15.  $x^2 e^{x^2}$

16.  $\dfrac{e^x}{x^2 + 1}$
17.  $\sqrt{e^x + 1}$
18.  $(e^x + 2)^8$

19.  $\sqrt{x} e^{\sqrt{x}}$
20.  $\sqrt{x^3 e^{3x}}$
21.  $\dfrac{x}{1 + e^x}$

22.  $\dfrac{x}{1 + e^{-x}}$
23.  $\dfrac{e^{2x} - e^{-2x}}{e^{2x} + e^{-2x}}$
24.  $(e^{3x} + 1)^4$

25.  $\sqrt{e^{3x} + 2}$
26.  $\sqrt{2 + e^{-3x}}$
27.  $5^x$

28.  $3^{3x}$
29.  $3^{\sqrt{x}}$
30.  $7^{x^7}$

31.  $x3^x$
32.  $3^x \cdot 5^x$
33.  $\ln |x^3 + x^2 + 1|$

34.  $\ln(x^4 + x^2 + 3)$
35.  $x^2 \ln (x^2)$
36.  $\dfrac{\ln(x^2)}{x + 1}$

**37.** $\sqrt{\ln|x^2 + x + 1|}$    **38.** $\sqrt{x^2 + 1} \ln(x + 1)$    **39.** $\ln\left|\dfrac{x}{x + 1}\right|$

**40.** $\ln\sqrt{\left|\dfrac{x}{x + 1}\right|}$    **41.** $e^{-x^2} \ln(x^2)$    **42.** $\dfrac{\ln(x^2)}{e^x + e^{-x}}$

In Exercises 43 through 52 take $x > 0$ and find the derivatives.

**43.** $x^2 \ln x$    **44.** $(x^3 + 2) \ln x$    **45.** $\dfrac{\ln x}{x^2 + 1}$

**46.** $\sqrt{\ln x}$    **47.** $(\ln x)^{11}$    **48.** $e^{x^2} \ln x$

**49.** $\ln(x^3)$    **50.** $\ln(x^7)$    **51.** $\ln(e^x + 1)$

**52.** $\ln(e^{3x} + x^2)$

In Exercises 53 through 64 find where the function is increasing and where it is decreasing. Find any relative minima and maxima. Find where the curve is concave up and where it is concave down. Graph. Assume $x > 0$ in 61 through 64.

**53.** $y = xe^{-3x}$    **54.** $y = xe^x$    **55.** $y = e^{-x^2}$

**56.** $y = e^{-x^4}$    **57.** $y = x^2e^x$    **58.** $y = x^4e^x$

**59.** $y = x^2e^{-x}$    **60.** $y = x^2e^{-0.5x^2}$    **61.** $y = x \ln x^2$

**62.** $y = x - \ln x$    **63.** $y = \dfrac{\ln x}{x^2}$    **64.** $y = (\ln x)^3$

**65.** Use the change of base theorem given in Exercise 50 of the Exercise Set 15.2 to show that

$$\frac{d}{dx}(\log_a |x|) = \frac{1}{x \ln a}$$

**66.** Use the formula in the previous exercise to find $\dfrac{d}{dx}(\log_{10} |x|)$.

## Applications

**67. Revenue.**   Suppose the price and demand of a commodity is related by $p(x) = e^{-2x}$. Find where the revenue $R(x)$ attains an absolute maximum (for $x$ nonnegative).

**68. Profits.**   Suppose the profit equation is given by $P(x) = xe^{-0.5x^2}$. Find where the profit $P(x)$ attains an absolute maximum (for $x$ nonnegative).

**69. Cost.**   If the cost function is given by $C(x) = \ln(x^2 + 5) + 100$, find the marginal cost. Is the cost increasing?

**70. Revenue.**   If the revenue function is given by

$$R(x) = \frac{\ln(x + 1)}{x + 1} - \frac{9}{x + 1},$$

find where the revenue is maximum.

71.  **Biology.**  The monomolecular growth assumption assumes that the rate of
     change $R$ of growth of a substance at time $t$ is given by

     $$R(t) = \frac{A_0}{b-1}(be^{-kt} - 1),$$

     where $b > 1$ and $k > 0$ are constants, and $A_0$ is the initial amount of the
     substance. Show that $R(t)$ is always decreasing.

72.  **Biology.**  The Gompertz growth curve is given by

     $$P(t) = ae^{-be^{-kt}},$$

     where $a$, $b$, and $k$ are positive constants. Find $P'(t)$.

73.  **Physical Chemistry.**  The amount of gas absorbed per unit area of absorbing
     surface $S$ is given by

     $$S(T) = kPe^{Q/RT},$$

     where $T$ is the temperature, $P$ is the constant gas pressure, $Q$ is the constant heat
     of absorption, and $R$ and $k$ are positive constants. What happens to the amount
     of gas absorbed when the temperature rises?

74.  **Biology.**  The concentration $C$ of oxygen in plant roots is given by

     $$C(r) = a + b\ln\frac{R}{r},$$

     where $R$ is the radius of the root, $r$ is the radius of the root plus the moisture
     film, and $a$ and $b$ are positive constants. Show that the concentration is a de-
     creasing function of $r$.

## Graphing Calculator Exercises

75.  On your graphing calculator, graph each of the functions in Exercises 53 through
     64.

## Solutions to Self-Help Exercise Set 15.3

1.  The domain of $f(x)$ is the entire real line. The intercept is $(0, 1)$. Notice that for
    all $x$ $f(x) = e^{-0.50x^2}$ is positive. Taking a derivative gives

    $$f'(x) = \frac{d}{dx}(e^{-0.50x^2}) = e^{-0.50x^2}\frac{d}{dx}(-0.50x^2) = -xe^{-0.50x^2}.$$

    Since $e^{-0.50x^2}$ is positive for all $x$, the sign of $f'(x)$ is the same as the sign of
    $-x$. Thus $f(x)$ is increasing on $(-\infty, 0)$ and decreasing on $(0, +\infty)$, with $x = 0$
    an absolute maximum.

    $$f''(x) = -e^{-0.50x^2} + x^2e^{-0.50x^2} = e^{-0.50x^2}(x^2 - 1)$$

This is negative if $|x| < 1$ and positive if $|x| > 1$. Thus the graph of $f(x)$ is concave up on $(-\infty, -1)$ and $(1, +\infty)$ and concave down on $(-1, 1)$. Also $\lim\limits_{x \to \pm\infty} e^{-0.50x^2} = 0$. The graph is shown in the figure.

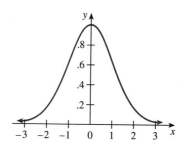

2.  First notice that the sign of $g(x) = x \ln x$ on $(0, +\infty)$ is the same as the sign of $\ln x$. Thus $x \ln x$ is negative on $(0, 1)$, zero at $x = 1$, and positive on $(1, +\infty)$. Taking $x = 0.1, 0.01, 0.001$, and $0.0001$ yields $x \ln x$ to be, respectively, $-0.2304, -0.0461, -0.0069$, and $-0.0009$. Thus we can assume (correctly) that $\lim\limits_{x \to 0+} x \ln x = 0$. We have

$$g'(x) = 1 + \ln x.$$

This is zero on $(0, +\infty)$ when $1 + \ln x = 0$ or when $\ln x = -1$. This implies $x = e^{-1}$. Since $1 + \ln x$ is an increasing function that is zero at $x = e^{-1}$, $g'(x) = 1 + \ln x$ is negative on $(0, e^{-1})$ and positive on $(e^{-1}, +\infty)$. Thus $g(x)$ is decreasing on $(0, e^{-1})$, zero at $x = e^{-1}$, and increasing on $(e^{-1}, +\infty)$. This implies that $x = e^{-1}$ is an absolute minimum. Taking the second derivative gives $g''(x) = 1/x$, which is positive on $(0, +\infty)$. Thus the graph of $g(x)$ is concave up for $x > 0$. The graph is shown in the figure.

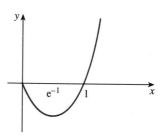

## 15.4   GROWTH AND DECAY

▶ *Introduction*

▶ *Population Growth*

▶ *Radioactive Decay*

▶ *Limited Growth*

**APPLICATION**

**Safe Time After Nuclear Accident**

Due to an unfortunate nuclear accident, radioactive iodine-131 has been released into the atmosphere and has been absorbed by the grass in some fields, which then has been ingested by cows. The milk from the affected cows has three times the maximum allowable amount of iodine-131. How long until it is safe to consume the milk from these animals? (Such an accident occurred in Michigan.) See Example 3 for the answer.

**APPLICATION**

**Carbon Dating**

An anthropologist discovers a bone fragment and learns from laboratory testing that exactly 20% of carbon-14 is missing. How old is the bone? See Example 4 for the answer.

**APPLICATION**

**Population of Earth**

What is an estimate of the long-term human population of the earth? See Example 7 for the answer.

*(© Eric L. Heyer/Grant Heilman, Inc.)*

## Introduction

We now study quantities that satisfy the equation $Q(t) = Q_0 e^{kt}$, where $Q(t)$ is the amount of the quantity. A graph is shown in Figure 15.16 when $k > 0$. We have already seen that the amount in an account that is compounding continuously satisfies such an equation. We will see in this section that a number of quantities in nature also follow such a model. If $k > 0$, we refer to $k$ as the **growth constant**.

Notice that since $Q(0) = Q_0 e^0 = Q_0$, $Q_0$ is the *initial* amount present.

First let us see how $Q'$ is related to $Q$. Differentiate $Q(t)$ and obtain

$$Q'(t) = Q_0 \frac{d}{dt}(e^{kt}) = Q_0 e^{kt} k = k[Q_0 e^{kt}] = kQ(t)$$

Thus we see that $Q'(t)$ is *proportional* to $Q(t)$ with proportionality constant $k$.

We shall often have a quantity $Q(t)$ that is initially present in an amount $Q_0$ and whose rate of change at any time $t$ is proportional to the amount of the quantity at time $t$ with a proportionality constant $k$. That is, the quantity satisfies

$$Q'(t) = kQ(t), \qquad Q(0) = Q_0. \tag{1}$$

We know from the above calculations that $Q(t) = Q_0 e^{kt}$ satisfies (1). We need to know if this is the *only* function that satisfies (1). The following theorem says that this is the case.

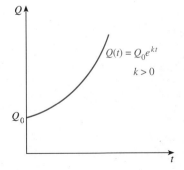

**Figure 15.16**

---

**Uniqueness Theorem**

If $Q(t)$ satisfies $\qquad Q'(t) = kQ(t), \qquad Q(0) = Q_0$

on the interval $[0, +\infty)$, then

$$Q(t) = Q_0 e^{kt}.$$

The uniqueness theorem can readily be established using integration theory from the next chapter.

Consider now what we would expect to happen merely by analyzing the equation $Q'(t) = kQ(t)$, $k > 0$ (without actually knowing what the quantity $Q(t)$ is). If $Q(t)$ is positive, $Q'(t)$ is positive. This implies that $Q(t)$ is increasing and therefore becoming larger. This in turn implies that the rate of change of $Q(t)$ becomes even larger and so on. One concludes that $Q(t)$ should be "exploding." This is indeed the case, since of course, $Q(t) = Q_0 e^{kt}$, and we know that the term $e^{kt}$ does become large quickly if $k > 0$.

## Population Growth

---

**Three More People Each Second**

Every 2 seconds, 9 babies are born and 3 people die. This increase of 3 people a second results in a growth of world population of 254,000 a day and 93 million a year.

---

Biologists argue that the rate of change of a population under a wide range of circumstances should be proportional to the population. Thus the population $P(t)$ should satisfy

$$P'(t) = kP(t), \qquad P(0) = P_0,$$

where $P_0$ is the initial population, $k$ is the growth constant, $k > 0$, and $t$ is time measured in some appropriate units. Thus from the uniqueness theorem

$$P(t) = P_0 e^{kt}.$$

Biologists have long known from actual data on many species that this is in fact the case in many important situations.

**E X A M P L E   1**     **Calculating the Future Population of the Earth**

According to the U.S. Bureau of Statistics, the population of the earth was 5 billion in 1986 and $5\frac{1}{3}$ billion in 1990. (a) If we assume exponential growth, what is the growth constant $k$? (b) What will be the population in the year 2000?

**Solutions**     (a)   Take $t = 0$ to be 1986 when the population is 5 billion. Then, in billions, $P(0) = P_0 = 5$. Since the population 4 years later is $5\frac{1}{3}$ billion, $P(4) = 5\frac{1}{3}$. Thus

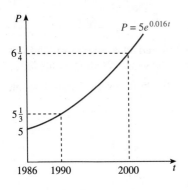

$$5\tfrac{1}{3} = P(4) = P_0 e^{kt} = 5e^{4k}$$

$$\tfrac{16}{15} = e^{4k} \qquad\qquad \tfrac{1}{5}\cdot 5\tfrac{1}{3} = \tfrac{1}{5}\cdot\tfrac{16}{3} = \tfrac{16}{15}$$

$$\ln \tfrac{16}{15} = \ln (e^{4k}) = 4k \qquad \ln e^z = z$$

$$k = \tfrac{1}{4}\ln\tfrac{16}{15}$$

$$\approx 0.016 \qquad\qquad \text{from a calculator}$$

(b)  Having found the growth constant $k$, we now have $P(t) = 5e^{.016t}$. The year 2000 corresponds to $t = 14$. Thus

$$P(14) = 5e^{.016(14)} \approx 5(1.25) \approx 6.25,$$

that is, 6.25 billion. Refer to Figure 15.17. ∎

**Figure 15.17**

### Radioactive Decay

Some elements exhibit radioactive decay. For these elements, on occasion, one nucleus will spontaneously divide into two or more other nuclei. Thus, over a period of time, the amount of the substance will decrease (decay). The rate at which decay takes place depends on the substance.

Physicists maintain that the rate of change of the amount $A(t)$ of the substance at any time is *negatively* proportional to $A(t)$. Thus $A(t)$ satisfies

$$A'(t) = -kA(t), \qquad A(0) = A_0,$$

where $k > 0$. The constant $k$ is referred to as the **decay constant**. Thus from the uniqueness theorem,

$$A(t) = A_0 e^{-kt}.$$

A graph is shown in Figure 15.18.

A common way of measuring how fast a radioactive substance is decaying is to use the **half-life**. This is the time that must elapse for half the material to decay. See Figure 15.19.

**Figure 15.18**

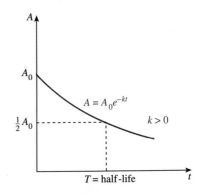

**Figure 15.19**

**E X A M P L E  2**     **Calculating Half-life**

A radioactive substance has been carefully monitored in a laboratory and has been observed to have decayed 0.4% in exactly four years. What is its half-life?

**Solution**     We are given that the amount after four years is 0.996 times the initial amount, that is, $A(4) = .996A_0$. Then

$$.996A_0 = A(4) = A_0e^{-k(4)} = A_0e^{-4k}$$

$$0.996 = e^{-4k} \qquad\qquad \text{cancel } A_0$$

$$\ln 0.996 = \ln (e^{-4k}) = -4k \qquad\qquad \ln e^z = z$$

$$k = -\frac{\ln .996}{4}$$

$$\approx .001002 \qquad\qquad \text{from a calculator}$$

Since we now have the decay constant $k$ we can write

$$A(t) = A_0e^{-0.001002t}.$$

Let $T$ be the time it takes for the substance to decay to half the original amount. Then $A(T) = \frac{1}{2}A_0$ and

$$\frac{1}{2} A_0 = A(T) = A_0e^{-0.001002T}$$

$$\frac{1}{2} = e^{-0.001002T} \qquad\qquad \text{cancel } A_0$$

$$\ln \left(\frac{1}{2}\right) = \ln (e^{-0.001002T})$$

$$= -0.001002T \qquad\qquad \ln e^z = z$$

$$T = -\frac{\ln \left(\frac{1}{2}\right)}{0.001002}$$

$$= \frac{\ln 2}{0.001002} \qquad\qquad \ln \left(\frac{1}{2}\right) = \ln 2^{-1} = -\ln 2$$

$$\approx 692 \qquad\qquad \text{from a calculator}$$

Thus the half-life of this substance is about 692 years. See Figure 15.20.  ∎

We can find the general relationship between the half-life and the decay constant. If we let $T$ be the half-life, then

$$\frac{1}{2} A_0 = A(T) = A_0e^{-kT}$$

$$\frac{1}{2} = e^{-kT} \qquad\qquad \text{cancel } A_0$$

$$\ln \left(\frac{1}{2}\right) = \ln (e^{-kT}) = -kT \qquad\qquad \ln e^z = z$$

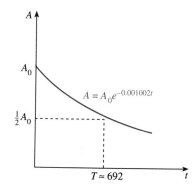

**Figure 15.20**

Since $\ln \left(\frac{1}{2}\right) = -\ln 2$, this yields

$$T = \frac{\ln 2}{k} \quad \text{or} \quad k = \frac{\ln 2}{T},$$

where $\ln 2 \approx 0.69315$.

---

**Connection between Half-life and Decay Constant**

If a quantity satisfies (1) and has a half-life of $T$ and a decay constant $k$, then

$$T = \frac{\ln 2}{k} \quad \text{or} \quad k = \frac{\ln 2}{T}.$$

---

**REMARK.** Notice that the half-life does *not* depend on the initial amount of the substance.

**E X A M P L E   3**   **Safe Time After Nuclear Accident**

Due to an unfortunate nuclear accident, radioactive iodine-131 has been released into the atmosphere and has been absorbed by the grass in some fields, which then has been ingested by cows. The milk from the affected cows has three times the maximum allowable amount of iodine-131. How long until it is safe to consume milk from these animals? The half-life of iodine-131 is eight days.

...unt of iodine-131 in the grass immediately after the accident.
...time $t$ is $P(t) = P_0 e^{-kt}$, where $t$ is given in days. We first
...ve formula gives

$$\quad \text{or} \quad k = \frac{\ln 2}{T} \approx \frac{.69315}{8} \approx .087.$$

...dine-131 returns to a safe level. A safe level in
...rd the initial unsafe amount. Then

$$-0.087T_s$$

ug of cold milk at 40
perature of the milk wih
**limited growth.**
temperature $T_0$ is placed in a
emperature of $C$. Then, according to
of the object is proportional to the difi
g medium and of the object. Thus there exi
that

$$= k[C - T(t)] \tag{2}$$

ire 15.21.  ∎

Radioactive decay of carbon-14 was first used by the Nobel Prize winner Willard Libby around about 1950 to date archeological finds. It is known that the ratio of radioactive carbon-14 to nonradioactive carbon-12 exists in all animals in a fixed constant percent while the animal is alive. When the animal dies, the carbon-14 then decays exponentially, while the amount of nonradioactive carbon-12 remains constant. From this and the known half-life of carbon-14, one can then estimate the date when the animal died.

**E X A M P L E  4**    **Carbon Dating**

An anthropologist discovers a bone fragment and learns from laboratory testing that exactly 20% of carbon-14 is missing. The half-life of carbon-14 is known to be 5570 years. How old is the bone?

**Solution**    We first need to find the decay constant for carbon-14. We use the above formula and obtain

$$k = \frac{0.69315}{T} = \frac{0.69315}{5570} \approx 0.000124.$$

Thus

$$A(t) = A_0 e^{-0.000124t},$$

where $A_0$ is the amount of carbon-14 present in the animal when alive and $t$ is the time since the death of the animal. The laboratory result indicates that $A(t) = .80A_0$. Then

$$.80A_0 = A(t)$$

$$.80 = e^{-0.0}$$

$$\ln .80 = \ln (e$$

$$t = -\frac{\ln}{.00}$$

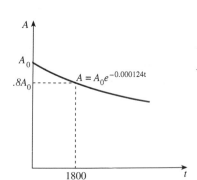

Thus the bone is about 1800 years

**Figure 15.22**

## Limited Growth

**Newton's Law of Cooling.**   If a j
a room kept at 70°F, then the tem
70°F. This is a familiar example o
Suppose an object with initi
medium that is kept at a constant t
law of cooling, the temperature $T($
in the temperature of the surroundi
a constant of proportionality, $k$, so

$$T'(t)$$

It is possible to show that the *only* function $T(t)$ that satisfies equation (2) is given by

$$T(t) = C - (C - T_0)e^{-kt} \tag{3}$$

You will be asked in the exercises to graph this function.

**The Logistic Equation.**   In Example 1 we noted that if the human population of the earth $P(t)$ in billions were growing at an exponential rate, then $P(t) = 5e^{0.016t}$, where $t = 0$ corresponds to 1986. But this implies that as $t \to +\infty$, $P(t)$ will increase without bound. In fact, using this model, the following table gives the population in trillions.

| Date            | 2386 | 2486 | 2586 | 2686 | 2786 |
|-----------------|------|------|------|------|------|
| $t$             | 400  | 500  | 600  | 700  | 800  |
| $P(t)$ in trillions | 3 | 15 | 74 | 366 | 1811 |

At this rate the population in the year 2786 will be 1811 trillion. But the surface of the earth has fewer than 2000 trillion square feet with 80% of this water! Clearly we have come to an absurd conclusion.

Biologists have noted that populations tend to grow exponentially when the population is small, but then level off in time due to the finite amount of available resources. If the population is larger than the available resources, then the population has been observed to decrease to some limiting value. The Dutch mathematical biologist P. Verhulst in 1838 replaced the growth model

$$P'(t) = kP(t) \tag{4}$$

with a more realistic model that incorporates this finite resource factor.

Notice that equation (4) says that the increase in the population will become larger as the population becomes larger. This is reasonable when the population is small, but as the population becomes large enough to be affected by the limited available resources, this increase should slow. Thus the equation should be modified to reflect this fact.

Verhulst suggested subtracting a positive term from the right-hand side of equation (4) that would become larger if the population became larger. The simplest such term is $cP^2$ where $c$ is a positive constant. Doing this gives

$$P'(t) = kP - cP^2. \tag{5}$$

The term $c$ is assumed very small compared to $k$. In fact for the population of the earth in billions $c \approx .003$. Thus if the population is less than a billion, the term $cP^2$ in equation (5) is much smaller than the term $kP$. In this case the right-hand side of equation (5) is approximately $kP$, which just gives equation (4) again. However as the population of $P$ increases the term $cP^2$ becomes larger and slows up the change in population.

It is instructive to factor the right-hand side of equation (5), obtaining

$$P'(t) = k\left(1 - \frac{c}{k}P\right)P.$$

It is furthermore useful to let $L = k/c$ and finally obtain

$$P'(t) = k\left(1 - \frac{P}{L}\right)P. \tag{6}$$

This equation is called the **logistic equation**. From this equation we can see that if $P < L$, then the right-hand side is *positive* and the population increases. Furthermore, as the population increases the term in front of $P$ in equation (6) becomes smaller, that is, the growth rate slows. Thus it is instructive to view the term

$$r = k\left(1 - \frac{P}{L}\right)$$

as a changing growth rate. As $P$ increases toward $L$, this growth rate $r$ slows to zero. Should $P$ ever exceed $L$, then this growth rate becomes *negative* and the population decreases.

Also notice that there are *two* constant functions that satisfy equation (6). The first is just $P = 0$. The second is $P = L$. These constant functions are also referred to as **equilibrium points** or **fixed points**.

It can be shown that the only function that satisfies the logistic equation (6) is

$$P(t) = \frac{LP_0}{P_0 + (L - P_0)e^{-kt}} \tag{7}$$

where $P_0 = P(0)$ is the initial population. Notice that

$$\lim_{t \to +\infty} P(t) = \lim_{t \to +\infty} \frac{LP_0}{P_0 + (L - P_0)e^{-kt}} = L.$$

Thus, no matter what the initial population $P_0$, the population always heads for the limiting value of $L$.

Also notice that

$$P'(t) = \frac{kLP_0(L - P_0)}{(P_0 + (L - P_0)e^{-kt})^2} e^{-kt} \tag{8}$$

and is positive if the initial population $P_0$ is less than the limiting population $L$. Thus, in this case, the population is always growing toward the limiting value $L$.

**E X A M P L E   6**     **Graphing the Logistic Curve When $P_0 < L$**

Graph the function $P(t)$ given in equation (7) when $P_0 < L$.

**Solution**     We already know that $P(t)$ increases steadily with a limiting value of $L$. To determine the concavity, we could calculate $P''(t)$ directly, but this is a bit tedious.

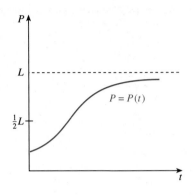

**Figure 15.23**

Since we are only interested in the *sign* of $P''(t)$, we can differentiate the expression in equation (6) and obtain

$$P''(t) = kP' - \frac{2k}{L} PP'$$

$$= kP' \left( 1 - \frac{2}{L} P \right).$$

Recall that when $P_0 < L$, $P' > 0$. Thus the sign of $P''$ is the same as the sign of the factor $\left( 1 - \frac{2}{L} P \right)$. This is positive if $P(t) < \frac{1}{2} L$ and negative if $P(t) > \frac{1}{2} L$. Thus the graph is concave up at first and then turns concave down. See Figure 15.23. ∎

In Figure 15.24, the population of the Northeast of the United States has been plotted. Notice how the curve looks like a logistic curve except perhaps for the period from 1930 to 1950. But consider that there was a major recession in the 1930s, a world war in the first half of the 1940s, and a baby boom after the war!

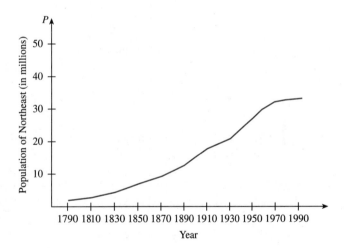

**Figure 15.24**

The logistic curve has been used very extensively in modeling. For example, M. B. Schaefer[1] used the logistic curve to give the total biomass of the Eastern Pacific yellowfin tuna with $L = 1.34 \times 10^8$ kg, and $k = 2.61$. In a different direction P. Pearse[2] gave the net stumpage value of an average tree in a typical

---

[1] "Fishery dynamics and the Present Status of the Yellowfin Tuna Population of the Eastern Pacific," *Bulletin of the Inter-American Tropical Tuna Commission*, vol. 12, no. 3 (1967).

[2] "The Optimal Forest Rotation," *Forest Chronicle*, vol. 43 (1967), 178–195.

stand of British Columbia Douglas fir trees. See Figure 15.25. (The trees have no commercial value until age 30 years.)

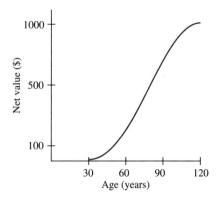

**Figure 15.25**

Because the logistic curve looks like an "S" it is sometimes called the S-curve. Notice that in the early growth stage, the curve is concave up, indicating that $P'$ is increasing. At this stage the population appears to be "exploding." Then the concavity turns negative and the growth slows.

Using historical records to estimate the past human population of the earth, biologists estimate the correct value of $k$ in the logistic equation to be $k = .03$. We are now prepared to do the following example.

**E X A M P L E   7**    **Estimating the Future Population of the World**

We noted above that biologists estimate that the correct value of $k$ in the logistic equation to be $k = 0.03$. When the human population of the earth was 5 billion, we estimated in Example 1 that the population was growing at about 1.6% per year. Estimate the constant $c$ in the logistic equation and obtain an estimate for the limiting value of the human population of the earth.

**Solution**    We rewrite the logistic equation (5) as

$$\frac{1}{P}\frac{dP}{dt} = k - cP.$$

If $\Delta t = 1$ year, then

$$\frac{1}{P}\frac{\Delta P}{\Delta t} = \frac{\Delta P}{P}$$

is the fractional change in the population, which has been given as $1.6/100$. But we can make the estimate

$$\frac{1}{P}\frac{dP}{dt} \approx \frac{1}{P}\frac{\Delta P}{\Delta t} \approx .016.$$

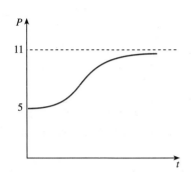

**Figure 15.26**

Putting this information in the above rewritten logistic equation gives

$$0.016 \approx \frac{1}{P}\frac{dP}{dt} = k - cP \approx 0.03 - c \cdot 5.$$

This yields

$$c = \frac{.014}{5} = .0028.$$

The limiting value of the population of the earth is

$$L = \frac{k}{c} = \frac{0.03}{.0028} = 10.7,$$

or about 11 billion. See Figure 15.26. ■

It is interesting to note that the change in concavity occurs at half this value, which is approximately the population at this time. Thus according to this model, the population explosion is over, and the population has just begun to start the leveling off phase in which the finite resource term comes into prominence.

The logistic equation can be used to model the spread of a technological innovation through an industry. Figure 15.27 shows the number of sales of mini-computers in the U.S. This is just one example of how the spread of a new technological innovation tends to explode initially and then level off.

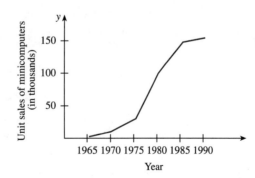

**Figure 15.27**

## SELF-HELP EXERCISE SET 15.4

1. A thousand dollars is deposited into an account growing at 6% per year compounded continuously. How long before the account has value $1500?

2. Plutonium-239 is a product of nuclear reactors with a half-life of about 24,000 years. What percentage of a given sample will remain after 10,000 years?

3. Let $P(t)$ denote the fraction of firms in an industry that are currently using a technological innovation. Assume that $P(t)$ satisfies a logistic curve model and that this innovation will eventually spread throughout the entire industry. Initially 10% of the firms used the innovation, and two years later 20% did. Find the logistic equation that $P(t)$ satisfies.

## EXERCISE SET 15.4

1. How long does it take an account growing at an annual rate of 7% compounded continuously to double?

2. How long does it take an account growing at an annual rate of 8% compounded continuously to double?

3. What annual rate $r$ when compounded continuously is necessary to double an account in 6 years?

4. What annual rate $r$ when compounded continuously is necessary to triple an account in 5 years?

5. Suppose a quantity satisfies $Q(t) = Q_0 e^{kt}$ with $k > 0$. If $T$ is the time it takes for the amount $Q$ to triple in size, show that $\ln 3 = kT$.

### Applications

6. **Population.**   According to the U.S. Bureau of Statistics, Nigeria is growing at a rate that will result in the population doubling in 23 years. Assuming that this population is growing at an exponential rate, what is the growth constant?

7. **Population.**   According to the U.S. Bureau of Statistics, Japan is growing at a rate that will result in the population doubling in 150 years. Assuming that this population is growing at an exponential rate, what is the growth constant?

8. **Population.**   According to the U.S. Bureau of Statistics, Sub-Saharan Africa has a growth constant of about 0.03. At this rate, how long will it take for this population to double?

9. **Population.**   According to the U.S. Bureau of Statistics, the developed world has a growth constant of about 0.005. At this rate, how long will it take for this population to double?

10. **Continuous Compounding.**   If $r$ is the annual rate of interest and $T$ is the time it takes for an account to increase by 50% in size and if the compounding is continuous, show that $\ln 1.5 = rT$.

11. **Population.**   According to the U.S. Bureau of Statistics, the world's population increased from 2 billion to 3 billion from the end of 1925 to the end of 1959. Assuming the population satisfied equation (1) of the text, what was the growth constant for this period?

12. **Population.**   According to the U.S. Bureau of Statistics, the world's population increased from 4 billion to 5 billion from the end of 1974 to the end of 1986.

Assuming the population satisfied equation (1) of the text, what was the growth constant for this period?

13. **Population.**   If the world's population from the beginning of 1960 to the end of 1986 had continued at the same rate as found in Exercise 11 and satisfies equation (1) of the text, what would have been the population in 1986?

14. **Population.**   If the world's population from the beginning of 1987 to the end of 2000 continues at the same rate as found in Exercise 12 and satisfies equation (1) of the text, what will be the population in 2000?

15. **Population.**   If a population grows according to $P(t) = P_0 e^{kt}$ and if the population at time $T$ is $P_1$, then show that

$$T = \frac{1}{k} \ln \frac{P_1}{P_0}$$

16. **Radioactive Decay.**   The decay constant for cobalt-60 is $k = 0.13$ when time is measured in years. Find the half-life of cobalt-60.

17. **Radioactive Decay.**   Radioactive potassium has a half-life of 1.3 billion years. Find its decay constant.

18. **Radioactive Decay.**   The decay constant for cobalt-60 is $k = 0.13$ when time is measured in years. Find the time it takes for this substance to decay to one third of its original amount.

19. **Radioactive Decay.**   Show that the time $T$ that it takes for a radioactive substance to decay to one third of its original amount is given by $T = \frac{1}{k} \ln 3$.

20. **Anthropology.**   An anthropologist discovers a bone fragment and learns from laboratory testing that exactly 10% of carbon-14 is missing. How old is the bone?

21. **Anthropology.**   An anthropologist discovers a bone fragment and learns from laboratory testing that exactly 90% of carbon-14 is missing. How old is the bone?

22. **Medicine.**   Mathium-106 is a radioactive substance with a half-life of only ten days. If an individual receives four times the maximum safe dose of mathium-106, how long must elapse before a safe level is reached?

23. **Public Health.**   In Example 3 of the text, suppose the milk from the cows had twenty times the maximum allowable amounts of iodine-131. How long until it is safe to consume the milk from these animals?

24. **Medicine.**   Radioactive tracers are used for medical diagnosis. For example iodine-131 with a half-life of 8 days is used in the diagnosis of the thyroid gland. Suppose 20 units of iodine-131 are shipped and take two days to arrive. How much of the substance arrives?

25. **Medicine.**   In the previous problem suppose 20 units are needed in two days. How much should be ordered?

26. **Biology.**   A population of mice was noted to quadruple in 3 years. What is the growth constant assuming the population satisfies equation (1) of the text?

27. **Population.**   Verhulst used the logistic model in 1845 to predict the population of the United States. He took $k = .03134$ and $c = (1.5887)10^{-10}$ and $P_0 =$

$(3.9)10^6$, where $t = 0$ corresponds to 1790. Using this model find Verhulst's projected population of the United States in 1890. Compare your answer to the actual population of 62 million.

28. **Population.**   Using the model in the previous example, find Verhulst's projected population in 1930. Compare your answer to the actual population of 124 million.

29. **Population.**   Estimate the limiting population of the United States if $k = .025$ in the logistic equation. The population in 1990 was 250 million and growing at 1% a year.

30. **Fishery.**   Use the logistic curve model of the Eastern Pacific yellowfin tuna given by M. B. Schaefer and the parameters given there to find the biomass at time $t = 10$ if $P_0 = 0.10\ L$.

31. **Population.**   If $P(t)$ is the solution of the logistic equation, show that $\lim\limits_{t \to +\infty} P'(t) = 0$. Draw a graph for the case $P_0 > L$.

32. **Technology.**   Suppose we assume in the discussion on the spread of technological innovation that eventually all the firms will institute the innovation. Suppose $T$ years after the introduction of the innovation it is observed that $P(T) = Q$. Show that

$$k = \frac{1}{T} \ln \frac{Q(1 - P_0)}{P_0(1 - Q)}.$$

33. **Physics.**   A jug of milk at a temperature of 40°F is placed in a room held at 70°F. Find the time it takes for the milk to reach 50°F. Assume $k = 0.40$ with $t$ in hours.

34. **Logistic Equation.**   Show that the function given by equation (7) of the text satisfies equation (6).

35. **Newton's Law of Heating.**   Show that the function given by equation (3) of the text satisfies equation (2). Draw a graph of $T(t)$.

36. **Spread of Rumors.**   In a corporation, rumor spreads of large impending layoffs. The rate of change of the percentage of the work force that has heard the rumor is proportional to the percentage that has not heard the rumor. Initially 0.1% of the workforce has heard the rumor. If 12.5% of the workforce has heard the rumor after the first day, what percentage will have heard the rumor after 6 days from the start of the rumor? *Hint:* If $P(t)$ is the percentage of the work force that has heard the rumor, then

$$\frac{d}{dt} P(t) = k(100 - P(t)),$$

and $P(0) = 0.1$.

37. **Spread of Disease.**   Jim Smith is a member of a small, completely isolated community of 100 residents. Jim leaves and returns with an infectious disease to which everyone in the community is susceptible. One day after returning, Jim gives the disease to one other resident. If the number of persons who get this disease satisfies a logistic equation, find how many of these residents will have contracted the disease 10 days after Jim returns.

38. **Profits.** The profits from the introduction of a new product often explode at first and then level off and thus can sometimes be modeled by a logistic equation. Suppose the annual profits $P(t)$ in millions of dollars $t$ years after the introduction of a certain new product satisfies

$$P'(t) = 1.2P - .05P^2,$$

where $P_0 = 1$. Graph $P(t)$. Find the point where the concavity changes. (This is called the **point of diminishing returns.**)

39. **Spread of News.** A person in a small community of 1000 inhabitants has just heard that he has won a $5 million lotto prize. He immediately tells 4 family members and the news spreads through the community at a rate that is proportional to the product of the number who know the news and the number who do not. In only 2 hours, 20 people have heard the news. If we assume that eventually all the inhabitants will hear the news, how long until half of them hear the news? *Hint:* If $P(t)$ is the number in the community that have heard the news, then $P(t)$ satisfies the logistic equation where $P(0) = 5$ and $P(2) = 20$.

40. **Spread of News.** Repeat the previous problem if instead of assuming that everyone in the community will eventually hear the news, assume that only 900 will eventually hear the news.

41. **Population.** Studies indicate that a small state forest with no deer could hold a maximum of 100 deer. Five deer are transplanted into this forest and one year later the population has increased to ten. Assuming the deer population satisfies a logistic equation, how long will it take for the deer population to reach 75?

42. **Profits.** What is the limiting value of the annual profits in Exercise 38?

## Graphing Calculator Exercise

Answer each of the remaining exercises by graphing two appropriate functions in the same viewing rectangle and use the zoom feature to find the point of intersection. Find the answer to two decimal places.

43. Find the time it takes for an account growing at an annual rate of 9% compounded continuously to double in value.

44. Find the time it takes for an account growing at an annual rate of 12% compounded continuously to increase by 50% in value.

45. Find the half-life of a radioactive substance with decay constant $k = .85$, where time is measured in years.

46. Find the time for a radioactive substance with decay constant $k = .85$ (time is measured in years) to decay to 25% of the original amount.

47. According to the U.S. Bureau of the Census, the population of Nigeria was 113 million in 1990 and growing at about 3% a year compounded continuously while the population of Japan was 123 million in 1990 and growing at about 0.5% a year compounded continuously. Find when the populations of these two countries will be the same, assuming the two growth rates continue at the given values.

**48.** According to the U.S. Bureau of the Census, the population of Indonesia was 181 million in 1990 and growing at about 1.9% a year compounded continuously while the population of the United States was 249 million in 1990 and growing at about 0.9% a year compounded continuously. Find when the populations of these two countries will be the same, assuming the two growth rates continue at the given values.

## Solutions to Self-Help Exercise Set 15.4

**1.** The amount in the account after $t$ years is $A(t) = 1000e^{0.06t}$. Let $t^*$ be the time needed for the amount in the account to become \$1500. Thus $A(t^*) = 1500$. Then

$$1500 = A(t^*) = 1000e^{0.060t^*}$$

$$1.5 = e^{0.06t^*}$$

$$\ln 1.5 = \ln e^{0.06t^*} = 0.06t^*$$

$$t^* = \frac{\ln 1.5}{.06} = \frac{.405465}{.06} \approx 6.76$$

or just short of 7 years.

**2.** We first have

$$k = \frac{\ln 2}{T} = \frac{0.69315}{24,000} \approx .000\ 0289.$$

Thus $Q(t) = Q_0 e^{-0.000\ 0289t}$, and the percentage of the given sample after 10,000 years is

$$100\ \frac{Q(10,000)}{Q_0} = 100e^{-0.000\ 0289(10,000)} \approx 75.$$

That is, 75% of this material will be left after 10,000 years.

**3.** Since eventually the entire industry uses the innovation, we must have $L = 1$. Also we have $P_0 = 0.10$. Thus equation (7) becomes

$$P(t) = \frac{.10}{.10 + .90e^{-kt}} = \frac{1}{1 + 9e^{-kt}}.$$

Also $P(2) = 0.20$, and this gives

$$0.20 = P(2) = \frac{1}{1 + 9e^{-2k}}$$

$$1 + 9e^{-2k} = 5$$

$$\frac{4}{9} = e^{-2k}$$

$$\ln \frac{4}{9} = \ln e^{-2k} = -2k$$

$$k = \frac{1}{2} \ln \frac{9}{4} \approx 0.4055$$

Thus

$$P(t) = \frac{1}{1 + 9e^{-0.4055t}}.$$

## SUMMARY OUTLINE OF CHAPTER 15

- An **exponential** function is a function of the form $y = a^x$.
- **Properties of $a^x$.**

| $a > 1$ | $0 < a < 1$ |
|---|---|
| domain $= (-\infty, +\infty)$ | domain $= (-\infty, +\infty)$ |
| range $= (0, +\infty)$ | range $= (0, +\infty)$ |
| increasing on $(-\infty, +\infty)$ | decreasing on $(-\infty, +\infty)$ |
| $a^x = a^y$ if and only if $x = y$ | $a^x = a^y$ if and only if $x = y$ |
| $\lim\limits_{x \to -\infty} a^x = 0, \ \lim\limits_{x \to +\infty} a^x = +\infty$ | $\lim\limits_{x \to -\infty} a^x = +\infty, \ \lim\limits_{x \to +\infty} a^x = 0$ |

- The number $e$ to five decimals is 2.71828.
- An account with an initial amount of $P$ that is growing at an annual rate of $r$ (expressed as a decimal) compounded continuously becomes $Pe^{rt}$ after $t$ years.
- If a sum of money is invested at an annual rate of $r$ expressed as a decimal and is compounded continuously, the **effective yield** is $r_{\text{eff}} = e^r - 1$.
- If an account earns an annual rate of $r$ expressed as a decimal and is compounded continuously, then the amount $P$, called the **present value**, needed in this account now so that a future balance of $A$ will be attained in $t$ years is $P = Ae^{-rt}$.
- **Natural Logarithm.**   If $x > 0$, the natural logarithm of $x$, denoted by $\ln x$, is defined as $y = \ln x$ if, and only if, $x = e^y$.
- **Properties of Natural Logarithms**

$$e^{\ln x} = x \text{ if } x > 0 \qquad \ln e^x = x \text{ for all } x$$

$$\ln 1 = 0 \qquad \ln xy = \ln x + \ln y$$

$$\ln \frac{x}{y} = \ln x - \ln y \qquad \ln x^a = a \ln x$$

- For any $a > 0$, $a^x = e^{kx}$ if $k = \ln a$
- **Derivatives of Exponential and Logarithmic Functions**

$$\frac{d}{dx}(e^x) = e^x \qquad \frac{d}{dx}(\ln x) = \frac{1}{x}$$

$$\frac{d}{dx}(e^{f(x)}) = e^{f(x)}f'(x) \qquad \frac{d}{dx}(a^{f(x)}) = a^{f(x)}f'(x) \ln a$$

$$\frac{d}{dx}(\ln f(x)) = \frac{f'(x)}{f(x)}$$

- If a quantity grows according to $Q(t) = Q_0 e^{kt}$ where $k > 0$, $k$ is called the **growth constant**.

- **Uniqueness Theorem.**   If $Q'(t) = kQ(t)$ and $Q(0) = Q_0$, then $Q(t) = Q_0 e^{kt}$.

- If a quantity decays according to $Q(t) = Q_0 e^{-kt}$ where $k > 0$, $k$ is called the **decay constant**.

- The **half-life** of a radioactive substance is the time that must elapse for half of the material to decay.

- **Connection between half-life and decay constant.**   If $k$ is the decay constant of a radioactive material and $T$ is the half-life, then $kT = \ln 2$.

- A **logistic equation** is any equation of the form $P'(t) = kP - cP^2$.

## Chapter 15 Review Exercises

In Exercise 1 through 6 graph.

**1.**  $y = 3^x$   **2.**  $y = 3^{-x}$   **3.**  $y = e^{2x}$

**4.**  $y = e^{-2x}$   **5.**  $y = e^{|x|}$   **6.**  $\ln |x|$

In Exercises 7 through 12 solve for $x$.

**7.**  $9^x = 27$   **8.**  $\left(\dfrac{1}{3}\right)^x = \dfrac{1}{27}$   **9.**  $\ln x = 1$

**10.**  $e^{4x} = e^{8x-2}$   **11.**  $\ln e^x = 4$   **12.**  $e^{\ln x} = 5$

In Exercises 13 through 16 find the limits.

**13.**  $\displaystyle \lim_{x \to +\infty} \frac{1}{1 + e^x}$   **14.**  $\displaystyle \lim_{x \to +\infty} (1 + e^{-2x})$

**15.**  $\displaystyle \lim_{x \to +\infty} \frac{1}{1 + \ln x}$   **16.**  $\displaystyle \lim_{x \to 0+} \frac{1}{1 + \ln x}$

In Exercises 17 through 22 differentiate.

**17.**  $e^{-7x}$   **18.**  $e^{2x^3+1}$   **19.**  $(x^2 + x + 2)e^x$

**20.**  $\ln(x^2 + 1)$   **21.**  $e^{x^2} \ln x$   **22.**  $\ln(x^2 + e^x)$

In Exercises 23 through 26 find where the function is increasing and where it is decreasing. Find where it is concave up and where it is concave down.

**23.**  $y = e^{-x^2-1}$   **24.**  $y = 1 - e^{-x^2}$

**25.**  $\ln(x^2 + 1)$   **26.**  $\ln(1 + e^x)$

**27.**  **Population.**   According to the U.S. Bureau of Statistics, the developing world is growing at a rate of 2% a year. If we assume exponential growth $P_0 e^{0.2t}$, how long until this population increases by 30 percent?

**28.**  **Population.**   According to the U.S. Bureau of Statistics, the population of India will double in 35 years. If we assume exponential growth $P_0 e^{kt}$, what must be the growth constant $k$?

29. **Radioactive Decay.**   A radioactive substance was noted to have decayed by 20% in 10 years. What is its half-life?

30. **Anthropology.**   An anthropologist discovers a bone fragment and is told from laboratory testing that 15% of carbon-14 is missing. How old is the bone?

31. **Supply and Demand.**   Suppose the rate of change of the supply of a certain commodity is proportional to the difference between the demand and supply so that

$$\frac{dS}{dt} = k(D - S) \qquad S(0) = S_0.$$

If the demand $D$ is a constant, find $S(t)$.

32. **Population.**   An isolated lake was stocked with 50 lake trout. Two years later it was estimated that there were 100 of these trout in this lake. It is estimated that this lake can support 500 of these trout. Assuming that the trout population satisfies a logistic equation, find how many lake trout will be in this lake in 4 years.

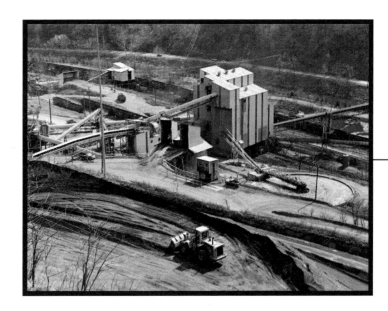

Integrals will be used to determine the life of the resource. (© Tcd Clutter/Photo Researchers, Inc.)

# Integration

16.1   Antiderivatives

16.2   Substitution

16.3   The Definite Integral

16.4   The Fundamental Theorem of Calculus

16.5   Area Between Two Curves

We now begin to study the second part of calculus. We will introduce the integral and see in what sense the integral is the opposite of the derivative. The integral will enable us to the find the areas between curves, which we will see, has many concrete applications. We will see how to recover the cost, revenue, and profit functions when given the rates of change of these quantities, how to find the population when given a measurement of the density, how to find important economic quantities such as consumers' and producers' surplus, and even how to measure to what extent an economic system is distributing income equitably.

## 16.1   ANTIDERIVATIVES

▶ *Antiderivatives*

▶ *Rules of Integration*

▶ *Applications*

What is the revenue function $R(x)$ if the marginal revenue is $2x - 10$? See Example 7 for the answer.

What is the cost function $C(x)$ if the marginal cost is $e^{0.01x} + .04x - 100$ and fixed costs are 1000? See Example 8 for the answer.

### Antiderivatives

Given a function $f(x) = x^2$, can we find some function $F(x)$ whose derivative is $x^2$? We know that in differentiating $x$ to a power, the power is reduced by one. Thus we might tentatively try the function $x^3$. But $\dfrac{d}{dx}(x^3) = 3x^2$. Thus we have *three* times our function $x^2$. Dividing by this number we obtain $\dfrac{d}{dx}\left(\dfrac{1}{3}x^3\right) = x^3$. This is one answer. But are there others? Recalling that the derivative of a constant is zero, we readily see that

$$\frac{d}{dx}\left(\frac{1}{3}x^3 + C\right) = x^2,$$

for any constant $C$. This gives us the family of answers

$$F(x) = \frac{1}{3}x^3 + C.$$

See Figure 16.1.

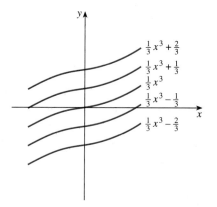

**Figure 16.1**

We must now address the following general problem. Given a function $f(x)$ we would like to find a function $F(x)$ such that $F'(x) = f(x)$. We call $F(x)$ an **antiderivative** of $f(x)$. We use the term *anti*derivative to refer to the fact that finding $F(x)$ is the *opposite* of differentiating.

---

**Definition of Antiderivative**

If $F'(x) = f(x)$, then $F(x)$ is called an antiderivative of $f(x)$.

---

We saw that $\frac{1}{3}x^3 + C$, where $C$ is any constant, are all antiderivatives of $x^2$. Are there still others? The following theorem indicates that the answer is no.

---

**Theorem 1**

If $F'(x) = G'(x)$ on $(a,b)$, then on $(a,b)$

$$F(x) = G(x) + C,$$

for some constant $C$.

---

That is, if two functions have the same derivative on $(a,b)$, then they differ by a constant on $(a,b)$.

**E X A M P L E   1**   **Finding all the Antiderivatives**

Find all antiderivatives of $x^2$.

**Solution**   From the above theorem and work we know that all antiderivatives of $x^2$ must be of the form

$$\frac{1}{3}x^3 + C,$$

where $C$ is any constant.   ■

When we include the arbitrary constant $C$, we refer to the function $\frac{1}{3}x^3 + C$ as the *general form* of the antiderivative. By this we mean that the collection of all antiderivatives of $x^2$ can be obtained by adding all possible constants to $\frac{1}{3}x^3$.

The collection of all antiderivatives of a function $f(x)$ is called the **indefinite integral** and is denoted by $\int f(x)\,dx$. This is read as *the integral of $f(x)$* or more carefully as *the integral of $f(x)$ with respect to $x$*. We refer to the symbol $\int$ as the **integral sign** and $f(x)$ as the **integrand**. The symbol $\int$ looks like an $S$, and in fact stands for *sum*. We shall later see the important connection between the integral and a certain sum.

If we know of one function $F(x)$ for which $F'(x) = f(x)$, then Theorem 1 indicates that the collection of all antiderivatives of $f(x)$ must be of the form $F(x) + C$, where $C$ is a constant. Thus we have the following.

---

**The Indefinite Integral**

The collection of all antiderivatives of a function $f(x)$ is called the **indefinite integral** and is denoted by $\int f(x)\, dx$.

    If we know one function $F(x)$ for which $F'(x) = f(x)$, then

$$\int f(x)\, dx = F(x) + C,$$

where $C$ is an arbitrary constant and called the **constant of integration**.

---

From Example 1 above we see that

$$\int x^2\, dx = \frac{1}{3} x^3 + C.$$

The following is also true

$$\int x^3\, dx = \frac{1}{4} x^4 + C.$$

This can be readily verified by differentiating the right-hand side to obtain the integrand. Thus

$$\frac{d}{dx}\left(\frac{1}{4} x^4 + C\right) = x^3.$$

## Rules of Integration

It would be convenient if we had some rules to find indefinite integrals. The following is the first such rule.

---

**Power Rule**

If $n \neq -1$, then

$$\int x^n\, dx = \frac{1}{n + 1} x^{n+1} + C.$$

---

In words, increase the power by one and divide by the resulting number, then add an arbitrary constant.

    To prove the power rule formula we need only differentiate the right-hand side and note that we obtain the integrand. Thus

$$\frac{d}{dx}\left(\frac{1}{n + 1} x^{n+1} + C\right) = x^n.$$

E X A M P L E  2     **Using the Power Rule**

Find (a) $\int x^5 \, dx$ and (b) $\int \dfrac{1}{\sqrt{x}} \, dx.$

**Solutions**

(a)   Here we set $n = 5$ in the power rule formula and obtain

$$\int x^5 \, dx = \frac{1}{5+1} x^{5+1} + C = \frac{1}{6} x^6 + C.$$

(b)   Here we first note that $\dfrac{1}{\sqrt{x}} = x^{-1/2}$ and set $n = -\frac{1}{2}$ in the power rule formula and obtain

$$\int \frac{1}{\sqrt{x}} \, dx = \int x^{-1/2} \, dx = \frac{1}{-\frac{1}{2}+1} x^{-1/2+1} + C = 2x^{1/2} + C \quad \blacksquare$$

R E M A R K .     As in the previous example, *always remove radical signs before integrating.*

Since $\dfrac{d}{dx}(kF(x)) = k \dfrac{d}{dx} F(x)$, we have the following rule.

---

**Constant Times Function Rule**

For any constant $k$

$$\int kf(x) \, dx = k \int f(x) \, dx.$$

---

In words, the integral of a constant times a function is the constant times the integral of the function.

E X A M P L E  3     **Using the Constant Times Function Rule**

Find (a) $\int 3t^9 \, dt$ and $\int \dfrac{\sqrt{x}}{\pi} \, dx.$

**Solutions**

(a)   Since 3 is a constant

$$\int 3t^9 \, dt = 3 \int t^9 \, dt = \frac{3}{10} t^{10} + C.$$

(b)   Since $\dfrac{1}{\pi}$ is a constant

$$\int \frac{\sqrt{x}}{\pi} \, dx = \int \frac{1}{\pi} \sqrt{x} \, dx = \frac{1}{\pi} \int \sqrt{x} \, dx = \frac{1}{\pi} \int x^{1/2} \, dx$$

$$= \frac{1}{\pi} \frac{1}{\frac{1}{2}+1} x^{1/2+1} + C$$

$$= \frac{2}{3\pi} x^{3/2} + C \quad \blacksquare$$

Since $\dfrac{d}{dx}(F(x) \pm G(x)) = \dfrac{d}{dx}F(x) \pm \dfrac{d}{dx}G(x)$, we have the following rule.

---

**Sum or Difference Rule**

$$\int [f(x) \pm g(x)] \, dx = \int f(x) \, dx \pm \int g(x) \, dx.$$

---

In words the integral of the sum (or difference) is the sum (or difference) of the integrals.

**E X A M P L E  4**     **Using the Sum and Difference Rules**

Find (a) $\displaystyle\int (u^2 + u^3) \, du$ and (b) $\displaystyle\int \left(2x^4 - \frac{1}{2}x^2\right) dx.$

**Solutions**     (a) $\displaystyle\int (u^2 + u^3) \, du$    $= \displaystyle\int u^2 \, du + \int u^3 \, du$

$$= \frac{1}{3}u^3 + \frac{1}{4}u^4 + C$$

(b) $\displaystyle\int \left(2x^4 - \frac{1}{2}x^2\right) dx = \int 2x^4 \, dx - \int \frac{1}{2}x^2 \, dx$

$$= 2\int x^4 \, dx - \frac{1}{2}\int x^2 \, dx$$

$$= \frac{2}{5}x^5 - \frac{1}{6}x^3 + C \quad \blacksquare$$

As we are seeing, for every rule on derivatives there is a corresponding rule on integrals. For example

$$\frac{d}{dx}(e^x) = e^x \qquad \text{implies} \qquad \int e^x \, dx = e^x + C.$$

Also for any constant $k \neq 0$

$$\frac{d}{dx}\left(\frac{1}{k}e^{kx}\right) = e^{kx} \qquad \text{implies} \qquad \int e^{kx} \, dx = \frac{1}{k}e^{kx} + C.$$

We then have the following rules for integration of the exponential function.

---

**Indefinite Integral of Exponential Functions**

$$\int e^x \, dx = e^x + C,$$

and for any constant $k \neq 0$,

$$\int e^{kx} \, dx = \frac{1}{k} e^{kx} + C.$$

---

**E X A M P L E   5**   **Integrating Exponential Functions**

Find $\int (e^{2x} - e^{-2x}) \, dx$.

**Solution**
$$\int (e^{2x} - e^{-2x}) \, dx = \int e^{2x} \, dx - \int e^{-2x} \, dx$$
$$= \frac{1}{2} e^{2x} + \frac{1}{2} e^{-2x} + C \quad \blacksquare$$

Since

$$\frac{d}{dx} (\ln |x|) = \frac{1}{x} \quad \text{implies} \quad \int \frac{1}{x} \, dx = \ln |x| + C,$$

we have the following rule.

---

**Indefinite Integral of $x^{-1} = \dfrac{1}{x}$**

$$\int \frac{1}{x} \, dx = \ln |x| + C.$$

---

Recall that in the power rule for integrating $x^n$ we had to restrict $n \neq -1$. (If we try to use this formula with $n = -1$ we obtain a zero in the denominator.) With this last rule we can now find the indefinite integral of $x^n$ for *any* number $n$.

**E X A M P L E   6**   **Integrating $x^{-1}$**

Find $\int \dfrac{2}{x} \, dx$.

**Solution**
$$\int \frac{2}{x} \, dx = 2 \int \frac{1}{x} \, dx$$
$$= 2 \ln |x| + C \quad \blacksquare$$

## Applications

E X A M P L E  7    **Finding Revenue Given Marginal Revenue**

Find the revenue function $R(x)$ if the marginal revenue is $2x - 10$.

**Solution**    We are given that $R'(x) = 2x - 10$. Thus

$$R(x) = \int (2x - 10)\, dx$$
$$= x^2 - 10x + C,$$

where $C$ is some constant not yet determined.

But we know that no sales result in no revenue, that is, $R(0) = 0$. Using this gives

$$0 = R(0) = (0)^2 - 10(0) + C = C.$$

Thus

$$R(x) = x^2 - 10x. \quad \blacksquare$$

E X A M P L E  8    **Finding Cost Given Marginal Cost**

Find the cost function $C(x)$ if the marginal cost is $e^{0.01x} + .04x - 100$ and fixed costs are 1000.

**Solution**    We are given that $C'(x) = e^{0.01x} + .04x - 100$. Thus

$$C(x) = \int (e^{0.01x} + .04x - 100)\, dx$$
$$= 100e^{0.01x} + 0.02x^2 - 100x + C^*,$$

where $C^*$ is some constant not yet determined.

If fixed costs are 1000, this translates into $C(0) = 1000$. Then

$$1000 = C(0) = 100e^0 + .02(0) - 100(0) + C^* = 100 + C^*.$$

Thus $C^* = 900$ and

$$C(x) = 100e^{0.01x} + 0.02x^2 - 100x + 900. \quad \blacksquare$$

## SELF-HELP EXERCISE SET 16.1

1.  Find $\int \left(1 + \sqrt[3]{x} + \dfrac{3}{x} - e^{2x}\right) dx$.

2.  If a ball is thrown upward with an initial velocity of 20 feet per second, then from physics it can be shown that the velocity (neglecting air resistance) is given by $v(t) = -32t + 20$. Find $s(t)$ if the ball is thrown from 6 feet above ground level and $s(t)$ measures the height of the ball in feet.

## EXERCISE SET 16.1

In Exercises 1 through 38 find the antiderivatives.

1. $\int x^{99}\, dx$

2. $\int x^{0.01}\, dx$

3. $\int x^{-99}\, dx$

4. $\int x^{-0.01}\, dx$

5. $\int 5\, dx$

6. $\int dx$

7. $\int \dfrac{5}{y^3}\, dy$

8. $\int \dfrac{1}{2y^5}\, dy$

9. $\int \dfrac{y^{3/2}}{\sqrt{2}}\, dy$

10. $\int \pi y^{2/3}\, dy$

11. $\int \sqrt[3]{u^2}\, du$

12. $\int \sqrt[5]{u^3}\, du$

13. $\int (6x^2 + 4x)\, dx$

14. $\int (8x^3 + 1)\, dx$

15. $\int (x^2 + x + 1)\, dx$

16. $\int (3u^2 + 4u^3)\, du$

17. $\int (\sqrt{2}u^{0.1} - 0.1u^{1.1})\, du$

18. $\int \left(\dfrac{3}{t^2} - 6t^2\right) dt$

19. $\int (t^2 - 1)\, dt$

20. $\int (t^{-2} + 3)\, dt$

21. $\int (\sqrt{t} - \sqrt[3]{t^5})\, dt$

22. $\int (t^3 + 4t^2 + 5)\, dt$

23. $\int (6t^5 - 4t^3 + 1)\, dt$

24. $\int \left(x + \dfrac{1}{x}\right) dx$

25. $\int \left(1 + \dfrac{3}{x}\right) dx$

26. $\int \dfrac{2}{\pi x}\, dx$

27. $\int \left(\pi + \dfrac{1}{x}\right) dx$

28. $\int \dfrac{t^4 + 3}{t^2}\, dt$

29. $\int \dfrac{t + 1}{\sqrt{t}}\, dt$

30. $\int \sqrt{t}(t + 1)\, dt$

31. $\int (e^x + e^{-x})\, dx$

32. $\int \left(e^x - \dfrac{1}{x}\right) dx$

33. $\int e^{5x}\, dx$

34. $\int e^{-6x}\, dx$

35. $\int (e^{5x} - e^{-5x})\, dx$

36. $\int (e^{5x} + e^{-5x})\, dx$

37. $\int \dfrac{x + 1}{x}\, dx$

38. $\int \dfrac{e^x + e^{-x}}{e^{2x}}\, dx$

## Applications

39. **Revenue.** Find the revenue function for a knife manufacturer if the marginal revenue, in dollars, is given by $30 - .5x$, where $x$ is the number of knives sold.

40. **Revenue.** Find the revenue function for a paper plate manufacturer if the marginal revenue, in dollars, is given by $200 - .26x$, where $x$ is the number of plates sold.

41. **Demand.** Find the demand function for a salt shaker manufacturer if marginal demand, in dollars, is given by $p'(x) = -x^{-3/2}$, where $x$ is the number of salt shakers sold and $p(1) = 3$.

42. **Demand.** Find the demand function for a table cloth manufacturer if marginal demand, in dollars, is given by $p'(x) = -x^{-3/2}$, where $x$ is the number of thousands of table cloths sold. Assume $p(1) = 100$.

**43.** **Cost.**   Find the cost function for an envelope manufacturer if the marginal cost, in dollars, is given by $100 - e^{.1x}$, where $x$ is the number of thousands of envelopes produced and fixed costs are $1000.

**44.** **Cost.**   Find the cost function for an adhesive tape manufacturer if the marginal cost, in dollars, is given by $150 - e^{.01x}$, where $x$ is the number of cases of tape produced. Assume $C(0) = 100$.

**45.** **Cost.**   Find the cost function for a computer disk manufacturer if the marginal cost, in hundreds of dollars, is given by $30\sqrt{x} - 6x^2$, where $x$ is the number of thousands of disks sold. Assume no fixed costs.

**46.** **Cost.**   Find the cost function for a spark plug manufacturer if the marginal cost, in dollars, is given by $30x - 4e^{2x}$, where $x$ is the number of thousands of plugs sold. Assume no fixed costs.

**47.** **Velocity.**   If a ball is thrown upward with an initial velocity of 30 ft per second, then from physics it can be shown that the velocity (neglecting air resistance) is given by $v(t) = -32t + 30$. Find $s(t)$ if the ball is thrown from 15 ft above ground level and $s(t)$ measures the height of the ball in feet.

**48.** **Acceleration.**   If a ball is thrown upward, then from physics it can be shown that the acceleration is given by $a(t) = -32$ ft per second. If a ball is thrown upward with an initial velocity of 10 ft per second, find $v(t)$.

**49.** **Velocity.**   If in the previous problem, the ball is at an initial height of 6 feet, find $s(t)$, the height of the ball after $t$ seconds.

**50.** **Depreciation.**   The value of a piece of machinery purchased for $20,000 is decreasing at the rate given by

$$V'(t) = -1000e^{-0.1t},$$

where $t$ is measured in years from purchase. How much will the machine be worth in 4 years?

**51.** **Radioactive Decay.**   A certain radioactive substance initially weighing 100 grams decays at a rate given by

$$A'(t) = -20e^{-0.5t},$$

where $t$ is measured in years. How much of the substance is there after 10 years?

**52.** **Population.**   The population of a certain city is increasing at the rate given by

$$P'(t) = 2000e^{0.04t},$$

where $t$ is time in years measured from the beginning of 1994 when the population was 200,000. Find the population at the end of 1996.

**53.** **Appreciation.**   A painting by one of the masters is purchased by a museum for $1,000,000 and increases in value at a rate given by

$$V'(t) = 10,000e^{0.1t},$$

where $t$ is measured in years from the time of purchase. What will the painting be worth in 10 years?

54. **Marginal Propensity to Consume.** Let $C(x)$ represent national consumption in trillions of dollars where $x$ is disposable national income in trillions of dollars. Suppose that the marginal propensity to consume, $C'(x)$, is given by

$$C'(x) = 1 + \frac{2}{\sqrt{x}},$$

and that consumption is \$0.1 trillion when disposable income is zero. Find the national consumption function.

55. **Marginal Propensity to Save.** Let $S$ represent national savings in trillions of dollars. We can assume that disposable national income equals national consumption plus national savings, that is, $x = C(x) + S(x)$. Thus $C'(x) = 1 - S'(x)$, where $S'(x)$ is the marginal propensity to save. If the marginal propensity to save is given by $0.5x$ and consumption is \$0.2 trillion dollars when disposable income is zero, find the consumption function.

56. **Sum-of-Years-Digit Depreciation.** The sum-of-years-digit depreciation method depreciates a piece of equipment more in the early years of its life and less in the later years. In this method the value decreases at a rate given by

$$\frac{d}{dt} V(t) = -k(T - t),$$

where $T$ is the useful life in years of the equipment, $t$ is the time in years since purchase, and $k$ is a positive constant. (Notice that as $t$ gets closer to $T$, $V'(t)$ becomes smaller and thus the value decreases less.) If a piece of machinery is purchased for $I$ dollars and has no salvage value, show that

$$k = \frac{2I}{T^2},$$

and write an expression for $V(t)$ in terms of $t$, $I$, and $T$.

## Solutions to Self-Help Exercise Set 16.1

1. First remove the radical and obtain

$$\int \left(1 + x^{1/3} + \frac{3}{x} - e^{2x}\right) dx = \int 1 \, dx + \int x^{1/3} \, dx + 3\int \frac{1}{x} dx - \int e^{2x} \, dx$$

$$= x + \frac{1}{\frac{4}{3}} x^{4/3} + 3 \ln |x| - \frac{1}{2} e^{2x} + C$$

$$= x + \frac{3}{4} x^{4/3} + 3 \ln |x| - \frac{1}{2} e^{2x} + C$$

2. Since $\dfrac{ds}{dt}(t) = v(t)$ and $s(0) = 6$, we have

$$s(t) = \int (-32t + 20) \, dt = -32\int t \, dt + 20 \int 1 \, dt = -16t^2 + 20t + C.$$

Then since

$$6 = s(0) = -16(0)^2 + 20(0) + C,$$

$C = 6$ and

$$s(t) = -16t^2 + 20t + 6.$$

## 16.2   SUBSTITUTION

▶ *The Method of Substitution*

▶ *Applications*

APPLICATION
**Production of Oil**

The rate at which a natural resource is extracted tends to increase initially and then fall off later after the easily accessible material has been extracted. Suppose the rate in hundreds of thousands of barrels per year at which oil is being extracted from a field during the early and intermediate stages of the life of the field is given by

$$P'(t) = \frac{t}{(t^2 + 1)^2} + 1,$$

where $t$ is in years. Find the amount of oil extracted during the first 7 years. See Example 7 for the answer.

### The Method of Substitution

The methods in the previous section are inadequate to find all the indefinite integrals that we will need. In this section we will develop the method of *substitution*.

The method of substitution is based on the chain rule. Let us consider the simplest case. Suppose $n$ is any real number with $n \neq -1$ and

$$F(x) = \frac{1}{n + 1} [f(x)]^{n+1}.$$

Then using the chain rule we know that $F'(x) = [f(x)]^n f'(x)$. Thus $F(x)$ is an antiderivative of $[f(x)]^n f'(x)$. Then

$$\int [f(x)]^n f'(x)\, dx = F(x) + C = \frac{1}{n + 1} [f(x)]^{n+1} + C.$$

This gives

$$\boxed{\int [f(x)]^n f'(x)\, dx = \frac{1}{n + 1} [f(x)]^{n+1} + C,\ n \neq -1.}$$

This is referred to as the **general power rule**.

We note that the above expression is the correct answer since differentiating the right-hand side gives the integrand on the left-hand side.

The way this works in practice is as follows. We let $u = f(x)$ and recall from the section on differentials that $du = f'(x) \, dx$. We then formally *substitute* all of this into the integral, obtaining

$$\int [f(x)]^n f'(x) \, dx = \int u^n \, du = \frac{1}{n+1} u^{n+1} + C$$

Now we recall that $u = f(x)$, and the last expression becomes

$$\frac{1}{n+1} [f(x)]^{n+1} + C,$$

which is the correct answer. Our only justification at this time of this formal procedure is that it always gives the correct answer. (In the next section we shall obtain insight into the meaning of the "*dx*" term in the indefinite integral that will then indicate why this formal procedure works.)

**E X A M P L E   1**   **Using the Method of Substitution**

Find $\int 2x(x^2 + 1)^{99} \, dx$.

**Solution**   We could actually solve this problem using the techniques of the last section by expanding the term $(x^2 + 1)^{99}$ by the binomial theorem and then multiply the result by $2x$. We would have an expression with 100 terms to integrate.

Instead we try the method of substitution. One should clearly understand that this method is in practice a trial-and-error method.

Suppose we try the substitution $u = x^2 + 1$. Then $du = 2x \, dx$. Substitute all of this into the integral and obtain

$$\int 2x(x^2 + 1)^{99} \, dx = \int (x^2 + 1)^{99} 2x \, dx$$

$$= \int u^{99} \, du$$

$$= \frac{1}{100} u^{100} + C$$

$$= \frac{1}{100} (x^2 + 1)^{100} + C \quad \blacksquare$$

How can we be certain we have the correct answer? Simply differentiate the alleged answer to see if we obtain the integrand. Since

$$\frac{d}{dx} \left( \frac{1}{100} (x^2 + 1)^{100} + C \right) = \frac{1}{100} (100)(x^2 + 1)^{99}(2x) + 0$$

$$= 2x(x^2 + 1)^{99},$$

our answer is correct.

How can one possibly know what to substitute? The above indicates that if you encounter an integrand that is of the form of a function raised to a power, then set $u$ equal to the function as a first try. It just might work! (Fortunately the above integrand has the factor $2x$ in it, otherwise we would be stuck and would have to resort to some other as yet unlearned technique.)

We now consider the case $n = 1$ that was avoided above. For the second substitution formula, we let $F(x) = \ln |f(x)|$. Then $F'(x) = \dfrac{f'(x)}{f(x)}$, and $F(x)$ is an antiderivative of $f'(x)/f(x)$. Thus

$$\int \frac{f'(x)}{f(x)}\, dx = F(x) + C = \ln |f(x)| + C.$$

This gives

$$\boxed{\int \frac{f'(x)}{f(x)}\, dx = \ln |f(x)| + C.}$$

In practice, if we let $u = f(x)$, then $du = f'(x)\, dx$. Formally substituting this into the integral yields

$$\int \frac{f'(x)}{f(x)}\, dx = \int \frac{1}{f(x)} f'(x)\, dx$$

$$= \int \frac{1}{u}\, du$$

$$= \ln |u| + C$$

$$= \ln |f(x)| + C,$$

which is the correct answer.

**E X A M P L E   2**    **Using the Method of Substitution**

Find $\displaystyle \int \frac{x + 1}{x^2 + 2x}\, dx.$

**Solution**    We let $u = x^2 + 2x$, then $du = (2x + 2)\, dx = 2\,(x + 1)\, dx$. Now formally substitute all of this into the integral and obtain

$$\int \frac{x + 1}{x^2 + 2x}\, dx = \int \frac{1}{x^2 + 2x}\,(x + 1)\, dx$$

$$= \int \frac{1}{u}\frac{1}{2}\, du \qquad\qquad (x + 1)\, dx = \frac{1}{2}\, du$$

$$= \frac{1}{2} \int \frac{du}{u}$$

$$= \frac{1}{2} \ln |u| + C$$

$$= \frac{1}{2} \ln |x^2 + 2x| + C \quad \blacksquare$$

As the last case of the use of substitution, let now $F(x) = e^{f(x)}$. Then $F'(x) = e^{f(x)}f'(x)$, and so $F(x)$ is an antiderivative of $e^{f(x)}f'(x)$. Thus

$$\int e^{f(x)} f'(x)\, dx = F(x) + C = e^{f(x)} + C.$$

This gives

$$\boxed{\int e^{f(x)}f'(x)\, dx = e^{f(x)} + C.}$$

This is referred to as the **general exponential rule**.

We know this must be the correct answer since differentiating the answer on the right-hand side gives the integrand on the left-hand side.

In practice we let $u = f(x)$ and then $du = f'(x)\, dx$. Formally substituting all of this into the integral gives

$$\int e^{f(x)}f'(x)\, dx = \int e^u\, du$$
$$= e^u + C$$
$$= e^{f(x)} + C,$$

which is the correct answer.

There are really only two cases in the method of substitution that we will consider. In the first case, one has $[f(x)]^n$ in the integrand. Then set $u = f(x)$ and $du = f'(x)\, dx$ and substitute. In the second case, one has $e^{f(x)}$ in the integrand. Then set $u = f(x)$ and $du = f'(x)\, dx$ and substitute.

---

**Procedure for Using the Method of Substitution**

1. Remove radicals and move any functions in the denominator to the numerator, eliminating any fractions (other than constants).
2. (a) If $[f(x)]^n$ is in the integrand, then set $u = f(x)$ and $du = f'(x)\, dx$ and substitute.
   (b) If $e^{f(x)}$ is in the integrand, then set $u = f(x)$ and $du = f'(x)\, dx$ and substitute.
3. With the above substitutions, eliminate $x$ and $dx$ from the integral so that only $u$ and $du$ appear.
4. Integrate with respect to $u$. If you cannot, then the method fails.
5. Now replace $u$ with $f(x)$.

---

**E X A M P L E   3**   **Using the General Power Rule**

Find $\int 3x^2\sqrt{x^3 + 3}\, dx$.

**Solution**   Following the first step of the above procedure, rewrite the integral as

$$\int 3x^2(x^3 + 3)^{1/2}\, dx.$$

We notice a function, $(x^3 + 3)$ raised to the power $\frac{1}{2}$. Thus we try $u = x^3 + 3$ and hope for the best. With $u$ defined in this way, we then have $du = 3x^2 \, dx$. Substitute all of this into the integral and obtain

$$\int 3x^2(x^3 + 3)^{1/2} \, dx = \int (x^3 + 3)^{1/2} 3x^2 \, dx$$

$$= \int u^{1/2} \, du$$

$$= \frac{2}{3} u^{3/2} + C$$

$$= \frac{2}{3} (x^3 + 3)^{3/2} + C \quad \blacksquare$$

To check this answer we simply differentiate and notice that it does indeed equal the integrand. That is,

$$\frac{d}{dx} \left( \frac{2}{3} (x^3 + 3)^{3/2} + C \right) = 3x^2(x^3 + 3)^{1/2}.$$

**E X A M P L E  4**    **Using the General Power Rule**

Find $\displaystyle\int \frac{1}{(2x + 1)^2} \, dx$.

**Solution**    Following the first step of the procedure rewrite the integral as

$$\int \frac{1}{(2x + 1)^2} \, dx = \int (2x + 1)^{-2} \, dx.$$

Then we see a function $(2x + 1)$ raised to the power $-2$. We then try the substitution $u = 2x + 1$ and then $du = 2 \, dx$. Making these substitutions yields

$$\int (2x + 1)^{-2} \, dx = \int u^{-2} \frac{1}{2} \, du \qquad\qquad dx = \frac{1}{2} \, du$$

$$= \frac{1}{2} \int u^{-2} \, du$$

$$= -\frac{1}{2} u^{-1} + C$$

$$= -\frac{1}{2} (2x + 1)^{-1} + C$$

$$= -\frac{1}{2(2x + 1)} + C \quad \blacksquare$$

E X A M P L E  5

**A Case When the Method of Substitution Fails**

Find $\int (x^5 + x^4 + 1)^5 (5x^3 + 4x^2) \, dx$.

Solution

We let $u = x^5 + x^4 + 1$ and $du = (5x^4 + 4x^3) \, dx = x(5x^3 + 4x^2) \, dx$. Substituting this, we obtain the integral

$$\int u^5 \frac{du}{x}.$$

We are unable to eliminate all $x$'s from the integrand and only obtain $u$'s. The method fails. ∎

E X A M P L E  6

**The General Exponential Rule**

Find $\int (x^2 + 1)e^{x^3 + 3x + 1} \, dx$.

Solution

Let $u = x^3 + 3x + 1$. Then $du = (3x^2 + 3) \, dx = 3(x^2 + 1) \, dx$. Then

$$\int (x^2 + 1)e^{x^3 + 3x + 1} \, dx = \int e^{x^3 + 3x + 1}(x^2 + 1) \, dx$$

$$= \int e^u \frac{1}{3} \, du \qquad (x^2 + 1) \, dx = \frac{1}{3} \, du$$

$$= \frac{1}{3} \int e^u \, du$$

$$= \frac{1}{3} e^u + C$$

$$= \frac{1}{3} e^{x^3 + 3x + 1} + C \quad ∎$$

This is the correct answer since

$$\frac{d}{dx}\left( \frac{1}{3} e^{x^3 + 3x + 1} + C \right) = (x^2 + 1)e^{x^3 + 3x + 1},$$

which is the integrand.

## Applications

**Historic Oil Well**

The first development oil well in Alaska's Prudhoe Bay, Number DS 1-1, produced over 19 million barrels of oil during its 19-year life. It went into production in December, 1970.

**E X A M P L E   7**    **Production of Oil**

The rate at which a natural resource is extracted tends to increase initially and then fall off later after the easily accessible material has been extracted. Suppose the rate in hundreds of thousands of barrels per year at which oil is being extracted during the early and intermediate stages of the life of a field is given by

$$P'(t) = \frac{t}{(t^2 + 1)^2} + 1,$$

where $t$ is in years. Find the amount of oil extracted during the first 7 years.

**Solution**    We must have

$$P(t) = \int \left( \frac{t}{(t^2 + 1)^2} + 1 \right) dt = \int (t^2 + 1)^{-2} t \, dt + \int dt.$$

The second integral is $t + C$. The first integral requires the substitution $u = t^2 + 1$. Then $du = 2t \, dt$ and

$$\int (t^2 + 1)^{-2} t \, dt + \int dt = \int u^{-2} \frac{1}{2} \, du + t + C$$

$$= -\frac{1}{2} u^{-1} + t + C$$

$$= -\frac{1}{2} \frac{1}{t^2 + 1} + t + C$$

There can be no production of oil in no time, thus

$$0 = P(0) = -\frac{1}{2} + C.$$

This implies that $C = \frac{1}{2}$. Thus

$$P(7) = -\frac{1}{2} \frac{1}{7^2 + 1} + 7 + \frac{1}{2} = 7.49.$$

Thus 749,000 barrels of oil are extracted in the first 7 years.    ■

## SELF-HELP EXERCISE SET 16.2

1.  Find $\int \frac{2x}{\sqrt[3]{x^2 + 4}} \, dx.$

2.  Find $\int \frac{e^{2x}}{e^{2x} + 1} \, dx.$

3.  Find $\int \frac{e^{1/x}}{x^2} \, dx.$

## EXERCISE SET 16.2

In Exercises 1 through 34 find the indefinite integral.

**1.** $\int 6(3x + 1)^{10} \, dx$

**2.** $\int 8(3 - 2x)^5 \, dx$

**3.** $\int x(3 - x^2)^7 \, dx$

**4.** $\int 10x(2x^2 + 1)^3 \, dx$

**5.** $\int 8(x + 1)(2x^2 + 4x - 1)^{3/2} \, dx$

**6.** $\int (x^3 + 2)(x^4 + 8x + 3)^{1/3} \, dx$

**7.** $\int (x^3 + x^2 + x + 1)(3x^4 + 4x^3 + 6x^2 + 12x + 1)^3 \, dx$

**8.** $\int (x^4 + x^2 + 2)(3x^5 + 5x^3 + 30x + 1)^5 \, dx$

**9.** $\int 2\sqrt{x + 1} \, dx$

**10.** $\int \sqrt[3]{x + 1} \, dx$

**11.** $\int x\sqrt{x^2 + 1} \, dx$

**12.** $\int x\sqrt{3x^2 + 1} \, dx$

**13.** $\int \dfrac{x}{\sqrt[3]{x^2 + 1}} \, dx$

**14.** $\int \dfrac{x}{\sqrt[5]{2x^2 + 5}} \, dx$

**15.** $\int \dfrac{\sqrt{x^{1/3} + 1}}{x^{2/3}} \, dx$

**16.** $\int \dfrac{\ln 2x}{x} \, dx$

**17.** $\int \dfrac{\sqrt{\ln x}}{x} \, dx$

**18.** $\int e^{2x+1} \, dx$

**19.** $\int e^{1-x} \, dx$

**20.** $\int 4xe^{x^2-1} \, dx$

**21.** $\int xe^{1-x^2} \, dx$

**22.** $\int \dfrac{e^{1/x^2}}{x^3} \, dx$

**23.** $\int \dfrac{e^{\sqrt{x}}}{\sqrt{x}} \, dx$

**24.** $\int \dfrac{1}{2x + 1} \, dx$

**25.** $\int \dfrac{1}{3x + 5} \, dx$

**26.** $\int \dfrac{x^3}{(x^4 + 4)^2} \, dx$

**27.** $\int \dfrac{x}{(x^2 + 3)^4} \, dx$

**28.** $\int \dfrac{e^x}{e^x + 1} \, dx$

**29.** $\int \dfrac{e^{-x}}{e^{-x} + 1} \, dx$

**30.** $\int \dfrac{e^x - e^{-x}}{e^x + e^{-x}} \, dx$

**31.** $\int \dfrac{e^{2x} - e^{-2x}}{e^{2x} + e^{-2x}} \, dx$

**32.** $\int \dfrac{1}{x \ln x} \, dx$

**33.** $\int \dfrac{1}{x \ln x^2} \, dx$

**34.** $\int \dfrac{1}{x \ln \sqrt{x}} \, dx$

## Applications

**35. Demand.** Find the demand function for a clock manufacturer if marginal demand, in dollars, is given by $p'(x) = -24x/(3x^2 + 1)^2$ and $p(1) = 10$, where $x$ is the number of thousands of clocks sold.

**36. Demand.** Find the demand function for a cardboard box manufacturer if marginal demand, in dollars, is given by $p'(x) = -4xe^{-x^2}$ and $p(1) = 10$, where $x$ is the number of thousands of boxes sold.

**37. Revenue.** Find the revenue function for a shoulder bag manufacturer if the marginal revenue, in dollars, is given by $4x(10 - x^2)$, where $x$ is the number of hundreds of bags sold.

**38. Revenue.** Find the revenue function for a hand purse manufacturer if the marginal revenue, in dollars, is given by $(1 - x)e^{2x-x^2}$, where $x$ is the number of thousands of purses sold.

39. **Cost.**   Find the cost function for a lipstick manufacturer if the marginal cost, in dollars, is given by $4x/(x^2 + 1)$, where $x$ is the number of cases of lipstick produced and fixed costs are $1000.

40. **Cost.**   Find the cost function for a hair brush manufacturer if the marginal cost, in dollars, is given by $x^3(100 + x^4)$, where $x$ is the number of cases of brushes produced and fixed costs are $1000.

41. **Production.**   Suppose the rate in tons per year at which copper is being extracted from a mine during the early and intermediate stages of the life of the mine is given by

$$P'(t) = \frac{-54t^2}{(t^3 + 1)^2} + 11.$$

Find the amount of copper extracted during the first 2 years.

42. **Depreciation.**   The value of a piece of machinery purchased for $10,000 is decreasing at the rate given by

$$V'(t) = -1000te^{-0.25t^2},$$

where $t$ is measured in years from purchase. How much will the machine be worth in 2 years?

43. **Population.**   The population of a certain city for the next several years is increasing at the rate given by

$$P'(t) = 2000(t + 10)e^{0.05t^2 + t},$$

where $t$ is time in years measured from the beginning of 1994 when the population was 100,000. Find the population at the end of 1995.

44. **Sales.**   After the introduction of a new camera, the rate of sales in thousands per month is given by

$$N'(t) = 10(1 + 2t)^{-5/2},$$

where $t$ is in months. Find the number sold in the first 4 months.

45. **Productivity.**   Suppose the number of items produced on a certain piece of machinery by an average employee is increasing at a rate given

$$N'(t) = \frac{1}{2t + 1},$$

where $t$ is measured in hours since placed on the machine for the first time. How many items are produced by the average employee in the first 3 hours?

46. **Memorization.**   Suppose the rate at which an average person can memorize a list of items is given by

$$N'(t) = \frac{15}{\sqrt{3t + 1}},$$

where $t$ is the number of hours spent memorizing. How many items does the average person memorize in 1 hour?

## Solutions to Self-Help Exercise Set 16.2

1.  Write the integral as $\int (x^2 + 4)^{-1/3} 2x \, dx$ and use the substitution $u = x^2 + 4$. Then $du = 2x \, dx$ and

$$\int (x^2 + 4)^{-1/3} 2x \, dx = \int u^{-1/3} \, du$$

$$= \frac{3}{2} u^{2/3} + C$$

$$= \frac{3}{2} (x^2 + 4)^{2/3} + C$$

2.  Write the integral as $\int (e^{2x} + 1)^{-1} e^{2x} \, dx$. This is a function raised to the power $n = -1$. Let $u = e^{2x} + 1$, then $du = 2e^{2x} \, dx$. Then

$$\int \frac{e^{2x}}{e^{2x} + 1} \, dx = \int \frac{\frac{1}{2} \, du}{u} = \frac{1}{2} \int \frac{du}{u}$$

$$= \frac{1}{2} \ln |u| + C$$

$$= \frac{1}{2} \ln(e^{2x} + 1) + C$$

3.  First write the integral as $\int e^{1/x} x^{-2} \, dx$. Let $u = 1/x = x^{-1}$. Then $du = -x^{-2} \, dx$ and

$$\int \frac{e^{1/x}}{x^2} \, dx = \int e^{1/x} x^{-2} \, dx$$

$$= \int e^u (-\, du) = -\int e^u \, du$$

$$= -e^u + C = -e^{1/x} + C$$

## 16.3   THE DEFINITE INTEGRAL

► *Sigma Notation*
► *Riemann Sums*
► *The Definite Integral*

APPLICATION

**Finding the Area Under a Curve**

Find the area between the graph of the curve $y = x^2$ and the $x$-axis on the interval $[0, 1]$. For the answer see Example 3 and the discussion that follows.

APPLICATION

**Method of Exhaustion**

In about 370 B.C., the Greek geometer Eudoxus made the method of exhaustion a rigorous procedure. In order to find the area of a circle, he first inscribed an equilateral triangle $P_0$ shown in Figure 16.2a and labeled $E$ the area outside the triangle but inside the circle. He then placed a regular hexagon $P_1$, as shown in Figure 16.2b, and noted that from simple geometrical principals the area outside $P_1$ and inside the circle was less than $\frac{1}{2} E$. Continuing in this manner, as shown in Figure 16.2c, the area outside the regular 12-sided polygon $P_2$ and inside the circle is less than $(\frac{1}{2})^2 E$. In such a manner, regular $3 \cdot 2^n$-sided polygons $P_n$ could be inscribed in a circle with the area outside the polygon $P_n$ and inside the circle less than $(\frac{1}{2})^n E$. Since $(\frac{1}{2})^n E \to 0$ as $n \to +\infty$, the areas of $P_n$ must approach the area of the circle. Since he knew the areas of $P_n$, he could find the area of the circle to any desired accuracy.

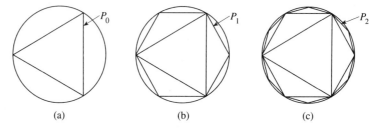

(a)                    (b)                    (c)

**Figure 16.2**

## Sigma Notation

We use the Greek letter $\Sigma$ (sigma) as a shorthand notation for summation. The expression

$$\sum_{i=1}^{5} 2i$$

means replace $i$ in the term $(2i)$ successively with the integers from 1 to 5 and then add the results. Thus

$$\sum_{i=1}^{5} 2i = 2(1) + 2(2) + 2(3) + 2(4) + 2(5) = 30.$$

**EXAMPLE 1**

Find (a) $\sum_{i=1}^{3} 4i^2$, (b) $\sum_{i=1}^{5} f(x_i)$, and (c) $\sum_{i=1}^{4} f(x_i)$ if $f(x) = 4x + 1$, $x_i = \frac{i}{4}$.

**Solutions**

(a)  $\sum_{i=1}^{3} 4i^2 = 4(1)^2 + 4(2)^2 + 4(3)^2 = 56$

(b)  $\sum_{i=1}^{5} f(x_i) = f(x_1) + f(x_2) + f(x_3) + f(x_4) + f(x_5)$

(c)   $\sum_{i=1}^{4} f(x_i) = \sum_{i=1}^{4} f(\frac{i}{4})$

$$= f(\tfrac{1}{4}) + f(\tfrac{2}{4}) + f(\tfrac{3}{4}) + f(\tfrac{4}{4})$$

$$= 2 + 3 + 4 + 5 = 14 \quad ■$$

We will need the following formulas, proofs of which are not given.

$$\sum_{i=1}^{n} i = 1 + 2 + 3 + \cdots + n = \frac{n(n + 1)}{2} \tag{1}$$

$$\sum_{i=1}^{n} i^2 = 1^2 + 2^2 + 3^2 + \cdots + n^2 = \frac{n(n + 1)(2n + 1)}{6}. \tag{2}$$

## Riemann Sums

**Bernhard Riemann, 1826–1866**

*(The Bettman Archive.)*

Riemann clarified the concept of the integral by developing the Riemann sums, which led later to the more important *Lebesgue* integral. He also developed a non-Euclidean geometry called Riemannian geometry which was used by Albert Einstein as a basis for Einstein's relativity theory. He also is known for the Riemann zeta function and the associated Riemann hypothesis, a celebrated unproven conjecture. His achievements were all the more remarkable, since he died from tuberculosis at the age of 39.

Assume we are given a nonnegative continuous function $y = f(x)$ on some finite interval $[a, b]$, such as shown in Figure 16.3. We have an intuitive idea of what is meant by the *area under the curve from a to b*. This is indicated by the shaded region in Figure 16.3. We wish to find a general procedure that will give this area. *This represents a fundamentally important problem in the calculus.*

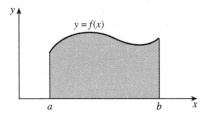

**Figure 16.3**

We shall see that the area under the graph of the curve can represent many important quantities in business, economics, biology, physics, and so on. Consider for example, a vehicle with the velocity given in Figure 16.4. Since distance is velocity multiplied by time, $d = vt$, and the velocity is constant during the last 2 minutes, one can readily calculate that the distance traveled during the second and third minute is $d = 60 \cdot \dfrac{2}{60} = 2$ miles. Notice that this is also the area under

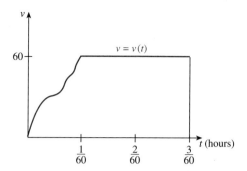

**Figure 16.4**

the curve on the interval $\left[\dfrac{1}{60}, \dfrac{3}{60}\right]$. However, during the first minute the velocity is changing. How can we calculate the distance traveled during this minute? We shall see in this section that this is also an area: the area under the curve on the interval $\left[0, \dfrac{1}{60}\right]$.

Returning to the general problem of finding the area indicated in Figure 16.3, we shall now see how to first approximate the area we are seeking as the finite sum of areas of "thin" rectangles with bases on the $x$-axis.

For any integer $n$ we divide the interval $[a, b]$ into $n$ subintervals, all of equal length. Each of these subintervals will form the base of a rectangle. If we designate the length of each of these subintervals as $\Delta x$, then $\Delta x = \dfrac{b-a}{n}$. Thus if we make $n$ large, $\Delta x$ will be small. Let us label the $n$ subintervals as $I_1, I_2, \ldots, I_n$. See Figure 16.5. We pick any point $x_1 \in I_1$ and take $f(x_1)$ as the height of the first rectangle. We also pick any point $x_2 \in I_2$ and take $f(x_2)$ as the height of the second rectangle. We continue in this manner obtaining $n$ rectangles $R_1$, $R_2, \ldots, R_n$, with areas equal, respectively, to $A_1 = f(x_1)\Delta x$, $A_2 = f(x_2)\Delta x, \ldots, A_n = f(x_n)\Delta x$. We designate $S_n$ as the sum of the areas of these $n$ rectangles and refer to $S_n$ as a Riemann sum. Thus

$$S_n = \sum_{i=1}^{n} f(x_i)\Delta x = f(x_1)\Delta x + f(x_2)\Delta x + \cdots + f(x_n)\Delta x.$$

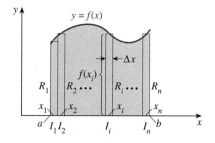

**Figure 16.5**

---

**Riemann Sum**

Suppose that $f(x)$ is continuous on the interval $[a, b]$. If this interval is divided into $n$ subintervals of equal length $\Delta x$ and $x_i$ is any point in the $i$th subinterval, then

$$S_n = \sum_{i=1}^{n} f(x_i)\Delta x$$

$$= f(x_1)\Delta x + f(x_2)\Delta x + \cdots + f(x_n)\Delta x$$

is called a **Riemann sum**.

---

**REMARK.** In general, for a fixed $n$, selecting different values of $x_i$ in $I_i$ will result in a different value for the Riemann sum $S_n$. Our notation does not indicate this. Thus keep in mind that for a fixed $n$ there is a Riemann sum $S_n$ for every possible selection of points $x_i \in I_i$.

Notice from Figure 16.6 that the Riemann sum $S_n$ seems to approximate the area under the curve we are seeking. It appears reasonable from these figures, that if $n$ is larger (or $\Delta x$ smaller), then the approximation will be better.

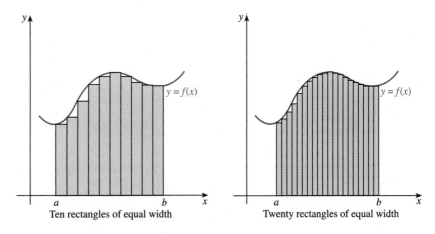

Ten rectangles of equal width      Twenty rectangles of equal width

**Figure 16.6**

EXAMPLE 2    **Finding Riemann Sums**

Let $f(x) = x^2$ on $[0, 1]$. If $x_i$ is always chosen to be the right endpoint of $I_i$, find the Riemann sums (a) $S_4$, (b) $S_8$. Show that with $x_i$ always chosen as the right endpoints that $S_4$ and $S_8$ are *larger* than the area and that $S_8 < S_4$. Thus $S_8$ is a better approximation to the area than $S_4$.

Solutions    (a) Since $f(x)$ is an increasing function, on any subinterval $[c, d]$ $f(x)$ must assume its maximum at the right endpoint. Thus from Figure 16.7 we see that the four rectangles lie *above* the curve. This implies in this case that $S_4$ is *larger* than the area under the curve. We have $\Delta x = \frac{1-0}{4} =$

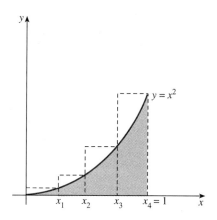

**Figure 16.7**

$\frac{1}{4}$ and the four subintervals are $I_1 = [0, \frac{1}{4}]$, $I_2 = [\frac{1}{4}, \frac{1}{2}]$, $I_3 = [\frac{1}{2}, \frac{3}{4}]$, and $I_4 = [\frac{3}{4}, 1]$. Therefore $x_1 = \frac{1}{4}$, $x_2 = \frac{1}{2}$, $x_3 = \frac{3}{4}$, and $x_4 = 1$. Then

$$S_4 = \Delta x[f(x_1) + f(x_2) + f(x_3) + f(x_4)]$$

$$= \frac{1}{4}\left[f\left(\frac{1}{4}\right) + f\left(\frac{2}{4}\right) + f\left(\frac{3}{4}\right) + f\left(\frac{4}{4}\right)\right]$$

$$= \frac{1}{4}\left[\left(\frac{1}{4}\right)^2 + \left(\frac{2}{4}\right)^2 + \left(\frac{3}{4}\right)^2 + \left(\frac{4}{4}\right)^2\right]$$

$$= \frac{1}{4^3}[1 + 2^2 + 3^2 + 4^2] = \frac{30}{64} = \frac{15}{32}$$

$$\approx .469.$$

(b)  Refer to Figure 16.8. Since the tops of the rectangles are above the curve, the Riemann sum found here will be greater than the area under the curve. Figure 16.9 overlays Figures 16.7 and 16.8, with the dotted line indicating the top of the rectangles used in Figure 16.7. Comparing Figure 16.8 with 16.7 in this manner, we see that our answer for part (b) should be definitely smaller than the Riemann sum found in (a). Thus the approximation is getting better. In this case $\Delta x = \frac{1}{8}$. The right endpoints $x_i$ of the eight subintervals are respectively $\frac{1}{8}$, $\frac{2}{8}$, $\frac{3}{8}$, ... , $\frac{8}{8}$. That is, $x_i = \frac{i}{8}$. See Figure 16.9. Then

$$S_8 = \Delta x[f(x_1) + f(x_2) + f(x_3) + \cdots + f(x_8)]$$

$$= \frac{1}{8}\left[f\left(\frac{1}{8}\right) + f\left(\frac{2}{8}\right) + f\left(\frac{3}{8}\right) + \cdots + f\left(\frac{8}{8}\right)\right]$$

$$= \frac{1}{8}\left[\left(\frac{1}{8}\right)^2 + \left(\frac{2}{8}\right)^2 + \left(\frac{3}{8}\right)^2 + \cdots + \left(\frac{8}{8}\right)^2\right]$$

$$= \frac{1}{8^3}[1 + 2^2 + 3^2 + \cdots + 8^2]$$

**Figure 16.8**

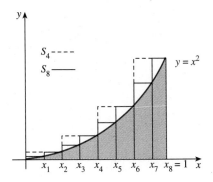

**Figure 16.9**

Now use formula (2) with $n = 8$ and obtain

$$\sum_{i=1}^{8} i^2 = 1^2 + 2^2 + 3^2 + \cdots + 8^2 = \frac{8(8 + 1)(16 + 1)}{6}.$$

Using this where we left off before yields

$$S_8 = \frac{1}{8^3} \left( \frac{8 \cdot 9 \cdot 17}{6} \right) = \frac{1}{64} \left( \frac{3 \cdot 17}{2} \right) = \frac{51}{128} \approx .398.$$

This is indeed smaller than the Riemann sum found in (a).  ∎

When $x_i$ is chosen, as in Example 2, so that $f(x_i)$ is the maximum value of $f(x)$ on $I_i$, we call the Riemann sum $S_n$ an **upper sum** (since the tops of all the rectangles are above the curve).

We now find $S_n$ for any $n$ when the $x_i$ are chosen to be the right endpoints of each subinterval.

**E X A M P L E  3**     **Finding a Riemann Sum in Closed Form**

For $y = f(x) = x^2$ on [0, 1], take $x_i$ to be the right endpoint of each subinterval $I_i$ and find $S_n$. Show that each $S_n$ is larger than the area and the $S_n$ will be smaller if $n$ is larger.

**Solution**     See Figure 16.10. Here $\Delta x = \dfrac{1}{n}$. Thus the right endpoints of the subintervals will

be $\dfrac{1}{n}, \dfrac{2}{n}, \dfrac{3}{n}, \ldots, \dfrac{n}{n}$, and $x_i = \dfrac{i}{n}$. Notice from the geometry that all the rectangles

lie *above* the graph of the curve $y = f(x) = x^2$ and thus $S_n$ must be *larger* than the area. We then have

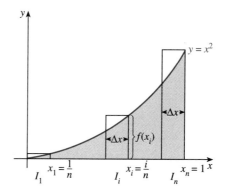

**Figure 16.10**

$$S_n = \Delta x[f(x_1) + f(x_2) + f(x_3) + \cdots + f(x_n)]$$

$$= \frac{1}{n}\left[f\left(\frac{1}{n}\right) + f\left(\frac{2}{n}\right) + f\left(\frac{3}{n}\right) + \cdots + f\left(\frac{n}{n}\right)\right]$$

$$= \frac{1}{n}\left[\left(\frac{1}{n}\right)^2 + \left(\frac{2}{n}\right)^2 + \left(\frac{3}{n}\right)^2 + \cdots + \left(\frac{n}{n}\right)^2\right]$$

$$= \frac{1}{n^3}[1 + 2^2 + 3^2 + \cdots + n^2]$$

$$= \frac{1}{n^3}\frac{n(n + 1)(2n + 1)}{6} \qquad \text{from (2)}$$

$$= \frac{2n^2 + 3n + 1}{6n^2}$$

$$= \frac{1}{3} + \frac{1}{2n} + \frac{1}{6n^2}$$

Notice from this formula that $S_n$ is smaller if $n$ is larger and $S_n > \frac{1}{3}$ for all $n$. ∎

**R E M A R K .**     Any *increasing* function $y = f(x)$, such as shown in Figure 16.10, will always attain its maximum on any $I_i$ at the right endpoint of this interval and attain its minimum at the left endpoint. Choosing $x_i$ always as the right endpoints of $I_i$ for such functions will always result in rectangles that are *above* the graph of $y = f(x)$ and a Riemann sum that is *greater* than the area under the curve. Choosing $x_i$ always as the left endpoints of $I_i$ for such functions will always result in rectangles that are *below* the graph of $y = f(x)$ and a Riemann sum that is *smaller* than the area under the curve.

## The Definite Integral

We have seen in the previous two examples that the Riemann sums $S_n$, which are all larger than the area, get smaller and closer to the actual area under the curve as $n$ gets larger.

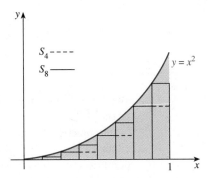

**Figure 16.11**

In the first Self-Help Exercise, the reader is asked to find the Riemann sums, which for purposes of comparison we denote by $S_n^*$, for the same function on the same interval, by taking $x_i$ always on the *left* of each subinterval. From Figure 16.11, it is apparent that $S_n^*$ will always be *less* than the area under the curve. The reader will find that

$$S_n^* = \frac{1}{3} - \frac{1}{2n} + \frac{1}{6n^2}.$$

Although not immediately apparent from this formula, it is true that as $n$ increase, $S_n^*$ increases. This can easily be seen geometrically by comparing $S_4^*$ with $S_8^*$, and so on, as in Figure 16.11. This is very similar to what Exodus accomplished with his inscribed regular polygons $P_n$. The areas of his inscribed polygons are all less than the area of the circle and steadily increase in area as $n$ gets larger with the area of $P_n$ approaching the area of the circle.

We have thus "trapped" the area above the $x$-axis and under the curve $y = f(x) = x^2$ on $[0, 1]$ between $S_n^*$ below and $S_n$ above. Table 16.1 gives some values of $S_n^*$ and $S_n$.

Looking at this limiting process from the geometric point of view, as $n$ becomes larger, $\Delta x$ becomes smaller, and by examining the previous figures, it appears that as $n \to +\infty$, $S_n$ will approach what we intuitively think of as the area under the curve from above while $S_n^*$ will approach the area from below. Using the closed forms of $S_n$ and $S_n^*$ and taking the limits we obtain

**Table 16.1**

| $n$ | $S_n^*$ | $S_n$ |
|---|---|---|
| 4 | 0.21875 | 0.46875 |
| 8 | 0.27344 | 0.39844 |
| 100 | 0.32835 | 0.33835 |
| 1000 | 0.33283 | 0.333834 |
| 10,000 | 0.33328 | 0.33338 |

$$\lim_{n \to +\infty} S_n = \lim_{n \to +\infty} \left( \frac{1}{3} + \frac{1}{2n} + \frac{1}{6n^2} \right) = \frac{1}{3}$$

$$\lim_{n \to +\infty} S_n^* = \lim_{n \to +\infty} \left( \frac{1}{3} - \frac{1}{2n} + \frac{1}{6n^2} \right) = \frac{1}{3}.$$

We thus conclude that the area above the $x$-axis and under the curve $y = f(x) = x^2$ on $[0, 1]$ is $\frac{1}{3}$.

In practice it is not practical to be able to pick the points $x_i$ to give rectangles either all above the graph of $y = f(x)$ or all below the graph. More typically, the points $x_i$ are conveniently selected to be the left endpoints, the right endpoints, or the midpoints of the subintervals, and we will be unable to determine if the tops of the rectangles are all above or all below the graph of the curve. We are therefore faced with two serious problems. If we calculate $S_n$ by picking the $x_i$ in some specific manner, can we be assured that the resulting Riemann sum $S_n$ is heading for some limit? And secondly, can we be assured that if we calculate $S_n$ by picking $x_i$ in some other way that these other Riemann sums will head for the very same limit?

The following result answers these questions in the affirmative if $f(x)$ is continuous.

---

**The Common Limit of Riemann Sums**

Suppose that $f(x)$ is continuous on the interval $[a, b]$. If this interval is divided into $n$ subintervals of equal length $\Delta x$ and $x_i$ is any point in the $i$th subinterval, then

$$\lim_{n \to +\infty} \sum_{i=1}^{n} f(x_i) \Delta x$$

exists and is the same limit independent of how the points $x_i$ are chosen in the $i$th subinterval.

---

With this we can now give the definition of the definite integral.

---

**Definition of Definite Integral**

Assume that $f(x)$ is continuous on $[a, b]$. The **definite integral** of $f$ over $[a, b]$, written $\int_a^b f(x)\, dx$ is defined to be the common limit of Riemann sums given above, that is,

$$\int_a^b f(x)\, dx = \lim_{n \to +\infty} \sum_{i=1}^{n} f(x_i) \Delta x.$$

---

The notation for the definite integral, $\int_a^b f(x)\, dx$, is very suggestive of a Riemann sum. One can think of the integral sign $\int$ as an elongated $S$ (or $\Sigma$) that

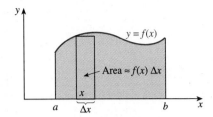

**Figure 16.12**

stands for "sum." Also, if $n$ is large, $dx \approx \Delta x$ and $f(x) \, dx$ is an area of a rectangle. See Figure 16.12. The notation for the definite integral thus suggests summing the area of such rectangles from $a$ to $b$. This interpretation is widely used in the applications and can yield significant insights.

Since the area above the $x$-axis and under the graph of the curve $y = f(x)$ when the function is nonnegative is intuitively the common limit of the Riemann sums, we *define* the term "area under the graph of the curve" to be this common limit or what is the same thing, the definite integral.

---

**Area Under the Curve**

If $f(x)$ is nonnegative and continuous on $[a, b]$, we say the area under the curve $y = f(x)$ and above the $x$-axis from $a$ to $b$ is given by the definite integral

$$\int_a^b f(x) \, dx.$$

---

Thus from the previous work the area under the graph of the curve $y = f(x) = x^2$ on $[0, 1]$ is $\frac{1}{3}$. There are no formulas from geometry that we could use to obtain this answer.

**E X A M P L E   4**   **Finding a Definite Integral**

For $y = x$ on $[1, 3]$, take $x_i$ to always be the left endpoint of $I_i$ and find

$$\lim_{n \to +\infty} S_n = \int_1^3 x \, dx.$$

**Solution**   First notice that this represents the area under the line $y = x$ from 1 to 3. See Figure 16.13. From the figure we readily see that the area is 4. Thus

$$\int_1^3 x \, dx = 4.$$

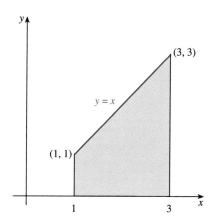

**Figure 16.13**

We will now verify this using Riemann sums. We first divide the interval $[1, 3]$ into $n$ equal subintervals each of length

$$\Delta x = \frac{3 - 1}{n} = \frac{2}{n}.$$

Since the point $x_i$ is chosen as the left endpoint of the $i$th subinterval, the $x_i$ are $1, 1 + \Delta x, 1 + 2\Delta x, \ldots, 1 + (n - 1)\Delta x$. See Figure 16.14. Thus $x_i = 1 + (i - 1)\Delta x$, and

$$S_n = \sum_{i=1}^{n} f(x_i)\Delta x$$

$$= \sum_{i=1}^{n} f(1 + (i - 1)\Delta x)\Delta x$$

$$= \sum_{i=1}^{n} (1 + (i - 1)\Delta x)\Delta x$$

$$= \Delta x \sum_{i=1}^{n} 1 + (\Delta x)^2 \sum_{i=1}^{n} (i - 1)$$

$$= (\Delta x)n + (\Delta x)^2 \frac{(n - 1)n}{2} \qquad \text{from (1)}$$

$$= \frac{2}{n} n + \left(\frac{2}{n}\right)^2 \frac{(n - 1)n}{2}$$

$$= 2 + 2 \frac{n - 1}{n}$$

$$= 4 - \frac{2}{n}$$

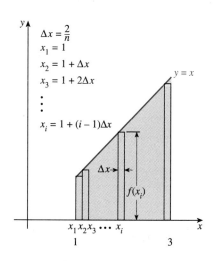

**Figure 16.14**

We see that as $n$ gets larger $S_n$ gets larger and

$$\int_1^3 x \, dx = \lim_{n \to +\infty} S_n$$

$$= \lim_{n \to +\infty} \left( 4 - \frac{2}{n} \right)$$

$$= 4. \quad \blacksquare$$

E X A M P L E   5    **Finding a Definite Integral Using Area**

Find

$$\int_0^2 \sqrt{4 - x^2} \, dx.$$

**Solution**    We know that this definite integral represents the area under the curve $y = \sqrt{4 - x^2}$ from 0 to 2. But this curve is just the portion of the circle of radius two centered at the origin that is in the first quadrant. See Figure 16.15. This area is just $\pi$. Thus

$$\int_0^2 \sqrt{4 - x^2} \, dx = \pi. \quad \blacksquare$$

In a later section we will take appropriate Riemann sums and trap this area of $\pi$ between two of them. This will provide another method of approximating $\pi$.

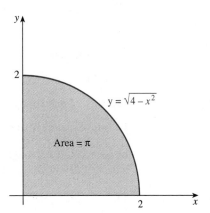

**Figure 16.15**

## SELF-HELP EXERCISE SET 16.3

1.  Let $f(x) = x^2$ on $[0, 1]$. Find the Riemann sum if $x_i$ is always chosen to be the left endpoint of $I_i$ and $n = 4$.

2.  For $y = f(x) = x^2$ on $[0, 1]$ take $x_i$ to be the left endpoint of each subinterval $I_i$ and find $S_n$. Show that each $S_n$ is smaller than the area under the curve.

3.  For $y = x$ on $[1, 3]$, take $x_i$ to always be the right endpoint of $I_i$ and find

$$\lim_{n \to +\infty} S_n = \int_1^3 x \, dx.$$

## EXERCISE SET 16.3

In Exercises 1 through 4 find the sum.

1.  $\sum_{i=1}^{4} (-3i)$                  2.  $\sum_{i=1}^{5} (i + 1)$

3.  $\sum_{i=1}^{10} 2$                     4.  $\sum_{i=1}^{5} (i - 2)$

In Exercises 5 through 12 find the requested Riemann sums for the indicated functions on the given intervals, using the indicated choices for $x_i$.

5.  (a) $S_5$, and (b) $S_{10}$, $f(x) = x$ on $[0, 1]$ with $x_i$ the right endpoint

6.  (a) $S_5$, and (b) $S_{10}$, $f(x) = x$ on $[0, 1]$ with $x_i$ the left endpoint

7.  (a) $S_4$, and (b) $S_8$, $f(x) = x^3$ on $[0, 1]$ with $x_i$ the right endpoint

8.  (a) $S_4$, and (b) $S_8$, $f(x) = x^3$ on $[0, 1]$ with $x_i$ the left endpoint

9.  (a) $S_5$, and (b) $S_{10}$, $f(x) = x$ on $[0, 1]$ with $x_i$ the midpoint

10.  (a) $S_4$, and (b) $S_8$, $f(x) = x^3$ on $[0, 1]$ with $x_i$ the midpoint

11.  (a) $S_3$, and (b) $S_6$, $f(x) = 4 - x$ on $[1, 4]$ with $x_i$ the right endpoint

12.  (a) $S_3$, and (b) $S_6$, $f(x) = 4 - x$ on $[0, 1]$ with $x_i$ the left endpoint

In Exercises 13 through 22 find $S_n$ using the indicated point $x_i$ and then find the indicated definite integral by finding $\lim_{n \to +\infty} S_n$.

13.  $\int_0^1 x \, dx$, $x_i$ the right endpoint         14.  $\int_1^5 x^2 \, dx$, $x_i$ the right endpoint

15.  $\int_0^1 x \, dx$, $x_i$ the left endpoint          16.  $\int_1^5 x^2 \, dx$, $x_i$ the left endpoint

17.  $\int_1^4 x \, dx$, $x_i$ the right endpoint         18.  $\int_1^5 x^2 \, dx$, $x_i$ the midpoint

19.  $\int_1^4 (4 - x) \, dx$, $x_i$ the right endpoint

20.  $\int_1^4 (4 - x) \, dx$, $x_i$ the left endpoint

**21.** $\int_2^4 (8 - 2x) \, dx$, $x_i$ the left endpoint

**22.** $\int_2^4 (8 - 2x) \, dx$, $x_i$ the right end point

In Exercises 23 through 32 find the definite integrals by finding the area of the appropriate geometric figure.

**23.** $\int_{-1}^3 5 \, dx$  **24.** $\int_{-10}^{-2} 4 \, dx$  **25.** $\int_0^2 3x \, dx$

**26.** $\int_1^3 2x \, dx$  **27.** $\int_1^4 (4 - x) \, dx$  **28.** $\int_{-1}^4 (4 - x) \, dx$

**29.** $\int_{-1}^1 \sqrt{1 - x^2} \, dx$  **30.** $\int_0^3 \sqrt{9 - x^2} \, dx$

**31.** $\int_0^2 \sqrt{1 - (x - 1)^2} \, dx$  **32.** $\int_1^2 \sqrt{1 - (x - 1)^2} \, dx$

It is not possible to find an antiderivative of the two functions $e^{-x^2}$ or $\dfrac{1}{\ln x}$ in terms of elementary known functions. Thus the *only* way of evaluating the definite integrals of such functions is by some approximation technique such as Riemann sums. We shall see later that the definite integral of $e^{-x^2}$ is of critical importance in the applications of probability.

**33.** Using the Riemann sum $S_4$ with $x_i$ the right endpoint of each subinterval, estimate $\int_0^1 e^{-x^2} \, dx$. Using a graph of $e^{-x^2}$ show why your answer must be *less* than the integral.

**34.** Using the Riemann sum $S_4$ with $x_i$ the left endpoint of each subinterval, estimate $\int_2^3 \dfrac{1}{\ln x} \, dx$. Using a graph of $1/\ln x$ show why your answer must be *greater* than the integral.

## Applications

Very often one does not know the values of $y = f(x)$ for all values of $x$ on the interval $[a, b]$, but rather for only a finite number of values from data that has been collected. In such a circumstance, it is impossible to know what the area under the curve $y = f(x)$ actually is. Nonetheless, one can use a Riemann sum to find an *approximation* to this area.

**35. Revenue.** In the following table the rate $R'(t)$ of revenue in millions of dollars per week is estimated in some manner for discrete equally spaced points $0 = t_0, t_1, \ldots, t_4 = 8$, where $t$ is in weeks. Find $S_4$ for the case when (a) $t_i$ is always taken as the left endpoint of each subinterval and (b) $t_i$ is always taken as the right endpoint of each subinterval. The answers will then be approximations to $\int_0^8 R'(t) \, dt$. (In the next section we will see that this integral equals $R(8) - R(0)$, which is the total revenue for the eight-week period.)

| $t_i$ | 0 | 2 | 4 | 6 | 8 |
|---|---|---|---|---|---|
| $R'(t_i)$ | 1.0 | 0.8 | 0.5 | 0.7 | 0.8 |

36. **Sales.**  In the following table the rate $R'(t)$ of sales for an entire industry in billions of dollars is given for discrete equally spaced points $0 = t_0, t_1, \ldots, t_{10} = 10$, where $t$ is in weeks. Find $S_{10}$ for the case when (a) $t_i$ is always taken as the left endpoint of each subinterval, and (b) $t_i$ is always taken as the right endpoint of each subinterval. The answers will then be approximations to $\int_0^{10} R'(t)\, dt$. (In the next section we will see that this integral equals $R(10) - R(0)$, which is the total sales for the ten-week period.)

| $t_i$ | 0 | 1 | 2 | 3 | 4 | 5 | 6 | 7 | 8 | 9 | 10 |
|---|---|---|---|---|---|---|---|---|---|---|---|
| $R'(t_i)$ | 10 | 12 | 15 | 16 | 20 | 18 | 16 | 15 | 12 | 10 | 6 |

37. **Population.**  In the following table the rate of growth $M'(t)$ of a bacterium has been estimated by measuring the rate of increase in mass at discrete equally spaced points $0 = t_0, t_1, \ldots, t_{10} = 20$, where $t$ is in hours. Find $M_{10}$ for the case when (a) $t_i$ is always taken as the left endpoint of each subinterval, and (b) $t_i$ is always taken as the right endpoint of each subinterval. The answers will then be approximations to $\int_0^{20} M'(t)\, dt$. (In the next section we will see that this integral equals $M(20) - M(0)$, which is the total growth for the twenty-hour period.)

| $t_i$ | 0 | 2 | 4 | 6 | 8 | 10 | 12 | 14 | 16 | 18 | 20 |
|---|---|---|---|---|---|---|---|---|---|---|---|
| $M'(t_i)$ | 10 | 14 | 20 | 30 | 50 | 100 | 100 | 80 | 50 | 20 | 0 |

38. **Velocity.**  In the following table the velocity $v(t) = s'(t)$ is given for discrete equally spaced points $0 = t_0, t_1, \ldots, t_5 = 1$. Find $S_5$ for the case when (a) $t_i$ is always taken as the left endpoint of each subinterval, and (b) $t_i$ is always taken as the right endpoint of each subinterval. The answers will then be approximations to $\int_0^1 v(t)\, dt$. (In the next section we will see that this integral equals $s(1) - s(0)$, which is the total distance traveled.)

| $t_i$ | 0 | 0.2 | 0.4 | 0.6 | 0.8 | 1.0 |
|---|---|---|---|---|---|---|
| $v(t_i)$ | 1.0 | 1.6 | 2.0 | 1.8 | 1.5 | 1.0 |

## Graphing Calculator Exercises

In Exercises 39 through 44 use program Riemann, found in Appendix C, to find the Riemann sums $\sum_{i=1}^{n} f(x_i)\, \Delta x$ when $x_i$ is taken on each subinterval to be at the left endpoint, the midpoint, the right endpoint for $n$ equal (a) 50, (b) 100, (c) 200. Compare your answers to the exact answer.

39. $f(x) = x^3$ on $[0, 1]$

40. $f(x) = x^3$ on $[1, 4]$

41. $f(x) = x(5 - x)$ on $[0, 5]$

42. $f(x) = \sqrt{x}$ on $[0, 4]$

43. $f(x) = \dfrac{1}{x}$ on $[1, 4]$

44. $f(x) = \sqrt{4 - x^2}$ on $[0, 2]$

## Solutions to Self-Help Exercise Set 16.3

1.  Since $f(x)$ is an increasing function, on any subinterval $[c, d]$ $f(x)$ must assume its minimum at the left endpoint. Thus we see from the figure, that the height of each rectangle is always the minimum value of the function on the subinterval. Thus the four rectangles lie *underneath* the curve, and the Riemann sum that is the sum of the areas of these four rectangles will be *less* than the area under the curve. In this case $\Delta x = \frac{1-0}{4} = \frac{1}{4}$. The four subintervals are $I_1 = [0, \frac{1}{4}]$, $I_2 = [\frac{1}{4}, \frac{1}{2}]$, $I_3 = [\frac{1}{2}, \frac{3}{4}]$, and $I_4 = [\frac{3}{4}, 1]$. Therefore $x_1 = 0$, $x_2 = \frac{1}{4}$, $x_3 = \frac{1}{2}$, and $x_4 = \frac{3}{4}$. Then

$$S_4 = f(x_1)\Delta x + f(x_2)\Delta x + f(x_3)\Delta x + f(x_4)\Delta x$$

$$= \Delta x[f(x_1) + f(x_2) + f(x_3) + f(x_4)]$$

$$= \frac{1}{4}\left[f(0) + f\left(\frac{1}{4}\right) + f\left(\frac{1}{2}\right) + f\left(\frac{3}{4}\right)\right]$$

$$= \frac{1}{4}\left[0 + \frac{1}{16} + \frac{1}{4} + \frac{9}{16}\right] = \frac{7}{32} \approx 0.219$$

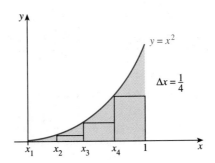

2.  See the figure. Here $\Delta x = \dfrac{1}{n}$. Thus the left endpoints of the subintervals will be $0, \dfrac{1}{n}, \dfrac{2}{n}, \ldots, \dfrac{n-1}{n}$, and $x_i = \dfrac{i-1}{n}$. Notice from the geometry that all the rectangles lie *below* the graph of the curve $y = f(x) = x^2$, and thus $S_n$ must be *smaller* than the area.

$$S_n = \Delta x[f(x_1) + f(x_2) + f(x_3) + \cdots + f(x_n)]$$

$$= \frac{1}{n}\left[f(0) + f\left(\frac{1}{n}\right) + f\left(\frac{2}{n}\right) + f\left(\frac{3}{n}\right) + \cdots + f\left(\frac{n-1}{n}\right)\right]$$

$$= \frac{1}{n}\left[0 + \left(\frac{1}{n}\right)^2 + \left(\frac{2}{n}\right)^2 + \left(\frac{3}{n}\right)^2 + \cdots + \left(\frac{n-1}{n}\right)^2\right]$$

$$= \frac{1}{n^3}[1 + 2^2 + 3^2 + \cdots + (n-1)^2]$$

$$= \frac{1}{n^3} \frac{(n-1)(n)(2n-1)}{6} \qquad \text{from (2)}$$

$$= \frac{2n^2 - 3n + 1}{6n^2}$$

$$= \frac{1}{3} - \frac{1}{2n} + \frac{1}{6n^2}$$

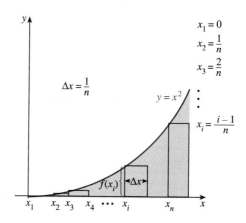

$$x_1 = 0$$
$$x_2 = \frac{1}{n}$$
$$x_3 = \frac{2}{n}$$
$$\Delta x = \frac{1}{n}$$
$$y = x^2$$
$$x_i = \frac{i-1}{n}$$

3. First divide the interval $[1, 3]$ into $n$ equal subintervals each of length $\Delta x = \frac{3-1}{n} = \frac{2}{n}$. Since the point $x_i$ is chosen as the right endpoint of the $i$th subinterval, the $x_i$ are $1 + \Delta x, 1 + 2\Delta x, \ldots, 1 + n\Delta x$. Thus $x_i = 1 + i\Delta x$. See the figure. Then

$$S_n = \sum_{i=1}^{n} f(x_i)\Delta x$$

$$= \sum_{i=1}^{n} f(1 + i\Delta x)\Delta x = \sum_{i=1}^{n} (1 + i\Delta x)\Delta x$$

$$= \Delta x \sum_{i=1}^{n} 1 + (\Delta x)^2 \sum_{i=1}^{n} i$$

$$= (\Delta x)n + (\Delta x)^2 \frac{n(n+1)}{2} \qquad \text{from (1)}$$

$$= \frac{2}{n}n + \left(\frac{2}{n}\right)^2 \frac{n(n+1)}{2}$$

$$= 2 + 2\frac{n+1}{n} = 4 + \frac{2}{n}$$

We see that as $n$ gets larger $S_n$ gets smaller, and

$$\lim_{n \to +\infty} S_n = \lim_{n \to +\infty} \left( 4 + \frac{2}{n} \right) = 4.$$

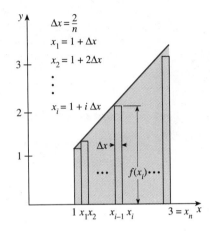

# 16.4 THE FUNDAMENTAL THEOREM OF CALCULUS

► *Properties of the Definite Integral*
► *The Fundamental Theorem of Calculus*
► *Average Value*
► *Density*

APPLICATION
**Extracting Natural Gas
from a Field**

The rate at which a natural resource is extracted often increases at first until the easily accessible part of the resource is exhausted. Then the rate of extraction tends to decline. Suppose natural gas is being extracted from a new field at a rate of $\dfrac{4t}{t^2 + 1}$ billions of cubic feet a year where $t$ is in years. If this field has an estimated 10 billion cubic feet, how long will it take to exhaust this field at the given rate of extraction? See Example 4 for the answer.

An industrial plant is located at the center of town. A straight highway goes through town and the concentration (density) of pollutants along this highway is given by

$$\delta(x) = 1000(24 - 3x^2)$$

in numbers of particles per mile per day, where $x$ is the distance from the center of town. What is the total number of particles of pollutant per day in town along this highway? See Example 6 for the answer.

## Properties of the Definite Integral

We state five properties for the definite integral that we defined in the last section. We assume that $f(x)$ and $g(x)$ are continuous on $[a, b]$.

---

**Properties of Definite Integrals**

1.  $\int_a^a f(x)\,dx = 0.$
2.  $\int_a^b kf(x)\,dx = k \int_a^b f(x)\,dx,\ k$ constant.
3.  $\int_a^b [f(x) \pm g(x)]\,dx = \int_a^b f(x)\,dx \pm \int_a^b g(x)\,dx.$
4.  $\int_a^b f(x)\,dx = \int_a^c f(x)\,dx + \int_c^b f(x)\,dx,\ a < c < b.$
5.  $\int_b^a f(x)\,dx = -\int_a^b f(x)\,dx,\ a < b.$

---

When both $f(x)$ and $g(x)$ are positive, the definite integrals are areas under curves. Thus Property 1 says that the area under a single point is zero. Property 2 says that the area under $kf$ is $k$ times the area under $f$. One part of Property 3 says that the area under $f + g$ is the area under $f$ plus the area under $g$. Property 4 says that, according to Figure 16.16, the area under the curve from $a$ to $b$ is the area under the curve from $a$ to $c$ plus the area under the curve from $c$ to $b$. The actual proofs require the use of Riemann sums and can be shown to be valid without the restriction that $f$ be nonnegative. The proofs are not given here. Property 5 is actually a *definition*.

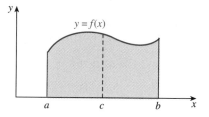

**Figure 16.16**

## The Fundamental Theorem of Calculus

In the last section we gave the definition of the definite integral $\int_a^b f(x)\, dx$ and interpreted it in the case that $f(x)$ was nonnegative as the area under the curve $y = f(x)$ from $a$ to $b$. In an earlier section we defined the indefinite integral $\int f(x)\, dx$ as the antiderivative of $f(x)$. The following theorem, called the fundamental theorem of calculus, connects these two integrals.

---

**Fundamental Theorem of Calculus**

Suppose $f(x)$ is continuous on $[a,b]$. If $F(x)$ is an antiderivative of $f(x)$, then

$$\int_a^b f(x)\, dx = F(b) - F(a).$$

---

For convenience we write

$$F(x)\big|_a^b = F(b) - F(a).$$

We know that if $F(x)$ is an antiderivative of $f(x)$, then any other antiderivative must be of the form $F(x) + C$ for some constant $C$. Let $G(x) = F(x) + C$. Then $G(x)$ is an antiderivative of $f(x)$. Let us then use $G(x)$ in the fundamental theorem of calculus to evaluate the definite integral. We obtain

$$\int_a^b f(x)\, dx = G(b) - G(a)$$
$$= [F(b) + C] - [F(a) + C]$$
$$= F(b) - F(a).$$

Thus *any* antiderivative will do.

Using the limit of Riemann sums we showed in the last section that

$$\int_0^1 x^2\, dx = \frac{1}{3}.$$

The fundamental theorem of calculus gives us a much easier way of obtaining this answer.

**E X A M P L E   1**      **Using the Fundamental Theorem**

---

Use the fundamental theorem of calculus to find $\int_0^1 x^2\, dx$.

**Solution**      Since $F(x) = x^3/3$ is an antiderivative of $x^2$, according to the fundamental theorem of calculus

$$\int_0^1 x^2\, dx = \frac{1}{3} x^3\big|_0^1 = \frac{1}{3}(1)^3 - \frac{1}{3}(0)^3 = \frac{1}{3}. \quad \blacksquare$$

**REMARK.**    The definite integral $\int_a^b f(x)\,dx$ is a *number* while the indefinite integral $\int f(x)\,dx$ is a collection of *functions*. For example, $\int_0^1 x^2\,dx = \frac{1}{3}$ while $\int x^2\,dx = \frac{1}{3}x^3 + C$.

We now give a sketch of the proof of the fundamental theorem of calculus when $f(x) > 0$. In this case the definite integral $\int_a^b f(x)\,dx$ represents the area under the curve $y = f(x)$ from $a$ to $b$.

We define the area function $A(t)$ by

$$A(t) = \int_a^t f(x)\,dx,$$

for $t \in [a, b]$. Thus $A(t)$ represents the area under the curve $y = f(x)$ from $a$ to $t$. See Figure 16.17. We will show that $A(t)$ is an antiderivative of $f(t)$, that is, we will show that

$$\frac{d}{dt}A(t) = f(t).$$

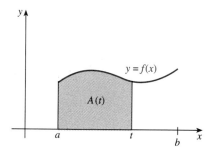

**Figure 16.17**

To show this we need to consider the quotient

$$\frac{A(t + \Delta t) - A(t)}{\Delta t}.$$

Property 4 above indicates that

$$A(t + \Delta t) = \int_a^{t+\Delta t} f(x)\,dx = \int_a^t f(x)\,dx + \int_t^{t+\Delta t} f(x)\,dx$$

$$= A(t) + \int_t^{t+\Delta t} f(x)dx.$$

Now $\int_t^{t+\Delta t} f(x)\,dx$ represents the area under the curve from $t$ to $t + \Delta t$ and is shown in Figure 16.18 (when $\Delta t > 0$). Now if $\Delta t$ is small, this area will be the area of a "thin" slice and approximately equal to the area of the rectangle with base $\Delta t$ and height $f(t)$. Thus the area under the curve from $t$ to $t + \Delta t$,

$$\int_t^{t+\Delta t} f(x)\,dx \approx f(t)\Delta t,$$

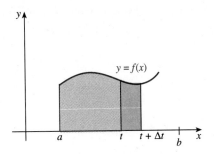

**Figure 16.18**

and this approximation improves as $\Delta t \to 0$. Then

$$\frac{A(t + \Delta t) - A(t)}{\Delta t} \approx \frac{1}{\Delta t} f(t)\Delta t = f(t).$$

Taking the limit as $\Delta t \to 0$ yields

$$A'(t) = f(t).$$

This shows that $A(t)$ is an antiderivative of $f(x)$. From Property 1 of definite integrals,

$$A(a) = \int_a^a f(x)\, dx = 0.$$

Also $A(b)$ is the area we are seeking, which we know must be $\int_a^b f(x)\, dx$ from the last section. Thus

$$\int_a^b f(x)\, dx = A(b) = A(b) - 0 = A(b) - A(a).$$

This concludes the proof.

In the process of establishing the fundamental theorem of calculus, we also established the following theorem.

---

**Theorem**

If $f(x)$ is continuous on the interval $[a, b]$, and if

$$F(x) = \int_a^x f(t)\, dt,$$

then for $x \in (a, b)$,

$$F'(x) = f(x).$$

---

**E X A M P L E   2**    **Finding a Definite Integral**

Find $\int_1^2 (1 + 6x^2)\, dx.$

**Solution**    Using Properties 2 and 3,

$$\int_1^2 (1 + 6x^2)\, dx = \int_1^2 dx + 6 \int_1^2 x^2\, dx$$

$$= x\big|_1^2 + 2x^3\big|_1^2$$

$$= [2 - 1] + 2[(2)^3 - (1)^3]$$

$$= 15 \quad \blacksquare$$

**E X A M P L E  3**    **Finding the Area Under a Curve**

Find the area above the $x$-axis and below the graph of the curve $y = x\sqrt{9 + x^2}$ on $[0, 4]$. See Figure 16.19.

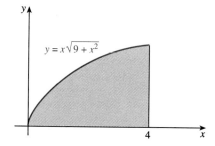

**Figure 16.19**

**Solution**    We are seeking $\int_0^4 x\sqrt{9 + x^2}\, dx$. Use the method of substitution. Let $u = 9 + x^2$. Then $du = 2x\, dx$. When the integration is with respect to $x$, then $x$ varies from 0 to 4. But if we are to integrate with respect to $u$, then we need to use limits of integration that correspond to the change in $u$. We have $u(0) = 9$ and $u(4) = 25$. Thus

$$\int_0^4 x\sqrt{9 + x^2}\, dx = \int_0^4 (9 + x^2)^{1/2} x\, dx$$

$$= \int_9^{25} u^{1/2}\, \frac{1}{2}\, du$$

$$= \frac{1}{2} \cdot \frac{2}{3}\, u^{3/2}\big|_9^{25}$$

$$= \frac{1}{3}\, (25^{3/2} - 9^{3/2})$$

$$= \frac{1}{3}\, (125 - 27)$$

$$= \frac{98}{3} \quad \blacksquare$$

EXAMPLE 4        **Extracting Natural Gas from a Field**

The rate at which a natural resource is extracted often increases at first until the easily accessible part of the resource is exhausted. Then the rate of extraction tends to decline. Suppose natural gas is being extracted from a new field at a rate of $\dfrac{4t}{t^2 + 1}$ billions of cubic feet a year where $t$ is in years. If this field has an estimated 10 billion cubic feet of gas, how long will it take to exhaust this field at the given rate of extraction?

Solution       See Figure 16.20. The amount extracted during the first $T$ years is $\displaystyle\int_0^T \dfrac{4t}{t^2 + 1}\, dt$.

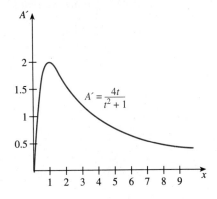

**Figure 16.20**

We are seeking the time $T$ for which this definite integral equals 10. Let $u = t^2 + 1$. Then $du = 2t\, dt$. As $t$ goes from 0 to $T$, $u$ goes from $u(0) = 1$ to $u(T) = T^2 + 1$. Then

$$
\begin{aligned}
10 &= \int_0^T \frac{4t}{t^2 + 1}\, dt \\[2mm]
&= 4 \int_1^{T^2+1} \frac{\tfrac{1}{2}\, du}{u} \\[2mm]
&= 2 \int_1^{T^2+1} \frac{du}{u} \\[2mm]
&= 2 \ln |u| \Big|_1^{T^2+1} \\[2mm]
&= 2(\ln(T^2 + 1) - \ln 1) \\[2mm]
&= 2 \ln(T^2 + 1) \\[2mm]
5 &= \ln(T^2 + 1) \\[2mm]
T^2 + 1 &= e^5 \\[2mm]
T &= \sqrt{e^5 - 1} \\[2mm]
&\approx 12.1
\end{aligned}
$$

or about 12.1 years.  ■

## Average Value

As mentioned in the last section, the indefinite integral represents many different quantities in business, economics, and science. In order to recognize these quantities as a definite integral, we must be able to write them as a limit of Riemann sums. We first consider the average of a function.

The average of a set of $n$ numbers $y_1, y_2, \ldots, y_n$, is

$$\frac{y_1 + y_2 + \cdots + y_n}{n}.$$

But what is the average value of a continuous function such as seen in Figure 16.21 on the *interval* $[a, b]$? Here there are an *infinite* number of values of $x$.

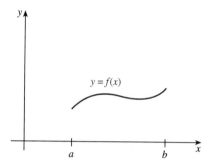

**Figure 16.21**

To answer this question let us start with an average such as

$$\frac{f(x_1) + f(x_2) + \cdots + f(x_n)}{n}$$

where $n$ is very large and the numbers $x_1, x_2, \ldots, x_n$, are distributed evenly throughout $[a, b]$. In fact let us divide the interval $[a, b]$ into $n$ equal subintervals $I_1, I_2, \ldots, I_n$, each of length $\Delta x = \dfrac{b - a}{n}$, and for each subinterval $I_i$, pick a point $x_i \in I_i$, say at the midpoint. Then if $n$ is very large

$$\frac{f(x_1) + f(x_2) + \cdots + f(x_n)}{n}$$

is an average of a very large number of values of $f(x)$ at points evenly distributed throughout $[a, b]$. Our strategy will be to let $n \to +\infty$ and see what happens to this average.

First however we multiply and divide each term in this last expression by $b - a$ and obtain

$$f(x_1) \frac{b-a}{n} \frac{1}{b-a} + f(x_2) \frac{b-a}{n} \frac{1}{b-a} + \cdots + f(x_n) \frac{b-a}{n} \frac{1}{b-a}.$$

Factoring out the term $\dfrac{1}{b-a}$ then gives

$$\frac{1}{b-a} \left[ f(x_1) \frac{b-a}{n} + f(x_2) \frac{b-a}{n} + \cdots + f(x_n) \frac{b-a}{n} \right].$$

Replacing $(b-a)/n$ with $\Delta x$ gives

$$\frac{1}{b-a} [f(x_1)\Delta x + f(x_2)\Delta x + \cdots + f(x_n)\Delta x],$$

which can be rewritten in the shorter form as

$$\frac{1}{b-a} \sum_{i=1}^{n} f(x_i)\Delta x.$$

We recognize this as $1/(b-a)$ times a Riemann sum of $f(x)$ over $[a, b]$. See Figure 16.22.

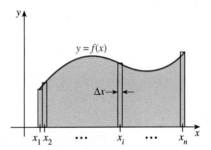

**Figure 16.22**

The Riemann sum approximates the area under the curve $f(x)$ on $[a, b]$. If we let $n \rightarrow +\infty$ this becomes

$$\frac{1}{b-a} \left[ \lim_{n \to +\infty} \sum_{i=1}^{n} f(x_i)\Delta x \right].$$

But since the term inside the square brackets is precisely the definition of the definite integral $\int_a^b f(x)\, dx$, this last term is

$$\frac{1}{b-a} \int_a^b f(x)\, dx.$$

This is what we define as the average of $f(x)$ over $[a, b]$.

---

**The Average Value of $f(x)$ over $[a, b]$**

If $f(x)$ is continuous on $[a, b]$, we define the average value of $f(x)$ on $[a, b]$ to be

$$\frac{1}{b - a} \int_a^b f(x) \, dx.$$

---

Notice that we do not require that $f(x)$ be a nonnegative function.

E X A M P L E  5    **Average Amount in an Account**

An account with an initial amount of $1000 is compounding continuously at an annual rate of 10% per year. What is the average amount of money in the account over the first 10 years?

**Solution**    The average amount on $[0, 10]$ is given by

$$\frac{1}{b - a} \int_a^b f(t) \, dt = \frac{1}{10 - 0} \int_0^{10} 1000 e^{.1t} \, dt$$

$$= 100 \int_0^{10} e^{.1t} \, dt$$

$$= 1000 e^{.1t} \big|_0^{10}$$

$$= 1000[e - 1]$$

$$\approx 1718$$

or $1718. Refer to Figure 16.23.  ■

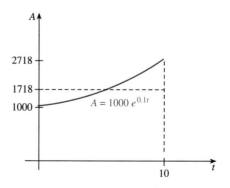

**Figure 16.23**

**Geometric Interpretation of Average Value.**   When the function $f(x)$ is non-negative on the interval $[a, b]$, one can give a simple geometric interpretation of the average value. Let us denote the average value of $f(x)$ on $[a, b]$ as $\bar{y}$. Then

$$\bar{y} = \frac{1}{b - a} \int_a^b f(x)\, dx,$$

and thus

$$\bar{y}(b - a) = \int_a^b f(x)\, dx.$$

Since $f(x)$ is nonnegative, the definite integral $\int_a^b f(x)\, dx$ is the area under the curve $y = f(x)$ on $[a, b]$. Thus if we construct a rectangle with base on the interval $[a, b]$ and height $\bar{y}$, then this rectangle will have area equal to $\bar{y}(b - a)$, which by the above formula is also the area under the curve $y = f(x)$ on $[a, b]$. See Figure 16.24.

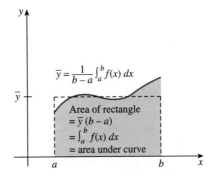

**Figure 16.24**

## Density

Suppose an industrial plant emits a certain amount of pollutants. We wish to determine the amount of pollutants falling to the ground along some straight line from the plant. At various distances from the plant along this line we could place instruments that count the number of pollutants per day falling on a one-foot stretch. We then would have data that give the number of pollutants per foot per day or the **density** at various points from the plant, as indicated in Figure 16.25. We could then determine a continuous function $\delta(x)$ whose graph goes through the data points as in Figure 16.26. The function $\delta(x)$ could be obtained

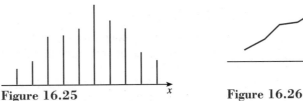

**Figure 16.25**                              **Figure 16.26**

just by drawing straight lines between the data points or by a more sophisticated curve-fitting scheme that would fit a higher degree polynomial to the data. Or we might have some theoretical means at our disposal that would give us $\delta(x)$. The function $\delta(x)$ is a **density function** and gives the number of pollutants per square foot (per day).

To find the total number of pollutants or the population on the interval $[a, b]$, proceed as follows. For any integer $n$, divide the interval $[a, b]$ into $n$ subintervals $I_1, I_2, \ldots, I_n$, each of length $\Delta x = \frac{b-a}{n}$. In a typical interval $I_i$, pick a point $x_i \in I_i$, say the midpoint to be specific.

As Figure 16.27 indicates, if $\Delta x$ is small then, since $\delta(x)$ is continuous, $\delta(x)$ will be approximately equal to $\delta(x_i)$ on the interval $I_i$. Thus the term $P_i = \delta(x_i)\Delta x$ is the number per square foot times the length of $I_i$ or the total number of pollutants on $I_i$, that is, the population on $I_i$. Of course, $P_i = \delta(x_i)\Delta x$ is also the area of the rectangle shown in Figure 16.27.

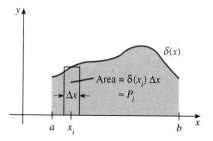

**Figure 16.27**

The total number of pollutants on $[a, b]$ is given by

$$P = P_1 + P_2 + \cdots + P_n = \sum_{i=1}^{n} P_i \approx \sum_{i=1}^{n} \delta(x_i)\Delta x.$$

As $n \to +\infty$, $\Delta x \to 0$, and this last approximation becomes more and more accurate. Also we notice that $\sum_{i=1}^{n} \delta(x_i)\Delta x$ is a Riemann sum $S_n$ of $\delta(x)$ on $[a, b]$, and by the very definition of the definite integral

$$\lim_{n \to +\infty} \sum_{i=1}^{n} \delta(x_i)\Delta x = \int_{a}^{b} \delta(x)\, dx.$$

Thus we have the following.

---

**Population Given the Density**

If $\delta(x)$ is a continuous function that gives the density in number per unit length, then the total population in the interval $[a, b]$ is

$$P = \int_{a}^{b} \delta(x)\, dx.$$

---

E X A M P L E  6    **Finding the Population Given the Density**

An industrial plant is located at the center of town. Let $x$ be the distance from the plant. A straight highway goes through town and the concentration (density) of pollutants along this highway is given by

$$\delta(x) = 1000(24 - 3x^2)$$

in numbers of particles per mile per day at the point $x$. See Figure 16.28. Find the total number of particles of pollutant per day in town along this highway.

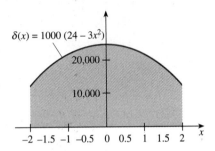

**Figure 16.28**

**Solution**    We have

$$P = \int_a^b \delta(x)\, dx$$

$$= \int_{-2}^{2} 1000(24 - 3x^2)\, dx$$

$$= 1000(24x - x^3)\big|_{-2}^{2}$$

$$= 1000[(48 - 8) - (-48 + 8)]$$

$$= 80,000 \quad \blacksquare$$

## SELF-HELP EXERCISE SET 16.4

1.  Evaluate $\int_{-1}^{0} 4xe^{x^2+2}\, dx$.

2.  Suppose the rate of sales of a certain brand of bicycle by a retailer in thousands of dollars per week is given by

$$\frac{d}{dt} S(t) = 10t - 0.30t^2, \qquad 0 \le t \le 5,$$

    where $t$ is the number of weeks after an advertising campaign has begun. Find the amount of sales during the third and fourth weeks.

3.  A study has indicated that the level $L(x)$ of air pollution a distance of $x$ miles from a certain factory is given by $L(x) = e^{-0.1x} + 1/x^2$. Find the average level of pollution between 10 and 20 miles from the factory.

## EXERCISE SET 16.4

In Exercises 1 through 24 evaluate the definite integrals.

**1.** $\int_1^2 4x^3\, dx$        **2.** $\int_{-2}^{-1} 6x^2\, dx$        **3.** $\int_{-2}^{-2} 3x^4\, dx$

**4.** $\int_{-2}^2 2x^5\, dx$        **5.** $\int_{-1}^0 (9x^2 - 1)\, dx$        **6.** $\int_{-1}^1 (1 - x^4)\, dx$

**7.** $\int_0^2 (4x^3 - 2x + 1)\, dx$        **8.** $\int_1^4 \left(\dfrac{3}{\sqrt{x}} - \dfrac{6}{\sqrt{x}}\right) dx$

**9.** $\int_1^2 (x^{-2} + 3x^{-4})\, dx$        **10.** $\int_{-2}^{-1} (x^{-5} + 1)\, dx$

**11.** $\int_{-2}^{-1} e^{2x}\, dx$        **12.** $\int_{-1}^0 e^{-x}\, dx$

**13.** $\int_1^3 \dfrac{1}{2x}\, dx$        **14.** $\int_2^4 \dfrac{3}{x}\, dx$        **15.** $\int_0^1 (2x - 1)^9\, dx$

**16.** $\int_{-1}^0 (1 + 2x)^5\, dx$        **17.** $\int_0^1 x(x^2 - 1)^7\, dx$        **18.** $\int_0^1 x\sqrt[3]{8 + x^2}\, dx$

**19.** $\int_0^4 \sqrt{2x + 1}\, dx$        **20.** $\int_0^4 \dfrac{1}{\sqrt{2x + 1}}\, dx$        **21.** $\int_{-1}^1 xe^{x^2+1}\, dx$

**22.** $\int_1^4 \dfrac{e^{\sqrt{x}}}{\sqrt{x}}\, dx$        **23.** $\int_1^2 \dfrac{1}{2x + 1}\, dx$        **24.** $\int_{-2}^0 \dfrac{x}{x^2 + 1}\, dx$

In Exercises 25 through 28 find the derivative of each function.

**25.** $\int_0^x e^{t^2}\, dt$        **26.** $\int_1^x \dfrac{e^t}{\sqrt{t}}\, dt$

**27.** $\int_2^x \dfrac{1}{\ln t}\, dt$        **28.** $\int_1^x \ln(t^2 + 1)\, dt$

In Exercises 29 through 38 find the average value of each of the given functions on the given interval.

**29.** $f(x) = 6$ on $[0, 10]$        **30.** $f(x) = -3$ on $[-10, -2]$

**31.** $f(x) = x$ on $[0, 10]$        **32.** $f(x) = x$ on $[-4, 4]$

**33.** $f(x) = 2x$ on $[-2, 2]$        **34.** $f(x) = x^2$ on $[0, 3]$

**35.** $f(x) = x^3$ on $[-1, 1]$        **36.** $f(x) = e^x$ on $[0, \ln 2]$

**37.** $f(x) = x(x - 1)$ on $[0, 2]$        **38.** $f(x) = x(x - 1)$ on $[-1, 2]$

## Applications

**39. Production.** Suppose oil is being extracted from a field at a rate given by

$$\frac{d}{dt} P(t) = 0.2e^{0.1t},$$

where $P(t)$ is measured in millions of barrels and $t$ in years. At this rate how much oil will be extracted during the second ten-year period?

40. **Production.** Suppose copper is being extracted from a certain mine at a rate given by

$$\frac{d}{dt} P(t) = 100e^{-0.2t},$$

where $P(t)$ is measured in tons of copper and $t$ in years. At this rate how much copper will be extracted during the third year?

41. **Sales.** Suppose the rate of sales of an item is given by

$$\frac{d}{dt} S(t) = -3t^2 + 36t,$$

where $t$ is the number of weeks after an advertising campaign has begun. How many items were sold during the third week?

42. **Sales.** A firm estimates that its sales will increase at a rate given by

$$\frac{d}{dt} S(t) = 10\sqrt{t},$$

where $t$ is measured in years. Find the total sales during the second and third years.

43. **Natural Resource Depletion.** Suppose a new oil reserve has been discovered with an estimated 10-billion-barrel capacity. Suppose production is given in billions of barrels and proceeds at a rate given by

$$\frac{d}{dt} P(t) = 1.5e^{-0.1t},$$

where $t$ is measured in years. At this rate, how long will it take to exhaust this resource?

44. **Natural Resource Depletion.** Suppose a resource such as oil, gas, or some mineral, has total reserves equal to $R$. Suppose the resource is being extracted at a rate given by

$$\frac{d}{dt} P(t) = ae^{-kt},$$

where $a$ and $k$ are positive constants and $t$ is time. Show that at this rate the resource will be exhausted in time

$$T = -\frac{\ln(1 - kR/a)}{k}.$$

45. **Chemical Spill.** A tank holding 10,000 gallons of a polluting chemical breaks at the bottom and spills out at the rate given by

$$f'(t) = 400e^{-0.01t},$$

where $t$ is measured in hours. How much spills during the third day?

**46. Chemical Spill.** In the previous problem how long before the tank is emptied?

**47. Sum-of-Years-Digit Depreciation.** The sum-of-years-digit depreciation method depreciates a piece of equipment more in the early years of its life and less in the later years. In this method the value changes at a rate given by

$$\frac{d}{dt} V(t) = -k(T - t),$$

where $T$ is the useful life in years of the equipment, $t$ is the time in years since purchase, and $k$ is a positive constant.

**a.** If the initial value of the equipment is $I$, show that the scrap value of the equipment is given by

$$I - kT^2/2.$$

**b.** Show that the number of years for the equipment to be worth half of its value is given by

$$T - \sqrt{T^2 - I/k}.$$

**48. Sum-of-Years-Digit Depreciation.** Suppose in the previous problem a certain computer with an initial value of $51,000 has $T = 10$ and $k = 1000$.
**a.** Find the scrap value of this computer.
**b.** Find the number of years it takes for the computer to be worth half its initial value using this method of depreciation.

**49. Sales.** From the time a firm introduced a new product, sales in thousands of this product have been increasing at a rate given by $S'(t) = 2 + t$, where $t$ is the time since the introduction of the new product. At the beginning of the third year, an advertising campaign was introduced and sales then increased at a rate given by $S_2'(t) = t^2$, where $t \geq 2$. Find the increase in sales due to the advertising campaign during the first two years of this campaign over what the sales would have been without the campaign.

**50. Production.** Suppose copper is being extracted from a certain mine at a rate given by

$$\frac{d}{dt} P(t) = 100e^{-0.2t},$$

where $P(t)$ is measured in tons of copper and $t$ in years from opening the mine. At the beginning of the sixth year, a mining innovation is introduced to boost production to a rate given by $Q'(t) = 500/t$. Find the increase in production of copper due to this innovation during the first five years of its use over what copper production would have been without its use.

**51. Average Amount of Money.** An account with an initial amount of $1000 is compounding continuously at an annual rate of 10% per year. What is the average amount of money in the account over the first 20 years?

**52. Average Price.** Given the demand equation $p = 100e^{-.2x}$, find the average price over the demand interval $[10, 20]$.

53. **Average Profit.**   The profits in millions of dollars of a certain firm is given by $P(t) = 6(t - 1)(t - 2)$, where $t$ is measured in years. Find the average profits per year over the period of time [0, 4].

54. **Average Revenue.**   The revenue in thousands of dollars is given by $R(x) = x^2 - 2x$, where $x$ is the number of thousands of items sold. Find the average revenue for the first 10,000 items sold.

55. **Average Air Pollution.**   A study has indicated that the level $L(x)$ of air pollution a distance of $x$ miles from a certain factory is given by $L(x) = e^{-0.1x} + 10$. Find the average level of pollution between 10 and 20 miles from the factory.

56. **Average Amount of a Drug.**   The amount of a drug in the body at time $t$ is given by $A(t) = 5e^{-.2t}$ where $t$ is in hours and $A$ is in milligrams. What is the average amount of drug in the body over the next 5-hour period?

57. **Average Blood Pressure.**   The blood pressure in an artery of a healthly individual can change substantially over only a few seconds. The average pressure is used in some studies of blood pressure. Suppose the function $p(t) = 40t^4 - 160t^3 + 160t^2 + 80$ gives the blood pressure in mm Hg in an artery over an interval [0, 2] of time measured in seconds. Find the average blood pressure in this artery over the two-second interval [0, 2].

58. **Density of Pollutants.**   Find the number of pollutants within one mile of the plant given in Example 6 of the text.

59. **Density of Pollutants.**   Find the number of pollutants between one and two miles from the plant in Example 6 of the text.

60. **Density of a Population.**   The density function in people per mile for the population of a small coastal town is given by

$$\delta(y) = \frac{32,000}{(2 + y)^3}$$

where $y$ is the distance in miles along a highway shown in the figure. If the town extends for 2 miles along this highway, what is the total population in the town along this road?

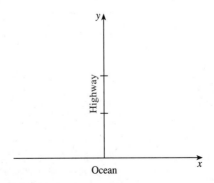

61. **Density of a Population.**   Find the total population in the previous problem along the highway in town within one mile of shore.

### Solutions to Self-Help Exercise Set 16.4

1. Let $u = x^2 + 2$. Then $du = 2x\,dx$. As $x$ goes from $-1$ to $0$, $u$ goes from $u(-1) = 3$ to $u(0) = 2$. Then

$$\int_{-1}^{0} 4xe^{x^2+2}\,dx = 2\int_{-1}^{0} e^{x^2+2}2x\,dx$$

$$= 2\int_{3}^{2} e^{u}\,du$$

$$= 2(e^2 - e^3)$$

2. We are seeking the definite integral $\int_{2}^{4} (10t - 0.30t^2)\,dt$. Then

$$\int_{2}^{4} (10t - 0.30t^2)\,dt = (5t^2 - 0.10t^3)|_{2}^{4}$$

$$= [80 - 6.4] - [20 - 0.80] = 54.4$$

or \$54,400.

3. The average on the interval $[10, 20]$ is

$$\frac{1}{b-a}\int_{a}^{b} f(x)\,dx = \frac{1}{20-10}\int_{10}^{20}\left(e^{-0.10x} + \frac{1}{x^2}\right)dx$$

$$= \frac{1}{10}\left(-10e^{-0.10x} - \frac{1}{x}\right)\Big|_{10}^{20}$$

$$= \frac{1}{10}\left[\left(-10e^{-2} - \frac{1}{20}\right) - \left(-10e^{-1} - \frac{1}{10}\right)\right] \approx 0.238$$

## 16.5     AREA BETWEEN TWO CURVES

▶ *Area Between Two Curves*

▶ *Consumers' and Producers' Surplus*

▶ *Lorentz's Curves*

**APPLICATION**

**Consumers' Surplus**

Let $p = D(x) = 45 - 0.18x^2$ be the demand equation where $p$ is the unit price of a commodity and $x$ is the quantity demanded by the consumers at that price. Suppose the current price is $p_0 = 27$. As can be seen in Figure 16.29, $D(x) > p_0$ when $x < x_0$. This means that there are consumers that are willing to pay a *higher* price for the commodity than $p_0$. These consumers are then actually experiencing a *savings*. What is the total amount of these savings? See Example 5 for the answer.

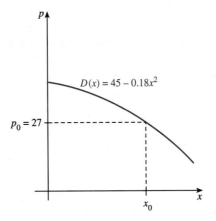

**Figure 16.29**

## Area Between Two Curves

Suppose we have two continuous functions $f(x)$ and $g(x)$ on the interval $[a, b]$ with $f(x) \geq g(x) \geq 0$. See Figure 16.30. We wish to find the area between the two curves.

Clearly the area between the two curves in Figure 16.30 is the area under $f$ less the area under $g$. That is, the area between the two curves is given by

$$\int_a^b f(x) \, dx - \int_a^b g(x) \, dx = \int_a^b [f(x) - g(x)] \, dx.$$

Consider now the more general situation shown in Figure 16.31. We still have $f(x) \geq g(x)$, but both functions can change sign.

It is apparent (see Figure 16.32) that there exists a positive constant large enough so that the graph of $y = g(x) + k$ remains *above* the $x$-axis. We also add

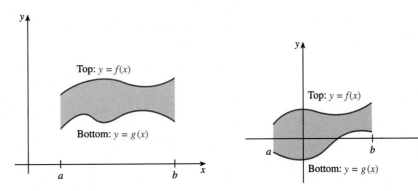

**Figure 16.30**                    **Figure 16.31**

$k$ to $f$ and let $F(x) = f(x) + k$ and $G(x) = g(x) + k$. The graphs of these functions are shown in Figure 16.32. But adding the very same constant to each function does not change the area between the resulting functions. Thus it is apparent that the area between $F$ and $G$ is the same as the area between $f$ and $g$.

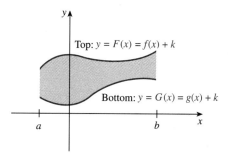

**Figure 16.32**

But now $F(x) \geq G(x) > 0$. Thus the area between $F$ and $G$ is just

$$\int_a^b [F(x) - G(x)]\, dx = \int_a^b [(f(x) + k) - (g(x) + k)]\, dx$$

$$= \int_a^b [f(x) - g(x)]\, dx$$

which is exactly the same formula as found before.

---

**Area Between Two Curves**

Let $y = f(x)$ and $y = g(x)$ be two continuous functions with $f(x) \geq g(x)$ on $[a, b]$. Then the area between the graphs of the two curves on $[a, b]$ is given by the definite integral

$$\int_a^b [f(x) - g(x)]\, dx.$$

---

We can think of this as the area between the *top* curve and the *bottom* curve or as

$$\int_a^b [\text{top} - \text{bottom}]\, dx.$$

**E X A M P L E 1** **Finding the Area Between Two Curves**

Find the area under the curve $y = \sqrt{x}$ and above the curve $y = x^2$ between $x = 0$ and $x = 1$.

**Solution** See Figure 16.33. The graph indicates that the top of the region is given by the function $y = \sqrt{x}$, and the bottom is given by $y = x^2$. According to the previous discussion, we are seeking

$$\int_a^b [\text{top} - \text{bottom}] \, dx = \int_0^1 [x^{1/2} - x^2] \, dx$$

$$= \left( \frac{2}{3} x^{3/2} - \frac{1}{3} x^3 \right) \Big|_0^1$$

$$= \frac{2}{3} - \frac{1}{3} = \frac{1}{3}. \quad \blacksquare$$

Notice that a graph is critical to determine exactly what is the top and what is the bottom curve and also to discern what is the interval of integration.

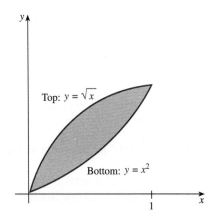

**Figure 16.33**

**E X A M P L E 2** **Finding the Area Between Two Curves**

Find the area between the two curves $y = x + 1$ and $y = x^2 - 1$.

**Solution** It is critical that we draw a graph. This is given in Figure 16.34. We see that the curve $y = x + 1$ is the top and the curve $y = x^2 - 1$ is the bottom. To determine where the two curves intersect, set $x^2 - 1 = y = x + 1$. This gives $0 = x^2 - x - 2 = (x + 1)(x - 2)$. Thus $x = -1$ and $x = 2$ are the $x$-coordinates of the points of intersection. Then

$$\int_a^b [\text{top} - \text{bottom}] \, dx = \int_{-1}^2 \left[ (x + 1) - (x^2 - 1) \right] dx$$

$$= \int_{-1}^2 [2 + x - x^2] \, dx$$

$$= \left( 2x + \frac{1}{2} x^2 - \frac{1}{3} x^3 \right) \Bigg|_{-1}^2$$

$$= \left[ 4 + 2 - \frac{8}{3} \right] - \left[ -2 + \frac{1}{2} + \frac{1}{3} \right]$$

$$= \frac{9}{2} \quad \blacksquare$$

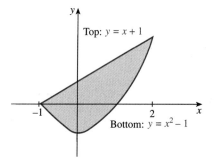

**Figure 16.34**

**E X A M P L E   3**    **Finding the Area of a Curve Below the x-axis**

Find the area of that part of the curve $y = x^2 - 1$ that is below the x-axis and in the fourth quadrant.

**Solution**    The region is shown in Figure 16.35. The curve $y = x^2 - 1$ is the bottom. The top is given by $y = 0$. The region of integration is easily seen to be $[0, 1]$. Then

$$\int_a^b [\text{top} - \text{bottom}] \, dx = \int_0^1 [(0) - (x^2 - 1)] \, dx$$

$$= \int_0^1 [1 - x^2] \, dx$$

$$= \left( x - \frac{1}{3} x^3 \right) \Bigg|_0^1$$

$$= \left( \frac{2}{3} \right) - (0) = \frac{2}{3}. \quad \blacksquare$$

This example illustrates how we can geometrically interpret $\int_a^b f(x) \, dx$

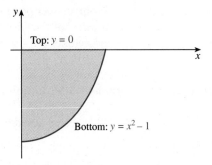

**Figure 16.35**

when $f(x) \leq 0$ on $[a, b]$. In Figure 16.36, we are given the graph of such a function. The area of the region between the $x$-axis and the curve $y = f(x)$ from $a$ to $b$ has as the top $y = 0$ and has as the bottom $y = f(x)$. Thus the area $A$ is just

$$A = \int_a^b [\text{top} - \text{bottom}] \, dx$$

$$= \int_a^b [0 - f(x)] \, dx$$

$$= -\int_a^b f(x) \, dx.$$

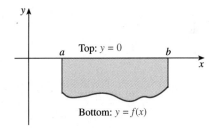

**Figure 16.36**

Therefore we have the following.

---

**Area of a Curve Under the $x$-axis**

If the graph of $y = f(x)$ is below the $x$-axis on $[a, b]$, then the area below the $x$-axis and above the graph of $y = f(x)$ on $[a, b]$ is

$$\text{area} = -\int_a^b f(x) \, dx.$$

---

From another point of view, when $f(x) \leq 0$ on $[a, b]$, $\int_a^b f(x)\, dx$ represents the *negative* of the area between the $x$-axis and the curve $y = f(x)$ from $a$ to $b$. Sometimes the region of integration must be divided into subregions.

**E X A M P L E  4**     **Graphs With Multiple Points of Intersection**

Find the area enclosed between the curves $y = -x$ and $y = 2 - x^2$ on $[-1, 3]$.

**Solution**     A careful graph is presented in Figure 16.37. Notice that the top of the region is *not* given by one single equation. (In fact neither is the bottom.) For example, the top of the region $R_1$ is given by $y = 2 - x^2$ while the top of the region $R_2$ is given by $y = -x$. In such a "bow tie" region, we must divide the region into two subregions in order to use our formulas.

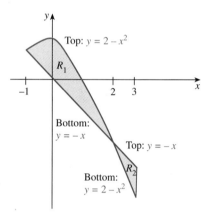

**Figure 16.37**

To do this we must find the point where the two functions cross. We then set $2 - x^2 = y = -x$ or $0 = x^2 - x - 2 = (x + 1)(x - 2)$. Thus the two curves cross when $x = -1$ and $x = 2$. So the two intervals of integration will be $[-1, 2]$ and $[2, 3]$. Let $A$ be the area we are seeking, $A_1$ the area of $R_1$, and $A_2$ the area of $R_2$, then

$$A = A_1 + A_2$$

$$= \int_{-1}^{2} [\text{top of } R_1 - \text{bottom of } R_1]\, dx + \int_{2}^{3} [\text{top of } R_2 - \text{bottom of } R_2]\, dx$$

$$= \int_{-1}^{2} [(2 - x^2) - (-x)]\, dx + \int_{2}^{3} [(-x) - (2 - x^2)]\, dx$$

$$= \int_{-1}^{2} [2 + x - x^2]\, dx + \int_{2}^{3} [x^2 - x - 2]\, dx$$

$$= \left( 2x + \frac{1}{2}x^2 - \frac{1}{3}x^3 \right)\Bigg|_{-1}^{2} + \left( \frac{1}{3}x^3 - \frac{1}{2}x^2 - 2x \right)\Bigg|_{2}^{3}$$

$$= \left( 4 + 2 - \frac{8}{3} \right) - \left( -2 + \frac{1}{2} + \frac{1}{3} \right) + \left( 9 - \frac{9}{2} - 6 \right) - \left( \frac{8}{3} - 2 - 4 \right)$$

$$= \frac{19}{3} \quad \blacksquare$$

### Consumers' and Producers' Surplus

Let $p = D(x)$ be the demand equation where $p$ is the unit price of a commodity and $x$ is the quantity demanded by the consumers at that price. As we know, this function must be decreasing. See Figure 16.38.

Let $p = S(x)$ be the supply equation where $p$ is the unit price of a commodity and $x$ is the quantity made available by the producers at that price. As we know, this function must be increasing. See Figure 16.38. We assume that the demand curve and the supply curve intersect at $x = x_0$. Thus $D(x_0) = p_0 = S(x_0)$. With $(x_0, p_0)$ the point of intersection, $p_0$ is the equilibrium price. At the equilibrium price consumers will purchase the same number of the commodity as the producers will supply.

As can be seen in Figure 16.38, $D(x) > p_0$ when $x < x_0$. This means that there are consumers who are willing to pay a *higher* price for the commodity than $p_0$. These consumers are then actually experiencing a *savings*. The total amount of such "savings" is called the **consumers' surplus**.

We shall now show that the consumer surplus is given by the area between the curve $p = D(x)$ and the straight line $p = p_0$ on the interval $[0, x_0]$, which is

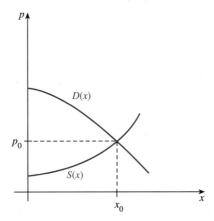

**Figure 16.38**

the definite integral

$$\int_0^{x_0} [D(x) - p_0] \, dx.$$

To see this we divide the interval $[0, x_0]$ into $n$ subintervals of equal length $\Delta x$. Thus $\Delta x$ is the number of consumers in each subinterval. In a typical interval $I_i$ we pick a point $x_i$, say the midpoint. Thus $x_i$ represents a "typical" consumer in the subinterval $I_i$. The number $p_i - p_0 = D(x_i) - p_0$ then represents the "typical" savings experienced by consumers on this interval. The total savings of consumers on this interval is then approximately the "typical" savings times the number in the interval or

$$[D(x_i) - p_0]\Delta x,$$

which is also the area of the rectangle in Figure 16.39.

To find an approximation of the total consumers' surplus, we sum $n$ such terms and obtain

$$\sum_{i=1}^{n} [D(x_i) - p_0]\Delta x.$$

This is also a Reimann sum of $[D(x) - p_0]$ over $[a, b]$. For $n$ larger and larger this approximation will become better and better. Taking the limit as $n \to +\infty$, that is,

$$\lim_{n \to +\infty} \sum_{i=1}^{n} [D(x_i) - p_0]\Delta x,$$

we obtain the exact consumers' surplus. But, by the definition of the definite integral, this limit is

$$\int_0^{x_0} [D(x) - p_0] \, dx.$$

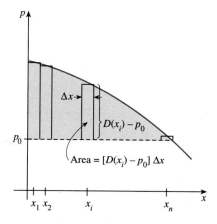

Figure 16.39

We thus have the following.

---

**Consumers' Surplus**

If $p = D(x)$ is the demand equation, $p_0$ is the current price of the commodity, and $x_0$ the current demand, then the consumers' surplus is given by

$$\int_0^{x_0} [D(x) - p_0] \, dx.$$

---

We also see from Figure 16.38 that $p = S(x) < p_0$ if $x < x_0$. This means that there are producers who are willing to sell the commodity for *less* than the going price. For these producers the current price of $p_0$ represents a *savings*. The total of all such savings is called the **producers' surplus**.

In exact analogy with consumers' surplus, the producers' surplus is given by the area between the straight line $p = p_0$ and the curve $p = S(x)$ on the interval $[0, x_0]$, which is the definite integral $\int_0^{x_0} [p_0 - S(x)] \, dx$. We then have the following.

---

**Producers' Surplus**

If $p = S(x)$ is the supply equation, $p_0$ is the current price of the commodity, and $x_0$ the current demand, then the producers' surplus is given by

$$\int_0^{x_0} [p_0 - S(x)] \, dx.$$

---

**E X A M P L E   5**     **Finding the Consumer's Surplus**

If the demand equation is given by $p = D(x) = 45 - 0.18x^2$ and the supply equation is given by $p = S(x) = 0.12x^2 + 1.5x$, find the consumers' surplus.

**Solution**   We need to first find the equilibrium quantity. To do this set $S(x) = D(x)$ and obtain

$$0.12x^2 + 1.5x = 45 - 0.18x^2$$

$$0 = .30x^2 + 1.5x - 45$$

$$= .30 \, [x^2 + 5x - 150]$$

$$= .30(x + 15)(x - 10).$$

Since the equilibrium quantity cannot be negative, we ignore the solution $x_0 = -15$. Thus the equilibrium quantity is $x_0 = 10$. For this value we have $p_0 = D(x_0) = D(10) = 27$. See Figure 16.40.

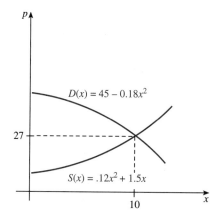

**Figure 16.40**

The consumers' surplus is then given by

$$\int_0^{x_0} [D(x) - p_0]\, dx = \int_0^{10} [45 - 0.18x^2 - 27]\, dx$$

$$= [18x - 0.06x^3]\Big|_0^{10}$$

$$= 120. \quad \blacksquare$$

## Lorentz's Curves

For a given population let $f(x)$ be the proportion of total income that is received by the lowest paid $100x\%$ of income recipients. Thus $f(.3) = .2$ means that the lowest paid 30% of income recipients receive 20% of the total income. Or if $f(.6) = .25$, this means that the lowest paid 60% receive 25% of the total income. We must have $0 \le x \le 1$ and $0 \le f(x) \le 1$.

We only include income recipients, thus $f(0) = 0$. Since all income is received by 100% of the recipients, $f(1) = 1$. Also $f(x) \le x$ since the lowest $100x\%$ of the income recipients cannot receive more than $100x\%$ of the total income. The graph of such an income distribution is called a **Lorentz curve**. See Figure 16.41 for a typical such curve.

Suppose a Lorentz curve is given by $f(x) = x^2$. Since $f(.5) = .25$, the lowest paid 50% receive 25% of the total income. Since $f(.1) = .01$, the lowest paid 10% receive 1% of the total income.

A perfect equality of income distribution is represented by the curve $f(x) = x$. This is true since in this case the lowest paid 1% would receive 1% of total income, the lowest paid 10% would receive 10% of total income, etc. Any deviation from this perfect equality represents an inequality of income distribution. See Figure 16.42.

A convenient and intuitive measure of this deviation is the *area* between the curve $y = x$ and $y = f(x)$. Economists then define the **coefficient of inequality**

**Figure 16.41**

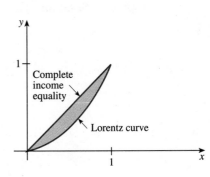

**Figure 16.42**

to be the ratio of this area to the area under the curve $y = x$ that represents perfect equality of income distribution. Since of course the area under $y = x$ on $[0, 1]$ is $\frac{1}{2}$, the ratio will be two times the area between $y = x$ and $y = f(x)$. We have the following.

---

**Coefficient of Inequality**

The coefficient of inequality $L$ of a Lorentz curve $y = f(x)$ is given by

$$L = 2 \int_0^1 [x - f(x)]\, dx.$$

---

**E X A M P L E   6**   **Calculating the Coefficient of Inequality**

Find the coefficient of inequality of the Lorentz curve $y = x^3$. Refer to Figure 16.43.

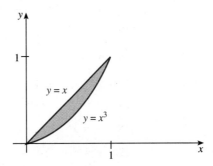

**Figure 16.43**

**Solution**

$$L = 2 \int_0^1 [x - f(x)]\, dx$$

$$= 2 \int_0^1 [x - x^3]\, dx$$

$$= 2 \left[ \frac{1}{2} x^2 - \frac{1}{4} x^4 \right] \Big|_0^1$$

$$= \frac{1}{2} \quad \blacksquare$$

## SELF-HELP EXERCISE SET 16.5

1. Find the area between the two curves $y = x + 2$ and $y = x^2$.

2. Find the area enclosed by the graphs of the two curves $y = x$ and $y = f(x) = \sqrt[3]{x}$.

3. If the demand equation is given by $p = D(x) = 45 - 0.18x^2$ and the supply equation is given by $p = S(x) = 0.12x^2 + 1.5x$, find the producers' surplus. (See Example 5.)

## EXERCISE SET 16.5

In Exercises 1 through 36 find the area enclosed by the given curves.

1. $y = x^3, y = 0, x = 1, x = 2$

2. $y = x^2, y = 0, x = -1, x = 2$

3. $y = e^{0.1x}, y = 1, x = 0, x = 10$

4. $y = e^{-x}, y = 3, x = -1, x = 0$

5. $y = x^2, y = x, x = 0, x = 1$

6. $y = \sqrt{x}, y = x, x = 0, x = 1$

7. $y = x, y = 1/x, x = e, x = e^2$

8. $y = 4, y = x^2$

9. $y = x^2 - 2x + 1, y = x + 1$

10. $y = x^2, y = 8 - x^2$

11. $y = x^3, y = -x^3, x = 0, x = 1$

12. $y = x, y = -1/x, x = 1/e, x = e$

13. $y = -x^2, y = -3x - 1, x = 0, x = 2$

14. $y = x^2 - 2x, y = x - 4, x = 0, x = 1$

15. $y = x^2 + 1, y = x, x = -1, x = 1$

16. $y = x + 2, y = \sqrt[3]{x}, x = -1, x = 1$

17. $y = x^2 + 4x + 4, y = 4 - x$

18. $y = x^2 - 2x + 1, y = 1 + 2x - x^2$

19. $y = x(x - 4), y = x$

20. $y = x(x - 2), y = x$

21. $y = e^{|x|}, y = 0, x = -1, x = 2$

22. $y = e^{-|x|}, y = 0, x = -1, x = 2$

**23.**  $y = x^2, y = 4, x = 0, x = 4$      **24.**  $y = x^3, y = 2, x = 0, x = 2$

**25.**  $y = x^3, y = -x, x = -1, x = 1$      **26.**  $y = e^x, y = 2, x = 0, x = 1$

**27.**  $y = -\sqrt[3]{x}, y = -x, x = -1, x = 1$

**28.**  $y = \sqrt[3]{x}, y = -x, x = -1, x = 1$

**29.**  $y = e^x, y = e^{-x}, x = -\ln 2, x = \ln 2$

**30.**  $y = e^{2x}, y = e^{-2x}, x = -1, x = 2$

**31.**  $y = x^3 - 3x, y = x$              **32.**  $y = x^3 - 3x, y = 2x^2$

**33.**  $y = |x|, y = 3, x = -1, x = 6$ (triple region)

**34.**  $y = |x|, y = \dfrac{1}{2}x + 2, x = -2, x = 4$ (triple region)

**35.**  $y = \sqrt[3]{x}, y = x, x = -8, x = 1$ (triple region)

**36.**  $y = -\sqrt[3]{x}, y = -x, x = -1, x = 8$ (triple region)

## Applications

In Exercises 37 through 40 find the consumers' surplus, using the given demand equations and the market price $p_0$.

**37.**  $D(x) = 20 - x^2, p_0 = 4$      **38.**  $D(x) = 30 - x^2, p_0 = 5$

**39.**  $D(x) = e^{-x}, p_0 = 0.1$      **40.**  $D(x) = 20 - x, p_0 = 10$

In Exercises 41 through 44 find the producers' surplus, using the given supply equation and the market price $p_0$.

**41.**  $S(x) = 100x, p_0 = 10$      **42.**  $S(x) = x^3, p_0 = 8$

**43.**  $S(x) = 9x^2, p_0 = 1$      **44.**  $S(x) = e^x, p_0 = e^2$

In Exercises 45 through 48 you are given the demand and supply equation. Find the equilibrium point and then calculate both the consumers' surplus and the producers' surplus.

**45.**  $D(x) = 12 - x$ and $S(x) = 2x$      **46.**  $D(x) = 50 - x^2$ and $S(x) = 5x$

**47.**  $D(x) = 20 - x^2$ and $S(x) = x$      **48.**  $D(x) = 64 - x^2$ and $S(x) = 3x^2$

In Exercises 49 through 56 show that the curves are Lorentz curves. (You will need to find $f''(x)$.) Then find the coefficient of inequality.

**49.**  $f(x) = x^2$      **50.**  $f(x) = x^5$

**51.**  $f(x) = \dfrac{1}{4}x^2 + \dfrac{3}{4}x$      **52.**  $f(x) = \dfrac{3}{4}x^2 + \dfrac{1}{4}x$

**53.**  $f(x) = 0.3x^2 + 0.7x$      **54.**  $f(x) = 0.7x^2 + 0.3x$

**55.**  $f(x) = 0.01x^2 + 0.99x$      **56.**  $f(x) = 0.99x^2 + 0.01x$

**57. Costs.** A firm has found that its costs in billions of dollars have been increasing (due primarily to inflation) at a rate given by $C_1'(t) = 1.2t$, where $t$ is the time measured in years since the item was begun being produced. At the beginning of the second year, an innovation in the production process has resulted in costs increasing at a rate given by $C_2'(t) = 0.9\sqrt{t}$. Find the total costs during the first four years. Assume $C_1(0) = 0$.

**58. Revenue.** A firm has found that its revenue in billions of dollars has been increasing at a rate given by $R'(t) = 3t^2$, where $t$ is the time measured in years since the item was introduced. At the beginning of the third year, new competition has resulted in revenue increasing only at a rate given by $R_1'(t) = 12t$. Find the total revenue during the first four years.

**59. Sales.** From the time a firm introduced a new product, sales in thousands of this product have been increasing at a rate given by $S'(t) = 2 + t$, where $t$ is the time since the introduction of the new product. At the beginning of the third year, an advertising campaign was introduced and sales then increased at a rate given by $S_2'(t) = t^2$. Find the increase in sales due to the advertising campaign during the first two years of this campaign over what the sales would have been without the campaign.

**60. Production.** Suppose copper is being extracted from a certain mine at a rate given by

$$\frac{d}{dt}P(t) = 100e^{-0.2t},$$

where $P(t)$ is measured in tons of copper and $t$ in years from opening the mine. At the beginning of the sixth year, a new mining innovation is introduced to boost production to a rate given by $Q'(t) = 500/t$. Find the increase in production of copper due to this innovation during the first five years of its use over what copper production would have been without its use.

## Graphing Calculator Exercises

In Exercises 61 through 64 graph the given pair of curves in the same viewing rectangle and use the zoom feature to find the points of intersection to two decimal places. Then use the program Riemann found in Appendix C to estimate the area enclosed by the given pairs of curves.

**61.** $y = x^5 + x^2 - 3x, \ y = x$     **62.** $y = x^5 + x^4 - 3x, \ y = x$

**63.** $y = x^5 + x^4 - 3x, \ y = 0.50x^3$

**64.** $y = x^5 + x^4 - 3x, \ y = 3x - x^2 - x^5$

## Solutions to Self-Help Exercise Set 16.5

**1.** A graph is given in the figure. In order to determine the interval of integration we need to determine the values of $x$ for which the two curves intersect. This will happen when $x + 2 = y = x^2$ or $0 = x^2 - x - 2 = (x + 1)(x - 2)$. Thus $x = -1$ and $x = 2$. Then

$$\int_a^b [\text{top} - \text{bottom}] \, dx = \int_{-1}^2 [(x + 2) - (x^2)] \, dx$$

$$= \left( \frac{1}{2} x^2 + 2x - \frac{1}{3} x^3 \right) \Bigg|_{-1}^2$$

$$= \left[ 2 + 4 - \frac{8}{3} \right] - \left[ \frac{1}{2} - 2 + \frac{1}{3} \right] = \frac{9}{2}$$

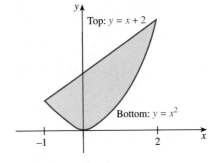

2. The region between the graphs of the 2 curves forms a ''bow tie'' region, so we must divide the region into two subregions in order to use our formulas. If we let $A$ be the area we are seeking, $A_1$ the area of $R_1$, and $A_2$ the area of $R_2$, then

$$A = A_1 + A_2$$

$$= \int_{-1}^0 [\text{top of } R_1 - \text{bottom of } R_1] \, dx + \int_0^1 [\text{top of } R_2 - \text{bottom of } R_2] \, dx$$

$$= \int_{-1}^0 [(x) - (x^{1/3})] \, dx + \int_0^1 [(x^{1/3}) - (x)] \, dx$$

$$= \left( \frac{1}{2} x^2 - \frac{3}{4} x^{4/3} \right) \Bigg|_{-1}^0 + \left( \frac{3}{4} x^{4/3} - \frac{1}{2} x^2 \right) \Bigg|_0^1$$

$$= -\left( \frac{1}{2} - \frac{3}{4} \right) + \left( \frac{3}{4} - \frac{1}{2} \right) = \frac{1}{2}$$

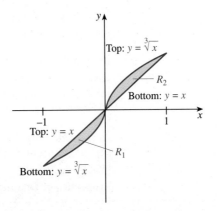

3. The equilibrium demand was already found to be $x_0 = 10$ with an equilibrium price of $p_0 = 27$ in Example 5. The producers' surplus is then given by

$$\int_0^{x_0} [p_0 - S(x)] \, dx = \int_0^{10} [27 - 0.12x^2 - 1.5x] \, dx$$
$$= [27x - 0.04x^3 - 0.75x^2]\big|_0^{10}$$
$$= 155.$$

## SUMMARY OUTLINE OF CHAPTER 16

- If $F'(x) = f(x)$, then $F(x)$ is called the **antiderivative** of $f(x)$.
- If two functions have the same derivative on $(a, b)$, then they differ by a constant on $(a, b)$.
- If $F'(x) = f(x)$, then the **indefinite integral** is $\int f(x) \, dx = F(x) + C$, where $C$ is an arbitrary constant called the **constant of integration**.
- **Rules of Integration**

$$\int x^n \, dx = \frac{1}{n+1} x^{n+1} + C, \, n \neq -1 \qquad \int k f(x) \, dx = k \int f(x) \, dx$$

$$\int [f(x) \pm g(x)] \, dx = \int f(x) \, dx \pm \int g(x) \, dx \qquad \int e^{kx} \, dx = \frac{1}{k} e^{kx} + C$$

$$\int \frac{1}{x} \, dx = \ln |x| + C$$

- **Method of Substitution**
  In the integral $\int [f(x)]^n g(x) \, dx$, substitute $u = f(x)$ and $du = f'(x) \, dx$
  In the integral $\int e^{f(x)} g(x) \, dx$, substitute $u = f(x)$ and $du = f'(x) \, dx$

- Divide the interval $[a, b]$ into $n$ subintervals of equal length $\Delta x$ and let $x_i$ by any point in the $i$th subinterval. Then $S_n = \Sigma_{i=1}^n f(x_i) \Delta x$ is called a **Riemann sum**.

- **Common Limit of Riemann Sums.** Suppose that $f(x)$ is continuous on the interval $[a, b]$. If this interval is divided into $n$ subintervals of equal length $\Delta x$ and $x_i$ is any point in the $i$th subinterval, then $\lim_{n \to +\infty} \Sigma_{i=1}^n f(x_i) \Delta x$ exists and is the same limit independent of how the points $x_i$ are chosen in the $i$th interval.

- **Definite Integral.** Assume that $f(x)$ is continuous on $[a, b]$. The **definite integral** of $f$ over $[a, b]$, written $\int_a^b f(x) \, dx$, is defined to be the common limit of Riemann sums given above, that is, $\int_a^b f(x) \, dx = \lim_{n \to +\infty} \Sigma_{i=1}^n f(x_i) \Delta x$.

- **Area Under the Curve.** If $f(x)$ is nonnegative and continuous on $[a, b]$, we say the **area under the curve** $y = f(x)$ from $a$ to $b$ is given by the indefinite integral $\int_a^b f(x) \, dx$.

- **Properties of Definite Integrals**

$$\int_a^a f(x)\, dx = 0$$

$$\int_a^b kf(x)\, dx = k \int_a^b f(x)\, dx,\ k \text{ constant}$$

$$\int_a^b [f(x) \pm g(x)]\, dx = \int_a^b f(x)\, dx \pm \int_a^b g(x)\, dx$$

$$\int_a^b f(x)\, dx = \int_a^c f(x)\, dx + \int_c^b f(x)\, dx,\ a < c < b$$

$$\int_b^a f(x)\, dx = -\int_a^b f(x)\, dx,\ a < b$$

- **Fundamental Theorem of Calculus.** Suppose $f(x)$ is continuous on $[a, b]$. If $F(x)$ is an antiderivative of $f(x)$, then $\int_a^b f(x)\, dx = F(b) - F(a)$.

- **Theorem.** If $f(x)$ is continuous on the interval $[a, b]$, and if $F(x) = \int_a^x f(t)\, dt$, then for $x \in (a, b)$, $F'(x) = f(x)$.

- If $f(x)$ is continuous on $[a, b]$ we define the **average value** of $f(x)$ on $[a, b]$ to be $\dfrac{1}{b - a} \int_a^b f(x)\, dx$.

- **Population Given the Density.** If $\delta(x)$ is a continuous function that gives the density in number per unit length, then the total population in the interval $[a, b]$ is $P = \int_a^b \delta(x)\, dx$.

- Let $y = f(x)$ and $y = g(x)$ be two continuous functions with $f(x) \geq g(x)$ on $[a, b]$. Then the **area between the graphs of the two curves** on $[a, b]$ is given by the definite integral $\int_a^b [f(x) - g(x)]\, dx$.

- If $p = D(x)$ is the demand equation, $p_0$ the current price of the commodity, and $x_0$ the current demand, then the **consumers' surplus** is given by $\int_0^{x_0} [D(x) - p_0]\, dx$.

- If $p = S(x)$ is the supply equation, $p_0$ the current price of the commodity, and $x_0$ the current demand, then the **producers' surplus** is given by $\int_0^{x_0} [p_0 - S(x)]\, dx$.

- A **Lorentz curve** is the graph of any function $y = f(x)$ defined on $[0, 1]$ for which $f(x) \leq x$, $f(0) = 0$, and $f(1) = 1$.

- The **coefficient of inequality** $L$ of a Lorentz curve $y = f(x)$ is given by $L = 2 \int_0^1 [x - f(x)]\, dx$.

## Chapter 16 Review Exercises

In Exercises 1 through 9 find the antiderivatives.

**1.** $\int x^9\, dx$

**2.** $\int x^{0.50}\, dx$

**3.** $\int \dfrac{2}{y^2}\, dy$

**4.** $\int \sqrt{3}y^2 \, dy$

**5.** $\int (6x^2 + 8x) \, dx$

**6.** $\int \left( 2x - \dfrac{3}{x} \right) dx$

**7.** $\int \dfrac{t^2 + 1}{\sqrt{t}} \, dt$

**8.** $\int e^{3x} \, dx$

**9.** $\int \dfrac{e^{2x} + e^{-2x}}{e^{3x}} \, dx$

In Exercises 10 through 18 use the method of substitution to evaluate the integrals.

**10.** $\int 4(2x + 1)^{20} \, dx$

**11.** $\int 60x^2(2x^3 + 1)^9 \, dx$

**12.** $\int \dfrac{x^2}{\sqrt{x^3 + 1}} \, dx$

**13.** $\int \dfrac{x + 1}{x^2 + 2x} \, dx$

**14.** $\int xe^{x^2 + 5} \, dx$

**15.** $\int \dfrac{10x}{(x^2 + 1)^8} \, dx$

**16.** $\int \dfrac{1}{\sqrt{xe^{\sqrt{x}}}} \, dx$

**17.** $\int \dfrac{e^x}{e^x + 1} \, dx$

**18.** $\int \dfrac{(\ln x)^3}{x} \, dx$

**19.** Find the Riemann sum $S_4$ where $f(x) = 2x^2 + 1$ on the interval $[0, 4]$, taking $x_i$ to be the left endpoints of the subintervals.

**20.** Find the Riemann sum $S_4$ where $f(x) = 2x^2 + 1$ on the interval $[0, 4]$, taking $x_i$ to be the right endpoints of the subintervals.

**21.** For any $n$ find the Riemann sum $S_n$ where $f(x) = 2x^2 + 1$ on the interval $[0, 1]$, taking $x_i$ to be the left endpoints of the subintervals. Find $\lim\limits_{n \to +\infty} S_n$ and relate this to some definite integral.

**22.** For any $n$ find the Riemann sum $S_n$ where $f(x) = 2x^2 + 1$ on the interval $[0, 1]$, taking $x_i$ to be the right endpoints of the subintervals. Find $\lim\limits_{n \to +\infty} S_n$ and relate this to some definite integral.

**23.** Find $\int_0^3 \sqrt{9 - x^2} \, dx$ by finding the area of an appropriate geometric figure.

**24.** Since $\displaystyle\int_1^2 \dfrac{1}{x} \, dx = \ln 2$ use the Riemann sum $S_4$ (with $x_i$ as the left endpoint of each subinterval) as an approximation to $\ln 2$. Using a graph of $y = \dfrac{1}{x}$, show why your answer must be greater than $\ln 2$.

**25.** Use the definite integral in Exercise 24 to approximate $\ln 2$ using the Riemann sum $S_4$ with $x_i$ as the right endpoint of each subinterval. Using a graph of $y = \dfrac{1}{x}$, show why your answer must be less than $\ln 2$.

In Exercises 26 through 33 evaluate the definite integrals.

**26.** $\int_1^3 6x^2 \, dx$

**27.** $\int_{-3}^{-1} 2x \, dx$

**28.** $\int_{-1}^1 (9x^2 - 2x + 1) \, dx$

**29.** $\displaystyle\int_1^2 \left( e^{2x} + \dfrac{1}{x} \right) dx$

**30.** $\displaystyle\int_0^1 4x(x^2 + 1)^9 \, dx$

**31.** $\displaystyle\int_0^2 x^2 e^{x^3 + 1} \, dx$

**32.** $\displaystyle\int_0^2 \frac{1}{2x+1}\,dx$                **33.** $\displaystyle\int_0^3 \frac{1}{\sqrt{2x+3}}\,dx$

**34.** Find the derivative of $\int_0^x e^{t^4}\,dt$.

**35.** Find the average value of $f(x) = x^3$ on $[-2, 2]$.

**36.** Find the average value of $f(x) = \dfrac{1}{x}$ on $[1, 3]$.

In Exercises 37 through 48 find the area enclosed by the given curves.

**37.**  $y = x^2$, $y = 0$, $x = 0$, $x = 3$          **38.**  $y = x^2 + 1$, $y = x$, $x = 0$, $x = 2$

**39.**  $y = 2\sqrt{x}$, $y = x$, $x = 0$, $x = 4$          **40.**  $y = \sqrt{x}$, $y = -x^2$, $x = 0$, $x = 1$

**41.**  $y = x^2 + 1$, $y = 1 - 2x$, $x = 0$, $x = 2$

**42.**  $y = x^2$, $y = 18 - x^2$

**43.**  $y = x^2 - 4x + 3$, $y = 2x - 2$

**44.**  $y = x^2 - 5x + 4$, $y = -x^2 + 5x - 4$.

**45.**  $y = \sqrt[5]{x}$, $y = x$

**46.**  $y = x^2 - 4x + 3$, $y = 2x - 2$, $x = 0$, $x = 5$

**47.**  $y = \dfrac{2}{x}$, $y = 1$, $x = 1$, $x = 3$

**48.**  $y = x^2 - 5x + 4$, $y = -x^2 + 5x - 4$, $x = 0$, $x = 4$

**49.** **Revenue.**  If marginal revenue is given by $2x - 10$, find the revenue function.

**50.** **Cost.**  If marginal cost is given by $3x^2 + 1$ and fixed costs are 1000, find the cost function.

**51.** **Depreciation.**  A piece of machinery purchased for \$100,000 is decreasing at a rate given by $V'(t) = -500e^{-0.20t}$, where $t$ is measured in years from date of purchase. How much will the machine be worth in 5 years?

**52.** **Sales.**  After an advertising campaign ends, the rate of change of sales (in thousands) of a product is given by $S'(t) = 4te^{-t^2}$, where $t$ is in years. How many will be sold in the next 2 years?

**53.** **Sales.**  In the following table the rate $R'(t)$ of sales for a firm in millions of dollars per month is given for discrete equally spaced points $0 = t_0, t_1, \ldots,$ $t_5 = 5$. Find the Riemann sum $S_5$ for the case when (a) $t_i$ is always taken as the left endpoint of each subinterval, and (b) $t_i$ is always taken as the right endpoint of each subinterval. The answers will then be approximations to $\int_0^5 R'(t)\,dt = R(5) - R(0)$, which is the total sales for the 5-month period.

| $t_i$      | 0 | 1 | 2 | 3 | 4  | 5 |
|------------|---|---|---|---|----|---|
| $R'(t_i)$  | 2 | 3 | 5 | 6 | 10 | 8 |

**54.** **Population.**  The population of a certain city is increasing at the rate given by

$$P'(t) = 2000e^{0.04t},$$

where $t$ is time in years measured from the beginning of 1992. Find the change in the population from the beginning of 1994 to the beginning of 1997.

55. **Medicine.**   A tumor is increasing in volume at a rate given by

$$V'(t) = 4e^{0.1t},$$

where $t$ is measured in months from the time of diagnosis. How much will the tumor increase in volume during the third and fourth months after diagnosis?

56. **Average Population.**   The population of a certain city is given $10,000e^{.1t}$ where $t$ is given in years. Find the average population during the next 10 years.

In Exercises 57 through 59 suppose the demand equation is given by $p = D(x) = 10 - x^2$ and the supply equation is given by $p = S(x) = 3x$.

57. **Supply and Demand.**   Find the equilibrium point.

58. **Consumers' Surplus.**   Find the consumers' surplus.

59. **Producers' Surplus.**   Find the producers' surplus.

60. **Lorentz's Curve.**   Show that $f(x) = x^4$ is a Lorentz's curve and find the coefficient of inequality.

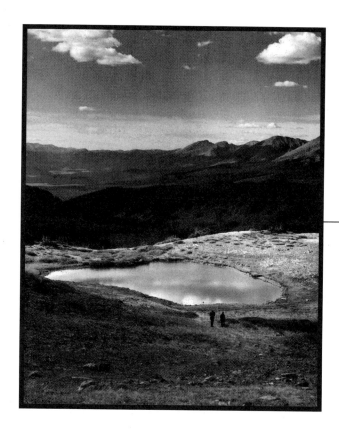

Numerical integration is useful in determining the approximate area of the lake. (Grant Heilman Photography, Inc.)

# 17

# Additional Topics in Integration

17.1  Integration by Parts

17.2  Integration Using Tables

17.3  Numerical Integration

17.4  Improper Integrals

17.5  Continuous Money Flow

**T**his chapter contains a collection of important additional topics in integration. In the first section we will learn a very important technique of integration and in the second section will see how to use a table of already calculated integrals, while in the third section we will see how to evaluate definite integrals using numerical techniques. In the fourth section we extend the notion of integration to unbounded intervals. The fifth section describes an important application of the integral to the theory of finance.

## 17.1  INTEGRATION BY PARTS

► *Integration by Parts*
► *Application*

**APPLICATION**

**Sales of a New Product**

The rate of sales of a new product will tend to increase rapidly initially and then decrease. Suppose the rate of sales of a new product is given by $S(t) = 1000t^2e^{-t}$ items per week, where $t$ is the number of weeks from the introduction of the product. How many items are sold in the first 3 weeks? The answer can be found in Example 4.

## Integration by Parts

If $f(x)$ and $g(x)$ are differentiable functions, then the product rule of differentiation says that

$$\frac{d}{dx} f(x)g(x) = f(x) \frac{d}{dx} g(x) + g(x) \frac{d}{dx} f(x).$$

Rearranging the terms yields

$$f(x)g'(x) = (f(x)g(x))' - g(x)f'(x).$$

Taking the indefinite integral gives

$$\int f(x)g'(x) \, dx = f(x)g(x) - \int g(x)f'(x) \, dx.$$

In practice this is written an alternate way. If we let $u = f(x)$ and $v = g(x)$, then $du = f'(x) \, dx$ and $dv = g'(x) \, dx$. The above formula then becomes $\int u \, dv = uv - \int v \, du$. This technique is called **integration by parts**.

---

**Integration by Parts**

$$\int u \, dv = uv - \int v \, du.$$

---

At first glance it is difficult to believe that anything has been accomplished. However, the goal is to define $u$ and $v$ so that $\int v \, du$ is easier to find than the original integral, $\int u \, dv$. The following examples will illustrate this. Keep in mind that your first choice of $u$ and $v$ may not work.

**EXAMPLE 1**   **Using Integration by Parts**

Find $\int xe^x \, dx$.

**Solution**

If we set

$$u = x \qquad dv = e^x \, dx,$$

then

$$du = dx \qquad v = e^x,$$

and

$$\int xe^x \, dx = \int u \, dv$$

$$= uv - \int v \, du$$

$$= xe^x - \int e^x \, dx$$

$$= xe^x - e^x + C = e^x (x - 1) + C \quad \blacksquare$$

Notice that $\int v \, du = \int e^x \, dx$ was simpler than the original integral. Had we made the substitution $u = e^x$ and $dv = x \, dx$, then we would have $\int v \, du = \int \frac{1}{2}x^2 e^x \, dx$, which is *more* complicated than the original integral.

EXAMPLE 2

## Using Integration by Parts

Find $\int x\sqrt{1 + x} \, dx$.

**Solution**    We might first think of letting $u = x$ since $du$ will be simple. Let us try this. Let

$$u = x \qquad dv = \sqrt{1 + x} \, dx.$$

Of course $du = dx$, and $v = \int \sqrt{1 + x} \, dx$. This last integral is similar to others we have done before and can be readily done by substitution. If we let $w = 1 + x$, then $dw = dx$ and

$$v = \int \sqrt{1 + x} \, dx = \int w^{1/2} \, dw = \frac{2}{3} w^{3/2} = \frac{2}{3} (1 + x)^{3/2}.$$

Then

$$\int x\sqrt{1 + x} \, dx = \int u \, dv$$

$$= uv - \int v \, du$$

$$= x \frac{2}{3} (1 + x)^{3/2} - \int \frac{2}{3} (1 + x)^{3/2} \, dx$$

Now we can evaluate the last integral in just the way we found $v$. Thus

$$\int x\sqrt{1 + x} \, dx = x \frac{2}{3} (1 + x)^{3/2} - \int \frac{2}{3} (1 + x)^{3/2} \, dx$$

$$= x \frac{2}{3} (1 + x)^{3/2} - \frac{2}{3} \cdot \frac{2}{5} (1 + x)^{5/2} + C$$

$$= \frac{2}{3} x (1 + x)^{3/2} - \frac{4}{15} (1 + x)^{5/2} + C \quad \blacksquare$$

**E X A M P L E   3**    **Using Integration by Parts**

Find (a) $\int x^3\sqrt{1 + x^2}\, dx$, (b) $\int_0^{\sqrt{3}} x^3\sqrt{1 + x^2}\, dx$.

**Solutions**    (a)    We might be tempted to set $u = x^3$ and $dv = \sqrt{1 + x^2}$, but then we could *not* find $v$ since we cannot find $\int \sqrt{1 + x^2}\, dx$. But we can integrate $\int x\sqrt{1 + x^2}\, dx$. Thus we set

$$u = x^2 \qquad dv = x\sqrt{1 + x^2}.$$

Then    $$du = 2x\, dx \qquad v = \frac{1}{3}(1 + x^2)^{3/2}.$$

(The reader should verify the correctness of $v$.) Thus

$$\int x^3\sqrt{1 + x^2}\, dx = \int u\, dv$$

$$= uv - \int v\, du$$

$$= x^2\, \frac{1}{3}(1 + x^2)^{3/2} - \int \frac{1}{3}(1 + x^2)^{3/2}\, 2x\, dx$$

$$= \frac{1}{3}x^2(1 + x^2)^{3/2} - \frac{2}{15}(1 + x^2)^{5/2} + C$$

(b)    To evaluate this definite integral we use the antiderivative found in (a).

$$\int_0^{\sqrt{3}} x^3\sqrt{1 + x^2}\, dx = \left(\frac{1}{3}x^2(1 + x^2)^{3/2} - \frac{2}{15}(1 + x^2)^{5/2}\right)\Bigg|_0^{\sqrt{3}}$$

$$= \left[4^{3/2} - \frac{2}{15}4^{5/2}\right] - \left[0 - \frac{2}{15}\right]$$

$$= \frac{58}{15} \quad \blacksquare$$

Sometimes one must use integration by parts *more than once*. Sometimes $\int v\, du$ is definitely simpler but requires another integration by parts. The following is such an example.

## Application

**E X A M P L E   4**    **Repeated Integration by Parts**

Suppose the rate of sales of a new product is given by $S'(x) = 1000x^2e^{-x}$ items per week, where $x$ is the number of weeks from the introduction of the product. Refer to Figure 17.1. How many items are sold in the first 3 weeks?

**Solution**    We are seeking $1000 \int_0^3 x^2e^{-x}\, dx$. We first find $\int x^2e^{-x}\, dx$.

We set $$u = x^2 \qquad dv = e^{-x}\,dx.$$

Then $$du = 2x\,dx \qquad v = -e^{-x}.$$

Then

$$\int x^2 e^{-x}\,dx = \int u\,dv$$

$$= uv - \int v\,du$$

$$= x^2(-e^{-x}) - \int -e^{-x}2x\,dx$$

$$= -x^2 e^{-x} + 2\int xe^{-x}\,dx$$

The integral we are left with, $\int xe^{-x}\,dx$, is definitely simpler, but to evaluate it requires another integration by parts. To do this we let

$$u = x \qquad dv = e^{-x}dx.$$

Then $$du = dx \qquad v = -e^{-x},$$

and

$$\int xe^{-x}\,dx = \int u\,dv$$

$$= uv - \int v\,du$$

$$= x(-e^{-x}) - \int -e^{-x}\,dx$$

$$= -xe^{-x} + \int e^{-x}\,dx$$

$$= -xe^{-x} - e^{-x} + C*$$

Substituting this back in the above equation yields

$$\int x^2 e^{-x}\,dx = -x^2 e^{-x} + 2\int xe^{-x}\,dx$$

$$= -x^2 e^{-x} + 2[-xe^{-x} - e^{-x} + C*]$$

$$= -e^{-x}[x^2 + 2x + 2] + C,$$

where we have set $C = 2C*$ for convenience. Then

$$1000\int_0^3 x^2 e^{-x}\,dx = -1000e^{-x}[x^2 + 2x + 2]\big|_0^3$$

$$= -1000e^{-3}[17] + 1000e^0[2]$$

$$= 1000(2 - 17e^{-3})$$

$$\approx 1153.6. \quad \blacksquare$$

$1000x^2 e^{-x}$

**Figure 17.1**

## SELF-HELP EXERCISE SET 17.1

1.  Find (a) $\int \ln x \, dx$, and (b) $\int_1^e \ln x \, dx$.

2.  Find $\int \dfrac{xe^x}{(x + 1)^2} \, dx$. *Hint:* You may need to try several choices of $u$ and $dv$ before one works.

## EXERCISE SET 17.1

In Exercises 1 through 20 evaluate.

**1.** $\int 2xe^{2x} \, dx$

**2.** $\int xe^{-2x} \, dx$

**3.** $\int_1^{e^2} x \ln x \, dx$

**4.** $\int_e^{e^2} x^2 \ln x \, dx$

**5.** $\int_1^e \ln x^2 \, dx$

**6.** $\int_1^e \ln x^3 \, dx$

**7.** $\int \dfrac{x}{\sqrt{1 + x}} \, dx$

**8.** $\int x(1 + x)^{3/2} \, dx$

**9.** $\int x(1 + x)^{10} \, dx$

**10.** $\int x(1 + x)^5 \, dx$

**11.** $\int x(x + 2)^{-2} \, dx$

**12.** $\int x(x + 2)^{-3} \, dx$

**13.** $\int x^5 \sqrt{1 + x^3} \, dx$

**14.** $\int (x + 1)^{2/3} x \, dx$

**15.** $\int x^3 (1 + x^2)^{10} \, dx$

**16.** $\int \dfrac{x^3}{\sqrt{1 + x^2}} \, dx$

**17.** $\int x^2 e^{2x} \, dx$

**18.** $\int x^2 e^{-2x} \, dx$

**19.** $\int_0^2 x^3 e^{x^2} \, dx$

**20.** $\int_1^2 (\ln x)^2 \, dx$

## Applications

**21. Sales.** The rate of sales of a new product will tend to increase rapidly initially and then fall off. Suppose the rate of sales of a new product is given by $S'(t) = 1000te^{-3t}$ items per week, where $t$ is in weeks from the introduction of the product. How many items are sold in the first four weeks? Assume $S(0) = 0$.

**22. Sales.** Suppose the rate of sales of a new product is given by $S'(t) = 1000t^2 e^{-4t}$ items per month, where $t$ is in months from the introduction of the product. How many items are sold in the first year? Assume $S(0) = 0$.

**23. Profits.** The marginal profits of an electric can opener manufacturer are given by $P'(x) = 0.1x\sqrt{x + 2}$, where $x$ is measured in thousands of can openers and $P(x)$ is measured in thousands of dollars. Find the total profits generated by increasing the number of can openers from 7000 to 14,000.

**24. Consumers' Surplus.** Find the consumers' surplus if $p = D(x) = 25 - x^3 \sqrt{5 + x^2}$ is the demand equation with $x$ measured in thousands of units and $p_0 = 1$ is the current price of the commodity. *Hint:* $D(2) = 1$.

25. **Costs.**   The cost function, in thousands of dollars, of a coat manufacturer is given by $x(1 + x)^4$, where $x$ is in thousands of coats. Find the average cost for the first 2000 coats.

26. **Price.**   The demand equation for a bath towel manufacturer is given by $p = p(x) = x(x + 1)^{-2}$, where $p$ is in dollars and $x$ is measured in thousands of towels. Find the average price over the interval $[1, 8]$.

27. **Population.**   The population of a new town is given by $P(t) = 1000 \ln t$, where $t \geq 1$ is measured in years. What is the average population over the time from $t = 1$ to $t = e$.

28. **Income Distribution.**   Find the coefficient of inequality for the Lorentz curve $y = xe^{x-1}$.

## Solutions to Self-Help Exercise Set 17.1

1.  **a.**   Our choices for $u$ are very limited. Let

$$u = \ln x \qquad dv = dx.$$

Then
$$du = \frac{1}{x} dx \qquad v = x,$$

and

$$\int \ln x \, dx = \int u \, dv$$

$$= uv - \int v \, du$$

$$= (\ln x)x - \int x \frac{1}{x} dx$$

$$= x \ln x - \int dx$$

$$= x \ln x - x + C$$

**b.**
$$\int_1^e \ln x \, dx = (x \ln x - x)\Big|_1^e$$

$$= (e \ln e - e) - (1 \cdot \ln 1 - 1)$$

$$= (0) - (-1) = 1$$

2.  Let
$$u = xe^x \qquad dv = \frac{1}{(1 + x)^2} dx.$$

Then
$$du = e^x(x + 1) \, dx \qquad v = \frac{-1}{x + 1}.$$

$$\int \frac{xe^x}{(1 + x)^2} \, dx = \int u \, dv$$

$$= uv - \int v \, du$$

$$= xe^x \left( \frac{-1}{x + 1} \right) - \int \left( -\frac{1}{x + 1} \right) e^x(x + 1) \, dx$$

$$= -\frac{xe^x}{x + 1} + \int e^x \, dx$$

$$= -\frac{xe^x}{x + 1} + e^x + C$$

$$= \frac{e^x}{x + 1} + C$$

## 17.2     INTEGRATION USING TABLES

▶ *Using Tables of Integrals*

▶ *Application*

**APPLICATION**

**Sales**

The rate of sales in millions of dollars per week of a toy is given by $\frac{e^x}{e^{2x} - 1}$ where $x$ is the number of weeks after the end of an advertising campaign. Find the total dollar sales during the second and third week. See Example 4 for the answer.

### Using Tables of Integrals

In previous sections we have seen how to integrate many expressions including $\int x^n \, dx$, $\int e^{kx} \, dx$, and so on. We have also been introduced to some techniques of integration such as substitution and integration by parts. A substantial number of different *techniques* of integration exist; many require considerable skill and ingenuity. Unfortunately there are not hard and fast rules on when to use each technique. In the face of these difficulties the most convenient technique is to refer to a table of integrals. We can always check an answer in any table by merely differentiating the answer in order to see if we obtain the integrand.

**E X A M P L E   1**     **Using a Table of Integrals**

Find $\int \frac{1}{x(2x + 3)} \, dx$.

**Solution** By looking through the table of integrals found in Appendix B.4, we notice that the integrand has the form of the formula in the sixth entry if we set $a = 2$ and $b = 3$. Thus we have

$$\int \frac{1}{x(2x + 3)} \, dx = \frac{1}{3} \ln \left| \frac{x}{2x + 3} \right| + C. \quad \blacksquare$$

**REMARK.** Notice we have added the arbitrary constant $C$. The table does not include this constant in order to save space.

EXAMPLE 2 **Using a Table of Integrals**

Find

$$\int \frac{1}{\sqrt{9x^2 + 1}} \, dx.$$

**Solution** If we look through the table we do not actually find this integral, but it appears to look somewhat like entry 22. Suppose we factor out the 9 from the integrand. Then we obtain

$$\frac{1}{3} \int \frac{1}{\sqrt{x^2 + \frac{1}{9}}} \, dx.$$

This integral is of the form of entry 22 with $a = \frac{1}{3}$. Thus

$$\int \frac{1}{\sqrt{9x^2 + 1}} \, dx = \frac{1}{3} \int \frac{1}{\sqrt{x^2 + \frac{1}{9}}} \, dx$$

$$= \frac{1}{3} \ln \left| x + \sqrt{x^2 + \frac{1}{9}} \right| + C. \quad \blacksquare$$

EXAMPLE 3 **Repeated Use of a Table of Integrals**

Find

$$\int x^2 e^{-x} \, dx.$$

**Solution** Although this integral can be done using integration by parts, one can also do it by using entry 38 in the table by taking $n = 2$ and $a = -1$. Doing this we obtain

$$\int x^2 e^{-x} \, dx = \frac{1}{(-1)} x^2 e^{-x} - \frac{2}{(-1)} \int x e^{-x} \, dx$$

$$= -x^2 e^{-x} + 2 \int x e^{-x} \, dx.$$

Now we are left to evaluate $\int x e^{-x} \, dx$, which we can do by using the very same entry but this time take $n = 1$ and $a = -1$. We then obtain for this latter integral

$$\int xe^{-x}\,dx = \frac{1}{(-1)}\,xe^{-x} - \frac{1}{(-1)}\int x^0 e^{-x}\,dx$$

$$= -xe^{-x} + \int e^{-x}\,dx$$

$$= -xe^{-x} - e^{-x} + C^*.$$

Now using this above yields

$$\int x^2 e^{-x}\,dx = -x^2 e^{-x} + 2\int xe^{-x}\,dx$$

$$= -x^2 e^{-x} + 2[-xe^{-x} - e^{-x} + C^*]$$

$$= -e^{-x}(x^2 + 2x + 2) + C,$$

where we have set $C = 2C^*$ for convenience.  ■

## Application

**E X A M P L E   4**    **Using Substitution in a Table of Integrals**

The rate of sales in millions of dollars per week of a toy is given by $\dfrac{e^x}{e^{2x} - 1}$ where $x$ is the number of weeks after the end of an advertising campaign. Find the total dollar sales during the second and third week.

**Solution**    We are seeking $\displaystyle\int_1^3 \frac{e^x}{e^{2x} - 1}\,dx$. The integrand involves exponentials, yet no formula in the table under the exponential heading applies. If we make the substitution $u = e^x$, then $du = e^x\,dx$ and the integral becomes

$$\int \frac{e^x}{e^{2x} - 1}\,dx = -\int \frac{e^x}{1 - e^{2x}}\,dx = -\int \frac{1}{1 - u^2}\,du.$$

This last integral can be found in entry 13 in the table with $a = 1$. Thus

$$\int \frac{e^x}{e^{2x} - 1} = -\int \frac{1}{1 - u^2}\,du$$

$$= -\frac{1}{2}\ln\left|\frac{u + 1}{u - 1}\right| + C$$

$$= \frac{1}{2}\ln\left|\frac{u - 1}{u + 1}\right| + C$$

$$= \frac{1}{2}\ln\left|\frac{e^x - 1}{e^x + 1}\right| + C$$

Then

$$\int_1^3 \frac{e^x}{e^{2x} - 1} \, dx = \frac{1}{2} \ln \frac{e^x - 1}{e^x + 1} \Big|_1^3 \approx 0.336,$$

or about \$336,000. See Figure 17.2.  ■

$$y = \frac{e^x}{e^{2x} - 1}$$

**Figure 17.2**

## SELF-HELP EXERCISE SET 17.2

**1.**  Find $\displaystyle\int \frac{4x}{\sqrt{x^4 + 4}} \, dx$.

## EXERCISE SET 17.2

In Exercises 1 through 26 evaluate using the tables.

**1.**  $\int x\sqrt{3x + 2} \, dx$          **2.**  $\displaystyle\int \frac{1}{x\sqrt{4 - x^2}} \, dx$          **3.**  $\int \sqrt{x^2 - 9} \, dx$

**4.**  $\displaystyle\int \frac{x}{2x - 1} \, dx$          **5.**  $\displaystyle\int \frac{1}{x\sqrt{1 - x}} \, dx$          **6.**  $\displaystyle\int \frac{\sqrt{4 - x^2}}{x} \, dx$

**7.**  $\displaystyle\int \frac{1}{2x^2 + 5x + 2} \, dx$          **8.**  $\displaystyle\int \frac{1}{x^2\sqrt{x^2 + 9}} \, dx$          **9.**  $\int x\sqrt{x^2 - 9} \, dx$

**10.**  $\displaystyle\int \frac{x^2}{(2x + 3)^2} \, dx$          **11.**  $\displaystyle\int \frac{1}{x^2(x - 3)} \, dx$          **12.**  $\displaystyle\int \frac{x}{(9 - x^2)^{3/2}} \, dx$

**13.**  $\displaystyle\int \frac{x^2}{(x^2 + 16)^{3/2}} \, dx$          **14.**  $\displaystyle\int \frac{\sqrt{x^2 + 16}}{x} \, dx$          **15.**  $\displaystyle\int \frac{1}{x(2x^3 + 1)} \, dx$

**16.**  $\displaystyle\int \frac{1}{2 + 3e^{4x}} \, dx$          **17.**  $\displaystyle\int \frac{1}{x^2\sqrt{1 - 4x^2}} \, dx$          **18.**  $\displaystyle\int \frac{1}{(1 - 4x^2)^{3/2}} \, dx$

**19.**  $\displaystyle\int \frac{x^2}{\sqrt{9x^2 - 1}} \, dx$          **20.**  $\displaystyle\int \frac{1}{(4x^2 + 1)^{3/2}} \, dx$          **21.**  $\int (\ln x)^3 \, dx$

**22.** $\int x^3 e^{2x}\, dx$

**23.** $\displaystyle\int \frac{e^{3x}}{e^x + 1}\, dx$

**24.** $\displaystyle\int \frac{e^{2x}}{(e^x + 1)^2}\, dx$

**25.** $\int x^2 e^{-2x}\, dx$

**26.** $\int (\ln(x + 1))^3\, dx$

## Applications

**27. Revenue.** Find the revenue function of a toaster manufacturer if marginal revenue, in hundreds of dollars, is given by $x\sqrt{x + 1}$, where $x$ is in hundreds of toasters.

**28. Sales.** The rate of change of the number of sales of refrigerators by a firm is given by

$$\frac{t}{(t^2 + 9)^{3/2}},$$

where $t$ is in weeks. How many refrigerators are sold in the first 4 weeks?

**29. Profit.** If the rate of change of profit in thousands of dollars per week is given by

$$P'(t) = \frac{10}{(t^2 + 9)^{3/2}},$$

where $t$ is measured in weeks and $P(0) = 0$, find $P = P(t)$.

**30. Costs.** The marginal cost, in thousands of dollars, of a brick manufacturer is given by

$$\frac{x}{\sqrt{x^2 + 9}},$$

where $x$ is in thousands of bricks. If fixed costs are $10,000, find $C = C(x)$.

**31. Consumers' Surplus.** Find the consumers' surplus if

$$p = D(x) = \frac{4e}{e + e^{0.10x}}$$

is the demand equation with $x$ measured in thousands of units and $p_0 = 2$ is the current price of the commodity.

**32. Producers' Surplus.** Find the producers' surplus if

$$p = S(x) = x\sqrt{x^2 + 5}$$

is the supply equation with $x$ measured in thousands of units and $p_0 = 6$ is the current price of the commodity. *Hint: $S(2) = 6$.*

**33. Pollutants.** The concentration (density) of pollutants, measured in thousands of particles per mile per day, at a distance of $x$ miles east of an industrial plant is given by

$$\delta(x) = \frac{1}{x^2 + 5x + 4}.$$

Find the amount of pollutants between $x = 0$ and $x = 5$ miles.

34. **Biology.** The rate of change of a certain population of plant in a small region is given by

$$P'(t) = \frac{1000}{(9t^2 + 1)^{3/2}},$$

where $t$ is in months. If the initial population is zero, find the population at any time. What happens as $t \to +\infty$?

## Solutions to Self-Help Exercise Set 17.2

Use entry 22 in the table of integrals after using the substitution $u = x^2$. Then $du = 2x\,dx$ and

$$\int \frac{4x}{\sqrt{x^4 + 4}}\,dx = 2\int \frac{2x}{\sqrt{x^4 + 4}}\,dx$$

$$= 2\int \frac{1}{\sqrt{u^2 + 4}}\,du$$

$$= 2\ln\left|u + \sqrt{u^2 + 4}\right| + C$$

$$= 2\ln\left|x^2 + \sqrt{x^4 + 4}\right| + C$$

## 17.3 NUMERICAL INTEGRATION

▶ *Introduction*
▶ *The Trapezoidal Rule*
▶ *Simpson's Rule*
▶ *Errors*

**APPLICATION**

**Finding Approximation of Definite Integrals**

Find approximations for $\int_0^2 \sqrt{4 - x^2}\,dx = \pi$ and $\int_1^2 \frac{1}{x}\,dx = \ln 2$. See Examples 3 and 4.

## Introduction

We really have only one method for evaluating the definite integral $\int_a^b f(x)\,dx$. This requires us to find an antiderivative $F(x)$ of $f(x)$, evaluate $F$ at $a$ and $b$, and use the formula

$$\int_a^b f(x)\,dx = F(b) - F(a).$$

If we cannot do this, our method fails.

But it turns out that this method *does* fail for many important and rather simple-looking integrals such as

$$\int_0^1 e^{-x^2}\, dx \qquad \text{and} \qquad \int_2^3 \frac{1}{\ln x}\, dx.$$

The first of these integrals plays a vital role in continuous probability theory, as we will see later.

Naturally $\int_0^1 e^{-x^2}\, dx$ exists since $e^{-x^2}$ is continuous on $[0, 1]$. Furthermore we know that in *theory* $\int_0^x e^{-t^2}\, dt$ is an antiderivative of $e^{-x^2}$ since we know from the fundamental theorem of calculus that if $f(x)$ is any continuous function on $[a, b]$ then

$$\frac{d}{dx} \int_a^x f(t)\, dt = f(x).$$

But it has been *proven* that $e^{-x^2}$ has no elementary antiderivative, that is, one that can be expressed as finite combinations of algebraic, exponential, logarithmic, and trigonometric functions. In effect this says that we cannot write down the antiderivative in a form that would permit us to evaluate the antiderivative at the two points $a$ and $b$.

This difficulty is not as disastrous as might first appear. There are two reasons for this. The first is the existence of *numerical* methods, which are the subject of this section, that will permit us to find definite integrals such as $\int_0^1 e^{-x^2}\, dx$ to (for all practical purposes) any desired accuracy. Thus knowing this integral to an accuracy of for example five or ten decimal places will be sufficient for any practical purpose.

The second reason is that many of the functions encountered in practice come from data collected in various ways. In this typical situation one does not know what the function $f(x)$ is at *every* point $x$ but rather only at the discrete points $x_1$, $x_2, \ldots, x_n$ where $f$ has been measured. See Figure 17.3. The numerical methods given in this section are well suited to give reasonable approximations to the definite integral $\int_a^b f(x)\, dx$ under such conditions.

We *defined* the definite integral $\int_a^b f(x)\, dx$ as the limit of Riemann sums. That is, we divided the interval $[a, b]$ into $n$ subintervals, $I_1, I_2, \ldots, I_n$ of equal length $\Delta x = \frac{b-a}{n}$, and

$$\int_a^b f(x)\, dx = \lim_{n \to +\infty} \sum_{i=1}^n f(x_i)\Delta x,$$

where $x_i \in I_i$ for each $i$. If $n$ is larger we expect, as in Figure 17.4, that the Riemann sum

$$\sum_{i=1}^n f(x_i)\Delta x,$$

to be close to the value of $\int_a^b f(x)\, dx$. This represents a *numerical* method of approximating the definite integral by using thin rectangles.

**Figure 17.3**

**Figure 17.4**

**Figure 17.5**

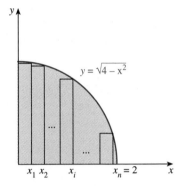

**Figure 17.6**

From Figure 17.5, $\int_0^2 \sqrt{4 - x^2}\, dx$ is the area under the circle $y = f(x) = \sqrt{4 - x^2}$ in the first quadrant. This circle is centered at the origin, has radius 2, and area equal to $\pi(2)^2 = 4\pi$. Therefore the part of the circle that is in the first quadrant has area equal to $\pi$. Thus

$$\int_0^2 \sqrt{4 - x^2}\, dx = \pi.$$

By approximating this integral, we will approximate $\pi$.

If we approximate this integral by the Riemann sum

$$S_n = \sum_{i=1}^n f(x_i^r)\Delta x,$$

where $x_i^r$ is the right endpoint of each interval, we see from Figure 17.6 that the rectangles are below the curve, and thus $S_n < \pi$ for all $n$.

Notice that $\Delta x = \dfrac{2}{n}$ and $x_i^r = \dfrac{2i}{n}$. Thus

$$S_n = \sum_{i=1}^n f(x_i^r)\Delta x = \sum_{i=1}^n \sqrt{4 - \left(\frac{2i}{n}\right)^2}\,\frac{2}{n} = \frac{4}{n}\sum_{i=1}^n \sqrt{1 - \left(\frac{i}{n}\right)^2}.$$

The table shows how this gets closer to $\pi$ as the values of $n$ get larger.

| $n$ | 4 | 8 | 50 | 500 |
|---|---|---|---|---|
| $S_n$ | 2.4957 | 2.8398 | 3.0983 | 3.1375 |

## The Trapezoid Rule

The Riemann sum approximates $\int_a^b f(x)\, dx$ by rectangles. In general, a better approximation can be obtained by using trapezoids as shown in Figure 17.7. Once again the interval $[a, b]$ has been divided into $n$ subintervals each of length $\Delta x = \dfrac{b - a}{n}$. Also $x_0 = a, x_1 = a + \Delta x, x_2 = a + 2\Delta x, \ldots, x_n = a + n\Delta x$.

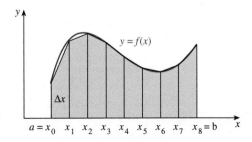

**Figure 17.7**

Recall that the area of a trapezoid is given by multiplying the length of the base times the average of the heights of the sides. Thus the areas of the $n$ trapezoids is given by

$$T_n = \frac{f(x_0) + f(x_1)}{2} \Delta x + \frac{f(x_1) + f(x_2)}{2} \Delta x + \frac{f(x_2) + f(x_3)}{2} \Delta x$$

$$+ \cdots + \frac{f(x_{n-2}) + f(x_{n-1})}{2} \Delta x + \frac{f(x_{n-1}) + f(x_n)}{2} \Delta x$$

$$= \Delta x \left( \frac{1}{2} f(x_0) + f(x_1) + f(x_2) + \cdots + f(x_{n-1}) + \frac{1}{2} f(x_n) \right)$$

We then have

---

**Trapezoidal Rule**

For any integer $n$ divide the interval $[a, b]$ into $n$ subintervals of equal length with endpoints $x_0, x_1, \ldots, x_n$. If

$$T_n = \frac{b - a}{n} \left[ \frac{1}{2} f(x_0) + f(x_1) + f(x_2) + \cdots + f(x_{n-1}) + \frac{1}{2} f(x_n) \right],$$

then

$$\int_a^b f(x) \, dx \approx T_n.$$

---

In general the trapezoidal rule will give much closer approximations than the rectangular rule using Riemann sums. We will return to this point later.

**E X A M P L E  1**    **Using the Trapezoidal Rule**

Approximate $\int_0^2 \sqrt{4 - x^2} \, dx = \pi$ by using $T_4$.

**Solution**    See Figure 17.8 where one can see that $T_4$ must be smaller than the area which is $\pi$. We have $\Delta x = \dfrac{2 - 0}{4} = 0.50$ and the points $x_0, x_1, x_2, x_3$, and $x_4$, are respectively 0, 0.50, 1.00, 1.50, and 2. We have from the above formula that

$$T_4 = \frac{2 - 0}{4} \left[ \frac{1}{2} f(0) + f(0.5) + f(1) + f(1.5) + \frac{1}{2} f(2) \right]$$

$$= \frac{1}{2} \left[ \frac{1}{2} \cdot 2 + \sqrt{4 - .25} + \sqrt{4 - 1} + \sqrt{4 - 2.25} + \frac{1}{2} \cdot 0 \right]$$

$$\approx 2.996 \quad \blacksquare$$

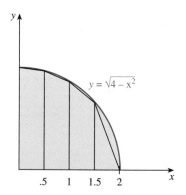

$y = \sqrt{4 - x^2}$

**Figure 17.8**

Now notice that

$$\int_1^2 \frac{1}{x}\, dx = \ln x \Big|_1^2 = \ln 2.$$

Thus we can approximate $\ln 2$ by approximating the integral $\int_1^2 \frac{1}{x}\, dx.$

### EXAMPLE 2  Using the Trapezoidal Rule

Use $T_4$ to find an approximation to $\int_1^2 \frac{1}{x}\, dx = \ln 2.$

**Solution**   See Figure 17.9, where one can see that $T_4$ must be larger than the area, that is, $T_4 > \ln 2$. We have $\Delta x = \frac{2-1}{4} = 0.25$ and the points $x_0$, $x_1$, $x_2$, $x_3$, and $x_4$, are, respectively, 1, 1.25, 1.5, 1.75, 2. Thus

$$T_4 = \frac{2-1}{4}\left[\frac{1}{2} f(1) + f(1.25) + f(1.5) + f(1.75) + \frac{1}{2} f(2)\right]$$

$$= \frac{1}{4}\left[\frac{1}{2} + \frac{4}{5} + \frac{2}{3} + \frac{4}{7} + \frac{1}{4}\right]$$

$$\approx 0.6970 \quad \blacksquare$$

Compare this to $\ln 2 = 0.6931$ to 4 decimal places.

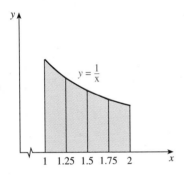

$y = \frac{1}{x}$

**Figure 17.9**

## Simpson's Rule

**James Gregory**
**1638–1675**

*(The Granger Collection, New York.)*

It was actually the English mathematician James Gregory who in 1668 discovered Simpson's rule. Thomas Simpson (1710–1761), after whom Simpson's rule is named, wrote a very popular calculus textbook.

When approximating the area under a curve using Riemann sums, we are using thin "slices" consisting of thin rectangles, that is, a slice with the top a horizontal line (**polynomial of degree zero**). When approximating using the trapezoidal rule we use thin slices with the top a line that connects the curve at two points (**polynomial of degree one**).

Intuitively we might expect that trapezoids would be the better approximation, in general, since the straight line forming the top of a trapezoid does hit the curve in at least *two* points (the endpoints of the subinterval), whereas the top of a rectangle need only hit the curve in *one* place. In general it is true that trapezoids will yield better approximations than rectangles.

However, it is easy to criticize the use of trapezoids, since the typical curve actually "bends," whereas the top of a trapezoid is a straight line (polynomial of degree one). We might then think of using for the top a convenient curve that "bends." The obvious next choice would be a polynomial of degree two, that is, a parabola. For technical reasons we shall need to divide that interval $[a, b]$ into an *even* number $n$ of subintervals and, as usual, of equal length.

Look at Figure 17.10 where the interval has been divided into four subintervals. We then wish to find a second degree polynomial $Ax^2 + Bx + C$ that goes through the three points $(x_0, f(x_0))$, $(x_1, f(x_1))$, and $(x_2, f(x_2))$, and use this as a top over the interval $[x_0, x_2]$.

In the same way we take the unique second degree polynomial $A_2x^2 + B_2x + C_2$ that goes through the three points $(x_2, f(x_2))$, $(x_3, f(x_3))$, and $(x_4, f(x_4))$ as the top over the subinterval $[x_2, x_4]$. We refer to the sum of the areas under these two polynomials as $S_4$, the four referring to the original four subintervals.

It turns out (no proof given) that the area under the first parabola from $x_0$ to $x_2$ is

$$\frac{\Delta x}{3} [f(x_0) + 4f(x_1) + f(x_2)],$$

where as usual $\Delta x = (b - a)/n$. In the same way, the area under the second parabola from $x_2$ to $x_4$ is

$$\frac{\Delta x}{3} [f(x_2) + 4f(x_3) + f(x_4)].$$

Thus we have that the sum of these two is

$$S_4 = \frac{\Delta x}{3} [f(x_0) + 4f(x_1) + 2f(x_2) + 4f(x_3) + f(x_4)].$$

In general we have the following.

---

**Simpson's Rule**

For any even integer $n$ divide the interval $[a, b]$ into $n$ equal subintervals with endpoints $a = x_0, x_1, \ldots, x_n = b$, and let

$$S_n = \frac{b - a}{3n} [f(x_0) + 4f(x_1) + 2f(x_2) + 4f(x_3) + \cdots$$

$$+ 2f(x_{n-2}) + 4f(x_{n-1}) + f(x_n)].$$

Then
$$S_n \approx \int_a^b f(x)\, dx.$$

---

Notice in this formula that, with the exceptions of the first and last terms, there is a 4 in front of a term $f(x_k)$ if $k$ is *odd* and a 2 in front of a term $f(x_k)$ if $k$ is *even*.

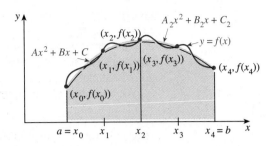

**Figure 17.10**

E X A M P L E 3

**Using Simpson's Rule**

Use Simpson's rule for $S_4$ to approximate $\int_0^2 \sqrt{4 - x^2}\, dx = \pi$.

**Solution**

The interval $[0, 2]$ is divided into 4 subintervals with $x_0 = 0$, $x_1 = 0.50$, $x_2 = 1$, $x_3 = 1.50$, $x_4 = 2$. We then have, using Simpson's rule, that

$$S_4 = \frac{2 - 0}{3 \cdot 4}\,[f(0) + 4f(0.5) + 2f(1) + 4f(1.5) + f(2)]$$

$$= \frac{1}{6}\,[2 + 4\sqrt{4 - .25} + 2\sqrt{4 - 1} + 4\sqrt{4 - 2.25} + 0]$$

$$\approx 3.084 \quad \blacksquare$$

This compares to $T_4 = 2.996$ for the trapezoidal rule and we see that $S_4$ is much closer to $\pi$ than $T_4$ is.

E X A M P L E 4

**Using Simpson's Rule**

Use $S_4$ to find an approximation to $\displaystyle\int_1^2 \frac{1}{x}\, dx = \ln 2$.

**Solution**

Here $x_0 = 1$, $x_1 = 1.25$, $x_2 = 1.50$, $x_3 = 1.75$, $x_4 = 2$ and

$$S_4 = \frac{2 - 1}{3 \cdot 4}\,[f(1) + 4f(1.25) + 2f(1.5) + 4f(1.75) + f(2)]$$

$$= \frac{1}{12}\left[1 + 4 \cdot \frac{4}{5} + 2 \cdot \frac{2}{3} + 4 \cdot \frac{4}{7} + \frac{1}{2}\right]$$

$$\approx 0.6933 \quad \blacksquare$$

Compare this to $T_4 = 0.6970$ for the trapezoidal rule, and we see that $S_4$ is much closer to $\ln 2$ than $T_4$ is since to four decimal places $\ln 2 = 0.6931$.

## Errors

An approximation is not worth much if we do not know how close it is to the correct answer. It is of course of vital importance to know just how accurate our approximations really are.

We now state two important results that give error estimates.

---

**Error in the Trapezoidal Rule**

The maximum possible error incurred in using the trapezoidal rule is

$$\frac{M(b - a)^3}{12n^2},$$

where $|f''(x)| \leq M$ on $[a, b]$.

---

Naturally we need to assume that $f''(x)$ exists and is bounded on $[a, b]$.

---

**Error in Simpson's Rule**

The maximum possible error incurred in using Simpson's rule is

$$\frac{M(b - a)^5}{180n^4},$$

where $|f^{(4)}(x)| \leq M$ on $[a, b]$.

---

Naturally we need to assume that $f^{(4)}(x)$ exists and is bounded on $[a, b]$.

**REMARK.**    Since these error formulas are correct, no matter what function $f$ is used, they are extremely conservative error bounds. That is, most likely the error in any particular problem will be much less.

---

**E X A M P L E   5**    **Finding Error Bounds**

Use the above error bound formulas to find the largest possible errors incurred in the previous calculations of $T_4$ and $S_4$ for the definite integral $\int_1^2 \frac{1}{x}\, dx$.

**Solutions**    *(a) Trapezoidal Rule Error.*   We first need to find a bound of $f''(x)$ on the interval $[1, 2]$. We have $f(x) = x^{-1}$, $f'(x) = -x^{-2}$, and $f''(x) = 2x^{-3}$. So we need to find an upper bound on $2x^{-3}$ on $[1, 2]$. Since the function $2x^{-3}$ is decreasing on this interval, it attains its maximum at the left endpoint $x = 1$ where we see that $f''(1) = 2$. Thus we take $M = 2$ in the error bound formula given above. Then

$$\frac{M(b - a)^3}{12n^2} = \frac{2(2 - 1)^3}{12(4)^2} \approx 0.01.$$

As we noted above the actual error was $T_4 - \ln 2 \approx .004$.

**(b) Simpson's Rule Error.** We first need to find a bound of $f^{(4)}(x)$ on the interval $[1, 2]$. We have $f''(x) = 2x^{-3}$, so $f^{(3)}(x) = -6x^{-4}$ and $f^{(4)}(x) = 24x^{-5}$. So we need to find an upper bound on $24x^{-5}$ on $[1, 2]$. Since the function $24x^{-5}$ is decreasing on this interval, it attains its maximum at the left endpoint $x = 1$, where we see that $f^{(4)}(1) = 24$. Thus we take $M = 24$ in the error bound formula given above. Then

$$\frac{M(b - a)^5}{180n^4} = \frac{24(2 - 1)^5}{180(4)^4} \approx 0.0005.$$

As we noted above, the actual error was $S_4 - \ln 2 \approx .0002$ ∎

If the original function $f(x)$ was a *first* degree polynomial, that is, $f(x) = Ax + B$, then the error bound formula for the trapezoidal rule indicates that there is no error, since in this case $f''(x) = 0$ and therefore $M = 0$. This is naturally what we expect, since the tops of the approximating trapezoids are themselves *first* degree polynomials.

But very surprisingly the Simpson's rule error bound formula indicates that Simpson's rule is exact not just for *second* degree polynomials, as you would be inclined to suspect, but also for third degree polynomials. This is true since the fourth derivative of a third degree polynomial is identically zero, and thus $M = 0$ in the above formula. For this reason Simpson's rule gives much better approximations than one might otherwise suspect.

| $f(x)$ on $[0, 2]$ | $x^2$ | $x^3$ | $1/(x + 1)$ | $\sqrt{1 + x^2}$ |
|---|---|---|---|---|
| Trapezoidal ($n = 2$) | 4.000 | 5.000 | 1.667 | 3.032 |
| Simpson's ($n = 2$) | 2.667 | 4.000 | 1.111 | 2.964 |
| Exact (to three decimals) | 2.667 | 4.000 | 1.099 | 2.958 |

## SELF-HELP EXERCISE SET 17.3

1. Use $T_8$ to approximate $\int_1^2 \frac{1}{x}\,dx$.

2. Use $S_8$ to approximate $\int_1^2 \frac{1}{x}\,dx$.

3. Using the error bound formulas, find the maximum error in the previous two exercises.

## EXERCISE SET 17.3

In Exercises 1 through 8 find $T_2$ and $S_2$ and for Exercises 1 through 6 compare your approximations to the correct answer.

1. $\int_0^2 x^3\,dx$

2. $\int_0^2 x^2\,dx$

3. $\int_1^3 \frac{1}{x}\,dx$

**4.** $\int_0^2 x^4 \, dx$

**5.** $\int_0^2 \sqrt{4 - x^2} \, dx$

**6.** $\int_0^2 \frac{1}{x+1} \, dx$

**7.** $\int_0^1 e^{x^2} \, dx$

**8.** $\int_2^3 \frac{1}{\ln x} \, dx$

In Exercises 9 through 16 find $T_4$ and $S_4$, and for Exercises 9 through 14 compare your approximations to the correct answer and to the approximations obtained above.

**9.** $\int_0^2 x^3 \, dx$

**10.** $\int_0^2 x^2 \, dx$

**11.** $\int_1^3 \frac{1}{x} \, dx$

**12.** $\int_0^2 x^4 \, dx$

**13.** $\int_0^2 \sqrt{4 - x^2} \, dx$

**14.** $\int_0^2 \frac{1}{x+1} \, dx$

**15.** $\int_0^1 e^{x^2} \, dx$

**16.** $\int_2^3 \frac{1}{\ln x} \, dx$

In Exercises 17 through 24 find $T_{10}$ and $S_{10}$, and for Exercises 17 through 22 compare your approximations to the correct answer and to the approximations obtained above.

**17.** $\int_0^2 x^3 \, dx$

**18.** $\int_0^2 x^2 \, dx$

**19.** $\int_1^3 \frac{1}{x} \, dx$

**20.** $\int_0^2 x^4 \, dx$

**21.** $\int_0^2 \sqrt{4 - x^2} \, dx$

**22.** $\int_0^2 \frac{1}{x+1} \, dx$

**23.** $\int_0^1 e^{x^2} \, dx$

**24.** $\int_2^3 \frac{1}{\ln x} \, dx$

## Applications

**25. Revenue.** In the following table the rate $R'(t)$ of revenue in millions of dollars per week is estimated in some manner for discrete equally spaced points $0 = t_0, t_1, \ldots, t_4 = 8$, where $t$ is in weeks. Find (a) $T_4$, and (b) $S_4$. The answers will then be approximations to $\int_0^8 R'(t) \, dt = R(8) - R(0) = R(8)$, which is the total revenue for the eight-week period.

| $t_i$ | 0 | 2 | 4 | 6 | 8 |
|---|---|---|---|---|---|
| $R'(t_i)$ | 1.0 | 0.8 | 0.5 | 0.7 | 0.8 |

**26. Sales.** In the following table the rate $R'(t)$ of sales for an entire industry in billions of dollars is given for discrete equally spaced points $0 = t_0, t_1, \ldots, t_{10} = 10$, where $t$ is in weeks. Find (a) $T_{10}$, and (b) $S_{10}$. The answers will then be approximations to $\int_0^{10} R'(t) \, dt = R(10) - R(0) = R(10)$, which is the total sales for the ten-week period.

| $t_i$ | 0 | 1 | 2 | 3 | 4 | 5 | 6 | 7 | 8 | 9 | 10 |
|---|---|---|---|---|---|---|---|---|---|---|---|
| $R'(t_i)$ | 10 | 12 | 15 | 16 | 20 | 18 | 16 | 15 | 12 | 10 | 6 |

27. **Production.** In the following table the rate $P'(t)$ of production of steel in thousands of tons is estimated in some manner for discrete equally spaced points $0 = t_0, t_1, \ldots, t_4 = 16$, where $t$ is in weeks. Find (a) $T_4$, and (b) $S_4$. The answers will then be approximations to $\int_0^8 P'(t)\, dt = P(8) - P(0) = P(8)$, which is the total production of steel for the eight-week period.

| $t_i$ | 0 | 4 | 8 | 12 | 16 |
|---|---|---|---|---|---|
| $P'(t_i)$ | 1 | 2 | 3 | 5 | 4 |

28. **Population.** In the following table the rate of growth $M'(t)$ of a strain of bacteria has been estimated by measuring the rate of increase in mass at discrete, equally spaced points $0 = t_0, t_1, \ldots, t_{10} = 20$, where $t$ is in hours. Find (a) $T_{10}$, and (b) $S_{10}$. The answers will then be approximations to $\int_0^{20} M'(t)\, dt = M(20) - M(0)$, which is the change in mass for the twenty-hour period.

| $t_i$ | 0 | 2 | 4 | 6 | 8 | 10 | 12 | 14 | 16 | 18 | 20 |
|---|---|---|---|---|---|---|---|---|---|---|---|
| $M'(t_i)$ | 10 | 14 | 20 | 30 | 50 | 100 | 100 | 80 | 50 | 20 | 0 |

29. **Velocity.** In the following table the velocity $v(t) = s'(t)$ is given for discrete, equally spaced points $0 = t_0, t_1, \ldots, t_6 = 3$. Find (a) $T_6$ and $S_6$. The answers will then be approximations to $\int_0^3 v(t)\, dt = s(3) - s(0)$, which is the total distance traveled.

| $t_i$ | 0 | 0.5 | 1 | 1.5 | 2 | 2.5 | 3 |
|---|---|---|---|---|---|---|---|
| $v(t_i)$ | 1.0 | 1.6 | 2.0 | 1.8 | 1.5 | 1.0 | 0.5 |

In Exercises 30 through 34 find the maximum error for $T_{10}$ and $S_{10}$ for the following.

30. $\int_0^2 x^3\, dx$

31. $\int_0^2 x^2\, dx$

32. $\int_1^3 \dfrac{1}{x}\, dx$

33. $\int_0^2 x^4\, dx$

34. $\int_0^2 \dfrac{1}{x+1}\, dx$

35. Can you use the trapezoidal rule error bound formula to find an error bound for $T_{10}$ for $\int_0^1 \sqrt{x}\, dx$? Why or why not?

36. For any $b > 0$ show directly (without using the Simpson's rule error bound formula) that $S_2$ gives the exact value for $\int_0^b x^3\, dx$.

## Graphing Calculator Exercises

In Exercises 37 through 41 use the three programs Riemann, Trapezoi, and Simpson, found in Appendix C to find the Riemann sums $\sum_{i=1}^n f(x_i)\Delta x$ with $x_i$ the left endpoint, the midpoint, and the right endpoint for $n = 50, 100, 250$, and also $S_n$ in Simpson's rule and $T_n$ in the trapezoidal rule for $n = 50, 100, 250$. Compare your answers with the exact answers where given.

**37.** $\displaystyle\int_0^2 \sqrt{4 - x^2}\, dx = \pi$

**38.** $\displaystyle\int_1^2 \frac{1}{x}\, dx = \ln 2$

**39.** $\displaystyle\int_0^1 e^{-0.5x^2}\, dx$

**40.** $\displaystyle\int_0^1 e^x\, dx = e^x - 1$

**41.** $\displaystyle\int_0^4 3\sqrt{x}\, dx = 16$

**42.** If $x$ is large, the logarithmic integral $li(x) = \displaystyle\int_2^x \frac{dt}{\ln t}$ approximates the number of prime numbers less than $x$. Find an approximation to $li(200)$. (There are 46 primes less than 200.)

## Solutions to Self-Help Exercise Set 17.3

**1.** We have $\Delta x = \frac{2-1}{8} = 0.125$, and the points $x_0, x_1, x_2, \ldots, x_8$, are, respectively, 1, 1.125, 1.25, ..., 1.875, 2. Thus

$$T_8 = \frac{1-0}{8}\left[\frac{1}{2}f(1) + f(1.125) + f(1.25) + f(1.375)\right.$$

$$\left. + f(1.50) + f(1.625) + f(1.75) + f(1.875) + \frac{1}{2}f(2)\right]$$

$$= \frac{1}{8}\left[\frac{1}{2} + \frac{1}{1.125} + \frac{1}{1.25} + \frac{1}{1.375}\right.$$

$$\left. + \frac{1}{1.5} + \frac{1}{1.625} + \frac{1}{1.75} + \frac{1}{1.875} + \frac{1}{4}\right]$$

$$\approx 0.6941$$

**2.** The points $x_0, x_1, \ldots, x_8$, are the same as in the previous exercise. Then

$$S_8 = \frac{2-1}{3 \cdot 8}\left[f(1) + 4f(1.125) + 2f(1.25) + 4f(1.375)\right.$$

$$+ 2f(1.5) + 4f(1.625) + 2f(1.75) + 4f(1.875) + f(2)]$$

$$= \frac{1}{24}\left[1 + 4\frac{1}{1.125} + 2\frac{1}{1.25} + 4\frac{1}{1.375}\right.$$

$$\left. + 2\frac{1}{1.5} + 4\frac{1}{1.625} + 2\frac{1}{1.75} + 4\frac{1}{1.875} + \frac{1}{2}\right]$$

$$\approx 0.6931545$$

**3.** Just as in Example 5 of the text, $|f''(x)| \le 2$, $|f^{(4)}(x)| \le 24$, and $b - a = 2 - 1 = 1$. For this exercise $n = 8$. Thus a bound on the trapezoid error is

$$\frac{M(b-a)^3}{12n^2} = \frac{2(2-1)^3}{12(8)^2} \approx .0026$$

Compare this with the actual error of 0.001. A bound on the Simpson's error is

$$\frac{M(b - a)^5}{180n^4} = \frac{24(2 - 1)^5}{180(8)^4} \approx .000\ 03$$

Compare this with the actual error of 0.000 007.

## 17.4   IMPROPER INTEGRALS

► *Improper Integrals*
► *Applications*

APPLICATION

**Long-Term Sales of an Engine Part**

In order to sell a large number of its most recent engines to the aircraft manufacturers of the world, an aircraft engine manufacturer had to guarantee to supply all spare parts for this engine in perpetuity. One year from now the company plans to stop manufacturing this engine and wishes in the next year to make all the spare parts they will ever need. They estimate that the rate at which engine manifolds will be needed per year is given by $r(t) = 200e^{-0.1t}$, where $t$ is in years from now. How many manifolds will be needed? See Example 4 for the answer.

### Improper Integrals

Up to now we have considered definite integrals $\int_a^b f(x)\ dx$ where the interval $[a, b]$ of integration was *bounded*, that is, both $a$ and $b$ are finite numbers. But in some important applications one must consider *infinite* intervals of integration.

If we ask for the area under a curve $y = f(x)$ from $a$ to $b$, where $b$ is any number such that $b > a$; this is given by $\int_a^b f(x)\ dx$. See Figure 17.11.

Suppose now that the function $f(x)$ is positive and that $f(x) \to 0$ as $x \to +\infty$ ($f(x) = 1/x^2$ is such an example). If we now ask for the area under the same curve to the right of $a$ (see Figure 17.12), we have an unbounded or "infinite"

**Figure 17.11**

**Figure 17.12**

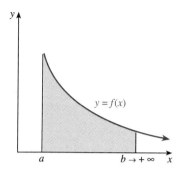

**Figure 17.13**

region. Can such a region have a *finite* area? What we need is a sensible and workable definition of "area" under the curve on $[a, +\infty)$. We want a "sensible" definition that agrees with our intuition and a "workable" definition that permits us to actually *find* the area.

We certainly know that for any $b > a$ the definite integral $\int_a^b f(x)\, dx$ is the area under the curve from $a$ to $b$. In this expression (or in Figure 17.11), it makes sense to let $b \to +\infty$. What we obtain we call the area under the curve on $[a, +\infty)$. This seems to be geometrically reasonable. See Figure 17.13.

This definition is workable since we understand what is meant by the area under the curve on $[a, b]$ when $b$ is *finite*, and we should be able to actually calculate this area in important examples. Furthermore we understand what is meant by taking the limit as $b$ goes to $+\infty$ and should be able to do this and obtain an answer. The formal definition is now given.

---

**Definition of** $\displaystyle\int_a^{+\infty} f(x)\, dx$

If $f(x)$ is continuous on $[a, +\infty)$, we define

$$\int_a^{+\infty} f(x)\, dx = \lim_{b \to +\infty} \int_a^b f(x)\, dx,$$

if this limit exists as a number. Otherwise we say that the integral *diverges*. If the limit exists as a number, we say that the integral *converges* to that number.

---

**REMARK.**   When the integral is over an infinite interval we call it an **improper integral**.

**E X A M P L E   1**   **Finding an Improper Integral**

Find

$$\int_1^{+\infty} \frac{1}{x^2}\, dx.$$

**Solution**   According to our definition we first need to evaluate $\int_1^b \frac{1}{x^2}\, dx$, for any $b > 1$. Of course, this just represents the area under the curve on the interval $[1, b]$. See Figure 17.14. We have

$$\int_1^b \frac{1}{x^2}\, dx = -\frac{1}{x}\Big|_1^b = -\frac{1}{b} + 1 = 1 - \frac{1}{b}.$$

Now we need to take the limit of this expression as $b \to +\infty$. Doing this we obtain

$$\int_1^{+\infty} \frac{1}{x^2}\, dx = \lim_{b \to +\infty} \int_1^b \frac{1}{x^2}\, dx = \lim_{b \to +\infty} \left(1 - \frac{1}{b}\right) = 1. \quad \blacksquare$$

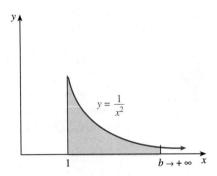

**Figure 17.14**

E X A M P L E   2

**Finding an Improper Integral**

Find

$$\int_{1}^{+\infty} \frac{1}{\sqrt{x}} \, dx.$$

**Solution**    According to our definition we first need to evaluate

$$\int_{1}^{b} \frac{1}{\sqrt{x}} \, dx,$$

for any $b > 1$. Of course this just represents the area under the curve on the interval $[1, b]$. See Figure 17.15. We have

$$\int_{1}^{b} x^{-1/2} \, dx = 2x^{1/2}\big|_{1}^{b} = 2(\sqrt{b} - 1).$$

Now we need to take the limit of this expression as $b \to +\infty$. See Figure 17.15. Doing this we obtain

**Figure 17.15**

**Figure 17.16**

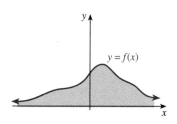

**Figure 17.17**

$$\int_1^{+\infty} x^{-1/2}\, dx = \lim_{b\to+\infty} \int_1^b x^{-1/2}\, dx = \lim_{b\to+\infty} 2(\sqrt{b} - 1),$$

which is unbounded and thus does not exist. This integral then diverges. ∎

Just as we defined an improper integral on the interval $[a, +\infty)$, so also we can define an improper integral on the interval $(-\infty, b]$. We now state this definition. See also Figure 17.16.

---

**Definition of** $\displaystyle\int_{-\infty}^b f(x)\, dx$

If $f(x)$ is continuous on $(-\infty, b]$, we define

$$\int_{-\infty}^b f(x)\, dx = \lim_{a\to-\infty} \int_a^b f(x)\, dx,$$

if this limit exists. If the limit does not exist, we say that the integral *diverges*.

---

Finally, we define an improper integral of the form $\int_{-\infty}^{+\infty} f(x)\, dx$. See Figure 17.17.

---

**Definition of** $\displaystyle\int_{-\infty}^{+\infty} f(x)\, dx$

If $f(x)$ is continuous on $(-\infty, +\infty)$, we define

$$\int_{-\infty}^{+\infty} f(x)\, dx = \int_{-\infty}^0 f(x)\, dx + \int_0^{+\infty} f(x)\, dx,$$

if both integrals on the right converge. If one or both of the integrals on the right diverges, then we say that $\displaystyle\int_{-\infty}^{+\infty} f(x)\, dx$ diverges.

---

**E X A M P L E   3**    **Finding an Improper Integral**

Find

$$\int_{-\infty}^{+\infty} 2xe^{-x^2}\, dx.$$

**Solution**    Refer to Figure 17.18. According to the previous definition we need to find both $\int_{-\infty}^0 2xe^{-x^2}\, dx$ and $\int_0^{+\infty} 2xe^{-x^2}\, dx$. To find the last integral we need to first find

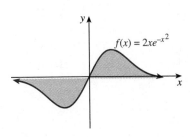

**Figure 17.18**

$\int_0^b 2xe^{-x^2}\, dx$ for every $b > 0$. Then

$$\int_0^b 2xe^{-x^2}\, dx = -e^{-x^2}\big|_0^b = 1 - e^{-b^2} = 1 - \frac{1}{e^{b^2}}.$$

Now taking the limit of this last expression as $b \to +\infty$, we obtain

$$\lim_{b \to +\infty} \int_0^b 2xe^{-x^2}\, dx = \lim_{b \to +\infty}\left(1 - \frac{1}{e^{b^2}}\right) = 1.$$

For the other integral we have for any $a < 0$,

$$\int_a^0 2xe^{-x^2}\, dx = -e^{-x^2}\big|_a^0 = -1 + e^{-a^2} = \frac{1}{e^{a^2}} - 1.$$

Now taking the limit of this last expression as $a \to -\infty$, we obtain

$$\lim_{a \to -\infty} \int_a^0 2xe^{-x^2}\, dx = \lim_{a \to -\infty}\left(\frac{1}{e^{a^2}} - 1\right) = -1.$$

Thus

$$\int_{-\infty}^{+\infty} 2xe^{-x^2}\, dx = \int_{-\infty}^0 2xe^{-x^2}\, dx + \int_0^{+\infty} 2xe^{-x^2}\, dx = -1 + 1 = 0 \quad \blacksquare$$

## Applications

The large aircraft manufacturers of the world do not produce the engines for their aircraft. Rather they purchase them from several large engine manufacturers. In this country General Electric and United Technologies dominate this fiercely competitive industry.

**E X A M P L E   4**

**Spare Parts Production**

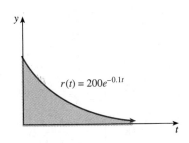

**Figure 17.19**

In order to sell a large number of its most recent engines to the aircraft manufacturers of the world, an aircraft engine manufacturer had to guarantee to supply all spare parts for this engine in perpetuity. One year from now the company plans to stop manufacturing this engine and wishes in the next year to make all the spare parts they will ever need. They estimate that the rate at which engine manifolds will be needed per year is given by

$$r(t) = 200e^{-0.1t},$$

where $t$ is in years from now. See Figure 17.19. How many manifolds will be needed?

**Solution**    Since the integral $\int_0^b r(t)\, dt$ is the amount needed from now to $b$ years from now, $\int_0^{+\infty} r(t)\, dt$ is the amount needed for all time. Then

$$\int_0^b r(t)\ dt = 200 \int_0^b e^{-0.10t}\ dt$$

$$= -2000e^{-0.10t}\big|_0^b$$

$$= -2000(e^{-0.10b} - 1)$$

$$\int_0^{+\infty} r(t)\ dt = \lim_{b\to +\infty} \int_0^b r(t)\ dt$$

$$= \lim_{b\to +\infty} -2000(e^{-0.10b} - 1)$$

$$= 2000$$

Thus the company needs to make 2000 engine manifolds.   ∎

## SELF-HELP EXERCISE SET 17.4

1.  Repeat Example 4 if $r(t) = 1000(t + 1)^{-3/2}$.

2.  Find $\int_{-\infty}^{+\infty} f(x)\ dx$ where $f(x) = 1$ if $|x| < 1$ and $f(x) = x^{-2}$ if $|x| \geq 1$.

## EXERCISE SET 17.4

In Exercises 1 through 20 evaluate.

**1.** $\displaystyle\int_1^{+\infty} \frac{1}{x^3}\ dx$

**2.** $\displaystyle\int_1^{+\infty} \frac{1}{\sqrt[3]{x}}\ dx$

**3.** $\displaystyle\int_1^{+\infty} \frac{1}{\sqrt[4]{x}}\ dx$

**4.** $\displaystyle\int_1^{+\infty} \frac{1}{x^4}\ dx$

**5.** $\displaystyle\int_1^{+\infty} \frac{1}{x^{1.01}}\ dx$

**6.** $\displaystyle\int_1^{+\infty} \frac{1}{x^{.99}}\ dx$

**7.** $\displaystyle\int_1^{+\infty} \frac{1}{x}\ dx$

**8.** $\displaystyle\int_{-\infty}^{-1} \frac{1}{x^2}\ dx$

**9.** $\displaystyle\int_{-\infty}^0 e^x\ dx$

**10.** $\int_0^{+\infty} e^{-x}\ dx$

**11.** $\displaystyle\int_{-\infty}^{-1} \frac{1}{x^3}\ dx$

**12.** $\displaystyle\int_{-\infty}^{-1} \frac{1}{\sqrt[3]{x}}\ dx$

**13.** $\int_0^{+\infty} x^2 e^{-x^3}\ dx$

**14.** $\int_{-\infty}^0 x^2 e^{x^3}\ dx$

**15.** $\displaystyle\int_{-\infty}^0 \frac{x}{(x^2 + 1)^2}\ dx$

**16.** $\displaystyle\int_{-\infty}^{+\infty} e^{-|x|}\ dx$

**17.** $\displaystyle\int_{-\infty}^{+\infty} \frac{x}{(x^2 + 1)^4}\ dx$

**18.** $\displaystyle\int_{-\infty}^{+\infty} \frac{x}{x^2 + 1}\ dx$

**19.** $\int_{-\infty}^{+\infty} x^3 e^{-x^4}\ dx$

**20.** $\displaystyle\int_{-\infty}^{+\infty} \frac{x^3}{(x^4 + 3)^2}\ dx$

## Applications

21. **Production.**   Repeat Example 4 if $r(t) = 1000(t + 1)^{-5/2}$.

22. **Mining.**   It is estimated that platinum is being extracted from a certain mine at the rate of $r(t) = 10 \dfrac{t}{(t^2 + 1)^2}$ tons per year where $t$ is in years. How much will be extracted from this mine in the long term?

23. **Sales.**   A company estimates that the rate of increase, in millions of dollars, of sales for a new product is given by $r(t) = 40e^{-2t}$, where $t$ is in years. If this rate continues forever, what will be the eventual sales?

24. **Profits.**   A company estimates that the rate of increase in millions of dollars of profits from a new product is given by $\dfrac{1}{t^{4/3}}$, where $t$ is in years. If this rate continues forever, what will be the eventual profits?

25. **Radioactive Waste.**   Suppose radioactive waste from a closed dump site is entering the atmosphere over an area at a rate given by $r(t) = 100e^{-kt}$ tons per year, where $t$ is measured in years. Assuming that this rate continues forever and $k = .05$, find the total amount of waste that will enter the atmosphere.

26. **Radioactive Waste.**   Repeat the previous exercise if $k = 0.10$.

27. **Disease.**   A contagious disease is spreading into the world's human population at the rate of $p'(t) = 10e^{-.05t}$, where $p(t)$ is measured in millions and $t$ in years. If this rate continues forever, what will be the total number of people that are infected with this disease?

28. **Biology.**   A population of mice in an area is growing at the rate of $p'(t) = \dfrac{1}{(t + 1)^{5/2}}$, where $p$ is measured in thousands and $t$ in months. If this rate continues forever, what will be the total number of mice in this area?

29. **Pollutants.**   Pollutants are leaking from a closed dumpsite into a small lake nearby. The number of thousands of particles of pollutants that is accumulating in the lake is given by $N(t) = \dfrac{1}{(t + 1)^2}$, where $t$ is measured in years. Assuming this equation holds forever, how many particles of pollutants will eventually enter this lake from the dump?

## Graphing Calculator Exercises

30. By making the change of variable $x = -\ln u$, show that the improper integral $\int_0^{+\infty} f(x)\, dx$ becomes $\int_0^1 u^{-1} f(-\ln u)\, du$. Use this and Simpson's rule for $n = 128$ to approximate

$$\int_0^{+\infty} \frac{e^{-x}}{1 + x^2}\, dx.$$

### Solutions to Self-Help Exercise Set 17.4

1. In the integral $\int_0^b (t + 1)^{-3/2} \, dx$ make the change of variable $u = t + 1$. Then $du = dt$ and

$$1000 \int_0^b (t + 1)^{-3/2} \, dt = 1000 \int_1^{b+1} u^{-3/2} \, du$$

$$= -2000u^{-1/2} \big|_1^{b+1}$$

$$= 2000 \left( 1 - \frac{1}{\sqrt{b + 1}} \right)$$

$$1000 \int_0^{+\infty} (t + 1)^{-3/2} \, dt = 1000 \lim_{b \to +\infty} \int_0^b (t + 1)^{-3/2} \, dt$$

$$= \lim_{b \to +\infty} 2000 \left( 1 - \frac{1}{\sqrt{b + 1}} \right) = 2000$$

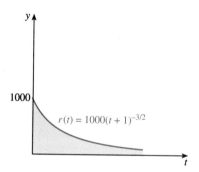

$r(t) = 1000(t + 1)^{-3/2}$

2. By symmetry, notice from the figure that $\int_{-\infty}^0 f(x) \, dx = \int_0^{+\infty} f(x) \, dx$. Then

$$\int_{-\infty}^{+\infty} f(x) \, dx = \int_{-\infty}^0 f(x) \, dx + \int_0^{+\infty} f(x) \, dx$$

$$= 2 \int_0^{+\infty} f(x) \, dx$$

$$= 2 \left( \int_0^1 dx + \int_1^{+\infty} x^{-2} \, dx \right)$$

$$= 2 + 2 \lim_{b \to +\infty} \int_1^b x^{-2} \, dx$$

$$= 2 + 2 \lim_{b \to +\infty} (-x^{-1} \big|_1^b)$$

$$= 2 + 2 \lim_{b \to +\infty} \left( 1 - \frac{1}{b} \right) = 4$$

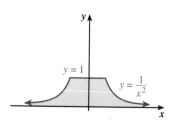

$y = 1$

$y = \dfrac{1}{x^2}$

**17.5**   **CONTINUOUS MONEY FLOW**

► *Continuous Money Flow*

► *Present Value of a Continuous Income Flow*

► *Continuous Reinvestment of Income*

► *Present Values for Infinite Intervals*

A P P L I C A T I O N
**Price of an Oil Well**

Suppose the rate of change in thousands of dollars of total income from an oil well is projected to be $f(t) = 100e^{-0.1t}$ where $t$ is measured in years. If the oil well is being offered for sale for $500,000 and has a useful life of ten years and current annual interest rates are at 10%, would you buy it? See Example 4 for the answer.

### Continuous Money Flow

We are already familiar with the idea of *continuous* compounding. When money is being compounded continuously, we are not suggesting that the bank is actually continuously placing money into our account. Rather we calculate what is in our account as if this were in fact happening. We then can think of our account as a **continuous flow of money**. This is not unlike the electric utility company that has a meter on our house continuously totaling the amount of electricity that they are selling to us. Since there has been a price set for the electricity this can be thought of as a continuous flow of income for the utility company.

We then assume that we have a continuous flow of money and that the positive function $f(t)$ represents the *rate of change* of this flow. (We could think of $f(t)$ as the velocity of the flow.) If we let $A(t)$ be the total amount of income obtained from this flow, then we are saying that $A'(t) = f(t)$. We will always assume that the amount of money from this flow at time zero is zero, that is, we will always assume that $A(0) = 0$. Then the fundamental theorem of calculus tells us that

$$A(t) = \int_0^t f(x)\, dx.$$

This is the area under the curve $f(x)$ on the interval $[0, t]$. See Figure 17.20.

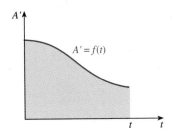

**Figure 17.20**

E X A M P L E   1   **Finding Income from an Income Flow**

Suppose an investment brings in an income of $1000 per year. How much income is obtained in three years.

Solution   Obviously the answer is $3000. But this can also be viewed as an income flow given by $f(t) = 1000$ dollars per year on the interval $[0, 3]$. The area under this curve is the definite integral

$$\int_0^3 1000 \, dx = 3000.$$

See Figure 17.21.  ∎

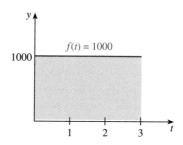

**Figure 17.21**

E X A M P L E  2    **Finding Income from an Income Flow**

Suppose the rate of change of income in thousands of dollars per year from an oil well is projected to be $f(t) = 100e^{-0.1t}$, where $t$ is measured in years. Find the total amount of money from this well during the first year. See Figure 17.22.

**Solution**    We have

$$A(1) = \int_0^1 f(x) \, dx$$

$$= \int_0^1 100e^{-0.1x} \, dx$$

$$= 100(-10)e^{-0.1t}\big|_0^1$$

$$= 1000[1 - e^{-0.1}]$$

$$\approx 95$$

or approximately $95,000.  ∎

## Present Value of a Continuous Income Flow

We will now take up the subject of **present value** of an income flow and find a formula that yields the present value. This will permit us to compare different income flows.

We again assume that we have some continuous flow of income, perhaps from oil or gas wells, accounts with continuous compounding of interest, or other accounts assumed to be compounding continuously. We assume that $f(t)$ gives

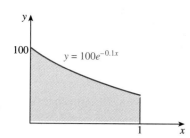

**Figure 17.22**

the rate of change of the total flow of funds with $t$ measured in some appropriate units of time that, for convenience, we will take to be years.

For any $T > 0$ we wish to define a number that we call the present value of this income flow on the interval $[0, T]$. *Intuitively we want this number to be the present amount of money needed to be able to generate the same income flow over the time interval $[0, T]$ as is generated by the given flow.* To do this we will have to assume that the present money can be invested at a certain given rate (the current available rate) over that time interval, say, a rate equal to $r$ compounded continuously.

Recall that if an amount $P$ is invested at an annual rate of $r$ (given as a decimal) and compounded continuously, then the amount becomes equal to $A = Pe^{rt}$ after $t$ years. Solving for $P$ gives the present value $P = Ae^{-rt}$. This is the present amount that is needed to attain the amount $A$ after $t$ years of continuous compounding at the annual rate of $r$.

To proceed, divide the interval $[0, T]$ into $n$ equal subintervals of length $\Delta t$ as in Figure 17.23. In a typical interval $I_i$ we pick some point $t_i$. The area under $f(t)$ on the interval $I_i$ is precisely the income generated by the given flow over the time interval $I_i$. This area $A_i$ can be approximated by the area of the rectangle given by $f(t_i)\Delta t$. The present value of this income $A_i$ is

$$P_i = A_i e^{-rt_i} \approx f(t_i)e^{-rt_i}\Delta t.$$

Summing all of these present values gives the total present value, $P$. This is also approximately the sum of all the areas of all $n$ rectangles or

$$P = \sum_{i=1}^{n} P_i \approx \sum_{i=1}^{n} f(t_i)e^{-rt_i}\Delta t,$$

which is a Riemann sum of $f(t)e^{-rt}$ on the interval $[0, T]$. If we take $n$ larger and larger, the approximations will be better and better. Then letting $n \to +\infty$, we should obtain the exact value. But this is

$$\lim_{n \to +\infty} \sum_{i=1}^{n} f(t_i)e^{-rt_i}\Delta t,$$

which is precisely the definition of the definite integral

$$\int_0^T f(t)e^{-rt}\, dt.$$

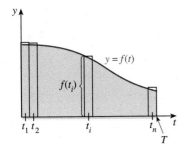

Figure 17.23

We thus have the following.

---

**Present Value of a Continuous Income Flow**

Let $f(t)$ be the continuous rate of change of total income on the interval $[0, T]$ and suppose that we can invest current money at an interest rate of $r$, where $r$ is a decimal, compounded continuously over the time interval $[0, T]$. Then the present value $P_V(T)$ of this continuous income flow over

[0, $T$] is given by

$$P_V(T) = \int_0^T f(t)e^{-rt}\, dt.$$

EXAMPLE 3     **Finding the Present Value of a Continuous Income Flow**

Suppose that the rate of change of an income flow is given by the constant (and continuous) function $f(t) = 1000$ dollars per year. If the rate $r$ is given by $r = .10$, find $P_V(3)$.

Solution

$$P_V(3) = \int_0^T f(t)e^{-rt}\, dt$$

$$= \int_0^3 1000e^{-.1t}\, dt$$

$$= 1000(-10)e^{-.1t}\big|_0^3$$

$$= 10,000[1 - e^{-.3}]$$

$$\approx 2592 \quad \blacksquare$$

EXAMPLE 4     **Determining the Price of an Oil Well**

If the oil well in Example 2 is being offered for sale for $500,000 and has a useful life of ten years, would you buy it if (a) $r = .1$, (b) $r = .05$?

Solutions      (a)   In this case

$$P_V(10) = \int_0^T f(t)e^{-rt}\, dt$$

$$= \int_0^{10} 100e^{-.1t}e^{-.1t}\, dt$$

$$= 100\int_0^{10} e^{-.2t}\, dt$$

$$= 100(-5)e^{-.2t}\big|_0^{10}$$

$$= 500[1 - e^{-2}]$$

$$\approx 432,$$

or $432,000. This is not a good deal since you can generate more income from $500,000 from an investment at current interest rates.

(b) In this case

$$P_V(10) = \int_0^T f(t)e^{-rt}\, dt$$

$$= \int_0^{10} 100e^{-.1t}e^{-.05t}\, dt$$

$$= 100 \int_0^{10} e^{-.15t}\, dt$$

$$= -\frac{100}{.15} e^{-.15t}\Big|_0^{10}$$

$$= \frac{100}{.15} [1 - e^{-1.5}]$$

$$\approx 518,$$

or \$518,000. This is a good deal since you can generate more income from this oil well than you can from \$500,000 from an investment at current interest rates. ■

**REMARK.** Naturally an important consideration in any investment is the riskiness of the investment. This has not been taken into account in the decision to buy in the previous example.

## Continuous Reinvestment of Income

As we noted before the formula $A(t) = Pe^{rt}$ relates the amount $A(t)$ in an account after $t$ years with initial amount $P$ and earning an annual rate $r$ compounded continuously. We now suppose that the continuous income stream from the flow is continuously invested and furthermore we assume that throughout the interval under consideration this income is always invested at the very same annual rate of $r$ compounded continuously. If we let $A^*(T)$ be the amount of money after $T$ years if all the income has been continuously invested at an annual rate of $r$ compounded continuously, then

$$A^*(T) = e^{rT}P_V(T).$$

To see why this is true, recall from the setup in Figure 17.23 that the amount of income obtained during the time interval $I_i$ was $A_i \approx f(t_i)\Delta t$. If this money is then invested at an annual rate of $r$ compounded continuously for the next $T - t_i$ years, this money will then become

$$A_i^*(T) = A_i e^{r(T-t_i)} \approx f(t_i)e^{r(T-t_i)}\Delta t,$$

or

$$A_i^*(T) \approx e^{rT}f(t_i)e^{-rt_i}\Delta t.$$

Summing all of these up gives the amount at time $T$, designated as $A^*(T)$, which according to this last formula, also is approximately equal to

$$\sum_{i=1}^{n} e^{rT}f(t_i)e^{-rt_i}\Delta t,$$

which can be written as

$$e^{rT}\sum_{i=1}^{n} f(t_i)e^{-rt_i}\Delta t.$$

Letting $n \to +\infty$ the latter becomes

$$e^{rT}\int_0^T f(t)e^{-rt}\, dt = e^{rT}P_V(T).$$

---

**Continuous Investment of Money**

Let $A^*(T)$ be the amount of money after $T$ years if all the income has been continuously invested at an annual rate of $r$ compounded continuously, then

$$A^*(T) = e^{rT}P_V(T) = e^{rT}\int_0^T f(t)e^{-rt}\, dt$$

---

E X A M P L E  5    **Finding the Total Amount When Continuously Reinvested**

Suppose in Example 3 that the income is continuously invested at the rate $r$ given there. How much money is there after 3 years?

**Solutions**    Using the formula for $A^*(T)$ and the result found in Example 3 we obtain

$$A^*(3) = e^{(.1)(3)}P_V(3) \approx e^{.3}(2592) \approx 3499,$$

that is, approximately \$3499.  ∎

Naturally the amount in Example 5 is greater than the amount in Example 1 since in Example 5 the money is being continuously reinvested as it comes in.

## Present Value for Infinite Intervals

If the income dispersement continues perpetually, then the function $f(t)$ must be given on the interval $[0, +\infty)$. The present value of all future income is denoted by $P_V(+\infty)$. We have the following.

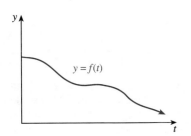

**Figure 17.24**

> **Present Value of a Continuous Income Flow on $[0, +\infty)$**
>
> Let $f(t)$ be the continuous rate of change of total income on the interval $[0, +\infty)$ and suppose that one can invest current money at an interest rate of $r$ in perpetuity, where $r$ is a decimal, compounded continuously, forever. Then the present value $P_V(+\infty)$ of this continuous income flow over $[0, +\infty)$ is given by
>
> $$P_V(+\infty) = \int_0^{+\infty} f(t)e^{-rt}\, dt$$
>
> $$= \lim_{T \to +\infty} \int_0^T f(t)e^{-rt}\, dt$$

**E X A M P L E   6**   **Present Value on an Infinite Interval**

What would be a fair initial price for the oil well in Example 2 as a function of $r$?

**Solution**

$$P_V(+\infty) = \int_0^{+\infty} f(t)e^{-rt}\, dt$$

$$= \lim_{T \to +\infty} \int_0^T f(t)e^{-rt}\, dt$$

$$= \lim_{T \to +\infty} \int_0^T 100e^{-.1t}e^{-rt}\, dt$$

$$= 100 \lim_{T \to +\infty} \int_0^T e^{-(r+.1)t}\, dt$$

$$= 100 \lim_{T \to +\infty} \frac{-1}{r+.1} e^{-(r+.1)t}\Big|_0^T$$

$$= \frac{100}{r+.1} \lim_{T \to +\infty} (1 - e^{-(r+.1)T})$$

$$= \frac{100}{r+.1}\quad \blacksquare$$

**R E M A R K .**   Notice that as $r$ increases the value of the oil well goes down. Thus if long-term interest rates go up, the value of the well will drop.

## SELF-HELP EXERCISE SET 17.5

1.  For the oil well in Example 2 find the total amount $A(t)$ of income over 10 years.

2.  If the oil well in Example 2 of the text is offered for sale for $500,000 and has a useful life of 8 years and $r = .05$, would you buy it?

3.   For the oil well in Example 2, find the total amount $A*(t)$ if all income had been continuously invested at an annual rate of 10% compounded continuously.

## EXERCISE SET 17.5

### Applications

In Exercises 1 through 10 $f(t)$ is the rate of change of total income per unit time. Find (a) the total amount of income at the given time $T$, (b) the present value $P_V(T)$ at the given time $T$ for the given interest rate $r$ that is compounding continuously, and (c) the amount at the given time $T$ if all income is reinvested continuously at the given annual rate of $r$ compounded continuously.

1.   $f(t) = 20, T = 5, r = 10\%$           2.   $f(t) = 30, T = 10, r = 10\%$

3.   $f(t) = 100e^{.05t}, T = 20, r = 5\%$

4.   $f(t) = 100e^{-.05t}, T = 40, r = 5\%$

5.   $f(t) = 20e^{-.05t}, T = 10, r = 10\%$

6.   $f(t) = 20e^{.05t}, T = 10, r = 10\%$

7.   $f(t) = t, T = 10, r = 10\%$           8.   $f(t) = t + 1, T = 10, r = 5\%$

9.   $f(t) = t^2, T = 10, r. = 10\%$           10.   $f(t) = t^2 + 1, T = 10, r = 8\%$

11.   **Investment Decision.**   An investment that has a continuous return of $150,000 per year for 10 years is being offered for sale for $850,000. Would you buy if the current interest rate $r$ that can be compounded continuously over the next 10 years is (a) 0.10?, (b) 0.15?

12.   **Investment Decision.**   An investment that has a continuous return of $240,000 per year for 10 years is being offered for sale for $1,500,000. Would you buy if the current interest rate $r$ that can be compounded continuously over the next 10 years is (a) 0.08?, (b) 0.12?

13.   **Oil Well.**   An oil well is being offered for sale for $310,000 and has a rate of change of income per year given by

$$f(t) = 100e^{-.1t},$$

in thousands of dollars and a useful life of 5 years. Would you buy if the current interest rate $r$ that can be compounded continuously over the next 5 years is (a) 0.10?, (b) 0.12?

14.   **Oil Well.**   An oil well is being offered for sale for $420,000 and has a rate of change of income per year given by

$$f(t) = 100e^{-.1t},$$

in thousands of dollars, and a useful life of 10 years. Would you buy if the current interest rate $r$ that can be compounded continuously over the next 10 years is (a) 0.10?, (b) 0.12?

In Exercises 15 through 18 $f(t)$ is the rate of change of total income per year and $r$ is the annual interest rate compounding continuously. Find $P_V(+\infty)$.

15. $f(t) = 100$, $r = 8\%$

16. $f(t) = 1000$, $r = 12\%$

17. $f(t) = 100e^{-.1t}$, $r = 8\%$

18. $f(t) = 1000e^{-.2t}$, $r = 10\%$

19. **Continuous Income Flow.** An investment with a continuous income flow of $10,000 per year for ever is being sold. Assume that the current annual interest rate of 10% compounded continuously will continue forever. Would you buy if the price was (a) $110,000?, (b) $90,000?

20. **Continuous Income Flow.** An investment with a continuous income flow of $16,000 per year forever is being sold. Assume that the current annual interest rate of 8% compounded continuously will continue forever. Would you buy if the price was (a) $250,000?, (b) $180,000?

21. **Investment Decision.** An investment with a continuous income flow of

$$24,000e^{.04t}$$

in dollars per year forever is being sold. Assume that the current annual interest rate of 8% compounded continuously will continue forever. Would you buy if the price was (a) $700,000?, (b) $550,000?

22. **Investment Decision.** An investment with a continuous income flow of

$$10,000e^{.05t}$$

in dollars per year forever is being sold. Assume that the current annual interest rate of 10% compounded continuously will continue forever. Would you buy if the price was (a) $300,000?, (b) $250,000?

## Solutions to Self-Help Exercise Set 17.5

1.

$$A(t) = \int_0^{10} f(t)\, dt$$

$$= \int_0^{10} 100e^{-.1t}\, dt$$

$$= 100(-10)e^{-.1t}\Big|_0^{10}$$

$$= 1000(1 - e^{-1}) \approx 632,$$

or $632,000.

2. In this case

$$P_V(8) = \int_0^T f(t)e^{-rt}\, dt$$

$$= \int_0^8 100e^{-.05t}e^{-.1t} \, dt$$

$$= 100 \int_0^8 e^{-.15t} \, dt$$

$$= -\frac{100}{.15} e^{-.15t}\Big|_0^8$$

$$= \frac{100}{.15} [1 - e^{-1.2}] \approx 466,$$

or $466,000. This is not a good deal since you can generate less income from this oil well than you can from $500,000 from an investment at current interest rates.

**3.** Recall from Example 4a that $P_V(10) \approx 432$, so

$$A^*(T) = e^{rT}P_V(T) = e^{(.1)(10)}P_V(10) \approx 1174,$$

or about $1,174,000.

## SUMMARY OUTLINE OF CHAPTER 17

- **Integration by parts:** $\int u \, dv = uv - \int v \, du$.

- **Trapezoidal Rule.**   For any integer $n$ divide the interval $[a, b]$ into $n$ subintervals of equal length with endpoints $a = x_0, x_1, \ldots, x_n = b$. If

$$T_n = \frac{b - a}{n}\left[\frac{1}{2}f(x_0) + f(x_1) + f(x_2) + \cdots + f(x_{n-1}) + \frac{1}{2}f(x_n)\right],$$

then

$$\int_a^b f(x) \, dx \approx T_n.$$

- **Simpson's Rule.**   For any even integer $n$ divide the interval $[a, b]$ into $n$ equal subintervals with endpoints $a = x_0, x_1, \ldots, x_n = b$, and let

$$S_n = \frac{b - a}{3n}[f(x_0) + 4f(x_1) + 2f(x_2) + 4f(x_3) + \cdots$$

$$+ 2f(x_{n-2}) + 4f(x_{n-1}) + f(x_n)].$$

Then

$$S_n \approx \int_a^b f(x) \, dx.$$

- **Error in the Trapezoidal Rule.**   The maximum possible error incurred in using the trapezoidal rule is $\dfrac{M(b - a)^3}{12n^2}$, where $|f''(x)| \leq M$ on $[a, b]$.

- **Error in Simpson's Rule.**   The maximum possible error incurred in using Simpson's rule is $\dfrac{M(b - a)^5}{180^4}$, where $|f^{(4)}(x)| \leq M$ on $[a, b]$.

- If $f(x)$ is continuous on $(a, +\infty)$, define

$$\int_a^{+\infty} f(x)\, dx = \lim_{b \to +\infty} \int_a^b f(x)\, dx,$$

if this limit exists as a finite number. Otherwise we say that the integral **diverges**. If the limit exists as a finite number we say that the integral **converges** to that number.

- If $f(x)$ is continuous on $(-\infty, b]$, define

$$\int_{-\infty}^b f(x)\, dx = \lim_{a \to -\infty} \int_a^b f(x)\, dx,$$

if this limit exists. If the limit does not exist we say that the integral **diverges**.

- If $f(x)$ is continuous on $(-\infty, +\infty)$ define

$$\int_{-\infty}^{+\infty} f(x)\, dx = \int_{-\infty}^0 f(x)\, dx + \int_0^{+\infty} f(x)\, dx,$$

if both integrals on the right exist (as finite numbers). If one or both of the integrals on the right diverges then we say that $\int_{-\infty}^{+\infty} f(x)$ **diverges**.

- Let $f(t)$ be the continuous rate of change of income on the interval $[0, T]$ and suppose that one can invest current money at an interest rate of $r$, where $r$ is a decimal, compounded continuously over the time interval $[0, T]$. Then the **present value** $P_V(T)$ of this continuous income flow over $[0, T]$ is given by

$$P_V(T) = \int_0^T f(t)e^{-rt}\, dt.$$

- Let $A^*(T)$ be the amount of money after $T$ years if all the income has been continuously invested at an annual rate of $r$ compounded continuously, then

$$A^*(T) = e^{rT}P_V(T) = e^{rT}\int_0^T f(t)e^{-rt}\, dt.$$

- Let $f(t)$ be the continuous rate of change of income on the interval $[0, +\infty)$ and suppose that one can invest current money at an interest rate of $r$ in perpetuity, where $r$ is a decimal, compounded continuously forever. Then the **present value** $P_V(+\infty)$ of this continuous income flow over $[0, +\infty)$ is given by

$$P_V(+\infty) = \int_0^{+\infty} f(t)e^{-rt}\, dt = \lim_{T \to +\infty} \int_0^T f(t)e^{-rt}\, dt.$$

## Chapter 17 Review Exercises

In Exercises 1 through 6 integrate by parts.

**1.** $\displaystyle\int xe^{5x}\, dx$

**2.** $\displaystyle\int \frac{x}{\sqrt{5 + 3x}}\, dx$

**3.** $\displaystyle\int x^5 \ln x\, dx$

**4.** $\displaystyle\int \frac{x}{(x + 1)^2}\, dx$

**5.** $\displaystyle\int x^3\sqrt{9 - x^2}\, dx$

**6.** $\displaystyle\int \frac{\ln x}{\sqrt{x}}\, dx$

In Exercises 7 through 12 evaluate using the table of integrals.

**7.** $\displaystyle\int \frac{x}{2x + 3}\, dx$

**8.** $\displaystyle\int \frac{1}{x(x + 4)^2}\, dx$

**9.** $\displaystyle\int \frac{1}{x^2\sqrt{4 - x^2}}\, dx$

**10.** $\displaystyle\int \frac{\sqrt{9 - x^2}}{x}\, dx$

**11.** $\displaystyle\int \sqrt{x^2 + 1}\, dx$

**12.** $\displaystyle\int \frac{1}{1 + e^x}\, dx$

In Exercises 13 through 16 evaluate the improper integrals.

**13.** $\displaystyle\int_{1}^{+\infty} x^{-1.1}\, dx$

**14.** $\displaystyle\int_{1}^{+\infty} x^{-.9}\, dx$

**15.** $\displaystyle\int_{-\infty}^{+\infty} x^2 e^{-|x^3|}\, dx$

**16.** $\displaystyle\int_{-\infty}^{+\infty} e^{-x}\, dx$

In Exercises 17 through 18 approximate using the trapezoidal rule and Simpson's rule for $n = 2$ and compare your answers to the correct answer.

**17.** $\displaystyle\int_{0}^{2} 5x^4\, dx$

**18.** $\displaystyle\int_{1}^{4} \frac{1}{x}\, dx$

**19.** Repeat Exercise 17 for $n = 4$.

**20.** Repeat Exercise 18 for $n = 4$.

In Exercises 21 through 23 assume $f(t) = 10e^{-.04t}$ is the rate of change of total income per year and the interest rate is 8% per year compounded continuously.

**21.** Find the total amount of income over the first 10 years.

**22.** Find the present value $P_V(10)$.

**23.** Find the amount in the account after 10 years if all income is continuously reinvested at the given interest rate.

**24.** A mine produces a rate of change of total income in thousands of dollars per year given by $f(t) = 100e^{-0.10t}$ and has a useful life of 5 years. Market interest rates are at 8% compounded continuously. What would be a reasonable price for this mine discounting any risk involved?

**25.** If the mine in the previous exercise had a life that went on forever, find $P_V(+\infty)$.

Integrals are used to
determine shelf life.
(Charles D.
Winters.)

# 18

# Probability and Calculus

18.1   Continuous Probability Density Functions

18.2   Expected Value and Variance

18.3   The Normal Distribution

**I**n this chapter we apply the integral calculus to continuous probability distributions.

## 18.1   CONTINUOUS PROBABILITY DENSITY FUNCTIONS

► *Continuous Probability Density Functions*

► *Applications*

**APPLICATION**

**The Life Span
of a Light Bulb**

Suppose that the life span of a certain light bulb has a probability density function given by $f(t) = 0.01e^{-.01t}$, where $t$ is given in hours. What is the probability that the light bulb will last at most 100 hours? See Example 6 for the answer.

## Continuous Probability Density Functions

Let $X$ be the discrete random variable giving the total number of dots on the top faces of two dice that have been tossed. Figure 18.1 shows a histogram for the probability distribution of this random variable. Since the base of each rectangle is one and the height gives the probability of each event, we then see that the *area* of each rectangle is also the probability. For example, the probability of four dots is 3/36, and this is also the area of the shaded rectangle shown in Figure 18.1 above the number 4.

In fact we can associate with the probability of any event an area of a histogram. For example, if $E$ is the event that the sum of dots on the top faces of the dice is 9, 10, or 11 when the dice are tossed, then the probability of $E$ is the area of the three shaded rectangles shown in Figure 18.2. Thus in discrete probability we became accustomed to associating probabilities with areas of rectangles. For a **continuous random variable** that can take any value in some interval $(a, b)$, we can associate with the probability of an event an area under a curve. This curve is called a **probability density function**.

**Figure 18.1**

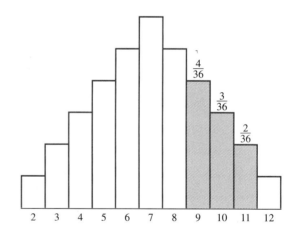

**Figure 18.2**

---

**Probability Density Function**

A function $f(x)$ defined on $(a, b)$ is called a probability density function on $(a, b)$ if the following are satisfied.

1.  For all $x \in (a, b)$, $f(x) \geq 0$.
2.  The area between the graph of $y = f(x)$ and the $x$-axis from $a$ and $b$ is one, i.e.,

$$\int_a^b f(x) \, dx = 1.$$

**Figure 18.3**

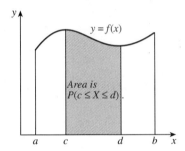

**Figure 18.4**

For the interval $(a, b)$, $a$ may be $-\infty$, $b$ may be $+\infty$, or both. See Figure 18.3.

We can now define the probability associated with a continuous random variable.

---

**Probability of a Continuous Random Variable**

Suppose $f(x)$ is a continuous probability density function on $(a, b)$ associated with the random variable $X$. The probability that the random variable $X$ assumes a value in the interval $(c, d) \subset (a, b)$, written $P(c < X < d)$, is defined to be the area between the graph of $y = f(x)$ and the $x$-axis from $c$ to $d$. That is,

$$P(c < X < d) = \int_c^d f(x)\ dx.$$

See Figure 18.4.

---

**REMARK.**     Since the area under a single point is zero, we must have

$$P(X = c) = P(c \le X \le c) = \int_c^c f(x)\ dx = 0.$$

We also must have

$$P(c < X < d) = P(c \le X \le d).$$

Sometimes we can obtain the probability density function by theoretical analysis. Other times we can obtain the probability density function (or a close approximation) from an experiment or from the collection and analysis of data. We now give an example of each instance.

**E X A M P L E   1**     **Determining a Probability Density Function Empirically**

---

We select children between one and two years old in this country at random and measure their weights in pounds. For any interval of weight such as [20, 21), we divide the numbers that have fallen into this interval by the total number measured to obtain the probability that the weight of such a child is between 20 and 21 pounds. This probability $p$ is used as the height of the rectangle shown in Figure 18.5 that lies above the interval [20, 21] so that the *area* of this rectangle is precisely the probability $p$. We do this for every one-pound interval from 0 to 45, as shown in Figure 18.5. Find a function $f(t)$ such that

$$P(c \le X \le d) = \int_c^d f(x)\ dx,$$

at least approximately.

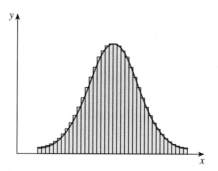

**Figure 18.5**

**Solution**     If we draw a function $f(t)$ that goes through the midpoint of the top of each rectangle such as indicated in Figure 18.5, the area under $f$ on any subinterval $[c, d]$ will be approximately the area of the rectangles enclosed by $c$ and $d$. Thus

$$P(c \le X \le d) \approx \int_c^d f(x) \, dx.$$

If this were not accurate enough we could subdivide the one-pound intervals further, perhaps into 16ths. Then take the heights of each very thin rectangle so that the *area* of the rectangle will be the probability that the weights of these children will lie in the one-ounce intervals. A continuous function smoothly connecting the midpoints of the tops of the very thin rectangles should give an even better approximation. ∎

**E X A M P L E   2**     **Determining a Probability Density Function by Theoretical Means**

**Figure 18.6**

Suppose we have a pointer attached to the center of a circle of circumference 10 inches and free to spin in a completely random manner, as shown in Figure 18.6. Let $X$ be the random variable given by the point on the interval $[0, 10]$ at which the pointer ends. (We assume that the pointer is perfectly sharp and does point at one single point.) Find the continuous function $f(x)$ that is defined on $[0, 10]$ such that

$$P(c \le X \le d) = \int_c^d f(x) \, dx.$$

**Solution**     From our knowledge of the physical setup we know that the probability that the pointer will give a value in any one interval is the same probability as any other interval of the *same* length. We will now show that the probability density function must be a constant.

From the physical setup we know that for any $s, t \in [0, 10]$ and for any $h > 0$,

$$P(s \le X \le s + h) = P(t \le X \le t + h),$$

since the intervals $[s, s + h]$ and $[t, t + h]$ have the same length $h$. If we furthermore assume that $f(x)$ is *continuous*, then for very small $h$, $f(x)$ will vary very little on $[s, s + h]$ and the average value of $f(x)$ on $[s, s + h]$ will be approximately equal to $f(s)$. Thus we have

$$f(s) \approx \text{average value of } f(x) \text{ on } [s, s + h]$$

$$= \frac{1}{h} \int_s^{s+h} f(x) \, dx$$

$$= \frac{1}{h} P(s \le X \le s + h)$$

$$= \frac{1}{h} P(t \le X \le t + h)$$

$$= \frac{1}{h} \int_t^{t+h} f(x) \, dx$$

$$= \text{average value of } f(x) \text{ on } [t, t + h]$$

$$\approx f(t)$$

Since this approximation improves as $h$ becomes smaller, we conclude that $f(s) = f(t)$. Since $s$ and $t$ were *any* points in $[0, 10]$, this means that $f(x)$ is a constant, say $C$.

Since $f(x) = C$, then

$$1 = P(0 \le X \le 10) = \int_0^{10} f(x) \, dx = \int_0^{10} C \, dx = 10C,$$

and $f(x) = C = \frac{1}{10}$. If an event $E$ is some interval $E = [c, d]$, then

$$P(E) = P(c \le X \le d) = \int_c^d f(x) \, dx = \int_c^d \frac{1}{10} \, dx = \frac{d - c}{10} = \frac{\text{size of E}}{\text{size of S}}.$$

This is intuitively what we expect, since this is analogous to a finite sample space with equally likely single events. ∎

EXAMPLE 3 | **Finding a Probability Density Function**

Find the only probability density function on $[a, b]$ that is a constant.

Solution | Set $f(x) = k$. Then the first condition in the definition of a probability density function requires that $k \ge 0$. The second condition requires that

$$1 = \int_a^b f(x) \, dx = \int_a^b k \, dx = k(b - a).$$

Thus $f(x) = k = 1/(b - a)$. See Figure 18.7. ∎

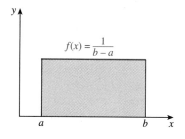

**Figure 18.7**

When the probability density function is constant as, in Figure 18.7, we say it is a *uniform* probability density function. Notice that in this case

$$P(c \le X \le d) = \int_c^d \frac{1}{b-a} \, dx = \frac{d-c}{b-a} = \frac{\text{length of } [c, d]}{\text{length of } [a, b]}.$$

This is an extension of the notion in a *finite* sample space $S$ with equally likely single events, since in this case the probability $P(E)$ of an event $E$ is given by

$$P(E) = \frac{\text{size of } E}{\text{size of } S}.$$

The uniform probability function will arise in any situation where *any two intervals of equal length give rise to the same probability*.

---

**Uniform Probability Density Function**

1. The probability distribution on $[a, b]$ with the constant probability density function $f(x) = \dfrac{1}{b-a}$ is called the **uniform probability density function**.

2. For the uniform probability density function on $[a, b]$

$$P(c \le X \le d) = \frac{d-c}{b-a}.$$

---

**E X A M P L E   4**    **Finding a Probability Density Function**

Find $k$ so that the function $f(x) = kx$ is a probability density function on $[0, 1]$.

**Solution**    The first condition requires that $k \ge 0$. The second condition requires that

$$1 = \int_a^b f(x) \, dx = \int_0^1 kx \, dx = k \frac{1}{2} x^2 \big|_0^1 = k \frac{1}{2}$$

Thus $k = 2$ and $f(x) = 2x$ is a probability density function on $[0, 1]$. See Figure 18.8. ∎

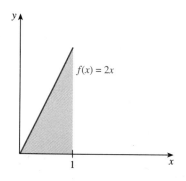

**Figure 18.8**

We have assumed that the interval $[a, b]$ is *finite*. But there are many situations when we need to consider infinite intervals. For example, we may wish to consider the probability that a light bulb will last for a certain amount of time. It is then convenient to take the interval under consideration to be $[0, +\infty)$ if we have no idea of how long the bulb may last. We naturally continue to require that $f(x) \ge 0$ on the interval $I$ under consideration and that the definite integral over $I$ still be one, where we interpret the integral as an improper integral when $I$ is infinite.

E X A M P L E 5  **Finding a Probability Density Function on an Infinite Interval**

Given a positive constant $k$, find $C$ so that the function $f(x) = Ce^{-kx}$ is a probability density function on $[0, +\infty)$.

**Solution**
Again the first condition in the definition of a probability density function requires that $C \geq 0$. The second condition requires that

$$1 = \int_0^{+\infty} f(x)\, dx$$

$$= \lim_{b \to +\infty} \int_0^b Ce^{-kx}\, dx$$

$$= \lim_{b \to +\infty} -\frac{C}{k} e^{-kx}\Big|_0^b$$

$$= \lim_{b \to +\infty} \frac{C}{k}(1 - e^{-kb})$$

$$= \frac{C}{k}.$$

$f(x) = ke^{-kx}$

**Figure 18.9**

Thus $C = k$ and $f(x) = ke^{-kx}$ is a probability density function on $[0, +\infty)$. We will refer to this as an **exponential probability density function**. See Figure 18.9. ∎

## Applications

The probability of the life of products often can be approximated by a probability with an exponential probability density function.

E X A M P L E 6  **The Life Span of a Light Bulb**

Suppose that the life span of a certain light bulb has a probability density function given by $f(t) = 0.01e^{-.01t}$, where $t$ is given in hours. Find the probability that the light bulb will last (a) at most 100 hours, (b) at least 100 hours, (c) between 100 and 200 hours.

**Solution**
First notice from the previous example that $f(t) = 0.01e^{-.01t}$ is in fact a probability density function where $k = 0.01$.

(a) Here the interval is $[0, 100]$. Thus (see Figure 18.10)

$$P(0 \leq X \leq 100) = \int_0^{100} f(x)\, dx = \int_0^{100} 0.01e^{-.01t}\, dt$$

$$= -e^{-.01t}\Big|_0^{100} = 1 - e^{-1} \approx 0.63.$$

(b) Here the interval is $[100, +\infty)$. Therefore we need to find

$$P(100 \leq X) = \int_{100}^{+\infty} f(x)\, dx,$$

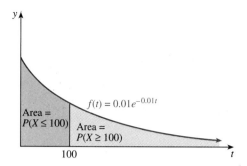

**Figure 18.10**

where $f(x) = 0.01e^{-.01t}$. This involves an improper integral. One can calculate the requested probability by avoiding this improper integral as follows.

$$P(100 \leq X) = 1 - P(X \leq 100)$$

$$= 1 - \int_0^{100} f(x)\, dx$$

$$\approx 1 - 0.63 \qquad \text{from (a)}$$

$$= .37$$

See Figure 18.10.

(c)    Here the interval is [100, 200]. Thus (see Figure 18.11),

$$P(100 \leq X \leq 200) = \int_{100}^{200} f(x)\, dx$$

$$= \int_{100}^{200} 0.01e^{-.01t}\, dt$$

$$= -e^{-.01t}\big|_{100}^{200}$$

$$= e^{-1} - e^{-2} \approx 0.23.$$

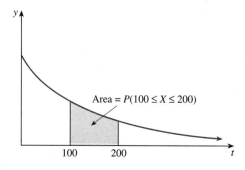

**Figure 18.11**

## SELF-HELP EXERCISE SET 18.1

1. What must be the value of $k$ if $f(x) = kx^2$ is to be a probability density function on $[1, 4]$?

2. The life in years of a certain automobile has a probability density function given by $f(t) = 0.1e^{-0.1t}$ on $[0, +\infty)$. Find the probability that the life of the vehicle is (a) at most 5 years, (b) at least 5 years, (c) between 5 and 10 years.

## EXERCISE SET 18.1

In Exercises 1 through 12 find $k$ such that the given function is a probability density function on the given interval.

1. $f(x) = kx$, $[1, 2]$
2. $f(x) = kx$, $[2, 3]$
3. $f(x) = kx^2$, $[0, 1]$
4. $f(x) = kx^2$, $[1, 2]$
5. $f(x) = kx^3$, $[0, 2]$
6. $f(x) = kx^4$, $[-1, 0]$
7. $f(x) = k/x$, $[1, e]$
8. $f(x) = k/x$, $[1, e^2]$
9. $f(x) = k/x^2$, $[1, +\infty)$
10. $f(x) = k/x^3$, $[1, +\infty)$
11. $f(x) = kx^2/(1 + x^3)^2$, $[0, +\infty)$
12. $f(x) = kx^2/(1 + x^3)^3$, $[0, +\infty)$

In Exercises 13 through 22 determine why each of the functions is *not* a probability density function on the given interval.

13. $f(x) = \frac{3}{2}x - 1$, $[0, 2]$
14. $f(x) = 3 - x^2$, $[0, 3]$
15. $f(x) = \frac{2}{3}(x^2 - 3x + 2)$, $[0, 3]$
16. $f(x) = \frac{1}{3}(x^2 - 4x)$, $[0, 3]$
17. $f(x) = 1/(x + 2)^2$, $[0, +\infty)$
18. $f(x) = 1/(x + 3)^2$, $[0, +\infty)$
19. $f(x) = 1/\sqrt{x}$, $[1, +\infty)$
20. $f(x) = 1/x$, $[1, +\infty)$
21. $f(x) = e^{-x}$, $(-\infty, +\infty)$
22. $f(x) = e^x$, $(-\infty, +\infty)$

23. Show that the function $f(x) = x/8$ is a probability density function on $[0, 4]$. Find (a) $P(0 \le X \le 1)$, (b) $P(1 \le X \le 2)$, (c) $P(0 \le X \le 4)$.

24. Show that the function $f(x) = 2x/15$ is a probability density function on $[1, 4]$. Find (a) $P(1 \le X \le 2)$, (b) $P(2 \le X \le 3)$, (c) $P(X = 3)$.

25. Show that the function $f(x) = 6x(1 - x)$ is a probability density function on $[0, 1]$. Find (a) $P(0 \le X \le 0.5)$, (b) $P(0.25 \le X \le 0.5)$, (c) $P(0.5 \le X \le 1)$.

26. Show that the function $f(x) = x^3/20$ is a probability density function on $[1, 3]$. Find (a) $P(1 \le X \le 2)$, (b) $P(2 \le X \le 3)$, (c) $P(X = 3)$.

## Applications

27. **Commuting Time.**   The time, measured in minutes, required by a certain individual to commute to work is a random phenomenon obeying a uniform probability law over the interval $[20, 25]$. If this person leaves home promptly at 7:06 A.M., what is the probability that he will arrive at work by 7:29 A.M.?

*(© William J. Kennedy/The Image Bank.)*

*(© Grant Heilman Photography, Inc.)*

**28. Roulette.** The circumference of a wheel is divided into 37 equal lengths (as in roulette) with numbers 0 through 36 marking the dividing lines. The wheel is twirled and after coming to rest, the point on the wheel located opposite a certain fixed marker is noted. Assuming the uniform probability, what is the probability that the point thus chosen will lie (a) between 0 and 10, (b) between two successive integers the first of which is odd, (c) on the number 7?

**29. Waiting Time.** A bus arrives at a certain stop every 30 minutes. A person arrives at this bus stop at random. What is the probability that this person (a) must wait no more than 5 minutes before the bus arrives, (b) must wait at least 12 minutes before the bus arrives, (c) must wait between 6 and 10 minutes, (d) must wait at least 30 minutes, (e) must wait at most 30 minutes?

**30. Waiting Time.** A geyser erupts every 60 minutes without fail. If a person arrives randomly to see this event, not knowing the schedule of the geyser, what is the probability that this person will have to wait (a) at least 15 minutes, (b) at most 35 minutes, (c) between 30 and 45 minutes, (d) at most one hour?

**31. Water Consumption.** The daily demand for water, in thousands of gallons, in a city is a continuous random variable with probability density function $f(x) = 1 - 0.50x$, on the interval $[0, 2]$. (a) What is the probability that this city will use less than 1500 gallons of water in one day? (b) What is the probability that this city will use more than 1500 gallons of water in one day? (c) What is the probability that this city will use between 500 and 1500 gallons of water in one day?

**32. Electricity Demand.** The daily demand for electricity, in thousands of kilowatt-hours, at an industrial plant is a continuous random variable with probability density function $f(x) = \frac{1}{21}(10 - x^2)$, on the interval $[0, 3]$. (a) What is the probability that this plant will use less than 2000 kilowatt-hours in one day? (b) What is the probability that this plant will use more than 1000 kilowatt-hours in one day? (c) What is the probability that this plant will use between 1000 and 2000 kilowatt-hours in one day?

**33. Battery Life.** The life in hours of a certain battery has a probability density function given by $f(t) = 0.15e^{-.15t}$ on $[0, +\infty)$. Find the probability that the life of the battery is (a) at most 10 hours, (b) at least 10 hours, (c) between 20 and 40 hours.

**34. Life of Part.** The life in days of a critical electronics component has a probability density function given by $f(t) = 0.001e^{-0.001t}$ on $[0, +\infty)$. Find the probability that the life of this component is (a) at most 1000 days, (b) at least 1000 days, (c) between 1000 and 2000 days.

**35. VCR Life.** The life in years of a certain VCR is given by the probability density function $f(t) = 1/(t + 1)^2$ on $[0, +\infty)$. Show that $f(t)$ is indeed a probability density function on $[0, +\infty)$. Find the probability that the life of the VCR is (a) at most two years, (b) at least two years, (c) between 1 and 4 years.

**36. Waiting Time.** The waiting time in hours at an auto muffler shop is a random variable with a probability density function given by $f(x) = \frac{4}{15}(x + 1)^3$ on $[0, 1]$. Find the probability that you will have to wait (a) at most 30 minutes, (b) at least 30 minutes, (c) between 15 and 45 minutes.

37. **Medicine.** The length of time in days for recovery from a certain disease is a random variable with probability density function $f(t) = .02e^{-.02t}$ on the interval $[0, +\infty)$. Find the probability that a person with this disease will recover in (a) less than 5 days, (b) more than 5 days, (c) between 5 and 10 days.

38. **Learning.** The number of hours for a new employee to learn a certain task is a random variable with probability density function $f(x) = \dfrac{1}{(x+1)^2}$, on the interval $[0, +\infty)$. Find the probability that a new employee will learn this task in (a) less than two hours, (b) more than two hours, (c) between one and two hours.

### Solutions to Self-Help Exercise Set 18.1

1. We must have

$$1 = \int_1^4 kx^2 \, dx = \frac{k}{3} x^3\Big|_1^4 = \frac{k}{3}(64 - 1) = 21k$$

Thus $k = \dfrac{1}{21}$.

2. **a.** We are seeking $P(X \le 5)$. Then

$$P(X \le 5) = \int_0^5 0.1e^{-0.1t} \, dt = -e^{-0.1t}\Big|_0^5 = 1 - e^{-0.5} \approx .39$$

**b.** We are seeking $P(X \ge 5)$. Then

$$P(X \ge 5) = 1 - P(X \le 5) \approx 1 - .39 = .61.$$

**c.** We are seeking $P(5 \le X \le 10)$. Then

$$P(5 \le X \le 10) = \int_5^{10} 0.1e^{-0.1t} \, dt$$

$$= -e^{-0.1t}\Big|_5^{10}$$

$$= e^{-0.5} - e^{-1} \approx .24$$

## 18.2    EXPECTED VALUE AND VARIANCE

▶ *Mean, Variance, Standard Deviation*
▶ *Median*

APPLICATION

**Mean and Variance for the Life of a Light Bulb**

Suppose that the life span of a certain light bulb is a random variable with a probability density function given by $f(t) = 0.01e^{-.01t}$, where $t$ is given in hours. What are the mean and variance? See Example 3 for the answer.

## Mean, Variance, Standard Deviation

Recall from discrete probability that if $x_1, x_2, \ldots, x_n$, are a *finite* number $n$ of possible values of a random variable $X$ with probabilities $p_1, p_2, \ldots, p_n$, respectively, the *mean* or *expected value* is defined as

$$E(X) = x_1 p_1 + x_2 p_2 + \cdots + x_n p_n.$$

We now would like to extend the notion of expected value to the case of a continuous probability distribution $f(x)$ on the interval $(a, b)$. But we can no longer talk about the probability that the random variable takes on *one single* value but rather must consider the probability that the random variable takes on values in some *interval*.

If the interval $(a, b)$ is finite, we divide the interval $[a, b]$ into $n$ subintervals, $I_1, I_2, \ldots, I_n$, each of length $\Delta x = \dfrac{b - a}{n}$. In each subinterval $I_i$ pick a point $x_i \in I_i$, say at the midpoint as shown in Figure 18.12. The probability that the random variable $X$ lies in $I_i$ is precisely the area under the curve $y = f(x)$ on the interval $I_i$. This area (see Figure 18.12) is approximately equal to the area of the rectangle $f(x_i)\Delta x$. Thus we have

$$E(X) = x_1 P(x \in I_1) + x_2 P(x \in I_2) + \cdots + x_n P(x \in I_n)$$

$$\approx x_1 f(x_1)\Delta x + x_2 f(x_2)\Delta x + \cdots + x_n f(x_n)\Delta x$$

$$= \sum_{i=1}^{n} x_i f(x_i)\Delta x,$$

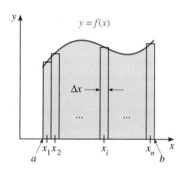

**Figure 18.12**

which is a Riemann sum for $xf(x)$ on $[a, b]$. The larger $n$ is the better the approximation. Now letting $n \to +\infty$ we obtain by definition the definite integral $\int_a^b xf(x)\, dx$. We have then motivated the following definition.

---

**Expected Value or Mean of a Continuous Random Variable**

The **expected value $E(X)$** (or **mean $\mu$**) **of a random variable $X$** with a continuous probability density function $f(x)$ on the interval $(a, b)$, is defined to be

$$E(X) = \mu = \int_a^b xf(x)\, dx.$$

---

Recall from discrete probability that if $x_1, x_2, \ldots, x_n$, are a *finite* number $n$ of possible values of a random variable $X$ with probabilities $p_1, p_2, \ldots, p_n$, respectively, the *variance* measures the spread of these values and is defined as

$$\text{Var}(X) = (x_1 - \mu)^2 p_1 + (x_2 - \mu)^2 p_2 + \cdots + (x_n - \mu)^2 p_n$$

$$= \sum_{i=1}^{n} (x_i - \mu)^2 p_i.$$

If we are to define a corresponding number in the case of a continuous random variable with continuous probability density function $f(x)$ on the interval $(a, b)$, we proceed as we did in defining the expected value. Then

$$\text{Var}(X) = \sum_{i=1}^{n} (x_i - \mu)^2 P(x_i \in I_i)$$

$$\approx \sum_{i=1}^{n} (x_i - \mu)^2 f(x_i) \Delta x,$$

which is a Riemann sum for $(x - \mu)^2 f(x)$ over $[a, b]$. Letting $n \to +\infty$ we obtain the definite integral $\int_a^b (x - \mu)^2 f(x) \, dx$. This motivates the following definition.

---

**Variance of a Continuous Random Variable**

The **variance** of a random variable $X$ with a continuous probability density function $f(x)$ on the interval $(a, b)$ is defined to be

$$\text{Var}(X) = \int_a^b (x - \mu)^2 f(x) \, dx.$$

---

This formula for variance can be simplified. First recall that

$$\mu = \int_a^b x f(x) \, dx \qquad \text{and} \qquad \int_a^b f(x) \, dx = 1.$$

Then

$$\text{Var}(X) = \int_a^b (x - \mu)^2 f(x) \, dx$$

$$= \int_a^b (x^2 - 2x\mu + \mu^2) f(x) \, dx$$

$$= \int_a^b x^2 f(x) \, dx - 2\mu \int_a^b x f(x) \, dx + \mu^2 \int_a^b f(x) \, dx$$

$$= \int_a^b x^2 f(x) \, dx - 2\mu\mu + \mu^2$$

$$= \int_a^b x^2 f(x) \, dx - \mu^2$$

---

**Alternate Form of Variance**

If $f(x)$ is a continuous probability density function on the interval $(a, b)$ with mean $\mu$, an alternative formula for the variance $\text{Var}(X)$ is given by

$$\text{Var}(X) = \int_a^b x^2 f(x) \, dx - \mu^2.$$

---

We also define the *standard deviation* $\sigma$ as the square root of the variance.

---

**Standard Deviation**

The *standard deviation*, denoted by $\sigma$, is the square root of the variance, that is,

$$\sigma = \sqrt{\text{Var}(X)}.$$

---

**E X A M P L E   1**     **Find the Mean and the Variance**

Find the mean and variance of the random variable with the constant or uniform probability density function $f(x) = 1/(b - a)$ on $[a, b]$.

**Solutions**        (a)   For the mean we have

$$\mu = \int_a^b xf(x)\, dx$$

$$= \frac{1}{b - a} \int_a^b x\, dx$$

$$= \frac{1}{b - a} \frac{1}{2} x^2 \Big|_a^b$$

$$= \frac{1}{b - a} \frac{1}{2} (b^2 - a^2)$$

$$= \frac{1}{b - a} \frac{1}{2} (b - a)(b + a)$$

$$= \frac{1}{2} (b + a).$$

(b)   We use the simplified formula for Var($x$) and first find $\int_a^b x^2 f(x)\, dx$. Thus

$$\int_a^b x^2 f(x)\, dx = \frac{1}{b - a} \int_a^b x^2\, dx$$

$$= \frac{1}{b - a} \frac{1}{3} x^3 \Big|_a^b$$

$$= \frac{1}{b - a} \frac{1}{3} (b^3 - a^3)$$

$$= \frac{1}{b - a} \frac{1}{3} (b - a)(b^2 + ab + a^2)$$

$$= \frac{1}{3} (b^2 + ab + a^2)$$

Now from (a) we have $\mu^2 = \frac{1}{4}(b + a)^2$, thus

$$\text{Var}(X) = \int_a^b x^2 f(x)\, dx - \mu^2$$

$$= \frac{1}{3}(b^2 + ab + a^2) - \frac{1}{4}(b + a)^2$$

$$= \frac{1}{12} b^2 - \frac{1}{6} ab + \frac{1}{12} a^2$$

$$= \frac{1}{12}(b^2 - 2ab + a^2)$$

$$= \frac{1}{12}(b - a)^2 \quad \blacksquare$$

---

**Mean, Variance, and Standard Deviation of a Random Variable with a Constant Probability Density Function**

Let $X$ be a random variable with constant probability density function $f(x) = 1/(b - a)$ on $[a, b]$. Then

$$\mu = \frac{1}{2}(b + a)$$

$$\text{Var}(X) = \frac{1}{12}(b - a)^2$$

$$\sigma = \frac{1}{2\sqrt{3}}(b - a)$$

---

**REMARK.** According to Figures 18.13 and 18.14, the probability density function spreads out as $b$ increases. But since $V(X) = \dfrac{(b - a)^2}{12}$, increasing $b$ is the same as increasing $V(X)$. This shows how the variance measures the spread

**Figure 18.13**

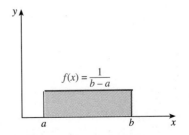

**Figure 18.14**

of a probability density function. A larger variance means a more spread-out function.

On unbounded intervals the formulas are the same but require improper integrals.

**E X A M P L E   2**      **Finding Mean and Variance on an Infinite Interval**

Find the mean and the variance of the random variable with probability density function given by

$$f(x) = ke^{-kx} \qquad k > 0,$$

on the unbounded interval $[0, +\infty)$. Refer to Figure 18.15.

(a)   We have

**Solutions**

$$\mu = \int_0^{+\infty} xf(x)\, dx$$

$$= \lim_{b \to +\infty} \int_0^b xf(x)\, dx$$

$$= \lim_{b \to +\infty} \int_0^b xke^{-kx}\, dx$$

We will need to integrate by parts. Let

$$u = x \qquad \text{and} \qquad dv = ke^{-kx}\, dx.$$

Then

$$du = dx \qquad \text{and} \qquad v = -e^{-kx}.$$

Thus

$$\int_0^b xke^{-kx}\, dx = \int_0^b u\, dv$$

$$= uv\big|_0^b - \int_0^b v\, du$$

$$= -xe^{-kx}\big|_0^b + \int_0^b e^{-kx}\, dx$$

$$= -be^{-kb} - \frac{1}{k}e^{-kx}\big|_0^b$$

$$= -b\frac{1}{e^{kb}} + \frac{1}{k} - \frac{1}{k}\frac{1}{e^{kb}}.$$

Since $k > 0$,

$$\frac{1}{k}\frac{1}{e^{kb}} \to 0 \qquad \text{as} \qquad b \to +\infty.$$

It can be shown that the first term $-b/e^{kb}$ also goes to zero as $b \to +\infty$. For example, using a calculator one can obtain in the special case that $k = 1$, that $10e^{-10} = .0005$ while your calculator will show that $20e^{-20}$ is already virtually zero. Thus

$$\int_0^b xke^{-k}\, dx \to \frac{1}{k} \qquad \text{as} \qquad b \to +\infty,$$

which means

$$\mu = \int_0^{+\infty} xf(x)\, dx = \frac{1}{k}.$$

(b)  To calculate the variance we first find $\int_0^{+\infty} x^2 ke^{-kx}\, dx$. In order to do this we first find $\int_0^b x^2 ke^{-kx}\, dx$ using integration by parts. Set

$$u = x^2 \qquad \text{and} \qquad dv = ke^{-kx}\, dx.$$

Then

$$du = 2x\, dx \qquad \text{and} \qquad v = -e^{-kx}.$$

Thus

$$\int_0^b x^2 ke^{-kx}\, dx = x^2(-e^{-kx})\big|_0^b + \int_0^b 2xe^{-kx}\, dx$$

$$= -\frac{b^2}{e^{kb}} + \frac{2}{k}\left[-b\frac{1}{e^{kb}} + \frac{1}{k} - \frac{1}{k}\frac{1}{e^{kb}}\right],$$

using the work in (a). But it can also be shown that

$$-\frac{b^2}{e^{kb}} \to 0 \qquad \text{as} \qquad b \to +\infty.$$

Therefore as $b \to +\infty$

$$\int_0^b x^2 ke^{-kx}\, dx \to \frac{2}{k^2}.$$

Thus since $\mu = 1/k$,

$$\text{Var}(X) = \int_0^{+\infty} x^2 ke^{-kx}\, dx - \mu^2$$

$$= \frac{2}{k^2} - \frac{1}{k^2}$$

$$= \frac{1}{k^2} \quad \blacksquare$$

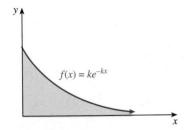

$f(x) = ke^{-kx}$

**Figure 18.15**

> **Mean, Variance, and Standard Deviation of a Random Variable with an Exponential Probability Density Function**
>
> Let $X$ be a random variable with an exponential probability density function $f(x) = ke^{-kx}$, $k > 0$, on $[0, +\infty)$. Then
>
> $$\mu = \frac{1}{k}$$
>
> $$\text{Var}(X) = \frac{1}{k^2}.$$
>
> $$\sigma = \frac{1}{k}$$

**E X A M P L E   3**    **Finding the Mean and Variance on an Infinite Interval**

Suppose that the life span of a certain light bulb is a random variable with a probability density function given by $f(t) = 0.01e^{-.01t}$, where $t$ is given in hours. Refer to Figure 18.16. Find the mean, variance, and standard deviation.

**Solution**    Substituting $k = .01$ in the previous example yields

$$\mu = \frac{1}{k} = \frac{1}{.01} = 100$$

$$\text{Var}(X) = \frac{1}{k^2} = \frac{1}{(.01)^2} = 10{,}000,$$

$$\sigma = \sqrt{V(x)} = 100 \quad \blacksquare$$

We then expect a typical light bulb from this group to last 100 hours, but the lifetimes will vary greatly by virtue of the large variance.

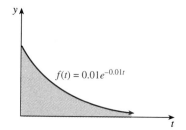

$f(t) = 0.01e^{-0.01t}$

**Figure 18.16**

## Median

For any continuous random variable, the **median outcome** $x_m$ is that outcome for which $P(X \le x_m) = \frac{1}{2}$.

> **The Median of a Continuous Random Variable**
>
> If $f(x)$ is the probability density function on $(a, b)$ associated with this continuous random variable, then the **median**, denoted by $x_m$, is that number for which
>
> $$\int_a^{x_m} f(x)\, dx = \frac{1}{2}.$$

**REMARK.**   As noted in Figure 18.17, the area under $f(x)$ on $[a, x_m]$ is exactly $\frac{1}{2}$, as is the area under $f(x)$ on $[x_m, b]$.

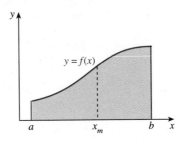

**Figure 18.17**

**E X A M P L E   4**     **Finding the Median**

Find the median of the random variable with probability density function given by

$$f(x) = ke^{-kx} \qquad k > 0,$$

on the unbounded interval $[0, +\infty)$. Refer to Figure 18.18.

**Solution**
$$\frac{1}{2} = \int_0^{x_m} ke^{-kx}\, dx$$

$$= -e^{-kx}\big|_0^{x_m}$$

$$= 1 - e^{-kx_m}$$

Thus

$$e^{-kx_m} = \frac{1}{2}$$

$$-kx_m = \ln\frac{1}{2} = -\ln 2$$

$$x_m = \frac{1}{k}\ln 2 \approx 0.693\,\frac{1}{k} \quad\blacksquare$$

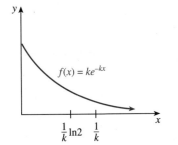

**Figure 18.18**

## SELF-HELP EXERCISE SET 18.2

1.  Find the mean, variance, and standard deviation of the random variable with probability density function $f(x) = 2x$ on $[0, 1]$.

2.  Find the mean, variance, and standard deviation of the random variable with exponential probability density function $f(x) = 0.5e^{-0.5x}$ on $[0, +\infty)$.

## EXERCISE SET 18.2

In Exercises 1 through 10 find the (a) mean $\mu$, (b) variance, (c) standard deviation $\sigma$ of the given probability density functions on the given interval.

1.  $f(x) = \dfrac{1}{4}$, [1, 5]

2.  $f(x) = \dfrac{1}{8}$, [2, 10]

3.  $f(x) = \dfrac{x}{8}$, [0, 4]

4.  $f(x) = \dfrac{2x}{15}$, [1, 4]

5.  $f(x) = 6x(1 - x)$, [0, 1]

6.  $f(x) = x^3/20$, [1, 3]

7.  $f(x) = .001e^{-.001x}$, [0, $+\infty$)

8.  $f(x) = .15e^{-.15x}$, [0, $+\infty$)

9.  $f(x) = 4/x^5$, [1, $+\infty$)

10.  $f(x) = 3/x^4$, [1, $+\infty$)

11.  Find the formula for the median of the uniform probability density function $f(x) = 1/(b - a)$ on [a, b]

12.  For the uniform probability density function $f(x) = 1/(b - a)$, on [a, b], show that

$$P(\mu - \sigma \leq X \leq \mu + \sigma) = \frac{1}{\sqrt{3}},$$

where $\mu$ and $\sigma$ are respectively the mean and standard deviation of the uniform distribution.

13.  For the probability density function $f(x) = ke^{-kx}$, $k > 0$, on [0, $+\infty$), show that

$$P(\mu - \sigma \leq X \leq \mu + \sigma) = 1 - e^{-2},$$

where $\mu$ and $\sigma$ are respectively the mean and standard deviation of the distribution.

14.  For the probability density function $f(x) = ke^{-kx}$, $k > 0$, on [0, $+\infty$), show that

$$P(X \leq \mu) = 1 - e^{-1},$$

where $\mu$ is the mean of the distribution.

15.  For the probability density function $f(x) = ke^{-kx}$, $k > 0$, on [0, $+\infty$), show that

$$P(\mu \leq X \leq +\infty) = e^{-1},$$

where $\mu$ is the mean of the distribution.

16.  Let $f(x) = xe^{-x}$. (a) Show $f(x)$ is a probability density function on [0, $+\infty$); (b) find the mean; (c) find the variance.

## Applications

17. **Waiting Time.** The waiting time, in minutes, in the speedy checkout line in a certain grocery store is a continuous random variable with a probability density function $f(x) = .20 - .02x$, on the interval $[0, 10]$. Find the expected waiting time, the variance, and the standard deviation.

18. **Product Life.** The time in years before a product wears out is a continuous random variable with a probability density function $f(x) = .15 - .0015x^2$ on the interval $[0, 10]$. Find the expected life of the product, the variance, and the standard deviation.

19. **Heating Oil Consumption.** The daily consumption for heating oil, in thousands of gallons, in a town is a continuous random variable with probability density function $f(x) = .06x - .006x^2$, on the interval $[0, 10]$. Find the expected consumption, the variance, and the standard deviation.

20. **Demand.** The daily demand for electricity in thousands of kilowatt-hours in a city is a continuous random variable with probability density function $f(x) = .012x^2 - .0012x^3$ on the interval $[0, 10]$. Find the expected demand, the variance, and the standard deviation.

21. **Water Consumption.** The daily demand for water, in thousands of gallons, in a city is a continuous random variable with probability density function $f(x) = 1 - 0.50x$ on the interval $[0, 2]$. Find the expected demand, the variance, and the standard deviation.

22. **Electricity Demand.** The daily demand for electricity, in thousands of kilowatt-hours, at an industrial plant is a continuous random variable with probability density function $f(x) = \frac{1}{21}(10 - x^2)$, on the interval $[0, 3]$. Find the expected demand, the variance, and the standard deviation.

23. **Revenue.** The daily revenue, in thousands of dollars, of a certain company is a continuous random variable with a probability density function $f(x) = 3/(x + 1)^4$, on the interval $[0, +\infty)$. Find the expected revenue, the variance, and the standard deviation.

24. **Profit.** The daily profit, in thousands of dollars, of a certain company is a continuous random variable with a probability density function $f(x) = 4/(x + 1)^5$, on the interval $[0, +\infty)$. Find the expected profit, the variance, and the standard deviation.

25. **Shelf-Life.** The shelf-life, in months, of a certain cereal is a continuous random variable with a probability density function $f(x) = \frac{3}{14}\sqrt{x + 1}$ on $[0, 3]$. Find the expected shelf-life, the variance, and the standard deviation. *Hint:* Use integration by parts.

26. **Waiting Time.** The waiting time in hours at an auto muffler shop is a random variable with a probability density function given by $f(x) = \frac{4}{15}(x + 1)^3$ on $[0, 1]$. What is the expected waiting time? *Hint:* Use integration by parts.

27. **Medicine.** The length of time in days for recovery from a certain disease is a random variable with probability density function $f(t) = .02e^{-.02t}$ on the interval $[0, +\infty)$. Find the expected recovery time, the variance, and the standard deviation.

28. **Learning.** The number of hours for a new employee to learn a certain task is a random variable with probability density function $f(x) = 3/x^4$, on the interval $[1, +\infty)$. Find the expected time of recovery, the variance, and the standard deviation.

## Solutions to Self-Help Exercise Set 18.2

1. $\mu = \displaystyle\int_a^b xf(x)\,dx = \int_0^1 x(2x)\,dx = \frac{2}{3}x^3\Big|_0^1 = \frac{2}{3}$

$\text{Var}(X) = \displaystyle\int_a^b x^2 f(x)\,dx - \mu^2 = \int_0^1 x^2(2x)\,dx - \left(\frac{2}{3}\right)^2 = \frac{1}{2}x^4\Big|_0^1 - \frac{4}{9} = \frac{1}{18}$

$\sigma = \sqrt{\text{Var}(X)} = \sqrt{\frac{1}{18}} = \frac{1}{3\sqrt{2}}$

2. According to the formulas developed in the text:

$$\mu = \frac{1}{k} = \frac{1}{.5} = 2$$

$$\text{Var}(X) = \frac{1}{k^2} = \frac{1}{(.5)^2} = 4$$

$$\sigma = \frac{1}{k} = 2$$

## 18.3    THE NORMAL DISTRIBUTION

▶ *The Standard Normal Distribution*

▶ *The Normal Distribution*

APPLICATION
**Weights of Adult Males**

Assume the weights of adult males in this country have a mean of 180 pounds and a standard deviation of 30 pounds. What is the probability that the weight of an adult male in this country is between 160 and 200 pounds? See Example 2 for the answer.

## The Standard Normal Distribution

The normal probability distribution is arguably the most important of the probability distributions. The reason is that many natural phenomena obey a normal distribution, and also, under various conditions, the normal distribution approximates many other probability laws and is easy to work with.

We first must consider the *standard* normal distribution.

---

**Standard Normal Distribution**

The random variable $Z$ with continuous probability density function given by

$$f(z) = \frac{1}{\sqrt{2\pi}} e^{-z^2/2},$$

on the interval $(-\infty, +\infty)$ is said to satisfy the **standard normal distribution** with mean $\mu = 0$ and stanard deviation $\sigma = 1$.

---

Our convention is to reserve the letter $Z$ for a random variable that has a standard normal distribution.

The graph of this function is shown in Figure 18.19. Notice its bell-shaped character and its symmetry about $z = 0$. The fact that it *is* a probability density function, that is,

$$\int_{-\infty}^{+\infty} \frac{1}{\sqrt{2\pi}} e^{-z^2/2}\, dz = 1,$$

can be shown using advanced techniques. Since this distribution is evenly spread about the origin, one suspects that $\mu = 0$, which is the case. Actually calculating the standard deviation is beyond the scope of this text. In fact $\sigma = 1$.

According to Figure 18.20, slightly more than 68% of the area under the curve is found over the interval $[-1, 1]$ while slightly more than 95% of the area under the curve is found over the interval $[-2, 2]$. Slightly more than 99.7% of the area under the curve is found over the interval $[-3, 3]$. This gives an indication of just how small the "tails" of the curve are.

Standard normal distribution

**Figure 18.19**

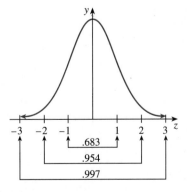

**Figure 18.20**

Since

$$P(c \le Z \le d) = \int_c^d \frac{1}{\sqrt{2\pi}} e^{-z^2/2} \, dz,$$

naturally, we will need to calculate this definite integral. However, recall that the function $e^{-z^2/2}$ is a function that can only be integrated by numerical means! But the importance of the normal distribution is so great that tables have been created that can be used to find such integrals. The tables give the values of

$$\int_{-\infty}^d \frac{1}{\sqrt{2\pi}} e^{-z^2/2} \, dz,$$

for various discrete values of $d$, which is the area under the curve to the left of $d$, shown as the shaded region in Figure 18.21. We designate this area as $\mathcal{A}(d)$. Use the table in Appendix B.1 to find this number.

If we are asked for $P(Z \le d)$, this is just $\mathcal{A}(d)$. Thus

$$P(Z \le d) = \mathcal{A}(d),$$

where one looks up $\mathcal{A}(d)$ in Appendix B.1.

If we are asked for $P(d \le Z)$, this is

$$\int_d^{+\infty} \frac{1}{\sqrt{2\pi}} e^{-z^2/2} \, dz,$$

which is also the unshaded area under the curve in Figure 18.21, that is, the area to the right of $d$. This area also is the entire area under the curve (which is one) less the area under the curve to the left of $d$. Thus

$$P(d \le Z) = 1 - \mathcal{A}(d).$$

Finally, if we are asked for $P(c \le Z \le d)$ this is

$$\int_c^d \frac{1}{\sqrt{2\pi}} e^{-z^2/2} \, dz,$$

which is the area under the curve from $c$ to $d$ shown in Figure 18.22. This can be obtained as the area to the left of $d$, $\mathcal{A}(d)$, less the area to the left of $c$, $\mathcal{A}(c)$.

**Figure 18.21**

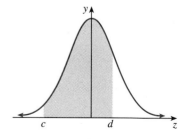

**Figure 18.22**

Thus

$$P(c \leq Z \leq d) = \mathcal{A}(d) - \mathcal{A}(c),$$

where both the numbers $\mathcal{A}(d)$ and $\mathcal{A}(c)$ are obtained in Appendix B.1.

---

**Finding the Probabilities of a Random Variable with a Standard Normal Distribution**

If $Z$ is a random variable with a standard normal probability density function, then

$$P(Z \leq d) = \int_{-\infty}^{d} \frac{1}{\sqrt{2\pi}} e^{-z^2/2} \, dz = \mathcal{A}(d)$$

$$P(d \leq Z) = \int_{d}^{+\infty} \frac{1}{\sqrt{2\pi}} e^{-z^2/2} \, dz = 1 - \mathcal{A}(d)$$

$$P(c \leq Z \leq d) = \int_{c}^{d} \frac{1}{\sqrt{2\pi}} e^{-z^2/2} \, dz = \mathcal{A}(d) - \mathcal{A}(c)$$

---

**E X A M P L E  1**  **Finding Probabilties of a Normally Distributed Random Variable**

Using the table of the normal distribution found in the Appendix find (a) $P(Z \leq 1.12)$, (b) $P(Z \geq 1.12)$, (c) $P(-.50 \leq Z \leq 0.71)$.

**Solutions**  (a)  Here

$$P(Z \leq 1.12) = \mathcal{A}(1.12),$$

which, according to the table in Appendix B.1, is .8686.

(b)  Here

$$P(Z \geq 1.12) = 1 - \mathcal{A}(1.12) = 1 - .8686 = .1314.$$

(c)  Here

$$P(-.50 \leq Z \leq 0.71) = \mathcal{A}(0.71) - \mathcal{A}(-0.50),$$

which, according to the table in Appendix B.1, is

$$.7611 - .3085 = .4526 \quad \blacksquare$$

## The Normal Distribution

We are now prepared to consider the normal distribution.

---

### The Normal Distribution

The continuous random variable $X$ is normally distributed with mean $\mu$ and standard deviation $\sigma$ if its probability density function is given by

$$f(x) = \frac{1}{\sigma\sqrt{2\pi}} \, e^{-0.5[(x-\mu)/\sigma]^2} \qquad \text{on } (-\infty, +\infty).$$

See Figure 18.23.

---

Showing that this function *is* a probability function, that is,

$$\int_{-\infty}^{+\infty} \frac{1}{\sigma\sqrt{2\pi}} \, e^{-0.5[(x-\mu)/\sigma]^2} \, dx = 1,$$

and that its mean is $\mu$, and that its standard deviation is $\sigma$ is beyond the scope of this text.

Notice that the standard normal distribution is obtained by setting $\mu = 0$ and $\sigma = 1$.

Figure 18.24 shows some normal density functions for $\sigma = 1$ and different values of $\mu$. The curves for different values of $\mu$ and the same $\sigma$ are simply translations of each other. Figure 18.25 shows several normal density functions all with $\mu = 0$ and different values of $\sigma$. The curves are all centered about the same point $\mu = 0$ but are more spread out for larger $\sigma$.

A normal probability density function has many properties that make it easy to manipulate. Thus for mathematical convenience we will sometimes assume that a random phenomenon has a normal density function if its actual density function has a shape similar to the bell shape of the normal density function. An example is the probability distribution function associated with the weight of a

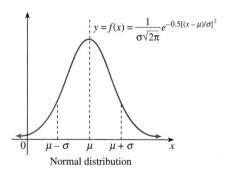

$$y = f(x) = \frac{1}{\sigma\sqrt{2\pi}} e^{-0.5[(x-\mu)/\sigma]^2}$$

Normal distribution

**Figure 18.23**

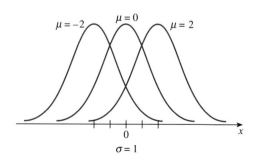

$\mu = -2$    $\mu = 0$    $\mu = 2$

$\sigma = 1$

**Figure 18.24**

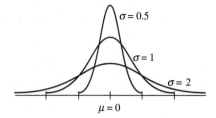

**Figure 18.25**

human being. However, in doing this one must exercise some care since the "tails" of the normal density function have small but finite areas. Thus we might conclude that there is *some* probability that a human being could have *negative* weight.

Also it may happen that a random phenomenon does not have a normal density function but that a simple numerical transformation of the measurement does.

Naturally, given specific values for $\mu$ and $\sigma$, we are interested in calculating

$$P(c \leq X \leq d) = \int_c^d \frac{1}{\sigma\sqrt{2\pi}} e^{-0.5[(x-\mu)/\sigma]^2} \, dx$$

for a whole variety of numbers $c$ and $d$. Will we need a complete table of values for every pair of numbers $\mu$ and $\sigma$? Fortunately not.

A simple substitution will change this definite integral to one that has the *standard* probability density function as the integrand. Let

$$z = \frac{x - \mu}{\sigma}.$$

Then $dz = dx/\sigma$. Also when $x = c$, $z = (c - \mu)/\sigma$, and when $x = d$, $z = (d - \mu)/\sigma$. Thus

$$P(c \leq X \leq d) = \int_c^d \frac{1}{\sigma\sqrt{2\pi}} e^{-0.5[(x-\mu)/\sigma]^2} \, dx$$

$$= \int_{(c-\mu)/\sigma}^{(d-\mu)/\sigma} \frac{1}{\sigma\sqrt{2\pi}} e^{-0.5z^2} \, \sigma \, dz$$

$$= \int_{(c-\mu)/\sigma}^{(d-\mu)/\sigma} \frac{1}{\sqrt{2\pi}} e^{-0.5z^2} \, dz.$$

This is the area under the *standard* normal density function on the interval $[(c - \mu)/\sigma, (d - \mu)/\sigma]$, which is $\mathcal{A}((d - \mu)/\sigma) - \mathcal{A}((c - \mu)/\sigma)$. This area can be found in the table in Appendix B.1 as we did before.

**Finding Normally Distributed Probabilities**

If $X$ is a normally distributed random variable with mean $\mu$ and standard deviation $\sigma$, then

$$P(X \le d) = \mathcal{A}\left(\frac{d - \mu}{\sigma}\right)$$

$$P(d \le X) = 1 - \mathcal{A}\left(\frac{d - \mu}{\sigma}\right)$$

$$P(c \le X \le d) = \mathcal{A}\left(\frac{d - \mu}{\sigma}\right) - \mathcal{A}\left(\frac{c - \mu}{\sigma}\right)$$

**E X A M P L E   2**    **Finding Normally Distributed Probabilities**

Assume the weights of adult males in this country have a mean of 180 pounds and a standard deviation of 30 pounds. Assuming a normal distribution, find the probability that the weight of an adult male in this country is (a) less than 160 pounds, (b) greater than 160 pounds, (c) between 160 and 200 pounds.

**Solutions**        (a)            $$P(X \le d) = \mathcal{A}\left(\frac{d - \mu}{\sigma}\right)$$

$$= \mathcal{A}\left(\frac{160 - 180}{30}\right)$$

$$\approx \mathcal{A}(-.67)$$

$$= .2514 \qquad \text{from Appendix B.1}$$

Refer to Figure 18.26.

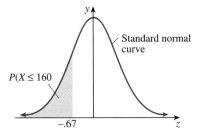

**Figure 18.26**

(b)   This is just one less the probability found in (a), or about .75.

(c)  $P(c \le X \le d) = \mathcal{A}\left(\dfrac{d - \mu}{\sigma}\right) - \mathcal{A}\left(\dfrac{c - \mu}{\sigma}\right)$

$$= \mathcal{A}\left(\dfrac{200 - 180}{30}\right) - \mathcal{A}\left(\dfrac{160 - 180}{30}\right)$$

$$\approx \mathcal{A}(.67) - \mathcal{A}(-.67)$$

$$= .7486 - .2514 \qquad \text{from (a) and Appendix B.1}$$

$$\approx .50 \quad \blacksquare$$

**R E M A R K .**    Notice that part (c) indicated that about 50% of the area under the standard normal distribution lies on the interval $[-.67, .67]$. Refer to Figure 18.27.

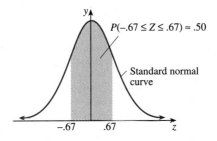

$P(-.67 \le Z \le .67) \approx .50$

Standard normal curve

$-.67 \qquad .67$

**Figure 18.27**

## SELF-HELP EXERCISE SET 18.3

In each of the following assume that $X$ is a normally distributed random variable with $\mu = 7$ and $\sigma = 10$. Find

**1.**  $P(X \le 5)$;  **2.**  $P(X \ge 5)$;  **3.**  $P(5 \le X \le 10)$.

## EXERCISE SET 18.3

In Exercises 1 through 12 find the indicated probabilities given that $Z$ is a random variable with a standard normal distribution.

**1.**  $P(Z \le 0.5)$  **2.**  $P(Z \le 0.75)$  **3.**  $P(Z \le -0.3)$

**4.**  $P(Z \le -1.5)$  **5.**  $P(Z \ge 1.5)$  **6.**  $P(Z \ge 0.6)$

**7.**  $P(Z \ge -0.8)$  **8.**  $P(Z \ge -1.42)$  **9.**  $P(1.0 \le Z \le 1.5)$

**10.**  $P(0 \le Z \le 1.5)$  **11.**  $P(-1.0 \le Z \le -0.5)$

**12.**  $P(-0.5 \le Z \le 1.0)$

In Exercises 13 through 20 find the indicated probabilities assuming that $x$ is a random variable with a normal distribution with the given mean and standard deviation.

13.  $P(x \le 50)$, $\mu = 38$, $\sigma = 8$        14.  $P(x \le 200)$, $\mu = 100$, $\sigma = 50$

15.  $P(x \ge 0.01)$, $\mu = 0.006$, $\sigma = 0.002$

16.  $P(x \ge -10)$, $\mu = 20$, $\sigma = 30$

17.  $P(100 \le x \le 150)$, $\mu = 20$, $\sigma = 100$

18.  $P(10 \le x \le 20)$, $\mu = 5$, $\sigma = 10$

19.  $P(0.01 \le x \le 0.02)$, $\mu = 0.005$, $\sigma = 0.01$

20.  $P(-3 \le x \le 10)$, $\mu = 4$, $\sigma = 10$

21.  Find the area under the standard normal density function on the interval $[-2, 2]$.

22.  Find the area under the standard normal density function on the interval $[-3, 3]$.

23.  Show that the function $f(x) = e^{-0.5x^2}$ is increasing on $(-\infty, 0)$, decreasing on $(0, +\infty)$, and has a maximum at $x = 0$.

24.  For any $\mu > 0$ and $\sigma > 0$, show that the function $f(x) = e^{-0.5[(x-\mu)/\sigma]^2}$ is increasing on $(-\infty, \mu)$, decreasing on $(\mu, +\infty)$, and has a maximum at $x = \mu$.

25.  Show that the function $f(x) = e^{-0.5x^2}$ is concave up on the intervals $(-\infty, -1)$ and $(1, +\infty)$ while concave down on $(-1, 1)$.

26.  Show that the function $f(x) = e^{-0.5[(x-\mu)/\sigma]^2}$ is concave up on the intervals $(-\infty, \mu - \sigma)$ and $(\mu + \sigma, +\infty)$ while concave down on $(\mu - \sigma, \mu + \sigma)$.

## Applications

27.  **Quality Control.**   A machine produces ball bearings with diameters normally distributed. The mean diameter is 3.50 cm and the standard deviation is 0.02 cm. Quality requirements demand a ball bearing to be rejected if the diameter is more than 0.05 cm different than the mean. What percentage of bearings will be rejected?

28.  **Quality Control.**   The length, in feet, of a certain structural steel beam is normally distributed with the mean 10 feet and standard deviation 1 inch. Quality requirements demand a beam to be rejected if the length is more than 2 inches different than the mean. What percentage of beams will be rejected?

29.  **Revenue.**   The weekly revenue, in thousands of dollars, of a certain retail store is normally distributed with mean 1000 and standard deviation 600. What percentage of weeks will revenue (a) exceed 1600, (b) be less than 600, (c) be between 800 and 1200?

30.  **Sales.**   The number of a particular item sold in a week is normally distributed with mean 1000 and standard deviation 400. What percentage of weeks will sales (a) exceed 1500, (b) be less than 600, (c) be between 800 and 1200?

31.  **Manufacturing.**   The number of ounces of detergent in an advertised one-pound box is normally distributed with mean 16.1 oz. and standard deviation 0.3. What percentage of boxes have at least one pound of detergent in them?

32. **Manufacturing.**  A company regularly orders 2000 copies of a newsletter from a printer. The number of copies delivered is normally distributed with mean 2000 and standard deviation 100. The delivery is not acceptable if the number delivered is less than 90% of the original order. What percentage of deliveries is not acceptable?

33. **Manufacturing.**  The amount of tape on a roll is normally distributed with mean 30 feet and standard deviation 2 feet. What percentage of rolls will have less than 27 feet?

34. **Manufacturing.**  The amount of soda in a 16-ounce can is normally distributed with the mean 16 ounces and standard deviation .50 ounce. What percentage of these cans will have less than 15 ounces?

35. **Medicine.**  The time, in hours to perform a certain operation is normally distributed with mean 3 hours and standard deviation 1 hour. What percentage of times will this operation be (a) less than 1 hour, (b) more than 3 hours, (c) betweeen 1 and 3 hours?

36. **Biology.**  The weight, in pounds, of a certain type of adult squirrel is normally distributed with mean 3 pounds and standard deviation .50 pound. What percentage of these squirrels have weight (a) less than 2 pounds, (b) greater than 4 pounds, (c) between 2 and 4 pounds?

37. **Learning Time.**  The learning time for a particular task on an assembly line is normally distributed with mean 5 hours and standard deviation 2 hours. If an employee is given an eight-hour shift in which to learn the task, what is the probability that the task will be learned before the shift ends?

38. **Testing.**  A corporation notes that scores on an intelligence test for newly hired employees is normally distributed with mean 100 and standard deviation 20. What percentage of new employees should score (a) above 110, (b) below 80, (c) between 95 and 105?

## Solutions to Self-Help Exercise Set 18.3

1.
$$P(X \leq d) = \mathcal{A}\left(\frac{d - \mu}{\sigma}\right)$$

$$= \mathcal{A}\left(\frac{5 - 7}{10}\right) = \mathcal{A}(-.20)$$

$$= .4207 \qquad \text{from Appendix B.1}$$

2. This is just one less the probability found in (1) or about .5793.

3.
$$P(c \leq X \leq d) = \mathcal{A}\left(\frac{d - \mu}{\sigma}\right) - \mathcal{A}\left(\frac{c - \mu}{\sigma}\right)$$

$$= \mathcal{A}\left(\frac{10 - 7}{10}\right) - \mathcal{A}\left(\frac{5 - 7}{10}\right)$$

$$= \mathcal{A}(.30) - \mathcal{A}(-.20)$$

$$= .6179 - .4207 \qquad \text{from (a) and Appendix B.1}$$

$$= .1972$$

## SUMMARY OUTLINE OF CHAPTER 18

- A function $f(x)$ defined on $(a, b)$ is called a **probability density function** on $(a, b)$ if the following are satisfied.

  1. For all $x \in (a, b)$, $f(x) \geq 0$.
  2. The area between the graph of $y = f(x)$ and the $x$-axis from $a$ and $b$ is one, i.e.,

  $$\int_a^b f(x)\, dx = 1.$$

  For the interval $(a, b)$, $a$ may be $-\infty$, $b$ may be $+\infty$, or both.

- Suppose $f(x)$ is a continuous probability density function on $(a, b)$ for a continuous random variable $X$. The **probability that the random variable $X$ assumes a value in the interval** $(c, d) \subset (a, b)$, written $P(c < X < d)$, is defined to be the area between the graph of $y = f(x)$ and the $x$-axis from $c$ to $d$. That is,

  $$P(c < X < d) = \int_c^d f(x)\, dx.$$

- The probability distribution on $[a, b]$ with the constant probability density function $f(x) = \dfrac{1}{b - a}$ is called the **uniform probability function**. For the uniform probability function

  $$P(c \leq X \leq d) = \frac{d - c}{b - a}.$$

- The probability density function given by $f(x) = ke^{-kx}$ on $[0, +\infty)$ is called the **exponential probability density function**.

- The **expected value** $E(X)$ (or **mean** $\mu$) of a random variable $X$ with a continuous probability density function $f(x)$ on the interval $(a, b)$ is defined to be

  $$E(X) = \mu = \int_a^b x f(x)\, dx.$$

- The **variance** of a random variable $X$ with a continuous probability density function $f(x)$ on the interval $(a, b)$ is defined to be

  $$V(X) = \int_a^b (x - \mu)^2 f(x)\, dx.$$

- **Alternate Form of Variance.**  If $f(x)$ is a continuous probability density function on the interval $(a, b)$ with mean $\mu$, the variance $V(X)$ is also given by

  $$\int_a^b x^2 f(x)\, dx - \mu^2$$

- The *standard deviation*, denoted by $\sigma$, is the square root of the variance, that is,

$$\sigma = \sqrt{V(X)}.$$

- The mean, variance, and standard deviation of a random variable with a constant probability density function $f(x) = \dfrac{1}{b-a}$ on $[a, b]$ is $\mu = \dfrac{1}{2}(b + a)$, $\text{Var}(X) = \dfrac{1}{12}(b - a)^2$, $\sigma = \dfrac{1}{2\sqrt{3}}(b - a)$.

- The mean, variance, and standard deviation of a random variable with an exponential probability density function $f(x) = ke^{-kx}$, $k > 0$, on $[0, +\infty)$ is $\mu = \dfrac{1}{k}$, $\text{Var}(X) = \dfrac{1}{k^2}$, $\sigma = \dfrac{1}{k}$.

- For any continuous random variable, the **median** outcome $x_m$ is that outcome for which $P(X \leq x_m) = \frac{1}{2}$.

- The random variable $Z$ with continuous probability distribution given by the function

$$f(z) = \frac{1}{\sqrt{2\pi}}\, e^{-z^2/2},$$

on the interval $(-\infty, +\infty)$ is said to satisfy the **standard normal distribution** with mean $\mu = 0$ and standard deviation $\sigma = 1$.

- If $Z$ is a random variable with a standard normal probability density function, then

$$P(Z \leq d) = \int_{-\infty}^{d} \frac{1}{\sqrt{2\pi}}\, e^{-z^2/2}\, dz = \mathcal{A}(d)$$

$$P(d \leq Z) = \int_{d}^{+\infty} \frac{1}{\sqrt{2\pi}}\, e^{-z^2/2}\, dz = 1 - \mathcal{A}(d)$$

$$P(c \leq Z \leq d) = \int_{c}^{d} \frac{1}{\sqrt{2\pi}}\, e^{-z^2/2}\, dz = \mathcal{A}(d) - \mathcal{A}(c)$$

- The continuous random variable $X$ is normally distributed with mean $\mu$ and standard deviation $\sigma$ if its probability density function is given by

$$f(x) = \frac{1}{\sigma\sqrt{2\pi}}\, e^{-0.5[(x-\mu)/\sigma]^2} \qquad \text{on } (-\infty, +\infty).$$

- If $X$ is a normally distributed random variable with mean $\mu$ and standard deviation $\sigma$, then

$$P(X \leq d) = \mathcal{A}\left(\frac{d - \mu}{\sigma}\right)$$

$$P(d \leq X) = 1 - \mathcal{A}\left(\frac{d - \mu}{\sigma}\right)$$

$$P(c \leq X \leq d) = \mathcal{A}\left(\frac{d - \mu}{\sigma}\right) - \mathcal{A}\left(\frac{c - \mu}{\sigma}\right)$$

## Chapter 18 Review Exercises

1. Find $k$ such that $kx$ is a probability density function on $[1, 3]$.

2. Find $k$ such that $ke^{-|x|}$ is a probability density function on $(-\infty, +\infty)$.

3. Find $k$ such that $2/x^2$ is a probability density function on $[1, k]$.

4. Show that $\frac{3}{8}x^2$ is a probability density function on $[0, 2]$. Find (a) $P(0 \le X \le 1)$, (b) $P(1 \le X)$, (c) $P(X = 1)$.

5. Find the mean of the random variable with probability density function $f(x) = \frac{1}{3}$ on $[0, 3]$.

6. Find the variance and standard deviation of the random variable with the probability density function given in the previous exercise.

7. Find the median of the random variable with the probability density function given in Exercise 5.

8. Find the mean of the random variable with probability density function $f(x) = \frac{1}{18}x$ on $[0, 6]$.

9. Find the variance and standard deviation of the random variable with the probability density function given in the previous exercise.

10. Find the median of the random variable with the probability density function given in Exercise 8.

11. Suppose $Z$ is a random variable with a standard normal probability density function. Find (a) $P(Z \le 1.25)$, (b) $P(Z \ge 1.25)$, (c) $P(1 \le Z \le 1.25)$.

12. Suppose $X$ is a random variable with a normal probability density function with mean $\mu = 2$ and standard deviation $\sigma = 5$. Find (a) $P(X \le 1)$, (b) $P(X \ge 1)$, (c) $P(1 \le X \le 1.5)$.

13. **Manufacturing.** The life in years of a certain automobile has a probability density function given by $f(t) = .20e^{-.20t}$. Find the probability that the life of this car is (a) at most 10 years, (b) at least 10 years, (c) between 5 and 10 years.

14. **Shelf-Life.** The shelf-life, in months, of a certain cereal is a continuous random variable with a probability density function $f(x) = \frac{3}{98}x\sqrt{9 + x^2}$, on $[0, 4]$. Find the probability that the shelf-life is (a) at most one month, (b) at least one month, (c) between one and two months.

15. **Manufacturing.** The length, in feet, of wire produced on a certain machine is normally distributed with the mean 10 feet and standard deviation .30 feet. What percentage of these will be cut less than 9.4 feet? More than 9.4 feet? Between 9.4 and 10.6 feet?

Partial derivatives, which measure the rate of change of height, depend on the direction, just as the skier's rate of descent depends on her direction down the slope. (© E. Gebhardt/ FPG International Corporation.)

# Functions of Several Variables

19.1   Functions of Several Variables

19.2   Partial Derivatives

19.3   Extrema of Functions of Two Variables

19.4   Lagrange Multipliers

19.5   Total Differentials and Approximations

19.6   Double Integrals

**W**e now consider functions of more than one variable. In the first section we will see that the graph of a function $z = f(x, y)$ of two variables represents a surface in three dimensions. In the second section we will see that the partial derivatives measure the rates of change of a function when moving in certain directions, just as your rate of change of height when moving in different directions from a point on the side of a mountain will differ depending on which direction you move. In the third and fourth sections we develop ways of finding the relative and absolute extrema of functions of several variables. In the fifth and sixth sections we extend the notions of differential and integral to functions of several variables.

## 19.1    FUNCTIONS OF SEVERAL VARIABLES

▶ *Functions of Several Variables*

▶ *Cartesian Coordinate System*

▶ *Level Curves*

▶ *Continuity*

APPLICATION

**Revenue**

Determine the revenue of a firm that sells $x$ thousands of boxes of regular corn flakes at a price of $p$ dollars per box and $y$ thousands of boxes of crunchy corn flakes at a price of $q$ dollars per box if $p = 4 - 2x + y$ and $q = 6 + x - 2y$. For the answer see Example 3.

### Functions of Several Variables

We have become familiar with functions of *one* variable. For example the area $A$ of a circle is a function of its radius $r$ and can be written as $A = A(r) = \pi r^2$. But to describe the area of a rectangle of width $w$ and length $l$ will require the function of *two* variables given by $A = A(w, l) = wl$. The notation $A(2, 3)$ means the area of a rectangle with width $w = 2$ and length $l = 3$, which is obtained by replacing $w$ with 2 and replacing $l$ with 3 in the formula for area; thus, $A(2, 3) = 2 \cdot 3 = 6$. In the same way $A(7, 4) = 7 \cdot 4 = 28$ is the area of a rectangle with width $w = 7$ and length $l = 4$.

If a company sells pens at \$2 each and sells $x$ pens, then the revenue is a function of *one* variable and is given by $R = R(x) = 2x$ dollars. If, in addition, the company sells $y$ pencils at 5 cents each, then the revenue, in dollars, now becomes a function of *two* variables and is given by $R = R(x, y) = 2x + .05y$. Thus if on one day 3000 pens and 4000 pencils are sold, the revenue is

$$R(3000, 4000) = 2(3000) + .05(4000) = 6200$$

dollars.

---

**Functions of Two Variables**

A **function of two variables**, denoted by $z = f(x, y)$, is a rule that associates to every point $(x, y)$ in some set $D$ called the **domain** a unique number denoted by $z = f(x, y)$.

The variables $x$ and $y$ are the **independent variables**, and $z$ is the **dependent variable**.

If a domain is not specified, we assume the domain is the largest possible set for which $f(x, y)$ makes sense.

E X A M P L E  1    **Evaluating Functions**

Let $f(x, y) = 2x - 3y$, $g(x, y) = \dfrac{1}{x + y}$, $h(x, y) = \sqrt{9 - x^2 - y^2}$. Find $f(5, 1), g(2, 3), h(1, 2)$.

Solutions    $f(5, 1) = 2(5) - 3(1) = 7$

$$g(2, 3) = \frac{1}{2 + 3} = \frac{1}{5}$$

$$h(1, 2) = \sqrt{9 - (1)^2 - (2)^2} = 2 \quad \blacksquare$$

E X A M P L E  2    **Finding the Domain**

Find the domain of each of the following functions: (a) $f(x, y) = 2x - 3y$, (b) $g(x, y) = \dfrac{1}{x + y}$, (c) $h(x, y) = \sqrt{9 - x^2 - y^2}$.

Solutions    (a)  One can calculate $2x - 3y$ for any $x$ and any $y$, so the domain is the entire $xy$-plane.

(b)  One can only calculate $\dfrac{1}{x + y}$ if the denominator is not zero. This happens when $y \neq -x$. Thus the domain of $g$ is all points in the $xy$-plane except for points on the line $y = -x$. See Figure 19.1a.

(c)  One can only form the square root of a nonnegative number. Thus the domain of $h$ is all points $(x, y)$ for which $9 - x^2 - y^2 \geq 0$, which is the same as $x^2 + y^2 \leq 9$. This is just all points inside and on the circle or radius 3 centered at the origin. See Figure 19.1b. $\quad \blacksquare$

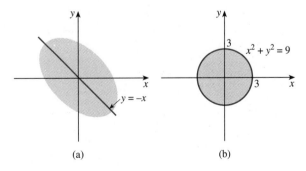

(a)                    (b)

**Figure 19.1**

E X A M P L E  3    **Revenue, Cost, and Profit Functions**

(a)  Suppose a firm produces two types of corn flakes, regular and crunchy. Naturally if the price of one of them increases, the demand for that one will decrease. But since these two types compete with each other, if the price of one of them increases, then demand for the other will *increase*. Let the number of thousands of boxes of regular and crunchy corn flakes sold by the manufacturer per week be $x$ and $y$, respectively, and the price per box be respectively $p$ and $q$. Suppose

$$p = 4 - 2x + y$$
$$q = 6 + x - 2y.$$

Find the weekly revenue function $R(x, y)$ and find the revenue when a box of regular corn flakes sells for \$1 and a box of crunchy corn flakes for \$2.

(b) If the weekly cost function of the firm is $C(x, y) = 1000(10 + 2x + 3y)$, find the weekly profit function $P(x, y)$ and $P(1, 2)$.

**Solutions**

(a) Since the revenue on the sale of regular corn flakes is

(number sold) $\times$ (price) $= (1000x)(p) = 1000x(4 - 2x + y)$,

and the revenue on the sale of crunchy corn flakes is

(number sold) $\times$ (price) $= (1000y)(q) = 1000y(6 + x - 2y)$,

the total revenue $R(x, y)$ is the sum of these two. Thus

$$R(x, y) = 1000x(4 - 2x + y) + 1000y(6 + x - 2y).$$

Then

$$R(1, 2) = 1000(1)(4 - 2(1) + (2)) + 1000(2)(6 + (1) - 2(2))$$
$$= 10,000.$$

Therefore, the weekly revenue is \$10,000.

(b) Since profit equals revenue less cost,

$$P(x, y) = R(x, y) - C(x, y)$$
$$= 1000x(4 - 2x + y) + 1000y(6 + x - 2y)$$
$$\quad - 1000(10 + 2x + 3y)$$
$$= 1000(2x - 2x^2 + 3y - 2y^2 + 2xy - 10).$$

Then

$$P(1, 2) = 1000(2(1) - 2(1)^2 + 3(2) - 2(2)^2 + 2(1)(2) - 10)$$
$$= -8000.$$

This represents a loss of \$8000 per week.   ■

Next we consider production functions, which play an important role in business. The total output of a firm $P$, measured by the total number of items produced per year, certainly depends on labor costs $L$ and capital investment $K$, with labor costs measured by dollars per year spent on wages and capital measured by dollars of capital investment per year. Assume that we explicitly know this relationship as $P = f(K, L)$. Then we call $f(K, L)$ the **production function** of the firm.

One important class of production functions are the **Cobb-Douglas production functions**. They are

$$f(K, L) = AL^b K^{1-b},$$

where $A$ and $b$ are positive constants with $0 < b < 1$.

**E X A M P L E  4**  **Evaluating a Cobb-Douglas Production Function**

Suppose the production function of a firm is given by the Cobb-Douglas production function

$$P(K, L) = 5L^{1/3}K^{2/3}$$

where $P$ is the number of thousands of items produced each year, $K$ is the capital investment per year in millions of dollars, and $L$ is the labor cost per year in millions of dollars. Find the number produced if $8 million is invested in capital and $27 million in labor.

**Solution**  Since

$$P(8, 27) = 5(27)^{1/3}(8)^{2/3} = 5(3)(4) = 60,$$

60,000 are produced.  ∎

## Cartesian Coordinate System

We are already familiar with a two-dimensional (Cartesian) coordinate system. This is necessary if we are to graph a function $y = f(x)$ of *one* variable since we need to plot the pair of points $(x, y) = (x, f(x))$ for all $x$ in the domain of $f$. In order to graph a function $z = f(x, y)$ of *two* variables we will need a three-dimensional (Cartesian) coordinate system.

A point in three-dimensional space can be uniquely represented by an ordered triple of numbers $(x, y, z)$, and every ordered triple $(x, y, z)$ can uniquely represent a point in three-dimensional space. The coordinate system is shown in Figure 19.2. The three axes shown in Figure 19.2, the $x$-axis, the $y$-axis, and the $z$-axis, are all perpendicular to each other.

**Figure 19.2**

**E X A M P L E  5**    **Using the Cartesian Coordinate System**

Locate $(1, -2, 5)$ in a three-dimensional Cartesian coordinate system.

**Solution**    To find the point $(1, -2, 5)$ start at the origin $(0, 0, 0)$ and move 1 unit in the positive $x$-direction along the $x$-axis, then move 2 units in the direction of the negative $y$-axis, then move 5 units vertically in the positive $z$-axis. See Figure 19.3.  ∎

We would like to find a formula for the distance $d$ between the two points $P(x_1, y_1, z_1)$ and $Q(x_2, y_2, z_2)$ shown in Figure 19.4. We construct a rectangular box so that the line from $P$ to $Q$ forms its diagonal, as shown in Figure 19.4. The bottom of the box lies in a plane, and so the distance $s$ can be found from the Pythagorean theorem to be

$$s = \sqrt{(x_2 - x_1)^2 + (y_2 - y_1)^2}.$$

**Figure 19.3**

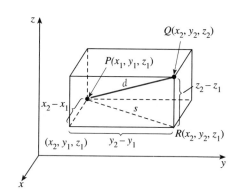

**Figure 19.4**

Now the triangle through the three points $R$, $P$, and $Q$ forms a right triangle, and again using the Pythagorean theorem, we have that

$$d^2 = s^2 + (z_2 - z_1)^2$$
$$= (x_2 - x_1)^2 + (y_2 - y_1)^2 + (z_2 - z_1)^2.$$

Thus we have the following.

---

**The Distance Formula**

The distance between the two points $(x_1, y_1, z_1)$ and $(x_2, y_2, z_2)$ is

$$d = \sqrt{(x_2 - x_1)^2 + (y_2 - y_1)^2 + (z_2 - z_1)^2}.$$

---

**E X A M P L E  6**

**Finding Distance Between Two Points**

Find the distance between the two points $(2, 3, -2)$ and $(1, -4, -1)$ shown in Figure 19.5.

**Solution**  Setting $(2, 3, -2) = (x_1, y_1, z_1)$ and $(1, -4, -1) = (x_2, y_2, z_2)$ yields

$$d = \sqrt{(x_2 - x_1)^2 + (y_2 - y_1)^2 + (z_2 - z_1)^2}$$
$$= \sqrt{[(1) - (2)]^2 + [(-4) - (3)]^2 + [(-1) - (-2)]^2}$$
$$= \sqrt{1 + 49 + 1} = \sqrt{51} \quad\blacksquare$$

We now define the **graph** of a function $z = f(x, y)$. A graph of a function will enable us to see a picture of the function. See Figure 19.6.

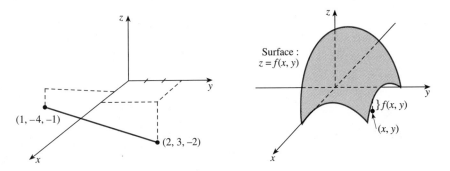

**Figure 19.5**                    **Figure 19.6**

---

**Graph of a Function of Two Variables**

Given a function $z = f(x, y)$ with domain $D$, the **graph** of $f$ is the set of points $(x, y, z)$ with $z = f(x, y)$ and $(x, y) \in D$. Since this is described by *two* variables, the result is a two-dimensional object called a **surface**.

---

**E X A M P L E  7**

**Graphing a Function**

Graph the function

$$z = f(x, y) = \sqrt{9 - x^2 - y^2}.$$

**Solutions**  In Example 2 the domain was found to be all points inside and on the circle in the $xy$-plane with center at $(0, 0)$ and radius 3. To graph the function $f(x, y)$, square each side of $z = \sqrt{9 - x^2 - y^2}$ and obtain

$$z^2 = 9 - x^2 - y^2,$$

which can be written as

$$x^2 + y^2 + z^2 = 9. \tag{1}$$

The left-hand side is the square of the distance from the point $(x, y, z)$ to $(0, 0, 0)$. Thus a point $(x, y, z)$ satisfies (1) if, and only if, this point is on the sphere with center at $(0, 0, 0)$ and radius 3. But since $z$ must be *positive*, the graph of $f(x, y)$ is the surface given by the top half of this sphere. See Figure 19.7.   ■

We can use the distance formula to find the equation of the sphere centered at the point $(x_0, y_0, z_0)$ with radius $r$ shown in Figure 19.8. The point $(x, y, z)$ will be on this sphere if, and only if, the distance from the point $(x, y, z)$ to the point $(x_0, y_0, z_0)$ is $r$, that is,

$$\sqrt{(x - x_0)^2 + (y - y_0)^2 + (z - z_0)^2} = r,$$

or

$$(x - x_0)^2 + (y - y_0)^2 + (z - z_0)^2 = r^2.$$

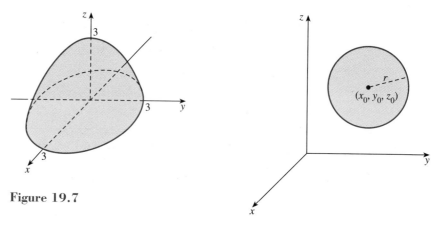

**Figure 19.7**

**Figure 19.8**

---

**Equation of Sphere**

The equation of the sphere centered at $(x_0, y_0, z_0)$ with radius $r$ is given by
$$(x - x_0)^2 + (y - y_0)^2 + (z - z_0)^2 = r^2.$$

---

It can be shown, but we do not do so here, that the graph of an equation of the form $Ax + By + Cz = D$ is a plane.

E X A M P L E  8    **Graphing a Function**

Graph $3x + 2y + z = 6$.

Solution    Since we know the graph is a plane, we can determine the plane by locating the intercepts. Setting $y = z = 0$ yields $x = 2$. Thus the point $(2, 0, 0)$ is on the plane. Also setting $x = z = 0$ yields $y = 3$. Thus $(0, 3, 0)$ is on the plane. Finally, setting $x = y = 0$ yields $z = 6$. Thus $(0, 0, 6)$ is on the plane. From this we can graph the plane. See Figure 19.9.  ∎

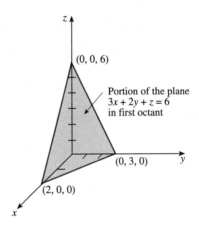

**Figure 19.9**

## Level Curves

*(© United States Geological Survey.)*

A topographic map conveys a description of a *three*-dimensional portion of the surface of the earth by using a *two*-dimension map. This is done by using **level curves**. If we intersect a surface given by the graph of $z = f(x, y)$ with a plane given by $z = z_0$, we obtain the equation $z_0 = f(x, y)$. The graph of this equation in the $xy$-plane is called a level curve for $f$. See Figure 19.10.

Thus in Figure 19.11, the curve with a 3000 next to it means that the surface above this curve has the constant height 3000. The next curve shown inside this one has the number 3500 next to it. This conveys the information that the surface is *rising* as we move from the level curve $z_0 = 3000$ to the level curve $z_1 = 3500$. This indicates that we are moving up a hill, as shown in Figure 19.11.

Figure 19.12 gives a typical weather chart. Here the temperature is $T = T(x, y)$. A level curve is also referred to as an **isobar**, that is, a curve on which the temperature is the same.

Figure 19.13 shows a typical chart of the barometric pressure. On each level curve the barometric pressure is the same.

The following example shows how we can exploit level curves to determine how a surface looks.

**Figure 19.10**

**Figure 19.11**

**Figure 19.12**

**Figure 19.13**

**E X A M P L E   9**    **Graphing Using Level Curves**

Graph $z = 4 - x^2 - y^2$ by using level curves.

**Solution**    To find the level curves we find the intersection of each horizontal plane $z = z_0$ with the surface and obtain

$$z_0 = 4 - x^2 - y^2$$

$$x^2 + y^2 = 4 - z_0.$$

Clearly we obtain the empty set if $z_0 > 4$. This conveys the information that no part of the surface extends above the plane $z = 4$. If $z_0 = 4$, we obtain $x^2 + y^2 = 0$, which is true only if $x = 0$ and $y = 0$. Thus the level curve for $z_0 = 4$ consists of the single point $(0, 0)$. If now $z_0 < 4$, we obtain a circle $x^2 + y^2 =$

$4 - z_0$ in the $xy$-plane with center at $(0, 0)$ and radius $\sqrt{4 - z_0}$. For example, if $z_0 = 1$ we have the level surface shown in Figure 19.14 with a ''1'' next to it. This indicates that along this level curve the surface is always 1 unit above the $xy$-plane. The level surface is a circle centered at $(0, 0)$ with radius $\sqrt{3}$. The level surface for $z_0 = 0$ is also shown in Figure 19.14 and is a circle centered at $(0, 0)$ with radius 2. This conveys the information that moving outward from $(0, 0)$ in any direction results in the height of the surface decreasing. Figure 19.15 then gives a three-dimensional picture of the graph. ■

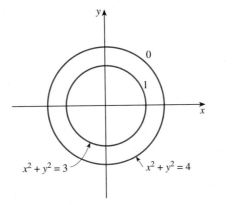

**Figure 19.14**                         **Figure 19.15**

## Continuity

We say that $f(x)$ is *continuous* at $x = a$ if $f(x)$ approaches $f(a)$, no matter how $x$ approaches $a$. A similar definition is used for a function of two variables.

---

**Continuity**

We say that $f(x, y)$ is **continuous** at $(a, b)$ if $f(x, y)$ approaches $f(a, b)$, no matter how $(x, y)$ approaches $(a, b)$.

---

Geometrically, a function $f(x)$ of a single variable is continuous on some interval if the curve that represents the graph of the function has no ''tears.'' In the same way a function $f(x, y)$ of two variables is continuous on some region if the surface that represents the graph of the function has no tears. Thus the function whose graph is shown in Figure 19.16 is continuous everywhere in $D$, whereas the graph of the function shown in Figure 19.17 has a tear and so is not continuous throughout $D$.

**Figure 19.16**

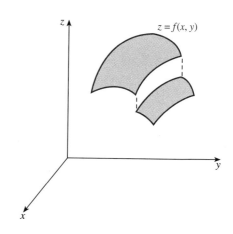

**Figure 19.17**

## SELF-HELP EXERCISE SET 19.1

1. The box shown in the figure has a square base and open top. Find the surface area $S(x, y)$ as a function of $x$ and $y$. Then find $S(5, 8)$.

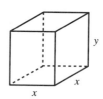

2. Plot the two points $(1, 2, -5)$ and $(0, 4, -3)$, and find the distance between them.

3. Use level curves to graph $z = x^2 + y^2 - 3$.

## EXERCISE SET 19.1

In Exercises 1 through 10 evaluate the function at the indicated points.

1. $f(x, ,y) = 3x + 4y + 2$, $(1, 2)$, $(0, 2)$, $(-2, 4)$

2. $f(x, y) = 2x - 5y$, $(0, 0)$, $(-2, 4)$, $(2, -3)$

3. $f(x, y) = 2x^2 + y^2 - 4$, $(1, 2)$, $(2, -3)$, $(-1, -2)$

4. $f(x, y) = x^2 + x - y^2$, $(2, 3)$, $(-1, 4)$, $(-2, -3)$

5. $f(x, y) = x/(x + y)$, $(1, 2)$, $(-1, 2)$, $(-2, -4)$

6. $f(x, y) = y/(x + y)$, $(2, 1)$, $(-1, 2)$, $(-2, -1)$

7.  $f(x, y) = \sqrt{8 - x^2 - y}$, $(1, 1)$, $(1, -2)$, $(-2, 0)$

8.  $f(x, y) = \sqrt{9 + x^2 + 2y^2}$, $(0, 0)$, $(1, 2)$, $(-1, -3)$

9.  $f(x, y) = 1/\sqrt{x - y}$, $(2, 1)$, $(5, 1)$, $(-1, -5)$

10. $f(x, y) = \sqrt[3]{xy}$, $(8, 1)$, $(9, 3)$, $(-1, -1)$

In Exercises 11 through 15 find the domain of the given function.

11. $f(x, y) = x + 3y$ 　　　　　　　　　　12. $f(x, y) = x - 3y$

13. $f(x, y) = x/(x + y)$ 　　　　　　　　14. $f(x, y) = y/(x - 2y)$

15. $f(x, y) = \sqrt{16 - x^2 - y^2}$ 　　　　16. $f(x, y) = 1/\sqrt{16 - x^2 - y^2}$

In Exercises 17 through 22 find the distance between the given two points.

17. $(1, 2, 3)$ and $(0, 4, 5)$ 　　　　　　　18. $(2, 4, 5)$ and $(3, 5, 8)$

19. $(0, 2, -4)$ and $(2, -3, 8)$ 　　　　　20. $(3, -1, -2)$ and $(2, -3, -5)$

21. $(-1, -2, -3)$ and $(-2, -3, -4)$

22. $(2, 0, -2)$ and $(1, -4, 3)$

In Exercises 23 through 32 graph or describe the surface.

23. $x + 2y + 3z = 12$ 　　　　　　　　　24. $3x - 2y + z = 6$

25. $2x + 5y + 10z = 20$ 　　　　　　　　26. $y = 4$

27. $z = 3$ 　　　　　　　　　　　　　　28. $x^2 + y^2 + z^2 = 25$

29. $x^2 + y^2 + z^2 = 36$ 　　　　　　　　30. $(x - 1)^2 + (y - 2)^2 + z^2 = 4$

31. $(x - 1)^2 + y^2 + (z - 3)^2 = 4$

32. $(x + 1)^2 + (y - 1)^2 + (z - 2)^2 = 9$

In Exercises 33 through 36 find the given level curves for the indicated functions and describe the surface.

33. $z = f(x, y) = x^2 + y^2$, $z_0 = 0$, $z_0 = 1$, $z_0 = 4$, $z_0 = 9$

34. $z = f(x, y) = 1 + x^2 + y^2$, $z_0 = 1, 2, 5, 10$

35. $z = f(x, y) = 1 - x^2 - y^2$, $z_0 = 1, 0, -3, -8$

36. $z = f(x, y) = 2 - x^2 - y^2$, $z_0 = 2, 1, -2, -7$

37. Find the surface area $S(x, y)$ of the box with square base and lid as a function of $x$ and $y$. Find $S(3, 5)$.

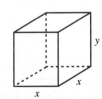

**38.** Find the surface area $S(r, h)$ of the cylinder as a function of $r$ and $h$. Find $S(5, 10)$.

## Applications

**39.  Demand.**   A company sells cameras and film to be used in the camera. Let $x$ and $y$ be, respectively, the number of cameras sold per day and the number of rolls of film sold per day and $p$ and $q$ the respective prices. Suppose $p = 1400 - 12x - y$ and $q = 802 - 3x - 0.5y$. Find the revenue function $R(x, y)$ and $R(100, 200)$.

**40.  Cost.**   Suppose the daily cost function $C(x, y)$ for the company in the previous exercise is $C(x, y) = 15{,}000 + 50x + 0.5y$. Find $C(100, 200)$.

**41.  Profits.**   Find the profit function $P(x, y)$ for the company discussed in the previous two exercises. Find $P(100, 200)$.

**42.  Interest.**   If $1000 is compounded continuously at an annual rate of $r$ and for $t$ years, write the amount $A(r, t)$ as a function of $r$ and $t$. Find $A(.10, 5)$.

**43.  Interest.**   If $1000 is compounding $m$ times a year at an annual rate of 8% for $t$ years, write the amount $A(m, t)$ as a function of $m$ and $t$. Find $A(12, 5)$.

**44.  Cobb-Douglas Production Function.**   A production function is given by $f(K, L) = 10K^{.25}L^{.75}$. Find $f(16, 81)$.

**45.  Medicine.**   In the Fick method for directly measuring cardiac output, the cardiac output $C$ is given by $C = 100x/y$. Here $x$ is the carbon dioxide, in cc per minute, released by the lungs and $y$ represents the change in carbon dioxide content of the blood leaving the lungs from when it entered, measured in cc of carbon dioxide per 100 cc of blood per minute. (a) What is the domain of this function? (b) Find $C(6, 3)$. (c) Find $C(1, 2)$.

**46.  Biology.**   The transfer of energy by convection from an animal results from a temperature difference between the animal's surface temperature and the surrounding air temperature. The convection coefficient is given by

$$h(V, D) = \frac{kV^{1/3}}{D^{2/3}},$$

where $k$ is a constant, $V$ is the wind velocity in cm/sec, and $D$ is the diameter of the animal's body in cm. (a) What is the domain of this function? (b) Find $h(8, 27)$. (c) Find $C(64, 8)$.

47. **Medicine.** The flow, $Q$ in cm$^3$/sec, of blood from a large vessel to a small capillary has been described by $Q(d, P) = 0.25C\pi d^2\sqrt{P}$, where $C$ is a constant, $d$ is the diameter of the capillary, and $P$ is the difference in pressure from the (large) vessel from that in the capillary. (a) What is the domain of this function? (b) Find $Q(4, 9)$ if $C = 2$. (c) Find $Q(2, 16)$ if $C = 2$.

48. **Biology.** The following formula relates the surface area $A$ in square meters of a human to the weight $w$ in kg and height $h$ in meters: $A(w, h) = 2.02w^{0.425}h^{0.725}$. What is the domain of this function? Estimate your surface area.

## Solutions to Self-Help Exercise Set 19.1

1. The area of the base is $x^2$, while the area of each of the 4 sides is $xy$. Thus the total surface area is $S(x, y) = x^2 + 4xy$. Then $S(5, 8) = 5^2 + 4(5)(8) = 185$.

2. The two points $(1, 2, -5)$ and $(0, 4, -3)$ are plotted in the figure.

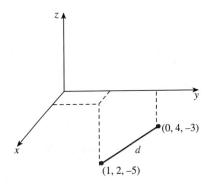

The distances between the 2 points is given by

$$d = \sqrt{(x_2 - x_1)^2 + (y_2 - y_1)^2 + (z_2 - z_1)^2}$$
$$= \sqrt{[0 - 1]^2 + [4 - 2]^2 + [(-3) - (-5)]^2}$$
$$= \sqrt{1 + 4 + 4} = 3$$

3. To find the level curves to $z = x^2 + y^2 - 3$, set $z = z_0$, $z_0 = x^2 + y^2 - 3$ or $x^2 + y^2 = z_0 + 3$. If $z_0 < -3$, this yields the empty set. If $z_0 = -3$, this gives one single point $(0, 0)$. If $z_0 > -3$, this gives a circle centered at $(0, 0)$ with radius $\sqrt{z_0 + 3}$. As $z_0$ increases, the radius increases. Thus for example, if $z_0 = 1$, the level curve is a circle with radius 2 shown in the figure. If $z_0 = 6$, the level curve is a circle with radius 3 shown in the figure. Moving away from $(0, 0)$ results in $z$ increasing.

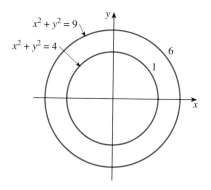

## 19.2 PARTIAL DERIVATIVES

▶ *Partial Derivatives*

▶ *Competitive and Complementary Demand Relations*

▶ *Second Order Partial Derivatives*

**APPLICATION**

**Determining if Two Commodities Compete with Each Other**

Suppose there are two commodities $A$ and $B$ with respective unit prices $p$ and $q$ and respective demand equations

$$x_A = 25 + \frac{q^3}{p^2} \quad \text{and} \quad x_B = 150 + \frac{2p^2}{q^3}.$$

Can you determine if these two commodities compete with each other? For the answer see Example 4.

### Partial Derivatives

Suppose we are given a function $z = f(x, y)$ with domain $D$. Given a point $(a, b)$ $\in D$, the plane $x = a$ that passes through this point also intersects the surface, as indicated in Figure 19.18a. The intersection of the plane with the surface forms a curve $z = f(a, y) = g(y)$ shown in Figure 19.18a. In Figure 19.18b the plane $x = a$ has been removed for better examination. Notice that with $x = a$ this curve is a function of *one* variable $y$. The slope of the tangent line to this curve is then given simply by

$$g'(b) = \lim_{\Delta y \to 0} \frac{f(a, b + \Delta y) - f(a, b)}{\Delta y}.$$

We refer to this as the **partial derivative of $f$ with respect to y** and denote it by

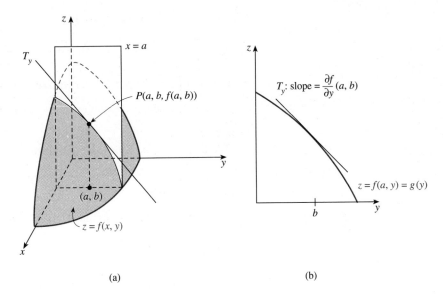

(a)                (b)

**Figure 19.18**

$\dfrac{\partial f}{\partial y}(a, b)$. The important practical point is that this derivative is calculated as an ordinary derivative with the $x$ variable treated as a *constant*.

In the same way the plane $y = b$ intersects the surface in a curve shown in Figure 19.19a. This curve is given by $z = f(x, b) = h(x)$, and the slope of the

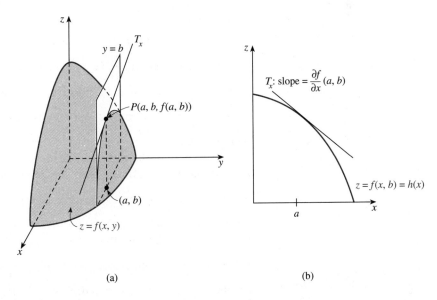

(a)                (b)

**Figure 19.19**

tangent line to this curve is given by

$$h'(a) = \lim_{\Delta x \to 0} \frac{f(a + \Delta x, b) - f(a, b)}{\Delta x}.$$

We refer to this as the **partial derivative of $f$ with respect to $x$** and denote it by $\frac{\partial f}{\partial x}(a, b)$. It is calculated as an ordinary derivative with the $y$ variable treated as a *constant.*

We then have the following definitions.

---

**Partial Derivatives**

If $z = f(x, y)$, then the **partial derivative of $f$ with respect to x** is

$$\frac{\partial f}{\partial x}(x, y) = \lim_{\Delta x \to 0} \frac{f(x + \Delta x, y) - f(x, y)}{\Delta x},$$

and the **partial derivative of $f$ with respect to y** is

$$\frac{\partial f}{\partial y}(x, y) = \lim_{\Delta y \to 0} \frac{f(x, y + \Delta y) - f(x, y)}{\Delta y},$$

if the limits exist.

---

**REMARK.**    If $z = f(x, y)$ we also denote $\frac{\partial f}{\partial x}(x, y)$ by $\frac{\partial z}{\partial x}$, $f_x$, or $f_x(x, y)$; and $\frac{\partial f}{\partial y}(x, y)$ by $\frac{\partial z}{\partial y}$, $f_y$, or $f_y(x, y)$.

If we move in the positive $x$-direction from the point $(x, y)$, the partial derivative with respect to $x$, $\frac{\partial f}{\partial x}(x, y)$, is the instantaneous rate of change of $f$ with respect to $x$. If we move in the positive $y$-direction from the point $(x, y)$, the partial derivative with respect to $y$, $\frac{\partial f}{\partial y}(x, y)$, is the instantaneous rate of change of $f$ with respect to $y$.

If we think of the surface of our function as the surface of a mountain and the point $(x, y, f(x, y))$ as a point on the surface that we are currently standing on, then as we move in the positive $x$-direction $\frac{\partial f}{\partial x}(x, y)$ measures our rate of ascent if this partial derivative is positive or rate of descent if the partial derivative is negative. See Figure 19.20. In the same way as we move in the positive $y$-direction $\frac{\partial f}{\partial y}(x, y)$ measures our rate of ascent if this partial derivative is positive or rate of descent if the partial derivative is negative. See Figure 19.21.

**Figure 19.20**

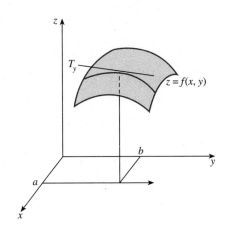

**Figure 19.21**

### E X A M P L E   1   **Finding Partial Derivatives**

Let $f(x, y) = x^2 + 2x^3y^4 + \frac{1}{3}x^3 - y$. Find (a) $\partial f/\partial x$ at $(1, 0)$, and (b) $\partial f/\partial y$ at $(1, 0)$. Interpret these numbers.

**Solutions**

(a)   In order to find $\partial f/\partial x$, treat $y$ as a *constant* and differentiate with respect to $x$ to obtain

$$\frac{\partial f}{\partial x}(x, y) = \frac{\partial}{\partial x}(x^2) + \frac{\partial}{\partial x}(2y^4x^3) + \frac{\partial}{\partial x}\left(\frac{1}{3}x^3\right) - \frac{\partial}{\partial x}(y)$$

$$= \frac{\partial}{\partial x}(x^2) + 2y^4\frac{\partial}{\partial x}(x^3) + \frac{1}{3}\frac{\partial}{\partial x}(x^3) - \frac{\partial}{\partial x}(y)$$

$$= 2x + 2y^43x^2 + x^2 + 0$$

$$= 2x + 6x^2y^4 + x^2.$$

Then

$$\frac{\partial f}{\partial x}(1, 0) = 2(1) + 6(1)^2(0)^4 + 1 = 3.$$

With $y$ kept at $y = 0$, the instantaneous rate of change of $f$ with respect to $x$ when $x = 1$ is 3. So as a point moves from $(1, 0)$ in the positive $x$-direction the function is increasing at 3 $z$-units per each $x$-unit. Refer to Figure 19.22.

(b)   In order to find $\partial f/\partial y$, treat $x$ as a *constant* and differentiate with respect to $y$ to obtain

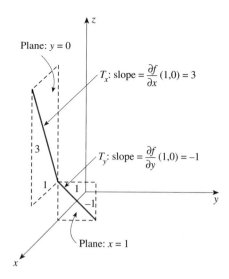

**Figure 19.22**

$$\frac{\partial f}{\partial y}(x, y) = \frac{\partial}{\partial y}(x^2) + \frac{\partial}{\partial y}(2x^3y^4) + \frac{\partial}{\partial y}\left(\frac{1}{3}x^3\right) - \frac{\partial}{\partial y}(y)$$

$$= \frac{\partial}{\partial y}(x^2) + 2x^3\frac{\partial}{\partial y}(y^4) + \frac{\partial}{\partial y}\left(\frac{1}{3}x^3\right) - \frac{\partial}{\partial y}(y)$$

$$= 0 + 2x^34y^3 + 0 - 1$$

$$= 8x^3y^3 - 1.$$

Then

$$\frac{\partial f}{\partial y}(1, 0) = -1.$$

With $x$ kept at $x = 1$, the instantaneous rate of change of $f$ with respect to $y$ when $y = 0$ is $-1$. So as a point moves from $(1, 0)$ in the positive $y$-direction, the function is decreasing at 1 $z$-unit per each $y$-unit.   ■

In the last section Cobb-Douglas production functions $f(K, L) = AL^bK^{1-b}$ were introduced. Recall that $P = f(K, L)$ is the total output of a firm, $L$ the cost of labor, and $K$ is capital investment.

The partial derivative $\partial f/\partial K$ is called the **marginal productivity of capital** while the partial derivative $\partial f/\partial L$ is called the **marginal productivity of labor**.

**E X A M P L E   2**    **Finding Marginal Productivities**

Suppose the production function of a firm is given by the Cobb-Douglas production function

$$P = f(K, L) = 27L^{1/3}K^{2/3}$$

where $P$ is the number of thousands of items produced each year, $K$ is the capital investment per year in millions of dollars, and $L$ is the labor cost per year in millions of dollars. Find the marginal productivities when $K = 8$ and $L = 27$ and interpret the result.

**Solution** The marginal productivity of capital and the marginal productivity of labor are given respectively by

$$\frac{\partial f}{\partial K}(K, L) = 18L^{1/3}K^{-1/3},$$

and

$$\frac{\partial f}{\partial L}(K, L) = 9L^{-2/3}K^{2/3}.$$

Substituting $K = 8$ and $L = 27$ yields

$$\frac{\partial f}{\partial K}(8, 27) = 18(27)^{1/3}(8)^{-1/3} = 18 \cdot 3 \cdot \frac{1}{2} = 27,$$

and

$$\frac{\partial f}{\partial L}(8, 27) = 9(27)^{-2/3}(8)^{2/3} = 9 \cdot \frac{1}{9} \cdot 4 = 4.$$

Thus with $L = 27$, the rate of change of production with respect to $K$ at $K = 8$ is 27. With $K = 8$, the rate of change of production with respect to $L$ at $L = 27$ is 4. We notice that production increases much more rapidly with an increase in capital spending, as compared with an increase in labor costs. Putting all of a (small) sum of money into capital investment will then lead to a much greater increase in production than putting all of this money into increasing labor from this point. ∎

## Competititive and Complementary Demand Relations

We are already familiar with a demand relation that relates the demand for a product to the price of that product. But the demand for a commodity such as home heating oil depends not only on the price of oil but also on the price of natural gas, and vice versa.

Suppose then that there are two commodities $A$ and $B$ such that the price of one affects the price of the other. Let the unit price of the first commodity be $p$ and the unit price of the second commodity be $q$. We assume that their respective demands $x_A$ and $x_B$ are functions of $p$ and $q$. Thus

$$x_A = f(p, q) \quad \text{and} \quad x_B = g(p, q).$$

We then have the following definitions

$$\frac{\partial x_A}{\partial p} = \text{marginal demand for } A \text{ with respect to } p$$

$$\frac{\partial x_A}{\partial q} = \text{marginal demand for } A \text{ with respect to } q$$

$$\frac{\partial x_B}{\partial p} = \text{marginal demand for } B \text{ with respect to } p$$

$$\frac{\partial x_B}{\partial q} = \text{marginal demand for } B \text{ with respect to } q.$$

If the price of the second commodity is held constant, then the demand $x_A$ should decrease if $p$ increases. But this implies[1] that

$$\frac{\partial x_A}{\partial p} < 0.$$

In the same way if the price of the first commodity is held constant, then the demand $x_B$ should decrease if $q$ increases. But this implies[1] that

$$\frac{\partial x_B}{\partial q} < 0.$$

Now suppose that the two commodities $A$ and $B$ are respectively home heating oil and natural gas. Then an increase in the price $q$ of natural gas will naturally lead to an increase in demand $x_A$ for home heating oil and an increase in the price $p$ of home heating oil will lead to an increase in demand $x_B$ for natural gas. This implies that

$$\frac{\partial x_A}{\partial q} > 0 \quad \text{and} \quad \frac{\partial x_B}{\partial p} > 0.$$

In this case we say that the two commodities are **competitive**.

On the other hand suppose that the two commodities $A$ and $B$ are automobiles and gasoline. Then an increase in the price of one will lead to a decrease in the demand for the other. This implies that

$$\frac{\partial x_A}{\partial q} < 0 \quad \text{and} \quad \frac{\partial x_B}{\partial p} < 0.$$

In this case we say that the two commodities are **complementary**. Figure 19.23 shows the graphs of typical demand functions when the two commodities are competitive and when they are complementary.

---

[1]Mathematically we could have the partial derivative equal to zero, but in practice this rarely happens.

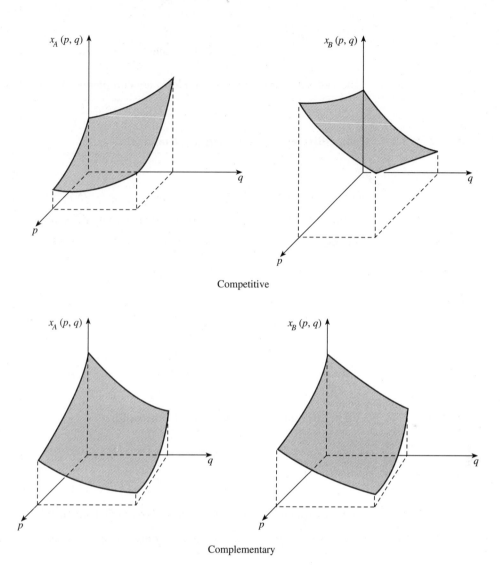

Competitive

Complementary

**Figure 19.23**

---

**Competitive and Complementary Demand Relations**

Suppose we have two commodities $A$ and $B$ with respective unit prices $p$ and $q$ and respective demand equations $x_A = f(p, q)$ and $x_B = g(p, q)$.

If
$$\frac{\partial x_A}{\partial q} > 0 \qquad \text{and} \qquad \frac{\partial x_B}{\partial p} > 0,$$

then we say that the two commodities are **competitive**.

If $$\frac{\partial x_A}{\partial q} < 0 \quad \text{and} \quad \frac{\partial x_B}{\partial p} < 0,$$

then we say that the two commodities are **complementary**.

## E X A M P L E  3    Competitive and Complementary Commodities

If

$$x_A = 50 - 2p^3 - q^2 \quad \text{and} \quad x_B = 75 - p^2 - 5q^3,$$

determine whether $A$ and $B$ are competitive or complementary.

**Solution**    First notice that these functions are reasonable demand functions since

$$\frac{\partial x_A}{\partial p} = -6p^2 < 0 \quad \text{and} \quad \frac{\partial x_B}{\partial q} = -15q^2 < 0.$$

Since

$$\frac{\partial x_A}{\partial q} = -2q < 0 \quad \text{and} \quad \frac{\partial x_B}{\partial p} = -2p < 0,$$

we see that the two commodities are complementary. ∎

## E X A M P L E  4    Competitive and Complementary Commodities

If

$$x_A = 25 + \frac{q^3}{p^2} \quad \text{and} \quad x_B = 150 + \frac{2p^2}{q^3},$$

determine whether $A$ and $B$ are competitive or complementary.

**Solution**    First notice that these functions are reasonable demand functions since

$$\frac{\partial x_A}{\partial p} = -2p^{-3}q^3 < 0 \quad \text{and} \quad \frac{\partial x_B}{\partial q} = -6q^{-4}p^2 < 0.$$

Since

$$\frac{\partial x_A}{\partial q} = \frac{3q^2}{p^2} > 0 \quad \text{and} \quad \frac{\partial x_B}{\partial p} = \frac{4p}{q^3} > 0,$$

we see that the two commodities are competitive. ∎

## Second Order Partial Derivatives

We know that the second derivative of a function of one variable is defined to be the derivative of the derivative. We do something very similar for partial derivatives.

---

**Second Order Partial Derivatives**

Given a function $z = f(x, y)$, we define the following four, second order partial derivatives.

$$\frac{\partial^2 f}{\partial x^2} = \frac{\partial}{\partial x}\left(\frac{\partial f}{\partial x}\right), \qquad \frac{\partial^2 f}{\partial y^2} = \frac{\partial}{\partial y}\left(\frac{\partial f}{\partial y}\right),$$

$$\frac{\partial^2 f}{\partial y \partial x} = \frac{\partial}{\partial y}\left(\frac{\partial f}{\partial x}\right), \qquad \frac{\partial^2 f}{\partial x \partial y} = \frac{\partial}{\partial x}\left(\frac{\partial f}{\partial y}\right).$$

---

**REMARK.**    We also use the notation

$$\frac{\partial^2 f}{\partial x^2} = f_{xx}, \qquad\qquad \frac{\partial^2 f}{\partial y^2} = f_{yy},$$

$$\frac{\partial^2 f}{\partial y \partial x} = (f_x)_y = f_{xy}, \qquad \frac{\partial^2 f}{\partial x \partial y} = (f_y)_x = f_{yx}.$$

**E X A M P L E   5**    **Finding Second Partial Derivatives**

Find all four, second order partial derivatives of $f(x, y) = e^{xy}$.

**Solution**    We first find the two first order partial derivatives:

$$f_x = ye^{xy}, \qquad f_y = xe^{xy}.$$

Then

$$f_{xx} = \frac{\partial}{\partial x}(f_x) = \frac{\partial}{\partial x}(ye^{xy}) = y^2 e^{xy}.$$

$$f_{xy} = \frac{\partial}{\partial y}(f_x) = \frac{\partial}{\partial y}(ye^{xy})$$

$$= y\frac{\partial}{\partial y}e^{xy} + e^{xy}\frac{\partial}{\partial y}y$$

$$= xye^{xy} + e^{xy}$$

$$f_{yy} = \frac{\partial}{\partial y}(f_y) = \frac{\partial}{\partial y}(xe^{xy}) = x^2 e^{xy}.$$

$$f_{yx} = \frac{\partial}{\partial x}(f_y) = \frac{\partial}{\partial x}(xe^{xy})$$

$$= x\frac{\partial}{\partial x}e^{xy} + e^{xy}\frac{\partial}{\partial x}x$$

$$= xye^{xy} + e^{xy} \quad \blacksquare$$

**R E M A R K .**   Notice that in the previous example $f_{xy} = f_{yx}$. This is no accident. *This will always happen if all second order partial derivatives are continuous.*

If $w = f(x, y, z)$ is a function of three variables, then the partial of $f$ with respect to one of the variables is obtained by taking the derivative with respect to that variable while treating the other two variables as constants.

**E X A M P L E   6**   **Finding Partial Derivatives of a Function of 3 Variables**

Find all three first order partial derivatives of $w = f(x, y, z) = x^2y^3z^4$.

**Solution**   We have

$$f_x = 2xy^3z^4 \qquad f_y = 3x^2y^2z^4 \qquad f_z = 4x^2y^3z^3 \quad \blacksquare$$

## SELF-HELP EXERCISE SET 19.2

1.   If $f(x, y) = x^2y^3$, find the two first partial derivatives, evaluate them at $(1, 1)$, and interpret this result.

2.   Find all second order partial derivatives of the function in the previous exercise.

## EXERCISE SET 19.2

In Exercises 1 through 20 find both first order partial derivatives. Then evaluate each partial derivative at the indicated point.

1.   $f(x, y) = x^2 + y^2$, $(1, 3)$

2.   $f(x, y) = x^2 - y^2$, $(1, 2)$

3.   $f(x, y) = x^2y - x^3y^2 + 10$, $(1, 2)$

4.   $f(x, y) = xy - x^2y^3 + y^4$, $(1, -1)$

5.   $f(x, y) = \sqrt{xy}$, $(1, 1)$

6.   $f(x, y) = \sqrt{1 + xy}$, $(0, 1)$

7.   $f(x, y) = \sqrt{1 + x^2y^2}$, $(1, 0)$

8.   $f(x, y) = \sqrt{1 + x^4y^3}$, $(1, 1)$

9.   $f(x, y) = e^{2x+3y}$, $(1, 1)$

10.   $f(x, y) = e^{x-y^2}$, $(0, 3)$

11.   $f(x, y) = xye^{xy}$, $(1, 1)$

12.   $f(x, y) = xye^{-xy}$, $(1, 1)$

13.   $f(x, y) = \ln(x + 2y)$, $(1, 0)$

14.   $f(x, y) = \ln(x^2 - y)$, $(1, e)$

15.   $f(x, y) = e^{xy} \ln x$, $(1, 0)$

16.   $f(x, y) = e^{xy} \ln y$, $(0, 1)$

17.   $f(x, y) = 1/xy$, $(1, 2)$

18.   $f(x, y) = y/x$, $(2, 1)$

19.   $f(x, y) = (x - y)/(x^2 + y^2)$, $(2, 1)$

20.   $f(x, y) = (x^2 - y^2)/(x^2 + y^2)$, $(2, 3)$

In Exercises 21 through 30 find all four of the second order partial derivatives. In each case check to see if $f_{xy} = f_{yx}$.

**21.** $f(x, y) = x^2 y^4$

**22.** $f(x, y) = x^2/y^3$

**23.** $f(x, y) = e^{2x - 3y}$

**24.** $f(x, y) = e^{x - y}$

**25.** $f(x, y) = \sqrt{xy}$

**26.** $f(x, y) = \ln(x + 2y)$

**27.** $f(x, y) = x^2 e^y$

**28.** $f(x, y) = x \ln y$

**29.** $f(x, y) = \sqrt{x^2 + y^2}$

**30.** $f(x, y) = xye^{xy}$

In Exercises 31 through 38 find all three first order partial derivatives.

**31.** $f(x, y, z) = xyz$

**32.** $f(x, y, z) = e^{xyz}$

**33.** $f(x, y, z) = \sqrt{x^2 + y^2 + z^2}$

**34.** $f(x, y, z) = x^2/(x^2 + y^2)$

**35.** $f(x, y, z) = e^{x + 2y + 3z}$

**36.** $f(x, y, z) = xe^y + ye^z$

**37.** $f(x, y, z) = \ln(x + 2y + 5z)$

**38.** $f(x, y, z) = xy \ln(y + 2z)$

In Exercises 39 through 44 determine if the demand relations are competitive or complementary.

**39.** $x_A = 100 - p - q,$ $\qquad x_B = 200 - 2p - 4q$

**40.** $x_A = 100 - p + 2q,$ $\qquad x_B = 50 + 2p - 4q$

**41.** $x_A = 20 - p^3 + q^3,$ $\qquad x_B = 10 + p - q^3$

**42.** $x_A = e^{-pq},$ $\qquad x_B = e^{-p-q}$

**43.** $x_A = 10 - \dfrac{p}{q},$ $\qquad x_B = 10 + p - q$

**44.** $x_A = 30 - \dfrac{p}{q},$ $\qquad x_B = 40 + \dfrac{p}{q}$

## Applications

**45.** **Compounding.** If \$1000 is invested at an annual rate of $r$ and compounded monthly, then the amount after $t$ years is given by

$$A(r, t) = 1000 \left(1 + \frac{r}{12}\right)^{12t}.$$

Find $\partial A(r, t)/\partial r$. Interpret your answer.

**46.** **Continuous Compounding.** If \$1000 is invested at an annual rate of $r$ and compounded continuously, the amount after $t$ years is given by $A(r, t) = 1000e^{rt}$. Find $\partial A(r, t)/\partial r$. Interpret your answer.

**47.** **Fishery.** Let $N$ denote the number of prey in a school of fish, $r$ the detection range of the school by a predator, and $c$ a constant that defines the average spacing of fish within the school. The volume of the region within which a predator can detect the school, called the *visual volume*, has been given by biologists as

$$V(r, N) = \frac{4\pi r^3}{3}\left(1 + \frac{c}{r}N^{1/3}\right)^3.$$

Find (a) $\partial V(r, N)/\partial r$, (b) $\partial V(r, N)/\partial N$. Interpret your answers.

48. **Biology.** The transfer of energy by convection from an animal results from a temperature difference between the animal's surface temperature and the surrounding air temperature. The convection coefficient is given by

$$h(V, D) = \frac{kV^{1/3}}{D^{2/3}},$$

where $k$ is a constant, $V$ is the wind velocity in cm/sec and $D$ is the diameter of the animal's body in cm. Find the two first order partial derivatives and interpret your answers.

49. **Biology.** If $p$ represents the total density of leaves in a tree per unit of ground area, the leaves are uniformly distributed among $n$ layers, and $r$ is the circular radius of the leaf, then the amount $L$ of light penetrating all $n$ layers is given approximately by

$$L(r, n, p) = \left(1 - \frac{p\pi r^2}{n}\right)^n.$$

Find the first order partial derivatives of $L$ with respect to $r$ and with respect to $p$ and state what each of these mean.

50. **Biology.** The concentration $C$ of oxygen at the surface of the roots of plants is given by

$$C(Q, r, R) = A + \frac{Qr^2}{2D}\ln\frac{r}{R},$$

where $A$ and $D$ are constants, $Q$ is the oxygen consumption of the roots, $r$ is the radius of the root, and $R$ is the radius of the root plus the moisture film. Find the three first order partial derivatives and interpret your answers.

## Solutions to Self-Help Exercise Set 19.2

1.   $\dfrac{\partial f}{\partial x}(x, y) = 2xy^3$          $\dfrac{\partial f}{\partial y}(x, y) = 3x^2y^2$

   $\dfrac{\partial f}{\partial x}(1, 1) = 2(1)(1)^3 = 2$     $\dfrac{\partial f}{\partial y}(1, 1) = 3(1)^2(1)^2 = 3$

With $y$ kept at $y = 1$, the instantaneous rate of change of $f$ with respect to $x$ when $x = 1$ is 2. Thus as a point $(x, y)$ moves in the positive $x$-direction, the function $f$ is increasing at 2 $z$-units per each $x$-unit. Also with $x$ kept at $x = 1$, the instantaneous rate of change of $f$ with respect to $y$ when $y = 1$ is 3. Thus as a point $(x, y)$ moves in the positive $y$-direction, the function $f$ is increasing at 3 $z$-units per each $y$-unit.

2.  $\dfrac{\partial^2 f}{\partial x^2} = \dfrac{\partial}{\partial x}\left(\dfrac{\partial f}{\partial x}\right) = \dfrac{\partial}{\partial x}(2xy^3) = 2y^3$

$\dfrac{\partial^2 f}{\partial y \partial x} = \dfrac{\partial}{\partial y}\left(\dfrac{\partial f}{\partial x}\right) = \dfrac{\partial}{\partial y}(2xy^3) = 6xy^2$

$\dfrac{\partial^2 f}{\partial y^2} = \dfrac{\partial}{\partial y}\left(\dfrac{\partial f}{\partial y}\right) = \dfrac{\partial}{\partial y}(3x^2y^2) = 6x^2y$

$\dfrac{\partial^2 f}{\partial x \partial y} = \dfrac{\partial}{\partial x}\left(\dfrac{\partial f}{\partial y}\right) = \dfrac{\partial}{\partial x}(3x^2y^2) = 6xy^2$

# 19.3   EXTREMA OF FUNCTIONS OF TWO VARIABLES

▶ *Relative Extrema*
▶ *Absolute Extrema*

A P P L I C A T I O N
**Finding an
Absolute Maximum**

A package mailed in this country must have length plus girth of no more than 108 inches. What are the dimensions of a rectangular package of greatest volume that can be mailed? See Example 4 for the answer.

A P P L I C A T I O N
**Finding an
Absolute Minimum**

What are the dimensions of the rectangular box of volume 8 cubic feet that minimizes the surface area? For the answer see Example 5.

## Relative Extrema

For a function of *one* variable we have already defined relative and absolute extrema. In the search for these extrema, we noted the need to find all critical points and found both the first and second derivative useful in determining the extrema. For a function of *two* variables we will in this section give the corresponding defintions of relative extrema, absolute extrema, and critical point and also give an appropriate second derivative test.

**Relative Maximum and Relative Minimum**

Suppose $z = f(x, y)$ is a function defined on some domain $D$.
    We say that $f$ has a **relative maximum** at $(a, b) \in D$ if there exists a circle centered at $(a, b)$ and entirely in $D$, such that

$$f(x, y) \le f(a, b)$$

for all points $(x, y)$ inside this circle.

We say that $f$ has a **relative minimum** at $(a, b) \in D$ if there exists a circle centered at $(a, b)$ and entirely in $D$, such that

$$f(x, y) \geq f(a, b)$$

for all points $(x, y)$ inside this circle.

We say that the function $z = f(x, y)$ has a **relative extremum** at $(a, b)$ $\in D$ if $f$ has either a relative maximum or a relative minimum at $(a, b)$.

See Figure 19.24 for examples.

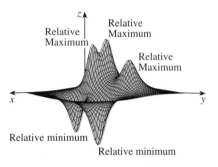

**Figure 19.24**

For a typical domain $D$ shown in Figure 19.25, we see that a point $(a, b) \in D$ will have a circle centered at $(a, b)$ and entirely in $D$ if, and only if, the point $(a, b)$ is not on the boundary of $D$. This is true in general and simply means that a relative extremum cannot occur at a boundary point of $D$. This is analogous to the one variable case, where the relative extrema were by definition excluded from occurring at the endpoints (boundary) of the interval $[a, b]$.

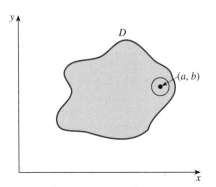

**Figure 19.25**

For the remainder of this chapter, *we always assume that the given function* $z = f(x, y)$ *has continuous first and second order partial derivatives in its domain of definition.*

Suppose we then have such a function $z = f(x, y)$ and that this function has a relative maximum at $(a, b)$. If we intersect the surface given by the graph of $z = f(x, y)$ with the plane $y = b$ as in Figure 19.26, then $f(a, b)$ is also a relative maximum of the curve $z = g(x) = f(x, b)$ of intersection. This implies from our knowledge of *one* variable that the slope $m_x$ of the tangent line at this point must be *zero*. But the slope of this tangent line is the partial derivative with respect to $x$. Thus $f_x(a, b) = 0$. In the same way we see from Figure 19.26 that at $(a, b)$ the slope of the tangent line to the curve of intersection of the surface with the plane $x = a$ is also zero and is just $f_y(a, b)$. Thus $f_y(a, b) = 0$. Similar remarks apply for the case of a relative minimum. We thus have the following important result.

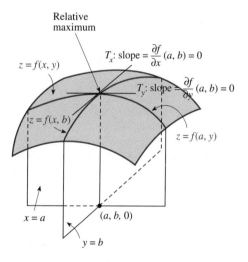

Figure 19.26

---

**Necessary Condition for Relative Extrema**

If $z = f(x, y)$ is defined and both first partial derivatives exist for all values of $(x, y)$ inside some circle about $(a, b)$ and $f$ assumes a relative extremum at $(a, b)$, then

$$\frac{\partial f}{\partial x}(a, b) = 0 \quad \text{and} \quad \frac{\partial f}{\partial y}(a, b) = 0.$$

---

A point not on the boundary of the domain of $f$ where *both* first partial derivatives are zero is called a **critical point**. (Recall that we are only dealing with functions whose first-order partial derivatives exist in their domains of definition.)

**Critical Point**

Assume that both partial derivatives of $z = f(x, y)$ exist in its domain of definition. **Critical points** of $f$ are points $(a, b)$, not on the boundary of the domain of $f$, for which

$$\frac{\partial f}{\partial x}(a, b) = 0 \qquad \text{and} \qquad \frac{\partial f}{\partial y}(a, b) = 0$$

In Figure 19.27 we see the three possibilities that can happen at a critical point. We can have a relative maximum, a relative minimum, or a **saddle point**. As Figure 19.27 indicates, at a saddle point both partial derivatives are zero, but the function does not attain a relative maximum nor a relative minimum.

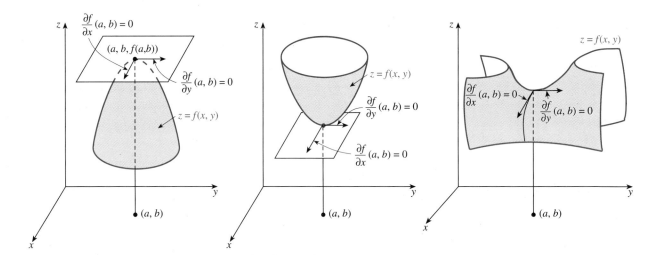

**Figure 19.27**

When dealing with a function of one variable, we know from the second derivative test that if the first derivative is zero at a point and the second derivative is positive at the same point, then the function has a relative minimum, with a corresponding result if the second derivative is negative. One might then suspect that if at a critical point for a function of *two* variables *all* the second order partial derivatives were positive, the function would have a relative minimum. *This is not true, in general.* The following example illustrates this.

**E X A M P L E   I**       **Function With Positive Second Derivatives But No Extrema**

For the function $f(x, y) = x^2 + 3xy + y^2$ show that

$$f_x(0, 0) = 0, \quad f_y(0, 0) = 0, \quad f_{xx}(0, 0) > 0, \quad f_{yy}(0, 0) > 0, \quad f_{xy}(0, 0) > 0,$$

but $f$ assumes no relative extrema at $(0, 0)$.

**Solution**       Since $f_x(x, y) = 2x + 3y$, $f_x(0, 0) = 0$, and since $f_y(x, y) = 3x + 2y$, $f_y(0, 0) = 0$. Also $f_{xx}(x, y) = 2 > 0$, $f_{yy}(x, y) = 2 > 0$, and $f_{xy}(x, y) = 3 > 0$. If now we move along the line $y = x$, we see that

$$f(x, x) = x^2 + 3xx + x^2 = 5x^2,$$

so that $f$ increases as we move away from $(0, 0)$ along this line. But along the line $y = -x$,

$$f(x, -x) = x^2 - 3xx + x^2 = -x^2,$$

so that $f$ decreases as we move away from $(0, 0)$ along this line.

In summary, moving away from the origin in one direction we ascend, but moving away from the origin in a different direction we descend. Thus we have neither a minimum nor a maximum. ■

The second derivative test for functions of more than one variable will then be more complicated. We now state this test.

---

**Second Derivative Test for Functions of Two Variables**

For $z = f(x, y)$ assume that $f_{xx}$, $f_{yy}$, and $f_{xy}$ all exist for every point inside a circle centered at $(a, b)$ and that $(a, b)$ is a critical point, that is,

$$f_x(a, b) = 0 \quad \text{and} \quad f_y(a, b) = 0.$$

Define the number $\Delta(a, b)$ by

$$\Delta(a, b) = f_{xx}(a, b)f_{yy}(a, b) - [f_{xy}(a, b)]^2.$$

Then

1.  $\Delta(a, b) > 0$ and $f_{xx}(a, b) < 0$ implies that $f$ has a relative maximum at $(a, b)$.
2.  $\Delta(a, b) > 0$ and $f_{xx}(a, b) > 0$ implies that $f$ has a relative minimum at $(a, b)$.
3.  $\Delta(a, b) < 0$ implies that $(a, b)$ is a saddle point, that is, $f$ has neither a relative minimum nor a relative maximum at $(a, b)$.
4.  If $\Delta(a, b) = 0$, $f$ can have a relative maximum, a relative minimum, or a saddle point.

Figure 19.28 summarizes this test.

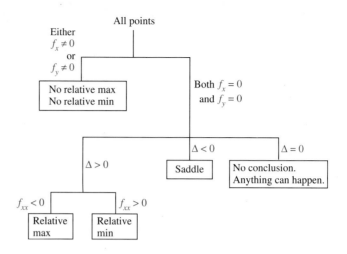

**Figure 19.28**

E X A M P L E   2      **Using the Second Derivative Test**

Find the relative extrema of

$$f(x, y) = -x^2 - y^2 + 2x + 4y + 5.$$

**Solution**     We first locate all critical points by finding both first partial derivatives and setting them both equal to zero. Thus

$$f_x = -2x + 2 = 0, \qquad f_y = -2y + 4 = 0.$$

The first equation implies that $x = 1$ and the second equation implies that $y = 2$; these are the only solutions and thus $(1, 2)$ is the only critical point. We note that

$$f_{xx} = -2, \qquad f_{yy} = -2, \qquad f_{xy} = 0.$$

Thus

$$\Delta(1, 2) = (-2)(-2) - (0)^2 = 4.$$

Since $\Delta(1, 2) > 0$ and $f_{xx}(1, 2) < 0$, the second derivative test indicates that $f$ has a relative maximum at $(1, 2)$.   ■

E X A M P L E   3      **Using the Second Derivative Test**

Find the relative extrema of

$$f(x, y) = x^2 - \hat{\phantom{x}} \; + \frac{1}{3}y^3 - 3y.$$

**Solution**  We first locate all critical points by finding both first partial derivatives and setting them both equal to zero. Thus

$$f_x = 2x - 2y = 0, \qquad f_y = -2x + y^2 - 3 = 0.$$

The first equation implies that $y = x$. Substituting this into the second equation yields $-2x + x^2 - 3 = 0$ or

$$0 = x^2 - 2x - 3 = (x - 3)(x + 1).$$

This has two solutions $x = -1$ and $x = 3$. Since $x = y$, this then indicates that the critical points are $(-1, -1)$ and $(3, 3)$. Notice that

$$f_{xx} = 2, \qquad f_{yy} = 2y, \qquad f_{xy} = -2.$$

Thus for the first critical point $(-1, -1)$, $f_{yy}(-1, -1) = -2$ and

$$\Delta(-1, -1) = (2)(-2) - (-2)^2 = -8.$$

Since $\Delta(-1, -1) < 0$, the second derivative test indicates that $f$ has a saddle point at $(-1, -1)$.

For the second critical point $(3, 3)$, $f_{yy}(3, 3) = 6$ and

$$\Delta(3, 3) = (2)(6) - (-2)^2 = 8.$$

Since $\Delta(3, 3) > 0$ and $f_{xx}(3, 3) > 0$, the second derivative test indicates that $f$ has a relative minimum at $(3, 3)$.  ∎

## Absolute Extrema

We have the following definitions for functions of two variables.

---

**Absolute Maximum and Absolute Minimum**

We say that $f$ has an **absolute maximum** on $D$ at $(a, b) \in D$ if for all $(x, y) \in D$

$$f(x, y) \le f(a, b).$$

We say that $f$ has an **absolute minimum** on $D$ at $(a, b) \in D$ if for all $(x, y) \in D$

$$f(x, y) \ge f(a, b).$$

Also we say that $f$ has an **absolute extremum** on $D$ at $(a, b) \in D$ if $f$ has either an absolute maximum or an absolute minimum on $D$ at $(a, b)$.

---

When we have a differentiable function of one variable $y = f(x)$ we know that a point at which the function attains an absolute extremum on a *bounded and closed* interval $[a, b]$ can occur at the critical points on the open interval $(a, b)$ or *can occur at the endpoints* where the derivative need not be zero. A similar situation exists for functions of two variables.

We call the interval $[a, b]$ *closed* since it includes its endpoints. In the same way we say that a region $D$ in the $xy$-plane is **closed** if it includes its boundary.

We know that a continuous function of one variable on a closed and bounded interval $[a, b]$ assumes its absolute maximum and its absolute minimum on the interval. We have a similar theorem for functions of two variables.

---

**Existence of Absolute Extrema on a Closed and Bounded Region**

If $z = f(x, y)$ is continuous on the closed and bounded region $D$, then $f$ assumes its absolute extrema in $D$, that is, there exist two points $(a, b) \in D$ and $(A, B) \in D$ such that

$$f(a, b) \leq f(x, y) \leq f(A, B) \qquad \text{for all } (x, y) \in D.$$

---

If $(x_c, y_c)$ is a critical point of $f(x, y)$, then it is a candidate for a relative extremum and therefore a candidate for an absolute extremum. But also *every point on the boundary of D is a candidate for an absolute extremum.*

**E X A M P L E    4**    **Finding an Absolute Maximum**

A package mailed in this country must have length plus girth of no more than 108 inches. Find the dimensions of a rectangular package of greatest volume that can be mailed.

**Solution**    Let $x$, $y$, and $z$ represent the dimensions of the package, with $z$ denoting the length. See Figure 19.29. We wish to find the maximum volume which is given by $V = xyz$. As Figure 19.29 indicates, the girth is $2x + 2y$ so that the condition that the girth plus the length be no more than 108 inches can be written as

$$\text{girth} + \text{length} = 2x + 2y + z \leq 108.$$

Naturally $x \geq 0$, $y \geq 0$, and $z \geq 0$. Since we wish the largest possible volume, we can assume that

$$2x + 2y + z = 108.$$

Solving this equation for $z$ and substituting into the equation of $V$ yields

$$V = xy(108 - 2x - 2y)$$
$$= 108xy - 2x^2y - 2xy^2$$

Since $2x + 2y + z = 108$ and $z \geq 0$, we must have

$$2x + 2y \leq 108.$$

Thus we are searching for the absolute maximum of $V(x, y) = 108xy - 2x^2y - 2xy^2$ on the region $D$ shown in Figure 19.30.

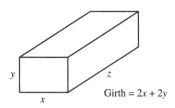

$y$
$z$
$x$
Girth $= 2x + 2y$

**Figure 19.29**

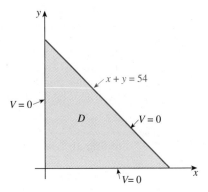

**Figure 19.30**

We first search for critical points inside $D$. Setting the two partial derivatives equal to zero, we obtain

$$0 = V_x = 108y - 4xy - 2y^2 = y(108 - 4x - 2y),$$

$$0 = V_y = 108x - 2x^2 - 4xy = x(108 - 2x - 4y).$$

The first equation implies that $y = 0$ or $108 - 4x - 2y = 0$. Since $y = 0$ is on the boundary of $D$, we discard this solution. The second equation implies that $x = 0$ or $108 - 2x - 4y = 0$. Since $x = 0$ is on the boundary of $D$, we discard this solution. Thus the only critical point in the interior of $D$ is given by the solution to

$$108 - 4x - 2y = 0 \quad \text{and} \quad 108 - 2x - 4y = 0.$$

Subtracting the second equation from twice the first yields $108 - 6x = 0$ or $x = 18$. Substituting this back into one of the two equations then yields $y = 18$. Thus $(18, 18)$ is the only critical point of $V(x, y)$ inside the region $D$, and thus it is certainly a candidate for the absolute maximum.

But all points of the boundary of $D$ are also candidates for an absolute maximum. However we notice that on the boundary of $D$ either $x = 0$ or $y = 0$ or $z = 0$, and all of these give $V = 0$, that is, $V = 0$ on the boundary of $D$. Thus $V$ does not assume its absolute maximum on the boundary of $D$.

We conclude that $V$ assumes its absolute maximum on $D$ at $(18, 18)$ *without even checking to see if $V(18, 18)$ is a relative maximum*, since the critical point $(18, 18)$ is now the only possible place in $D$ where $V$ could assume its absolute maximum. Finally, when $x = 18$ and $y = 18$ are substituted into the equation $2x + 2y + z = 108$, we obtain $z = 36$ for the final dimension of the box.

Let us double check our work to see if $V(18, 18)$ is indeed a relative maximum. We have

$$V_{xx} = -4y, \quad V_{yy} = -4x, \quad V_{xy} = 108 - 4x - 4y,$$

and

$$V_{xx}(18, 18) = -72, \quad V_{yy}(18, 18) = -72, \quad V_{xy}(18, 18) = -36.$$

Thus

$$\Delta(18, 18) = (-72)(-72) - (-36)^2 > 0,$$

and $V_{xx}(18, 18) < 0$, and so by the second derivative test $V$ has a relative maximum at this point.   ∎

**E X A M P L E   5**      **Finding an Absolute Minimum**

Find the dimensions of the rectangular box of volume 8 cubic feet that minimizes the surface area.

**Solution**

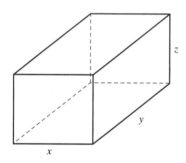

**Figure 19.31**

If as in Figure 19.31 the dimensions of the box are given by $x$, $y$, and $z$, the volume is $V = xyz$, and also, according to the problem, $V = 8$. Thus $8 = xyz$. The surface area $S$ is given by $S = 2xy + 2xz + 2yz$. Since $8 = xyz$, we have $z = 8/xy$ and

$$S = 2xy + 2x\,\frac{8}{xy} + 2y\,\frac{8}{xy}$$

$$= 2xy + \frac{16}{y} + \frac{16}{x}$$

We thus wish to minimize $S$ over the region $D$ given by $x > 0$ and $y > 0$.

We find the critical points by taking both first partial derivatives and setting them equal to zero, obtaining

$$0 = S_x = 2y - \frac{16}{x^2} = \frac{2}{x^2}\,(x^2 y - 8),$$

$$0 = S_y = 2x - \frac{16}{y^2} = \frac{2}{y^2}\,(xy^2 - 8).$$

The first of these equation gives $x^2 y = 8$ and the second gives $xy^2 = 8$. Then $x^2 y = xy^2$, and since $x > 0$ and $y > 0$, this implies that $x = y$. Then $8 = x^2 y = x^3$, and so $x = 2$ and also $y = 2$. Thus $(2, 2)$ is the only critical point of $S(x, y)$ in the region $D$.

If we now look carefully at the formula $S(x, y) = 2xy + 16/y + 16/x$, we see that since $x > 0$ and $y > 0$, $S(x, y)$ becomes unbounded if either $x$ or $y$ become large without bound or if either $x$ or $y$ go to zero. Thus we can visualize the surface $S(x, y)$ as extremely high near either axis and also for large $x$ and $y$. Refer to Figure 19.32. We conclude from this that $S(x, y)$ cannot have a minimum for any small value of either $x$ or $y$ or that $S(x, y)$ cannot have a minimum for any large value of either $x$ or $y$.

Thus the single critical point $(2, 2)$ is the *only* candidate for a point at which $S$ assumes an absolute minimum, and, therefore, $S$ *must* assume its absolute min-

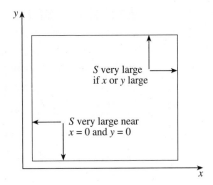

**Figure 19.32**

imum at this point. Substituting $x = 2$ and $y = 2$ into the equation $8 = xyz$ yields $z = 2$. Thus the box must be a cube with each side equal to 2 feet.

Let us double check our work to see if $(2, 2)$ is indeed a relative minimum. We have

$$S_{xx} = \frac{32}{x^3}, \qquad S_{yy} = \frac{32}{y^3}, \qquad S_{xy} = 2,$$

and

$$S_{xx}(2, 2) = 4, \qquad S_{yy}(2, 2) = 4, \qquad S_{xy}(2, 2) = 2.$$

Thus

$$\Delta(2, 2) = (4)(4) - (2)^2 = 12 > 0,$$

and $S_{xx}(2, 2) > 0$, and so by the second derivative test $V$ has a relative minimum at this point. ∎

## SELF-HELP EXERCISE SET 19.3

1. Determine the relative extrema, if any, for the function $f(x) = -x^3 - y^3 - 3xy$.

2. Suppose a firm has two products $X$ and $Y$ that compete with each other. Let the unit prices for the products $X$ and $Y$ be, respectively, $x$ and $y$ and the demand equations, respectively,

$$p(x, y) = 4 - x + y$$

$$q(x, y) = 8 + x - 2y$$

where $p(x, y)$ and $q(x, y)$ are the number of items of $X$ and $Y$, respectively, produced and sold. Find where the revenue attains a relative maximum.

## EXERCISE SET 19.3

In Exercises 1 through 20 find all critical points and determine whether each point is a relative minimum, relative maximum, or a saddle point.

1.  $f(x, y) = x^2 + xy + 2y^2 - 8x + 3y$

2.  $f(x, y) = x^2 + 2y^2 - xy + 3y$

3.  $f(x, y) = -x^2 + xy - y^2 + 3x + 8$

4.  $f(x, y) = -5x^2 - xy - y^2 - 4x - 8y$

5.  $f(x, y) = -x^2 + 2xy + 3y^2 - 8y$

6.  $f(x, y) = 3x^2 - 2xy + y^2 + x$

7.  $f(x, y) = -3x^2 + xy - y^2 - 4x - 3y$

8.  $f(x, y) = -x^2 - xy - 3y^2 + 4x + 2y$

9.  $f(x, y) = 2x^2 + xy + y^2 + 4$

10. $f(x, y) = 2x^2 - xy + y^2 - 8x + 7$

11. $f(x, y) = -x^2 + 2xy + 2y^2 - 4x - 8y$

12. $f(x, y) = -2x^2 + 2xy - y^2 + 2x + 4y$

13. $f(x, y) = x^3 + y^3 - 3xy - 1$

14. $f(x, y) = 2x^3 - 3x^2 - 12x + y^2 - 2y$

15. $f(x, y) = xy - x^3 - y^2$        16. $f(x, y) = -x^2 + y^3 + 6x - 12y$

17. $f(x, y) = 3/xy - 1/x^2y - 1/xy^2$        18. $f(x, y) = e^{xy}$

19. $f(x, y) = e^{x^2 - y^2}$        20. $f(x, y) = xy + x/y^2$

21. Show that $f(x, y) = x^4y^4$ has a relative minimum at $(0, 0)$ and that $\Delta(0, 0) = 0$.

22. Show that $f(x, y) = -x^4y^4$ has a relative maximum at $(0, 0)$ and that $\Delta(0, 0) = 0$.

23. Show that $f(x, y) = x^3y^3$ has a saddle point at $(0, 0)$ and that $\Delta(0, 0) = 0$.

24. Show that if $\Delta(a, b) > 0$ and $f_{xx}(a, b) > 0$, then $f_{yy}(a, b) > 0$.

In Exercises 25 through 28 determine whether the given function is increasing or decreasing at the point (a) $(1, 0)$ as $y$ increases, (b) $(0, 1)$ as $x$ increases, (c) $(1, 5)$ as $y$ increases, (d) $(1, 5)$ as $x$ increases.

25. $f(x, y) = x^2 + y^2 + xy + x + y$

26. $f(x, y) = -x^2 - y^2 + xy + x + y$

27. $f(x, y) = x^2 + y^2 - xy - x - y$

28. $f(x, y) = -x^2 - y^2 - xy - x - y$

## Applications

In Exercises 29 through 30 do the following.

(a)   Find the single critical point in the interior of the first quadrant.

(b)   Show that the region over which the function is to be maximized is a triangular region and that the function is zero on the boundary of this triangle. Conclude from this alone that the function assumes its maximum at the critical point found in (a).

(c)   Check your work by using the second derivative test to verify that the critical point found in (a) is a point at which the function attains a relative maximum.

29.   **Construction.**   An architect is attempting to fit a rectangular closet of maximal volume into a chopped-off corner of a building. She determines that the available space is as shown in the figure where the constricting plane is given by $4x + 2y + z = 12$ with the dimensions given in feet. What are the dimensions of the closet of maximum volume?

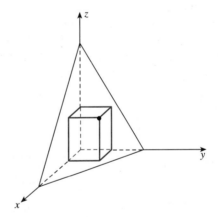

30.   **Number Theory.**   Find three positive numbers whose sum is 48 and whose product is as large as possible.

In Exercises 31 and 32 do the following.

(a)   Find the single critical point in the first quadrant.

(b)   Show that the function to be minimized is large for large $x$ and $y$ and also for small $x$ and $y$. Conclude from this alone that the function assumes its minimum at the critical point found in (a).

(c)   Check your work by using the second derivative test to verify that the critical point found in (a) is a point at which the function attains a relative minimum.

31.   **Cost.**   Find the dimensions of the cheapest rectangular box with open top and volume of 96 cubic feet with the cost in square feet of the base three times that of the cost of each side.

32.   **Number Theory.**   Find three positive numbers whose product is 64 and whose sum is as large as possible.

In the remaining problems (except for Exercise 36), there is only one critical point in the domain under consideration. Locate this critical point and use the second derivative test. Assume that the relative extrema are the absolute extrema.

33. **Cost.**  A firm has two separate plants producing the same item. Let $x$ and $y$ be the amounts produced in the respective plants with respective cost functions given by $C_1 = 100 - 12x + 3x^2/100$ and $C_2 = 100 - 10y + y^3/3000$. What allocation of production in the two plants will minimize the firm's cost?

34. **Profit.**  A firm manufactures and sells two products, $X$ and $Y$, that sell for \$15 and \$10 each, respectively. The cost of producing $x$ units of $X$ and $y$ units of $Y$ is

$$C(x, y) = 400 + 7x + 4y + 0.01(3x^2 + xy + 3y^2).$$

Find the values of $x$ and $y$ that maximize the firm's profit.

35. **Competitive Pricing.**  If the demand equations are given by $f(x, y) = 4 - x + 3y$ and $g(x, y) = 8 + x - 5y$, find where the revenue attains an absolute maximum for $x \geq 0$ and $y \geq 0$.

36. **Competitive Pricing.**  If the demand equations are given by $f(x, y) = 4 - x + 2y$ and $g(x, y) = 8 + x - 2y$, show that the revenue does not attain an absolute maximum for $x \geq 0$ and $y \geq 0$. Do this by showing that along the line $y = \dfrac{3}{4}x$

$$R(x, y) = R\left(x, \frac{3}{4}x\right) = 10x + \frac{1}{8}x^2,$$

which is not bounded.

37. **Competitive Pricing.**  If the demand equations are the same as in Exercise 35 and the cost function is given by $C(x, y) = 2x + 4y + 10,000$, find the point at which the profit attains a maximum for $x \geq 0$ and $y \geq 0$.

38. **Geometry.**  A rectangular box without a top is to have a surface area of 48 square feet. What dimensions will yield the maximum volume?

39. **Cost.**  A firm has three separate plants producing the same product and has a contract to sell 1000 units of its product. Each plant has a different cost function. Let $x$, $y$, and $z$ be the number of items produced at the three plants and assume the respective cost functions are

$$C_1(x) = 100 + .03x^2, \quad C_2(y) = 100 + 2y + .05y^2, \quad C_3(z) = 100 + 12z.$$

Find the allocation of production in the three plants that will minimize the firm's total cost.

40. **Facility Location.**  A company has three stores. The first is located 4 miles east and 2 miles north of the center of town; the second is located 2 miles east and 1 mile north of the center of town; and the third store is located 1 mile east and 6 miles south of the center of town. A warehouse is to be located convenient to these three stores. Find the location of the warehouse if the sum of the squares of the distances from each store to the warehouse is to be minimized.

41. **Profit.** A firm sells its product in the United States and in Japan. The cost function given in dollars is

$$C(z) = 100,000 + 80z + .05z^2,$$

where $z$ is the amount produced. Let $x$ be the amount sold in the United States and $y$ the amount sold in Japan. Then $x \geq 0$ and $y \geq 0$. Considering the two different cultures, we assume the following two different demand equations (given in dollars)

$$p_U(x) = 500 - .05x \quad \text{and} \quad p_J(y) = 400 - .03y.$$

Suppose that at most 3000 units of the product can be sold in each country. What are the values of $x$ and $y$ that will maximize the firm's profit $P(x, y)$?

42. **Profit.** Suppose in Exercise 41 the production facilities are located in the United States and the transportation cost of moving the product to Japan is $110 per item shipped to Japan. With this additional cost now find the values of $x$ and $y$ that will maximize the firm's profit.

## Solutions to Self-Help Exercise Set 19.3

1. Taking the first partial derivatives and setting them equal to zero gives

$$f_x = -3x^2 - 3y = 0, \qquad f_y = -3y^2 - 3x = 0.$$

From the first of these, we obtain $y = -x^2$. Substituting this into the second then gives $3x^4 + 3x = 3x(x^3 + 1)$. Therefore $x = 0$ or $x = -1$. The critical points are thus $(0, 0)$ and $(-1, -1)$. Also

$$f_{xx} = -6x, \qquad f_{yy} = -6y, \qquad f_{xy} = -3.$$

At $(0, 0)$,

$$\Delta(0, 0) = f_{xx}(0, 0)f_{yy}(0, 0) - (f_{xy}(0, 0))^2 = -9 < 0,$$

and therefore $(0, 0)$ is a saddle point.

At $(-1, -1)$,

$$\Delta(-1, -1) = f_{xx}(-1, -1)f_{yy}(-1, -1) - (f_{xy}(-1, -1))^2$$
$$= (6)(6) - (-3)^2 = 27,$$

and $f_{xx}(-1, -1) = 6 > 0$. Therefore $f(-1, -1)$ is a relative minimum.

2. The revenue $R(x, y)$ is

$$R(x, y) = xp(x, y) + yq(x, y)$$
$$= 4x + 8y + 2xy - x^2 - 2y^2.$$

Take

$$D = \{(x, y) : x \geq 0, y \geq 0\}.$$

We then are seeking to maximize $R(x, y)$ over the region $D$.

We first look for critical points interior to $D$ by setting the first partial derivatives equal to zero.

$$0 = R_x = 4 + 2y - 2x$$

$$0 = R_y = 8 + 2x - 4y$$

This yields the two linear equations

$$x - y = 2$$

$$x - 2y = -4$$

Subtracting the two equations yields $y = 6$. This gives $x = 8$.

Using the second derivative test to classify this critical point we have

$$R_{xx} = -2, \qquad R_{yy} = -4, \qquad R_{xy} = 2.$$

Thus

$$\Delta(8, 6) = (-2)(-4) - (2)^2 = 4 > 0$$

and $R_{xx}(8, 6) < 0$, which implies that the $R(x, y)$ has a relative maximum at $(8, 6)$. (It can be shown that this is also an absolute maximum.)

## 19.4   LAGRANGE MULTIPLIERS

▶ *The Method of Lagrange Multipliers*

▶ *Applications*

**APPLICATION**

**Maximizing Production**

Suppose we have a Cobb-Douglass production function $f(x, y) = 15x^{1/3}y^{2/3}$, where $x$ is the number of units of labor, $y$ is the number of units of capital, and $f$ is the number of units of a certain product that is produced. If each unit of labor costs \$200 and each unit of capital costs \$100, and the total expense for both is limited to \$7,500,000, what is the number of units of labor and capital needed to maximize production? The answer can be found in Example 3.

### The Method of Lagrange Multipliers

There is no maximum volume of a rectangular box of dimensions $x$, $y$, and $z$, since $V(x, y, z) = xyz$ is obviously unbounded in the first octant. However, we were asked in Example 4 of the last section to maximize the volume of such a box subject to the **constraint** that the girth plus the length be 108, that is, subject to $g(x, y, z) = 2x + 2y + z - 108 = 0$. Subject to the given constraint, a maximum existed and we found it. The minimum surface area of a rectangular box is clearly zero. However, in the last section we were asked in Example 5 to minimize the surface area $S(x, y, z) = 2xy + 2xz + 2yz$ of a box subject to the *constraint* that the volume be 8, that is, $g(x, y, z) = xyz - 8 = 0$. Subject to the given constraint, a *positive* minimum existed and we found it. Consider now a third example. The maximum of $f(x, y) = 4 - x^2 - y^2$ is clearly 4. However, the maximum of $f$ subject to the constraint that $g(x, y) = y - 1 = 0$, is 3 and therefore different. See Figure 19.33. Notice that the maximum or minimum of a function can be very different than the maximum or minimum subject to a *constraint*.

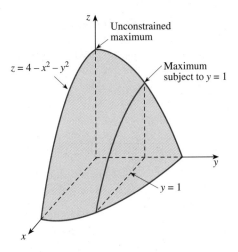

**Figure 19.33**

The above are examples of a situation that arises often. We wish to maximize or minimize some function $f(x, y, z)$ subject to a constraint $g(x, y, z) = 0$, or we wish to maximize or minimize some function of two variables $f(x, y)$ subject to a constraint $g(x, y) = 0$. When solving such problems in the last section, we merely solved the constraint equation $g(x, y, z) = 0$ for $z$ and substituted the result in $f(x, y, z)$, obtaining a function of only *two* variables and then found the extrema. In general, this may be extremely difficult, if not impossible, to do. But even if this can be done, the resulting equation can sometimes be unnecessarily complicated. There is an alternate method called the **method of Lagrange multipliers** that avoids these difficulties. We state the method for three independent variables.

---

**Method of Lagrange Multipliers**

Candidates $(x_c, y_c, z_c)$ for a relative extrema of $f(x, y, z)$ subject to the constraint $g(x, y, z) = 0$ can be found among the critical points $(x_c, y_c, z_c, \lambda)$ of the auxiliary function

$$F(x, y, z, \lambda) = f(x, y, z) + \lambda g(x, y, z),$$

that is, among the solutions of the equations

$$F_x = 0, \qquad F_y = 0, \qquad F_z = 0, \qquad g = 0. \qquad (1)$$

It is important to realize that this method only yields candidates for the extrema. The test to determine whether the relative extrema are minima or maxima is too complicated to be given here. However we noted in the last section that often there is only one candidate and that we can argue in some way that this candidate is the one we are seeking.

## Applications

We first revisit a "fence problem" from an earlier chapter. We do this in order to gain some geometric insight into Lagrange multipliers.

**E X A M P L E   1**    **Using Lagrange Multipliers**

What are the dimensions of the rectangular pasture of largest area that can be enclosed by a fence of fixed perimeter $P$?

**Solution**    Let $x$ and $y$ be the dimensions of the rectangular pasture. Then we are seeking to maximize the area $A(x, y) = xy$ subject to the constraint $2x + 2y = P$. Let us see if we can solve this problem geometrically. Figure 19.34 shows several level curves for the function $A(x, y)$. Notice from the figure that the maximum cannot occur on the level curve $A(x, y) = A_1$, since there are other level curves (such as $A(x, y) = A_2$) on which $A(x, y)$ is larger and that also intersect the constraint equation given by $g(x, y) = 2x + 2y - P = 0$. Notice that *the maximum must occur at a point where a level curve is tangent to the constraint curve.* Because of the symmetry of the curves, this must be where $x = y$.

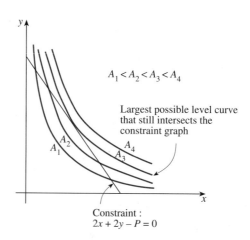

**Figure 19.34**

Now let us solve using Lagrange multipliers. Form the function

$$F(x, y) = A(x, y) + \lambda g(x, y) = xy + \lambda(2x + 2y - P),$$

and solve the equations

$$F_x = y + 2\lambda = 0$$

$$F_y = x + 2\lambda = 0$$

$$g = 2x + 2y - P = 0$$

The first two equations yield $x = y$. Substituting this into the constraint equation then yields $x = y = P/4$. ∎

It turns out in general (though not shown here) that *the critical points of the auxiliary function F in the Lagrange multipliers method occur at points where a level curve of f is tangent to the constraint curve g.* This is much less useful when the dimension of the problem is higher because of the difficulty of visualizing the graphs involved.

Let us redo Example 4 of the last section using this new method.

**E X A M P L E   2**    **Using Lagrange Multipliers**

A package mailed in this country must have length plus girth of no more than 108 inches. Find the dimensions of a rectangular package of greatest volume that can be mailed.

**Solution**    Let $x$, $y$, and $z$ represent the dimensions of the package, with $z$ denoting the length. See Figure 19.35. We wish to maximize the volume $V = xyz$. The condition that the girth plus the length be no more than 108 inches can be written as

$$\text{girth} + \text{length} = 2x + 2y + z \leq 108.$$

Naturally $x \geq 0$, $y \geq 0$, and $z \geq 0$. Since we wish the largest possible volume, we can assume that

$$2x + 2y + z = 108.$$

Thus our constraint equation is given by $g(x, y, z) = 2x + 2y + z - 108 = 0$. We then form the the auxiliary function $F(x, y, z\ \lambda)$ given by

$$F(x, y, z, \lambda) = xyz + \lambda(2x + 2y + z - 108),$$

and solve the equations

$$F_x = yz + 2\lambda = 0,$$

$$F_y = xz + 2\lambda = 0,$$

$$F_z = xy + \lambda = 0,$$

$$g(x, y, z) = 2x + 2y + z - 108 = 0.$$

**Figure 19.35**

Solving the first three equations for $-\lambda$ and setting them equal yields

$$-\lambda = \frac{yz}{2} = \frac{xz}{2} = xy.$$

This immediately yields $x = y$ and then $z = 2y = 2x$. Substituting this into the constraint equation then yields

$$2x + 2x + 2x = 108,$$

or $x = 18$, and then $y = x = 18$ and $z = 2x = 36$. We then must argue as we did in the last section that this point must be where the function attains its maximum volume. ∎

### E X A M P L E   3     Using Lagrange Multipliers

Suppose we have a Cobb–Douglass production function

$$f(x, y) = 15x^{1/3}y^{2/3},$$

where $x$ is the number of units of labor, $y$ is the number of units of capital, and $f$ is the number of units of a certain product that is produced. If each unit of labor costs \$200 and each unit of capital costs \$100, and the total expense for both is limited to \$7,500,000, find the number of units of labor and capital needed to maximize production.

**Solution**     We have that $200x + 100y = 7,500,000$. Thus the constraint equation is

$$g(x, y) = 200x + 100y - 7,500,000 = 0.$$

Form the auxiliary function $F(x, y, \lambda)$ given by

$$F(x, y, \lambda) = 15x^{1/3}y^{2/3} + \lambda(200x + 100y - 7,500,000),$$

and solve the equations

$$F_x = 5x^{-2/3}y^{2/3} + 200\lambda = 0,$$

$$F_y = 10x^{1/3}y^{-1/3} + 100\lambda = 0,$$

$$g(x, y) = 200x + 100y - 7,500,000 = 0.$$

Solving the first two equations for $-\lambda$ and setting them equal yields

$$-\lambda = \frac{x^{-2/3}y^{2/3}}{40} = \frac{x^{1/3}y^{-1/3}}{10}.$$

Multiplying each side of this equation by $x^{2/3}y^{1/3}$ yields $y = 4x$. Substituting this into the constraint equation yields

$$200x + 100(4x) = 7,500,000$$

$$x = 12,500,$$

and $y = 50,000$. The relative maximum production is then

$$f(12\ 500,\ 50\ 000) = 15(12\ 500)^{1/3}(50\ 000)^{2/3} \approx 472,500$$

units. It turns out that this is also the maximum production. ∎

Every plane has an equation of the form $ax + by + cz = d$. There are a number of ways of finding the perpendicular distance from a point to a plane. The following way uses Lagrange multipliers.

**E X A M P L E  4**   **Using Lagrange Multipliers**

Find the shortest distance from the point $(x_0,\ y_0,\ z_0)$ to the plane $ax + by + cz = E$.

**Solution**   We will minimize the square of the distance $d$ to avoid working with square roots. Thus we wish to minimize

$$d^2 = f(x, y, z) = (x - x_0)^2 + (y - y_0)^2 + (z - z_0)^2$$

subject to the constraint

$$g(x, y, z) = ax + by + cz - E = 0.$$

We form the auxiliary function $F(x, y, z, \lambda)$ given by

$$F(x, y, z, \lambda) = (x - x_0)^2 + (y - y_0)^2 + (z - z_0)^2$$
$$+ \lambda(ax + by + cz - E),$$

and solve the equations

$$F_x = 2(x - x_0) + a\lambda = 0,$$
$$F_y = 2(y - y_0) + b\lambda = 0,$$
$$F_z = 2(z - z_0) + c\lambda = 0,$$
$$g(x, y, z) = ax + by + cz - E = 0.$$

Solving the first three equations for $-\lambda$ and setting them equal yields

$$-\lambda = 2\frac{x - x_0}{a} = 2\frac{y - y_0}{b} = 2\frac{z - z_0}{c}.$$

This yields

$$y - y_0 = \frac{b}{a}(x - x_0),$$

$$z - z_0 = \frac{c}{b}(y - y_0) = \frac{c}{a}(x - x_0).$$

For convenience let

$$E_0 = ax_0 + by_0 + cz_0 \quad \text{and} \quad A = a^2 + b^2 + c^2.$$

Now substituting back into the constraint equation yields

$$
\begin{aligned}
E &= ax + by + cz \\
&= ax + by + cz - E_0 + E_0 \\
&= ax + by + cz - ax_0 - by_0 - cz_0 + E_0 \\
&= a(x - x_0) + b(y - y_0) + c(z - z_0) + E_0 \\
&= a(x - x_0) + \frac{b^2}{a}(x - x_0) + \frac{c^2}{a}(x - x_0) + E_0 \\
&= \frac{1}{a}[a^2 + b^2 + c^2](x - x_0) + E_0 \\
&= \frac{1}{a}A(x - x_0) + E_0
\end{aligned}
$$

Thus $x - x_0 = a \dfrac{E - E_0}{A}$ and then $y - y_0 = b \dfrac{E - E_0}{A}$ and $z - z_0 = c \dfrac{E - E_0}{A}$.

Substituting this into $f$ yields

$$d^2 = (a^2 + b^2 + c^2)\frac{(E - E_0)^2}{A^2} = A\frac{(E - E_0)^2}{A^2} = \frac{(E - E_0)^2}{A}.$$

Thus

$$d = \frac{|E - ax_0 - by_0 - cz_0|}{\sqrt{a^2 + b^2 + c^2}}.$$

From geometric considerations, we know this must be the absolute minimum. ∎

**Joseph Louis Lagrange, 1736–1813**

After Euler, Lagrange is considered the greatest mathematician of the eighteenth century. Lagrange had wide interests and had fundamental accomplishments in the theory of equations, differential equations, and number theory. He was the first great mathematician to recognize the unsatisfactory state of the foundations of analysis and calculus and to introduce more rigor into these subjects.

*(The Bettmann Archive.)*

## SELF-HELP EXERCISE SET 19.4

1. Find the dimensions of the rectangular box with lid of volume 8 cubic feet that minimizes the surface area using Lagrange multipliers. (This was done in Example 5 of the last section without using Lagrange multipliers.)

## EXERCISE SET 19.4

In Exercises 1 through 31 solve using Lagrange multipliers.

1.  Minimize $f(x, y) = 3x^2 + y^2$ subject to the constraint $x + y - 4 = 0$.

2.  Minimize $f(x, y) = x^2 + 3y^2$ subject to the constraint $x - y + 4 = 0$.

3.  Maximize $f(x, y) = -x^2 + xy - 4y^2$ subject to the constraint $x + y + 4 = 0$.

4.  Maximize $f(x, y) = -5x^2 - xy - y^2$ subject to the constraint $x + y + 20 = 0$.

5.  Minimize $f(x, y) = 2x^2 + xy + y^2 + x$ subject to the constraint $2x + y - 3 = 0$.

6.  Minimize $f(x, y) = 2x^2 - xy + y^2 + 7y$ subject to the constraint $x - 2y + 1 = 0$.

7.  Maximize $f(x, y) = -2x^2 + xy - y^2 + 3x + y$ subject to the constraint $2x + 3y + 11 = 0$.

8.  Maximize $f(x, y) = -x^2 - xy - 3y^2 + x - y$ subject to the constraint $-x + y + 2 = 0$.

9.  Maximize $f(x, y) = xy$ in the first quadrant subject to the constraint $x^2 + y^2 - 8 = 0$.

10.  Minimize $f(x, y) = xy^2$ in the first quadrant subject to the constraint $x^2 + y^2 - 8 = 0$,

11.  Maximize $f(x, y) = xy$ in the first quadrant subject to the constraint $x + y - 8 = 0$.

12.  Maximize $f(x, y) = x^2y$ in the first quadrant subject to the constraint $x + y - 9 = 0$.

13.  Minimize $f(x, y) = x^2 + 2y^2 + 4z^2$ subject to the constraint $x + y + z - 7 = 0$.

14.  Minimize $f(x, y) = x + 2y - 3z$ subject to the constraint $x^2 + 4y^2 - z = 0$.

15.  Maximize $f(x, y) = xyz$ in the first quadrant subject to the constraint $x^2 + y^2 + 4z^2 - 12 = 0$.

16.  Maximize $f(x, y) = xy + yz$ subject to the constraint $x^2 + y^2 + z^2 - 16 = 0$.

17.  Find two positive numbers whose sum is 20 and whose product is as large as possible.

18.  Find two positive numbers whose sum is 40 such that the sum of their squares is as small as possible.

19.  Find three positive numbers whose sum is 36 and whose product is as large as possible.

20.  Find three positive numbers whose product is 64 and whose sum is as small as possible.

## Applications

**21. Construction.** Find the most economical production for a closed cylindrical can that will hold $16\pi$ cubic inches.

**22. Construction.** A manufacturer wishes to construct an open rectangular box by removing squares from the corners of a square piece of cardboard and bending up sides as shown in the figure. What are the dimensions of the box of maximum volume?

**23. Fencing.** A person wishes to enclose a rectangular parking lot, using a building as one boundary and adding fence for the other boundaries. If there are 400 feet of fencing available, find the dimensions of the largest parking lot that can be enclosed.

**24. Cost.** What are the dimensions of the rectangular field of 20,000 square feet that will minimize the cost of fencing if one side cost three times as much as the other three.

**25. Cost.** A closed rectangular box of volume 324 cubic inches is to be made with a square base. If the material for the bottom costs twice as much per square foot as the material for the sides and top, find the dimensions of the box that minimize the cost of materials.

**26. Production.** Suppose a Cobb-Douglass production function is given by $f(x, y) = 100x^{.75}y^{.25}$, where $x$ is the number of units of labor, $y$ is the number of units of capital, and $f$ is the number of units of a certain product that is produced. If each unit of labor costs \$100 and each unit of capital costs \$200, and the total expense for both is limited to \$1,000,000, find the number of units of labor and capital needed to maximize production.

**27. Production.** Suppose a Cobb-Douglass production function is given by $f(x, y) = 100x^{.20}y^{.80}$, where $x$ is the number of units of labor, $y$ is the number of units of capital, and $f$ is the number of units of a certain product that is produced. If each unit of labor costs \$100 and each unit of capital costs \$200, and the total expense for both is limited to \$1,000,000, find the number of units of labor and capital needed to maximize production.

**28. Geometry.** A rectangular box without top is to have a surface area of 48 square feet. What dimensions will yield the maximum volume?

29. **Construction.**   Solve Exercise 29 from the last section using Lagrange multipliers.

30. **Cost.**   Solve Exercise 31 from the last section using Lagrange multipliers.

31. **Cost.**   Solve Exercise 39 from the last section using Lagrange multipliers.

### Solutions to Self-Help Exercise Set 19.4

1. Let the dimensions of a box be given by the three numbers $x$, $y$, and $z$. Then the volume is $V = xyz$, and also, according to the problem, $V = 8$. Thus $8 = xyz$. From this we obtain the constraint equation $g(x, y, z) = xyz - 8 = 0$. The surface area $S$ is given by $S(x, y, z) = 2xy + 2xz + 2yz$. It is this surface area that we wish to minimize.

   We then form the auxiliary function $F(x, y, z, \lambda)$ given by

   $$F(x, y, z, \lambda) = 2xy + 2xz + 2yz + \lambda(xyz - 8),$$

   and solve the equations

   $$F_x = 2y + 2z + \lambda yz = 0,$$

   $$F_y = 2x + 2z + \lambda xz = 0,$$

   $$F_z = 2x + 2y + \lambda xy = 0,$$

   $$g(x, y, z) = xyz - 8 = 0.$$

   Solving the first three equations for $-\lambda$ and setting them equal yields

   $$-\lambda = 2\,\frac{y + z}{yz} = 2\,\frac{x + z}{xz} = 2\,\frac{x + y}{xy}.$$

   Setting the first two fractions equal yields $xy + xz = xy + yz$, giving $x = y$, while the other equation yields $xy + yz = xz + yz$, which implies that $y = z$. Thus $x = y = z$. Substituting this into the constraint equation yields $x^3 - 8 = 0$ or $x = 2$. Then $y = 2$ and $z = 2$. We then must argue as we did in the last section that this point must be where the function attains its minimum surface.

**19.5**   ## TOTAL DIFFERENTIALS AND APPROXIMATIONS

▶ *Total Differentials and Approximations*

▶ *Applications*

**APPLICATION**

**Finding Easy-To-Calculate Approximations**

Suppose we have a Cobb-Douglass production function $f(x, y) = 81x^{1/3}y^{2/3}$, where $x$ is the number of units of labor, $y$ is the number of units of capital, and $f$ is the number of units of a certain product that is produced. Find an approximation for the change in production given changes in labor $\Delta x$ and changes in capital $\Delta y$ as a *linear* function of $\Delta x$ and $\Delta y$. For the answer see Example 2 and the discussion that follows.

### Total Differentials and Approximations

Recall that for a function of *one* variable, $y = f(x)$, the actual change in $y$ as $x$ changes from $a$ to $a + \Delta x$ is given by

$$\Delta y = f(a + \Delta x) - f(a).$$

Recall also that if $\Delta x = dx$ is small, this actual change in $y$ can be approximated by the differential $dy$, which is the change along the tangent line, that is,

$$\Delta y \approx dy = f'(a) \cdot dx.$$

Thus for $\Delta x$ small, the graph of $y = f(x)$ is approximately the same as the graph of the tangent line. See Figure 19.36. We shall now extend these ideas to functions of *two* variables.

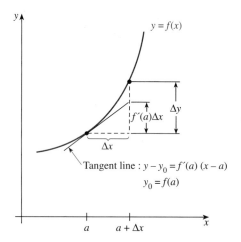

**Figure 19.36**

We have already seen that $\dfrac{\partial f}{\partial x}(a, b)$ gives the slope of the line $T_x$ tangent to the curve in which the plane $y = b$ intersects the surface $z = f(x, y)$ and that $\dfrac{\partial f}{\partial y}(a, b)$ gives the slope of the line $T_y$ tangent to the curve in which the plane $x = a$ intersects the surface $z = f(x, y)$. The lines $T_x$ and $T_y$ determine a plane. This plane is called the **tangent plane** to the surface at the point $P = (a, b, z_0)$ where $z_0 = f(a, b)$. See Figure 19.37. It can be shown, but we do not do so here, that this tangent plane is given by the equation

$$z - z_0 = f_x(a, b)(x - a) + f_y(a, b)(y - b).$$

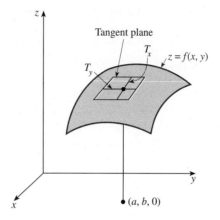

**Figure 19.37**

Notice carefully in the last equation that the point $(a, b, z_0)$ is on the *surface*, whereas the point $(x, y, z)$ is on the *tangent plane*. If we let

$$dx = \Delta x = x - a \qquad \text{and} \qquad dy = \Delta y = y - b,$$

and denote the corresponding change in $z$ as the **differential** $dz = z - z_0$, then the equation of the tangent plane can be written as

$$dz = f_x(a, b)\Delta x + f_y(a, b)\Delta y = f_x(a, b) \cdot dx + f_y(a, b) \cdot dy.$$

This indicates how the changes $\Delta x = dx$ and $\Delta y = dy$ produce the change $dz$ *on the tangent plane.* Since the tangent plane *is tangent* to the surface, we fully expect from the geometry of the situation (see Figure 19.37) that if the changes in $\Delta x$ and $\Delta y$ are small, the actual change in $z$, which is $f(a + \Delta x, b + \Delta y) - f(a, b)$, denoted by $\Delta z$, will be approximately the change on the tangent plane $dz$. Thus we have the following.

---

**Differentials and Approximations**

Suppose $f_x(a, b)$ and $f_y(a, b)$ exist.
  If $dx$ and $dy$ are any numbers, then the **total differential** is

$$dz = f_x(a, b) \cdot dx + f_y(a, b) \cdot dy.$$

If $dx = \Delta x$ and $dy = \Delta y$ are small, then

$$\Delta z \approx dz,$$

that is,

$$f(a + \Delta x, b + \Delta y) - f(a, b) \approx f_x(a, b) \cdot dx + f_y(a, b) \cdot dy.$$

---

**E X A M P L E   1**    **Using Differentials for an Approximation**

Use differentials to approximate $\sqrt[3]{(2.06)^2 + (1.97)^2}$.

**Solution**    Since 2.06 and 1.97 are very close to 2 and we can easily take the cube root of $2^2 + 2^2 = 8$, we set $a = 2 = b$ and $dx = \Delta x = 0.06$ and $dy = \Delta y = -0.03$. We take

$$f(x, y) = \sqrt[3]{x^2 + y^2}.$$

Then $f(a, b) = f(2, 2) = \sqrt[3]{8} = 2$ and

$$f_x(x, y) = \frac{2x}{3(x^2 + y^2)^{2/3}}, \qquad f_y(x, y) = \frac{2y}{3(x^2 + y^2)^{2/3}},$$

and

$$f_x(2, 2) = \frac{1}{3}, \qquad f_y(2, 2) = \frac{1}{3}.$$

Thus

$$
\begin{aligned}
\sqrt[3]{(2.06)^2 + (1.97)^2} &= f(a + \Delta x, b + \Delta y) \\
&\approx f(a, b) + \Delta z \\
&= f(a, b) + f_x(a, b) \cdot dx + f_y(a, b) \cdot dy \\
&= 2 + \frac{1}{3} dx + \frac{1}{3} dy \\
&= 2 + \frac{1}{3}(.06) + \frac{1}{3}(-.03) \\
&= 2.01.
\end{aligned}
$$

The actual value of $\sqrt[3]{(2.06)^2 + (1.97)^2}$ is 2.0103 to four decimal places. Thus, in this case, the error using the tangent plane approximation is very small.  ∎

**R E M A R K .**    Notice that for $x_0 = 2$ and $y_0 = 2$, we have the formula

$$\Delta z \approx \frac{1}{3} \cdot dx + \frac{1}{3} \cdot dy,$$

which can be used to give quick estimates of $\Delta z$ for a variety of values of $dx$ and $dy$. More generally, for any $(x, y)$ we have

$$\Delta z \approx \frac{2x}{3(x^2 + y^2)^{2/3}} \cdot dx + \frac{2y}{3(x^2 + y^2)^{2/3}} \cdot dy$$

where $\Delta z = f(x + \Delta x, y + \Delta y) - f(x, y)$, $\Delta x = dx$, and $\Delta y = dy$.

## Applications

We now give a number of applications.

**E X A M P L E   2**

**Using Differentials for an Approximation**

Suppose we have a Cobb-Douglass production function

$$f(x, y) = 81x^{1/3}y^{2/3},$$

where $x$ is the number of units of labor, $y$ is the number of units of capital, and $f$ is the number of units of a certain product that is produced. Using the tangent plane, find an approximation for the change in production as $x$ changes from 27 to 29 and as $y$ changes from 125 to 122.

**Solution**

We take $a = 27$ and $dx = \Delta x = 2$. Also $b = 125$ and $dy = \Delta y = -3$. We first need to calculate the first order partial derivatives. Thus

$$f_x(x, y) = 27x^{-2/3}y^{2/3} \quad \text{and} \quad f_y(x, y) = 54x^{1/3}y^{-1/3},$$

and

$$f_x(27, 125) = 27(27)^{-2/3}(125)^{2/3} = 27(9)^{-1}(25) = 75,$$

and

$$f_y(27, 125) = 54(27)^{1/3}(125)^{-1/3} = 54(3)(5)^{-1} = 32.4.$$

Then

$$\begin{aligned}
\Delta z &= f(27 + 2, 125 - 3) - f(27, 125) \\
&= f(a + \Delta x, b + \Delta y) - f(a, b) \\
&\approx f_x(a, b) \cdot dx + f_y(a, b) \cdot dy \\
&= f_x(27, 125)(2) + f_y(27, 125)(-3) \\
&= (75)(2) + (32.4)(-3) \\
&= 52.8.
\end{aligned}$$

The actual change in $z$ is

$$\Delta z = f(29, 122) - f(27, 125) \approx 46.5. \quad \blacksquare$$

**R E M A R K .**   Notice that for $(a, b) = (27, 125)$ we have the formula

$$\Delta z \approx (75) \cdot dx + (32.4) \cdot dy.$$

This formula can be used to give quick estimates of the change in production given various changes in $x$ and $y$. More generally for any $(x, y)$ we have

$$\Delta z \approx 27x^{-2/3}y^{2/3} \cdot dx + 54x^{1/3}y^{-1/3} \cdot dy.$$

where $\Delta z = f(x + \Delta x, y + \Delta y) - f(x, y)$, $\Delta x = dx$, and $\Delta y = dy$.

**E X A M P L E   3**

**Using Differentials for an Approximation**

(a)   Use the tangent line approximation to estimate the change in volume of a cylindrical can of inside radius $r$ and height $h$ for any small changes in $r$ and $h$ when $r = 2$ and $h = 5$. See Figure 19.38.

(b)  Suppose the side of the can is 0.02 inches thick and the top and bottom is 0.03 inches thick. Use the formula in (a) to estimate the volume of the metal in the can.

**Solutions**          (a)  The volume $V(r, h)$ is given by

$$V(r, h) = \pi r^2 h.$$

The first order partial derivatives are

$$V_r(r, h) = 2\pi rh, \qquad V_h(r, h) = \pi r^2.$$

Then

$$\begin{aligned}
\Delta V &= V(r + \Delta r, h + \Delta h) - V(r, h) \\
&\approx V_r(r, h) \cdot dr + V_h(r, h) \cdot dh \\
&= 2\pi rh \cdot dr + \pi r^2 \cdot dh \\
&= 2\pi(2)(5) \cdot dr + \pi(2)^2 \cdot dh \\
&= 20\pi \cdot dr + 4\pi \cdot dh
\end{aligned}$$

**Figure 19.38**

(b)  Take $dr = \Delta r = 0.02$. Since the top and bottom are each .03 inches thick, $dh = \Delta h = 2(.03)$. Then we have

$$\Delta V \approx 20\pi \cdot dr + 4\pi \cdot dh = 20\pi(.02) + 4\pi(.06) \approx 2.01,$$

or 2.01 cubic inches.

## SELF-HELP EXERCISE SET 19.5

1.  A firm has a cost function given by

$$C(x, y) = x^3 + 3xy + y^2.$$

Currently $x = 20$ and $y = 100$. Use the tangent line approximation to approximate the change in costs $\Delta C$ as a function of the change in $x$ and the change in $y$.

2.  Take $\Delta x = \Delta y = 1$ in the answer in the previous problem to estimate the change in costs of producing one more of each item. Compare your approximation to the actual answer.

## EXERCISE SET 19.5

In Exercises 1 through 16 use the tangent plane approximation to estimate

$$\Delta z = f(a + \Delta x, b + \Delta y) - f(a, b)$$

for the given function at the given point and for the given values of $\Delta x$ and $\Delta y$.

1.  $f(x, y) = x^2 + y^3$, $(a, b) = (2, 1)$ $\Delta x = 0.1$, $\Delta y = 0.2$

2.  $f(x, y) = 2x^3 + xy$, $(a, b) = (1, 2)$, $\Delta x = 0.1$, $\Delta y = 0.3$

**3.** $f(x, y) = x^2y^3$, $(a, b) = (3, 2)$, $\Delta x = 0.1$, $\Delta y = -0.1$

**4.** $f(x, y) = x^3y^4$, $(a, b) = (-1, 1)$, $\Delta x = 0.1$, $\Delta y = -0.2$

**5.** $f(x, y) = \frac{x - y}{x + y}$, $(a, b) = (1, 2)$, $\Delta x = 0.2$, $\Delta y = 0.1$

**6.** $f(x, y) = \frac{x + y}{x - y}$, $(a, b) = (3, 4)$, $\Delta x = -0.1$, $\Delta y = 0.2$

**7.** $f(x, y) = \frac{x}{x^2 + y^2}$, $(a, b) = (3, 4)$, $\Delta x = -0.1$, $\Delta y = 0.2$

**8.** $f(x, y) = \frac{y}{x^2 + y^2}$, $(a, b) = (3, 4)$, $\Delta x = 0.02$, $\Delta y = -0.03$

**9.** $f(x, y) = \ln(x + 2y)$, $(a, b) = (2, 4)$, $\Delta x = 0.1$, $\Delta y = 0.2$

**10.** $f(x, y) = \ln(x^2 + y^2)$, $(a, b) = (1, 2)$, $\Delta x = 0.5$, $\Delta y = -0.5$

**11.** $f(x, y) = \sqrt{x + y}$, $(a, b) = (4, 5)$, $\Delta x = -0.1$, $\Delta y = -0.2$

**12.** $f(x, y) = \sqrt{x^2 + y}$, $(a, b) = (2, 5)$, $\Delta x = 1/30$, $\Delta y = 1/15$

**13.** $f(x, y) = e^{x+y}$, $(a, b) = (1, -1)$, $\Delta x = .02$, $\Delta y = .03$

**14.** $f(x, y) = e^{xy}$, $(a, b) = (2, 0.5)$, $\Delta x = .02$, $\Delta y = -.01$

**15.** $f(x, y) = xe^y$, $(a, b) = (0, 1)$, $\Delta x = 0.1$, $\Delta y = 0.2$

**16.** $f(x, y) = xye^{xy}$, $(a, b) = (1, 2)$, $\Delta x = 0.02$, $\Delta y = 0.01$

In Exercises 17 through 28 use the tangent line approximation to estimate the given quantities.

**17.** $\sqrt{4.04} \sqrt{9.06}$          **18.** $\sqrt{4.08} \sqrt{8.94}$

**19.** $\sqrt{(3.02)^2 + (4.03)^2}$      **20.** $\sqrt{(3.01)^2 + (3.97)^2}$

**21.** $\sqrt{(2.06)^3 + (.97)^2}$      **22.** $\sqrt{(1.94)^3 + (1.03)^2}$

**23.** $(1.02)^7 + (1.98)^3$    **24.** $(1.02)^7(1.98)^3$    **25.** $(.99)^7(2.98)^3$

**26.** $(.97)^7 + (1.98)^3$    **27.** $e^{(1.02)^2 - (.98)^2}$    **28.** $e^{(2.01)^3 - 7.97}$

## Applications

**29. Cost.** A cost function in thousands of dollars for a firm producing two products $X$ and $Y$ is given by

$$C(x, y) = 1000 + x^3 + xy + y^2,$$

where $x$ and $y$ are the respective number in thousands of $X$ and $Y$ produced. If currently 1000 of $X$ and 2000 of $Y$ are being produced, use the tangent plane approximation to estimate the change in cost if the firm produces an additional 100 of the $X$ items and 300 of the $y$ items.

**30. Revenue.** Suppose the revenue function for the firm in the previous problem is given by

$$R(x, y) = \sqrt{xy} (10 - xy),$$

and the firm is currently selling what they produce. Use the tangent plane approximation to estimate the change in revenue if the firm produces and sells an additional 100 of the $X$ items and 300 of the $Y$ items.

31. **Production.**   Suppose we have a Cobb-Douglass production function

$$f(x, y) = 100x^{1/4}y^{3/4},$$

where $x$ is the number of units of labor, $y$ is the number of units of capital, and $f$ is the number of units of a certain product that is produced. Using the tangent plane, find an approximation for the change in production as $x$ changes from 16 to 18 and as $y$ changes from 81 to 80.

32. **Manufacturing.**   Use the tangent plane approximation to estimate the volume of metal in a closed cylindrical can of inner radius 3 inches and inner height 6 inches if the metal is 0.03 inches thick.

33. **Manufacturing.**   Use the tangent plane approximation to estimate the volume of metal in a closed rectangular metal box with square bottom and top with each side of inner length one foot and inner height 3 feet if the metal is 0.05 feet thick.

34. **Biology.**   In measuring the volume of a rod-shaped bacterium (right circular cylinder), suppose an error of 2% is made measuring its radius and a 1% error in measuring its length. Using the tangent line approximation, estimate the greatest relative error ($|\Delta V|/V$) in calculating its volume when $r = 1$ unit and $h = 10$ units.

35. **Biology.**   The following formula relates the surface area $A$ in square meters of a human to the weight $w$ in kg and height $h$ in meters:

$$A(w, h) = 2.02w^{0.425}h^{0.725}.$$

Use the tangent plane approximation to estimate the difference in surface area of two humans, one weighing 100 kg and 2 meters tall and the other weighing 101 kg and 2.1 meters tall.

36. **Medicine.**   Poiseville's law states that the resistance of a blood vessel of length $L$ and radius $r$ is given by

$$R = \frac{kL}{r^4},$$

where $k$ is a constant. Use the tangent plane approximation to show how small changes in $L$ and $r$ affect the resistance $R$.

## Solutions to Self-Help Exercise Set 19.5

1.   We have $C_x(x, y) = 3x^2 + 3y$ and $C_y(x, y) = 3x + 2y$. Thus $C_x(20, 100) = 1500$, $C_y(20, 100) = 260$, and

$$\Delta C \approx 1500\Delta x + 260\Delta y.$$

2.   Setting $\Delta x = \Delta y = 1$ in the previous formula yields

$$\Delta C \approx 1500(1) + 260(1) = 1760.$$

The actual change in $C$ is given by

$$C(21, 101) - C(20, 100) = 25{,}825 - 24{,}000 = 1825.$$

## **19.6**  DOUBLE INTEGRALS

▶ *Riemann Sums and Integrals*

▶ *Iterated Integrals*

▶ *Applications*

APPLICATION
**Concentration of Pollutants**

An industrial plant is located at the precise center of a town shaped as a square with each side of length 4 miles. If the plant is placed at the point $(0, 0)$ of the $xy$-plane, then certain pollutants are dispersed in such a manner that the concentration at any point $(x, y)$ in town is given by $C(x, y) = 1000(24 - 3x^2 - 3y^2)$, where $C(x, y)$ is the number of particles of pollutants per square mile of surface per day at a point $(x, y)$ in town. What is the average concentration and number of these pollutants in this town? For the answer see Examples 6 and 7.

### Riemann Sums and Integrals

In this section we shall first be interested in finding the volume $V$ between a surface given by $z = f(x, y)$ and a region $D$ in the $xy$-plane. We will then see that this can be used for many applications. We will assume that $f(x, y)$ is nonnegative on $D$. See Figure 19.39.

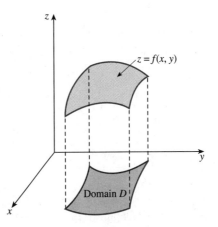

**Figure 19.39**

The domain $D$ is shown in the $xy$-plane in Figure 19.40. In this figure the domain $D$ has been encased with a rectangle $R$ as shown where

$$R(x, y) = \{(x, y) : a \le x \le b, c \le y \le d\}.$$

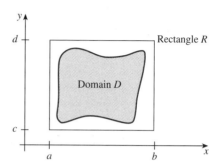

**Figure 19.40**

For any integer $n$ we divide both intervals $[a, b]$ and $[c, d]$ in $n$ equal subintervals and define

$$\Delta x = \frac{b - a}{n}, \qquad \Delta y = \frac{d - c}{n}.$$

This gives us a collection of subrectangles shown in Figure 19.41. We label all subrectangles that lie *entirely inside D* in some manner by going from left to right and top to bottom, as indicated in Figure 19.41. Let the total number of subrectangles be $N = N(n)$ and label them as $R_1, R_2, \ldots, R_N$. (If the region $D$ were the rectangle $R$, then $N = N(n) = n^2$.) In each $R_i$ pick some point $(x_i, y_i) \in R_i$.

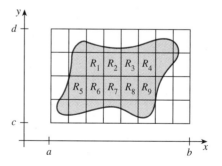

**Figure 19.41**

Now for each $i = 1, 2, \ldots, N$, construct a rectangular box as in Figure 19.42 of height $f(x_i, y_i)$ and base $R_i$. If $n$ is large, the volume

$$V_i = f(x_i, y_i)\Delta x \Delta y$$

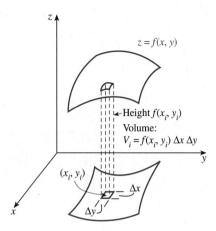

**Figure 19.42**

of this rectangular box will approximate the area under the surface $z = f(x, y)$ and over $R_i$. The sum of all such volumes will then approximate the volume we are seeking. This sum is called a **Riemann sum of $f$ on $D$** and is denoted by $S_n$. We thus have

$$V \approx S_n = V_1 + V_2 + \cdots + V_N$$

$$= f(x_1, y_1)\Delta x \Delta y + f(x_2, y_2)\Delta x \Delta y + \cdots + f(x_N, y_N)\Delta x \Delta y$$

$$= \sum_{i=1}^{N} f(x_i, y_i)\Delta x \Delta y.$$

**E X A M P L E  1**  **Finding a Riemann Sum**

Find the Riemann sum of $f(x, y) = 6xy^2$ on the rectangle

$$R = \{(x, y) : 0 \le x \le 2, 0 \le y \le 1\},$$

for $n = 2$ by taking the points $(x_i, y_i)$ at the top right of each subrectangle.

**Solution**   Each interval [0, 2] and [0, 1] is divided into two parts and

$$\Delta x = \frac{2 - 0}{2} = 1, \qquad \Delta y = \frac{1 - 0}{2} = \frac{1}{2}.$$

The four subrectangles are obtained, as shown in Figure 19.43 where the four points $(x_1, y_1)$, $(x_2, y_2)$, $(x_3, y_3)$, $(x_4, y_4)$ are, respectively, (1, 1), (2, 1), (1, 0.50), (2, 0.50). Note that the function increases as either $x$ or $y$ increases. Thus, on any subrectangle $R_i$, the function is largest at the top right point $(x_i, y_i)$. Thus the top

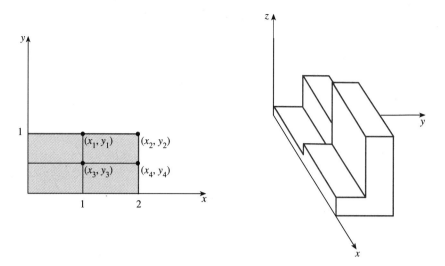

**Figure 19.43**

of each of the rectangular boxes will be above the surface and the Riemann sum will be larger than the volume. Then

$$S_2 = f(x_1, y_1)\Delta x\Delta y + f(x_2, y_2)\Delta x\Delta y + f(x_3, y_3)\Delta x\Delta y + f(x_4, y_4)\Delta x\Delta y$$

$$= f(1, 1)\Delta x\Delta y + f(2, 1)\Delta x\Delta y + f(1, 0.50)\Delta x\Delta y + f(2, 0.50)\Delta x\Delta y$$

$$= (6)\frac{1}{2} + (12)\frac{1}{2} + (1.50)\frac{1}{2} + (3)\frac{1}{2}$$

$$= 11.25 \quad \blacksquare$$

If we take the points $(x_i, y_i)$ at the lower left of $R_i$, then the tops of the rectangular boxes will be under the surface and the corresponding Riemann sum $S_n^*$ will be less than the volume.

The following table shows the results of calculating the Riemann sums taking $(x_i, y_i)$ at the lower left and the upper right.

| $n$ | $n^2$ | $S_n^*$ | $S_n$ |
|-----|-------|---------|-------|
| 4 | 16 | 1.969 | 7.031 |
| 8 | 64 | 2.871 | 5.379 |
| 100 | 10,000 | 3.901 | 4.100 |
| 500 | 250,000 | 3.980 | 4.020 |

If, in the previous discussion, we then let $n$ become larger and larger, we expect the Riemann sum to become a better and better approximation to the volume $V$. By letting $n \to +\infty$, we intuitively see that we should obtain the volume $V$. According to the table, the volume appears to be 4. (In Example 2, we shall see that this is the case.)

For a function $f(x, y)$ that is continuous on $D$, the limit $\lim\limits_{n \to +\infty} S_n$ will exist and be the same limit no matter how the points $(x_i, y_i)$ in $R_i$ are chosen. The result also holds whether or not $f(x, y)$ is nonnegative. We have the following theorem.

---

**Definite Integral**

Let $f(x, y)$ be continuous on the bounded region $D$. Then the **definite integral of $f$ over $D$** is

$$\iint_D f(x, y)\, dA = \lim_{n \to +\infty} \sum_{i=1}^{N} f(x_i, y_i)\Delta x \Delta y,$$

where the limit exists and is the same no matter how the points $(x_i, y_i)$ are chosen in $R_i$.

---

When $f(x, y) \geq 0$, then the indefinite integral is intuitively the volume under the surface and over the domain $D$. We then *define* the volume to be this Riemann integral.

---

**Volume as a Definite Integral**

Let $f(x, y)$ be continuous on the bounded region $D$ and nonnegative there. Then the volume under the surface $z = f(x, y)$ and over the domain $D$ is the definite integral of $f$ over $D$ or

$$V = \iint_D f(x, y)\, dA.$$

---

## Iterated Integrals

Suppose the domain $D$ of $z = f(x, y)$ has a "top" and a "bottom" given respectively by $y = g_2(x)$ and $y = g_1(x)$ as in Figure 19.44. For any fixed $\overline{x} \in [a, b]$, draw a straight line connecting the top and the bottom of the domain $D$, as shown in Figure 19.44. As shown in Figure 19.45, the plane $x = \overline{x}$ cuts out a plane region with top $z = z(\overline{x}, y)$. We can readily find the area $A(\overline{x})$ on the plane $x = \overline{x}$ under the curve $z = f(\overline{x}, y)$ as

$$A(\overline{x}) = \int_{g_1(\overline{x})}^{g_2(\overline{x})} f(\overline{x}, y)\, dy.$$

This is the cross-sectional area of a slice of the volume.

**Figure 19.44**

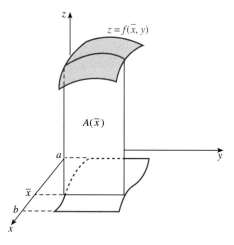

**Figure 19.45**

For any integer $n$, we now divide the interval $[a, b]$ into $n$ equal subintervals, each with base of length $\Delta x = \dfrac{b - a}{n}$. The $i$th such interval $I_i$ is shown in Figure 19.46. Each interval then slices out a thin volume under the surface as shown in Figure 19.47. For each $i$, we pick any point $x_i \in I_i$. By virtue of continuity and

**Figure 19.46**

**Figure 19.47**

the smallness of $\Delta x$, the volume $V_i$ of this slice is approximately the thickness, $\Delta x$, times the surface area of either side. The surface area of either side is approximately $A(x_i)$, given before. Thus

$$V_i \approx A(x_i)\Delta x.$$

If we sum all $n$ of these up, we will approximately obtain the volume $V$ under the surface $z = f(x, y)$ over the domain $D$. We have

$$V = V_1 + V_2 + \cdots + V_n$$

$$\approx A(x_1)\Delta x + A(x_2)\Delta x + \cdots + A(x_n)\Delta x$$

$$= \sum_{i=1}^{n} A(x_i)\Delta x.$$

But we recognize this as a Riemann sum of $A(x)$ on the interval $[a, b]$. Thus as $n \to +\infty$,

$$\sum_{i=1}^{n} A(x_i)\Delta x \to \int_a^b A(x)\ dx.$$

Thus we have

$$V = \int_a^b A(x)\ dx,$$

where

$$A(x) = \int_{g_1(x)}^{g_2(x)} f(x, y)\ dy.$$

We must understand that in this last integral for $A(x)$, $y$ *is held constant* during the integration. Putting this together we have

$$V = \int_a^b \left( \int_{g_1(x)}^{g_2(x)} f(x, y) \, dy \right) dx.$$

We thus have the following result.

---

**Iterated Integral**

If the domain $D$ is given as in Figure 19.44, and $z = f(x, y) \geq 0$ is continuous on $D$, then the volume $V$ between the surface $z = f(x, y)$ and the domain $D$ is

$$\iint_D f(x, y) \, dA = \int_a^b \left( \int_{g_1(x)}^{g_2(x)} f(x, y) \, dy \right) dx.$$

---

**R E M A R K S .**    The following two remarks are critical.

1.  In the above, $\displaystyle\int_{g_1(x)}^{g_2(x)} f(x, y) \, dy$ is to be done *first*, that is, we integrate with respect to $y$ *first*.

2.  When calculating $\displaystyle\int_{g_1(x)}^{g_2(x)} f(x, y) \, dy$, treat $x$ as a *constant*. Thus we integrate first with respect to $y$, while holding $x$ constant.

Since there will be two basic types of regions to master, we must be able to distinguish between the two. Toward this end we should get in the habit when taking a double integral over a region, such as shown in Figure 19.44, to take the following steps.

---

**Procedure for Setting Up the Iterated Integral**

1.  Clearly label the "top" and the "bottom" of the domain $D$.
2.  Draw the slice shown in Figure 19.46 so that one realizes that $x$ is being held constant while the $y$ is varying from the "bottom" to the "top." This gives the limits of integration for the *inner* integral.
3.  Now "sum" these slices "on $x$," that is, $x$ goes from $a$ to $b$. This gives the limits of integration on the *outer* integral.

E X A M P L E   2    **An Iterated Integral Over a Rectangle**

Find the volume under the surface given by $z = 6xy^2$ over the domain given by the rectangle

$$D = \{(x, y) : 0 \le x \le 2, 0 \le y \le 1\},$$

that is, find

$$\iint_D 6xy^2 \, dA.$$

**Solution**    The region $D$ has a bottom given by $y = g_1(x) = 0$ and a top given by $y = g_2(x) = 1$. See Figure 19.48. The slice shown in Figure 19.48 or 19.49 indicates that while holding $x$ constant, we integrate with respect to $y$ from the bottom to the top. Thus the limits of integration on the inner integral go from $y = 0$ to $y = 1$. Now the slices are "summed" on $x$ from 0 to 2. This gives the limits of integration on the outer integral. Thus

$$\iint_D f(x, y) \, dA = \int_a^b \left( \int_{g_1(x)}^{g_2(x)} f(x, y) \, dy \right) dx$$

$$= \int_0^2 \left( \int_0^1 6xy^2 \, dy \right) dx$$

$$= \int_0^2 \left( 2xy^3 \big|_{y=0}^{y=1} \right) dx$$

$$= \int_0^2 2x \, dx$$

$$= x^2 \big|_0^2$$

$$= 4 \quad \blacksquare$$

**Figure 19.48**

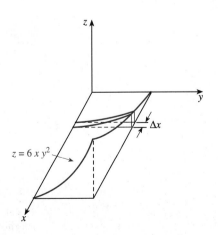

**Figure 19.49**

**E X A M P L E   3**    **Finding an Iterated Integral**

Find the volume under the surface given by $z = 48xy$ over the domain given by the region $D$ in the first quadrant of the $xy$-plane bounded by $y = x$ and $y = x^2$.

**Solution**    The region $D$ has a bottom given by $y = g_1(x) = x^2$ and a top given by $y = g_2(x) = x$. See Figure 19.50. The slice shown in Figure 19.50 or 19.51 indicates that while holding $x$ constant, we integrate with respect to $y$ from the bottom to the top. Thus the limits of integration on the inner integral go from $y = x^2$ to $y = x$. Now the slices are "summed" on $x$ from 0 to 1. This gives the limits of integration on the outer integral.

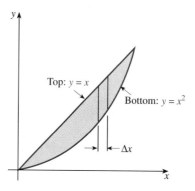

**Figure 19.50**                    **Figure 19.51**

Thus

$$\iint_D f(x, y)\, dA = \int_a^b \left( \int_{g_1(x)}^{g_2(x)} f(x, y)\, dy \right) dx$$

$$= \int_0^1 \left( \int_{x^2}^{x} 48xy\, dy \right) dx$$

$$= \int_0^1 (24xy^2 \big|_{y=x^2}^{y=x})\, dx$$

$$= \int_0^1 (24x^3 - 24x^5)\, dx$$

$$= (6x^4 - 4x^6) \big|_0^1$$

$$= 2 \quad \blacksquare$$

Now let us consider the case in which the domain $D$ of $z = f(x, y)$ has a "left" and a "right" given respectively by $x = g_1(y)$ and $x = g_2(y)$ as in Figure 19.52. We proceed in a manner similar to what we did before, this time summing thin slices of volumes shown in Figure 19.53. This leads to the following iterated integral.

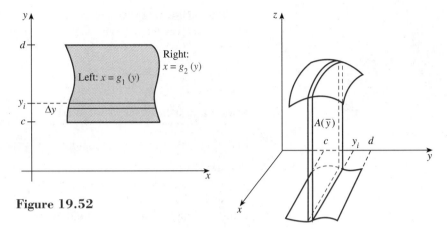

**Figure 19.52**

**Figure 19.53**

---

**Iterated Integral**

If the domain $D$ is given as in Figure 19.52, and $z = f(x, y) \geq 0$ is continuous on $D$, then the volume $V$ between the surface $z = f(x, y)$ and the domain $D$ is

$$\iint_D f(x, y) \, dA = \int_c^d \left( \int_{g_1(y)}^{g_2(y)} f(x, y) \, dx \right) dy.$$

---

In this iterated integral we say we are *integrating with respect to x first*, whereas in the previous iterated integral we say that we are *integrating with respect to y first*.

**REMARKS.**     The following two remarks are critical.

1.  Notice that $\displaystyle\int_{g_1(y)}^{g_2(y)} f(x, y) \, dx$ is to be done *first*.

2.  When calculating $\displaystyle\int_{g_1(y)}^{g_2(y)} f(x, y) \, dx$, *treat y as a constant*.

One should get in the habit when taking a double integral over a region such as shown in Figure 19.52 of following the procedures listed below.

---

**Procedure for Setting Up an Iterated Integral**

1.  Clearly label the "left" and the "right" of the domain $D$.
2.  Draw the slice shown in Figure 19.52 so that one realizes that $y$ is being held constant while $x$ is varying from the "left" to the "right." This gives the limits of integration for the *inner* integral.

3.   Now sum these slices "on $y$," that is, the slices go from $c$ to $d$. This gives the limits of integration on the *outer* integral.

EXAMPLE 4     **Integrating with Respect to $x$ First**

Redo Example 3 by integrating first with respect to $x$.

**Solution**     This region has a left side and a right side. The left side is given by $x = g_1(y) = y$, and the right side is given by $x = g_2(y) = \sqrt{y}$. See Figure 19.54. The slice shown in the figure indicates we will integrate from the left to the right while holding $y$ constant. Thus the limits of integration on the inner integral go from $x = y$ to $x = \sqrt{y}$. Now the slices are summed on $y$ from 0 to 1. This gives the limits of integration on the outer integral. Then

$$\int\int_D f(x,\ y)\ dA \ = \ \int_c^d \left( \int_{g_1(y)}^{g_2(y)} f(x,\ y)\ dx \right) dy$$

$$= \ \int_0^1 \left( \int_y^{\sqrt{y}} 48xy\ dx \right) dy$$

$$= \ \int_0^1 \left( [24x^2y]\big|_{x=y}^{x=\sqrt{y}} \right) dy$$

$$= \ \int_0^1 (24y^2 - 24y^3)\ dy$$

$$= \ (8y^3 - 6y^4)\big|_0^1$$

$$= \ 2 \quad \blacksquare$$

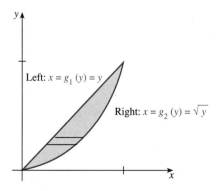

**Figure 19.54**

**E X A M P L E   5**      **Changing the Order of Integration**

Find

$$\int_0^1 \left( \int_y^1 2e^{x^2}\, dx \right) dy.$$

**Solution**      We run into an immediate difficulty since we cannot find an antiderivative for
$e^{x^2}$. We need to see what the region of integration is in the $xy$-plane and to see if
integrating with respect to $y$ first will help.

When doing the inner integral, $y$ is being held constant and $x$ is going from
(the left) $x = y$ to (the right) $x = 1$. Then the outer integral indicates that $y$ is
going from 0 to 1. Thus we have the region shown in Figure 19.55, with the slice
indicating how the above integral is to be done.

In order to *switch the order of integration*, we look at the region as in Figure
19.56, with a bottom given by $y = 0$ and a top given by $y = x$. Thus we first
integrate with respect to $y$ from $y = 0$ to $y = x$. Then sum on $x$ from 0 to 1.
Thus

$$\iint_D f(x, y)\, dA = \int_a^b \left( \int_{g_1(x)}^{g_2(x)} f(x, y)\, dy \right) dx$$

$$= \int_0^1 \left( \int_0^x 2e^{x^2}\, dy \right) dx$$

$$= \int_0^1 \left( 2e^{x^2} y \big|_{y=0}^{y=x} \right) dx$$

$$= \int_0^1 2x e^{x^2}\, dx$$

$$= e^{x^2} \big|_0^1$$

$$= e - 1 \quad \blacksquare$$

**Figure 19.55**

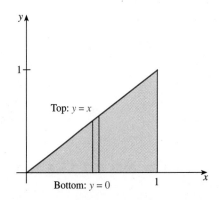

**Figure 19.56**

## Applications

We have already defined the average value of a continuous function $f(x)$ of *one* variable on the interval $[a, b]$ to be

$$\text{average value} = \frac{1}{b - a} \int_a^b f(x) \, dx.$$

For a nonnegative function $f$, this is the same as the area under the function $y = f(x)$ on the interval $[a, b]$, divided by the length of this interval.

There is a similar definition for the **average value** of a function $z = f(x, y)$ such as shown in Figure 19.57 of *two* variables on a domain $D$ in the $xy$-plane.

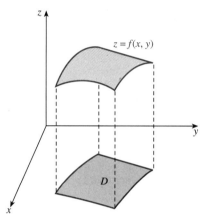

**Figure 19.57**

---

**Average Value**

If $f(x, y)$ is a continuous function on a domain $D$, the average value of $f$ on $D$ is

$$\frac{1}{A(D)} \int \int_D f(x, y) \, dA,$$

where $A(D)$ is the area of the domain $D$.

---

Notice that for a nonnegative function $f$, this is the same as the volume under the surface $z = f(x, y)$ and above the domain $D$ divided by the area of the domain.

**E X A M P L E   6**    **Average Concentration of Pollutants**

An industrial plant is located at the precise center of a town shaped as a square with each side of length 4 miles. If the plant is placed at the point $(0, 0)$ of the $xy$-plane (see Figure 19.58), then certain pollutants are dispersed in such a manner that the concentration at any point $(x, y)$ in town is given by

$$C(x, y) = 1000(24 - 3x^2 - 3y^2),$$

where $C(x, y)$ is the number of particles of pollutants per square mile of surface per day at a point $(x, y)$ in town. What is the average concentration of these pollutants in this town?

**Solution**

This concentration varies over the area of the town being largest naturally at the center of town where the plant is located and smallest at the four corners of town. In fact $C(0, 0) = 24,000$ while $C(2, 2) = C(2, -2) = C(-2, -2) = C(-2, 2) = 0$. The concentration function $C(x, y)$ could be obtained by taking samples at various points around the town and then "fitting a surface to the data" in a fashion analogous to the way we fitted a straight line to data points by the method of least squares in Section 1.4 when dealing with functions of one variable. Each sample could be a pollutant count over, say, a single square foot of surface. Multiplying this number by the number of square feet in a square mile then gives the number of particles per square mile. This last number is then $C(x, y)$ at the center of the 1 foot by 1 foot square.

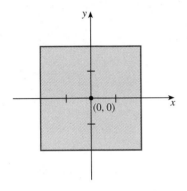

**Figure 19.58**

The average value is the double integral $\displaystyle\iint_D C(x, y)\, dA$, divided by the area of $D$ where $D$ is the square shown in Figure 19.58. If the domain $D$ were some irregularly shaped region, we would have to resort to Riemann sums to approximate this double integral. But the region $D$ has a "bottom" and "top" and also a "left" and "right." Thus we can evaluate this double integral as an iterated integral and also can then integrate first with respect to $x$ or first with respect to $y$. We integrate first with respect to $x$, as indicated by the slice in Figure 19.59. The left is given by $x = -2$ and the right by $x = 2$. The slices are summed from $y = -2$ to $y = 2$. Thus

**Figure 19.59**

$$\iint_D C(x, y)\, dA = 1000 \int_{-2}^{2} \left( \int_{-2}^{2} [24 - 3x^2 - 3y^2]\, dx \right) dy$$

$$= 1000 \int_{-2}^{2} ([24x - x^3 - 3xy^2]\big|_{x=-2}^{x=2})\, dy$$

$$= 1000 \int_{-2}^{2} ([48 - 8 - 6y^2] - [-48 + 8 + 6y^2])\, dy$$

$$= 1000 \int_{-2}^{2} (80 - 12y^2)\, dy$$

$$= 1000(80y - 4y^3)\big|_{-2}^{2}$$

$$= 1000[(160 - 32) - (-160 + 32)]$$

$$= 256,000.$$

The area of $D$ is the area of the square and is 16. Thus the average concentration of pollutants is $256,000/16 = 16,000$ particles per square mile or about 0.0006 particles per square foot per day. ∎

In Example 6 it is natural to ask just how many particles of pollutants there are in town. The concentration $C(x, y)$ is an example of a **density** function that we usually denote in general by $\delta(x, y)$. The density function gives the number per unit area. In the case of the pollutants we counted the number of particles per square mile. In general, we have the following.

---

**Density**

If $\delta(x, y)$ is a continuous function that gives the density in number per unit of area, then the total number in $D$ is given by

$$\iint_D \delta(x, y) \, dA.$$

---

**E X A M P L E   7**    **Amount of Pollutants**

Find the total number of particles of pollutant in town per day in the previous example.

**Solution**    Since the total number is given by $\iint_D \delta(x, y) \, dA$ and $\delta(x, y) = C(x, y)$, we have from Example 6 that

$$\iint_D \delta(x, y) \, dA = \iint_D C(x, y) \, dA = 256{,}000. \quad \blacksquare$$

## SELF-HELP EXERCISE SET 19.6

1.  Find the volume under the surface given by $z = 10 - x^2 - y^2$ and over the domain given by the rectangle

$$D = \{(x, y) : 0 \le x \le 1, 0 \le y \le 3\};$$

that is, find

$$\iint_D (10 - x^2 - y^2) \, dA.$$

2.  Find the volume under the surface given by $z = 8xy$ over the domain given by the triangular region $D$ in the $xy$-plane bounded by $y = x$, $y = 0$, and $x = 1$ by integrating first with respect to $y$.

3.  Redo the previous exercise by integrating first with respect to $x$.

# EXERCISE SET 19.6

In Exercises 1 through 4 find the Riemann sum $S_n$ for the given $n$ for $f(x, y) = 6xy^2$ on the rectangle $R = \{(x, y) : 0 \le x \le 2, 0 \le y \le 1\}$, by taking the points $(x_i, y_i)$ at the indicated points of each subrectangle. (Compare with Example 1 of the text.)

1.   $S_2$, $(x_i, y_i)$ at the center of each subrectangle

2.   $S_3$, $(x_i, y_i)$ at the center of each subrectangle

3.   $S_2$, $(x_i, y_i)$ at the lower left of each subrectangle

4.   $S_4$, $(x_i, y_i)$ at the center of each subrectangle

5.   Find $S_2$ for $z = x^2 y$ for $D = \{(x, y) | 0 \le x \le 4, 0 \le y \le 2\}$ where $(x_i, y_i)$ are taken at the lower left of each subrectangle.

6.   Repeat the previous exercise for $S_4$.

In Exercises 7 through 28 find the double integral over the indicated region $D$ in two ways. (a) Integrate first with respect to $x$; (b) integrate first with respect to $y$.

7.   $\int\int_D (2x + 3y) \, dA$, $D = \{(x, y): 0 \le x \le 1, 0 \le y \le 1\}$

8.   $\int\int_D (x^2 + y) \, dA$, $D = \{(x, y): 0 \le x \le 1, 0, \le y \le 1\}$

9.   $\int\int_D xy \, dA$, $D = \{(x, y): 0 \le x \le 2, 0 \le y \le 1\}$

10.  $\int\int_D xy \, dA$, $D = \{(x, y): 0 \le x \le 1, 0 \le y \le 3\}$

11.  $\int\int_D 6x^2 y \, dA$, $D = \{(x, y): 1 \le x \le 2, 2 \le y \le 3\}$

12.  $\int\int_D 6xy^2 \, dA$, $D = \{(x, y): 1 \le x \le 2, 2 \le y \le 3\}$

13.  $\int\int_D y\sqrt{y^2 + x} \, dA$, $D = \{(x, y): 0 \le x \le 1, 0 \le y \le 1\}$

14.  $\int\int_D x\sqrt{x^2 + y} \, dA$, $D = \{(x, y): 0 \le x \le 1, 0 \le y \le 4\}$

15.  $\int\int_D \frac{1}{xy} \, dA$, $D = \{(x, y): 1 \le x \le 2, 1 \le y \le 3\}$

16.  $\int\int_D (1 + \frac{x}{y}) \, dA$, $D = \{(x, y): 1 \le x \le 2, 1 \le y \le 3\}$

17.  $\int\int_D e^{x+y} \, dA$, $D = \{(x, y): 0 \le x \le 2, 0 \le y \le 3\}$

18.  $\int\int_D xye^{0.5(x^2+y^2)} \, dA$, $D = \{(x, y): 0 \le x \le 2, 0 \le y \le 3\}$

19.  $\int\int_D 1/(1 + x + y)^3 \, dA$, $D = \{(x, y): 0 \le x \le 2, 0, \le y \le 3\}$

20.  $\int\int_D 1/(1 + x + y)^2 \, dA$, $D = \{(x, y): 1 \le x \le 2, 1 \le y \le 3\}$

21.  $\int\int_D xe^{x^2} \, dA$, $D = \{(x, y): 0 \le x \le 2, 0 \le y \le 1\}$

22.  $\int\int_D ye^{y^2} \, dA$, $D = \{(x, y): 0 \le x \le 2, 0 \le y \le 1\}$

23.  $\int\int_D (x + y) \, dA$, $D$ is the triangular region with vertices at $(0, 0)$, $(1, 0)$, $(1, 1)$

24.  $\int\int_D (x + y) \, dA$, $D$ is the triangular region with vertices at $(0, 0)$, $(0, 1)$, $(1, 1)$

25. $\int \int_D (x^2 + y^2) \, dA$, $D$ is the triangular region with vertices at $(0, 0)$, $(1, 0)$, $(1, 1)$

26. $\int \int_D (x^2 + y^2) \, dA$, $D$ is the triangular region with vertices at $(0, 0)$, $(0, 1)$, $(1, 1)$

27. $\int \int_D xy \, dA$, $D$ is the triangular region with vertices at $(0, 0)$, $(1, 0)$, $(1, 1)$

28. $\int \int_D x^2y \, dA$, $D$ is the triangular region with vertices at $(1, 0)$, $(2, 0)$, $(2, 1)$

In Exercises 30 through 34 find the volume under the surface of the given function and over the indicated region.

29. $f(x, y) = 1$, $D$ is the region in the first quadrant bounded by the curves $y = x$ and $y = x^3$.

30. $f(x, y) = 1$, $D$ is the region in the first quadrant bounded by the curves $y^2 = x^3$, $y = x$.

31. $f(x, y) = xy^2$, $D$ is the region in the first quadrant bounded by the curves $y = x$ and $y = x^2$.

32. $f(x, y) = x/y^2$, $D$ is the region in the first quadrant bounded by the curves $x = 0$, $x = y^{3/2}$, $y = 1$, $y = 2$.

33. $f(x, y) = xy$, $D$ is the region in the first quadrant bounded by the curves $y = x$ and $x = y^2$.

34. $f(x, y) = xe^y$, $D$ is the region in the first quadrant bounded by the curves $y = 0$, $y = x^2$, $x = 0$, $x = 1$.

In Exercises 35 through 42 integrating first with respect to one of $x$ or $y$ is difficult while integrating first with respect to the other is not. Find $\int \int_D f(x, y) \, dA$ the easier way.

35. $f(x, y) = x$, $D$ is the region bounded by the curves $y = x^2 + 1$, $y = x$, $x = -1$, $x = 1$.

36. $f(x, y) = x$, $D$ is the region in the fourth quadrant bounded by the curves $y = -x^2$, $y = -3x - 1$, $x = 0$, $x = 2$.

37. $f(x, y) = 1$, $D$ is the region in the first and fourth quadrants bounded by the curves $x = 1 - y$, $y = x^2 - 1$, $x = 0$.

38. $f(x, y) = 1$, $D$ is the region in the first quadrant bounded by the curves $x = y - 2$, $x = y^3$, $x = 0$, $x = 1$.

39. $f(x, y) = 1$, $D$ is the region in the first quadrant bounded by the curves $xy = 1$, $y = x$, $y = 2$.

40. $f(x, y) = y$, $D$ is the region in the first quadrant bounded by the curves $x = 1$, $x = y^2$.

41. $f(x, y) = y$, $D$ is the region in the first quadrant bounded by the curves $y = 2x$, $y = x$, $x = 2$.

42. $f(x, y) = y$, $D$ is the region in the first quadrant bounded by the curves $y = \frac{1}{2}x$, $y = x$, $y = 2$.

## Applications

43. **Concentration of Pollutants.** In Example 6 of the text, what is the average concentration of pollutants in the region

$$D_1 = \{(x, y): 0 \leq x \leq 1, 0 \leq y \leq 2\}?$$

44. **Concentration of Pollutants.** If $D$ is the region given in Example 6 and $D_1$ is the region given in the previous problem, find the average concentration of pollutants over the region consisting $D$ less $D_1$.

45. **Cobb-Douglas Production Function.** What is the average value of the Cobb-Douglass production function $f(x, y) = x^{1/3}y^{2/3}$ for the range of $x$ and $y$ given by

$$D = \{(x, y): 8 \leq x \leq 27, 1 \leq y \leq 8\}.$$

46. **Cobb-Douglas Production Function.** What is the average value of the Cobb-Douglass production function $f(x, y) = x^{1/3}y^{2/3}$ for the range of $x$ and $y$ given by

$$D = \{(x, y): 8 \leq x \leq 27, 1 \leq y \leq 8\}.$$

47. **Amount of Pollutants.** In Example 6 of the text what is the total amount of pollutants in the region $D_1$, given in Exercise 43.

48. **Amount of Pollutants.** In Example 6 of the text, what is the total amount of pollutants in the region

$$D_2 = \{(x, y): 0 \leq x \leq 2, 0 \leq y \leq 1\}.$$

49. **Population.** The density function for a coastal town is given by

$$\delta(x, y) = \frac{120,000}{(2 + x + y)^3}$$

in people per square mile where $x$ and $y$ are in miles and where the town is the square

$$D = \{(x, y): 0 \leq x \leq 2, 0 \leq y \leq 2\}.$$

(The shore runs along the $x$-axis.) Find the total population of this town.

50. **Average Population.** Find the average density of population in the town of the previous problem.

51. **Population.** Find the total population living in the town given in the previous problem that are within one mile of shore.

52. **Population.** Find the total population living in the town given in the previous problem that are at least one mile from shore.

## Solutions to Self-Help Exercise Set 19.6

1. The region $D$ has a bottom given by $y = g_1(x) = 0$ and a top given by $y = g_2(x) = 3$. See the figure. The slice shown in the figure indicates that while holding $x$ constant, we integrate with respect to $y$ from the bottom to the top.

Thus the limits of integration on the inner integral go from $y = 0$ to $y = 3$. Now the slices are summed on $x$ from 0 to 1. This gives the limits of integration on the outer integral. Thus

$$\int\int_D f(x, y)\, dA = \int_a^b \left( \int_{g_1(x)}^{g_2(x)} f(x, y)\, dy \right) dx$$

$$= \int_0^1 \left( \int_0^3 (10 - x^2 - y^2)\, dy \right) dx$$

$$= \int_0^1 \left( \left[ 10y - x^2 y - \frac{1}{3} y^3 \right] \Big|_{y=0}^{y=3} \right) dx$$

$$= \int_0^1 (30 - 3x^2 - 9)\, dx$$

$$= \int_0^1 (21 - 3x^2)\, dx$$

$$= (21x - x^3)\big|_0^1 = 20.$$

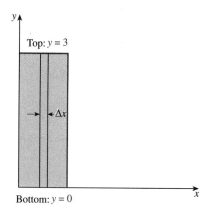

Top: $y = 3$

$\leftarrow \Delta x$

Bottom: $y = 0$

2.  The region $D$ has a bottom given by $y = g_1(x) = 0$ and a top given by $y = g_2(x) = x$. See the figure. The slice shown in the figure indicates that while holding $x$ constant, we integrate with respect to $y$ from the bottom to the top. Thus the limits of integration on the inner integral go from $y = 0$ to $y = x$. Now the slices are summed on $x$ from 0 to 1. This gives the limits of integration on the outer integral. Thus

$$\int\int_D f(x, y)\, dA = \int_a^b \left( \int_{g_1(x)}^{g_2(x)} f(x, y)\, dy \right) dx$$

$$= \int_0^1 \left( \int_0^x 8xy\, dy \right) dx$$

$$= \int_0^1 \left( [4xy^2] \big|_{y=0}^{y=x} \right) dx$$

$$= \int_0^1 4xx^2 \, dx$$

$$= x^4 \big|_0^1 = 1$$

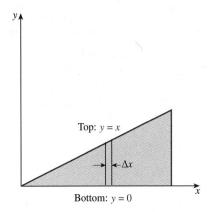

Top: $y = x$

$\Delta x$

Bottom: $y = 0$

3. We notice that this region does have a left side and a right side. The left side is given by $x = g_1(y) = y$ and the right side is given by $x = g_2 = 1$. See the figure. The slice shown in the figure indicates we will integrate from the left to the right while holding $y$ constant. Thus the limits of integration on the inner integral go from $x = y$ to $x = 1$. Now the slices are summed on $y$ from 0 to 1. This gives the limits of integration on the outer integral. Then

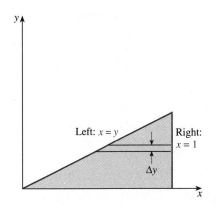

Left: $x = y$

Right: $x = 1$

$\Delta y$

$$\iint_D f(x, y) \, dA = \int_c^d \left( \int_{g_1(y)}^{g_2(y)} f(x, y) \, dx \right) dy$$

$$= \int_0^1 \left( \int_y^1 8xy \, dx \right) dy$$

$$= \int_0^1 \left( 4x^2 y \big|_{x=y}^{x=1} \right) dy$$

$$= \int_0^1 (4y - 4y^3) \, dy$$

$$= (2y^2 - y^4) \big|_0^1 = 1 \quad \blacksquare$$

## SUMMARY OUTLINE OF CHAPTER 19

- A **function of two variables**, denoted by $z = f(x, y)$, is a rule that associates to every point $(x, y)$ in some set $D$ (called the **domain**) a unique point denoted by $z = f(x, y)$.

- The **distance between the two points** $(x_1, y_1, z_1)$ and $(x_2, y_2, z_2)$ is

$$d = \sqrt{(x_2 - x_1)^2 + (y_2 - y_1)^2 + (z_2 - z_1)^2}.$$

- Given a function $z = f(x, y)$ with domain $D$, the **graph** is the set of points $(x, y, z)$ with $z = f(x, y)$ and $(x, y) \in D$. Since this is described by *two* variables, the result is a two-dimensional object called a **surface**.

- The equation of the **sphere** centered at $(x_0, y_0, z_0)$ with radius $r$ is given by

$$(x - x_0)^2 + (y - y_0)^2 + (z - z_0)^2 = r^2.$$

- The graph of an equation of the form $Ax + By + Cz = D$ is a **plane**.

- The graph of $f(x, y) = z_0$ in the xy-plane is called a **level curve**.

- We say that $f(x, y)$ is **continuous** at $(a, b)$ if $f(x, y)$ approaches $f(a, b)$, no matter how $(x, y)$ approaches $(a, b)$.

- If $z = f(x, y)$, then the **partial derivative of $f$ with respect to $x$** is

$$\frac{\partial f}{\partial x}(x, y) = \lim_{\Delta x \to 0} \frac{f(x + \Delta x, y) - f(x, y)}{\Delta x},$$

and the **partial derivative of $f$ with respect to $y$** is

$$\frac{\partial f}{\partial y}(x, y) = \lim_{\Delta y \to 0} \frac{f(x, y + \Delta y) - f(x, y)}{\Delta y},$$

if the limits exist.

- If $\dfrac{\partial x_A}{\partial q} > 0$     and     $\dfrac{\partial x_B}{\partial p} > 0,$

then we say that the two commodities are **competitive**.

If $\dfrac{\partial x_A}{\partial q} < 0$     and     $\dfrac{\partial x_B}{\partial p} < 0,$

then we say that the two commodities are **complementary**.

- Given a function $z = f(x, y)$, we define the following four, second order partial derivatives.

$$\frac{\partial^2 f}{\partial x^2} = \frac{\partial}{\partial x}\left(\frac{\partial f}{\partial x}\right), \qquad \frac{\partial^2 f}{\partial y^2} = \frac{\partial}{\partial y}\left(\frac{\partial f}{\partial y}\right),$$

$$\frac{\partial^2 f}{\partial y \partial x} = \frac{\partial}{\partial y}\left(\frac{\partial f}{\partial x}\right), \qquad \frac{\partial^2 f}{\partial x \partial y} = \frac{\partial}{\partial x}\left(\frac{\partial f}{\partial y}\right).$$

- We say that $f$ has a **relative maximum** at $(a, b) \in D$ if there exists a circle centered at $(a, b)$ and entirely in $D$, such that

$$f(x, y) \leq f(a, b)$$

for all points $(x, y)$ inside this circle.

  We say that $f$ has a **relative minimum** at $(a, b) \in D$ if there exists a circle centered at $(a, b)$ and entirely within $D$, such that

$$f(x, y) \geq f(a, b)$$

for all points $(x, y)$ inside this circle.

  We say that the function $z = f(x, y)$ has a **relative extremum** at $(a, b) \in D$ if $f$ has either a relative maximum or a relative minimum at $(a, b)$.

- **Necessary Condition for Relative Extrema.**   If $z = f(x, y)$ is defined, and both first partial derivatives exist for all values of $(x, y)$ inside some circle about $(a, b)$, and $f$ assumes a relative extremum at $(a, b)$, then

$$\frac{\partial f}{\partial x}(a, b) = 0 \quad \text{and} \quad \frac{\partial f}{\partial y}(a, b) = 0.$$

- Suppose that both partial derivatives of $z = f(x, y)$ exist in its domain of definition. **Critical points** are points $(a, b)$, not on the boundary of the domain of $f$, for which

$$\frac{\partial f}{\partial x}(a, b) = 0 \quad \text{and} \quad \frac{\partial f}{\partial y}(a, b) = 0.$$

- **Second Derivative Test for Functions of Two Variables.** For $z = f(x, y)$ assume that $f_{xx}, f_{yy}$, and $f_{xy}$ all exist for every point near $(a, b)$. Suppose $(a, b)$ is a critical point, that is, assume that $f_x(a, b) = 0$ and $f_y(a, b) = 0$. Define

$$\Delta(a, b) = f_{xx}(a, b)f_{yy}(a, b) - [f_{xy}(a, b)]^2.$$

Then

1. $\Delta(a, b) > 0$ and $f_{xx}(a, b) < 0$ implies that $f(a, b)$ is a relative maximum.
2. $\Delta(a, b) > 0$ and $f_{xx}(a, b) > 0$ implies that $f(a, b)$ is a relative minimum.
3. $\Delta(a, b) < 0$ implies $(a, b)$ is a saddle point, that is, $(a, b)$ is neither a relative minimum nor a relative maximum.
4. $\Delta(a, b) = 0$ implies that the test fails, that is, when $\Delta(a, b) = 0$, there can be a relative maximum, a relative minimum, or a saddle.

- We say that $f$ has an **absolute maximum** on $D$ at $(a, b) \in D$ if for all $(x, y) \in D$, $f(x, y) \le f(a, b)$.

  We say that $f$ has an **absolute minimum** on $D$ at $(a, b) \in D$ if for all $(x, y) \in D$, $f(x, y) \ge f(a, b)$.

  Also we say that $f$ has an **absolute extremum** on $D$ at $(a, b) \in D$ if $f$ has either an absolute maximum or an absolute minimum on $D$ at $(a, b)$.

- **Existence of Absolute Extrema on a Closed and Bounded Region.** If $z = f(x, y)$ is continuous on the closed and bounded region $D$, then $f$ assumes its absolute extrema in $D$, that is, there exists two points $(a, b) \in D$ and $(A, B) \in D$ such that

$$f(a, b) \le f(x, y) \le f(A, B) \quad \text{for all } (x, y) \in D.$$

- **Method of Lagrange Multipliers.** Candidates $(x_c, y_c, z_c)$ for a relative extremum of $f(x, y, z)$, subject to the constraint $g(x, y, z) = 0$, can be found among the critical points $(x_c, y_c, z_c, \lambda)$ of the auxiliary function

$$F(x, y, z, \lambda) = f(x, y, z) + \lambda g(x, y, z),$$

that is, among the solutions of the equations

$$F_x = 0, \quad F_y = 0, \quad F_z = 0, \quad g = 0.$$

- **Equation of Tangent Plane.** The plane which is tangent to the surface $z = f(x, y)$ at the point $P = (a, b, z_0)$ is given by the equation

$$z - z_0 = f_x(a, b)(x - a) + f_y(a, b)(y - b),$$

where $z_0 = f(a, b)$.

- **Differential.** Assume $f_x(a, b)$ and $f_y(a, b)$ exist. If $dx$ and $dy$ are any numbers, then the total differential is

$$dz = f_x(a, b) \cdot dx + f_y(a, b) \cdot dy$$

- **Tangent Plane Approximations.** If the first partial derivatives of $f(x, y)$ are continuous at $(a, b)$ and if $\Delta x$ and $\Delta y$ are small then $\Delta z \approx dz$, that is,

$$f(a + \Delta x, b + \Delta y) - f(a, b) \approx f_x(a, b)\Delta x + f_y(a, b)\Delta y.$$

- **Riemann Sum.** For any integer $n$ divide both intervals $[a, b]$ and $[c, d]$ into $n$ equal subintervals and define $\Delta x = \dfrac{b - a}{n}$ and $\Delta y = \dfrac{d - c}{n}$. This gives a collection of subrectangles. Label all subrectangles that lie *entirely inside D* (the domain of $f(x, y)$) in some manner. Let the total number of subrectangles be $N$. In each subrectangle $R_i$ pick some point $(x_i, y_i) \in R_i$. The Riemann sum is defined to be

$$\sum_{i=1}^{N} f(x_i, y_i)\Delta x \Delta y.$$

- Let $f(x, y)$ be continuous on the bounded region $D$. Then the **definite integral of f over D** is

$$\iint_D f(x, y)\, dA = \lim_{n \to +\infty} \sum_{i=1}^{N} f(x_i, y_i)\Delta x \Delta y,$$

where the limit exists and is the same no matter how the points $(x_i, y_i)$ are chosen in $R_i$.

- Let $f(x, y)$ be continuous on the bounded region $D$ and nonnegative there. Then the volume under the surface $z = f(x, y)$ and over the domain $D$ is the definite integral of $f$ over $D$ or

$$V = \iint_D f(x, y)\, dA$$

- If the domain $D$ is given as in the figure and $z = f(x, y) \geq 0$ is continuous on $D$, then the volume $V$ between the surface $z = f(x, y)$ and the domain $D$ is

$$\iint_D f(x, y)\, dA = \int_a^b \left( \int_{g_1(x)}^{g_2(x)} f(x, y)\, dy \right) dx.$$

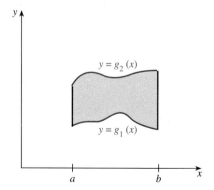

- If the domain $D$ is given as in the figure and $z = f(x, y) \geq 0$ is continuous on $D$, then the volume $V$ between the surface $z = f(x, y)$ and the domain $D$ is

$$\iint_D f(x, y)\, dA = \int_c^d \left( \int_{g_1(y)}^{g_2(y)} f(x, y)\, dx \right) dy.$$

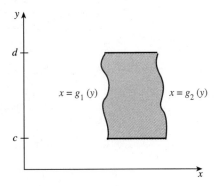

## Chapter 19 Review Exercises

1.  Evaluate the function $z = f(x, y) = 2x^2 + 3y - 1$ at the points $(0, 0)$, $(1, 2)$, $(2, -1)$.

2.  Evaluate the function $z = f(x, y) = \dfrac{2x^2}{5y}$ at the points $(0, 1)$, $(1, 2)$, $(2, -1)$.

3.  Determine the domains of the functions in the previous two exercises.

4.  Find the distance between the two points $(1, -2, 5)$ and $(3, -4, -1)$.

5.  For the function $z = f(x, y) = 16 - x^2 - y^2$, find the level surfaces corresponding to $z_0 = 16$, $z_0 = 12$, $z_0 = 7$. Describe the surface $z = f(x, y)$.

6.  Find both first order partial derivatives to the function $z = f(x, y) = x + x^2 y^5 + 10$.

7.  Evaluate the two first order partial derivatives found in the previous exercise at the point $(2, 1)$.

8.  Find both first order partial derivatives to the function $z = f(x, y) = e^{-xy} + \ln xy$.

9.  Find all three second order partial derivatives of the function in Exercise 6.

10. Find all three second order partial derivatives of the function in Exercise 8.

11. Find all three first order partial derivatives to the function $z = f(x, y, z) = e^{-xyz} + xy^2 z^3$.

In Exercises 12 through 19 find all the critical points and determine whether each critical point is a relative minimum, relative maximum, or a saddle point.

12. $f(x, y) = x^2 - xy + 4y^2$

13. $f(x, y) = -x^2 + 2xy - 8y^2$

14. $f(x, y) = -x^2 + xy - y^2 - 3x$

15. $f(x, y) = (x - 2)^2 + (y + 3)^2$

16. $f(x, y) = x^3 + y^3 + 3xy$

17. $f(x, y) = 3x^2 + y^3 - 6xy - 9y$

18. $f(x, y) = y^3 - 3y^2 - x^2 - 9y + 10x$

19. $f(x, y) = 12xy - x^3 - 36y^2$

20. Let $z = f(x, y) = ax^2 + 2bxy + cy^2$, where $a$, $b$, and $c$ are constants. Show that $(0, 0)$ is a critical point. Show that the point $(0, 0)$ is a relative minimum if $ac - b^2 > 0$ and $a > 0$, and a relative maximum if $ac - b^2 > 0$ and $a < 0$.

21. Using Lagrange multipliers, minimize $z = 4x^2 + 2xy - 3y^2$ subject to the constraint $x + y = 1$.

22. Using Lagrange multipliers, maximize $z = -2x^2 - xy - y^2 - x$ subject to the constraint $2x + y - 1 = 0$.

23. Use the tangent line approximation to estimate

$$\Delta z = f(x_0 + \Delta x, y_0 + \Delta y) - f(x_0, y_0)$$

if $f(x, y) = 2x^2 + 3y^2$, $(x_0, y_0) = (1, 2)$, $\Delta x = 0.1$, $\Delta y = 0.2$.

24. Repeat the previous exercise if $f(x, y) = x^2 y^4$, $(x_0, y_0) = (1, 2)$, $\Delta x = 0.2$, $\Delta y = -0.1$.

25. Use the tangent line approximation to estimate

$$\sqrt{3.01^2 + 3.98^2}.$$

In Exercises 26 through 31 find the double integral over the indicated region $D$ in two ways. (a) Integrate first with respect to $x$; (b) integrate first with respect to $y$.

26. $\int\int_D (3x + 6y)\, dA$, $D = \{(x, y)|0 \le x \le 2, 1 \le y \le 2\}$.

27. $\int\int_D (12x^2 y^3)\, dA$, $D = \{(x, y)|0 \le x \le 1, 1 \le y \le 3\}$.

28. $\int\int_D (x^2 + y)\, dA$, where $D$ is the triangular region with vertices at $(0, 0)$, $(1, 0)$, $(1, 1)$.

29. $\int\int_D (x + xy)\, dA$, where $D$ is the triangular region with vertices at $(0, 0)$, $(0, 1)$, $(1, 1)$.

30. $\int\int_D (x + y)\, dA$, where $D$ is the region between the two curves $y = 1$ and $y = x^2$.

31. $\int\int_D (x + y)\, dA$, where $D$ is the region between the two curves $y = x$ and $y = \sqrt{x}$.

In Exercises 32 through 35 change the order of integration and then evaluate.

32. $\displaystyle\int_0^1 \int_0^x (x + y)\, dy\, dx$

33. $\displaystyle\int_0^1 \int_0^{\sqrt{x}} (x + y)\, dy\, dx$

34. $\displaystyle\int_0^1 \int_0^{\sqrt{y}} (x + y)\, dx\, dy.$

35. $\displaystyle\int_0^1 \int_y^{\sqrt{y}} dx\, dy$

36. Find the average value of $f(x, y) = 8xy^3$ on the region $D = \{(x, y)|0 \le x \le 2, 1 \le y \le 3\}$.

37. **Cobb-Douglas Production Function.**  A production function is given by $f(K, L) = 10K^{.20}L^{.80}$. Find $f(243, 32)$.

38. **Cobb-Douglas Production Function.**  Find the first partial derivatives of the production function given in the previous exercise.

39. **Cobb-Douglas Production Function.** Find all the second partial derivatives of the production function given in the previous exercise.

40. **Cost.** A cost function is given by $C(x, y) = 1000 + x^2 + y^2 - xy$. Find $C(10, 1)$.

41. **Cost.** Find the first partial derivatives of the cost function given in the previous exercise.

42. **Cost.** Find all the second partial derivatives of the cost function given in the previous exercise.

43. **Continuous Compounding.** If \$P is invested at an annual rate of $r$ and compounded continuously, the amount after $t$ years is $A(P, r, t) = Pe^{rt}$. Find the three first partial derivatives of this function.

44. **Packaging.** Find the dimensions of the cheapest rectangular box of volume 16 cubic feet for which the cost per square foot of the top and bottom is twice that of the sides.

45. **Minimizing Production Costs.** The daily cost function of a firm that produces $x$ units of a product $X$ and $y$ units of another product $Y$ is given by $C(x, y) = 244 + x^2 - 20x + y^2 - 24y$. Find the number of units of each of the products that will minimize the cost.

46. **Maximizing Profit.** The firm in the previous exercise can sell each unit of product $X$ for \$2 and each unit of product $Y$ for \$4. Find the levels of production that will maximize the profits.

47. **Competititve Pricing.** A firm makes $x$ thousand units of regular toothpaste and $y$ thousand units of mint. Suppose the prices are $p$ and $q$ respectively and the demand equations are given by

$$p = 4 - x + y$$
$$q = 14 + x - 2y$$

Find the number of units of each that will maximize revenue.

48. **Fishery.** A small lake is to be stocked with 2 types of fish. There are $x$ thousand of the first fish and $y$ thousand of the second; the average weight of the first is $(3 - 2x - y)$ pounds and of the second is $(4 - x - 3y)$. Find the numbers of each fish that maximize the total weight of these fish.

49. **Measurements.** A rectangular cardboard box of dimensions 10 in. by 10 in. by 20 in. is being produced. Use the tangent plane approximation to approximate the volume of cardboard if the sides are 0.15 inches thick and the top and bottom are 0.25 inches thick.

# A

# Review

A.1  Exponents and Roots

A.2  Polynomials and Rational Expressions

A.3  Equations

A.4  Inequalities

## A.1  EXPONENTS AND ROOTS

► *Exponents*

► *Roots*

► *Rational Exponents*

► *Rationalizing*

### EXPONENTS

The symbol $x^2$ denotes $x \cdot x$, the symbol $x^3$ denotes $x \cdot x \cdot x$, and, in general, we have the following.

**Exponents**

If $n$ is a positive integer

$$x^n = \underbrace{x \cdot x \cdot \cdots \cdot x}_{n \text{ factors}}.$$

The number $n$ is called the **exponent**, and $x$ is called the **base**.

**Examples**

$$2^4 = 2 \cdot 2 \cdot 2 \cdot 2 = 16$$

$$\left(-\frac{1}{2}\right)^3 = \left(-\frac{1}{2}\right)\left(-\frac{1}{2}\right)\left(-\frac{1}{2}\right) = -\frac{1}{8}$$

**Zero Exponent**

If $x \neq 0$, then

$$x^0 = 1.$$

The expression $0^0$ is not defined.

**Examples**

$$2^0 = 1 \qquad \pi^0 = 1 \qquad (a + b)^0 = 1$$

**Negative Exponent**

If $n$ is a positive integer, then

$$x^{-n} = \frac{1}{x^n}.$$

**Examples**

$$2^{-1} = \frac{1}{2} \qquad x^{-3} = \frac{1}{x^3} \qquad \left(\frac{2}{3}\right)^{-2} = \frac{1}{\left(\frac{2}{3}\right)^2} = \frac{1}{\frac{4}{9}} = \frac{9}{4}$$

**Properties of Exponents**

If $m$ and $n$ are integers, the following properties are true.

| Property | Example |
|---|---|
| 1.  $x^m x^n = x^{m+n}$ | $x^2 \cdot x^3 = x^{2+3} = x^5$ |
| 2.  $(x^m)^n = x^{mn}$ | $(x^2)^3 = x^{2\cdot3} = x^6$ |
| 3.  $(xy)^n = x^n y^n$ | $(2 \cdot 4)^3 = 2^3 \cdot 4^3 = 8 \cdot 64 = 512$ |
| 4.  $\left(\dfrac{x}{y}\right)^n = \dfrac{x^n}{y^n}$ if $y \neq 0$ | $\left(\dfrac{2}{3}\right)^2 = \dfrac{2^2}{3^2} = \dfrac{4}{9}$ |
| 5.  $\dfrac{x^m}{x^n} = x^{m-n}$ if $x \neq 0$ | $\dfrac{x^5}{x^2} = x^{5-2} = x^3$ |
|     $\dfrac{x^m}{x^n} = \dfrac{1}{x^{n-m}}$ if $x \neq 0$ | $\dfrac{x^2}{x^5} = \dfrac{1}{x^{5-2}} = \dfrac{1}{x^3}$ |

**Examples**

$$(2x^2y^3)(3x^3y^4) = (2)(3)(x^2x^3)(y^3y^4) = 6x^5y^7$$

$$(2x^2y^3z)^4 = 2^4(x^2)^4(y^3)^4(z)^4 = 16x^8y^{12}z^4$$

$$\left(\frac{2x^2}{3y^3}\right)^3 = \frac{(2x^2)^3}{(3y^3)^3} = \frac{8x^6}{27y^9}$$

$$(a^4b^{-2})^{-3} = (a^4)^{-3}(b^{-2})^{-3} = (a^{-12})(b^6) = \frac{b^6}{a^{12}}$$

$$(a^2b^3)^2(a^4b)^{-1} = (a^2b^3)^2\frac{1}{a^4b} = \frac{a^4b^6}{a^4b} = b^5.$$

We can also do the last example as follows.

$$(a^2b^3)^2(a^4b)^{-1} = (a^4b^6)(a^{-4}b^{-1}) = a^{4-4}b^{6-1} = b^5.$$

## Roots

**Roots**

- If $n$ is a positive integer and $y^n = x$, then we say that $y$ is an $n$th root of $x$.
- If $n$ is odd, there is exactly one $n$th root, which we denote by $y = \sqrt[n]{x} = x^{1/n}$.
- If $n$ is even and $x$ is negative, there is no (real) root.
- If $n$ is even and $x$ is positive, we use the symbols $y = \sqrt[n]{x} = x^{1/n}$ to denote the positive $n$th root.
- If $n = 2$, we write $\sqrt[2]{x} = \sqrt{x}$.

The symbol $\sqrt{\phantom{x}}$ is called a **radical sign**.

**Examples**

$$\sqrt{9} = 3 \qquad \sqrt[3]{-64} = -4 \qquad \sqrt[4]{\dfrac{1}{16}} = \dfrac{1}{2}$$

---

**Properties of Roots**

| Radical Formulation | Exponent Formulation |
|---|---|
| 1. $\sqrt[n]{xy} = \sqrt[n]{x}\,\sqrt[n]{y}$ | $(xy)^{1/n} = x^{1/n}y^{1/n}$ |
| 2. $\sqrt[n]{\dfrac{x}{y}} = \dfrac{\sqrt[n]{x}}{\sqrt[n]{y}}$ | $\left(\dfrac{x}{y}\right)^{1/n} = \dfrac{x^{1/n}}{y^{1/n}},\ y \neq 0$ |
| 3. $\sqrt[m]{\sqrt[n]{x}} = \sqrt[mn]{x}$ | $(x^{1/n})^{1/m} = x^{1/mn}$ |

---

**Examples**

$$\sqrt{40} = \sqrt{4 \cdot 10} = \sqrt{4}\sqrt{10} = 2\sqrt{10}$$

$$\sqrt[3]{-24} = \sqrt[3]{(-8)(3)} = \sqrt[3]{-8}\sqrt[3]{3} = -2\sqrt[3]{3}$$

$$\sqrt{\sqrt[3]{64}} = \sqrt[6]{64} = 2$$

$$\sqrt[3]{8x^6y^{10}} = \sqrt[3]{8}\sqrt[3]{x^6}\sqrt[3]{y^{10}} = 2\sqrt[3]{(x^2)^3}\sqrt[3]{(y^3)^3y} = 2x^2\sqrt[3]{(y^3)^3}\sqrt[3]{y} = 2x^2y^3\sqrt[3]{y}$$

## Rational Exponents

**Rational Exponents**

If $m$ and $n$ are integers, $x \neq 0$, and $x^{1/n}$ exists, then

$$x^{m/n} = (x^{1/n})^m = (\sqrt[n]{x})^m$$

or, equivalently,

$$x^{m/n} = (x^m)^{1/n} = \sqrt[n]{x^m}.$$

**Examples**

$$4^{3/2} = (\sqrt{4})^3 = 2^3 = 8 \qquad \text{or} \qquad 4^{3/2} = \sqrt{4^3} = \sqrt{64} = 8$$

$$(-8)^{2/3} = (\sqrt[3]{-8})^2 = (-2)^2 = 4 \quad \text{or} \quad (-8)^{2/3} = \sqrt[3]{(-8)^2} = \sqrt[3]{64} = 4$$

*The properties of exponents are also true for rational exponents.*

**Example**

$$x^{3/2}y^{2/3}(\sqrt[5]{4}x^{1/3}y^{1/2})^{-5/2} = 4x^{3/2}y^{2/3}(4^{1/5})^{-5/2}(x^{1/3})^{-5/2}(y^{1/2})^{-5/2}$$

$$= 4x^{3/2}y^{2/3}4^{-1/2}x^{-5/6}y^{-5/4}$$

$$= 4x^{3/2-5/6}\frac{1}{2}y^{2/3-5/4}$$

$$= 2x^{4/6}y^{-7/12} = 2\frac{x^{2/3}}{y^{7/12}}$$

## Rationalizing

To remove a radical from the denominator of a quotient, multiply both the numerator and denominator of the quotient by an appropriate expression.

**Examples**

$$\frac{1}{\sqrt{3}} = \frac{1}{\sqrt{3}}\frac{\sqrt{3}}{\sqrt{3}} = \frac{\sqrt{3}}{(\sqrt{3})^2} = \frac{\sqrt{3}}{3}$$

$$\sqrt[5]{\frac{x}{y^2}} = \sqrt[5]{\frac{x}{y^2}\frac{y^3}{y^3}} = \frac{\sqrt[5]{xy^3}}{\sqrt[5]{y^5}} = \frac{\sqrt[5]{xy^3}}{y}$$

We can also rationalize the numerator.

**Example**

$$2 + \sqrt{3} = \frac{2 + \sqrt{3}}{1}\frac{2 - \sqrt{3}}{2 - \sqrt{3}} = \frac{(2)^2 - (\sqrt{3})^2}{2 - \sqrt{3}} = \frac{1}{2 - \sqrt{3}}$$

## EXERCISE SET A.1

In Exercises 1 through 64 simplify. Write the answers without using negative exponents.

**1.** $x^2x^8$             **2.** $y^3y^{11}$             **3.** $(x^3)^5$

**4.** $(y^4)^3$             **5.** $(x + 2)^4(x + 2)$             **6.** $(y + 3)^5(y + 3)^2$

**7.** $\dfrac{x^9}{x^4}$             **8.** $\dfrac{y^3}{y^7}$             **9.** $\dfrac{(x + 1)^5}{(x + 1)^2}$

**10.** $\dfrac{(x^2y)^2}{(xy^3)^3}$             **11.** $8^0$             **12.** $(2^0 + 2)^0$

**13.** $2^{-1}$             **14.** $4^{-2}$             **15.** $\left(\dfrac{1}{10}\right)^{-1}$

**16.** $\left(\dfrac{4}{9}\right)^{-2}$             **17.** $\left(\dfrac{xy^2}{z}\right)^0$             **18.** $(a^3bc^0)^{-2}$

**19.** $(a^2b)^2(ab^3)^{-1}$      **20.** $(ab^2c^{-1})^{-2}$      **21.** $\left(\dfrac{x^{-1}y^{-1}}{z}\right)^{-1}$

**22.** $\left(\dfrac{x^2y^{-2}z}{x^3yz^{-2}}\right)^{-2}$      **23.** $\left(\dfrac{x^2y^{-5}}{x^{-2}y^{-3}}\right)^{-3}$      **24.** $\dfrac{(x+1)^3(x+2)^4}{[(x+1)(x+2)]^3}$

**25.** $\sqrt[3]{-64}$      **26.** $\sqrt[5]{32}$      **27.** $\sqrt[5]{-32}$

**28.** $\sqrt[3]{-\dfrac{1}{9}}$      **29.** $\sqrt[4]{(-7)^4}$      **30.** $\sqrt[4]{7^4}$

**31.** $\sqrt[3]{(-3)^3}$      **32.** $\sqrt[3]{3^3}$      **33.** $\sqrt{\dfrac{25}{4}}$

**34.** $\sqrt[3]{\dfrac{8}{27}}$      **35.** $\sqrt[3]{-\dfrac{8}{125}}$      **36.** $\sqrt[5]{\dfrac{243}{32}}$

**37.** $\sqrt{\sqrt{16}}$      **38.** $\sqrt[3]{\sqrt{64}}$      **39.** $8^{2/3}$

**40.** $64^{2/3}$      **41.** $(0.001)^{1/3}$      **42.** $\left(-\dfrac{1}{125}\right)^{1/3}$

**43.** $36^{-3/2}$      **44.** $125^{2/3}$      **45.** $32^{4/5}$

**46.** $32^{-4/5}$      **47.** $64^{4/3}$      **48.** $\left(\dfrac{9}{16}\right)^{-5/2}$

**49.** $\left(\dfrac{1000}{27}\right)^{4/3}$      **50.** $(-1)^{3/5}$      **51.** $\left(\dfrac{8}{27}\right)^{-2/3}$

**52.** $100^{3/2}$      **53.** $\sqrt{4x^4}$      **54.** $\sqrt[4]{16y^8}$

**55.** $\sqrt[3]{ab^2}\sqrt[3]{a^2b}$      **56.** $\sqrt[3]{8x^6}$      **57.** $\sqrt{72a^4b^3c^5}$

**58.** $\sqrt[3]{8x^4b^8}$      **59.** $\sqrt[5]{32a^{10}b^{15}}$      **60.** $\sqrt{18a^4b^5}$

**61.** $\sqrt[3]{\dfrac{8a^9b^6}{c^{12}}}$      **62.** $\sqrt[5]{\dfrac{4x^6}{x^{-4}y^5}}$      **63.** $\sqrt{ab}\sqrt{a^3b^3}$

**64.** $\sqrt[3]{ab}\sqrt[3]{a^2b^2}$

In Exercises 65 through 72 write the expressions using rational exponents rather than radicals.

**65.** $\dfrac{1}{\sqrt{x}} + \sqrt{x}$      **66.** $\sqrt{\sqrt{x}}$      **67.** $\sqrt[3]{(x+1)^2}$      **68.** $(\sqrt[3]{(x+1)})^2$

**69.** $\sqrt[3]{x^2}$      **70.** $\sqrt{x^3}$      **71.** $\sqrt{\dfrac{x}{y}}$      **72.** $\sqrt[3]{\dfrac{x}{y}}$

In Exercises 73 through 84 simplify.

**73.** $(x^2+1)^{3/4}(x^2+1)^{2/3}$      **74.** $\dfrac{(x^2+1)^{1/3}}{(x^2+1)^{-3/4}}$

**75.** $\dfrac{x^{2/3}x^{-1/3}}{x^{2/3}}$      **76.** $\dfrac{y^{-1/4}y^{-1/2}}{y^{-3/4}}$

77. $\dfrac{a^{2/3}b^{3/4}}{a^{-1/3}b^{-1/4}}$

78. $\dfrac{a^{1/3}b^{-2/5}}{a^{-2/3}b^{1/5}}$

79. $\dfrac{a^{2/3}b^{4/3}}{a^{2/3}b^{1/3}}$

80. $\dfrac{a^{-1/3}b^{-3/4}}{a^{-2/3}b^{1/4}}$

81. $\dfrac{(x+1)^{2/3}(x+2)^{1/3}}{(x+1)^{-1/3}(x+2)^{-2/3}}$

82. $\dfrac{(x+1)^{-1/3}(y+1)^{-2/3}}{(x+1)^{2/3}(y+1)^{-5/3}}$

83. $[(x+1)^3]^{5/3}$

84. $[(y-2)^{1/3}]^6$

In Exercises 85 through 92 rationalize the denominator.

85. $\dfrac{1}{\sqrt{5}}$

86. $\dfrac{1}{\sqrt{7}}$

87. $\dfrac{1}{\sqrt{2}+1}$

88. $\dfrac{1}{3-\sqrt{5}}$

89. $\dfrac{1}{\sqrt[3]{5}}$

90. $\dfrac{3}{\sqrt[4]{3}}$

91. $\dfrac{1+\sqrt{3}}{1-\sqrt{3}}$

92. $\dfrac{2-\sqrt{5}}{2+\sqrt{5}}$

In Exercises 93 through 96 rationalize the numerator.

93. $\dfrac{\sqrt{5}-\sqrt{2}}{3}$

94. $\dfrac{\sqrt{7}-\sqrt{3}}{4}$

95. $\dfrac{\sqrt{10}-2}{6}$

96. $\sqrt{10}-3$

## A.2   POLYNOMIALS AND RATIONAL EXPRESSIONS

▶ *Polynomials*

▶ *Factoring*

▶ *Rational Expressions*

### Polynomials

We begin with the following definition.

---

**Polynomial**

A polynomial is an expression of the form

$$a_nx^n + a_{n-1}x^{n-1} + \cdots + a_1x + a_0,$$

where $n$ is a nonnegative integer and each coefficient $a_i$ is a real number.

---

If $a_n \neq 0$, the polynomial is said to have **degree n**.

Lower order polynomials have special names. A polynomial such as 3 has degree equal to *zero* and is called a **constant**. A polynomial such as $x - \frac{1}{2}$ has degree equal to *one* and is called linear. A polynomial such as $4x^2 - 2x + 4$

has degree equal to *two* and is called a quadratic. A polynomial such as $2x^3 - 5x^2 - \frac{3}{2}x - 6$ has degree equal to *three* and is called a **cubic**.

Since for any real number $x$ a polynomial is a real number, polynomials must follow the same rules as any real number.

**E X A M P L E   1**

Simplify the following. (a) $(x^3 + 2x^2 + 4x - 7) + (2x^3 - x^2 + 2x + 10)$, (b) $(x^4 - 2x^2 + 3) - (x^4 + x^3 - x - 4)$, (c) $(x^2 - 2)(3x^3 - x + 1)$

**Solutions**

$(a)$   $(x^3 + 2x^2 + 4x - 7) + (2x^3 - x^2 + 2x + 10)$

$$= (1 + 2)x^3 + (2 - 1)x^2 + (4 + 2)x + (-7 + 10)$$

$$= 3x^3 + x^2 + 6x + 3$$

$(b)$   $(x^4 - 2x^2 + 3) - (x^4 + x^3 - x - 4)$

$$= (1 - 1)x^4 - x^3 - 2x^2 + x + (3 + 4)$$

$$= -x^3 - 2x^2 + x + 7$$

$(c)$   $(x^2 - 2)(3x^3 - x + 1) = x^2(3x^3 - x + 1) - 2(3x^3 - x + 1)$

$$= (3x^5 - x^3 + x^2) + (-6x^3 + 2x - 2)$$

$$= 3x^5 + (-1 - 6)x^3 + x^2 + 2x - 2$$

$$= 3x^5 - 7x^3 + x^2 + 2x - 2 \quad\blacksquare$$

We now list some products that occur frequently. They can be checked by multiplying the terms on the left side.

---

**Special Products**

1.  $(a + b)(a - b) = a^2 - b^2$
2.  $(a + b)^2 = a^2 + 2ab + b^2$
3.  $(a - b)^2 = a^2 - 2ab + b^2$
4.  $(ax + b)(cx + d) = acx^2 + (ad + bc)x + bd$
5.  $(x + y)^3 = x^3 + 3x^2y + 3xy^2 + y^3$
6.  $(x - y)^3 = x^3 - 3x^2y + 3xy^2 - y^3$
7.  $(a - b)(a^2 + ab + b^2) = a^3 - b^3$
8.  $(a + b)(a^2 - ab + b^2) = a^3 + b^3$

---

## Factoring

We will often find it useful to **factor** a polynomial, that is, write the polynomial as a product of two or more nonconstant polynomials, each of which is as low an order as possible.

For example, using the first special product listed above, we see that

$$x^2 - 1 = (x + 1)(x - 1).$$

Although, in general, factoring a polynomial can be extremely difficult, many polynomials can be factored using the other special products listed above. The following are some examples.

**E X A M P L E   2**
_____

Factor (a) $4x^2 - 9$, (b) $8x^3 - 27$, (c) $81x^4 - 16$, (d) $4x^{-4} - 9x^{-6}$.

**Solutions**

(a)   Use Formula 1 with $a = 2x$ and $b = 3$. Then

$$4x^2 - 9 = (2x)^2 - 3^2$$
$$= (2x + 3)(2x - 3)$$

(b)   Use Formula 7 with $a = 2x$ and $b = 3$.

$$8x^3 - 27 = (2x)^3 - (3)^3$$
$$= (2x - 3)[(2x)^2 + (2x)(3) + (3)^2]$$
$$= (2x - 3)(4x^2 + 6x + 9)$$

(c)   Use Formula 1 twice as follows.

$$81x^4 - 16 = (9x^2)^2 - (4)^2$$
$$= (9x^2 + 4)(9x^2 - 4)$$
$$= (9x^2 + 4)(3x + 2)(3x - 2)$$

(d)   There are two ways of factoring this expression. In calculus one normally factors out the highest negative power of $x$ first. Doing this gives

$$4x^{-4} - 9x^{-6} = x^{-6}(4x^2 - 9)$$
$$= x^{-6}(2x + 3)(2x - 3) \quad ■$$

## Rational Expressions

A **rational expression** is a sum, difference, product, or quotient of terms, each of which is the quotient of polynomials. Thus the following are rational expressions.

$$\frac{x^2 + 1}{x^3 - x - 3} \qquad 2x^3 + \frac{5x + 1}{x^2 - 2}$$

Rational expressions must satisfy the same rules as real numbers.

---

**Rules for Rational Expressions**

Let $a(x)$, $b(x)$, $c(x)$, $d(x)$ be rational expressions with $b(x) \neq 0$ and $d(x) \neq 0$.
Also $c(x) \neq 0$ in Rule 5. Then

*Rule 1*
$$\frac{a(x)}{b(x)} + \frac{c(x)}{b(x)} = \frac{a(x) + c(x)}{b(x)}$$

*Rule 2*
$$\frac{a(x)}{b(x)} - \frac{c(x)}{b(x)} = \frac{a(x) - c(x)}{b(x)}$$

*Rule 3*
$$\frac{a(x)d(x)}{b(x)d(x)} = \frac{a(x)}{b(x)}$$

*Rule 4*
$$\frac{a(x)}{b(x)} \cdot \frac{c(x)}{d(x)} = \frac{a(x)c(x)}{b(x)d(x)}$$

*Rule 5*
$$\frac{a(x)}{b(x)} \div \frac{c(x)}{d(x)} = \frac{a(x)}{b(x)} \cdot \frac{d(x)}{c(x)}$$

---

**E X A M P L E   3**     **Rule 1**

Simplify $\dfrac{x^2 + x + 1}{x^2 + 1} - \dfrac{x + 1}{x^2 + 1}$.

**Solution**
$$\frac{x^2 + x + 1}{x^2 + 1} - \frac{x + 1}{x^2 + 1} = \frac{(x^2 + x + 1) - (x + 1)}{x^2 + 1} = \frac{x^2}{x^2 + 1} \quad \blacksquare$$

**E X A M P L E   4**     **Rule 3**

Simplify $\dfrac{x^2 - 1}{x^2 + 4x + 3}$.

**Solution**
$$\frac{x^2 - 1}{x^2 + 4x + 3} = \frac{(x + 1)(x - 1)}{(x + 1)(x + 3)} = \frac{x - 1}{x + 3} \quad \blacksquare$$

**E X A M P L E   5**     **Rule 4**

Simplify $\dfrac{x^2 + 3x + 2}{2x^2 - x - 1} \dfrac{x - 1}{x + 1}$.

**Solution**
$$\frac{x^2 + 3x + 2}{2x^2 - x - 1} \cdot \frac{x - 1}{x + 1} = \frac{(x + 1)(x + 2)(x - 1)}{(x - 1)(2x + 1)(x + 1)} = \frac{x + 2}{2x + 1} \quad \blacksquare$$

**E X A M P L E  6**     **Rule 5**

Simplify $\dfrac{x^3 - 2x^2 + x}{x^2 + 2x} \div \dfrac{x - 1}{x^2 - 4}$.

**Solution**

$$\dfrac{x^3 - 2x^2 + x}{x^2 + 2x} \div \dfrac{x - 1}{x^2 - 4} = \dfrac{x^3 - 2x^2 + x}{x^2 + 2x} \cdot \dfrac{x^2 - 4}{x - 1}$$

$$= \dfrac{x(x^2 - 2x + 1)}{x(x + 2)} \cdot \dfrac{x^2 - 4}{x - 1}$$

$$= \dfrac{(x - 1)^2(x - 2)(x + 2)}{(x + 2)(x - 1)}$$

$$= (x - 1)(x - 2) \quad \blacksquare$$

**Least Common Multiple.**   In order to combine the two fractions

$$\dfrac{1}{x^2 - 2x + 1} - \dfrac{1}{x^3 - x},$$

we must first factor, obtaining

$$\dfrac{1}{(x - 1)^2} - \dfrac{1}{x(x - 1)(x + 1)}.$$

We now take as the least common denominator, the product of each of the factors, taking the power of each factor as the highest power that appears. Thus the least common denominator for the above example is $x(x - 1)^2(x + 1)$. Thus

$$\dfrac{1}{(x - 1)^2} - \dfrac{1}{x(x - 1)(x + 1)} = \dfrac{1}{(x - 1)^2}\dfrac{x(x + 1)}{x(x + 1)} - \dfrac{1}{x(x - 1)(x + 1)}\dfrac{x - 1}{x - 1}$$

$$= \dfrac{x(x + 1) - (x - 1)}{x(x - 1)^2(x + 1)}$$

$$= \dfrac{x^2 + x - x + 1}{x(x - 1)^2(x + 1)}$$

$$= \dfrac{x^2 + 1}{x(x - 1)^2(x + 1)}$$

The following is a problem that comes up in calculus.

**E X A M P L E  7**

Simplify $\dfrac{\dfrac{1}{x + h} - \dfrac{1}{x}}{h}$.

**Solution**

$$\frac{\dfrac{1}{x+h} - \dfrac{1}{x}}{h} = \frac{1}{h}\left(\frac{1}{x+h} - \frac{1}{x}\right)$$

$$= \frac{1}{h}\left(\frac{1}{x+h}\frac{x}{x} - \frac{1}{x}\frac{x+h}{x+h}\right)$$

$$= \frac{1}{h}\frac{x-(x+h)}{x(x+h)}$$

$$= \frac{1}{h}\frac{-h}{x(x+h)}$$

$$= \frac{-1}{x(x+h)} \quad \blacksquare$$

**Rationalizing the Numerator.** In calculus one often must rationalize the numerator.

**E X A M P L E  8**

Rationalize $\dfrac{\sqrt{x}-2}{x-4}$.

**Solution**

$$\frac{\sqrt{x}-2}{x-4} = \frac{\sqrt{x}-2}{x-4}\frac{\sqrt{x}+2}{\sqrt{x}+2}$$

$$= \frac{(\sqrt{x})^2 - (2)^2}{(x-4)(\sqrt{x}+2)}$$

$$= \frac{(x-4)}{(x-4)(\sqrt{x}+2)}$$

$$= \frac{1}{\sqrt{x}+2} \quad \blacksquare$$

## EXERCISE SET A.2

In Exercises 1 through 14 simplify.

**1.** $(x^2 + 2x - 1) + (2x^2 + x + 3)$      **2.** $(x^2 - x + 2) + (-x^2 - x - 1)$

**3.** $(-2x^2 + x - 2) + (-x^2 - 2x + 1)$

**4.** $(-3x^2 - x + 1) + (x^2 + x - 1)$

**5.** $(x^4 + x^2 + x + 1) - (x^3 + 2x - 3)$

**6.** $(x^4 - 2x^3 - 1) - (2x^4 + x^2 + x - 1)$

7.  $(2x^3 - 2x^2 + 1) - (x^3 - x^2 - x - 1)$

8.  $(-x^3 - x + 1) - (x^3 - x^2 + 2)$

9.  $(x + 1)(x^2 + 2x + 3)$            10.  $(x + 2)(x^2 - x - 2)$

11.  $(2x - 1)(x^2 - x + 3)$           12.  $(2x - 3)(x^2 - 2)$

13.  $(x^2 + 3)(x^2 + 1)$              14.  $(x^2 + 2)(x^2 + x + 2)$

In Exercises 15 through 58 factor.

15.  $x^3 - 27$        16.  $x^3 - 8$        17.  $x^3 - \dfrac{1}{8}$        18.  $x^3 - \dfrac{1}{64}$

19.  $27x^3 - 8$      20.  $8x^3 - 125$     21.  $\dfrac{1}{8} x^3 - 27$      22.  $\dfrac{1}{64} x^3 - 27$

23.  $\dfrac{1}{64} x^3 - \dfrac{1}{27}$    24.  $\dfrac{1}{125} x^3 - \dfrac{1}{8}$    25.  $a^3 x^3 - b^3$    26.  $a^3 x^3 - \dfrac{1}{b^3}$

27.  $x^2 - 16$       28.  $x^2 - 81$       29.  $16x^2 - 1$       30.  $81x^2 - 1$

31.  $x^4 - \dfrac{1}{16}$    32.  $x^4 - \dfrac{1}{81}$    33.  $\dfrac{1}{16} x^4 - 1$    34.  $\dfrac{1}{81} x^4 - 1$

35.  $x^4 - 4$        36.  $x^4 - 9$        37.  $\dfrac{1}{4} x^4 - 1$    38.  $\dfrac{1}{9} x^4 - 1$

39.  $\dfrac{x^2 - x - 2}{x^2 - 3x + 2}$              40.  $\dfrac{x^2 + 3x + 2}{x^2 + 5x + 6}$

41.  $\dfrac{x^3 + x^2}{x - x^3}$                    42.  $\dfrac{x^2 - 4}{x^2 + x - 6}$

43.  $\dfrac{x^2 - 1}{x^2 - 4} \cdot \dfrac{x - 2}{x + 1}$        44.  $\dfrac{x^2 - 2x + 1}{x^2 + 3x + 2} \cdot \dfrac{x + 1}{x - 1}$

45.  $\dfrac{x^3 - 1}{x^2 - 16} \cdot \dfrac{x - 4}{x^2 + x + 1}$    46.  $\dfrac{x^4 - 1}{x + 2} \cdot \dfrac{x}{x^2 - 1}$

47.  $\dfrac{x^2 - 1}{x + 1} \div \dfrac{x - 1}{2x + 1}$          48.  $\dfrac{2x^2 + x - 1}{x^2 - 2x + 1} \div \dfrac{2x - 1}{x - 1}$

49.  $\dfrac{x^2 + 1}{x + 1} \div \dfrac{x}{x + 1}$              50.  $\dfrac{a^3 - b^3}{a^2 - b^2} \div \dfrac{1}{a + b}$

51.  $\dfrac{x}{x + 1} - \dfrac{1}{x + 1}$              52.  $\dfrac{x + 1}{x^2 + 1} - \dfrac{2x + 3}{x^2 + 1}$

53.  $\dfrac{x}{x^2 - 1} - \dfrac{1}{x^2 - 1}$          54.  $\dfrac{x^2}{x^4 - 8} + \dfrac{4}{x^4 - 8}$

55.  $\dfrac{x^2}{x^2 - 1} - \dfrac{x}{x - 1}$          56.  $\dfrac{1}{x^2 + x - 2} - \dfrac{1}{x^2 - 4}$

**57.** $\dfrac{x^2}{x^4 - 16} - \dfrac{1}{x^2 + 4}$     **58.** $\dfrac{1}{x - 3} + \dfrac{1}{x^2 - 9} - \dfrac{1}{x^2 - 2x - 3}$

In Exercises 59 through 66 rationalize the denominator.

**59.** $\dfrac{1}{\sqrt{x} + 1}$   **60.** $\dfrac{1}{\sqrt{x} - 1}$   **61.** $\dfrac{1}{\sqrt{x} + \sqrt{y}}$   **62.** $\dfrac{1}{\sqrt{x} - \sqrt{3}}$

**63.** $\dfrac{y - 9}{\sqrt{y} - 3}$   **64.** $\dfrac{a - b}{\sqrt{a} - \sqrt{b}}$   **65.** $\dfrac{a^2 - 1}{\sqrt{a} - 1}$   **66.** $\dfrac{x^2 - y^2}{\sqrt{x} - \sqrt{y}}$

In Exercises 67 through 70 rationalize the numerator.

**67.** $\dfrac{\sqrt{x} - 1}{x - 1}$     **68.** $\dfrac{\sqrt{x} - \sqrt{y}}{x - y}$

**69.** $\dfrac{\sqrt{x + h} - \sqrt{x}}{h}$     **70.** $\dfrac{\sqrt{x} - \sqrt{x - h}}{h}$

## A.3   EQUATIONS

- ▶ *Linear Equations*
- ▶ *Quadratic Equations*
- ▶ *Polynomial Equations*
- ▶ *Linear Equations with Absolute Values*

### Linear Equations

The polynomial $ax + b$ is called a **linear** expression and the equation $ax + b = 0$ is called a **linear equation**.

If $a \neq 0$, the linear equation $ax + b = 0$ can always be solved as follows.

$$ax + b = 0$$

$$ax = -b$$

$$x = -\frac{b}{a}$$

**EXAMPLE 1**

Solve $\dfrac{2x}{3} - \dfrac{x - 1}{3} = \dfrac{1}{4} - \dfrac{1}{2}\left(3x - \dfrac{3x - 2}{3}\right)$.

**Solution**   First remove the parenthesis.

$$\frac{2x}{3} - \frac{x - 1}{3} = \frac{1}{4} - \frac{3}{2}x + \frac{3x - 2}{6}$$

Remove the fractions by multiplying by 12.

$$4(2x) - 4(x - 1) = 3 - 6(3x) + 2(3x - 2)$$
$$8x - 4x + 4 = 3 - 18x + 6x - 4$$

Move all $x$-terms to the left and all constant terms to the right.

$$8x - 4x + 18x - 6x = 3 - 4 - 4$$
$$16x = -5$$
$$x = -\frac{5}{16} \quad \blacksquare$$

## Quadratic Equations

If $a \neq 0$, the polynomial $ax^2 + bx + c$ is called a **quadratic expression**. The equation

$$ax^2 + bx + c = 0, \qquad a \neq 0,$$

is called a **quadratic equation**. We shall always assume that $a$, $b$, $c$, and $x$ are *real* numbers.

Some quadratic equations can be solved easily by factoring.

**EXAMPLE 2** _____

Solve (a) $x^2 - 4 = 0$, (b) $9x^2 - 16 = 0$, (c) $3x^2 + 7x - 6 = 0$.

**Solutions**

(a)  To solve $x^2 - 4 = 0$, first recognize the expression $x^2 - 4$ as a difference of squares $a^2 - b^2$ with $a = x$ and $b = 2$. Then factoring gives

$$0 = x^2 - 4 = x^2 - 2^2 = (x + 2)(x - 2).$$

The only way the product of two expressions can be zero is if at least one of them is zero. Thus either $x + 2 = 0$ or $x - 2 = 0$. This in turn implies that $x = -2$ or $x = 2$. Thus the quadratic equation $x^2 - 4 = 0$ has two solutions, $x = -2$ and $x = 2$.

(b)  To solve $9x^2 - 16 = 0$ first recognize the expression $9x^2 - 16$ as a difference of squares $a^2 - b^2$ with $a = 3x$ and $b = 4$. Factoring then gives

$$0 = 9x^2 - 16 = (3x)^2 - 4^2 = (3x + 4)(3x - 4)$$
$$\text{implies } 3x + 4 = 0 \qquad \text{or} \quad 3x - 4 = 0$$
$$x = -\frac{4}{3} \quad \text{or} \qquad x = \frac{4}{3}$$

(c)  To solve $3x^2 + 7x - 6 = 0$, first factor $3x^2 + 7x - 6$ into a product of two linear expressions. Since the first term is $3x^2$ and the last term is negative, we try $(3x + \quad)(x - \quad)$ or $(3x - \quad)(x + \quad)$. Now 6 can be written as $6 = 3 \cdot 2$ or $6 = 6 \cdot 1$. Using trial and error with the first way gives the following.

$$(3x - 3)(x + 2) = 3x^2 + 3x - 6 \qquad \text{wrong}$$

$$(3x + 3)(x - 2) = 3x^2 - 3x - 6 \qquad \text{wrong}$$

$$(3x + 2)(x - 3) = 3x^2 - 7x - 6 \qquad \text{wrong}$$

$$(3x - 2)(x + 3) = 3x^2 + 7x - 6 \qquad \text{correct}$$

Thus we have $3x^2 + 7x - 6 = (3x - 2)(x + 3)$. Therefore

$$0 = 3x^2 + 7x - 6 \quad = \quad (3x - 2)(x + 3)$$

$$\text{implies } 3x - 2 = 0 \quad \text{or} \quad x + 3 = 0$$

$$x = \frac{2}{3} \quad \text{or} \qquad x = -3 \quad \blacksquare$$

The procedures used in the last example do not always work. A procedure that does always work is to use the quadratic formula.

---

**Solutions of the Quadratic Equation**

The solutions of the quadratic equation

$$ax^2 + bx + c = 0, \qquad a \neq 0$$

are given by the quadratic formula

$$x = \frac{-b \pm \sqrt{b^2 - 4ac}}{2a}$$

The term $b^2 - 4ac$ is called the **discriminant**.

---

The quadratic formula breaks down into three cases.

**Case 1.** If $b^2 - 4ac > 0$, there are the two solutions

$$\frac{-b \pm \sqrt{b^2 - 4ac}}{2a}.$$

**Case 2.** If $b^2 - 4ac = 0$, there is one solution given by

$$-\frac{b}{2a}.$$

**Case 3.** If $b^2 - 4ac < 0$, there are no solutions in the set of real numbers.

**E X A M P L E   3**

Solve $2x^2 - 3x - 1 = 0$.

**Solution**    Use the quadratic formula with $a = 2$, $b = -3$, $c = -1$ and obtain

$$x = \frac{-(-3) \pm \sqrt{(-3)^2 - 4(2)(-1)}}{2(2)}$$

$$= \frac{3 \pm \sqrt{17}}{4} \quad \blacksquare$$

## Polynomial Equations

We now consider solving polynomial equations of any order. We have just seen how any quadratic equation can be solved using the quadratic formula. It would be convenient if any polynomial equation such as $x^4 + 3x^3 - 2x^2 - 8 = 0$ could be solved using some corresponding "polynomial formula." Unfortunately, no such formula exists for polynomials of degree four or higher except for very special cases. We can solve polynomial equations in the cases when the polynomial can be completely factored. We now consider such cases.

**E X A M P L E   4**

Solve (a) $4x^4 - 36x^2 = 0$, (b) $2x^5 - 2x = 0$, (c) $10x^3 - 5x^2 - 5x = 0$.

**Solutions**

(a)  $\begin{aligned} 0 = 4x^4 - 36x^2 &= 4x^2(x^2 - 9) \qquad\qquad \text{common factor} \\ &= 4x^2(x + 3)(x - 3) \quad\;\; \text{difference of squares} \end{aligned}$

From this factorization we can immediately obtain the solutions $x = 0$, $x = -3$, and $x = 3$.

(b)  $\begin{aligned} 0 = 2x^5 - 2x &= 2x(x^4 - 1) \qquad\qquad\qquad\; \text{common factor} \\ &= 2x(x^2 + 1)(x^2 - 1) \qquad\quad\; \text{difference of squares} \\ &= 2x(x^2 + 1)(x + 1)(x - 1) \;\; \text{difference of squares} \end{aligned}$

From this factorization we can immediately obtain the solutions $x = 0$, $x = -1$, and $x = 1$.

(c)  $\begin{aligned} 0 = 10x^3 - 5x^2 - 5x &= 5x(2x^2 - x - 1) \qquad \text{common factor} \\ &= 5x(2x + 1)(x - 1) \;\; \text{factoring a quadratic} \end{aligned}$

From this factorization we can immediately obtain the solutions $x = 0$, $x = 1$, and $x = -1/2$.  $\blacksquare$

## Linear Equations with Absolute Value

We now give the definition of absolute value.

---

**Absolute Value**

$$|x| = \begin{cases} x & \text{if } x \geq 0 \\ -x & \text{if } x < 0 \end{cases}$$

---

**REMARK.**  You may have learned to find the absolute value by "dropping the sign." Doing this for algebraic expressions will be disastrous. For example $|-x|$ is $x$ if $x$ is positive but $-x$ if $x$ is negative. Thus $|-x|$ is sometimes equal to $x$ and sometimes to $-x$.

It follows immediately that the absolute value has the following property.

---

**A Property of Absolute Value**

For any $a \geq 0$,

$$|x| = a \qquad \text{if, and only if,} \qquad x = a \qquad \text{or} \qquad x = -a.$$

---

**E X A M P L E  5**

Solve $|2x - 3| = 2$.

**Solution**  This will be true if $2x - 3 = 2$ or if $2x - 3 = -2$. Thus

$$2x - 3 = 2 \qquad \text{or} \qquad 2x - 3 = -2$$
$$2x = 5 \qquad \text{or} \qquad 2x = 1$$
$$x = \frac{5}{2} \qquad \text{or} \qquad x = \frac{1}{2} \quad \blacksquare$$

## EXERCISE SET A.3

In Exercises 1 through 10 solve.

1. $x - 2(x - 3) = 4 - 2(x + 1)$

2. $3(x - 1) - 4(x - 2) - 3 = -(3x - 1)$

3. $-(2x - 1) + 3(x + 4) - 4 = 2(4x - 1)$

4. $(4x - 5) - 3(2x + 1) - 4 = 3 + 2(x - 1) - 3(2x + 1)$

**5.** $\dfrac{1}{2}x - \dfrac{x-2}{4} - 1 = 2(x - 2)$

**6.** $\dfrac{x-1}{2} - \dfrac{2x-3}{4} - \dfrac{1}{2} = \dfrac{1}{2}(3x - 1)$

**7.** $\dfrac{2x-3}{3} - \dfrac{2x+1}{2} - \dfrac{2}{3} = \dfrac{1}{3}(4x + 3)$

**8.** $\dfrac{7}{12}x - \dfrac{2x-1}{3} + \dfrac{5}{6} = \dfrac{1}{2}(5x - 2)$

**9.** $\dfrac{1}{2}\left(3x - \dfrac{1}{2}\dfrac{x-1}{2}\right) = 2x - \dfrac{3}{2}$

**10.** $\dfrac{1}{2}\left(\dfrac{3x-2}{3} - \dfrac{1}{3}\dfrac{2x-1}{2}\right) = \dfrac{1}{3}\left(\dfrac{2x-3}{2} + \dfrac{3}{2}\right)$

In Exercises 11 through 50 set the quadratic expression equal to zero and solve by factoring.

**11.** $y^2 - 4$      **12.** $x^2 - 9$      **13.** $4y^2 - 9$      **14.** $36x^2 - 25$

**15.** $\dfrac{1}{4}x^2 - 1$      **16.** $\dfrac{1}{9}y^2 - 1$      **17.** $x^2 - \dfrac{1}{4}$      **18.** $y^2 - \dfrac{1}{9}$

**19.** $25x^2 - 9$      **20.** $16x^2 - 25$      **21.** $\dfrac{1}{4}y^2 - 9$      **22.** $\dfrac{1}{16}y^2 - \dfrac{1}{9}$

**23.** $4y^2 - \dfrac{1}{9}$      **24.** $9y^2 - \dfrac{1}{4}$      **25.** $a^2x^2 - b^2$      **26.** $a^2y^2 - \dfrac{1}{b^2}$

**27.** $x^2 + 4x + 4$      **28.** $x^2 + 4x + 3$      **29.** $x^2 - 5x + 4$

**30.** $x^2 + 2x - 8$      **31.** $2x^2 + 3x - 2$      **32.** $2x^2 - 3x + 1$

**33.** $3x^2 + 7x + 2$      **34.** $6x^2 + 5x + 1$      **35.** $6x^2 + x - 1$

**36.** $6x^2 + x - 1$      **37.** $10x^2 - 12x + 2$      **38.** $10x^2 - 11x - 6$

**39.** $6x^2 + 13x + 6$      **40.** $6x^2 - 5x - 6$      **41.** $12x^2 - 11x + 2$

**42.** $12x^2 - x - 6$      **43.** $9x^2 - 6x + 1$      **44.** $x^2 - 6x - 9$

**45.** $9x^2 + 12x + 4$      **46.** $4x^2 + 20x + 25$      **47.** $x^2 - x + \dfrac{1}{4}$

**48.** $x^2 - \dfrac{1}{2}x + \dfrac{1}{16}$      **49.** $4x^2 + 2x + \dfrac{1}{4}$      **50.** $9x^2 - 3x + \dfrac{1}{4}$

In Exercises 51 through 58 set the polynomial expression equal to zero and solve by factoring.

**51.** $x^4 - 16$               **52.** $x^4 - 81$            **53.** $16x^4 - 81$

**54.** $81x^4 - 16$        **55.** $x^5 + x^4 - 2x^3$        **56.** $x^7 + x^6 - 6x^5$

**57.** $2x^6 - 162x^2$       **58.** $3x^7 - 48x^3$

In Exercises 59 through 72 set each of quadratic expressions equal to zero. If there is a solution, solve by using the quadratic formula.

**59.** $2x^2 + 4x + 2$       **60.** $2x^2 - 5x + 3$       **61.** $5x^2 - 7x + \dfrac{1}{5}$

**62.** $4x^2 - 9x + 2$       **63.** $-x^2 + x + 1$       **64.** $-2x^2 - 3x + 2$

**65.** $-3x^2 - 4x + 5$      **66.** $-2x^2 - 3x + 4$      **67.** $-x^2 + 4x - 1$

**68.** $-2x^2 + 5x - 2$      **69.** $-3x^2 + 7x - \dfrac{2}{3}$      **70.** $-x^2 + 3x - 2$

**71.** $x^2 + x + 1$        **72.** $x^2 + 2x - 2$

In Exercises 73 through 84 solve.

**73.** $|2x - 1| = 1$       **74.** $|2x - 3| = 2$       **75.** $|3x - 1| = 5$

**76.** $|5x - 2| = 4$       **77.** $|2 - 3x| = 2$       **78.** $|3 - 2x| = 1$

**79.** $|4 - x| = 5$        **80.** $|3 - 5x| = 3$       **81.** $\left|\dfrac{1}{2} - x\right| = \dfrac{2}{3}$

**82.** $\left|x - \dfrac{1}{3}\right| = \dfrac{1}{4}$       **83.** $\left|\dfrac{1}{2}x - \dfrac{2}{3}\right| = \dfrac{1}{6}$       **84.** $\left|\dfrac{1}{3}x - \dfrac{2}{5}\right| = \dfrac{1}{2}$

## A.4   INEQUALITIES

▶ *Linear Inequalities*

▶ *Inequalities with Absolute Values*

▶ *Nonlinear Inequalities*

### Linear Inequalities

We begin this section with some fundamental properties of inequalities.

---

**Properties of Inequalities**

   1.  $a > b$ implies $a + c > b + c$.

   2.  $a > b$, $c > 0$ implies $ac > bc$.

   3.  $a > b$, $c < 0$ implies $ac < bc$.

---

**E X A M P L E  1**

Solve the inequality $5x - 4 < 2x + 5$.

**Solution**

$$5x - 4 < 2x + 5$$
$$5x - 2x < 5 + 4$$
$$3x < 9$$
$$x < 3$$

**Figure A.1**

Thus all numbers *less* than 3 is the solution set. This set can also be represented using the interval notation $(-\infty, 3)$. See Figure A.1.  ■

**E X A M P L E  2**

Solve the inequality $-2 \leq 6 - 2x < 8$.

**Solution**

$$-2 \leq 6 - 2x \quad \text{and} \quad 6 - 2x < 8$$
$$2x \leq 6 + 2 \qquad\qquad -2x < 8 - 6 = 2$$
$$2x \leq 8 \qquad\qquad 2x > -2$$
$$x \leq 4 \qquad\qquad x > -1$$

**Figure A.2**

Thus $x$ must simultaneously satisfy $x \leq 4$ and $x > -1$. We can describe this solution set by the interval notation $(-1, 4]$. See Figure A.2.  ■

## Inequalities with Absolute Values

---

**Properties of Inequalities with Absolute Values**

For any $a > 0$,

1.  $|x| < a$ if and only if $-a < x < a$.
2.  $|x| > a$ if and only if either $x > a$ or $x < -a$.

---

**E X A M P L E  3**

Solve the inequality $|2x + 4| < 2$.

**Solution**

According to Property 1, this is equivalent to

$$-2 < 2x + 4 < 2.$$
$$-2 < 2x + 4 \quad \text{and} \quad 2x + 4 < 2$$
$$-2x < 6 \qquad\qquad 2x < -2$$
$$x > -3 \qquad\qquad x < -1$$

**Figure A.3**

In interval notation, the solution set is $(-3, -1)$. See Figure A.3.  ■

**E X A M P L E  4**

Solve the inequality $|3 - 4x| > 5$.

**Solution**  According to Property 2, this is equivalent to

$$3 - 4x > 5 \quad \text{or} \quad 3 - 4x < -5$$
$$-4x > 2 \qquad\qquad -4x < -8$$
$$x < -\frac{1}{2} \qquad\qquad x > 2$$

In interval notation, the solution set is $(-\infty, -\frac{1}{2})$ together with $(2, +\infty)$. See Figure A.4.  ∎

**Figure A.4**

## Nonlinear Inequalities

**E X A M P L E  5**

Solve the inequality $x^2 - 3x < -2$.

**Solution**  This is equivalent to

$$x^2 - 3x + 2 < 0.$$

Factoring the left side, this is equivalent to

$$(x - 1)(x - 2) < 0.$$

The only way the product of the two terms $(x - 1)$ and $(x - 2)$ can be negative is for one to be positive and the other negative. Thus we need to carefully look at the signs of these two factors. Clearly $x - 1 < 0$ if $x < 1$ and $x - 1 > 0$ if $x > 1$. In the same way $x - 2 < 0$ if $x < 2$ and $x - 2 > 0$ if $x > 2$. We then indicate this with a **sign chart** of these two factors as shown in Figure A.5. The sign of $(x - 1)(x - 2)$ is then obtained by taking the product of the signs of the individual factors. Thus on the interval $(-\infty, 1)$ the sign of both of the factors is negative, so the sign of the product is positive. On the interval $(1, 2)$ one factor is negative while the other factor is positive. Thus the sign of the product is negative. On the interval $(2, +\infty)$ both factors are positive so that the sign of the product is also positive.

From this we see that the product $(x - 1)(x - 2)$ is negative only when $1 < x < 2$. Thus the solution set is $(1, 2)$.  ∎

**Figure A.5**

**E X A M P L E  6**

Solve the inequality $2x^2 - x \geq 3$.

**Solution**  This is equivalent to

$$2x^2 - x - 3 \geq 0.$$

Factoring the left side, this is equivalent to

$$(x + 1)(2x - 3) \geq 0.$$

**Figure A.6**

Figure A.6 gives the sign chart of the two factors $(x + 1)$ and $(2x - 3)$. It is suggested that the first factor be the one with the least zero. This seems to be the least confusing and most systematic, but is not mandatory. In the last line is given the product of the signs in the previous two lines. In this way we find all places where the product $(x + 1)(2x - 3)$ is greater or equal to zero. This is the set $(-\infty, -1]$ together with the set $[\frac{3}{2}, +\infty)$. ∎

**E X A M P L E   7**

Solve the inequality $\dfrac{x^2 - 2}{x} < 1$.

**Solution**    Notice that we cannot have $x = 0$.

We must reduce the inequality to one with zero on the right. Thus

$$\frac{x^2 - 2}{x} < 1$$

$$\frac{x^2 - 2}{x} - 1 < 0$$

$$\frac{x^2 - 2}{x} - \frac{x}{x} < 0$$

$$\frac{x^2 - x - 2}{x} < 0$$

$$\frac{(x + 1)(x - 2)}{x} < 0$$

The sign of the quotient will be determined by the signs of the three factors $(x + 1)$, $x$, and $(x - 2)$. We note the signs of the three factors in the sign chart given in Figure A.7. The sign chart of the quotient given in the last line in Figure A.7 is obtained by multiplying the individual signs on the previous three lines. Thus since the sign of all three factors is negative on $(-\infty, -1)$, the product is negative. On the interval $(-1, 0)$, two of the factors have negative signs and one has positive sign, thus the product is positive. We see from Figure A.7 that the sign of the quotient is negative on the interval $(-\infty, -1)$ together with $(0, 2)$. The "N" in the sign chart indicates that the quotient is not defined at $x = 0$. ∎

```
                      -1        0            2
        ────────────────┼────────┼────────────┼──────────
(x + 1)       – – –    + + +     + + +        + + +
      x       – – –    – – –     + + +        + + +
(x - 2)       – – –    – – –     – – –        + + +
(x + 1)(x - 2)
──────────    – – –    + + +     – – –        + + +
      x
```

**Figure A.7**

**E X A M P L E   8**

Solve the inequality $\dfrac{x^2 - 4}{x^2 + 1} \leq 0$.

**Solution**   We factor and obtain

$$\frac{x^2 - 4}{x^2 + 1} \leq 0$$

$$\frac{(x + 2)(x - 2)}{x^2 + 1} \leq 0$$

We notice that the factor $x^2 + 1$ is *always* positive. Figure A.8 shows the sign chart of the three factors. The sign of the quotient is shown on the last line, where we see that the quotient is less than or equal to zero on $[-2, 2]$. ■

$$
\begin{array}{l}
\qquad\qquad\quad -2 \qquad 2 \\
\qquad\qquad\; \rule{5cm}{0.4pt} \\
(x + 2) \quad - - - \quad + + + \quad + + + \\
(x - 2) \quad - - - \quad - - - \quad + + + \\
(x^2 + 1) \;\; + + + \quad + + + \quad + + + \\
\dfrac{(x + 2)(x - 2)}{x^2 + 1} \;\; + + + \quad - - - \quad + + +
\end{array}
$$

**Figure A.8**

## EXERCISE SET A.4

Solve the following.

1.  $4x - 1 < x + 3$
2.  $9x + 3 > 5x + 7$
3.  $3x - 5 > 5x + 7$

4.  $3x - 2 < 7x - 10$
5.  $5x - 3 \leq 2x - 6$
6.  $2x - 7 \geq 4x + 3$

7.  $3x + 7 \geq 5x - 7$
8.  $3 - x \leq 2 - 3x$
9.  $\dfrac{5x - 1}{3} < \dfrac{2x + 1}{6}$

10. $\dfrac{2x - 1}{2} > -\dfrac{x - 1}{4}$
11. $1 - x \geq \dfrac{x - 1}{2}$
12. $\dfrac{x}{2} - \dfrac{1}{3} \leq \dfrac{x + 3}{4}$

13. $|x + 1| < 3$
14. $|x + 2| \leq 4$
15. $|2x - 1| \leq 1$

16. $|3 - 2x| < 6$
17. $|x - 1| > 2$
18. $|x + 2| \geq 3$

19. $|3x - 2| \geq 6$
20. $|6 - x| > 3$
21. $|2x + 3| < \dfrac{1}{6}$

22. $\left|\dfrac{x}{2} - 3\right| \leq 4$
23. $\dfrac{|2x - 5|}{3} \geq 5$
24. $\left|\dfrac{1}{2} - 2x\right| > 4$

25. $(x - 1)(x - 3) < 0$
26. $(x + 1)(x - 2) > 0$
27. $(x + 2)(x - 3) \geq 0$

28. $(x + 2)(x + 1) \leq 0$     29. $x^2 - 4 < 0$     30. $x^2 - x - 6 > 0$

31. $x^2 + 2x - 3 \geq 0$     32. $x^2 + 2x \leq 0$     33. $2x^2 - 5x < -2$

34. $2x^2 + x > 3$     35. $3x^2 + 7x \geq 6$     36. $6x^2 + 3 \leq 11x$

37. $\dfrac{x - 1}{x + 2} < 0$     38. $\dfrac{x + 2}{x + 3} > 0$     39. $\dfrac{x - 1}{x - 2} \geq 0$

40. $\dfrac{x - 2}{x + 3} \leq 0$     41. $\dfrac{x}{x + 2} > \dfrac{1}{x + 2}$     42. $\dfrac{2x}{x + 3} < \dfrac{3}{x + 3}$

43. $\dfrac{1}{x - 1} \leq \dfrac{x}{1 - x}$     44. $\dfrac{2x}{x - 4} \geq \dfrac{1}{x - 4}$     45. $x \geq \dfrac{1}{x}$

46. $x \leq \dfrac{4}{x}$     47. $\dfrac{x^2 + 2}{x} \geq 3x$     48. $\dfrac{x^2 - 3}{x} < 2x$

49. $\dfrac{(x - 2)(x - 3)}{x - 1} < 0$     50. $\dfrac{x^2 - 1}{x - 3} > 0$

51. $\dfrac{x^2 - 1}{x - 2} \geq 0$     52. $\dfrac{x^2 - 2x + 1}{x} \leq 0$     53. $\dfrac{x^2 + 2x + 1}{x} \leq 0$

54. $\dfrac{x^2 + 1}{x - 2} \geq 0$     55. $\dfrac{x^2 - x - 2}{x^2} < 0$     56. $\dfrac{x^2 - 1}{x^2 + 1} > 0$

a p p e n d i x   b

# Tables

B.1   Area Under a Normal Curve

B.2   Financial Tables

B.3   Basic Geometric Formulas

B.4   Table of Integrals

**Table B.1**

**Area Under Standard Normal Curve to the Left of** $z = \dfrac{x - \mu}{\sigma}$

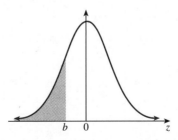

| $z$ | $.00$ | .01 | .02 | .03 | .04 | .05 | .06 | .07 | .08 | .09 |
|---|---|---|---|---|---|---|---|---|---|---|
| $-3.4$ | .0003 | .0003 | .0003 | .0003 | .0003 | .0003 | .0003 | .0003 | .0003 | .0002 |
| $-3.3$ | .0005 | .0005 | .0005 | .0004 | .0004 | .0004 | .0004 | .0004 | .0004 | .0003 |
| $-3.2$ | .0007 | .0007 | .0006 | .0006 | .0006 | .0006 | .0006 | .0005 | .0005 | .0005 |
| $-3.1$ | .0010 | .0009 | .0009 | .0009 | .0008 | .0008 | .0008 | .0008 | .0007 | .0007 |
| $-3.0$ | .0013 | .0013 | .0013 | .0012 | .0012 | .0011 | .0011 | .0011 | .0010 | .0010 |

**Table B.1** *(continued)*

| z | .00 | .01 | .02 | .03 | .04 | .05 | .06 | .07 | .08 | .09 |
|---|-----|-----|-----|-----|-----|-----|-----|-----|-----|-----|
| −2.9 | .0019 | .0018 | .0017 | .0017 | .0016 | .0016 | .0015 | .0015 | .0014 | .0014 |
| −2.8 | .0026 | .0025 | .0024 | .0023 | .0023 | .0022 | .0021 | .0021 | .0020 | .0019 |
| −2.7 | .0035 | .0034 | .0033 | .0032 | .0031 | .0030 | .0029 | .0028 | .0027 | .0026 |
| −2.6 | .0047 | .0045 | .0044 | .0043 | .0041 | .0040 | .0039 | .0038 | .0037 | .0036 |
| −2.5 | .0062 | .0060 | .0059 | .0057 | .0055 | .0054 | .0052 | .0051 | .0049 | .0048 |
| −2.4 | .0082 | .0080 | .0078 | .0075 | .0073 | .0071 | .0069 | .0068 | .0066 | .0064 |
| −2.3 | .0107 | .0104 | .0102 | .0099 | .0096 | .0094 | .0091 | .0089 | .0087 | .0084 |
| −2.2 | .0139 | .0136 | .0132 | .0129 | .0125 | .0122 | .0119 | .0116 | .0113 | .0110 |
| −2.1 | .0179 | .0174 | .0170 | .0166 | .0162 | .0158 | .0154 | .0150 | .0146 | .0143 |
| −2.0 | .0228 | .0222 | .0217 | .0212 | .0207 | .0202 | .0197 | .0192 | .0188 | .0183 |
| −1.9 | .0287 | .0281 | .0274 | .0268 | .0262 | .0256 | .0250 | .0244 | .0239 | .0233 |
| −1.8 | .0359 | .0352 | .0344 | .0336 | .0329 | .0322 | .0314 | .0307 | .0301 | .0294 |
| −1.7 | .0446 | .0436 | .0427 | .0418 | .0409 | .0401 | .0392 | .0384 | .0375 | .0367 |
| −1.6 | .0548 | .0537 | .0526 | .0516 | .0505 | .0495 | .0485 | .0475 | .0465 | .0455 |
| −1.5 | .0668 | .0655 | .0643 | .0630 | .0618 | .0606 | .0594 | .0582 | .0571 | .0559 |
| −1.4 | .0808 | .0793 | .0778 | .0764 | .0749 | .0735 | .0722 | .0708 | .0694 | .0681 |
| −1.3 | .0968 | .0951 | .0934 | .0918 | .0901 | .0885 | .0869 | .0853 | .0838 | .0823 |
| −1.2 | .1151 | .1131 | .1112 | .1093 | .1075 | .1056 | .1038 | .1020 | .1003 | .0985 |
| −1.1 | .1357 | .1335 | .1314 | .1292 | .1271 | .1251 | .1230 | .1210 | .1190 | .1170 |
| −1.0 | .1587 | .1562 | .1539 | .1515 | .1492 | .1469 | .1446 | .1423 | .1401 | .1379 |
| −0.9 | .1841 | .1814 | .1788 | .1762 | .1736 | .1711 | .1685 | .1660 | .1635 | .1611 |
| −0.8 | .2119 | .2090 | .2061 | .2033 | .2005 | .1977 | .1949 | .1922 | .1894 | .1867 |
| −0.7 | .2420 | .2389 | .2358 | .2327 | .2296 | .2266 | .2236 | .2206 | .2177 | .2148 |
| −0.6 | .2743 | .2709 | .2676 | .2643 | .2611 | .2578 | .2546 | .2514 | .2483 | .2451 |
| −0.5 | .3085 | .3050 | .3015 | .2981 | .2946 | .2912 | .2877 | .2843 | .2810 | .2776 |
| −0.4 | .3446 | .3409 | .3372 | .3336 | .3300 | .3264 | .3228 | .3192 | .3156 | .3121 |
| −0.3 | .3821 | .3783 | .3745 | .3707 | .3669 | .3632 | .3594 | .3557 | .3520 | .3483 |
| −0.2 | .4207 | .4168 | .4129 | .4090 | .4052 | .4013 | .3974 | .3936 | .3897 | .3859 |
| −0.1 | .4602 | .4562 | .4522 | .4483 | .4443 | .4404 | .4364 | .4325 | .4286 | .4247 |
| −0.0 | .5000 | .4960 | .4920 | .4880 | .4840 | .4801 | .4761 | .4721 | .4681 | .4641 |
| 0.0 | .5000 | .5040 | .5080 | .5120 | .5160 | .5199 | .5239 | .5279 | .5319 | .5359 |
| 0.1 | .5398 | .5438 | .5478 | .5517 | .5557 | .5596 | .5636 | .5675 | .5714 | .5753 |
| 0.2 | .5793 | .5832 | .5871 | .5910 | .5948 | .5987 | .6026 | .6064 | .6103 | .6141 |
| 0.3 | .6179 | .6217 | .6255 | .6293 | .6331 | .6368 | .6406 | .6443 | .6480 | .6517 |
| 0.4 | .6554 | .6591 | .6628 | .6664 | .6700 | .6736 | .6772 | .6808 | .6844 | .6879 |
| 0.5 | .6915 | .6950 | .6985 | .7019 | .7054 | .7088 | .7123 | .7157 | .7190 | .7224 |
| 0.6 | .7257 | .7291 | .7324 | .7357 | .7389 | .7422 | .7454 | .7486 | .7517 | .7549 |
| 0.7 | .7580 | .7611 | .7642 | .7673 | .7704 | .7734 | .7764 | .7794 | .7823 | .7852 |
| 0.8 | .7881 | .7910 | .7939 | .7967 | .7995 | .8023 | .8051 | .8078 | .8106 | .8133 |
| 0.9 | .8159 | .8186 | .8212 | .8238 | .8264 | .8289 | .8315 | .8340 | .8365 | .8389 |
| 1.0 | .8413 | .8438 | .8461 | .8485 | .8508 | .8531 | .8554 | .8577 | .8599 | .8621 |
| 1.1 | .8643 | .8665 | .8686 | .8708 | .8729 | .8749 | .8770 | .8790 | .8810 | .8830 |
| 1.2 | .8849 | .8869 | .8888 | .8907 | .8925 | .8944 | .8962 | .8980 | .8997 | .9015 |
| 1.3 | .9032 | .9049 | .9066 | .9082 | .9099 | .9115 | .9131 | .9147 | .9162 | .9177 |
| 1.4 | .9192 | .9207 | .9222 | .9236 | .9251 | .9265 | .9278 | .9292 | .9306 | .9319 |

**Table B.1** *(continued)*

| z | .00 | .01 | .02 | .03 | .04 | .05 | .06 | .07 | .08 | .09 |
|---|-----|-----|-----|-----|-----|-----|-----|-----|-----|-----|
| 1.5 | .9332 | .9345 | .9357 | .9370 | .9382 | .9394 | .9406 | .9418 | .9429 | .9441 |
| 1.6 | .9452 | .9463 | .9474 | .9484 | .9495 | .9505 | .9515 | .9525 | .9535 | .9545 |
| 1.7 | .9554 | .9564 | .9573 | .9582 | .9591 | .9599 | .9608 | .9616 | .9625 | .9633 |
| 1.8 | .9641 | .9649 | .9656 | .9664 | .9671 | .9678 | .9686 | .9693 | .9699 | .9706 |
| 1.9 | .9713 | .9719 | .9726 | .9732 | .9738 | .9744 | .9750 | .9756 | .9761 | .9767 |
| 2.0 | .9772 | .9778 | .9783 | .9788 | .9793 | .9798 | .9803 | .9808 | .9812 | .9817 |
| 2.1 | .9821 | .9826 | .9830 | .9834 | .9838 | .9842 | .9846 | .9850 | .9854 | .9857 |
| 2.2 | .9861 | .9864 | .9868 | .9871 | .9875 | .9878 | .9881 | .9884 | .9887 | .9890 |
| 2.3 | .9893 | .9896 | .9898 | .9901 | .9904 | .9906 | .9909 | .9911 | .9913 | .9916 |
| 2.4 | .9918 | .9920 | .9922 | .9925 | .9927 | .9929 | .9931 | .9932 | .9934 | .9936 |
| 2.5 | .9938 | .9940 | .9941 | .9943 | .9945 | .9946 | .9948 | .9949 | .9951 | .9952 |
| 2.6 | .9953 | .9955 | .9956 | .9957 | .9959 | .9960 | .9961 | .9962 | .9963 | .9964 |
| 2.7 | .9965 | .9966 | .9967 | .9968 | .9969 | .9970 | .9971 | .9972 | .9973 | .9974 |
| 2.8 | .9974 | .9975 | .9976 | .9977 | .9977 | .9978 | .9979 | .9979 | .9980 | .9981 |
| 2.9 | .9981 | .9982 | .9982 | .9983 | .9984 | .9984 | .9985 | .9985 | .9986 | .9986 |
| 3.0 | .9987 | .9987 | .9987 | .9988 | .9988 | .9989 | .9989 | .9989 | .9990 | .9990 |
| 3.1 | .9990 | .9991 | .9991 | .9991 | .9992 | .9992 | .9992 | .9992 | .9993 | .9993 |
| 3.2 | .9993 | .9993 | .9994 | .9994 | .9994 | .9994 | .9994 | .9995 | .9995 | .9995 |
| 3.3 | .9995 | .9995 | .9995 | .9996 | .9996 | .9996 | .9996 | .9996 | .9996 | .9997 |
| 3.4 | .9997 | .9997 | .9997 | .9997 | .9997 | .9997 | .9997 | .9997 | .9997 | .9998 |

**Table B.2**

$i$ = **rate of interest per period**     $n$ = **number of periods**     $i = \frac{1}{4}\%$

| $n$ | $(1 + i)^n$ | $(1 + i)^{-n}$ | $s_{\overline{n}\rvert i}$ | $a_{\overline{n}\rvert i}$ | $\dfrac{1}{s_{\overline{n}\rvert i}}$ | $\dfrac{1}{a_{\overline{n}\rvert i}}$ |
|---|---|---|---|---|---|---|
| 1 | 1.0025 0000 | 0.9975 0623 | 1.0000 0000 | 0.9975 0623 | 1.0000 0000 | 1.0025 0000 |
| 2 | 1.0050 0625 | 0.9950 1869 | 2.0025 0000 | 1.9925 2492 | 0.4993 7578 | 0.5018 7578 |
| 3 | 1.0075 1877 | 0.9925 3734 | 3.0075 0625 | 2.9850 6227 | 0.3325 0139 | 0.3350 0139 |
| 4 | 1.0100 3756 | 0.9900 6219 | 4.0150 2502 | 3.9751 2446 | 0.2490 6445 | 0.2515 6445 |
| 5 | 1.0125 6266 | 0.9875 9321 | 5.0250 6258 | 4.9627 1766 | 0.1990 0250 | 0.2015 0250 |
| 6 | 1.0150 9406 | 0.9851 3038 | 6.0376 2523 | 5.9478 4804 | 0.1656 2803 | 0.1681 2803 |
| 7 | 1.0176 3180 | 0.9826 7370 | 7.0527 1930 | 6.9305 2174 | 0.1417 8928 | 0.1442 8928 |
| 8 | 1.0201 7588 | 0.9802 2314 | 8.0703 5110 | 7.9107 4487 | 0.1239 1035 | 0.1264 1035 |
| 9 | 1.0227 2632 | 0.9777 7869 | 9.0905 2697 | 8.8885 2357 | 0.1100 0462 | 0.1125 0462 |
| 10 | 1.0252 8313 | 0.9753 4034 | 10.1132 5329 | 9.8638 6391 | 0.0988 8015 | 0.1013 8015 |
| 11 | 1.0278 4634 | 0.9729 0807 | 11.1385 3642 | 10.8367 7198 | 0.0897 7840 | 0.0922 7840 |
| 12 | 1.0304 1596 | 0.9704 8187 | 12.1663 8277 | 11.8072 5384 | 0.0821 9370 | 0.0846 9370 |
| 13 | 1.0329 9200 | 0.9680 6171 | 13.1967 9872 | 12.7753 1555 | 0.0757 7595 | 0.0782 7595 |
| 14 | 1.0355 7448 | 0.9656 4759 | 14.2297 9072 | 13.7409 6314 | 0.0702 7510 | 0.0727 7510 |
| 15 | 1.0381 6341 | 0.9632 3949 | 15.2653 6520 | 14.7042 0264 | 0.0655 0777 | 0.0680 0777 |

**Table B.2** (*continued*)

$i = \frac{1}{4}\%$

| $n$ | $(1 + i)^n$ | $(1 + i)^{-n}$ | $s_{\overline{n}|i}$ | $a_{\overline{n}|i}$ | $\dfrac{1}{s_{\overline{n}|i}}$ | $\dfrac{1}{a_{\overline{n}|i}}$ |
|---|---|---|---|---|---|---|
| 16 | 1.0407 5882 | 0.9608 3740 | 16.3035 2861 | 15.6650 4004 | 0.0613 3642 | 0.0638 3642 |
| 17 | 1.0433 6072 | 0.9584 4130 | 17.3442 8743 | 16.6234 8133 | 0.0576 5587 | 0.0601 5587 |
| 18 | 1.0459 6912 | 0.9560 5117 | 18.3876 4815 | 17.5795 3250 | 0.0543 8433 | 0.0568 8433 |
| 19 | 1.0485 8404 | 0.9536 6700 | 19.4336 1727 | 18.5331 9950 | 0.0514 5722 | 0.0539 5722 |
| 20 | 1.0512 0550 | 0.9512 8878 | 20.4822 0131 | 19.4844 8828 | 0.0488 2288 | 0.0513 2288 |
| 21 | 1.0538 3352 | 0.9489 1649 | 21.5334 0682 | 20.4334 0477 | 0.0464 3947 | 0.0489 3947 |
| 22 | 1.0564 6810 | 0.9465 5011 | 22.5872 4033 | 21.3799 5488 | 0.0442 7278 | 0.0467 7278 |
| 23 | 1.0591 0927 | 0.9441 8964 | 23.6437 0843 | 22.3241 4452 | 0.0422 9455 | 0.0447 9455 |
| 24 | 1.0617 5704 | 0.9418 3505 | 24.7028 1770 | 23.2659 7957 | 0.0404 8121 | 0.0429 8121 |
| 25 | 1.0644 1144 | 0.9394 8634 | 25.7645 7475 | 24.2054 6591 | 0.0388 1298 | 0.0413 1298 |
| 26 | 1.0670 7247 | 0.9371 4348 | 26.8289 8619 | 25.1426 0939 | 0.0372 7312 | 0.0397 7312 |
| 27 | 1.0697 4015 | 0.9348 0646 | 27.8960 5865 | 26.0774 1585 | 0.0358 4736 | 0.0383 4736 |
| 28 | 1.0724 1450 | 0.9324 7527 | 28.9657 9880 | 27.0098 9112 | 0.0345 2347 | 0.0370 2347 |
| 29 | 1.0750 9553 | 0.9301 4990 | 30.0382 1330 | 27.9400 4102 | 0.0332 9093 | 0.0357 9093 |
| 30 | 1.0777 8327 | 0.9278 3032 | 31.1133 0883 | 28.8678 7134 | 0.0321 4059 | 0.0346 4059 |
| 31 | 1.0804 7773 | 0.9255 1653 | 32.1910 9210 | 29.7933 8787 | 0.0310 6449 | 0.0335 6449 |
| 32 | 1.0831 7892 | 0.9232 0851 | 33.2715 6983 | 30.7165 9638 | 0.0300 5569 | 0.0325 5569 |
| 33 | 1.0858 8687 | 0.9209 0624 | 34.3547 4876 | 31.6375 0262 | 0.0291 0806 | 0.0316 0806 |
| 34 | 1.0886 0159 | 0.9186 0972 | 35.4406 3563 | 32.5561 1234 | 0.0282 1620 | 0.0307 1620 |
| 35 | 1.0913 2309 | 0.9163 1892 | 36.5292 3722 | 33.4724 3126 | 0.0273 7533 | 0.0298 7533 |
| 36 | 1.0940 5140 | 0.9140 3384 | 37.6205 6031 | 34.3864 6510 | 0.0265 8121 | 0.0290 8121 |
| 37 | 1.0967 8653 | 0.9117 5445 | 38.7146 1171 | 35.2982 1955 | 0.0258 3004 | 0.0283 3004 |
| 38 | 1.0995 2850 | 0.9094 8075 | 39.8113 9824 | 36.2077 0030 | 0.0251 1843 | 0.0276 1843 |
| 39 | 1.1022 7732 | 0.9072 1272 | 40.9109 2673 | 37.1149 1302 | 0.0244 4335 | 0.0269 4335 |
| 40 | 1.1050 3301 | 0.9049 5034 | 42.0132 0405 | 38.0198 6336 | 0.0238 0204 | 0.0263 0204 |
| 41 | 1.1077 9559 | 0.9026 9361 | 43.1182 3706 | 38.9225 5697 | 0.0231 9204 | 0.0256 9204 |
| 42 | 1.1105 6508 | 0.9004 4250 | 44.2260 3265 | 39.8229 9947 | 0.0226 1112 | 0.0251 1112 |
| 43 | 1.1133 4149 | 0.8981 9701 | 45.3365 9774 | 40.7211 9648 | 0.0220 5724 | 0.0245 5724 |
| 44 | 1.1161 2485 | 0.8959 5712 | 46.4499 3923 | 41.6171 5359 | 0.0215 2855 | 0.0240 2855 |
| 45 | 1.1189 1516 | 0.8937 2281 | 47.5660 6408 | 42.5108 7640 | 0.0210 2339 | 0.0235 2339 |
| 46 | 1.1217 1245 | 0.8914 9407 | 48.6849 7924 | 43.4023 7047 | 0.0205 4022 | 0.0230 4022 |
| 47 | 1.1245 1673 | 0.8892 7090 | 49.8066 9169 | 44.2916 4137 | 0.0200 7762 | 0.0225 7762 |
| 48 | 1.1273 2802 | 0.8870 5326 | 50.9312 0842 | 45.1786 9463 | 0.0196 3433 | 0.0221 3433 |
| 49 | 1.1301 4634 | 0.8848 4116 | 52.0585 3644 | 46.0635 3580 | 0.0192 0915 | 0.0217 0915 |
| 50 | 1.1329 7171 | 0.8826 3457 | 53.1886 8278 | 46.9461 7037 | 0.0188 0099 | 0.0213 0099 |
| 51 | 1.1358 0414 | 0.8804 3349 | 54.3216 5449 | 47.8266 0386 | 0.0184 0886 | 0.0209 0886 |
| 52 | 1.1386 4365 | 0.8782 3790 | 55.4574 5862 | 48.7048 4176 | 0.0180 3184 | 0.0205 3184 |
| 53 | 1.1414 9026 | 0.8760 4778 | 56.5961 0227 | 49.5808 8953 | 0.0176 6906 | 0.0201 6906 |
| 54 | 1.1443 4398 | 0.8738 6312 | 57.7375 9252 | 50.4547 5265 | 0.0173 1974 | 0.0198 1974 |
| 55 | 1.1472 0484 | 0.8716 8391 | 58.8819 3650 | 51.3264 3656 | 0.0169 8314 | 0.0194 8314 |

**Table B.2** *(continued)*
$i = \frac{1}{4}\%$

| $n$ | $(1 + i)^n$ | $(1 + i)^{-n}$ | $s_{\overline{n}\rvert i}$ | $a_{\overline{n}\rvert i}$ | $\dfrac{1}{s_{\overline{n}\rvert i}}$ | $\dfrac{1}{a_{\overline{n}\rvert i}}$ |
|---|---|---|---|---|---|---|
| 56 | 1.1500 7285 | 0.8695 1013 | 60.0291 4135 | 52.1959 4669 | 0.0166 5858 | 0.0191 5858 |
| 57 | 1.1529 4804 | 0.8673 4178 | 61.1792 1420 | 53.0632 8847 | 0.0163 4542 | 0.0188 4542 |
| 58 | 1.1558 3041 | 0.8651 7883 | 62.3321 6223 | 53.9284 6730 | 0.0160 4308 | 0.0185 4308 |
| 59 | 1.1587 1998 | 0.8630 2128 | 63.4879 9264 | 54.7914 8858 | 0.0157 5101 | 0.0182 5101 |
| 60 | 1.1616 1678 | 0.8608 6911 | 64.6467 1262 | 55.6523 5769 | 0.0154 6869 | 0.0179 6869 |
| 61 | 1.1645 2082 | 0.8587 2230 | 65.8083 2940 | 56.5110 7999 | 0.0151 9564 | 0.0176 9564 |
| 62 | 1.1674 3213 | 0.8565 8085 | 66.9728 5023 | 57.3676 6083 | 0.0149 3142 | 0.0174 3142 |
| 63 | 1.1703 5071 | 0.8544 4474 | 68.1402 8235 | 58.2221 0557 | 0.0146 7561 | 0.0171 7561 |
| 64 | 1.1732 7658 | 0.8523 1395 | 69.3106 3306 | 59.0744 1952 | 0.0144 2780 | 0.0169 2780 |
| 65 | 1.1762 0977 | 0.8501 8848 | 70.4839 0964 | 59.9246 0800 | 0.0141 8764 | 0.0166 8764 |
| 66 | 1.1791 5030 | 0.8480 6831 | 71.6601 1942 | 60.7726 7631 | 0.0139 5476 | 0.0164 5476 |
| 67 | 1.1820 9817 | 0.8459 5343 | 72.8392 6971 | 61.6186 2974 | 0.0137 2886 | 0.0162 2886 |
| 68 | 1.1850 5342 | 0.8438 4382 | 74.0213 6789 | 62.4624 7355 | 0.0135 0961 | 0.0160 0961 |
| 69 | 1.1880 1605 | 0.8417 3947 | 75.2064 2131 | 63.3042 1302 | 0.0132 9674 | 0.0157 9674 |
| 70 | 1.1909 8609 | 0.8396 4037 | 76.3944 3736 | 64.1438 5339 | 0.0130 8996 | 0.0155 8996 |
| 71 | 1.1939 6356 | 0.8375 4650 | 77.5854 2345 | 64.9813 9989 | 0.0128 8902 | 0.0153 8902 |
| 72 | 1.1969 4847 | 0.8354 5786 | 78.7793 8701 | 65.8168 5774 | 0.0126 9368 | 0.0151 9368 |
| 73 | 1.1999 4084 | 0.8333 7442 | 79.9763 3548 | 66.6502 3216 | 0.0125 0370 | 0.0150 0370 |
| 74 | 1.2029 4069 | 0.8312 9618 | 81.1762 7632 | 67.4815 2834 | 0.0123 1887 | 0.0148 1887 |
| 75 | 1.2059 4804 | 0.8292 2312 | 82.3792 1701 | 68.3107 5146 | 0.0121 3898 | 0.0146 3898 |
| 76 | 1.2089 6191 | 0.8271 5523 | 83.5851 6505 | 69.1379 0670 | 0.0119 6385 | 0.0144 6385 |
| 77 | 1.2119 8532 | 0.8250 9250 | 84.7941 2797 | 69.9629 9920 | 0.0117 9327 | 0.0142 9327 |
| 78 | 1.2150 1528 | 0.8230 3491 | 86.0061 1329 | 70.7860 3411 | 0.0116 2708 | 0.0141 2708 |
| 79 | 1.2180 5282 | 0.8209 8246 | 87.2211 2857 | 71.6070 1657 | 0.0114 6511 | 0.0139 6511 |
| 80 | 1.2210 9795 | 0.8189 3512 | 88.4391 8139 | 72.4259 5169 | 0.0113 0721 | 0.0138 0721 |
| 81 | 1.2241 5070 | 0.8168 9289 | 89.6602 7934 | 73.2428 4458 | 0.0111 5321 | 0.0136 5321 |
| 82 | 1.2272 1108 | 0.8148 5575 | 90.8844 3004 | 74.0577 0033 | 0.0110 0298 | 0.0135 0298 |
| 83 | 1.2302 7910 | 0.8128 2369 | 92.1116 4112 | 74.8705 2402 | 0.0108 5639 | 0.0133 5639 |
| 84 | 1.2333 5480 | 0.8107 9670 | 93.3419 2022 | 75.6813 2072 | 0.0107 1330 | 0.0132 1330 |
| 85 | 1.2364 3819 | 0.8087 7476 | 94.5752 7502 | 76.4900 9548 | 0.0105 7359 | 0.0130 7359 |
| 86 | 1.2395 2928 | 0.8067 5787 | 95.8117 1321 | 77.2968 5335 | 0.0104 3714 | 0.0129 3714 |
| 87 | 1.2426 2811 | 0.8047 4600 | 97.0512 4249 | 78.1015 9935 | 0.0103 0384 | 0.0128 0384 |
| 88 | 1.2457 3468 | 0.8027 3915 | 98.2938 7060 | 78.9043 3850 | 0.0101 7357 | 0.0126 7357 |
| 89 | 1.2488 4901 | 0.8007 3731 | 99.5396 0527 | 79.7050 7581 | 0.0100 4625 | 0.0125 4625 |
| 90 | 1.2519 7114 | 0.7987 4046 | 100.7884 5429 | 80.5038 1627 | 0.0099 2177 | 0.0124 2177 |
| 91 | 1.2551 0106 | 0.7967 4859 | 102.0404 2542 | 81.3005 6486 | 0.0098 0004 | 0.0123 0004 |
| 92 | 1.2582 3882 | 0.7947 6168 | 103.2955 2649 | 82.0953 2654 | 0.0096 8096 | 0.0121 8096 |
| 93 | 1.2613 8441 | 0.7927 7973 | 104.5537 6530 | 82.8881 0628 | 0.0095 6446 | 0.0120 6446 |
| 94 | 1.2645 3787 | 0.7908 0273 | 105.8151 4972 | 83.6789 0900 | 0.0094 5044 | 0.0119 5044 |
| 95 | 1.2676 9922 | 0.7888 3065 | 107.0796 8759 | 84.4677 3966 | 0.0093 3884 | 0.0118 3884 |
| 96 | 1.2708 6847 | 0.7868 6349 | 108.3473 8681 | 85.2546 0315 | 0.0092 2957 | 0.0117 2957 |
| 97 | 1.2740 4564 | 0.7849 0124 | 109.6182 5528 | 86.0395 0439 | 0.0091 2257 | 0.0116 2257 |
| 98 | 1.2772 3075 | 0.7829 4388 | 110.8923 0091 | 86.8224 4827 | 0.0090 1776 | 0.0115 1776 |
| 99 | 1.2804 2383 | 0.7809 9140 | 112.1695 3167 | 87.6034 3967 | 0.0089 1508 | 0.0114 1508 |
| 100 | 1.2836 2489 | 0.7790 4379 | 113.4499 5550 | 88.3824 8346 | 0.0088 1446 | 0.0113 1446 |

**Table B.2** *(continued)*

$i = \frac{1}{2}\%$

| $n$ | $(1 + i)^n$ | $(1 + i)^{-n}$ | $s_{\overline{n}|i}$ | $a_{\overline{n}|i}$ | $\dfrac{1}{s_{\overline{n}|i}}$ | $\dfrac{1}{a_{\overline{n}|i}}$ |
|---|---|---|---|---|---|---|
| 1 | 1.0050 0000 | 0.9950 2488 | 1.0000 0000 | 0.9950 2488 | 1.0000 0000 | 1.0050 0000 |
| 2 | 1.0100 2500 | 0.9900 7450 | 2.0050 0000 | 1.9850 9938 | 0.4987 5312 | 0.5037 5312 |
| 3 | 1.0150 7513 | 0.9851 4876 | 3.0150 2500 | 2.9702 4814 | 0.3316 7221 | 0.3366 7221 |
| 4 | 1.0201 5050 | 0.9802 4752 | 4.0301 0013 | 3.9504 9566 | 0.2481 3279 | 0.2531 3279 |
| 5 | 1.0252 5125 | 0.9753 7067 | 5.0502 5063 | 4.9258 6633 | 0.1980 0997 | 0.2030 0997 |
| 6 | 1.0303 7751 | 0.9705 1808 | 6.0755 0188 | 5.8963 8441 | 0.1645 9546 | 0.1695 9546 |
| 7 | 1.0355 2940 | 0.9656 8963 | 7.1058 7939 | 6.8620 7404 | 0.1407 2854 | 0.1457 2854 |
| 8 | 1.0407 0704 | 0.9608 8520 | 8.1414 0879 | 7.8229 5924 | 0.1228 2886 | 0.1278 2886 |
| 9 | 1.0459 1058 | 0.9561 0468 | 9.1821 1583 | 8.7790 6392 | 0.1089 0736 | 0.1139 0736 |
| 10 | 1.0511 4013 | 0.9513 4794 | 10.2280 2641 | 9.7304 1186 | 0.0977 7057 | 0.1027 7057 |
| 11 | 1.0563 9583 | 0.9466 1487 | 11.2791 6654 | 10.6770 2673 | 0.0886 5903 | 0.0936 5903 |
| 12 | 1.0616 7781 | 0.9419 0534 | 12.3355 6237 | 11.6189 3207 | 0.0810 6643 | 0.0860 6643 |
| 13 | 1.0669 8620 | 0.9372 1924 | 13.3972 4018 | 12.5561 5131 | 0.0746 4224 | 0.0796 4224 |
| 14 | 1.0723 2113 | 0.9325 5646 | 14.4642 2639 | 13.4887 0777 | 0.0691 3609 | 0.0741 3609 |
| 15 | 1.0776 8274 | 0.9279 1688 | 15.5365 4752 | 14.4166 2465 | 0.0643 6436 | 0.0693 6436 |
| 16 | 1.0830 7115 | 0.9233 0037 | 16.6142 3026 | 15.3399 2502 | 0.0601 8937 | 0.0651 8937 |
| 17 | 1.0884 8651 | 0.9187 0684 | 17.6973 0141 | 16.2586 3186 | 0.0565 0579 | 0.0615 0579 |
| 18 | 1.0939 2894 | 0.9141 3616 | 18.7857 8791 | 17.1727 6802 | 0.0532 3173 | 0.0582 3173 |
| 19 | 1.0993 9858 | 0.9095 8822 | 19.8797 1685 | 18.0823 5624 | 0.0503 0253 | 0.0553 0253 |
| 20 | 1.1048 9558 | 0.9050 6290 | 20.9791 1544 | 18.9874 1915 | 0.0476 6645 | 0.0526 6645 |
| 21 | 1.1104 2006 | 0.9005 6010 | 22.0840 1101 | 19.8879 7925 | 0.0452 8163 | 0.0502 8163 |
| 22 | 1.1159 7216 | 0.8960 7971 | 23.1944 3107 | 20.7840 5896 | 0.0431 1380 | 0.0481 1380 |
| 23 | 1.1215 5202 | 0.8916 2160 | 24.3104 0322 | 21.6756 8055 | 0.0411 3465 | 0.0461 3465 |
| 24 | 1.1271 5978 | 0.8871 8567 | 25.4319 5524 | 22.5628 6622 | 0.0393 2061 | 0.0443 2061 |
| 25 | 1.1327 9558 | 0.8827 7181 | 26.5591 1502 | 23.4456 3803 | 0.0376 5186 | 0.0426 5186 |
| 26 | 1.1384 5955 | 0.8783 7991 | 27.6919 1059 | 24.3240 1794 | 0.0361 1163 | 0.0411 1163 |
| 27 | 1.1441 5185 | 0.8740 0986 | 28.8303 7015 | 25.1980 2780 | 0.0346 8565 | 0.0396 8565 |
| 28 | 1.1498 7261 | 0.8696 6155 | 29.9745 2200 | 26.0676 8936 | 0.0333 6167 | 0.0383 6167 |
| 29 | 1.1556 2197 | 0.8653 3488 | 31.1243 9461 | 26.9330 2423 | 0.0321 2914 | 0.0371 2914 |
| 30 | 1.1614 0008 | 0.8610 2973 | 32.2800 1658 | 27.7940 5397 | 0.0309 7892 | 0.0359 7892 |
| 31 | 1.1672 0708 | 0.8567 4600 | 33.4414 1666 | 28.6507 9997 | 0.0299 0304 | 0.0349 0304 |
| 32 | 1.1730 4312 | 0.8524 8358 | 34.6086 2375 | 29.5032 8355 | 0.0288 9453 | 0.0338 9453 |
| 33 | 1.1789 0833 | 0.8482 4237 | 35.7816 6686 | 30.3515 2592 | 0.0279 4727 | 0.0329 4727 |
| 34 | 1.1848 0288 | 0.8440 2226 | 36.9605 7520 | 31.1955 4818 | 0.0270 5586 | 0.0320 5586 |
| 35 | 1.1907 2689 | 0.8398 2314 | 38.1453 7807 | 32.0353 7132 | 0.0262 1550 | 0.0312 1550 |
| 36 | 1.1966 8052 | 0.8356 4492 | 39.3361 0496 | 32.8710 1624 | 0.0254 2194 | 0.0304 2194 |
| 37 | 1.2026 6393 | 0.8314 8748 | 40.5317 8549 | 33.7025 0372 | 0.0246 7139 | 0.0296 7139 |
| 38 | 1.2086 7725 | 0.8273 5073 | 41.7354 4942 | 34.5298 5445 | 0.0239 6045 | 0.0289 6045 |
| 39 | 1.2147 2063 | 0.8232 3455 | 42.9441 2666 | 35.3530 8900 | 0.0232 8607 | 0.0282 8607 |
| 40 | 1.2207 9424 | 0.8191 3886 | 44.1588 4730 | 36.1722 2786 | 0.0226 4552 | 0.0276 4552 |

**Table B.2** *(continued)*

$i = \frac{1}{4}\%$

| $n$ | $(1 + i)^n$ | $(1 + i)^{-n}$ | $s_{\overline{n}\rceil i}$ | $a_{\overline{n}\rceil i}$ | $\dfrac{1}{s_{\overline{n}\rceil i}}$ | $\dfrac{1}{a_{\overline{n}\rceil i}}$ |
|---|---|---|---|---|---|---|
| 41 | 1.2268 9821 | 0.8150 6354 | 45.3796 4153 | 36.9872 9141 | 0.0220 3631 | 0.0270 3631 |
| 42 | 1.2330 3270 | 0.8110 0850 | 46.6065 3974 | 37.7982 9991 | 0.0214 5622 | 0.0264 5622 |
| 43 | 1.2391 9786 | 0.8069 7363 | 47.8395 7244 | 38.6052 7354 | 0.0209 0320 | 0.0259 0320 |
| 44 | 1.2453 9385 | 0.8029 5884 | 49.0787 7030 | 39.4082 3238 | 0.0203 7541 | 0.0253 7541 |
| 45 | 1.2516 2082 | 0.7989 6402 | 50.3241 6415 | 40.2071 9640 | 0.0198 7117 | 0.0248 7117 |
| 46 | 1.2578 7892 | 0.7949 8907 | 51.5757 8497 | 41.0021 8547 | 0.0193 8894 | 0.0243 8894 |
| 47 | 1.2641 6832 | 0.7910 3390 | 52.8336 6390 | 41.7932 1937 | 0.0189 2733 | 0.0239 2733 |
| 48 | 1.2704 8916 | 0.7870 9841 | 54.0978 3222 | 42.5803 1778 | 0.0184 8503 | 0.0234 8503 |
| 49 | 1.2768 4161 | 0.7831 8250 | 55.3683 2138 | 43.3635 0028 | 0.0180 6087 | 0.0230 6087 |
| 50 | 1.2832 2581 | 0.7792 8607 | 56.6451 6299 | 44.1427 8635 | 0.0176 5376 | 0.0226 5376 |
| 51 | 1.2896 4194 | 0.7754 0902 | 57.9283 8880 | 44.9181 9537 | 0.0172 6269 | 0.0222 6269 |
| 52 | 1.2960 9015 | 0.7715 5127 | 59.2180 3075 | 45.6897 4664 | 0.0168 8675 | 0.0218 8675 |
| 53 | 1.3025 7060 | 0.7677 1270 | 60.5141 2090 | 46.4574 5934 | 0.0165 2507 | 0.0215 2507 |
| 54 | 1.3090 8346 | 0.7638 9324 | 61.8166 9150 | 47.2213 5258 | 0.0161 7686 | 0.0211 7686 |
| 55 | 1.3156 2887 | 0.7600 9277 | 63.1257 7496 | 47.9814 4535 | 0.0158 4139 | 0.0208 4139 |
| 56 | 1.3222 0702 | 0.7563 1122 | 64.4414 0384 | 48.7377 5657 | 0.0155 1797 | 0.0205 1797 |
| 57 | 1.3288 1805 | 0.7525 4847 | 65.7636 1086 | 49.4903 0505 | 0.0152 0598 | 0.0202 0598 |
| 58 | 1.3354 6214 | 0.7488 0445 | 67.0924 2891 | 50.2391 0950 | 0.0149 0481 | 0.0199 0481 |
| 59 | 1.3421 3946 | 0.7450 7906 | 68.4278 9105 | 50.9841 8855 | 0.0146 1392 | 0.0196 1392 |
| 60 | 1.3488 5015 | 0.7413 7220 | 69.7700 3051 | 51.7255 6075 | 0.0143 3280 | 0.0193 3280 |
| 61 | 1.3555 9440 | 0.7376 8378 | 71.1188 8066 | 52.4632 4453 | 0.0140 6096 | 0.0190 6096 |
| 62 | 1.3623 7238 | 0.7340 1371 | 72.4744 7507 | 53.1972 5824 | 0.0137 9796 | 0.0187 9796 |
| 63 | 1.3691 8424 | 0.7303 6190 | 73.8368 4744 | 53.9276 2014 | 0.0135 4337 | 0.0185 4337 |
| 64 | 1.3760 3016 | 0.7267 2826 | 75.2060 3168 | 54.6543 4839 | 0.0132 9681 | 0.0182 9681 |
| 65 | 1.3829 1031 | 0.7231 1269 | 76.5820 6184 | 55.3774 6109 | 0.0130 5789 | 0.0180 5789 |
| 66 | 1.3898 2486 | 0.7195 1512 | 77.9649 7215 | 56.0969 7621 | 0.0128 2627 | 0.0178 2627 |
| 67 | 1.3967 7399 | 0.7159 3544 | 79.3547 9701 | 56.8129 1165 | 0.0126 0163 | 0.0176 0163 |
| 68 | 1.4037 5785 | 0.7123 7357 | 80.7515 7099 | 57.5252 8522 | 0.0123 8366 | 0.0173 8366 |
| 69 | 1.4107 7664 | 0.7088 2943 | 82.1553 2885 | 58.2341 1465 | 0.0121 7206 | 0.0171 7206 |
| 70 | 1.4178 3053 | 0.7053 0291 | 83.5661 0549 | 58.9394 1756 | 0.0119 6657 | 0.0169 6657 |
| 71 | 1.4249 1968 | 0.7017 9394 | 84.9839 3602 | 59.6412 1151 | 0.0117 6693 | 0.0167 6693 |
| 72 | 1.4320 4428 | 0.6983 0243 | 86.4088 5570 | 60.3395 1394 | 0.0115 7289 | 0.0165 7289 |
| 73 | 1.4392 0450 | 0.6948 2829 | 87.8408 9998 | 61.0343 4222 | 0.0113 8422 | 0.0163 8422 |
| 74 | 1.4464 0052 | 0.6913 7143 | 89.2801 0448 | 61.7257 1366 | 0.0112 0070 | 0.0162 0070 |
| 75 | 1.4536 3252 | 0.6879 3177 | 90.7265 0500 | 62.4136 4543 | 0.0110 2214 | 0.0160 2214 |
| 76 | 1.4609 0069 | 0.6845 0923 | 92.1801 3752 | 63.0981 5466 | 0.0108 4832 | 0.0158 4832 |
| 77 | 1.4682 0519 | 0.6811 0371 | 93.6410 3821 | 63.7792 5836 | 0.0106 7908 | 0.0156 7908 |
| 78 | 1.4755 4622 | 0.6777 1513 | 95.1092 4340 | 64.4569 7350 | 0.0105 1423 | 0.0155 1423 |
| 79 | 1.4829 2395 | 0.6743 4342 | 96.5847 8962 | 65.1313 1691 | 0.0103 5360 | 0.0153 5360 |
| 80 | 1.4903 3857 | 0.6709 8847 | 98.0677 1357 | 65.8023 0538 | 0.0101 9704 | 0.0151 9704 |

**Table B.2** *(continued)*

$i = \frac{1}{4}\%$

| $n$ | $(1 + i)^n$ | $(1 + i)^{-n}$ | $s_{\overline{n}|i}$ | $a_{\overline{n}|i}$ | $\dfrac{1}{s_{\overline{n}|i}}$ | $\dfrac{1}{a_{\overline{n}|i}}$ |
|---|---|---|---|---|---|---|
| 81 | 1.4977 9026 | 0.6676 5022 | 99.5580 5214 | 66.4699 5561 | 0.0100 4439 | 0.0150 4439 |
| 82 | 1.5052 7921 | 0.6643 2858 | 101.0558 4240 | 67.1342 8419 | 0.0098 9552 | 0.0148 9552 |
| 83 | 1.5128 0561 | 0.6610 2346 | 102.5611 2161 | 67.7953 0765 | 0.0097 5028 | 0.0147 5028 |
| 84 | 1.5203 6964 | 0.6577 3479 | 104.0739 2722 | 68.4530 4244 | 0.0096 0855 | 0.0146 0855 |
| 85 | 1.5279 7148 | 0.6544 6248 | 105.5942 9685 | 69.1075 0491 | 0.0094 7021 | 0.0144 7021 |
| 86 | 1.5356 1134 | 0.6512 0644 | 107.1222 6834 | 69.7587 1135 | 0.0093 3513 | 0.0143 3513 |
| 87 | 1.5432 8940 | 0.6479 6661 | 108.6578 7968 | 70.4066 7796 | 0.0092 0320 | 0.0142 0320 |
| 88 | 1.5510 0585 | 0.6447 4290 | 110.2011 6908 | 71.0514 2086 | 0.0090 7431 | 0.0140 7431 |
| 89 | 1.5587 6087 | 0.6415 3522 | 111.7521 7492 | 71.6929 5608 | 0.0089 4837 | 0.0139 4837 |
| 90 | 1.5665 5468 | 0.6383 4350 | 113.3109 3580 | 72.3312 9958 | 0.0088 2527 | 0.0138 2527 |
| 91 | 1.5743 8745 | 0.6351 6766 | 114.8774 9048 | 72.9664 6725 | 0.0087 0493 | 0.0137 0493 |
| 92 | 1.5822 5939 | 0.6320 0763 | 116.4518 7793 | 73.5984 7487 | 0.0085 8724 | 0.0135 8724 |
| 93 | 1.5901 7069 | 0.6288 6331 | 118.0341 3732 | 74.2273 3818 | 0.0084 7213 | 0.0134 7213 |
| 94 | 1.5981 2154 | 0.6257 3464 | 119.6243 0800 | 74.8530 7282 | 0.0083 5950 | 0.0133 5950 |
| 95 | 1.6061 1215 | 0.6226 2153 | 121.2224 2954 | 75.4756 9434 | 0.0082 4930 | 0.0132 4930 |
| 96 | 1.6141 4271 | 0.6195 2391 | 122.8285 4169 | 76.0952 1825 | 0.0081 4143 | 0.0131 4143 |
| 97 | 1.6222 1342 | 0.6164 4170 | 124.4426 8440 | 76.7116 5995 | 0.0080 3583 | 0.0130 3583 |
| 98 | 1.6303 2449 | 0.6133 7483 | 126.0648 9782 | 77.3250 3478 | 0.0079 3242 | 0.0129 3242 |
| 99 | 1.6384 7611 | 0.6103 2321 | 127.6952 2231 | 77.9353 5799 | 0.0078 3115 | 0.0128 3115 |
| 100 | 1.6466 6849 | 0.6072 8678 | 129.3336 9842 | 78.5426 4477 | 0.0077 3194 | 0.0127 3194 |

$i = \frac{3}{4}\%$

| $n$ | $(1 + i)^n$ | $(1 + i)^{-n}$ | $s_{\overline{n}|i}$ | $a_{\overline{n}|i}$ | $\dfrac{1}{s_{\overline{n}|i}}$ | $\dfrac{1}{a_{\overline{n}|i}}$ |
|---|---|---|---|---|---|---|
| 1 | 1.0075 0000 | 0.9925 5583 | 1.0000 0000 | 0.9925 5583 | 1.0000 0000 | 1.0075 0000 |
| 2 | 1.0150 5625 | 0.9851 6708 | 2.0075 0000 | 1.9777 2291 | 0.4981 3200 | 0.5056 3200 |
| 3 | 1.0226 6917 | 0.9778 3333 | 3.0225 5625 | 2.9555 5624 | 0.3308 4579 | 0.3383 4579 |
| 4 | 1.0303 3919 | 0.9705 5417 | 4.0452 2542 | 3.9261 1041 | 0.2472 0501 | 0.2547 0501 |
| 5 | 1.0380 6673 | 0.9633 2920 | 5.0755 6461 | 4.8894 3961 | 0.1970 2242 | 0.2045 2242 |
| 6 | 1.0458 5224 | 0.9561 5802 | 6.1136 3135 | 5.8455 9763 | 0.1635 6891 | 0.1710 6891 |
| 7 | 1.0536 9613 | 0.9490 4022 | 7.1594 8358 | 6.7946 3785 | 0.1396 7488 | 0.1471 7488 |
| 8 | 1.0615 9885 | 0.9419 7540 | 8.2131 7971 | 7.7366 1325 | 0.1217 5552 | 0.1292 5552 |
| 9 | 1.0695 6084 | 0.9349 6318 | 9.2747 7856 | 8.6715 7642 | 0.1078 1929 | 0.1153 1929 |
| 10 | 1.0775 8255 | 0.9270 0315 | 10.3443 3940 | 9.5995 7958 | 0.0966 7123 | 0.1041 7123 |

**Table B.2** *(continued)*

$i = \frac{3}{4}\%$

| $n$ | $(1 + i)^n$ | $(1 + i)^{-n}$ | $s_{\overline{n}|i}$ | $a_{\overline{n}|i}$ | $\dfrac{1}{s_{\overline{n}|i}}$ | $\dfrac{1}{a_{\overline{n}|i}}$ |
|---|---|---|---|---|---|---|
| 11 | 1.0856 6441 | 0.9210 9494 | 11.4219 2194 | 10.5206 7452 | 0.0875 5094 | 0.0950 5094 |
| 12 | 1.0938 0690 | 0.9142 3815 | 12.5075 8636 | 11.4349 1267 | 0.0799 5148 | 0.0874 5148 |
| 13 | 1.1020 1045 | 0.9074 3241 | 13.6013 9325 | 12.3423 4508 | 0.0735 2188 | 0.0810 2188 |
| 14 | 1.1102 7553 | 0.9006 7733 | 14.7034 0370 | 13.2430 2242 | 0.0680 1146 | 0.0755 1146 |
| 15 | 1.1186 0259 | 0.8939 7254 | 15.8136 7923 | 14.1369 9495 | 0.0632 3639 | 0.0707 3639 |
| 16 | 1.1269 9211 | 0.8873 1766 | 16.9322 8183 | 15.0243 1261 | 0.0590 5879 | 0.0665 5879 |
| 17 | 1.1354 4455 | 0.8807 1231 | 18.0592 7394 | 15.9050 2492 | 0.0553 7321 | 0.0628 7321 |
| 18 | 1.1439 6039 | 0.8741 5614 | 19.1947 1849 | 16.7791 8107 | 0.0520 9766 | 0.0595 9766 |
| 19 | 1.1525 4009 | 0.8676 4878 | 20.3386 7888 | 17.6468 2984 | 0.0491 6740 | 0.0566 6740 |
| 20 | 1.1611 8414 | 0.8611 8985 | 21.4912 1897 | 18.5080 1969 | 0.0465 3063 | 0.0540 3063 |
| 21 | 1.1698 9302 | 0.8547 7901 | 22.6524 0312 | 19.3627 9870 | 0.0441 4543 | 0.0516 4543 |
| 22 | 1.1786 6722 | 0.8484 1589 | 23.8222 9614 | 20.2112 1459 | 0.0419 7748 | 0.0494 7748 |
| 23 | 1.1875 0723 | 0.8421 0014 | 25.0009 6336 | 21.0533 1473 | 0.0399 9846 | 0.0474 9846 |
| 24 | 1.1964 1353 | 0.8358 3140 | 26.1884 7059 | 21.8891 4614 | 0.0381 8474 | 0.0456 8474 |
| 25 | 1.2053 8663 | 0.8296 0933 | 27.3848 8412 | 22.7187 5547 | 0.0365 1650 | 0.0440 1650 |
| 26 | 1.2144 2703 | 0.8234 3358 | 28.5902 7075 | 23.5421 8905 | 0.0349 7693 | 0.0424 7693 |
| 27 | 1.2235 3523 | 0.8173 0380 | 29.8046 9778 | 24.3594 9286 | 0.0335 5176 | 0.0410 5176 |
| 28 | 1.2327 1175 | 0.8112 1966 | 31.0282 3301 | 25.1707 1251 | 0.0322 2871 | 0.0397 2871 |
| 29 | 1.2419 5709 | 0.8051 8080 | 32.2609 4476 | 25.9758 9331 | 0.0309 9723 | 0.0384 9723 |
| 30 | 1.2512 7176 | 0.7991 8690 | 35.5029 0184 | 26.7750 8021 | 0.0298 4816 | 0.0373 4816 |
| 31 | 1.2606 5630 | 0.7932 3762 | 34.7541 7361 | 27.5683 1783 | 0.0287 7352 | 0.0362 7352 |
| 32 | 1.2701 1122 | 0.7873 3262 | 36.0148 2991 | 28.3556 5045 | 0.0277 6634 | 0.0352 6634 |
| 33 | 1.2796 3706 | 0.7814 7158 | 37.2849 4113 | 29.1371 2203 | 0.0268 2048 | 0.0343 2048 |
| 34 | 1.2892 3434 | 0.7756 5418 | 38.5645 7819 | 29.9127 7621 | 0.0259 3053 | 0.0334 3053 |
| 35 | 1.2989 0359 | 0.7698 8008 | 39.8538 1253 | 30.6826 5629 | 0.0250 9170 | 0.0325 9170 |
| 36 | 1.3086 4537 | 0.7641 4896 | 41.1527 1612 | 31.4468 0525 | 0.0242 9973 | 0.0317 9973 |
| 37 | 1.3184 6021 | 0.7584 6051 | 42.4613 6149 | 32.2052 6576 | 0.0235 5082 | 0.0310 5082 |
| 38 | 1.3283 4866 | 0.7528 1440 | 43.7798 2170 | 32.9580 8016 | 0.0228 4157 | 0.0303 4157 |
| 39 | 1.3383 1128 | 0.7472 1032 | 45.1081 7037 | 33.7052 9048 | 0.0221 6893 | 0.0296 6893 |
| 40 | 1.3483 4861 | 0.7416 4796 | 46.4464 8164 | 34.4469 3844 | 0.0215 3016 | 0.0290 3016 |
| 41 | 1.3584 6123 | 0.7361 2701 | 47.7948 3026 | 35.1830 6545 | 0.0209 2276 | 0.0284 2276 |
| 42 | 1.3686 4969 | 0.7306 4716 | 49.1532 9148 | 35.9137 1260 | 0.0203 4452 | 0.0278 4452 |
| 43 | 1.3789 1456 | 0.7252 0809 | 50.5219 4117 | 36.6389 2070 | 0.0197 9338 | 0.0272 9338 |
| 44 | 1.3892 5642 | 0.7198 0952 | 51.9008 5573 | 37.3587 3022 | 0.0192 6751 | 0.0267 6751 |
| 45 | 1.3996 7584 | 0.7144 5114 | 53.2901 1215 | 38.0731 8136 | 0.0187 6521 | 0.0262 6521 |
| 46 | 1.4101 7341 | 0.7091 3264 | 54.6897 8799 | 38.7823 1401 | 0.0182 8495 | 0.0257 8495 |
| 47 | 1.4297 4971 | 0.7038 5374 | 56.0999 6140 | 39.4861 6775 | 0.0178 2532 | 0.0253 2532 |
| 48 | 1.4314 0533 | 0.6986 1414 | 57.5207 1111 | 40.1847 8189 | 0.0173 8504 | 0.0248 8504 |
| 49 | 1.4421 4087 | 0.6934 1353 | 58.9521 1644 | 40.8781 9542 | 0.0169 6292 | 0.0244 6292 |
| 50 | 1.4529 5693 | 0.6882 5165 | 60.3942 5732 | 41.5664 4707 | 0.0165 5787 | 0.0240 5787 |

Appendix B Tables

**Table B.2** *(continued)*

$i = \frac{3}{4}\%$

| $n$ | $(1 + i)^n$ | $(1 + i)^{-n}$ | $s_{\overline{n}|i}$ | $a_{\overline{n}|i}$ | $\dfrac{1}{s_{\overline{n}|i}}$ | $\dfrac{1}{a_{\overline{n}|i}}$ |
|---|---|---|---|---|---|---|
| 51 | 1.4638 5411 | 0.6831 2819 | 61.8472 1424 | 42.2495 7525 | 0.0161 6888 | 0.0236 6888 |
| 52 | 1.4748 3301 | 0.6780 4286 | 63.3110 6835 | 42.9276 1812 | 0.0157 9503 | 0.0232 9503 |
| 53 | 1.4858 9426 | 0.6729 9540 | 64.7859 0136 | 43.6006 1351 | 0.0154 3546 | 0.0229 3546 |
| 54 | 1.4970 3847 | 0.6679 8551 | 66.2717 9562 | 44.2685 9902 | 0.0150 8938 | 0.0225 8938 |
| 55 | 1.5082 6626 | 0.6630 1291 | 67.7688 3409 | 44.9316 1193 | 0.0147 5605 | 0.0222 5605 |
| 56 | 1.5195 7825 | 0.6580 7733 | 69.2771 0035 | 45.5896 8926 | 0.0144 3478 | 0.0219 3478 |
| 57 | 1.5309 7509 | 0.6531 7849 | 70.7966 7860 | 46.2428 6776 | 0.0141 2496 | 0.0216 2496 |
| 58 | 1.5424 5740 | 0.6483 1612 | 72.3276 5369 | 46.8911 8388 | 0.0138 2597 | 0.0213 2597 |
| 59 | 1.5540 2583 | 0.6434 8995 | 73.8701 1109 | 47.5346 7382 | 0.0135 3727 | 0.0210 3727 |
| 60 | 1.5656 8103 | 0.6386 9970 | 75.4241 3693 | 48.1733 7352 | 0.0132 5836 | 0.0207 5836 |
| 61 | 1.5774 2363 | 0.6339 4511 | 76.9898 1795 | 48.8073 1863 | 0.0129 8873 | 0.0204 8873 |
| 62 | 1.5892 5431 | 0.6292 2592 | 78.5672 4159 | 49.4365 4455 | 0.0127 2795 | 0.0202 2795 |
| 63 | 1.6011 7372 | 0.6245 4185 | 80.1564 9590 | 50.0610 8640 | 0.0124 7560 | 0.0199 7560 |
| 64 | 1.6131 8252 | 0.6198 9266 | 81.7576 6962 | 50.6809 7906 | 0.0122 3127 | 0.0197 3127 |
| 65 | 1.6252 8139 | 0.6152 7807 | 83.3708 5214 | 51.2962 5713 | 0.0119 9460 | 0.0194 9460 |
| 66 | 1.6374 7100 | 0.6106 9784 | 84.9961 3353 | 51.9069 5497 | 0.0117 6524 | 0.0192 6524 |
| 67 | 1.6497 5203 | 0.6061 5170 | 86.6336 0453 | 52.5131 0667 | 0.0115 4286 | 0.0190 4286 |
| 68 | 1.6621 2517 | 0.6016 3940 | 88.2833 5657 | 53.1147 4607 | 0.0113 2716 | 0.0188 2716 |
| 69 | 1.6745 9111 | 0.5971 6070 | 89.9454 8174 | 53.7119 0677 | 0.0111 1785 | 0.0186 1785 |
| 70 | 1.6871 5055 | 0.5927 1533 | 91.6200 7285 | 54.3046 2210 | 0.0109 1464 | 0.0184 1464 |
| 71 | 1.6998 0418 | 0.5883 0306 | 93.3072 2340 | 54.8929 2516 | 0.0107 1728 | 0.0182 1728 |
| 72 | 1.7125 5271 | 0.5839 2363 | 95.0070 2758 | 55.4768 4880 | 0.0105 2554 | 0.0180 2554 |
| 73 | 1.7253 9685 | 0.5795 7681 | 96.7195 8028 | 56.0564 2561 | 0.0103 3917 | 0.0178 3917 |
| 74 | 1.7383 3733 | 0.5752 6234 | 98.4449 7714 | 56.6316 8795 | 0.0101 5796 | 0.0176 5796 |
| 75 | 1.7513 7486 | 0.5709 7999 | 100.1833 1446 | 57.2026 6794 | 0.0099 8170 | 0.0174 8170 |
| 76 | 1.7645 1017 | 0.5667 2952 | 101.9346 8932 | 57.7693 9746 | 0.0098 1020 | 0.0173 1020 |
| 77 | 1.7777 4400 | 0.5625 1069 | 103.6991 9949 | 58.3319 0815 | 0.0096 4328 | 0.0171 4328 |
| 78 | 1.7910 7708 | 0.5583 2326 | 105.4769 4349 | 58.8902 3141 | 0.0094 8074 | 0.0169 8074 |
| 79 | 1.8045 1015 | 0.5541 6701 | 107.2680 2056 | 59.4443 9842 | 0.0093 2244 | 0.0168 2244 |
| 80 | 1.8180 4398 | 0.5500 4170 | 109.0725 3072 | 59.9944 4012 | 0.0091 6821 | 0.0166 6821 |
| 81 | 1.8316 7931 | 0.5459 4710 | 110.8905 7470 | 60.5403 8722 | 0.0090 1790 | 0.0165 1790 |
| 82 | 1.8454 1691 | 0.5418 8297 | 112.7222 5401 | 61.0822 7019 | 0.0088 7136 | 0.0163 7136 |
| 83 | 1.8592 5753 | 0.5378 4911 | 114.5676 7091 | 61.6201 1930 | 0.0087 2847 | 0.0162 2847 |
| 84 | 1.8732 0196 | 0.5338 4527 | 116.4269 2845 | 62.1539 6456 | 0.0085 8908 | 0.0260 8908 |
| 85 | 1.8872 5098 | 0.5298 7123 | 118.3001 3041 | 62.6838 3579 | 0.0084 5308 | 0.0159 5308 |
| 86 | 1.9014 0536 | 0.5259 2678 | 120.1873 8139 | 63.2097 6257 | 0.0083 2034 | 0.0158 2034 |
| 87 | 1.9156 6590 | 0.5220 1169 | 122.0887 8675 | 63.7317 7427 | 0.0081 9076 | 0.0156 9076 |
| 88 | 1.9300 3339 | 0.5181 2575 | 124.0044 5265 | 64.2499 0002 | 0.0080 6423 | 0.0155 6423 |
| 89 | 1.9445 0865 | 0.5142 6873 | 125.9344 8604 | 64.7641 6875 | 0.0079 4064 | 0.0154 4064 |
| 90 | 1.9590 9246 | 0.5104 4043 | 127.8789 9569 | 65.2746 0918 | 0.0078 1989 | 0.0153 1989 |

**Table B.2** *(continued)*

$i = \frac{3}{4}\%$

| $n$ | $(1 + i)^n$ | $(1 + i)^{-n}$ | $s_{\overline{n}|i}$ | $a_{\overline{n}|i}$ | $\dfrac{1}{s_{\overline{n}|i}}$ | $\dfrac{1}{a_{\overline{n}|i}}$ |
|---|---|---|---|---|---|---|
| 91 | 1.9737 8565 | 0.5066 4063 | 129.8380 8715 | 65.7812 4981 | 0.0077 0190 | 0.0152 0190 |
| 92 | 1.9885 8905 | 0.5028 6911 | 131.8118 7280 | 66.2841 1892 | 0.0075 8657 | 0.0150 8657 |
| 93 | 2.0035 0346 | 0.4991 2567 | 133.8004 6185 | 66.7832 4458 | 0.0074 7382 | 0.0149 7382 |
| 94 | 2.0185 2974 | 0.4954 1009 | 135.8039 6531 | 67.2786 5467 | 0.0073 6356 | 0.0148 6356 |
| 95 | 2.0336 6871 | 0.4917 2217 | 137.8224 9505 | 67.7703 7685 | 0.0072 5571 | 0.0147 5571 |
| 96 | 2.0489 2123 | 0.4880 6171 | 139.8561 6377 | 68.2584 3856 | 0.0071 5020 | 0.0146 5020 |
| 97 | 2.0642 8814 | 0.4844 2850 | 141.9050 8499 | 68.7428 6705 | 0.0070 4696 | 0.0145 4696 |
| 98 | 2.0797 7030 | 0.4808 2233 | 143.9693 7313 | 69.2236 8938 | 0.0069 4592 | 0.0144 4592 |
| 99 | 2.0953 6858 | 0.4772 4301 | 146.0491 4343 | 69.7009 3239 | 0.0068 4701 | 0.0143 4701 |
| 100 | 2.1110 8384 | 0.4736 9033 | 148.1445 1201 | 70.1746 2272 | 0.0067 5017 | 0.0142 5017 |

$i = 1\%$

| $n$ | $(1 + i)^n$ | $(1 + i)^{-n}$ | $s_{\overline{n}|i}$ | $a_{\overline{n}|i}$ | $\dfrac{1}{s_{\overline{n}|i}}$ | $\dfrac{1}{a_{\overline{n}|i}}$ |
|---|---|---|---|---|---|---|
| 1 | 1.0100 0000 | 0.9900 9901 | 1.0000 0000 | 0.9900 9901 | 1.0000 0000 | 1.0100 0000 |
| 2 | 1.0201 0000 | 0.9802 9605 | 2.0100 0000 | 1.9703 9506 | 0.4975 1244 | 0.5075 1244 |
| 3 | 1.0303 0100 | 0.9705 9015 | 3.0301 0000 | 2.9409 8521 | 0.3300 2211 | 0.3400 2211 |
| 4 | 1.0406 0401 | 0.9609 8034 | 4.0604 0100 | 3.9019 6555 | 0.2462 8109 | 0.2562 8109 |
| 5 | 1.0510 1005 | 0.9514 6569 | 5.1010 0501 | 4.8534 3124 | 0.1960 3980 | 0.2060 3980 |
| 6 | 1.0615 2015 | 0.9420 4524 | 6.1520 1506 | 5.7954 7647 | 0.1625 4837 | 0.1725 4837 |
| 7 | 1.0721 3535 | 0.9327 1805 | 7.2135 3521 | 6.7281 9453 | 0.1386 2828 | 0.1486 2828 |
| 8 | 1.0828 5671 | 0.9234 8322 | 8.2856 7056 | 7.6516 7775 | 0.1206 9029 | 0.1306 9029 |
| 9 | 1.0936 8527 | 0.9143 3982 | 9.3685 2727 | 8.5660 1758 | 0.1067 4036 | 0.1167 4036 |
| 10 | 1.1046 2213 | 0.9052 8695 | 10.4622 1254 | 9.4713 0453 | 0.0955 8208 | 0.1055 8208 |
| 11 | 1.1156 6835 | 0.8963 2372 | 11.5668 3467 | 10.3676 2825 | 0.0864 5408 | 0.0964 5408 |
| 12 | 1.1268 2503 | 0.8874 4923 | 12.6825 0301 | 11.2550 7747 | 0.0788 4879 | 0.0888 4879 |
| 13 | 1.1380 9328 | 0.8786 6260 | 13.8093 2804 | 12.1337 4007 | 0.0724 1482 | 0.0824 1482 |
| 14 | 1.1494 7421 | 0.8699 6297 | 14.9474 2132 | 13.0037 0304 | 0.0669 0117 | 0.0769 0117 |
| 15 | 1.1609 6896 | 0.8613 4947 | 16.0968 9554 | 13.8650 5252 | 0.0621 2378 | 0.0721 2378 |
| 16 | 1.1725 7864 | 0.8528 2126 | 17.2578 6449 | 14.7178 7378 | 0.0579 4460 | 0.0679 4460 |
| 17 | 1.1843 0443 | 0.8443 7749 | 18.4304 4314 | 15.5622 5127 | 0.0542 5806 | 0.0642 5806 |
| 18 | 1.1961 4748 | 0.8360 1731 | 19.6147 4757 | 16.3982 6858 | 0.0509 8205 | 0.0609 8205 |
| 19 | 1.2081 0895 | 0.8277 3992 | 20.8108 9504 | 17.2260 0850 | 0.0480 5175 | 0.0580 5175 |
| 20 | 1.2201 9004 | 0.8195 4447 | 22.0190 0399 | 18.0455 5297 | 0.0454 1531 | 0.0554 1531 |

**Table B.2** *(continued)*

$i = 1\%$

| $n$ | $(1 + i)^n$ | $(1 + i)^{-n}$ | $s_{\overline{n}|i}$ | $a_{\overline{n}|i}$ | $\dfrac{1}{s_{\overline{n}|i}}$ | $\dfrac{1}{a_{\overline{n}|i}}$ |
|---|---|---|---|---|---|---|
| 21 | 1.2323 9194 | 0.8114 3017 | 23.2391 9403 | 18.8569 8313 | 0.0430 3075 | 0.0530 3075 |
| 22 | 1.2447 1586 | 0.8033 9621 | 24.4715 8598 | 19.6603 7934 | 0.0408 6372 | 0.0508 6372 |
| 23 | 1.2571 6302 | 0.7954 4179 | 25.7163 0183 | 20.4558 2113 | 0.0388 8584 | 0.0488 8584 |
| 24 | 1.2697 3465 | 0.7875 6613 | 26.9734 6485 | 21.2433 8726 | 0.0370 7347 | 0.0470 7347 |
| 25 | 1.2824 3200 | 0.7797 6844 | 28.2431 9950 | 22.0231 5570 | 0.0354 0675 | 0.0454 0675 |
| 26 | 1.2952 5631 | 0.7720 4796 | 29.5256 3150 | 22.7952 0366 | 0.0338 6888 | 0.0438 6888 |
| 27 | 1.3082 0888 | 0.7644 0392 | 30.8208 8781 | 23.5596 0759 | 0.0324 4553 | 0.0424 4553 |
| 28 | 1.3212 9097 | 0.7568 3557 | 32.1290 9669 | 24.3164 4316 | 0.0311 2444 | 0.0411 2444 |
| 29 | 1.3345 0388 | 0.7493 4215 | 33.4503 8766 | 25.0657 8530 | 0.0298 9502 | 0.0398 9502 |
| 30 | 1.3478 4892 | 0.7419 2292 | 34.7848 9153 | 25.8077 0822 | 0.0287 4811 | 0.0387 4811 |
| 31 | 1.3613 2740 | 0.7345 7715 | 36.1327 4045 | 26.5422 8537 | 0.0276 7573 | 0.0376 7573 |
| 32 | 1.3749 4068 | 0.7273 0411 | 37.4940 6785 | 27.2695 8947 | 0.0266 7089 | 0.0366 7089 |
| 33 | 1.3886 9009 | 0.7201 0307 | 38.8690 0853 | 27.9896 9255 | 0.0257 2744 | 0.0357 2744 |
| 34 | 1.4025 7699 | 0.7129 7334 | 40.2576 9862 | 28.7026 6589 | 0.0248 3997 | 0.0348 3997 |
| 35 | 1.4166 0276 | 0.7059 1420 | 41.6602 7560 | 29.4085 8009 | 0.0240 0368 | 0.0340 0368 |
| 36 | 1.4307 6878 | 0.6989 2495 | 43.0768 7836 | 30.1075 0504 | 0.0232 1431 | 0.0332 1431 |
| 37 | 1.4450 7647 | 0.6920 0490 | 44.5076 4714 | 30.7995 0994 | 0.0224 6805 | 0.0324 6805 |
| 38 | 1.4595 2724 | 0.6851 5337 | 45.9527 2361 | 31.4846 6330 | 0.0217 6150 | 0.0317 6150 |
| 39 | 1.4741 2251 | 0.6783 6967 | 47.4122 5085 | 32.1630 3298 | 0.0210 9160 | 0.0310 9160 |
| 40 | 1.4888 6373 | 0.6716 5314 | 48.8863 7336 | 32.8346 8611 | 0.0204 5560 | 0.0304 5560 |
| 41 | 1.5037 5237 | 0.6650 0311 | 50.3752 3709 | 33.4996 8922 | 0.0198 5102 | 0.0298 5102 |
| 42 | 1.5187 8989 | 0.6584 1892 | 51.8789 8946 | 34.1581 0814 | 0.0192 7563 | 0.0292 7563 |
| 43 | 1.5339 7779 | 0.6518 9992 | 53.3977 7936 | 34.8100 0806 | 0.0187 2737 | 0.0287 2737 |
| 44 | 1.5493 1757 | 0.6454 4546 | 54.9317 5715 | 35.4554 5352 | 0.0182 0441 | 0.0282 0441 |
| 45 | 1.5648 1075 | 0.6390 5492 | 56.4810 7472 | 36.0945 0844 | 0.0177 0505 | 0.0277 0505 |
| 46 | 1.5804 5885 | 0.6327 2764 | 58.0458 8547 | 36.7272 3608 | 0.0172 2775 | 0.0272 2775 |
| 47 | 1.5962 6344 | 0.6264 6301 | 59.6263 4432 | 37.3536 9909 | 0.0167 7111 | 0.0267 7111 |
| 48 | 1.6122 2608 | 0.6202 6041 | 61.2226 0777 | 37.9739 5949 | 0.0163 3384 | 0.0263 3384 |
| 49 | 1.6283 4834 | 0.6141 1921 | 62.8348 3385 | 38.5880 7871 | 0.0159 1474 | 0.0259 1474 |
| 50 | 1.6446 3182 | 0.6080 3882 | 64.4631 8218 | 39.1961 1753 | 0.0155 1273 | 0.0255 1273 |
| 51 | 1.6610 7814 | 0.6020 1864 | 66.1078 1401 | 39.7981 3617 | 0.0151 2680 | 0.0251 2680 |
| 52 | 1.6776 8892 | 0.5960 5806 | 67.7688 9215 | 40.3941 9423 | 0.0147 5603 | 0.0247 5603 |
| 53 | 1.6944 6581 | 0.5901 5649 | 69.4465 8107 | 40.9843 5072 | 0.0143 9956 | 0.0243 9956 |
| 54 | 1.7114 1047 | 0.5843 1336 | 71.1410 4688 | 41.5686 6408 | 0.0140 5658 | 0.0240 5658 |
| 55 | 1.7285 2457 | 0.5785 2808 | 72.8524 5735 | 42.1471 9216 | 0.0137 2637 | 0.0237 2637 |
| 56 | 1.7458 0982 | 0.5728 0008 | 74.5809 8192 | 42.7199 9224 | 0.0134 0824 | 0.0234 0824 |
| 57 | 1.7632 6792 | 0.5671 2879 | 76.3267 9174 | 43.2871 2102 | 0.0131 0156 | 0.0231 0156 |
| 58 | 1.7809 0060 | 0.5615 1365 | 78.0900 5966 | 43.8486 3468 | 0.0128 0573 | 0.0228 0573 |
| 59 | 1.7987 0960 | 0.5559 5411 | 79.8709 6025 | 44.4045 8879 | 0.0125 2020 | 0.0225 2020 |
| 60 | 1.8166 9670 | 0.5504 4962 | 81.6696 6986 | 44.9550 3841 | 0.0122 4445 | 0.0222 4445 |

**Table B.2** *(continued)*

$i = 1\%$

| $n$ | $(1 + i)^n$ | $(1 + i)^{-n}$ | $s_{\overline{n}|i}$ | $a_{\overline{n}|i}$ | $\dfrac{1}{s_{\overline{n}|i}}$ | $\dfrac{1}{a_{\overline{n}|i}}$ |
|---|---|---|---|---|---|---|
| 61 | 1.8348 6367 | 0.5449 9962 | 83.4863 6655 | 45.5000 3803 | 0.0119 7800 | 0.0219 7800 |
| 62 | 1.8532 1230 | 0.5396 0358 | 85.3212 3022 | 46.0396 4161 | 0.0117 2041 | 0.0217 2041 |
| 63 | 1.8717 4443 | 0.5342 6097 | 87.1744 4252 | 46.5739 0258 | 0.0114 7125 | 0.0214 7125 |
| 64 | 1.8904 6187 | 0.5289 7126 | 89.0461 8695 | 47.1028 7385 | 0.0112 3013 | 0.0212 3013 |
| 65 | 1.9093 6649 | 0.5237 3392 | 90.9366 4882 | 47.6266 0777 | 0.0109 9667 | 0.0209 9667 |
| 66 | 1.9284 6015 | 0.5185 4844 | 92.8460 1531 | 48.1451 5621 | 0.0107 7052 | 0.0207 7052 |
| 67 | 1.9477 4475 | 0.5134 1429 | 94.7744 7546 | 48.6585 7050 | 0.0105 5136 | 0.0205 5136 |
| 68 | 1.9672 2220 | 0.5083 3099 | 96.7222 2021 | 49.1669 0149 | 0.0103 3889 | 0.0203 3889 |
| 69 | 1.9868 9442 | 0.5032 9801 | 98.6894 4242 | 49.6701 9949 | 0.0101 3280 | 0.0201 3280 |
| 70 | 2.0067 6337 | 0.4983 1486 | 100.6763 3684 | 50.1685 1435 | 0.0099 3282 | 0.0199 3282 |
| 71 | 2.0268 3100 | 0.4933 8105 | 102.6831 0021 | 50.6618 9539 | 0.0097 3870 | 0.0197 3870 |
| 72 | 2.0470 9931 | 0.4884 9609 | 104.7099 3121 | 51.1503 9148 | 0.0095 5019 | 0.0195 5019 |
| 73 | 2.0675 7031 | 0.4836 5949 | 106.7570 3052 | 51.6340 5097 | 0.0093 6706 | 0.0193 6706 |
| 74 | 2.0882 4601 | 0.4788 7078 | 108.8246 0083 | 52.1129 2175 | 0.0091 8910 | 0.0191 8910 |
| 75 | 2.1091 2847 | 0.4741 2949 | 110.9128 4684 | 52.5870 5124 | 0.0090 1609 | 0.0190 1609 |
| 76 | 2.1302 1975 | 0.4694 3514 | 113.0219 7530 | 53.0564 8638 | 0.0088 4784 | 0.0188 4784 |
| 77 | 2.1515 2195 | 0.4647 8726 | 115.1521 9506 | 53.5212 7364 | 0.0086 8416 | 0.0186 8416 |
| 78 | 2.1730 3717 | 0.4601 8541 | 117.3037 1701 | 53.9814 5905 | 0.0085 2488 | 0.0185 2488 |
| 79 | 2.1947 6754 | 0.4556 2912 | 119.4767 5418 | 54.4370 8817 | 0.0083 6983 | 0.0183 6983 |
| 80 | 2.2167 1522 | 0.4511 1794 | 121.6715 2172 | 54.8882 0611 | 0.0082 1885 | 0.0182 1885 |
| 81 | 2.2388 8237 | 0.4466 5142 | 123.8882 3694 | 55.3348 5753 | 0.0080 7179 | 0.0180 7179 |
| 82 | 2.2612 7119 | 0.4422 2913 | 126.1271 1931 | 55.7770 8666 | 0.0079 2851 | 0.0179 2851 |
| 83 | 2.2838 8390 | 0.4378 5063 | 128.3883 9050 | 56.2149 3729 | 0.0077 8887 | 0.0177 8887 |
| 84 | 2.3067 2274 | 0.4335 1547 | 130.6722 7440 | 56.6484 5276 | 0.0076 5273 | 0.0176 5273 |
| 85 | 2.3297 8997 | 0.4292 2324 | 132.9789 9715 | 57.0776 7600 | 0.0075 1998 | 0.0175 1998 |
| 86 | 2.3530 8787 | 0.4249 7350 | 135.3087 8712 | 57.5026 4951 | 0.0073 9050 | 0.0173 9050 |
| 87 | 2.3766 1875 | 0.4207 6585 | 137.6618 7499 | 57.9234 1535 | 0.0072 6418 | 0.0172 6418 |
| 88 | 2.4003 8494 | 0.4165 9985 | 140.0384 9374 | 58.3400 1520 | 0.0071 4089 | 0.0171 4089 |
| 89 | 2.4243 8879 | 0.4124 7510 | 142.4388 7868 | 58.7524 9030 | 0.0070 2056 | 0.0170 2056 |
| 90 | 2.4486 3267 | 0.4083 9119 | 144.8632 6746 | 59.1608 8148 | 0.0069 0306 | 0.0169 0306 |
| 91 | 2.4731 1900 | 0.4043 4771 | 147.3119 0014 | 59.5652 2919 | 0.0067 8832 | 0.0167 8832 |
| 92 | 2.4978 5019 | 0.4003 4427 | 149.7850 1914 | 59.9655 7346 | 0.0066 7624 | 0.0166 7624 |
| 93 | 2.5228 2869 | 0.3963 8046 | 152.2828 6933 | 60.3619 5392 | 0.0065 6673 | 0.0165 6673 |
| 94 | 2.5480 5698 | 0.3924 5590 | 154.8056 9803 | 60.7544 0982 | 0.0064 5971 | 0.0164 5971 |
| 95 | 2.5735 3755 | 0.3885 7020 | 157.3537 5501 | 61.1429 8002 | 0.0063 5511 | 0.0163 5511 |
| 96 | 2.5992 7293 | 0.3847 2297 | 159.9272 9256 | 61.5277 0299 | 0.0062 5284 | 0.0162 5284 |
| 97 | 2.6252 6565 | 0.3809 1383 | 162.5265 6548 | 61.9086 1682 | 0.0061 5284 | 0.0161 5284 |
| 98 | 2.6515 1831 | 0.3771 4241 | 165.1518 3114 | 62.2857 5923 | 0.0060 5503 | 0.0160 5503 |
| 99 | 2.6780 3349 | 0.3734 0832 | 167.8033 4945 | 62.6591 6755 | 0.0059 5936 | 0.0159 5936 |
| 100 | 2.7048 1383 | 0.3697 1121 | 170.4813 8294 | 63.0288 7877 | 0.0058 6574 | 0.0158 6574 |

**Table B.2** *(continued)*

$i = 1\frac{1}{2}\%$

| $n$ | $(1 + i)^n$ | $(1 + i)^{-n}$ | $s_{\overline{n}|i}$ | $a_{\overline{n}|i}$ | $\dfrac{1}{s_{\overline{n}|i}}$ | $\dfrac{1}{a_{\overline{n}|i}}$ |
|---|---|---|---|---|---|---|
| 1 | 1.0150 0000 | 0.9852 2167 | 1.0000 0000 | 0.9852 2167 | 1.0000 0000 | 1.0150 0000 |
| 2 | 1.0302 2500 | 0.9706 6175 | 2.0150 0000 | 1.9558 8342 | 0.4962 7792 | 0.5112 7792 |
| 3 | 1.0456 7838 | 0.9563 1699 | 3.0452 2500 | 2.9122 0042 | 0.3283 8296 | 0.3433 8296 |
| 4 | 1.0613 6355 | 0.9421 8423 | 4.0909 0338 | 3.8543 8465 | 0.2444 4479 | 0.2594 4479 |
| 5 | 1.0772 8400 | 0.9282 6033 | 5.1522 6693 | 4.7826 4497 | 0.1940 8932 | 0.2090 8932 |
| 6 | 1.0934 4326 | 0.9145 4219 | 6.2295 5093 | 5.6971 8717 | 0.1605 2521 | 0.1755 2521 |
| 7 | 1.1098 4491 | 0.9010 2679 | 7.3229 9419 | 6.5982 1396 | 0.1356 5616 | 0.1515 5616 |
| 8 | 1.1264 9259 | 0.8877 1112 | 8.4328 3911 | 7.4959 2508 | 0.1185 8402 | 1.1335 8402 |
| 9 | 1.1433 8998 | 0.8745 9224 | 9.5593 3169 | 8.3605 1732 | 0.1046 0982 | 0.1196 0982 |
| 10 | 1.0605 4083 | 0.8616 6723 | 10.7027 2167 | 9.2221 8455 | 0.0934 3518 | 0.1084 3418 |
| 11 | 1.1779 4894 | 0.8489 3323 | 11.8632 6249 | 10.0711 1779 | 0.0842 9384 | 0.0992 9384 |
| 12 | 1.1956 1817 | 0.8363 8742 | 13.0412 1143 | 10.9075 0521 | 0.0766 7999 | 0.0916 7999 |
| 13 | 1.2135 5244 | 0.8240 2702 | 14.2368 2960 | 11.7315 3222 | 0.0702 4036 | 0.0852 4036 |
| 14 | 1.2317 5573 | 0.8118 4928 | 15.4503 8205 | 12.5433 8150 | 0.0647 2332 | 0.0797 2332 |
| 15 | 1.2502 3207 | 0.7998 5150 | 16.6821 3778 | 13.3432 3301 | 0.0599 4436 | 0.0749 4436 |
| 16 | 1.2689 8555 | 0.7880 3104 | 17.9323 6984 | 14.1312 6405 | 0.0557 6508 | 0.0707 6508 |
| 17 | 1.2880 2033 | 0.7763 8526 | 19.2013 5539 | 14.9076 4931 | 0.0520 7966 | 0.0670 7966 |
| 18 | 1.3073 4064 | 0.7649 1159 | 20.4893 7572 | 15.6725 6089 | 0.0488 0578 | 0.0638 0578 |
| 19 | 1.3269 5075 | 0.7536 0747 | 21.7967 1636 | 16.4261 6837 | 0.0458 7847 | 0.0608 7847 |
| 20 | 1.3468 5501 | 0.7424 7042 | 23.1236 6710 | 17.4686 3879 | 0.0432 4574 | 0.0582 4574 |
| 21 | 1.3670 5783 | 0.7314 9795 | 24.4705 2211 | 17.9001 3673 | 0.0408 6550 | 0.0558 6550 |
| 22 | 1.3875 6370 | 0.7206 8763 | 25.8375 7994 | 18.6208 2437 | 0.0387 0332 | 0.0537 0332 |
| 23 | 1.4083 7715 | 0.7100 3708 | 27.2251 4364 | 19.3308 6145 | 0.0367 3075 | 0.0517 3075 |
| 24 | 1.4295 0281 | 0.6995 4392 | 28.6335 2080 | 20.0304 0537 | 0.0349 2410 | 0.0499 2410 |
| 25 | 1.4509 4535 | 0.6892 0583 | 30.0630 2361 | 20.7196 1120 | 0.0332 6345 | 0.0482 6345 |
| 26 | 1.4727 0953 | 0.6790 2052 | 31.5139 6896 | 21.3986 3172 | 0.0317 3196 | 0.0467 3196 |
| 27 | 1.4948 0018 | 0.6689 8574 | 32.9866 7850 | 22.0676 1746 | 0.0303 1527 | 0.0453 1527 |
| 28 | 1.5172 2218 | 0.6590 9925 | 34.4814 7867 | 22.7267 1671 | 0.0290 0108 | 0.0440 0108 |
| 29 | 1.5399 8051 | 0.6493 5887 | 35.9987 0085 | 23.3760 7558 | 0.0277 7878 | 0.0427 7878 |
| 30 | 1.5630 8022 | 0.6397 6243 | 37.5386 8137 | 24.0158 3801 | 0.0266 3919 | 0.0416 3919 |
| 31 | 1.5865 2642 | 0.6303 0781 | 39.1017 6159 | 24.6461 4582 | 0.0255 7430 | 0.0405 7430 |
| 32 | 1.6103 2432 | 0.6209 9292 | 40.6882 8801 | 25.2671 3874 | 0.0245 7710 | 0.0395 7710 |
| 33 | 1.6344 7918 | 0.6118 1568 | 42.2986 1233 | 25.8789 5442 | 0.0236 4144 | 0.0386 4144 |
| 34 | 1.6589 9637 | 0.6027 7407 | 43.9330 9152 | 26.4817 2849 | 0.0227 6189 | 0.0377 6189 |
| 35 | 1.6838 8132 | 0.5938 6608 | 45.5920 8789 | 27.0755 9458 | 0.0219 3363 | 0.0369 3363 |
| 36 | 1.7091 3954 | 0.5850 8974 | 47.2759 6921 | 27.6606 8431 | 0.0211 5240 | 0.0361 5240 |
| 37 | 1.7347 7663 | 0.5764 4309 | 48.9851 0874 | 28.2371 2740 | 0.0204 1437 | 0.0354 1437 |
| 38 | 1.7607 9828 | 0.5679 2423 | 50.7198 8538 | 28.8050 5163 | 0.0197 1613 | 0.0347 1613 |
| 39 | 1.7872 1025 | 0.5595 3126 | 52.4806 8366 | 29.3645 8288 | 0.0190 5463 | 0.0340 5463 |
| 40 | 1.8140 1841 | 0.5512 6232 | 54.2678 9391 | 29.9158 4520 | 0.0184 2710 | 0.0334 2710 |

**Table B.2** *(continued)*

$i = 1\frac{1}{2}\%$

| $n$ | $(1 + i)^n$ | $(1 + i)^{-n}$ | $s_{\overline{n}|i}$ | $a_{\overline{n}|i}$ | $\dfrac{1}{s_{\overline{n}|i}}$ | $\dfrac{1}{a_{\overline{n}|i}}$ |
|---|---|---|---|---|---|---|
| 41 | 1.8412 2868 | 0.5431 1559 | 56.0819 1232 | 30.4589 6079 | 0.0178 3106 | 0.0328 3106 |
| 42 | 1.8688 4712 | 0.5350 8925 | 57.9231 4100 | 30.9940 5004 | 0.0172 6426 | 0.0322 6426 |
| 43 | 1.8968 7982 | 0.5271 8153 | 59.7919 8812 | 31.5212 3157 | 0.0167 2465 | 0.0317 2465 |
| 44 | 1.9253 3302 | 0.5193 9067 | 61.6888 6794 | 32.0406 2223 | 0.0162 1038 | 0.0312 1038 |
| 45 | 1.9542 1301 | 0.5117 1494 | 63.6142 0096 | 32.5523 3718 | 0.0157 1976 | 0.0307 1976 |
| 46 | 1.9835 2621 | 0.5041 5265 | 65.5684 1398 | 33.0564 8983 | 0.0152 5125 | 0.0302 5125 |
| 47 | 2.0132 7910 | 0.4967 0212 | 67.5519 4018 | 33.5531 9195 | 0.0148 0342 | 0.0298 0342 |
| 48 | 2.0434 7829 | 0.4893 6170 | 69.5652 1929 | 34.0425 5365 | 0.0143 7500 | 0.0293 7500 |
| 49 | 2.0741 3046 | 0.4821 2975 | 71.6086 9758 | 34.5246 8339 | 0.0139 6478 | 0.0289 6478 |
| 50 | 2.1052 4242 | 0.4750 0468 | 73.6828 2804 | 34.9996 8807 | 0.0135 7168 | 0.0285 7168 |
| 51 | 2.1368 2106 | 0.4679 8491 | 75.7880 7046 | 35.4676 7298 | 0.0131 9469 | 0.0281 9469 |
| 52 | 2.1688 7337 | 0.4610 6887 | 77.9248 9152 | 35.9287 4185 | 0.0128 3287 | 0.0278 3287 |
| 53 | 2.2014 0647 | 0.4542 5505 | 80.0937 6489 | 36.3829 9690 | 0.0124 8537 | 0.0274 8537 |
| 54 | 2.2344 2757 | 0.4475 4192 | 82.2951 7136 | 36.8305 3882 | 0.0121 5138 | 0.0271 5138 |
| 55 | 2.2679 4398 | 0.4409 2800 | 84.5295 9893 | 37.2714 6681 | 0.0118 3018 | 0.0268 3018 |
| 56 | 2.3019 6314 | 0.4344 1182 | 86.7975 4292 | 37.7058 7863 | 0.0115 2106 | 0.0265 2106 |
| 57 | 2.3364 9259 | 0.4279 9194 | 89.0995 0606 | 38.1338 7058 | 0.0112 2341 | 0.0262 2341 |
| 58 | 2.3715 3998 | 0.4216 6694 | 91.4359 9865 | 38.5555 3751 | 0.0109 3661 | 0.0259 3661 |
| 59 | 2.4071 1308 | 0.4154 3541 | 93.8075 3863 | 38.9709 7292 | 0.0106 6012 | 0.0256 6012 |
| 60 | 2.4432 1978 | 0.4092 9597 | 96.2146 5171 | 39.3802 6889 | 0.0103 9343 | 0.0253 9343 |
| 61 | 2.4798 6807 | 0.4032 4726 | 98.6578 7149 | 39.7835 1614 | 0.0101 3604 | 0.0251 3604 |
| 62 | 2.5170 6609 | 0.3972 8794 | 101.1377 3956 | 40.1808 0408 | 0.0098 8751 | 0.0248 8751 |
| 63 | 2.5548 2208 | 0.3914 1669 | 103.6548 0565 | 40.5722 2077 | 0.0096 4741 | 0.0246 4741 |
| 64 | 2.5931 4442 | 0.3856 3221 | 106.2096 2774 | 40.9578 5298 | 0.0094 1534 | 0.0244 1534 |
| 65 | 2.6320 4158 | 0.3799 3321 | 108.8027 7215 | 41.3377 8618 | 0.0091 9094 | 0.0241 9094 |
| 66 | 2.6715 2221 | 0.3743 1843 | 111.4348 1374 | 41.7121 0461 | 0.0089 7386 | 0.0239 7386 |
| 67 | 2.7115 9504 | 0.3687 8663 | 114.1063 3594 | 42.0808 9125 | 0.0087 6376 | 0.0237 6376 |
| 68 | 2.7522 6896 | 0.3633 3658 | 116.8179 3098 | 42.4442 2783 | 0.0085 6033 | 0.0235 6033 |
| 69 | 2.7935 5300 | 0.3579 6708 | 119.5701 9995 | 42.8021 9490 | 0.0083 6329 | 0.0233 6329 |
| 70 | 2.8354 5629 | 0.3526 7692 | 122.3637 5295 | 43.1548 7183 | 0.0081 7235 | 0.0231 7235 |
| 71 | 2.8779 8814 | 0.3474 6495 | 125.1992 0924 | 43.5023 3678 | 0.0079 8727 | 0.0229 8727 |
| 72 | 2.9211 5796 | 0.3423 3000 | 128.0771 9738 | 43.8446 6677 | 0.0078 0779 | 0.0228 0779 |
| 73 | 2.9649 7533 | 0.3372 7093 | 130.9983 5534 | 44.1819 3771 | 0.0076 3368 | 0.0226 3368 |
| 74 | 3.0094 4996 | 0.3322 8663 | 133.9633 3067 | 44.5142 2434 | 0.0074 6473 | 0.0224 6473 |
| 75 | 3.0545 9171 | 0.3273 7599 | 136.9727 8063 | 44.8416 0034 | 0.0073 0072 | 0.0223 0072 |
| 76 | 3.1004 1059 | 0.3225 3793 | 140.0273 7234 | 45.1641 3826 | 0.0071 4146 | 0.0221 4146 |
| 77 | 3.1469 1674 | 0.3177 7136 | 143.1277 8292 | 45.4819 0962 | 0.0069 8676 | 0.0219 8676 |
| 78 | 3.1941 2050 | 0.3130 7523 | 146.2746 9967 | 45.7949 8485 | 0.0068 3645 | 0.0218 3645 |
| 79 | 3.2420 3230 | 0.3084 4850 | 149.4688 2016 | 46.1034 3335 | 0.0066 9036 | 0.0216 9036 |
| 80 | 3.2906 6279 | 0.3038 9015 | 152.7108 5247 | 46.4073 2349 | 0.0065 4832 | 0.0215 4832 |

**Table B.2** *(continued)*

$i = 1\frac{1}{2}\%$

| $n$ | $(1+i)^n$ | $(1+i)^{-n}$ | $s_{n\rceil i}$ | $a_{n\rceil i}$ | $\dfrac{1}{s_{n\rceil i}}$ | $\dfrac{1}{a_{n\rceil i}}$ |
|---|---|---|---|---|---|---|
| 81 | 3.3400 2273 | 0.2993 9916 | 156.0015 1525 | 46.7067 2265 | 0.0064 1019 | 0.0214 1019 |
| 82 | 3.3901 2307 | 0.2949 7454 | 159.3415 3798 | 47.0016 9720 | 0.0062 7583 | 0.0212 7583 |
| 83 | 3.4409 7492 | 0.2906 1531 | 162.7316 6105 | 47.2923 1251 | 0.0061 4509 | 0.0211 4509 |
| 84 | 3.4925 8954 | 0.2863 2050 | 166.1726 3597 | 47.5786 1784 | 0.0060 1784 | 0.0210 1784 |
| 85 | 3.5449 7838 | 0.2820 8917 | 169.6652 2551 | 47.8607 2218 | 0.0058 9396 | 0.0208 9396 |
| 86 | 3.5981 5306 | 0.2779 2036 | 173.2101 0389 | 48.1386 4254 | 0.0057 7333 | 0.0207 7333 |
| 87 | 3.6521 2535 | 0.2738 1316 | 176.8083 5695 | 48.4124 5571 | 0.0056 5584 | 0.0206 5584 |
| 88 | 3.7069 0723 | 0.2697 6666 | 180.4604 8230 | 48.6822 2237 | 0.0055 4138 | 0.0205 4138 |
| 89 | 3.7625 1084 | 0.2657 7996 | 184.1673 8954 | 48.9480 0234 | 0.0054 2984 | 0.0204 2984 |
| 90 | 3.8189 4851 | 0.2618 5218 | 187.9299 0038 | 49.2098 5452 | 0.0053 2113 | 0.0203 2113 |
| 91 | 3.8762 3273 | 0.2579 8245 | 191.7488 4889 | 49.4678 3696 | 0.0052 1516 | 0.0202 1516 |
| 92 | 3.9343 7622 | 0.2541 6990 | 195.6250 8162 | 49.7220 0686 | 0.0051 1182 | 0.0201 1182 |
| 93 | 3.9933 9187 | 0.2504 1369 | 199.5594 5784 | 49.9724 2055 | 0.0050 1104 | 0.0200 1104 |
| 94 | 4.0532 9275 | 0.2467 1300 | 203.5528 4971 | 50.2191 3355 | 0.0049 1273 | 0.0199 1273 |
| 95 | 4.1140 9214 | 0.2430 6699 | 207.6061 4246 | 50.4622 0054 | 0.0048 1681 | 0.0198 1681 |
| 96 | 4.1758 0352 | 0.2394 7487 | 211.7202 3459 | 50.7016 7541 | 0.0047 2321 | 0.0197 2321 |
| 97 | 4.2384 4057 | 0.2359 3583 | 215.8960 3811 | 50.9376 1124 | 0.0046 3186 | 0.0196 3186 |
| 98 | 4.3020 1718 | 0.2324 4909 | 220.1344 7868 | 51.1700 6034 | 0.0045 4268 | 0.0195 4268 |
| 99 | 4.3665 4744 | 0.2290 1389 | 224.4364 9586 | 51.3990 7422 | 0.0044 5560 | 0.0194 5560 |
| 100 | 4.4320 4565 | 0.2256 2944 | 228.8030 4330 | 51.6247 0367 | 0.0043 7057 | 0.0193 7057 |

$i = 2\%$

| $n$ | $(1+i)^n$ | $(1+i)^{-n}$ | $s_{n\rceil i}$ | $a_{n\rceil i}$ | $\dfrac{1}{s_{n\rceil i}}$ | $\dfrac{1}{a_{n\rceil i}}$ |
|---|---|---|---|---|---|---|
| 1 | 1.0200 0000 | 0.9803 9216 | 1.0000 0000 | 0.9803 9216 | 1.0000 0000 | 1.0200 0000 |
| 2 | 1.0404 0000 | 0.9611 6878 | 2.0200 0000 | 1.9415 6094 | 0.4950 4950 | 0.5150 4950 |
| 3 | 1.0612 0800 | 0.9423 2233 | 3.0604 0000 | 2.8838 8327 | 0.3267 5467 | 0.3467 5467 |
| 4 | 1.0824 3216 | 0.9238 4543 | 4.1216 0800 | 3.8077 2870 | 0.2426 2375 | 0.2626 2375 |
| 5 | 1.1040 8080 | 0.9057 3081 | 5.2040 4016 | 4.7134 5951 | 0.1921 5839 | 0.2121 5839 |
| 6 | 1.1261 6242 | 0.8879 7138 | 6.3081 2096 | 5.6014 3089 | 0.1585 2581 | 0.1785 2581 |
| 7 | 1.1486 8567 | 0.8705 6018 | 7.4342 8338 | 6.4719 9107 | 0.1345 1196 | 0.1545 1196 |
| 8 | 1.1716 5938 | 0.8534 9037 | 8.5829 6905 | 7.3254 8144 | 0.1165 0980 | 0.1365 0980 |
| 9 | 1.1950 9257 | 0.8367 5527 | 9.7546 2843 | 8.1622 3671 | 0.1025 1544 | 0.1225 1544 |
| 10 | 1.2189 9442 | 0.8203 4830 | 10.9497 2100 | 8.9825 8501 | 0.0913 2653 | 0.1113 2653 |
| 11 | 1.2433 7431 | 0.8042 6304 | 12.1687 1542 | 9.7868 4805 | 0.0821 7794 | 0.1021 7794 |
| 12 | 1.2682 4179 | 0.7884 9318 | 13.4120 8973 | 10.5753 4122 | 0.0745 5960 | 0.0945 5960 |
| 13 | 1.2936 0663 | 0.7730 3253 | 14.6803 3152 | 11.3483 7375 | 0.0681 1835 | 0.0881 1835 |
| 14 | 1.3194 7876 | 0.7578 7502 | 15.9739 3815 | 12.1062 4877 | 0.0626 0197 | 0.0826 0197 |
| 15 | 1.3458 6834 | 0.7430 1473 | 17.2934 1692 | 12.8492 6350 | 0.0578 2547 | 0.0778 2547 |

**Table B.2** *(continued)*

$i = 1\frac{1}{2}\%$  $2\%$

| $n$ | $(1 + i)^n$ | $(1 + i)^{-n}$ | $s_{\overline{n}\mid i}$ | $a_{\overline{n}\mid i}$ | $\dfrac{1}{s_{\overline{n}\mid i}}$ | $\dfrac{1}{a_{\overline{n}\mid i}}$ |
|---|---|---|---|---|---|---|
| 16 | 1.3727 8571 | 0.7284 4581 | 18.6392 8525 | 13.5777 0931 | 0.0536 5013 | 0.0736 5013 |
| 17 | 1.4002 4142 | 0.7141 6256 | 20.0120 7096 | 14.2918 7188 | 0.0499 6984 | 0.0699 6984 |
| 18 | 1.4282 4625 | 0.7001 5937 | 21.4123 1238 | 14.9920 3125 | 0.0467 0210 | 0.0667 0210 |
| 19 | 1.4568 1117 | 0.6864 3076 | 22.8405 5863 | 15.6784 6201 | 0.0437 8177 | 0.0637 8177 |
| 20 | 1.4859 4740 | 0.6729 7133 | 24.2973 6980 | 16.3514 3334 | 0.0411 5672 | 0.0611 5672 |
| 21 | 1.5156 6634 | 0.6597 7582 | 25.7833 1719 | 17.0112 0916 | 0.0387 8477 | 0.0587 8477 |
| 22 | 1.5459 7967 | 0.6468 3904 | 27.2989 8354 | 17.6580 4820 | 0.0366 3140 | 0.0566 3140 |
| 23 | 1.5768 9926 | 0.6341 5592 | 28.8449 6321 | 18.2922 0412 | 0.0346 6810 | 0.0546 6810 |
| 24 | 1.6084 3725 | 0.6217 2149 | 30.4218 6247 | 18.9139 2560 | 0.0328 7110 | 0.0528 7110 |
| 25 | 1.6406 0599 | 0.6095 3087 | 32.0302 9972 | 19.5234 5647 | 0.0312 2044 | 0.0512 2044 |
| 26 | 1.6734 1811 | 0.5975 7928 | 33.6709 0572 | 20.1210 3576 | 0.0296 9923 | 0.0496 9923 |
| 27 | 1.7068 8648 | 0.5858 6204 | 35.3443 2383 | 20.7068 9780 | 0.0282 9309 | 0.0482 9309 |
| 28 | 1.7410 2421 | 0.5743 7455 | 37.0512 1031 | 21.2812 7236 | 0.0269 8967 | 0.0469 8967 |
| 29 | 1.7758 4469 | 0.5631 1231 | 38.7922 3451 | 21.8443 8466 | 0.0257 7836 | 0.0457 7836 |
| 30 | 1.8113 6158 | 0.5520 7089 | 40.5680 7921 | 22.3964 5555 | 0.0246 4992 | 0.0446 4992 |
| 31 | 1.8475 8882 | 0.5412 4597 | 42.3794 4079 | 22.9377 0152 | 0.0235 9635 | 0.0435 9635 |
| 32 | 1.8845 4059 | 0.5306 3330 | 44.2270 2961 | 23.4683 3482 | 0.0226 1061 | 0.0426 1061 |
| 33 | 1.9222 3140 | 0.5202 2873 | 46.1115 7020 | 23.9885 6355 | 0.0216 8653 | 0.0416 8653 |
| 34 | 1.9606 7603 | 0.5100 2817 | 48.0338 0160 | 24.4985 9172 | 0.0208 1867 | 0.0408 1867 |
| 35 | 1.9998 8955 | 0.5000 2761 | 49.9944 7763 | 24.9986 1933 | 0.0200 0221 | 0.0400 0221 |
| 36 | 2.0398 8734 | 0.4902 2315 | 51.9943 6719 | 25.4888 4248 | 0.0192 3285 | 0.0392 3285 |
| 37 | 2.0806 8509 | 0.4806 1093 | 54.0342 5453 | 25.9694 5341 | 0.0185 0678 | 0.0385 0678 |
| 38 | 2.1222 9879 | 0.4711 8719 | 56.1149 3962 | 26.4406 4060 | 0.0178 2057 | 0.0378 2057 |
| 39 | 2.1647 4477 | 0.4619 4822 | 58.2372 3841 | 26.9025 8883 | 0.0171 7114 | 0.0371 7114 |
| 40 | 2.2080 3966 | 0.4528 9042 | 60.4019 8318 | 27.3554 7924 | 0.0165 5575 | 0.0365 5575 |
| 41 | 2.2522 0046 | 0.4440 1021 | 62.6100 2284 | 27.7994 8945 | 0.0159 7188 | 0.0359 7188 |
| 42 | 2.2972 4447 | 0.4353 0413 | 64.8622 2330 | 28.2347 9358 | 0.0154 1729 | 0.0354 1729 |
| 43 | 2.3431 8936 | 0.4267 6875 | 67.1594 6777 | 28.6615 6233 | 0.0148 8993 | 0.0348 8993 |
| 44 | 2.3900 5314 | 0.4184 0074 | 69.5026 5712 | 29.0799 6307 | 0.0143 8794 | 0.0343 8794 |
| 45 | 2.4378 5421 | 0.4101 9680 | 71.8927 1027 | 29.4901 5987 | 0.0139 0962 | 0.0339 0962 |
| 46 | 2.4866 1129 | 0.4021 5373 | 74.3305 6447 | 29.8923 1360 | 0.0134 5342 | 0.0334 5342 |
| 47 | 2.5363 4352 | 0.3942 6836 | 76.8171 7576 | 30.2865 8196 | 0.0130 1792 | 0.0330 1792 |
| 48 | 2.5870 7039 | 0.3865 3761 | 79.3535 1927 | 30.6731 1957 | 0.0126 0184 | 0.0326 0184 |
| 49 | 2.6388 1179 | 0.3789 5844 | 81.9405 8966 | 31.0520 7801 | 0.0122 0396 | 0.0322 0396 |
| 50 | 2.6915 8803 | 0.3715 2788 | 84.5794 0145 | 31.4236 0589 | 0.0118 2321 | 0.0318 2321 |
| 51 | 2.7454 1979 | 0.3642 4302 | 87.2709 8948 | 31.7878 4892 | 0.0114 5856 | 0.0314 5856 |
| 52 | 2.8003 2819 | 0.3571 0100 | 90.0164 0927 | 32.1449 4992 | 0.0111 0909 | 0.0311 0909 |
| 53 | 2.8563 3475 | 0.3500 9902 | 92.8167 3746 | 32.4950 4894 | 0.0107 7392 | 0.0307 7392 |
| 54 | 2.9134 6144 | 0.3432 3433 | 95.6730 7221 | 32.8382 8327 | 0.0104 5226 | 0.0304 5226 |
| 55 | 2.9717 3067 | 0.3365 0425 | 98.5865 3365 | 33.1747 8752 | 0.0101 4337 | 0.0301 4337 |

**Table B.2** *(continued)*

$i = 2\%$

| $n$ | $(1 + i)^n$ | $(1 + i)^{-n}$ | $s_{n|i}$ | $a_{n|i}$ | $\dfrac{1}{s_{n|i}}$ | $\dfrac{1}{a_{n|i}}$ |
|---|---|---|---|---|---|---|
| 56 | 3.0311 6529 | 0.3299 0613 | 101.5582 6432 | 33.5046 9365 | 0.0098 4656 | 0.0298 4656 |
| 57 | 3.0917 8859 | 0.3234 3738 | 104.5894 2961 | 33.8281 3103 | 0.0095 6120 | 0.0295 6120 |
| 58 | 3.1536 2436 | 0.3170 9547 | 107.6812 1820 | 34.1452 2650 | 0.0092 8667 | 0.0292 8667 |
| 59 | 3.2166 9685 | 0.3108 7791 | 110.8348 4257 | 34.4561 0441 | 0.0090 2243 | 0.0290 2243 |
| 60 | 3.2810 3079 | 0.3047 8227 | 114.0515 3942 | 34.7608 8668 | 0.0087 6797 | 0.0287 6797 |
| 61 | 3.3466 5140 | 0.2988 0614 | 117.3325 7021 | 35.0596 9282 | 0.0085 2278 | 0.0285 2278 |
| 62 | 3.4135 8443 | 0.2929 4720 | 120.6792 2161 | 35.3526 4002 | 0.0082 8643 | 0.0282 8643 |
| 63 | 3.4818 5612 | 0.2872 0314 | 124.0928 0604 | 35.6398 4316 | 0.0080 5848 | 0.0280 5848 |
| 64 | 3.5514 9324 | 0.2815 7170 | 127.5746 6216 | 35.9214 1486 | 0.0078 3855 | 0.0278 3855 |
| 65 | 3.6225 2311 | 0.2760 5069 | 131.1261 5541 | 36.1974 6555 | 0.0076 2624 | 0.0276 2624 |
| 66 | 3.6949 7357 | 0.2706 3793 | 134.7486 7852 | 36.4681 0348 | 0.0074 2122 | 0.0274 2122 |
| 67 | 3.7688 7304 | 0.2653 3130 | 138.4436 5209 | 36.7334 3478 | 0.0072 2316 | 0.0272 2316 |
| 68 | 3.8442 5050 | 0.2601 2873 | 142.2125 2513 | 36.9935 6351 | 0.0070 3173 | 0.0270 3173 |
| 69 | 3.9211 3551 | 0.2550 2817 | 146.0567 7563 | 37.2485 9168 | 0.0068 4665 | 0.0268 4665 |
| 70 | 3.9995 5822 | 0.2500 2761 | 149.9779 1114 | 37.4986 1929 | 0.0066 6765 | 0.0266 6765 |
| 71 | 4.0795 4939 | 0.2451 2511 | 153.9774 6937 | 37.7437 4441 | 0.0064 9446 | 0.0264 9446 |
| 72 | 4.1611 4038 | 0.2403 1874 | 158.0570 1875 | 37.9840 6314 | 0.0063 2683 | 0.0263 2683 |
| 73 | 4.2443 6318 | 0.2356 0661 | 162.2181 5913 | 38.2196 6975 | 0.0061 6454 | 0.0261 6454 |
| 74 | 4.3292 5045 | 0.2309 8687 | 166.4625 2231 | 38.4506 5662 | 0.0060 0736 | 0.0260 0736 |
| 75 | 4.4158 3546 | 0.2264 5771 | 170.7917 7276 | 38.6771 1433 | 0.0058 5508 | 0.0258 5508 |
| 76 | 4.5041 5216 | 0.2220 1737 | 175.2076 0821 | 38.8991 3170 | 0.0057 0751 | 0.0257 0751 |
| 77 | 4.5942 3521 | 0.2176 6408 | 179.7117 6038 | 39.1167 9578 | 0.0055 6447 | 0.0255 6447 |
| 78 | 4.6861 1991 | 0.2133 9616 | 184.3059 9558 | 39.3301 9194 | 0.0054 2576 | 0.0254 2576 |
| 79 | 4.7798 4231 | 0.2092 1192 | 188.9921 1549 | 39.5394 0386 | 0.0052 9123 | 0.0252 9123 |
| 80 | 4.8754 3916 | 0.2051 0973 | 193.7719 5780 | 39.7445 1359 | 0.0051 6071 | 0.0251 6071 |
| 81 | 4.9729 4794 | 0.2010 8797 | 198.6473 9696 | 39.9456 0156 | 0.0050 3405 | 0.0250 3405 |
| 82 | 5.0724 0690 | 0.1971 4507 | 203.6203 4490 | 40.1427 4663 | 0.0049 1110 | 0.0249 1110 |
| 83 | 5.1738 5504 | 0.1932 7948 | 208.6927 5180 | 40.3360 2611 | 0.0047 9173 | 0.0247 9173 |
| 84 | 5.2773 3214 | 0.1894 8968 | 213.8666 0683 | 40.5255 1579 | 0.0046 7581 | 0.0246 7581 |
| 85 | 5.3828 7878 | 0.1857 7420 | 219.1439 3897 | 40.7112 8999 | 0.0045 6321 | 0.0245 6321 |
| 86 | 5.4905 3636 | 0.1821 3157 | 224.5268 1775 | 40.8934 2156 | 0.0044 5381 | 0.0244 5381 |
| 87 | 5.6003 4708 | 0.1785 6036 | 230.0173 5411 | 41.0719 8192 | 0.0043 4750 | 0.0243 4750 |
| 88 | 5.7123 5402 | 0.1750 5918 | 235.6177 0119 | 41.2470 4110 | 0.0042 4416 | 0.0242 4416 |
| 89 | 5.8266 0110 | 0.1716 2665 | 241.3300 5521 | 41.4186 6774 | 0.0041 4370 | 0.0241 4370 |
| 90 | 5.9431 3313 | 0.1682 6142 | 247.1566 5632 | 41.5869 2916 | 0.0040 4602 | 0.0240 4602 |
| 91 | 6.0619 9579 | 0.1649 6217 | 253.0997 8944 | 41.7518 9133 | 0.0039 5101 | 0.0239 5101 |
| 92 | 6.1832 3570 | 0.1617 2762 | 259.1617 8523 | 41.9136 1895 | 0.0038 5859 | 0.0238 5859 |
| 93 | 6.3069 0042 | 0.1585 5649 | 265.3450 2093 | 42.0721 7545 | 0.0037 6868 | 0.0237 6868 |
| 94 | 6.4330 3843 | 0.1554 4754 | 271.6519 2135 | 42.2276 2299 | 0.0036 8118 | 0.0236 8118 |
| 95 | 6.5616 9920 | 0.1523 9955 | 278.0849 5978 | 42.3800 2254 | 0.0035 9602 | 0.0235 9602 |

**Table B.2** *(continued)*

$i = 2\%$

| $n$ | $(1 + i)^n$ | $(1 + i)^{-n}$ | $s_{n\rceil i}$ | $a_{n\rceil i}$ | $\dfrac{1}{s_{n\rceil i}}$ | $\dfrac{1}{a_{n\rceil i}}$ |
|---|---|---|---|---|---|---|
| 96 | 6.6929 3318 | 0.1494 1132 | 284.6466 5898 | 42.5294 3386 | 0.0035 1313 | 0.0235 1313 |
| 97 | 6.8267 9184 | 0.1464 8169 | 291.3395 9216 | 42.6759 1555 | 0.0034 3242 | 0.0234 3242 |
| 98 | 6.9633 2768 | 0.1436 0950 | 298.1663 8400 | 42.8195 2505 | 0.0033 5383 | 0.0233 5383 |
| 99 | 7.1025 9423 | 0.1407 9363 | 305.1297 1168 | 42.9603 1867 | 0.0032 7729 | 0.0232 7729 |
| 100 | 7.2446 4612 | 0.1380 3297 | 312.2323 0591 | 43.0983 5164 | 0.0032 0274 | 0.0232 0274 |

$i = 2\frac{1}{2}\%$

| $n$ | $(1 + i)^n$ | $(1 + i)^{-n}$ | $s_{n\rceil i}$ | $a_{n\rceil i}$ | $\dfrac{1}{s_{n\rceil i}}$ | $\dfrac{1}{a_{n\rceil i}}$ |
|---|---|---|---|---|---|---|
| 1 | 1.0250 0000 | 0.9756 0976 | 1.0000 0000 | 0.9756 0976 | 1.0000 0000 | 1.0250 0000 |
| 2 | 1.0506 2500 | 0.9518 1440 | 2.0250 0000 | 1.9274 2415 | 0.4938 2716 | 0.5188 2716 |
| 3 | 1.0768 9063 | 0.9285 9941 | 3.0756 2500 | 2.8560 2356 | 0.3251 3717 | 0.3501 3717 |
| 4 | 1.1038 1289 | 0.9059 5064 | 4.1525 1563 | 3.7619 7421 | 0.2408 1788 | 0.2658 1788 |
| 5 | 1.1314 0821 | 0.8838 5429 | 5.2563 2852 | 4.6458 2850 | 0.1902 4686 | 0.2152 4686 |
| 6 | 1.1596 9342 | 0.8622 9687 | 6.3877 3673 | 5.5081 2536 | 0.1565 4997 | 0.1815 4997 |
| 7 | 1.1886 8575 | 0.8412 6524 | 7.5474 3015 | 6.3493 9060 | 0.1324 9543 | 0.1574 9543 |
| 8 | 1.2184 0290 | 0.8207 4657 | 8.7361 1590 | 7.1701 3717 | 0.1144 6735 | 0.1394 6735 |
| 9 | 1.2488 6297 | 0.8007 2836 | 9.9545 1880 | 7.9708 6553 | 0.1004 5689 | 0.1254 5689 |
| 10 | 1.2800 8454 | 0.7811 9840 | 11.2033 8177 | 8.7520 6393 | 0.0892 5876 | 0.1142 5876 |
| 11 | 1.3120 8666 | 0.7621 4478 | 12.4834 6631 | 9.5142 0871 | 0.0801 0596 | 0.1051 0596 |
| 12 | 1.3448 8882 | 0.7435 5589 | 13.7955 5297 | 10.2577 6460 | 0.0724 8713 | 0.0974 8713 |
| 13 | 1.3785 1104 | 0.7254 2038 | 15.1404 4179 | 10.9831 8497 | 0.0660 4827 | 0.0910 4827 |
| 14 | 1.4129 7382 | 0.7077 2720 | 16.5189 5284 | 11.6909 1217 | 0.0605 3652 | 0.0855 3652 |
| 15 | 1.4482 9817 | 0.6904 6556 | 17.9319 2666 | 12.3813 7773 | 0.0557 6646 | 0.0807 6646 |
| 16 | 1.4845 0562 | 0.6736 2493 | 19.3802 2483 | 13.0550 0266 | 0.0515 9899 | 0.0765 9899 |
| 17 | 1.5216 1826 | 0.6571 9506 | 20.8647 3045 | 13.7121 9772 | 0.0479 2777 | 0.0729 2777 |
| 18 | 1.5596 5872 | 0.6411 6591 | 22.3863 4871 | 14.3533 6363 | 0.0446 7008 | 0.0696 7008 |
| 19 | 1.5986 5019 | 0.6255 2772 | 23.9460 0743 | 14.9788 9134 | 0.0417 6062 | 0.0667 6062 |
| 20 | 1.6386 1644 | 0.6102 7094 | 25.5446 5761 | 15.5891 6229 | 0.0391 4713 | 0.0641 4713 |
| 21 | 1.6795 8185 | 0.5953 8629 | 27.1832 7405 | 16.1845 4857 | 0.0367 8733 | 0.0617 8733 |
| 22 | 1.7215 7140 | 0.5808 6467 | 28.8628 5590 | 16.7654 1324 | 0.0346 4661 | 0.0596 4661 |
| 23 | 1.7646 1068 | 0.5666 9724 | 30.5844 2730 | 17.3321 1048 | 0.0326 9638 | 0.0576 9638 |
| 24 | 1.8087 2595 | 0.5528 7535 | 32.3490 3798 | 17.8849 8583 | 0.0309 1282 | 0.0559 1282 |
| 25 | 1.8539 4410 | 0.5393 9059 | 34.1577 6393 | 18.4243 7642 | 0.0292 7592 | 0.0542 7592 |
| 26 | 1.9002 9270 | 0.5262 3472 | 36.0117 0803 | 18.9506 1114 | 0.0277 6875 | 0.0527 6875 |
| 27 | 1.9478 0002 | 0.5133 9973 | 37.9120 0073 | 19.4640 1087 | 0.0263 7687 | 0.0513 7687 |
| 28 | 1.9964 9502 | 0.5008 7778 | 39.8598 0075 | 19.9648 8866 | 0.0250 8793 | 0.0500 8793 |
| 29 | 2.0464 0739 | 0.4886 6125 | 41.8562 9577 | 20.4535 4991 | 0.0238 9127 | 0.0488 9127 |
| 30 | 2.0975 6758 | 0.4767 4269 | 43.9027 0316 | 20.9302 9259 | 0.0227 7764 | 0.0477 7764 |

**Table B.2** *(continued)*

$i = 2\frac{1}{2}\%$

| $n$ | $(1 + i)^n$ | $(1 + i)^{-n}$ | $s_{n|i}$ | $a_{n|i}$ | $\dfrac{1}{s_{n|i}}$ | $\dfrac{1}{a_{n|i}}$ |
|---|---|---|---|---|---|---|
| 31 | 2.1500 0677 | 0.4651 1481 | 46.0002 7074 | 21.3954 0741 | 0.0217 3900 | 0.0467 3900 |
| 32 | 2.2037 5694 | 0.4537 7055 | 48.1502 7751 | 21.8491 7796 | 0.0207 6831 | 0.0457 6831 |
| 33 | 2.2588 5086 | 0.4427 0298 | 50.3540 3445 | 22.2918 8094 | 0.0198 5938 | 0.0448 5938 |
| 34 | 2.3153 2213 | 0.4319 0534 | 52.6128 8531 | 22.7237 8628 | 0.0190 0675 | 0.0440 0675 |
| 35 | 2.3732 0519 | 0.4213 7107 | 54.9282 0744 | 23.1451 5734 | 0.0182 0558 | 0.0432 0558 |
| 36 | 2.4325 3532 | 0.4110 9372 | 57.3014 1263 | 23.5562 5107 | 0.0174 5158 | 0.0424 5158 |
| 37 | 2.4933 4870 | 0.4010 6705 | 59.7339 4794 | 23.9573 1812 | 0.0167 4090 | 0.0417 4090 |
| 38 | 2.5556 8242 | 0.3912 8492 | 62.2272 9664 | 24.3486 0304 | 0.0160 7012 | 0.0410 7012 |
| 39 | 2.6195 7448 | 0.3817 4139 | 64.7829 7906 | 24.7303 4443 | 0.0154 3615 | 0.0404 3615 |
| 40 | 2.6850 6384 | 0.3724 3062 | 67.4025 5354 | 25.1027 7505 | 0.0148 3623 | 0.0398 3623 |
| 41 | 2.7521 9043 | 0.3633 4695 | 70.0876 1737 | 25.4661 2200 | 0.0142 6786 | 0.0392 6786 |
| 42 | 2.8209 9520 | 0.3544 8483 | 72.8398 0781 | 25.8206 0683 | 0.0137 2876 | 0.0387 2876 |
| 43 | 2.8915 2008 | 0.3458 3886 | 75.6608 0300 | 26.1664 4569 | 0.0132 1688 | 0.0382 1688 |
| 44 | 2.9638 0808 | 0.3374 0376 | 78.5523 2308 | 26.5038 4945 | 0.0127 3037 | 0.0377 3037 |
| 45 | 3.0379 0328 | 0.3291 7440 | 81.5161 3116 | 26.8330 2386 | 0.0122 6751 | 0.0372 6751 |
| 46 | 3.1138 5086 | 0.3211 4576 | 84.5540 3443 | 27.1541 6962 | 0.0118 2676 | 0.0368 2676 |
| 47 | 3.1916 9713 | 0.3133 1294 | 87.6678 8530 | 27.4674 8255 | 0.0114 0669 | 0.0364 0669 |
| 48 | 3.2714 8956 | 0.3056 7116 | 90.8595 8243 | 27.7731 5371 | 0.0110 0599 | 0.0360 0599 |
| 49 | 3.3532 7680 | 0.2982 1576 | 94.1310 7199 | 28.0713 6947 | 0.0106 2348 | 0.0356 2348 |
| 50 | 3.4371 0872 | 0.2909 4221 | 97.4843 4879 | 28.3623 1168 | 0.0102 5806 | 0.0352 5806 |
| 51 | 3.5230 3664 | 0.2838 4606 | 100.9214 5751 | 28.6461 5774 | 0.0099 0870 | 0.0349 0870 |
| 52 | 3.6111 1235 | 0.2769 2298 | 104.4444 9395 | 28.9230 8072 | 0.0095 7446 | 0.0345 7446 |
| 53 | 3.7013 9016 | 0.2701 6876 | 108.0556 0629 | 29.1932 4948 | 0.0092 5449 | 0.0342 5449 |
| 54 | 3.7939 2491 | 0.2635 7928 | 111.7569 9645 | 29.4568 2876 | 0.0089 4799 | 0.0339 4799 |
| 55 | 3.8887 7303 | 0.2571 5052 | 115.5509 2136 | 29.7139 7928 | 0.0086 5419 | 0.0336 5419 |
| 56 | 3.9859 9236 | 0.2508 7855 | 119.4396 9440 | 29.9648 5784 | 0.0083 7243 | 0.0333 7243 |
| 57 | 4.0856 4217 | 0.2447 5956 | 123.4256 8676 | 30.2096 1740 | 0.0081 0204 | 0.0331 0204 |
| 58 | 4.1877 8322 | 0.2387 8982 | 127.5113 2893 | 30.4484 0722 | 0.0078 4244 | 0.0328 4244 |
| 59 | 4.2924 7780 | 0.2329 6568 | 131.6991 1215 | 30.6813 7290 | 0.0075 9307 | 0.0325 9307 |
| 60 | 4.3997 8975 | 0.2272 8359 | 135.9915 8995 | 30.9086 5649 | 0.0073 5340 | 0.0323 5340 |
| 61 | 4.5097 8449 | 0.2217 4009 | 140.3913 7970 | 31.1303 9657 | 0.0071 2294 | 0.0321 2294 |
| 62 | 4.6225 2910 | 0.2163 3179 | 144.9011 6419 | 31.3467 2836 | 0.0069 0126 | 0.0319 0126 |
| 63 | 4.7380 9233 | 0.2110 5541 | 149.5236 9330 | 31.5577 8377 | 0.0066 8790 | 0.0316 8790 |
| 64 | 4.8565 4464 | 0.2059 0771 | 154.2617 8563 | 31.7636 9148 | 0.0064 8249 | 0.0314 8249 |
| 65 | 4.9779 5826 | 0.2008 8557 | 159.1183 3027 | 31.9645 7705 | 0.0062 8463 | 0.0312 8463 |
| 66 | 5.1024 0721 | 0.1959 8593 | 164.0962 8853 | 33.1605 6298 | 0.0060 9398 | 0.0310 9398 |
| 67 | 5.2299 6739 | 0.1912 0578 | 169.1986 9574 | 32.3517 6876 | 0.0059 1021 | 0.0309 1021 |
| 68 | 5.3607 1658 | 0.1865 4223 | 174.4286 6314 | 32.5383 1099 | 0.0057 3300 | 0.0307 3300 |
| 69 | 5.4947 3449 | 0.1819 9241 | 179.7893 7971 | 32.7203 0340 | 0.0055 6206 | 0.0305 6206 |
| 70 | 5.6321 0286 | 0.1775 5358 | 185.2841 1421 | 32.8978 5698 | 0.0053 9712 | 0.0303 9712 |

**Table B.2** *(continued)*

$i = 2\frac{1}{2}\%$

| $n$ | $(1 + i)^n$ | $(1 + i)^{-n}$ | $s_{\overline{n}|i}$ | $a_{\overline{n}|i}$ | $\dfrac{1}{s_{\overline{n}|i}}$ | $\dfrac{1}{a_{\overline{n}|i}}$ |
|---|---|---|---|---|---|---|
| 71 | 5.7729 0543 | 0.1732 2300 | 190.9162 1706 | 33.0710 7998 | 0.0052 3790 | 0.0302 3790 |
| 72 | 5.9172 2806 | 0.1689 9805 | 196.6891 2249 | 33.2400 7803 | 0.0050 8417 | 0.0300 8417 |
| 73 | 6.0651 5876 | 0.1648 7615 | 202.6063 5055 | 33.4049 5417 | 0.0049 3568 | 0.0299 3568 |
| 74 | 6.2167 8773 | 0.1608 5478 | 208.6715 0931 | 33.5658 0895 | 0.0047 9222 | 0.0297 9222 |
| 75 | 6.3722 0743 | 0.1569 3149 | 214.8882 9705 | 33.7227 4044 | 0.0046 5358 | 0.0296 5358 |
| 76 | 6.5315 1261 | 0.1531 0389 | 221.2605 0447 | 33.8758 4433 | 0.0045 1956 | 0.0295 1956 |
| 77 | 6.6948 0043 | 0.1493 6965 | 227.7920 1709 | 34.0252 1398 | 0.0043 8997 | 0.0293 8997 |
| 78 | 6.8621 7044 | 0.1457 2649 | 234.4868 1751 | 34.1709 4047 | 0.0042 6463 | 0.0292 6463 |
| 79 | 7.0337 2470 | 0.1421 7218 | 241.3489 8795 | 34.3131 1265 | 0.0041 4338 | 0.0291 4338 |
| 80 | 7.2095 6782 | 0.1387 0457 | 248.3827 1265 | 34.4518 1722 | 0.0040 2605 | 0.0290 2605 |
| 81 | 7.3898 0701 | 0.1353 2153 | 255.5922 8047 | 34.5871 3875 | 0.0039 1248 | 0.0289 1248 |
| 82 | 7.5745 5219 | 0.1320 2101 | 262.9820 8748 | 34.7191 5976 | 0.0038 0254 | 0.0288 0254 |
| 83 | 7.7639 1599 | 0.1288 0098 | 270.5566 3966 | 34.8479 6074 | 0.0036 9608 | 0.0286 9608 |
| 84 | 7.9580 1389 | 0.1256 5949 | 278.3205 5566 | 34.9736 2023 | 0.0035 9298 | 0.0285 9298 |
| 85 | 8.1569 6424 | 0.1225 9463 | 286.2785 6955 | 35.0962 1486 | 0.0034 9310 | 0.0284 9310 |
| 86 | 8.3608 8834 | 0.1196 0452 | 294.4355 3379 | 35.2158 1938 | 0.0033 9633 | 0.0283 9633 |
| 87 | 8.5699 1055 | 0.1166 8733 | 302.7964 2213 | 35.3325 0671 | 0.0033 0255 | 0.0283 0255 |
| 88 | 8.7841 5832 | 0.1138 4130 | 311.3663 3268 | 35.4463 4801 | 0.0032 1165 | 0.0282 1165 |
| 89 | 9.0037 6228 | 0.1110 6468 | 320.1504 9100 | 35.5574 1269 | 0.0031 2353 | 0.0281 2353 |
| 90 | 9.2288 5633 | 0.1083 5579 | 329.1542 5328 | 35.6657 6848 | 0.0030 3809 | 0.0280 3809 |
| 91 | 9.4595 7774 | 0.1057 1296 | 338.3831 0961 | 35.7714 8144 | 0.0029 5523 | 0.0279 5523 |
| 92 | 9.6960 6718 | 0.1031 3460 | 347.8426 8735 | 35.8746 1604 | 0.0028 7486 | 0.0278 7486 |
| 93 | 9.9384 6886 | 0.1006 1912 | 357.5387 5453 | 35.9752 3516 | 0.0027 9690 | 0.0277 9690 |
| 94 | 10.1869 3058 | 0.0981 6500 | 367.4772 2339 | 36.0734 0016 | 0.0027 2126 | 0.0277 2126 |
| 95 | 10.4416 0385 | 0.0957 7073 | 377.6641 5398 | 36.1691 7089 | 0.0026 4786 | 0.0276 4786 |
| 96 | 10.7026 4395 | 0.0934 3486 | 388.1057 5783 | 36.2626 0574 | 0.0025 7662 | 0.0275 7662 |
| 97 | 10.9702 1004 | 0.0911 5596 | 398.8084 0177 | 36.3537 6170 | 0.0025 0747 | 0.0276 0747 |
| 98 | 11.2444 6530 | 0.0889 3264 | 409.7786 1182 | 36.4426 9434 | 0.0024 4034 | 0.0274 4034 |
| 99 | 11.5255 7693 | 0.0867 6355 | 421.0230 7711 | 36.5294 5790 | 0.0023 7517 | 0.0273 7517 |
| 100 | 11.8137 1635 | 0.0846 4737 | 432.5486 5404 | 36.6141 0526 | 0.0023 1188 | 0.0273 1188 |

$i = 3\%$

| $n$ | $(1 + i)^n$ | $(1 + i)^{-n}$ | $s_{\overline{n}|i}$ | $a_{\overline{n}|i}$ | $\dfrac{1}{s_{\overline{n}|i}}$ | $\dfrac{1}{a_{\overline{n}|i}}$ |
|---|---|---|---|---|---|---|
| 1 | 1.0300 0000 | 0.9708 7379 | 1.0000 0000 | 0.9708 7379 | 1.0000 0000 | 1.0300 0000 |
| 2 | 1.0609 0000 | 0.9425 9591 | 2.0300 0000 | 1.9134 6970 | 0.4926 1084 | 0.5226 1084 |
| 3 | 1.0927 2700 | 0.9151 4166 | 3.0909 0000 | 2.8286 1135 | 0.3235 3036 | 0.3535 3036 |
| 4 | 1.1255 0881 | 0.8884 8705 | 4.1836 2700 | 3.7170 9840 | 0.2390 2705 | 0.2690 2705 |
| 5 | 1.1592 7407 | 0.8626 0878 | 5.3091 3581 | 4.5797 0719 | 0.1883 5457 | 0.2183 5457 |

**Table B.2** *(continued)*

$i = 3\%$

| $n$ | $(1 + i)^n$ | $(1 + i)^{-n}$ | $s_{\overline{n}|i}$ | $a_{\overline{n}|i}$ | $\dfrac{1}{s_{\overline{n}|i}}$ | $\dfrac{1}{a_{\overline{n}|i}}$ |
|---|---|---|---|---|---|---|
| 6 | 1.1940 5230 | 0.8374 8426 | 6.4684 0988 | 5.4171 9144 | 0.1545 9750 | 0.1845 9750 |
| 7 | 1.2298 7837 | 0.8130 9151 | 7.6624 6218 | 6.2302 8296 | 0.1305 0635 | 0.1605 0635 |
| 8 | 1.2667 7008 | 0.7894 0923 | 8.8923 3605 | 7.0196 9219 | 0.1124 5639 | 0.1424 5639 |
| 9 | 1.3047 7318 | 0.7664 1673 | 10.1591 0613 | 7.7861 0892 | 0.0984 3386 | 0.1284 3386 |
| 10 | 1.3439 1638 | 0.7440 9391 | 11.4638 7931 | 8.5302 0284 | 0.0872 3051 | 0.1172 3051 |
| 11 | 1.3842 3387 | 0.7224 2128 | 12.8077 9569 | 9.2526 2411 | 0.0780 7745 | 0.1080 7745 |
| 12 | 1.4257 6089 | 0.7013 7988 | 14.1920 2956 | 9.9540 0399 | 0.0704 6209 | 0.1004 6209 |
| 13 | 1.4685 3371 | 0.6809 5134 | 15.6177 9045 | 10.6349 5533 | 0.0640 2954 | 0.0940 2954 |
| 14 | 1.5125 8972 | 0.6611 1781 | 17.0863 2416 | 11.2960 7314 | 0.0585 2634 | 0.0885 2634 |
| 15 | 1.5579 6742 | 0.6418 6195 | 18.5989 1389 | 11.9379 3509 | 0.0537 6658 | 0.0837 6658 |
| 16 | 1.6047 0644 | 0.6231 6694 | 20.1568 8130 | 12.5611 0203 | 0.0496 1085 | 0.0796 1085 |
| 17 | 1.6528 4763 | 0.6050 1645 | 21.7615 8774 | 13.1661 1847 | 0.0459 5253 | 0.0759 5253 |
| 18 | 1.7024 3306 | 0.5873 9461 | 23.4144 3537 | 13.7535 1308 | 0.0427 0870 | 0.0727 0870 |
| 19 | 1.7535 0605 | 0.5702 8603 | 25.1168 6844 | 14.3237 9911 | 0.0398 1388 | 0.0698 1388 |
| 20 | 1.8061 1123 | 0.5536 7575 | 26.8703 7449 | 14.8774 7486 | 0.0372 1571 | 0.0672 1571 |
| 21 | 1.8602 9456 | 0.5375 4928 | 28.6764 8572 | 15.4150 2414 | 0.0348 7178 | 0.0648 7178 |
| 22 | 1.9161 0341 | 0.5218 9520 | 30.5367 8030 | 15.9369 1664 | 0.0327 4739 | 0.0627 4739 |
| 23 | 1.9735 8651 | 0.5066 9175 | 32.4528 8370 | 16.4436 0839 | 0.0308 1390 | 0.0608 1390 |
| 24 | 2.0327 9411 | 0.4919 3374 | 34.4264 7022 | 16.9355 4212 | 0.0290 4742 | 0.0590 4742 |
| 25 | 2.0937 7793 | 0.4776 0557 | 36.4592 6432 | 17.4131 4769 | 0.0274 2787 | 0.0574 2787 |
| 26 | 2.1565 9127 | 0.4636 9473 | 38.5530 4225 | 17.8678 4242 | 0.0259 3829 | 0.0559 3829 |
| 27 | 2.2212 8901 | 0.4501 8906 | 40.7096 3352 | 18.3270 3147 | 0.0245 6421 | 0.0545 6421 |
| 28 | 2.2879 2768 | 0.4370 7675 | 42.9309 2252 | 18.7641 0823 | 0.0232 9323 | 0.0532 9323 |
| 29 | 2.3565 6551 | 0.4243 4636 | 45.2188 5020 | 19.1884 5459 | 0.0221 1467 | 0.0524 1467 |
| 30 | 2.4272 6247 | 0.4119 8676 | 47.5754 1571 | 19.6004 4135 | 0.0210 1926 | 0.0510 1926 |
| 31 | 2.5000 8035 | 0.3999 8715 | 50.0026 7818 | 20.0004 2849 | 0.0199 9893 | 0.0499 9893 |
| 32 | 2.5750 8276 | 0.3883 3703 | 52.5027 5852 | 20.3887 6553 | 0.0190 4662 | 0.0490 4662 |
| 33 | 2.6523 3524 | 0.3770 2625 | 55.0778 4128 | 20.7657 9178 | 0.0181 5612 | 0.0481 5612 |
| 34 | 2.7319 0530 | 0.3660 4490 | 57.7301 7652 | 21.1318 3668 | 0.0173 2196 | 0.0473 2196 |
| 35 | 2.8138 6245 | 0.3553 8340 | 60.4620 8181 | 21.4872 2007 | 0.0165 3929 | 0.0465 3929 |
| 36 | 2.8982 7833 | 0.3450 3243 | 63.2759 4427 | 21.8322 5250 | 0.0158 0379 | 0.0458 0379 |
| 37 | 2.9852 2668 | 0.3349 8294 | 66.1742 2259 | 22.1672 3544 | 0.0151 1162 | 0.0451 1162 |
| 38 | 3.0747 8348 | 0.3252 2615 | 69.1594 4927 | 22.4924 6159 | 0.0144 5934 | 0.0444 5934 |
| 39 | 3.1670 2698 | 0.3157 5355 | 72.2342 3275 | 22.8082 1513 | 0.0138 4385 | 0.0428 4385 |
| 40 | 3.2620 3779 | 0.3065 5684 | 75.4012 5973 | 23.1147 7197 | 0.0132 6238 | 0.0432 6238 |
| 41 | 3.3598 9893 | 0.2976 2800 | 78.6632 9753 | 23.4123 9997 | 0.0127 1241 | 0.0427 1241 |
| 42 | 3.4606 9589 | 0.2889 5922 | 82.0231 9645 | 23.7013 5920 | 0.0121 9167 | 0.0421 9167 |
| 43 | 3.5645 1677 | 0.2805 4294 | 85.4838 9234 | 23.9819 0213 | 0.0116 9811 | 0.0416 9811 |
| 44 | 3.6714 5227 | 0.2723 7178 | 89.0484 0911 | 24.2542 7392 | 0.0112 2985 | 0.0412 2985 |
| 45 | 3.7815 9584 | 0.2644 3862 | 92.7198 6139 | 24.5187 1254 | 0.0107 8518 | 0.0407 8518 |

**Table B.2** *(continued)*

$i = 3\%$

| $n$ | $(1 + i)^n$ | $(1 + i)^{-n}$ | $s_{\overline{n}|i}$ | $a_{\overline{n}|i}$ | $\dfrac{1}{s_{\overline{n}|i}}$ | $\dfrac{1}{a_{\overline{n}|i}}$ |
|---|---|---|---|---|---|---|
| 46 | 3.8950 4372 | 0.2567 3653 | 96.5014 5723 | 24.7754 4907 | 0.0103 6254 | 0.0403 6254 |
| 47 | 4.0118 9503 | 0.2492 5876 | 100.3965 0095 | 25.0247 0783 | 0.0099 6051 | 0.0399 6051 |
| 48 | 4.1322 5188 | 0.2419 9880 | 104.4083 9598 | 25.2667 0664 | 0.0095 7777 | 0.0395 7777 |
| 49 | 4.2562 1944 | 0.2349 5029 | 108.5406 4785 | 25.5016 5693 | 0.0092 1314 | 0.0392 1314 |
| 50 | 4.3839 0602 | 0.2281 0708 | 112.7968 6729 | 25.7297 6401 | 0.0088 6549 | 0.0388 6549 |
| 51 | 4.5154 2320 | 0.2214 6318 | 117.1807 7331 | 25.9512 2719 | 0.0085 3382 | 0.0385 3382 |
| 52 | 4.6508 8590 | 0.2150 1280 | 121.6961 9651 | 26.1662 3999 | 0.0082 1718 | 0.0382 1718 |
| 53 | 4.7904 1247 | 0.2087 5029 | 126.3470 8240 | 26.3749 9028 | 0.0079 1471 | 0.0349 1471 |
| 54 | 4.9341 2485 | 0.2026 7019 | 131.1374 9488 | 26.5776 6047 | 0.0076 2558 | 0.0376 2558 |
| 55 | 5.0821 4859 | 0.1967 6717 | 136.0716 1972 | 26.7744 2764 | 0.0073 4907 | 0.0373 4907 |
| 56 | 5.2346 1305 | 0.1910 3609 | 141.1537 6831 | 26.9654 6373 | 0.0070 8447 | 0.0370 8447 |
| 57 | 5.3916 5144 | 0.1854 8193 | 146.3883 8136 | 27.1509 3566 | 0.0068 3114 | 0.0368 3114 |
| 58 | 5.5534 0098 | 0.1800 6984 | 151.7800 3280 | 27.3310 0549 | 0.0065 8848 | 0.0365 8848 |
| 59 | 5.7200 0301 | 0.1748 2508 | 157.3334 3379 | 27.5058 3058 | 0.0063 5593 | 0.0363 5593 |
| 60 | 5.8916 0310 | 0.1697 3309 | 163.0534 3680 | 27.6755 6367 | 0.0061 3296 | 0.0361 3296 |
| 61 | 6.0683 5120 | 0.1647 8941 | 168.9450 3991 | 27.8403 5307 | 0.0059 1908 | 0.0359 1908 |
| 62 | 6.2504 0173 | 0.1599 8972 | 175.0133 9110 | 28.0003 4279 | 0.0057 1385 | 0.0357 1385 |
| 63 | 6.4379 1379 | 0.1553 2982 | 181.2637 9284 | 28.1556 7261 | 0.0055 1682 | 0.0355 1682 |
| 64 | 6.6310 5120 | 0.1508 0565 | 187.7017 0662 | 28.3064 7826 | 0.0053 2760 | 0.0353 2760 |
| 65 | 6.8299 8273 | 0.1464 1325 | 194.3327 5782 | 28.4528 9152 | 0.0051 4581 | 0.0351 4581 |
| 66 | 7.0348 8222 | 0.1421 4879 | 201.1627 4055 | 28.5950 4031 | 0.0049 7110 | 0.0349 7110 |
| 67 | 7.2459 2868 | 0.1380 0853 | 208.1976 2277 | 28.7330 4884 | 0.0048 0313 | 0.0348 0313 |
| 68 | 7.4633 0654 | 0.1339 8887 | 215.4435 5145 | 28.8670 3771 | 0.0046 4159 | 0.0346 4159 |
| 69 | 7.6872 0574 | 0.1300 8628 | 222.9068 5800 | 28.9971 2399 | 0.0044 8618 | 0.0344 8618 |
| 70 | 7.9178 2191 | 0.1262 9736 | 230.5940 6374 | 29.1234 2135 | 0.0043 3663 | 0.0343 3663 |
| 71 | 8.1553 5657 | 0.1226 1880 | 238.5118 8565 | 29.2460 4015 | 0.0041 9266 | 0.0341 9266 |
| 72 | 8.4000 1727 | 0.1190 4737 | 246.6672 4222 | 29.3650 8752 | 0.0040 5404 | 0.0340 5404 |
| 73 | 8.6520 1778 | 0.1155 7998 | 255.0672 5949 | 29.4806 6750 | 0.0039 2053 | 0.0339 2053 |
| 74 | 8.9115 7832 | 0.1122 1357 | 263.7192 7727 | 29.5928 8107 | 0.0037 9191 | 0.0337 9191 |
| 75 | 9.1789 2567 | 0.1089 4521 | 272.6308 5559 | 29.7018 2628 | 0.0036 6796 | 0.0336 6796 |
| 76 | 9.4542 9344 | 0.1057 7205 | 281.8097 8126 | 29.8075 9833 | 0.0035 4849 | 0.0335 4849 |
| 77 | 9.7379 2224 | 0.1026 9131 | 291.2640 7469 | 29.9102 8964 | 0.0034 3331 | 0.0334 3331 |
| 78 | 10.0300 5991 | 0.0997 0030 | 301.0019 9693 | 30.0099 8994 | 0.0033 2224 | 0.0333 2224 |
| 79 | 10.3309 6171 | 0.0967 9641 | 311.0320 5684 | 30.1067 8635 | 0.0032 1510 | 0.0335 1510 |
| 80 | 10.6408 9056 | 0.0939 7710 | 321.3630 1855 | 30.2007 6345 | 0.0031 1175 | 0.0331 1175 |
| 81 | 10.9601 1727 | 0.0912 3990 | 332.0039 0910 | 30.2920 0335 | 0.0030 1201 | 0.0330 1201 |
| 82 | 11.2889 2079 | 0.0885 8243 | 342.9640 2638 | 30.3805 8577 | 0.0029 1576 | 0.0329 1576 |
| 83 | 11.6275 8842 | 0.0860 0236 | 354.2529 4717 | 30.4665 8813 | 0.0028 2284 | 0.0328 2284 |
| 84 | 11.9764 1607 | 0.0834 9743 | 365.8805 3558 | 30.5500 8556 | 0.0027 3313 | 0.0327 3313 |
| 85 | 12.3357 0855 | 0.0810 6547 | 377.8569 5165 | 30.6311 5103 | 0.0026 4650 | 0.0326 4650 |

**Table B.2** *(continued)*

$i = 3\%$

| $n$ | $(1+i)^n$ | $(1+i)^{-n}$ | $s_{\overline{n}|i}$ | $a_{\overline{n}|i}$ | $\dfrac{1}{s_{\overline{n}|i}}$ | $\dfrac{1}{a_{\overline{n}|i}}$ |
|---|---|---|---|---|---|---|
| 86 | 12.7057 7981 | 0.0787 0434 | 390.1926 6020 | 30.7098 5537 | 0.0025 6284 | 0.0325 6284 |
| 87 | 13.0869 5320 | 0.0764 1198 | 402.8984 4001 | 30.7862 6735 | 0.0024 8202 | 0.0324 8202 |
| 88 | 13.4795 6180 | 0.0741 8639 | 415.9853 9321 | 30.8604 5374 | 0.0024 0393 | 0.0324 0393 |
| 89 | 13.8839 4865 | 0.0720 2562 | 429.4649 5500 | 30.9324 7936 | 0.0023 2848 | 0.0323 2848 |
| 90 | 14.3004 6711 | 0.0699 2779 | 443.3489 0365 | 31.0024 0714 | 0.0022 5556 | 0.0322 5556 |
| 91 | 14.7294 8112 | 0.0678 9105 | 457.6493 7076 | 31.0702 9820 | 0.0021 8508 | 0.0321 8508 |
| 92 | 15.1713 6556 | 0.0659 1364 | 472.3788 5189 | 31.1362 1184 | 0.0021 1694 | 0.0321 1694 |
| 93 | 15.6265 0652 | 0.0639 9383 | 487.5502 1744 | 31.2002 0567 | 0.0020 5107 | 0.0320 5107 |
| 94 | 16.0953 0172 | 0.0621 2993 | 503.1767 2397 | 31.2623 3560 | 0.0019 8737 | 0.0319 8737 |
| 95 | 16.5781 6077 | 0.0603 2032 | 519.2720 2568 | 31.3226 5592 | 0.0019 2577 | 0.0319 2577 |
| 96 | 17.0755 0559 | 0.0585 6342 | 535.8501 8645 | 31.3812 1934 | 0.0018 6619 | 0.0318 6619 |
| 97 | 17.5877 7076 | 0.0568 5769 | 552.9256 9205 | 31.4380 7703 | 0.0018 0856 | 0.0318 0856 |
| 98 | 18.1154 0388 | 0.0552 0164 | 570.5134 6281 | 31.4932 7867 | 0.0017 5281 | 0.0317 5281 |
| 99 | 18.6588 6600 | 0.0535 9383 | 588.6288 6669 | 31.5468 7250 | 0.0016 9886 | 0.0316 9886 |
| 100 | 19.2186 3198 | 0.0520 3284 | 607.2877 3269 | 31.5989 0534 | 0.0016 4667 | 0.0316 4667 |

$i = 3\frac{1}{2}\%$

| $n$ | $(1+i)^n$ | $(1+i)^{-n}$ | $s_{\overline{n}|i}$ | $a_{\overline{n}|i}$ | $\dfrac{1}{s_{\overline{n}|i}}$ | $\dfrac{1}{a_{\overline{n}|i}}$ |
|---|---|---|---|---|---|---|
| 1 | 1.0350 0000 | 0.9661 8357 | 1.0000 0000 | 0.9661 8357 | 1.0000 0000 | 1.0350 0000 |
| 2 | 1.0712 2500 | 0.9335 1070 | 2.0350 0000 | 1.8996 9428 | 0.4914 0049 | 0.5264 0049 |
| 3 | 1.1087 1788 | 0.9019 4271 | 3.1062 2500 | 2.8016 3698 | 0.3219 3418 | 0.3569 3418 |
| 4 | 1.1475 2300 | 0.8714 4223 | 4.2149 4288 | 3.6730 7921 | 0.2372 5114 | 0.2722 5114 |
| 5 | 1.1876 8631 | 0.8419 7317 | 5.3624 6588 | 4.5150 5238 | 0.1864 8137 | 0.2214 8137 |
| 6 | 1.2292 5533 | 0.8135 0064 | 6.5501 5218 | 5.3285 5302 | 0.1526 6821 | 0.1876 6821 |
| 7 | 1.2722 7926 | 0.7859 9096 | 7.7794 0751 | 6.1145 4398 | 0.1285 4449 | 0.1635 4449 |
| 8 | 1.3168 0904 | 0.7594 1156 | 9.0516 8677 | 6.8739 5554 | 0.1104 7665 | 0.1454 7665 |
| 9 | 1.3628 9735 | 0.7337 3097 | 10.3684 9581 | 7.6076 8651 | 0.0964 4601 | 0.1314 4601 |
| 10 | 1.4105 9876 | 0.7089 1881 | 11.7313 9316 | 8.3166 0532 | 0.0852 4137 | 0.1202 4137 |
| 11 | 1.4599 6972 | 0.6849 4571 | 13.1419 9192 | 9.0015 5104 | 0.0760 9197 | 0.1110 9197 |
| 12 | 1.5110 6866 | 0.6617 8330 | 14.6019 6164 | 9.6633 3433 | 0.0684 8395 | 0.1034 8395 |
| 13 | 1.5639 5606 | 0.6394 0415 | 16.1130 3030 | 10.3027 3849 | 0.0620 6157 | 0.0970 6157 |
| 14 | 1.6186 9452 | 0.6177 8179 | 17.6769 8636 | 10.9205 2028 | 0.0565 7073 | 0.0915 7073 |
| 15 | 1.6753 4883 | 0.5968 9062 | 19.2956 8088 | 11.5174 1090 | 0.0518 2507 | 0.0868 2507 |
| 16 | 1.7339 8604 | 0.5767 0591 | 20.9710 2971 | 12.0941 1681 | 0.0476 8483 | 0.0826 8483 |
| 17 | 1.7946 7555 | 0.5572 0378 | 22.7050 1575 | 12.6513 2059 | 0.0440 4313 | 0.0790 4313 |
| 18 | 1.8574 8920 | 0.5383 6114 | 24.4996 9130 | 13.1896 8173 | 0.0408 1684 | 0.0758 1684 |
| 19 | 1.9225 0132 | 0.5201 5569 | 26.3571 8050 | 13.7098 3742 | 0.0379 4033 | 0.0729 4033 |
| 20 | 1.9897 8886 | 0.5025 6588 | 28.2796 8181 | 14.2124 0330 | 0.0353 6108 | 0.0703 6108 |

**Table B.2** *(continued)*

$i = 3\frac{1}{2}\%$

| $n$ | $(1 + i)^n$ | $(1 + i)^{-n}$ | $s_{\overline{n}|i}$ | $a_{\overline{n}|i}$ | $\dfrac{1}{s_{\overline{n}|i}}$ | $\dfrac{1}{a_{\overline{n}|i}}$ |
|---|---|---|---|---|---|---|
| 21 | 2.0594 3147 | 0.4855 7090 | 30.2694 7068 | 14.6979 7420 | 0.0330 3659 | 0.0680 3659 |
| 22 | 2.1315 1158 | 0.4691 5063 | 32.3289 0215 | 15.1671 2484 | 0.0309 3207 | 0.0659 3207 |
| 23 | 2.2061 1448 | 0.4532 8563 | 34.4604 1373 | 15.6204 1047 | 0.0290 1880 | 0.0640 1880 |
| 24 | 2.2833 2849 | 0.4379 5713 | 36.6665 2821 | 16.0583 6760 | 0.0272 7283 | 0.0622 7283 |
| 25 | 2.3632 4498 | 0.4231 4699 | 38.9498 5669 | 16.4815 1459 | 0.0256 7404 | 0.0606 7404 |
| 26 | 2.4459 5856 | 0.4088 3767 | 41.3131 0168 | 16.8903 5226 | 0.0242 0540 | 0.0592 0540 |
| 27 | 2.5315 6711 | 0.3950 1224 | 43.7590 6024 | 17.2853 6451 | 0.0228 5241 | 0.0578 5241 |
| 28 | 2.6201 7196 | 0.3816 5434 | 46.2906 2734 | 17.6670 1885 | 0.0216 0265 | 0.0566 0265 |
| 29 | 2.7118 7798 | 0.3687 4815 | 48.9107 9930 | 18.0357 6700 | 0.0204 4538 | 0.0554 4538 |
| 30 | 2.8067 9370 | 0.3562 7841 | 51.6226 7728 | 18.3920 4541 | 0.0193 7133 | 0.0543 7133 |
| 31 | 2.9050 3148 | 0.3442 3035 | 54.4294 7098 | 18.7362 7576 | 0.0183 7240 | 0.0533 7240 |
| 32 | 3.0067 0759 | 0.3325 8971 | 57.3345 0247 | 19.0688 6547 | 0.0174 4150 | 0.0524 4150 |
| 33 | 3.1119 4235 | 0.3213 4271 | 60.3412 1005 | 19.3902 0818 | 0.0165 7242 | 0.0515 7242 |
| 34 | 3.2208 6033 | 0.3104 7605 | 63.4531 5240 | 19.7006 8423 | 0.0157 5966 | 0.0507 5966 |
| 35 | 3.3335 9045 | 0.2999 7586 | 66.6740 1274 | 20.0006 6110 | 0.0149 9835 | 0.0499 9835 |
| 36 | 3.4502 6611 | 0.2898 3272 | 70.0076 0318 | 20.2904 9381 | 0.0142 8416 | 0.0492 8416 |
| 37 | 3.5710 2543 | 0.2800 3161 | 73.4578 6930 | 20.5705 2542 | 0.0136 1325 | 0.0486 1325 |
| 38 | 3.6960 1132 | 0.2705 6194 | 77.0288 9472 | 20.8410 8736 | 0.0129 8214 | 0.0479 8214 |
| 39 | 3.8253 7171 | 0.2614 1250 | 80.7249 0604 | 21.1024 9987 | 0.0123 8775 | 0.0473 8775 |
| 40 | 3.9592 5972 | 0.2525 7247 | 84.5502 7775 | 21.3550 7234 | 0.0118 2728 | 0.0468 2728 |
| 41 | 4.0978 3381 | 0.2440 3137 | 88.5095 3747 | 21.5991 0371 | 0.0112 9822 | 0.0462 9822 |
| 42 | 4.2412 5799 | 0.2357 7910 | 92.6073 7128 | 21.8348 8281 | 0.0107 9828 | 0.0457 9828 |
| 43 | 4.3897 0202 | 0.2278 0590 | 96.8486 2928 | 22.0626 8870 | 0.0103 2539 | 0.0453 2539 |
| 44 | 4.5433 4160 | 0.2201 0231 | 101.2383 3130 | 22.2827 9102 | 0.0098 7768 | 0.0448 7768 |
| 45 | 4.7023 5855 | 0.2126 5924 | 105.7816 7290 | 22.4954 5026 | 0.0094 5343 | 0.0444 5343 |
| 46 | 4.8669 4110 | 0.2054 6787 | 110.4840 3145 | 22.7009 1813 | 0.0090 5108 | 0.0440 5108 |
| 47 | 5.0372 8404 | 0.1985 1968 | 115.3509 7255 | 22.8994 3780 | 0.0086 6919 | 0.0436 6919 |
| 48 | 5.2135 8898 | 0.1918 0645 | 120.3882 5659 | 23.0912 4425 | 0.0083 0646 | 0.0433 0646 |
| 49 | 5.3960 6459 | 0.1853 2024 | 125.6018 4557 | 23.2765 6450 | 0.0079 6167 | 0.0429 6167 |
| 50 | 5.5849 2686 | 0.1790 5337 | 130.9979 1016 | 23.4556 1787 | 0.0076 3371 | 0.0426 3371 |
| 51 | 5.7803 9930 | 0.1729 9843 | 136.5828 3702 | 23.6286 1630 | 0.0073 2156 | 0.0423 2156 |
| 52 | 5.9827 1327 | 0.1671 4824 | 142.3632 3631 | 23.7957 6454 | 0.0070 2429 | 0.0420 2429 |
| 53 | 6.1921 0824 | 0.1614 9589 | 148.3459 4958 | 23.9572 6043 | 0.0067 4100 | 0.0417 4100 |
| 54 | 6.4088 3202 | 0.1560 3467 | 154.5380 5782 | 24.1132 9510 | 0.0064 7090 | 0.0414 7090 |
| 55 | 6.6331 4114 | 0.1507 5814 | 160.9468 8984 | 24.2640 5323 | 0.0062 1323 | 0.0412 1323 |
| 56 | 6.8653 0108 | 0.1456 6004 | 167.5800 3099 | 24.4097 1327 | 0.0059 6730 | 0.0409 6730 |
| 57 | 7.1055 8662 | 0.1407 3433 | 174.4453 3207 | 24.5504 4760 | 0.0057 3245 | 0.0407 3245 |
| 58 | 7.3542 8215 | 0.1359 7520 | 181.5509 1869 | 24.6864 2281 | 0.0055 0810 | 0.0405 0810 |
| 59 | 7.6116 8203 | 0.1313 7701 | 188.9052 0085 | 24.8177 9981 | 0.0052 9366 | 0.0402 9366 |
| 60 | 7.8780 9090 | 0.1269 3431 | 196.5168 8288 | 29.9447 3412 | 0.0050 8862 | 0.0400 8862 |

**Table B.2** *(continued)*

$i = 3\frac{1}{2}\%$

| $n$ | $(1 + i)^n$ | $(1 + i)^{-n}$ | $s_{\overline{n}|i}$ | $a_{\overline{n}|i}$ | $\dfrac{1}{s_{\overline{n}|i}}$ | $\dfrac{1}{a_{\overline{n}|i}}$ |
|-----|-------------|----------------|----------------------|----------------------|---------------------------------|---------------------------------|
| 61  | 8.1538 2408  | 0.1226 4184 | 204.3949 7378 | 25.0673 7596 | 0.0048 9249 | 0.0398 9249 |
| 62  | 8.4392 0793  | 0.1184 9453 | 212.5487 9786 | 25.1858 7049 | 0.0047 0480 | 0.0397 0480 |
| 63  | 8.7345 8020  | 0.1144 8747 | 220.9880 0579 | 25.3003 5796 | 0.0045 2513 | 0.0395 2513 |
| 64  | 9.0402 9051  | 0.1106 1591 | 229.7225 8599 | 25.4109 7388 | 0.0043 5308 | 0.0393 5308 |
| 65  | 9.3567 0068  | 0.1068 7528 | 238.7628 7650 | 25.5178 4916 | 0.0041 8826 | 0.0391 8826 |
| 66  | 9.6841 8520  | 0.1032 6114 | 248.1195 7718 | 25.6211 1030 | 0.0040 3031 | 0.0390 3031 |
| 67  | 10.0231 3168 | 0.0997 6922 | 257.8037 6238 | 25.7208 7951 | 0.0038 7892 | 0.0388 7892 |
| 68  | 10.3739 4129 | 0.0963 9538 | 267.8268 9406 | 25.8172 7489 | 0.0037 3375 | 0.0387 3375 |
| 69  | 10.7370 2924 | 0.0931 3563 | 278.2008 3535 | 25.9104 1052 | 0.0035 9453 | 0.0385 9453 |
| 70  | 11.1128 2526 | 0.0899 8612 | 288.9378 6459 | 26.0003 9664 | 0.0034 6095 | 0.0384 6095 |
| 71  | 11.5017 7414 | 0.0869 4311 | 300.0506 8985 | 26.0873 3975 | 0.0033 3277 | 0.0383 3277 |
| 72  | 11.9043 3624 | 0.0840 0300 | 311.5524 6400 | 26.1713 4275 | 0.0032 0973 | 0.0382 0973 |
| 73  | 12.3209 8801 | 0.0811 6232 | 323.4568 0024 | 26.2525 0508 | 0.0030 9160 | 0.0380 9160 |
| 74  | 12.7522 2259 | 0.0784 1770 | 335.7777 8824 | 26.3309 2278 | 0.0029 7816 | 0.0379 7816 |
| 75  | 13.1985 5038 | 0.0757 6590 | 348.5300 1083 | 26.4066 8868 | 0.0028 6919 | 0.0378 6919 |
| 76  | 13.6604 9964 | 0.0732 0376 | 361.7285 6121 | 26.4798 9244 | 0.0027 6450 | 0.0377 6450 |
| 77  | 14.1386 1713 | 0.0707 2827 | 375.3890 6085 | 26.5506 2072 | 0.0026 6390 | 0.0376 6390 |
| 78  | 14.6334 6873 | 0.0683 3650 | 389.5276 7798 | 26.6189 5721 | 0.0025 6721 | 0.0375 6721 |
| 79  | 15.1456 4013 | 0.0660 2560 | 404.1611 4671 | 26.6849 8281 | 0.0024 7426 | 0.0374 7426 |
| 80  | 15.6757 3754 | 0.0637 9285 | 419.3067 8685 | 26.7487 7567 | 0.0023 8489 | 0.0373 8489 |
| 81  | 16.2243 8835 | 0.0616 3561 | 434.9825 2439 | 26.8104 1127 | 0.0022 9894 | 0.0372 9894 |
| 82  | 16.7922 4195 | 0.0595 5131 | 451.2069 1274 | 26.8699 6258 | 0.0022 1628 | 0.0372 1628 |
| 83  | 17.3799 7041 | 0.0575 3750 | 467.9991 5469 | 26.9275 0008 | 0.0021 3676 | 0.0371 3676 |
| 84  | 17.9882 6938 | 0.0555 9178 | 485.3791 2510 | 26.9830 9186 | 0.0020 6025 | 0.0370 6025 |
| 85  | 18.6178 5881 | 0.0537 1187 | 503.3673 9448 | 27.0368 0373 | 0.0019 8662 | 0.0369 8662 |
| 86  | 19.2694 8387 | 0.0518 9553 | 521.9852 5329 | 27.0886 9926 | 0.0019 1576 | 0.0369 1576 |
| 87  | 19.9439 1580 | 0.0501 4060 | 541.2547 3715 | 27.1388 3986 | 0.0018 4756 | 0.0368 4756 |
| 88  | 20.6419 5285 | 0.0484 4503 | 561.1986 5295 | 27.1872 8489 | 0.0017 8190 | 0.0367 8190 |
| 89  | 21.3644 2120 | 0.0468 0679 | 581.8406 0581 | 27.2340 9168 | 0.0017 1868 | 0.0367 1868 |
| 90  | 22.1121 7595 | 0.0452 2395 | 603.2050 2701 | 27.2793 1564 | 0.0016 5781 | 0.0366 5781 |
| 91  | 22.8861 0210 | 0.0436 9464 | 625.3172 0295 | 27.3230 1028 | 0.0015 9919 | 0.0365 9919 |
| 92  | 23.6871 1568 | 0.0422 1704 | 648.2033 0506 | 27.3652 2732 | 0.0015 4273 | 0.0365 4273 |
| 93  | 24.5161 6473 | 0.0407 8941 | 671.8904 2073 | 27.4060 1673 | 0.0014 8834 | 0.0364 8834 |
| 94  | 25.3742 3049 | 0.0349 1006 | 696.4065 8546 | 27.4454 2680 | 00.014 3594 | 0.0364 3594 |
| 95  | 26.2623 2856 | 0.0380 7735 | 721.7808 1595 | 27.4835 0415 | 0.0013 8546 | 0.0363 8546 |
| 96  | 27.1815 1006 | 0.0367 8971 | 748.0431 4451 | 27.5202 9387 | 0.0013 3692 | 0.0363 3682 |
| 97  | 28.1328 6291 | 0.0355 4562 | 775.2246 5457 | 27.5558 3948 | 0.0012 8995 | 0.0362 8995 |
| 98  | 29.1175 1311 | 0.0343 4359 | 803.3575 1748 | 27.5901 8308 | 0.0012 4478 | 0.0362 4478 |
| 99  | 30.1366 2607 | 0.0331 8221 | 832 4750 3059 | 27.6233 6529 | 0.0012 0124 | 0.0362 0124 |
| 100 | 31.1914 0798 | 0.0320 6011 | 862.6116 5666 | 27.6554 2540 | 0.0011 5927 | 0.0361 5927 |

**Table B.2** *(continued)*

$i = 4\%$

| $n$ | $(1 + i)^n$ | $(1 + i)^{-n}$ | $s_{\overline{n}|i}$ | $a_{\overline{n}|i}$ | $\dfrac{1}{s_{\overline{n}|i}}$ | $\dfrac{1}{a_{\overline{n}|i}}$ |
|---|---|---|---|---|---|---|
| 1 | 1.0400 0000 | 0.9615 3846 | 1.0000 0000 | 0.9615 3846 | 1.0000 0000 | 1.0400 0000 |
| 2 | 1.0816 0000 | 0.9245 5621 | 2.0400 0000 | 1.8860 9467 | 0.4901 9608 | 0.5301 9608 |
| 3 | 1.1248 6400 | 0.8889 9636 | 3.1216 0000 | 2.7750 9103 | 0.3203 4854 | 0.3603 4854 |
| 4 | 1.1698 5856 | 0.8548 0419 | 4.2464 6400 | 3.6298 9522 | 0.2354 9005 | 0.2754 9005 |
| 5 | 1.2166 5290 | 0.8219 2711 | 5.4163 2256 | 4.4518 2233 | 0.1846 2711 | 0.2246 2711 |
| 6 | 1.2653 1902 | 0.7903 1453 | 6.6329 7546 | 5.2421 3686 | 0.1507 6190 | 0.1907 6190 |
| 7 | 1.3159 3178 | 0.7599 1781 | 7.8982 9448 | 6.0020 5467 | 0.1266 0961 | 0.1666 0961 |
| 8 | 1.3685 6905 | 0.7306 9021 | 9.2142 2626 | 6.7327 4487 | 0.1085 2783 | 0.1485 2783 |
| 9 | 1.4233 1181 | 0.7025 8674 | 10.5827 9531 | 7.4353 3161 | 0.0944 9299 | 0.1344 9299 |
| 10 | 1.4802 4428 | 0.6755 6417 | 12.0061 0712 | 8.1108 9578 | 0.0832 9094 | 0.1232 9094 |
| 11 | 1.5394 5406 | 0.6495 8093 | 13.4863 5141 | 8.7604 7671 | 0.0741 4904 | 0.1141 4904 |
| 12 | 1.6010 3222 | 0.6245 9705 | 15.0258 0546 | 9.3850 7376 | 0.0665 5217 | 0.1065 5217 |
| 13 | 1.6650 7351 | 0.6005 7409 | 16.6268 3768 | 9.9856 4785 | 0.0601 4373 | 0.1001 4373 |
| 14 | 1.7316 7645 | 0.5774 7508 | 18.2919 1119 | 10.5631 2293 | 0.0546 6897 | 0.0946 6897 |
| 15 | 1.8009 4351 | 0.5552 6450 | 20.0235 8764 | 11.1183 8743 | 0.0499 4110 | 0.0899 4110 |
| 16 | 1.8729 8125 | 0.5339 0818 | 21.8245 3114 | 11.6522 9561 | 0.0458 2000 | 0.0858 2000 |
| 17 | 1.9479 0050 | 0.5133 7325 | 23.6975 1239 | 12.1656 6885 | 0.0421 9852 | 0.0821 9852 |
| 18 | 2.0258 1652 | 0.4936 2812 | 25.6454 1288 | 12.6592 9697 | 0.0389 9333 | 0.0789 9333 |
| 19 | 2.1068 4918 | 0.4746 4242 | 27.6712 2940 | 13.1339 3940 | 0.0361 3862 | 0.0761 3862 |
| 20 | 2.1911 2314 | 0.4563 8695 | 29.7780 7858 | 13.5903 2634 | 0.0335 8175 | 0.0735 8175 |
| 21 | 2.2787 6807 | 0.4388 3360 | 31.9692 0172 | 14.0291 5995 | 0.0312 8011 | 0.0712 8011 |
| 22 | 2.3699 1879 | 0.4219 5539 | 34.2479 6979 | 14.4511 1533 | 0.0291 9881 | 0.0691 9881 |
| 23 | 2.4647 1554 | 0.4057 2633 | 36.6178 8858 | 14.8568 4167 | 0.0273 0906 | 0.0673 0906 |
| 24 | 2.5633 0416 | 0.3901 2147 | 39.0826 0412 | 15.2469 6314 | 0.0255 8683 | 0.0655 8683 |
| 25 | 2.6658 3633 | 0.3751 1680 | 41.6459 0829 | 15.6220 7994 | 0.0240 1196 | 0.0640 1196 |
| 26 | 2.7724 6978 | 0.3606 8923 | 44.3117 4462 | 15.9827 6918 | 0.0225 6738 | 0.0625 6738 |
| 27 | 2.8833 6858 | 0.3468 1657 | 47.0842 1440 | 16.3295 8575 | 0.0212 3854 | 0.0612 3854 |
| 28 | 2.9987 0332 | 0.3334 7747 | 49.9675 8298 | 16.6630 6322 | 0.0200 1298 | 0.0600 1298 |
| 29 | 3.1186 5145 | 0.3206 5141 | 52.9662 8630 | 16.9837 1463 | 0.0188 7993 | 0.0588 7993 |
| 30 | 3.2433 9751 | 0.3083 1867 | 56.0849 3775 | 17.2920 3330 | 0.0178 3010 | 0.0578 3010 |
| 31 | 3.3731 3341 | 0.2964 6026 | 59.3283 3526 | 17.5884 9356 | 0.0168 5535 | 0.0568 5535 |
| 32 | 3.5080 5875 | 0.2850 5794 | 62.7014 6867 | 17.8735 5150 | 0.0159 4859 | 0.0559 4859 |
| 33 | 3.6483 8110 | 0.2740 9417 | 66.2095 2742 | 18.1476 4567 | 0.0151 0357 | 0.0551 0357 |
| 34 | 3.7943 1634 | 0.2635 5209 | 69.8579 0851 | 18.4111 9776 | 0.0143 1477 | 0.0543 1477 |
| 35 | 3.9460 8899 | 0.2534 1547 | 73.6522 2486 | 18.6646 1323 | 0.0135 7732 | 0.0535 7732 |
| 36 | 4.1039 3255 | 0.2436 6872 | 77.5983 1385 | 18.9082 8195 | 0.0128 8688 | 0.0528 8688 |
| 37 | 4.2680 8986 | 0.2342 9685 | 81.7022 4640 | 19.1425 7880 | 0.0122 3957 | 0.0522 3957 |
| 38 | 4.4388 1345 | 0.2252 8543 | 85.9703 3626 | 19.3678 6423 | 0.0116 3192 | 0.0516 3192 |
| 39 | 4.6163 6599 | 0.2166 2061 | 90.4091 4971 | 19.5844 8484 | 0.0110 6083 | 0.0510 6083 |
| 40 | 4.8010 2063 | 0.2082 8904 | 95.0255 1570 | 19.7927 7388 | 0.0105 2349 | 0.0505 2349 |

**Table B.2** *(continued)*

$i = 4\%$

| $n$ | $(1 + i)^n$ | $(1 + i)^{-n}$ | $s_{\overline{n}|i}$ | $a_{\overline{n}|i}$ | $\dfrac{1}{s_{\overline{n}|i}}$ | $\dfrac{1}{a_{\overline{n}|i}}$ |
|---|---|---|---|---|---|---|
| 41 | 4.9930 6145 | 0.2002 7793 | 99.8265 3633 | 19.9930 5181 | 0.0100 1738 | 0.0500 1738 |
| 42 | 5.1927 8391 | 0.1925 7493 | 104.8195 9778 | 20.1856 2674 | 0.0095 4020 | 0.0495 4020 |
| 43 | 5.4004 9527 | 0.1851 6820 | 110.0123 8169 | 20.3707 9494 | 0.0090 8989 | 0.0490 8989 |
| 44 | 5.6165 1508 | 0.1780 4635 | 115.4128 7696 | 20.5488 4129 | 0.0086 6454 | 0.0486 6454 |
| 45 | 5.8411 7568 | 0.1711 9841 | 121.0293 9204 | 20.7200 3970 | 0.0082 6246 | 0.0482 6246 |
| 46 | 6.0748 2271 | 0.1646 1386 | 126.8705 6772 | 20.8846 5356 | 0.0078 8205 | 0.0478 8205 |
| 47 | 6.3178 1562 | 0.1582 8256 | 132.9453 9043 | 21.0429 3612 | 0.0075 2189 | 0.0475 2189 |
| 48 | 6.5705 2824 | 0.1521 9476 | 139.2632 0604 | 21.1951 3088 | 0.0071 8065 | 0.0471 8065 |
| 49 | 6.8333 4937 | 0.1463 4112 | 145.8337 3429 | 21.3414 7200 | 0.0068 5712 | 0.0468 5712 |
| 50 | 7.1066 8335 | 0.1407 1262 | 152.6670 8366 | 21.4821 8462 | 0.0065 5020 | 0.0465 5020 |

$i = 4\frac{1}{2}\%$

| $n$ | $(1 + i)^n$ | $(1 + i)^{-n}$ | $s_{\overline{n}|i}$ | $a_{\overline{n}|i}$ | $\dfrac{1}{s_{\overline{n}|i}}$ | $\dfrac{1}{a_{\overline{n}|i}}$ |
|---|---|---|---|---|---|---|
| 1 | 1.0450 0000 | 0.9569 3780 | 1.0000 0000 | 0.9569 3780 | 1.0000 0000 | 1.0450 0000 |
| 2 | 1.0920 2500 | 0.9157 2995 | 2.0450 0000 | 1.8726 6775 | 0.4889 9756 | 0.5339 9756 |
| 3 | 1.1411 6613 | 0.8762 9660 | 3.1370 2500 | 2.7489 6435 | 0.3187 7336 | 0.3637 7336 |
| 4 | 1.1925 1860 | 0.8385 6134 | 4.2781 9113 | 3.5875 2570 | 0.2337 4365 | 0.2787 4365 |
| 5 | 1.2461 8194 | 0.8024 5105 | 5.4707 0973 | 4.3899 7674 | 0.1827 9164 | 0.2277 9164 |
| 6 | 1.3022 6012 | 0.7678 9574 | 6.7168 9166 | 5.1578 7248 | 0.1488 7839 | 0.1938 7839 |
| 7 | 1.3608 6183 | 0.7348 2846 | 8.0191 5179 | 5.8927 0094 | 0.1247 0147 | 0.1697 0147 |
| 8 | 1.4221 0061 | 0.7031 8513 | 9.3800 1362 | 6.5958 8607 | 0.1066 0965 | 0.1516 0965 |
| 9 | 1.4860 9514 | 0.6729 0443 | 10.8021 1423 | 7.2687 9050 | 0.0925 7447 | 0.1375 7447 |
| 10 | 1.5529 6942 | 0.6439 2768 | 12.2882 0937 | 7.9127 1818 | 0.0813 7882 | 0.1263 7882 |
| 11 | 1.6228 5305 | 0.6161 9874 | 13.8411 7879 | 8.5289 1692 | 0.0722 4818 | 0.1172 4818 |
| 12 | 1.6958 8143 | 0.5896 6386 | 15.4640 3184 | 9.1185 8078 | 0.0646 6619 | 0.1096 6619 |
| 13 | 1.7721 9610 | 0.5642 7164 | 17.1599 1327 | 9.6828 5242 | 0.0582 7535 | 0.1032 7535 |
| 14 | 1.8519 4492 | 0.5399 7286 | 18.9321 0937 | 10.2228 2528 | 0.0528 2032 | 0.0978 2032 |
| 15 | 1.9352 8244 | 0.5167 2044 | 20.7840 5429 | 10.7395 4573 | 0.0481 1381 | 0.0931 1381 |
| 16 | 2.0223 7015 | 0.4944 6932 | 22.7193 3673 | 11.2340 1505 | 0.0440 1537 | 0.0890 1537 |
| 17 | 2.1133 7681 | 0.4731 7639 | 24.7417 0689 | 11.7071 9143 | 0.0404 1758 | 0.0854 1758 |
| 18 | 2.2084 7877 | 0.4528 0037 | 26.8550 8370 | 12.1599 9180 | 0.0372 3690 | 0.0822 3690 |
| 19 | 2.3078 6031 | 0.4333 0179 | 29.0635 6246 | 12.5932 9359 | 0.0344 0734 | 0.0794 0734 |
| 20 | 2.4117 1402 | 0.4146 4286 | 31.3714 2277 | 13.0079 3645 | 0.0318 7614 | 0.0768 7614 |
| 21 | 2.5202 4116 | 0.3967 8743 | 33.7831 3680 | 13.4047 2388 | 0.0296 0057 | 0.0746 0057 |
| 22 | 2.6336 5201 | 0.3797 0089 | 36.3033 7795 | 13.7844 2476 | 0.0275 4565 | 0.0725 4565 |
| 23 | 2.7521 6635 | 0.3633 5013 | 38.9370 2996 | 14.1477 7489 | 0.0256 8249 | 0.0706 8249 |
| 24 | 2.8760 1383 | 0.3477 0347 | 41.6891 9631 | 14.4954 7837 | 0.0239 8703 | 0.0689 8703 |
| 25 | 3.0054 3446 | 0.3327 3060 | 44.5652 1015 | 14.8282 0896 | 0.0224 3903 | 0.0674 3903 |

**Table B.2** *(continued)*

$i = 4\frac{1}{2}\%$

| $n$ | $(1 + i)^n$ | $(1 + i)^{-n}$ | $s_{\overline{n}|i}$ | $a_{\overline{n}|i}$ | $\dfrac{1}{s_{\overline{n}|i}}$ | $\dfrac{1}{a_{\overline{n}|i}}$ |
|---|---|---|---|---|---|---|
| 26 | 3.1406 7901 | 0.3184 0248 | 47.5706 4460 | 15.1466 1145 | 0.0210 2137 | 0.0660 2137 |
| 27 | 3.2820 0956 | 0.3046 9137 | 50.7113 2361 | 15.4513 0282 | 0.0197 1946 | 0.0647 1946 |
| 28 | 3.4296 9999 | 0.2915 7069 | 53.9933 3317 | 15.7428 7351 | 0.0185 2081 | 0.0635 2081 |
| 29 | 3.5840 3649 | 0.2790 1502 | 57.4230 3316 | 16.0218 8853 | 0.0174 1461 | 0.0624 1461 |
| 30 | 3.7453 1813 | 0.2670 0002 | 61.0070 6966 | 16.2888 8854 | 0.0163 9154 | 0.0613 9154 |
| 31 | 3.9138 5745 | 0.2555 0241 | 64.7523 8779 | 16.5443 9095 | 0.0154 4345 | 0.0604 4345 |
| 32 | 4.0899 8104 | 0.2444 9991 | 68.6662 4524 | 16.7888 9086 | 0.0145 6320 | 0.0595 6320 |
| 33 | 4.2740 3018 | 0.2339 7121 | 75.7562 2628 | 17.0228 6207 | 0.0137 4453 | 0.0587 4453 |
| 34 | 4.4663 6154 | 0.2238 9589 | 77.0302 5646 | 17.2467 5796 | 0.0129 8191 | 0.0579 8191 |
| 35 | 4.6673 4781 | 0.2142 5444 | 81.4966 1800 | 17.4610 1240 | 0.0122 7045 | 0.0572 7045 |
| 36 | 4.8773 7846 | 0.2050 2817 | 86.1639 6581 | 17.6660 4058 | 0.0116 0578 | 0.0566 0578 |
| 37 | 5.0968 6049 | 0.1961 9921 | 91.0413 4427 | 17.8622 3979 | 0.0109 8402 | 0.0559 8402 |
| 38 | 5.3262 1921 | 0.1877 5044 | 96.1382 0476 | 18.0499 9023 | 0.0104 0169 | 0.0554 0169 |
| 39 | 5.5658 9908 | 0.1796 6549 | 101.4644 2398 | 18.2296 5572 | 0.0098 5567 | 0.0548 5567 |
| 40 | 5.8163 6454 | 0.1719 2870 | 107.0303 2306 | 18.4015 8442 | 0.0093 4315 | 0.0543 4315 |
| 41 | 6.0781 0094 | 0.1645 2507 | 112.8466 8760 | 18.5661 0949 | 0.0088 6158 | 0.0538 6158 |
| 42 | 6.3516 1548 | 0.1574 4026 | 118.9247 8854 | 18.7235 4975 | 0.0084 0868 | 0.0534 0868 |
| 43 | 6.6374 3818 | 0.1506 6054 | 125.2764 0402 | 18.8742 1029 | 0.0079 8235 | 0.0529 8235 |
| 44 | 6.9361 2290 | 0.1441 7276 | 131.9138 4220 | 19.0183 8305 | 0.0075 8071 | 0.0525 8071 |
| 45 | 7.2482 4843 | 0.1379 6437 | 138.8499 6510 | 19.1563 4742 | 0.0072 0202 | 0.0522 0202 |
| 46 | 7.5744 1961 | 0.1320 2332 | 146.0982 1353 | 19.2883 7074 | 0.0068 4471 | 0.0518 4471 |
| 47 | 7.9152 6849 | 0.1263 3810 | 153.6726 3314 | 19.4147 0884 | 0.0065 0734 | 0.0515 0734 |
| 48 | 8.2714 5557 | 0.1208 9771 | 161.5879 0163 | 19.5356 0654 | 0.0061 8858 | 0.0511 8858 |
| 49 | 8.6436 7107 | 0.1156 9158 | 169.8593 5720 | 19.6512 9813 | 0.0058 8722 | 0.0508 8722 |
| 50 | 9.0326 3627 | 0.1107 0965 | 178.5030 2828 | 19.7620 0778 | 0.0056 0215 | 0.0506 0215 |

$i = 5\%$

| $n$ | $(1 + i)^n$ | $(1 + i)^{-n}$ | $s_{\overline{n}|i}$ | $a_{\overline{n}|i}$ | $\dfrac{1}{s_{\overline{n}|i}}$ | $\dfrac{1}{a_{\overline{n}|i}}$ |
|---|---|---|---|---|---|---|
| 1 | 1.0500 0000 | 0.9523 8095 | 1.0000 0000 | 0.9523 8095 | 1.0000 0000 | 1.0500 0000 |
| 2 | 1.1025 0000 | 0.9070 2948 | 2.0500 0000 | 1.8594 1043 | 0.4878 0488 | 0.5378 0488 |
| 3 | 1.1576 2500 | 0.8638 3760 | 3.1525 0000 | 2.7232 4803 | 0.3172 0856 | 0.3672 0856 |
| 4 | 1.2155 0625 | 0.8227 0247 | 4.3101 2500 | 3.5459 5050 | 0.2320 1183 | 0.2820 1183 |
| 5 | 1.2762 8156 | 0.7835 2617 | 5.5256 3125 | 4.3294 7667 | 0.1809 7480 | 0.2309 7480 |
| 6 | 1.3400 9564 | 0.7462 1540 | 6.8019 1281 | 5.0756 9207 | 0.1470 1747 | 0.1970 1747 |
| 7 | 1.4071 0042 | 0.7106 8133 | 8.1420 0845 | 5.7863 7340 | 0.1228 1982 | 0.1728 1982 |
| 8 | 1.4774 5544 | 0.6768 3936 | 9.5491 0888 | 6.4632 1276 | 0.1047 2181 | 0.1547 2181 |
| 9 | 1.5513 2822 | 0.6446 0892 | 11.0265 6432 | 7.1078 2168 | 0.0906 9008 | 0.1406 9008 |
| 10 | 1.6288 9463 | 0.6139 1325 | 12.5778 9254 | 7.7217 3493 | 0.0795 0457 | 0.1295 0457 |

**Table B.2** *(continued)*
$i = 5\%$

| $n$ | $(1 + i)^n$ | $(1 + i)^{-n}$ | $s_{n\rceil i}$ | $a_{n\rceil i}$ | $\dfrac{1}{s_{n\rceil i}}$ | $\dfrac{1}{a_{n\rceil i}}$ |
|---|---|---|---|---|---|---|
| 11 | 1.7103 3936 | 0.5846 7929 | 14.2067 8716 | 8.3064 1422 | 0.0703 8889 | 0.1203 8889 |
| 12 | 1.7958 5633 | 0.5568 3742 | 15.9171 2652 | 8.8632 5164 | 0.0628 2541 | 0.1128 2541 |
| 13 | 1.8856 4914 | 0.5303 2135 | 17.7129 8285 | 9.3935 7299 | 0.0564 5577 | 0.1064 5577 |
| 14 | 1.9799 3160 | 0.5050 6795 | 19.5986 3199 | 9.8986 4094 | 0.0510 2397 | 0.1010 2397 |
| 15 | 2.0789 2818 | 0.4810 1710 | 21.5785 6359 | 10.3796 5804 | 0.0463 4229 | 0.0963 4229 |
| 16 | 2.1828 7459 | 0.4581 1152 | 23.6574 9177 | 10.8377 6956 | 0.0422 6991 | 0.0922 6991 |
| 17 | 2.2920 1832 | 0.4362 9669 | 25.8403 6636 | 11.2740 6625 | 0.0386 9914 | 0.0886 9914 |
| 18 | 2.4066 1923 | 0.4155 2065 | 28.1323 8467 | 11.6895 8690 | 0.0355 4622 | 0.0855 4622 |
| 19 | 2.5269 5020 | 0.3957 3396 | 30.5390 0391 | 12.0853 2086 | 0.0327 4501 | 0.0827 4501 |
| 20 | 2.6532 9771 | 0.3768 8948 | 33.0659 5410 | 12.4622 1034 | 0.0302 4259 | 0.0802 4259 |
| 21 | 2.7859 6259 | 0.3589 4236 | 35.7192 5181 | 12.8211 5271 | 0.0279 9611 | 0.0779 9611 |
| 22 | 2.9252 6072 | 0.3418 4987 | 38.5052 1440 | 13.1630 0258 | 0.0259 7051 | 0.0759 7051 |
| 23 | 3.0715 2376 | 0.3255 7131 | 41.4304 7512 | 13.4885 7388 | 0.0241 3682 | 0.0741 3682 |
| 24 | 3.2250 9994 | 0.3100 6791 | 44.5019 9887 | 13.7986 4179 | 0.0224 7090 | 0.0724 7090 |
| 25 | 3.3863 5494 | 0.2953 0277 | 47.7270 9882 | 14.0939 4457 | 0.0209 5246 | 0.0709 5246 |
| 26 | 3.5556 7269 | 0.2812 4073 | 51.1134 5376 | 14.3751 8530 | 0.0195 6432 | 0.0695 6432 |
| 27 | 3.7334 5632 | 0.2678 4832 | 54.6691 2645 | 14.6430 3362 | 0.0182 9186 | 0.0682 9186 |
| 28 | 3.9201 2914 | 0.2550 9364 | 58.4025 8277 | 14.8981 2726 | 0.0171 2253 | 0.0671 2253 |
| 29 | 4.1161 3560 | 0.2429 4632 | 62.3227 1191 | 15.1410 7358 | 0.0160 4551 | 0.0660 4551 |
| 30 | 4.3219 4238 | 0.2313 7745 | 66.4388 4750 | 15.3724 5103 | 0.0150 5144 | 0.0650 5144 |
| 31 | 4.5380 3949 | 0.2203 5947 | 70.7607 8988 | 15.5928 1050 | 0.0141 3212 | 0.0641 3212 |
| 32 | 4.7649 4147 | 0.2098 6617 | 75.2988 2937 | 15.8026 7667 | 0.0132 8042 | 0.0632 8042 |
| 33 | 5.0031 8854 | 0.1998 7254 | 80.0637 7084 | 16.0025 4921 | 0.0124 9004 | 0.0624 9004 |
| 34 | 5.2533 4797 | 0.1903 5480 | 85.0669 5938 | 16.1929 0401 | 0.0117 5545 | 0.0617 5545 |
| 35 | 5.5160 1537 | 0.1812 9029 | 90.3203 0735 | 16.3741 9429 | 0.0110 7171 | 0.0610 7171 |
| 36 | 5.7918 1614 | 0.1726 5741 | 95.8363 2272 | 16.5468 5171 | 0.0104 3446 | 0.0604 3446 |
| 37 | 6.0814 0694 | 0.1644 3563 | 101.6281 3886 | 16.7112 8734 | 0.0098 3979 | 0.0598 3979 |
| 38 | 6.3854 7729 | 0.1566 0536 | 107.7095 4580 | 16.8678 9271 | 0.0092 8423 | 0.0592 8423 |
| 39 | 6.7047 5115 | 0.1491 4797 | 114.0950 2309 | 17.0170 4067 | 0.0087 6462 | 0.0587 6462 |
| 40 | 7.0399 8871 | 0.1420 4568 | 120.7997 7424 | 17.1590 8635 | 0.0082 7816 | 0.0582 7816 |
| 41 | 7.3919 8815 | 0.1352 8160 | 127.8397 6295 | 17.2943 6796 | 0.0078 2229 | 0.0578 2229 |
| 42 | 7.7615 8756 | 0.1288 3962 | 135.2317 5110 | 17.4232 0758 | 0.0073 9471 | 0.0573 9471 |
| 43 | 8.1496 6693 | 0.1227 0440 | 142.9933 3866 | 17.5459 1198 | 0.0069 9333 | 0.0569 9333 |
| 44 | 8.5571 5028 | 0.1168 6133 | 151.1430 0559 | 17.6627 7331 | 0.0066 1625 | 0.0566 1625 |
| 45 | 8.9850 0779 | 0.1112 9651 | 159.7001 5587 | 17.7740 6982 | 0.0062 6173 | 0.0562 6173 |
| 46 | 9.4342 5818 | 0.1059 9668 | 168.6851 6366 | 17.8800 6650 | 0.0059 2820 | 0.0559 2820 |
| 47 | 9.9059 7109 | 0.1009 4921 | 178.1194 2185 | 17.9810 1571 | 0.0056 1421 | 0.0556 1421 |
| 48 | 10.4012 6965 | 0.0961 4211 | 188.0253 9294 | 18.0771 5782 | 0.0053 1843 | 0.0553 1843 |
| 49 | 10.9213 3313 | 0.0915 6391 | 198.4266 6259 | 18.1687 2173 | 0.0050 3965 | 0.0550 3965 |
| 50 | 11.4673 9979 | 0.0872 0373 | 209.3479 9572 | 18.2559 2546 | 0.0047 7674 | 0.0547 7674 |

**Table B.2** *(continued)*
$i = 6\%$

| $n$ | $(1 + i)^n$ | $(1 + i)^{-n}$ | $s_{\overline{n}|i}$ | $a_{\overline{n}|i}$ | $\dfrac{1}{s_{\overline{n}|i}}$ | $\dfrac{1}{a_{\overline{n}|i}}$ |
|---|---|---|---|---|---|---|
| 1 | 1.0600 0000 | 0.9433 9623 | 1.0000 0000 | 0.9433 9623 | 1.0000 0000 | 1.0600 0000 |
| 2 | 1.1236 0000 | 0.8899 9644 | 2.0600 0000 | 1.8333 9267 | 0.4854 3689 | 0.5454 3689 |
| 3 | 1.1910 1600 | 0.8396 1928 | 3.1836 0000 | 2.6730 1195 | 0.3141 0981 | 0.3741 0981 |
| 4 | 1.2624 7696 | 0.7920 9366 | 4.3746 1600 | 0.4651 0561 | 0.2285 9149 | 0.2885 9149 |
| 5 | 1.3382 2558 | 0.7472 5817 | 5.6370 9296 | 4.2123 6379 | 0.1773 9640 | 0.2373 9640 |
| 6 | 1.4185 1911 | 0.7049 6054 | 6.9753 1854 | 4.9173 2433 | 0.1433 6463 | 0.2033 6263 |
| 7 | 1.5036 3026 | 0.6650 5711 | 8.3938 3765 | 5.5823 8144 | 0.1191 3502 | 0.1791 3502 |
| 8 | 1.5938 4807 | 0.6274 1237 | 9.8974 6791 | 6.2097 9381 | 0.1010 3594 | 0.1610 3594 |
| 9 | 1.6894 7896 | 0.5918 9846 | 11.4913 1598 | 6.8016 9227 | 0.0870 2224 | 0.1470 2224 |
| 10 | 1.7908 4770 | 0.5583 9478 | 13.1807 9494 | 7.3600 8705 | 0.0758 6796 | 0.1358 6796 |
| 11 | 1.8982 9856 | 0.5267 8753 | 14.9716 4264 | 7.8868 7458 | 0.0667 9294 | 0.1267 9294 |
| 12 | 2.0121 9647 | 0.4969 6936 | 16.8699 4120 | 8.3838 4394 | 0.0592 7703 | 0.1192 7703 |
| 13 | 2.1329 2826 | 0.4688 3902 | 18.8821 3767 | 8.8526 8296 | 0.0529 6011 | 0.1129 6011 |
| 14 | 2.2609 0396 | 0.4423 0096 | 21.0150 6593 | 9.2949 8393 | 0.0475 8491 | 0.1075 8491 |
| 15 | 2.3965 5819 | 0.4172 6506 | 23.2759 6988 | 9.7122 4899 | 0.0429 6276 | 0.1029 6276 |
| 16 | 2.5403 5168 | 0.3936 4628 | 25.6725 2808 | 10.1058 9527 | 0.0389 5214 | 0.0989 5214 |
| 17 | 2.6927 7279 | 0.3713 6442 | 28.2128 7976 | 10.4772 5969 | 0.0354 4480 | 0.0954 4480 |
| 18 | 2.8543 3915 | 0.3503 4379 | 30.9056 5255 | 10.8276 0348 | 0.0323 5654 | 0.0923 5654 |
| 19 | 3.0255 9950 | 0.3305 1301 | 33.7599 9170 | 11.1581 1649 | 0.0296 2086 | 0.0896 2086 |
| 20 | 3.2071 3547 | 0.3118 0473 | 36.7855 9120 | 11.4699 2122 | 0.0271 8456 | 0.0871 8456 |
| 21 | 3.3995 6360 | 0.2941 5540 | 39.9927 2668 | 11.7640 7662 | 0.0250 0455 | 0.0850 0455 |
| 22 | 3.6035 3742 | 0.2775 0510 | 43.3922 9028 | 12.0415 8172 | 0.0230 4557 | 0.0830 4557 |
| 23 | 3.8197 4966 | 0.2617 9726 | 46.9958 2769 | 12.3033 7898 | 0.0212 7848 | 0.0812 7848 |
| 24 | 4.0489 3464 | 0.2469 7855 | 50.8155 7735 | 12.5503 5753 | 0.0196 7900 | 0.0796 7900 |
| 25 | 4.2918 7072 | 0.2329 9863 | 54.8645 1200 | 12.7833 5616 | 0.0182 2672 | 0.0782 2672 |
| 26 | 4.5493 8296 | 0.2198 1003 | 59.1563 8272 | 13.0031 6619 | 0.0169 0435 | 0.0769 0435 |
| 27 | 4.8223 4594 | 0.2073 6795 | 63.7057 6568 | 13.2105 3414 | 0.0156 9717 | 0.0756 9717 |
| 28 | 5.1116 8670 | 0.1956 3014 | 68.5281 1162 | 13.4061 6428 | 0.0145 9255 | 0.0745 9255 |
| 29 | 5.4183 8790 | 0.1845 5674 | 73.6397 9832 | 13.5907 2102 | 0.0135 7961 | 0.0735 7961 |
| 30 | 5.7434 9117 | 0.1741 1013 | 79.0581 8622 | 13.7648 3115 | 0.0126 4891 | 0.0726 7891 |
| 31 | 6.0881 0064 | 0.1642 5484 | 84.8016 7739 | 13.9290 8599 | 0.0117 9222 | 0.0717 9222 |
| 32 | 6.4533 8668 | 0.1549 5740 | 90.8897 7803 | 14.0840 4339 | 0.0110 0234 | 0.0710 0234 |
| 33 | 6.8405 8988 | 0.1461 8622 | 97.3431 6471 | 14.2302 2961 | 0.0102 7293 | 0.0702 7293 |
| 34 | 7.2510 2528 | 0.1379 1153 | 104.1837 5460 | 14.3681 4114 | 0.0095 9843 | 0.0695 9843 |
| 35 | 7.6860 8679 | 0.1301 0522 | 111.4347 7987 | 14.4982 4636 | 0.0089 7386 | 0.0689 7386 |
| 36 | 8.1472 5200 | 0.1227 4077 | 119.1208 6666 | 14.6209 8713 | 0.0083 9483 | 0.0683 9483 |
| 37 | 8.6360 8712 | 0.1157 9318 | 127.2681 1866 | 14.7367 8031 | 0.0078 5743 | 0.0678 5743 |
| 38 | 9.1542 5235 | 0.1092 3885 | 135.9042 0578 | 14.8460 1916 | 0.0073 5812 | 0.0673 5812 |
| 39 | 9.7035 0749 | 0.1030 5552 | 145.0584 5813 | 14.9490 7468 | 0.0068 9377 | 0.0668 9377 |
| 40 | 10.2857 1794 | 0.0972 2219 | 154.7619 6562 | 15.0462 9687 | 0.0064 6154 | 0.0664 6154 |

**Table B.2** *(continued)*

$i = 6\%$

| $n$ | $(1 + i)^n$ | $(1 + i)^{-n}$ | $s_{\overline{n}|i}$ | $a_{\overline{n}|i}$ | $\dfrac{1}{s_{\overline{n}|i}}$ | $\dfrac{1}{a_{\overline{n}|i}}$ |
|---|---|---|---|---|---|---|
| 41 | 10.9028 6101 | 0.0917 1905 | 165.0476 8356 | 15.1380 1592 | 0.0060 5886 | 0.0660 5886 |
| 42 | 11.5570 3267 | 0.0865 2740 | 175.9505 4457 | 15.2245 4332 | 0.0056 8342 | 0.0656 8342 |
| 43 | 12.2504 5463 | 0.0816 2962 | 187.5075 7724 | 15.3061 7294 | 0.0053 3312 | 0.0653 3312 |
| 44 | 12.9854 8191 | 0.0770 0908 | 199.7580 3188 | 15.3831 8202 | 0.0050 0606 | 0.0650 0606 |
| 45 | 13.7646 1083 | 0.0726 5007 | 212.7435 1379 | 15.4558 3209 | 0.0047 0050 | 0.0647 0050 |
| 46 | 14.5904 8748 | 0.0685 3781 | 226.5081 2462 | 15.5243 6990 | 0.0044 1485 | 0.0644 1485 |
| 47 | 15.4659 1673 | 0.0646 5831 | 241.0986 1210 | 15.5890 2821 | 0.0041 4768 | 0.0641 4768 |
| 48 | 16.3938 7173 | 0.0609 9840 | 256.5645 2882 | 15.6500 2661 | 0.0038 9765 | 0.0638 9765 |
| 49 | 17.3775 0403 | 0.0575 4566 | 272.9584 0055 | 15.7075 7227 | 0.0036 6356 | 0.0636 6356 |
| 50 | 18.4201 5427 | 0.0542 8836 | 290.3359 0458 | 15.7618 6064 | 0.0034 4429 | 0.0634 4429 |

$i = 7\%$

| $n$ | $(1 + i)^n$ | $(1 + i)^{-n}$ | $s_{\overline{n}|i}$ | $a_{\overline{n}|i}$ | $\dfrac{1}{s_{\overline{n}|i}}$ | $\dfrac{1}{a_{\overline{n}|i}}$ |
|---|---|---|---|---|---|---|
| 1 | 1.0700 0000 | 0.9345 7944 | 1.0000 0000 | 0.9345 7944 | 1.0000 0000 | 1.0700 0000 |
| 2 | 1.1449 0000 | 0.8734 3873 | 2.0700 0000 | 1.8080 1817 | 0.4830 9179 | 0.5530 9179 |
| 3 | 1.2250 4300 | 0.8162 9788 | 3.2149 0000 | 2.6243 1604 | 0.3110 5167 | 0.3810 5167 |
| 4 | 1.3107 9601 | 0.7628 9521 | 4.4399 4300 | 3.3872 1126 | 0.2252 2812 | 0.2952 2812 |
| 5 | 1.4025 5173 | 0.7129 8618 | 5.7507 3901 | 4.1001 9744 | 0.1738 9069 | 0.2438 9069 |
| 6 | 1.5007 3035 | 0.6663 4222 | 7.1532 9074 | 4.7665 3966 | 0.1397 9580 | 0.2097 9580 |
| 7 | 1.6057 8148 | 0.6227 4974 | 8.6540 2109 | 5.3892 8940 | 0.1155 5322 | 0.1855 5322 |
| 8 | 1.7181 8618 | 0.5820 0910 | 10.2598 0257 | 5.9712 9851 | 0.0974 6776 | 0.1674 6776 |
| 9 | 1.8384 5921 | 0.5439 3374 | 11.9779 8875 | 6.5152 3225 | 0.0834 8647 | 0.1534 8647 |
| 10 | 1.9671 5136 | 0.5083 4929 | 13.8164 4796 | 7.0235 8154 | 0.0723 7750 | 0.1423 7750 |
| 11 | 2.1048 5195 | 0.4750 9280 | 15.7835 9932 | 7.4986 7434 | 0.0633 5690 | 0.1333 5690 |
| 12 | 2.2521 9159 | 0.4440 1196 | 17.8884 5127 | 7.9426 8630 | 0.0559 0199 | 0.1259 0199 |
| 13 | 2.4098 4500 | 0.4149 6445 | 20.1406 4286 | 8.3576 5074 | 0.0496 5085 | 0.1196 5085 |
| 14 | 2.5785 3415 | 0.3878 1724 | 22.5504 8786 | 8.7454 6799 | 0.0443 4494 | 0.1143 4494 |
| 15 | 2.7590 3154 | 0.3624 4602 | 25.1290 2201 | 9.1079 1401 | 0.0397 9462 | 0.1097 9462 |
| 16 | 2.9521 6375 | 0.3387 3460 | 27.8880 5355 | 9.4466 4860 | 0.0358 5765 | 0.1058 5765 |
| 17 | 3.1588 1521 | 0.3165 7439 | 30.8402 1730 | 9.7632 2299 | 0.0324 2519 | 0.1024 2519 |
| 18 | 3.3799 3228 | 0.2958 6392 | 33.9990 3251 | 10.0590 8691 | 0.0294 1260 | 0.0994 1260 |
| 19 | 3.6165 2754 | 0.2765 0833 | 37.3789 6479 | 10.3355 9524 | 0.0267 5301 | 0.0967 5301 |
| 20 | 3.8696 8446 | 0.2584 1900 | 40.9954 9232 | 10.5940 1425 | 0.0243 9293 | 0.0943 9293 |
| 21 | 4.1405 6237 | 0.2415 1309 | 44.8651 7678 | 10.8355 2733 | 0.0222 8900 | 0.0922 8900 |
| 22 | 4.4304 0174 | 0.2257 1317 | 49.0057 3916 | 11.0612 4050 | 0.0204 0577 | 0.0904 0577 |
| 23 | 4.7405 2986 | 0.2109 4688 | 53.4361 4090 | 11.2721 8738 | 0.0187 1393 | 0.0887 1393 |
| 24 | 5.0723 6695 | 0.1971 4662 | 58.1766 7076 | 11.4693 3400 | 0.0171 8902 | 0.0871 8902 |
| 25 | 5.4274 3264 | 0.1842 4918 | 63.2490 3772 | 11.6535 8318 | 0.0158 1052 | 0.0858 1052 |

**Table B.2** *(continued)*

$i = 7\%$

| $n$ | $(1 + i)^n$ | $(1 + i)^{-n}$ | $s_{\overline{n}|i}$ | $a_{\overline{n}|i}$ | $\dfrac{1}{s_{\overline{n}|i}}$ | $\dfrac{1}{a_{\overline{n}|i}}$ |
|---|---|---|---|---|---|---|
| 26 | 5.8073 5292 | 0.1721 9549 | 68.6764 7036 | 11.8257 7867 | 0.0145 6103 | 0.0845 6103 |
| 27 | 6.2138 6763 | 0.1609 3037 | 74.4838 2328 | 11.9867 0904 | 0.0134 2573 | 0.0834 2573 |
| 28 | 6.6488 3836 | 0.1504 0221 | 80.6976 9091 | 12.1371 1125 | 0.0123 9193 | 0.0823 9193 |
| 29 | 7.1142 5705 | 0.1405 6282 | 87.3465 2927 | 12.2776 7407 | 0.0114 4865 | 0.0814 4865 |
| 30 | 7.6122 5504 | 0.1313 6712 | 94.4607 8632 | 12.4090 4118 | 0.0105 8640 | 0.0805 8640 |
| 31 | 8.1451 1290 | 0.1227 7301 | 102.0730 4137 | 12.5318 1419 | 0.0097 9691 | 0.0797 9691 |
| 32 | 8.7152 7080 | 0.1147 4113 | 110.2181 5426 | 12.6465 5532 | 0.0090 7292 | 0.0790 7292 |
| 33 | 9.3253 3975 | 0.1072 3470 | 118.9334 2506 | 12.7537 9002 | 0.0084 0807 | 0.0784 0807 |
| 34 | 9.9781 1354 | 0.1002 1934 | 128.2587 6481 | 12.8540 0936 | 0.0077 9674 | 0.0777 9674 |
| 35 | 10.6765 8148 | 0.0936 6294 | 138.2368 7835 | 12.9476 7230 | 0.0072 3396 | 0.0772 3396 |
| 36 | 11.4239 4219 | 0.0875 3546 | 148.9134 5984 | 13.0352 0776 | 0.0067 1531 | 0.0767 1531 |
| 37 | 12.2236 1814 | 0.0818 0884 | 160.3374 0202 | 13.1170 1660 | 0.0062 3685 | 0.0762 3685 |
| 38 | 13.0792 7141 | 0.0764 5686 | 172.5610 2017 | 13.1934 7345 | 0.0057 9505 | 0.0757 9505 |
| 39 | 13.9948 2041 | 0.0714 5501 | 185.6402 9158 | 13.2649 2846 | 0.0053 8676 | 0.0753 8676 |
| 40 | 14.9744 5784 | 0.0667 8038 | 199.6351 1199 | 13.3317 0884 | 0.0050 0914 | 0.0750 0914 |
| 41 | 16.0226 6989 | 0.0624 1157 | 214.6095 6983 | 13.3941 2041 | 0.0046 5962 | 0.0746 5962 |
| 42 | 17.1442 5678 | 0.0583 2857 | 230.6322 3972 | 13.4524 4898 | 0.0043 3591 | 0.0743 3591 |
| 43 | 18.3443 5475 | 0.0545 1268 | 247.7764 9650 | 13.5069 6167 | 0.0040 3590 | 0.0740 3590 |
| 44 | 19.6284 5959 | 0.0509 4643 | 266.1208 5125 | 13.5579 0810 | 0.0037 5769 | 0.0737 5769 |
| 45 | 21.0024 5176 | 0.0476 1349 | 285.7493 1084 | 13.6055 2159 | 0.0034 9957 | 0.0734 9957 |
| 46 | 22.4726 2338 | 0.0444 9859 | 306.7517 6260 | 13.6500 2018 | 0.0032 5996 | 0.0732 5996 |
| 47 | 24.0457 0702 | 0.0415 8747 | 329.2243 8598 | 13.6916 0764 | 0.0030 3744 | 0.0730 3744 |
| 48 | 25.7289 0651 | 0.0388 6679 | 353.2700 9300 | 13.7304 7443 | 0.0028 3070 | 0.0728 3070 |
| 49 | 27.5299 2997 | 0.0363 2413 | 378.9989 9951 | 13.7667 9853 | 0.0026 3853 | 0.0726 3853 |
| 50 | 29.4570 2506 | 0.0339 4776 | 406.5289 2947 | 13.8007 4629 | 0.0024 5985 | 0.0724 5985 |

$i = 8\%$

| $n$ | $(1 + i)^n$ | $(1 + i)^{-n}$ | $s_{\overline{n}|i}$ | $a_{\overline{n}|i}$ | $\dfrac{1}{s_{\overline{n}|i}}$ | $\dfrac{1}{a_{\overline{n}|i}}$ |
|---|---|---|---|---|---|---|
| 1 | 1.0800 0000 | 0.9259 2593 | 1.0000 0000 | 0.9259 2593 | 1.0000 0000 | 1.0800 0000 |
| 2 | 1.1664 0000 | 0.8573 3882 | 2.0800 0000 | 1.7832 6475 | 0.4807 6923 | 0.5607 6923 |
| 3 | 1.2597 1200 | 0.7938 3224 | 3.2464 0000 | 2.5770 9699 | 0.3080 3351 | 0.3880 3351 |
| 4 | 1.3604 8896 | 0.7350 2985 | 4.5061 1200 | 3.3121 2684 | 0.2219 2080 | 0.3019 2080 |
| 5 | 1.4693 2808 | 0.6805 8320 | 5.8666 0096 | 3.9927 1004 | 0.1704 5645 | 0.2504 5645 |
| 6 | 1.5868 7432 | 0.6301 6963 | 7.3359 2904 | 4.6228 7966 | 0.1363 1539 | 0.2163 1539 |
| 7 | 1.7138 2427 | 0.5834 9040 | 8.9228 0336 | 5.2063 7006 | 0.1120 7240 | 0.1920 7240 |
| 8 | 1.8509 3021 | 0.5402 6888 | 10.6366 2763 | 5.7466 3894 | 0.0940 1476 | 0.1740 1476 |
| 9 | 1.9990 0463 | 0.5002 4897 | 12.4875 5784 | 6.2468 8791 | 0.0800 7971 | 0.1600 7971 |
| 10 | 2.1589 2500 | 0.4631 9349 | 14.4865 6247 | 6.7100 8140 | 0.0690 2949 | 0.1490 2949 |

**Table B.2** *(continued)*

$i = 8\%$

| $n$ | $(1 + i)^n$ | $(1 + i)^{-n}$ | $s_{\overline{n}|i}$ | $a_{\overline{n}|i}$ | $\dfrac{1}{s_{\overline{n}|i}}$ | $\dfrac{1}{a_{\overline{n}|i}}$ |
|---|---|---|---|---|---|---|
| 11 | 2.3316 3900 | 0.4288 8286 | 16.6454 8746 | 7.1389 6426 | 0.0600 7634 | 0.1400 7634 |
| 12 | 2.5181 7012 | 0.3971 1376 | 18.9771 2646 | 7.5360 7802 | 0.0526 9502 | 0.1326 9502 |
| 13 | 2.7196 2373 | 0.3676 9792 | 21.4952 9658 | 7.9037 7594 | 0.0465 2181 | 0.1265 2181 |
| 14 | 2.9371 9362 | 0.3404 6104 | 24.2149 2030 | 8.2442 3698 | 0.0412 9685 | 0.1212 9685 |
| 15 | 3.1721 6911 | 0.3152 4170 | 27.1521 1393 | 8.5594 7869 | 0.0368 2954 | 0.1168 2954 |
| 16 | 3.4259 4264 | 0.2918 9047 | 30.3242 8304 | 8.8513 6916 | 0.0329 7687 | 0.1129 7687 |
| 17 | 3.7000 1805 | 0.2702 6895 | 33.7502 2569 | 9.1216 3811 | 0.0296 2943 | 0.1096 2943 |
| 18 | 3.9960 1950 | 0.2502 4903 | 37.4502 4374 | 9.3718 8714 | 0.0267 0210 | 0.1067 0210 |
| 19 | 4.3157 0106 | 0.2317 1206 | 41.4462 6324 | 9.6035 9920 | 0.0241 2763 | 0.1041 2763 |
| 20 | 4.6609 5714 | 0.2145 4821 | 45.7619 6430 | 9.8181 4741 | 0.0218 5221 | 0.1018 5221 |
| 21 | 5.0338 3372 | 0.1986 5575 | 50.4229 2144 | 10.0168 0316 | 0.0198 3225 | 0.0998 3225 |
| 22 | 5.4365 4041 | 0.1839 4051 | 55.4567 5516 | 10.2007 4366 | 0.0180 3207 | 0.0980 3207 |
| 23 | 5.8714 6365 | 0.1703 1528 | 60.8932 9557 | 10.3710 5895 | 0.0164 2217 | 0.0964 2217 |
| 24 | 6.3411 8074 | 0.1576 9934 | 66.7647 5922 | 10.5287 5828 | 0.0149 7796 | 0.0949 7796 |
| 25 | 6.8484 7520 | 0.1460 1790 | 73.1059 3995 | 10.6747 7619 | 0.0136 7878 | 0.0936 7878 |
| 26 | 7.3963 5321 | 0.1352 0176 | 79.9544 1515 | 10.8099 7795 | 0.0125 0713 | 0.0925 0713 |
| 27 | 7.9880 6147 | 0.1251 8682 | 87.3507 6836 | 10.9351 6477 | 0.0114 4810 | 0.0914 4810 |
| 28 | 8.6271 0639 | 0.1159 1372 | 95.3388 2983 | 11.0510 7849 | 0.0104 8891 | 0.0904 8891 |
| 29 | 9.3172 7490 | 0.1073 2752 | 103.9659 3622 | 11.1584 0601 | 0.0096 1854 | 0.0896 1854 |
| 30 | 10.0626 5689 | 0.0993 7733 | 113.2832 1111 | 11.2577 8334 | 0.0088 2743 | 0.0888 2743 |
| 31 | 10.8676 6944 | 0.0920 1605 | 123.3458 6800 | 11.3497 9939 | 0.0081 0728 | 0.0881 0728 |
| 32 | 11.7370 8300 | 0.0852 0005 | 134.2135 3744 | 11.4349 9944 | 0.0074 5081 | 0.0874 5081 |
| 33 | 12.6760 4964 | 0.0788 8893 | 145.9506 2044 | 11.5138 8837 | 0.0068 5163 | 0.0868 5163 |
| 34 | 13.6901 3361 | 0.0730 4531 | 158.6266 7007 | 11.5869 3367 | 0.0063 0411 | 0.0863 0411 |
| 35 | 14.7853 4429 | 0.0676 3454 | 172.3168 0368 | 11.6545 6822 | 0.0058 0326 | 0.0858 0326 |
| 36 | 15.9681 7184 | 0.0626 2458 | 187.1021 4797 | 11.7171 9279 | 0.0053 4467 | 0.0853 4467 |
| 37 | 17.2456 2558 | 0.0579 8572 | 203.0703 1981 | 11.7751 7851 | 0.0049 2440 | 0.0849 2440 |
| 38 | 18.6252 7563 | 0.0536 9048 | 220.3159 4540 | 11.8288 6899 | 0.0045 3894 | 0.0845 3894 |
| 39 | 20.1152 9768 | 0.0497 1341 | 238.9412 2103 | 11.8785 8240 | 0.0041 8513 | 0.0841 8513 |
| 40 | 21.7245 2150 | 0.0460 3093 | 259.0565 1871 | 11.9246 1333 | 0.0038 6016 | 0.0838 6016 |
| 41 | 23.4624 8322 | 0.0426 2123 | 280.7810 4021 | 11.9672 3457 | 0.0035 6149 | 0.0835 6149 |
| 42 | 25.3394 8187 | 0.0394 6411 | 304.2435 2342 | 12.0066 9867 | 0.0032 8684 | 0.0832 8684 |
| 43 | 27.3666 4042 | 0.0365 4084 | 329.5830 0530 | 12.0432 3951 | 0.0030 3414 | 0.0830 3414 |
| 44 | 29.5559 7166 | 0.0338 3411 | 356.9496 4572 | 12.0770 7362 | 0.0028 0152 | 0.0828 0152 |
| 45 | 31.9204 4939 | 0.0313 2788 | 386.5056 1738 | 12.1084 0150 | 0.0025 8728 | 0.0825 8728 |
| 46 | 34.4740 8534 | 0.0290 0730 | 418.4260 6677 | 12.1374 0880 | 0.0023 8991 | 0.0823 8991 |
| 47 | 37.2320 1217 | 0.0268 5861 | 452.9001 5211 | 12.1642 6741 | 0.0022 0799 | 0.0822 0799 |
| 48 | 40.2105 7314 | 0.0248 6908 | 490.1321 6428 | 12.1891 3649 | 0.0020 4027 | 0.0820 4027 |
| 49 | 46.4274 1899 | 0.0230 2693 | 530.3427 3742 | 12.2121 6341 | 0.0018 8557 | 0.0818 8557 |
| 50 | 43.9016 1251 | 0.0213 2123 | 573.7701 5642 | 12.2334 8464 | 0.0017 4286 | 0.0817 4286 |

## B.3    BASIC GEOMETRIC FORMULAS

- **Pythagorean Theorem**
  $$c^2 = a^2 + b^2$$

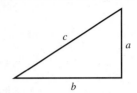

- **Rectangle**
  Area $= ab$
  Perimeter $= 2a + 2b$

- **Parallelogram**
  Height $= h$
  Area $= ah$
  Perimeter $= 2a + 2b$

- **Triangle**
  Height $= h$
  Area $= \frac{1}{2}hc$
  Perimeter $= a + b + c$

- **Trapezoid**
  Area $= \frac{1}{2}(a + b)h$

- **Circle**
  Radius $= r$
  Diameter $= d = 2r$
  Area $= \pi r^2$
  Circumference $= 2\pi r$
  $\pi = 3.14159$

- **Rectangular Solid**
  Volume $= abc$
  Total surface area $= 2ab + 2ac + 2bc$

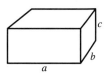

- **Right Circular Cylinder**
  Radius of base $= r$
  Height $= h$
  Volume $= \pi r^2 h$
  Total surface area $= 2\pi r^2 + 2\pi rh$

- **Sphere**
  Radius $= r$
  Diameter $= d = 2r$
  Volume $= \frac{4}{3}\pi r^3$
  Total surface area $= 4\pi r^2$

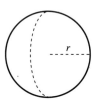

- **Right Circular Cone**
  Radius of base $= r$
  Height $= h$
  Volume $= \frac{1}{3}\pi r^2 h$
  Total surface area $= \pi r^2 + \pi r\sqrt{r^2 + h^2}$

## B.4   TABLE OF INTEGRALS

### Integrands Containing $(ax + b)$

**1.** $\displaystyle \int x(ax + b)^n \, dx = \frac{(ax + b)^{n+1}}{a^2}\left[\frac{ax + b}{n + 2} - \frac{b}{n + 1}\right]$
$(n \neq -1, -2)$

**2.** $\displaystyle \int \frac{x}{ax + b}\, dx = \frac{x}{a} - \frac{b}{a^2}\ln|ax + b|$

**3.** $\displaystyle \int \frac{x}{(ax + b)^2}\, dx = \frac{1}{a^2}\left[\ln|ax + b| + \frac{b}{ax + b}\right]$

4. $\displaystyle\int \frac{x^2}{ax + b}\, dx = \frac{1}{a^3}\left[\frac{1}{2}(ax + b)^2 - 2b(ax + b) + b^2 \ln |ax + b|\right]$

5. $\displaystyle\int \frac{x^2}{(ax + b)^2}\, dx = \frac{1}{a^3}\left[ax + b - \frac{b^2}{ax + b} - 2b \ln |ax + b|\right]$

6. $\displaystyle\int \frac{1}{x(ax + b)}\, dx = \frac{1}{b} \ln \left|\frac{x}{ax + b}\right| \qquad (b \neq 0)$

7. $\displaystyle\int \frac{1}{x^2(ax + b)}\, dx = -\frac{1}{bx} + \frac{a}{b^2} \ln \left|\frac{ax + b}{x}\right| \qquad (b \neq 0)$

8. $\displaystyle\int \frac{1}{x(ax + b)^2}\, dx = \frac{1}{b(ax + b)} - \frac{1}{b^2} \ln \left|\frac{ax + b}{x}\right| \qquad (b \neq 0)$

## Integrands Containing $\sqrt{ax + b}$

9. $\displaystyle\int x\sqrt{ax + b}\, dx = \frac{2}{a^2}\left[\frac{(ax + b)^{5/2}}{5} - \frac{b(ax + b)^{3/2}}{3}\right]$

10. $\displaystyle\int x^2\sqrt{ax + b}\, dx$
$$= \frac{2}{a^3}\left[\frac{(ax + b)^{7/2}}{7} - \frac{2b(ax + b)^{5/2}}{5} + \frac{b^2(ax + b)^{3/2}}{3}\right]$$

11. $\displaystyle\int \frac{x}{\sqrt{ax + b}}\, dx = \frac{2ax - 4b}{3a^2}\sqrt{ax + b}$

12. $\displaystyle\int \frac{1}{x\sqrt{ax + b}}\, dx = \frac{1}{\sqrt{b}} \ln \left|\frac{\sqrt{ax + b} - \sqrt{b}}{\sqrt{ax + b} + \sqrt{b}}\right| \qquad (b > 0)$

## Integrands Containing $a^2 \pm x^2$

13. $\displaystyle\int \frac{1}{a^2 - x^2}\, dx = \frac{1}{2a} \ln \left|\frac{x + a}{x - a}\right|$

14. $\displaystyle\int \frac{x}{a^2 \pm x^2}\, dx = \pm\frac{1}{2} \ln |a^2 \pm x^2|$

15. $\displaystyle\int \frac{1}{x(a^2 \pm x^2)}\, dx = \frac{1}{2a^2} \ln \left|\frac{x^2}{a^2 \pm x^2}\right|$

## Integrands Containing $\sqrt{a^2 - x^2}$

16. $\displaystyle\int \frac{x}{\sqrt{a^2 - x^2}}\, dx = -\sqrt{a^2 - x^2}$

17. $\displaystyle\int \frac{1}{x\sqrt{a^2 - x^2}}\, dx = -\frac{1}{a}\ln\left|\frac{a + \sqrt{a^2 - x^2}}{x}\right|$

18. $\displaystyle\int \frac{1}{x^2\sqrt{a^2 - x^2}}\, dx = -\frac{\sqrt{a^2 - x^2}}{a^2 x}$

19. $\displaystyle\int \frac{1}{(a^2 - x^2)^{3/2}}\, dx = \frac{1}{a^2}\frac{x}{\sqrt{a^2 - x^2}}$

20. $\displaystyle\int \frac{x}{(a^2 - x^2)^{3/2}}\, dx = \frac{1}{\sqrt{a^2 - x^2}}$

21. $\displaystyle\int \frac{\sqrt{a^2 - x^2}}{x}\, dx = \sqrt{a^2 - x^2} - a\ln\left|\frac{a + \sqrt{a^2 - x^2}}{x}\right|$

## Integrands Containing $\sqrt{x^2 \pm a^2}$

22. $\displaystyle\int \frac{1}{\sqrt{x^2 \pm a^2}}\, dx = \ln\left|x + \sqrt{x^2 \pm a^2}\right|$

23. $\displaystyle\int \frac{x}{\sqrt{x^2 \pm a^2}}\, dx = \sqrt{x^2 \pm a^2}$

24. $\displaystyle\int \frac{x^2}{\sqrt{x^2 \pm a^2}}\, dx = \frac{1}{2}x\sqrt{x^2 \pm a^2} \mp \frac{1}{2}a^2\ln\left|x + \sqrt{x^2 \pm a^2}\right|$

25. $\displaystyle\int \frac{1}{x\sqrt{x^2 + a^2}}\, dx = -\frac{1}{a}\ln\left|\frac{a + \sqrt{x^2 + a^2}}{x}\right|$

26. $\displaystyle\int \frac{1}{x^2\sqrt{x^2 \pm a^2}}\, dx = \mp\frac{\sqrt{x^2 \pm a^2}}{a^2 x}$

27. $\displaystyle\int \frac{1}{(x^2 \pm a^2)^{3/2}}\, dx = \pm\frac{1}{a^2}\frac{x}{\sqrt{x^2 \pm a^2}}$

28. $\displaystyle\int \frac{x}{(x^2 \pm a^2)^{3/2}}\, dx = -\frac{1}{\sqrt{x^2 \pm a^2}}$

29. $\displaystyle\int \frac{x^2}{(x^2 \pm a^2)^{3/2}}\, dx = -\frac{x}{\sqrt{x^2 \pm a^2}} + \ln\left|x + \sqrt{x^2 \pm a^2}\right|$

30. $\displaystyle\int x\sqrt{x^2 \pm a^2}\, dx = \frac{1}{3}(x^2 \pm a^2)^{3/2}$

31. $\displaystyle\int \sqrt{x^2 \pm a^2}\, dx = \frac{1}{2}x\sqrt{x^2 \pm a^2} \pm \frac{1}{2}a^2\ln\left|x + \sqrt{x^2 \pm a^2}\right|$

32. $\displaystyle\int \frac{\sqrt{x^2 + a^2}}{x}\, dx = \sqrt{x^2 + a^2} - a\ln\left|\frac{a + \sqrt{x^2 + a^2}}{x}\right|$

**33.** $\displaystyle \int \frac{\sqrt{x^2 \pm a^2}}{x^2}\, dx = -\frac{\sqrt{x^2 \pm a^2}}{x} + \ln\left|x + \sqrt{x^2 \pm a^2}\right|$

## Integrands Containing $(ax^2 + bx + c)$

**34.** $\displaystyle \int \frac{1}{ax^2 + bx + c}\, dx$

$$= \frac{1}{\sqrt{b^2 - 4ac}} \ln\left|\frac{2ax + b - \sqrt{b^2 - 4ac}}{2ax + b + \sqrt{b^2 - 4ac}}\right| \quad (b^2 - 4ac > 0)$$

## Integrands Containing $(ax^n + b)$

**35.** $\displaystyle \int \frac{1}{x(ax^n + b)}\, dx = \frac{1}{nb} \ln\left|\frac{x^n}{ax^n + b}\right| \qquad (n \neq 0,\ b \neq 0)$

**36.** $\displaystyle \int \frac{1}{x\sqrt{ax^n + b}}\, dx = \frac{1}{n\sqrt{b}} \ln\left|\frac{\sqrt{ax^n + b} - \sqrt{b}}{\sqrt{ax^n + b} + \sqrt{b}}\right| \qquad (b > 0)$

## Integrands Containing Exponentials and Logarithms

**37.** $\displaystyle \int x e^{ax}\, dx = \frac{1}{a^2}(ax - 1)e^{ax}$

**38.** $\displaystyle \int x^n e^{ax}\, dx = \frac{1}{a} x^n e^{ax} - \frac{n}{a}\int x^{n-1} e^{ax}\, dx$

**39.** $\displaystyle \int \frac{1}{b + ce^{ax}}\, dx = \frac{1}{ab}\left[ax - \ln(b + ce^{ax})\right] \qquad (ab \neq 0)$

**40.** $\displaystyle \int x^n \ln|x|\, dx = \frac{1}{n+1} x^{n+1}\left[\ln|x| - \frac{1}{n+1}\right] \qquad (n \neq -1)$

**41.** $\displaystyle \int (\ln x)^n\, dx = x\,(\ln x)^n - n\int (\ln x)^{n-1}\, dx$

## Miscellaneous Integrals

**42.** $\displaystyle \int \sqrt{\frac{x + a}{x + b}}\, dx = \sqrt{(x + b)(x + a)} + (a - b)\ln\left|\sqrt{x + b} + \sqrt{x + a}\right|$

# The Graphing Calculator

C.1    The Graphing Calculator

C.2    Programs

## C.1    THE GRAPHING CALCULATOR

► *Introduction*
► *Graphing*
► *Matrices and the Gauss–Jordan Method*
► *Matrix Inverses*
► *Statistical Calculations*

### Introduction

Inexpensive hand-held graphing calcualtors that are available today have a number of powerful and useful features. First of all, these graphing calculators can graph functions on a small screen. They also have a zoom feature that permits the user to magnify a particular part of the graph to essentially any desired amount. One considerable advantage of this zoom feature is the ability to find, to a considerable degree of accuracy, the zeros of a function or the solution of two linear or nonlinear equations.

Another important feature of these calculators is their ability to manipulate matrices of orders as large as $6 \times 6$. The calculator can not only add, subtract, and multiply matrices of these orders, but can also find the inverses. These calculators also have the capability to perform elementary row operations on a matrix, permitting the user to apply the Gauss–Jordan method to systems of equations.

These calculators also can perform a variety of statistical calculations, such as finding the mean, standard deviation, linear and nonlinear regression, least square equations, and correlation coefficients.

Finally, and very important, these calculators have a programing feature that gives the user the flexibility to create individual programs much like those of a small personal computer. Many such programs are given in the second section of this appendix.

The most popular graphing calculators are the Texas Instruments TI-81 graphing calculator (or the more powerful TI-82 and TI-85), the Casio power graphic fx-7700G, the Sharp EL-5200, and the Hewlett-Packard HP-28S and HP-48. In this appendix we will demonstrate the features of the TI-81. Since all of the calculators have similar features, you should be able to translate the instructions to the other calculators by carefully reading the appropriate manual.

## Graphing

We will demonstrate the graphing capabilities of the TI-81 by showing how to graph each of the equations in the system

$$3x + 9y = 45$$

$$2x + y = 10$$

that we considered at the beginning of Section 2.1. In order to graph these equations on the TI-81, we first need to solve each of the linear equations for $y$ and obtain

$$y = -\frac{1}{3}x + 5$$

$$y = -2x + 10$$

After turning on the TI-81, press the $\boxed{Y=}$ key and the first line will read

$$: Y_1 =$$

with a blinking cursor set right after the $=$ sign. To obtain the term $-\dfrac{x}{3} + 5$, depress the following keys

$$\boxed{(-)}\ \boxed{X|T}\ \boxed{\div}\ \boxed{3}\ \boxed{+}\ \boxed{5}\ \boxed{ENTER}$$

The $\boxed{X|T}$ key is the same as $x$. Now move the cursor to the point after the equal sign of the expression

$$: Y_2 =$$

on the second line. To input the expression $-2x + 10$, depress the keys

$$\boxed{(-)}\ \boxed{2}\ \boxed{X|T}\ \boxed{+}\ \boxed{10}\ \boxed{ENTER}$$

Now press the $\boxed{\text{ZOOM}}$ key and then $\boxed{6}$. A graph of each of the two equations will appear on the screen.

In order to find the solution of the sytem, we need to find the coordinates of the point of intersection of the two lines. To do this depress the $\boxed{\text{ZOOM}}$ key again and then depress $\boxed{2}$. The graph will reappear on the screen with a flashing cursor. By using the arrow buttons near the top right of the calculator move the cursor to the point of intersection and again depress the $\boxed{\text{ENTER}}$ key. The graph reappears, but this time in magnified form. The flashing cursor should be near the point of intersection. Place the cursor at the point of intersection. The numbers at the bottom of the screen are the coordinates of the cursor and represent an approximation of the solution to the system. Depress the $\boxed{\text{ENTER}}$ key again and repeat this process until you have obtained the coordinates to sufficient accuracy. After only three or four repetitions, the coordinates should already be approximately

$$x = 3.001 \qquad y = 4.001.$$

After only several more repetitions, you should obtain an answer something like

$$x = 3.00000064 \qquad y = 4.0000019$$

After several more repetitions, you should finally obtain

$$x = 3 \qquad y = 4.$$

Notice that the graph was drawn on a screen with the $x$-axis and $y$-axis going from $-10$ to $+10$. Often the range of the $x$- or $y$-axis needs to be adjusted. To change the range, depress the $\boxed{\text{RANGE}}$ key and set the range as desired.

## Matrices and the Gauss–Jordan Method

We will now see how to use the TI-81 to perform the steps of the Gauss–Jordan method. Consider the system in Example 1 of Section 2.2

$$-3x + 6y = 12$$
$$2x - 4y = -8.$$

Our first step is replace this system with the augmented matrix

$$\begin{bmatrix} -3 & 6 & | & 12 \\ 2 & -4 & | & -8 \end{bmatrix}.$$

To enter this matrix into the calculator first depress the two keys

$$\boxed{\text{MATRX}} \ \boxed{\triangleright}.$$

Now depress the $\boxed{1}$ key to create the matrix $A$. The cursor is now flashing over the first number, designating the order of the matrix. Since our matrix has order $2 \times 3$, depress $\boxed{2}$ $\boxed{\text{ENTER}}$ and then $\boxed{3}$ $\boxed{\text{ENTER}}$. Now the cursor asks for $a_{11}$, so depress the keys $\boxed{(-)}$ $\boxed{3}$ $\boxed{\text{ENTER}}$. The cursor now asks for $a_{12}$. So depress the keys $\boxed{6}$ $\boxed{\text{ENTER}}$, and so forth, until all the entries of the matrix have been inputted. Now depress the keys 2nd CLEAR and return to the home screen. To display your matrix depress 2nd 1 ENTER.

The first step in the Gauss–Jordan method is to divide the first row by $-3$. To do this on the calculator depress the keys MATRX 3. On the screen you will see

$$*\text{Row}($$

with a flashing cursor after the ( symbol. Now depress the following keys:

(−) 1 ÷ 3 ALPHA · 2nd 1 ALPHA · 1 )

The line should read

$$*\text{Row}(-1/3,[A],1)$$

This indicates that we are multiplying the first row of the matrix $A$ by $-1/3$.
Now depress ENTER and the screen should read

$$\begin{bmatrix} 1 & -2 & | & -4 \\ 2 & -4 & | & -8 \end{bmatrix}$$

This answer is stored in ANS and by pressing the keys 2nd (−) we can access this answer.

The next step in the Gauss–Jordan method is $R_2 - 2R_1 \to R_2$. To do this on the calculator depress the keys MATRX 4. On the screen you will see

$$*\text{Row}+($$

with a flashing cursor after the ( symbol. Now depress the following keys

(−) 2 ALPHA · 2nd (−) ALPHA · 1 ALPHA · 2 )

The line should read

$$*\text{Row} + (-2,\text{Ans},1,2)$$

This indicates that we are multiplying the first row of the matrix $A$ by $-2$ and adding the result to the second row.
Now depress ENTER and the screen should read

$$\begin{bmatrix} 1 & -2 & | & -4 \\ 0 & 0 & | & 0 \end{bmatrix}$$

This is the final step in the Gauss–Jordan method.
Swapping two rows is done in a similar fashion.


## Matrix Inverses

Finding the inverse of a square matrix on the calculator is particularly easy. We shall find the inverse of the matrix

$$\begin{bmatrix} 1 & 1 & 2 \\ 2 & 3 & 2 \\ 1 & 1 & 3 \end{bmatrix}$$

found in Example 2 of Section 2.5.

Enter this $3 \times 3$ matrix as the $A$ matrix in your calculator. Return to the home screen and simply depress the keys

| 2nd | | 1 | | $x^{-1}$ | | ENTER |

The inverse of the matrix will appear on the screen.

## Statistical Calculations

The calculator can easily calculate the means and standard deviations requested in Chapter 8. For example, let us calculate the mean found in Example 1 of Section 8.2. This data is as follows:

| Number Sold in One Quarter | 1 | 2 | 3 | 4 | 5 |
|---|---|---|---|---|---|
| Frequency of Occurrence | 10 | 15 | 16 | 7 | 2 |

To begin, first clear the statistical section of all old data by depressing the keys

| 2nd | | MATRX | | ▷ | | ▷ | | 2 | | ENTER |

Now, to enter the data, first depress the keys

| 2nd | | MATRX | | ▷ | | ▷ | | 1 |

Now enter 1 for $x_1$, 10 for $y_1$, and so forth, until all the data is entered. Then depress the keys

| 2nd | | MATRX | | 1 | | ENTER |

The first term $\bar{x} = 2.52$ is the mean.

To find the standard deviation of the data 1.020, .980, 1.030, .970, given for the first assembly line at the beginning of Section 8.5, first clear the statistical section of all data as before. Now enter this data as before, except that you do not change the $y_i$'s which are already set equal to 1. Repeat the previous steps and you will obtain the standard deviation as $\sigma_x = .0254950976$.

## C.2   PROGRAMS

► *Limit*

► *SecantN*

► *SecantG*

► *Tangent*

► *Riemann*

► *Trapezoi*

► *Simpson*

The preceding list cites all the programs that are in this section. The programs are written for the TI-81.


## Program Limit

This program evaluates a function at the points $a \pm 0.1$, $a \pm 0.01$, ..., and stores the answers in the data section.

:Disp "A="
:Input X
:Disp "N DECIMAL
S"
:Input N
:Sci
:ClrStat
:1→Arow
:2→Acol
:1→I
:.1→D
:X→A
:Lbl 1
:A-D→X
:$Y_1$→{x}($I$)
:$Y_1$→[A](1,1)
:A+D→X
:$Y_1$→{y}($I$)
:$Y_1$→[A](1,2)
:Disp [A]
:Pause
:D/10→D
:$I$+1→$I$
:If I<$N$+.5
:Goto 1

## Program SecantN

This program evalautes the slope of the secant line $\dfrac{f(a + h) - f(a)}{h}$ for values of $h$ approaching zero and stores them in the data section

```
:Disp "X="
:Input X
:Disp "N DECIMAL
S"
:Input N
:Sci
:ClrStat
:1→Arow
:2→Acol
:1→I
:.1→H
:X→A
:Y₁→B
:Lbl 1
:A-H→X
:(B − Y₁)/H→{x}(I)
:(B − Y₁)/H→[A](1,
1)
:A+H→X
:(Y₁ − B)/H→{ y}(I)
:(Y₁ − B)/H→[A](1,
2)
:Disp [A]
:Pause
:H/10→H
:1+I→I
:If I<N+.5
:Goto 1
```

## Program SecantG

This program first draws a graph of the function $y = f(x)$ and a tangent line at an inputed point. The program then graphs a secant line and successive secant lines for values of $h$ approaching zero. One then sees in the viewing rectangle the secant lines approaching the tangent line.

```
:ClrDraw
:Disp "X="
:Input X
:Disp "N STEP"
:Input N
:X→A
:Y₁→B
:Xmax-X→W
:W/100→E
:NDeriv(Y₁,E)→M
:Xmax→C
:B+WM→D
:Line(A,B,C,D)
:DispGraph
:Pause
:1→H
:1→J
:Lbl 1
:A+HW→X
:Y₁→D
:(D-B)/(HW)→M
:B+WM→D
:Line(A,B,C,D)
:Pause
:H/2^.5→H
:J+1→J
:If J<N+.5
:Goto 1
:Disp "END"
```

## Program Tangent

This program plots graphs of tangent lines along the graph of the curve.

```
:ClrDraw
:DispGraph
:Pause
```

:Disp ''N POINTS''

:Input N

:$(X_{max}-X_{min})/N{\to}S$

:$X_{min}+S/2{\to}X$

:$(X_{max}-X_{min})/100$

${\to}E$

:Lbl 1

:NDeriv($Y_1$,E)${\to}M$

:PT-On(X,M)

:$Y_1{\to}Y$

:X-10E${\to}A$

:Y-10EM${\to}B$

:X+10E${\to}C$

:Y+10EM${\to}D$

:Line(A,B,C,D)

:Pause

:X+S${\to}X$

:If $X{<}X_{max}$

:Goto 1

:Disp ''END''


## Program Riemann

This program evaluates three Riemann sums $\Sigma_{i=1}^{n}\ f(x_i)\ \Delta x$, taking $x_i$ at the left endpoint, the midpoint, and the right endpoint of the subintervals.

:All-Off

:Disp ''LOWER LIM
IT''

:Input A

:Disp ''UPPER LIM
IT''

:Input b

:Disp ''N DIVISIO
NS''

:Input N

:1${\to}J$

:0${\to}S$

:0${\to}T$

:0→U
:(B-A)/N→W
:Lbl 1
:A+(J-1)W→L
:A+JW→R
:(L+R)/2→M
:L→X
:$Y_1$→L
:S+LW→S
:M→X
:$Y_1$→M
:T+MW→T
:R→X
:$Y_1$→R
:U+RW→U
:IS>(J,N)
:Goto 1
:Disp "LEFT ENDP
T"
:Disp S
:Disp "MID PT"
:Disp T
:Disp "RIGHT END
PT"
:Disp U

## Program Trapezoi

This program uses the trapezoidal rule to approximate $\int_a^b f(x)\ dx$.

:Disp "LOWER LIM
IT"
:Input A
:Disp "UPPER LIM
IT"
:Input B
:Disp "DIVISION
S"
:Input N

:$0 \rightarrow S$

:$(B-A)/N \rightarrow W$

:$1 \rightarrow J$

:Lbl 1

:$A + JW \rightarrow X$

:$Y_1 \rightarrow Y$

:$S + Y_1 \rightarrow S$

:$J + 1 \rightarrow J$

:If $J < N$

:Goto 1

:$A \rightarrow X$

:$S + Y_1/2 \rightarrow S$

:$B \rightarrow X$

:$S + Y_1/2 \rightarrow S$

:$SW \rightarrow S$

:Disp ''AREA = ''

:Disp S


## Program Simpson

This program uses Simpson's rule to approximate $\int_a^b f(x)\ dx$.

:Disp ''LOWER LIM

IT''

:Input A

:Disp ''UPPER LIM

IT''

:Input B

:Disp ''DIVISION

S''

:Input D

:$0 \rightarrow S$

:$(B - A)/2D \rightarrow W$

:$1 \rightarrow J$

:Lbl 1

:$A + 2(J - 1)W \rightarrow L$

:$A + 2JW \rightarrow R$

:$(L + R)/2 \rightarrow M$

:L→X
:$Y_1$→L
:M→X
:$Y_1$→M
:R→X
:$Y_1$→R
:W(L+4M+R)/3+S→S
:IS>(J,D)
:Goto 1
:Disp ''AREA=''
:Disp S

# Answers

---

**Section 1.1**

**1.** Yes, No    **3.** Yes, No

**5.** $\{x|1 \leq x \leq 15, x \text{ odd integer}\}$, $\{x|3 \leq x \leq 21, x \text{ a multiple of } 3\}$

**7.** $\{x|2 \leq x \leq 16, x \text{ an even integer}\}$, $\{x|x \text{ an odd positive integer}\}$

**9.** $\{5, 6, 7, 8, 9, 10, 11\}$, $\{2, 4, 6, 8, 10, 12\}$

**11.** $\{4, 5, 6, 7, 8, 9, 10\}$, $\{1, 3, 5, 7, 9, 11, 13\}$    **13.** F, T, T, F

**15.**

$\longleftarrow \quad\quad\quad\quad\quad\quad \longrightarrow$
$\quad -1 \quad\quad\quad\quad\quad\quad\quad 3$

**17.**

$\longleftarrow \quad\quad\quad\quad\quad\quad \dashv$
$\quad 2 \quad\quad\quad\quad\quad\quad\quad 5$

**19.**

$\longleftarrow \quad\quad\quad\quad\quad\quad \longrightarrow$
$\quad -2$

**21.**

$\longleftarrow \quad\quad\quad\quad\quad\quad \dashv$
$\quad\quad\quad\quad\quad\quad\quad\quad 1.5$

**23–33.**

**35.**

**37.**

**39.**

**41.**

**43.**

**45.**

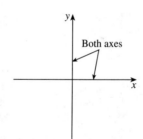

**47.** $\sqrt{5}$

**49.** $\sqrt{37}$　　**51.** $5\sqrt{2}$　　**53.** $\sqrt{2}|b - a|$

**55.**

**57.**

**59.**

**61.**

**63.**

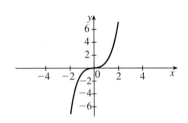

**65.**  25 miles        **67.**  13 miles

**69.**  10 days        **71.**  $100x + 150y = 3000$

**73.**  $125x + 190y = 800$        **75.**  $c = \sqrt{a^2 - b^2}$

**Section 1.2**

**1.**  4        **3.**  $-3$        **5.**  0        **7.**  Undefined        **9.**  $2, \frac{1}{2}, -1$        **11.**  $-1, 1, 1$

**13.**  $\frac{1}{2}, -4, 2$        **15.**  Undefined, 3, no $y$-intercept        **17.**  0, no $x$-intercept, 4

**19.**  $\frac{1}{2}, 1, -\frac{1}{2}$        **21.**  $y = -3x + 7$        **23.**  $y = -3$        **25.**  $y = -x + 2$

**27.**  $x = -2$        **29.**  $y = -\frac{b}{a}x + b$        **31.**  $y = -\frac{2}{5}x + \frac{11}{5}$        **33.**  $y = 0$

**35.**  $x = 1$        **37.**  $y = 2x + 5$        **39.**  $y = -\frac{5}{3}x + 4$

**41.**  $y = -\frac{2}{5}x + 4$        **43.**  No        **45.**  $10x + 25y = 235$

**47.**  $.08x + .05y = 576$, \$160        **49.**  $3x + 4y = 45, 6$

**51.**  $m = 89.83$. The length of the twig increases 89.83 times as much as the diameter does.

**Section 1.3**

**1.**  $C = 3x + 10{,}000$        **3.**  $C = .02x + 1000$        **5.**  $R = 5x$        **7.**  $R = 0.1x$

**9.**  $P = 2x - 10{,}000$        **11.**  $P = .08x - 1000$        **13.**  $x = 2$        **15.**  $x = 20$

**17.**  $(1.5, 4.5)$        **19.**  $(1, 15)$        **21.**  $V = -5000t + 50{,}000$, \$45,000, \$25,000

**23.**  $p = -.005x + 22.5$        **25.**  (a) $3x + 4y = 120$ (b) $m = -\frac{3}{4}$, making 4 more bookcases requires making 3 fewer desks.

**27.**  (a) $y - 1000 + .05x, y = 1500 + .04x$ (b) \$50,000

**29.** $t = \frac{25}{23}$

**31.** $y = 6x + 23$, \$4.5 million.

**33.** Outside  **35.** (a) Manual (b) Automatic

| | Total Costs | |
|---|---|---|
| **Volume** | Manual | Automatic |
| 1,000 | \$  2600 | \$8020 |
| 10,000 | \$17,000 | \$8200 |

## Section 1.4

**1.** $y = .50x + .50$  **3.** $y = 1.1x + 0.1$  **5.** $y = -0.9x + 4.5$
**7.** $y = -0.7x + 3.4$  **9.** $y = 5x + 5.4$, 35.4 (or rounded to 35)
**11.** $y = x + 2.5$, \$5.5 million.  **13.** $y = .1006x - 167.12$

## Chapter 1 Review

**1.** $\{x|x$ is an odd positive integer$\}$  **2.** (a) $\{m, i, s, p\}$ (b) $\{-1, 3\}$

**3–6.**

**7.**

**8.** $\sqrt{34}$

**9.**

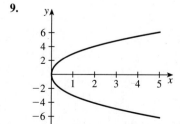

**10.** $-\frac{7}{3}$   **11.** $\frac{3}{2}$   **12.** Does not exist   **13.** $3, -\frac{2}{3}, 2$

**14.** $0$, no $x$-intercept, $-2$   **15.** No slope, 3, no $y$-intercept

**16.** $y - 5 = -2(x + 2)$   **17.** $y = -x - 2$   **18.** $y + 5 = 3(x + 3)$

**19.** $y - 7 = \frac{1}{4}(x + 3)$   **20.** $C = 6x + 2000$   **21.** $R = 10x$

**22.** $P = 4x - 2000, 500$   **23.** $x = 150$   **24.** $x = 1000$

**25.** 150 miles   **26.** 42 minutes   **27.** $p - 1 = -\frac{1}{200,000}(x - 100,000)$

**28.** 12.000   **29.** $4.5x + 5.5y = 15$

**30.** (a) Boston (b) Houston   **31.** $y = \frac{5}{7}x + \frac{1}{7}$

### Section 2.1

**1.** $(2, 5)$   **3.** $(1, -2)$   **5.** $(-1, -1)$   **7.** $(-2, 1)$   **9.** No solution

**11.** No solution   **13.** No solution   **15.** No solution   **17.** $(1, 2, 0)$

**19.** $(5, 0, 5)$   **21.** No solution   **23.** $d = 20, q = 5$

**25.** $n = 10, d = 20, q = 6$   **27.** 10 of A, 6 of B

**29.** \$700 in first bank, \$300 in second   **31.** 160 of style A and 300 of style B

**33.** 3 of type A, 5 of type B, 4 of type C   **35.** Since the solution requires a *negative* amount (-\$300,000) in low-risk stocks, the client's demands are impossible to meet.

### Section 2.2

**1.** Yes   **3.** Yes   **5.** No   **7.** $\left.\begin{array}{l} x = 2 \\ y = 3 \end{array}\right\}, x = 2, y = 2$

**9.** $\left.\begin{array}{l} x + 2y = 4 \\ 0 = 0 \end{array}\right\}, x = 4 - 2t, y = t$, $t$ a parameter

**11.** $\left.\begin{array}{l} x + 2z + 3u = 4 \\ y + 2z + 3u = 5 \end{array}\right\}, x = 4 - 2s - 3t, y = 5 - 2s - 3t, z = s, u = t,$

$s$ and $t$ parameters

**13.** $\left.\begin{array}{l} x + 2z + 4u = 6 \\ y + 2z + 3u = 1 \\ 0 = 0 \end{array}\right\}, x = 6 - 2s - 4t, y = 1 - 2s - 3t, z = s, u = t,$

$s$ and $t$ parameters

**15.** $\left.\begin{array}{l} y + 2z + 2v = 1 \\ u + v = 2 \\ w = 3 \end{array}\right\}, x = r, y = 1 - 2s - 2t, z = s, u = 2 - t, v = t, w = 3,$

$r$ and $s$ and $t$ are parameters   **17.** $(1, 2)$   **19.** $(1, 2, 3)$   **21.** $(\frac{3}{2}, \frac{3}{2}, -\frac{1}{2})$

**23.** $(-1, 1, 1)$   **25.** $(6 - \frac{7}{3}t, 10 - \frac{8}{3}t, t)$   **27.** $(-6 + 3s + 2t, s, t)$

**29.** $(-7 + 5s - 3t, s, t)$   **31.** $(1.5, -.5 - t, t)$   **33.** No solution

**35.** $(1, t, t)$   **37.** $(2, 1, 1, 2)$   **39.** $(1, -1, 2, 3)$   **41.** $(1, 3)$

**43.** No solution   **45.** Solution is $(d - 2b + (2a - c)t, \frac{3b - d}{2} - (\frac{3a - c}{2})t, t)$

**47.** $(n, d, q) = (-8 + 3t, 44 - 4t, t), 3 \le t \le 11$

**49.** 9 of type A and 4 of type B or 3 of type A, 5 of type B, and 4 of type C

**51.** 90,000 − 2*t* dollars in low risk stocks, *t* dollars in medium risk stocks, and *t* dollars in bonds, where $0 \le t \le 45,000$     **53.** 100 of style A, 150 of B, and 50 of C
**55.** Fired. Solution is $(150, -50 - 2s, s, 500)$ requiring a negative number of Y
**57.** 3 oranges, 2 cups strawberries, 1 cup blackberries
**59.** System is $x_1 + x_4 = 700$, $x_1 + x_2 = 600$, $x_2 + x_3 = 800$, $x_3 + x_4 = 900$. Solution is $x_1 = 700 - t$, $x_2 = -100 + t$, $x_3 = 900 - t$, $x_4 = t$, where $100 \le t \le 700$.

**Section 2.3**

**1.** $4 \times 2$     **3.** $1 \times 2$     **5.** 1     **7.** 8     **9.** −2     **11.** 6     **13.** 2

**15.** $\begin{bmatrix} 5 & 5 & 5 \\ 5 & 5 & 5 \end{bmatrix}$     **17.** $\begin{bmatrix} 1 & 0 & 0 \\ 0 & 1 & 0 \\ 0 & 0 & 1 \end{bmatrix}$     **19.** $\begin{bmatrix} 0 & 0 & 0 \\ 0 & 0 & 0 \\ 0 & 0 & 0 \end{bmatrix}$

**21.** $x = 3, y = 0, z = 4$     **23.** $\begin{bmatrix} 4 & 5 \\ 1 & 10 \\ -3 & 10 \\ 5 & 6 \end{bmatrix}$     **25.** $[0 \quad 6]$

**27.** $\begin{bmatrix} 2 & -3 \\ -5 & -2 \\ 3 & -4 \\ -1 & 10 \end{bmatrix}$     **29.** $[-2 \quad -2]$     **31.** $\begin{bmatrix} 2 & 8 \\ 6 & 12 \\ -6 & 14 \\ 6 & -4 \end{bmatrix}$     **33.** $[4 \quad 16]$

**35.** $\begin{bmatrix} 7 & 6 \\ -1 & 14 \\ -3 & 13 \\ 7 & 14 \end{bmatrix}$     **37.** $[1 \quad 16]$     **39.** Not defined     **41.** $\begin{bmatrix} 0 & 0 \\ 0 & 0 \\ 0 & 0 \\ 0 & 0 \end{bmatrix}$

**43.** The element in the *i*th row and *j*th column of $A + B$ is $a_{ij} + b_{ij}$. But this equals $b_{ij} + a_{ij}$ which is the element in the *i*th row and *j*th column of $B + A$.
**45.** The element in the *i*th row and *j*th column of $A + O$ is $a_{ij} + 0$. But this equals $a_{ij}$ which is the element in the *i*th row and *j*th column of $A$.
**47.** The element in the *i*th row and *j*th column of $A - A$ is $a_{ij} - a_{ij}$. But this equals 0, which is the element in the *i*th row and *j*th column of $O$.
**49.** $X + A = (B - A) + A = B + (-A + A) = B + O = B$

**51.** $M + A = \begin{bmatrix} 520 & 310 & 100 \\ 365 & 310 & 730 \end{bmatrix}$     **53.** $.9A = \begin{bmatrix} 270 & 180 & 90 \\ 180 & 180 & 360 \end{bmatrix}$

**55.** $N = \begin{bmatrix} 250 & 300 & 350 & 300 \\ 300 & 200 & 250 & 150 \\ 200 & 240 & 320 & 220 \end{bmatrix}, L = \begin{bmatrix} 300 & 350 & 450 & 400 \\ 320 & 240 & 280 & 250 \\ 250 & 260 & 420 & 280 \end{bmatrix},$

$$N + L = \begin{bmatrix} 550 & 650 & 800 & 700 \\ 620 & 440 & 530 & 400 \\ 450 & 500 & 740 & 500 \end{bmatrix}, L - N = \begin{bmatrix} 50 & 50 & 100 & 100 \\ 20 & 40 & 30 & 100 \\ 50 & 20 & 100 & 60 \end{bmatrix},$$

$$1.2N = \begin{bmatrix} 300 & 360 & 420 & 360 \\ 360 & 240 & 300 & 180 \\ 240 & 288 & 384 & 264 \end{bmatrix}, 0.5(N + L) = \begin{bmatrix} 275 & 325 & 400 & 350 \\ 310 & 220 & 265 & 200 \\ 225 & 250 & 370 & 250 \end{bmatrix}$$

**Section 2.4**

**1.** 16    **3.** 23    **5.** Not defined

**7.** Order of $AB$ is $2 \times 4$. Order of $BA$ is undefined.

**9.** Order of $AB$ is undefined. Order of $BA$ is $5 \times 4$

**11.** Order of $AB$ is $3 \times 5$. Order of $BA$ is undefined.

**13.** Order of $AB$ and $BA$ are $6 \times 6$.

**15.** Order of $AB$ is $20 \times 30$. Order of $BA$ is undefined.    **17.** $\begin{bmatrix} 29 \\ 23 \end{bmatrix}$

**19.** $[43 \quad 29]$    **21.** $\begin{bmatrix} .9 & .8 \\ 3.3 & 2.2 \end{bmatrix}$    **23.** $\begin{bmatrix} 36 \\ 18 \end{bmatrix}$    **25.** $\begin{bmatrix} .8 & 2 \\ 1.3 & 2.8 \end{bmatrix}$

**27.** $\begin{bmatrix} 3 & 30 & 31 & 14 \\ 8 & 46 & 26 & 26 \end{bmatrix}$    **29.** $[.07 \quad .18 \quad .05]$    **31.** $\begin{bmatrix} 26 \\ 10 \\ 12 \end{bmatrix}$

**33.** $[-4 \quad 4 \quad 21]$    **35.** $\begin{bmatrix} 5 & 10 & 21 \\ 0 & 4 & 8 \\ 16 & 5 & 15 \end{bmatrix}$    **37.** Not defined    **39.** $\begin{bmatrix} 6 & 10 \\ 15 & 25 \end{bmatrix}$

**41.** Not defined    **47.** $(A + B)^2 = \begin{bmatrix} 8 & 8 \\ 32 & 40 \end{bmatrix}, A^2 + 2AB + B^2 = \begin{bmatrix} 13 & 13 \\ 32 & 35 \end{bmatrix}$

**49.** $A(B + C) = AB + AC = \begin{bmatrix} 7 & 6 \\ 10 & 9 \end{bmatrix}, A(BC) = (AB)C = \begin{bmatrix} 1 & 1 \\ 1 & 1 \end{bmatrix}$

**53.** $A = \begin{bmatrix} 2 & 3 \\ 3 & -5 \end{bmatrix} X = \begin{bmatrix} x_1 \\ x_2 \end{bmatrix}, B = \begin{bmatrix} 5 \\ 7 \end{bmatrix}$

**55.** $A = \begin{bmatrix} 2 & 5 & 3 \\ 4 & -7 & -2 \\ 5 & -2 & 6 \end{bmatrix}, X = \begin{bmatrix} x_1 \\ x_2 \\ x_3 \end{bmatrix}, B = \begin{bmatrix} 16 \\ 12 \\ 24 \end{bmatrix}$

**57.** $A = \begin{bmatrix} 2 & -3 & 3 \\ 5 & 6 & -2 \end{bmatrix}, X = \begin{bmatrix} x_1 \\ x_2 \\ x_3 \end{bmatrix}, B = \begin{bmatrix} 3 \\ 1 \end{bmatrix}$

**59.** $A = [2 \quad 3], X = \begin{bmatrix} x_1 \\ x_2 \end{bmatrix}, B = [7]$

**61.** $NR = \begin{bmatrix} 27 & 5.3 & 42 \\ 25 & 4.4 & 38 \\ 21 & 3.4 & 35 \end{bmatrix}$, $QC = \begin{bmatrix} 2954 \\ 2714 \\ 2271 \end{bmatrix}$ The number 27 in the first row and first

column indicates that 27 hundred tires are needed in LA. The number 38 in the second row and third column indicates that 38 hundred plugs are needed in NYC, and so forth.

**63.** (a) $B = \begin{bmatrix} .4 & 0 & 0 \\ .6 & 1 & .5 \\ 0 & 0 & .5 \end{bmatrix}$ (b) $C = \begin{bmatrix} 1 & .2 & 0 \\ 0 & .3 & 0 \\ 0 & .2 & .4 \\ 0 & .3 & .6 \end{bmatrix}$ (c) $W = CZ = C(BY) = (CB)Y =$

$(CB)(AX) = (CBA)X$ (d) $w_1 = 1930$, $w_2 = 1695$, $w_3 = 2550$, $w_4 = 2825$.

**Section 2.5**

**1.** Yes **3.** Yes **5.** Yes **7.** Yes **9.** Yes **11.** $\begin{bmatrix} 10 & -3 \\ -3 & 1 \end{bmatrix}$

**13.** $\frac{1}{7}\begin{bmatrix} 2 & 1 \\ 3 & 5 \end{bmatrix}$ **15.** No inverse **17.** No inverse **19.** $\begin{bmatrix} 1 & -.5 & 0 \\ -1 & 1 & -1 \\ 1 & -.5 & 1 \end{bmatrix}$

**21.** $\frac{1}{3}\begin{bmatrix} 1 & 2 & 4 \\ 1 & -1 & -2 \\ -1 & 1 & -1 \end{bmatrix}$ **23.** $\frac{1}{2}\begin{bmatrix} -2 & 0 & 2 \\ 7 & -1 & -3 \\ -3 & 1 & 1 \end{bmatrix}$

**25.** $\frac{1}{2}\begin{bmatrix} 3 & -1 & -1 & -1 \\ 1 & 1 & -1 & -1 \\ -1 & 1 & 1 & -1 \\ -1 & -1 & 1 & 3 \end{bmatrix}$ **27.** $A^{-1} = \frac{1}{2}\begin{bmatrix} -1 & 4 \\ -1 & 2 \end{bmatrix}$, $x = -9, y = -7$

**29.** $A^{-1} = \begin{bmatrix} 3 & -1 \\ -2 & 1 \end{bmatrix}$, $x = 9, y = -3$

**31.** $A^{-1} = \frac{1}{2}\begin{bmatrix} -5 & 4 \\ 3 & -2 \end{bmatrix}$, $x = -7, y = 5$ **33.** $x = 11, y = -3$

**35.** $x = \frac{11}{7}, y = \frac{27}{7}$ **37.** $x = 3, y = -1, z = -1$

**39.** $x = 2, y = -\frac{7}{2}, z = \frac{3}{2}$ **41.** $x_1 = -1, x_2 = -1, x_3 = 0, x_4 = 3$

**43.** $\begin{bmatrix} a & b \\ c & d \end{bmatrix}\frac{1}{D}\begin{bmatrix} d & -b \\ -c & a \end{bmatrix} = \frac{1}{D}\begin{bmatrix} ad - bc & -ab + ba \\ cd - dc & -bc + da \end{bmatrix} = \begin{bmatrix} 1 & 0 \\ 0 & 1 \end{bmatrix}$

$\frac{1}{D}\begin{bmatrix} d & -b \\ -c & a \end{bmatrix}\begin{bmatrix} a & b \\ c & d \end{bmatrix} = \frac{1}{D}\begin{bmatrix} da - bc & db - bd \\ -ca + ac & -cb + ad \end{bmatrix} = \begin{bmatrix} 1 & 0 \\ 0 & 1 \end{bmatrix}$

**45.** (a) $75,000 in bond fund, $25,000 in stock fund (b) $25,000 in bond fund, $25,000 in stock fund

**47.** (a) 80 of style A and 40 of style B (b) 10 of style A and 130 of style B.

## Section 2.6

**1.** $\dfrac{1}{41}\begin{bmatrix} 1,100,000,000 \\ 1,800,000,000 \end{bmatrix}$    **3.** $\begin{bmatrix} 375,000,000 \\ 437,500,000 \end{bmatrix}$    **5.** $\dfrac{1}{41}\begin{bmatrix} 21,000,000,000 \\ 23,000,000,000 \end{bmatrix}$

**7.** $\dfrac{1}{48}\begin{bmatrix} 1,550,000,000 \\ 2,700,000,000 \\ 1,600,000,000 \end{bmatrix}$    **9.** $\dfrac{1}{871}\begin{bmatrix} 32,500,000,000 \\ 33,100,000,000 \\ 30,600,000,000 \end{bmatrix}$    **11.** $p_1 = \frac{2}{7}t,\ p_2 = t.$

**13.** $p_1 = \frac{1}{6}t,\ p_2 = t.$    **15.** $p_1 = \frac{14}{39}t,\ p_2 = \frac{20}{39}t,\ p_3 = t.$

**17.** $p_1 = \frac{28}{85}t,\ p_2 = \frac{82}{85}t,\ p_3 = t.$

## Chapter 2 Review

**1.** $x = 1,\ y = 2$    **2.** No solution    **3.** $x = 1 - 3t,\ y = t$    **4.** $x = 3,\ y = 4$

**5.** $x = 1.6 - .2t,\ y = .8 + .4t,\ z = t$    **6.** $x = 10 - t,\ y = t,\ z = 0$

**7.** $x = 1,\ y = 3,\ z = 1$    **8.** $x = -2 + 2t,\ y = 7 - 3t,\ z = t$

**9.** $\begin{bmatrix} 3 & 9 \\ 6 & 3 \\ -3 & 9 \end{bmatrix}$    **10.** $\begin{bmatrix} -4 & -10 \\ -2 & 0 \\ -6 & 4 \end{bmatrix}$    **11.** $\begin{bmatrix} -1 & -1 \\ 4 & 3 \\ -9 & 13 \end{bmatrix}$    **12.** 14

**13.** Not defined    **14.** Not defined    **15.** $\begin{bmatrix} 4 \\ 2 \end{bmatrix}$    **16.** Not defined

**17.** $\begin{bmatrix} 11 & 3 & 5 \\ 7 & 1 & 0 \\ 7 & 3 & 7 \end{bmatrix}$    **18.** $\begin{bmatrix} 3 & 3 \\ 3 & 16 \end{bmatrix}$    **19.** Not defined    **20.** 12

**21.** Part (a) follows from part (b) by taking $B = A$. To show part (b) notice that

$$AB = \begin{bmatrix} a_{11} & 0 & 0 \\ 0 & a_{22} & 0 \\ 0 & 0 & a_{33} \end{bmatrix}\begin{bmatrix} b_{11} & 0 & 0 \\ 0 & b_{22} & 0 \\ 0 & 0 & b_{33} \end{bmatrix} = \begin{bmatrix} a_{11}b_{11} & 0 & 0 \\ 0 & a_{22}b_{22} & 0 \\ 0 & 0 & a_{33}b_{33} \end{bmatrix}$$

**22.** $\dfrac{1}{3}\begin{bmatrix} 1 & -1 \\ -2 & 5 \end{bmatrix}$    **23.** Not defined    **24.** $-\dfrac{1}{7}\begin{bmatrix} 1 & 3 \\ 3 & 2 \end{bmatrix}$

**25.** $\dfrac{1}{3}\begin{bmatrix} -12 & 6 & -3 \\ 5 & -2 & 1 \\ -1 & 1 & 1 \end{bmatrix}$    **26.** Does not exist    **27.** $\dfrac{1}{3}\begin{bmatrix} 2 & -1 & 0 \\ 6 & 0 & -3 \\ -5 & 1 & 3 \end{bmatrix}$

**28.** $x = 1,\ y = -4$    **29.** $x = -17,\ y = -16$    **30.** $x = 8,\ y = -3,\ z = 2$

**31.** $x = 23,\ y = 60,\ z = -53$    **32.** $\dfrac{1}{21}\begin{bmatrix} 550,000,000 \\ 1,150,000,000 \end{bmatrix}$

**33.** $\dfrac{1}{1597}\begin{bmatrix} 42,000,000,000 \\ 68,300,000,000 \\ 76,800,000,000 \end{bmatrix}$    **34.** $p_1 = \frac{3}{8}t,\ p_2 = t.$

**35.** $p_1 = \frac{37}{85}t,\ p_2 = \frac{78}{85}t,\ p_3 = t$

**36.** $1 + t$ three-par holes, $8 - 2t$ four-par holes, $t$ five-par holes, $0 \le t \le 4$

**37.** 7 standard, 5 deluxe   **38.** 8 of the first and 6 of the second
**39.** 20 hams and 30 beefs   **40.** 90, 60, 150 liters
**41.** Plant 1 should produce 100, Plant 2 should produce 200, and Plant 3 should produce 50.

## Section 3.1

**1.**

**3.**

**5.**

**7.**

**9.**

**11.**

**13.**

**15.**

**17.**

**19.**

**21.**

**23.**

**25.**

**27.**

**29.**

**31.**

**33.**

**35.**

**37.**

**39.**

**41.**

**43.** $x + y \leq 200$

$\qquad x \geq 100$

$\qquad y \geq 50$

$\qquad x, y \geq 0$

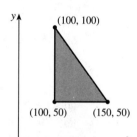

**45.** $x + 2y \leq 20$

$\quad 2x + 2y \leq 22$

$\quad 2x + y \leq 15$

$\qquad x, y \geq 0$

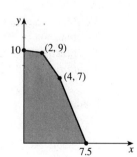

**47.**
$$x + 3y \leq 45$$
$$10x + 20y \leq 350$$
$$2x + y \leq 55$$
$$x, y \geq 0$$

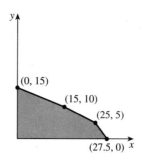

**49.**
$$x + 4y \leq 800$$
$$x + 2y \leq 500$$
$$2x + y \leq 700$$
$$x, y \geq 0$$

## Section 3.2

**1.** Max = 12, min = 3     **3.** Max = 16, min = 7     **5.** No max, min = 6

**7.** Max = 1400 at (10, 40)     **9.** Max = 180 at (30, 40)

**11.** Min = 7 at (3, 4)     **13.** Min = 2 at (1, 0)     **15.** Max = 48 at (8, 4)

**17.** Min = 20 at (5, 0)     **19.** No solution     **21.** No solution

**23.** (a) $a \geq -0.5$ (b) $-1 \leq a \leq -0.5$ (c) $-2 \leq a \leq -1$ (d) $a \leq -2$

**25.** (a) A (b) B (c) C (d) D (e) E

**27.** With the single exception of the point $A$, starting at any point in the feasible region, we can move a little either in the positive $x$ or positive $y$ direction and still remain in the feasible region and increase $z$.

**29.** Min 5000 at $x = 100$, $y = 50$     **31.** Max 530 at (2, 9)

**33.** Max 115 at (25, 5) with labor in excess

**35.** Max 400 at (300, 100) with food I in excess

**37.** Min 39,000 at (5, 12) or (3, 15)     **39.** Max 1150 at (200, 150)

**41.** Max 20 at (5, 15), (6, 14), (7, 13), (8, 12), (9, 11), (10, 10).

## Section 3.3

**1.** Maximize $C = .06x_1 + .08x_2 + .10x_3 + .13x_4$

Subject to
$$x_1 + x_2 + x_3 + x_4 \leq 100{,}000$$
$$-x_1 \phantom{+x_2 +x_3} + x_4 \leq 0$$
$$-x_2 + x_3 \phantom{+x_4} \leq 0$$
$$x_1 \phantom{+x_2 +x_3 +x_4} \geq 10{,}000$$
$$x_1 \phantom{+x_2 +x_3 +x_4} \leq 40{,}000$$
$$x_3 \phantom{+x_4} \geq 10{,}000$$
$$x_1, x_2, x_3, x_4 \geq 0$$

**3.** Minimize $C = 20x_1 + 40x_2 + 50x_3 + 20x_4$

Subject to
$$x_1 + x_2 \qquad\qquad \leq 500$$
$$x_3 + x_4 \leq 100$$
$$x_1 \qquad + x_3 \qquad \geq 150$$
$$x_2 \qquad + x_4 \geq 250$$
$$x_1, x_2, x_3, x_4 \geq 0$$

**5.** Minimize $\quad C = 4x_1 + 3x_2 + x_3$

Subject to
$$x_1 \quad + x_2 \quad + x_3 \leq 20$$
$$30x_1 + 30x_2 + 10x_3 \geq 500$$
$$20x_1 + 20x_2 + 10x_3 \geq 329$$
$$35x_1 + 50x_2 + 20x_3 \geq 425$$
$$x_1, x_2, x_3 \geq 0$$

**7.** Minimize $\quad C = 20x_1 + 30x_2 + 20x_3$

Subject to
$$40x_1 + 20x_2 \qquad\qquad \geq 4400$$
$$30x_1 + 20x_2 + 30x_3 \geq 1800$$
$$10x_1 + 60x_2 + 40x_3 \geq 2800$$
$$x_1, x_2, x_3 \geq 0$$

**9.** Minimize $\quad C = 40{,}000x_1 + 30{,}000x_2 + 45{,}000x_3$

Subject to
$$8000x_1 + 4000x_2 + 4500x_3 \geq 40{,}000$$
$$3000x_1 + 6000x_2 + 6500x_3 \geq 24{,}000$$
$$x_1, x_2, x_3 \geq 0$$

**11.** Minimize $\quad C = 2000x_1 + 3000x_2 + 4000x_3$

Subject to
$$2x_1 \quad + x_2 \quad + x_3 \geq 21$$
$$x_1 \quad + x_2 + 3x_3 \geq 19$$
$$x_1 + 4x_2 + 3x_3 \geq 25$$
$$x_1, x_2, x_3 \geq 0$$

**13.** Minimize $\quad C = 4x_1 + 2x_2 + 2x_3 + 4x_4$

Subject to
$$x_1 + x_2 \qquad\qquad \leq 200 \qquad x_1 \text{ is number from SW to NW}$$
$$x_3 + x_4 \leq 600 \qquad x_2 \text{ is number from SE to NW}$$
$$x_1 \qquad + x_3 \qquad \geq 400 \qquad x_3 \text{ is number from SW to NE}$$
$$x_2 \qquad + x_4 \geq 300 \qquad x_4 \text{ is number from SE to NE}$$
$$x_1, x_2, x_3, x_4 \geq 0$$

**15.**  Maximize          $R = 50(x_1 + x_3 + x_5) + 65(x_2 + x_4 + x_6)$

Subject to

$$x_1 + x_2 \leq 200$$
$$x_3 + x_4 \leq 30$$
$$x_5 + x_6 \leq 60$$
$$-.25x_1 + .75x_3 - .25x_5 \geq 0$$
$$-.15x_2 + .85x_4 - .15x_6 \geq 0$$
$$-.10x_2 - .10x_4 + .90x_6 \geq 0$$
$$x_1, x_2, x_3, x_4, x_5, x_6 \geq 0$$

$x_1$ = oz of sugar in first candy, $x_2$ = oz of sugar in second candy, $x_3$ = oz of nuts in first candy, $x_4$ = oz of nuts in second candy, $x_5$ = oz of chocolate in first candy, $x_6$ = oz of chocolate in second candy.

**17.**  Minimize          $C = .50x_1 + .40x_2 + .60x_3 + .75x_4$

Subject to

$$x_1 + x_2 + x_3 + x_4 \leq 18$$
$$x_1 + 2x_2 \geq 10$$
$$100x_1 + 50x_2 + 10x_4 \geq 800$$
$$100x_1 + 200x_3 + 20x_4 \geq 500$$
$$50x_3 + 70x_4 \geq 200$$
$$200x_1 + 150x_2 + 75x_3 + 250x_4 \leq 3000$$
$$2000x_2 + 5000x_4 \leq 25000$$
$$5x_1 + 20x_2 + 30x_3 + 70x_4 \leq 130$$
$$x_1, x_2, x_3, x_4 \geq 0$$

**19.**  Minimize  $z = x_1 + x_2 + x_3 + x_4 + x_5 + x_6 + x_7$

Subject to

$$x_1 + x_4 + x_5 + x_6 + x_7 \geq 82 \text{ Monday constraint}$$
$$x_1 + x_2 + x_5 + x_6 + x_7 \geq 87 \text{ Tuesday constraint}$$
$$x_1 + x_2 + x_3 + x_6 + x_7 \geq 77 \text{ Wednesday constraint}$$
$$x_1 + x_2 + x_3 + x_4 + x_7 \geq 73 \text{ Thursday constraint}$$
$$x_1 + x_2 + x_3 + x_4 + x_5 \geq 75 \text{ Friday constraint}$$
$$x_2 + x_3 + x_4 + x_5 + x_6 \geq 42 \text{ Saturday constraint}$$
$$x_3 + x_4 + x_5 + x_6 + x_7 \geq 23 \text{ Sunday constraint}$$
$$x_1, x_2, x_3, x_4, x_5, x_6, x_7 \geq 0$$

where $x_1$ is the number of employees who begin their work week on Monday, $x_2$ the number who begin their work week on Tuesday, and so on.

**21.**  25 tables and 5 desks. Excess of labor.

**23.**  300 smallmouth and 100 largemouth. Excess of food I.

**25.**  200 smallmouth and 150 largemouth. Excess of food III.

**27.**  (a) Labor (b) Wood

## Chapter 3 Review

1. 10 sacks of each at a cost of $350
2. 3 of the first and 2 of the second at a cost of $5 per 10 yards
3. 26 chairs and 8 sofas with income $9200
4. (a) Wood in excess by 10 feet (b) 42 chairs and no sofas with income $16,800 and wood in excess by 90 feet and material in excess by 80 square yards
5. 3 of first and 6 of second with profit of $150
6. (a) Labor in excess by 3 hours (b) 6 of the first and 3 of the second with wood in excess by 6 square feet
7. North 2 days and South 3 days with cost $18,000
8. 30 acres in corn and 210 acres in soybeans, with revenue $59,040
9. 12 pounds of the first and 21 pounds of the second with profit of $144

## Section 4.1

1. (a)
$$x_1 + x_2 + s_1 = 6$$
$$2x_1 + x_2 + s_2 = 8$$

(b)

| $x_1$ | $x_2$ | $s_1$ | $s_2$ | $z$ | |
|---|---|---|---|---|---|
| 1 | 1 | 1 | 0 | 0 | 6 |
| 2 | 1 | 0 | 1 | 0 | 8 |
| $-3$ | $-2$ | 0 | 0 | 1 | 0 |

(c)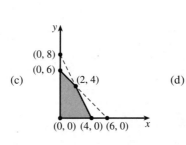

(d)

| $x_1$ | $x_2$ | $s_1$ | $s_2$ | $(x_1, x_2)$ | |
|---|---|---|---|---|---|
| 0 | 0 | 6 | 8 | (0, 0) | corner |
| 0 | 6 | 0 | 2 | (0, 6) | corner |
| 0 | 8 | $-2$ | 0 | (0, 8) | not corner |
| 6 | 0 | 0 | $-4$ | (6, 0) | not corner |
| 4 | 0 | 2 | 0 | (4, 0) | corner |
| 2 | 4 | 0 | 0 | (2, 4) | corner |

3. (a)
$$x_1 + 3x_2 + s_1 = 30$$
$$x_1 + x_2 + s_2 = 12$$
$$4x_1 + x_2 + s_3 = 24$$

(b)

| $x_1$ | $x_2$ | $s_1$ | $s_2$ | $s_3$ | $z$ | |
|---|---|---|---|---|---|---|
| 1 | 3 | 1 | 0 | 0 | 0 | 30 |
| 1 | 1 | 0 | 1 | 0 | 0 | 12 |
| 4 | 1 | 0 | 0 | 1 | 0 | 24 |
| $-20$ | $-40$ | 0 | 0 | 0 | 1 | 0 |

(c)

| $x_1$ | $x_2$ | $s_1$ | $s_2$ | $s_3$ | $(x_1, x_2)$ | |
|---|---|---|---|---|---|---|
| 0 | 0 | 30 | 12 | 24 | (0, 0) | corner |
| 0 | 10 | 0 | 2 | 14 | (0, 10) | corner |
| 0 | 12 | −6 | 0 | 12 | (0, 12) | not corner |
| 0 | 24 | −42 | −12 | 0 | (0, 24) | not corner |
| 30 | 0 | 0 | −18 | −96 | (30, 0) | not corner |
| 12 | 0 | 18 | 0 | −24 | (12, 0) | not corner |
| 6 | 0 | 24 | 6 | 0 | (6, 0) | corner |
| 3 | 9 | 0 | 0 | 3 | (3, 9) | corner |
| $\frac{42}{11}$ | $\frac{96}{11}$ | 0 | $-\frac{6}{11}$ | 0 | $(\frac{42}{11}, \frac{96}{11})$ | not corner |
| 4 | 8 | 2 | 0 | 0 | (4, 8) | corner |

(d)

**5.**

| $x_1$ | $x_2$ | $s_1$ | $s_2$ | $z$ | |
|---|---|---|---|---|---|
| 1 | 2 | .50 | 0 | 0 | 4 |
| 0 | −9 | −2.5 | 1 | 0 | 0 |
| 0 | 6 | 2 | 0 | 1 | 16 |

**7.**

| $x_1$ | $x_2$ | $s_1$ | $s_2$ | $z$ | |
|---|---|---|---|---|---|
| 0 | 3.6 | 1 | −.40 | 0 | 0 |
| 1 | .20 | 0 | .20 | 0 | 4 |
| 0 | −1.2 | 0 | .80 | 1 | 16 |

**9.**

| $x_1$ | $x_2$ | $s_1$ | $s_2$ | $s_3$ | $z$ | |
|---|---|---|---|---|---|---|
| 1 | 2 | .50 | 0 | 0 | 0 | 4 |
| 0 | −9 | −2.5 | 1 | 0 | 0 | 5 |
| 0 | 0 | −1.5 | 0 | 1 | 0 | 3 |
| 0 | 4 | 1.5 | 0 | 0 | 1 | 12 |

**11.**

| $x_1$ | $x_2$ | $s_1$ | $s_2$ | $s_3$ | $z$ | |
|---|---|---|---|---|---|---|
| 0 | 3.6 | 1 | $-.40$ | 0 | 0 | $-2$ |
| 1 | .20 | 0 | .20 | 0 | 0 | 5 |
| 0 | 5.4 | 0 | $-.60$ | 1 | 0 | 0 |
| 0 | $-1.4$ | 0 | .60 | 0 | 1 | 15 |

**13.**

| $x_1$ | $x_2$ | $s_1$ | $s_2$ | $s_3$ | $z$ | |
|---|---|---|---|---|---|---|
| 0 | 0 | 1 | 0 | $-\frac{2}{3}$ | 0 | $-2$ |
| 0 | $-9$ | 0 | 1 | $-\frac{5}{3}$ | 0 | 0 |
| 1 | 2 | 0 | 0 | $\frac{1}{3}$ | 0 | 5 |
| 0 | 4 | 0 | 0 | 1 | 1 | 15 |

**15.** Nonbasic: $x_1, x_2$. Basic: $s_1, s_2$. $x_1 = x_2 = 0$, $s_1 = 10$, $s_2 = 20$. Corner.

**17.** Nonbasic: $x_1, s_1$. Basic: $x_2, s_2$. $x_1 = s_1 = 0$, $x_2 = 5$, $s_2 = 10$. Corner.

**19.** Nonbasic: $x_2, s_2$. Basic: $x_1, s_1$. $x_2 = s_2 = 0$, $x_1 = 4$, $s_1 = -2$. Not corner.

**21.** Nonbasic: $x_1, s_3$. Basic: $x_2, s_1, s_2$. $x_1 = s_3 = 0$, $x_2 = 2$, $s_1 = 21$, $s_2 = 30$. Corner.

**23.** Nonbasic: $x_2, s_2$. Basic: $x_1, s_1, s_3$. $x_2 = s_2 = 0$, $x_1 = 2$, $s_1 = 0$, $s_3 = -5$. Not corner.

## Section 4.2

**1.**

| $x_1$ | $x_2$ | $s_1$ | $s_2$ | $z$ | | |
|---|---|---|---|---|---|---|
| 1 | 2 | .5 | 0 | 0 | 4 | $x_1 = 4$ |
| 0 | $-9$ | $-2.5$ | 1 | 0 | 5 | $x_2 = 0$ |
| 0 | 6 | 2 | 0 | 1 | 16 | max $= 16$ |

**3.**

| $x_1$ | $x_2$ | $s_1$ | $s_2$ | $z$ | | |
|---|---|---|---|---|---|---|
| 0 | 1 | 3 | $-.4$ | 0 | 1 | $x_1 = 2$ |
| 1 | 0 | 1 | .2 | 0 | 2 | $x_2 = 1$ |
| 0 | 0 | 4 | 1.4 | 1 | 14 | max $= 14$ |

**5.**

| $x_1$ | $x_2$ | $s_1$ | $s_2$ | $z$ | |
|---|---|---|---|---|---|
| $\frac{4}{3}$ | 0 | 1 | 6 | 0 | 6 |
| $\frac{1}{3}$ | 1 | 0 | 1 | 0 | 1 |
| $\frac{5}{3}$ | 0 | 0 | 3 | 1 | 5 |

$x_1 = 0$
$x_2 = 1$
max = 5

**7.**

| $x_1$ | $x_2$ | $s_1$ | $s_2$ | $s_3$ | $z$ | |
|---|---|---|---|---|---|---|
| 1 | 2.5 | 1 | 0 | 22 | 0 | 26 |
| 0 | .5 | 0 | 1 | 5 | 0 | 31 |
| 1 | .5 | 0 | 0 | 4 | 0 | 1 |
| 0 | 5 | 0 | 0 | 80 | 1 | 40 |

$x_1 = 1$
$x_2 = 0$
max = 40

**9.**

| $x_1$ | $x_2$ | $s_1$ | $s_2$ | $s_3$ | $z$ | |
|---|---|---|---|---|---|---|
| 0 | 5 | 1 | 0 | 0 | 0 | 1 |
| 0 | $\frac{4}{3}$ | 0 | 1 | 0 | 0 | $\frac{1}{3}$ |
| 1 | 1 | 0 | 0 | 0 | 0 | 2 |
| 0 | $\frac{5}{3}$ | 0 | 0 | 1 | 0 | $\frac{11}{12}$ |
| 0 | $\frac{14}{3}$ | 0 | 0 | 0 | 1 | $\frac{26}{3}$ |

$x_1 = 2$
$x_2 = 0$
max = $\frac{26}{3}$

**11.**

**13.**

**15.**

**17.**

**19.**

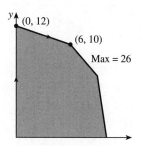

**21.** Max = 24 at (0, 12, 0).

**23.** Max = 8 at (0, 0, 4).     **25.** Max = 40 at (0, 2, 0).

**27.** Max = 450 at (0, 100, 50).     **29.** Max = 8 at (0, 1, 2).

**31.** No solution.     **33.** Max = 260 at (0, 4, 2).     **35.** Max = 210 at (0, 15, 50).

**37.** Max = 10,700 at (21, 0, 20).     **39.** Max = 164 at (20, 5, 16).

**41.** For Exercise 35: $s_1$ is the number of pounds of excess cashews, $s_2$ is the number of pounds of excess almonds, and $s_3$ is the number of pounds of excess peanuts. There is an excess of $s_2 = 3$ pounds of almonds. For Exercise 37: $s_1$ is the number of feet of excess wood, $s_2$ is the number of pounds of excess foam rubber, and $s_3$ is the number of square yards of excess material. There is an excess of $s_1 = 120$ feet of wood. For Exercise 39: $s_1$ is the number of board-feet of excess wood, $s_2$ is the number of pounds of excess iron, and $s_3$ is the number of hours of excess labor. No excess resources.

## Section 4.3

**1.** Minimum 4.5 at (1.5, 0)     **3.** Minimum 13 at (3, 1)

**5.** Minimum 70 at (1, 3) and (3, 2)     **7.** Minimum 11 at (4, 1)

**9.** Minimum 16 at (4, 1)     **11.** Minimum 2 at (0, 1, 0)

**13.** Minimum 472 at (10, 12, 8)     **15.** Minimum 3 at (0, 1, 2)     **17.** No solution

**19.** The matrix form of each of these four problems is

$$
\begin{aligned}
\text{Minimize} \qquad & w = B^T Y \\
\text{Subject to} \qquad & A^T Y \geq C^T \\
& Y \geq 0.
\end{aligned}
$$

and the dual is

$$
\begin{aligned}
\text{Maximize} \qquad & z = CX \\
\text{Subject to} \qquad & AX \leq B \\
& X \geq 0.
\end{aligned}
$$

For Exercise 7

$$
B = \begin{bmatrix} 2 \\ 3 \end{bmatrix}, \quad A = \begin{bmatrix} 4 & 1 \\ 1 & 1 \\ 1 & 2 \end{bmatrix}, \quad Y = \begin{bmatrix} y_1 \\ y_2 \end{bmatrix}, \quad C = \begin{bmatrix} 8 & 5 & 6 \end{bmatrix}
$$

For Exercise 8

$$B = \begin{bmatrix} 20 \\ 30 \end{bmatrix}, \quad A = \begin{bmatrix} 3 & 1 \\ 1 & 1 \\ 0 & 1 \end{bmatrix}, \quad Y = \begin{bmatrix} y_1 \\ y_2 \end{bmatrix}, \quad C = \begin{bmatrix} 8 & 4 & 1 \end{bmatrix}$$

For Exercise 9

$$B = \begin{bmatrix} 3 \\ 4 \end{bmatrix}, \quad A = \begin{bmatrix} 3 & 2 \\ 1 & 0 \\ 0 & 1 \end{bmatrix}, \quad Y = \begin{bmatrix} y_1 \\ y_2 \end{bmatrix}, \quad C = \begin{bmatrix} 14 & 2 & 1 \end{bmatrix}$$

For Exercise 10

$$B = \begin{bmatrix} 1 \\ 2 \end{bmatrix}, \quad A = \begin{bmatrix} -1 & 1 \\ 1 & 1 \\ 0 & 1 \end{bmatrix}, \quad Y = \begin{bmatrix} y_1 \\ y_2 \end{bmatrix}, \quad C = \begin{bmatrix} 0 & 4 & 1 \end{bmatrix}$$

**21.** Minimum of $4.50 with 4 oranges and 2 cups strawberries

**23.** First plant 4 days, second plant 2 days. Minimum cost is $230,000.

**25.** $2250 in first, $8000 in second, none in third

**27.** 90 bags of type A, 20 of type B, none of type C. Minimum cost is $2400.

**29.** North mine 8 days, Center 2 days, South 3. Minimum cost is $34,000.

**31.** 200 pink azaleas and no red from the East, 50 pink and 300 red from the West. Minimum cost is $2650.

**33.** None from SW to NW, 400 from SW to NE, 200 from SE to NW, 100 from SE to NE. Minimum cost per week is $1600.

**35.** The primal problem is: Maximize $P = 26x_1 + 24x_2 + 30x_3$ subject to:

$$2x_1 + x_2 + x_3 \le 20$$

$$x_1 + x_2 + 4x_3 \le 30$$

$$x_1 + x_2 + 3x_3 \le 40$$

$$x_1, x_2, x_3 \ge 0$$

The dual is: Minimize $z = 20y_1 + 30y_2 + 40y_3$ subject to:

$$2y_1 + y_2 + y_3 \ge 26$$

$$y_1 + y_2 + 3y_3 \ge 24$$

$$y_1 + 4y_2 + y_3 \ge 30$$

$$y_1, y_2, y_3 \ge 0$$

We think of $y_1, y_2, y_3$ as the respective prices of one unit of each of the ingredients. The cost of the ingredients in the first product is $2y_1 + y_2 + y_3$, which must be at least the profit of 26 dollars. This gives the first problem constraint in the dual problem. In the same way we obtain the other two problem constraints. Our cost is $w = 20y_1 + 30y_2 + 40y_3$, which we wish to minimize.

**37.** The primal problem is: Minimize $C = 75y_1 + 75y_2$

$$\text{Subject to} \quad y_1 + 2y_2 \geq 8$$
$$75y_1 + 60y_2 \geq 420$$
$$y_1, y_2 \geq 0$$

The dual is: Maximize $z = 8x_1 + 420x_2$

$$\text{Subject to} \quad x_1 + 75x_2 \leq 75$$
$$2x_1 + 60x_2 \leq 75$$
$$x_1, x_2 \geq 0$$

We think of $x_1$ as the price of one gm of fiber (in pure form) and $x_2$ as one mg of vitamin C (in pure form). If we think of selling fiber and vitamin C separately, then we cannot obtain more for them than their cost combined in oranges and strawberries. This gives the two problem constraints in the dual problem. Our income is $z = 8x_1 + 420x_2$, which we wish to maximize.

**39.** The primal problem is: Minimize $C = 20y_1 + 30y_2 + 20y_3$

$$\text{Subject to} \quad 40y_1 + 20y_2 \geq 4000$$
$$30y_1 + 20y_2 + 30y_3 \geq 2000$$
$$10y_1 + 55y_2 + 40y_3 \geq 2000$$
$$y_1, y_2, y_3 \geq 0$$

The dual is: Maximize $z = 4000x_1 + 2000x_2 + 2000x_3$

$$\text{Subject to} \quad 40x_1 + 30x_2 + 10x_3 \leq 20$$
$$20x_1 + 20x_2 + 55x_3 \leq 30$$
$$30x_2 + 40x_3 \leq 20$$
$$x_1, x_2, x_3 \geq 0$$

We think of $x_1$ as the price of one pound of nitrogen (in pure form), $x_2$ as one pound of phosphoric acid (in pure form), and $x_3$ as one pound of potash (in pure form). If we think of selling nitrogen, phosphoric acid, and potash separately, then we cannot obtain more for them than their cost combined in each of the three types of fertilizers. This gives the three problem constraints in the dual problem. Our income is $z = 4000x_1 + 2000x_2 + 2000x_3$, which we wish to maximize.

**41.** The primal problem is: Minimize $C = 2000y_1 + 3000y_2 + 4000y_3$

$$\text{Subject to} \quad 2y_1 + y_2 + y_3 \geq 21$$
$$y_1 + y_2 + 3y_3 \geq 19$$
$$y_1 + 4y_2 + 3y_3 \geq 25$$
$$y_1, y_2, y_3 \geq 0$$

The dual is: Maximize $z = 21x_1 + 19x_2 + 25x_3$

$$\text{Subject to} \quad 2x_1 + x_2 + x_3 \leq 2000$$
$$x_1 + x_2 + 4x_3 \leq 3000$$
$$x_1 + 3x_2 + 3x_3 \leq 4000$$
$$x_1, x_2, x_3 \geq 0$$

We think of $x_1$ as the price of one ton of low-grade ore, $x_2$ as one ton of medium-grade ore, and $x_3$ as one ton of low-grade ore. If we think of selling the three grades of ore, then we cannot obtain more for them than their cost combined from each of the three mines. This gives the three problem constraints in the dual problem. Our income is $z = 4000x_1 + 2000x_2 + 2000x_3$, which we wish to maximize.

## Section 4.4

**1.** Maximum of 6 at $(2, 2)$    **3.** Maximum of 10 at $(2, 4)$

**5.** Maximum of 16 at $(8, 0)$    **7.** Maximum of 13 at $(2, 3)$

**9.** Maximum of 1280 at $(180, 220, 260)$ and $(620, 0, 40)$

**11.** Minimum of 4.5 at $(1.5, 0)$    **13.** Minimum of 13 at $(3, 1)$

**15.** Minimum of 70 at $(1, 3)$ or $(3, 2)$    **17.** Minimum of 11 at $(4, 1)$

**19.** Minimum of 16 at $(4, 1)$    **21.** Minimum of 2 at $(0, 1, 0)$

**23.** Minimum of 472 at $(10, 12, 8)$    **25.** Minimum of 3 at $(0, 1, 2)$

**27.** Minimum cost of \$11,000 with 150 shipped from Stockpile I to Plant A, 150 shipped from Stockpile I to Plant B, none shipped from Stockpile II to Plant A, and 100 shipped from Stockpile II to Plant B.    **29.** Maximum of 8.2 at $(30, 30, 40)$

**31.** Minimum cost of \$2400 with 90 bags of Type A, 20 bags of Type B, none of Type C

**33.** The first plant 4 days, the second 2 days with minimum cost of \$220,000.

**35.** The North mine 8 days, the Center 2, and the South 3. \$34,000

## Section 4.5

**1.** (a) The new last column is

$$20 - 2h_2$$

$$30 + h_1 - h_2 - h_3$$

$$25 - h_2 + 3h_3$$

$$40 + 4h_2 + 3h_3$$

(b) $20 - 2h_2 \geq 0$, $30 + h_1 - h_2 - h_3 \geq 0$, $25 - h_2 + 3h_3 \geq 0$

(c) $(20 - 2h_2, 25 - h_2 + 3h_3)$ (d) $40 + 4h_2 + 3h_3$ (e) 10 (f) (14,16), 46

(g) 0, 4, and 3. The optimal $z$-value is increased by 0 if the first constraint is increased by 1, by 4 if the second constraint is increased by 1, and by 3 if the third constraint is increased by 1.

**3.** (a) The new last column is

$$20 + h_1 + 4h_2 + h_3 + 5h_4$$

$$30 + 3h_1 + 2h_2 + h_4$$

$$25 + 2h_1 + h_2 + 2h_4$$

$$40 + h_1 + 3h_2 + 6h_4$$

$$90 + 2h_1 + 4h_2 + 5h_4$$

(b) $20 + h_1 + 4h_2 + h_3 + 5h_4 \geq 0$, $30 + 3h_1 + 2h_2 + h_4 \geq 0$
$25 + 2h_1 + h_2 + 2h_4 \geq 0$, $40 + h_1 + 3h_2 + 6h_4 \geq 0$

(c) $(30 + 3h_1 + 2h_2 + h_4, 25 + 2h_1 + h_2 + 2h_4, 40 + h_1 + 3h_2 + 6h_4)$

(d) $90 + 2h_1 + 4h_2 + 5h_4$ (e) 20 (f) (43, 34, 57), 111

(g) 2, 4, 0, and 5. The optimal $z$-value is increased by 2 if the first constraint is increased by 1, by 4 if the second constraint is increased by 1, by 0 if the third constraint is increased by 1, and by 5 if the fourth constraint is increased by 1.
**5.** (24, 0) with maximum of 5.      **7.** (0, 15) with maximum of 80.

## Chapter 4 Review

**1.** (500, 0), 500      **2.** (0, 100), 400      **3.** (0, 1), 2      **4.** (35, 0), 70
**5.** (6, 0), 18      **6.** $(\frac{32}{7}, \frac{25}{7})$, 29      **7.** (7.5, 0), 15      **8.** (27, 2), 60
**9.** No solution      **10.** (4, 1), 21
**11.** 100 acres of the first field in wheat, 120 acres of the first field in corn, 100 acres of the second field in wheat, no acres of the second field in corn
**13.** (a) 50 purses, 55 belts, new maximum is 90 (b) 2, 0, 4. The optimal $z$-value is increased by 2 if the first constraint is increased by 1, by 0 if the second constraint is increased by 1, by 4 if the third constraint is increased by 1.

## Section 5.1

**1.** $26.67      **3.** $480      **5.** $2046.67      **7.** $1120.00      **9.** 8.57%
**11.** 3.5%      **13.** $1980.20      **15.** $5357.14      **17.** 5.11%      **19.** 7.12%
**21.** $16.67, $983.33      **23.** $500, $4500      **25.** 2.5 years      **27.** 12.5 years.
**29.** $1666.67, $48,333.33

## Section 5.2

**1.** $1100.00      **3.** (a) $1080.00 (b) $1082.43 (c) $1083.00 (d) $1083.22
(e) $1083.28
**5.** (a) $21,724.52 (b) $23,769.91 (c) $24,273.39 (d) $24,472.28 (e) $24,523.94
**7.** (a) $3262.04 (b) $7039.99 (c) $14,974.46 (d) $31,409.42 (e) $93,050.97
(f) $276,863.55
**9.** (a) $1538.62 (b) $2367.36 (c) $3642.48 (d) $13,267.68
**11.** (a) 8.16% (b) 8.24% (c) 8.30% (d) 8.32% (e) 8.33%
**13.** (a) $1784.31 (b) $1664.13 (c) $1655.56
**15.** (a) $3183.76 (b) $2769.32 (c) $2740.89
**17.** $31,837.58      **19.** $26,897.08
**21.** About $3 trillion      **23.** 9.57%      **25.** 81.3 months, 351.6 weeks
**27.** 128.3 months, 3894.3 days      **29.** $1104.79 yes      **31.** 6.875%
**33.** 8.26%, 8.32%, second      **35.** $70,485.40

## Section 5.3

**1.** $17,383.88      **3.** $18,120.60      **5.** $98,925.54      **7.** $15,645.49
**9.** $2,045,953.85      **11.** $1316.40      **13.** $148.36      **15.** $7.23
**17.**

$$FV = PMT \, \frac{(1 + i)^n - 1}{i}$$

$$\frac{iFV}{PMT} + 1 = (1 + i)^n$$

$$\log\left(\frac{iFV}{PMT} + 1\right) = n\log(1 + i)$$

$$n = \frac{\log\left(\dfrac{iFV}{PMT} + 1\right)}{\log(1 + i)}$$

**19.** 34.2 years   **21.** $675,764.89   **23.** $206,348.77

**25.** $1210.61, $28,209.02, $18,392.71, $1418.71

**27.** $265.17. Equity: $3316.64, $6944.40, $10,912.47, $15,252.77, $20,000.00

**29.** $14,836.23. Equity: $129,611.03, $287,529.47, $479.937.77, $714,368.74, $1,000,000.00. Interest earned is $406,550.80. Interest earned during 18th period is $7738.86.

## Section 5.4

**1.** $339,758.16   **3.** $286,748.03   **5.** $245,453.69   **7.** $2296.07, $3776.42

**9.** $2642.37, $5854.22   **11.** $665.30, $139,508.00   **13.** $804.62, $189,663.20

**15.** $898.83, $61,789.40   **17.** $1014.27, $82,568.60   **19.** $253.63, $2174.24

**21.** $263.34, $2640.32   **23.** $34.18, $230.48   **25.** $36.15, $301.40

**27.** $85,812.06, $88.92, $588.80   **29.** $77,412.95, $332.45, $750.54

**31.**

| End of Period | Repayment Made | Interest Charged | Payment Toward Principal | Outstanding Principal |
|---|---|---|---|---|
| 0 | | | | 10,000 |
| 1 | 2296.07 | 1000.00 | 1296.07 | 8703.93 |
| 2 | 2296.07 | 870.39 | 1425.68 | 7278.25 |
| 3 | 2296.07 | 727.83 | 1568.24 | 5710.01 |
| 4 | 2296.07 | 571.00 | 1725.07 | 3984.94 |
| 5 | 2296.07 | 398.49 | 1897.58 | 2087.36 |
| 6 | 2296.10 | 208.74 | 2087.36 | 0000.00 |

**33.**

| End of Period | Repayment Made | Interest Charged | Payment Toward Principal | Outstanding Principal |
|---|---|---|---|---|
| 0 | | | | 10,000 |
| 1 | 2642.37 | 1500.00 | 1142.37 | 8857.63 |
| 2 | 2642.37 | 1328.64 | 1313.73 | 7543.90 |
| 3 | 2642.37 | 1131.59 | 1510.78 | 6033.12 |
| 4 | 2642.37 | 904.97 | 1737.40 | 4295.72 |
| 5 | 2642.37 | 644.36 | 1998.01 | 2297.71 |
| 6 | 2642.37 | 344.66 | 2297.71 | 0000.00 |

**35.** (a)

$$PV = PMT \frac{[1 - (1 + i)^{-n}]}{i}$$

$$\frac{PV(i)}{PMT} = 1 - (1 + i)^{-n}$$

$$(1 + i)^{-n} = 1 - \frac{PV(i)}{PMT} = \frac{PMT - PV(i)}{PMT}$$

$$-n \log(1 + i) = \log \frac{PMT - PV(i)}{PMT}$$

$$n \log(1 + i) = \log \frac{PMT}{PMT - PV(i)}$$

$$n = \frac{\log \frac{PMT}{PMT - PV(i)}}{\log(1 + i)}$$

(b) 70 payments.

**37.** The first. The present value of the first is $67,100.81, which is higher than that of the second, which is $63,817.96

**39.** The present value of the first option is only $320,882.88. Accept the lump sum.

**41.** (a) The first saves the most, $24,781.30. (b) The second saves the most, $35,383.35.

**43.** Your monthly payment in the first case will be $295.24, while using the rebate and the credit union will give payments per month of $292.56. The second is better.

**Chapter 5 Review**

1. (a) $1469.33 (b) $1485.95 (c) $1489.85 (d) $1491.37 (e) $1491.76
2. (a) $1610.51 (b) $1638.62 (c) $1645.31 (d) $1647.93 (e) $1648.61
3. (a) 9.20% (b) 9.31% (c) 9.38% (d) 9.41% (e) 9.42%
4. (a) 11.30% (b) 11.46% (c) 11.57% (d) 11.62% (e) 11.63%
5. (a) $4631.93 (b) $4505.23 (c) $4496.05
6. (a) $4224.11 (b) $4106.46 (c) $4066.15     7. $214,548.21     8. $6499.31
9. 9.05%     **10.**     $(1 + r)^T = 3$

$$T \log(1 + r) = \log 3$$

$$T = \frac{\log 3}{\log(1 + r)}$$

11. $408,922.62     12. $33,111.50; $675,540.00; $14,179.87
13. The present value of the 19 payments of $100,000 each, plus the immediate disbursement of $100,000 is $1,060,359.90.
14. $839.20, $1014.27, $151,760.00, $82,568.60, $95.40, $282.65, $217.03, $642.99

**Section 6.1**

1. (a) False (b) False.     3. (a) True (b) False.

5. (a) False (b) True (c) False (d) True.     7. (a) $\phi$, {3} (b) $\phi$, {3}, {4}, {3, 4}

9.

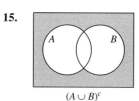

11. (a) 2 (b) 3 (c) 4 (d) 1

13.

$A \cup B^c$

15.

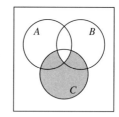

$(A \cup B)^c$

17. (a) I (b) V (c) II (d) VI (e) VIII (f) IV (g) III (h) VII

19.

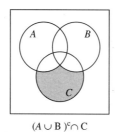

$(A \cup B) \cap C^c$

21.

$(A \cap B)^c \cap C$

23.

$(A \cup B)^c \cap C$

25. (a) {4, 5, 6} (b) {1, 2, 3, 4, 5, 6, 7, 8}

27. (a) {1, 2, 3}, (b) {9, 10}     29. (a) {5, 6} (b) {1, 2, 3, 4, 7, 8, 9, 10}

31. (a) $\phi$ (b) $\phi$

33. (a) People in your state who do not own an auto (b) People in your state who own an auto or a house (c) People in your state who own an auto or not a house

35. (a) People in your state who own an auto but not a house (b) People in your state who do not own an auto and do not own a house (c) People in your state who do not own an auto or do not own a house

37. (a) People in your state who own an auto and a house and a piano (b) People in

your state who own an auto or a house or a piano (c) People in your state who own both an auto and a house or else own a piano

**39.** (a) People in your state who do not own both an auto and a house but do own a piano (b) People in your state who do not own an auto, nor a house, nor a piano (c) People in your state who own a piano, but do not own a car or a house

**41.** (a) $N \cap F$ (b) $N \cap H^c$    **43.** (a) $N \cup S$ (b) $N^c \cap S^c$

**45.** (a) $(N \cap H) \cup (S \cap H)$ (b) $N \cap F \cap H^c$

**47.** (a) $F \cap H \cap (N \cup S)^c$ (b) $F \cap H^c \cap (N \cup S)^c$

## Section 6.2

**1.** 135    **3.** 70    **5.** 60    **7.** 150    **9.** (a) 1100 (b) 750 (c) 100

**11.** (a) 100 (b) 500    **13.** 110    **15.** 90    **17.** 15    **19.** 7    **21.** 3

**23.** 56    **27.** (a) 30 (b) 360 (c) 70    **29.** (a) 20 (b) 25 (c) 345    **31.** 30

## Section 6.3

**1.** $5 \cdot 4 \cdot 3 = 60$    **3.** $8 \cdot 7 \cdot 6 \cdot 5 \cdot 4 = 6720$    **5.** $7! = 5040$

**7.** $9 \cdot 8 \cdot 7 \cdot 6 \cdot 5 \cdot 4 \cdot 3 \cdot 2 = 362,880$    **9.** $4 \cdot 30 = 120$

**11.** $4 \cdot 10 \cdot 5 = 200$    **13.** $10 \cdot 10 \cdot 10 \cdot 10 \cdot 10 \cdot 10 = 1,000,000$

**15.** $2 \cdot 10 \cdot 5 \cdot 3 = 300$    **17.** $(26)^3(10)^3 = 17,576,000$    **19.** 96

**21.** $5 \cdot 4 \cdot 4 \cdot 3 \cdot 3 \cdot 2 \cdot 2 = 2880$    **23.** $5! = 120$    **25.** $2! \, 4! \, 6! = 34,560$

**27.** $9! 5! = 43,545,600$    **29.** $3! 4! = 144$    **31.** $12 \cdot 11 \cdot 10 \cdot 9 = 11,880$

**33.** $P(10, 8) = 1,814,400$    **35.** $(10 \cdot 9 \cdot 8)(8 \cdot 7 \cdot 6) = 241,920$

**37.** $P(11, 4)P(9, 3)P(5, 2) = 79,833,600$

## Section 6.4

**1.** $(8 \cdot 7 \cdot 6)/(3 \cdot 2 \cdot 1) = 56$    **3.** $(8 \cdot 7 \cdot 6 \cdot 5 \cdot 4)/(5 \cdot 4 \cdot 3 \cdot 2 \cdot 1) = 56$

**5.** 12    **7.** 35    **9.** 105    **11.** 1    **13.** 1

**15.**

**17.** 10    **19.** $C(46, 6) = 9,366,819$    **21.** $C(20, 5) = 15,504$

**23.** $C(20, 4) = 4845$    **25.** $C(40, 5) = 658,008$    **27.** $C(11, 3)C(7, 2) = 3465$

**29.** $C(20, 7)C(6, 3) = 1,550,400$    **31.** $C(12, 3)P(9, 3) = 110,880$

**33.** $C(11, 4)P(7, 2) = 13,860$    **35.** $C(9, 5) = 126$    **37.** $C(8, 5) = 56$

**39.** $2^8 - C(8, 1) - C(8, 0) = 247$    **41.** $C(12, 3)C(8, 3)C(4, 2)2 = 147,840$

**43.** $C(4, 3) \cdot 48 \cdot 44/2 = 4224$    **45.** $C(4, 2)C(4, 2)C(13, 2) \cdot 44 = 123,552$

## Section 6.5

1. $a^5 - 5a^4b + 10a^3b^2 - 10a^2b^3 + 5ab^4 - b^5$
3. $32x^5 + 240x^4y + 720x^3y^2 + 1080x^2y^3 + 810xy^4 + 243y^5$
5. $1 - 6x + 15x^2 - 20x^3 + 15x^4 - 6x^5 + x^6$
7. $16 - 32x^2 + 24x^4 - 8x^6 + x^8$
9. $s^{12} + 6s^{10}t^2 + 15s^8t^4 + 20s^6t^6 + 15s^4t^8 + 6s^2t^{10} + t^{12}$
11. $a^{10} - 10a^9b + 45a^8b^2, 45a^2b^8 - 10ab^9 + b^{10}$
13. $x^{11} + 11x^{10}y + 55x^9y^2, 55x^2y^9 + 11xy^{10} + y^{11}$
15. $1 - 12z + 66z^2, 66z^{10} - 12z^{11} + z^{12}$
17. $1 - 12x^3 + 66x^6, 66x^{30} - 12x^{33} + x^{36}$     **19.** 210     **21.** 1260
23. 113,400     **25.** 5040     **27.** 1, 7, 21, 35, 35, 21, 7, 1     **29.** 420
31. 34,650     **33.** 2520     **35.** (a) 6 (b) 3 (c) 24 (d) 12 (e) 6
37. (a) 34,650 (b) 3780

## Chapter 6 Review

1. (a) Yes (b) No (c) Yes     **2.** $\{x \mid x = 5n, n = 1, 2, \ldots, 8\}$
3. $\{0, \sqrt{2}, -\sqrt{2}\}$     **4.** $\phi, \{A\}, \{B\}, \{C\}, \{A, B\}, \{A, C\}, \{B, C\}, \{A, B, C\}$

5.

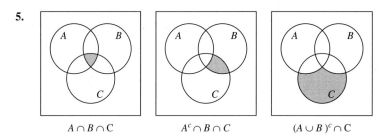

$A \cap B \cap C$     $A^c \cap B \cap C$     $(A \cup B)^c \cap C$

6. $A \cup B = \{1, 2, 3, 4\}, A \cap B = \{2, 3\}, B^c = \{1, 5, 6\}, A \cap B \cap C = \phi$,
$(A \cup B) \cap C = \{4\}, A \cap B^c \cap C = \phi$

7. (a) My current instructors who are less than 6 feet tall. (b) My current instructors who are at least 6 feet tall or are male. (c) My current instructors who are female and weigh at most 180. (d) My current male instructors who are at least 6 feet tall and weigh more than 180 pounds. (e) My current male instructors who are less than 6 feet tall and weigh more than 180 pounds. (f) My current female instructors who are at least 6 feet tall and weigh more than 180 pounds.

8. (a) $M^c$ (b) $M^c \cap W^c$ (c) $M \cap (H \cup W)$

10.

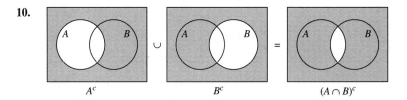

$A^c$     $\cup$     $B^c$     $=$     $(A \cap B)^c$

11. 120     **12.** 10     **13.** 50     **14.** (a) 40, (b) 45, (c) 19
15. $P(10, 4) = 10 \cdot 9 \cdot 8 \cdot 7 = 5040, C(10, 4) = 10 \cdot 9 \cdot 8 \cdot 7/(4 \cdot 3 \cdot 2) = 210$

**16.** 6    **17.** 5!6! = 86,400    **18.** $P(6, 3)P(5, 3) = 7200$
**19.** $C(5, 3)C(6, 4) = 150$    **20.** 44
**21.** $C(10, 7) = 120$    **22.** $32 - 80x + 80x^2 - 40x^3 + 10x^4 - x^5$
**23.** $-36x^7 + 9x^8 - x^9$    **24.** 3150    **25.** 7560    **26.** 369,600

## Section 7.1

**1.** $\{a\}, \{b\}, \{c\}, \{a, b\}, \{a, c\}, \{b, c\}, \{a, b, c\}, \phi$
**3.** $\{HHH, HHT, HTH, HTT, THH, THT, TTH, TTT\}, \{HHH, HHT, HTH, THH\}$
**5.** (a) $\{111, 110, 101, 100, 011, 010, 001, 000\}$, (b) $\{111, 110, 101, 011\}$
**7.** (a) {white, black, red}, (b) {black, red}
**9.** $\{ABC, ABD, ABF, ACD, ACF, ADF, BCD, BCF, BDF, CDF\}$,
$\{ABC, ABD, ABF, ACD, ACF, ADF\}$
**11.** (a) $\{0, 1, 2, 3, 4, 5, 6, 7, 8, 9, 10\}$ (b) $E = \{6, 7, 8, 9, 10\}$
(c) $F = \{0, 1, 2, 3, 4\}$ (d) $E \cup F = \{0, 1, 2, 3, 4, 6, 7, 8, 9, 10\}$. $E \cap F = \phi$.
$E^c = \{0, 1, 2, 3, 4, 5\}$. $E \cap F^c = \{6, 7, 8, 9, 10\}$. $E^c \cap F^c = \{5\}$
(e) $E \cup F$ and $E^c \cap F^c$. $E^c$ and $E \cap F^c$. $E \cap F^c$ and $E^c \cap F^c$.
**13.** (a) $\{SSS, SSF, SFS, SFF, FSS, FSF, FFS, FFF\}$ (b) $E = \{SSS, SSF, SFS, FSS\}$
(c) $G = \{SSS, SSF, SFS, SFF\}$ (d) $E \cup G = \{SSS, SSF, SFS, SFF, FSS\}$. $E \cap G = $
$\{SSS, SSF, SFS\}$. $G^c = \{FSS, FSF, FFS, FFF\}$. $E^c \cap G = \{SFF\}$. $(E \cup G)^c = $
$\{FSF, FFS, FFF\}$ (e) $E \cup G$ and $(E \cup G)^c$. $E \cap G$ and $G^c$. $E \cap G$ and $E^c \cap G$. $G^c$
and $E^c \cap G$. $E^c \cap G$ and $(E \cup G)^c$. $E \cap G$ and $(E \cup G)^c$.
**15.** $E \cap F^c$    **17.** $E^c \cap F^c$    **19.** $E \cap F \cap G^c$
**21.** $E \cup F \cup G = $ all 26 letters. $E^c \cap F^c \cap G^c = \phi$. $E \cap F \cap G = \phi$. $E \cup F^c \cup G$
$= \{a, b, c, d, e, i, o, u\}$

## Section 7.2

**1.** (a) 23 (b) 232 (c) 2317    **3.** .40    **5.** .15    **7.** (a) .12 (b) .80 (c) .08
**9.** .20    **11.** .56    **13.** A: .125, B: .175, C: .40, D: .20, F: .10
**15.** .15, .20, .14, .25, .16, .10    **17.** .30    **19.** .72    **21.** .55    **23.** .15
**25.** Since $f(E) + f(E^c) = f(E \cup E^c) = f(S) = N$,

$$\frac{f(E)}{N} + \frac{f(E^c)}{N} = \frac{N}{N} = 1.$$

for large $N$, $\dfrac{f(E)}{N} \approx p(E)$ and $\dfrac{f(E^c)}{N} \approx p(E^c)$. Thus $p(E) + p(E^c) = 1$
**27.** Since $f(S) = f(S_1) + \cdots + f(S_n)$,

$$\frac{f(S_1)}{N} + \cdots + \frac{f(S_n)}{N} = \frac{f(S)}{N} = \frac{N}{N} = 1.$$

For large $N$, $\dfrac{f(S_1)}{N} \approx p_n$. This implies

$$p_1 + \cdots + p_n = 1.$$

## Section 7.3

1. .50, .90   3. .18, .63, .62   5. .25   7. $\frac{1}{6}$   9. $\frac{7}{30}$   11. .60, .80, 0
13. .70, .70, 0, .30, 1   15. .60, .10   17. .20, .10   19. .20
21. .20, .10, .15   23. .85   25. (a) 3 : 17 (b) 1 : 3   27. (a) 2 : 3 (b) 1 : 4
29. 7 : 5   31. .20   33. Since $E \cup E^c = S$ and $E \cap E^c = \phi$,

$$1 = p(S) = p(E \cup E^c) = p(E) + p(E^c).$$

The result then follows.

35. Using the fact that $E_1$, $E_2$, and $E_3$ are pairwise mutually disjoint,

$$p(E_1 \cup E_2 \cup E_3) = p((E_1 \cup E_2) \cup E_3)$$
$$= p(E_1 \cup E_2) + p(E_3) - p((E_1 \cup E_2) \cap E_3)$$
$$= p(E_1) + p(E_2) + p(E_3) - p((E_1 \cap E_3) \cup (E_2 \cap E_3))$$
$$= p(E_1) + p(E_2) + p(E_3) - p(\phi \cup \phi)$$
$$= p(E_1) + p(E_2) + p(E_3)$$

37. Since $E \cap F^c$, $E^c \cap F$, and $E \cap F$ are pairwise mutually disjoint and their union is $E \cup F$, we have

$$p((E \cap F^c) \cup (E^c \cap F)) + p(E \cap F) = p((E \cap F^c) \cup (E^c \cap F) \cup (E \cap F))$$
$$= p(E \cup F)$$
$$= p(E) + p(F) - p(E \cap F)$$

The result then follows.

39. Since $E^c \cap F$ and $E \cap F$ are disjoint and their union is $F$,

$$p(E^c \cap F) + p(E \cap F) = p((E^c \cap F) \cup (E \cap F)) = p(F).$$

The result then follows.

41. .04, .96   43. .039, .049, .951   45. (a) .009 (b) .001 (c) .006 (d) .974

## Section 7.4

1. $\frac{3}{8}, \frac{7}{8}$   3. $\frac{1}{12}$   5. $\frac{1}{36}, \frac{1}{12}, \frac{5}{36}$   7. $\frac{3}{10}, \frac{1}{3}$   9. $6 \cdot 7 \cdot 8/1330$   11. $\frac{20}{323}$
13. $\frac{6}{1326}$   15. $\frac{78}{1326}$   17. $\frac{208}{1326}$   19. $4/2,598,960$   21. $624/2,598,960$
23. $10,200/2,598,960$   25. $123,552/2,598,960$   27. .0833; .2361; .4271; .6181
29. 1   31. $\frac{6}{45}$   33. $56/2^{10} \approx .0547$   35. $\frac{5}{11}$

## Section 7.5

1. $\frac{3}{7}, \frac{3}{5}$   3. $\frac{2}{3}, \frac{4}{5}$   5. 1, 0   7. $\frac{2}{3}, \frac{1}{2}$   9. $\frac{1}{3}, 1$   11. $0, \frac{1}{2}$   13. No
15. Yes   17. No   19. Yes   21. $\frac{2}{11}$   23. $\frac{1}{7}$

25. (a) $\frac{10,200}{132,600} \approx .077$ (b) $\frac{49}{25 \cdot 33} \approx .059$   27. .12, .64, .60   29. $\frac{1}{3}, \frac{3}{10}$   31. No

33. .65   35. .72   37. First assume that $E$ and $F$ are independent. Then

$$p_2 = p(E \cap F) = p(E)p(F)$$
$$= (p_1 + p_2)(p_2 + p_3)$$

$$= p_1p_2 + p_1p_3 + p_2^2 + p_2p_3$$

$$= p_2(p_1 + p_2 + p_3) + p_1p_3$$

$$= p_2(1 - p_4) + p_1p_3$$

$$= p_2 - p_2p_4 + p_1p_3$$

$$p_2p_4 = p_1p_3$$

Now suppose $p_1p_3 = p_2p_4$, then

$$p(E)p(F) = (p_1 + p_2)(p_2 + p_3)$$

$$= p_1p_3 + p_1p_2 + p_2p_3 + p_2^2$$

$$= p_2p_4 + p_1p_2 + p_2p_3 + p_2^2$$

$$= p_2[p_4 + p_1 + p_3 + p_2]$$

$$= p_2$$

$$= p(E \cap F)$$

**39.** If $E$ and $F$ are independent, then $p(E \cap F) = p(E)p(F)$ and

$$p(E^c)p(F^c) = [1 - p(E)][1 - p(F)]$$

$$= 1 - p(E) - p(F) + p(E)p(F)$$

$$= 1 - p(E) - p(F) + p(E \cap F)$$

$$= 1 - p(E \cup F)$$

$$= p((E \cup F)^c)$$

$$= p(E^c \cap F^c)$$

**41.** Since $E \cap F = \phi$, $p(E \cap F) = 0$. But since $p(E) > 0$ and $p(F) > 0$, $p(E)p(F) > 0$, and therefore $p(E)p(F) > 0 = p(E \cap F)$.

**43.**

$$P(E^c|F) = \frac{P(E^c \cap F)}{p(F)}$$

$$= \frac{p(F) - p(E \cap F)}{p(F)}$$

$$= 1 - p(E|F)$$

**45.** $p(E|F) = \dfrac{p(E \cap F)}{p(F)} = \dfrac{p(F)}{p(F)} = 1$

**47.** $p(E|F) + p(E^c|F) = \dfrac{p(E \cap F)}{p(F)} + \dfrac{p(E^c \cap F)}{p(F)} = \dfrac{p(E \cap F) + p(E^c \cap F)}{p(F)} =$

$\dfrac{p(F)}{p(F)} = 1$.     **49.** .02, .017, $\frac{6}{17}$     **51.** $\frac{8}{26}$     **53.** $\frac{5}{7}, \frac{5}{21}, \frac{1}{21}$     **55.** Yes

**57.** No     **59.** .000001

## Section 7.6

**1.** $\frac{3}{7}, \frac{57}{93}$    **3.** $\frac{3}{19}, \frac{1}{3}$    **5.** $\frac{4}{11}$    **7.** $\frac{2}{17}$    **9.** (a) $\frac{2}{23}$ (b) $\frac{21}{23}$    **11.** (a) $\frac{1}{56}$ (b) $\frac{5}{28}$

**13.** $\frac{100}{136} \approx .74$    **15.** $\frac{396}{937} \approx .42, \frac{5}{937} \approx .005$    **17.** $\frac{5}{7}$    **19.** $\frac{10}{19}$    **21.** (a) $\frac{1}{12}$ (b) $\frac{1}{4}$

**23.** $P(1|N) = \frac{6}{20}, P(2|N) = \frac{4}{20}, P(3|N) = \frac{6}{20}, P(4|N) = \frac{4}{20}$

## Section 7.7

**1.** $5(.2)^4(.8) = .0064$    **3.** $35(.5)^7 = .273$    **5.** $70(.25)^4(.75)^4 = .087$

**7.** $20(.5)^6 = .3125$    **9.** $35(.1)^4(.9)^3 = .00255$    **11.** $10(.1)^3(.9)^2 = .0081$

**13.** $45(.5)^{10} = .0439$    **15.** $56(.5)^{10} = .0547$    **17.** $11(.5)^{10} = .0107$

**19.** $7(.6)^6(.4) = .1306$    **21.** $7(.6)^6(.4) + (.6)^7 = .1586$

**23.** $(.4)^7 + 7(.6)(.4)^6 + 21(.6)^2(.4)^5 = .0962$    **25.** $1 - (.633)^4 = .839$

**27.** $(.839)^{10} = .173$    **29.** $6(.085)^2(.915)^2 + 4(.085)^3(.915) + (0.85)^4 = .03859$

**31.** $(.03859)^3 = .00006$    **33.** $12(.05)(.95)^{11} = .341$

**35.** $(.95)^{12} + 12(.05)(.95)^{11} + 66(.05)^2(.95)^{10} = .980$

**37.** $C(20, 10)(.8)^{10}(.2)^{10} = .002$

**39.** $190(.8)^{18}(.2)^2 + 20(.8)^{19}(.2) + (.8)^{20} = .206$    **41.** $190(.05)^2(.95)^{18} = .189$

**43.** $(.95)^{20} + 20(.05)(.95)^{19} + 190(.05)^2(.95)^{18} + 1140(.05)^3(.95)^{17} = .984$    -

## Chapter 7 Review

**1.** (a) .016, .248, .628, .088, .016, .044 (b) .892 (c) .264    **2.** (a) $\frac{5}{30}$ (b) $\frac{15}{30}$ (c) $\frac{20}{30}$

**3.** $p(E \cup F) = .60, p(E \cap F) = 0, p(E^c) = .75$

**4.** $p(E \cup F) = .55, p(E^c \cap F) = .35, p((E \cup F)^c) = .45$    **5.** $p(b) = \frac{1}{6}$

**6.** .75    **7.** (a) .09 (b) .91    **8.** (a) .13 (b) .09 (c) .55

**9.** $\dfrac{5 \cdot 10 \cdot 15}{c(30, 3)} \approx .185, \frac{1}{6}$

**10.** $C(10, 3)C(5, 4)C(15, 2)/C(30, 9), C(9, 3)C(6, 4)(\frac{10}{30})^3(\frac{5}{30})^4(\frac{15}{30})^2$

**11.** $p(E|F) = .40, p(E^c|F) = .60, p(F^c|E^c) = \frac{4}{7}$    **12.** Not independent

**13.** $(.05)^3$    **14.** .254    **15.** (a) $\frac{2}{31}$ (b) $\frac{4}{31}$    **16.** $C(6, 3)(.3)^3(.7)^3$

**17.** $1140(.55)^{17}(.45)^3 + 190(.55)^{18}(.45)^2 + 20(.55)^{19}(.45) + (.55)^{20} = .0049$. Yes.

**18.** .50

## Section 8.1

**1.** $\{2, 3, 4, \ldots\}$ infinite discrete    **3.** Continuous

**5.**

| A | B | C | D | F |
|---|---|---|---|---|
| .12 | .20 | .38 | .22 | .08 |

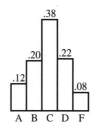

**7.**

| 1 | 2 | 3 | 4 | 5 | 6 | 7 | 8 | 9 | 10 |
|---|---|---|---|---|---|---|---|---|---|
| .10 | .04 | .06 | 0 | .08 | .20 | .20 | .22 | .06 | .04 |

**9.**

| Random variable $x$ | 0 | 1 | 2 | 3 | 4 | 5 |
|---|---|---|---|---|---|---|
| $P(X = x)$ | $\frac{6}{36}$ | $\frac{10}{36}$ | $\frac{8}{36}$ | $\frac{6}{36}$ | $\frac{4}{36}$ | $\frac{2}{36}$ |

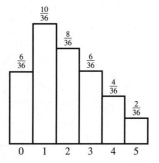

**11.**

| Random variable $x$ | 10 | 7 | 4 | 1 | $-2$ | $-5$ |
|---|---|---|---|---|---|---|
| $P(X = x)$ | $\frac{1}{32}$ | $\frac{5}{32}$ | $\frac{10}{32}$ | $\frac{10}{32}$ | $\frac{5}{32}$ | $\frac{1}{32}$ |

**13.**

| 0 | 1 | 2 |
|---|---|---|
| $\frac{21}{45}$ | $\frac{21}{45}$ | $\frac{3}{45}$ |

**15.** $z = .30$     **17.** (a) .15 (b) 0 (c) .40 (d) .23

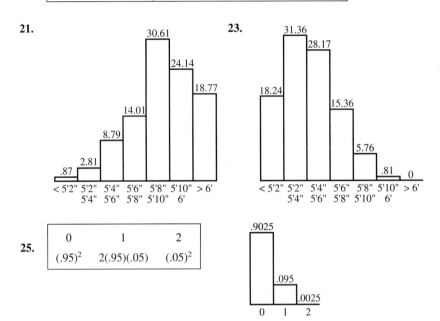

**19.**

| Random variable $y$ | 0 | 1 | 2 | 3 | 4 | 5 |
|---|---|---|---|---|---|---|
| $P(Y = y)$ | .20 | .15 | .05 | .35 | .15 | .10 |

**21.**

**23.**

**25.**

| 0 | 1 | 2 |
|---|---|---|
| $(.95)^2$ | $2(.95)(.05)$ | $(.05)^2$ |

Section 8.2

**1.**

| 0 | 2 | 4 | 6 | 8 | 10 |
|---|---|---|---|---|---|
| .50 | .05 | .10 | .05 | .20 | .10 |

$E(X) = 3.4$

**3.** 0    **5.** .847    **7.** 0, $-25\%$    **9.** $-\$.19$    **11.** $-\$1/19$

**13.** $-\$1/19$    **15.** $.80 or $-.20 if you account for the cost of the $1.00 ticket.

**17.** $\$1/6$    **19.** Yes. $\dfrac{\dfrac{x_1 + \cdots + x_{100}}{100} + \dfrac{x_{101} + \cdots + x_{200}}{100}}{2} = \dfrac{x_1 + \cdots x_{200}}{200}$

**21.** Yes since

$$\frac{1234 + \cdots + 1333}{100} = \frac{(1234 - 1234) + (1235 - 1234) + \cdots + (1333 - 1234) + 100(1234)}{100}$$

$$= \frac{0 + 1 + \cdots + 99}{100} + 1234$$

**23.** (a) $E(Y) = (x_1 - c)p_1 + \cdots + (x_n - c)p_n$

$$= [x_1 p_1 + \cdots x_n p_n] - c[p_1 + \cdots + p_n]$$

$$= E(X) - c$$

(b) 13.1 (c) .1    **25.** .09    **27.** .83    **29.** $1800, $100    **31.** 2.5
**33.** $300, $255, the first    **35.** 156    **37.** 1.29    **39.** 6.6

## Section 8.3

**1.** (a) $\mu = 3$, var $= 2$, $\sigma = \sqrt{2}$ (b) $\mu = 3$, var $= 1.2$, $\sigma = \sqrt{1.2}$
**3.** Geddes: $69\frac{1}{3}$, 1.556, 1.247; Schreger: $69\frac{2}{3}$, .222, .471; Alcott: $69\frac{2}{3}$, 2.889, 1.700: Geddes has lowest average, Schreger the most consistent    **5.** $-.30$, 2.01, 1.418
**7.** 1.5, 3.850, 1.962    **9.** 1, second

**11.** (a) $\text{Var}(cX) = (cx_1)^2 p_1 + \cdots + (cx_n)^2 p_n - (E(cX))^2$

$$= c^2(x_1^2 p_1 + \cdots + x_n^2 p_n) - (cE(X))^2$$

$$= c^2(x_1^2 p_1 + \cdots + x_n^2 p_n - (E(X))^2)$$

$$= c^2 \text{Var}(X)$$

(b) 3.24 (c) $\text{Var}(\frac{1}{3}X) = .36$    **13.** .96    **15.** $c = 10$
**17.** Bank: 0, 0, 0. Bond: $-.16$, .034, .185. The bond varies more and carries more risk.
**19.** U.I. 2.33, U.S.S. 12.834. U.S.S. more risky.
**21.** A: .83, 1.121, 1.059. B: 1.3, 1.810, 1.345. Salesman B sells the most, but Salesman A is more consistent.
**23.** Gonorrhea: 789.333, 7531.5, 86.8. Syphilis: 95.17, 560.5, 23.7. The number of gonorrhea cases is varying the most.    **25.** San Diego: 71, 21.5, 4.6. Chicago: 58.75, 375.2, 19.4. Chicago.
**27.** Marihuana: 23, 52.3, 7.23. Cocaine: 5.67, 6.56, 2.56. Marihuana.
**29.** Males: 431.8, 13.36, 3.655. Females: 421, 6.000, 2.449. Males.
**31.** Carbon monoxide: 6.48, .162, .402. Ozone: .124, .0001, .010. Carbon monoxide.
**33.** Viral hepatitis: .5, .015, $\sqrt{.015}$. Meningitis: .6, .015, $\sqrt{.015}$. They are varying the same.    **35.** $1 - 1/(1.5)^2 = .56$    **37.** .75

## Section 8.4

**1.** .7580    **3.** .3085    **5.** .1151    **7.** .7257    **9.** .1359    **11.** .0919
**13.** .9772    **15.** .0228    **17.** .383    **19.** .2417    **21.** (a) .6826 (b) .0062
**23.** 71.165    **25.** .0228    **27.** (a) 15.87% (b) 6.68% (c) 53.28%
**29.** 116.45 tons    **31.** A 94.2, B 82.9, C 67.1, D 55.8

## Section 8.5

**1.** (a) .1102 (b) .7977    **3.** .0192    **5.** (a) .9901 (b) .9307
**7.** (a) .9744 (b) .8793    **9.** (a) .1539 (b) .7891    **11.** (a) .9969 (b) .9326
**13.** .9872
**15.** Yes, since under the old system the probability of 18 or less defectives is .0166.

**17.** Yes, since the probability of this response before the change was .0000 to four decimal places.

**19.** (a) 1139 (b) 113, 906     **21.** (a) 6765 (b) 676,506

**Chapter 8 Review**

**1.**

| $x$ | 0 | 1 | 2 | 3 | 4 | 5 |
|---|---|---|---|---|---|---|
| $p$ | .168 | .360 | .309 | .132 | .028 | .002 |

**2.** (b) .20 (c) .35 (d) .30 (e) .70     **3.** 2.97
**4.** $1.30 or $.30 if the cost of the $1.00 ticket is subtracted.     **5.** $400
**6.** Wins: $\mu = 28$, $\sigma^2 = 21.2$, $\sigma = 4.60$. Loses: $\mu = 12.6$, $\sigma^2 = 13.44$, $\sigma = 3.67$
**7.** 13–29: 3862.4, 501,978.64, 708.5; 30–39: 9338.2, 4,050,740.6, 2012.6. The 30–39 year old group     **8.** 3.709, 1.926     **9.** .96
**10.** (a) .8078 (b) .1922 (c) .4993     **11.** $c = 2.33$     **12.** .0228     **13.** 38.3%
**14.** .0107     **15.** .1539

**Section 9.1**

**1.** No    **3.** No    **5.** No    **7.** No    **9.** No    **11.** Yes

**13.** $X_1 = \begin{bmatrix} .5 \\ .5 \end{bmatrix}$, $X_2 = \begin{bmatrix} .45 \\ .55 \end{bmatrix}$, $X_4 = \begin{bmatrix} .4445 \\ .5555 \end{bmatrix}$

**15.** $X_1 = \begin{bmatrix} .25 \\ .75 \end{bmatrix}$, $X_2 = \begin{bmatrix} .225 \\ .775 \end{bmatrix}$, $X_4 = \begin{bmatrix} .22225 \\ .77775 \end{bmatrix}$

**17.** $X_1 = \begin{bmatrix} .5 \\ .5 \\ 0 \end{bmatrix}$, $X_2 = \begin{bmatrix} .25 \\ .5 \\ .25 \end{bmatrix}$, $X_4 = \begin{bmatrix} .3125 \\ .3125 \\ .375 \end{bmatrix}$     **19.** $\begin{bmatrix} .7 & .5 \\ .3 & .5 \end{bmatrix}$, .64

**21.** $\begin{bmatrix} .3 & .8 \\ .7 & .2 \end{bmatrix}$, .65     **23.** .62     **25.** .625     **27.** $\begin{bmatrix} .6 & .1 & .1 \\ .2 & .8 & .2 \\ .2 & .1 & .7 \end{bmatrix}$, .44

**29.** .28 in first field, .50 in second, .22 in third.

**Section 9.2**

**1.** Yes    **3.** No    **5.** Yes    **5.** Yes    **7.** No    **9.** Yes    **11.** $\begin{bmatrix} \frac{5}{7} \\ \frac{2}{7} \end{bmatrix}$

**13.** $\begin{bmatrix} .40 \\ .60 \end{bmatrix}$    **15.** $\begin{bmatrix} \frac{4}{7} \\ \frac{3}{7} \end{bmatrix}$    **17.** $\begin{bmatrix} \frac{1}{3} \\ \frac{4}{9} \\ \frac{2}{9} \end{bmatrix}$    **19.** $\begin{bmatrix} \frac{3}{11} \\ \frac{5}{11} \\ \frac{3}{11} \end{bmatrix}$    **21.** $\begin{bmatrix} \frac{5}{8} \\ \frac{3}{8} \end{bmatrix}$    **23.** $\begin{bmatrix} \frac{8}{15} \\ \frac{7}{15} \end{bmatrix}$

**25.** $\begin{bmatrix} .20 \\ .50 \\ .30 \end{bmatrix}$    **27.** $\begin{bmatrix} .30 \\ .60 \\ .10 \end{bmatrix}$

**Section 9.3**

**1.** Yes   **3.** No   **5.** $\begin{bmatrix} 1 & 0 & \frac{1}{3} \\ 0 & 1 & \frac{2}{3} \\ 0 & 0 & 0 \end{bmatrix}$   **7.** $\begin{bmatrix} 1 & 0 & 0 & \frac{1}{6} \\ 0 & 1 & 0 & \frac{1}{3} \\ 0 & 0 & 1 & \frac{1}{2} \\ 0 & 0 & 0 & 0 \end{bmatrix}$

**9.** $\begin{bmatrix} 1 & 1 & 1 \\ 0 & 0 & 0 \\ 0 & 0 & 0 \end{bmatrix}$   **11.** $\begin{bmatrix} 1 & 0 & \frac{17}{64} & \frac{18}{64} \\ 0 & 1 & \frac{47}{64} & \frac{46}{64} \\ 0 & 0 & 0 & 0 \\ 0 & 0 & 0 & 0 \end{bmatrix}$

**13.** (a) $\begin{bmatrix} 1 & .25 \\ 0 & .75 \end{bmatrix}$ (b) State 1 is absorbing and one can go from state 2 to state 1.

(c) $\begin{bmatrix} 1 & 1 \\ 0 & 0 \end{bmatrix}$; The second 1 in the first row indicates that the probability is 1 that the task will be learned in the long run.

**15.** The probability is $\frac{15}{19}$ that the person goes from \$1 to \$0. The probability is $\frac{9}{19}$ that the person goes from \$2 to \$0. The probability is $\frac{4}{19}$ that the person goes from \$1 to \$3. The probability is $\frac{10}{19}$ that the person goes from \$2 to \$3.

**17.** (a) $\begin{bmatrix} 1 & 0 & \frac{1}{6} & \frac{1}{6} \\ 0 & 1 & 0 & \frac{2}{3} \\ 0 & 0 & \frac{1}{6} & 0 \\ 0 & 0 & \frac{2}{3} & \frac{1}{6} \end{bmatrix}$ (b) The first two states, completion and scrap, are absorbing.

(c) The probability is $\frac{16}{25}$ of going from first manufacturing state to completion. The probability is $\frac{4}{5}$ of going from second manufacturing state to completion. The probability is $\frac{9}{25}$ of going from first manufacturing state to scrap. The probability is $\frac{1}{5}$ of going from second manufacturing state to scrap.

**19.** Probability of $\frac{5}{6}$ of < 30 becoming paid. Probability of $\frac{2}{3}$ of < 60 becoming paid. Probability of $\frac{1}{6}$ of < 30 becoming bad. Probability of $\frac{1}{3}$ of < 60 becoming bad.

**Chapter 9 Review**

**1.** None   **2.** Regular stochastic   **3.** Regular stochastic
**4.** Nonabsorbing stochastic   **5.** Nonabsorbing stochastic   **6.** Absorbing stochastic

**7.** $A^2 = \begin{bmatrix} .55 & .30 \\ .45 & .70 \end{bmatrix}$   $A^4 = \begin{bmatrix} .4375 & .3750 \\ .5625 & .6250 \end{bmatrix}$

The elements in $A^2$ and $A^4$ represent the probabilities of going from one state to another in two stages and four stages, respectively.

**8.** $A^2 = \begin{bmatrix} .19 & .16 & .16 \\ .31 & .34 & .31 \\ .50 & .50 & .53 \end{bmatrix}$   $A^4 = \begin{bmatrix} .1657 & .1648 & .1648 \\ .3193 & .3202 & .3193 \\ .5150 & .5150 & .5159 \end{bmatrix}$

The elements in $A^2$ and $A^4$ represent the probabilities of going from one state to another in two stages and four stages, respectively. **9.** $\begin{bmatrix} .40 \\ .60 \end{bmatrix}$ **10.** $\begin{bmatrix} \frac{16}{97} \\ \frac{31}{97} \\ \frac{50}{97} \end{bmatrix}$

**11.** $\begin{bmatrix} 1 & 0 & .4 \\ 0 & 1 & .6 \\ 0 & 0 & 0 \end{bmatrix}$ **12.** $\begin{bmatrix} 1 & 1 & 1 \\ 0 & 0 & 0 \\ 0 & 0 & 0 \end{bmatrix}$

**13.** 40% in the first location and 60% in the second.

**14.** In the long run $\frac{9}{35}$ in the upper class, $\frac{19}{35}$ in middle class, and $\frac{1}{5}$ in lower class.

**15.** 50% of the mechanics leave as master mechanics and 50% leave as mechanics.

### Section 10.1

**1.** Value is 2. Row picks second row. Column picks first column.

**3.** Value is $-2$. Row picks second row. Column picks second column.

**5.** Value is 5. Row picks third row. Column picks first column.

**7.** Value is $-1$. Row picks second row. Column picks second or third column.

**9.** No **11.** Value is 0. Row picks third row. Column picks second column.

**13.** No matter what $a$ is, 1 is a saddle point.

**15.** $\begin{matrix} p \\ n \\ d \end{matrix} \begin{bmatrix} 0 & 5 & 10 \\ -5 & 0 & 10 \\ -10 & -10 & 0 \end{bmatrix}$; Yes with value 0. Each player plays a penny.

**17.** $\begin{matrix} p \\ n \\ d \end{matrix} \begin{bmatrix} 1 & -4 & -9 \\ -4 & 5 & -5 \\ -9 & -5 & 10 \end{bmatrix}$; No.

### Section 10.2

**1.** .625 **3.** $-.52$ **5.** 2.1 **7.** 1.56

**9.** $v = 0, p^T = [.50 \quad .50], q^T = [\frac{2}{3} \quad \frac{1}{3}]$

**11.** $v = .50, p^T = [.50 \quad .50], q^T = [.75 \quad .25]$

**13.** $v = .50, p^T = [.75 \quad .25], q^T = [.50 \quad 0 \quad .50]$

**15.** $v = -.25, p^T = [.75 \quad .25 \quad 0], q^T = [.25 \quad .75]$

**17.** $v = .50, p^T = [0 \quad .75 \quad .25], q^T = [0 \quad .50 \quad .50]$

**19.** $v = -\frac{4}{7}, p^T = [\frac{6}{7} \quad \frac{1}{7}]$ **21.** $v = \frac{5}{3}, q^T = [\frac{1}{3} \quad \frac{2}{3}]$

**23.** $v = \frac{19}{30}$. The first should have a 20% sale $\frac{1}{6}$ of the time and a 30% sale $\frac{5}{6}$ of the time. The second should have a 20% sale $\frac{2}{3}$ of the time and a 30% sale $\frac{1}{3}$ of the time.

**25.** $v = .50$. The second should advertise one half the time on radio and one half the time in the newspaper. The first should advertise three quarters of the time on TV and one quarter of the time in the newspaper.

### Section 10.3

**1.** $v = -.50, p^T = [\frac{5}{8} \quad \frac{3}{8}], q^T = [.50 \quad .50]$

**3.** $v = .60, p^T = [\frac{2}{5} \quad \frac{3}{5}], q^T = [\frac{4}{5} \quad \frac{1}{5}]$ **5.** $v = -\frac{1}{3}, p^T = [\frac{1}{3} \quad \frac{2}{3}], q^T = [\frac{1}{3} \quad \frac{2}{3}]$

**7.** $v = -.25, p^T = [.75 \quad .25], q^T = [.25 \quad .75]$ **9.** $v = -\frac{4}{7}, p^T = [\frac{6}{7} \quad \frac{1}{7}]$.

**11.** $v = \frac{5}{3}, q^T = [\frac{1}{3} \quad \frac{2}{3}]$    **13.** $v = -.50, p^T = [\frac{5}{8} \quad \frac{3}{8}], q^T = [.50 \quad .50]$

**15.** $v = .60, p^T = [\frac{2}{5} \quad \frac{3}{5}], q^T = [\frac{4}{5} \quad \frac{1}{5}]$

**17.** Solving the maximum problem for column (after adding 6 to the payoff matrix) gives the following last line in the final tableau

| $x_1$ | $x_2$ | $x_3$ | $s_1$ | $s_2$ | $z$ | |
|-------|-------|-------|-------|-------|-----|-----|
| 0 | $\frac{3}{38}$ | $\frac{4}{38}$ | $\frac{6}{38}$ | $\frac{1}{38}$ | 1 | $\frac{7}{38}$ |

The value is then $\dfrac{1}{\frac{7}{38}} - 6 = -\frac{4}{7}$. The strategy for column is

$$q^T = \frac{1}{\frac{7}{38}} [0 \quad \tfrac{3}{38} \quad \tfrac{4}{38}] = [0 \quad \tfrac{3}{7} \quad \tfrac{4}{7}]$$

By using the dual theory we can read off the strategy for row by using the numbers under the slack variable and obtain

$$p^T = \frac{1}{\frac{7}{38}} [\tfrac{6}{38} \quad \tfrac{1}{38}] = [\tfrac{6}{7} \quad \tfrac{1}{7}]$$

This agrees with the answer found in Example 9.

**19.** Solving the minimum problem for row (after adding 3 to the payoff matrix) gives the following last line in the final tableau:

| $y_1$ | $y_2$ | $y_3$ | $s_1$ | $s_2$ | $z$ | |
|-------|-------|-------|-------|-------|-----|-----|
| 0 | $\frac{2}{14}$ | $\frac{1}{14}$ | $\frac{1}{14}$ | $\frac{2}{14}$ | 1 | $\frac{3}{14}$ |

The value is then $\dfrac{1}{\frac{3}{14}} - 3 = \frac{5}{3}$. The strategy for row is

$$p^T = \frac{1}{\frac{3}{14}} [0 \quad \tfrac{2}{14} \quad \tfrac{1}{14}] = [0 \quad \tfrac{2}{3} \quad \tfrac{1}{3}]$$

By using dual theory we can read off the strategy for column by using the numbers under the slack variable and obtain

$$q^T = \frac{1}{\frac{3}{14}} [\tfrac{1}{14} \tfrac{2}{14}] = [\tfrac{1}{3} \quad \tfrac{2}{3}]$$

This agrees with the answer found in Example 11.

**21.** Each picks even and odd half the time.

**23.** $v = \frac{19}{30}$. The first should have a 20% sale $\frac{1}{6}$ of the time and a 30% sale $\frac{5}{6}$ of the time. The second should have a 20% sale of $\frac{2}{3}$ of the time and a 30% sale $\frac{1}{3}$ of the time.

**25.** $v = .50$. The second should advertise $\frac{1}{2}$ the time on radio and $\frac{1}{2}$ the time in the newspaper. The first should advertise $\frac{3}{4}$ of the time on TV and $\frac{1}{4}$ of the time in the newspaper.

## Chapter 10 Review

1. $v = 0$, $p^T = [1 \quad 0]$, $q^T = [0 \quad 0 \quad 1]$
2. $v = 1$, $p^T = [1, 0]$, $q^T = [0 \quad 1 \quad 0 \quad 0]$
3. $v = \frac{5}{3}$, $p^T = [0 \quad \frac{2}{3} \quad \frac{1}{3}]$, $q^T = [\frac{2}{3} \quad 0 \quad \frac{1}{3}]$
4. $v = \frac{3}{2}$, $p^T = [\frac{1}{2} \quad 0 \quad \frac{1}{2}]$, $q^T = [0 \quad \frac{1}{2} \quad \frac{1}{2}]$    5. $v = \frac{1}{5}$, $p^T = [\frac{2}{5} \quad \frac{3}{5}]$, $q^T = [\frac{2}{5} \quad \frac{3}{5}]$
6. $v = \frac{3}{4}$, $p^T = [\frac{3}{4} \quad \frac{1}{4}]$, $q^T = [\frac{5}{8} \quad \frac{3}{8}]$    7. $v = -\frac{1}{4}$, $p^T = [\frac{3}{4} \quad \frac{1}{4}]$, $q^T = [\frac{1}{4} \quad \frac{3}{4} \quad 0]$
8. $v = \frac{1}{2}$, $p^T = [\frac{1}{6} \quad \frac{5}{6}]$, $q^T = [\frac{1}{2} \quad \frac{1}{2} \quad 0]$    9. $v = -\frac{1}{4}$, $p^T = [\frac{3}{4} \quad \frac{1}{4}]$
10. $v = \frac{25}{6}$, $q^T = [\frac{1}{6} \quad \frac{5}{6}]$    11. $v = -\frac{1}{4}$, $q^T = [0 \quad \frac{3}{4} \quad \frac{1}{4}]$
12. $v = \frac{25}{6}$, $p^T = [0 \quad \frac{5}{6} \quad \frac{1}{6}]$,    13. $v = \frac{1}{3}$, $p^T = [\frac{1}{3} \quad \frac{1}{3} \quad \frac{1}{3}]$, $q^T = [\frac{1}{3} \quad \frac{1}{3} \quad \frac{1}{3}]$
14. If $a \geq b$ or $a \geq c$, then $a$ is a saddle. If $a < b \leq c$, then $b$ is a saddle. If $a < c \leq b$, then $c$ is a saddle.

## Section 11.1

1. Yes    3. No    5. Yes
7. (1) domain = $\{1, 4\}$, range = $\{2, 7\}$ (2) domain = $\{3, 5\}$, range = $\{7\}$
9. All real numbers $x$ except $x = 1$    11. All $x \geq 0$
13. All $x > 1$    15. All real numbers $x$ except $x = 2, 3$
17. All real numbers    19. $\{x | x \geq -2, x \neq 0\}$    21. All real numbers
23. All real numbers    25. Domain = $[0, 2]$, range = $[0, 1]$
27. Domain = $[0, 2]$, range = $\{1, 2\}$    29. Domain = $[0, 6]$, range = $[0, 1]$
31. $5, 1, 3, 6\sqrt{3} + 3, 2x^6 + 3, 2/x^3 + 3$
33. $1/2, 2/3, -1, 1/\sqrt{2}, 1/(x + 3), (x + 1)/(x + 2)$
35. $3, 0, \sqrt{2x^2 - 2}, \sqrt{2/x - 1}, \sqrt{-2x - 1}, -\sqrt{2x - 1}$    37. $1, -1, |x|, x^2$
39. $1, 2, 2, 2$    41. $0, 1, 0, 1$    43. Yes    45. No    47. No    49. Yes
51. $y = -\frac{2}{5}x + 3$    53. No    55. $y = \sqrt[3]{x}$    57. No

59.

61.

63.

65.

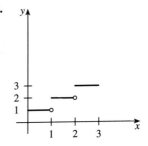

**67.** 3     **69.** $6x + 3h$     **71.** $-\dfrac{1}{x(x+h)}$

**73.** $f(a) + f(b) = ma + mb = m(a + b) = f(a + b)$

**75.** $f(x + 1) = 3^{x+1} = 3 \cdot 3^x = 3f(x)$     $f(a + b) = 3^{a+b} = 3^a \cdot 3^b = f(a) \cdot f(b)$

**77.** $A(r) = \pi r^2$     **79.** $V = 3x^3$     **81.** $d(t) = 60t$     **83.** $R(x) = 60x$

**85.** $C(x) = \begin{cases} x & \text{if } 0 \le x \le 500 \\ 500 + 2(x - 500) & \text{if } x > 500 \end{cases}$

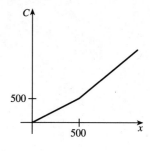

**87.** $T(x) = \begin{cases} 0 & \text{if } 0 \le x \le 50 \\ .01(x - 50) & \text{if } x > 50 \end{cases}$

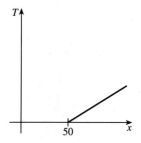

**89.** Domain $= (-\infty, +\infty)$, range $= (-\infty, 8]$

**91.** Domain $= (-\infty, +\infty)$, range $= [2, +\infty)$

**93.** Domain $= (-\infty, -3] \cup [3, +\infty)$, range $= [0, +\infty)$

**95.** Domain $= (-\infty, -4) \cup (-4, 1]$, range $= (-\infty, +\infty)$

**97.** (a) Domain $= (-\infty, +\infty)$, range $= (-\infty, 1]$, (b) Domain $= (-\infty +\infty)$, range $= [1, +\infty)$

**Section 11.2**

**1.**

**3.**

**5.**

(1, 0)

**7.**

(−2, 0)

**9.**

(1, 2)

**11.**

(1, −2)

**13.**

(−2, −3)

**15.**

(−2, 4)

**17.**

(1, 3)

**19.**

(2, −4)

**21.**

**23.**

**25.**

**27.**

**29.**

**31.**

**33.**

**35.**

**37.**          **39.**

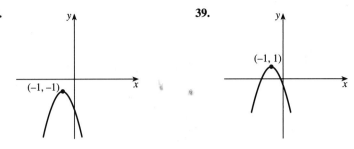

**41.** $x = 5$     **43.** $x = 1.5$     **45.** $x = 500$

**47.** $P(x) = -2x^2 + 20x - 42$, maximum when $x = 5$, break-even points are $x = 3, 7$

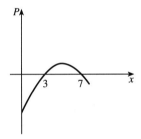

**49.** $P(x) = -x^2 + 15x - 36$, maximum when $x = 15/2$, break-even points are $x = 3, 12$

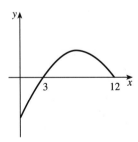

**51.** $P(x) = -3x^2 + 4x - 3$, maximum when $x = 2/3$, no break-even points

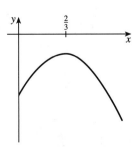

**53.** Degree $= 8$, domain $= (-\infty, +\infty)$  **55.** Degree $= 9$, domain $= (-\infty, +\infty)$

**57.** $(-\infty, 2) \cup (2, +\infty)$  **59.** $(-\infty, +\infty)$  **61.** All real numbers except zero.

**63.** 20.25  **65.** $50 \times 50$  **67.** 110  **69.** $r = 0.50$

**71.** Reaches maximum height in 2 seconds; hits the ground in 4.25 seconds.

### Section 11.3

**1.** (a) 11, (b) 3, (c) 28, (d) 7/4

**3.** $x^2 + 2x + 4$, $-x^2 + 2x + 2$, $(2x + 3)(x^2 + 1)$, $(2x + 3)/(x^2 + 1)$. All the domains are $(-\infty, +\infty)$.

**5.** $3x + 4$, $x + 2$, $2x^2 + 5x + 3$, $(2x + 3)/(x + 1)$. Domain is $(-\infty, +\infty)$ in first three cases and $(-\infty, -1) \cup (-1, +\infty)$ in last case.

**7.** $\sqrt{x + 1} + x + 2$, $\sqrt{x + 1} - x - 2$, $\sqrt{x + 1}(x + 2)$, $\sqrt{x + 1}/(x + 2)$. Domains are all $[-1, +\infty)$.

**9.** $2x + 1 + 1/x$, $2x + 1 - 1/x$, $(2x + 1)/x$, $x(2x + 1)$. Domains are all $(-\infty, 0) \cup (0, +\infty)$.  **11.** $125, -1, 5, -13$

**13.** $2x^3 + 3$, $(2x + 3)^3$. Domains: $(-\infty, +\infty)$, $(-\infty, +\infty)$

**15.** $x, x$. Domains: $(-\infty, +\infty)$, $(-\infty, +\infty)$

**17.** $\sqrt{x^4 + 1}$, $x^2 + 1$. Domains: $(-\infty, +\infty)$, $[0, +\infty)$

**19.** $4x^3 + 1$, $2(2x + 1)^3$, $4x + 3$, $16x^9$  **21.** $f(x) = x^5$, $g(x) = x + 5$

**23.** $f(x) = \sqrt[3]{x}$, $g(x) = x + 1$  **25.** $f(x) = |x|$, $g(x) = x^2 - 1$

**27.** $f(x) = 1/x$, $g(x) = x^2 + 1$  **29.** $9t + 10$

**31.** $R(x) - C(x)$ which is the profit $P(x)$  **33.** $V(t) = 4\pi(30 - 2t)^3/3$

**35.** $g(x) = 40x$, $f(r) = 5.5r$, $y = f(g(x)) = (f \circ g)(x) = 40(5.5)x = 220x$. There are 220 yards in a furlong.

### Chapter 11 Review

**1.** Yes  **2.** No  **3.** Yes  **4.** No  **5.** No  **6.** Yes  **7.** $\{x | x \neq 3\}$

**8.** $\{x | x > 5\}$  **9.** $1, 1/3, 1/(a^2 - 2a + 2), 1/(x^2 + 2x + 2)$

**10.**

**11.**

**12.**

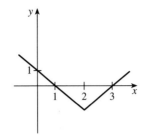

**13.** $x^3 + 2x + 3$. Domain $= (-\infty, +\infty)$

**14.** $x^3 - 2x + 5$. Domain $= (-\infty, +\infty)$
**15.** $(x^3 + 4)(2x - 1)$. Domain $= (-\infty, +\infty)$
**16.** $(x^3 + 4)/(2x - 1)$. Domain $= (-\infty, .50) \cup (.50, +\infty)$
**17.** $9x^2 + 6x + 3, 3x^2 + 7$. Domains: $(-\infty, +\infty), (-\infty, +\infty)$
**18.** $\sqrt{x}, \sqrt{x - 1} + 1$. Domains: $[0, +\infty), [1, +\infty)$     **19.** $10x^2$

**20.**

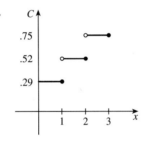

**21.** $4\pi t^2$

## Section 12.1

**1.** (a) 2, 2, 2 (b) 0, does not exist, does not exist (c) 2, 2, 2 (d) 2, 2, 2 (e) 1, 2, does not exist (f) 3, 4 does not exist

**3.**

| $x$ | 1.9 | 1.99 | 1.999 | $\rightarrow$ | 2 | $\leftarrow$ | 2.001 | 2.01 | 2.1 |
|---|---|---|---|---|---|---|---|---|---|
| $f(x)$ | 2.859 | 3.881 | 3.988 | $\rightarrow$ | 4 | $\leftarrow$ | 4.012 | 4.121 | 5.261 |

**5.**

| $x$ | 1.9 | 1.99 | 1.999 | $\rightarrow$ | 2 | $\leftarrow$ | 2.001 | 2.01 | 2.1 |
|---|---|---|---|---|---|---|---|---|---|
| $f(x)$ | $-10$ | $-100$ | $-1000$ | $\rightarrow$ | No | $\leftarrow$ | 1000 | 100 | 10 |

**7.**

| $x$ | 3.9 | 3.99 | 3.999 | $\rightarrow$ | 4 | $\leftarrow$ | 4.001 | 4.01 | 4.1 |
|---|---|---|---|---|---|---|---|---|---|
| $f(x)$ | 3.9748 | 3.9975 | 3.9997 | $\rightarrow$ | 4 | $\leftarrow$ | 4.0002 | 4.0025 | 4.0248 |

**9.** 0, 0, 0

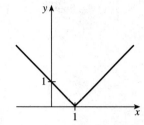

**11.** 1, 0, does not exist

**13.** 1, 1, 1

**15.** 0, does not exist, does not exist

**17.** $-1$, 1, does not exist

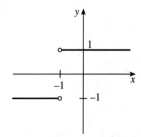

**19.** None of the limits exist

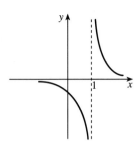

**21.** 128    **23.** 2    **25.** −2    **27.** −6
**29.** 0    **31.** 0    **33.** 13    **35.** 5    **37.** 13    **39.** 0    **41.** −3
**43.** −35    **45.** 0    **47.** 1    **49.** 2    **51.** 4    **53.** 2    **55.** −2
**57.** Does not exist    **59.** 1/2    **61.** 12

**63.** (a) .29, .52, does not exist (b) .52, .52, .52

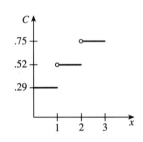

**65.** .1, 1, 10. $C(x)$ becomes large without bound.

**Section 12.2**

**1.** 0, 2, 3, 4    **3.** Nowhere    **5.** 0    **7.** 1    **9.** 0    **11.** −1    **13.** 1
**15.** Everywhere except $x = 0, -1$    **17.** Everywhere except $x = \pm 1$
**19.** Everywhere    **21.** [−1, 1]    **23.** Positive    **25.** Negative
**27.** Negative on (0, 2), positive on (2, 4)

**29.** Not continuous at $x = 50$

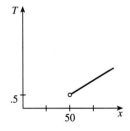

**31.** Not continuous at $x = 100{,}000$ and $x = 1{,}000{,}000$

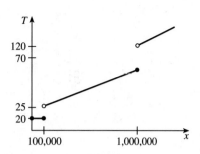

**Section 12.3**

1. (a) $-2t + 8 - h$, 5, 0, $-4$ (b) $v(t) = -2t + 8$ (c) 2, 0, $-2$
3. (a) 2 (b) 2    5. (a) $2x - 4 + h$ (b) $2x - 4$    7. (a) $6x + 3h$ (b) $6x$
9. $-2, 0, 2$    11. 2    13. $2x$    15. $-4x$    17. $2x + 2$
19. $-4t - 4$    21. $3u^2$    23. $3t^2$    25. $0, -8.5, 7/3, 12$
27. $15.9, 13/15, -327/15$    29. $4, 0, -4$    31. $8, 0, -2$
33. $-400, 0, 400$    35. $0, 4, 16$    37. $0.8$

**Section 12.4**

1. 5    3. $2x$    5. $6x + 3$    7. $-6x^2 + 1$    9. $-1/(x + 2)^2$
11. $-2/(2x - 1)^2$    13. $\frac{1}{2\sqrt{t+1}}$    15. $\frac{1}{\sqrt{2t + 5}}$
17. (1) $y - 2 = 5(x - 1)$; (3) $y - 5 = 2(x - 1)$; (5) $y - 5 = 9(x - 1)$
19. Everywhere except $x = 0$    21. Everywhere except $x = 1$
23. Everywhere except $x = 1$    25. Everywhere in $(-1, 1)$ except $x = 0$
27. Everywhere in $(1, 3)$ except $x = 2$    29. $f'(0)$ does not exist.
31. $f'(0)$ does not exist.

**33.**

**35.**

**37.**

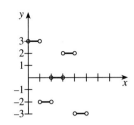

**39.** $2/3, 2/5$

**41.** $-25, -4$    **43.** $-.15, -3/125$

## Section 12.5

**1.** $0$   **3.** $0$   **5.** $23x^{22}$   **7.** $1.4x^{.4}$   **9.** $0.4x^{-0.6}$   **11.** $\frac{4}{3}x^{1/3}$

**13.** $-5.1x^{-2.7}$   **15.** $.12/\sqrt{x}$   **17.** $2$   **19.** $\pi$   **21.** $-1 - 2x$

**23.** $\frac{2}{\pi}x + \pi$   **25.** $-1 - 2x - 3x^2$   **27.** $5x^4 - 1$   **29.** $0.7x - 0.3$

**31.** $-.02x - .09x^2$   **33.** $(x - 1)/3$   **35.** $9x^2 - 9/x^4$   **37.** $2x + 1/x^2$

**39.** $9x^{-4} + 8x$   **41.** $-81t^2$   **43.** $-t^{-2} - 2t^{-3} - 3t^{-4}$

**45.** $\frac{1}{3}t^{-2/3} + \frac{1}{3}t^{-4/3}$   **47.** $5u^{2/3} - 2u^{-1/3}$   **49.** $4u + 3$

**51.** $\frac{1}{2}u^{-1/2} - \frac{9}{2}u^{-5/2}$   **53.** $x^{-1/2} + \frac{1}{6}x^{-3/2}$   **55.** $\pi^2 x^{\pi - 1}$

**57.** $\sqrt{2}x^{\sqrt{2}-1} + \sqrt{2}x^{-\sqrt{2}-1}$   **59.** $-\frac{3}{2}x^{-5/2} - \frac{4}{5}x^{-9/5}$   **61.** $\frac{2}{3}x^{-1/3}$

**63.** $-\frac{2}{5}x^{-7/5}$   **65.** $y - 7 = 5(x - 2)$   **67.** $y - 2 = -3(x - 1)$

**69.** $y = -3(x + 1)$   **71.** $x = 0$   **73.** $x = \pm\sqrt{3}$   **75.** $x = -2, 1$

**77.** $.60$   **79.** $10$   **81.** $30$   **83.** $3$   **85.** $83$   **87.** $2$

**89.** $-4x + 24$   **91.** $-8x + 24, 2, 4$   **93.** $6t^2 - 4$   **95.** $130$

**97.** $-4al/243$

## Section 12.6

**1.** $(2x + 1)3 + 2(3x - 2)$   **3.** $(4 - x) - (3 + x)$

**5.** $(x + 1)2x + (x^2 - 3)$   **7.** $(4 - x^5)(-3) - 5x^4(2 - 3x)$

**9.** $(x^7 - 3)5x^4 + 7x^6(x^5 + 5)$

**11.** $(x^2 + x + 1)(2x - 1) + (2x + 1)(x^2 - x - 3)$

**13.** $\dfrac{x + 1}{2\sqrt{x}} + \sqrt{x} + 1$   **15.** $(2x - 3)(1 - x^{-2}) + 2(x + x^{-1})$

**17.** $-1/(x + 2)^2$   **19.** $-3x^2/(x^3 + 1)^2$   **21.** $2/(x + 2)^2$

**23.** $8/(x + 5)^2$   **25.** $4/(x - 2)^2$   **27.** $-1/(2x - 1)^2$   **29.** $-6u/(u^2 + 1)^2$

**31.** $(2 - 2u - u^2)/(u^2 + 2)^2$   **33.** $(u^4 + 9u^2 + 2u)/(u^2 + 3)^2$

**35.** $\dfrac{-8u^3 + 2u - 1}{3u^{2/3}(u^3 - u - 1)^2}$   **37.** $(1 - x^2)/(1 + x^2)^2$   **39.** $0.1/(10 - 0.1x)^2$

**41.** $\dfrac{2}{5} \cdot \dfrac{I^2 + 100{,}000I}{(I + 50{,}000)^2}$   **43.** $\dfrac{15 - 6t^2}{(2t^2 + 5)^2}$   **45.** $4000\dfrac{t^3}{(1 + t^4)^2}$

## Section 12.7

**1.** $14(2x + 1)^6$   **3.** $8x(x^2 + 1)^3$   **5.** $3(2x - 3)^{1/2}$   **7.** $3.3x^2(x^3 + 1)^{0.1}$

**9.** $1/\sqrt{2x + 1}$   **11.** $4x(x^2 + 1)^{-2/3}$   **13.** $-1/(x + 1)^2$

**15.** $6x^2/(x^3 + 1)^2$   **17.** $-x(x^2 + 1)^{-3/2}$

**19.** $15(2x + 1)(3x - 2)^4 + 2(3x - 2)^5$

**21.** $16x(1 - 4x)(x^2 + 1)^7 - 4(x^2 + 1)^8$

**23.** $10(1 - x)^4(2x - 1)^4 - 4(1 - x)^3(2x - 1)^5$

**25.** $12\sqrt{x}(4x + 3)^2 + \frac{1}{2}x^{-1/2}(4x + 3)^3$ **27.** $-12(x + 1)^{-5}$

**29.** $\dfrac{-2x}{(x + 3)^4}$ **31.** $\frac{1}{4}(\sqrt{x} + 1)^{-1/2}x^{-1/2}$

**33.** $\frac{1}{4}(\sqrt{x} - 1)(\sqrt{x} + 3)^{-1/2}x^{-1/2} + \frac{1}{2}x^{-1/2}(\sqrt{x} + 3)^{1/2}$

**35.** $\left(\dfrac{x + 1}{x - 1}\right)^{1/2}\dfrac{1}{(x + 1)^2}$ **37.** $-x(1 + x^2)^{-3/2}$ **39.** $4000(1 + \frac{1}{12}i)^{47}$

**41.** $\dfrac{4\pi cr^2}{3}\left(1 + \dfrac{c}{r}n^{1/3}\right)^2 n^{-2/3}$ **43.** $\dfrac{an}{2\sqrt{n - b}} + a\sqrt{n - b}$

## Chapter 12 Review

**1.** $x = -1$: 1, 1, 1; $x = 0$: 2, 1, Does not exist; $x = 1$: 2, Does not exist, Does not exist; $x = 2$: 1, 1, 1. **2.** $-1, 0, 1$ **3.** $-1, 0, 1, 2$ **4.** 12.7 **5.** $\pi^3$

**6.** 0 **7.** $-3$ **8.** $-4$ **9.** 1 **10.** $-7$ **11.** 3 **12.** 7

**13.** Does not exist **14.** Does not exist **15.** 2 **16.** 3 **17.** 1.5

**18.** Does not exist **19.** $4x + 1$ **20.** $2x + 5$ **21.** $-2/x^3$

**22.** $-2/(2x + 5)^2$ **23.** $\dfrac{1}{2\sqrt{x + 7}}$ **24.** $1/\sqrt{2x - 1}$

**25.** $y - 9 = 12(x - 2)$ **26.** $y - 10 = 3(x - 4)$ **27.** $y - 6 = 11(x - 1)$

**28.** $y - \frac{1}{3} = -\frac{1}{54}(x - 9)$ **29.** $6x^5 - 6x$ **30.** $\frac{24}{5}x^{1/5} + 2x^{-3/5}$

**31.** $-6x^{-4} - 2x^{-5/3}$ **32.** $2.6x^{0.3} - 7.2x^{-3.4}$ **33.** $2\pi^2 x + 2$

**34.** $3\pi^3 x^2 + 2e^2 x$ **35.** $\frac{2}{3}x^{-1/3}$ **36.** $-\frac{5}{3}x^{-8/3}$

**37.** $(x + 3)(3x^2 + 1) + (x^3 + x + 1)$

**38.** $(x^2 + 5)(5x^4 - 3) + 2x(x^5 - 3x + 2)$

**39.** $\sqrt{x}(4x^3 - 4x) + \frac{1}{2}x^{-1/2}(x^4 - 2x^2 + 3)$

**40.** $\sqrt[3]{x}(3x^2 - 3) + \frac{1}{3}x^{-2/3}(x^3 - 3x + 5)$

**41.** $x^{3/2}(2x + 4) + \frac{3}{2}x^{1/2}(x^2 + 4x + 7)$

**42.** $x^{-7/2}(3x^2 + 2) - \frac{7}{2}x^{-9/2}(x^3 + 2x + 1)$ **43.** $2/(x + 1)^2$

**44.** $\dfrac{(2x + 1)\frac{1}{2}x^{-1/2} - 2x^{1/2}}{(2x + 1)^2}$ **45.** $\dfrac{x^4 + 3x^2}{(x^2 + 1)^2}$ **46.** $\dfrac{3x^2 - 4x - 5}{(3x - 2)^2}$

**47.** $\dfrac{-2x^5 + 4x^3 + 2x}{(x^4 + 1)^2}$ **48.** $30(3x + 1)^9$ **49.** $40x(x^2 + 4)^{19}$

**50.** $\frac{5}{4}(\sqrt{x} - 5)^{3/2}x^{-1/2}$ **51.** $\dfrac{2x + 1}{2\sqrt{x^2 + x + 1}}$

**52.** $5x^3(x^2 + x + 2)^4(2x + 1) + 3x^2(x^2 + x + 2)^5$

**53.** $5(x^2 + 3)(x^4 + x + 1)^4(4x^3 + 1) + 2x(x^4 + x + 1)^5$

**54.** $-4(x^2 + 1)(x^4 + x + 1)^{-5}(4x^3 + 1) + 2x(x^2 + x + 1)^{-4}$

**55.** $\frac{3}{4}(x^{3/2} + 1)^{-1/2}x^{1/2}$ **56.** $\frac{1}{6}(x^{1/2} + 1)^{-2/3}x^{-1/2}$ **57.** Yes **58.** No

**59.** Large without bound **60.** Large without bound **61.** 0

**62.** Continuous everywhere except 1, 2, 3, ... **63.** 200, 0, $-200$ **64.** $-2, 0, 2$

**65.** 25 **66.** $-900/(3x + 1)^2$

## Section 13.1

1. Critical value at $x = 2$
   Increasing on $(0, 2)$
   Decreasing on $(2, 4)$

3. Critical value at $x = 2$
   Increasing on $(0, 4)$

5. Critical value at $x = 3$
   Increasing on $(3, 4)$
   Decreasing on $(0, 3)$

7. Critical value at $x = 3$
   Increasing on $(0, 4)$

9. Critical value at $x = 2$
   Increasing on $(2, +\infty)$
   Decreasing on $(-\infty, 2)$

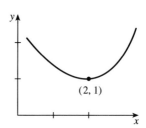

11. Critical values at $x = 1, 3$
    Increasing on $(-\infty, 1)$ and $(3, +\infty)$
    Decreasing on $(1, 3)$

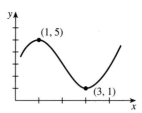

13. Critical values at $x = 1, 3, 5$
    Increasing on $(1, 3)$ and $(5, +\infty)$
    Decreasing on $(-\infty, 1)$ and $(3, 5)$

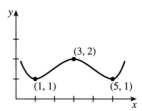

15. Critical values at $x = -2, 1, 3$
    Increasing on $(-\infty, -2)$ and $(1, 3)$
    Decreasing on $(-2, 1)$ and $(3, +\infty)$

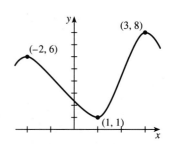

**17.** Critical values at $x = -1, 2, 4$
Increasing on $(-\infty, 2)$ and $(4, +\infty)$
Decreasing on $(2, 4)$

**19.** Critical values at $x = 0, 2, 4$
Increasing on $(2, +\infty)$
Decreasing on $(-\infty, 2)$

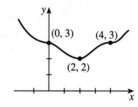

**21.** No critical values
Increasing on $(-\infty, +\infty)$

**23.** No critical values
Decreasing on $(-\infty, +\infty)$

**25.** Critical value at $x = -3/2$
Increasing on $(-3/2, +\infty)$
Decreasing on $(-\infty, -3/2)$

**27.** Critical value at $x = 1$
Increasing on $(-\infty, 1)$
Decreasing on $(1, +\infty)$

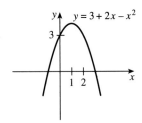

$y = 3 + 2x - x^2$

**29.** Critical values at $x = 1, 2$
Increasing on $(-\infty, 1)$ and $(2, +\infty)$
Decreasing on $(1, 2)$

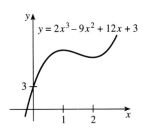

$y = 2x^3 - 9x^2 + 12x + 3$

**31.** Critical values at $x = -\frac{1}{2}, \frac{1}{2}$
Increasing on $(-\frac{1}{2}, \frac{1}{2})$
Decreasing on $(-\infty, -\frac{1}{2})$ and $(\frac{1}{2}, +\infty)$

$y = -4x^3 + 3x - 2$

**33.** Critical values at $x = 0, 1$
Increasing on $(1, +\infty)$
Decreasing on $(-\infty, 1)$

$y = 3x^4 - 4x^3 + 1$

**35.** Critical values at $x = 0, 4$
Increasing on $(-\infty, 0)$ and $(4, +\infty)$
Decreasing on $(0, 4)$

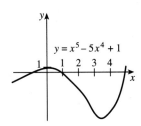

$y = x^5 - 5x^4 + 1$

**37.** Critical values at $x = -1, 0, 1$
Increasing on $(-1, 0)$ and $(1, +\infty)$
Decreasing on $(-\infty, -1)$ and $(0, 1)$

**39.** Critical values at $x = -4, 0, \frac{1}{2}$
Increasing on $(-\infty, -4)$ and $(0, \frac{1}{2})$
Decreasing on $(-4, 0)$ and $(\frac{1}{2}, +\infty)$

**41.** Critical value at $x = 0$
Increasing on $(0, +\infty)$
Decreasing on $(-\infty, 0)$

**43.** Critical value at $x = 0$
Increasing on $(0, +\infty)$
Decreasing on $(-\infty, 0)$

**45.** Critical value at $x = -1$
Increasing on $(-\infty, +\infty)$

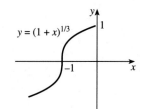

**47.** Critical value at $x = 1$
Increasing on $(1, +\infty)$
Decreasing on $(-\infty, 1)$

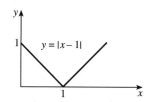

**49.** $f'(x) = 3ax^2 + b$ is positive if both $a$ and $b$ are positive, and negative if both $a$ and $b$ are negative.

**51.** Increasing on $(0, 50)$, decreasing on $(50, +\infty)$

**53.** Increasing on $(0, 6)$, decreasing on $(6, +\infty)$

**55.** Increasing on $(0, 200/3)$, decreasing on $(200/3, 100)$

**57.** Decreasing on $(0, +\infty)$

**Section 13.2**

**1.** Critical value at $x = 2$
Relative maximum at $x = 2$

**3.** Critical values at $x = 2, 4$
Relative maximum at $x = 4$
Relative minimum at $x = 2$

**5.** Critical values at $x = 1, 2, 3, 4$
Relative maxima at $x = 1, 4$
Relative minimum at $x = 3$

**7.** Critical values at $x = 8, 10$
Relative maximum at $x = 8$
Relative minimum at $x = 10$

**9.** Critical value at $x = 3$
Increasing on $(3, +\infty)$
Decreasing on $(-\infty, 3)$
No relative maxima
Relative minimum at $x = 3$

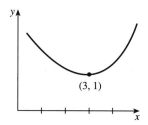

**11.** Critical values at $x = 2, 4$
Increasing on $(-\infty, 2), (4, +\infty)$
Decreasing on $(2, 4)$
Relative maximum at $x = 2$
Relative minimum at $x = 4$

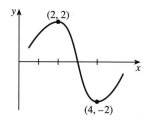

**13.** Critical values at $x = 1, 3, 5$
Increasing on $(1, 3), (5, +\infty)$
Decreasing on $(-\infty, 1), (3, 5)$
Relative maximum at $x = 3$
Relative minima at $x = 1, 5$

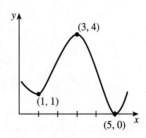

**15.** Critical values at $x = 1, 3, 5$
Increasing on $(-\infty, 1), (3, +\infty)$
Decreasing on $(1, 3)$
Relative maximum at $x = 1$
Relative minimum at $x = 3$

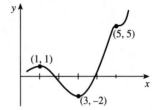

**17.** Critical values at $x = -1, 1, 2$
Increasing on $(-\infty, 1), (2, +\infty)$
Decreasing on $(1, 2)$
Relative maximum at $x = 1$
Relative minimum at $x = 2$

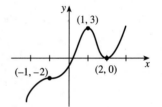

**19.** Critical values at $x = -1, 1, 2$
Increasing on $(-1, 1), (2, +\infty)$
Decreasing on $(-\infty, -1), (1, 2)$
Relative maximum at $x = 1$
Relative minima at $x = -1, 2$

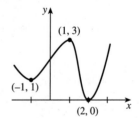

**21.** Critical value at $x = 2$
Increasing on $(2, +\infty)$
Decreasing on $(-\infty, 2)$
No relative maxima
Relative minimum at $x = 2$

**23.** No critical values
Increasing on $(-\infty, +\infty)$
Never decreasing
No relative maxima
No relative minima

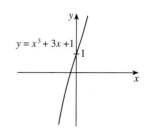

$y = x^3 + 3x + 1$

**25.** Critical value at $x = 0$
Increasing on $(0, +\infty)$
Decreasing on $(-\infty, 0)$
No relative maxima
Relative minimum at $x = 0$

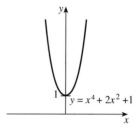

$y = x^4 + 2x^2 + 1$

**27.** Critical value at $x = 1$
Increasing on $(-\infty, 1)$
Decreasing on $(1, +\infty)$
Relative maximum at $x = 1$
No relative minima

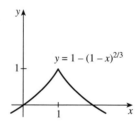

$y = 1 - (1 - x)^{2/3}$

**29.** Critical values at $x = -1, 1$
Increasing on $(-\infty, -1), (1, +\infty)$
Decreasing on $(-1, 1)$
Relative maximum at $x = -1$
Relative minimum at $x = 1$

$y = x^3 - 3x + 1$

**31.** Critical values at $x = \frac{1}{2}, \frac{3}{2}$
Increasing on $(-\infty, \frac{1}{2}), (\frac{3}{2}, +\infty)$
Decreasing on $(\frac{1}{2}, \frac{3}{2})$
Relative maximum at $x = \frac{1}{2}$
Relative minimum at $x = \frac{3}{2}$

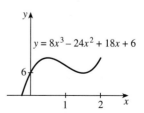

$y = 8x^3 - 24x^2 + 18x + 6$

**33.** Critical values at $x = -2, 0, 2$
Increasing on $(-2, 0), (2, +\infty)$
Decreasing on $(-\infty, -2), (0, 2)$
Relative maximum at $x = 0$
Relative minima at $x = -2, 2$

**35.** Critical values at $x = -1, 1$
Increasing on $(-1, 1)$
Decreasing on $(-\infty, -1), (1, +\infty)$
Relative maximum at $x = 1$
Relative minimum at $x = -1$

**37.** Critical values at $x = -2, 0, 2$
Increasing on $(-\infty, -2), (2, \infty)$
Decreasing on $(-2, 2)$
Relative maximum at $x = -2$
Relative minimum at $x = 2$

**39.** Critical values at $x = 0, 1$
Increasing on $(-\infty, 0), (1, +\infty)$
Decreasing on $(0, 1)$
Relative maximum at $x = 0$
Relative minimum at $x = 1$

**41.** Critical values at $x = 0, 1$
Increasing on $(0, 1)$
Decreasing on $(-\infty, 0), (1, +\infty)$
Relative maximum at $x = 1$
Relative minimum at $x = 0$

**43.** $x = 500$     **45.** $x = 20$     **47.** $x = 3$

**51.** 16, 6912 hundred thousand pounds     **53.** $1/2$     **55.** $\sqrt{b}$

### Section 13.3

**1.** $80x^3 + 2$     **3.** $2(x + 1)^{-3}$     **5.** $-(2x + 1)^{-3/2}$     **7.** $240x^2, 480x$

**9.** $-6x^{-4}, 24x^{-5}$     **11.** $\dfrac{8}{27}x^{-7/3}, -\dfrac{56}{81}x^{-10/3}$     **13.** No inflection value.
Concave down on $(0, 4)$     **15.** 2. Concave down on $(0, 2)$, concave up on $(2, 4)$

**17.** 2. Concave down on $(0, 2)$, concave up on $(2, 4)$

**19.** No inflection value. Concave down on $(0, 2)$, concave up on $(2, 4)$

**21.** 2. Concave up on $(2, +\infty)$, concave down on $(-\infty, 2)$

**23.** 1, 3. Concave up on $(-\infty, 1)$ and $(3, +\infty)$, concave down on $(1, 3)$

**25.** 1, 3, 5. Concave up on $(1, 3)$ and $(5, +\infty)$, concave down on $(-\infty, 1)$ and $(3, 5)$

**27.** 0, 4. Concave up on $(-\infty, 0)$ and $(4, +\infty)$, concave down on $(0, 4)$

**29.** $-2$. Concave up on $(-2, +\infty)$, concave down on $(-\infty, -2)$

**31.** $-2, 0, 2$. Concave up on $(-2, 0)$ and $(2, +\infty)$, concave down on $(-\infty, -2)$ and $(0, 2)$

**33.**   Critical value at $x = 0$.
Inflection value at $x = 0$
Increasing on $(-\infty, +\infty)$
Decreasing nowhere
Concave up on $(0, +\infty)$
Concave down on $(-\infty, 0)$

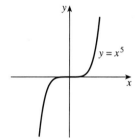

**35.**   Critical value at $x = 0$
Relative maximum at $x = 0$
No inflection values
Increasing on $(-\infty, 0)$
Decreasing on $(0, +\infty)$
Concave up nowhere
Concave down on $(-\infty, +\infty)$

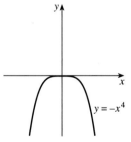

**37.**   Critical value at $x = 4$
Relative minimum at $x = 4$
No inflection values
Increasing on $(4, +\infty)$
Decreasing on $(-\infty, 4)$
Concave up on $(-\infty, +\infty)$
Concave down nowhere

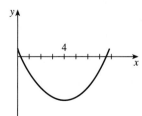

**39.** Critical value at $x = \frac{1}{2}$
Relative maximum at $x = \frac{1}{2}$
No inflection values
Increasing on $(-\infty, \frac{1}{2})$
Decreasing on $(\frac{1}{2}, +\infty)$
Concave up nowhere
Concave down on $(-\infty, +\infty)$

**41.** Critical values at $x = 0, 2$
Relative maximum at $x = 0$
Relative minimum at $x = 2$
Inflection value at $x = 1$
Increasing on $(-\infty, 0), (2, +\infty)$
Decreasing on $(0, 2)$
Concave up on $(1, +\infty)$
Concave down on $(-\infty, 1)$

$y = x^3 - 3x^2 + 2$

**43.** Critical values at $x = 1, 3$
Relative maximum at $x = 3$
Relative minimum at $x = 1$
Inflection value at $x = 2$
Increasing on $(1, 3)$
Decreasing on $(-\infty, 1), (3, +\infty)$
Concave up on $(-\infty, 2)$
Concave down on $(2, +\infty)$

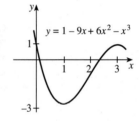

$y = 1 - 9x + 6x^2 - x^3$

**45.** Critical values at $x = 0, 1$
Relative maximum at $x = 0$
Relative minimum at $x = 1$
Inflection value at $x = \frac{3}{4}$
Increasing on $(-\infty, 0), (1, +\infty)$
Decreasing on $(0, 1)$
Concave up on $(\frac{3}{4}, +\infty)$
Concave down on $(-\infty, \frac{3}{4})$

$y = 4x^5 - 5x^4 + 1$

**47.** No critical values
No inflection values
Increasing on $(0, +\infty)$
Decreasing nowhere
Concave up nowhere
Concave down on $(0, +\infty)$

$y = \sqrt{x}$

**49.** No critical values
No inflection values
Increasing nowhere
Decreasing on $(-\infty, 1)$
Concave up nowhere
Concave down on $(-\infty, 1)$

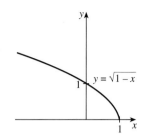

**51.** Critical value at $x = 2$
Inflection value at $x = 2$
Increasing on $(-\infty, \infty)$
Decreasing nowhere
Concave up on $(2, +\infty)$
Concave down $(-\infty, 2)$

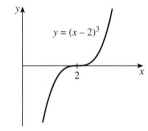

**53.** Critical value at $x = \frac{1}{2}$
Relative minimum at $x = \frac{1}{2}$
No inflection values
Increasing on $(\frac{1}{2}, +\infty)$
Decreasing on $(-\infty, \frac{1}{2})$
Concave up on $(-\infty, +\infty)$
Concave down nowhere

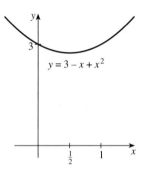

**55.** Critical value at $x = 0$
Relative minimum at $x = 0$
No inflection values
Increasing on $(0, +\infty)$
Decreasing on $(-\infty, 0)$
Concave up on $(-\infty, +\infty)$
Concave down nowhere

**57.** Critical value at $x = 0$
Relative minimum at $x = 0$
No inflection values
Increasing on $(0, +\infty)$
Decreasing $(-\infty, 0)$
Concave up on $(-\infty, +\infty)$
Concave down nowhere

**59.** Critical value at $x = 3$
Relative minimum at $x = 3$
No inflection values
Increasing on $(3, \infty)$
Decreasing on $(-\infty, 3)$
Concave up on $(-\infty, +\infty)$
Concave down nowhere

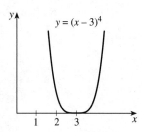

**61.** Critical values at $x = 0, 2$
Relative maximum at $x = 2$
Inflection values at $x = 0, \frac{4}{3}$
Increasing on $(-\infty, 2)$
Decreasing on $(2, +\infty)$
Concave up on $(0, \frac{4}{3})$
Concave down on $(-\infty, 0), (\frac{4}{3}, +\infty)$

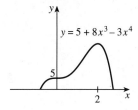

**63.** Since $f''(x) = 2a$, the sign of $f''(x)$ and $a$ are the same.
**65.** $f''(x) = 12ax^2 + 2b$
**67.** Since $f'(x) = 3x^2 - 3b$, the critical values are $x = \pm\sqrt{b}$. Since $f''(x) = 6x$, $f(-\sqrt{b})$ is a relative maximum and $f(\sqrt{b})$ is a relative minimum. **69.** 6
**71.** (a) Increasing on $(0, +\infty)$, (b) Concave up on $(0, 10)$,
concave down on $(10, +\infty)$, inflection point at $(10, 25)$,
(c)

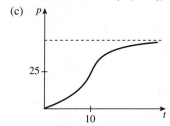

**73.** Increasing on $(0, +\infty)$
Concave up on $(0, 3)$
Concave down on $(3, +\infty)$
Inflection point at $(3, 103/1080)$

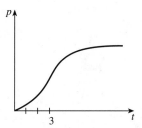

**75.** Positive on $\left(-\infty, -\dfrac{\sqrt{6}}{6}\right)$ and $\left(\dfrac{\sqrt{6}}{6}, +\infty\right)$, negative on $\left(-\dfrac{\sqrt{6}}{6}, \dfrac{\sqrt{6}}{6}\right)$
**77.** $r = \frac{1}{2}$

**Section 13.4**

**1.** $\frac{2}{3}$     **3.** $\frac{4}{3}$     **5.** 0     **7.** $+\infty$     **9.** $+\infty$     **11.** 0     **13.** 2

**15.** 2     **17.** 0

**19.** Critical values at $x = 0, -3$
Inflection values at $x = 0, -2$

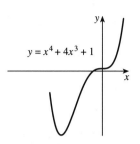

$y = x^4 + 4x^3 + 1$

**21.** Critical values at $x = -4, 0, 4$
Inflection values at $\pm 4/\sqrt{3}$

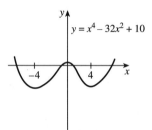

$y = x^4 - 32x^2 + 10$

**23.** Critical values at $x = -2\sqrt{2}, 0, 2\sqrt{2}$
Inflection values at $x = -2, 0, 2$

$y = 3x^5 - 40x^3 + 20$

**25.** No critical values
No inflection values
Asymptotes: $x = 1, y = 0$

$y = \dfrac{1}{x-1}$

**27.** No critical values
No inflection values
Asymptotes: $x = -2, y = 1$

$y = \dfrac{x+1}{x+2}$

29. Critical value at $x = 0$
    Inflection values at $x = \pm 1/\sqrt{3}$
    Asymptote: $y = 0$

$y = \dfrac{1}{x^2+1}$

31. Critical values at $x = \pm 3$
    No inflection values
    Asymptote: $x = 0$

$y = x + \dfrac{9}{x}$

33. Critical values at $x = \pm 1$
    No inflection values
    Asymptote: $x = 0$

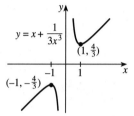

$y = x + \dfrac{1}{3x^3}$

$\left(1, \frac{4}{3}\right)$

$\left(-1, -\frac{4}{3}\right)$

35. Critical values at $x = 0, 2$
    No inflection values
    Asymptote: $x = 1$

$y = \dfrac{x^2}{x-1}$

37. Critical values at $x = \pm 1$
    Inflection values at $x = 0, \pm\sqrt{3}$
    Asymptote: $y = 0$

$y = \dfrac{x}{x^2+1}$

**39.** Critical value at $x = 2$
Inflection value at $x = 3$
Asymptote: $x = 0, y = 0$.

$$y = \frac{x-1}{x^2}$$

**41.** Critical value at $x = 4$
Inflection value at $x = 6$

$$y = \frac{x^2 + 2x - 4}{x^2}$$

**43.** No critical values
No inflection values
Asymptote: $x = 100$

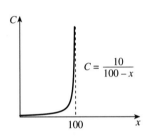

$$C = \frac{10}{100 - x}$$

**45.** No critical values
No inflection values
Asymptote: $n = 9$

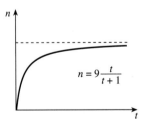

$$n = 9\frac{t}{t + 1}$$

**47.** Critical value at $x = \sqrt{a/b}$
Asymptote: $x = 0$

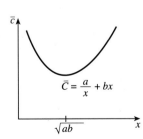

$$\bar{C} = \frac{a}{x} + bx$$

**49.** $c \to 0$     **51.** $p(t) \to 100$

## Section 13.5

1. Absolute minimum at 0, absolute maximum at 4
3. Absolute minimum at 0, absolute maximum at 3
5. Absolute minimum at 3, absolute maximum at 1
7. Absolute minimum at 1 and 3, absolute maximum at 0 and 4
9. Absolute minimum at $x = 2$, absolute maximum at $x = 4$
11. Absolute minimum at $x = 1$, absolute maximum at $x = 2$
13. Absolute minimum at $x = -2, 1$, absolute maximum at $x = -1$ and 2
15. Absolute minimum at $x = 0$, absolute maximum at $x = -2$
17. Absolute minimum at $x = 2$, absolute maximum at $x = 1$
19. Absolute minimum at $x = 2$, absolute maximum at $x = -2$
21. Absolute minimum at $x = 1$, absolute maximum at $x = -1$
23. Absolute minimum at $x = 0$, absolute maximum at $x = 2$
25. Absolute minimum at $x = 1$, absolute maximum at $x = 3$
27. Absolute minimum at $x = -3, 0$, absolute maximum at $x = 2$
29. Absolute minimum at $x = -2, 2$, absolute maximum at $x = -3, 3$
31. Absolute minimum at $x = 2$, absolute maximum at $x = 8$
33. Absolute minimum at $x = \frac{1}{2}$, no absolute maximum
35. Absolute minimum at $x = 0$, no absolute maximum
37. Absolute minimum at $x = 0$, no absolute maximum
39. Absolute minimum at $x = 1$, absolute maximum at $x = \frac{1}{2}$
41. No absolute minimum, no absolute maximum
43. Absolute minimum at $x = 1$, no absolute maximum
45. No absolute minimum, absolute maximum at $x = -1$
47. msy $= 243$, mvp $\approx 2.09$, cc $\approx 12.91$     49. Use all the wire to form the circle.
51. Use all the fence for the square.

## Section 13.6

1. 8, 8     3. 10, 10     5. 2, 8     7. 10, $-10$     9. $2a^2$
11. $(c/2a, c/2b)$     13. $200 \times 200$     15. $x = 1.5$     17. $10 \times 20$
19. $100 \times 200$     21. 40     23. $450     25. $4 \times 4 \times 2$     27. $6 \times 6 \times 9$
29. C is $10\sqrt{3}$ feet to the right of D.
31. Cut at $\dfrac{48}{3\sqrt{3} + 4} \approx 5.22$ and use the shorter piece for the square.
33. $\dfrac{500}{9 + \sqrt{3}} \approx 46.6$ yds.     35. 2 runs of 5000 each

## Chapter 13 Review

1. Critical values at 3, 5, 7, 11, 13, 15. Increasing on (3, 7), (11, 13), (15, 18).
   Decreasing on (1, 3), (7, 11), (13, 15)
2. Critical values at $-2$, 1. Increasing on $(-\infty, -2)$, $(1, +\infty)$. Decreasing on
   $(-2, 1)$. Relative maximum at $x = -2$. Relative minimum at $x = 1$.
3. Critical values at $-4$, 2, 5. Increasing on $(-\infty, -4)$, $(5, +\infty)$. Decreasing on
   $(-4, 5)$. Relative maximum at $x = -4$. Relative minimum at $x = 5$.

**4.** Critical value at $x = -\frac{3}{4}$
Relative minimum at $x = -\frac{3}{4}$
Increasing on $(-.75, +\infty)$
Decreasing on $(-\infty, -.75)$
Concave up on $(-\infty, +\infty)$
Concave down nowhere

$y = 2x^2 + 3x - 4$

**5.** Critical value at $x = 1$
Relative maximum at $x = 1$
Increasing on $(-\infty, 1)$
Decreasing on $(1, +\infty)$
Concave up nowhere
Concave down on $(-\infty, +\infty)$

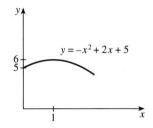

$y = -x^2 + 2x + 5$

**6.** Critical value at $x = -1$
Relative maximum at $x = -1$
Increasing on $(-\infty, -1)$
Decreasing on $(-1, +\infty)$
Concave up nowhere
Concave down on $(-\infty, +\infty)$

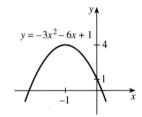

$y = -3x^2 - 6x + 1$

**7.** Critical value at $x = -2$
Relative minimum at $x = -2$
Increasing on $(-2, +\infty)$
Decreasing on $(-\infty, -2)$
Concave up on $(-\infty, +\infty)$
Concave down nowhere

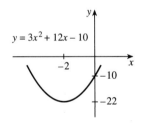

$y = 3x^2 + 12x - 10$

**8.** Critical values at $x = -1, 1$
Relative maximum at $x = -1$
Relative minimum at $x = 1$
Increasing on $(-\infty, -1), (1, +\infty)$
Decreasing on $(-1, 1)$
Concave up on $(0, +\infty)$
Concave down on $(-\infty, 0)$
Inflection value at $x = 0$

$y = \frac{1}{3}x^3 - x + 1$

9. Critical value at $x = \pm 1$
   Relative maximum at $x = 1$
   Relative minimum at $x = -1$
   Increasing on $(-1, 1)$
   Decreasing on $(-\infty, -1)$, $(1, +\infty)$
   Concave up on $(-\infty, 0)$
   Concave down on $(0, +\infty)$
   Inflection value at $x = 0$

10. Critical values at $x = 0, 3$
    Relative minimum at $x = 3$
    Increasing on $(3, +\infty)$
    Decreasing on $(-\infty, 3)$
    Concave up on $(-\infty, 0)$, $(2, +\infty)$
    Concave down on $(0, 2)$
    Inflection values at $x = 0, 2$

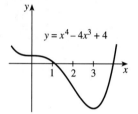

11. Critical values at $x = 0, 12$
    Relative maximum at $x = 0$
    Relative minimum at $x = 12$
    Increasing on $(-\infty, 0)$, $(12, +\infty)$
    Decreasing on $(0, 12)$
    Concave up on $(9, +\infty)$
    Concave down on $(-\infty, 9)$
    Inflection value at $x = 9$

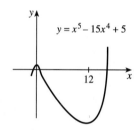

12. Critical values at $x = -3, 0, 3$
    Relative maximum at $x = 0$
    Relative minimum at $x = -3, 3$
    Increasing on $(-3, 0)$, $(3, +\infty)$
    Decreasing on $(-\infty, -3)$, $(0, 3)$
    Concave up on $(-\infty, -\sqrt{3})$, $(\sqrt{3}, +\infty)$
    Concave down on $(-\sqrt{3}, \sqrt{3})$
    Inflection values at $x = \pm\sqrt{3}$

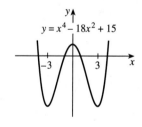

13. Critical values at $x = -3, 0, 3$
    Relative maximum at $x = 3$
    Relative minimum at $x = -3$
    Increasing on $(-3, 3)$,
    Decreasing on $(-\infty, -3)$, $(3, +\infty)$
    Concave up on $(-\infty, -3/\sqrt{2})$, $(0, 3/\sqrt{2})$
    Concave down on $(-3/\sqrt{2}, 0)$, $(3/\sqrt{2}, +\infty)$
    Inflection values at $x = 0, \pm 3/\sqrt{2}$

14. Critical value at $x = 2$
Relative minimum at $x = 2$
Increasing on $(-\infty, 0), (2, +\infty)$
Decreasing on $(0, 2)$
Concave up on $(-\infty, 0), (0, +\infty)$
Concave down nowhere
No inflection values
Asymptote: $x = 0$

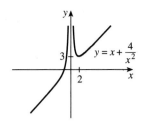

15. Critical values at $x = -1, 1$
Relative minimum at $x = -1, 1$
Increasing on $(-1, 0), (1, +\infty)$
Decreasing on $(-\infty, -1), (0, 1)$
Concave up on $(-\infty, 0), (0, +\infty)$
Concave down nowhere
No inflection values
Asymptote: $x = 0$

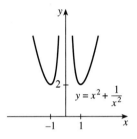

16. Critical value at $x = -\sqrt[3]{2}$
Relative maximum at $x = -\sqrt[3]{2}$
Increasing on $(-\infty, -\sqrt[3]{2}), (0, +\infty)$
Decreasing on $(-\sqrt[3]{2}, 0)$
Concave up nowhere
Concave down on $(-\infty, 0), (0, +\infty)$
No inflection values.
Asymptote: $x = 0$

17. No critical values
Increasing on $(0, +\infty)$
Decreasing on $(-\infty, 0)$
Concave up on $(-\infty, -3), (3, +\infty)$
Concave down on $(-3, 0), (0, 3)$
Inflection values at $x = \pm 3$
Asymptote: $x = 0$

18. No critical values
Increasing nowhere
Decreasing on $(-\infty, 5), (5, +\infty)$
Concave up on $(5, +\infty)$
Concave down on $(-\infty, 5)$
No inflection values
Asymptotes: $x = 5, y = 1$

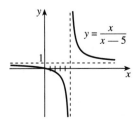

**19.** No critical values
Increasing nowhere
Decreasing on $(-\infty, 5)$, $(5, +\infty)$
Concave up on $(5, +\infty)$
Concave down on $(-\infty, 5)$
No inflection values

$$y = \frac{x-1}{x-5}$$

**20.** Critical value at $x = 0$
Relative minimum at $x = 0$
Increasing $(0, +\infty)$
Decreasing on $(-\infty, 0)$
Concave up on $(-\infty, +\infty)$
Concave down nowhere
No inflection value

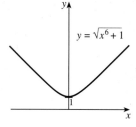

$$y = \sqrt{x^6 + 1}$$

**21.** Critical value at $x = 9$
Increasing nowhere
Decreasing on $(-\infty, +\infty)$
Concave up on $(9, +\infty)$
Concave down on $(-\infty, 9)$
Inflection value at $x = 9$

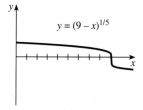

$$y = (9 - x)^{1/5}$$

**22.** Absolute maximum at $x = 1$, absolute minimum at $x = -1$
**23.** Absolute maximum at $x = -2$, absolute minimum at $x = 2$
**24.** No absolute maximum, absolute minimum at $x = 1$
**25.** Absolute maximum at $x = 1$, no absolute minimum
**26.** Absolute maximum at $x = 2$, absolute minimum at $x = -1$
**27.** Absolute maximum at $x = 3$, absolute minimum at $x = 0$
**28.** Absolute maximum at $x = 0$, absolute minimum at $x = 2$
**29.** Absolute maximum at $x = 1$, absolute minimum at $x = 2$
**30.** Absolute maximum at $x = 1$, absolute minimum at $x = 2$
**31.** Absolute maximum at $x = 4$, absolute minimum at $x = 1$
**33.** Relative minimum at $x = a$, relative maximum at $x = b$
**34.** Increasing on $(\sqrt{a/b}, +\infty)$, decreasing on $(0, \sqrt{a/b})$     **35.** 10
**38.** $\dfrac{N}{k} \dfrac{1-a}{2-a}$     **39.** $\sqrt[3]{a/2b}$     **40.** Only construct a square     **41.** $70
**42.** $72     **43.** $r = 3/\sqrt[3]{\pi}$, $h = 12/\sqrt[3]{\pi}$     **44.** 4 runs of 250 cases each

## Section 14.1

**1.** 2.025    **3.** 1.975    **5.** 2.01    **7.** 1.98    **9.** 2.02    **11.** 1.2

**13.** 38    **15.** 95    **17.** (a) $dy = (2x + 1)\, dx$ (b) $dy = 7\, dx$ (c) $dy = 0.7$

**19.** (a) $dy = (6x^2 + 1)\, dx$ (b) $dy = 25\, dx$ (c) $dy = -2.5$

**21.** (a) $dy = \frac{3}{2} x^{1/2}\, dx$ (b) $dy = 3\, dx$ (c) $dy = 0.03$

**23.** (a) $dy = \dfrac{1 - x^2}{(1 + x^2)^2}\, dx$ (b) $dy = 0$ (c) $dy = 0$    **25.** $.8\pi$    **27.** $-120$

**29.** $-.015$    **31.** 288,000    **33.** $1.6\pi$    **35.** 30

## Section 14.2

**1.** $-2$    **3.** $-\frac{3}{5}$    **5.** $\frac{9}{2}$    **7.** $-\frac{2}{7}$    **9.** $-2$    **11.** $-\frac{1}{2}$    **13.** $\frac{16}{31}$

**15.** $-\frac{1}{2}$    **17.** $-3$    **19.** $-\frac{15}{7}$    **21.** $-\frac{3}{4}$    **23.** $-\frac{5}{4}$    **25.** $-\frac{1}{4}\sqrt[4]{al}$

**27.** 6

## Section 14.3

**1.** 6    **3.** 16    **5.** 12    **7.** $-1$    **9.** $-1$    **11.** $-2$    **13.** $-200$

**15.** 400    **17.** $-1.6$ mi/hr    **19.** $-3$ ft/sec    **21.** $\dfrac{125}{16\pi}$ yds/min

**23.** 40 sq ft/sec    **25.** (a) $\frac{1}{7}$ inelastic (b) 1 unit (c) 3 elastic

**27.** (a) $\frac{1}{3}$ inelastic (b) $\frac{1}{2}$ inelastic (c) $\frac{10}{11}$ inelastic

**29.** (a) $\dfrac{2}{1.001}$ elastic (b) $\frac{2}{11}$ inelastic    **31.** (a) 2, elastic (b) 2, elastic

**33.** (a) 15 (b) 15    **35.** (a) None (b) None

## Chapter 14 Review

**1.** $-\frac{11}{30}$    **2.** $-\frac{5}{16}$    **3.** $-1$    **4.** 0    **5.** $dy = \frac{1}{5} x^{-4/5}\, dx$

**6.** $dy = \frac{2}{3} x^{-1/3}\, dx$    **7.** $dy = -3x^{-4}\, dx$    **8.** $dy = \frac{1}{2}(x^{-1/2} - x^{-3/2})\, dx$

**9.** 1.01    **10.** 3.9    **11.** .11    **12.** 2.503

**13.** (a) $\frac{2}{3}$ inelastic (b) 1 unit (c) $\frac{3}{2}$ elastic    **14.** $4\pi$ cu microns    **15.** 4.5 ft/sec

## Section 15.1

**1.**

**3.**

**5.**

**7.**

**9.** 2     **11.** $\frac{3}{4}$     **13.** $\frac{3}{4}$     **15.** $-\frac{45}{23}$

**17.**

**19.**

**21.**

**23.**

**25.** $-1$     **27.** $-3$     **29.** $-\frac{1}{2}$

**31.** $-\frac{1}{2}$     **33.** 1     **35.** $-1$     **37.** 0, 1     **39.** 1     **41.** 5     **43.** 1

**45.** 4     **47.** \$1083.29     **49.** \$24,532.53

**51.** (a) \$1822.12 (b) \$2718.28 (c) \$4055.20 (d) \$6049.65 (e) \$11,023.18
(f) \$20,085.54

**53.** (a) \$1568.31 (b) \$2459.60 (c) \$3857.43 (d) \$14,879.73     **55.** 8.329%

**57.** \$1652.99

**Section 15.2**

**1.** $e^2$     **3.** $e^{-\pi}$     **5.** $e^{5/3}$     **7.** 4     **9.** $-1$     **11.** $2\pi$     **13.** 8

**15.** 81     **17.** $2\ln x + \frac{1}{2}\ln y + \ln z$     **19.** $\frac{1}{2}\ln x + \frac{1}{2}\ln y - \ln z$

**21.** $\frac{1}{2}(\ln x + \ln y - \ln z)$    **23.** $\ln x^2 y$    **25.** $\ln \dfrac{\sqrt{x}}{\sqrt[3]{y}}$    **27.** $\ln \dfrac{x^3 y}{\sqrt[3]{z}}$

**29.** $0.2 \ln .3$    **31.** $-0.5 \ln 3$    **33.** $\frac{1}{3}(1 - \ln 2)$    **35.** $\pm\sqrt{\ln 4}$

**37.** $e^{-1.5} - 7$    **39.** $e/4$    **41.** $\ln 10 \approx 2.3026$    **43.** $\ln 2 \approx .6931$

**45.** $\log_{10} x = \dfrac{\ln x}{\ln 10}$    **47.** $\dfrac{\ln 2}{.08} \approx 8.7$ years    **49.** $\dfrac{\ln 2}{4 \ln (1.015)} \approx 11.6$

**51.** $\dfrac{\ln 1.8}{10} \approx .0588$    **53.** $\dfrac{\ln 1.5}{10} \approx .0405$    **55.** $17.3$ years

## Section 15.3

**1.** $4e^{4x}$    **3.** $-3e^{-3x}$    **5.** $4xe^{2x^2+1}$    **7.** $\dfrac{e^{\sqrt{x}}}{2\sqrt{x}}$    **9.** $x^2 e^x + 2xe^x$

**11.** $x^3 e^x + 3x^2 e^x$    **13.** $-xe^{-x} + e^{-x}$    **15.** $2x^3 e^{x^2} + 2xe^{x^2}$

**17.** $\dfrac{e^x}{2\sqrt{e^x + 1}}$    **19.** $\frac{1}{2}e^{\sqrt{x}} + \dfrac{e^{\sqrt{x}}}{2\sqrt{x}}$    **21.** $(1 + e^x - xe^x)/(1 + e^x)^2$

**23.** $\dfrac{8}{(e^{2x} + e^{-2x})^2}$    **25.** $\dfrac{3e^{3x}}{2\sqrt{e^{3x} + 2}}$    **27.** $5^x \ln 5$    **29.** $\dfrac{3^{\sqrt{x}} \ln 3}{2\sqrt{x}}$

**31.** $3^x + x3^x \ln 3$    **33.** $(3x^2 + 2x)/(x^3 + x^2 + 1)$    **35.** $4x \ln |x| + 2x$

**37.** $\dfrac{2x + 1}{2(x^2 + x + 1)\sqrt{\ln |x^2 + x + 1|}}$    **39.** $\dfrac{1}{x(x + 1)}$

**41.** $-4xe^{-x^2} \ln |x| + \dfrac{2e^{-x^2}}{x}$    **43.** $2x \ln x + x$

**45.** $\dfrac{-2x \ln x}{(x^2 + 1)^2} + \dfrac{1}{x(x^2 + 1)}$    **47.** $\dfrac{11}{x}(\ln x)^{10}$    **49.** $3/x$    **51.** $e^x/(e^x + 1)$

**53.** Increasing on $(-\infty, 1/3)$
Decreasing on $(1/3, +\infty)$
Concave up on $(2/3, +\infty)$
Concave down on $(-\infty, 2/3)$

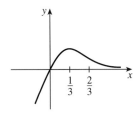

**55.** Increasing on $(-\infty, 0)$
Decreasing on $(0, +\infty)$
Concave up on $(-\infty, -\sqrt{2}/2), (\sqrt{2}/2, +\infty)$
Concave down on $(-\sqrt{2}/2, \sqrt{2}/2)$

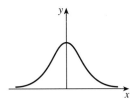

**57.** Increasing on $(-\infty, -2)$, $(0, +\infty)$
Decreasing on $(-2, 0)$
Concave up on $(-\infty, -2 - \sqrt{2})$,
$(-2 + \sqrt{2}, +\infty)$
Concave down on $(-2 - \sqrt{2}, -2 + \sqrt{2})$

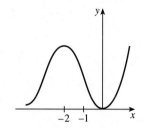

**59.** Increasing on $(0, 2)$
Decreasing on $(-\infty, 0)$, $(2, +\infty)$
Concave up on $(-\infty, 2 - \sqrt{2})$, $(2 + \sqrt{2}, +\infty)$
Concave down on $(2 - \sqrt{2}, 2 + \sqrt{2})$

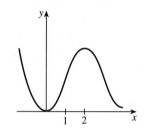

**61.** Increasing on $(e^{-1}, +\infty)$
Decreasing on $(0, e^{-1})$
Concave up on $(0, +\infty)$
Concave down nowhere

**63.** Increasing on $(0, \sqrt{e})$
Decreasing on $(\sqrt{e}, +\infty)$
Concave up on $(e^{5/6}, +\infty)$
Concave down on $(0, e^{5/6})$

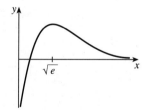

**65.** $\dfrac{d}{dx}(\log_a |x|) = \dfrac{d}{dx}\left(\dfrac{\ln |x|}{\ln a}\right)$

$$= \dfrac{1}{\ln a}\dfrac{d}{dx}(\ln |x|)$$

$$= \dfrac{1}{\ln a}\dfrac{1}{x}$$

**67.** $x = .5$    **69.** $\dfrac{2x}{x^2 + 5}$, Yes    **71.** $R'(t) = \dfrac{A_0}{b - 1}(-kbe^{-kt}) < 0$

**73.** Decreases since $\dfrac{dS(T)}{dT} < 0$

### Section 15.4

**1.** $\dfrac{\ln 2}{.07} \approx 9.9$    **3.** $\dfrac{\ln 2}{6} \approx .116$

**5.** $3Q_0 = Q(T) = Q_0 e^{kT}$

$\qquad 3 = e^{kT}$

$\qquad \ln 3 = \ln e^{kT} = kT$

**7.** $\dfrac{\ln 2}{150} \approx .0046$    **9.** $\dfrac{\ln 2}{.005} \approx 139$    **11.** $\dfrac{\ln 1.5}{34} \approx .0119$    **13.** 4.14 billion.

**15.** $\qquad P_1 = Q(T) = P_0 e^{kT}$

$\qquad \dfrac{P_1}{P_0} = e^{kT}$

$\qquad \ln \dfrac{P_1}{P_0} = \ln e^{kT} = kT$

$\qquad T = \dfrac{1}{k} \ln \dfrac{P_1}{P_0}$

**17.** $5.3 \cdot 10^{-10}$

**19.** $\dfrac{1}{3} P_0 = P(T) = P_0 e^{-kT}$

$\qquad \dfrac{1}{3} = e^{-kT}$

$\qquad \ln \dfrac{1}{3} = \ln e^{-kT} = -kT$

$\qquad T = -\dfrac{1}{k} \ln \dfrac{1}{3} = \dfrac{1}{k} \ln 3$

**21.** $\dfrac{\ln 10}{.000124} \approx 18{,}569$ years old    **23.** $\dfrac{\ln 20}{.087} \approx 34.4$ days

**25.** 23.8 units    **27.** 62.4 million    **29.** 417 million

**31.** $\displaystyle\lim_{t \to +\infty} P'(t) = \lim_{t \to +\infty} \dfrac{kLP_0(L - P_0)}{(P_0 - (L - P_0)e^{-kt})^2} e^{-kt} = \dfrac{kLP_0(L - P_0)}{(P_0 + (L - P_0)(0))^2}(0) = 0$

**33.** 1.01 hours

**35.** $T'(t) = \dfrac{d}{dt}[C - (C - T_0)e^{-kt}]$

$\phantom{T'(t)} = k(C - T_0)e^{-kt}$

$\phantom{T'(t)} = kC - kC + k(C - T_0)e^{-kt}$

$\phantom{T'(t)} = k[C - T(t)]$

**37.** 92    **39.** 7.6 hours    **41.** 5.4 years

**Chapter 15 Review**

**1.**

**2.**

**3.**

**4.**

**5.**

**6.**

**7.** $\frac{3}{2}$   **8.** 3   **9.** $e$

**10.** .50   **11.** 4   **12.** 5   **13.** 0   **14.** 1   **15.** 0   **16.** 0

**17.** $-7e^{-7x}$   **18.** $6x^2e^{2x^3+1}$   **19.** $e^x(x^2 + 3x + 3)$   **20.** $2x/(x^2 + 1)$

**21.** $e^{x^2}/x + 2xe^{x^2}\ln x$   **22.** $(2x + e^x)/(x^2 + e^x)$

**23.** Increasing on $(-\infty, 0)$, decreasing on $(0, +\infty)$, concave up on $(-\infty, -\sqrt{2}/2)$ and $(\sqrt{2}/2, +\infty)$, concave down on $(-\sqrt{2}/2, \sqrt{2}/2)$

**24.** Increasing on $(0, +\infty)$, decreasing on $(-\infty, 0)$, concave up on $(-\sqrt{2}/2, \sqrt{2}/2)$, concave down on $(-\infty, -\sqrt{2}/2)$ and $(\sqrt{2}/2, +\infty)$

**25.** Increasing on $(0, +\infty)$, decreasing on $(-\infty, 0)$, concave up on $(-1, 1)$, concave down on $(-\infty, -1)$ and $(1, +\infty)$

**26.** Increasing on $(-\infty, +\infty)$, decreasing nowhere, concave up on $(-\infty, +\infty)$, concave down nowhere

**27.** $\dfrac{\ln 1.3}{.02} \approx 13$   **28.** $\dfrac{\ln 2}{35} \approx .0198$   **29.** 31.1 years   **30.** 1311 years

**31.** $S(t) = D - (D - S_0)e^{-kt}$   **32.** 180

## Section 16.1

**1.** $\frac{1}{100}x^{100} + C$   **3.** $-\frac{1}{98}x^{-98} + C$   **5.** $5x + C$   **7.** $-\frac{5}{2}y^{-2} + C$

**9.** $\dfrac{\sqrt{2}}{5}y^{5/2} + C$   **11.** $\frac{3}{5}u^{5/3} + C$   **13.** $2x^3 + 2x^2 + C$

**15.** $\frac{1}{3}x^3 + \frac{1}{2}x^2 + x + C$   **17.** $\frac{10\sqrt{2}}{11}u^{1.1} - \frac{1}{21}u^{2.1} + C$   **19.** $\frac{1}{3}t^3 - t + C$

**21.** $\frac{2}{3}t^{3/2} - \frac{3}{8}t^{8/3} + C$   **23.** $t^6 - t^4 + t + C$   **25.** $x + 3\ln|x| + C$

**27.** $\pi x + \ln|x| + C$   **29.** $\frac{2}{3}t^{3/2} + 2t^{1/2} + C$   **31.** $e^x - e^{-x} + C$

**33.** $\frac{1}{5}e^{5x} + C$   **35.** $\frac{1}{5}e^{5x} + \frac{1}{5}e^{-5x} + C$   **37.** $x + \ln|x| + C$

**39.** $R(x) = 30x - 0.25x^2$   **41.** $p(x) = 2x^{-1/2} + 1$

**43.** $C(x) = 100x - 10e^{0.1x} + 1010$   **45.** $C(x) = 20x^{3/2} - 2x^3$

**47.** $-16t^2 + 30t + 15$   **49.** $-16t^2 + 10t + 6$   **51.** 60.27 grams

**53.** \$1,171,828   **55.** $x - .25x^2 + 0.2$

## Section 16.2

**1.** $\frac{2}{11}(3x + 1)^{11} + C$   **3.** $-\frac{1}{16}(3 - x^2)^8 + C$   **5.** $\frac{4}{5}(2x^2 + 4x - 1)^{5/2} + C$

**7.** $(3x^4 + 4x^3 + 6x^2 + 12x + 1)^4/48 + C$   **9.** $\frac{4}{3}(x + 1)^{3/2} + C$

**11.** $\frac{1}{3}(x^2 + 1)^{3/2} + C$   **13.** $\frac{3}{4}(x^2 + 1)^{2/3} + C$   **15.** $2(x^{1/3} + 1)^{3/2} + C$

**17.** $\frac{2}{3}(\ln x)^{3/2} + C$   **19.** $-e^{1-x} + C$   **21.** $-\frac{1}{2}e^{1-x^2} + C$

**23.** $2e^{\sqrt{x}} + C$   **25.** $\frac{1}{3}\ln|3x + 5| + C$   **27.** $-\frac{1}{6}(x^2 + 3)^{-3} + C$

**29.** $-\ln(e^{-x} + 1) + C$    **31.** $\frac{1}{2}\ln(e^{2x} + e^{-2x}) + C$    **33.** $\frac{1}{2}\ln\left|\ln|x|\right| + C$

**35.** $p(x) = \dfrac{4}{3x^2 + 1} + 9$    **37.** $R(x) = 20x^2 - x^4$

**39.** $C(x) = 2\ln(x^2 + 1) + 1000$    **41.** 6 tons

**43.** $20{,}000e^{2.2} + 80{,}000 \approx 260{,}500$    **45.** $\dfrac{\ln 7}{2} \approx 0.97$

### Section 16.3

**1.** $-30$    **3.** 20    **5.** (a) $\frac{3}{5}$ (b) $\frac{11}{20}$    **7.** (a) $\frac{25}{64} \approx .391$ (b) $\frac{81}{256} \approx .316$
**9.** (a) 1/2 (b) 1/2    **11.** (a) 3 (b) 15/4    **13.** $1/2 + 1/2n \to 1/2$
**15.** $1/2 - 1/2n \to 1/2$    **17.** $15/2 + 9/2n \to 15/2$
**19.** $9/2 - 9/2n \to 9/2$    **21.** $4 + 4/\,n \to 4$    **23.** 20    **25.** 6
**27.** 9/2    **29.** $\frac{1}{2}\pi$    **31.** $\frac{1}{2}\pi$    **33.** .6640    **35.** (a) 6 (b) 5.6
**37.** (a) 948 (b) 928

### Section 16.4

**1.** 15    **3.** 0    **5.** 2    **7.** 14    **9.** 11/8    **11.** $(e^{-2} - e^{-4})/2$
**13.** $\frac{1}{2}\ln 3$    **15.** 0    **17.** $-1/16$    **19.** $\frac{26}{3}$    **21.** 0    **23.** $\frac{1}{2}\ln\frac{5}{3}$
**25.** $e^{x^2}$    **27.** $1/\ln x$    **29.** 6    **31.** 5    **33.** 0    **35.** 0    **37.** $\frac{1}{3}$
**39.** $2(e^2 - e) \approx 9.34$ million barrels    **41.** 71    **43.** About 11 years
**45.** About 5281 gal    **49.** $\frac{26}{3}$ thousands    **51.** \$3194.53    **53.** \$8 million
**55.** About 10.23    **57.** $101\frac{1}{3}$    **59.** 34,000    **61.** 2222

### Section 16.5

**1.** $\frac{15}{4}$    **3.** $10e - 20$    **5.** $\frac{1}{6}$    **7.** $\frac{1}{2}(e^4 - e^2) - 1$    **9.** $\frac{9}{2}$    **11.** $\frac{1}{2}$
**13.** $\frac{16}{3}$    **15.** $\frac{8}{3}$    **17.** $\frac{125}{6}$    **19.** $\frac{125}{6}$    **21.** $e^2 + e - 2$    **23.** 16
**25.** $\frac{3}{2}$    **27.** $\frac{1}{2}$    **29.** 1    **31.** 8    **33.** $\frac{23}{2}$    **35.** $\frac{83}{4}$    **37.** $\frac{128}{3}$
**39.** $0.9 - 0.1\ln 10$    **41.** $\frac{1}{2}$    **43.** $\frac{2}{9}$    **45.** 8, 16    **47.** $\frac{128}{3}$, 8
**49.** $\frac{1}{3}$    **51.** $\frac{1}{12}$    **53.** $\frac{1}{10}$    **55.** $\frac{1}{300}$    **57.** \$4.8 billion    **59.** $\frac{26}{3}$

### Chapter 16 Review

**1.** $\frac{1}{10}x^{10} + C$    **2.** $\frac{2}{3}x^{3/2} + C$    **3.** $-2y^{-1} + C$    **4.** $\dfrac{\sqrt{3}y^3}{3} + C$

**5.** $2x^3 + 4x^2 + C$    **6.** $x^2 - 3\ln|x| + C$    **7.** $\frac{2}{5}t^{5/2} + 2t^{1/2} + C$
**8.** $\frac{1}{3}e^{3x} + C$    **9.** $-e^{-x} - \frac{1}{5}e^{-5x} + C$    **10.** $\frac{2}{21}(2x + 1)^{21} + C$
**11.** $(2x^3 + 1)^{10} + C$    **12.** $\frac{2}{3}(x^3 + 1)^{1/2} + C$    **13.** $.5\ln|x^2 + 2x| + C$
**14.** $\frac{1}{2}e^{x^2+5} + C$    **15.** $-\frac{5}{7}(x^2 + 1)^{-7} + C$    **16.** $-2e^{-\sqrt{x}} + C$
**17.** $\ln(e^x + 1) + C$    **18.** $\frac{1}{4}(\ln x)^4 + C$    **19.** 32    **20.** 64
**21.** $\dfrac{5}{3} - \dfrac{1}{n} + \dfrac{1}{3n^2} \to \dfrac{5}{3}$    **22.** $5/3 + 1/n + 1/3n^2 \to 5/3$    **23.** $9\pi/4$
**24.** $\frac{1}{4}(1 + 4/5 + 2/3 + 4/7)$    **25.** $\frac{1}{4}(4/5 + 2/3 + 4/7 + 1/2)$    **26.** 52
**27.** $-8$    **28.** 8    **29.** $(e^4 - e^2)/2 + \ln 2$    **30.** $(2^{10} - 1)/5$
**31.** $(e^9 - e)/3$    **32.** $(\ln 5)/2$    **33.** $3 - \sqrt{3}$    **34.** $e^{x^4}$    **35.** 0
**36.** $.50\ln 3$    **37.** 9    **38.** 8/3    **39.** 8/3    **40.** 1    **41.** 20/3
**42.** 72    **43.** 32/3    **44.** 9    **45.** 2/3    **46.** 13    **47.** $4\ln 2 - 2\ln 3$

**48.** $38/3$     **49.** $x^2 - 10x$     **50.** $x^3 + x + 1000$     **51.** $\$98,419.70$

**52.** $1.96$ if $s(0) = 0$     **53.** (a) 26 (b) 32     **54.** 6906     **55.** 10.82

**56.** 17,183     **57.** $x_0 = 2$, $p_0 = 6$     **58.** $16/3$     **59.** 6     **60.** $L = 3/5$

## Section 17.1

**1.** $xe^{2x} - \frac{1}{2}e^{2x} + C$     **3.** $\frac{3}{4}e^4 + \frac{1}{4}$     **5.** 2

**7.** $2x(1 + x)^{1/2} - \frac{4}{3}(1 + x)^{3/2} + C$     **9.** $\frac{x}{11}(1 + x)^{11} - \frac{1}{132}(1 + x)^{12} + C$

**11.** $-\frac{x}{x+2} + \ln|x + 2| + C$     **13.** $\frac{2}{9}x^3(1 + x^3)^{3/2} - \frac{4}{45}(1 + x^3)^{5/2} + C$

**15.** $\frac{x^2}{22}(1 + x^2)^{11} - \frac{1}{264}(1 + x^2)^{12} + C$     **17.** $\frac{1}{2}x^2e^{2x} - \frac{1}{2}xe^{2x} + \frac{1}{4}e^{2x} + C$

**19.** $\frac{3}{2}e^4 + \frac{1}{2}$     **21.** 111     **23.** $\$26,306.67$     **25.** $\$36.47$

**27.** $1000/(e - 1) \approx 582$

## Section 17.2

**1.** $\dfrac{2}{9}\left[\dfrac{(3x + 2)^{5/2}}{5} - \dfrac{2(3x + 2)^{3/2}}{3}\right] + C$

**3.** $\dfrac{x}{2}\sqrt{x^2 - 9} - \dfrac{9}{2}\ln|x + \sqrt{x^2 - 9}| + C$

**5.** $\ln\left|\dfrac{\sqrt{1 - x} - 1}{\sqrt{1 - x} + 1}\right| + C$     **7.** $\dfrac{1}{3}\ln\left|\dfrac{2x + 1}{2x + 4}\right| + C$     **9.** $\dfrac{1}{3}(x^2 - 9)^{3/2} + C$

**11.** $\dfrac{1}{3x} + \dfrac{1}{9}\ln\left|\dfrac{x - 3}{x}\right| + C$

**13.** $-\dfrac{x}{\sqrt{x^2 + 16}} + \ln|x + \sqrt{x^2 + 16}| + C$     **15.** $\dfrac{1}{3}\ln\left|\dfrac{x^3}{2x^3 + 1}\right| + C$

**17.** $-\dfrac{\sqrt{1 - 4x^2}}{x} + C$     **19.** $\dfrac{1}{18}x\sqrt{9x^2 - 1} + \dfrac{1}{54}\ln|x + \dfrac{1}{3}\sqrt{9x^2 - 1}| + C$

**21.** $x[(\ln x)^3 - 3(\ln x)^2 + 6\ln x - 6] + C$

**23.** $\dfrac{1}{2}(e^x + 1)^2 - 2(e^x + 1) + \ln(e^x + 1) + C$

**25.** $-e^{-2x}\left(\dfrac{1}{2}x^2 + \dfrac{1}{2}x + \dfrac{1}{4}\right) + C$     **27.** $R(x) = \dfrac{2(x + 1)^{5/2}}{5} - \dfrac{2(x + 1)^{3/2}}{3} +$

$\dfrac{4}{15}$     **29.** $\dfrac{10t}{9\sqrt{t^2 + 9}}$     **31.** 4.8     **33.** 327

## Section 17.3

**1.** $T_2 = 5$, $S_2 = 4$, exact $= 4$

**3.** To 4 decimal places $T_2 = 1.1667$, $S_2 = 1.1111$, exact $= 1.0986$

**5.** To 4 decimal places $T_2 = 2.7321$, $S_2 = 2.9761$, exact $= 3.1416$

**7.** To 6 decimal places $T_2 = 1.571583$, $S_2 = 1.475731$, exact $= 1.462652$

**9.** $T_4 = 4.25$, $S_4 = 4$, exact $= 4$

**11.** To 4 decimal places $T_4 = 1.1167$, $S_4 = 1.1$, exact $= 1.0986$

**13.** To 4 decimal places $T_4 = 2.9957$, $S_4 = 3.0836$, exact $= 3.1416$

**15.** To 6 decimal places $T_4 = 1.490679$, $S_4 = 1.463711$, exact $= 1.462652$

17. $T_{10} = 4.04$, $S_{10} = 4$, exact $= 4$
19. To 6 decimal places $T_{10} = 1.101562$, $S_{10} = 1.098661$, exact $= 1.098612$
21. To 4 decimal places $T_{10} = 3.1045$, $S_{10} = 3.1270$, exact $= 3.1416$
23. To 6 decimal places $T_{10} = 1.467175$, $S_{10} = 1.462681$, exact $= 1.462652$
25. (a) $T_4 = 5.8$ (b) $S_4 \approx 5.87$  27. (a) $T_4 = 50$ (b) $S_4 = 52$
29. (a) $T_6 = 4.325$ (b) $S_6 = 4.35$  31. (a) $\frac{1}{75}$ (b) 0  33. (a) .32 (b) .00043
35. No, since $f''(x) = -\frac{1}{4}x^{-3/2}$ is unbounded on $(0, 1]$.

**Section 17.4**

1. $\frac{1}{2}$  3. diverges  5. 100  7. diverges  9. 1  11. $-\frac{1}{2}$  13. $\frac{1}{3}$
15. $-\frac{1}{2}$  17. 0  19. 0  21. Approximately 667 tons  23. $20 million
25. 2000 tons  27. 200 million  29. 1 thousand

**Section 17.5**

1. (a) $100 (b) $200(1 - e^{-.50}) = $78.69$ (c) $200(e^{.50} - 1) = $129.74$
3. (a) $2000(e - 1) = $3436.56$ (b) $2000 (c) $2000e = $5436.56$
5. (a) $400(1 - e^{-.50}) = $157.39$ (b) $2000(1 - e^{-1.5})/15 = $103.58$
(c) $2000(e - e^{-.50})/15 = $281.57$
7. (a) $50 (b) $(100 - 200e^{-1}) = $26.42$ (c) $(100e - 200) = $71.83$
9. (a) $333.33 (b) $(2000 - 5000e^{-1}) = $160.60$ (c) $(2000e - 5000) = $436.56$
11. (a) Yes, since $P_V(10) = $948,180.84$ (b) No, since $P_V(10) = $776,869.84$
13. (a) Yes, since $P_V(5) \approx $316$ thousand (b) No, since $P_V(5) \approx $303$ thousand
15. $1250  17. $555.56
19. (a) No, since $P_V(+\infty) = $100,000$ (b) Yes, since $P_V(+\infty) = $100,000$
21. (a) No, since $P_V(+\infty) = $600,000$ (b) Yes, since $P_V(+\infty) = $600,000$

**Chapter 17 Review**

1. $\frac{1}{25}(5x - 1)e^{5x} + C$  2. $\frac{2}{27}\sqrt{5 + 3x}\,[3x - 10] + C$
3. $\frac{1}{6}x^6(\ln x - \frac{1}{6}) + C$  4. $-\frac{x}{1+x} + \ln|1 + x| + C$
5. $-\frac{1}{5}(9 - x^2)^{3/2}[x^2 + 6] + C$  6. $2\sqrt{x}(\ln x - 2) + C$
7. $\frac{1}{2}x - \frac{3}{4}\ln|2x + 3| + C$  8. $\frac{1}{4(x + 4)} - \frac{1}{16}\ln\left|\frac{x + 4}{x}\right| + C$
9. $-\frac{\sqrt{4 - x^2}}{4x} + C$  10. $\sqrt{9 - x^2} - 3\ln\left|\frac{3 + \sqrt{9 - x^2}}{x}\right| + C$
11. $\frac{1}{2}x\sqrt{x^2 + 1} + \frac{1}{2}\ln|x + \sqrt{x^2 + 1}| + C$  12. $x - \ln(1 + e^x) + C$
13. 10  14. Diverges  15. 2/3  16. Diverges
17. (a) 45 (b) 100/3 (c) Correct $= 32$
18. To 4 decimal places (a) 1.5375 (b) 1.425 (c) Correct is 1.3863
19. To 4 decimal places (a) 35.3125 (b) 32.0833 (c) Correct $= 32$
20. To 4 decimal places (a) 1.4281 (b) 1.3916 (c) Correct is 1.3863
21. $250(1 - e^{-.4}) = $82.42$  22. $\dfrac{250}{3}(1 - e^{-1.2}) = $58.23$
23. $\dfrac{250}{3}(e^{.8} - e^{-.4}) = $129.59$  24. $\dfrac{5000}{9}(1 - e^{-.9}) = $329.68$ thousand
25. $555.6 thousand

## Section 18.1

**1.** $\frac{2}{3}$   **3.** 3   **5.** .25   **7.** 1   **9.** 1   **11.** 3

**13.** Not always positive   **15.** Not always positive

**17.** $\int_1^{+\infty} f(x)\, dx = .50 \neq 1$   **19.** $\int_1^{+\infty} f(x)\, dx$ diverges

**21.** $\int_{-\infty}^{+\infty} f(x)\, dx$ diverges   **23.** (a) $\frac{1}{16}$ (b) $\frac{3}{16}$ (c) 1   **25.** (a) $\frac{1}{2}$ (b) $\frac{11}{32}$ (c) $\frac{1}{2}$

**27.** 0.6   **29.** (a) $\frac{1}{6}$ (b) $\frac{18}{30}$ (c) $\frac{4}{30}$ (d) 0 (e) 1   **31.** (a) $\frac{15}{16}$ (b) $\frac{1}{16}$ (c) $\frac{1}{2}$

**33.** (a) $1 - e^{-1.5} \approx .777$ (b) $e^{-1.5} \approx .223$ (c) $e^{-3} - e^{-6} \approx .0473$

**35.** (a) $\frac{2}{3}$ (b) $\frac{1}{3}$ (c) $\frac{3}{10}$

**37.** (a) $1 - e^{-.1} \approx .095$ (b) $e^{-.1} \approx .905$ (c) $e^{-.1} - e^{-.2} \approx .086$

## Section 18.2

**1.** (a) 3 (b) $\frac{4}{3}$ (c) $2/\sqrt{3}$   **3.** (a) $\frac{8}{3}$ (b) $\frac{8}{9}$ (c) $2\sqrt{2}/3$   **5.** (a) $\frac{1}{2}$ (b) $\frac{1}{20}$ (c) $\dfrac{1}{2\sqrt{5}}$

**7.** (a) 1000 (b) 1,000,000 (c) 1000   **9.** (a) $\frac{4}{3}$ (b) $\frac{2}{9}$ (c) $\sqrt{2}/3$   **11.** $\frac{1}{2}(a + b)$

**13.** Since $\mu = \sigma = \dfrac{1}{k}$,

$$\int_{\mu-\sigma}^{\mu+\sigma} ke^{-kx}\, dx = \int_0^{2/k} ke^{-kx}\, dx = -e^{-kx}\big|_0^{2/k} = 1 - e^{-2}$$

**15.** Since $\mu = \dfrac{1}{k}$,

$$\int_\mu^{+\infty} ke^{-kx}\, dx = \lim_{b \to +\infty} \int_{1/k}^b ke^{-kx}\, dx$$

$$= \lim_{b \to +\infty} -e^{-kx}\big|_{1/k}^b = \lim_{b \to +\infty} (e^{-1} - e^{-kb}) = e^{-1}$$

**17.** (a) $\frac{10}{3}$ minutes (b) $\frac{50}{9}$ (c) $\frac{5}{3}\sqrt{2}$   **19.** (a) 5 thousand gallons (b) 5 (c) $\sqrt{5}$

**21.** (a) $\frac{2}{3}$ thousand gallons (b) $\frac{2}{9}$ (c) $\sqrt{2}/3$

**23.** (a) .50 thousand dollars (b) .75 (c) $\sqrt{3}/2$

**25.** (a) $\frac{58}{35} = 1.6571$ months (b) $\frac{848}{245} - \left(\frac{58}{35}\right)^2 \approx .7151$ (c) .8456

**27.** (a) 50 days (b) 2500 (c) 50

## Section 18.3

**1.** .6915   **3.** .3821   **5.** .0668   **7.** .7881   **9.** .0919   **11.** .1498

**13.** .9332   **15.** .0228   **17.** .1151   **19.** .2417   **21.** .9544

**27.** 1.24%   **29.** (a) 15.87% (b) 25.14% (c) 25.86%   **31.** 62.93%

**33.** 6.68%   **35.** (a) 2.28% (b) 50% (c) 47.72%   **37.** .9332

## Chapter 18 Review

**1.** .25   **2.** .50   **3.** 2   **4.** (a) $\frac{1}{8}$ (b) $\frac{7}{8}$ (c) 0   **5.** 1.5

**6.** $\sigma^2 = .75, \sigma = \sqrt{3}/2$   **7.** 1.5   **8.** 4   **9.** (a) $\sigma^2 = 2$ (b) $\sigma = \sqrt{2}$

**10.** $3\sqrt{2}$   **11.** (a) .8944 (b) .1056 (c) .0531   **12.** (a) .4207 (b) .5793 (c) .0395

**13.** (a) $1 - e^{-2} \approx .8647$ (b) $e^{-2} \approx .1353$ (c) $e^{-1} - e^{-2} \approx .2325$

**14.** (a) .0472 (b) .9528 (c) .1556   **15.** (a) 2.28% (b) 97.72% (c) 95.44%

## Section 19.1

1. (a) 13 (b) 10 (c) 12     3. (a) 2 (b) 13 (c) 2     5. (a) $\frac{1}{3}$ (b) $-1$ (c) $\frac{1}{3}$

7. (a) $\sqrt{6}$ (b) 3 (c) 2     9. (a) 1 (b) $\frac{1}{2}$ (c) $\frac{1}{2}$     11. All $x$ and $y$

13. $\{(x, y)|x \neq -y\}$     15. $\{(x, y)|x^2 + y^2 \leq 16\}$     17. 3     19. $\sqrt{173}$

21. $\sqrt{3}$     23. Plane with intercepts $(0, 0, 4)$, $(0, 6, 0)$, $(12, 0, 0)$

25. Plane with intercepts $(0, 0, 2)$, $(0, 4, 0)$, $(10, 0, 0)$

27. Horizontal plane 3 units above $xy$-plane

29. Sphere of radius 6 centered at $(0, 0, 0)$

31. Sphere of radius 2 centered at $(1, 0, 3)$

33. (a) $z_0 = 0$: $x = 0 = y$ (b) $z_0 = 1$: $x^2 + y^2 = 1$ (c) $z_0 = 4$: $x^2 + y^2 = 4$
(d) $z_0 = 9$: $x^2 + y^2 = 9$ (e) Surface is a bowl

35. (a) $z_0 = 1$: $x = 0 = y$ (b) $z_0 = 0$: $x^2 + y^2 = 1$ (c) $z_0 = -3$: $x^2 + y^2 = 4$
(d) $z_0 = -8$: $x^2 + y^2 = 9$ (e) Surface is a mountain

37. $S(x, y) = 2x^2 + 4xy$, $S(3, 5) = 78$

39. $R(x, y) = 1400x + 802y - 4xy - 12x^2 - 0.5y^2$, 80,400

41. $P(x, y) = -15,000 + 1350x + 801.5y - 4xy - 12x^2 - 0.5y^2$, 60,300

43. $A(m, t) = 1000(1 + .08/m)^{mt}$, \$1489.85

45. (a) $\{(x, y)|y \neq 0\}$ (b) 200 (c) 50     47. (a) $\{(d, P)|P \geq 0\}$ (b) $24\pi$ (c) $8\pi$

## Section 19.2

1. $f_x = 2x$, $f_y = 2y$, $f_x(1, 3) = 2$, $f_y(1, 3) = 6$

3. $f_x = 2xy - 3x^2y^2$, $f_y = x^2 - 2x^3y$, $f_x(1, 2) = -8$, $f_y(1, 2) = -3$

5. $f_x = \dfrac{\sqrt{y}}{2\sqrt{x}}$, $f_y = \dfrac{\sqrt{x}}{2\sqrt{y}}$, $1/2$, $1/2$

7. $f_x = xy^2/\sqrt{1 + x^2y^2}$, $f_y = x^2y/\sqrt{1 + x^2y^2}$, 0, 0

9. $f_x = 2e^{2x+3y}$, $f_y = 3e^{2x+3y}$, $2e^5$, $3e^5$

11. $f_x = ye^{xy} + xy^2e^{xy}$, $f_y = xe^{xy} + x^2ye^{xy}$, $2e$, $2e$

13. $f_x = 1/(x + 2y)$, $f_y = 2/(x + 2y)$, 1, 2

15. $f_x = e^{xy}/x + ye^{xy} \ln x$, $f_y = xe^{xy} \ln x$, 1, 0

17. $f_x = -1/x^2y$, $f_y = -1/xy^2$, $-1/2$, $-1/4$

19. $f_x = (y^2 - x^2 + 2xy)/(x^2 + y^2)^2$, $f_y = (y^2 - x^2 - 2xy)/(x^2 + y^2)^2$, $1/25$,
$-7/25$     21. $f_{xx} = 2y^4$, $f_{yy} = 12x^2y^2$, $f_{xy} = f_{yx} = 8xy^3$

23. $f_{xx} = 4e^{2x-3y}$, $f_{yy} = 9e^{2x-3y}$, $f_{xy} = f_{yx} = -6e^{2x-3y}$

25. $f_{xx} = -\frac{1}{4}y^{1/2}x^{-3/2}$, $f_{yy} = -\frac{1}{4}y^{-3/2}x^{1/2}$, $f_{xy} = f_{yx} = \frac{1}{4}x^{-1/2}y^{-1/2}$

27. $f_{xx} = 2e^y$, $f_{yy} = x^2e^y$, $f_{xy} = f_{yx} = 2xe^y$

29. $f_{xx} = y^2/(x^2 + y^2)^{3/2}$, $f_{yy} = x^2/(x^2 + y^2)^{3/2}$, $f_{xy} = f_{yx} = -xy/(x^2 + y^2)^{3/2}$

31. $f_x = yz$, $f_y = xz$, $f_z = xy$

33. $f_x = x/\sqrt{x^2 + y^2 + z^2}$, $f_y = y/\sqrt{x^2 + y^2 + z^2}$, $f_z = z/\sqrt{x^2 + y^2 + z^2}$

35. $f_x = e^{x+2y+3z}$, $f_y = 2e^{x+2y+3z}$, $f_z = 3e^{x+2y+3z}$

37. $f_x = 1/(x + 2y + 5z)$, $f_y = 2/(x + 2y + 5z)$, $f_z = 5/(x + 2y + 5z)$

39. Complementary     41. Competitive     43. Competitive

45. $A_r = 1000t(1 + r/12)^{12t-1}$. This gives the rate of change of the amount with respect to changes in the interest rate $r$.

**47.** $\dfrac{\partial V}{\partial r} = 4\pi r^2 \left(1 + \dfrac{c}{r} N^{1/3}\right)^2$. This gives the rate of change of the detection volume

with respect to changes in the detection radius $r$. $\dfrac{\partial V}{\partial N} = \dfrac{4\pi c r^2}{3} \left(1 + \dfrac{c}{r} N^{1/3}\right)^2 N^{-2/3}$.

This gives the rate of change of the detection volume with respect to changes in the number of prey $N$.

**49.** $\dfrac{\partial L}{\partial r} = -2p\pi r \left(1 - \dfrac{p\pi r^2}{n}\right)^{n-1}$. This gives the rate of change of the amount of

light with respect to changes in the radius $r$ of the leaves.

$\dfrac{\partial L}{\partial p} = -\pi r^2 \left(1 - \dfrac{p\pi r^2}{n}\right)^{n-1}$. This gives the rate of change of the amount of light with

respect to changes in the total density $p$ of the leaves.

## Section 19.3

**1.** Relative minimum at $(5, -2)$     **3.** Relative maximum at $(2, 1)$
**5.** Saddle at $(1, 1)$     **7.** Relative maximum at $(-1, -2)$
**9.** Relative minimum at $(0, 0)$     **11.** Saddle at $(0, 2)$
**13.** Saddle at $(0, 0)$, relative minimum at $(1, 1)$
**15.** Saddle at $(0, 0)$ relative maximum at $(1/6, 1/12)$
**17.** Relative maximum at $(1, 1)$     **19.** Saddle at $(0, 0)$
**21.** Since $f(x, y) = x^4 y^4 \geq 0$ and $f(0, 0) = 0$, $f$ has a relative minimum at $(0, 0)$.
Since $f_{xx} = 12x^2 y^4$, $f_{yy} = 12x^4 y^2$, and $f_{xy} = 16x^3 y^3$, $\Delta(0, 0) = 0$.
**23.** If $y = x$, then $f(x, y) = f(x, x) = x^3 x^3 = x^6$ and $f(x, x)$ increases as $x$ increases.
But if $y = -x$, $f(x, y) = f(x, -x) = -x^6$ and $f(x, -x)$ decreases as $x$ increases.
Since $f_{xx} = 6xy^3$, $f_{yy} = 6x^3 y$, and $f_{xy} = 9x^2 y^2$, $\Delta(0, 0) = 0$.
**25.** Increasing in all cases
**27.** Increasing in the third case and decreasing in all the others
**29.** $x = 1, y = 2, z = 4$     **31.** Base $4 \times 4$ and height 6
**33.** $x = 200, y = 100$     **35.** $x = 18, y = 8$     **37.** $x = 9, y = 4$
**39.** $x = 200, y = 100, z = 700$     **41.** $x = 1600, y = 1000$

## Section 19.4

**1.** Minimum at $x = 1, y = 3$     **3.** Maximum at $x = -3, y = -1$
**5.** Minimum at $x = 1, y = 1$     **7.** Maximum at $x = -1, y = -3$
**9.** Minimum at $x = 2, y = 2$     **11.** Maximum at $x = 4, y = 4$
**13.** Minimum at $x = 4, y = 2, z = 1$     **15.** Maximum at $x = 2, y = 2, z = 1$
**17.** 10 and 10     **19.** 12, 12, and 12     **21.** $r = 2, h = 4$
**23.** $x = 200, y = 100$     **25.** $6 \times 6 \times 9$     **27.** $x = 2000, y = 4000$

## Section 19.5

**1.** 1    **3.** $-6$    **5.** $\frac{1}{15}$    **7.** $-\frac{11}{1250}$    **9.** $\frac{1}{20}$    **11.** $-\frac{1}{20}$    **13.** .05
**15.** $e/10 \approx .2718$    **17.** 6.05    **19.** 5.036    **21.** 3.11    **23.** 8.9
**25.** 24.57    **27.** 1.08    **29.** $2000    **31.** 118.75    **33.** .70 cu ft
**35.** .9573 sq m

## Section 19.6

1. $\frac{15}{4}$     3. $\frac{3}{4}$     5. 8     7. 2.5     9. 1     11. 35     13. $\frac{4}{15}(2\sqrt{2} - 1)$

15. ln 2 ln 3     17. $(e^2 - 1)(e^3 - 1)$     19. $\frac{7}{24}$     21. $(e^4 - 1)/2$     23. $\frac{1}{2}$

25. $\frac{1}{3}$     27. $\frac{1}{8}$     29. $\frac{1}{4}$     31. $\frac{1}{40}$     33. $\frac{1}{24}$     35. $-\frac{2}{3}$     37. $\frac{7}{6}$

39. $\frac{1}{2} + \ln 2$     41. 4     43. 19,000     45. $\frac{3627}{532}$     47. 38,000

49. 10,000     51. 7000

## Chapter 19 Review

1. $f(0, 0) = -1, f(1, 2) = 7, f(2, -1) = 4$

2. $f(0, 0) = 0, f(1, 2) = 1/5, f(2, -1) = -8/5$

3. (1) all $x$ and $y$ (2) $\{(x, y)|y \neq 0\}$     4. $2\sqrt{11}$

5. $z_0 = 16$: $x^2 + y^2 = 0$ or the single point $(0, 0)$, $z_0 = 12$: $x^2 + y^2 = 4$, circle of radius 2 centered at $(0, 0)$, $z_0 = 7$: $x^2 + y^2 = 9$, circle of radius 3 centered at $(0, 0)$

6. $f_x = 1 + 2xy^5, f_y = 5x^2y^4$     7. $f_x(2, 1) = 5, f_y(2, 1) = 20$

8. $f_x = -ye^{-xy} + \dfrac{1}{x}, f_y = -xe^{-xy} + \dfrac{1}{y}$,

9. $f_{xx} = 2y^5, f_{xy} = f_{yx} = 10xy^4, f_{yy} = 20x^2y^3$

10. $f_{xx} = y^2e^{-xy} - \dfrac{1}{x^2}, f_{xy} = f_{yx} = e^{-xy}(xy - 1), f_{yy} = x^2e^{-xy} - \dfrac{1}{y^2}$

11. $f_x = -yze^{-xyz} + y^2z^3, f_y = -xze^{-xyz} + 2xyz^3, f_z = -xye^{-xyz} + 3xy^2z^2$

12. Relative minimum at $(0, 0)$     13. Relative maximum at $(0, 0)$

14. Relative maximum at $(-2, -1)$     15. Relative minimum at $(2, -3)$

16. Saddle at $(0, 0)$, relative maximum at $(-1, -1)$

17. Saddle at $(-1, -1)$, relative minimum at $(3, 3)$

18. Saddle at $(5, 3)$, relative maximum at $(5, -1)$

19. Saddle at $(0, 0)$, relative maximum at $(\frac{2}{3}, \frac{1}{9})$

20. Since $f_{xx}(0, 0) = 2a$ and $\Delta(0, 0) = 4(ac - b^2)$, the result follows.

21. Minimum at $(4, -3)$     22. Maximum at $(\frac{1}{4}, \frac{1}{2})$     23. 2.8     24. 3.2

25. 4.99     26. 24     27. 80     28. $\frac{5}{12}$     29. $\frac{7}{24}$     30. $\frac{4}{5}$     31. $\frac{3}{20}$

32. $\frac{1}{2}$     33. $\frac{13}{20}$     34. $\frac{13}{20}$     35. $\frac{1}{6}$     36. 80     37. 480

38. $f_K = 2K^{-.80}L^{.80}, f_L = 8K^{.20}L^{-.20}$

39. $f_{KK} = -1.6K^{-1.8}L^{.80}, f_{KL} = 1.6K^{-.80}L^{-.20}, f_{LL} = -1.6K^{.20}L^{-1.20}$

40. 1091     41. $C_x = 2x - y, C_y = 2y - x$

42. $C_{xx} = 2, C_{xy} = C_{yx} = -1, C_{yy} = 2$     43. $A_P = e^{rt}, A_r = Pte^{rt}, A_t = Pre^{rt}$

44. $2 \times 2 \times 4$     45. $x = 10, y = 12$     46. $x = 11, y = 14$

47. 11,000 of regular and 9000 of mint     48. 500 of each

49. 170 cubic inches

## Appendix A.1

1. $x^{10}$     3. $x^{15}$     5. $(x + 2)^5$     7. $x^5$     9. $(x + 1)^3$     11. 1

13. $1/2$     15. 10     17. 1     19. $a^3/b$     21. $xyz$     23. $y^6/x^{12}$

25. $-4$     27. $-2$     29. 7     31. $-3$     33. 5/2     35. $-2/5$     37. 2

39. 4     41. 0.1     43. 1/216     45. 16     47. 256     49. 10,000/81

51. 9/4     53. $2x^2$     55. $ab$     57. $6\sqrt{2}a^2b^{3/2}c^{5/2}$     59. $2a^2b^3$

61. $2a^3b^2/c^4$     63. $a^2b^2$     65. $x^{-1/2} + x^{1/2}$     67. $(x + 1)^{2/3}$

69. $x^{2/3}$     71. $(x/y)^{1/2}$     73. $(x^2 + 1)^{17/12}$     75. $1/x^{1/3}$     77. $ab$

**79.** $b$    **81.** $(x + 1)(x + 2)$    **83.** $(x + 1)^5$    **85.** $\sqrt{5}/5$    **87.** $\sqrt{2} - 1$

**89.** $\sqrt[3]{25}/5$    **91.** $-(1 + \sqrt{3})^2/2$    **93.** $1/(\sqrt{5} + \sqrt{2})$

**95.** $1/(\sqrt{10} + 2)$

## Appendix A.2

**1.** $3x^2 + 3x + 2$    **3.** $-3x^2 - x - 1$    **5.** $x^4 - x^3 + x^2 - x + 4$

**7.** $x^3 - x^2 + x + 2$    **9.** $x^3 + 3x^2 + 5x + 3$    **11.** $2x^3 - 3x^2 + 7x - 3$

**13.** $x^4 + 4x^2 + 3$    **15.** $(x - 3)(x^2 + 3x + 9)$    **17.** $(x - \frac{1}{2})(x^2 + \frac{1}{2}x + \frac{1}{4})$

**19.** $(3x - 2)(9x^2 + 6x + 4)$    **21.** $(\frac{1}{2}x - 3)(\frac{1}{4}x^2 + \frac{3}{2}x + 9)$

**23.** $(\frac{1}{4}x - \frac{1}{3})(\frac{1}{16}x^2 + \frac{1}{12}x + \frac{1}{9})$    **25.** $(ax - b)(a^2x^2 + abx + b^2)$

**27.** $(x - 4)(x + 4)$    **29.** $(4x - 1)(4x + 1)$    **31.** $(x^2 + \frac{1}{4})(x + \frac{1}{2})(x - \frac{1}{2})$

**33.** $(\frac{1}{4}x^2 + 1)(\frac{1}{2}x + 1)(\frac{1}{2}x - 1)$    **35.** $(x^2 + 2)(x + \sqrt{2})(x - \sqrt{2})$

**37.** $(\frac{1}{2}x^2 + 1)(\frac{1}{\sqrt{2}}x + 1)(\frac{1}{\sqrt{2}}x - 1)$    **39.** $\frac{x+1}{x-1}$    **41.** $\frac{x}{1-x}$    **43.** $\frac{x-1}{x+2}$

**45.** $\frac{x-1}{x+4}$    **47.** $2x + 1$    **49.** $\frac{x^2+1}{x}$    **51.** $\frac{x-1}{x+1}$    **53.** $\frac{1}{x+1}$    **55.** $-\frac{x}{(x-1)(x+1)}$

**57.** $\frac{4}{(x^2+4)(x+2)(x-2)}$    **59.** $\frac{\sqrt{x} - 1}{x - 1}$    **61.** $\frac{\sqrt{x} - \sqrt{y}}{x-y}$    **63.** $\sqrt{y} + 3$

**65.** $(a + 1)(\sqrt{a} + 1)$    **67.** $\frac{1}{\sqrt{x} + 1}$    **69.** $\frac{1}{\sqrt{x+h} + \sqrt{x}}$

## Appendix A.3

**1.** $-4$    **3.** $\frac{11}{7}$    **5.** $2$    **7.** $-\frac{19}{10}$    **9.** $\frac{13}{5}$    **11.** $\pm 2$    **13.** $\pm\frac{3}{2}$

**15.** $\pm 2$    **17.** $\pm\frac{1}{2}$    **19.** $\pm\frac{3}{5}$    **21.** $\pm 6$    **23.** $\pm\frac{1}{6}$    **25.** $\pm\frac{b}{a}$

**27.** $-2$    **29.** $1, 4$    **31.** $\frac{1}{2}, -2$    **33.** $-\frac{1}{3}, -2$    **35.** $\frac{1}{3}, -\frac{1}{2}$    **37.** $\frac{1}{5}, 1$

**39.** $-\frac{2}{3}, -\frac{3}{2}$    **41.** $\frac{1}{4}, \frac{2}{3}$    **43.** $\frac{1}{3}$    **45.** $-\frac{2}{3}$    **47.** $\frac{1}{2}$    **49.** $-\frac{1}{4}$

**51.** $\pm 2$    **53.** $\pm\frac{3}{2}$    **55.** $0, 1, -2$    **57.** $x = 0, \pm 3$    **59.** $-1$

**61.** $\frac{7}{10} \pm \frac{3}{10}\sqrt{5}$    **63.** $\frac{1}{2} \pm \frac{1}{2}\sqrt{5}$    **65.** $-\frac{2}{3} \pm \frac{\sqrt{19}}{3}$    **67.** $2 \pm \sqrt{3}$

**69.** $\frac{7}{6} \pm \frac{\sqrt{41}}{6}$    **71.** No real solution    **73.** $1, 0$    **75.** $2, -\frac{4}{3}$

**77.** $0, \frac{4}{3}$    **79.** $-1, 9$    **81.** $-\frac{1}{6}, \frac{7}{6}$    **83.** $\frac{5}{3}, 1$

## Appendix A.4

**1.** $x < \frac{4}{3}$    **3.** $x < -6$    **5.** $x \le -1$    **7.** $x \le 7$    **9.** $x < \frac{3}{8}$

**11.** $x \le 1$    **13.** $-4 < x < 2$    **15.** $0 \le x \le 1$    **17.** $x < -1$ or $x > 3$

**19.** $x \le -\frac{4}{3}$ or $x \ge \frac{8}{3}$    **21.** $-\frac{19}{12} < x < -\frac{17}{12}$    **23.** $x \le -5$ or $x \ge 10$

**25.** $1 < x < 3$    **27.** $x \le -2$ or $x \ge 3$    **29.** $-2 < x < 2$

**31.** $x \le -3$ or $x \ge 1$    **33.** $\frac{1}{2} < x < 2$    **35.** $x \le -3$ or $x \ge \frac{2}{3}$

**37.** $-2 < x < 1$    **39.** $x \le 1$ or $x > 2$    **41.** $x < -2$ or $x > 1$

**43.** $-1 \le x < 1$    **45.** $-1 \le x < 0$ or $x \ge 1$    **47.** $x < -1$ or $0 < x < 1$

**49.** $x < 1$ or $2 < x < 3$    **51.** $-1 \le x \le 1$ or $x > 2$    **53.** $x < 0$

**55.** $-1 < x < 2$ with $x \ne 0$

# Index

Abortions, 484–485
Absolute extrema, 787, 1087
Absolute maximum, 787, 1087
Absolute minimum, 787, 1087
Absolute value, A-18
Absorbing Markov process, 537
Absorbing stochastic matrix, 537
Acceleration, 765, 773, 902
Accidental deaths, 507
Adhesive tape manufacturing, 902
Advertising, 359, 396, 419, 508, 570,
    582
  campaign, 967
  promotion, 465
Africa, population of, 884
AIDS, 511
Air pollutant concentration, 486
Air pollution, average, 947
Airline scheduling, 106
Airline tours, 181–182
Alcohol, 485
Alcott, Amy, 480
Algebra of functions, 619
Allocation of funds, 167
Allocation of machine time, 59
Amortization, 302
Amortization schedule, 303

Amount of pollutants, 1128
Animal nutrition, 246
Animal ranching, 219
Annual rate, 271
Annuity, 289
  amortization, 302
  amortization schedule, 303
  equity in a sinking fund, 293
  future amount, 270
  future value, 276
  maturity value, 273
  ordinary, 289
  outstanding principal, 304–305
  payment toward principal, 305
  present value, 271, 284, 848
  sinking fund, 292, 312
  sinking fund payment, 292
  term of, 289
Antarctic blue whale, 792
  carrying capacity, 792
  measuring, 37
  minimum viable population level,
    792
Anthropology, 885, 891
Antiderivatives, 894
Apple orchard, 807
Appreciation, 741, 902

Approximations, 1106–1107
Area, between two curves, 949–950
  under a curve, 923
  under normal curve, B-1–B-3
Arrests, 507
Assembly line, 439
Asteroid, 40
Asymptote(s), horizontal, 776
  vertical, 646
Atlantic City Golf Classic, 480
Attack equation, 813
Augmented matrix, 67
Auto-catalytic reaction, 757
Automobile, accidents, 487
  life, 1050
  selection, 526
Average, 457
Average air pollution, 947
Average amount of drug, 947
Average amount of money, 946
Average blood pressure, 947
Average concentration of pollutants,
  1126
Average cost, 653, 708, 812–813
  function, 783
  of long-distance call, 726–727
Average population, 1131

Average price, 946
Average profit, 947
Average rate of change, 666
Average revenue, 947
    function, 757
Average value, 1126
    of a function, 938, 941
Average velocity, 640–641, 666
Award selection, 341
Awarding of contracts, 405, 420

Bacteria, 703, 823, 838, 928, 993,
    1112
Ball bearings, 472, 495
    quality control, 1046
Base $e$, 844, 846
Baseball, 341, 350, 450, 506, 511
    batting average, 450
    slugging average, 465
Baseball players, Cobb, Ty, 434
    Griffey, Ken Jr., 432
    Kiner, Ralph, 434
    Maris, Roger, 452
    Ruth, Babe, 434, 465
    Williams, Ted, 450, 452
Basic geometric formulas, B-35–B-36
Basic variables, 190
Basketball, 341, 431, 438, 453, 507
Bath towel manufacturing, 977
Battery life, 1026
Battery reliability, 438
Batting average, 450
Bayes, Thomas, 422
Bayes' theorem, 422
Behavior, long-term, 539
Bernoulli, James, 386
Bernoulli trials, 430
Bicycle sales, 943
Bidding on contracts, 393, 395
Big-M method, 240
Binomial distribution, 448, 459, 462
Binomial theorem, 351–352
Biology, 13, 527, 535, 615, 703–704,
    714, 727, 742, 757, 773,
    783–784, 983, 1001, 1066–1067,
    1080
Biomass, 751
Birth rate, 487
Birthday problem, 402
Birthdays, 404–405

Bladder cancer, 507
Blending, 170
Blood flow, 742, 828
Blood pressure, average, 947
Blood type, 429
Bone measurement, 823
Boole, George, 316, 327
Bounded interval, 1087
Box dimensions, 1090, 1093, 1102,
    1104
Boxed trifecta, 349
Bradley, Pat, 481
Break-even point, 30
Brick manufacturing, 982
Bronchi, 757, 829
Bus company revenue, 806, 813

Calculus, fundamental theorem of, 933
Camera sales, 912
Cancer, 383, 419, 499, 625, 676,
    816–817
    bladder, 507
    carcinogen, 383
    cervical, 499
    lung, 407
    thyroid, 507
Capital, expenditure, 308
    analysis, 301
    marginal productivity, 1072
Car loan, 302–303, 308
Car rental, 549
Car sales, 552
Carbon-14, 878
Carbon dating, 878
Carbon monoxide, 486
Carcinogen, 383
Card game(s), 340, 345, 350, 370,
    401–402, 404, 407–408, 417
Cardboard box manufacturing, 911
Cardiac output, 1066
Carrying capacity, 751
    Antarctic blue whale, 792
    mice, 776
Cartesian coordinate system, 4
    coordinate axes, 1057
    coordinates of a point, 1057
    ordered triple, 1057
    three-dimensional, 1057
    $x$-axis, 5
    $x$-coordinate, 5

$x$-intercept, 21
$xy$-plane, 5
$y$-axis, 5
$y$-coordinate, 5
Casino gambling, 381
Cauchy, Augustin Louis, 634
Cayley, Arthur, 109
Central limit theorem, 490
Cereal packaging, 479
Cereal profits, 1055–1056
Ceres, 40
Certainty event, 370
Cervical cancer, 499
Chain rule, 716
Charge accounts, 545–546
Chebyshev's inequality, 478
Chemical spill, 945–946
Chemical toxin, 654
    cost, 783
Chemistry, 714, 784
    physical, 871
Childbirth, 484–485
Chinese farming, 78
Chinese way of calculating with
    arrays, 78
Cigarettes, 485
Circle, formula, B-35
Clock manufacturing, 911
Closed Leontif input-output model,
    126, 129
Closed region, 1088
Closet construction, 1093
Coastal shark fishery, 754
Cobb, Ty, 434
Cobb-Douglass production function,
    1057, 1100, 1104, 1109, 1112,
    1131, 1138-1139
Cocaine, 485, 507
Coefficient matrix, 100
Coefficient of inequality, 958–959
Coin appreciation, 287
Coin tossing, 350, 367, 371, 378, 382,
    401, 403, 417, 452, 502–503,
    505–506, 512
College instruction, 439
Column matrix, 83
Column player, 553
Combinations, 343
Committee(s), 405
Committee selection, 342, 344,
    346–347, 349, 359

Commuting time, 1025
Company contracts, 438
Comparing bank rates, 287–288
Comparing investments, 301, 468–469
Comparing present values, 307–308
Competitive demand relation,
    1073–1075
Competitive pricing, 1094, 1139
Complement of a set, 317
Complement of an event, 369
Complementary demand relation,
    1073–1075
Complementary slackness, 260
Completing the square, 608
Component selection, 359
Composite function, 620
Compound interest, 278, 722
    effective annual yield, 283
    effective rate, 274
Compounding, continuous, 844, 859,
    1079
Computer chips, 433
Computer failure, 415
Concave down, 760
Concave up, 760
Concavity, test for, 761
Concentration, of drug, 714, 727, 784
    of pollutants, 1128, 1131
Conditional probability, 407, 409
Cone, right circular, formula, B-36
Constant sign theorem, 661
Constraints, 150, 1096
Consumer advertising, 362
Consumer survey, 362, 382
Consumers' surplus, 955, 957, 961,
    976, 982
Container construction, 799–801, 806
Continuity, 1063
    at a point, 657
    of polynomial functions, 658
    of rational functions, 658
    on an open interval, 657
    rules, 658
Continuous compounding, 844, 859,
    1079
Continuous money flow, 1003
    continuous reinvestment, 1007–1008
    on infinite intervals, 1008–1009
    present value, 1004–1005
    present value on infinite intervals,
        1008–1009

rate, 1003
    total income, 1003
Continuous random variable, 447
Contract awarding, 405, 420
Contract bidding, 393, 395
Contractor costs, 74
Coordinate plane, 5
Copper mine, 912
Copper production, 945, 962
Corner of the graph of a function,
    682–683
Corporate farming, 178
Corporate success determination, 372
Cost(s), 13, 29, 43–44, 1066, 1094,
    1111
    airline, 175, 234–235, 251
    fencing, 615
    fixed, 29
    of long-distance call, 664
    oil refinery, 175, 234, 251
    postage, 628
    variable, 29
Cost equation, 676, 703, 822, 835–836
Cost function, 625, 690, 741, 756, 772,
    830–831, 870, 977, 1139
Cournot, Augustin, 28
Course registration, 331–333
Course selection, 349
Credit card account, 270
Crime, 512
Critical point, 1083–1084
Critical value, 732
Crown's method, 240
Crude oil blending, 170
Customer selection, 350
Cylinder, right circular, formula, B-36


Daily numbers game, 466
Daniel, Beth, 481
Danzig, George, 138, 149, 166, 185,
    201, 239
de Fermat, Pierre, 373
de Mere, Chevalier, 373
De Morgan, Augustus, 327
De Morgan laws, 320
Death rate, 487
Decay constant, 875
Decreasing function, 731
Deer, 714

antlers, 26
    population, 887
Defective component, 438
Defective flashlights, 506, 508–509
Defective transistors, 399–400, 415
Definite integral, area, 923, 950
    average value of a function, 938,
        941
    definition, 922
    double, 1117
    fundamental theorem of calculus,
        933
    numerical approximation, 985–986
    properties, 932
    Riemann sum, 917
Degree of polynomial, A-7
Delivery fees, 603, 662
Demand, 36, 44, 48
    and supply, 79
    competitive, 1073–1075
    complementary, 1073–1075
    for cameras and film, 1066
    for electricity, 1037
Demand curve, 31
Demand equation, 28, 31, 703, 713,
    721, 756, 822, 828, 835, 863,
    870, 960
Demand function, 741, 828
Denominators, rationalizing, 648
Density, 1128
    of pollutants, 947
Density function, 941
Dependent variable, 591
Depreciation, 902–903, 912, 946,
    967
Derivative, definition of, 679
    implicit, 824
    marginal analysis, 696
    marginal average cost, 708
    marginal cost, 698, 967
    marginal profit, 698
    marginal propensity to consume,
        903
    marginal propensity to save, 903,
        994
    marginal revenue, 698, 721, 967
    nonexistence, 682
    related rates, 830
    rules, chain rule, 715
        constant, 692
        constant times a function, 695

Derivative, definition of (*continued*)
    general power rule, 717
    power rule, 693
    product rule, 705
    quotient rule, 707
    sum and differences, 695
    second, 759
Descartes, Renee, 2
Descriptive statistics, 443
Detection range of a school of fish, 722
Detergent box weight distribution, 1046
Developed world, population of, 884
Diagonal matrix, 89
Dice rolling, 367, 376–377, 381–382, 394–395, 401, 403, 416, 449–450, 453, 466, 506
Diet, 163, 173, 229
    health food mix, 183
    meal preparation, 177
    vitamin selection, 174
    vitamin supplements, 147, 162, 235
Differential, 819, 1106–1107
Diminishing returns, 772–773, 887
Dirichlet, Lejeune, 590
Discount, 272
Discounted note, 277
Discriminant, A-16
Disease(s), 1001
    cancer, 383, 419, 499, 625, 676, 816–817
        bladder, 507
        carcinogen, 383
        cervical, 499
        lung, 407
        thyroid, 507
    gonorrhea, 484
    heart, 414, 420, 507
    meningitis, 486
    plague, 485
    polio, 485
    spread of, 886
    syphilis, 484
    tuberculosis, 424
    venereal, 484
    viral hepatitis, 486
Disjoint sets, 319
Disk manufacturing, 902
Disposable diaper market, 713
Distance, between two points, 5–6
    between two ships, 625

Distance formula, 1058
Distribution of population by height, 454–455
Distribution of production, 135
Divorce rate, 487–488, 493
Domain, 591
Domination, 567–568
Double integral, average value, 1126
    changing order of integration, 1125
    definite, 1117
    density, 1128
    iterated, 1117, 1120, 1123
    Riemann sum, 1115
    volume, 1117
Doubling time, 276, 282, 312, 857, 859, 884
Douglas fir, 881–882
Drug(s), average amount, 947
    cocaine, 485, 507
    concentration, 714, 727, 784
    dosage, 654
    effectiveness, 508
    marijuana, 485, 507
    testing, 439
    use, 485
Dual problem, 219

$e$, 844, 846
Earth satellite, 420
Eastern Pacific yellowfin tuna, 881
Economics, 425, 428
Education, 333
Education fund, 296–297
Effective annual yield, 283
Effective rate, 274
Effective yield, 847
Einstein, Albert, 915
Einstein's relativity theory, 915
Elastic demand, 833
Elasticity of demand, 832
Elections, 37
Electric can opener manufacturing, 976
Electric utility average cost, 773
Electricity, cost, 812–813
    current intensity, 727
    demand, 1026, 1037
    tax, 663–664
Element, 3
Elementary event, 368
Elementary operations, 57
Elementary row operations, 67

Ellipsoid algorithm, 186
Empirical probability, 374
Employee, absences, 484
    assignments, 350
    attendance, 468
    bonus, 496
    productivity, 713, 912
    selection, 350
Empty set, 316
Entertainment, 333
Envelope manufacturing, 902
Equality of matrices, 83
Equality of sets, 3
Equation, linear, A-14
    polynomial, A-17
    quadratic, A-15–A-16
Equilibrium point, 32, 556, 880, 955, 961
    Equilibrium price, 32, 955
Equilibrium quantity, 32, 955
Equipment fund, 297
Equity in a sinking fund, 293
Equivalent systems of linear equations, 54, 58
Estes learning model, 544
European travel, 329–331
Event(s), 368
    certainty, 370
    complement, 369
    elementary, 368
    impossible, 370
    independent, 412
    independent events theorem, 413
    intersection, 369
    mutually exclusive, 370
    union, 369
Excess resources, 166, 211
Executive option, 354
Executive scheduling, 341
Exodus, 914
Expected return on lottery ticket, 466
Expected value, 459, 1028
    of a game, 563
Experiment, 367
Exponent, A-2
    negative, A-2
    properties, A-3
    rational, A-3–A-4
    zero, A-3
Exponential function, 842
    derivative, 861
    in terms of natural exponent, 856
    integral, 899

Exponential probability density
    function, 1023
  mean, 1034
  median, 1034
  standard deviation, 1034
  variance, 1034
Extended simplex method, 243
Extreme value theorem, 788

Facility location, 49, 1094
Factorial, 337
Factoring, A-8–A-9
Fair game, 555
False negative, 424
False positive, 424
Family composition, 407, 409
Family organization, 350
Family planning, 403, 418
Farming, 176–177, 181–182, 266–267
Fast food restaurants, 520, 539
Feasible region, 142, 189
Federal employees, 469
Federal tax revenue, 736, 742
Fence construction, 780, 797,
    805–806, 808, 813, 1098, 1104
Fermat, Pierre de, 373
Fick method, 1066
Financial tables, B-3–B-4
Finite discrete random variable, 447
Finite stochastic process, 410
First derivative test, 746
First quadrant, 5
Fishery, 163–164, 751, 791, 795, 886,
    1079, 1139
  Antarctic blue whale, 792
  bass, 147, 163, 179
  biomass, 751
  carrying capacity, 752
  coastal shark, 754
  harvesting, 751
  maximum sustainable yield, 751
  minimum viable biomass level, 791
  overfishing, 791
  Pacific halibut, 753
  Pacific salmon, 753
  renewable natural resources, 751
  sharks, 754
  trout, 216
Fixed costs, 29
Fixed point, 880
Flight control, 415

Flight safety, 415
Food survey, 332
Football, 341, 342
Foreign visits, 329–331
Forest fire, 623, 701
Forestry, 722, 881–882
Fourth quadrant, 5
Free variable, 70
Free water vapor diffusion coefficient,
    704
Function, 590
  absolute extrema, 787
  absolute maximum, 787
  absolute minimum, 787
  algebra of, 619
  composite, 620
  concave down, 760
  concave up, 760
  continuous, 657
  critical value, 732
  decreasing, 731
  density, 941
  dependent variable, 519
  domain, 591
  exponential, 842
  graph, 594
  increasing, 731
  independent variable, 591
  inflection point, 761
  inflection value, 761
  limit of, 636
  logarithmic, 853
  piecewise-defined, 594
  polynomial, 612
  power, 613
  quadratic, 605–606
  range, 591
  rational, 612
  relative extrema, 745
  relative maximum, 745
  relative minimum, 745
  several variables, 1054
  vertical line test, 596
Function of several variables, 1054
  absolute extrema, 1087
  continuity, 1063
  critical point, 1083–1084
  definition, 1054
  differential, 1106–1107
  domain, 1054
  evaluation, 1055
  first order partial derivative, 1068

  graph, 1059
  independent variables, 1054
  level curve, 1061
  relative extrema, 1081–1082
  relative maxima, 1081–1082
  relative minima, 1081–1082
  second order derivative test, 1085
  second order partial derivatives,
    1076–1077
  tangent plane, 1106
Fundamental theorem of calculus, 933
Fundamental theorem of duality, 222
Fundamental theorem of game theory,
    564
Furniture manufacturing, 119, 147,
    162, 178, 182, 215–216, 260
Future amount, 270
Future value, 276
  of an annuity, 289
  of an ordinary annuity, 291

Gambler's ruin, 538, 544
Game, 552
  column player, 553
  domination, 567–568
  expected value, 563
  fair, 555
  fundamental theorem, 564
  mixed strategy, 560
  nonstrictly determined, 559
  payoff matrix, 554
  payoff value, 555
  pure strategy, 563
  row player, 553
  saddle point, 555
  strictly determined, 552, 556
  translated, 567
  zero-sum, 553
Game show, 439–440
Game theory, fundamental theorem of,
    564
  linear programming and, 573,
    577–578
  simplex method in, 579
Gasoline blending, 170
Gauss, Carl Friedrich, 40, 52
Gauss-Jordan method, 55
  elementary operations, 57
  elementary row operations, 67
  free variable, 70
  general solution, 68

Gauss-Jordan method (*continued*)
   leading one, 69
   leading variable, 70
   parameter, 68
   particular solution, 68
   reduced matrix, 68
   with graphing calculator, C-3–C-4
Geddes, Jane, 480
General power rule, 717
General solution, 68
Geometric formulas, basic, B-35–B-36
Golf, 135, 341
Golf tournament scores, 437
Golfers, 480–481
   Daniel, Beth, 481
   Geddes, Jane, 480
   Little, Sally, 481
   McGann, M., 481
   Okamoto, A., 481
   Schreger, Cindy, 480
Gompertz growth curve, 871
Gonorrhea, 484
Grade distribution, 383–384, 452
Grade point average, 465
Graph of a function, 594
Graph of an equation, 7–9
Graphing calculator, C-1–C-5
   programs for, C-5–C-12
Graunt, John, 375
Gregory, James, 987
Griffey, Ken Jr., 432
Guessing on a test, 453

Hair brush manufacturing, 912
Half-life, 875, 887, 891
Hand purse manufacturing, 911
Health, 485
Health care unit location, 470
Health food mix, 183
Heart disease, 414, 420, 507
Heart operations, 468
Heating oil consumption, 1037
Herschel, Sir William, 40
Highway speed measurement, 382
Histogram, 445
Homeowners survey, 438
Horizontal asymptotes, 776
Horizontal rule, 17
Horse racing, 392
Hotel profits, 807

Hotel revenue, 807
Hours studying, 452
House down payment, 297
House selection, 341
Household survey, 328
Huygens, Christian, 386

Identity matrix, 97
Immigration, 507
Implicit differentiation, 824
Implicit function theorem, 828
Impossible event, 370
Improper integral, 995
   convergent, 996
   divergent, 996
Income, continuous flow, 1011
   distribution, 958
   maximization, corporate farming, 178
   survey, 372
Income tax cliffs, 660–661
Increasing function, 731
Indefinite integral, 895
   general form, 895
   integrand, 895
   integration by parts, 972
   rules, constant times function, 897
      exponential function, 899
      general exponential, 907
      general power, 904
      power rule, 896
      sum or difference, 898
   substitution, 904
Independent events, 412
Independent events theorem, 413
Independent set of events, 414
Independent variable, 591
India, population of, 890
Individual retirement account, 296
Indonesia, population of, 890
Industrial pollutants, 941, 943
Inelastic demand, 833
Inequalities, linear, A-20
   nonlinear, A-22
Infestation, spread of, 628
Infinite discrete random variable, 447
Inflation, 962
Inflection point, 761
Inflection value, 761
Input-output analysis, closed Leontief
   model, 126, 129

input-output matrix, 123
   open Leontief model, 123
Input-output matrix, 123
Insect temperature, 44
Instantaneous rate of change, 667, 669
Instantaneous velocity, 640, 666, 668
Insurance, 467
Integral(s), definite, 922
   improper, 995
   indefinite, 895
   iterated, 1117, 1120, 1123
   tables of, 978–980, B-36–B-39
Integration, by parts, 972
   by substitution, 904
   numerical, 985–986
   using tables, 978–980
Intelligence test (IQ) scores, 498,
   1047
Interest, 270, 278, 722
   compound, 278, 722
   doubling time, 276, 282, 312, 857,
      859, 884
   per period, 278
   present value, 271, 284
   simple, 270
   simple discount, 272, 277
   tripling time, 277, 312
Interest rate(s), 270
   nominal, 283
   tables of, B-3–B-34
International Whaling Commission,
   792
Intersection of events, 369
Intersection of sets, 319
Intuitive probability, 378
Inventory, 36
Inventory control, 808, 813
Inventory cost, 802
Inventory cost model, 802
Inverse of a matrix, 109
Investment(s), 25, 62, 63
   and risk, 174–175, 235, 251, 477,
      483–484
Investment advisor, 468
Investment decision, 1010–1011
Investment mix, 115, 119, 173–174,
   250, 332
Investment returns, 465, 468
Irrational powers, 842
Isobar, 1061
Isoprofit lines, 150

Iterated integral, 1117, 1120, 1123
Itinerary selection, 342

Japan, population of, 884
Jordan, Wilhelm, 52

Karmarkar, Narenda, 186
Karmarkar's algorithm, 186
Kennedy administration tax policy, 738
Keogh account, 312
Khachiyan, L. G., 186
Kiner, Ralph, 434
Knife manufacturing revenue, 901
Koopmans, Tjalling, 239

Labor, allocation, 36
  marginal productivity, 1072
Labor relations, 363
Laboratory management, 363
Laffer, Arthur, 736
Laffer curve, 736
Lagrange, Joseph Louis, 1102
Lagrange multipliers, 1096–1097
Laplace, Pierre Simon, 386
Law enforcement officers killed, 675
Law of large numbers, 504–505, 508
Leading one, 69
Leading variable, 70
Leaf area, 742
Learning, 823, 1027, 1038
Learning time, 722, 1047
Leasing, 308
Least common multiple, A-11
Leibniz, Gottfried, 632, 634
Leontief, Wassily, 123
Level curve, 1061
Libby, Willard, 878
License plate(s), number of, 339
  selection, 340
Life insurance, 511
Life insurance policy, 461
Life of a part, 1026
Light bulb life span, 1023–1024, 1034
Light intensity, 757
Limit(s), at infinity, 775
  definition of, 636
  existence, 636

from the left, 636
from the right, 636
polynomial function, 647
program, for graphing calculator, C-6–C-7
replacement theorem, 638
rules, 646
Limited growth, 878
Linear cost model, 29
Linear equation, 20, A-14
  horizontal rule, 17
  parallel lines, 21
  perpendicular lines, 22
  point-slope form, 19
  slope, 15
  slope-intercept form, 20
  vertical line, 19
Linear inequalities, 139, A-20
  properties, A-20–A-21
Linear models, 28
Linear programming, and game theory, 573, 577–578
  big-M method, 240
  constraints, 150
  Crown's method, 240
  dual problem, 219
  ellipsoid algorithm, 186
  feasible region, 142
  fundamental theorem of duality, 222
  Karmarkar's algorithm, 186
  method of corners, 155
  nonnegativity constraint, 139
  objective function, 150
  post-optimal analysis, 166
  problem constraints, 150
  sensitivity analysis, 254
  sensitivity theorem, 258
  shadow prices, 228, 258
  standard maximum problems, 201
Linear revenue model, 29
Lines, 15
Lipstick manufacturing, 912
Little, Sally, 481
Lloyd's of London, 374
Logarithmic function, derivative, 865
  properties of, 853
Logistic equation, 879–880
  equilibrium point, 880
  fixed point, 880
London life table, 375, 382
Long run, 528

Long-term behavior, 539
Longcor, Willard H., 377
Lorentz curve, 958
Lot acceptance sampling, 504
Lottery, 381, 466, 511
Lottery prize, 312
Lotto game, 343, 349, 403
Louisiana purchase, 287
Lung(s), 757, 829
Lung cancer, 407

Machine allocation, 36
Make or buy decision, 37
Malfunctioning machine, 438
Management personnel selection, 363
Manufacturing, 147, 162–163, 254, 396, 428–429, 438, 512
  adhesive tape, 902
  bath towel, 977
  brick, 982
  cardboard box, 911
  clock, 911
  defective parts, 419
  electric can opener, 976
  envelope, 902
  furniture, 119, 147, 162, 178, 182, 215–216, 260
  hair brush, 912
  hand purse, 911
  lipstick, 912
  metal connectors, 145, 148, 157, 215, 240
  park benches, 142, 182
  pen, 754
  piston rings, 498
  purses, 267
  salt shaker, 901
  shoulder bag, 911
  spark plug, 902
  table cloth, 901
  tape, 1047
  toaster, 982
  wire, 1050
  wooden boat, 138
Marginal, analysis, 696
  average cost, 708
  cost, 698, 967
  productivity of capital, 1072
  productivity of labor, 1072
  profit, 698

Marginal, analysis (*continued*)
  propensity to consume, 903
  propensity to save, 903, 994
  revenue, 698, 721, 967
Marijuana, 485, 507
Maris, Roger, 452
Market competition, 570
Market share, 654, 783
Marketing, 396, 526–527, 535, 571, 582
Markov process, 516
  absorbing, 537
  absorbing stochastic matrix, 537
  gambler's ruin, 538, 544
  random walk, 538
  regular, 532
  regular stochastic matrix, 529
  regular transition matrix, 530
  state of a, 516
  steady-state distribution, 531
  steady-state matrix, 531
  stochastic matrix, 524
Marriage rate, 487–488
Matrix(ces), 67
  addition, 84
  augmented, 67
  coefficient, 100
  column, 83
  diagonal, 89
  equality, 83
  identity, 97
  inverse(s), 109
    with graphing calculator,
    C-4–C-5
  multiplication, 91
  order, 82
  production, 86, 89–90
  reduced, 68
  regular stochastic, 529
  regular transition, 530
  row, 83
  square, 83
  steady-state, 528
  stochastic, 524
  strictly upper triangular, 105
  subtraction, 84
  transition, 517
  transpose, 230
  upper triangular, 89, 104
  with graphing calculator, C-3–C-4
  zero, 85
Maturity value, 273

Maximum sustainable yield, 751
McDonald's Golf Championship,
  481
McGann, M., 481
Meal selection, 340, 349
Mean, 457, 1028
Measurement(s), 1139
Measurement conversions, 623, 625
Median, 463, 1034
Median age, 44, 730–731
Medical diagnosis, 429, 885
Medical operation time, 1047
Medical team rescue, 804
Medicine, 37, 396, 419, 424–425, 428,
  435, 468, 625, 654, 676,
  703–704, 714, 727, 757, 784,
  828–829, 836, 885, 968,
  1066–1067, 1112
Member, 3
Memorization, 912
Meningitis, 486
Menu selection, 342
Mere, Chevalier de, 373
Metal connector manufacturing, 145,
  148, 157, 215, 240
Method of corners, 155
Method of exhaustion, 914
Method of least squares, 41
Mice, carrying capacity, 776
  population, 764, 1001
Minimum viable biomass level, 791
Mining, 182, 1014
  copper, 945
  platinum, 1001
Mixed strategy, 560
Mixture, 62, 262–263
  acid, 135
  alcohol, 116
  candy, 176
  fruit drinks, 180
  fruit juices, 262–263
  nuts, 147, 162, 178–179, 216
  tea, 119
Mode, 463
Monomolecular growth, 871
Morgan, Augustus de, 327
Mortgage payments, 312
Mosteller, Frederick, 377
Multiplication principle, 334–335
Mutual fund(s), 359, 405, 444, 452
  selection, 347

Mutually exclusive events, 370


Napier, John, 851, 855
National Marine Fisheries Service,
  754
Natural gas extraction, 937
Natural logarithm, 851, 853
Natural resource depletion, 945
Navigation, 12, 603, 831–832, 836
Negative exponent, 1141
Networks, 106
Neumann, John von, 123, 552
New York Giants, 341
News, spread of, 887
Newsletter copies, 1047
Newspaper readers, 331
Newton, Isaac, 632, 634
Newton's law of cooling, 878
Newton's law of heating, 886
Nigeria, population of, 884, 887
Nitrogen dioxide, 486
Nominal rate, 283
Nonbasic variables, 190
Nonlinear inequalities, A-22
Nonnegativity constraint, 139
Nonstrictly determined game, 557
Normal curve, 490
  area under, B-1–B-3
Normal distribution, 1042
  mean, 1042–1044
  standard, 1038–1039
  standard deviation, 1042–1044
  variance, 1042–1044
Normal probability density function,
  1042–1044
Normal probability distribution, 493,
  495
Nuclear accident, 877
Nuclear reactors, 883
Number $e$, 844, 846
Number line, 4
Number theory, 1093
Numerator, rationalizing, 648, 1051
Numerical approximation, of definite
  integral, 985–986
Numerical integration, Simpson's rule,
  988
  errors in, 991
  trapezoidal rule, 985–986
  errors in, 990

Nutrition, 13, 49, 62–63, 78
  meal selection, 234

Objective function, 150
Octane rating of gasoline, 172
Odds, 391
Office scheduling, 77
Oil drilling, 434
Oil pollution, 469–470
Oil production, 909, 944
Oil refinery, 170, 234, 251
  blending, 176
Oil slick, 12, 831
Oil well income, 1004, 1010–1011
Oil well price, 1006
Okamoto, A., 481
Open Leontief input-output model, 123
Optical lens, 420
Optimal strategies, 565, 575–576
Order of a matrix, 82
Ordered partitions, 355–358
Ordinary annuity, 289
Origin, 4, 5
Outcome of an experiment, 367
Outstanding principal, 304–305
Overfishing, 791
Ozone, 486

Pacific halibut fishery, 753
Pacific salmon fishery, 753
Package construction, 799–801, 806
Package dimensions, 1088, 1090, 1099
Packaging, 497, 603, 628, 1139
Pain medication, 331
Paper plate manufacturing revenue,
  901
Parabola, 609
Parallel lines, 21
Parallelogram, formula, B-35
Parameter, 68
Parental determination, 429
Park bench manufacturing, 142, 182
Partial derivative, first order, 1068
  Lagrange multipliers, 1096–1097
  second derivative test, 1085
  second order, 1076–1077
Particular solution, 68
Pascal, Blaise, 373
Pascal's triangle, 354

Payment toward principal, 305
Payoff matrix, 554
Payoff value, 555
Pen manufacturer, 754
Pension fund investment selection, 77
Perfecta, 335
Perfume production, 813
Permutation, 337
Perpendicular lines, 22
Personnel, 435
Personnel retention, 549
Pet food manufacturer survey, 332
Petroleum consumption, 675
Photosynthesis, 742, 757, 783, 829
Physical chemistry, 871
Physics, 836, 838, 886
Piazzi, Giuseppe, 40
Piecewise-defined functions, 594
Pine disease, 628
Pipe construction, 807
Piston ring manufacturing, 498
Pivot, column, 204
  element, 206
  operation, 193
  row, 206
  selection, 206
Pivoting, 193
Plagues, 485, 765
Plane, equation of, 1061
Plant fertilizer, 144, 156
Plant nutrition, 175, 235, 237, 251
Plant productivity, 783
Plant roots, 871, 1080
Plant scheduling, 179
Platinum mining, 1001
Plutonium-239, 883
Point-slope equation of a line, 19
Poiseuille's law, 704, 742, 828
Polio, 485
Politics, 526, 535
Pollutants, 982, 1001
  amount of, 1128
  average, 947, 1126
  concentration of, 486, 1126
  density of, 947
Pollution, 512
Polynomial, A-7
  degree, A-7
  equations, A-17
  factoring, A-8–A-9
  function, 612

Population, of farmers, 730
  of United States, 730
Population explosion, 883
Population theory, 902
  Africa, 884
  average, 968, 1131
  bacteria, 993
  biomass, 751
  birth rate, 487
  carrying capacity, 752
  death rate, 487
  deer, 887
  density, 942, 947
  developed world, 884
  Eastern Pacific yellowfin tuna, 881
  exploding, 764
  fishery, 751, 791
  growth, 874
  harvesting, 751
  human, 874
  India, 890
  Indonesia, 888
  Japan, 884
  lake trout, 891
  limited resources, 763
  limiting human population, 882
  logistics equation, 879
  maximum sustainable yield, 751
  mice, 764, 776, 1001
  minimum viable biomass level, 791
  Nigeria, 884, 887
  of Northeast, 881
  predator, 615
  prey, 615
  renewable natural resources, 751
  S-curve, 763
  United States, 885–886
  world, 884–885
Portfolio management, 167
Post-optimal analysis, 166
Postage cost, 628
Postal rates, 653
Power function, 613
Power rule, 693
Predators, 615, 773
Present value, 271, 284, 848
Present value of an ordinary annuity, 300
Prey, 615, 773
Price, average, 946
Price function, 690
Price wars, 570, 581

Pricing, competitive, 1139
Primal problem, 219
Principal, 270
Prize selection, 350
Probability, 378, 387
    central limit theorem, 490
    Chebyshev's inequality, 478
    combinations, 343
    conditional, 407, 409
    density function, 1018–1019
    distribution, 448
    empirical, 374
    expected value, 459, 1028
    intuitive, 378
    law of large numbers, 504–505, 508
    London life table, 375, 382
    mean, 1028
    median, 1034
    multiplication principle, 334–335
    odds, 391
    ordered partition, 355–358
    outcome of an experiment, 367
    permutation, 337
    product rule, 409
    random variable, probability density
        function, 1018–1019
    relative frequency, 374
    sample space, 367
    standard deviation, 1030
    tree, 410
    trial, 367
    uniform sample space, 398
    union of events, 369
    universal set, 317
    variance, 1029
        alternate form, 1029
Problem constraints, 150
Proceeds, 273
Process selection and capacity, 37
Producers' surplus, 957, 961, 982
Product, defects, 419
    life, 372, 498, 1037
    mix, 581
    reliability, 420
    selection, 340, 342, 349
    testing, 372, 383, 405
Product rule, for derivative, 705
    of probability, 409
Production, 15, 23
    costs, 1139
    function, 1056

line, 544, 545
matrix, 86, 89–90
of perishable items, 77–78
planning, 98, 101
scheduling, 63, 77–78, 147, 163,
    235, 237, 251–252
    mining, 175, 235–236, 251–252
Profit(s), 13, 29, 49, 237, 756, 1001,
    1037, 1066, 1094–1095
    average, 947
    equation, 703, 822
    function, 690, 741, 870
    orchard, 615
    sleds, 216, 237
Promotion, 570, 581
Prudhoe Bay, Alaska, 909
Psychology, 722
    and sales, 419
    of success, 415
Public expenditure on education, 471
Public health, 885
Purchasing, flowers, 236
Purchasing decision, 73
Purchasing mix, 176
Pure strategy, 563
Purse manufacturing, 267, 911
Pythagorean theorem, formula, 1199


Quadrant, 5
Quadratic equation, A-15–A-16
    completing the square, 608
    discriminant, A-16
    solutions of, A-16
    standard form, 609
Quadratic function, 605–606
    parabola, 609
    vertex, 609
Quality, 487
Quality control, 396, 405, 419,
    428–429, 455, 468, 495, 498,
    507, 512, 1046
Quinella, 348
Quotient rule, 707–708


Radioactive decay, 875, 885, 902
Radioactive tracers, 885
Radioactive waste, 1001
Random phenomena, 365
Random selection, 561

Random variable, 447
    continuous, expected value, 1028
        mean, 1028
        median, 1034
        probability density function,
            1018–1019
            exponential, 1023
            normal, 1042–1044
            standard, 1038–1039
            uniform, 1022
        standard deviation, 1030
        variance, 1029
            alternate form, 1029
Random walk, 538
Range of a function, 591
Rate of sales, 974, 976
Rational exponent, A-3–A-4
Rational expression, A-9–A-10
Rational function, 612
Rationalizing, 1044
    denominators, 648
    numerators, 648, 1051
Reagan administration tax policy, 736
Real estate appreciation, 287
Recovery time, 1027, 1037
Rectangle, formula, 1199
Rectangular solid, formula, 1200
Reduced matrix, 68
Refrigerator sales, 982
Regular Markov process, 532
Regular stochastic matrix, 529
Regular transition matrix, 530
Related rates, 830
Relative extrema, 745, 1081–1082
Relative frequency, 374
Relative maximum, 745, 1081–1082
Relative minimum, 745, 1081–1082
Renewable natural resources, 751, 791
Replacement theorem for limits, 638
Retirement and taxes, 296
Retirement fund, 296
Retirement packages, comparing of,
    308
Revenue, 13, 927, 962, 992, 1037,
    1046, 1111
    average, 947
    bus company, 613, 618
    candy, 261–262
    equation, 676, 703, 822
    function, 625, 690, 756–757, 772,
        870

Revenue (*continued*)
   orchard, 615
   shoes, 603
Riemann, Bernhard, 915
   program, for graphing calculator,
     C-9–C-10
Riemann sum, 917, 1113
Right circular cone, formula, B-36
Right circular cylinder, formula, B-36
Risk, in investments, 174–175, 235,
     251, 477, 483–484
Room temperatures, 497
Roots, A-3
   properties, A-4
Roster notation, 3
Roulette, 381, 464, 466, 1026
Row matrix, 83
Row player, 553
Rumors, spread of, 886
Russell, Bertrand, 316
Ruth, Babe, 434, 465

S-curve, 763
Saddle point, 555
Safety, 507
Salary equation, 664
Salary incentives, 498
Sales, 43, 63, 396, 467, 484, 487, 497,
     928, 945, 967, 992, 1001, 1046
   car dealership, 455
   commissions, 36, 603
   condominiums, 135
   equation, 676
   function, 690
   machine tools, 420
   of new product, 962
   records, 383
   restaurant, 135
   selection, 333
   sweaters, 77
   sweepers, 135
Salt shaker manufacturing, 901
Salt water concentration, 714, 784
Sample space, 367
Saving for machinery, 288
Savings, 312
Savings certificate interest, 659–660
Scatter diagram, 40
Scheduling, 26, 62, 63
Scholarships, 359

Scholastic Aptitude Test scores, 486
School rezoning, 176, 236
Schools of fish, 722, 1079
Schreger, Cindy, 480
Scrap value, 33
Secant(s), program, for graphing
   calculator, C-7–C-8
Secant line, 671
Second derivative, 759
Second derivative test, 766
Second quadrant, 5
Selections for promotion, 371
Sensitivity analysis, 254
Sensitivity theorem, 258
Set(s), 2
   complement of a, 317
   disjoint, 319
   element, 3
   empty, 316
   equality of, 3
   intersection, 319
   member, 3
   roster notation, 3
   set-builder notation, 3
   subset, 316
   union, 318
   universal, 317
   Venn diagram, 317
Set-builder notation, 3
Setting prices, 570, 581
Shadow prices, 228, 258
Sharks, 754
Shelf-life, 1037, 1050
Shipping, 147, 162
Shoulder bag manufacturing, 911
Sigma notation, 914
Sign chart, 661, A-22
Simple discount, 272, 277
Simple interest, 270
Simplex method, 203
   basic variables, 190
   extended, 243
   in game theory, 579
   nonbasic variables, 190
   pivot column, 204
   pivot row, 204
   pivoting, 193
   primal problem, 219
   simplex tableau, 191
   slack variables, 187
   unit column, 192

Simplex tableau, 191
Simpson, program, for graphing
   calculator, C-11–C-12
Simpson's rule, 988
   errors in, 991
Sinking fund, 292, 312
Sinking fund payment, 292
Site selection, 349
Slack variables, 187
Slope, 15
   tangent line, 642, 671
Slope-intercept equation of a line, 20
Slugging average, 465
Smoking, 414, 420
Soccer, 512
Social interaction, 107
Social mobility, 549
Social security, 363
Soda production, 1047
Solution of a system of linear
   equations, 54
Solution set, 54
Space craft, 438
Spare parts, planning, 105
   production, 999
Spark plug manufacturing, 902
Sphere, equation of, 1060
   formula, B-36
Spherical tumor, 703
Spread of disease, 886
Spread of infestation, 628
Spread of news, 887
Spread of rumors, 615, 773, 886
Spread of technology, 713, 764, 772,
     784, 883, 886
Square matrix, 83
Squirrel weight distribution, 1047
Standard deck of cards, 345
Standard deviation, 472–474, 1030
Standard form of quadratic equation,
   609
Standard maximum problems, 201
Standard normal distribution, 490–491,
   1038–1039
Standard normal probability density
   function, 1038–1039
State of a Markov process, 516
Statistical calculations, with graphing
   calculator, C-5
Statistics, 443
   descriptive, 443

Statistics (*continued*)
lot acceptance sampling, 504
mean, 457
median, 463
mode, 463
standard deviation, 472–474
variance, 472, 474
alternate form, 476
Steady-state distribution, 531
Steady-state matrix, 528
Steel beam quality control, 1046
Steel production, 993
Stochastic matrix, 524
absorbing, 537
regular, 528
Stock(s), 483
Stock indices, 420
Stock selection, 349, 358, 405
Straight-line depreciation, 33, 36
Strictly determined games, 552, 556
Strictly upper triangle matrix, 105
Student ratings, 453
Stumpage value, 881–882
Subset, 316
Successful marriage odds, 395
Sulfur content of oil, 171
Sulfur dioxide, 486
Superfecta, 342
Suppliers, 419
Supply and demand, 891
equilibrium, 955
Supply curve, 32
Supply equation, 32
Supreme Court cases, 469
Supreme Court decisions, 350
Survey, 327
Sylvester, James Joseph, 81, 109
Syphilis, 484
Systems of linear equations, 53
equivalent, 54, 58
solution, 54
solution set, 54

Table cloth manufacturing, 901
Tables of integrals, 978–980,
B-36–B-39
Tangent, program, for graphing
calculator, C-8–C-9
Tangent line, 642, 673
approximation, 816, 818

equation of, 681
vertical, 684–685
Tangent plane, 1106
Tape manufacturing, 1047
Taxes, 605
Teaching methods, 439
Technological innovation, spread of,
713, 764, 772, 784, 883, 886
Telephone company charges, 599
Temperature, 37, 485
Tennis, 350
Term of an annuity, 289
Test grades, 498
Testing a product, 405
Tetrahedron dice, 371, 403
Theater revenue, 799
Third quadrant, 5
Threshold strength, 727, 784
Thyroid cancer, 507
Thyroid gland, 885
Tine test, 424
Toaster manufacturing, 982
Toothpaste market, 783
Tornadoes, 471
Toxic emissions costs, 713
Toy sales, 980
Trachea, 757, 829
Traffic flow, 78–79, 106
Transition matrix, 517
Translated game, 567
Transportation, 2, 11–12, 48, 174, 236,
244, 250
air cargo, 147, 162
charges, 12
costs, 26, 168–169
selection, 333
Trapezoi, program, for graphing
calculator, C-10–C-11
Trapezoid, formula, B-35
Trapezoidal rule, 985–986
errors in, 990
Tree diagram, 334
Tree leaves, 1080
Trial, 367
Triangle, formula, B-35
Trifecta, 336
Tripling time, 277, 312, 859
Tuberculosis, 424
Tumor, 835
cancerous, 625, 676
Twigs, 26

Uniform probability density function,
1022
mean, 1031
standard deviation, 1031
variance, 1031
Uniform sample space, 398
Union of events, 369
Union of sets, 318
Unit column, 192
Unit elasticity, 833
Universal set, 317
Upper sum, Riemann, 917
Upper triangle matrix, 89, 104
Useful life, 487

Vacation selection, 527
Value of a game, 564
Variable costs, 29
Variance, 472, 474, 1029
alternate form, 476, 1029
VCR life, 1026
VCR profits, 738
Velocity, 603, 615–616, 666, 703,
727, 900, 902, 928, 993
Venereal disease, 484
Venn diagram, 317
Verhulst, 763, 879
Vertex of quadratic function, 609
Vertical line, 19
Vertical line test, 596
Vertical tangent, 684–685
Viral hepatitis, 486
Vitamin supplements, 147, 162
Volume, 1117
von Neumann, John, 123, 552
Voter turnout, 469

Waiting time, 1026, 1037
Warrantee period, 496
Water allocation, 178
Water consumption, 1026, 1037
Water vapor, 829
Weather records, 382
Weight of top soil, 487
Whales, 37
Wiess' law, 727, 742, 784
Williams, Ted, 450, 452
Windpipe, 757, 829
Wire manufacturing, 1050

Wooden boat manufacturing, 138
Work scheduling, 177–178
Work stoppages, 675
World, population of, 884
World Series odds, 395

$x$-axis, 5
$x$-coordinate, 5

$x$-intercept, 21
$xy$-plane, 5

$y$-axis, 5
$y$-coordinate, 5
Yacht sales, 460
Yogurt selection, 334
Young, Cy, 511

Zero exponent, A-3
Zero matrix, 85
Zero-sum game, 553